BASIC ELECTRICAL AND ELECTRONICS ENGINEERING

J.B. Gupta

BASIC ELECTRICAL AND ELECTRONICS ENGINEERING

By

J.B. Gupta

B.Sc. (Elec. Engg.); C. Eng. (I);
F.I.E. (India); M.I.S.T.E.

S.K. KATARIA & SONS®

Publisher of Engineering Books

4885/109, Prakash Mahal, Dr. Subhash Bhargav Lane,
Opposite Delhi Medical Association, Daryaganj, New Delhi–110002 (INDIA)
Phones: +91-11-43551243; Mobile: +91-9871775858, 7678425209
e-mail: katariabook@gmail.com; katariabooks@yahoo.com
Website: www.skkatariaandsons.com

(e-books, subscriptions and bulk buying also available on our website)

Published by:

S.K. Kataria & Sons®

4885/109, Prakash Mahal, Dr. Subhash Bhargav Lane,
Opposite Delhi Medical Association, Daryaganj, New Delhi-110002 (INDIA)
Phone: +91-11-43551243; Mobile: +91-9871775858, 7678425209
e-mail: katariabook@gmail.com; katariabooks@yahoo.com
Website: www.skkatariaandsons.com

Edited by: ***Rajeev Manglik***
B.E. (Elec.) Pb. Univ.

Rohit Manglik
B.E. (Elec.) Univ. of Roorkee

First Edition: February, 1999; Sixth Edition: July, 2011; Reprint: 2026

Price : ₹ 795/-

Laser Typesetting by : Plus Computers, Delhi

Printed at : New A.S. Printing Press, Delhi

PREFACE TO THE SIXTH EDITION

The author feels great pleasure in presenting the Sixth edition of the book titled **Basic Electrical and Electronics Engineering** with marked improvement over the previous one. The book has been thoroughly revised, rewritten and redesigned keeping in view the latest revised syllabus (effective from session 2011-12) prescribed by Pb. Technical University (P.T.U.) Jalandhar. The matter has been reorganised accordingly so as to facilitate the students as well as teachers. A large number of solved examples, exercises, short answer type questions with answers and numerical problems from latest examination papers of P.T.U. and other Universities/Technical Universities have been included in the text.

The book has been divided into two parts viz. "Electrical Engineering" and "Electronics Engineering" according to syllabus prescribed by P.T.U. The book consists of 18 chapters in all (11 chapters of Electrical Engineering and 7 chapters of Electronics Engineering).

The solution to illustrative examples covering almost all topics and sub-topics of various chapters makes the entire presentation easy to follow. Selected exercises and numerical problems with answers given at the end of each chapter help the students in evaluating their understanding of the subject. Highlights given at the end of almost each chapter help the students in refreshing their knowledge while short answer type questions with answers help them in preparing for viva-voce/practical/ theory examinations.

It is earnestly hoped that with the extensive revision, modifications, and improvement, the book will prove a "complete textbook" for the First Year students of Degree Engineering studying in Engineering Colleges affiliated to Pb. Technical University (P.T.U.) Jalandhar.

Every care has been taken to eliminate misprints, omissions and errors but it is too much to expect that no inaccuracy/error/misprint has crept in and the author would be grateful to the readers for bringing to his notice any of such errors/misprints/omissions they may come across while going through the book.

The author will like to thank the readers (both teachers and students) who sent their valuable suggestions that became the basis for the revision of the book.

The author shall feel satisfied if the book meets the needs of the students for whom it is meant. Suggestions and criticism for the further improvement of the book shall be welcomed.

The author will like to appreciate the sincere efforts and contribution made by his sons Rajeev Manglik and Rohit Manglik in presentation of the new edition in this form.

The author will like to thank all the scientists, engineers, authors and publishers whose works and text have been the source of enlightenment, inspiration and guidance in presenting this new edition.

Lastly, but not the least, the author is thankful to his publisher, Sanjeev Kumar Kataria for bringing out the book in such a short time and for pricing it moderately despite heavy cost of paper and printing.

July 14, 2011

J.B. GUPTA

PREFACE TO THE FIRST EDITION

The primary aim of presenting this book is to meet the needs of first year students of degree engineering (all branches) studying in Engineering Colleges affiliated to the Panjab Technical University. The subject matter has been presented in an easy comprehensive manner, logical sequence using simple language and to give a detailed exposition of the fundamentals.

The book consists of 10 chapters providing a comprehensive coverage of course as prescribed by Pb. Technical University for first year students. A large number of solved examples, theoretical exercises and numerical problems, all of degree standards, have been included in the text. The solution to illustrative examples covering almost all topics and subtopics of various chapters makes the entire presentation easy to follow. Exercises and numerical problems with answers given at the end of each chapter helps the students in evaluating their understanding of the subject. A large number of reasoning questions (What, Why, Where and How type) given at the end of each chapter will make the readers to think deep into the subject matter, while answering. This will improve their reasoning capability, consequently, they shall be in a position to respond in a better way while appearing in the viva-voce/practical examination and facing the Selection Boards.

A large number of objective questions (multiple-choice, true/false and fill-in the blank types) has been given at the end of each chapter. These questions have been designed to bring out the hidden aspects and also to test the knowledge, understanding and application aspect of learning. These questions will certainly help the students to score better in their regular as well as competitive examinations.

Every care has been taken to eliminate misprints, omissions and errors but it is too much to expect that no inaccuracy/misprint/error has crept in and the author would be grateful to the readers for bringing to his notice any such errors/misprints/omissions they may come across while going through the book.

The author shall feel satisfied if the book meets the needs of the students for whom it is meant. Suggestions and criticism for the improvement of the book shall be welcomed.

The author will like to appreciate sincere efforts and contribution made by his son Mr. Rajeev Manglik BE (Electrical) and Er. (Mrs.) Shalini Manglik in presenting the book in this form.

The author will like to thank all the Scientists, Engineers, Authors and Publishers whose works and text have been the source of enlightenment, inspiration and guidance to me in the presentation.

Lastly, but not the least, the author is thankful to his publisher for bringing out the book in such a short time and for pricing it moderately despite heavy cost of paper and printing.

February 11, 1999 **J.B. GUPTA**

OTHER USEFUL BOOKS

J.B. Gupta *A Course in Electrical Technology*
J.B. Gupta *Basic Electrical Engineering [A Course in Electrical Technology Vol. 1]*
J.B. Gupta *Electrical Machines [A Course in Electrical Technology Vol. 2]*
J.B. Gupta *Electronic Devices and Circuits [A Course in Electrical Technology Vol. 3]*
J.B. Gupta *Theory and Performance of Electrical Machines [DC Machines, AC Machines and Polyphase Circuits]*
J.B. Gupta *Electrical Machines-I [DC Machines]*
J.B. Gupta *Electrical Machines-II [AC Machines]*
J.B. Gupta *An Integrated Course in Electronics Engineering*
J.B. Gupta *Fundamentals of Electrical and Electronics Engineering*
J.B. Gupta *A Course in Electrical Power [Generation & Economic Considerations; Transmission & Distribution; Switchgear & Protection; Utilization of Electrical Power & Electric Traction]*
J.B. Gupta *Generation and Economic Considerations [Electrical Power Systems Vol. 1]*
J.B. Gupta *Transmission and Distribution [Electrical Power Systems Vol. 2]*
J.B. Gupta *Switchgear and Protection [Electrical Power Systems Vol. 3]*
J.B. Gupta *Electrical Power Systems [Vol. 1 of A Course in Electrical Power]*
J.B. Gupta *Utilization of Electrical Power and Electric Traction [Vol. 2 of A Course in Electrical Power]*
J.B. Gupta *A Course in Electrical and Electronic Measurements and Instrumentation*
J.B. Gupta *Question Bank in Electrical Engineering*
J.B. Gupta *An Integrated Course in Electrical Engineering [Objective Type Questions With Answers & Notes]*
J.B. Gupta *Electrical Installation Estimating & Costing*
J.B. Gupta *Electrical and Electronics Engineering [Second Semester R.T.U.]*
J.B. Gupta *Electrical Engineering [G.B.T.U.; M.M.T.U. and U.T.U.]*
J.B. Gupta *Electrical Science [G.G.S.I.P. Univ.]*
J.B. Gupta *Basic Electrical and Electronics Engineering [Pb. Technical University]*
J.B. Gupta *A Textbook of Electrical and Electronics Engineering [RGTU, Bhopal]*
J.B. Gupta *Electrical Technology [M.D. Univ. and Kurukshetra Univ.]*
J.B. Gupta *Electrical Science [AMITY Univ.]*
J.B. Gupta *Electronics Engineering [G.B.; M.M. and Uttarakhand Technical Universities]*
J.B. Gupta *Electrical Measurements and Measuring Instruments [M.D.U., P.T.U.; G.B.; M.M. and Uttarakhand Technical Universities]*
J.B. Gupta *Electronic Measurements and Instrumentation [3rd Semester Pb. Technical Univ.]*
J.B. Gupta *Electronic Instrumentation and Measurements [4th Semester G.B. and M.M. Technical Universities]*
J.B. Gupta *Electronic Devices and Circuits [3rd Semester Pb. Technical Univ.]*
J.B. Gupta *Electrical Machines*
J.B. Gupta *Electrical and Electronics Engineering Materials [4th Semester G.B. and M.M. Technical Universities]*
J.B. Gupta *Electromechanical Energy Conversion-I [4th Semester G.B. and M.M. Technical Universities]*
J.B. Gupta *Analog Electronics [4th Semester Pb. Technical Univ.]*
J.B. Gupta *Pulse and Digital Switching Circuits [5th Semester Pb. Technical Univ.]*
J.B. Gupta *Linear Integrated Circuits [5th Semester Pb. Technical Univ.]*
J.B. Gupta *Electromechanical Energy Conversion-II [5th Semester G.B. and M.M. Technical Universities]*
J.B. Gupta *Electronics Measurements and Instrumentation [5th Semester M.D. Univ.]*
J.B. Gupta *Electromechanical Energy Conversion [3rd Semester M.D.U.]*
J.B. Gupta *Analog Electronics [4th Semester M.D. Univ.]*
J.B. Gupta *Analog Electronic Circuits [5th Semester M.D. Univ.]*
J.B. Gupta *Basic Electronics [M.D. Univ.]*
J.B. Gupta *Fundamentals of Electronic Devices [3rd Semester G.B. and M.M. Technical Universities]*
J.B. Gupta *A Text Book of Analog Electronics [4th Semester R.T.U.]*
J.B. Gupta *Elements of Electrical Engineering [Gujarat Technical Univ.]*
J.B. Gupta *Electronic Devices and Circuits [3rd Semester M.D. Univ.]*
J.B. Gupta *Electrical Machines–I [3rd Semester M.D. Univ.]*
J.B. Gupta *Analog Electronics [3rd Semester G.G.S.I.P. Univ. Delhi]*
J.B. Gupta *Elements of Power System [5th Semester G.B. and M.M. Technical Universities]*
J.B. Gupta *Elements of Electronic Devices and Circuits [3rd Semester R.T.U.]*
J.B. Gupta *Power System Analysis [6th Semester G.B. and M.M. Technical Universities]*
J.B. Gupta *Switchgear and Protection [7th Semester G.B. and M.M. Technical Universities]*
J.B. Gupta *Electrical Machines-II [5th Semester M.D. University]*

CONTENTS

PART–A

ELECTRICAL ENGINEERING

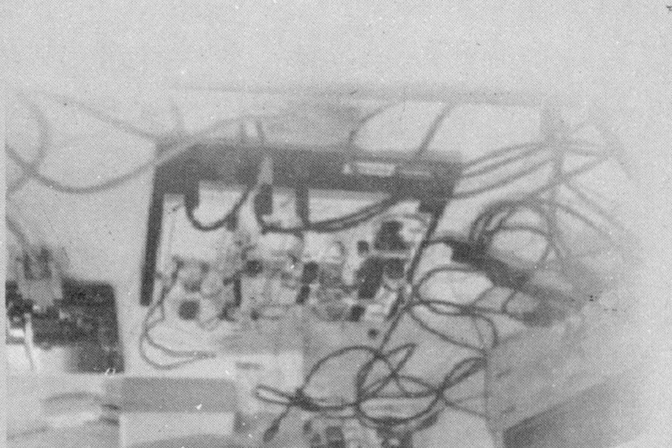

1 CHAPTER

Direct Current (DC) Circuits

1.1. NATURE OF ELECTRICITY

It is comparatively easy to describe electricity by its action than by its nature. Electricity is convenient form of energy and have got numerous applications, such as for lighting, transportation, communication, heating, elevators, and for driving of various types of machine tools owing to its inherent numerous advantages.

Although no one knows precisely what electricity is, it has been possible to develop theories about electricity through experiments and observations of its behaviour. As a result, it is now believed that all matters are essentially electrical in nature.

In order to appreciate the nature of electricity, it is necessary to consider the structure of matter. Matter has been defined any thing that occupies space and has weight such as copper, iron, water, air, smoke, aluminium, silver etc. Matter is made up of extremely small particles known as *molecules*. A molecule is the smallest particle into which a given substance may be divided without changing its identity. Molecules can further be subdivided chemically into still minute particles known as *atoms*. An atom may be regarded as the smallest particle of an element that can exist. The substances, whose molecules consist of similar atoms, are called the *elements* whereas the substances whose molecules consist of dissimilar atoms are called the *compounds*.

According to modern electron theory, the atoms of all elements are composed of two parts, namely, the central *nucleus* and the surrounding or *orbital electrons* (Fig. 1.1).

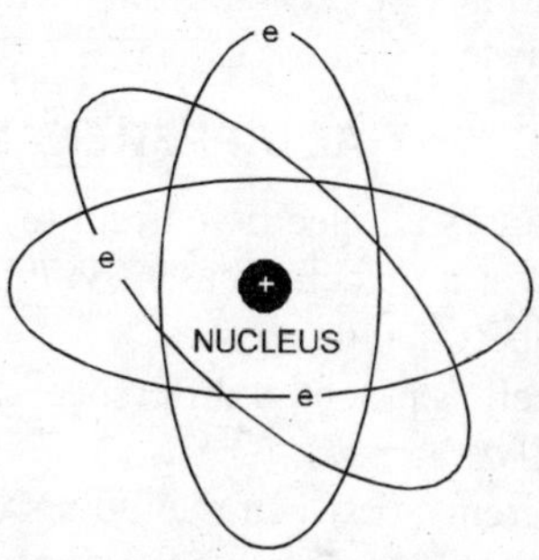

Fig. 1.1. ***Rutherford's Nuclear Model of Atomic Structure***

INSIDE THIS CHAPTER

The electrons each have a negative electric charge of 1.602×10^{-19} coulomb and some particles within the nucleus have a positive charge of the same magnitude. Since opposite charges attract each other, a force of attraction exists between the oppositely charged electrons and nucleus. Compared to the mass of nucleus, electrons are relatively tiny particles of almost negligible mass (each electron weighing 9.107×10^{-31} kg).

The nucleus of an atom is largely a cluster of two types of particles, called the *protons* and *neutrons.* Protons each have a positive electrical charge, equal in magnitude (but opposite in polarity) to the negative charge of an electron. A neutron has no charge at all. Protons and neutrons, each have mass about 1800 times the mass of an electron. For a given atom, the number of protons in the nucleus normally is equal to the number of orbiting electrons. Thus an atom is electrically neutral. If an atom loses an electron, it loses the negative charge and becomes a *positive ion.* Similarly if an atom gains an additional electron, it becomes a *negative ion.* However, nearly all atoms have some electrons which are loosely bound to their nuclei. These electrons are known as *free electrons* and may be dislodged by one means or another and transferred from one atom to another. The body, that contains unequal number of electrons and protons is said to be *electrically charged.* If a body contains electrons more than its normal number then the body is said to be – *vely charged.* Similarly a body containing electrons less than its normal number is said to be + *vely charged.*

The atoms of some materials such as silver, copper, aluminium and zinc, known as *conductors,* have many free electrons. These free electrons of a metal move about haphazardly in all directions from atom to atom but when a certain electrical pressure or potential is applied to such metals at the two ends, the electrons move only in one direction. The drift of electrons in a conductor in one direction is known as the *electric current i.e.* the flow of electric current takes place by the movement of the electrons in a conductor. Since electrons are negatively charged, the direction of their motion is opposite to the direction of conventional current which is from higher potential to the lower potential. In non-metallic materials, such as glass, mica, slate and porcelain, the electrons are very closely bound to the nucleus and it is difficult to remove the electrons from the atoms. Such materials are known as *non-conductors or insulators.*

1.2. ELECTRIC CURRENT

Electric current may be defined as the time rate of net motion of an electric charge across a cross-sectional boundary (Fig. 1.2). A random motion of electrons in a metal does not constitute a current unless there is a net transfer of charge with time

i.e. electric current,

$$i = \text{Rate of transfer of electric charge}$$

$$= \frac{\text{Quantity of electric charge transferred during a given time duration}}{\text{Time duration}} = \frac{dQ}{dt} \quad ...(1.1)$$

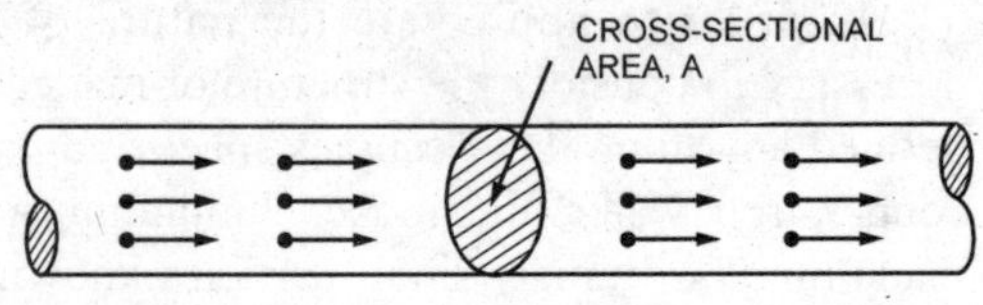

Fig. 1.2

Coulomb is the practical as well as SI unit for measurement of electric charge. One coulomb is approximately equal to 624×10^{16} electrons.

Since current is the rate of flow of electric charge through a conductor and coulomb is the unit of electric charge, the current may be specified in coulombs per second. In practice the term coulomb per second is seldom used, a shorter term, *ampere* is used instead.

1.3. ELECTROMOTIVE FORCE AND POTENTIAL DIFFERENCE

Electromotive force (emf) is the force that causes an electric current to flow in an electric circuit while the *potential difference* (pd) between two points in an electric circuit is that difference in their electrical state which tends to cause flow of electric current between them.

Volt is a unit of electromotive force as well as potential difference in practical as well as in SI system of units.

The volt is defined as that potential difference between two points of a conductor carrying a current of one ampere when the power dissipated between these points is equal to one watt.

1.4. RESISTANCE

Resistance may be defined as that property of a substance which opposes (or restricts) the flow of an electric current (or electrons) through it.

The practical as well as mks (or SI) unit of resistance is ohm (Ω), which is defined as that resistance between two points of a conductor when a potential difference of one volt, applied between these points, produces in this conductor a current of one ampere, the conductor not being a source of any emf.

For insulators having high resistance, much bigger units kilo-ohm or kΩ (10^3 ohm) and mega ohm or MΩ (10^6 ohm) are used. In case of very small resistances smaller units like milli-ohm (10^{-3} ohm) or micro-ohm (10^{-6} ohm) are employed.

1.5. OHM's LAW

As the rate of flow of water through a pipe is directly proportional to the effective pressure (*i.e.* difference of pressure at two ends) and inversely proportional to the frictional resistance, similarly the current flowing through a conductor is directly proportional to the potential difference across the ends of the conductor and inversely proportional to the conductor resistance. This relation was discovered by Georg Simon Ohm and so it is known as Ohm's law.

If I is the current flowing through a conductor of resistance R across which a potential difference V is applied then according to Ohm's law

$$I \propto V \text{ and } I \propto \frac{1}{R} \quad \text{or} \quad I \propto \frac{V}{R} \quad \text{or} \quad I = \frac{V}{R}$$ ...(1.2) where V is in volts, R is in ohms and I is in amperes.

Ohm's law may be defined as follows:

Physical state i.e. temperature etc. remaining the same, the current flowing through a conductor is directly proportional to the potential difference applied across its ends.

Or

The ratio of potential difference applied across a conductor and current flowing through it remains constant provided physical state i.e. temperature etc. of the conductor remains unchanged.

$$i.e. \quad \frac{V}{I} = \text{Constant} = R \qquad ...(1.3)$$

where R is known as the resistance of the conductor.

Ohm's law may be alternatively expressed as

$$V = I R \qquad ...(1.4)$$

Equations (1.2), (1.3) and (1.4) give Ohm's law in three forms with which the student should be familiar.

Limitations. Ohm's law cannot be applied to circuits consisting of electronic tubes or transistors because such elements are not bilateral *i.e.* they behave in different way when the direction of flow of current is reversed as in case of a diode. Ohm's law also cannot be applied to circuits consisting of nonlinear elements such as powdered carbon, thyrite, electric arc etc. For example, for silicon carbide, the relationship between applied voltage (or potential difference) V and current flowing I is given as $V = K I^m$ where K and m are constants and m is less than unity.

1.6. LAWS OF RESISTANCE

The resistance of a wire depends upon its length, area of x-section, type of material, purity and hardness of material of which it is made of and the operating temperature.

Resistance of a wire is

(*a*) directly proportional to its length, l *i.e.* $R \propto l$

(*b*) inversely proportional to its area of x-section, a *i.e.* $R \propto 1/a$

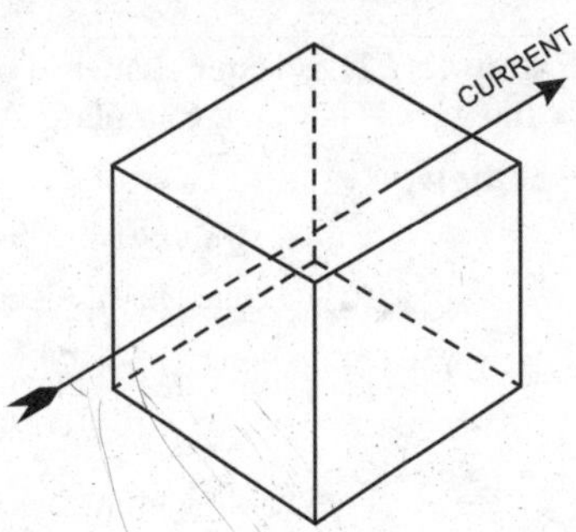

Fig. 1.3

Combining above two facts we have R α l/a

$$\text{or } R = \rho \frac{l}{a} \qquad ...(1.5)$$

where ρ (rho) is a constant depending upon the nature of the material and is known as the *specific resistance or resistivity of* the material of the wire.

To determine the nature of the constant ρ let us imagine a conductor of unit length and unit cross-sectional area, for example a cube whose edges are each of length one unit, and let the current flow into the cube at right angles to one face and out at the other face. Then putting $l = 1$ and $a = 1$ in Eq. (1.5) we have R = ρ. *Hence resistance of a material of unit length having unit cross-sectional area is defined as the resistivity or specific resistance of the material.*

Specific resistance or resistivity of a material is also defined as the resistance between opposite faces of a unit cube of that material.

Resistivity is measured in ohm-metres (Ω-*m*) or ohms per metre cube in mks (or SI) system and ohm-cm (Ω-cm) or ohm per cm cube in cgs system.

$$1\ \Omega\text{-m} = 100\ \Omega\text{-cm}$$

1.7. CONDUCTANCE AND CONDUCTIVITY

The reciprocal of resistance *i.e.* 1/R is called the *conductance* and is denoted by English letter G. It is defined as the inducement offered by the conductor to the flow of current and is measured in siemens (S). Earlier, the unit of conductance was mho (℧).

1 siemen = 1 mho

From Eq. (1.5)

$$G = \frac{1}{R} = \frac{1}{\rho \frac{l}{a}} = \frac{1}{\rho} \cdot \frac{a}{l} = \sigma \frac{a}{l} \qquad ...(1.6)$$

where $\sigma = \frac{1}{\rho}$ and is known as *specific conductance* or *conductivity* of the material. Hence *conductivity* is the reciprocal of the resistivity and is defined as the conductance between the two opposite faces of a unit cube. The unit of conductivity is siemens/metre (S/m).

Example 1.1. A coil consists of 2,000 turns of copper wire having a cross-sectional area of 0.8 mm^2. The mean length per turn is 80 cm, and the resistivity of copper is 0.02 μ Ω-m. Find the resistance of the coil.

Solution : Length of the coil, l = Number of turns × mean length per turn = 2,000 × 0.8 = 1,600 m

Cross-sectional area of wire, a = 0.8 mm^2 = 0.8×10^{-6} m^2

Resistivity of copper, ρ = 0.02 μΩ-m = $2 \times 10^{-2} \times 10^{-6}$ Ω-m = 2×10^{-8} Ω-m

Resistance of the coil, $R = \rho \frac{l}{a} = 2 \times 10^{-8} \times \frac{1,600}{0.8 \times 10^{-6}} = 40\ \Omega$ **Ans.**

Example 1.2. A heater element is made of nichrome wire having resistivity equal to 100 × 10^{-8} Ω-m. The diameter of the wire is 0.4 mm. Calculate the length of the wire required to get a resistance of 40 Ω.

Solution : Resistance of nichrome wire, R = 40 Ω

Resistivity of nichrome wire, ρ = 100×10^{-8} Ω-m

Diameter of nichrome wire, d = 0.4 mm = 4×10^{-4} m

Cross-sectional area of nichrome wire, $a = \frac{\pi}{4} d^2 = \frac{\pi}{4}\left(4 \times 10^{-4}\right)^2 = 4\pi \times 10^{-8}\ \text{m}^2$

Length of nichrome wire required, $l = \frac{R \times a}{\rho} = \frac{40 \times 4\pi \times 10^{-8}}{100 \times 10^{-8}} = 5.03$ m **Ans.**

Example 1.3. The resistance of a conductor 1 mm² in cross section and 20 m long is 0.346 Ω. Determine the specific resistance of the conductor material.

Solution :

Resistance of conductor, R = 0.346 Ω

Length of conductor, l = 20 m

Cross-sectional area of conductor, a = 1 mm² = 1×10^{-6} m²

$$\text{Specific resistance of conductor material, } \rho = \frac{R \times a}{l} = \frac{0.346 \times 1 \times 10^{-6}}{20} = 1.73 \times 10^{-8}\ \Omega\text{-m} \quad \textbf{Ans.}$$

Example 1.4. A wire of length 1 m has a resistance of 2 Ω. Obtain the resistance if specific resistance is doubled, diameter is doubled and the length is made three times of the first.

[*Pb. Technical Univ. Basic Electrical and Electronics Engineering Second Semester 2004-05; May 2010*]

Solution :

New specific resistance, $\rho_2 = 2\rho_1$

New length of wire, $l_2 = 3\, l_1$

$$\text{New cross-sectional area, } a_2 = \frac{\pi}{4}\left(d_2^2\right) = \frac{\pi}{4}(2d_1)^2 = 4\, a_1$$

$$\text{New resistance, } R_2 = \frac{\rho_2 l_2}{a_2} = \frac{2\rho_1 \times 3l_1}{4a_1} = 1.5\,\frac{\rho_1 l_1}{a_1} = 1.5\, R_1 = 1.5 \times 2 = 3\ \Omega \quad \textbf{Ans.}$$

Example 1.5. Determine the resistance of the material in terms of volume v, length l and resistivity ρ. Hence determine the resistivity of the material of a conductor of volume 0.05 m³, length 300 metres and resistance 0.0306 Ω.

Solution : Cross-sectional area of material of volume v units and length l units,

$$a = \frac{v}{l} \quad \text{...}(i)$$

$$\text{Resistance of material, } R = \rho\frac{l}{a} = \rho\frac{l}{v/l} = \rho\frac{l^2}{v} \quad \textbf{Ans.} \quad \because \text{ from Eq. } (i)\ a = \frac{v}{l}$$

Substituting v = 0.05 m³, l = 300 m and R = 0.0306 Ω in above equation we get

$$0.0306 = \rho\,\frac{(300)^2}{0.05}$$

$$\text{or } \rho = \frac{0.0306 \times 0.05}{(300)^2} = 1.7 \times 10^{-8}\ \Omega\text{-m} \quad \textbf{Ans.}$$

Example 1.6. Determine (*i*) length and (*ii*) diameter of copper cylinder of volume v, resistance R and specific resistance ρ in terms of v, R and ρ.

Solution :

$$\text{Volume of cylinder, } v = a \times l \quad \text{...}(i)$$

$$\text{Resistance of cylinder, } R = \rho\,\frac{l}{a} \quad \text{...}(ii)$$

Multiplying Eqs (*i*) and (*ii*) we get

$$vR = a\, l \times \rho\,\frac{l}{a} = \rho\, l^2$$

$$\text{or } l = \sqrt{\frac{vR}{\rho}} \quad \textbf{Ans.}$$

Dividing Eq. (*i*) by Eq. (*ii*) we get

$$\frac{v}{R} = \frac{al}{\rho l/a} = \frac{a^2}{\rho}$$

$$\text{or } a = \sqrt{\frac{v\rho}{R}}$$

$$\text{or } \frac{\pi}{4}d^2 = \sqrt{\frac{v\rho}{R}}$$

$$\text{or diameter of cylinder, } d = \left(\frac{16\, v\rho}{\pi^2 R}\right)^{1/4} \quad \textbf{Ans.}$$

Example 1.7. One km of wire having a diameter of 11.7 mm and of resistance 0.031 Ω is drawn so that its diameter is 5.0 mm. What does its resistance become ?

Solution :

Resistance of old wire, $R_1 = 0.031\ \Omega$

Diameter of old wire, $d_1 = 11.7 \text{ mm} = 11.7 \times 10^{-3}$ m

Diameter of new wire, $d_2 = 5.0 \text{ mm} = 5 \times 10^{-3}$ m

$$\text{Area of x-section of old conductor, } a_1 = \frac{\pi}{4}d_1^2 = \frac{\pi}{4}(11.7 \times 10^{-3})^2$$

$$\text{Area of x-section of new conductor, } a_2 = \frac{\pi}{4}d_2^2 = \frac{\pi}{4}(5 \times 10^{-3})^2$$

$$\text{Resistance of a wire, } R = \rho\frac{l}{a} = \rho\frac{v/a}{a} = \rho\frac{v}{a^2},$$ where v is the volume of wire

$\therefore\ R \propto \frac{1}{a^2}$ if wire is of constant volume and of same material.

$$\therefore \quad \text{Resistance of new wire, } R_2 = R_1\left(\frac{a_1}{a_2}\right)^2 = 0.031\frac{\left[\frac{\pi}{4}\left(11.7\times10^{-3}\right)^2\right]^2}{\left[\frac{\pi}{4}\left(5\times10^{-3}\right)^2\right]^2} = 0.929\ \Omega \quad \textbf{Ans.}$$

1.8. CONDUCTORS, INSULATORS AND SEMICONDUCTORS

As already said in some materials such as silver, copper, aluminium, atoms possess a relatively appreciable number of free electrons and, therefore, these materials offer least resistance to the flow of electric current. Such materials are known as *conductors.*

Silver is the best conductor of electricity but its use is limited due to its high cost. **Copper** is the next best conductor of electricity and is most commonly used in electrical engineering. Aluminium is third in rank in electrical conductivity after silver and copper and is used in manufacturing of wires, round or square bus-bars, capacitor foils and cable sheaths.

On the other hand in some materials such as mica, glass, slate and porcelain, the electrons are very closely bound to the nucleus and it is difficult to remove the electrons from the atoms. Such materials offer very high resistance to the flow of electric current and are known as *insulators.*

Materials lying between conductors and insulators, such as germanium, silicon, sulphur, carbon, are known as *semiconductors.*

Though there is no rigid line separating the conductors from semiconductors and semiconductors from insulators, but still according to resistivity the materials of resistivity of the order from 10^{-8} to 10^{-3}, 10^{-3} to 10^{6} and 10^{6} to 10^{18} may be classified as conductors, semiconductors and insulators respectively.

The resistance of the conductors (except of carbon) increases with the increase in temperature, therefore, on high temperature good conductors become bad conductors even.

The insulators lose their insulating power with the increase in temperature and some may even become good conductors if they are strongly heated.

Insulators are also affected by the moisture, which often converts the insulators into poor conductors and poor conductors into medium conductors.

Conductors except metallic and carbon ones get chemically decomposed when current is passed through them.

1.9. RESISTANCE VARIATION WITH TEMPERATURE

As already mentioned in Art 1.8, the resistance of all pure metallic conductors increases with the increase in temperature but the resistance of the insulators and non-metallic materials generally decreases with the increase in temperature.

If the resistance of any pure metal is plotted on a temperature base, it is found that over the range of temperature from 0 to 100 °C the graph is practically a straight line, as illustrated in Fig. 1.4. If this straight line is extended, it cuts the temperature axis at some temperature, $-t_0$ °C, known as *inferred zero resistance temperature.* This does not mean that the resistance of the metal is actually zero at that temperature, but $-t_0$ °C is the temperature at which the resistance would be zero if the rate of decrease between 100 and 0 °C were maintained constant at all temperatures. From the similarity of the triangles in Fig. 1.4.

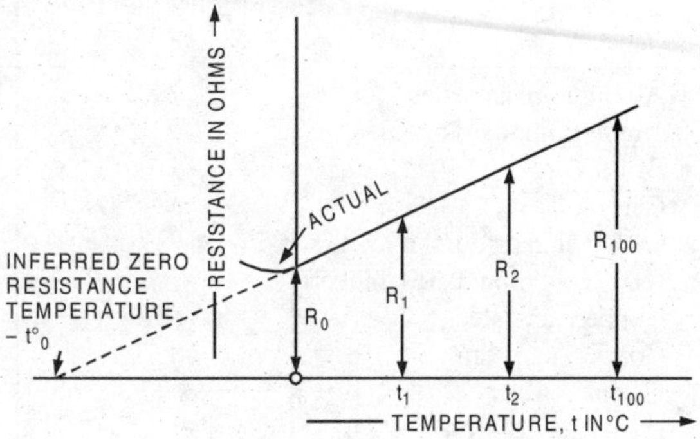

Fig. 1.4

$$\frac{R_2}{R_1} = \frac{t_0 + t_2}{t_0 + t_1} \qquad ...(1.7)$$

where R_1 and R_2 are the resistances at temperatures t_1 °C and t_2 °C respectively. Thus if the resistance R_1 for any temperature t_1 °C is known, then resistance for any other temperature t_2 can be computed from above equation provided that t_0 for that particular material is known. Inferred zero resistance temperatures, t_0 for various materials are given in Table 1.1.

The variation of resistance with temperature is often utilized in determining temperature variations. For example, in testing of an electric machine, the resistance of the coil is measured both before and after the test run, and the increase in resistance is a measure of the rise in temperature. For computation of temperature rise Eq. (1.7) may be transposed to the following form

$$t_2 - t_1 = \frac{R_2 - R_1}{R_1}(t_0 + t_1) \qquad ...(1.8)$$

A temperature of 20 °C has been adopted as the *standard reference temperature* for measurement of resistance, and the hand books give the resistance of the various materials at that temperature. Consequently, when a designer is computing the resistance of any conductor from its dimensions, the initial temperature t, at which the resistance is known, is generally 20 °C.

1.10. TEMPERATURE COEFFICIENT OF RESISTANCE

Let a metallic conductor having a resistance of R_0 at 0 °C be heated to t °C and let its resistance at this temperature be R_t. From Eq. (1.7)

$$\frac{R_t}{R_0} = \frac{t_0 + t}{t_0 + 0}$$

$$\text{or } R_t = R_0 + \frac{1}{t_0} R_0\, t$$

$$\text{or change in resistance, } \Delta R = R_t - R_0 = \frac{1}{t_0} R_0\, t = \alpha_0\, R_0\, t \qquad ...(1.9)$$

TABLE 1.1

Resistivities, Temperature Coefficients and Inferred Zero Resistance Temperatures

Material	Resistivity in Ω-m at 20 °C (× 10^{-8})	Temperature Coefficient of Resistance in ohms per ohm per °C rise in temperature at 20 °C	Inferred Zero Resistance Temperature in °C
Aluminium, annealed	2.82	0.0039	– 236
Aluminium, hard drawn	2.92	0.0038	– 243
Brass	6 – 8	0.002	– 480
Carbon	3,000 – 7,000	– 0.0005	—
Constantan or Eureka (60% copper and 40% nickel)	49	0.000008	– 1,25,000
Copper, annealed	1.724	0.00393	– 234.5
Copper, hard drawn	1.79	0.00382	– 242
German Silver (80% Cu, 12% Ni, 4% Zn)	33	0.0004	– 2,480
Gold	2.44	0.0034	– 274
Iron	12	0.005	– 180
Lead	22	0.0039	– 236
Manganin (84% Cu, 12% Mn, 4% Ni)	42 to 74	0.00003	– 33,000
Mercury	96	0.00089	– 1,100
Molybdenum	5.7	0.0033	– 280
Nichrome (60% Cu, 25% Fe,15 % Cr)	100	0.00044	– 2,250
Nickel	7.8	0.006	– 147
Phosphor, Bronze	5–10	0.0035	– 266
Platinum	10	0.003	– 310
Platinum–iridium	24.6	0.0012	– 814
Silver (99.98% pure)	1.6	0.0038	– 243
Tungsten	5.51	0.0045	– 200

Note : *It has been found that the resistivity of the metallic conductors is significantly affected by variations in temperature but it is also affected by other factors like tension and pressure to some extent. For most metals except lithium and calcium resistivity decreases with the increase in pressure. However, resistivity increases with increase in tension.*

where $\alpha_0 = \dfrac{1}{t_0}$ and is called the temperature coefficient of resistance of the material at 0 °C

From Eq. (1.9) it may be concluded that change in resistance due to change in temperature

(*a*) varies directly as its initial resistance,

(*b*) varies directly as rise in temperature and

(*c*) depends on the nature of the material of the conductor.

The Eq. (1.9) may be rewritten as

$$\alpha_0 = \frac{\Delta R}{R_0\, t} \qquad ...(1.10)$$

So the temperature coefficient of resistance may be defined as the ratio of increase in resistance per degree rise of temperature to the original resistance.

If R_0 is the resistance of any conductor at 0 °C and α_0 is the temperature coefficient of resistance at 0°, then resistance at t° C is given as

R_t = Original resistance + increase in resistance $= R_0 + R_0 \alpha_0 t = R_0 (1 + \alpha_0 t)$ ...(1.11)

The above expression holds good for both increase as well as decrease in temperature.

It is to be noted that

(*i*) temperature coefficient of resistance for all pure metallic conductors is positive *i.e.* the resistance of all pure metallic conductors increases with the increase in temperature, that of non-metallic materials such as of carbon is negative *i.e.* the resistance of non-metallic materials such as of carbon decreases with the increase in temperature. The temperature coefficient of resistance of alloys like constantan and manganin is negligible.

(*ii*) temperature coefficient of resistance is not constant but depends on the initial temperature on which the increment in resistance is based. When the increment is based on the resistance measured at 0 °C, then the temperature coefficient of resistance has the value of α_0. At any other temperature t °C, value of temperature coefficient of resistance is α_t and so on. For any material the temperature coefficient of resistance at 0 °C *i.e.* α_0 has the maximum value.

The temperature coefficient of resistance at any temperature t_1 is given as

$$\alpha_1 = \frac{1}{\frac{1}{\alpha_0} + t_1} \quad \text{...(1.12) (refer to Example 1.8)}$$

The temperature coefficient of resistance at temperature t_2 in terms of temperature coefficient of resistance at temperature t_1 is given as

$$\alpha_2 = \frac{1}{\frac{1}{\alpha_1} + (t_2 - t_1)} \quad \text{...(1.13) (refer to Example 1.9)}$$

If R_1 is the resistance of any conductor at t_1 °C and α_1 is the temperature coefficient of resistance at t_1 °C, then resistance of the conductor at t_2 °C is given as

$$R_2 = R_1 [1 + \alpha_1 (t_2 - t_1)] \quad \text{...(1.14)}$$

(*iii*) As the resistance of the material changes with the change in temperature so it is obvious that resistivity of the material depends on temperature.

Knowledge of temperature coefficient of resistance is utilized in determining the temperature rise of electrical machines, as already described in Art 1.9.

Values of resistivity, temperature coefficient of resistance at 20 °C and inferred zero resistance temperature for various materials are given in Table 1.1.

Example 1.8. Show that if α_0 be the resistance temperature coefficient of a conductor at 0 °C, then resistance tempreature coefficient at t_1 °C is given as

$$\alpha_1 = \frac{1}{\frac{1}{\alpha_0} + t_1}$$

Solution : Let R_0, R_1 and R_2 be the resistances of a conductor at 0°C, t_1 °C and t_2 °C temperatures respectively, then from Eq. (1.11)

$$\text{Resistance at } t_1 \text{ °C, } R_1 = R_0 (1 + \alpha_0 t_1) \quad \text{...}(i)$$

$$\text{Resistance at } t_2 \text{ °C, } R_2 = R_0 (1 + \alpha_0 t_2) \quad \text{...}(ii)$$

$$\text{Also } R_2 = R_1 [1 + \alpha_1 (t_2 - t_1)] \quad \text{(Refer to Eq. 1.14)}$$

or

$$\frac{R_2}{R_1} = 1 + \alpha_1 (t_2 - t_1) \quad \text{...}(iii)$$

Dividing Eq. (*ii*) by Eq. (*i*), we get

$$\frac{R_2}{R_1} = \frac{1+\alpha_0 t_2}{1+\alpha_0 t_1} = \frac{1+\alpha_0 t_2 + \alpha_0 t_1 - \alpha_0 t_1}{1+\alpha_0 t_1}$$

$$= \frac{1+\alpha_0 t_1}{1+\alpha_0 t_1} + \frac{\alpha_0 t_2 - \alpha_0 t_1}{1+\alpha_0 t_1} = 1 + \frac{\alpha_0 (t_2 - t_1)}{1+\alpha_0 t_1} = 1 + \frac{\alpha_0}{1+\alpha_0 t_1}(t_2 - t_1) \quad ...(iv)$$

Comparing Eqs. (*iii*), and (*iv*) we get

$$1 + \alpha_1 (t_2 - t_1) = 1 + \frac{\alpha_0}{1+\alpha_0 t_1}(t_2 - t_1)$$

$$\text{or } \alpha_1 = \frac{\alpha_0}{1+\alpha_0 t_1} = \frac{1}{\frac{1}{\alpha_0} + t_1} \quad \text{Proved.}$$

Example 1.9. Show that if α_1 is the resistance temperature coefficient of a conductor at t_1 °C then resistance tempreature coefficient at t_2 °C is given by

$$\frac{1}{\frac{1}{\alpha_1} + (t_2 - t_1)}$$

Solution : Let R_1, R_2 and R_3 be the resistances of a conductor at temperatures t_1, t_2 and t_3°C respectively.

Then $\quad R_2 = R_1 \, [1 + \alpha_1 (t_2 - t_1)] \quad ...(i)$

and $\quad R_3 = R_1 \, [1 + \alpha_1 (t_3 - t_1)] \quad ...(ii)$

Also $\quad R_3 = R_2 \, [1 + \alpha_2 (t_3 - t_2)]$

or $\quad \frac{R_3}{R_2} = 1 + \alpha_2 (t_3 - t_2) \quad ...(iii)$

Dividing Eq. (*ii*) by Eq. (*i*) we get

$$\frac{R_3}{R_2} = \frac{1+\alpha_1 (t_3 - t_1)}{1+\alpha_1 (t_2 - t_1)} = \frac{1+\alpha_1 (t_2 - t_1) + \alpha_1 (t_3 - t_1) - \alpha_1 (t_2 - t_1)}{1+\alpha_1 (t_2 - t_1)} = 1 + \frac{\alpha_1 (t_3 - t_2)}{1+\alpha_1 (t_2 - t_1)} \quad ...(iv)$$

Comparing Eqs. (*iii*) and (*iv*) we get

$$1 + \alpha_2 (t_3 - t_2) = 1 + \frac{\alpha_1 (t_3 - t_2)}{1+\alpha_1 (t_2 - t_1)}$$

$$\text{or } \alpha_2 = \frac{\alpha_1}{1+\alpha_1 (t_2 - t_1)} = \frac{1}{\frac{1}{\alpha_1} + (t_2 - t_1)} \quad \text{Proved.}$$

Example 1.10. Find the temperature at which resistance of the conductor becomes double to that at 0°C.

[*Pb. Technical Univ. Basic Electrical and Electronics Engineering May 2008*]

Solution : Resistance of a conductor at t°C is given by equation

$$R_t = R_0(1 + \alpha_0 t)$$

where R_0 is temperature at 0°C and α_0 is temperature coefficient of resistance at 0°C

Substituting $R_t = 2R_0$ in above equation, we have

$$2R_0 = R_0(1 + \alpha_0 t)$$

$$\text{or } t = \frac{1}{\alpha_0} \quad \textbf{Ans.}$$

Example 1.11. A motor winding has a resistance of 80 Ω at room temperature of 20°C before switching on to a 230 V. After 4 hour run the winding resistance is 100 Ω. Find the temperature rise if the material temperature coefficient is 1/234.5/°C. *[Pb. Technical Univ. Basic Electrical and Electronics Engineering May 2006]*

Solution : If the temperature coefficient of resistance of winding material at room temperature of 20°C is α_1 and resistance of winding material at 20°C is R_1, then resistance of winding material at t°C is given by the equation

$$R_2 = R_1[1 + \alpha_1(t_2 - t_1)]$$

or
$$100 = 80\left[1 + \frac{1}{234.5}(t_2 - 20)\right]$$

$$\because\ R_2 = 100\ \Omega,\ R_1 = 80\ \Omega,\ t_1 = 20°C \text{ and } \alpha_1 = \frac{1}{234.5}\ /°C$$

or
$$\text{Temperature rise, } (t_2 - 20) = \left(\frac{100}{80} - 1\right) \times 234.5 = 58.625°C \ \textbf{Ans.}$$

Example 1.12. A coil of relay is made of copper wire. At a temperature of 20 °C, the resistance of the coil is 400 Ω. Calculate the resistance of the coil at temperature of 80 °C. The temperature coefficient of copper is 0.0038 Ω/Ω/ °C at 0 °C.

Solution :

$$\text{Temperature coefficient of resistance at 0 °C, } \alpha_0 = 0.0038\ \Omega/\Omega/°C$$

$$\text{Temperature coefficient of resistance at 20 °C, } \alpha_{20} = \frac{1}{\frac{1}{\alpha_0} + t} = \frac{1}{\frac{1}{0.0038} + 20} = 0.003532$$

$$\text{Resistance of wire at 20 °C, } R_1 = 400\ \Omega$$

$$\text{Resistance of wire at 80 °C, } R_2 = R_1\ [1 + \alpha_{20}(t_2 - t_1)]$$
$$= 400\ [1 + 0.003532\ (80 - 20)] = 484.8\ \Omega \ \textbf{Ans.}$$

Example 1.13. A platinum coil has a resistance of 3.146 Ω at 40 °C and 3.767 Ω at 100 °C. Find the resistance at 0 °C and the temperature coefficient of resistance at 40 °C.

Solution : Let the temperature coefficient of resistance of platinum at 0 °C be α_0 and resistance of platinum coil at 0 °C be R_0, then

$$\text{Resistance at 40 °C, } R_1 = R_0\ (1 + 40\ \alpha_0) \quad ...(i)$$

$$\text{Resistance at 100 °C, } R_2 = R_0\ (1 + 100\ \alpha_0) \quad ...(ii)$$

Dividing Eq. (*ii*) by Eq. (*i*) we have

$$\frac{R_2}{R_1} = \frac{1 + 100\ \alpha_0}{1 + 40\ \alpha_0}$$

or
$$\frac{3.767}{3.146} = \frac{1 + 100\ \alpha_0}{1 + 40\ \alpha_0}$$

or $\alpha_0 = 0.00379$

Temperature coefficient of resistance at 40 °C,

$$\alpha_{40} = \frac{1}{\frac{1}{\alpha_0} + 40} = \frac{1}{\frac{1}{0.00379} + 40} = 0.00329\ \Omega/\Omega/\ °C \quad \textbf{Ans.}$$

Substituting $R_1 = 3.146$ and $\alpha_0 = 0.00379$ in Eq. (*i*) we have

$$3.146 = R_0\ (1 + 40 \times 0.00379)$$

or
$$R_0 = \frac{3.146}{1 + 40 \times 0.00379} = 2.732\ \Omega \ \textbf{Ans.}$$

Example 1.14. A coil has resistance of 18 Ω when its mean temperature is 20°C and of 20 Ω when its mean temperature is 50 °C. Find its mean temperature rise when its resistance is 21 Ω and the surrounding temperature is 15 °C.

Solution : Let R_0 be the resistance of the coil at 0°C and α_0 be its temperature coefficient of resistance at 0 °C

$$\text{Resistance at 20 °C, } 18 = R_0 (1 + 20 \alpha_0) \quad ...(i)$$

$$\text{Resistance at 50 °C, } 20 = R_0 (1 + 50 \alpha_0) \quad ...(ii)$$

Dividing Eq. (*ii*) by Eq. (*i*) we have

$$\frac{20}{18} = \frac{1 + 50\alpha_0}{1 + 20\alpha_0}$$

$$\text{or} \quad \alpha_0 = \frac{1}{250} = 0.004$$

If t °C is the temperature of coil when its resistance is 21 Ω, then

$$21 = R_0 (1 + 0.004\, t) \quad ...(iii)$$

Dividing Eq. (*iii*) by Eq. (*ii*) we have

$$\frac{21}{20} = \frac{1 + 0.004t}{1 + 50 \times 0.004}$$

$$\text{or} \quad t = 65 \text{ °C}$$

Temperature rise = t – surrounding temperature = 65 – 15 = 50 °C **Ans.**

Example 1.15. A specimen of copper wire has a specific resistance of 1.7 × 10^{-8} Ω-mm at 0°C, and has a temperature coefficient of $\frac{1}{254.5}$ per °C at 20°C. Find the specific resistance and temperature coefficient at 70°C. [*Pb. Technical Univ. Basic Electrical and Electronics Engineering May 2007*]

Solution : Temperature coefficient of resistance at 20°C

$$\alpha_{20} = \frac{1}{254.5} \ \Omega/\Omega/°\text{C}$$

Temperature coefficient of resistance at 70 °C,

$$\alpha_{70} = \frac{1}{\frac{1}{\alpha_{20}} + (t_2 - t_1)} = \frac{1}{254.5 + (70 - 20)} = \frac{1}{304.5} \ \Omega/\Omega/°\text{C} \quad \textbf{Ans.}$$

If α_0 is the temperature coefficient of resistance at 0 °C, then temperature coefficient of resistance at 20 °C

$$\alpha_{20} = \frac{1}{\frac{1}{\alpha_0} + 20}$$

$$\text{or} \quad \alpha_0 = \frac{1}{\frac{1}{\alpha_{20}} - 20} = \frac{1}{254.5 - 20} = \frac{1}{234.5} \ \Omega/\Omega/°\text{C}$$

Specific resistance at 70 °C, $\rho_{70} = \rho_o (1 + \alpha_o t) = 1.7 \times 10^{-8} \left(1 + \frac{1}{234.5} \times 70\right) = 2.207 \times 10^{8}$ Ω-mm **Ans.**

Note : *From the above example it is obvious that the specific resistance of a given material does not remain constant, but changes with the change in temperature.*

Example 1.16. A potential difference of 200 V is applied to a copper field coil at a temperature of 15 °C and the current is 10 A. What will be the mean temperature of the coil when the current has fallen to 8 A ? Applied voltage being the same. Given α = $\frac{1}{234.5}$ per °C at 0 °C.

Solution :

$$\text{Resistance at 15 °C, } R_1 = \frac{V}{I_1} = \frac{200}{10} = 20 \ \Omega$$

Resistance at final temperature of t_2°C, $R_2 = \frac{V}{I_2} = \frac{200}{8} = 25\ \Omega$

Temperature coefficient of copper at 0 °C, $\alpha_0 = \frac{1}{234.5}$

$$\text{Since} \quad \frac{R_2}{R_1} = \frac{1 + \alpha_0 t_2}{1 + \alpha_0 t_1}$$

$$\text{or} \quad \frac{25}{20} = \frac{1 + \frac{t_2}{234.5}}{1 + \frac{15}{234.5}}$$

$$\text{or } t_2 = 77.38\ °\text{C} \quad \textbf{Ans.}$$

Example 1.17. Discuss the temperature dependence of resistance.

The filament of 60 W, 230 V lamp has a normal working temperature of 2,000°C. Take the temperature coefficient of the material to be 0.005. Find the approximate current which flows at the instant of switching on the supply to the cold lamp. [*Pb. Technical Univ. Basic Electrical and Electronics Engineering Dec. 2005*]

Solution : Resistance of 60 W metal filament, $R_{2,000} = \frac{V^2}{W} = \frac{(230)^2}{60} = 881.67\ \Omega$

Let the resistance of 60 W metal filament at room temperature (say 20°C) be R_2.

Temperature coefficient of resistance (say at room temperature of 20°C) be $\alpha_{20} = 0.005$

$$\text{So} \quad R_{2,000} = R_{20}[1 + \alpha_{20}(t_2 - t_1)]$$

$$\text{or} \quad R_{20} = \frac{R_{2,000}}{1 + \alpha_{20}(t_2 - t_1)} = \frac{881.67}{1 + 0.005(2,000 - 20)} = \frac{881.67}{10.9} = 80.887\ \Omega$$

Current flowing at the instant of switching on the bulb

$$I_{20} = \frac{V}{R_{20}} = \frac{230}{80.887} = 2.843\ \text{A} \quad \textbf{Ans.}$$

Example 1.18. Find the current flowing at the instant of switching on a 100 watt metal filament lamp on to a 220 V circuit. Given that the incandescent filament temperature is 2,000 °C and the resistance temperature coefficient at room temperature of 20 °C is 0.005. [*Pb. Univ. Applied Electricity 1973*]

Solution : Resistance of 100 watt metal filament, $R_{100} = \frac{V^2}{W} = \frac{(220)^2}{100} = 484\ \Omega$

Let the resistance of 100 watt metal filament at room temperature be R_{20}

Temperature coefficient of resistance at 20 °C, $\alpha_{20} = 0.005$

$$\because \quad R_{100} = R_{20}\ [1 + \alpha_{20}(t_2 - t_1)]$$

$$\therefore \quad R_{20} = \frac{R_{100}}{1 + \alpha_{20}(t_2 - t_1)} = \frac{484}{1 + 0.005 \times (2,000 - 20)} = \frac{484}{10.9} = 44.4\ \Omega$$

Current flowing at the instant of switching on a 100 W metal filament lamp,

$$I_{20} = \frac{V}{R_{20}} = \frac{220}{44.4} = 4.95\ \text{A} \quad \textbf{Ans.}$$

1.11. SI SYSTEM OF UNITS

SI is the latest form of metric system and absorbs in it the rationalized MKSA system. SI stands for "Systeme International d' Unites" in French. This abbreviation is now adopted by the International Standardising Organisation as the abbreviated name of this new system of units in all languages.

The SI system is, in fact, simply the RMKSA* system expanded by adding the degree Kelvin, Candela and mole as basic units of temperature, luminous intensity and amount of substance respectively. The SI system is a comprehensive, logical and coherent system, designed for use in all branches of science, engineering and technology.

This system derives all the units from the following seven base units.

Quantity	*Unit*	*Symbol*
Length	metre	m
Mass	kilogram	kg
Time	second	s
Intensity of electric current	ampere	A
Thermodynamic temperature	Kelvin	K
Luminous intensity	candela	cd
Amount of substance	mole	mol

Recommended prefixes for formation of multiples and sub-multiples of units are given below:

Multiple	Prefix	Symbol	Fraction	Prefix	Symbol
10	deca	da*	0.1	deci	d*
100	hecto	h*	0.01	centi	c*
1000	kilo	k	0.001	milli	m
1000 000	mega	M	0.000 001	micro	μ
1000 000 000	giga	G	0.000 000 001	nano	n
1000 000 000 000	tera	T	0.000 000 000 001	pico	p
			0.000 000 000 000 001	fento	f
			0.000 000 000 000 000 001	atto	a

*Should be restricted as much as possible.

1.12. WORK, POWER AND ENERGY

Work is said to be done by or against a force, when its point of application moves in or opposite to the direction of the force and is measured by the product of the force and the displacement of the point of application in the direction of force.

i.e. Work done, W = Force [F] × distance [d]

The SI or MKS unit of work is the joule, which is defined as the work done when a force of one newton acts through a distance of one metre in the direction of the force. Hence if a force F acts through distance d in its own direction

$$W = F \text{ [newtons]} \times d \text{ [metres]} = Fd \text{ joules}$$

Power is defined as the rate of doing work or the amount of work done in unit time.

The MKS or SI unit of power is the joule/second or watt. In practice, the watt is often found to be inconveniently small and so a bigger unit, the kilowatt is frequently used.

1 kilowatt = 1,000 watts

The bigger unit of power, most commonly used in engineering practice (not at all in SI system) is Horse power defined as below:

Metric Horse Power : It is the practical unit of power in MKS system (not in SI system) which according to ISI specifications is equal to 75 kgf-m of work done per second.

* RMKSA stands for rationalized metre, kilogram, second, ampere.

Solution :

$$\text{Output of motor} = 10 \text{ hp}$$

$$\text{Speed of motor, N} = 600 \text{ rpm}$$

$$\text{Torque developed in N-m, T} = \frac{\text{Output in hp} \times 735.5}{2\pi N/60} = \frac{10 \times 735.5}{2\pi \times \frac{600}{60}} = 117 \text{ N-m } \textbf{Ans.}$$

$\because$ 1 hp (metric) = 735.5 N-m/s

$$\text{Motor input} = \frac{\text{Motor output}}{\text{Efficiency}} = \frac{10 \times 735.5}{0.85} = 8{,}653 \text{ N-m/s or watts}$$

$$\text{Current drawn from 220 V main, I} = \frac{\text{Motor input}}{\text{V}} = \frac{8{,}653}{220} = 39.33 \text{ A } \textbf{Ans.}$$

$$\text{Units consumed per hour} = \frac{\text{Motor input in watts}}{1{,}000} = \frac{8{,}653}{1{,}000} = 8.653 \text{ kWh}$$

$$\text{Units consumed per day (6 hours)} = 8.653 \times 6 = 52 \text{ units } \textbf{Ans.}$$

$$\text{Units consumed per month} = 52 \times 30 = 1{,}560$$

$$\text{Monthly bill} = \text{Rs } 4.5 \times 1{,}560 = \text{Rs. } 7{,}020 \textbf{ Ans.}$$

Example 1.21. How much heat is produced by 2 kW electric heater when it is operated for 30 minutes ?

[Pb. Technical Univ. Basic Electrical and Electronics Engineering May 2005; May 2010]

Solution : Heat produced = Rating in kW × Time of operation

$$= 2 \text{ kW} \times \frac{1}{2} \text{ hour} = 1 \text{ kWh} = 36 \times 10^5 \text{ joules or } 860 \text{ k cal. } \textbf{Ans.}$$

Example 1.22. An immersion heater takes 1 hour to heat 50 kg of water from 20°C to boiling point. Obtain power rating of the heater. (Assume efficiency of heating equipment = 90%).

[Pb. Technical Univ. Basic Electrical and Electronics Engineering Dec. 2004]

Solution : Useful energy required to raise the temperature of water

$$= ms(t_2 - t_1) = 50\ (100 - 20) = 4{,}000 \text{ kcals}$$

$$\text{Energy input} = \frac{\text{Useful energy required}}{\text{Efficiency of heating equipment}} = \frac{4{,}000}{90/100} = 4{,}444.4 \text{ k cals}$$

$$\text{Energy input in kWh} = \frac{\text{Energy input in k calories}}{860} = \frac{4{,}444.4}{860} = 5.186 \text{ kWh}$$

$$\text{Power rating of heater, P} = \frac{\text{Energy input in kWh}}{\text{Time of operation}} = \frac{5.168}{1} \simeq 5.2 \text{ kW } \textbf{Ans.}$$

Example 1.23. A current was sent through a resistance wire of 10 ohms which is fully immersed in 2 kg of water. At the end of 15 minutes, the rise in temperature was observed to be 60 °C. Determine the value of current.

Solution :

$$\text{Water to be heated, } m = 2 \text{ kg}$$

$$\text{Rise in temperature, } (t_2 - t_1) = 60\ ^\circ\text{C}$$

$$\text{Specific heat of water, } s = 4{,}200 \text{ J/kg/}^\circ\text{C}$$

Energy utilized in raising the temperature of water,

$$= ms\ (t_2 - t_1) = 2 \times 4{,}200 \times 60 = 504{,}000 \text{ J}$$

$$\text{Energy input} = \frac{\text{Energy utilized}}{\text{Heating efficiency}} = \frac{504{,}000}{0.9} = 560{,}000$$

Assuming heating efficiency to be 90 percent

$$\text{Power drawn} = \frac{\text{Energy input in joules}}{\text{Time taken in seconds}} = \frac{560{,}000}{15 \times 60} = 622 \text{ watts}$$

$$\text{Current drawn, I} = \sqrt{\frac{P}{R}} = \sqrt{\frac{622}{10}} = 7.9 \text{ A } \textbf{Ans.} \qquad \because \text{ power developed, } P = I^2 R$$

Example 1.24. An electric kettle contains 1.5 kg of water at 15 °C. It takes 15 minutes to raise the temperature of water to 95 °C. Assuming the heat losses due to radiations and heating the kettle to be 15 k calories, find the current taken. The supply voltage is 100 V.

Solution : Quantity of water to be heated, m = 1.5 kg

Rise of temperature = $t_2 - t_1 = 95 - 15 = 80$ °C

Heat losses = 15 kilo calories

Useful energy required to raise the temperature of water

$$= ms\,(t_2 - t_1) = 1.5 \times 1.0 \times 80 = 120 \text{ k. calories}$$

$$\text{Energy input} = ms\,(t_2 - t_1) + \text{heat losses} = 120 + 15 = 135 \text{ k.calories} = 135 \times 4{,}200 \text{ J}$$

$$\text{Rating of kettle} = \frac{\text{Energy input in joules}}{\text{Operating time in seconds}} = \frac{135 \times 4{,}200}{15 \times 60} = 630 \text{ watts}$$

$$\text{Current drawn, I} = \frac{\text{Rating of kettle}}{\text{Supply voltage}} = \frac{630}{100} = 6.3 \text{ A } \textbf{Ans.}$$

Example 1.25. The following are the details of domestic electric supply for a month :

(*i*) 5 lamps of 60 watts each working for 6 hours per day.

(*ii*) 4 fluorescent tubes each of 40 watts working for 2.5 hours per day.

(*iii*) One 1.5 kW water heater working for 2 hours per day.

If the cost of each unit of energy is Rs. 4.00, what will be the cost of electrical energy in this month.

Solution : (*i*) Lamp load, P_L = Wattage of each lamp × Number of lamps

= 60 × 5 = 300 watts

Number of units consumed by lamp load, each day

$$= \frac{P_L \times \text{number of working hours per day}}{1{,}000} = \frac{300 \times 6}{1{,}000} = 1.8$$

(*ii*) Fluorescent tube Load, P_T = 40 × 4 = 160 watts

$$\text{Number of units consumed by tube load, each day} = \frac{160 \times 2.5}{1{,}000} = 0.4$$

(*iii*) Number of units consumed by water heater, each day = 1.5 × 2 = 3

Number of units consumed per day = 1.8 + 0.4 + 3 = 5.2

Units consumed in a month (30 days) = 5.2 × 30 = 156

Cost of electrical energy consumed in a month = Cost of each unit × units consumed per month

= Rs. 4.00 × 156 = Rs. 624.00 **Ans.**

1.17. CIRCUIT ELEMENTS

An *electric circuit* (or network) is an interconnection of physical electrical devices such as an *energy source* (or sources), an *energy convertor* or convertors (load or loads), and *conductors* that connect them.

An *energy source* (or source), such as a primary or secondary cell, a generator, and the like, is a device that converts chemical, mechanical, thermal or some other form of energy into electrical energy.

An *energy convertor,* also called the *load*, (such as lamp, heating appliance, or an electric motor) converts electrical energy into light, heat, mechanical work and so on.

Events in an electrical circuit may be defined in terms of emf (or voltage) and current.

When electrical energy is generated, transmitted and converted under conditions such that the currents and voltages involved remain constant with time, the electric circuit is identified as *direct current* (dc) *circuit.* If the currents and voltages do change with time, the circuit is defined as *alternating current* (ac) *circuit.*

A graphic representation of an electric circuit is called a *circuit diagram* (Fig. 1.5). Such a diagram consists of interconnected symbols called *circuit elements or circuit parameters.* Two elements are necessary to represent processes in a dc circuit. These are source of emf E_s and of internal (or source) resistance R_s and the load resistance (which includes the resistance of the conductors) R.

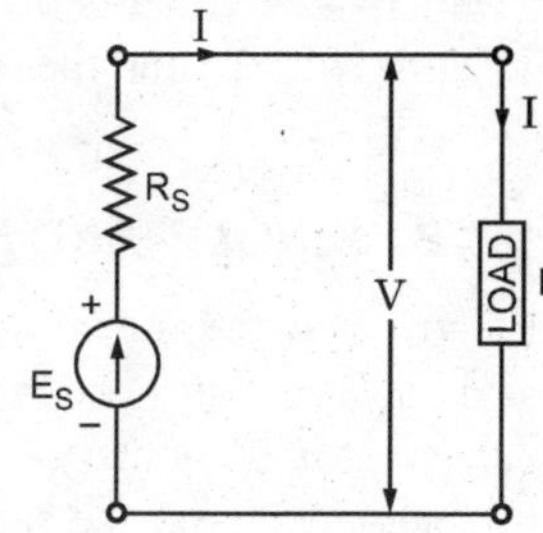

Fig. 1.5

In any electric circuit the energy convertor (or load) and the conductors connecting it to the source make up the *external circuit* in which current flows from the + ve side to the – ve side of the source whereas inside the source, current flows in the opposite direction, *i.e.* from the – ve side to the + ve side. The source emf is directed from the terminal at a lower potential to that at a higher one. In diagrams this is shown by arrows.

The *source emf* (or open-circuit voltage) is the voltage that appears across the source when no load is connected across it.

When a load is connected to the source terminals and the circuit is closed, an electric current starts flowing through the circuit. Now voltage across source terminals (called the *terminal voltage*) is not equal to source emf. It is due to voltage drop inside the source, *i.e.* across the source resistance.

$$\text{Voltage drop inside the source} = I\,R_s$$

The relationship between the current through a resistance and the voltage across the same resistance is called its *volt-ampere (or voltage-current) characteristic.* When represented graphically, voltages are laid off as abscissae and currents as ordinates.

There are two types of volt-ampere characteristics– straight line and nonlinear (curve), as shown in Figs. 1.6 (*a*) and 1.6 (*b*) respectively.

Resistive elements for which the volt-ampere characteristic is a straight line [Fig. 1.6 (*a*)] are called *linear*, and the electric circuits containing only linear resistances are called *linear circuits.*

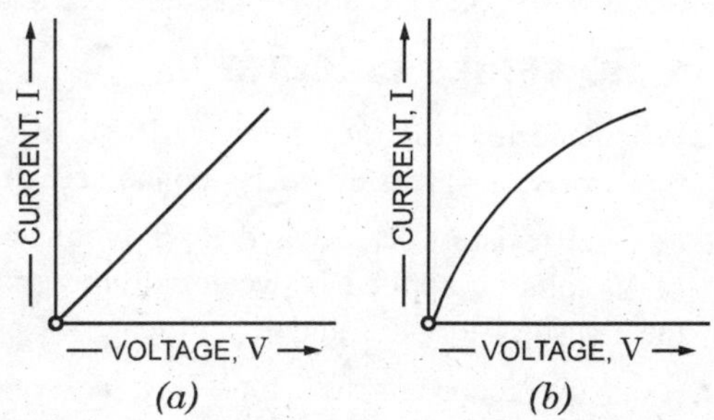

Fig. 1.6

Resistive elements for which the volt-ampere characteristic is other than a straight line are termed *nonlinear,* and so the electric circuits containing them are called *nonlinear circuits.* Examples of nonlinear elements are tungsten lamps, vacuum tubes and transistors, etc.

An electric circuit, whose characteristics or properties are same in either direction (e.g. a distribution or transmission line), is called the *bilateral circuit.* The distribution or transmission line can be made to perform its function equally well in either direction.

An electric circuit, whose characteristics or properties change with the direction of its operation (*e.g.* a diode rectifier), is called the *unilateral circuit.* A diode rectifier cannot perform rectification in both directions.

A network is said to be *passive* if it contains no source of emf in it. The *equivalent resistance* between any two terminals of a passive network is the ratio of potential difference across the two terminals to the current flowing into (or out of) the network. When a network contains one or more sources of emf and/or current, it is said to be *active*.

In case, a branch is removed from an electric network, the remainder of the network is left with a pair of terminals. The part of the network, which is considered with respect to the removed branch or terminal pair or port is termed as *one-port network.* When two branches are removed so that the network is left with four terminals or two pairs of terminals, the remainder network is called the *two-port network.* Usually one port accepts a source and the other port is coupled to a load, so that there is an input port and an output port in any two-port system.

1.18. SINGLE- AND MULTI-MESH NETWORK

Electric network may provide a single closed path (known as a *mesh or loop)* or several closed paths for the flow of current. An elementary *single-mesh* network in which all the elements carry the same current is shown in Fig. 1.5. An elementary multi-mesh network is shown in Fig. 1.7. It has 6 nodes, seven branches, three loops and two meshes. A *junction* (*or node**) is a point in a network where two or more branches meet. A *branch* is any section of a network which joins two nodes directly, that is, without passing through a third node. A *loop* is a closed path in a network formed by a number of connected branches. Mesh is a loop that contains no other loop within it.

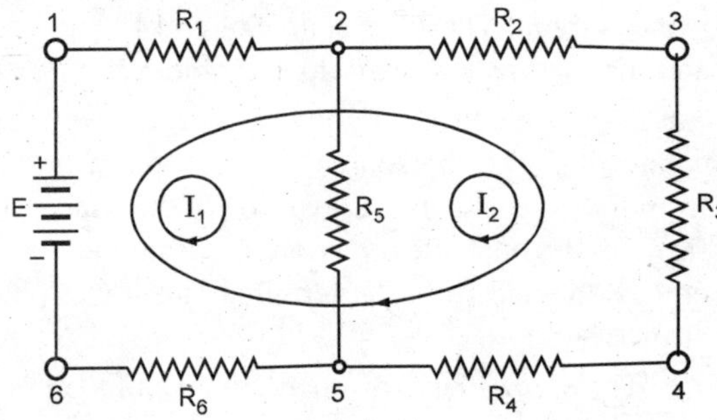

Fig. 1.7

If in the circuit diagram, as in Fig. 1.8 (*a*), there is a bold dot at the intersection of two branches, these branches are electrically connected and have a common node. Otherwise they simply cross, as in Fig. 1.8. (*b*), and are not connected electrically.

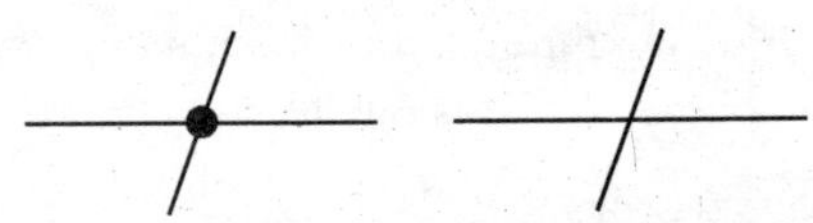

(*a*) (*b*)

Fig. 1.8

1.19. SERIES CIRCUITS

When the resistors are connected end to end, so that they form only one path for the flow of current, then resistors are said to be connected in series and such circuits are known as *series circuits.*

Let resistors R_1, R_2 and R_3 be connected in series as shown in Fig. 1.9, and the potential difference of V volts be applied between extreme ends A and D to cause flow of current of I amperes through all the resistors R_1 , R_2 and R_3.

Now according to Ohm's law

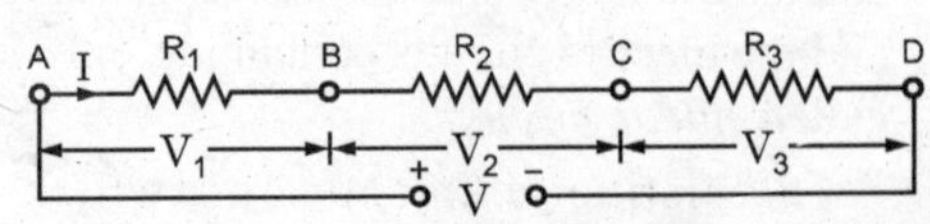

Fig. 1.9. ***Series Circuit***

Voltage drop across resistor R_1, $V_1 = I R_1$

Voltage drop across resistor R_2, $V_2 = I R_2$

Voltage drop across resistor R_3, $V_3 = I R_3$

Voltage drop across whole circuit, V = Voltage drop across resistor R_1 + voltage drop across resistor R_2 + voltage drop across resistor R_3

i.e. $V = I R_1 + I R_2 + I R_3 = I (R_1 + R_2 + R_3)$

$$\text{or} \quad \frac{V}{I} = R_1 + R_2 + R_3 \qquad ...(1.17)$$

and according to Ohm's law $\frac{V}{I}$ gives the whole circuit resistance, say R

* Nodes are considered to be resistanceless and infinitesimal.

$\therefore$ Effective resistance of the series circuit, $R = R_1 + R_2 + R_3$ ...(1.18)

Thus when a number of resistors are connected in series, the equivalent resistance is given by the arithmetic sum of their individual resistances.

$$i.e.\ R = R_1 + R_2 + R_3 + \dots R_n \qquad ...(1.19)$$

From the above discussions for a series circuit we conclude that

1. same current flows through all parts of the circuit,
2. applied voltage is equal to the sum of voltage drops across the different parts of the circuit,
3. different resistors have their individual voltage drops,
4. voltage drop across individual resistor is directly proportional to its resistance, current being the same in each resistor,
5. voltage drops are additive,
6. resistances are additive,
7. powers are additive.

Series circuits are common in electrical equipment. The tube filaments in small radios are usually in series. Current controlling devices are wired in series with the controlled equipment. Fuses are in series with the equipment they protect. A thermostat switch is in series with the heating element in an electric iron. Automatic house-heating equipment has a thermostat, electromagnet coils, and safety cut-outs in series with a voltage source. Rheostats are placed in series with the coils in large motors for motor current control.

1.20. PARALLEL CIRCUITS

When a number of resistors are connected in such a way that one end of each of them is joined to a common point and the other ends being joined to another common point, as shown in Fig. 1.10 then resistors are said to be connected in parallel and such circuits are known as *parallel circuits*. In these circuits current is divided into as many paths as the number of resistances.

Let the resistors R_1, R_2 and R_3 be connected in parallel, as shown in Fig. 1.10, and the potential difference of V volts be applied across the circuit.

Since potential difference across each resistor is same and equal to potential difference applied to the circuit *i.e.* V

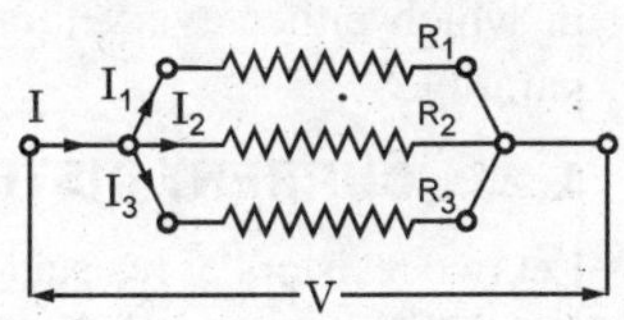

Fig. 1.10. Parallel Circuit

$\therefore$ According to Ohm's law

$$\text{Current in resistor } R_1,\ I_1 = \frac{V}{R_1} \qquad ....(1.20)$$

$$\text{Current in resistor } R_2,\ I_2 = \frac{V}{R_2} \qquad ....(1.21)$$

$$\text{Current is resistor } R_3,\ I_3 = \frac{V}{R_3} \qquad ...(1.22)$$

Adding Eqs. (1.20), (1.21) and (1.22) we have

$$I_1 + I_2 + I_3 = \frac{V}{R_1} + \frac{V}{R_2} + \frac{V}{R_3} = V\left(\frac{1}{R_1} + \frac{1}{R_2} + \frac{1}{R_3}\right) \qquad ...(1.23)$$

and since $I_1 + I_2 + I_3 = I$, the total current flowing through the circuit

$$\text{so } I = I_1 + I_2 + I_3 = V\left(\frac{1}{R_1} + \frac{1}{R_2} + \frac{1}{R_3}\right)$$

$$\text{or} \quad \frac{I}{V} = \frac{1}{R_1} + \frac{1}{R_2} + \frac{1}{R_3}$$

and since $\frac{I}{V} = \frac{1}{R}$ where R is the equivalent resistance of the whole circuit

$$\therefore \quad \frac{1}{R} = \frac{1}{R_1} + \frac{1}{R_2} + \frac{1}{R_3} \qquad ...(1.24)$$

Thus when a number of resistors are connected in parallel, the reciprocal of the equivalent resistance is given by the arithmetic sum of the reciprocals of their individual resistances.

In general if n resistors of resistances R_1, R_2, R_3, ... R_n are connected in parallel, then equivalent resistance R of the circuit is given by the Eq.

$$\frac{1}{R} = \frac{1}{R_1} + \frac{1}{R_2} + \frac{1}{R_3} + ... + \frac{1}{R_n} \qquad ...(1.25)$$

$$\text{Also} \quad G = G_1 + G_2 + G_3 +, ... G_n \qquad ...(1.26)$$

where $G = \frac{1}{R}$, $G_1 = \frac{1}{R_1}$, $G_2 = \frac{1}{R_2}$, $G_3 = \frac{1}{R_3}$ and so on.

From the above discussions for a parallel circuit we conclude that

1. same voltage acts across all parts of the circuit,
2. different resistors (or branches) have their individual currents,
3. total circuit current is equal to the sum of individual currents through the various resistors (or branches),
4. branch currents are additive,
5. conductances are additive,
6. powers are additive,
7. the reciprocal of the equivalent or combined resistance is equal to the sum of the reciprocals of the resistances of the individual branches.

Parallel circuits are very common in use. Various lamps and appliances in a house are connected in parallel, so that each one can be operated independently. A series circuit is an "all or none" circuit, in which either every thing operates or nothing operates. For individual control, devices are wired in parallel.

1.21. CURRENT DISTRIBUTION IN PARALLEL CIRCUITS

Let two resistors of resistances R_1 and R_2 be connected in parallel across a pd of V volts. According to Ohm's law

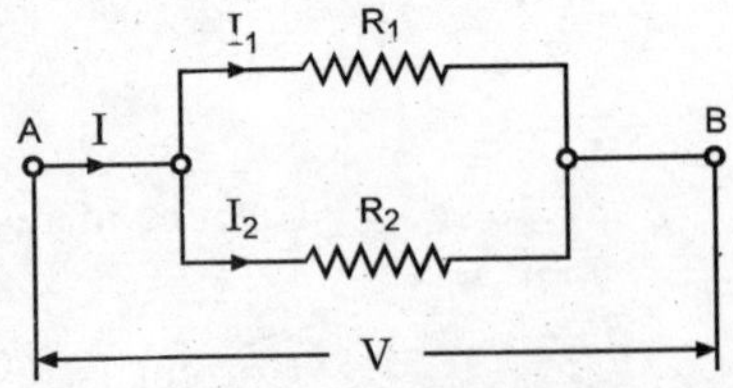

Fig. 1.11

$$\text{Current flowing through resistor } R_1, I_1 = \frac{V}{R_1} \qquad ...(1.27)$$

$$\text{Current flowing through resistor } R_2, I_2 = \frac{V}{R_2} \qquad ...(1.28)$$

Dividing Eq. (1.27) by Eq. (1.28) we have

$$\frac{I_1}{I_2} = \frac{V/R_1}{V/R_2} = \frac{R_2}{R_1} \qquad ...(1.29)$$

Hence current flowing through each resistor, when connected in parallel, is inversely proportional to their respective resistances.

Since conductance is reciprocal of resistance and if G_1 and G_2 are the respective conductances of resistors R_1 and R_2 then

$$\frac{1}{G_1} = R_1 \text{ and } \frac{1}{G_2} = R_2$$

Substituting the above values of R_1 and R_2 in Eq. (1.29) we have

$$\frac{I_1}{I_2} = \frac{1/G_2}{1/G_1} = \frac{G_1}{G_2} \quad ...(1.30)$$

Adding 1 on both sides of the above expression we have

$$\frac{I_1}{I_2} + 1 = \frac{G_1}{G_2} + 1$$

$$\text{or} \quad \frac{I_1 + I_2}{I_2} = \frac{G_1 + G_2}{G_2}$$

$$\text{or} \quad I_2 = \frac{I_1 + I_2}{G_1 + G_2} \times G_2$$

$$\text{Since } I_1 + I_2 = I \text{ and } G_1 + G_2 = \frac{1}{R_1} + \frac{1}{R_2} = \frac{1}{R} = G$$

$$\therefore \quad I_2 = I \times \frac{G_2}{G} \quad ...(1.31)$$

$$\text{and } I_1 = I \times \frac{G_1}{G} \quad ...(1.32)$$

Hence current in any branch of a parallel circuit is directly proportional to its respective conductance and is equal to the total current flowing through the circuit multiplied by the ratio of the conductance of the branch to that of the circuit.

The same relation holds good for parallel circuit consisting of more than two resistors and is very useful for its solution.

Let us consider a circuit consisting of resistances R_1, R_2, R_3 and R_4 ohms respectively connected in parallel across a potential difference of V volts, as shown in Fig. 1.12.

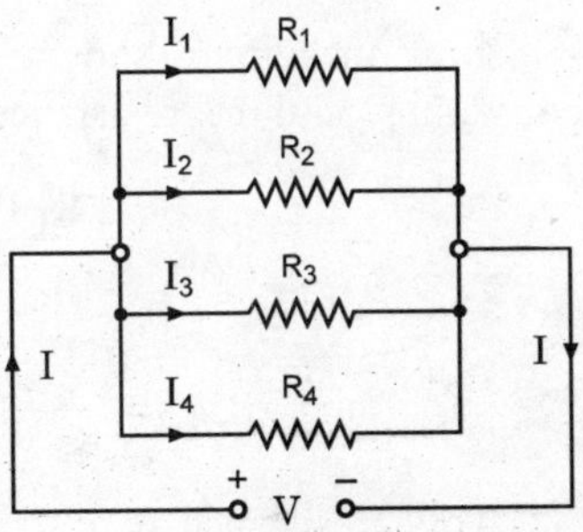

Fig. 1.12

$$\text{Conductance of first resistor, } G_1 = \frac{1}{R_1}$$

$$\text{Conductance of second resistor, } G_2 = \frac{1}{R_2}$$

$$\text{Conductance of third resistor, } G_3 = \frac{1}{R_3}$$

$$\text{Conductance of fourth resistor, } G_4 = \frac{1}{R_4}$$

Total conductance of the whole circuit is given by

$$G = \frac{1}{R} = \frac{1}{R_1} + \frac{1}{R_2} + \frac{1}{R_3} + \frac{1}{R_4} = G_1 + G_2 + G_3 + G_4$$

Total current flowing through the circuit,

$$I = \frac{V}{R} = V \times G$$

$$\text{or } V = \frac{I}{G} \quad ...(1.33)$$

$$\text{Current flowing through resistance } R_1, I_1 = \frac{V}{R_1} = \frac{I}{G} \times G_1 = \frac{G_1}{G} I$$

Similarly current flowing through resistance R_2, $I_2 = V \times G_2 = I \times \frac{G_2}{G}$

Current flowing through resistance R_3, $I_3 = V \times G_3 = I \times \frac{G_3}{G}$

and current flowing through resistance R_4, $I_4 = V \times G_4 = I \times \frac{G_4}{G}$.

1.22. SERIES-PARALLEL CIRCUITS

So far, only simple series and simple parallel circuits have been considered. Practical electric circuits very often consist of combinations of series and parallel resistances. Such circuits may be solved by the proper application of Ohm's law and the rules for series and parallel circuits to the various parts of the complex circuit. There is no definite procedure to be followed in solving complex circuits, the solution depends on the known facts concerning the circuit and the quantities which one desires to find. One simple rule may usually be followed, however, reduce the parallel branches to an equivalent series branch and then solve the circuit as a simple series circuit.

For example consider a series-parallel circuit shown in Fig. 1.13 for solution.

First of all equivalent resistances of all parallel branches are determined separately, *e.g.* of branches AB and CD by the laws of parallel circuits discussed in Art. 1.20.

Equivalent resistance of parallel branch AB, $R_{AB} = \frac{1}{\frac{1}{R_1} + \frac{1}{R_2}} = \frac{R_1 R_2}{R_1 + R_2}$

and equivalent resistance of parallel branch CD, $R_{CD} = \frac{1}{\frac{1}{R_4} + \frac{1}{R_5} + \frac{1}{R_6}}$

$$= \frac{R_4 R_5 R_6}{R_5 R_6 + R_4 R_6 + R_4 R_5}$$

Now the circuit shown in Fig. 1.13 gets reduced to a simple series circuit shown in Fig. 1.14 consisting of three resistances,

$$R_{AB} = \frac{R_1 R_2}{R_1 + R_2}, \quad R_{BC} = R_3$$

and $R_{CD} = \frac{R_4 R_5 R_6}{R_5 R_6 + R_4 R_6 + R_4 R_5}$

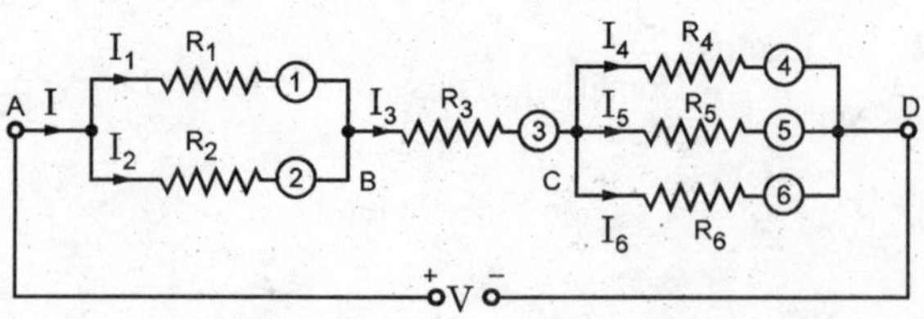

Fig. 1.13

Total resistance of circuit,

$$R_T = R_{AB} + R_{BC} + R_{CD}$$

Now circuit current may be determined from the relation

$$I = \frac{V}{R_T}$$

After knowing I, potential differences across branches AB, BC and CD are determined from the relations

PD across branch AB, $V_{AB} = I.R_{AB} = \frac{V}{R_T} R_{AB}$

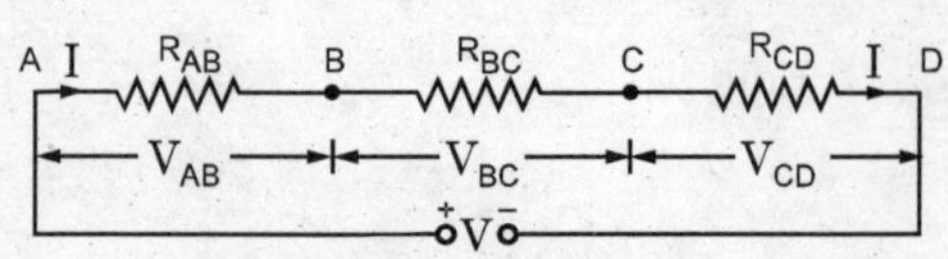

Fig. 1.14

PD across branch BC, $V_{BC} = I.R_{BC} = \frac{V}{R_T} R_{BC}$

and PD across branch CD, $V_{CD} = IR_{CD} = \frac{V}{R_T} R_{CD}$.

After determination of potential difference across each parallel branch, the currents in the various resistances are determined from the relations

$$\text{Current in resistance } R_1 = I_1 = \frac{V_{AB}}{R_1}$$

$$\text{Current in resistance } R_2 = I_2 = \frac{V_{AB}}{R_2}$$

$$\text{Current in resistance } R_3 = I_3 = I$$

$$\text{Current in resistance } R_4 = I_4 = \frac{V_{CD}}{R_4}$$

$$\text{Current in resistance } R_5 = I_5 = \frac{V_{CD}}{R_5}$$

$$\text{Current in resistance } R_6 = I_6 = \frac{V_{CD}}{R_6}$$

Thus equivalent resistance of the whole circuit, voltage drop across each branch and currents in the various resistors may be determined.

Example 1.26. When a resistor is placed across 230 V supply (dc) the current is 12 A. What is the value of resistor that must be connected in parallel to increase the load current to 16 A ?

[Pb. Technical Univ. Basic Electrical and Electronics Engineering Dec. 2005]

Solution : Resistance of the given resistor, $R_1 = \frac{V}{I} = \frac{230}{12} = 19.167\ \Omega$

Equivalent resistance of parallel combination.

$$R_P = \frac{V}{I_P} = \frac{230}{16} = 14.375\ \Omega$$

Let the value of the resistor connected in parallel be R_2 Ω then

$$R_P = \frac{R_1 R_2}{R_1 + R_2}$$

$$\text{or } R_P(R_1 + R_2) = R_1 R_2$$

$$\text{or } R_2 = \frac{R_P R_1}{R_1 - R_P} = \frac{14.375 \times 19.167}{19.167 - 14.375} = 57.5\ \Omega \textbf{ Ans.}$$

Example 1.27. A wire of 100 Ω resistance is cut into how many pieces so that when they are connected in parallel resultant is 1 ohm. *[Pb. Technical Univ. Basic Electrical and Electronics Engineering May 2008]*

Solution : Let the wire be cut into n pieces of equal length.

Now the resistance of each piece of wire, $R = \frac{100}{n}\ \Omega$ $\quad \because R \alpha l$

Equivalent resistance when the pieces are connected in parallel,

$$R_{eq} = \frac{\text{Resistance of each path}}{\text{Number of parallel paths}} = \frac{100/n}{n} = \frac{100}{n^2}.$$

Substituting $R_{eq} = 1\ \Omega$, we have

$$1 = \frac{100}{n^2} \quad \text{or} \quad n^2 = 100 \quad \text{or} \quad n = \sqrt{100} = 10 \textbf{ Ans.}$$

Example 1.28. Two resistances of 20 Ω and 30 Ω respectively are connected in parallel. These two parallel resistances are further connected in series with a resistance of 15 Ω. If the current through the 15 Ω resistance is 3 A find (*a*) the currents through the 20 Ω and 30 Ω resistances respectively (*b*) the voltage across the whole circuit (*c*) the total power consumed.

Solution : Equivalent resistance of branch AB, $R_{AB} = \frac{1}{\frac{1}{20}+\frac{1}{30}} = 12\ \Omega$

Effective resistance of the circuit, $R_{eff} = R_{AB} + R_{BC} = 12 + 15 = 27\ \Omega$

Circuit current, I = Current through 15 Ω resistance = 3 A

(*b*) Voltage across the whole circuit, $V = I R_{eff} = 3 \times 27 = 81$ V **Ans.**

(*c*) Total power consumed, $P = VI = 81 \times 3 = 243$ watts **Ans.**

(*a*) Voltage drop across branch AB, $V_{AB} = I R_{AB} = 3 \times 12 = 36$ V

Current through 20 Ω resistance, $I_1 = \frac{V_{AB}}{R_1} = \frac{36}{20} = 1.8$ A **Ans.**

Current through 30 Ω resistance, $I_2 = \frac{V_{AB}}{R_2} = \frac{36}{30} = 1.2$ A **Ans.**

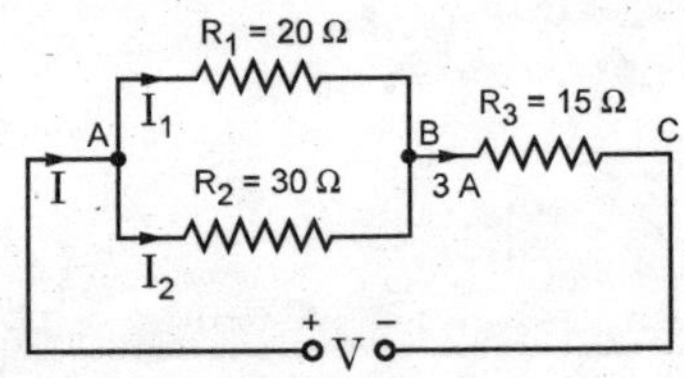

Fig. 1.15

Example 1.29. A dc circuit comprises two resistors; resistor A of value 25 Ω and resistor B of unknown value connected in parallel, together with a third resistor C of value of 5 Ω connected in series with the parallel branch. Find the voltage to be applied across the whole circuit and the value of resistor B if the pd across C is 90 V and the total power consumed is 4,320 W.

Solution : The circuit diagram is shown in Fig. 1.16.

Current flowing through the circuit,

I = Current flowing through resistor C

$$= \frac{\text{Voltage drop across resistor C}}{\text{Value of resistor C in ohms}} = \frac{90}{5} = 18 \text{ A}$$

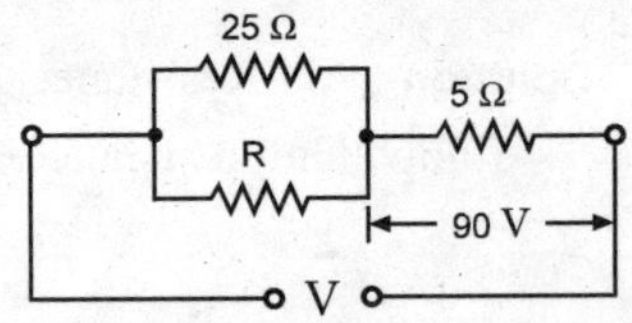

Fig. 1.16

Voltage applied across the whole circuit

$$= \frac{\text{Total power consumed}}{\text{Current through the circuit}} = \frac{4{,}320}{18} = 240 \text{ V } \textbf{Ans.}$$

Voltage drop across parallel combination = V – 90 = 240 – 90 = 150 V

Current through resistor of 25 Ω = $\frac{150}{25}$ = 6 A

Current through unknown resistor, R = 18 – 6 = 12 A

Unknown resistor, R = $\frac{150}{12}$ = 12.5 Ω **Ans.**

1.23. LAWS GOVERNING DISTRIBUTION OF ELECTRIC CURRENTS IN VARIOUS BRANCHES OF A NETWORK

The basic laws, that electric circuits follow rationally from the nature of the electrical quantities, lead directly to methods for the systematic analysis of electric circuits. These laws are known as *Kirchhoff's laws*, and they describe the relationships among circuit voltages and circuit currents that must be satisfied. These laws are very helpful in determining the equivalent resistance (or impedance) of a complex network and the current flowing in the various branches of the network.

Gustov Robert Kirchhoff derived two basic laws governing networks, one commonly known as *first law* or *current law* (KCL) or *point law,* whereas the second is called *second law* or *voltage law* (KVL) or *mesh law.*

1. Kirchhoff's First Law or Current Law (KCL) or Point Law. According to this law in any network of wires carrying currents, the algebraic sum of all currents meeting at a point (or junction) is zero or the sum of incoming currents towards any point is equal to the sum of outgoing currents away from that point. If I_1, I_2, I_3, I_4, I_5, and I_6 are the currents meeting at junction O, flowing in the directions of arrow heads marked on them (Fig. 1.17), taking incoming currents as positive and outgoing currents as negative, according to Kirchhoff's first law (KCL)

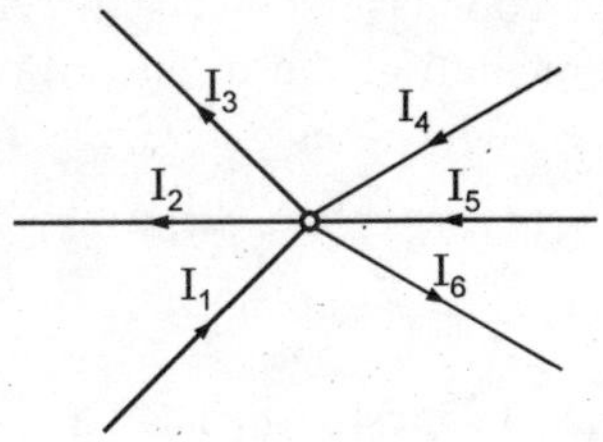

Fig. 1.17

$$I_1 - I_2 - I_3 + I_4 + I_5 - I_6 = 0$$
$$\text{or} \quad I_1 + I_4 + I_5 = I_2 + I_3 + I_6$$

2. Kirchhoff's Second Law or Voltage Law (KVL) or Mesh Law. According to this law in any closed circuit or mesh the algebraic sum of emfs acting in that circuit or mesh is equal to the algebraic sum of the products of the currents and resistances of each part of the circuit.

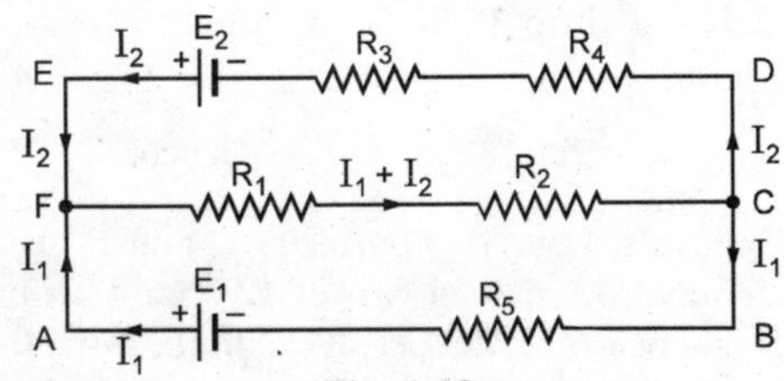

Fig. 1.18

If the circuit shown in Fig. 1.18 is considered, then according to Kirchhoff's second law (or KVL)

In mesh AFCBA, $E_1 = R_1 (I_1 + I_2) + R_2 (I_1 + I_2) + R_5 I_1$

or $E_1 = (R_1 + R_2) (I_1 + I_2) + R_5 I_1$

In mesh FEDCF, $-E_2 = -R_3 I_2 - R_4 I_2 - R_2 (I_1 + I_2) - R_1 (I_1 + I_2)$

or $E_2 = R_3 I_2 + R_4 I_2 + (R_1 + R_2) (I_1 + I_2)$

and in mesh AFEDCBA, $E_1 - E_2 = -R_3 I_2 - R_4 I_2 + R_5 I_1$

or $E_1 - E_2 = R_5 I_1 - (R_3 + R_4) I_2$

1.23.1. Application of Kirchhoff's Laws To Circuits

First of all the current distribution in various branches of the circuit is made with directions of their flow complying with first law of Kirchhoff. Then Kirchhoff's second law is applied to each mesh (one by one) separately and algebraic equations are obtained by equating the algebraic sum of emfs acting in a mesh equal to the algebraic sum of respective drops in the same mesh. By solving the equations so obtained unknown quantities can be determined. While applying Kirchhoff's second law, the question of algebraic signs may be troublesome and is a frequent source of error. If, however, the following rules are kept in mind, no difficulty should occur.

The resistive drops in a mesh due to current flowing in clockwise direction must be taken positive drops.

The resistive drops in a mesh due to current flowing in counter-clockwise direction must be taken as negative drops.

Similarly the battery emf causing current to flow in clockwise direction in a mesh must be taken as positive emf and the battery emf causing current to flow in counter-clockwise direction in a mesh must be taken as negative emf.

For example for the circuit shown in Fig. 1.18 let the current distribution be made as shown, which satisfy Kirchhoff's first law fully.

Taking first, mesh AFCBA for the application of Kirchhoff's second law, we see that there is only one emf acting in the mesh (E_1) and since it tries to send current in clockwise direction, E_1 be taken as positive, similarly all the resistive drops *i.e.* $R_1(I_1+I_2)$, $R_2(I_1+I_2)$ and $R_5 I_1$ are clockwise, these must be taken as positive.

∴ According to Kirchhoff 's second law in mesh AFCBA

$$E_1 = R_1(I_1+I_2) + R_2(I_1+I_2) + R_5 I_1$$

In mesh FEDCF, there is only one emf acting in the mesh (E_2) and since it tries to send current in counter-clockwise direction through the mesh under consideration, it may be taken as negative. Since all of the resistive drops $R_3 I_2$, $R_4 I_2$, $R_2(I_1+I_2)$ and $R_1(I_1+I_2)$ are counter-clockwise, these may be taken as negative.

Hence according to Kirchhoff 's second law in mesh FEDCF we get,

$$-E_2 = -R_3 I_2 - R_4 I_2 - R_2(I_1+I_2) - R_1(I_1+I_2)$$

$$\text{or } E_2 = (R_3+R_4) I_2 + (R_1+R_2)(I_1+I_2)$$

In mesh AFEDCBA, emf E_1 tries to cause current in clockwise direction so be taken as positive and emf E_2 tries to cause current in counter-clockwise direction so be taken as negative. Similarly resistive drop $R_5 I_1$ being clockwise be taken as positive and resistive drops $R_3 I_2$ and $R_4 I_2$ being counter-clockwise be taken as negative.

$$\text{Hence } E_1 - E_2 = -R_3 I_2 - R_4 I_2 + R_5 I_1 = R_5 I_1 - (R_3+R_4) I_2$$

Example 1.30. Two batteries A and B are connected in parallel and a load of 10 Ω is connected across their terminals. A has an emf of 12 V and an internal resistance of 2 Ω and B has an emf of 8 V and an internal resistance of 1 Ω. Use Kirchhoffs laws to determine the magnitude of currents and also the directions in each of the batteries. Also determine the pd across external resistance.

Solution : Let the current distribution in the circuit be as shown in Fig. 1.19.

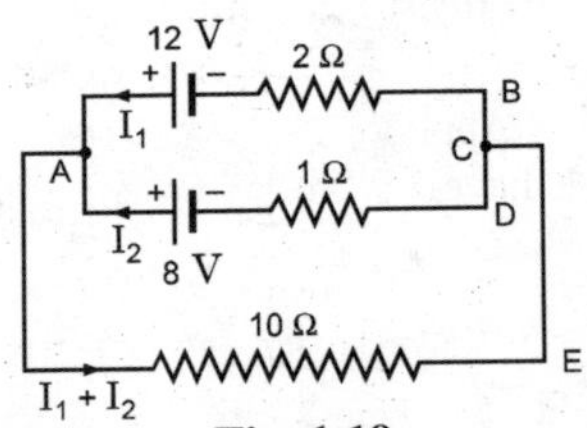

Fig. 1.19

Applying Kirchhoff's voltage law (KVL) to meshes ABCDA and ADCEA we have

$$2 I_1 - I_2 = 12 - 8$$

$$\text{and } I_2 + 10(I_1 + I_2) = 8$$

$$\text{or } 2 I_1 - I_2 = 4 \quad ...(i)$$

$$\text{and } 10 I_1 + 11 I_2 = 8 \quad ...(ii)$$

Solving equations (*i*) and (*ii*) we have

$I_1 = 1.625$ A **Ans.** and $I_2 = -0.75$ A **Ans.**

The negative sign with current I_2 means that the current I_2 is flowing into the 8-V battery and not out of it.

Current flowing through the load resistance of 10 Ω = $I_1 + I_2$ = 1.625 + (– 0.75) = 0.875 A

PD across load resistance = $(I_1 + I_2)$ R = 0.875 × 10 = 8.75 V **Ans.**

Example 1.31. Find the value of R and current through each branch if current in branch AO is zero.

[*Pb. Technical Univ. Basic Electrical and Electronics Engineering Dec. 2007*]

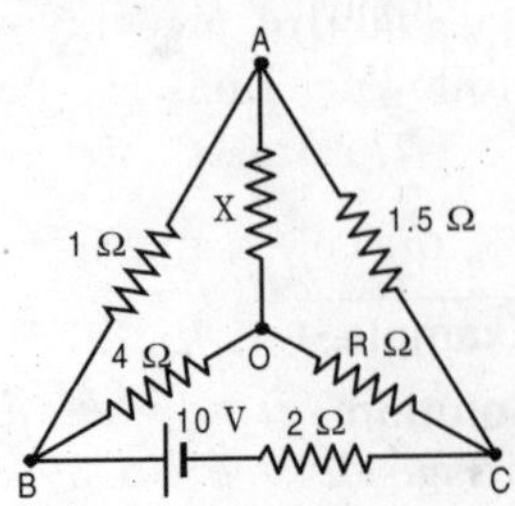

Fig. 1.20

Solution : Let the current through branches BA and BO of the circuit be I_1 and I_2 amperes respectively. Applying KCL to junctions B, A and O, we have

Current supplied by battery or current in branch CB = $I_1 + I_2$

Current through branch AC = I_1 ∵ Current through branch AO is zero

Current through branch OC = I_2

Now applying Kirchhoff 's voltage law to meshes ABOA, CAOC and BCOB respectively we have

$$-I_1 + 4I_2 = 0$$

$$\text{or } I_1 = 4I_2 \quad ...(i)$$

$$1.5I_1 - RI_2 = 0$$

$$\text{or } R = \frac{1.5I_1}{I_2} = 1.5 \times 4 = 6\ \Omega \ \textbf{Ans.}$$

$$\text{and } 4I_2 + RI_2 + 2(I_1 + I_2) = 10$$
$$\text{or } 4I_2 + 6I_2 + 2(4I_2 + I_2) = 10$$
$$20I_2 = 10$$
$$\text{or } I_2 = 0.5 \text{ A}$$

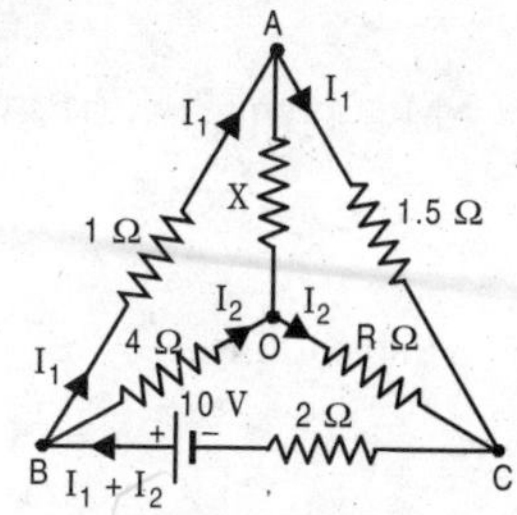

Fig. 1.21

Now Current in branch BA = $I_1 = 4I_2 = 4 \times 0.5 = 2$ A **Ans.**

Current in branch AC = $I_1 = 2$ A **Ans.**

Current in branch BO = $I_2 = 0.5$ A **Ans.**

Current in branch OC = $I_2 = 0.5$ A **Ans.**

Current in branch CB = $I_1 + I_2 = 2 + 0.5 = 2.5$ A **Ans.**

Example 1.32. In Fig. 1.22 the potential of point A = – 30 V. Using Kirchhoff 's laws find (*a*) value of V and (*b*) power dissipated by 5 Ω resistance.

Solution : The potential of point A is –30 V and potential of point G is zero, being grounded. So the *pd* across 12 Ω resistor connected between terminals A and H is 30 V and, therefore, current flowing through 12 Ω resistor is $\frac{30}{12}$ *i.e.* 2.5 A from terminal H to A.

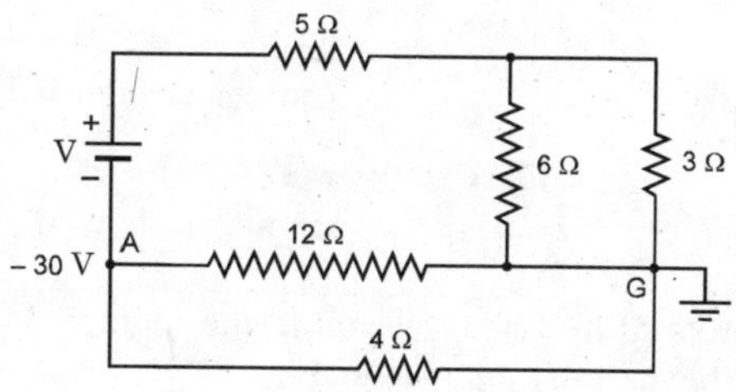

Fig. 1.22

Let the current distribution in the other branches of the circuit be, as shown in Fig. 1.23. Applying Kirchhoff 's current law (KCL) to node H we have

$$I_2 + I_3 = 2.5 \quad ...(i)$$

Applying Kirchhoff's voltage law to meshes CHGDC and AFEGHA we have

$$6\ I_2 - 3\ (I_1 - I_2) = 0$$

$$\text{or } I_2 = \frac{I_1}{3} \quad ...(ii)$$

$$\text{and } 4\ (I_1 - I_2 - I_3) - 12 \times 2.5 = 0$$
$$\text{or } I_1 - (I_2 + I_3) = 7.5$$

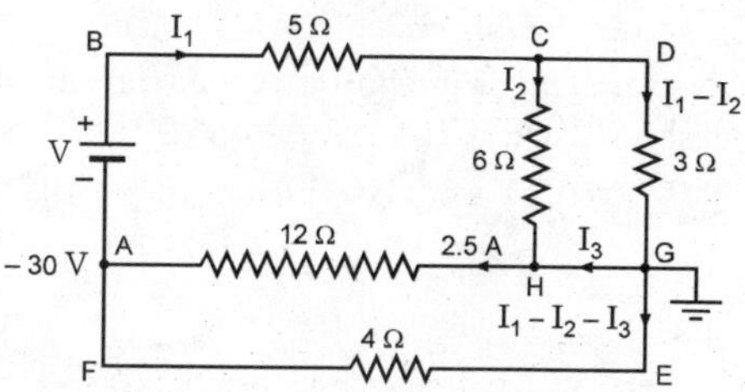

Fig. 1.23

Substituting $I_2 + I_3 = 2.5$ from Eq. (*i*) in the above equation we have

$$I_1 - 2.5 = 7.5$$
$$\text{or } I_1 = 10 \text{ A}$$

$$\text{and } I_2 = \frac{I_1}{3} = \frac{10}{3} \text{ A}$$

Applying KVL to mesh ABCHA we have

$$5\ I_1 + 6\ I_2 + 12 \times 2.5 = V$$

$$(a) \text{ or } V = 5\ I_1 + 6\ I_2 = 5 \times 10 + 6 \times \frac{10}{3} - 30 = 40 \text{ V } \textbf{Ans.}$$

(*b*) Power dissipated in 5-Ω resistor = $I_1^2 \times 5 = 10^2 \times 5 = 500$ W **Ans.**

Example 1.33. Find the current distribution in the network shown in Fig. 1.24.

Solution : Let the current distribution in the network be as shown in Fig. 1.24.

Applying Kirchhoff's second law to meshes OBAO and BCAB we have

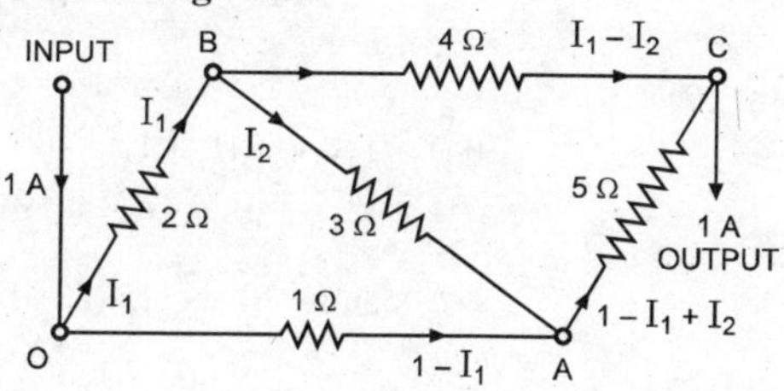

Fig. 1.24

$$2\ I_1 + 3\ I_2 - (1 - I_1) \times 1 = 0$$

$$\text{and } 4\ (I_1 - I_2) - 5\ (1 - I_1 + I_2) - 3\ I_2 = 0$$

$$\text{or } 3\ I_1 + 3\ I_2 = 1 \quad ...(i)$$

$$\text{and} \quad 9 I_1 - 12 I_2 = 5 \qquad ...(ii)$$

Multiplying Eq. (*i*) by 3 and subtracting it from Eq. (*ii*) we have

$$-21 I_2 = 2 \quad \text{or} \quad I_2 = \frac{-2}{21}$$

Substituting $I_2 = \frac{-2}{21}$ in Eq. (*i*) we have

$$I_1 = \frac{1}{3} - I_2 = \frac{1}{3} - \left(\frac{-2}{21}\right) = \frac{9}{21} \text{ or } \frac{3}{7} \text{ A}$$

So current in branch OB $= I_1 = \frac{3}{7}$ A **Ans.**

Current in branch OA $= 1 - I_1 = 1 - \frac{3}{7} = \frac{4}{7}$ A **Ans.**

Current in branch BA $= I_2 = \frac{-2}{21}$ A *i.e.* $\frac{2}{21}$ A from point A to point B **Ans.**

Current in branch BC $= I_1 - I_2 = \frac{3}{7} - \left(\frac{-2}{21}\right) = \frac{11}{21}$ A **Ans.**

Current in branch AC $= 1 - I_1 + I_2 = 1 - \frac{3}{7} + \left(\frac{-2}{21}\right) = \frac{10}{21}$ A **Ans.**

Example 1.34. Determine the voltage V_B and V_C in the network shown in Fig. 1.25.

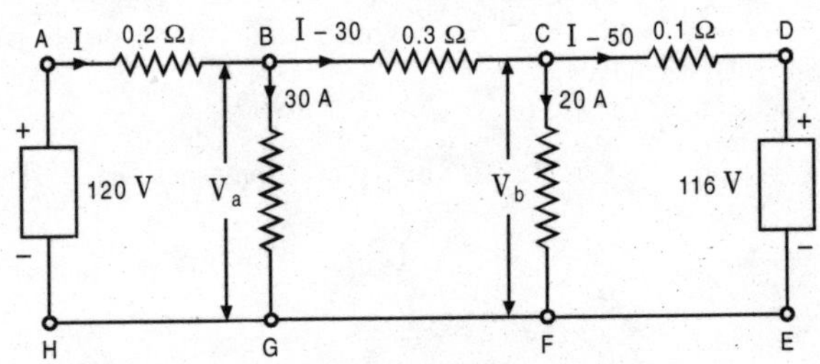

Fig. 1.25

Solution : Let the current distribution in the network be as shown in Fig. 1.25.

Applying Kirchhoff's second law to the outermost mesh ABCDEFGHA, we have,

$$0.2\,I + 0.3\,(I - 30) + 0.1\,(I - 50) = 120 - 116$$

$$\text{or} \quad 0.6\,I = 18$$

$$\text{or} \quad I = 30 \text{ A}$$

Voltage, $V_B = V_A - 0.2I_1 = 120 - 0.2\,I = 120 - 0.2 \times 30 = 114$ V **Ans.**

Voltage, $V_C = V_D + 0.1(I_1 - 50) = 116 + 0.1\,(I - 50) = 116 + 0.1\,(30 - 50) = 114$ V **Ans.**

Example 1.35. A bridge network ABCD is arranged as follows :

Resistances between terminals A-B, B-C, C-D, D-A and B-D are 10, 30, 15, 20 and 40 Ω respectively. A 2-V battery of negligible internal resistance is connected between terminals A and C. Determine the value and direction of the current in 40-Ω resistor.

Solution : Assume current distribution in the network as shown in Fig. 1.26.

Applying Kirchhoff's second law to meshes ABDA, BCDB and ADCA, we have

$$10\,I_1 + 40\,I_3 - 20\,I_2 = 0$$

$$\text{or} \quad I_1 + 4\,I_3 - 2\,I_2 = 0 \qquad ...(i)$$

$$30\,(I_1 - I_3) - 15\,(I_2 + I_3) - 40\,I_3 = 0$$

$$\text{or} \quad 6\,I_1 - 17\,I_3 - 3\,I_2 = 0 \qquad ...(ii)$$

$$\text{and} \quad 20\,I_2 + 15\,(I_2 + I_3) = 2$$

$$\text{or} \quad 7\,I_2 + 3\,I_3 = 0.4 \qquad ...(iii)$$

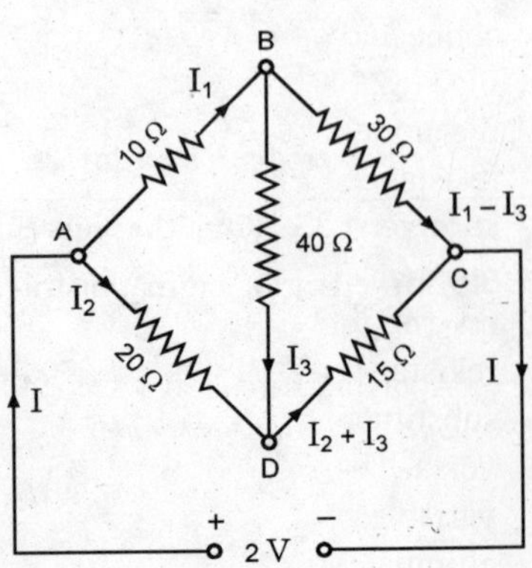

Fig. 1.26

Multiplying Eq. (*i*) by 6 and subtracting it from Eq. (*ii*) we have

$$41\,I_3 - 9\,I_2 = 0 \quad \text{or} \quad I_2 = \frac{41}{9}\,I_3$$

Substituting $I_2 = \frac{41}{9} I_3$ in Eq. (*iii*) we have

$$7 \times \frac{41}{9} I_3 + 3\,I_3 = 0.4 \quad \text{or} \quad I_3 = \frac{0.4 \times 9}{314}\ \text{A} = 11.465\ \text{m A}$$

i.e. Current flowing through a resistor of 40-Ω resistor is 11.465 mA from point B to point D. **Ans.**

Example 1.36. Find current I in the following circuit using KVL. (Fig. 1.27)

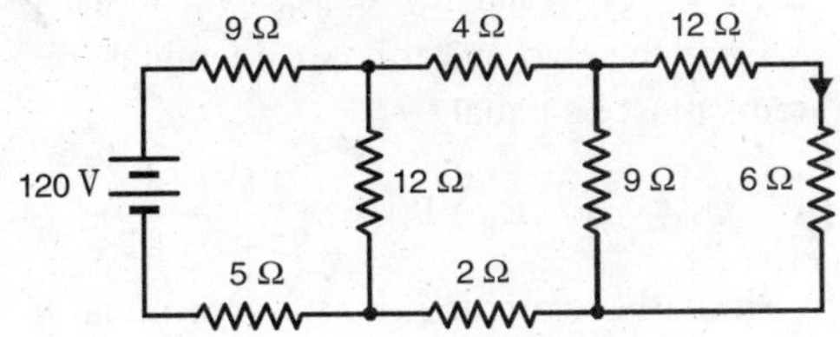

Fig. 1.27

Solution : The given network is redrawn as illustrated in Fig. 1.28. There are three independent loops. The loop currents have been taken clockwise as marked in Fig. 1.28. The individual branch currents along with their direction of flow are also shown in the circuit diagram. Applying KVL to meshes I, II and III respectively, we have

$$9I_1 + 12(I_1 - I_2) + 5I_1 = 120$$

$$\text{or} \quad 13I_1 - 6I_2 = 60 \qquad ...(i)$$

$$4I_2 + 9(I_2 - I) + 2I_2 + 12(I_2 - I_1) = 0$$

$$\text{or} \quad 9I_2 - 4I_1 - 3I = 0 \qquad ...(ii)$$

$$12I + 6I - 9(I_2 - I) = 0$$

$$\text{or} \quad I_2 = 3I \qquad ...(iii)$$

Substituting $I_2 = 3I$ in Eq. (*ii*), we have

$$9 \times 3I - 4I_1 - 3I = 0$$

$$\text{or} \quad I_1 = 6I \qquad ...(iv)$$

Substituting $I_1 = 6I$ and $I_2 = 3I$ in Eq. (*i*), we have

$$13 \times 6I - 6 \times 3I = 60$$

or $I = 1$ A **Ans.**

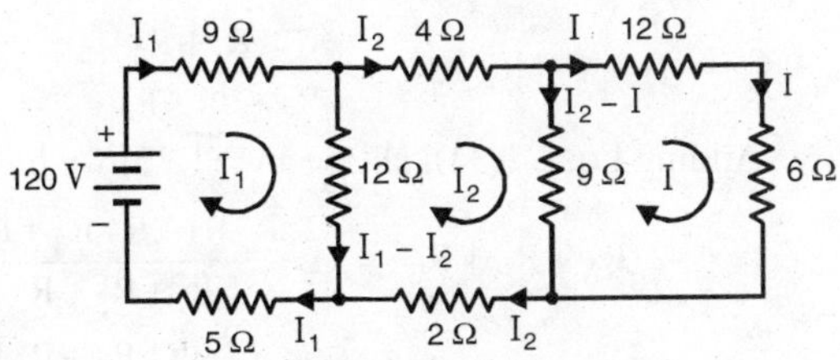

Fig. 1.28

1.24. NETWORK REDUCTION BY DELTA-STAR TRANSFORMATION OR VICE VERSA

Three resistances connected nose-to-tail as shown in Fig. 1.29 (*a*), are said to be *delta-* (or Δ) or *mesh-connected* (they form a mesh). Three resistances connected together at a common point O, as shown in Fig. 1.29 (*b*), are said to *be star* – (or Y) *connected.* If the nodes (A, B, and C) to which the two sets of resistances are connected are part of a larger network, it is possible to assign values to the two sets of resistances so that they have exactly the same effect on the network. If, therefore, delta-connected resistances are part of a network it is possible to substitute them by the star-connected ones and vice versa. The obvious advantages are that a delta-star transformation eliminates a mesh which reduces by one the variables and equations necessary to solve a network by mesh analysis whereas star-delta transformation eliminates a node (node O) which reduces by one the variables and equations necessary to solve a network by node analysis.

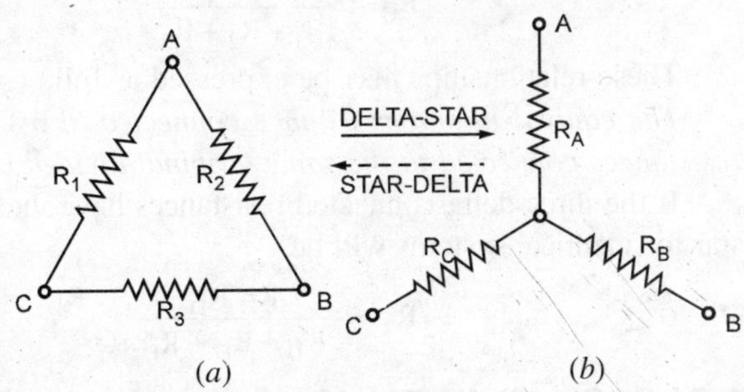

Fig. 1.29. ***Delta-Star Transformation***

1.24.1. Delta-Star Transformation

The replacement of delta or mesh by equivalent star system is known as delta-star transformation.

The two systems will be equivalent or identical if the resistances measured between any pair of lines is same in both of the systems, when the third line is open.

Hence resistances between terminals B and C,

$$R_{BC} = R_3 \parallel (R_1 + R_2) = \frac{R_3 (R_1 + R_2)}{R_1 + R_2 + R_3} \quad \text{in delta system}$$

$$\text{and } R_{BC} = R_B + R_C \quad \text{in star system}$$

Since the two systems are identical, resistances measured between terminals B and C in both of the systems must be equal.

$$\text{So } R_B + R_C = \frac{R_3 (R_1 + R_2)}{R_1 + R_2 + R_3} \qquad \text{...(1.34)}$$

Similarly resistances between terminals C and A being equal in the two systems

$$R_C + R_A = \frac{R_1 (R_2 + R_3)}{R_1 + R_2 + R_3} \qquad \text{...(1.35)}$$

And resistance between terminals A and B

$$R_A + R_B = \frac{R_2 (R_1 + R_3)}{R_1 + R_2 + R_3} \qquad \text{...(1.36)}$$

Adding Eqs. (1.34), (1.35) and (1.36) we have

$$2 (R_A + R_B + R_C) = \frac{2 (R_1 R_3 + R_2 R_3 + R_1 R_2)}{R_1 + R_2 + R_3}$$

$$\text{or } R_A + R_B + R_C = \frac{R_1 R_3 + R_2 R_3 + R_1 R_2}{R_1 + R_2 + R_3} \qquad \text{...(1.37)}$$

Subtracting Eqs. (1.34), (1.35) and (1.36) from Eq. (1.37) we have respectively

$$R_A = \frac{R_1 R_2}{R_1 + R_2 + R_3} \qquad \text{...(1.38)}$$

$$R_B = \frac{R_2 R_3}{R_1 + R_2 + R_3} \qquad \text{...(1.39)}$$

$$R_C = \frac{R_3 R_1}{R_1 + R_2 + R_3} \qquad \text{...(1.40)}$$

These relationships may be expressed as follows:

The equivalent star resistance connected to a given terminal is equal to the product of the two delta resistances connected to the same terminal divided by the sum of the delta connected resistances.

If the three delta-connected resistances have the same value R_D, the three resistances in the equivalent star for identical systems will be

$$R_S = \frac{R_D R_D}{R_D + R_D + R_D} = \frac{R_D}{3} \qquad \text{...(1.41)}$$

1.24.2. Star-Delta Transformation

Multiplying Eqs. (1.38) and (1.39), (1.39) and (1.40) and (1.40) and (1.38) and then adding them we get

$$R_A R_B + R_B R_C + R_C R_A$$

$$= \frac{R_1 R_2^2 R_3 + R_1 R_3^2 R_2 + R_1^2 R_2 R_3}{(R_1 + R_2 + R_3)^2} = \frac{R_1 R_2 R_3 (R_2 + R_3 + R_1)}{(R_1 + R_2 + R_3)^2} = \frac{R_1 R_2 R_3}{R_1 + R_2 + R_3} \qquad \text{...(1.42)}$$

Dviding Eq. (1.42) by Eqs. (1.38), (1.39) and (1.40) separately we have

$$R_3 = \frac{R_A R_B + R_B R_C + R_C R_A}{R_A} = R_B + R_C + \frac{R_B R_C}{R_A} \quad ...(1.43)$$

$$R_1 = \frac{R_A R_B + R_B R_C + R_C R_A}{R_B} = R_A + R_C + \frac{R_A R_C}{R_B} \quad ...(1.44)$$

$$\text{and } R_2 = \frac{R_A R_B + R_B R_C + R_C R_A}{R_C} = R_A + R_B + \frac{R_A R_B}{R_C} \quad ...(1.45)$$

The above relationship may be expressed as below:

The equivalent delta resistance between two terminals is the sum of the two star resistances connected to those terminals plus the product of the same divided by the third star resistance.

If the three star-connected resistances have the same value, say R_S, the three resistances of the equivalent delta for identical systems will be

$$R_D = R_S + R_S + \frac{R_S R_S}{R_S} = 3 R_S \quad ...(1.46)$$

The advantage of delta-star transformation may be shown by reference to network of Fig. 1.30. Fig. 1.30 (*a*) illustrates the network before conversion, where the dotted lines are drawn around the delta connection to be transformed into a star. Fig. 1.30 (*b*) illustrates the same network after transformation. The currents in the transformed form [Fig. 1.30 (*b*)] are much simpler to determine.

The advantage of star-delta transformation can be illustrated by reference to network of Fig. 1.31. Figure 1.31 (*a*) illustrates the network prior to transformation, with the dotted lines around the star to be transformed. Fig. 1.31 (*b*) illustrates the same network after transformation. The original network is now reduced to a simple series parallel connection of resistances.

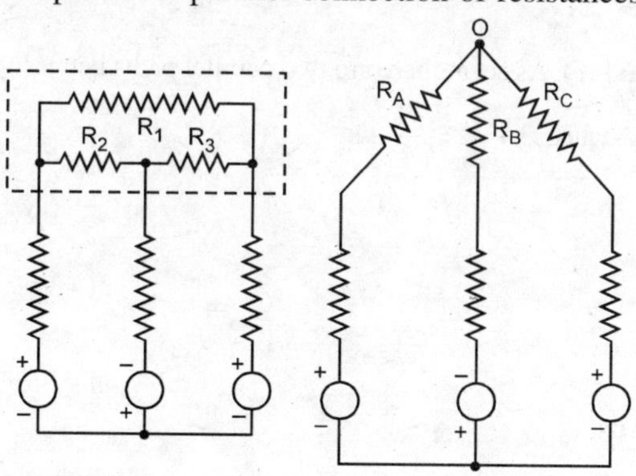

(*a*) (*b*)

Fig. 1.30

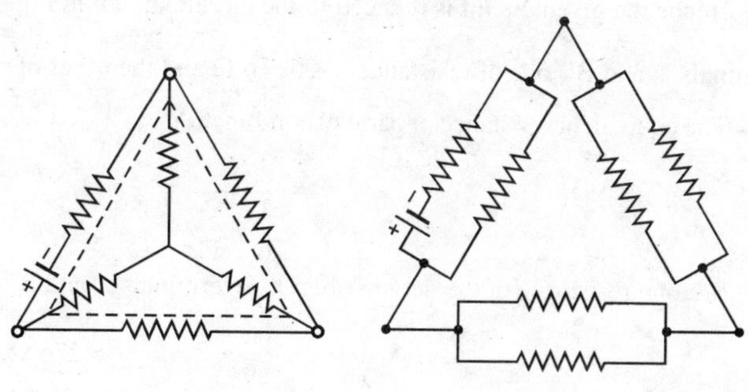

(*a*) (*b*)

Fig. 1.31

Example 1.37. Three resistances *r*, 2 *r*, and 3 *r* are connected in delta. Determine the resistances for an equivalent star connection.

Solution : Resistances for the equivalent star-connection (Fig. 1.32) are worked out as below:

$$R_A = \frac{R_1 R_2}{R_1 + R_2 + R_3} = \frac{r \times 2r}{r + 2r + 3r} = \frac{r}{3} \text{ Ans.}$$

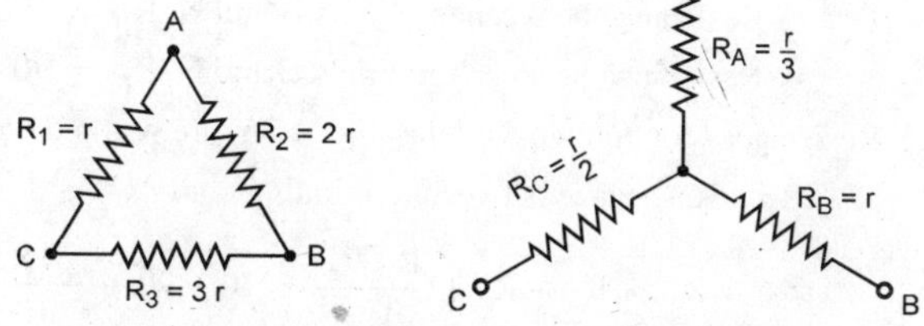

(*a*) *Delta-Connected Resistors* (*b*) *Equivalent Star-Connected Resistors*

Fig. 1.32

$$R_B = \frac{R_2 R_3}{R_1 + R_2 + R_3} = \frac{2r \times 3r}{r + 2r + 3r} = r \ \textbf{Ans.}$$

$$R_C = \frac{R_3 R_1}{R_1 + R_2 + R_3} = \frac{3r \times r}{r + 2r + 3r} = \frac{r}{2} \ \textbf{Ans.}$$

Example 1.38. Find the resistance between AB of the circuit shown in Fig. 1.33. Use Y-Δ (star-delta) transformation.

Solution : As shown in Fig. 1.34 (*a*), delta ACD has been reduced to equivalent star shown in Fig. 1.34 (*b*).

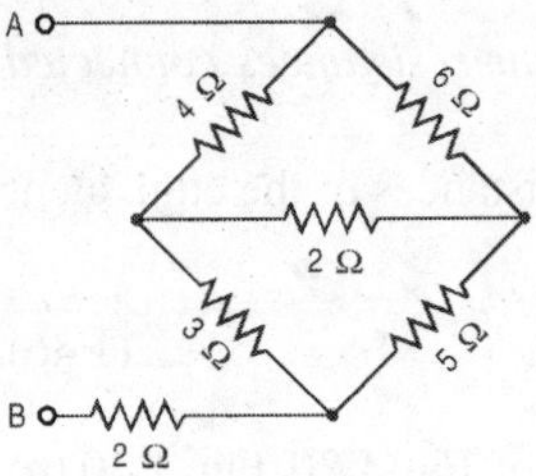

Fig. 1.33

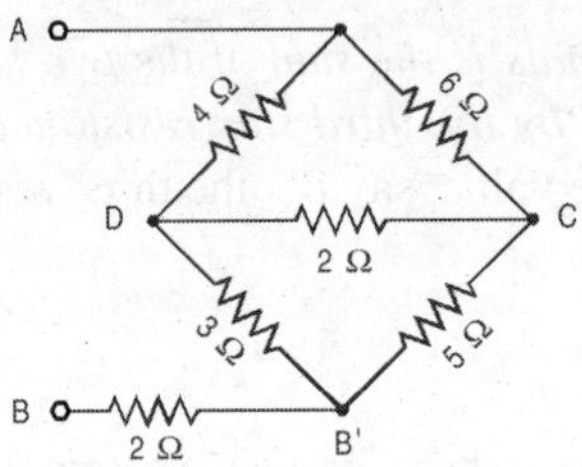

Fig. 1.34 (*a*)

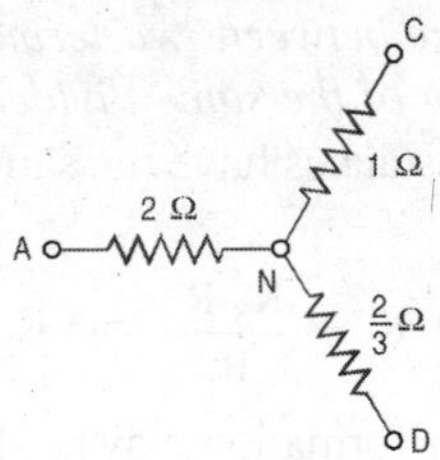

Fig. 1.34 (*b*)

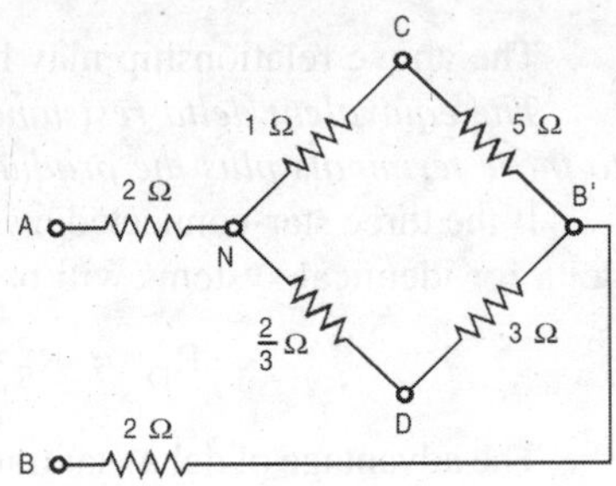

Fig. 1.34 (*c*)

$$R_A = \frac{R_{AC} R_{AD}}{R_{AC} + R_{CD} + R_{DA}} = \frac{6 \times 4}{6 + 2 + 4} = 2\ \Omega$$

$$\text{Similarly, } R_C = \frac{6 \times 2}{12} = 1\ \Omega$$

$$\text{and } R_D = \frac{4 \times 2}{12} = \frac{2}{3}\ \Omega$$

Hence the given circuit is reduced to the circuit shown in Fig. 1.34 (*c*). As seen, there are two parallel paths between terminals N and B', one of resistance 1 + 5 = 6 Ω and the other of resistance $3 + \frac{2}{3} = \frac{11}{3}\ \Omega$.

The equivalent resistance of parallel combination

$$= \frac{1}{\frac{1}{6} + \frac{3}{11}} = \frac{66}{29}\ \Omega$$

So total resistance of the network between terminals A and B

$$= \frac{66}{29} + 2 + 2 = 6.276\ \Omega \ \textbf{Ans.}$$

Example 1.39. A star-network in which N is star point made up as follows:

AN = 70 Ω, BN = 100 Ω, CN = 90 Ω. Find an equivalent delta network.

If the above star-delta network are superimposed, what would be measured resistance between A and C ?

Solution: Resistance between terminals A and N, $R_A = 70\ \Omega$

Resistance between terminals B and N, $R_B = 100\ \Omega$

Resistance between terminals C and N, $R_C = 90\ \Omega$

Resistances for equivalent delta (Fig. 1.35) are worked out as below:

Resistance between terminals C and A,

$$R_1 = R_A + R_C + \frac{R_A R_C}{R_B} = 70 + 90 + \frac{70 \times 90}{100} = 223\ \Omega \ \textbf{Ans.}$$

Resistance between terminals A and B,

$$R_2 = R_A + R_B + \frac{R_A R_B}{R_C} = 70 + 100 + \frac{70 \times 100}{90} = 247.78\ \Omega \quad \textbf{Ans.}$$

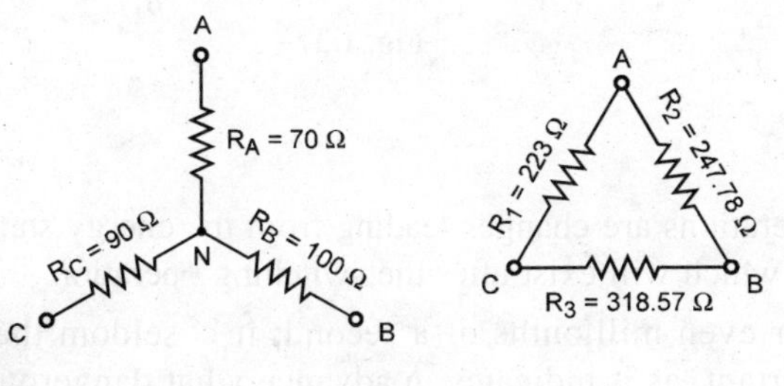

Fig. 1.35

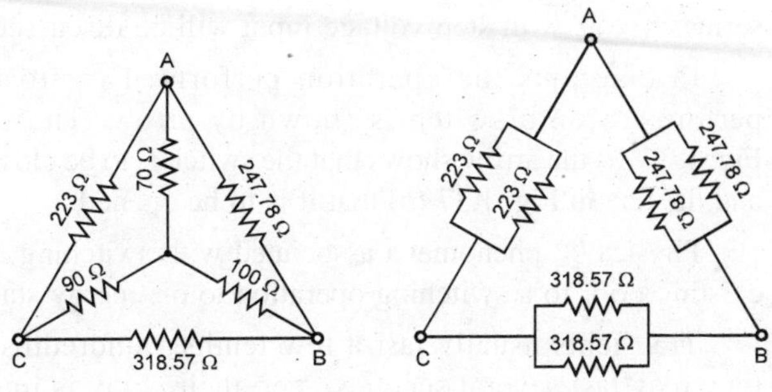

(*a*) (*b*)

Fig. 1.36

Resistance between terminals B and C,

$$R_3 = R_B + R_C + \frac{R_B R_C}{R_A} = 100 + 90 + \frac{100 \times 90}{70} = 318.57\ \Omega \quad \textbf{Ans.}$$

Now, when star and delta networks are superimposed, the equivalent network becomes as shown in Figs. 1.36 (*b*) and (*c*) by converting star into equivalent delta.

From network shown in Fig. 1.36 (*c*) resistance measured between terminals A and C,

$$R_{AC} = R_1 \parallel (R_2 + R_3)$$
$$= 111.5 \parallel (123.89 + 159.285)$$
$$= 111.5 \parallel 283.175 = \frac{111.5 \times 283.175}{111.5 + 283.175} = 80\ \Omega \quad \textbf{Ans.}$$

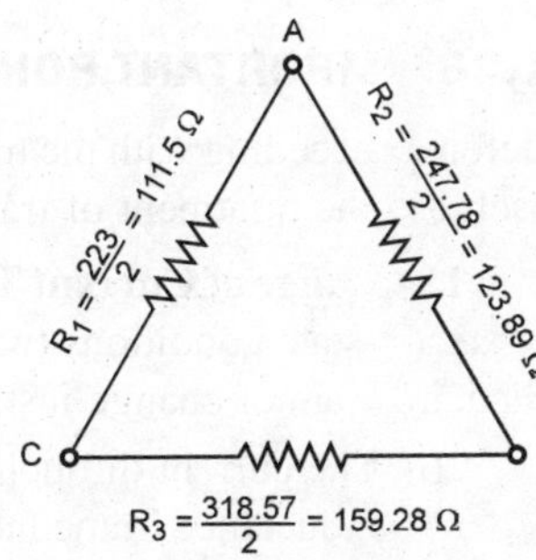

Fig. 1.36 (*c*)

1.25. TRANSIENT RESPONSE

The transient electric phenomena occur between two steady-state conditions; for instance when a load is suddenly changed, an appreciable time elapse before the generators and the line adapt themselves to the new conditions, and the currents and the voltages during the intermediate period are called the *transients.*

The transient currents are not caused by any part of the applied voltage, but are entirely associated with the changes in the stored energy in inductors and capacitors.

Since there is no stored energy in resistors hence there are *no transients in pure resistive circuits.* In inductive circuits the energy is stored in the form of electromagnetic energy and in capacitive circuits the energy is stored in the form of electrostatic energy. When there are changes in only one form of the stored energy (electromagnetic or electrostatic), as in R-L or R-C circuits, the transients are called the *single-energy transients*. When both inductance and capacitance are present, as in R-L-C circuits, the energy of the circuit is stored in two forms, and there is a possibility of periodic transformation of the magnetic energy into the dielectric energy and vice versa, electric oscillations, surges and waves are set up and the transients are called the *double-energy transients*. There is also a possibility of a *triple-energy transients*, when, for instance a synchronous motor is hunting at the end of a long transmission line which possesses inductance and capacitance, because in this case the total energy of the system is stored not only in the magnetic and electric forms but in mechanical form as well.

In general transient phenomena occur whenever:

(*i*) load circuit is suddenly connected or disconnected to/from the supply.

(*ii*) there is a sudden change in the applied voltage from one finite value to another and

(*iii*) the circuits are short circuited.

Here only transient response of R-L series and R-C series circuits with step voltage input will be discussed.

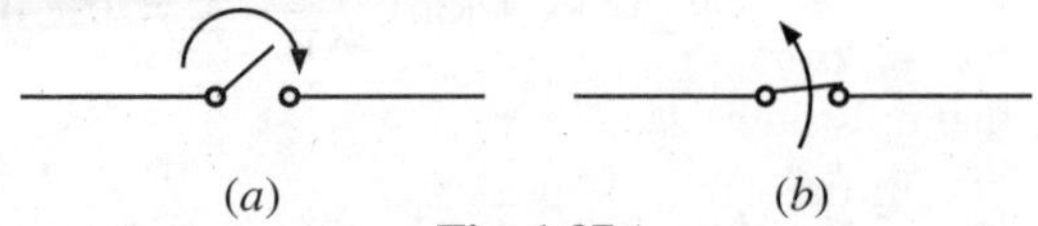

Fig. 1.37

In diagrams the operation performed (or to be performed) on a switch is shown by arrow. Thus in Fig. 1.37 (*a*) the arrow shows that the switch is to be closed and the one in Fig. 1.37 (*b*) that it is to be opened.

Physically, phenomena associated with switching operations are changes leading from the energy state existing prior to a switching operation to on energy state which will exist after the switching operation.

Transients usually last a few tenths, hundredths or even millionths of a second; it is seldom that they may last several seconds. Yet, their study is important, as it indicates in advance what dangerous rises in voltage or current, many times their steady-state values, may happen in the individual sections of a circuit.

1.26. IMPORTANT POINTS AND DEFINITIONS

Before proceeding with the transient analysis it is necessary to introduce some important points and definitions useful in the treatment of transients.

1. Change of Current Through an Inductance and Voltage Across a Capacitance. Under any transient or steady-state conditions two basic points hold: The current through an inductance and the voltage across a capacitor cannot change instantaneously. There are two corollaries to these two points as mentioned below:

(*i*) The current through any inductance just prior to switching is equal to the current through the same inductance immediately after switching *i.e.* $i_L(0^-) = i_L(0^+)$ where $t = 0^-$ is the instant immediately before switching; $t = 0^+$ is the instant just after switching.

(*ii*) The voltage across a capacitance just before switching is equal to the voltage across the same capacitance immediately after switching *i.e.* $v_C(0^-) = v_C(0^+)$

2. Response of Single Circuit Elements. The circuit response of single elements is given below in tabular form.

TABLE 1.2

Element	Voltage Across Element	Current in Element
Resistance, R	$v(t) = Ri(t)$	$i(t) = \frac{1}{R} v(t)$
Inductance, L	$v(t) = L\frac{di}{dt}$	$i(t) = \frac{1}{L}\int v\,dt$
Capacitance, C	$v(t) = \frac{1}{C}\int i\,dt$	$i(t) = C\frac{dv}{dt}$

3. Initial Conditions Before and After Switching. As already mentioned, currents through inductances and voltage across capacitances immediately before and after switching are respectively equal. As to other quantities, such as voltage across inductances and resistances, currents through capacitances and resistances, they can change instantaneously, and so their values immediately after switching may differ from those before switching. Hence, a distinction should be made between initial conditions before and after switching.

Initial conditions before switching are those which exist at time $t = 0^-$ and initial conditions after switching are those which exist at time $t = 0^+$.

4. Independent and Dependent Initial Values After Switching. For any circuit the currents and voltages that exist in or across any element at time $t = 0^+$ (after switching) can be determined by Kirchhoff's laws.

In the equations written according to these laws the current through the inductances and the voltage across capacitances are assigned the same values as before switching. Now that some of the terms in the equations are known, the remaining currents and voltages can be readily determined.

The currents through inductances and the voltages across capacitances known from the state of the circuit prior to switching are referred to as *independent initial values*. The remaining currents and voltages at $t = 0^+$ determined by Kirchhoff's laws, or otherwise, from the independent initial values are referred to as *dependent initial values*.

5. Zero and Non-Zero Initial Values. If by the time a transient is initiated, *i.e.*, just before switching, all currents through and all voltages across passive elements have become equal to zero, the circuit is said to be in the state of *zero initial conditions* and the respective values are termed *zero initial conditions*. If, however, by the time a transient is initiated the currents and voltages in a circuit are other than zero, the circuit is said to be in the state of *non-zero initial conditions* and the respective values are termed *non-zero initial values*. In the former case the currents through inductances and the voltages across the capacitances will vary starting from zero values, in the later, starting from non-zero values.

6. Initial and Final Conditions in Elements.

(a) Resistor. Current through a resistor changes instantaneously, if the voltage changes instantaneously and vice-versa.

(b) Inductor. As already mentioned, in case of an inductor current through it does not change instantaneously. If the initial conditions are zero, at the time of closing the switch for connecting an inductor to an energy source, the inductor will behave like an open circuit *i.e.* no current will flow at $t = 0^+$. If at the switching instant a current I_0 was flowing in the inductor, the inductor will behave as a current source of I_0 amperes.

(c) Capacitor. As already mentioned, in case of a capacitor voltage across it does not change instantaneously. If an uncharged capacitor is connected to an energy source, at the time of switching, the capacitor will behave like a short circuit. If at the time of switching the capacitor was charged with an initial charge of Q_0, the capacitor will behave as a voltage source of v_0 volts where $v_0 = Q_0/C$.

Initial and final conditions in various elements are given below:

Elements	Initial Conditions	Equivalent Circuit At $t = 0^+$	Equivalent Circuit At $t = \infty$
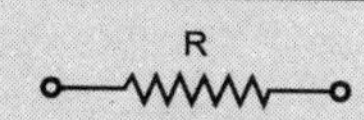	—	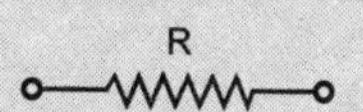	
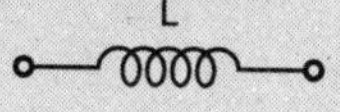	Zero	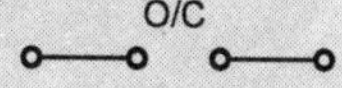	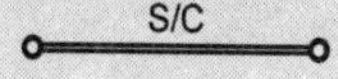
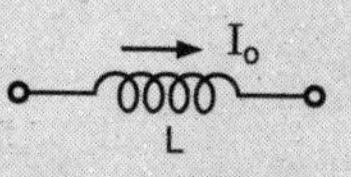	I_0		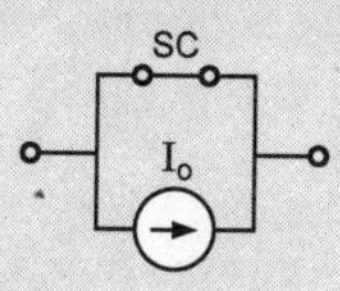
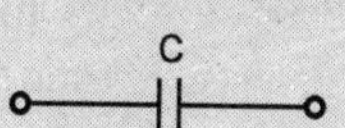	Zero		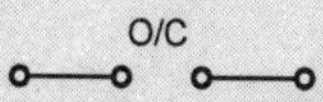
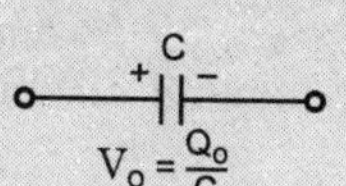	$V_0 = \dfrac{Q_0}{C}$	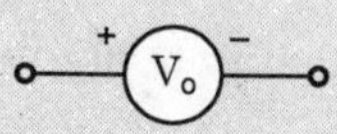	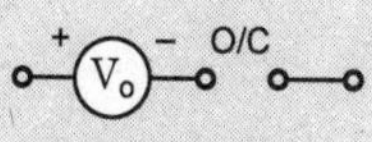

1.27. DC TRANSIENTS FOR R-L SERIES CIRCUIT (OR STEP VOLTAGE RESPONSE OF R-L SERIES CIRCUIT)

Consider an inductive circuit consisting of resistance R(ohms) and inductance L(henries) with switch S, as shown in Fig. 1.38. The switch S has been open for a long period of time, and is closed at $t = 0$.

Just before application of step voltage the current through the inductor

$$i_L(0^{-1}) = 0 \quad \because \text{ the circuit was open before switching in operation}$$

Applying Kirchhoff's voltage law to the circuit shown in Fig. 1.38 after switching-in operation

$$R\,i + L\frac{di}{dt} = V \text{ for } t > 0 \qquad \text{...(1.47)}$$

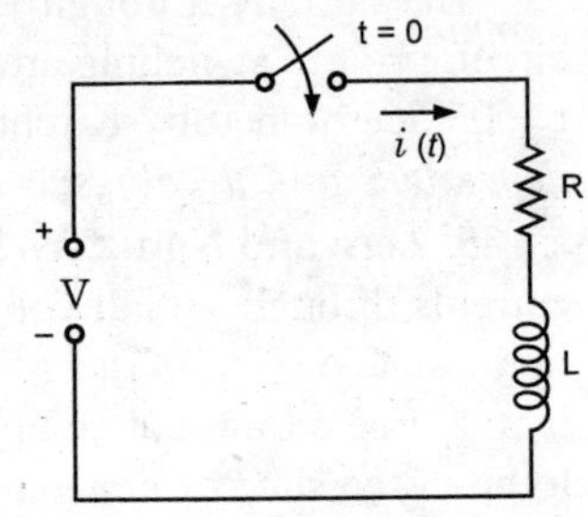

Fig. 1.38

The above equation is a non-homogeneous linear differential (first order) equation with the excitation term appearing on its RHS. The complete solution of above equation consists of two parts which are called the particular integral and complementary function. The particular integral (forced response i_f) is the solution corresponding to the steady-state conditions

$$i.e., \quad i_f = \frac{V}{R}$$

The above solution is obtained without any constant of integration and cannot, therefore, represent initial conditions. Complementary function is, therefore, necessary to represent the initial conditions. This is the solution of the Eq. (1.47) with zero substituted for V namely the solution of $R\,i + L\frac{di}{dt} = 0$ which is $A\,e^{\frac{-R}{L}t}$ and thus complete solution of Eq. (1.47) is given as

$$i = \frac{V}{R} + A\,e^{\frac{-R}{L}t} \qquad \text{...(1.48)}$$

in which A is an arbitrary constant and it must satisfy the initial condition on inductance current. As already stated

$$i(0^+) = i(0^-) = 0$$

Substituting in Eq. (1.48), we have

$$0 = \frac{V}{R} + A$$

$$\text{or} \quad A = -\frac{V}{R}$$

$$\text{Hence } i(t) = \frac{V}{R} - \frac{V}{R}e^{\frac{-R}{L}t}$$

$$= \frac{V}{R}\left(1 - e^{\frac{-R}{L}t}\right) \qquad \text{...(1.49)}$$

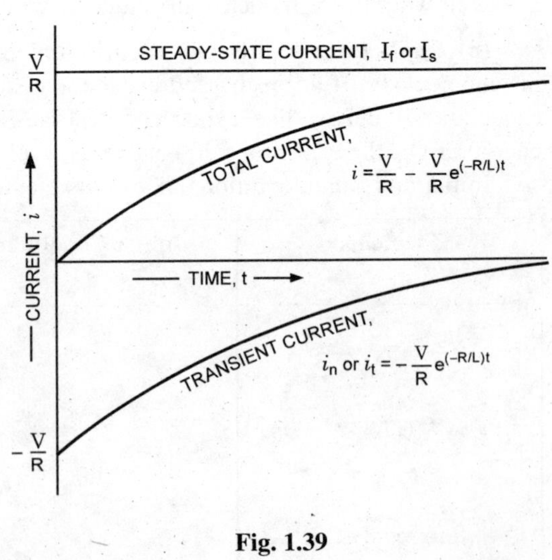

Fig. 1.39

where t is the time in seconds and e is the base of natural (hyperbola) logarithms. This expression is known as Helmholtz expression.

Now the first term, *i.e.*, V/R in the above expression for $i(t)$ is the current corresponding to the steady-state condition, *i.e.*, forced response, i_f and the second term $\frac{-V}{R}e^{\frac{-R}{L}t}$ is a transient (natural response, i_n) which approaches zero as t approaches infinity.

Thus the resultant current may be assumed to consist of a final steady-state current (or forced response, i_f) of value $\frac{V}{R}$ and a transient current (natural response, i_n) of value $\frac{-V}{R} e^{\frac{-R}{L}t}$ superimposed on the steady-state current.

The manner in which the transient current dies away, depends upon the value of ratio of R to L. The ratio $\frac{L}{R}$ is called the *time constant* and denoted by τ. The reciprocal of the time constant (*i.e.* $\frac{R}{L}$) is called the *damping coefficient* of the circuit.

Curves for steady-state current I_s and the transient current i_t are shown in Fig. 1.39. The curve for the resultant current has been obtained by superimposition of steady-state current I_f and transient current i_n where

$$i_s \text{ or } i_f = \frac{V}{R} \quad ...(1.50)$$

and

$$i_t \text{ or } i_n = \frac{-V}{R} e^{\frac{-R}{L}t} \quad ...(1.51)$$

Theoretically, the transient current takes infinite time to die away, but in practice it disappears in a very short time.

The values of resultant, steady-state and transient voltage across the resistor R can be determined by multiplying the Eqs. (1.49), (1.50) and (1.51) respectively by R. Thus

$$v_R(t) = i(t)\,R = V\left(1 - e^{\frac{-R}{L}t}\right) \quad ...(1.52)$$

The voltage across the inductance is given as

$$v_L(t) = L\frac{di(t)}{dt} = L \times \frac{V}{R}\left[0 - \left(\frac{-R}{L}\right)e^{\frac{-R}{L}t}\right] = V\,e^{\frac{-R}{L}t} \quad ...(1.53)$$

Since just prior to switch closure the voltage across the inductance was zero, and at the instant of closure of switch it was $v_L = V$, the total forced and free voltages across the inductance changes respectively.

The curves for resultant, steady-state and transient voltages across the resistor and transient voltage across the inductor are illustrated in Fig. 1.40.

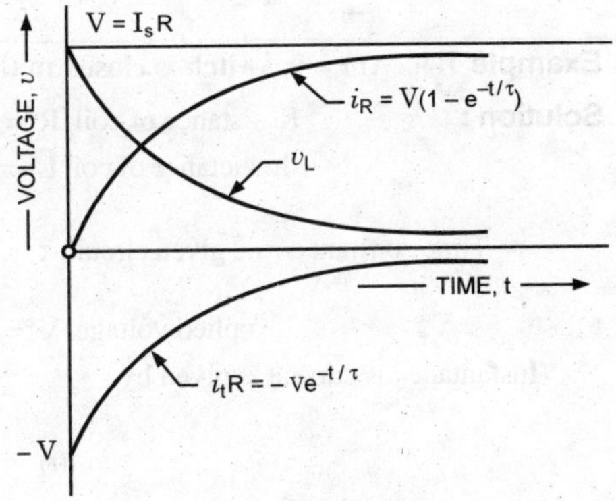

Fig. 1.40

The response to step voltage given by Eq. (1.49) rewritten here

$$i(t) = \frac{V}{R}\left(1 - e^{\frac{-R}{L}t}\right)$$

$$= I_s\left(1 - e^{-t/\tau}\right) \quad ...(1.54)$$

is plotted in Fig. 1.41 where the current rises exponentially from 0 to I_s or $I_f = \frac{V}{R}$ in accordance with the time constant (or natural frequency).

Immediately after the switch is closed, the rate of rise of current *i.e.* initial rate of rise of current can be determined by differentiating Eq. (1.49) w.r.t. for $t = 0$. Hence we get

$$\frac{di(t)}{dt} = \frac{V}{R \times \frac{L}{R}} e^{\frac{-R}{L}t} = \frac{V}{L} e^0 = \frac{V}{L} \quad \because t = 0$$

If this rate of rise were maintained then time required to reach steady-state value of current $I_s = \frac{V}{R}$ would have been $\frac{V/R}{V/L} = \frac{L}{R}$. This time duration is called the time constant of the circuit.

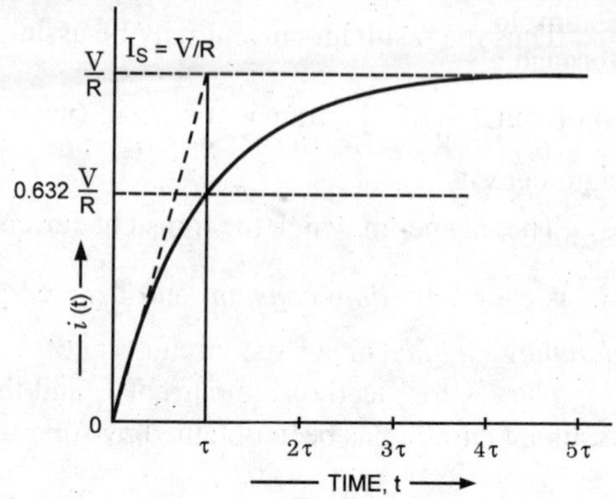

Fig. 1.41

Hence the time constant, τ may be defined as the time required for the current to rise to its final steady-state value if it is continued rising at its initial rate $\left(i.e. \frac{V}{L}\right)$

If we substitute $t = \tau = \frac{L}{R}$ in Eq. (1.54), we have

$$i = I_s(1 - e^{-1}) = 0.632\, I_s \text{ or } 0.632 \frac{V}{R}$$

So the *time constant* may be defined as the time required for the current to reach 0.632 of its final steady-state value (*i.e.*, 0.632 I_s).

Thus inductance behaviour at $t = 0^+$ and $t = \infty$ stated earlier can be confirmed by response of Eq. (1.54) from which it follows that

$$i_L(0^+) = I_s(1 - e^0) = 0 \; i.e., \text{ inductance behaves as an open circuit at } t = 0^+$$

and $$i_L(\infty) = I_s(1 - e^{-\infty}) = I_s = \frac{V}{R}, \; i.e., \text{ inductance behaves as a short circuit at } t = \infty$$

(for dc excitation).

Example 1.40. At $t = 0$, switch is closed in the circuit shown in Fig. 1.42. Find $i(t)$.

Solution :

Resistance of coil, R = 5 Ω

Inductance of coil L = 2 H

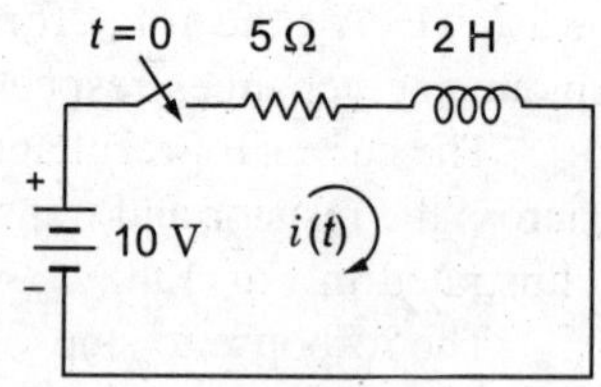

Fig. 1.42

Time constant of the given circuit, $\tau = \frac{L}{R} = \frac{2}{5} = 0.4$ second

Applied voltage, V = 10 V

Instantaneous current is given by

$$i(t) = \frac{V}{R}(1 - e^{-t/\tau}) = \frac{10}{5}(1 - e^{-t/0.4}) = 2(1 - e^{-2.5t}) \quad \textbf{Ans.}$$

Example 1.41. If a coil has a resistance of 10 ohms and inductance 1 H, estimate the value of current after 0.1 second of switching on 100 V dc supply.

Solution :

Time constant of coil, $\tau = \frac{L}{R} = \frac{1}{10} = 0.1$ second

Instantaneous current is given by $i(t) = \frac{V}{R}(1 - e^{-t/\tau}) = \frac{100}{10}(1 - e^{0.1/0.1}) = 6.32$ A **Ans.**

Example 1.42. A choke coil with an inductance of 20 H and resistance 10 Ω is connected across 200 V dc supply through a switch.

(*i*) What is the initial current when switch is closed.

(*ii*) What is final steady current ?

(*iii*) What is current at any instant of 1.25 seconds after switching.

Solution : Applied voltage, V = 200 V

$$\text{Time constant of choke coil, } \tau = \frac{L}{R} = \frac{20}{10} = 2 \text{ second}$$

Instantaneous current is given by

$$i(t) = \frac{V}{R}(1-e^{-t/\tau}) = \frac{200}{10}(1-e^{-t/2}) = 20(1-e^{-0.5t}) \quad ...(i)$$

(*i*) Initial current on closing of switch can be obtained by substituting $t = 0$ in Eq. (*i*) so we have

$$I(0) = 20(1 - e^{-0}) = 0 \quad \textbf{Ans.}$$

(*ii*) Final steady-state current, $I_s = \frac{V}{R} = \frac{200}{10} = 20$ A

(*iii*) Current after 1.25 s of switching can be had by substituting $t = 1.25$ s in Eq. (*i*) so

Current after 1.25 s of switching $= 20(1 - e^{-0.5 \times 1.25t}) = 20(1 - 0.53526) = 9.2948$ A **Ans.**

Example 1.43. A coil of resistance 20 Ω and inductance 0.8 H is connected to a 200 V dc supply. Find (*i*) rate of change of current at the instant of closing switch (*ii*) final steady value of current (*iii*) time constant of the circuit and (*iv*) time taken for the current to rise to half its final value.

Solution : (*i*) Rate of change of current at the instant of switching, *i.e.* at $t = 0^+$,

$$\frac{di(t)}{dt} = \frac{V}{L} = \frac{200}{0.8} = 250 \text{ A/s } \textbf{Ans.}$$

$$\text{(ii) Final steady value of current, } I_s = \frac{V}{R} = \frac{200}{20} = 10 \text{ A } \textbf{Ans.}$$

$$\text{(iii) Time constant of circuit, } \tau = \frac{L}{R} = \frac{0.8}{20} = 0.04 \text{ second } \textbf{Ans.}$$

(*iv*) Expression for instantaneous current while current is rising may be written as

$$i(t) = I_s(1 - e^{-t/\tau})$$

$$\text{when } i(t) = \frac{1}{2} I_s$$

$$\frac{1}{2} I_s = I_s(1 - e^{-t/0.04})$$

$$\text{or } e^{-25t} = 0.5$$

$$\text{or } t = -0.04 \log_e 0.5 = 0.028 \text{ second } \textbf{Ans.}$$

Example 1.44. A direct voltage of 200 V is suddenly applied to a series R-L circuit having R = 20 Ω and inductance 0.2 H. Determine the voltage drop across the inductor at the instant of switching and at 0.02 s later.

Solution: Since inductor initially behaves as an open circuit, *i.e.*, at $t = 0^+$, $i(0^+) = 0$

Hence, at the instant of switching, the voltage drop across inductor is 200 V

Instantaneous current is given as

$$i(t) = \frac{V}{R}\left(1-e^{\frac{-R}{L}t}\right) = \frac{200}{20}\left(1-e^{\frac{-20}{0.2}t}\right) = 10(1 - e^{-100t})$$

$$\text{at } t = 0.02 \text{ second, } i(t) = 10(1 - e^{-100 \times 0.02}) = 10(1 - e^{-2}) = 8.647 \text{ A}$$

$$\text{Since} \quad R\,i(t) + L\,\frac{di(t)}{dt} = 200$$

$$\text{or} \quad L\,\frac{di(t)}{dt} = 200 - Ri(t) = 200 - 20 \times 8.647 = 27 \text{ V} \quad \textbf{Ans.}$$

Example 1.45. A 100 V dc is applied to a circuit consisting of a resistance of 100 Ω in series with an inductance of 10 H through the switch. If the switch is closed at $t = 0$, find out (*i*) the expressions for $i(t)$, $v_R(t)$ and $v_L(t)$, (*ii*) the value of $i(t)$ for $t = 0.2$ second and (*iii*) time at which $v_R(t) = v_L(t)$.

Solution:

Resistance, R = 100 Ω

Inductance, L = 10 H

Voltage, V = 100 V

$$\text{Time constant, } \tau = \frac{L}{R} = \frac{10}{100} = 0.1 \text{ second}$$

(*i*) The expression for the current $i(t)$ is given as

$$i(t) = \frac{V}{R}\left(1 - e^{-t/\tau}\right) = \frac{100}{100}\left(1 - e^{-t/0.1}\right) = \left(1 - e^{-10t}\right) \text{ A } \textbf{Ans.}$$

The expression for voltage drop across resistor R is given as

$$v_R(t) = R \times i(t) = 100\left(1 - e^{-10t}\right) \text{ volts. } \textbf{Ans.}$$

The expression for voltage drop across inductor L is given as

$$v_L(t) = -L\,\frac{d\,i(t)}{d\,t} = -10 \times \frac{d}{d\,t}\left(1 - e^{-10t}\right) = 100\,e^{-10t} \text{ volts. } \textbf{Ans.}$$

(*ii*) When $t = 0.2$ second

$$\text{Current } i(t) = \left(1 - e^{-10 \times 0.2}\right) = 1 - e^{-2} = 0.8647 \text{ A } \textbf{Ans.}$$

(*iii*) Let the voltage drops across resistance and inductance be equal at t seconds after closure of switch, then

$$100\left(1 - e^{-10t}\right) = 100\,e^{-10t}$$

$$\text{or } t = \frac{-1}{10}\log_e 0.5 = 0.0693 \text{ second } \textbf{Ans.}$$

1.28. DC TRANSIENTS FOR R-C SERIES CIRCUITS (OR STEP VOLTAGE RESPONSE OF R-C SERIES CIRCUIT)

Consider an R-C series circuit shown in Fig. 1.43. The switch S is closed at time $t = 0$.

Let at any instant pd across the capacitor be $v_C(t)$, charge acquired by the capacitor be $q(t)$ coulombs and charging current be $i(t)$ amperes.

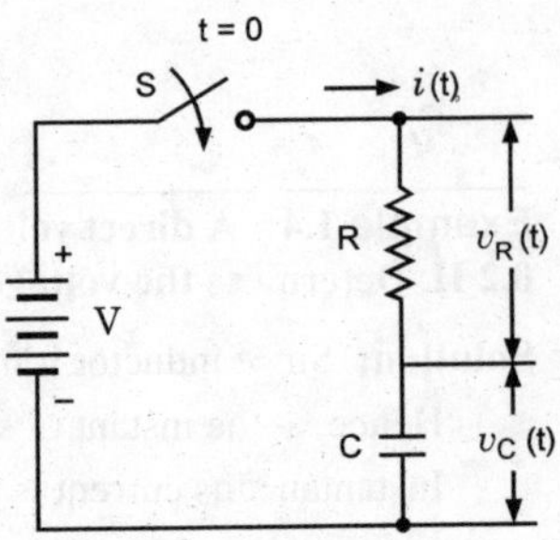

Fig. 1.43

$$\text{Voltage drop across resistor, } v_R(t) = R\,i(t)$$

$$\text{Voltage drop across capacitor, } v_C(t) = \frac{q(t)}{C}\; t = \frac{1}{C}\int_{-\infty}^{t} i(t)\,dt$$

$$\text{Since applied voltage, } V = v_R(t) + v_C(t)$$

$$\text{So} \quad R\,i(t) + \frac{1}{C}\int_{-\infty}^{t} i(t)\,dt = V$$

$$\text{or} \quad R\,i(t) + \frac{1}{C}\int_{0}^{t} i(t)\,dt + v_C(0^+) = V \qquad \text{...(1.55)}$$

Assuming capacitor initially uncharged, we have

$$Ri(t) + \frac{1}{C}\int_{0}^{t} i(t)\,dt = V \qquad \text{...(1.56)}$$

Differentiating above Eq. (1.56), we have

$$\frac{d\,i(t)}{dt} + \frac{1}{RC}\,i(t) = 0 \qquad \text{...(1.57)}$$

General solution of the above differential equation is

$$i(t) = A\, e^{\frac{-1}{RC}t} \qquad \text{...(1.58)}$$

At $t = 0^+$, $i(0^+) = \frac{V}{R}$ since capacitor behaves as a short circuit at $t = 0^+$

$$\text{So} \quad \frac{V}{R} = A\,e^0 \quad \text{or} \quad A = \frac{V}{R}$$

$$\text{and} \quad i(t) = \frac{V}{R}\, e^{\frac{-1}{RC}t} = \frac{V}{R} e^{-t/\tau} \qquad \text{...(1.59)}$$

where τ is the time constant of the circuit and is equal to RC

$$\text{Voltage across resistor, } v_R(t) = R\,i(t) = V\,e^{-t/\tau} \qquad \text{...(1.60)}$$

$$\text{Voltage across capacitor } v_C(t) = V - v_R(t) = V - V\,e^{-t/\tau} = V(1 - e^{-t/\tau}) \qquad \text{...(1.61)}$$

Voltage across capacitor can also be given as

$$v_C(t) = \frac{1}{C}\int_0^t i(t)\,dt = \frac{1}{C}\cdot\frac{V}{R}\left[e^{\frac{-t}{CR}}\right]_0^t \times(-RC) = -V\,e^{\frac{-t}{CR}}\Big|_0^t = -V\,(e^{\frac{-t}{CR}} - 1) = V\,(1 - e^{-t/CR}) = V\,(1 - e^{-t/\tau}) \qquad \text{...(1.62)}$$

At $t = 0^+$; $i(t) = \frac{V}{R}$; $v_R(t) = V$ and $v_C(t) = 0$

At $t = \infty$; $i(t) = 0$; $v_R(t) = 0$ and $v_C(t) = V$

At $t = \tau = CR$; $i(t) = \frac{V}{R} e^{-1} = 0.368\frac{V}{R}$; $v_R(t) = V\,e^{-1} = 0.368$ V and $v_C(t) = 0.632$ V

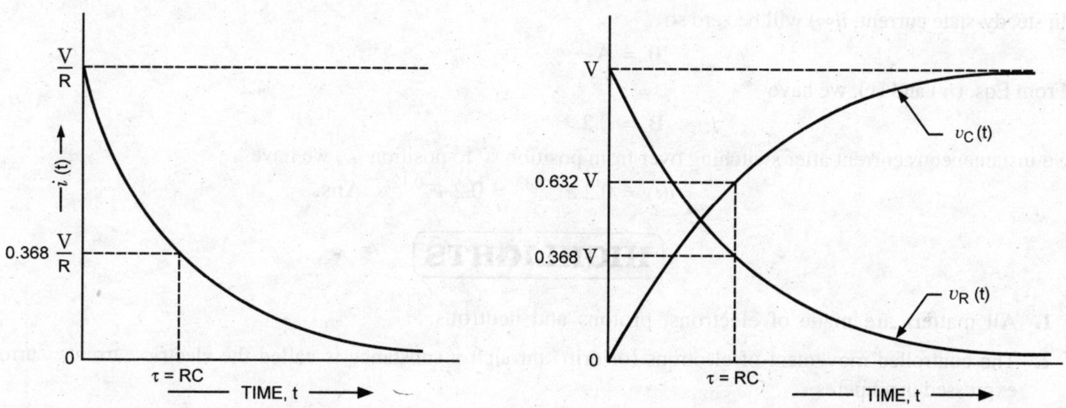

(a) For i (t) *(b) For v_R (t) and v_C (t)*

Fig. 1.44

The curve for $i(t)$ is shown in Fig. 1.44 (*a*) while for $v_R(t)$ and $v_C(t)$ are shown in Fig. 1.44 (*b*).

Example 1.46. A resistance R and 5 μF capacitor are connected in series across a 200 V dc supply. Calculate the value of R such that the voltage across the capacitor becomes 100 V in 5 seconds after the circuit is switched on.

Solution : Voltage across the capacitor is given as

$$v_C(t) = V(1 - e^{\frac{-t}{CR}}) \quad \text{... refer to Eq. (1.61)}$$

So $$100 = 200\left(1 - e^{\frac{-5}{R \times 5 \times 10^{-6}}}\right) = 200\left(1 - e^{\frac{-10^6}{R}}\right)$$

or $$e^{\frac{-10^6}{R}} = 0.5$$

or $R = 1.45 \times 10^6\ \Omega$ or $1.45\ \text{M}\Omega$ **Ans.**

Example 1.47. In the circuit shown, find $i(t)$ at $t = 0^+$ following switching of S from (1) to (2). Assume steady state of the circuit while S was at (1).

Solution : When switch is in position ①. At steady-state, capacitor acts as an open circuit, so current

$$I_S = \text{Current through } 10\ \Omega \text{ resistor}$$

$$= \frac{12}{50 + 10} = 0.2\ \text{A}$$

Voltage drop across capacitor,

$$V_{CS} = \text{Voltage drop across } 10\ \Omega \text{ resistor}$$

$$= I_S \times R = 0.2 \times 10 = 2\ \text{V}$$

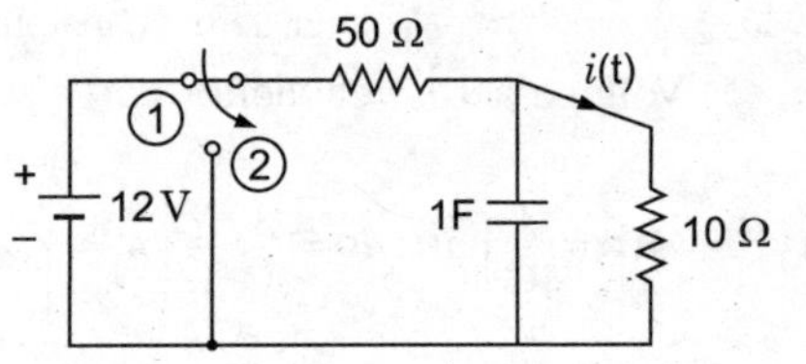

Fig. 1.45

After switching over from position ① to position ②

$$I(0^+) = I_S = 0.2\ \text{A} \quad \text{...(i)}$$

and $$V_C(0) = 2\ \text{V} \quad \text{...(ii)}$$

Instantaneous current i is given as

$$i(t) = A + B\, e^{-t/RC}$$

$$= A + B\, e^{-t/8.33} \quad \text{...(iii)} \qquad \because \text{RC} = \frac{50 \times 10}{50 + 10} \times 1 = 8.33\ s$$

Current just after switching over from position ① to position ② can be had by substituting $t = 0$ so, we have

$$0.2 = A + B \quad \text{...(iv)}$$

In steady-state current, $i(\infty)$ will be zero so

$$0 = A \quad \text{...(v)}$$

From Eqs. (*iv*) and (*v*), we have

$$B = 0.2$$

So instantaneous current after switching over from position ① to position ②, we have

$$i(t) = 0.2\, e^{-t/8.33} = 0.2\, e^{-0.12t} \quad \textbf{Ans.}$$

HIGHLIGHTS

1. All matters are made of electrons, protons and neutrons.
2. The controlled movement of electrons (or drift) through a substance is called the electric current. Current is expressed in amperes.
3. Electromotive force (emf) or potential difference is that which causes the current to flow in a closed circuit. The unit of potential difference in SI system is volt.
4. Resistance may be defined as that property of substance which opposes (or restricts) the flow of an electric current (or electrons) through it. The unit of resistance in SI system is ohm (Ω).

5. Ohm's law may be given as

$$\frac{V}{I} = \text{Constant} = R \qquad \text{for no change in physical state}$$

6. Laws of resistance are summed up as

$$R = \rho \frac{l}{a}$$

where ρ is the specific resistance or resistivity of the material of wire, which may be defined as the resistance between opposite faces of a unit cube of that material.

Resistivity is measured in ohm-metres (Ω-m) or ohms per metre cube.

7. Conductance G is reciprocal of resistance *i.e.* $G = \frac{1}{R}$. The unit of conductance is siemens (S).

8. Conductivity σ is reciprocal of resistivity ρ *i.e.* $\sigma = \frac{1}{\rho}$. The unit of conductivity is siemens per metre.

9. A current (or voltage) is called alternating if it periodically changes its direction and magnitude whereas direct currents are steady and flow in one direction only.

10. Temperature coefficient of resistance may be defined as the ratio of increase in resistance per degree of rise of temperature to the original resistance *i.e.*

$$\alpha_0 = \frac{\Delta R}{R_0 t}$$

It is expressed in Ω/Ω/°C.

11. In case of good conductors, especially metals, the variations in resistance with temperature (within normal range) are given by

$$R_t = R_0 (1 + \alpha_0 t)$$

Similarly, resistivity of given material varies as

$$\rho_t = \rho_0 (1 + \alpha_0 t)$$

12. Temperature coefficient of resistance is not constant but depends upon the initial temperature on which the increment in resistance is based. The values of temperature coefficient for different temperatures can be had from the relations

$$\alpha_1 = \frac{1}{\frac{1}{\alpha_0} + t_1} \quad \text{and} \quad \alpha_2 = \frac{1}{\frac{1}{\alpha_1} + (t_2 - t_1)}$$

13. Thermal unit of energy, kilocalorie (kcal), is the amount of heat required to raise the temperature of 1 kg of water through 1 °C

$$1 \text{ kcal} = 1{,}000 \text{ calories} = 4{,}180 \text{ J} \simeq 4{,}200 \text{ J}$$

14. Electrical unit of power is watt and is given as

$$1 \text{ W} = 1 \text{ J/s}$$

15. Electrical unit of energy in kWh is given as

$$1 \text{ kWh} = 36 \times 10^5 \text{ J or } 3.6 \text{ MJ}$$

16. Conversion efficiency is given by

$$\text{Conversion efficiency, } \eta = \frac{\text{Energy utilised or output energy}}{\text{Total input energy}}$$

17. Current through a resistor changes instantaneously if the voltage changes instantaneously and vice-versa.

18. In case of an inductor, current through it cannot change instantaneously.

19. In case of a capacitor, the voltage across it cannot change instantaneously.

20. Step Voltage Response of R-L Series Circuit

Time constant, $\tau = \frac{L}{R}$

$$i(t) = \frac{V}{R}(1 - e^{-t/\tau})$$

$$v_R(t) = V(1 - e^{-t/\tau})$$

$$v_L(t) = V e^{-t/\tau}$$

21. Step Voltage Response of R-C Series Circuit

Time constant, $\tau = RC$

$$i(t) = \frac{V}{R} e^{-t/\tau}$$

$$v_R(t) = V e^{-t/\tau}$$

$$v_C(t) = V(1 - e^{-t/\tau})$$

EXERCISES

1. State Ohm's law and give its limitations.
[Pb. Technical Univ. Basic Electrical and Electronics Engineering December 2008]
2. What is Ohm's law ? State its limitations.
[Pb. Technical Univ. Basic Electrical and Electronics Engineering Dec. 2005]
3. State Ohm's law and give its significance.
[Pb. Technical Univ. Basic Electrical and Electronics Engineering May 2007]
4. State laws of resistance and hence define specific resistance.
5. Discuss the temperature dependence of resistance.
[Pb. Technical Univ. Basic Electrical and Electronics Engineering Dec. 2005]
6. Derive an expression for resistance of a conductor material and explain the effect of temperature on it.
[Pb. Technical Univ. Basic Electrical and Electronics Engineering May 2008]
7. Define resistance temperature coefficient. How does its value change with temperature ?
8. Define temperature coefficient of resistance and give its units.
[Pb. Technical Univ. Basic Electrical and Electronics Engineering December 2009]
9. Define temperature coefficient of resistance. Write down the names of the materials having negative temperature coefficient.
10. Give concept of work and energy.
[Pb. Technical Univ. Basic Electrical and Electronics Engineering Dec. 2006]
11. How is energy related to power ? State the SI units of work, power and energy. What are the practical units of electrical energy and mechanical power ?
12. Explain the terms watt, joule, watt-hour and kWh.
13. State and explain Joule's law of electric heating.
14. Compute equivalent resistance of three resistances R_1, R_2 and R_3 connected in (*i*) series (*ii*) parallel.
15. State the two fundamental laws of circuit analysis.
[Pb. Technical Univ. Electrical Engineering May 2002]
16. State and explain Kirchhoff's laws. **[Pb. Technical Univ. Basic Electrical and Electronics Engg. Dec. 2006; Dec. 2007; Dec. 2008; May 2009]**
17. State KCL and KVL.
[Pb. Technical Univ. Basic Electrical and Electronics Engineering December 2009]
18. Derive the necessary equations for converting a delta network into an equivalent star network.
[Pb. Technical Univ. Basic Electrical and Electronics Engineering May 2006]

19. What are transients? Why are they produced? Explain with an example.
20. Determine expressions for $i(t)$, $v_R(t)$, and $v_L(t)$ when an R-L series circuit is switched on to a step voltage of V volts.
21. What are transients ? Discuss the step voltage response of R-L series circuit with step voltage as input.
22. Determine expressions for $i(t)$, $v_R(t)$ and $v_C(t)$ when an R-C series circuit is switched on to a step voltage of V volts.

SHORT ANSWER TYPE QUESTIONS WITH ANSWERS

Q. 1. What is electric current ?

Ans. The controlled movement of electrons (or drift) through a substance is called the electric current.

Q. 2. What is difference between emf and potential difference ?

Ans. EMF (electromotive force) is the force that causes a current to flow in an electric circuit while the potential difference between any two points in an electric circuit is that difference in their electrical state which tends to cause flow of electricity from one point to the other.

Q. 3. What is meant by electrical resistance ?

Ans. Resistance may be defined as that property of a substance which opposes (or restricts) the flow of electric current (or electrons) through it.

Q. 4. What is Ohm's law ? **[P.T.U. B.E.E.E. May 2006]**

Ans. Physical state i.e. temperature rise etc. remaining the same the current flowing through a conductor is directly proportional to the potential difference applied across its ends.

Q. 5. What are the limitations of Ohm's law ?

Ans. Ohm's law cannot be applied to the circuits consisting of (*i*) electronic tubes or transistors (because such elements are not bilateral) and (*ii*) nonlinear elements such as powdered iron, electric arc etc.

Q. 6. How does the resistance of a homogeneous material having constant length vary with the changing cross-sectional area ?

Ans. Resistance of a homogeneous material having constant length is inversely proportional to its area of x-section *i.e.* $R \propto \frac{1}{a}$.

Q. 7. What is meant by specific resistance of material ?

Ans. The specific resistance, also called the resistivity, is equal to the resistance of the material of unit length having unit cross-sectional area.

Q. 8. The length of a conductor is doubled and its area of x-section is also doubled, then the resistance will

(*i*) increase four times (*ii*) remain unchanged

(*iii*) decrease to four times (*iv*) change at random. **[P.T.U. B.E.E.E. May 2006]**

Ans. Resistance remain unchanged because $R_2 = \rho\frac{l_2}{a_2} = \rho \times \frac{2l_1}{2a_1} = \rho\frac{l_1}{a_1} = R.$

Q. 9. How does the resistance of metals and semiconductors varies with temperature ?

Ans. The resistance of metals increases with the increase in temperature while the resistance of semiconductors decreases with the increase in temperature.

Q. 10. Define temperature coefficient of resistance and give its units. **[P.T.U. December 2009]**

Ans. The temperature coefficient of resistance may be defined as the ratio of increase in resistance per degree rise of temperature to the original resistance. Its unit is Ω/Ω/°C.

Q. 11. What is the effect of temperature on the resistance of the conductor ? **[P.T.U. B.E.E.E. Dec. 2007]**

Ans. Please refer to Art. 1.9.

Q. 12. What is the effect of temperature on resistivity ? **[P.T.U. B.E.E.E. May 2007]**

Ans. Please refer to Art. 1.9.

Q. 13. State factors affecting the resistivity of conductors.

Ans. Resistivity of a material depends mainly on 3 factors viz. temperature, alloying and mechanical stressing.

The resistivity of all pure metallic conductors increases with the increase in temperature.

Alloying is another factor which affects the resistivity of a material. By adding traces of some impurities to a metal, its resistivity increases.

The resistivity of a material also changes under the influence of mechanical treatment. Although mechanical stressing increases the resistivity, annealing restores the electrical conductivity by establishing regularity in crystal structure.

Q. 14. Give the relations between (*i*) kWh and k calories and (*ii*) kWh and mega joules.

Ans. (*i*) 1 kWh = 860 K calories (*ii*) 1 kWh = 3.6 MJ

Q. 15. What do you mean by heating effect of electricity ? **[P.T.U. B.E.E.E. Dec. 2004]**

Ans. According to the *law of conservation of energy* the electrical energy expended must be converted in some other form of energy and that other form is heat *i.e.* electrical energy expended is converted into heat energy and conversion of electrical energy into heat energy is called the *heating or thermal effect of electric current*.

Q. 16. What are linear and nonlinear circuits ? **[P.T.U. B.E.E.E. Dec. 2007]**

Ans. Resistive elements for which the volt-ampere characteristic is a straight line [Fig. 1 (*a*)] are called *linear*, and the electric circuits containing only linear resistances are called *linear circuits.*

Resistive elements for which the volt-ampere characteristic is other than a straight line are termed *nonlinear,* and so the electric circuits containing them are called *nonlinear circuits.* Examples of nonlinear elements are tungsten lamps, vacuum tubes and transistors, etc.

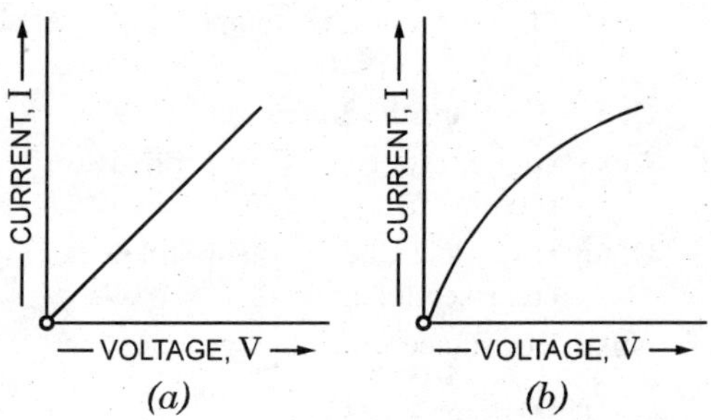

Fig. 1

Q. 17. Give reasons why are all equipments connected in parallel to the supply. **[P.T.U. B.E.E. May 2006]**

Ans. All equipments are connected in parallel so that each one have same voltage across it and can be operated independently.

PROBLEMS

1. A gas-filled lamp has a resistance of 18 Ω when not burning. When it is burning with a voltage of 120 V, its resistance is 200 Ω. Calculate the current (*i*) at the instant of switching on and (*ii*) normal current when burning. [**Ans.** (*i*) 6.667 A (*ii*) 0.6 A]
2. Calculate the length of copper wire, 1.5 mm in diameter, to have a resistance of 0.3 ohm, given that resistivity of copper is 0.017 microohm-m. [**Ans.** 31.185 m]
3. Find the ratio of cross-sectional area of copper wire to that of aluminium wire to have the same resistance and length if their resistivities are 1.72 μΩ-cm and 2.8 μΩ-cm respectively. [**Ans.** 1 : 0.6143]
4. A conductor has a cross section of 10 cm^2 and specific resistance of 7.5 μΩ-cm at 0 °C. What will be its resistance in ohms per km when the temperature is 40 °C ? Take temperature coefficient of the material 0.005 per °C. [**Ans.** 0.09 Ω]
5. The shunt field coil of a generator has a resistance of 200 Ω at 20 °C. Find its resistance at 80° C, if the temperature coefficient of resistance of the material of wire at 20 °C is 0.004. [**Ans.** 248 Ω]

6. It is found that the resistance of a coil of wire increases from 40 Ω at 15 °C to 50 Ω at 60 °C. Calculate the resistance temperature coefficient of the material and the resistance of the coil at 0 °C.

[**Ans.** $\frac{1}{165}$ per °C; 36.667Ω]

7. The resistance of the field coils with copper conductors of a dynamo is 120 Ω at 25 °C. After working for 6 hours on full load, the resistance of the coil increases to 140 Ω. Calculate the mean temperature rise of the field coil. Take the temperature coefficient of the conductor material as 0.0042 at 0 °C.

[**Ans.** 43.8 °C]

8. The output of an electric motor is 12.5 kW and its efficiency is 90%. Determine the input in kW. Also determine the torque in (*i*) N-m and (*ii*) kg-m when it is running at a speed of 600 rpm.

[**Ans.** 13.89 kW; 221 N-m; 22.53 kg-m]

9. A 220 V dc motor develops 8 kW at 750 rpm. Determine the torque developed in N-m. If the efficiency of the motor is 88%, calculate the current drawn from the mains and the number of units consumed if the motor operates for 6 hours. [**Ans.** 115.75 N-m; 41.32 A; 54.55 units]

10. What is the resistance of a 100 W, 200 V bulb ? [**Ans.** 400 Ω]

11. An electric kettle was marked 500 W, 230 V and was found to take 15 minutes to bring 1 kg of water at 15 °C to the boiling point. Determine the heating efficiency of the kettle. [**Ans.** 78.95%]

12. A 500 watts, 240 V electric kettle takes 20 minutes to bring 1.5 kg of water from 20 °C to boiling point. Determine the heating efficiency of the kettle. Also determine the cost of operating the kettle for 4 hours if the energy is available at Rs. 4.25 per unit. [**Ans.** 83.72%; Rs. 8.50]

13. Calculate the total number of units consumed in operating a 100 watt lamp for 5 hours, 500 W electric iron for 3 hours and 1 kW heater for 4 hours. Find also the cost of energy consumed if energy is supplied at the rate of Rs. 5.50 paise per unit. [**Ans.** 6 units, Rs. 33]

14. Four resistors A, B, C and D are connected in parallel. If A is of 56 Ω; B is of 50 Ω; C is of 40 Ω and takes a current of 2.5 A and the total current drawn from the supply is 9.0 A, determine the resistance of D in ohms. [**Ans.** 36.84 Ω]

15. Two resistors of equal values when placed in parallel gives an effective resistance equal to R. What will be the effective resistance if the two resistors are connected in series. [**Ans.** 4 R]

16. A resistor of value 1.6 Ω is connected in series with two resistors of value 4 Ω and 6 Ω which are connected in parallel. Find the total resistance of the series-parallel combination. [**Ans.** 4 Ω]

17. Two resistances of 4 Ω and 6 Ω in parallel are in series with another resistor of 12 Ω. If the current flowing in 12-Ω resistor is 2 A, determine (*i*) the current flowing through 4 Ω and 6 Ω resistors, (*ii*) voltage across the whole circuit and (*iii*) power drawn from the supply.

[**Ans.** (*i*) 1.2 A and 0.8 A; (*ii*) 28.8 V; and (*iii*) 57.6 watts]

18. In the circuit shown in Fig. 2, find the voltage across and the current in each element.

[**Ans.** Current through 2 Ω and 4 Ω resistors = 2 A and voltage across these resistors are 4 V and 8 V respectively. Current through 4 Ω and 8 Ω resistors = 1 A and voltage across them are 4 V and 8 V respectively]

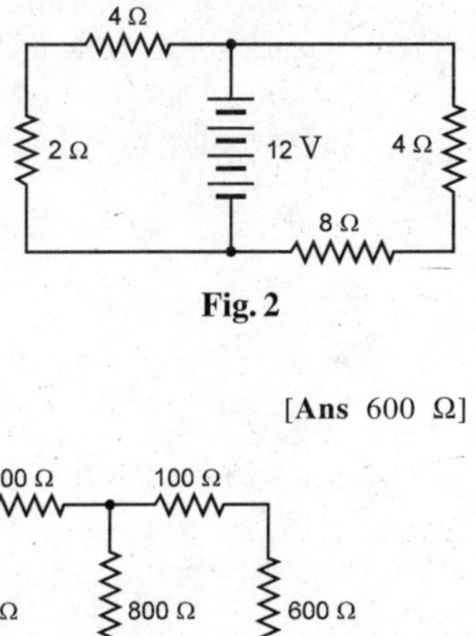

Fig. 2

19. How much current a battery of 8 volts will give when it is connected across AB in the circuit diagram given below (Fig. 3) [**Ans.** 1 A]

20. What is the equivalent resistance of the network shown in Fig. 4. [**Ans** 600 Ω]

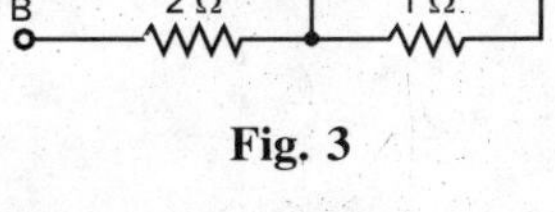

Fig. 3

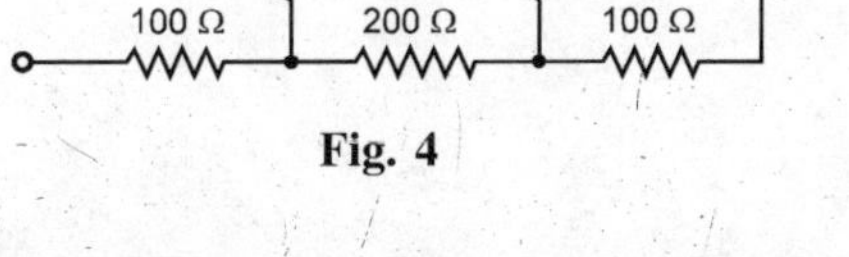

Fig. 4

21. In the given circuit (Fig. 5) find the current in each branch and direction of current.

[**Ans.** $\frac{96}{41}$ A, $\frac{60}{41}$ A, $\frac{36}{41}$ A]

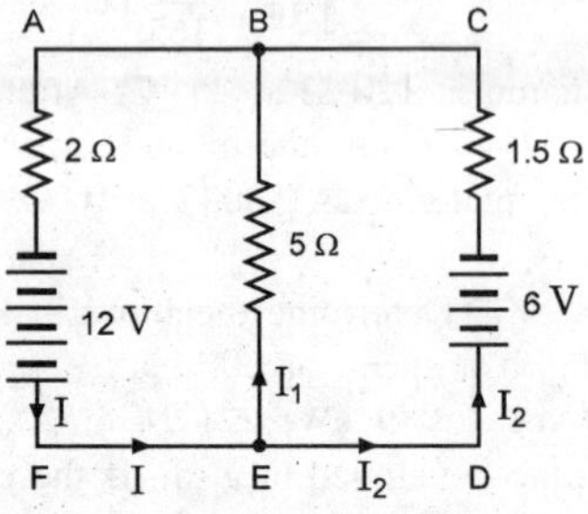

Fig. 5

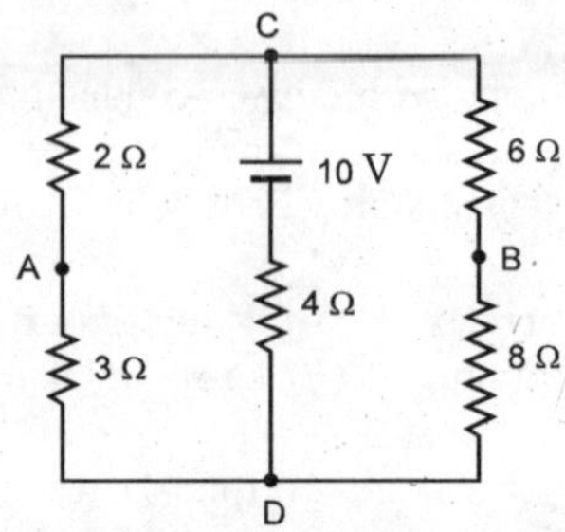

Fig. 6

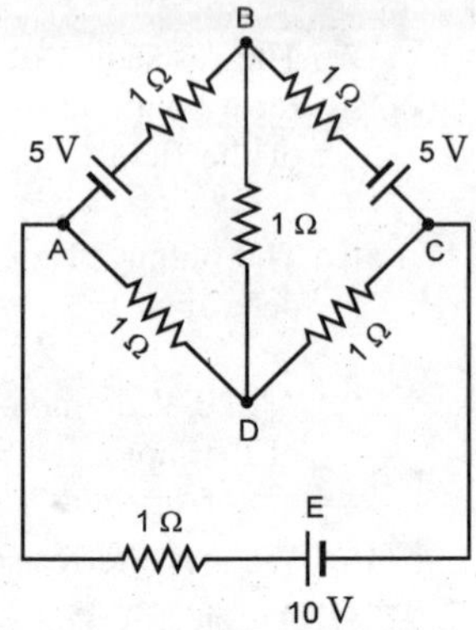

Fig. 7

22. Calculate the current in the battery, the current in each branch and pd across AB of the network shown in Fig. 6.

[**Ans.** $\frac{95}{73}$ A, $\frac{70}{73}$ A, $\frac{25}{73}$ A, $\frac{10}{73}$ A, $\frac{10}{73}$ V]

23. Determine the branch currents in the network of Fig. 7 when the value of each branch resistance is one ohm.

[**Ans.** 6.25 A, 6.25 A, 0, 1.25 A,1.25 A,7.5 A]

24. A Wheatstone bridge consists of AB = 400 Ω, BC = 220 Ω, CD = 225 Ω and DA = 100 Ω. A 2-V cell is connected between A and C and a galvanometer of 200 Ω resistance between B and D. Estimate the current through the galvanometer by applying Kirchhoff's laws.

[**Pb. Univ. Electrical Engineering-I Nov. 1983**]

[**Ans.** 2/703 A from D to B]

25. The step voltage applied to a coil of resistance 25 Ω and inductance 0.5 H is 220 V. Calculate (*i*) initial rate of rise of current (*ii*) time constant and (*iii*) steady-state current.

[**Ans.** (*i*) 440 A/s (*ii*) 0.02 s (*iii*) 8.8 A]

26. The step voltage applied to a series R-L circuit with R = 15 Ω is 30 V. Determine the value of L required to make the current of 1.0 A at 250 microseconds. Assume that the initial current is zero.

[**Ans.** 5.41 mH]

27. The step voltage applied to a series R-C circuit with R = 0.5 MΩ and C = 8 μF is 200 V. Determine (*i*) time constant of the circuit (*ii*) the initial charging current (*iii*) time taken for the pd across the capacitor to grow to 100 V and (*iv*) the current and the pd across the capacitor 4 *s* after application of step voltage. [**Ans.** (*i*) 4 s (*ii*) 0.4 m A (*iii*) 6.438 s (*iv*) 0.147 mA ; 126.42 V]

28. Determine $v_C(t)$ for $t > 0$ in the circuit shown in Fig. 8. Assume initial condition $v_C(0^-) = 9$ V.

[**Ans.** $v_C(t) = 1 + 8e^{-4t}$ volts]

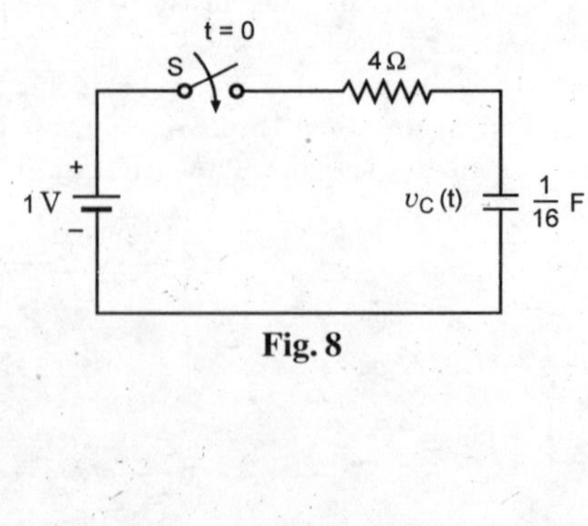

Fig. 8

□□□

2

CHAPTER

Alternating (Sinusoidal) Voltage and Current

2.1. INTRODUCTION

A current (or voltage) is called *alternating* if it reverses periodically in direction, and its magnitude undergoes a definite cycle of changes in definite intervals of time. Each cycle of alternating current (or voltage) consists of two half cycles, during one of which the current (or voltage) acts in one direction; while during the other in opposite direction. In more restricted sense, alternating current is a periodically varying current, the average value of which, over a period, is zero. The direct current always flows in one direction, and its magnitude remains unaltered. In order to produce an alternating current through an electric circuit, a source capable of reversing the emf periodically (ac generator) is required while for generating dc in an electric circuit, a source capable of developing a constant emf is required such as a battery or dc generator. The graphical representations of direct current and alternating current are given in Figs. 2.1(*a*) and 2.1 (*b*) respectively.

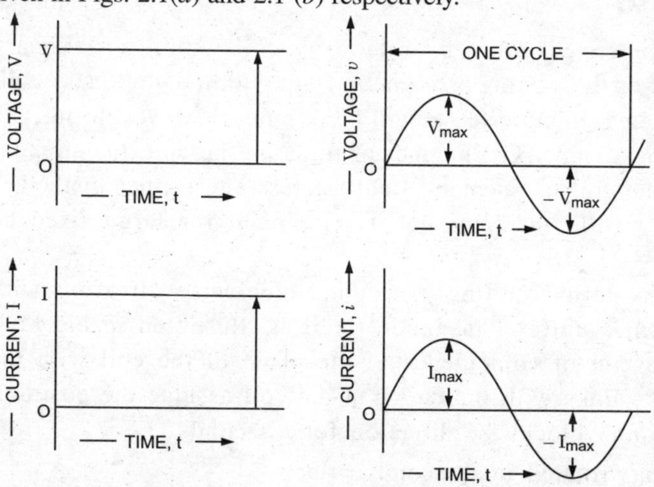

(*a*) *DC Voltage and Current* (*b*) *AC Voltage and Current*

Fig. 2.1

INSIDE THIS CHAPTER

At present a large percentage of the electrical energy (nearly all) being used for domestic and commercial purposes is generated as alternating current. In fact, almost the whole of the vast amount of electrical energy used throughout the world for every imaginable purpose is generated by alternating current generators. This is not due to any superiority of alternating current over direct current in the sphere of applicability to industrial and domestic use. In fact there are certain types of works for which alternating current is unsuitable and, therefore, direct current is absolutely necessary such as for electroplating, charging of storage batteries, refining of copper, refining of aluminium, electrotyping, production of industrial gases by electrolysis, municipal traction etc. In some power applications, the ac motor is unsatisfactory such as for metal rolling mills, paper making machines, high speed gearless elevators, automatic machine tools and high-speed printing presses. Direct current required for these applications is nowadays derived from an ac supply by the use of suitable convertors or rectifiers. For lighting and heating dc and ac are equally useful. The reasons for generation of electrical energy in the form of alternating current are given below:

1. AC generators have no commutator and can, therefore, be built in very large units to run at high speeds producing high voltages (as high as 11,000 volts), so that the construction and operating cost per kW is low, whereas dc generator capacities and voltages are limited to comparatively low values.
2. Alternating current can be generated at comparatively high voltages and can be raised and lowered readily by a static machine called the *transformer* which makes the transmission and distribution of electrical energy economical. In direct current use of transformers is not possible.
3. AC induction motor is cheaper in initial cost as well as in maintenance since it has got no commutator and is more efficient than dc motor for constant speed work, so it is desirable to generate power as alternating current.
4. The high transmission efficiency in ac makes the generation of electrical energy economical by generating it in large quantities in a single station and distributing over a large territory.
5. The switchgear (*e.g.* switches, circuit breakers etc.) for ac system is simpler than that required in a dc system.
6. The maintenance cost of ac equipment is less.

2.2. GENERATION OF ALTERNATING EMF

We know that an alternating emf can be generated either by rotating a coil within a stationary magnetic field, as illustrated in Fig. 2.2 (*a*) or by rotating a magnetic field within a stationary coil, as illustrated in Fig. 2.2 (*b*). The emf generated, in either case, will be of sinusoidal waveform. The magnitude of emf generated in the coil depends upon the number of turns on the coil, the strength of magnetic field and the speed at which the coil or magnetic field rotates. The former method is employed in case of small ac generators while the later one is employed for large sized ac generators.

Now consider a rectangular coil of N turns rotating in counter-clockwise direction with angular velocity of ω radians per second in a uniform magnetic field, as illustrated in Fig. 2.3.

Let the time be measured from the instant of coincidence of the plane of the coil with the X-axis. At this instant maximum flux, Φ_{max} links with the coil. Let the coil assume the position, as shown in Fig. 2.3, after moving in counter clockwise direction for t seconds.

The angle θ through which the coil has rotated in t seconds $= \omega t$

In this position, the component of flux along perpendicular to the plane of coil $= \Phi_{max} \cos \omega t$.

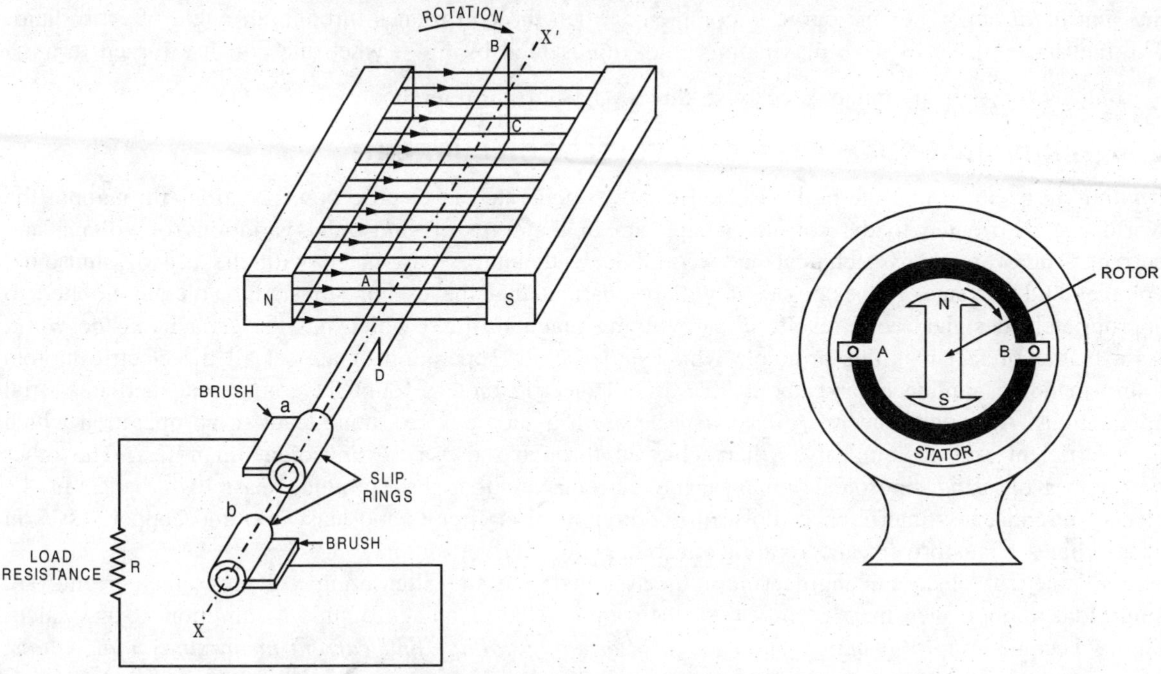

(a) *(b)*

Fig. 2.2. ***Generation of Alternating EMF***

Hence flux linkage of the coil at this instant = Number of turns on coil × linking flux

i.e., instantaneous flux linkage = N Φ_{max} cos ωt

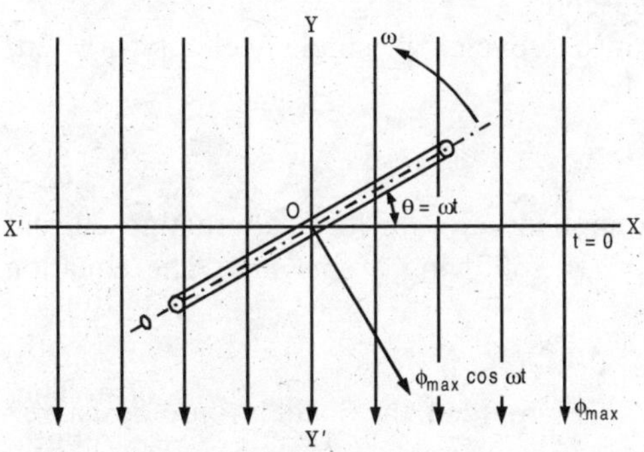

Fig. 2.3

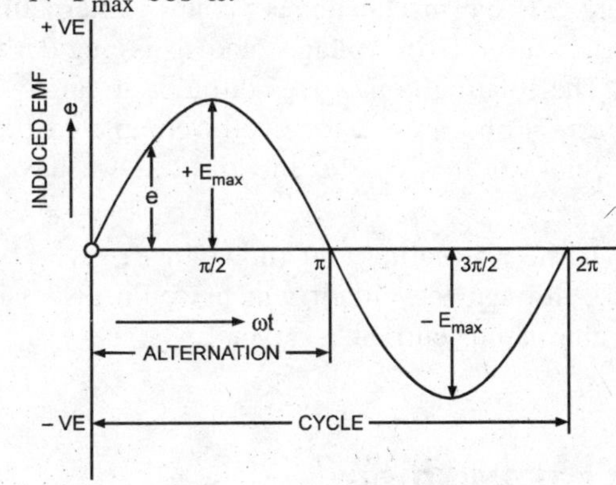

Fig. 2.4

Since emf induced in a coil is equal to the rate of change of flux linkage with minus sign, EMF induced at any instant,

$$e = -\frac{d}{dt}\left[N\Phi_{max}\cos\omega t\right] = \Phi_{max}\,N\,\frac{d}{dt}\left[-\cos\omega t\right] = \Phi_{max}\,N\omega\sin\omega t \qquad ...(2.1)$$

when $\omega t = 0$, $\sin\omega t = 0$, therefore, induced emf is zero, when $\omega t = \frac{\pi}{2}$, $\sin\frac{\pi}{2} = 1$, therefore, induced emf is maximum, which is denoted by E_{max} and is equal to Φ_{max} Nω. Substituting Φ_{max} Nω = E_{max} in Eq. (2.1) we have

$$\text{Instantaneous emf, } e = E_{max}\sin\omega t \qquad ...(2.2)$$

So the emf induced varies as the sine function of the time angle, ωt, and if emf induced is plotted against time, a curve of sine wave shape is obtained as illustrated in Fig. 2.4. Such an emf is called the *sinusoidal* emf. The sine curve is completed when the coil rotates through an angle of 2π radians. The induced emf e will have maximum value, represented by E_{max}, when the coil has turned through $\frac{\pi}{2}$ radians (or 90°) in counter-clockwise direction from the reference axis (*i.e.* OX axis).

2.3. SINUSOIDAL QUANTITIES (EMF, VOLTAGE OR CURRENT)

It is not an accident that the bulk of electric power generated in electric power stations throughout the world and distributed to the consumers appears in the form of sinusoidal variations of voltage and current. There are many technical and economical advantages associated with the use of sinusoidal voltages and currents. For example, it will be learned that the use of sinusoidal voltages applied to appropriately designed coils results in a revolving magnetic field which has the capacity to do work. As a matter of fact it is this principle which underlies the operation of almost all the electric motors found in home appliances and about 90% of all electric motors found in commercial and industrial applications. Although other waveforms can be used in such devices, none leads to an operation which is as efficient and economical as that achieved through the use of sinusoidal quantities. The other advantages of using sinusoidal voltages and currents are that they produce least disturbance in the electric circuits, less interference (noise) to communication circuits and least iron and copper losses in ac machines (transformers and rotating machines) for a given output.

In practical electrical engineering it is assumed that the alternating voltages and currents are sinusoidal, though they may slightly deviate from it. The advantage of this assumption is that calculations become simple. *It may be noted that alternating voltage and current mean sinusoidal voltage and current unless stated otherwise.*

Alternating emf following sine law (*i.e.* sinusoidal emf) is illustrated in Fig. 2.4 and is expressed in the form

$$e = E_{max} \sin \omega t \qquad ...(2.3)$$

where e is the instantaneous value of alternating emf (or voltage), E_{max} is the maximum value of the alternating emf (or voltage) and ω is angular velocity of the coil.

The rotating coil moves through an angle of 2π radians in one cycle, so angular velocity $\omega = 2\pi f$ where f is the number of cycles completed per second.

Substituting $\omega = 2\pi f$ in Eq. (2.3) we have

$$e = E_{max} \sin 2\pi f t \qquad ...(2.4)$$

If the alternating emf (or voltage) given by Eq. (2.3) is applied across a load, alternating current flows through the circuit which would also vary sinusoidally *i.e.* following a sine law. The equation for alternating current is given as

$$i = I_{max} \sin \omega t = I_{max} \sin 2\pi f t \qquad ...(2.5)$$

provided the load is pure resistive*

2.4. TERMINOLOGY

An *alternating quantity* (voltage or current) is one which changes continuously in magnitude and alternates in direction at regular intervals of time. It rises from zero to maximum positive value, falls to zero, increases to a maximum in the reverse direction and falls back to zero again, as illustrated in Fig. 2.4. The emf (or voltage) and current repeat the procedure.

The important ac terms are defined below:

2.4.1. Waveform

The shape of the curve of the voltage or current when plotted against time as abcissa (base) is called the *waveform*. The waveform of an alternating voltage varying sinusoidally is shown in Fig. 2.4.

* The load may be resistive, inductive or capacitive. It will be shown in chapter 3 that if the load is inductive or capacitive the current equation differs in time angle.

The waveform of the induced emf in an alternator differs slightly from that of sine wave but for calculation purposes it is treated as such.

2.4.2. Instantaneous Value

The value of alternating quantity (emf, voltage or current) at any particular instant is called the *instantaneous value* and is designated by a small italic letter (*e* for emf, *v* for voltage and *i* for current). The instantaneous values of an alternating quantity can be determined either from the curve or from an equation of the alternating quantity. For example, the instantaneous values of emf represented by the curve shown in Fig. 2.4 at 0, $\frac{\pi}{2}$, π and $\frac{3\pi}{2}$ are zero, $+E_{max}$, zero and $-E_{max}$ respectively.

2.4.3. Alternation and Cycle

When a periodic wave, such as sinusoidal wave, goes through one complete set of positive or negative values, it completes one *alternation* and when it goes through one complete set of positive and negative values it is said to have completed one *cycle*.

Alternation and cycle can also be defined in terms of angular measure. One alternation corresponds to 180°(or π radians) while one cycle corresponds to 360° (or 2π radians).

2.4.4. Time Period and Frequency

The time taken in seconds by an alternating quantity to complete one cycle is known as *time period* or *periodic time* and is denoted by T.

The number of cycles completed per second by an alternating quantity is known as *frequency* and is denoted by *f*. In SI system the frequency is expressed in hertz (pronounced as hurts). One hertz (or Hz) is equal to one cycle per second.

The number of cycles completed per second $= f$

$$\text{Time period, T} = \text{Time taken in completing one cycle} = 1/f$$

$$\text{or } f = 1/\text{T} \qquad ...(2.6)$$

In India the standard frequency for power supply is 50 Hz.

2.4.5. Angular Velocity and Frequency

A glance at Fig. 2.4 indicates that each cycle spans 2π radians. Hence, if this quantity is divided by the time period, *angular velocity* of the sine function is obtained. It is denoted by ω and is expressed in radians per second.

$$\therefore \text{ Angular velocity, } \omega = \frac{\text{Angle turned}}{\text{Time taken}} = \frac{2\pi}{\text{T}} = \frac{2\pi}{1/f}$$

$$= 2\pi f \text{ radians per second} \qquad ...(2.7)$$

2.4.6. Electrical Time Degrees and Mechanical Degrees

It is seen that the coil must revolve past a pair of poles in order to carry the generated emf through one complete cycle. In circuit work, one complete cycle of voltage or current is designated as 360 electrical degrees or 2π electrical radians. To correlate with this designation the arc through which a coil of dynamo must rotate in order to generate one cycle of emf is called 360 electrical degrees. In a 2-pole machine one complete revolution of coil produces one cycle of emf. But in a multipolar machine, such as four or six or eight pole machine, the emf completes one cycle or 360 electrical degrees or 2π electrical radians as soon as the coil passes a pair of poles and a mechanical degree will be equal to as many electrical degrees as there are pairs of poles in the dynamo structure. In a multipolar machine, the number of cycles completed per second by generated emf,

f = Pair of poles × number of revolutions made per second

$$= \frac{P}{2} \times \frac{N}{60} = \frac{PN}{120} \quad ...(2.8)$$

where N is the speed of rotation of the coil in rpm.

2.4.7. Amplitude

The maximum value, positive or negative, which an alternating quantity attains during one cycle is called the *amplitude* of the alternating quantity. The amplitude of an alternating emf (or voltage) and current is designated by E_{max} (or V_{max}) and I_{max} respectively.

Example 2.1. What is the time period of the wave produced by a 6-pole alternator which is driven at 1,000 rpm ?

[*Pb. Technical Univ. Basic Electrical Engineering, June 2003*]

Solution : Number of poles on alternator, P = 6

Speed of alternator, N = 1,000 rpm

Number of cycles completed per second by generated emf,

$$f = \frac{PN}{120} = \frac{6 \times 1{,}000}{120} = 50$$

$$\text{Time period, T} = \frac{1}{f} = \frac{1}{50} = 0.02 \text{ second} \quad \textbf{Ans.}$$

2.5. DETERMINATION OF MAXIMUM VALUE AND FREQUENCY FROM EMF OR CURRENT EQUATIONS

From Eq. (2.4) $e = E_{max} \sin 2\pi ft$

From above equation we observe that

(i) the maximum value of an alternating emf is given by the coefficient of the sine of the time angle.

(ii) the frequency is given by coefficient of time t divided by 2π

$$i.e. \quad f = \frac{\text{Coefficient of time, } t}{2\pi} \quad ...(2.9)$$

Similarly we can also find the maximum value and frequency of the current from the equation of instantaneous values of current.

Example 2.2. An ac voltage of 50 Hz frequency has a peak value of 220 V. Write down the expression for the instantaneous value of this voltage. [*Pb Technical Univ. Electrical Engineering January 2000*]

Solution : Supply frequency, f = 50 Hz

Peak value of ac voltage, V_{max} = 220 V

Expression for instantaneous value of ac voltage (assumed sinusoidal) with θ as zero when time is zero, is given as

$$v = V_{max} \sin \omega t = V_{max} \sin 2\pi ft = 220 \sin 2\pi \times 50\, t = 220 \sin 314t \quad \textbf{Ans.}$$

Example 2.3. An alternating current of frequency 60 Hz has a maximum value of 120 A. Write down the equation for its instantaneous value. Find the time taken to reach 96 A for the first time.

Solution : Maximum value of current, I_{max} = 120 A

Frequency of alternating current, f = 60 Hz

The equation for instantaneous value of alternating current (assumed sinusoidal) with θ as zero when time is zero, is given by the expression

$$i = I_{max} \sin 2\pi ft = 120 \sin 2\pi \times 60t = 120 \sin 120\pi t \quad \textbf{Ans.}$$

Let the instantaneous value of current be 96 A for the first time, t seconds after the instant the current is zero and becoming positive.

$$\therefore\ 96 = 120 \sin 120\pi t$$

$$\text{or } t = \frac{1}{120\pi} \sin^{-1} \frac{96}{120} = \frac{1}{120\pi} \sin^{-1} 0.8 = 0.00246 \text{ second or } 2.46 \text{ ms } \textbf{Ans.}$$

Example 2.4. An alternating current of frequency 50 Hz has a maximum value of 100 A. Calculate (*a*) its value $\frac{1}{600}$ second after the instant the current is zero and its value decreasing there afterwards (*b*) how many seconds after the instant the current is zero (increasing there afterwards) will the current attain the value of 86.6 A ?

Solution : The current wave form is shown in Fig. 2.5.

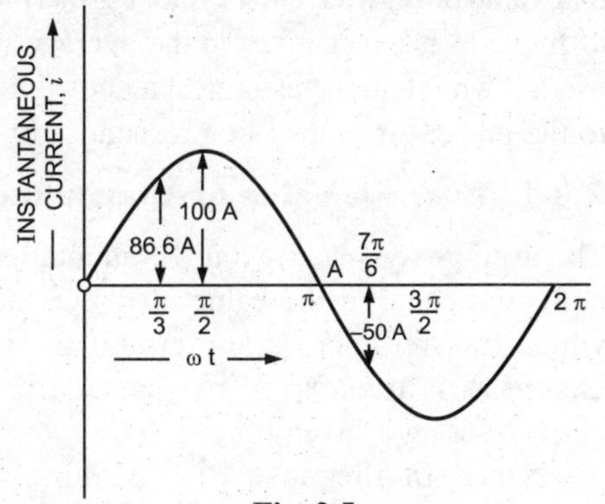

Fig. 2.5

The equation of the alternating current (assumed sinusoidal) with respect to the origin O is

$$i = 100 \sin 100\,\pi\, t \qquad ...(i)$$

(*a*) Since the current is measured from the instant the current is zero and is decreasing there afterwards (*i.e.* from point A in Fig. 2.5), the equation for the alternating current with respect to the point A becomes

$$i = 100 \sin (100\,\pi\, t + \pi) = -100 \sin 100\,\pi t \qquad ...(ii)$$

Substituting $t = \frac{1}{600}$ second in above equation we get the instantaneous value of current $\frac{1}{600}$ second after the instant the current is zero and decreasing there afterward.

$$\text{So } i = -100 \sin 100\,\pi \times \frac{1}{600} = -100 \sin \frac{\pi}{6} = -50 \text{ A } \textbf{Ans.}$$

(*b*) Let the current attain the value of 86.6 A, *t* seconds after the zero value of the current. Now substituting $i = 86.6$ A in Eq. (*i*) we get

$$86.6 = 100 \sin 100\,\pi\, t$$

$$\text{or } t = \frac{1}{100\,\pi} \sin^{-1} \frac{86.6}{100} = \frac{1}{100\,\pi} \times \frac{\pi}{3} = \frac{1}{300} \text{ second } \textbf{Ans.}$$

2.6. PLOTTING OF SINE WAVEFORM

Sine curve may be graphically drawn, as illustrated in Fig. 2.6. Draw a circle of radius equal to the maximum value of sinusoidal quantity. Divide the circumference of the circle drawn so into any number of equal parts, say 12, and draw a horizontal line AB (the base on which the sine wave is to be drawn) passing through the centre of the circle. Divide the line AB into the same number of equal parts *i.e.* 12 and number the points correspondingly. Draw perpendicular ordinates from each point. Project the points on the circle horizontally to meet the perpendicular ordinates having corresponding numbers. Draw smooth curve through these points. Curve so drawn will be of sine waveform.

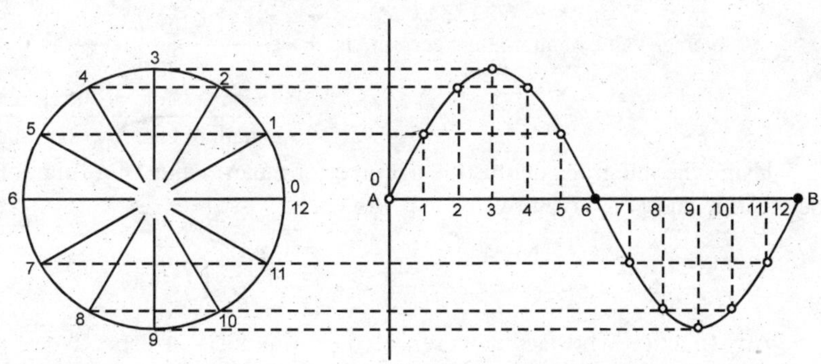

Fig. 2.6

2.7. AVERAGE AND EFFECTIVE (RMS) VALUES OF ALTERNATING VOLTAGE AND CURRENT

In a dc system, the voltage and current are constant and, therefore, there is no problem in specifying their magnitude. But in case of ac system, an alternating voltage or current varies from instant to instant and so poses a problem how to specify the magnitude of an alternating voltage or current. An alternating voltage or current may possibly be expressed in terms of peak (maximum) value, average (mean) value or effective (rms) value.

In specifying an alternating voltage or current, its peak or maximum value is rarely used because it has that value only twice each cycle. Furthermore, the average or mean value cannot be used because it is positive as much as it is negative, so the average value is zero. Although the average value over half cycle might be used, it would not be as logical a choice as what we shall find *effective* (*virtual or rms*) *value* which is related to the power developed in a resistance by the alternating voltage or current.

2.7.1. Average Value of Alternating Current

The average (or mean) value of an alternating current is equal to the value of direct current which transfers across any circuit the same charge as is transferred by that alternating current during a given time.

Since in the case of a symmetrical alternating current (*i.e.* one whose two half cycles are exactly similar, whether sinusoidal or non-sinusoidal) the average or mean value over a complete cycle is zero hence for such alternating quantities average or mean value means the value determined by taking the average of instantaneous values during half cycle or one alternation only. However, for unsymmetrical alternating current, as half wave rectified current, the average value means the value determined by taking the mean of instantaneous values over the whole cycle.

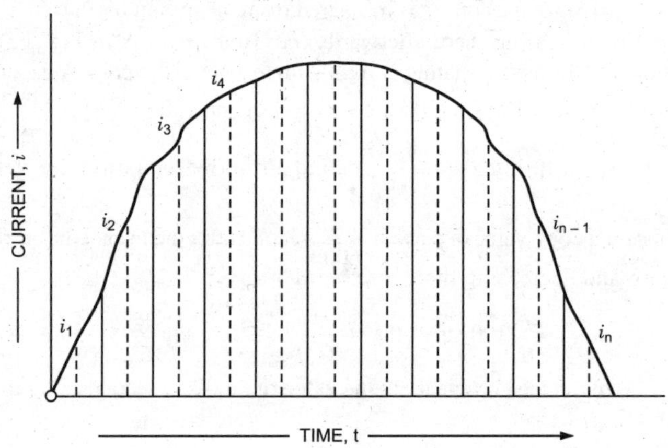

Fig. 2.7

The average value is determined by measuring the lengths of a number of equidistant ordinates and then taking their mean *i.e.* of $i_1, i_2, i_3 \ldots i_n$ etc. which are mid-ordinates.

$$\because \text{ Average value of alternating current, } I_{av} = \frac{i_1 + i_2 + i_3 + i_4 \ldots + i_n}{n}$$

$$= \frac{\text{Area of one alternation (or half cycle)}}{\text{Length of base over one alternation (or half cycle)}}$$

Using the integral calculus the average (or mean) value of a function $f(t)$ over a specific interval of time between t_1 and t_2 is given by

$$F_{av} = \frac{1}{t_2 - t_1} \int_{t_1}^{t_2} f(t)\, dt \qquad \ldots(2.10)$$

Any function whose cycle is repeated continuously, irrespective of its wave shape, is termed as *periodic function*, such as sinusoidal function, and its average value is given by

$$F_{av} = \frac{1}{T} \int_0^T f(t)\, dt \qquad \ldots(2.11)$$

where T is time period of periodic function.

In case of a symmetrical alternating current, whether sinusoidal or non-sinusoidal the average value is determined by taking average of one half cycle or one alternation only.

i.e., for symmetrical waveforms, $$F_{av} = \frac{2}{T}\int_0^{T/2} f(t)\, dt \qquad ...(2.12)$$

2.7.2. RMS Value or Effective Value of Alternating Current

The rms or effective value of an alternating current or voltage is given by that steady current or voltage which when flows or applied to a given resistance for a given time produces the same amount of heat as when the alternating current or voltage is flowing or applied to the same resistance for the same time.

Consider an alternating current of waveform shown in Fig. 2.8 flowing through a resistor of R ohms. Divide the base of one alternation into n equal parts and let the mid-ordinates be $i_1, i_2, i_3 ... i_n$ etc.

Heat produced during 1st interval $= i_1^2 \, R \times \frac{T}{n}$ joules

Heat produced during 2nd interval $= i_2^2 \, R \times \frac{T}{n}$ joules

Heat produced during 3rd interval $= i_3^2 \, R \times \frac{T}{n}$ joules

Heat produced during nth interval $= i_n^2 \, R \times \frac{T}{n}$ joules

Total heat produced in time T

$$= RT \left(\frac{i_1^2 + i_2^2 + i_3^2 + ... i_n^2}{n} \right) \text{ joules}$$

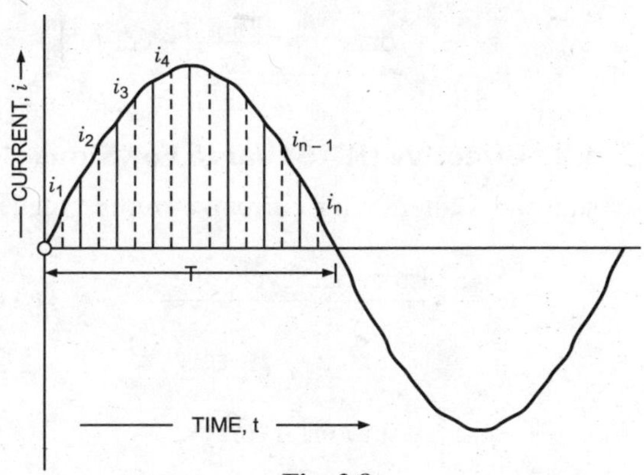

Fig. 2.8

Now if I_{eff} is the effective current, then heat produced by this current in time T = $I_{eff}^2 RT$ joules. By definition these two expressions are equal

$$\therefore \quad I_{eff}^2 \, RT = RT \left(\frac{i_1^2 + i_2^2 + i_3^2 + ... i_n^2}{n} \right)$$

$$\text{or} \quad I_{eff}^2 = \frac{i_1^2 + i_2^2 + i_3^2 + ... i_n^2}{n}$$

$$\text{or} \quad I_{eff} = \sqrt{\frac{i_1^2 + i_2^2 + i_3^2 + ... i_n^2}{n}}$$

Hence the *effective* or *virtual value* of alternating current or voltage is equal to the square root of the mean of the squares of successive ordinates and that is why it is known as *root-mean-square (rms) value.*

Using the integral calculus the root mean square (rms) or effective value of an alternating quantity over a time period is given by

$$F_{rms} = \sqrt{\frac{1}{T}\int_0^T f^2(t)\, dt} \qquad ...(2.13)$$

2.8. AVERAGE AND EFFECTIVE (RMS) VALUES OF SINUSOIDAL CURRENT AND VOLTAGE

2.8.1. Average Value For Sinusoidal Current or Voltage

The average value of a sine wave over a complete cycle is zero. Therefore, the half cycle average value is intended.

Instantaneous value of sinusoidal current is given by

$$i = I_{max} \sin \omega t$$

Considering first half cycle *i.e.* when ωt varies from 0 to π we get,

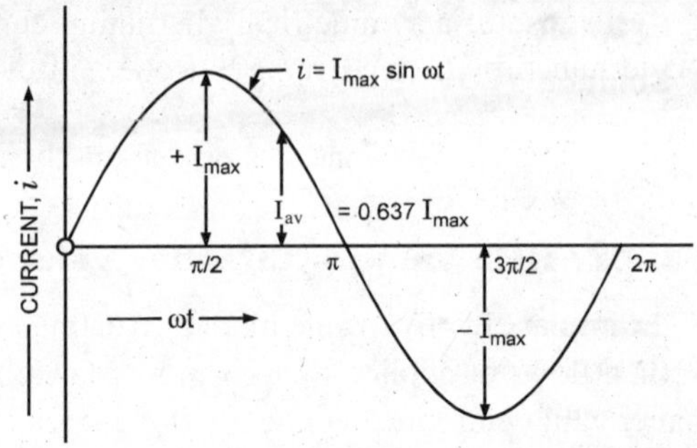

Fig. 2.9

$$I_{av} = \frac{\text{Area of first half cycle}}{\pi}$$

$$= \frac{1}{\pi}\int_0^{\pi} i\,d(\omega t)$$

$$= \frac{1}{\pi}\int_0^{\pi} I_{max} \sin \omega t\, d(\omega t)$$

$$\text{or } I_{av} = \frac{I_{max}}{\pi}\left[-\cos \omega t\right]_0^{\pi} = \frac{2}{\pi} I_{max} = 0.637\ I_{max} \quad ...(2.14)$$

Similarly $E_{av} = 0.637\ E_{max}$

2.8.2. Effective (RMS) Value For Sinusoidal Current or Voltage

A sinusoidal alternating current is represented by $i = I_{max} \sin \omega t$

$$I_{rms}^2 = \frac{\text{Area of first half cycle of } i^2}{\pi} = \frac{1}{\pi}\int_0^{\pi} i^2\, d(\omega t)$$

$$= \frac{1}{\pi}\int_0^{\pi} I_{max}^2 \sin^2 \omega t\, d(\omega t)$$

$$= \frac{I_{max}^2}{2\pi}\int_0^{\pi} (1 - \cos 2\omega t)\, d(\omega t)$$

$$= \frac{I_{max}^2}{2\pi}\left[\omega t - \frac{1}{2}\sin 2\omega t\right]_0^{\pi}$$

$$= \frac{I_{max}^2}{2\pi} \times \pi = \frac{I_{max}^2}{2}$$

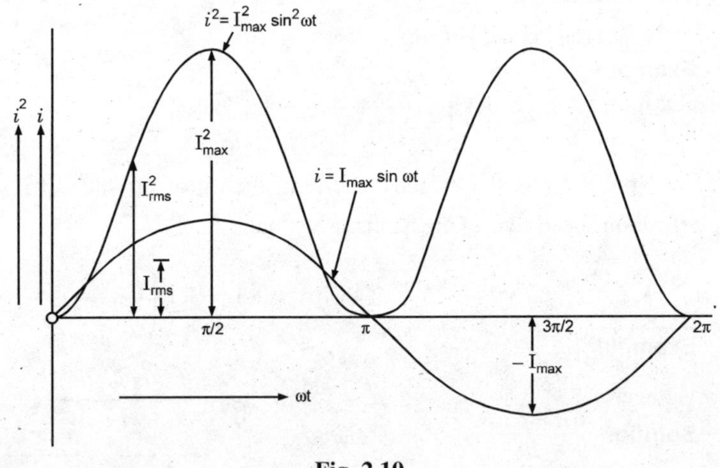

Fig. 2.10

$$\text{or } I_{rms} = \sqrt{\frac{I_{max}^2}{2}} = \frac{I_{max}}{\sqrt{2}} \quad ...(2.15)$$

$$\text{Similarly} \quad E_{rms} = \frac{E_{max}}{\sqrt{2}}$$

Example 2.5. Write the mathematical expression for a 50 Hz sinusoidal voltage supplied for domestic purpose at 230 V. [*Pb. Technical Univ. Basic Electrical and Electronics Engineering Dec. 2005*]

Solution : RMS value for a 50 Hz sinusoidal voltage supplied for domestic purposes,

$$V_{rms} = 230 \text{ V}$$

$$\text{Maximum value, } V_{max} = \sqrt{2} \times V_{rms} = \sqrt{2} \times 230 = 325.27 \text{ V}$$

Expression for instantaneous value with θ as zero when time is zero,

$$v = V_{max} \sin \omega t = V_{max} \sin 2\pi f t = 325.27 \sin 2\pi \times 50t = 325.27 \sin 314t \textbf{ Ans.}$$

Example 2.6. The ac supply at a house is 230 V, 50 Hz. Find maximum, average and rms values of voltage.

Solution : As mentioned in Art. 2.7 the alternating voltage is specified in terms of its effective or rms value.

So rms value of supply voltage, $V_{rms} = 230 \text{ V}$

Maximum value of supply voltage, $V_{max} = \sqrt{2}\, V_{rms} = \sqrt{2} \times 230 = 325.27 \text{ V}$ **Ans.**

Average value of supply voltage, $V_{av} = 0.637 \times V_{max} = 0.637 \times 325.27 = 207.2 \text{ V}$ **Ans.**

Example 2.7. An alternating current when passed through a resistance immersed in water for 5 minutes, just raised the temperature of water to boiling point. When a direct current of 4 amperes was passed through the same resistance under identical conditions it took 8 minutes to boil the water. Find the rms value of the alternating current.

Solution : Let the rms value of alternating current passed through the resistance be I_{rms} amperes.

Heat produced when an alternating current of I_{rms} amperes is passed through a resistance R immersed in water for 5 minutes

$$= (I_{rms})^2 \times R \times 5 \times 60 = 300\, I_{rms}^2 \text{ R joules} \quad ...(i)$$

Heat produced when a direct current of 4 A is passed through the same resistance R immersed in the water for 8 minutes

$$= (4)^2 \times R \times 8 \times 60 = 7{,}680 \text{ R joules} \quad ...(ii)$$

Since heat produced in both of the cases is same, therefore, equating Eq. (*i*) and (*ii*) we get

$$300\, I_{rms}^2 R = 7{,}680 \text{ R}$$

$$\text{or} \quad I_{rms} = \sqrt{\frac{7{,}680}{300}} = \sqrt{25.6} = 5.06 \text{ A} \quad \textbf{Ans.}$$

Example 2.8. What is the equation for a sinusoidal current of 25 Hz frequency having an rms value of 40 A ?

Solution :

Frequency of current, $f = 25 \text{ Hz}$

RMS value of alternating current, $I_{rms} = 40 \text{ A}$

Peak value of sinusoidal current, $I_{max} = \sqrt{2}\, I_{rms} = 40\sqrt{2} \text{ A}$

Expression for 25 Hz sinusoidal current of peak value $40\sqrt{2}$ A is given by the expression

$$i = I_{max} \sin \omega t = I_{max} \sin 2\pi f t$$

$$= 40\sqrt{2} \sin 2\pi \times 25\, t = 56.57 \sin 157t \quad \textbf{Ans.}$$

Example 2.9. An ac voltage wave is expressed as $e = 300 \sin 314t$. Find the peak voltage, frequency, rms voltage and instantaneous voltage at $t = \frac{1}{600}$ second.

Solution : Instantaneous value of alternating voltage is given as

$$e = 300 \sin 314t$$

(*i*) Peak value of alternating voltage,

V_{max} = Coefficient of the sine of the time angle = 300 V **Ans.**

(*ii*) Frequency, $f = \frac{314}{2\pi} = \frac{314}{2 \times 3.14} = 50 \text{ Hz}$ **Ans.** $\quad \because f = \frac{\text{Coefficient of time, } t}{2\pi}$

(*iii*) RMS voltage, $V_{rms} = \frac{V_{max}}{\sqrt{2}} = \frac{300}{\sqrt{2}} = 212.1 \text{ V}$ **Ans.**

(*iv*) Instantaneous voltage at $t = \frac{1}{600}$ second,

$$v = 300 \sin 314 \times \frac{1}{600} = 300 \sin 100\pi \times \frac{1}{600} = 300 \sin \frac{\pi}{6} = 150 \text{ V} \quad \textbf{Ans.}$$

Example 2.10. An ac has frequency 50 Hz and rms current 25 amp. Write equation of instantaneous current and find (*i*) current at time 0.0025 second, (*ii*) Time at which current is 14.14 amp.

[*Pb. Technical Univ. Basic Electrical and Electronics Engineering May 2008*]

Solution : Peak value of current, $I_{max} = \sqrt{2}\, I_{rms} = \sqrt{2} \times 25 = 35.36$ A

Supply frequency, $f = 50$ Hz

The equation of alternating current is given by expression

(*i*) $i = I_{max} \sin 2\pi ft = 35.36 \sin 2\pi \times 50t = 35.36 \sin 314t$ **Ans.**

Substituting $i = 14.14$ A in above equation, we have

$$14.14 = 35.36 \times \sin 100\pi t$$

or Time at which current is 14.14 A is given by the expression

(*ii*) $t = \frac{1}{100\pi} \sin^{-1} \frac{14.14}{35.36} = \frac{1}{100\pi} \sin^{-1} 0.4 = \frac{1}{100\pi} \times 0.131\pi = 0.00131$ second or 1.31 ms **Ans.**

Example 2.11. An ac sinusoidal current has rms value of 40 A at 50 Hz frequency. Write expression for instantaneous current and obtain its value after passing through maximum positive value 0.002 second.

[*Pb. Technical Univ. Basic Electrical and Electronics Engineering Dec. 2004*]

Solution : Expression for instantaneous value of an ac sinusoidal current with θ as zero when time is zero is given as

$$i = I_{max} \sin \omega t = I_{max} \sin 2\pi ft = 40\sqrt{2} \sin 2\pi \times 50t = 56.57 \sin 314t.$$

When time is measured from the positive maximum value, the above equation gets modified into

$$v = 56.57 \sin\left(314t + \frac{\pi}{2}\right) = 56.57 \cos 314t$$

$$= 56.57 \cos 314 \times 0.002$$

$$= 56.57 \cos 0.628 = 56.57 \cos \frac{\pi}{5} = 45.76 \text{ A } \textbf{Ans.}$$

Example 2.12. Instantaneous current is given by the relation $i = 20 \sin 314t$. Find rms and average value of ac.

[*Pb. Technical Univ. Basic Electrical and Electronics Engineering Dec. 2007*]

Solution : Instantaneous value of alternating current is given by the expression $i = 20 \sin 314t$

Maximum value of alternating current is given by the coefficient of the sine of the time angle

i.e. $I_{max} = 20$ A

RMS value of the alternating current, $I_{rms} = \frac{I_{max}}{\sqrt{2}} = \frac{20}{\sqrt{2}} = 14.14$ A **Ans.**

Average value of alternating current, $I_{av} = 0.637\, I_{max} = 0.637 \times 20 = 12.74$ A **Ans.**

Example 2.13. An alternating current varying sinusoidally with a frequency of 50 Hz has an rms value of 20 A. Write down the equation for the instantaneous value and find this value (*a*) 0.0025 second (*b*) 0.0125 second after passing through a positive maximum value. At what time, measured from a positive maximum value, will the instantaneous current be 14.14 A ?

Solution : The current waveform is shown in Fig. 2.11

Peak value of current, $I_{max} = \sqrt{2} I_{rms} = \sqrt{2} \times 20 = 28.28$ A

The equation of alternating current with respect to origin O is

$$i = I_{max} \sin 2\pi ft = 28.28 \sin 2\pi \times 50t = 28.28 \sin 100\,\pi t$$

The above equation is valid when time is measured from the instant the current is zero and increasing in positive direction. Since the time is measured from the positive maximum value (point A in Fig. 2.11), the above equation is modified to

$$i = 28.28 \sin\left(100\pi t + \frac{\pi}{2}\right)$$

$$= 28.28 \cos 100\pi t \quad \textbf{Ans.}$$

(*a*) When $t = 0.0025$ s, the instantaneous value of current,

$$i_1 = 28.28 \cos 100\,\pi \times 0.0025$$

$$= 28.28 \cos \frac{\pi}{4} = 28.28 \times \frac{1}{\sqrt{2}} = 20 \text{ A } \textbf{Ans.}$$

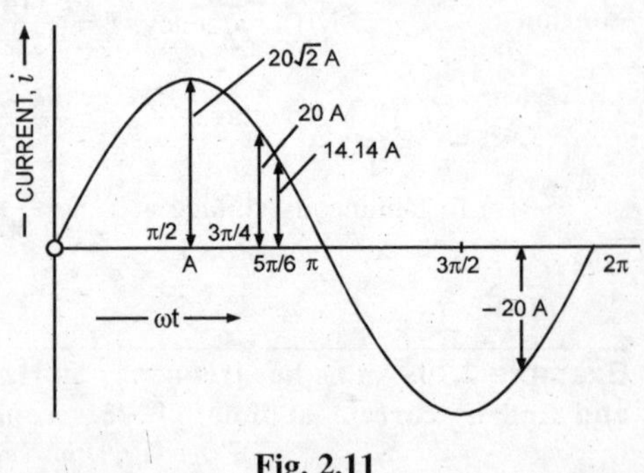

Fig. 2.11

(*b*) When t = 0.0125 s, the instantaneous value of current,

$$i_2 = 28.28 \cos 100\pi \times 0.0125$$

$$= 28.28 \cos \frac{5\pi}{4} = 28.28\left(\frac{-1}{\sqrt{2}}\right) = -20 \text{ A } \textbf{Ans.}$$

(*c*) Substituting i = 14.14 A in expression $i = 28.28 \cos 100\,\pi t$ we have

$$14.14 = 28.28 \cos 100\,\pi t$$

$$\text{or } t = \frac{1}{100\pi} \cos^{-1} \frac{14.14}{28.28} = \frac{1}{100\pi} \cos^{-1} 0.5 = \frac{1}{100\pi} \times \frac{\pi}{3} = \frac{1}{300} \text{ second } \textbf{Ans.}$$

2.9. FORM FACTOR AND PEAK FACTOR

2.9.1. Form Factor

In certain cases it is convenient to have calculations at first upon the mean value of the emf over half a period, therefore, it becomes essential to have some means of connecting this mean value with the effective or rms value. The knowledge of form factor, which is defined as the ratio of effective value to the average or mean value of periodic wave is, therefore, necessary.

Mathematically form factor is given by the relation

$$\text{Form factor} = \frac{\text{Effective value}}{\text{Average value}}$$

$$\text{Form factor for sinusoidal wave, } K_f = \frac{E_{rms}}{E_{av}} = \frac{\dfrac{E_{max}}{\sqrt{2}}}{\dfrac{E_{max}}{\pi/2}} = 1.11$$

2.9.2. Peak Factor

Knowledge of peak factor of an alternating voltage is very essential in connection with determining the dielectric strength since the dielectric stress developed in any insulating material is proportional to the maximum value of the voltage applied to it.

Peak or crest or amplitude factor of a periodic wave is defined as the ratio of maximum or peak to the effective or rms value of the wave

$$\textit{i.e. } \text{Peak factor, } K_p = \frac{\text{Maximum value}}{\text{Effective value}}$$

$$\text{Peak factor for sinusoidal wave, } K_p = \frac{E_{max}}{E_{rms}} = \frac{E_{max}}{E_{max}/\sqrt{2}} = 1.414$$

Example 2.14. The equation of an alternating current $i = 42.42 \sin 628t$. Determine (*i*) its maximum value (*ii*) frequency (*iii*) rms value (*iv*) average value (*v*) form factor.

Solution : Instantaneous value of alternating current is given by the expression

$$i = 42.42 \sin 628t$$

(*i*) Maximum value of an alternating current is given by the coefficient of the sine of the time angle

$$\textit{i.e. } I_{max} = 42.42 \text{ A } \textbf{Ans.}$$

(*ii*) Frequency is given by coefficient of time, t divided by 2π

$$\therefore \text{ Frequency, } f = \frac{628}{2\pi} = 100 \text{ Hz } \textbf{Ans.}$$

(*iii*) RMS value of the alternating current,

$$I_{rms} = \frac{I_{max}}{\sqrt{2}} = \frac{42.42}{\sqrt{2}} = 30 \text{ A } \textbf{Ans.}$$

(*iv*) Average value of the alternating current,

$$I_{av} = 0.637\ I_{max} = 0.637 \times 42.42 = 27\text{A}\ \textbf{Ans.}$$

$$(v)\ \text{Form factor } K_f = \frac{I_{rms}}{I_{av}} = \frac{30}{27} = 1.11\ \textbf{Ans.}$$

Example 2.15. Calculate the average and root mean square values, the form factor and peak factor of a periodic current wave having the following values for equal time intervals over half cycle, changing suddenly from one value to the next: 0, 40, 60, 80, 100, 80, 60, 40, 0.

Solution : Since the wave is symmetrical, so considering one half cycle only,

$$\text{Average value, } I_{av} = \frac{i_1 + i_2 + i_3 + i_4 + i_5 + i_6 + i_7 + i_8}{8} = \frac{0 + 40 + 60 + 80 + 100 + 80 + 60 + 40}{8} = 57.5\ \text{A}\ \textbf{Ans.}$$

RMS value, I_{rms}

$$= \sqrt{\frac{i_1^2 + i_2^2 + i_3^2 + i_4^2 + i_5^2 + i_6^2 + i_7^2 + i_8^2}{8}}$$

$$= \sqrt{\frac{0 + 1{,}600 + 3{,}600 + 6{,}400 + 10{,}000 + 6{,}400 + 3{,}600 + 1{,}600}{8}}$$

$= 64.42$ A **Ans.**

$$\text{Peak value, } I_{max} = 100\ \text{A}$$

$$\text{Form factor} = \frac{I_{rms}}{I_{av}} = \frac{64.42}{57.5} = 1.12\ \textbf{Ans.}$$

$$\text{Peak factor} = \frac{I_{max}}{I_{rms}} = \frac{100}{64.42} = 1.552\ \textbf{Ans.}$$

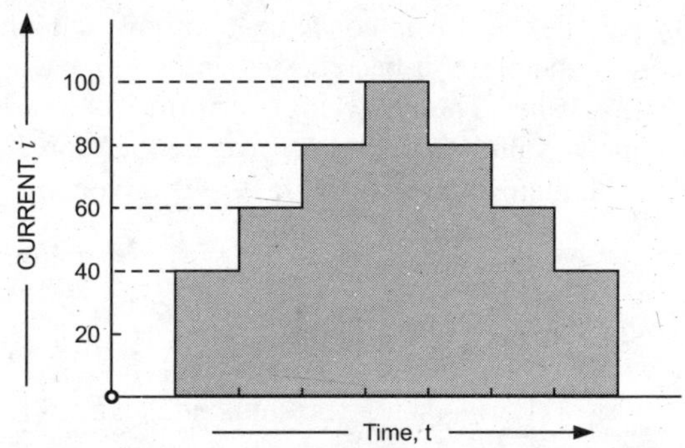

Fig. 2.12

2.10. RMS VALUE, AVERAGE VALUE, PEAK FACTOR AND FORM FACTOR OF HALF-WAVE RECTIFIED ALTERNATING (SINUSOIDAL) CURRENT

Half wave rectified alternating current is one whose one-half cycle has been suppressed *i.e.* one which flows for half the time during one cycle. It is illustrated in Fig. 2.13 where suppressed half cycle is shown dotted.

As mentioned earlier, for determining rms and average values of such an alternating current summation would be carried over the period for which current actually flows *i.e.* 0 to π but would be averaged for the whole cycle *i.e.* from 0 to 2π.

∴ RMS value of current,

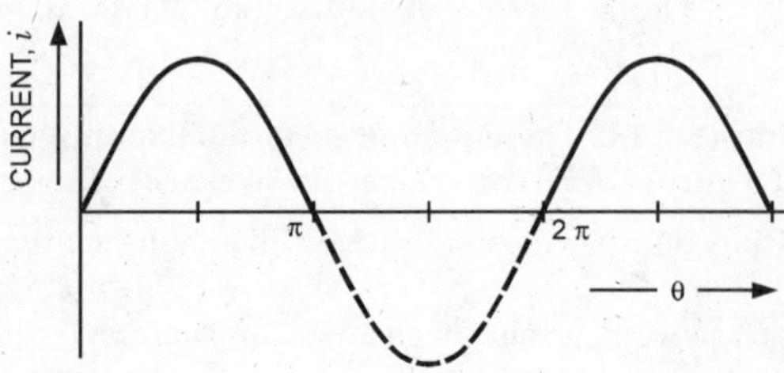

Fig. 2.13

$$I_{rms} = \sqrt{\frac{1}{2\pi}\int_0^{\pi} i^2\, d\theta}$$

$$= \sqrt{\frac{1}{2\pi}\int_0^{\pi} I^2_{max} \sin^2\theta\, d\theta}$$

$$= \frac{I_{max}}{\sqrt{2\pi}}\sqrt{\int_0^{\pi} \sin^2\theta\, d\theta}$$

$$= \frac{I_{max}}{\sqrt{4\pi}}\sqrt{\int_0^{\pi}(1 - \cos 2\theta)\, d\theta} = \frac{I_{max}}{\sqrt{4\pi}}\sqrt{\left[\theta - \frac{1}{2}\sin 2\theta\right]_0^{\pi}} = \frac{I_{max}}{\sqrt{4\pi}} \times \sqrt{\pi} = \frac{I_{max}}{2}$$

$$\text{Average value of current, } I_{av} = \frac{1}{2\pi}\int_0^{\pi} i\,d\theta = \frac{1}{2\pi}\int_0^{\pi} I_{max}\sin\theta\,d\theta = \frac{I_{max}}{2\pi}\left[-\cos\theta\right]_0^{\pi} = \frac{I_{max}}{\pi}$$

$$\text{Peak Factor} = \frac{I_{max}}{I_{rms}} = \frac{I_{max}}{I_{max}/2} = 2$$

$$\text{Form factor} = \frac{I_{rms}}{I_{av}} = \frac{I_{max}/2}{I_{max}/\pi} = \frac{\pi}{2} = 1.57$$

Example 2.16. Find the average and effective values of voltage for sinusoidal waveform shown in Fig. 2.14.

Solution : The equation of the given sinusoidal waveform is $v = 100 \sin\theta$

$$V_{av} = \frac{1}{2\pi}\int_{\pi/4}^{\pi} 100\sin\theta\,d\theta$$

$$= \frac{100}{2\pi}\left[-\cos\theta\right]_{\pi/4}^{\pi} = 27.17 \text{ V } \textbf{Ans.}$$

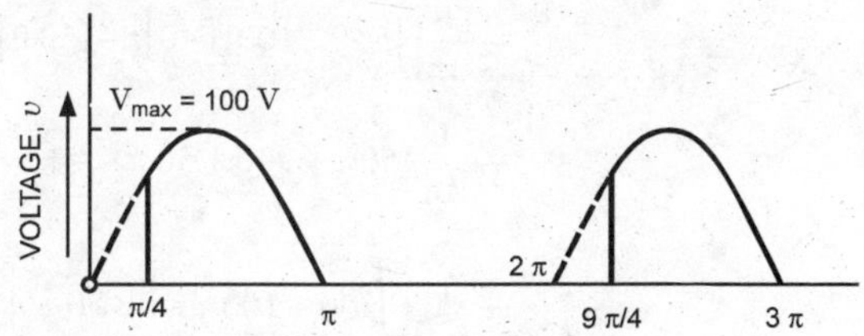

Fig. 2.14

$$V_{rms}^2 = \frac{1}{2\pi}\int_{\pi/4}^{\pi} 100^2\sin^2\theta\,d\theta$$

$$= \frac{10{,}000}{4\pi}\int_{\pi/4}^{\pi}(1-\cos 2\theta)\,d\theta = \frac{2{,}500}{\pi}\left[\theta - \frac{\sin 2\theta}{2}\right]_{\pi/4}^{\pi} = \frac{2{,}500}{\pi}\left[\pi - \frac{\pi}{4} + \frac{1}{2}\right] = 2{,}273$$

$$\text{or } V_{rms} = \sqrt{2{,}273} = 47.67 \text{ V } \textbf{Ans.}$$

Example 2.17. A voltage $v(t) = 220\sqrt{2}\sin 100t$ is applied to the circuit shown. What is the rms value of current through the resistor R of 100 Ω ? Derive the formula used. *[Pb. Technical Univ. Electrical Engineering May 2002]*

Solution : The maximum value of voltage

$$\text{applied to the circuit } V_{max} = 220\sqrt{2} \text{ V}$$

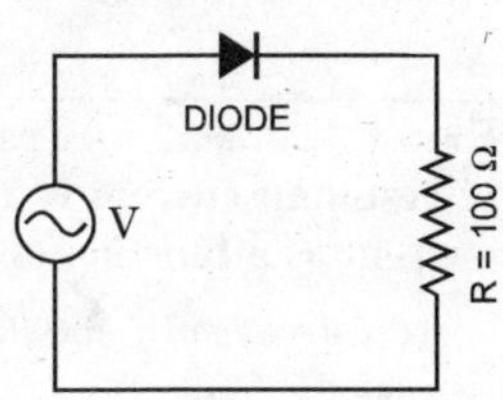

Fig. 2.15

$$\text{Maximum value of circuit current, } I_{max} = \frac{V_{max}}{R}$$

$$= \frac{220\sqrt{2}}{100} = 3.11 \text{ A}$$

$$\text{RMS value of circuit current, } I_{rms} = \frac{I_{max}}{2} = \frac{3.11}{2} = 1.55 \text{ A } \textbf{Ans.}$$

Example 2.18. Define average value of alternating current. Find this value for the waveform shown in Fig. 2.16. *[Pb Technical Univ. Basic Electrical Engineering December 2002]*

Solution : Maximum value of alternating current, $I_{max} = 10$ A

$$\text{Average value, } I_{av} = \frac{I_{max}}{\pi}$$

$$= \frac{10}{\pi} = 3.183 \text{ A } \textbf{Ans.}$$

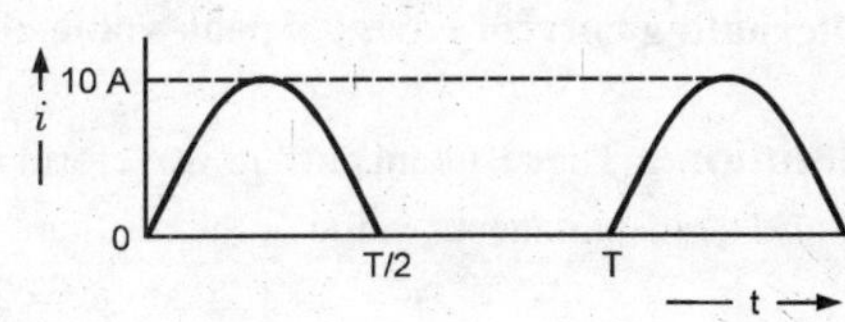

Fig. 2.16

Example 2.19. A transmission line carries a dc voltage of 50 V and a half-wave rectified sinusoidal voltage as shown in Fig. 2.17. Calculate (*i*) rms value, (*ii*) the average value, and (*iii*) form factor.

[Pb. Technical Univ. Basic Electrical and Electronics Engineering Dec. 2004]

Solution : The instantaneous value of the voltage wave is given by the expression

$$v = 50 + 100 \sin\theta \text{ for } 0 < \theta < \pi$$

$$v = 50 \quad \text{for } \pi < \theta < 2\pi$$

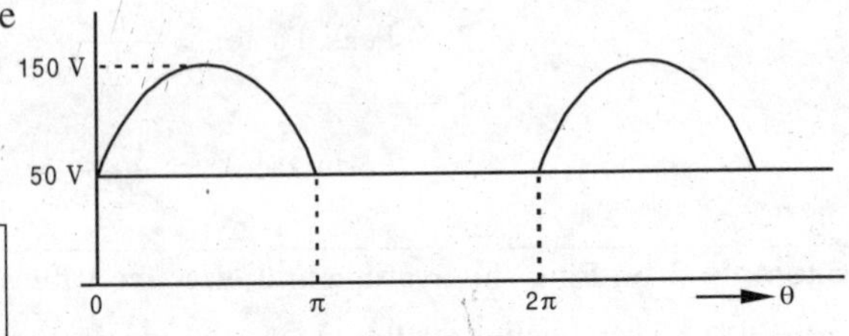

Fig. 2.17

(*ii*)
$$V_{Av} = \frac{1}{2\pi}\left[\int_0^{\pi}(50 + 100 \sin\theta)\, d\theta + \int_{\pi}^{2\pi} 50\, d\theta\right]$$

$$= \frac{1}{2\pi}\left[[50\theta - 100\cos\theta]_0^{\pi} + [50\,\theta]_{\pi}^{2\pi}\right]$$

$$= \frac{1}{2\pi}[50\pi + 200 + 50\,\pi] = \frac{200 + 100\pi}{2\pi} = 50 + \frac{100}{\pi} = 81.83 \text{ V } \textbf{Ans.}$$

(*i*)
$$V_{rms}^2 = \frac{1}{2\pi}\left[\int_0^{\pi}(50 + 100 \sin\theta)^2 d\theta + \int_{\pi}^{2\pi} 2{,}500\, d\theta\right]$$

$$= \frac{1}{2\pi}\left[\int_0^{\pi}[2{,}500 + 10{,}000 \sin\theta + 5{,}000\,(1 - \cos 2\theta)\right] d\theta + \int_{\pi}^{2\pi} 2{,}500\, d\theta$$

$$= \frac{1}{2\pi}[(7{,}500\theta - 10{,}000 \cos\theta - 2{,}500 \sin 2\theta]_0^{\pi} + [2{,}500\theta]_{\pi}^{2\pi}$$

$$= \frac{1}{2\pi}[7{,}500\pi + 20{,}000 - 0 + 2{,}500\pi] = \frac{10{,}000\pi + 20{,}000}{2\pi} = 5{,}000 + \frac{10{,}000}{\pi} = 8{,}183$$

$$V_{rms} = \sqrt{8{,}183} = 90.46 \text{ V } \textbf{Ans.}$$

(*iii*)
$$\text{Form factor} = \frac{V_{rms}}{V_{av}} = \frac{90.46}{81.83} = 1.105 \textbf{ Ans.}$$

Example 2.20. A wire carries both dc and ac current with maximum value of ac 20 A and 5 A of dc. The frequency of resultant current is 25 Hz. Write mathematical expression for the wire current as a function of time and draw the wave shape.

[Pb. Technical Univ. Basic Electrical and Electronics Engineering Dec. 2005]

Solution : The resultant current waveform is shown in Fig. 2.18. The instantaneous value of the resultant current is given by the expression

$$i = I_{dc} + I_{max} \sin 2\pi ft$$

$$= 5 + 20 \sin 2\pi \times 25t$$

$$= 5 + 20 \sin 157t \textbf{ Ans.}$$

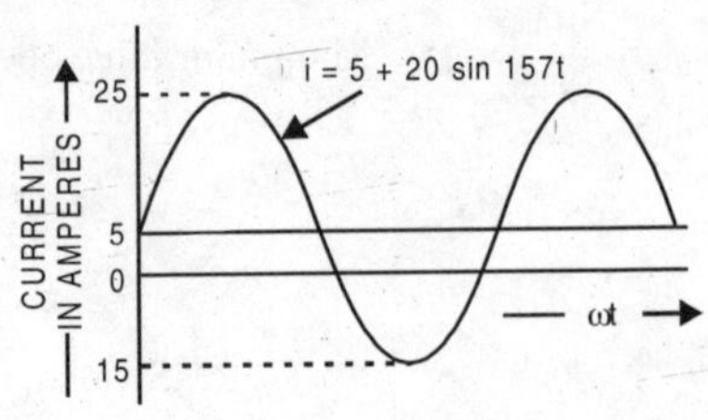

Fig. 2.18

Example 2.21. A circuit carries a current which is the resultant of direct current of 20 A, and a sinusoidal alternating current having a peak value of 20 A. Find the rms value of current in the circuit.

[Pb. Technical Univ. Basic Electrical Engineering June 2003]

Solution : The resultant current waveform is shown in Fig. 2.19.

The instantaneous value of the resultant current wave is given by the expression

$$i = 20 + 20 \sin \omega t = 20(1 + \sin \omega t) = 20(1 + \sin\theta)$$

and $$i^2 = 400(1 + \sin\theta)^2 = 400 + 800 \sin\theta + 400 \sin^2\theta$$

or $i^2 = 400 + 800 \sin\theta + 200\,(1 - \cos 2\theta)$

$= 600 + 800 \sin\theta - 200 \cos 2\theta$

$$I^2_{rms} = \frac{1}{2\pi}\int_0^{2\pi} i^2\,d\theta \text{ considering complete cycle}$$

$$= \frac{1}{2\pi}\int_0^{2\pi} (600 + 800\sin\theta - 200\cos 2\theta)\,d\theta$$

$$= \frac{1}{2\pi} \times 600 \times 2\pi = 600$$

or $I_{rms} = \sqrt{600} = 24.5$ A **Ans.**

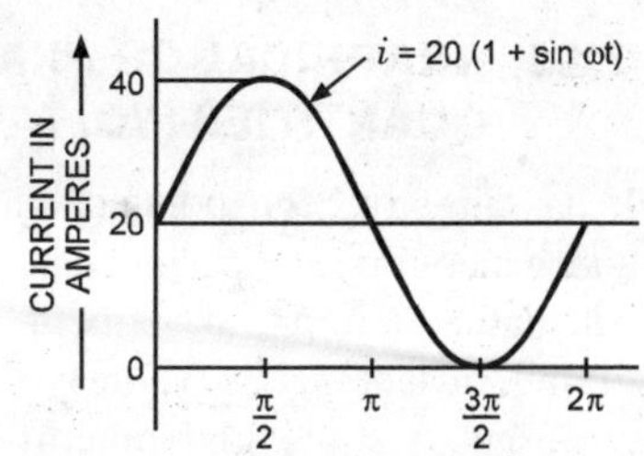

Fig. 2.19

Example 2.22. Find the average and rms values of the current
***i(t)* = 10 + 10 sin 314*t*.** [*Pb. Technical Univ. Basic Electrical and Electronics Engineering May 2007*]

Solution : The instantaneous value of current wave is given by the expression

$$i = 10 + 10 \sin 314t = 10\,(1 + \sin 314t) = 10(1 + \sin\theta) \quad \text{where } \theta = 314t$$

Average value, $I_{av} = \frac{1}{2\pi}\int_0^{2\pi} i\,d\theta$ considering the complete cycle

$$= \frac{1}{2\pi}\int_0^{2\pi} 10(1 + \sin\theta)\,d\theta = \frac{1}{2\pi}\int_0^{2\pi} 10\theta - 10\cos\theta = \frac{10\times 2\pi}{2\pi} - 0 = 10 \text{ A } \textbf{Ans.}$$

$i^2 = 100(1 + \sin\theta)^2 = 100 + 200\sin\theta + 100\sin^2\theta$

or $i^2 = 100 + 200\sin\theta + 50(1 - \cos 2\theta) = 150 + 200\sin\theta - 50\cos 2\theta$

$$I^2_{rms} = \frac{1}{2\pi}\int_0^{2\pi} i^2\,d\theta \text{ considering complete cycle}$$

$$= \frac{1}{2\pi}\int_0^{2\pi} 150 + 200\sin\theta - 50\cos 2\theta = \frac{1}{2\pi}\times 150 \times 2\pi = 150$$

or $I_{rms} = \sqrt{150} = \sqrt{150} = 12.25$ A **Ans.**

2.11. RMS VALUE, AVERAGE VALUE, FORM FACTOR AND PEAK OF A TRIANGULAR WAVEFORM

Let the maximum value of the current be I_{max} amperes.

Since $i = I_{max}$ when $\theta = \pi$. Hence, expression for the instantaneous current can be written as

$$i = \frac{I_{max}}{\pi}\theta \text{ for } 0 < \theta < \pi$$

RMS value of the current wave,

$$I_{rms} = \sqrt{\frac{1}{\pi}\int_0^{\pi} i^2\,d\theta} = \sqrt{\frac{1}{\pi}\int_0^{\pi}\frac{I^2_{max}}{\pi^2}\theta^2\,d\theta} = \frac{I_{max}}{\sqrt{3}}$$

Average value of the current wave,

$$I_{av} = \frac{1}{\pi}\int_0^{\pi} i\,d\theta = \frac{1}{\pi}\int_0^{\pi}\frac{I_{max}}{\pi}\theta\,d\theta = \frac{I_{max}}{2}.$$

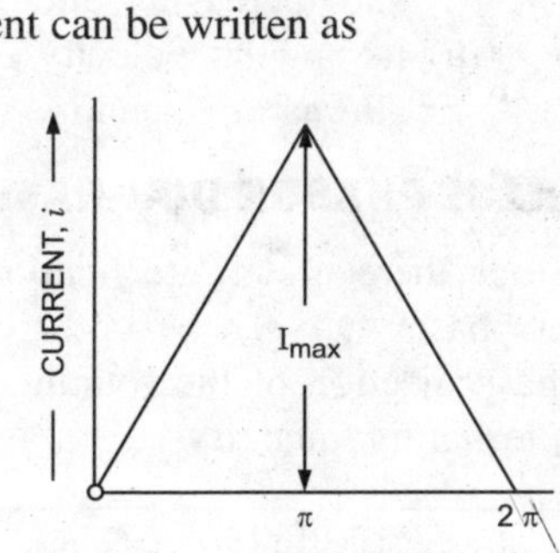

Fig. 2.20

$$\text{Form factor, } K_f = \frac{I_{rms}}{I_{av}} = \frac{I_{max}/\sqrt{3}}{I_{max}/2} = \frac{2}{\sqrt{3}} = 1.155$$

$$\text{Peak factor, } K_p = \frac{I_{max}}{I_{rms}} = \frac{I_{max}}{I_{max}/\sqrt{3}} = \sqrt{3}$$

2.12. GRAPHICAL OR PHASOR REPRESENTATION OF ALTERNATING (SINUSOIDAL) QUANTITIES (VOLTAGES AND CURRENTS)

It has already been pointed out that it is assumed that alternating voltages and currents follow sine law and generators are designed to give emfs having sine waveform. The above said assumption makes the calculations simple. The method of representing alternating quantities by waveform or by the equations giving instantaneous values is quite cumbersome. For solution of ac problems it is advantageous to represent a sinusoidal quantity (voltage or current) by a line of definite length rotating in counter-clockwise* direction with the same angular velocity as that of the sinusoidal quantity. Such a rotating line is called the *phasor***.

Consider a line OA (or phasor as it is called) representing to scale the maximum value of an alternating quantity, say emf *i.e.* OA = E_{max} and rotating in counter-clockwise direction at an angular velocity ω radians/second about the point O, as shown in Fig. 2.21. An arrowhead is put at the outer end of the phasor, partly to indicate which end is assumed to move and partly to indicate the precise length of the phasor when two or more phasors happen to coincide.

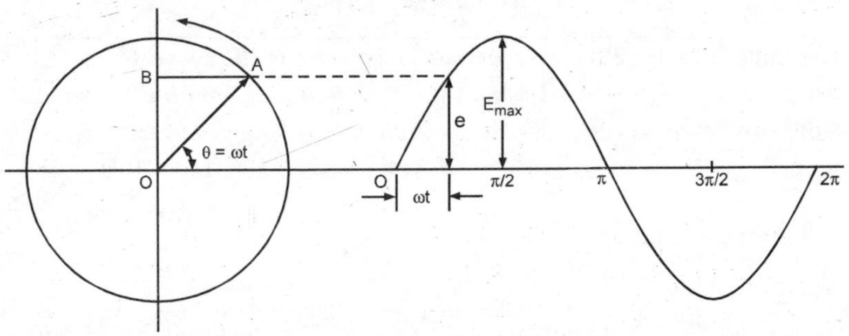

Fig. 2.21. *Phasor Representation of An Alternating Quantity*

Figure 2.21 shows OA when it has rotated through an angle θ, being equal to ω*t*, from the position occupied when the emf was passing through its zero value. The projection of OA on Y-axis,

$$OB = OA \sin\theta = E_{max} \sin \omega t = e, \text{ the value of the emf at that instant.}$$

Thus the projection of OA on the vertical axis represents to scale the instantaneous value of emf.

It will be seen that the phasor OA rotating in counter-clockwise direction will represent a sinusoidal quantity (voltage or current) if

(*i*) its length is equal to the peak or maximum value of the sinusoidal voltage or current to a suitable scale.

(*ii*) it is in horizontal position at the instant the alternating quantity (voltage or current) is zero and increasing and

(*iii*) its angular velocity is such that it completes one revolution in the same time as taken by the alternating quantity (voltage or current) to complete one cycle.

2.13. PHASOR DIAGRAM USING RMS VALUES

Since there is definite relation between maximum value and rms value ($E_{max} = \sqrt{2}\, E_{rms}$), the length of phasor OA can be taken equal to rms value if desired. But it should be noted that in such cases, the projection of the rotating phasor on the vertical axis will not give the instantaneous value of that alternating quantity.

---

* It is a standard convention that the phasor is rotated in counter-clockwise direction—a convention that is in harmony with the general use of polar coordinates.

** Strictly speaking, sinusoidal quantities are scalar quantities varying periodically with time and, according to the definition of vector quantities, they are not true vectors. Voltage is simply energy or work per coulomb and cannot be classified as a vector. Current is also not a vector quantity because it is merely the flow of electrons through a wire. But keeping in mind that any sinusoidal quantity at a given frequency is completely specified by its amplitude and phase angle, its similarity to a vector quantity is evident, since the amplitude may be considered as the magnitude and the phase angle as the direction of a vector. To account for the difference the term *phasor* has been adopted, instead of the term vector, for representing graphically the magnitude and phase of a sinusoidal current or voltage.

The phasor diagram drawn in rms values of the alternating quantities helps in understanding the behaviour of the ac machines under different loading conditions.

2.14. PHASE AND PHASE ANGLE

By phase of an alternating current is meant the fraction of the time period of that alternating current that has elapsed since the current last passed through the zero position of reference. The phase angle of any quantity means the angle the phasor representing the quantity makes with the reference line (which is taken to be at zero degrees or radians). For example the phase angle of current I_2 in Fig. 2.22 is ($-\Phi$).

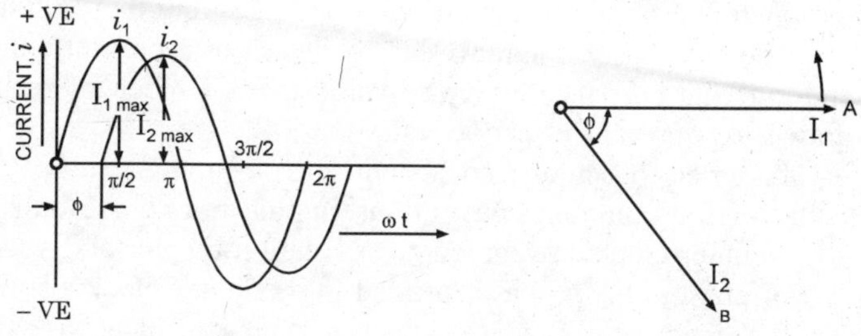

(a) (b)

Fig. 2.22

2.15. PHASE DIFFERENCE

When two alternating quantities, say, two emfs, or two currents or one voltage and one current are considered simultaneously, the frequency being the same, they may not pass through a particular point at the same instant. One may pass through its maximum value at the instant when the other passes through the value other than its maximum one. These two quantities are said to have a *phase difference*. Phase difference is always given either in degrees or in radians.

The phase difference is measured by the angular distance between the points where the two curves cross the base or reference line in the same direction.

The quantity ahead in phase is said to lead the other quantity while the second quantity is said to lag behind the first one. In Fig. 2.22 (*b*) current I_1 represented by phasor OA leads the current I_2 represented by phasor OB by Φ or current I_2 lags behind the current I_1 by Φ. The leading current I_1 goes through its zero and maximum values first and the current I_2 goes through its zero and maximum values after time angle Φ. The two waves representing these two currents are shown in Fig. 2.22 (*a*). If I_1 is taken as reference phasor, the two currents can be expressed as

$$i_1 = I_{1\,max} \sin \omega t$$

$$\text{and} \quad i_2 = I_{2max} \sin (\omega t - \Phi)$$

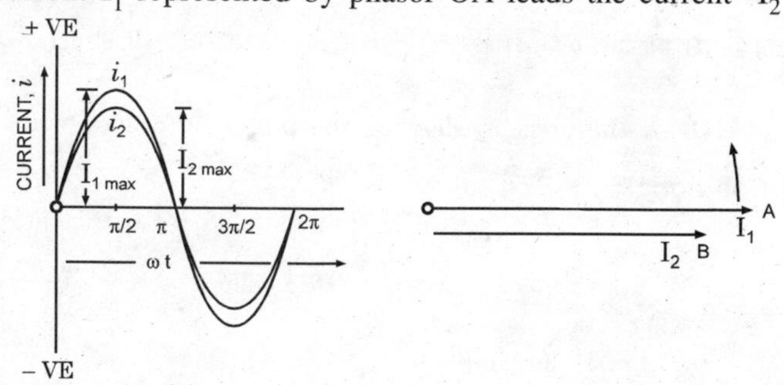

(a) (b)

Fig. 2.23

The two quantities are said to be in phase with each other if they pass through zero values at the same instant and rise in the same direction, as shown in Fig. 2.23, while the two quantities passing through zero values at the same instant but rising in opposite directions, as shown in Fig. 2.24 are said to be in phase opposition, *i.e.* phase difference is 180°. When the two alternating quantities have a phase difference of 90° or $\frac{\pi}{2}$ radians, they are said to be in quadrature.

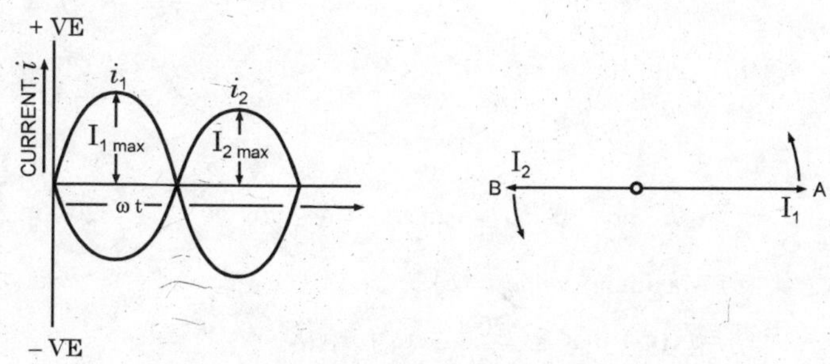

(a) (b)

Fig. 2.24

2.16. CONVENTIONS FOR DRAWING PHASOR DIAGRAMS

As already mentioned the alternating quantities (voltages and currents) in practice are represented by straight lines having definite direction and length. Such lines are called the *phasors* and the diagrams in which phasors represent currents, voltages and their phase difference are known as *phasor diagrams.*

Though phasor diagrams can be drawn to represent either maximum or effective values of voltages and currents but since effective values are of much more importance, the phasor diagrams are mostly drawn to represent effective values.

In order to achieve consistent and accurate results it is essential to follow certain conventions. Some of the common conventions in this regard are enlisted below:

1. Counter-clockwise direction of rotation of phasors is usually taken as positive direction of rotation of phasors *i.e.* a phasor rotated in a counter-clockwise direction from a given phasor is said to lead the given phasor while a phasor rotated in clockwise direction is said to lag the given phasor.
2. For series circuits, in which the current is common to all parts of the circuit, the current phasor is usually taken as reference phasor for other phasors in the same diagram and drawn on horizontal line.
3. In parallel circuits in which the voltage is common to all branches of the circuit, the voltage phasor is usually taken as reference phasor and drawn on the horizontal line. Other phasors are referred to the common voltage phasor.
4. It is not necessary that current and voltage phasors are drawn to the same scale; in fact it is often desirable to draw the current phasor to a larger scale than the voltage phasor when the values of currents being represented are small. However, if several voltage phasors are to be used in the same phasor diagram, they should all be drawn to the same scale. Likewise all current phasors in the same diagram should be drawn to the same scale.

Example 2.23. Calculate (*a*) the maximum value and (*b*) the root mean square values of the following quantities:

(*i*) 40 sin ω*t* (*ii*) B sin $\left(\omega t - \frac{\pi}{2}\right)$ and (*iii*) 10 sin ω*t* –17.3 cos ω*t*.

Draw the phasors showing the phase difference with respect to A sin $\left(\omega t - \frac{\pi}{6}\right)$.

Solution : (*i*) For 40 sin ω*t*

(*a*) Maximum value = Coefficient of the sine of the time angle = 40 **Ans.**

(*b*) RMS value $= \frac{\text{Maximum value}}{\sqrt{2}} = \frac{40}{\sqrt{2}} = 28.28$ **Ans.**

(*ii*) For B sin $\left(\omega t - \frac{\pi}{2}\right)$

(*a*) Maximum value = B **Ans.**

(*b*) RMS value $= \frac{B}{\sqrt{2}}$ **Ans.**

(*iii*) For 10 sin ω*t* – 17.3 cos ω*t*, which may be written as

$$20\left(\frac{1}{2}\sin\omega t - \frac{\sqrt{3}}{2}\cos\omega t\right)$$

$$\text{or}\quad 20\left(\cos\frac{\pi}{3}\sin\omega t - \sin\frac{\pi}{3}\cos\omega t\right)\quad\text{or}\quad 20\sin\left(\omega t - \frac{\pi}{3}\right)$$

(*a*) Maximum value = 20

(*b*) RMS value $= \frac{20}{\sqrt{2}} = 14.14$ **Ans.**

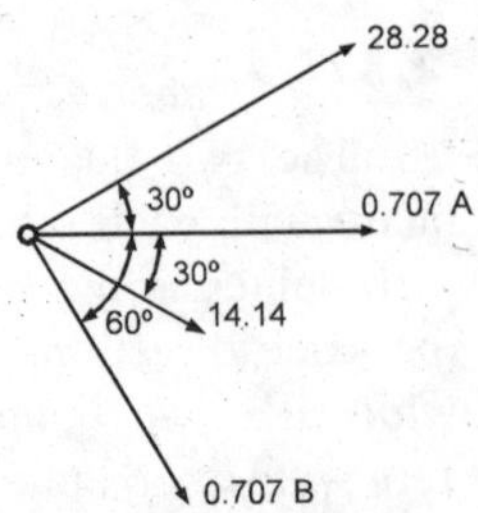

Fig. 2.25

Phasors showing the phase difference with respect to A sin $\left(\omega t - \frac{\pi}{6}\right)$ are shown in Fig. 2.25.

Example 2.24. Draw to scale a phasor diagram showing the following voltages:

$$v_1 = 100 \sin 500\, t$$

$$v_2 = 200 \sin \left(500\, t + \frac{\pi}{3}\right)$$

$$v_3 = -\ 50 \cos 500\, t.$$

$$v_4 = 150 \sin \left(500\, t - \frac{\pi}{4}\right)$$

Solution : The given voltages are

$$v_1 = 100 \sin 500t$$

$$v_2 = 200 \sin \left(500\, t + \frac{\pi}{3}\right)$$

$$v_3 = -\,50 \cos 500t = 50 \sin \left(500\, t - \frac{\pi}{2}\right)$$

and $$v_4 = 150 \sin \left(500t - \frac{\pi}{4}\right)$$

The phasor diagram is shown in Fig. 2.26.

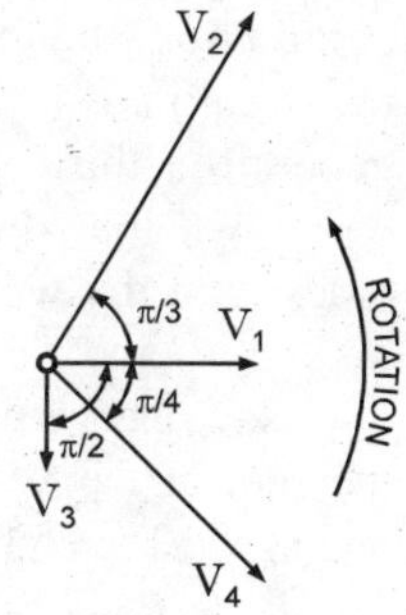

Fig. 2.26

Example 2.24. Alternating voltage and current are given by the following equations:

$$i = 10 \sin 314t \text{ and } e = 100 \sin \left(314\, t + \frac{\pi}{6}\right)$$

Find out : (1) Does the current lead or lag the voltage, find the phase difference between them. (2) Show them graphically and vectorially (3) Find the frequency.

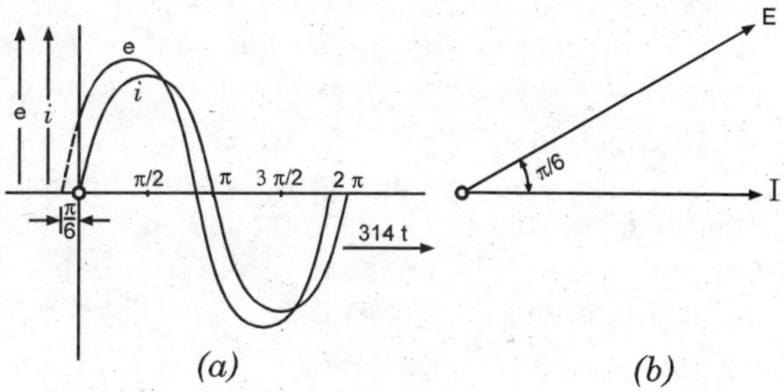

Fig. 2.27

Solution : (1) The voltage leads the current by $\frac{\pi}{6}$ radians or 30° or the current lags behind the voltage by 30°. **Ans.**

(2) Alternating voltage and current are shown graphically and vectorially in Figs. 2.27 (*a*) and 2.27 (*b*) respectively.

(3) Frequency, $f = \frac{314}{2\pi} = 50$ Hz. **Ans.**

2.17. ADDITION AND SUBTRACTION OF ALTERNATING QUANTITIES

In most of the ac circuits, it is necessary to consider the combined action of several emfs or voltages acting in a series circuit and action of several currents flowing through the different branches of a parallel circuit.

Let it be required to add two currents given by the equations

$$i_1 = I_{1\,max} \sin \omega t \text{ and } i_2 = I_{2\,max} \sin(\omega t - \Phi)$$

The resultant sum may be expressed as

$$i_r = i_1 + i_2 = I_{1\,max} \sin \omega t + I_{2\,max} \sin (\omega t - \Phi)$$

but it is too awkward and gives no idea of the peak value and phase angle of the resultant current.

The currents may be added graphically by plotting their curves in the same system of coordinates and then adding the ordinates of i_1 and i_2 point by point, according to the equation $i_r = i_1 + i_2$. Evidently this method is also too cumbersome and unwieldy to be practical. This is particularly so when situation arise where more than two sinusoidal quantities are to be added.

A simpler and more direct method consists in adding the sinusoidal quantities as phasors.

Consider the phasors $I_{1\,max}$ and $I_{2\,max}$ that would generate the two curves i_1 and i_2 and let them be in a position, as shown in Fig. 2.28 at one particular instant of time. If we now add $I_{1\,max}$ and $I_{2\,max}$ by completing the parallelogram as shown, the diagonal $I_{r\,max}$ will, when rotated, generate a third sine curve. It remains to be shown that this third sine curve coincides with the waveform of i_r obtained by adding i_1 and i_2 point by point.

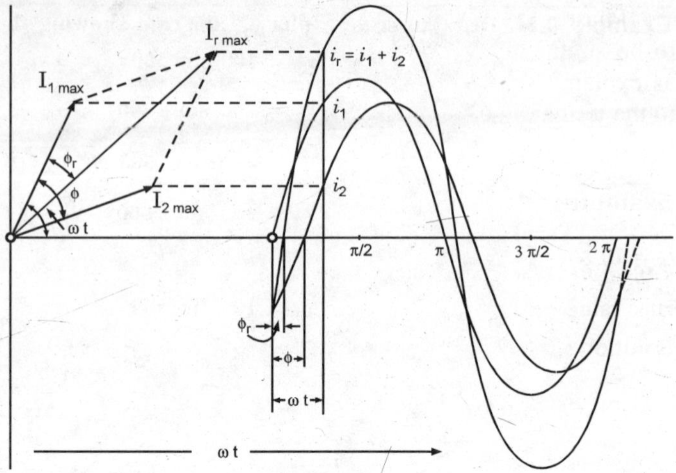

Fig. 2.28

Now the vertical component of $I_{r\,max}$ is the sum of the vertical components of $I_{1\,max}$ and $I_{2\,max}$. Therefore, the waveform of i_r is the graph generated by rotating $I_{r\,max}$ in counter-clockwise direction.

It follows, therefore, that two or more alternating quantities may be added in the same way as forces are added, namely by constructing parallelograms or closed polygons and either measuring or calculating the lengths of the diagonals or closing sides and the magnitude of the phase angles.

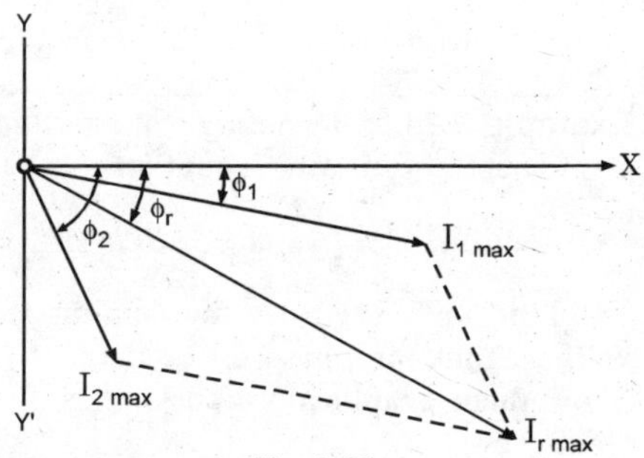

Fig. 2.29

Illustrations: The way in which the two given currents i_1 and i_2 can be added by the parallelogram rule of phasor addition is illustrated in Fig. 2.29 where the currents are shown as phasor drawn from the origin O of the system of coordinates. The resultant phasor is the diagonal of the parallelogram formed by the phasors $I_{1\,max}$ and $I_{2\,max}$.

This method is more convenient when more than two phasors are to be added, as shown in Fig. 2.30. From the end point of $I_{1\,max}$ a phasor is constructed parallel to $I_{2\,max}$ of the same magnitude and direction as the latter; then from the end point of $I_{2\,max}$ a phasor is constructed parallel to $I_{3\,max}$ and so on. Phasor $I_{r\,max}$ from the origin of the first phasor ($I_{1\,max}$) to the end point of the last phasor ($I_{5\,max}$) represents the sum of all the phasors.

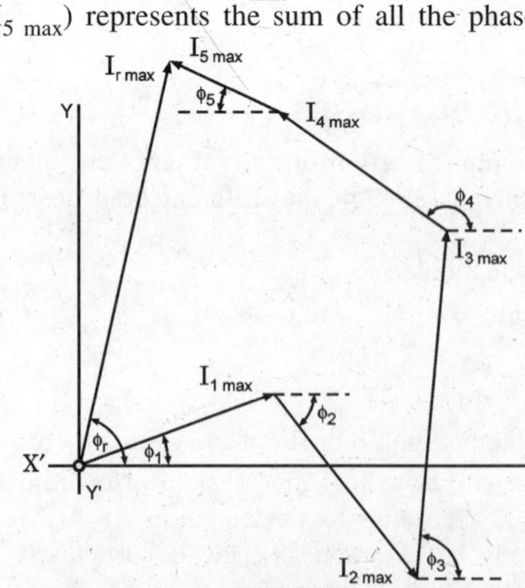

Fig. 2.30

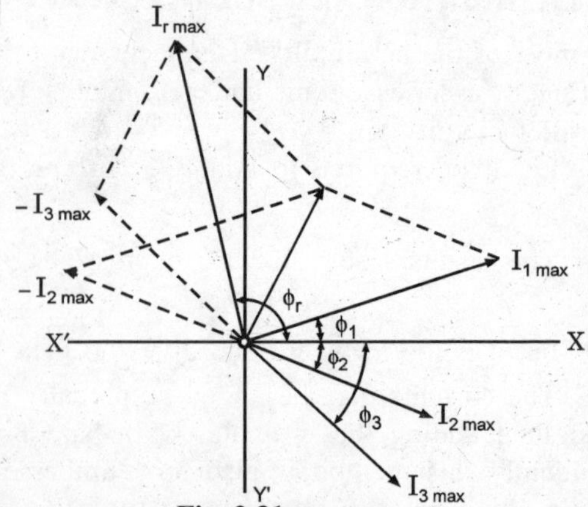

Fig. 2.31

Phasors may also be subtracted by the above method. For example if phasors $I_{2\,max}$ and $I_{3\,max}$ are to be subtracted from phasor $I_{1\,max}$ each of the two phasors be reversed in direction and then added as explained above [Fig. 2.31]. This time, the phasor drawn from the origin of the first phasor $I_{1\,max}$ to the terminal point of the last phasor $I_{3\,max}$ gives the difference of phasors $I_{2\,max}$ and $I_{3\,max}$ from phasor $I_{1\,max}$.

Example 2.25. Two alternating currents represented by the equations $i_1 = 7 \sin \omega t$ and $i_2 = 10 \sin\left(\omega t + \frac{\pi}{3}\right)$ are fed into a common conductor. Find the equation for the resultant current and its rms value.

Solution :

$$i_1 = 7 \sin \omega t$$

$$i_2 = 10 \sin\left(\omega t + \frac{\pi}{3}\right)$$

Resultant current, $i_r = i_1 + i_2 = 7 \sin \omega t + 10 \sin\left(\omega t + \frac{\pi}{3}\right)$

$$= 7 \sin \omega t + 10 \sin \omega t \cos \frac{\pi}{3} + 10 \cos \omega t \sin \frac{\pi}{3}$$

$$= 12 \sin \omega t + 8.66 \cos \omega t$$

Multiplying and dividing RHS of the above expression by $\sqrt{(12)^2 + (8.66)^2}$ *i.e.* 14.8, we get

$$i_r = 14.8\left(\sin \omega t \times \frac{12}{14.8} + \cos \omega t \times \frac{8.66}{14.8}\right)$$

$$= 14.8 (\sin \omega t \cos \Phi_r + \cos \omega t \sin \Phi_r)$$

where $\cos \Phi_r = \frac{12}{14.8}$ and $\sin \Phi_r = \frac{8.66}{14.8}$

or $i_r = 14.8 \sin (\omega t + \Phi_r)$ where $\Phi_r = \text{Tan}^{-1} \frac{8.66}{12} = 0.199\pi$ radians

or $i_r = 14.8 \sin (\omega t + 0.199\pi)$ **Ans.**

Peak value of resultant current, $I_{r\,max} = 14.8$ A

RMS value of resultant current, $I_{r\,rms} = \frac{I_{r\,max}}{\sqrt{2}} = \frac{14.8}{\sqrt{2}} = 10.46$ A **Ans.**

***Analytical Method*:** Resolving currents $I_{1\,max}$ and $I_{2\,max}$ along X-axis and Y-axis we get

Algebraic sum of X-components $= 7 \cos 0^\circ + 10 \cos \frac{\pi}{3} = 7 + 5 = 12$ A

Algebraic sum of Y-components $= 7 \sin 0^\circ + 10 \sin \frac{\pi}{3} = 8.66$ A

Maximum value of resultant current,

$$I_{r\,max} = \sqrt{(\text{X-components})^2 + (\text{Y-components})^2}$$

$$= \sqrt{12^2 + (8.66)^2} = 14.8 \text{ A}$$

and phase angle, $\Phi_r = \text{Tan}^{-1} \frac{\text{Y-components}}{\text{X-components}} = \text{Tan}^{-1} \frac{8.66}{12} = 0.199\pi$

$\therefore$ Resultant current, $i_r = I_{r\,max} \sin (\omega t + \Phi_r) = 14.8 \sin (\omega t + 0.199\pi)$ **Ans.**

Graphical Method: Take OX as reference phasor. Draw phasor OA 3.5 cm in length along OX so as to represent current $I_{1\ max}$ of 7 amperes (to the scale 1 cm = 2 A) in magnitude as well as in direction. Then draw phasor AB = 5 cm in length making an angle of $\frac{\pi}{3}$ radians or 60° with the reference phasor so as to represent current $I_{2\ max}$ of 10 amperes in magnitude as well as in direction. The phasor OB, obtained by joining points O and B, will represent the resultant current in magnitude as well as in direction. On measurement from Fig. 2.33

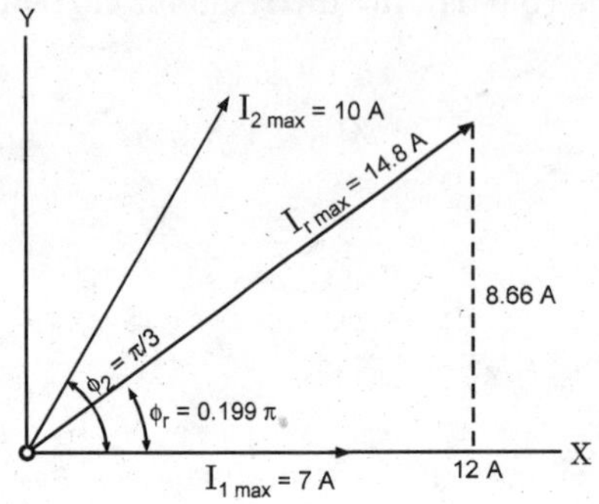

Fig. 2.32

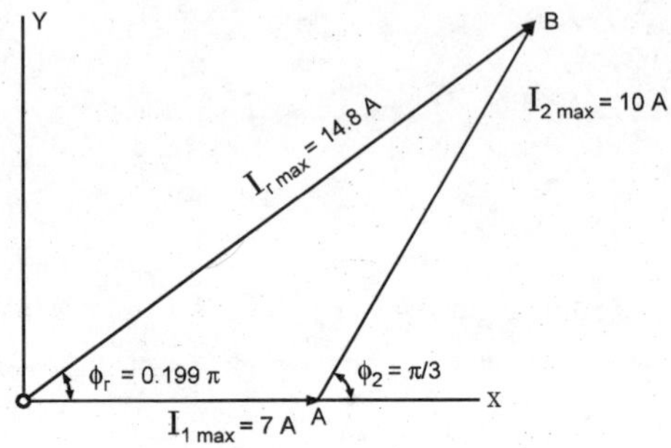

Fig. 2.33

Maximum value of resultant current, $I_{r\ max}$ = OB × 2 = 7.4 × 2 = 14.8 A

and phase angle, Φ_r = 35.8° or 0.199 π radians

Hence the resultant current may be expressed as

$$i_r = 14.8 \sin(\omega t + 0.199\,\pi) \text{ **Ans.**}$$

Example 2.26. Three sinusoidal voltages acting in series are given by

$$\mathbf{V_1 = 10 \sin 440\,t}$$
$$\mathbf{V_2 = 10\sqrt{2} \sin (440\,t - 45^\circ)}$$
$$\mathbf{V_3 = 20 \cos 440\,t}$$

Determine

(*i*) an expression for the resultant voltage,

(*ii*) the frequency and rms value of the resultant voltage.

Solution :

$$v_1 = 10 \sin 440\,t$$
$$v_2 = 10\sqrt{2} \sin (440\,t - 45^\circ)$$
$$v_3 = 20 \cos 440\,t = 20 \sin \left(440\,t + \frac{\pi}{2}\right)$$

Resolving voltages V_{1max}, V_{2max} and V_{3max} along X-axis and Y-axis we have

$$\text{Algebraic sum of X-components} = 10 \cos 0^\circ + 10\sqrt{2} \cos(-45^\circ) + 20 \cos 90^\circ$$
$$= 10 + 10\sqrt{2} \times \frac{1}{\sqrt{2}} + 0 = 20$$

$$\text{Algebraic sum of Y-components} = 10 \sin 0^\circ + 10\sqrt{2} \sin(-45^\circ) + 20 \sin 90^\circ$$
$$= 0 - 10\sqrt{2} \times \frac{1}{\sqrt{2}} + 20 = 10$$

$$\text{Maximum value of voltage, } V_{r\,max} = \sqrt{(\text{X-components})^2 + (\text{Y-components})^2} = \sqrt{20^2 + 10^2} = 22.36 \text{ V}$$

$$\text{Phase angle, } \phi_r = \text{Tan}^{-1} \frac{\text{Y-component}}{\text{X-component}} = \text{Tan}^{-1} \frac{10}{20} = 26.56^\circ \text{ or } 0.1476\,\pi \text{ radian}$$

(*i*) So expression for resultant voltage, v_r = 22.36 sin (440 t + 0.1476 π) **Ans.**

$$(ii)\ \text{Frequency, } f = \frac{440}{2\pi} = 70 \text{ Hz } \textbf{Ans.}$$

$$\text{RMS value of resultant voltage} = \frac{V_{max}}{\sqrt{2}} = \frac{22.36}{\sqrt{2}} = 15.81 \text{ volts } \textbf{Ans.}$$

HIGHLIGHTS

1. An *alternating quantity* (voltage or current) is one which changes continuously in magnitude and alternates in direction at regular intervals of time.

An alternating quantity that varies sinusoidally (*i.e.* according to the sine of angle θ) is called the *sinusoidal quantity.*

Sinusoidal quantities are expressed as

$$e = E_{max} \sin\theta = E_{max} \sin \omega t = E_{max} \sin 2\pi f t$$

$$\text{and } i = I_{max} \sin\theta = I_{max} \sin \omega t = I_{max} \sin 2\pi f t$$

2. The shape of the curve of the voltage or current when plotted against time as abscissa (base) is called the *waveform.*

3. When a periodic wave, such as sinusoidal wave, goes through one complete set of positive and negative values it is said to have completed one *cycle*. One cycle corresponds to 360º or 2π radians.

Alternation is one half of cycle and corresponds to 180º or π radians.

4. The time taken in seconds by an alternating quantity to complete one cycle is known as *time period* or *periodic time* (T), while the number of cycles completed per second by an alternating quantity is known as *frequency* (*f*).

$$f = \frac{1}{T}$$

In India the standard frequency for power supply is 50 Hz (cycles per second).

5. Angular velocity, ω = 2π*f* radians per second.

6. In a multipolar machine, the number of cycles completed per second by generated emf,

$$f = \text{Pairs of poles} \times \text{number of revolutions made per second}$$

$$= \frac{P}{2} \times \frac{N}{60} = \frac{PN}{120}$$

7. The maximum value, positive or negative, which an alternating quantity attains during one cycle is called the *amplitude* of the alternating quantity.

8. The general expressions for an alternating voltage and current are given as

$$v = V_{max} \sin(\omega t \pm \phi)\ ; \quad i = I_{max} \sin(\omega t \pm \phi)$$

where ϕ is the phase angle in degrees or radians.

The coefficient of the sine of the time angle gives maximum or peak value of the alternating quantity (emf, voltage or current)

Coefficient of time *t* divided by 2π gives the frequency of the periodic wave.

9. The *average* (*or mean*) *value* of an alternating current is expressed by that steady current (dc) which transfers across any circuit the same charge as transferred by that alternating current

Average or mean value of alternating current is given as

$$I_{av} = \frac{i_1 + i_2 + i_3 + \dots\dots + i_n}{n} = \frac{\text{Area of one alternation}}{\text{Length of base over one alternation}}$$

For sinusoidal current $I_{av} = \frac{2}{\pi} I_{max} = 0.637\, E_{max}$.

10. *Effective* or *virtual value* of alternating current or voltage is equal to the square root of the mean of the squares of successive ordinates and that is why it is known as *root-mean square* (rms) value.

$$i.e.\ I_{eff} \text{ or } I_{rms} = \sqrt{\frac{i_1^2 + i_2^2 + i_3^2 + \dots + i_n^2}{n}}$$

The *rms* or *effective value* of an alternating current or voltage is given by that steady current or voltage which when flows or applied to a given resistance for a given time produces the same amount of heat as when the alternating current or voltage flowing or applied to the same resistance for the same time.

For sinusoidal current, $I_{rms} = \dfrac{I_{max}}{\sqrt{2}} = 0.707\, I_{max}$

11. The ratio of rms value to the average value of a waveform is known as *form factor*. For sinusoidal wave its value is 1.11.

12. The ratio of maximum value to the rms value of a waveform is known as peak factor. For sinusoidal wave its value is $\sqrt{2}$ or 1.414.

13. By *phase* of an alternating quantity is meant the fraction of time period of that alternating quantity that has elapsed since the quantity last passed through zero position of reference.

The angular displacement between the maximum positive values of two alternating quantities having the same frequency is called the phase difference between them.

An alternating quantity that attains its positive maximum value prior to the other is called the *leading quantity*.

An alternating quantity that attains its positive maximum value after the other is called the *lagging quantity*.

14. The resultant current of two currents repersented by

$$i_1 = I_{max_1} \sin \omega t \text{ and } i_2 = I_{max_2} \sin (\omega t + \theta)$$

flowing through a common conductor is given as

$$i_r = I_{r\,max} \sin (\omega t \pm \phi)$$

$$\text{where } I_{r\,max} = \sqrt{I^2_{max_1} + I^2_{max_2} + 2 I_{max_1} I_{max_2} \cos \theta}$$

$$\text{and } \phi = \text{Tan}^{-1} \frac{I_{max_2} \sin \theta}{I_{max_1} + I_{max_2} \cos \theta}.$$

The sign of the phase angle ϕ of the resultant current will depend upon its position w.r.t. the reference phasor.

EXERCISES

1. Explain why a large percentage of electrical energy used for commercial purposes is generated as ac.

2. Describe briefly the generation of alternating voltage and current and deduce the equation for an alternating emf, $e = E_m \sin \theta$.

3. Define the terms: Instantaneous value, Average value, RMS value, frequency and phase.
[Pb. Technical Univ. Electrical Engineering June 2000; December 2000; May 2001]

4. Define and explain the following:
(*i*) Frequency (*ii*) Time period (*iii*) Instantaneous value (*iv*) Average value (*v*) RMS value with respect to alternating quantities and (*vi*) Amplitude.

5. Define rms, average values and form factor of sinusoidal alternating voltage.
[Pb. Technical Univ. Basic Electrical and Electronics Engineering Dec. 2004; May 2010]

6. What do you understand by rms value and average value as applied to an ac circuit ?
[Pb. Technical Univ. Basic Electrical and Electronics Engineering May 2007]

7. From fundamental, derive expression for form factor of a sinusoidal voltage.
[Pb. Technical Univ. Basic Electrical and Electronics Engineering May 2007]

8. Derive expressions for average and rms values of a sinusoidally varying quantity.

9. Explain the terms rms value and average value of ac sinusoidal current. Also derive the rms value of ac sinusoidal current having I_m as maximum value.
[Pb. Univ. Elements of Elec. Engg. II December 1988]

10. Derive the values of form factor and peak factor of a sinusoidal varying quantity.

11. Differentiate between form factor and peak factor.

[Pb. Technical Univ. Basic Electrical Engineering June 2003]

12. Derive expressions for the rms and average values of the half-wave rectified alternating (sinusoidal) current in terms of its amplitude.

13. Obtain the values of the (*i*) form factor and (*ii*) peak factor of half-wave rectified sinusoidal wave.

14. Explain the need and method of representing alternating quantities as phasor quantity.

15. What do you understand by phasor representation of alternating quantity ? What is an alternating quantity ? When is an alternating quantity said to be lagging or leading with respect to another quantity ?

16. Explain with diagrams what do you understand by (*i*) in-phase (*ii*) lagging and (*iii*) leading as applied to sinusoidal ac quantities.

17. Explain the difference between ac and dc quantities. Draw the waveform and phasor diagrams of two alternating quantities having a phase difference of 90º.

18. Explain the component method for the addition and subtraction of alternating quantities.

SHORT ANSWER TYPE QUESTIONS WITH ANSWERS

Q. 1. What is the difference between ac and dc ?

Ans. A current is called alternating if it periodically changes its direction and magnitude whereas direct currents are steady and flow in one direction only.

Q. 2. Why sinusoidal wave shape is insisted for voltages and currents while generating, transmitting and utilising ac electric power ? **[P.T.U. January 2000]**

Ans. Sinusoidal wave shape is insisted for voltages and currents while generating, transmitting and utilising ac electrical power because it has following advantages:

(*i*) Sinusoidal voltages and currents produce minimum disturbance in the electrical circuits during operation.

(*ii*) Sinusoidal voltages and currents cause less interference (noise) to nearby communication circuits (telephone lines etc.).

(*iii*) Sinusoidal voltages and currents result in low iron as well as low copper losses in transformers, and ac rotating machines for a given output. So ac machines with sinusoidal voltages and currents operate at higher efficiency.

Q. 3. What do you understand by ω ?

Ans. Each cycle of a sinusoidal wave spans 2π radians. Hence, if this quantity is divided by the time period, angular velocity of the sinusoidal wave is obtained. It is denoted by ω and is expressed in radians per second.

Q. 4. Why the rms value of an alternating current or voltage is used to denote its amplitude ?

Ans. RMS value of an alternating current or voltage is used to denote its amplitude because it is related to the power developed in a resistance by the alternating current or voltage.

Q. 5. What is rms value of an alternating current ? **[P.T.U. December 2010]**

OR

Define rms value of ac. **[P.T.U. December 2009]**

OR

Define root mean square value of a sinusoidal wave. **[P.T.U. December 2008]**

Ans. The effective or rms value of an alternating current is given by that steady current which when flows through a given resistance for a given time produces the same amount of heat as when the alternating current is flowing through the same resistance for the same time duration.

For sinusoidal current, $I_{rms} = \dfrac{I_{max}}{\sqrt{2}}$

Q. 6. Do waveshapes other than sine wave have effective value ? Explain. **[P.T.U. Dec. 2000; May 2010]**

Ans. Yes, all the wave shapes including the sine wave have effective value since in each half cycle of the wave, work is being done. Actually the effective value of an alternating wave (sinusoidal or non-sinusoidal) is defined as on next page :

The rms or effective value of an alternating current or voltage is given by that steady current or voltage which when flows or applied to a given resistance for a given time produces the same amount of heat as when the alternating current or voltage is flowing or applied to the same resistance for the same time.

Q. 7. Define average value of a sinusoidal wave. **[P.T.U. May 2009]**

Ans. The *average (or mean) value* of an alternating current is expressed by that steady current (dc) which transfers across any circuit the same charge as transferred by that alternating current

Average or mean value of alternating current is given as

$$I_{av} = \frac{i_1 + i_2 + i_3 + ... + i_n}{n} = \frac{\text{Area of one alternation}}{\text{Length of base over one alternation}}$$

$$\text{For sinusoidal current } I_{av} = \frac{2}{\pi} I_{max} = 0.637\ E_{max}.$$

Q. 8. Differentiate between form factor and peak factor. **[P.T.U. June 2003]**

Ans. Form factor is defined as the ratio of the effective value to the average or mean value of the periodic wave while peak factor is defined as the ratio of peak or maximum value to the effective or rms value of the periodic wave.

$$\textit{i.e. } K_f = \frac{\text{RMS value}}{\text{Average value}} \text{ while } K_p = \frac{\text{Peak value}}{\text{RMS value}}$$

Q. 9. What is significance of form factor ?

Ans. Form factor is a means of relating the mean value with the effective or rms value of alternating quantity and it is useful in determination of effective or rms values of the alternating quantities whose mean or average values over half a period can be determined conveniently.

Q. 10. What is the significance of peak factor ?

Ans. Knowledge of peak factor of an alternating voltage is very essential in connection with determining the dielectric strength since the dielectric stress developed in an insulating material is proportional to the peak value of the voltage applied to it.

Q. 11. Mention the values of form factor and peak factor for half-wave rectified sinusoidal current.

Ans. For half wave rectified sinusoidal current

$$\text{Form factor} = \frac{\pi}{2} = 1.57$$

$$\text{Peak factor} = 2$$

Q. 12. What is the difference between a vector and a phasor ? **[P.T.U. January 2000]**

Ans. *Vector* is a quantity that has magnitude as well as direction. A sinusoidal quantity (voltage or current) is represented by a line of definite length rotating in counter-clockwise direction with the same angular velocity as that of the sinusoidal quantity. Such a rotating line is called the phasor. Any sinusoidal quantity (current or voltage) at a given frequency is completely specified by its amplitude and phase angle, its similarity to a vector quantity is evident, since the amplitude may be considered as the magnitude and the phase angle as the direction of a vector. To account for the difference the term *phasor* has been adopted, instead of the term vector, for representing graphically the magnitude and phase of a sinusoidal current or voltage.

Q. 13. What is the significance of the phasor representation of an alternating quantity ?

Ans. The phasor representation of an alternating quantity enables us to understand its magnitude and position with respect to reference line.

Q. 14. What do you mean by phase and phase difference ?

Ans. The phase of an alternating quantity (voltage or current) at any instant is defined as the fractional part of a cycle through which the quantity has advanced while the phase difference may be defined as the angular displacement between the maximum positive values of the two phasors representing the two quantities having the same frequency.

PROBLEMS

1. For ac single phase 230 V, 50 Hz, assuming sinusoidal waveform find (*i*) maximum value (*ii*) rms value (*iii*) frequency (*iv*) instantaneous value voltage equation (*v*) voltage 0.003 second after starting from zero and rising in + ve direction. [**Ans.** (*i*) 325.3 V (*ii*) 230 V (*iii*) 50 Hz (*iv*) 325.3 sin 314*t* (*v*) 263.17 V]
2. Calculate the maximum value, and the root mean square value of the following quantities:
   (*i*) 40 sin ωt (*ii*) (A + B) sin $\left(\omega t - \frac{\pi}{2}\right)$; (*iii*) 10 sin ωt – 17.3 cos ωt.
   [**Ans.** (*i*) 40, $20\sqrt{2}$ (*ii*) A + B, $\frac{A+B}{\sqrt{2}}$, (*iii*) 20, $10\sqrt{2}$]
3. e = 300 sin 628 t volts. Find maximum value, rms value, average value and frequency. Find the instantaneous value of voltage at t = 0.001 second after e is zero and rising in positive direction.
   [**Ans.** 300 V, 212.13 V, 190.98 V, 100 Hz, 176.26 V]
4. An alternating current is represented by $i = 400 \sin\left(157t + \frac{\pi}{6}\right)$ amperes.

   Determine: (*i*) Peak value (*ii*) Average value (*iii*) RMS value (*iv*) Frequency (*v*) Time period (*vi*) Form factor (*vii*) Peak factor, and (*viii*) Phase angle of the current represented.
   [**Ans.** (*i*) 400 A (*ii*) 254.65 A (*iii*) 282.84 A, (*iv*) 25 Hz (*v*) 0.04 second (*vi*) 1.11 (*vii*) 1.414 (*viii*) 30°]
5. A voltage waveform is given by the expression
   v = 200 sin (628 t + 60°) volts. Determine:
   (*i*) maximum value (*ii*) rms value (*iii*) frequency (*iv*) time period and (*v*) phase angle of the voltage.
   [**Ans.** (*i*) 200 V (*ii*) 141.4 V (*iii*) 100 Hz (*iv*) 0.01s (*v*) 60°]
6. (*i*) Find the rms value and frequency for an ac voltage given by v = 100 sin 314 t volts. (*ii*) The power losses, if this voltage is applied to a 20 Ω resistance. [**Ans.** (*i*) 70.71 V, 50 Hz (*ii*) 250 W]
7. The currents flowing in the three branches of an ac parallel circuit are :
   $i_1 = 20 \sin \omega t$; $i_2 = 25 \sin\left(\omega t - \frac{\pi}{6}\right)$; $i_3 = 15 \sin\left(\omega t + \frac{\pi}{6}\right)$. Determine the total current drawn from the supply. [**Ans.** 54.87 sin (ωt – 0.029π)]
8. Three sinusoidal voltages acting in series are given by :

$$v_1 = 10 \sin 440\, t$$
$$v_2 = 10\sqrt{2} \sin (440\, t - 45°)$$
$$v_3 = 20 \cos 440\, t$$

   Determine :
   (*i*) an expression for the resultant voltage,
   (*ii*) the frequency and rms value of the resultant voltage.
   [**Ans.** (*i*) v = 22.36 sin (440 t + 0.1476 π), (*ii*) 70 Hz; 15.81 V]

□□□

3

CHAPTER

Single Phase AC Circuits

3.1. INTRODUCTION

INSIDE THIS CHAPTER

In a dc circuit the relationship between the applied voltage V and current flowing through the circuit I is a simple one and is given by the expression $I = \frac{V}{R}$ but in an ac circuit this simple relationship does not hold good. Variations in current and applied voltage set up magnetic and electrostatic effects respectively and these must be taken into account with the resistance of the circuit while determining the quantitative relations between current and applied voltage. With comparatively low-voltage, heavy-current circuits magnetic effects may be very large, but electrostatic effects are usually negligible. On the other hand with high-voltage circuits electrostatic effects may be of appreciable magnitude, and magnetic effects are also present.

In this chapter we will discuss how do the magnetic effects due to variations in current and electrostatic effects due to variations in the applied voltage affect the relationship between the applied voltage and current.

3.2. PURELY RESISTIVE CIRCUITS

A purely resistive or a non-inductive circuit is a circuit which has inductance so small that at normal frequency its reactance is negligible as compared to its resistance. Ordinary filament lamps, water resistances etc. are the examples of non-inductive resistances. If the circuit is purely non-inductive, no reactance emf (*i.e.* self-induced or back emf) is set up and whole of the applied voltage is utilised in overcoming the ohmic resistance of the circuit.

Consider an ac circuit containing a non-inductive resistance of R ohms connected across a sinusoidal voltage represented by $v = V_{max} \sin \omega t$, as shown in Fig. 3.1 (*a*).

As already said, when the current flowing through a pure resistance changes, no back emf is set up, therefore, applied voltage has to overcome the ohmic drop of iR only

$$i.e. \quad iR = v$$

$$\text{or} \quad i = \frac{v}{R} = \frac{V_{max}}{R} \sin \omega t$$

Current will be maximum when $\omega t = \frac{\pi}{2}$ or $\sin \omega t = 1$

$$\therefore \quad I_{max} = \frac{V_{max}}{R}$$

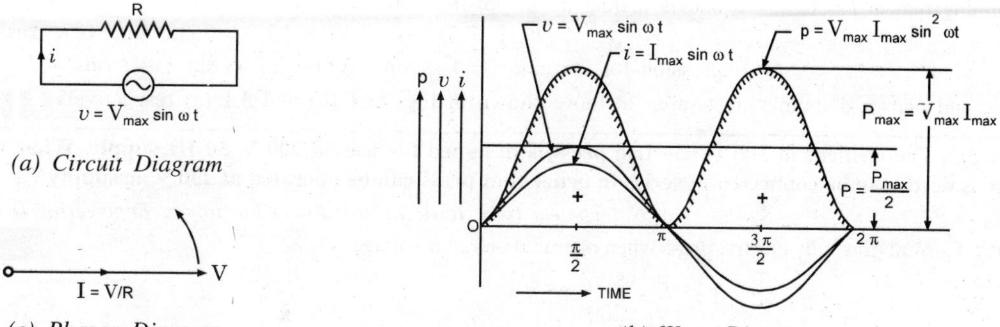

Fig. 3.1. ***Purely Resistive Circuit***

and instantaneous current may be expressed as

$$i = I_{max} \sin \omega t$$

From the expressions of instantaneous applied voltage and instantaneous current, it is evident that in a pure resistive circuit, the applied voltage and current are in phase with each other, as shown by wave and phasor diagrams in Figs. 3.1 (*b*) and 3.1 (*c*) respectively.

3.2.1. Power in a Purely Resistive Circuit

The instantaneous power delivered to the circuit in question is the product of the instantaneous values of applied voltage and current.

$$\text{i.e.} \quad p = vi = V_{max} \sin \omega t \, I_{max} \sin \omega t = V_{max} \, I_{max} \sin^2 \omega t$$

$$\text{or} \quad p = \frac{V_{max} \, I_{max}}{2} (1 - \cos 2 \, \omega t) \quad \text{Since } \sin^2 \omega t = \frac{1 - \cos 2\omega t}{2}$$

$$= \frac{V_{max} \, I_{max}}{2} - \frac{V_{max} \, I_{max}}{2} \cos 2\omega t$$

$$\text{Average power, } P = \text{Average of } \frac{V_{max} \, I_{max}}{2} - \text{Average of } \frac{V_{max} \, I_{max}}{2} \cos 2\omega t$$

Since average of $\frac{V_{max} \, I_{max}}{2} \cos 2\omega t$ over a complete cycle is zero

$$P = \frac{V_{max} \, I_{max}}{2} = \frac{V_{max}}{\sqrt{2}} \cdot \frac{I_{max}}{\sqrt{2}} = VI \text{ watts}$$

where V and I are the rms values of applied voltage and current respectively.

Thus for purely resistive circuits, the expression for power is the same as for d c circuits. From the power curve for a purely resistive circuit shown in Fig. 3.1 (*b*) it is evident that power is always positive. This is so because the instantaneous values of voltage and current are always either both positive or negative and, therefore, the product is always positive. This means that the voltage source constantly delivers power to the circuit and the circuit consumes it.

Example 3.1. A 250 V (rms), 50 Hz voltage is applied across a circuit consisting of a pure (non-inductive) resistance of 20 Ω. Determine (*i*) the current flowing through the circuit and (*ii*) power absorbed by the circuit. Give the expressions for the voltage and current. Give the wave and phasor diagrams.

Solution : (*i*) Current flowing through the circuit, $I = \frac{V}{R} = \frac{250}{20} = 12.5$ A **Ans.**

(*ii*) Power absorbed, $P = VI = 250 \times 12.5 = 3{,}125$ W or 3.125 kW **Ans.**

Peak value of applied voltage, $V_{max} = \sqrt{2}\, V_{rms} = \sqrt{2} \times 250 = 353.6$ V

Peak value of current, $I_{max} = \sqrt{2}\, I_{rms} = \sqrt{2} \times 12.5 = 17.68$ A

Expression for voltage, $v = V_{max} \sin 2\pi f t$

Assuming that at $t = 0$, $v = 0$ and is increasing in positive direction

$$v = 353.6 \sin 2\pi \times 50\, t = 353.6 \sin 314\, t \textbf{ Ans.}$$

Expression for current, $i = I_{max} \sin 2\pi f t = 17.68 \sin 314\, t$ **Ans.**

Wave and phasor diagrams are similar to those shown in Figs. 3.1 (*b*) and 3.1 (*c*) respectively.

Example 3.2. The element of 500 W electric press is designed for use on 200 V, 50 Hz supply. What value of resistance is needed to be connected in series in order that press can be operated at 230 V ac supply ?

[*Pb. Technical Univ. Basic Electrical and Electronics Engineering Dec. 2004*]

Solution : Current drawn by electric press when operated on rated voltage.

$$I = \frac{P}{V} = \frac{500}{200} = 2.5 \text{ A}$$

Voltage drop across resistance R connected in series with the electric press,

$$V_R = 230 - 200 = 30 \text{ V}$$

$$\text{Resistance, } R = \frac{V_R}{I} = \frac{30}{2.5} = 12\ \Omega \textbf{ Ans.}$$

3.3. PURELY INDUCTIVE CIRCUITS

An inductive circuit is a coil with or without an iron core having negligible resistance. Practically pure inductance can never be had as the inductive coil has always small resistance. However, a coil of thick copper wire wound on a laminated iron core has negligible resistance and is known as a *choke coil.**

When an alternating voltage is applied to a purely inductive coil, an emf, known as self induced emf, is induced in the coil which opposes the applied voltage. Since coil has no resistance, therefore, at every instant applied voltage has to overcome this self-induced emf only.

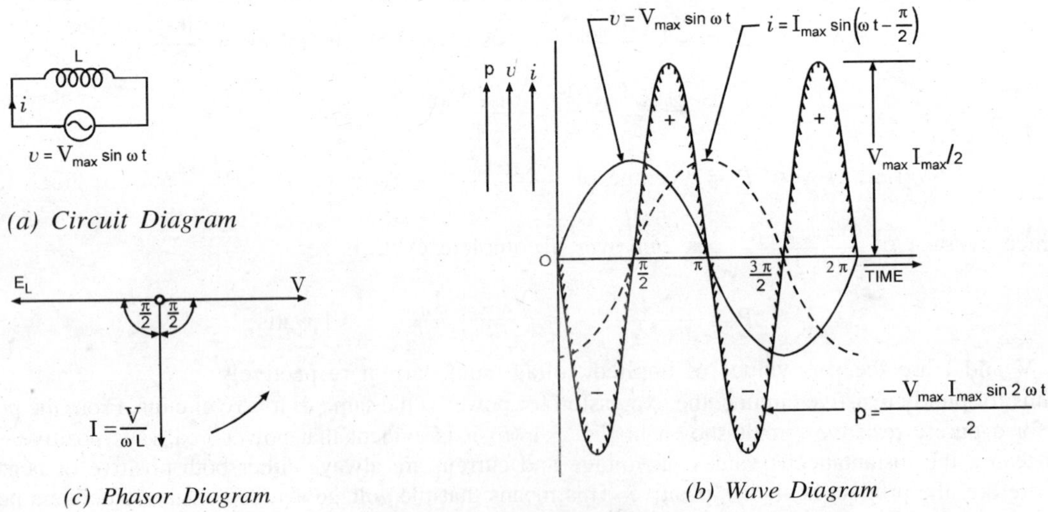

(*a*) *Circuit Diagram*

(*b*) *Wave Diagram*

(*c*) *Phasor Diagram*

Fig. 3.2. ***Purely Inductive Circuit***

---

* Inductors or choke coils are made of iron core because large valued flux densities can be produced in iron cores and so inductances of large values can be had. Air-cored inductors become too much bulky to give an inductance of a given value.

Let the applied voltage $v = V_{max} \sin \omega t$
and self-inductance of coil = L henry

Self induced emf in the coil, $e_L = -L \frac{di}{dt}$

Since applied voltage at every instant is equal and opposite to the self-induced emf *i.e.* $v = -e_L$

$$\therefore \quad V_{max} \sin \omega t = -\left(-L \frac{di}{dt}\right)$$

$$\text{or } di = \frac{V_{max}}{L} \sin \omega t \, dt$$

Integrating both sides we get

$$i = \frac{V_{max}}{L} \int \sin \omega t \, dt$$

$$\text{or } i = \frac{V_{max}}{\omega L} (-\cos \omega t) + A$$

where A is a constant of integration, which is found to be zero from initial conditions

$$i.e. \quad i = \frac{-V_{max}}{\omega L} \cos \omega t$$

$$\text{or} \quad i = \frac{V_{max}}{\omega L} \sin \left(\omega t - \frac{\pi}{2}\right)$$

Current will be maximum when $\sin \left(\omega t - \frac{\pi}{2}\right) = 1$, hence, value of maximum current,

$$I_{max} = \frac{V_{max}}{\omega L}$$

and instantaneous current may be expressed as

$$i = I_{max} \sin \left(\omega t - \frac{\pi}{2}\right)$$

From the expressions of instantaneous applied voltage and instantaneous current flowing through a pure inductive coil it is observed that the current lags behind the applied voltage by $\frac{\pi}{2}$, as shown in Fig. 3.2 (*b*) by wave diagram and in Fig. 3.2 (*c*) by phasor diagram.

3.3.1. Inductive Reactance

ωL in the expression $I_{max} = \frac{V_{max}}{\omega L}$ is known as *inductive reactance* and is denoted by X_L *i.e.* $X_L = \omega L$

If L is in henry and ω is in radians per second then X_L will be in ohms.

3.3.2. Power in a Purely Inductive Circuit

$$\text{Instantaneous power, } p = v \times i = V_{max} \sin \omega t \, I_{max} \sin \left(\omega t - \frac{\pi}{2}\right)$$

$$\text{or} \quad p = -V_{max} I_{max} \sin \omega t \cos \omega t = -\frac{V_{max} I_{max}}{2} \sin 2\omega t$$

The power measured by wattmeter is the average value of p which is zero since average of a sinusoidal quantity of double frequency over a complete cycle is zero. *Hence in a pure inductive circuit power absorbed is zero.*

Physically the above fact can be explained as below:

During the second quarter of a cycle the current and the magnetic flux of the coil increases and the coil draws power from the supply source to build up the magnetic field (the power drawn is positive and the energy drawn by the coil from the supply source is represented by the area between the curve *p* and the time axis). The energy stored in the magnetic field during build up is given as $W_{max} = \frac{1}{2} L I_{max}^2$.

In the next quarter the current decreases. The emf of self induction will, however, tend to oppose its decrease. The coil acts as a generator of electrical energy, returning the stored energy in the magnetic field to the supply source (now the power drawn by the coil is negative and the curve *p* lies below the time axis).

The chain of events repeats itself during the next half cycles. Thus, a proportion of power is continually exchanged between the field and the inductive circuit and the power consumed by a purely inductive coil is zero.

Example 3.3. A 0.014 H choke coil with negligible resistance is connected to a 220 $V_{(rms)}$, 50 Hz supply. Determine (*i*) inductive reactance of the coil (*ii*) current flowing through the coil (*iii*) power consumed by the coil and (*iv*) maximum energy stored in the coil. Write down the expressions for voltage and current and give the wave and phasor diagrams.

Solution : (*i*) Inductive reactance of coil, $X_L = 2\pi f L = 2\pi \times 50 \times 0.014 = 4.4\ \Omega$ **Ans.**

(*ii*) Current flowing through the coil, $I = \frac{V}{X_L} = \frac{220}{4.4} = 50$ A **Ans.**

(*iii*) Power consumed by the coil, P = zero **Ans.**

(*iv*) Maximum energy stored in the coil, $W_{max} = \frac{1}{2} L I_{max}^2 = \frac{1}{2} \times 0.014 \times (\sqrt{2} \times 50)^2 = 35$ J **Ans.**

Assuming that at time $t = 0$, $v = 0$ and is increasing in positive direction

Expression for the voltage, $v = V_{max} \sin 2\pi f t$

$= 220\sqrt{2} \sin 2\pi \times 50t = 311 \sin 314t$ **Ans.**

Expression for the current, $i = I_{max} \sin\left(2\pi f t - \frac{\pi}{2}\right) = 50\sqrt{2} \sin\left(314\,t - \frac{\pi}{2}\right)$ **Ans.**

Wave and phasor diagrams are similar to those shown in Figs. 3.2 (*b*) and 3.2 (*c*) respectively.

Example 3.4. The voltage and current through a circuit element are

$$v = 100 \sin (314\,t + 45°) \text{ volts}$$
$$i = 10 \sin (314\,t + 315°) \text{ amperes.}$$

(*i*) Identify the circuit element (*ii*) Find the value (*iii*) Obtain expression for power.

Solution : $v = 100 (\sin 314\,t + 45°)$ volts

$i = 10 \sin (314\,t + 315°) = 10 \sin (314\,t + 315° - 360°) = 10 \sin (314\,t - 45°)$ amperes

(*i*) From the equations for voltage and current it is observed that the circuit current *i* lags behind the applied voltage *v* by 90°. It means the circuit element is an *inductor.* **Ans.**

(*ii*) Inductive reactance of the inductor, $X_L = \frac{V_{rms}}{I_{rms}} = \frac{100/\sqrt{2}}{10/\sqrt{2}} = 10\ \Omega$ **Ans.**

and $L = \frac{X_L}{\omega} = \frac{10}{314} = 0.0318$ H or 31.8 mH **Ans.**

(*iii*) Expression for power is given as

$$p = \frac{-V_{max}\, I_{max}}{2} \sin 2\omega t = \frac{-100 \times 10}{2} \sin (2 \times 314t) = -500 \sin 628t$$ **Ans.**

Example 3.5. A coil of inductance 0.05 H is connected to a supply of 220 V, 50 Hz. Calculate the current in the coil.

[*Pb. Technical Univ. Basic Electrical and Electronics Engineering May 2007*]

Solution : Inductive reactance of the coil $X_L = 2\pi f L = 2\pi \times 50 \times 0.05 = 15.7\ \Omega$

Current in coil, $I = \frac{V}{X_L} = \frac{220}{15.7} = 14.0$ A **Ans.**

3.4. PURELY CAPACITIVE CIRCUITS

When a dc voltage is impressed across the plates of perfect condenser, it will become charged to full voltage almost instantaneously. The charging current will flow only during the period of "build up" and will cease to flow as soon as the capacitor has attained the steady voltage of the source. This implies that for a direct current, a capacitor is a break in the circuit or an infinitely high resistance.

When a capacitor is connected to an alternating current circuit the capacitor is charged and discharged during alternate quarter cycles.

Let an alternating voltage represented by $v = V_{max} \sin \omega t$ be applied across a capacitor of capacitance C farads. The expression for instantaneous charge is given as

$$q = C V_{max} \sin \omega t$$

Since the capacitor current is equal to the rate of change of charge, therefore, the capacitor current may be obtained by differentiating the above equation

$$i = \frac{dq}{dt} = \frac{d}{dt}[CV_{max} \sin \omega t]$$

$$= \omega C V_{max} \cos \omega t = \frac{V_{max}}{1/\omega C} \sin\left(\omega t + \frac{\pi}{2}\right)$$

Current is maximum when $t = 0$

$$\therefore \quad I_{max} = \frac{V_{max}}{1/\omega C}$$

Substituting $\frac{V_{max}}{1/\omega C} = I_{max}$ in the above equation for instantaneous current, we get

$$i = I_{max} \sin\left(\omega t + \frac{\pi}{2}\right)$$

From the equations of instantaneous applied voltage and instantaneous current flowing through capacitance, it is observed that the current leads the applied voltage by $\frac{\pi}{2}$, as shown in Figs. 3.3 (*b*) and 3.3 (*c*) by wave and phasor diagrams respectively.

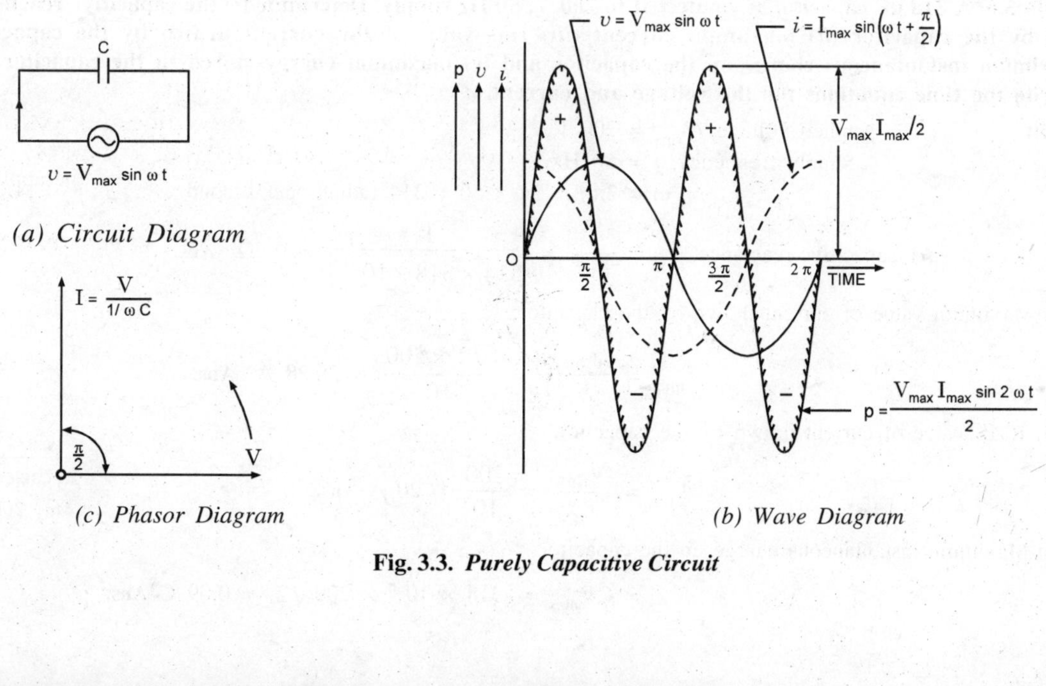

Fig. 3.3. *Purely Capacitive Circuit*

3.4.1. Capacitive Reactance

$\frac{1}{\omega C}$ in the expression $I_{max} = \frac{V_{max}}{1/\omega C}$ is known as *capacitive reactance* and is denoted by X_C

$$i.e \quad X_C = \frac{1}{\omega C}$$

If C is in farads and ω is in radians/s, X_C will be in ohms.

3.4.2. Power in a Purely Capacitive Circuit

$$\text{Instantaneous power, } p = v\,i = V_{max} \sin \omega t .\ I_{max} \sin\left(\omega t + \frac{\pi}{2}\right)$$

$$= V_{max}\, I_{max} \sin \omega t \cos \omega t = \frac{V_{max}\, I_{max}}{2} \sin 2\omega t$$

$$\text{Average power, } P = \frac{V_{max}\, I_{max}}{2} \times \text{average of } \sin 2\omega t \text{ over a complete cycle} = 0.$$

Hence power absorbed in a pure capacitive circuit is zero. The same is shown graphically in Fig. 3.3 (*b*). The energy taken from the supply circuit is stored in the capacitor during the first quarter-cycle and returned during the next.

The energy stored by a capacitor at maximum voltage across its plates is given by the expression

$$W_C = \frac{1}{2} CV^2_{max}$$

This can be realized when it is recalled that no heat is produced and no work is done while current is flowing through a capacitor. As a matter of fact, in commercial capacitors, there is a slight energy loss in the dielectric in addition to a minute I^2R loss due to flow of current over the plates having definite ohmic resistance. Although it raises the power factor from zero to 0.002 or even a little more, but for ordinary purposes the power factor is taken to be zero. Obviously the phase angle due to dielectric and ohmic losses decreases slightly.

Example 3.6. A 318 μF capacitor is connected to 200 V, 50 Hz supply. Determine (*i*) the capacitive reactance offered by the capacitor (*ii*) maximum current (*iii*) rms value of the current drawn by the capacitor (*iv*) maximum instantaneous charge on the capacitor and (*v*) maximum energy stored in the capacitor.

Write the time equations for the voltage and current.

Solution : Applied voltage, $V_{rms} = 200\ V$

Supply frequency, $f = 50$ Hz

$$\omega = 2\pi f = 2\pi \times 50 = 314 \text{ radians per second}$$

(*i*) Capacitive reactance, $X_C = \frac{1}{\omega C} = \frac{1}{314 \times 318 \times 10^{-6}} = 10\ \Omega$ **Ans.**

(*ii*) Maximum value of current drawn by the capacitor,

$$I_{max} = \frac{V_{max}}{X_C} = \frac{\sqrt{2} \times 200}{10} = 28.28 \text{ A} \quad \textbf{Ans.}$$

(*iii*) RMS value of current drawn by the capacitor,

$$I_{rms} = \frac{V_{rms}}{X_C} = \frac{200}{10} = 20 \text{ A} \quad \textbf{Ans.}$$

(*iv*) Maximum instantaneous charge on the capacitor

$$= CV_{max} = 318 \times 10^{-6} \times 200\sqrt{2} = 0.09 \text{ C} \quad \textbf{Ans.}$$

(v) Maximum energy stored, $W_C = \frac{1}{2} C (V_{max})^2 = \frac{1}{2} \times 318 \times 10^{-6} \times (200\sqrt{2})^2 = 12.72$ J **Ans.**

Expression for the voltage, $v = V_{max} \sin \omega t$ assuming that at time $t = 0$. $v = 0$ and is increasing in positive direction

$$= 200\sqrt{2} \sin 314t \text{ Ans.}$$

Expression for current, $i = I_{max} \sin\left(\omega t + \frac{\pi}{2}\right) = 28.28 \sin\left(314t + \frac{\pi}{2}\right)$ **Ans.**

Example 3.7. A capacitor of 25 μF is connected to a supply of 200 V, 50 Hz. What will be current flowing through the capacitor ? *[Pb. Technical Univ. Basic Electrical and Electronics Engineering Dec. 2006]*

Solution : Current flowing through the capacitor,

$$I = 2\pi f CV = 2\pi \times 50 \times 25 \times 10^{-6} \times 200 = 1.57 \text{ A} \textbf{ Ans.}$$

3.5. RESISTANCE AND INDUCTANCE IN SERIES

This is the case most generally met within practice, nearly all circuits contain both resistance and inductance.

Consider an ac circuit consisting of resistance of R ohms and inductance of L henrys connected in series, as shown in Fig. 3.4 (*a*).

Let the supply frequency be f and current flowing through the circuit be of I amperes (rms value).

Now voltage drop across resistance, $V_R = IR$ in phase with the current.

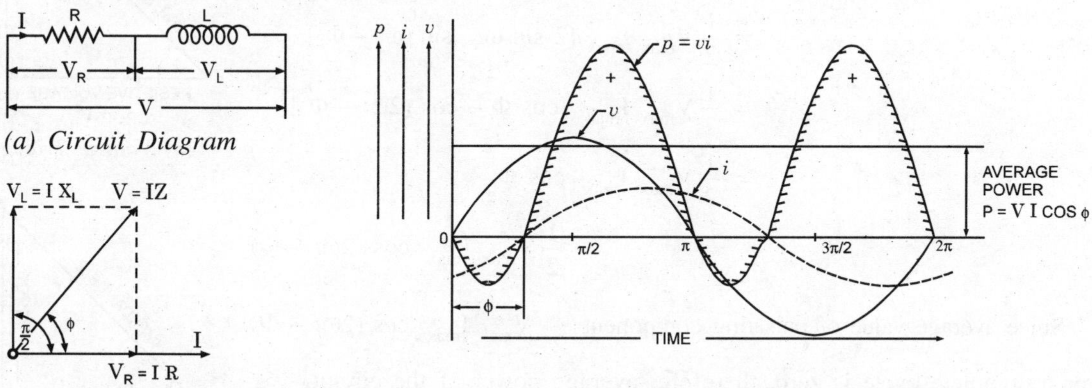

(*a*) *Circuit Diagram*

(*c*) *Phasor Diagram*

(*b*) *Wave Diagram*

Fig. 3.4 . ***Resistance and Inductance in Series***

Voltage drop across inductance, $V_L = IX_L = I\omega L$ leading I by $\frac{\pi}{2}$ radians, as shown in Fig. 3.4 (*c*)

The applied voltage, being equal to the phasor sum of V_R and V_L, will be given by the diagonal of the parallelogram.

$$\text{Applied voltage, } V = \sqrt{(V_R)^2 + (V_L)^2} = \sqrt{(IR)^2 + (IX_L)^2} = I\sqrt{R^2 + X_L^2} = IZ$$

where $X_L = \omega L = 2\pi f L$

Quantity $\sqrt{R^2 + X_L^2}$ is known as *impedance*, denoted by Z and is expressed in ohms.

From phasor diagram it is also evident that the current lags behind the applied voltage V by angle Φ, which is given by

$$\text{Tan } \Phi = \frac{V_L}{V_R} = \frac{IX_L}{IR} = \frac{X_L}{R}$$

$$\text{or} \quad \Phi = \text{Tan}^{-1} \frac{X_L}{R}$$

Since X_L and R are known, the value of phase angle Φ can be computed. If the applied voltage $v = V_{max} \sin \omega t$, then expression for the circuit current will be

$$i = I_{max} \sin (\omega t - \Phi)$$

where $I_{max} = \dfrac{V_{max}}{Z}$ and $\Phi = \text{Tan}^{-1} \dfrac{X_L}{R}$.

3.5.1. Impedance Triangle

If a triangle ABC is drawn so that AB = $\dfrac{V_R}{I}$ = R, BC = $\dfrac{V_L}{I} = X_L$ and AC = $\dfrac{V}{I}$ = Z, it is a triangle similar to voltage triangle [Fig. 3.5 (*a*)]. Such a triangle is called an *impedance triangle*. The angle between Z and R sides of the impedance triangle is known as *phase angle of the circuit* and cosine of this angle is known as *power factor of the circuit*.

$$\text{Power factor} = \text{Cos } \Phi = \frac{R}{Z}$$

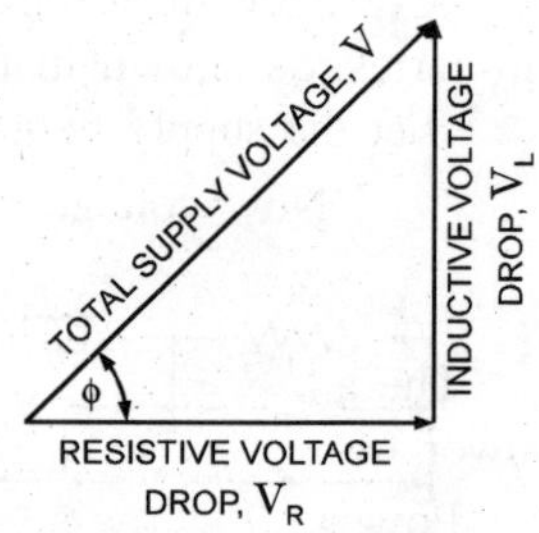

(*a*) *Voltage Triangle*

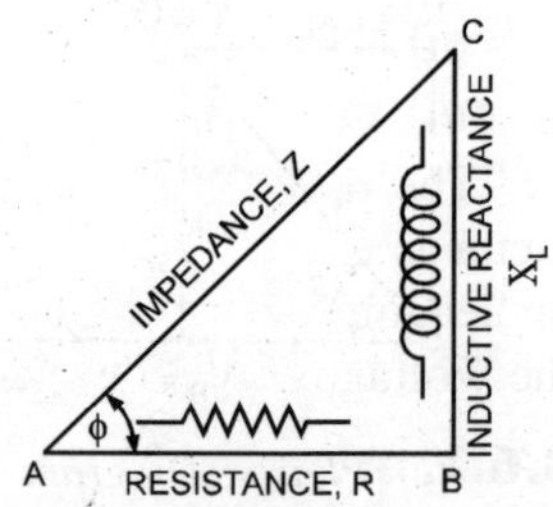

(*b*) *Impedance Triangle*

Fig. 3.5

3.5.2. Power in Resistance - Inductance Circuit

Instantaneous power, $p = vi = V_{max} \sin \omega t \times I_{max} \sin (\omega t - \Phi)$

$$= \frac{1}{2} V_{max} I_{max} [2 \sin \omega t \, . \, \sin (\omega t - \Phi)]$$

$$= \frac{1}{2} V_{max} I_{max} [\cos \Phi - \cos (2\omega t - \Phi)]$$

$$= \frac{1}{2} V_{max} I_{max} \cos \Phi - \frac{1}{2} V_{max} I_{max} \cos (2\omega t - \Phi)$$

Since average value of pulsating component $\dfrac{1}{2} V_{max} I_{max} \cos (2\omega t - \Phi)$ over a complete cycle is zero, therefore, average power of the circuit,

$$P = \text{Constant component, } \frac{1}{2} V_{max} I_{max} \cos \Phi$$

$$= \frac{V_{max}}{\sqrt{2}} . \frac{I_{max}}{\sqrt{2}} \cos \Phi = V \, I \cos \Phi$$

where V and I are the rms values of voltage and current and Φ is the phase angle between applied voltage V and circuit current I.

Alternatively power consumed by the circuit,

P = Power consumed by resistance

$\because$ power consumed by inductance is zero

$$= I^2 R = I(IR) = \frac{V}{Z} . IR = VI \frac{R}{Z} = VI \cos \Phi$$

Since from impedance triangle $\cos \Phi = \dfrac{R}{Z}$.

So the power in an ac circuit is given by the product of rms values of current, and voltage and cosine of the phase angle between voltage and current. Cosine of the phase angle between the voltage and current, cos Φ is known as the *power factor* of the circuit, and is equal to $\frac{R}{Z}$ which is obvious from impedance triangle [Fig. 3.5 (*b*)]

3.6. APPARENT POWER, TRUE POWER, REACTIVE POWER AND POWER FACTOR

The product of rms values of current and voltage, VI is called the *apparent power* and is measured in volt-amperes or kilovolt amperes (kVA).

The *true power* in an ac circuit is obtained by multiplying the apparent power by the power factor and is expressed in watts or kilowatts (kW).

The product of apparent power, VI and the sine of the angle between voltage and current, sin Φ is called the *reactive power*. This is also known as *wattless power* and is expressed in reactive volt-amperes or kilovolt amperes reactive (kVAR).

i.e. Apparent power, $S = VI$ volt-amperes or $\frac{VI}{1,000}$ kVA

True power, $P = VI \cos \Phi$ watts or $\frac{VI \cos \Phi}{1,000}$ kW

Reactive power, $Q = VI \sin \Phi$ VAR or $\frac{VI \sin \Phi}{1,000}$ kVAR

and $KVA = \sqrt{(kW)^2 + (kVAR)^2}$

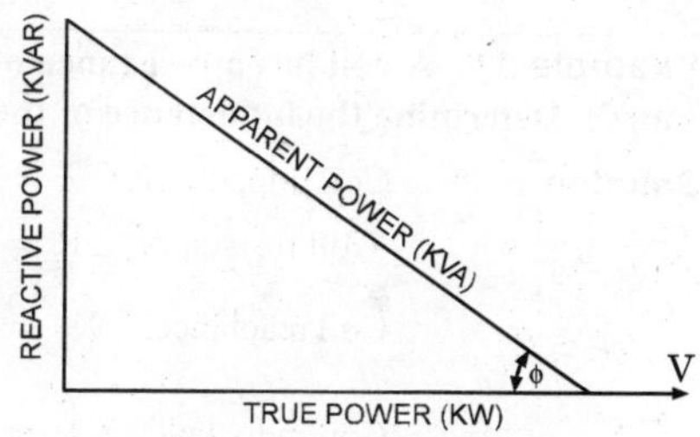

(*a*) *Power Triangle*

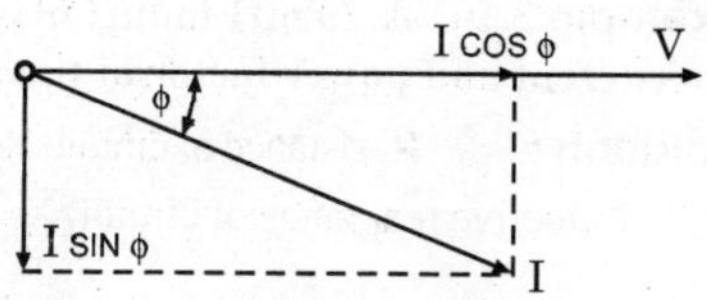

(*b*)

Fig. 3.6

The above relations can easily be followed by referring to the power triangle shown in Fig. 3.6 (*a*).

Power factor may be defined as

(*i*) cosine of the phase angle between voltage and current or

(*ii*) the ratio of the resistance to impedance or

(*iii*) the ratio of true power to apparent power.

The power factor can never be greater than unity. The power factor is expressed either as fraction or as a percentage. It is usual practice to attach the word 'lagging' or 'leading' with the numerical value of power factor to signify whether the current lags behind or leads the voltage.

3.6.1. Active Component of Current

The current component which is in phase with circuit voltage (*i.e.* I cos Φ) and contributes to active or true power of the circuit is called the *active* (*wattful* or *in-phase*) *component* of current.

3.6.2. Reactive Component of Current

The current component which is in quadrature (or 90° out of phase) to circuit voltage (*i.e* I sin Φ) and contributes to reactive power of the circuit, is called the *reactive* (or *wattless*) *component* of current.

3.7. Q-FACTOR OF COIL

Reciprocal of power factor is known as Q-factor of the coil. It is also called the quality factor or figure of merit of a coil.

$$\text{Mathematically Q-factor} = \frac{1}{\text{Power factor}} = \frac{1}{\text{Cos } \Phi} = \frac{Z}{R}$$

If R is very small in comparison to inductive reactance X_L, the

$$\text{Q-factor} = \frac{X_L}{R} = \frac{\omega L}{R}$$

$$\text{Also } Q = 2\pi \times \frac{\text{Maximum energy stored}}{\text{Energy dissipated per cycle}}$$

Example 3.8. A coil has a resistance of 10 Ω and draws a current of 5 A when connected across a 100 V, 50 Hz source. Determine the power factor of the circuit. *[Pb. Technical Univ. Electrical Engineering June 2000]*

Solution : Coil resistance, $R = 10\ \Omega$

$$\text{Coil impedance, } Z = \frac{\text{Voltage applied across the coil}}{\text{Current flowing through the coil}} = \frac{100}{5} = 20\ \Omega$$

$$\text{Power factor of the circuit, } \cos\phi = \frac{R}{Z} = \frac{10}{20} = 0.5 \text{ (lagging) } \textbf{Ans.}$$

Since the coil is inductive, the *pf* is always lagging.

Example 3.9. A coil has a resistance of 10 Ω and draws a current of 5 A when connected across a 100 V, 50 Hz source. Determine the inductance of the coil. *[Pb. Technical Univ. Electrical Engineering December 2000]*

Solution : Coil impedance, $Z = 20\ \Omega$, as determined above in Example 3.8.

Coil resistance, $R = 10\ \Omega$

$$\text{Coil reactance, } X_L = \sqrt{Z^2 - R^2} = \sqrt{20^2 - 10^2} = 17.32\ \Omega$$

$$\text{Coil inductance, } L = \frac{X_L}{2\pi f} = \frac{17.32}{2\pi \times 50} = 0.05513 \text{ H or } 55.13 \text{ mH } \textbf{Ans.}$$

Example 3.10. A 20 mH inductor is connected in series with a 10 Ω resistor across a 220 V, 50 Hz supply. Find the current and power factor of the circuit. *[Pb. Technical Univ. Electrical Engineering December 2001]*

Solution : Resistance of circuit, $R = 10\ \Omega$

$$\text{Inductive reactance of circuit, } X_L = 2\pi f L = 2\pi \times 50 \times 20 \times 10^{-3} = 6.28\ \Omega$$

$$\text{Impedance of the circuit, } Z = \sqrt{R^2 + X_L^2} = \sqrt{10^2 + (6.28)^2} = 11.81\ \Omega$$

$$\text{Circuit current, } I = \frac{V}{Z} = \frac{220}{11.81} = 18.628 \text{ A } \textbf{Ans.}$$

$$\text{Circuit power factor, } \cos\phi = \frac{R}{Z} = \frac{10}{11.81} = 0.8467 \text{ (lagging) } \textbf{Ans.}$$

Example 3.11. A coil has resistance of 10 Ω and draws a current of 5 A when connected across 100 V, 50 Hz source. Determine the reactive power of the circuit. *[Pb. Technical Univ. Electrical Engineering May 2001]*

Solution : Coil resistance, $R = 10\ \Omega$

$$\text{Coil impedance, } Z = \frac{V}{I} = \frac{100}{5} = 20\ \Omega$$

$$\text{Power factor of the circuit, } \cos\phi = \frac{R}{Z} = \frac{10}{20} = 0.5 \text{ (lagging)}$$

$$\sin\phi = \sqrt{1 - \cos^2\phi} = \sqrt{1 - (0.5)^2} = 0.866$$

$$\text{Reactive power} = VI\sin\phi = 100 \times 5 \times 0.866 = 433 \text{ VAR } \textbf{Ans.}$$

Example 3.12. A 120 V ac circuit contains 10 Ω resistance and 30 Ω reactance in series. What would be the average power in the circuit ?

Solution : Impedance of the circuit, $Z = \sqrt{R^2 + X^2} = \sqrt{10^2 + 30^2} = 31.62\ \Omega$

$$\text{Circuit current, } I = \frac{V}{Z} = \frac{120}{31.62} = 3.8 \text{ A}$$

$$\text{Power factor of the circuit,} \cos\Phi = \frac{R}{Z} = \frac{10}{31.62} = 0.3162$$

Average power in the circuit, $P = VI \cos\Phi = 120 \times 3.8 \times 0.3162 = 144$ watts **Ans.**

Example 3.13. A voltage $e = 200 \sin 100\pi t$ is applied to a coil having R = 200 Ω and L = 0.38 H. Find the expression for the current and power taken by the coil.

Solution : Peak value of applied voltage, $E_{max} = 200$ V

$$\omega = 100\pi$$

Circuit resistance, $R = 200\ \Omega$

Circuit reactance, $X_L = \omega L = 100\pi \times 0.38 = 119.38\ \Omega$

$$\text{Circuit impedance, } Z = \sqrt{R^2 + X_L^2} = \sqrt{(200)^2 + (119.38)^2} = 232.92\ \Omega$$

$$\text{Peak value of current, } I_{max} = \frac{V_{max}}{Z} = \frac{200}{232.92} = 0.859 \text{ A}$$

$$\text{Phase angle, } \Phi = \tan^{-1}\frac{X_L}{R} = \tan^{-1}\frac{119.38}{200} = 0.171\pi \text{ radians or } 30.8^\circ$$

∴ Expression for current, $i = I_{max} \sin(\omega t - \Phi) = 0.859 \sin(100\pi t - 0.171\pi)$ **Ans.**

$$\text{Power taken by the coil, } P = VI\cos\Phi = \frac{200}{\sqrt{2}} \times \frac{0.859}{\sqrt{2}} \cos 30.8^\circ = 73.8 \text{ W}$$ **Ans.**

Example 3.14. In a circuit containing a resistance of 5,000 Ω and an inductance of 1.0 H in series, a voltage of 150 V (rms) at 400 Hz is applied.

Determine the magnitude and phase of the current and the voltages across the resistance and the inductance. Calculate the power taken from the supply.

Solution : Inductive reactance, $X_L = 2\pi f L = 2\pi \times 400 \times 1.0 = 2{,}513\ \Omega$

$$\text{Impedance of the circuit, } Z = \sqrt{R^2 + X_L^2} = \sqrt{(5{,}000)^2 + (2{,}513)^2} = 5{,}600\ \Omega$$

$$\text{Circuit current, } I = \frac{V}{Z} = \frac{150}{5{,}600} = 0.0268 \text{ ampere}$$ **Ans.**

$$\text{Phase angle, } \Phi = \tan^{-1}\frac{X_L}{R} = \tan^{-1}\frac{2{,}513}{5{,}000} = \tan^{-1} 0.5026 = 26.7^\circ \text{ (lag)}$$ **Ans.**

Voltage across the resistance, $V_R = IR = 0.0268 \times 5{,}000 = 134$ V **Ans.**

Voltage across the inductance, $V_L = IX_L = 0.0268 \times 2{,}513 = 67.35$ V **Ans.**

Power taken from the supply, $P = VI\cos\Phi = 150 \times 0.0268 \times \cos 26.7^\circ = 3.6$ watts **Ans.**

Example 3.15. A current of 10 A flows in a circuit lagging behind the applied voltage of 100 V, 50 Hz, by 30°. Determine the values of resistance, reactance and impedance of the circuit.

Solution :

$$\text{Circuit impedance, } Z = \frac{V}{I} = \frac{100}{10} = 10\ \Omega$$ **Ans.**

Circuit phase angle, $\Phi = 30^\circ$ (lag)

$$\text{Circuit resistance, } R = Z \cos \Phi = 10 \cos 30° = 10 \frac{\sqrt{3}}{2} = 8.66\ \Omega$$

$$\text{Circuit reactance, } X_L = Z \sin \Phi = 10 \sin 30° = 10 \times \frac{1}{2} = 5\ \Omega \ \textbf{Ans.}$$

Note : Since phase angle is of lag, reactance will be inductive.

Example 3.16. In an a c circuit the current is given by *i* = 22 sin (314 – π/6) A. If the voltage of the circuit is the reference quantity, determine (*i*) power factor (*ii*) rms value of current, (*iii*) frequency of current.

Solution : From the equation of circuit current it is seen that it lags behind the reference quantity by $\frac{\pi}{6}$ radian and circuit voltage is the reference quantity, so

$$\text{Phase angle, } \phi = \frac{\pi}{6} \text{ radian or } \frac{180}{6} \text{ i.e. } 30° \text{ (lagging)}$$

$$(i) \text{ Power factor} = \cos \phi = \cos 30° = 0.866 \text{ (lagging) } \textbf{Ans.}$$

$$(ii) \text{ Peak value of current, } I_{max} = \text{Coefficient of the sine of the time angle} = 22 \text{ A}$$

$$\text{So rms value of current, } I_{rms} = \frac{I_{max}}{\sqrt{2}} = \frac{22}{\sqrt{2}} = 15.56 \text{ A } \textbf{Ans.}$$

$$(iii) \text{ Frequency, } f = \frac{\text{Coefficient of time, } t}{2\pi} = \frac{314}{2\pi} = 50 \text{ Hz } \textbf{Ans.}$$

Example 3.17. A resistance and an inductance are connected in series across a voltage *v* = 283 sin 314 *t*. The current expression is found to be 4 sin (314 *t* – π/4). Find the values of resistance, inductance and power factor.

Solution :

$$\text{Circuit voltage, } v = 283 \sin 314t \text{ volts}$$

$$\text{Circuit current, } i = 4 \sin \left(314\,t - \frac{\pi}{4}\right)$$

From the equations of circuit voltage and circuit current it is seen that current I lags behind the applied voltage V by $\frac{\pi}{4}$ radian or 45°

$$\text{So phase angle, } \phi = 45° \text{ (lagging)}$$

$$\text{Power factor of the circuit} = \cos \phi = \cos 45° = 0.707 \text{ (lagging) } \textbf{Ans.}$$

$$\text{Circuit impedance, } Z = \frac{V_{max}}{I_{max}} = \frac{283}{4} = 70.75\ \Omega$$

$$\text{Circuit resistance, } R = Z \cos \phi = 70.75 \times 0.707 = 50\ \Omega \ \textbf{Ans.}$$

$$\text{Circuit reactance, } X_L = Z \sin \phi = 70.75 \times \sin 45° = 70.75 \times 0.707 = 50\ \Omega$$

$$\text{Frequency, } f = \frac{314}{2\pi} = 50 \text{ Hz}$$

$$\text{Inductance, } L = \frac{X_L}{2\pi f} = \frac{50}{2\pi \times 50} = 0.1592 \text{ H } \textbf{Ans.}$$

Example 3.18. A current of 5 A flows through a non-inductive resistance in series with a choking coil when supplied at 250 V, 50 Hz. If the voltage across the resistance is 125 V and across the coil is 200 V, calculate impedance, reactance and resistance of the coil.

Solution : Circuit current, I = 5 A

Voltage across non-inductive resistance,

$$V_R = 125 \text{ V}$$

Impedance of choking coil, $Z = \dfrac{\text{Voltage across choking coil}}{\text{Circuit current, I}}$

$= \dfrac{200}{5} = 40\ \Omega$ **Ans.**

Draw phasor OA representing voltage across non-inductive resistance R (*i.e.* 125 V) and complete triangle OAB by taking AB = 200 V (voltage across the choking coil) and OB = 250 V (supply voltage).

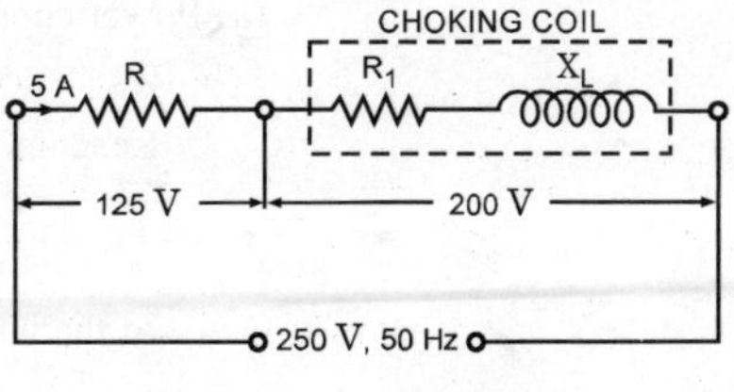

(a) Circuit Diagram

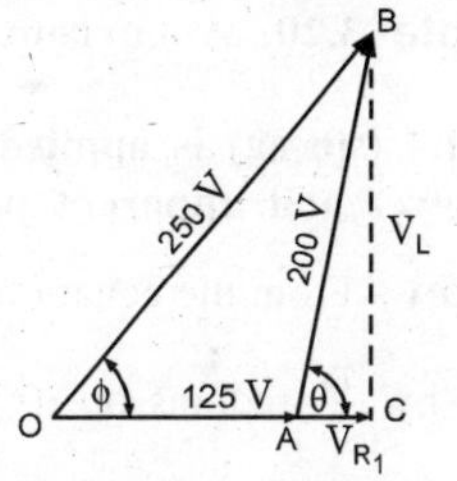

(b) Phasor Diagram

Fig. 3.7

Now from Δ OAB [Fig. 3.7 (*b*)]

$$OB^2 = OA^2 + AB^2 + 2\ OA \times AB\ \cos\theta$$

or $250^2 = 125^2 + 200 + 2 \times 125 \times 200 \cos\theta$

or $\cos\theta = \dfrac{250^2 - 200^2 - 125^2}{2 \times 200 \times 125} = 0.1375$

Resistance of choking coil, $R_1 = Z\cos\theta = 40 \times 0.1375$

$= 5.5\ \Omega$ **Ans.**

Reactance of choking coil, $X_L = \sqrt{Z^2 - (R_1)^2} = \sqrt{40^2 - 5.5^2}$

$= 39.62\ \Omega$ **Ans.**

Example 3.19. A coil is connected in series with a non-inductive resistance of 30 Ω across 240 V, 50 Hz supply. The reading of a voltmeter across the coil is 180 V and across the resistance is 130 V. Calculate (*a*) power consumed by the coil, and (*b*) pf of the whole circuit.

[*Pb. Technical Univ. Basic Electrical and Electronics Engineering May 2005*]

Solution : Circuit current, $I = \dfrac{\text{Voltage across non-inductive resistance}}{\text{Magnitude of non-inductive resistance}} = \dfrac{V_R}{R} = \dfrac{130}{30} = 4.333$ A

Impedance of choke coil, $Z = \dfrac{\text{Voltage across choke coil}}{\text{Circuit current, I}} = \dfrac{180}{4.333} = 41.54\ \Omega$

From ΔOAB shown in Fig. 3.8(*b*)

$$OB^2 = OA^2 + AB^2 + 2OA \times AB \times \cos\theta$$

or $\cos\theta = \dfrac{OB^2 - OA^2 - AB^2}{2 \times OA \times OB} = \dfrac{240^2 - 130^2 - 180^2}{2 \times 130 \times 180} = 0.17735.$

Resistance of coil, $R_1 = Z\cos\theta = 41.54 \times 0.17735 = 7.367\ \Omega$

Reactance of coil, $X_L = \sqrt{Z^2 - R_1^2} = \sqrt{(41.54)^2 - (7.367)^2} = 40.88\ \Omega.$

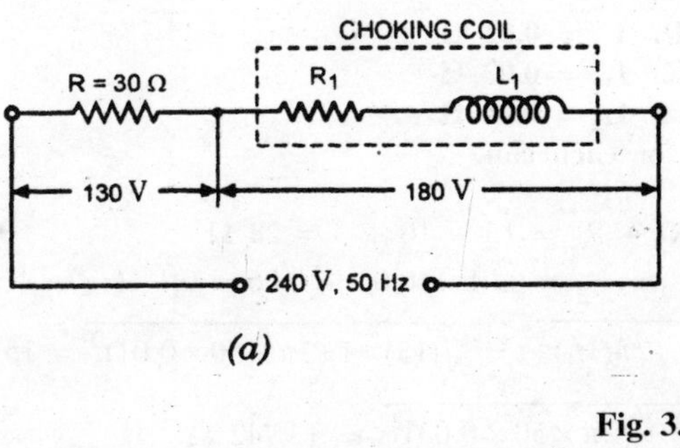

(a)

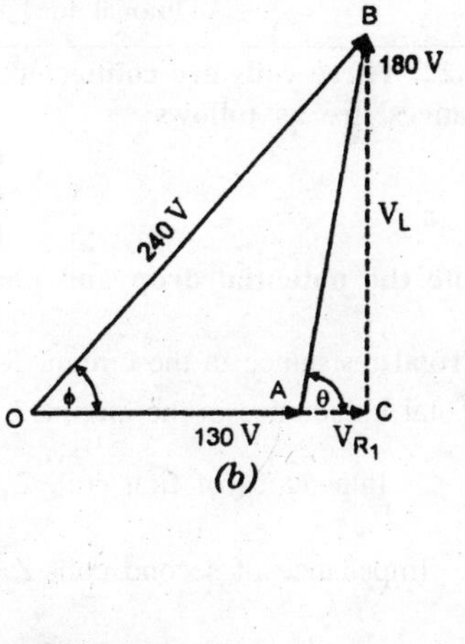

(b)

Fig. 3.8

(*a*) Power consumed by the coil $= I^2R_1 = (4.333)^2 \times 7.367 = 138.34$ W **Ans.**

(*b*) Resistance of whole circuit $= R + R_1 = 30 + 7.367 = 37.367\ \Omega$

Reactance of whole circuit = Reactance of coil = 40.88 Ω

Impedance of whole circuit $= \sqrt{(37.367)^2 + (40.88)^2} = 55.38\ \Omega$

Power factor of the whole circuit, $\cos\theta = \dfrac{37.367}{55.38} = 0.6747$ (lagging) **Ans.**

Example 3.20. A current, $i = 14.14 \sin\left(\omega t - \dfrac{\pi}{6}\right)$ flows in an electric circuit when a voltage of $v = 141.4 (\sin \omega t)$ is applied to it. Determine the power factor of the circuit. Determine also the values of true power and apparent power and of circuit components.

Solution : From the equations of applied voltage and circuit current, it is seen that current I lags behind the applied voltage by $\dfrac{\pi}{6}$ radians or 30°.

Thus phase angle, $\phi = 30°$ (lagging)

Power factor of the circuit $= \cos\phi = \cos 30° = 0.866$ (lagging) **Ans.**

Circuit voltage, $V = \dfrac{V_{max}}{\sqrt{2}} = \dfrac{141.4}{\sqrt{2}} = 100$ V

Circuit current, $I = \dfrac{I_{max}}{\sqrt{2}} = \dfrac{14.14}{\sqrt{2}} = 10$ A

True power, $P = VI\cos\phi = 100 \times 10 \times 0.866 = 866$ watts **Ans.**

Apparent power, $S = VI = 100 \times 10 = 1{,}000$ volt-amperes **Ans.**

Circuit impedance, $Z = \dfrac{V}{I} = \dfrac{100}{10} = 10\ \Omega$

Circuit resistance, $R = Z\cos\phi = 10 \times 0.866 = 8.66\ \Omega$ **Ans.**

Circuit reactance, $X_L = \sqrt{Z^2 - R^2} = \sqrt{10^2 - 8.66^2} = 5\ \Omega$ **Ans.**

Example 3.21. An alternator is supplying a load of 300 kW at a pf of 0.6 lagging. If the power factor is raised to unity, how many more kW can the alternator supply for the same kVA loading ?

Solution : Load supplied, P = 300 kW

Load power factor, $\cos\Phi = 0.6$ lagging

KVA loading $= \dfrac{P}{\cos\Phi} = \dfrac{300}{0.6} = 500$

For unity pf load that can be supplied, $P' =$ KVA loading $\times 1.0 = 500 \times 1.0 = 500$ kW

∴ Additional load supplied $= P' - P = 500 - 300 = 200$ kW **Ans.**

Example 3.22. Three coils are connected in series across a 100 V, 50 Hz supply. Values of their resistances and inductances are as follows:

$$R_1 = 15\ \Omega,\ L_1 = 0.01\text{ H}$$
$$R_2 = 10\ \Omega,\ L_2 = 0.03\text{ H}$$
$$R_3 = 3\ \Omega,\ L_3 = 0.06\text{ H}$$

Calculate the potential drop and phase angle for each coil.

Solution :

Total resistance in the circuit, $R = R_1 + R_2 + R_3 = 15 + 10 + 3 = 28\ \Omega$

Total inductance in the circuit, $L = L_1 + L_2 + L_3 = 0.01 + 0.03 + 0.06 = 0.1$ H

Impedance of first coil, $Z_1 = \sqrt{R_1^2 + (2\pi f L_1)^2} = \sqrt{(15)^2 + (2\pi \times 50 \times 0.01)^2} = 15.325\ \Omega$

Impedance of second coil, $Z_2 = \sqrt{(10)^2 + (2\pi \times 50 \times 0.03)^2} = 13.742\ \Omega$

$$\text{Impedance of third coil, } Z_3 = \sqrt{(3)^2 + (2\pi \times 50 \times 0.06)^2} = 19.087\ \Omega$$

$$\text{Impedance of whole circuit, } Z = \sqrt{R^2 + (2\pi f L)^2} = \sqrt{(28)^2 + (2\pi \times 50 \times 0.1)^2} = 42.083\ \Omega$$

$$\text{Current through the circuit, } I = \frac{V}{Z} = \frac{100}{42.083} = 2.376\ \text{A}$$

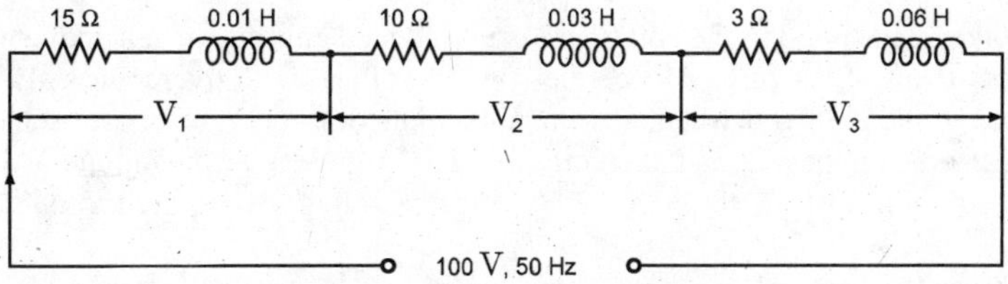

Fig. 3.9. ***Circuit Diagram***

Potential drop across first coil, $V_1 = I Z_1 = 2.376 \times 15.325 = 36.417$ V **Ans.**

Potential drop across second coil, $V_2 = I Z_2 = 2.376 \times 13.742 = 32.65$ V **Ans.**

Potential drop across third coil, $V_3 = I Z_3 = 2.376 \times 19.087 = 45.35$ V **Ans.**

Phase angle of first coil, $\Phi_1 = \text{Cos}^{-1} \dfrac{R_1}{Z_1} = \text{Cos}^{-1} \dfrac{15}{15.325} = 11°51'$ **Ans.**

Phase angle of second coil, $\Phi_2 = \text{Cos}^{-1} \dfrac{R_2}{Z_2} = \text{Cos}^{-1} \dfrac{10}{13.742} = 43°18'$ **Ans.**

Phase angle of third coil, $\Phi_3 = \text{Cos}^{-1} \dfrac{R_3}{Z_3} = \text{Cos}^{-1} \dfrac{3}{19.087} = 80°57'$ **Ans.**

3.8. POWER IN IRON-CORED CHOKING COIL

The power drawn by a choking coil is used up to supply the iron losses in the core in addition to that used up for heating the coil.

Power consumed by a choking coil,

P = Iron losses (hysteresis and eddy current loss, P_i + power loss in ohmic resistance *i.e.* $I^2 R$) R being the dc resistance or true resistance of the coil

$$= P_i + I^2 R$$

$$\text{or} \quad \frac{P}{I^2} = \frac{P_i}{I^2} + R$$

$\dfrac{P}{I^2}$ is called the *effective* resistance of the choking and it is equal to the true resistance plus resistance representing iron loss $\dfrac{P_i}{I^2}$

$$\therefore \quad R_{eff} = \frac{P}{I^2} = R + \frac{P_i}{I^2}$$

Example 3.23. A direct current of 1 A passes through a series circuit of 1 Ω resistor and 1 mH inductor. What is the voltage across the circuit ? If the current was a 50 Hz current, then what would be the voltage ?

[Pb. Technical Univ. Electrical Engineering May 2002]

Solution : For direct current

Voltage across the circuit, $V = IR = 1 \times 1 = 1$ V **Ans.**

$\because$ in a dc circuit the flow of current is opposed by only resistance.

For AC With 50 Hz Frequency

Voltage across the circuit, $V = IZ = I\sqrt{R^2 + (2\pi + fL)^2} = 1 \times \sqrt{1^2 + (2\pi \times 50 \times 1 \times 10^{-3})^2}$

$= 1.048$ V **Ans.**

Example 3.24. A coil takes 4 A when connected to 24 V, dc supply. If this coil is connected to 40 V, 50 Hz ac supply then same amount of power is consumed. Explain why are different voltages required for same power consumption. Calculate inductance of the coil and the phase angle between voltage and current ?

Solution : When dc voltage is applied to the coil, there is no induced emf due to self-inductance of the coil because the current is steady and frequency is zero. Hence the applied voltage is to overcome only the resistive drop $I \cdot R$, where R is the resistance of the coil. But when ac voltage is applied, the coil offers resistance as well as reactance, the combination of the two *i.e.* impedance of the coil is certainly greater in magnitude as compared to resistance. Hence to cause the same current to flow through the coil greater voltage is required to overcome the impedance of the coil.

When dc voltage is applied

Applied voltage, $V_{dc} = 24$ V

Current through coil, $I_{dc} = 4$ A

Resistance of coil, $R = \dfrac{V_{dc}}{I_{dc}} = \dfrac{24}{4} = 6\ \Omega$

When ac voltage is applied

Applied voltage, $V_{ac} = 40$ V

Resistance of coil, R = The same as when dc is applied *i.e.* 6 Ω

Current, $I_{ac} = 4$ A

Coil impedance, $Z = \dfrac{V_{ac}}{I_{ac}} = \dfrac{40}{4} = 10\ \Omega$

Inductive reactance of coil, $X_L = \sqrt{Z^2 - R^2} = \sqrt{(10)^2 - (6)^2} = 8\ \Omega$

Inductance of the coil, $L = \dfrac{X_L}{2\pi f} = \dfrac{8}{2\pi \times 50} = 0.0255$ H **Ans.**

The angle between the applied pd and current,

$$\Phi = \text{Tan}^{-1}\frac{X_L}{R} = \text{Tan}^{-1}\frac{8}{6} = 53.13^\circ \text{ (lag)} \quad \textbf{Ans.}$$

3.9. RESISTANCE AND CAPACITANCE IN SERIES

Consider an ac circuit consisting of resistance of R ohms and capacitance of C farads connected in series, as shown in Fig. 3.10 (*a*).

Let the supply frequency be of *f* Hz and current flowing through the circuit be of I amperes (rms value).

Voltage drop across resistance,

$V_R = IR$ in phase with the current.

Voltage drop across capacitance,

$V_C = IX_C$ lagging behind I by $\frac{\pi}{2}$ radians or 90°, as shown in Fig. 3.10 (*b*).

The applied voltage, being equal to phasor sum of V_R and V_C, is given in magnitude by

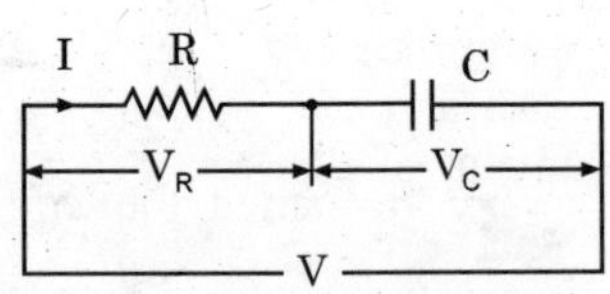

(a) Circuit Diagram

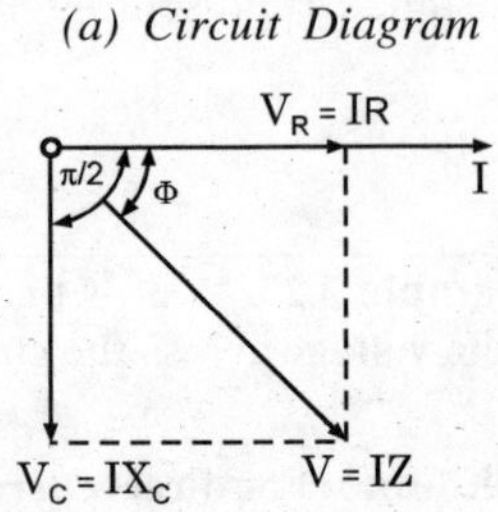

(b) Phasor Diagram

Fig. 3.10

$$V = \sqrt{(V_R)^2 + (V_C)^2} = \sqrt{(IR)^2 + (IX_C)^2}$$

$$= I\sqrt{R^2 + X_C^2} = IZ$$

where $Z^2 = R^2 + X_C^2$

The applied voltage lags behind the current by an angle Φ

$$\text{where } \tan\Phi = \frac{V_C}{V_R} = \frac{IX_C}{IR} = \frac{X_C}{R} = \frac{1}{\omega RC}$$

$$\text{or } \Phi = \tan^{-1}\frac{1}{\omega RC}$$

$$\text{Power factor, } \cos\Phi = \frac{R}{Z}$$

If instantaneous voltage is represented by

$$v = V_{max} \sin \omega t$$

then instantaneous current will be expressed as

$$i = I_{max} \sin(\omega t + \Phi)$$

and power consumed by the circuit is given by

$$P = VI\cos\Phi$$

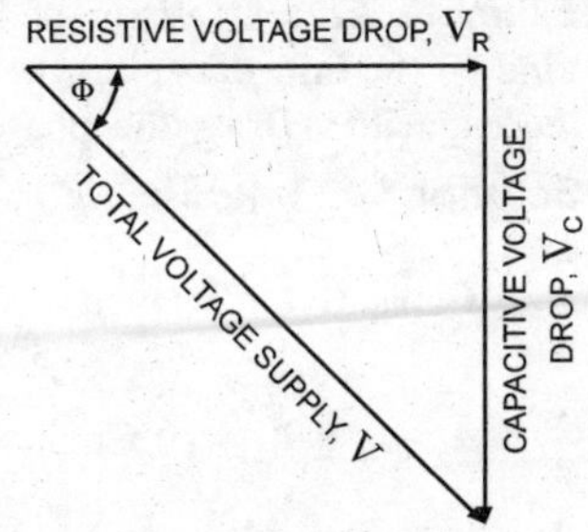

(a) Voltage Triangle

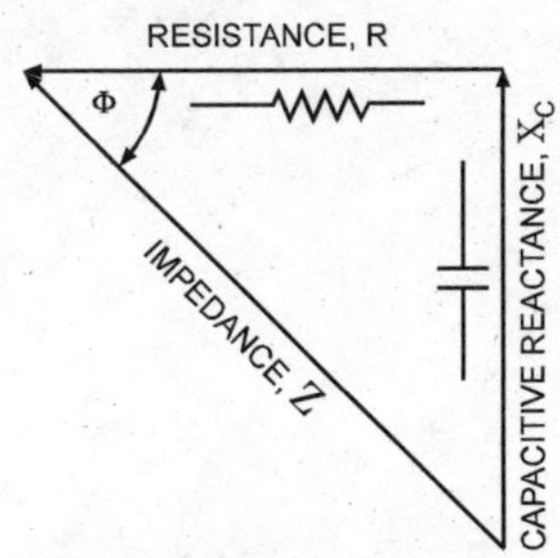

(b) Impedance Triangle

Fig. 3.11

Voltage triangle and impedance triangle are shown in Figs. 3.11 (*a*) and 3.11 (*b*) respectively.

Example 3.25. A circuit consists of 20 Ω resistance and in series with capacitance of 200 μF, connected across 50 Hz supply. The current through the circuit is 10.8 sin 314*t* amperes. Determine the voltage across each component and across the circuit. [*Pb. Technical Univ. Basic Electrical Engineering December 2002*]

Solution : Maximum value of circuit current, I_{max} = Coefficient of the sine of the time angle = 10.8 A

$$\text{RMS value of circuit current, } I_{rms} = \frac{I_{max}}{\sqrt{2}} = \frac{10.8}{\sqrt{2}} = 7.637 \text{ A}$$

Voltage across 20 Ω resistance $= I_{rms} \times R = 7.637 \times 20 = 152.74$ V **Ans.**

$$\text{Voltage across 200 μF capacitor} = I_{rms} X_C = \frac{7.637}{2\pi \times 50 \times 200 \times 10^{-6}} = 121.55 \text{ V } \textbf{Ans.}$$

$$\text{Voltage across the circuit, } V = \sqrt{(152.74)^2 + (121.55)^2} = 195.2 \text{ V } \textbf{Ans.}$$

Example 3.26. A voltage of 125 V at 60 Hz is applied across a non-inductive resistor connected in series with a condenser. The current in the circuit is 2.2 A. The power loss in the resistor is 96.8 watts and that in the condenser is negligible. Calculate the resistance and capacitance.

Solution : Applied voltage, V = 125 volts

Circuit current, I = 2.2 amperes

$$\text{Circuit impedance, } Z = \frac{V}{I} = \frac{125}{2.2} = 56.82\ \Omega$$

Power loss in the resistor $= I^2R = 96.8$ watts

$$\text{Resistance of the circuit, } R = \frac{I^2R}{I^2} = \frac{96.8}{(2.2)^2} = 20\ \Omega \textbf{ Ans.}$$

Since power loss in the condenser is negligible, resistance of the capacitor is zero.

$$\text{Capacitive reactance of the circuit, } X_C = \sqrt{Z^2 - R^2} = \sqrt{(56.82)^2 - (20)^2} = 53.2\ \Omega$$

$$\text{Capacitance of the circuit, } C = \frac{1}{\omega X_C} = \frac{1}{2\pi f X_C} = \frac{1}{2\pi \times 60 \times 53.2} = 50 \text{ μF } \textbf{Ans.}$$

Example 3.27. A resistor of 100 Ω is connected in series with a 56 μ F capacitor to a supply at 230 V, 50 Hz. Find (*i*) the impedance (*ii*) the current (*iii*) the phase angle (*iv*) the pf (*v*) the voltage across the resistor and across the capacitor. Draw the phasor diagram.

Solution : Resistance of the circuit, R = 100 Ω

$$\text{Capacitive reactance of the circuit, } X_C = \frac{1}{2\pi f C} = \frac{1}{2\pi \times 50 \times 56 \times 10^{-6}} = 56.84\ \Omega$$

(*i*) Circuit impedance, $Z = \sqrt{R^2 + X_C^2}$

$$= \sqrt{(100)^2 + (56.84)^2}$$

$= 115\ \Omega$ **Ans.**

(*ii*) Circuit current, $I = \frac{V}{Z} = \frac{230}{115} = 2$ A **Ans.**

(*iii*) Phase angle, $\Phi = \text{Tan}^{-1} \frac{X_C}{R}$

$= \text{Tan}^{-1} \frac{56.84}{100} = 29.6°$ **Ans.**

(*iv*) Power factor, cos Φ = Cos 29.6° = 0.87 (lead) **Ans.**

(*v*) Voltage across resistor, $V_R = IR = 2 \times 100 = 200$ V **Ans.**

(*vi*) Voltage across capacitor, $V_C = I X_C = 2 \times 56.84 = 113.7$ V **Ans.**

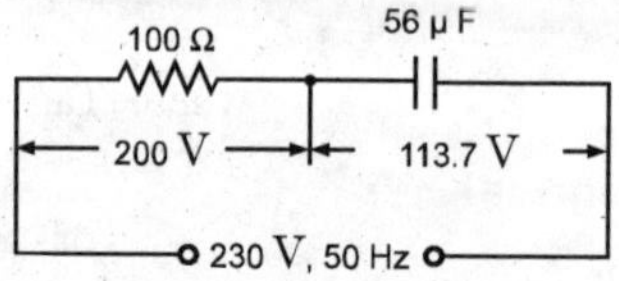

(*a*) *Circuit Diagram*

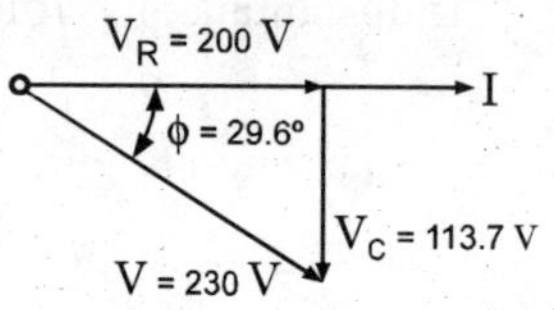

(*b*) *Phasor Diagram*

Fig. 3.12

Example 3.28. A series R-C circuit takes a power of 7,000 W when connected to 200 V, 50 Hz supply. The voltage across the resistor is 130 V. Calculate (*i*) the resistance (R), current (I), pf (cos ϕ), capacitance (C) and impedance (Z) and (*ii*) Also write equations for *v*(*t*) and *i*(*t*).

[*Pb. Technical Univ. Basic Electrical and Electronics Engineering Dec. 2005*]

Solution : Supply voltage, V = 200 V

Voltage across resistor, $V_R = 130$ V

$$\text{Power factor, } \cos\phi = \frac{R}{Z} = \frac{I \times R}{I \times Z} = \frac{V_R}{V_Z} = \frac{130}{200} = 0.65 \text{ (leading) } \textbf{Ans.}$$

∵ Circuit is capacitive.

$$\text{Current, } I = \frac{\text{Power drawn, P}}{V \times \cos\phi} = \frac{7{,}000}{200 \times 0.65} = 53.846 \text{ A } \textbf{Ans.}$$

$$\text{Resistance, } R = \frac{V_R}{I} = \frac{130}{53.846} = 2.41\ \Omega \textbf{ Ans.}$$

$$\text{Impedance, } Z = \frac{V}{I} = \frac{200}{53.846} = 3.714\ \Omega \textbf{ Ans.}$$

$$\text{Capacitive reactance, } X_C = \sqrt{Z^2 - R^2} = \sqrt{(3.714)^2 - (2.41)^2} = 2.826\ \Omega$$

$$\text{Capacitance, } C = \frac{1}{2\pi f X_C} = \frac{1}{2\pi \times 50 \times 2.826} = 1126\ \mu\text{F } \textbf{Ans.}$$

$$\text{Phase angle, } \phi = \text{Cos}^{-1} \text{ pf} = \text{Cos}^{-1}\ 0.65 = 49.46° \text{ or } 0.275\pi \text{ radian}$$

Equation for *v* (*t*) may be given as

$$v(t) = V_{max} \sin 2\pi f t \quad \text{assuming } \theta \text{ as zero when } t = 0$$

$$= 200\sqrt{2} \sin (2\pi \times 50)t = 282.84 \sin 314t \textbf{ Ans.}$$

Equation for *i*(*t*) may be given as

$$i(t) = I_{max} \sin (2\pi f t + \phi)$$

$$= 53.846 \times \sqrt{2} \sin (100\pi t + 0.275\pi) = 76.15 \sin (100\pi t + 0.275\pi) \textbf{ Ans.}$$

Example 3.29. A two-element series circuit is connected across an ac source $e = 200\sqrt{2}\sin(314t + 20°)$ V. The current in the circuit is then found to be $i = 10\sqrt{2}\cos(314t - 25°)$ A. Determine the parameters of the circuit.

Solution : $e = 200\sqrt{2}\sin(314t + 20°)$ volts

$$i = 10\sqrt{2}\cos(314t - 25°) = 10\sqrt{2}\sin(314t - 25° + 90°) = 10\sqrt{2}\sin(314t + 65°)$$

From the equations for emf and current it is seen that the applied emf e leads by 20° whereas current i leads by 65° with regard to the reference quantity, the phase angle being 65° – 20° = 45° (leading). It means the given circuit is capacitive or R-C circuit.

$$\text{Circuit impedance, } Z = \frac{E_{max}}{I_{max}} = \frac{200\sqrt{2}}{10\sqrt{2}} = 20\ \Omega$$

$$\text{Circuit resistance, } R = Z\cos\phi = 20\cos 45° = 20 \times \frac{1}{\sqrt{2}} = 14.14\ \Omega \quad \textbf{Ans.}$$

$$\text{Circuit reactance, } X = Z\sin\phi = 20\sin 45° = 20 \times \frac{1}{\sqrt{2}} = 14.14\ \Omega \text{ (capacitive)}$$

$$\text{Supply frequency, } f = \frac{\text{Coefficient of time, } t}{2\pi} = \frac{314}{2\pi} = 50 \text{ Hz}$$

$$\text{Circuit capacitance, } C = \frac{1}{2\pi f X_C} = \frac{1}{2\pi \times 50 \times 14.14} = 225\ \mu\text{F} \quad \textbf{Ans.}$$

Example 3.30. A voltage $v = 100\sin 314t - 50\cos 314t$, is applied to a circuit having R = 20 Ω in series with C = 100 μF. Obtain expression for instantaneous current, rms value of current and the power in the circuit.

[Pb. Technical Univ. Electrical Engineering 1999]

Solution : Applied voltage $v = 100\sin 314t - 50\cos 314t$

Multiplying and dividing RHS of the above expression by $\sqrt{100^2 + (-50)^2}$ *i.e.* 111.8, we have

$$v = 111.8\left(\sin 314t \times \frac{100}{111.8} - \cos 314t \times \frac{50}{111.8}\right)$$

$$= 111.8\,(\sin 314t\cos\theta - \cos 314t\sin\theta)$$

$$\text{where} \quad \cos\theta = \frac{100}{111.8} \quad \text{and} \quad \sin\theta = \frac{50}{111.8}$$

$$= 111.8\sin(314t - \theta) \text{ where } \theta = \text{Tan}^{-1}\frac{50}{100} = 26.56°$$

$$= 111.8\sin(314t - 26.56°)$$

$$\text{Circuit resistance, } R = 20\ \Omega$$

$$\text{Capacitive reactance of circuit, } X_C = \frac{1}{\omega C} = \frac{1}{314 \times 100 \times 10^{-6}} = 31.83\ \Omega$$

$$\text{Circuit impedance, } Z = \sqrt{R^2 + X_C^2} = \sqrt{20^2 + (31.83)^2} = 37.59\ \Omega$$

$$\text{Phase angle, } \phi = \text{Tan}^{-1}\frac{X_C}{R} = \text{Tan}^{-1}\frac{31.83}{20} = 57.86°\text{(leading)}$$

Hence expression for instantaneous current is given as

$$i = I_{max}\sin(314t - 26.56° + 57.86°)$$

$$= \frac{V_{max}}{Z}\sin(314t + 31.3°) = \frac{111.8}{37.59}\sin(314t + 31.3°)$$

$$= 2.974\sin(314t + 31.3°) \quad \textbf{Ans.}$$

$$\text{RMS value of current, } I = \frac{I_{max}}{\sqrt{2}} = \frac{2.974}{\sqrt{2}} = 2.1 \text{ A} \quad \textbf{Ans.}$$

$$\text{Power in the circuit, } P = VI\cos\phi = \frac{111.8}{\sqrt{2}} \times 2.1\cos 57.86° = 88.32 \text{ W} \quad \textbf{Ans.}$$

Example 3.31. A 120 V, 60 W lamp is to be operated on 220 V, 50 Hz supply mains. Calculate what value of (*i*) non-inductive resistance (*ii*) pure inductance and (*iii*) pure capacitance would be required in order that lamp is run on correct voltage. Which method is preferable and why ? Draw relevant phasor diagram.

Solution : Compared with the resistance of the filament of a lamp its inductance is negligible. As such a lamp is considered to be a non-inductive resistance.

$$\text{Rated power of lamp, P} = 60 \text{ W}$$

$$\text{Rated voltage of lamp, V} = 120 \text{ V}$$

Current drawn by the lamp when operated on rated voltage

$$I = \frac{P}{V} = \frac{60}{120} = 0.5 \text{ A}$$

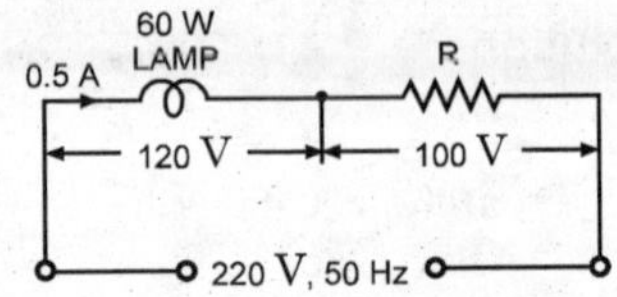

(*a*)

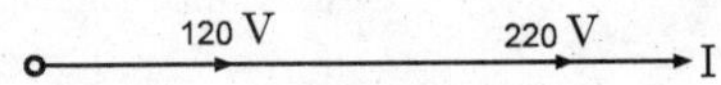

(*b*)

Fig. 3.13

(*i*) Let a non-inductive resistance of R ohms be connected in series with the lamp, as shown in Fig. 3.13 (*a*)

Since in this case voltage across the lamp and that across the resistance V_R, both are in phase, their arithmetic sum is equal to applied voltage *i.e.*

$$V_R + 120 = 220$$

$$\text{or } V_R = 220 - 120 = 100 \text{ V}$$

$$\text{Resistance, R} = \frac{V_R}{I} = \frac{100}{0.5} = 200\ \Omega \ \textbf{Ans.}$$

(*ii*) Let pure inductance of L henry be connected in series with the lamp, as shown in Fig. 3.14 (*a*)

Since in this case voltage drop across the inductance leads the current by 90° or π /2 radians while that across the lamp will be in phase with it, from phasor diagram shown in Fig. 3.14 (*b*) we have

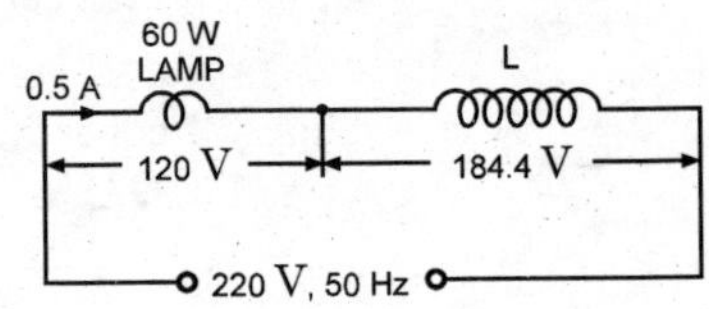

(*a*)

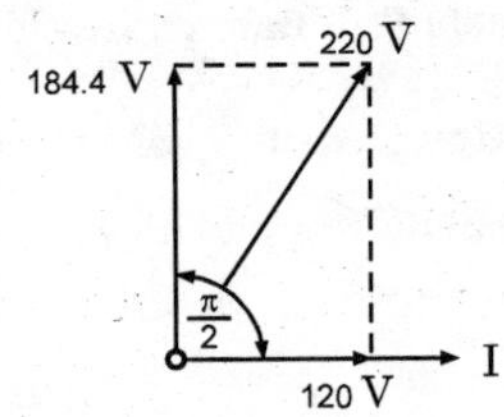

(*b*)

Fig. 3.14

$$120^2 + V_L^2 = 220^2$$

$$\text{or } V_L = \sqrt{220^2 - 120^2} = 184.4 \text{ volts}$$

$$\text{or } 2\pi f L I = 184.4 \text{ V}$$

$$\text{or } L = \frac{184.4}{2\pi \times 50 \times 0.5} = 1.174 \text{ H} \ \textbf{Ans.}$$

(*iii*) Let the pure capacitance of C farads be connected in series with the lamp, as shown in Fig. 3.15 (*a*).

Since in this case voltage drop across the capacitor lags behind the current by 90° or π/2 radians while that across the lamp will be in phase with it, from phasor diagram shown in Fig. 3.15 (*b*), we have

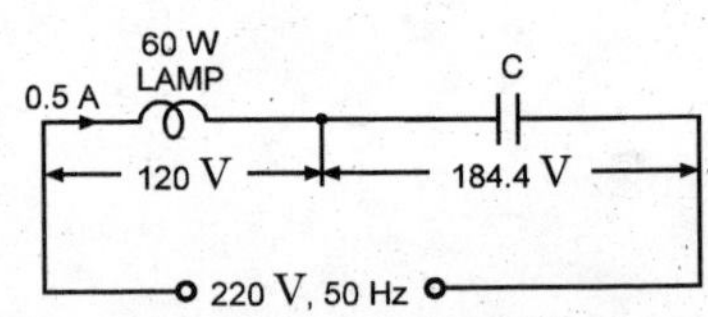

(*a*) *Circuit Diagram*

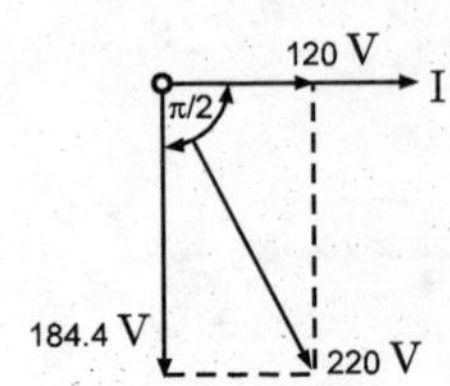

(*b*) *Phasor Diagram*

Fig. 3.15

$$(120)^2 + (V_C)^2 = 220^2$$

$$\text{or } V_C = \sqrt{(220)^2 - (120)^2} = 184.4 \text{ V}$$

$$\text{or } \frac{I}{2\pi f C} = 184.4$$

$$\text{or } C = \frac{I}{2\pi f \times 184.4} = \frac{0.5}{2\pi \times 50 \times 184.4} = 8.63\ \mu\text{F} \ \textbf{Ans.}$$

In first case there will be a power loss of $0.5^2 \times 200$ *i.e.* 50 watts, in second case though there will be no power loss but power factor will be lagging, but in third case there will be no power loss and the power factor will be leading, which is desirable. Hence third alternative will be preferred.

3.10. RESISTANCE, INDUCTANCE AND CAPACITANCE IN SERIES

Consider an ac circuit containing resistance of R ohms, inductance of L henries and capacitance of C farads connected in series, as shown in Fig. 3.16 (*a*).

Let the current flowing through the circuit be of I amperes and supply frequency be *f* Hz.

Voltage drop across resistance, V_R = IR in phase with I

Voltage drop across inductance, $V_L = I\omega L$ leading I by $\pi/2$ radians or 90°

Voltage drop across capacitance, $V_C = \frac{I}{\omega C}$ or $I X_C$ lagging behind I by $\pi/2$ radians or 90°

V_L and V_C are 180° out of phase with each other (or reverse in phase), therefore, when combined by parallelogram they cancel each other. The circuit can either be effectively inductive or capacitive depending upon which voltage drop (V_L or V_C) is predominant. Let us consider the case when V_L is greater than V_C.

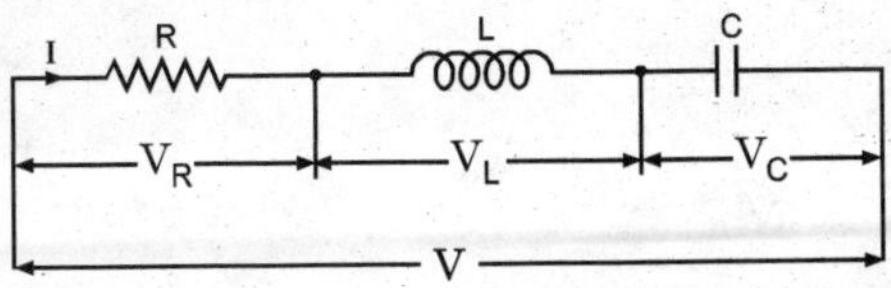

(a) Circuit Diagram

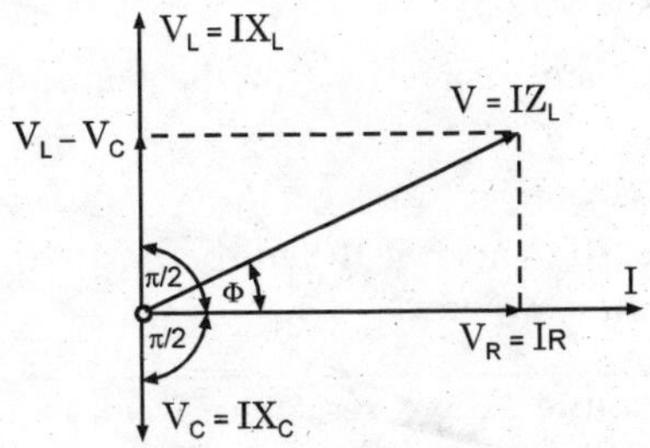

(b) Phasor Diagram

Fig. 3.16

The applied voltage V, being equal to the phasor sum of V_R, V_L and V_C is given in magnitude by

$$V = \sqrt{(V_R)^2 + (V_L - V_C)^2} = \sqrt{(IR)^2 + (IX_L - IX_C)^2} = I\sqrt{(R)^2 + (X_L - X_C)^2}$$

The term $\sqrt{R^2 + (X_L - X_C)^2}$ is known as impedance of the circuit and is represented by Z. Its unit is ohm.

Phase angle Φ between voltage and current is given by

$$\Phi = \text{Tan}^{-1}\frac{V_L - V_C}{V_R} = \text{Tan}^{-1}\frac{IX_L - IX_C}{IR} = \text{Tan}^{-1}\frac{X_L - X_C}{R} = \text{Tan}^{-1}\frac{X}{R}$$

Φ will be + ve *i.e.* applied voltage will lead the current if $X_L > X_C$ and Φ will be –ve *i.e.* applied voltage will be behind the current if $X_L < X_C$.

Power factor of the circuit is given by

$$\text{Cos } \Phi = \frac{R}{Z} = \frac{R}{\sqrt{R^2 + (X_L - X_C)^2}}$$

Power consumed in the circuit, $P = I^2 R$ or VI cos Φ

3.10.1. Reactance

Inductive reactance, X_L is directly proportional to frequency being equal to ωL or $2\pi f L$ and capacitive reactance, X_C is inversely proportional to frequency being equal to $\frac{1}{\omega C}$ or $\frac{1}{2\pi f C}$.

Inductive reactance causes the current to lag behind the applied voltage, while the capacitive reactance causes the current to lead the voltage. So when inductance and capacitance are connected in series, their effects neutralize each other and their combined effect is then their difference. The combined effect of inductive reactance and capacitive reactance is called the reactance and is found by subtracting the capacitive reactance from the inductive reactance or according to equation

$$X = X_L - X_C$$

When $X_L > X_C$ *i.e.* $X_L - X_C$ is positive, the circuit is inductive and phase angle Φ is positive.
When $X_L < X_C$ *i.e.* $X_L - X_C$ is negative, the circuit is capacitive and phase angle Φ is negative.
When $X_L = X_C$ *i.e.* $X_L - X_C = 0$, the circuit is purely resistive and phase angle Φ is zero.
If the expression for applied voltage is taken as

$$v = V_{max} \sin \omega t$$

then expression for the current will be

$$i = I_{max} \sin (\omega t \pm \Phi) \text{ where } I_{max} = \frac{V_{max}}{\text{Circuit impedance, Z}}$$

The value of Φ will be positive when current leads *i.e.* when $X_C > X_L$ and negative when current lags *i.e.* when $X_L > X_C$.

Example 3.32. A resistance R, an inductance L = 0.01 H and a capacitance C are connected in series. When a voltage v = 400 cos (3,000 t – 10°) V is applied to the series combination, the current flowing is $10\sqrt{2}$ cos (3,000 t – 55°) A. Find R and C.

Solution :

$$v = 400 \cos (3{,}000t - 10°)$$
$$= 400 \sin (3{,}000t - 10° + 90°) = 400 \sin (3{,}000\, t + 80°)$$
$$i = 10\sqrt{2} \cos(3{,}000t - 55°)$$
$$= 10\sqrt{2} \sin (3{,}000t - 55° + 90°) = 10\sqrt{2} \sin (3{,}000\, t + 35°)$$

Phase angle, $\Phi = 80° - 35° = 45°$ *i.e.* circuit is inductive and current lags behind voltage

$$\omega = 3{,}000$$

Inductive reactance, $X_L = \omega L = 3{,}000 \times 0.01 = 30\ \Omega$

Circuit impedance, $Z = \dfrac{V_{max}}{I_{max}} = \dfrac{400}{10\sqrt{2}} = 20\sqrt{2} = 28.28\ \Omega$

Net reactance, $X = X_L - X_C = Z \sin \Phi = 28.28 \sin 45° = 28.28 \times \dfrac{1}{\sqrt{2}} = 20\ \Omega$

or Capacitive reactance $X_C = X_L - X = 30 - 20 = 10\ \Omega$

Circuit Capacitance, $C = \dfrac{1}{\omega X_C} = \dfrac{1}{3{,}000 \times 10} = 33.33\ \mu F$ **Ans.**

Circuit resistance, $R = Z \cos \Phi = 28.28 \cos 45° = 20\ \Omega$ **Ans.**

Example 3.33. Find an expression for the instantaneous value of the current through an R-L-C series circuit in which R = 80 Ω, L = 41.3 mH and C = 0.797 μF, when applied voltage is 1 sin 2000 πt volt.

Solution : Supply frequency, $f = \dfrac{2{,}000\,\pi}{2\pi} = 1{,}000$ Hz

Circuit resistance, $R = 80\ \Omega$

Inductive reactance of the circuit, $X_L = 2\pi f L = 2\pi \times 1{,}000 \times 41.3 \times 10^{-3} = 259.5\ \Omega$

Capacitive reactance of the circuit, $X_C = \dfrac{1}{2\pi f C} = \dfrac{1}{2\pi \times 1{,}000 \times 0.797 \times 10^{-6}} = 199.7\ \Omega$

Circuit impedance, $Z = \sqrt{R^2 + (X_L - X_C)^2} = \sqrt{(80)^2 + (259.5 - 199.7)^2} = 99.88\ \Omega$

Maximum value of current, $I_{max} = \frac{V_{max}}{Z} = \frac{1}{99.88} = 0.01$ A

Phase angle of the circuit, $\Phi = \text{Tan}^{-1} \frac{X_L - X_C}{R} = \text{Tan}^{-1} \frac{259.5 - 199.7}{80} = 36.8°$ or $0.204\ \pi$ rad

Hence instantaneous value of current will be given by the expression

$i = 0.01 \sin (2{,}000\ \pi t - 0.204\pi)$ since circuit is inductive. **Ans.**

Example 3.34. A resistance of 12 Ω, inductance of 0.1 H and a capacitance of 100 μF are connected in series across ac 220 V, 100 Hz supply. Calculate the current and its power factor, power consumed and phase angle mentioning whether it is leading or lagging.

[Pb. Technical Univ. Basic Elec. Engineering December 2003; May 2010]

Solution :

Circuit resistance, $R = 12\ \Omega$

Supply frequency, $f = 100$ Hz

Inductive reactance of the circuit, $X_L = 2\pi f L$

$= 2\pi \times 100 \times 0.1 = 62.83\ \Omega$

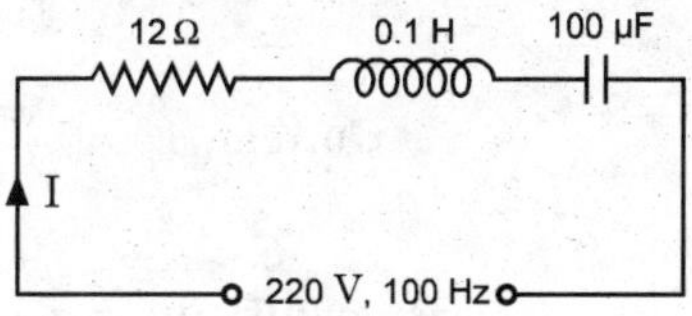

Fig. 3.17

Capacitive reactance of the circuit, $X_C = \frac{1}{2\pi f C}$

$= \frac{1}{2\pi \times 100 \times 100 \times 10^{-6}} = 15.92\ \Omega$

Circuit impedance, $Z = \sqrt{R^2 + (X_L - X_C)^2} = \sqrt{12^2 + (62.83 - 15.92)^2}$

$= 48.42\ \Omega$ (inductive)

∵ Inductive reactance exceeds capacitive reactance

Circuit current, $I = \frac{V}{Z} = \frac{220}{48.42} = 4.54$ A **Ans.**

Circuit power factor $= \text{Cos}\ \phi = \frac{R}{Z} = \frac{12}{48.42} = 0.2478$ (lagging) **Ans.**

Power consumed, $P = I^2R = (4.54)^2 \times 12 = 247.7$ watts **Ans.**

Phase angle, $\phi = \text{Cos}^{-1}\ 0.2478 = 75.65°C$ (lagging) **Ans.** ∵ circuit is inductive

3.11. INTRODUCTION TO PARALLEL CIRCUITS

In practice we come across the parallel circuits more often because multiple system of transmission and distribution is extensively used nowadays.

Practically all lighting and power circuits are constant voltage circuits with the loads connected in parallel. As in parallel dc circuits, the voltage is the same across each branch of a parallel ac circuit. The total line current supplied to the circuit is equal to the phasor sum of the branch currents. The total line current supplied to a parallel circuit may be determined by determining the currents in each branch and adding them, taking into account their phase relations, or by determining an equivalent impedance and dividing the applied voltage by the equivalent impedance.

Parallel ac circuits can be solved by the following methods:

1. Phasor method.
2. Admittance method.
3. Symbolic method or j-method.

The application of a particular method will depend upon the conditions of the problem. However, in general, that method should be used which gives quick results.

3.12. PHASOR METHOD

Consider a parallel circuit consisting of three branches of impedances Z_1 (R_1, L_1 C_1), Z_2 (R_2, L_2, C_2) and Z_3 (R_3, L_3, C_3) respectively and connected in parallel across an alternating voltage of *V* volts (rms), as shown in Fig. 3.18 (*a*).

Since all the three branches are connected in parallel, the voltage across each branch is the same and equal to supply voltage V, but currents through them will be different.

Branch 1

$$\text{Impedance, } Z_1 = \sqrt{(R_1)^2 + \left(X_{C_1} - X_{L_1}\right)^2}$$

$$\text{Current, } I_1 = \frac{V}{Z_1}$$

$$\text{Phase angle, } \Phi_1 = \text{Tan}^{-1} \frac{X_{C_1} - X_{L_1}}{R_1}$$

Current I_1 will lead or lag behind the applied voltage if Φ_1 is + ve or – ve respectively.

Similarly in Branch 2

$$\text{Impedance, } Z_2 = \sqrt{(R_2)^2 + \left(X_{C_2} - X_{L_2}\right)^2}$$

$$\text{Current, } I_2 = \frac{V}{Z_2}$$

$$\text{Phase angle, } \Phi_2 = \text{Tan}^{-1} \frac{X_{C_2} - X_{L_2}}{R_2}$$

and in Branch 3

$$\text{Impedance, } Z_3 = \sqrt{(R_3)^2 + \left(X_{C_3} - X_{L_3}\right)^2}$$

$$\text{Current, } I_3 = \frac{V}{Z_3}$$

$$\text{Phase angle, } \Phi_3 = \text{Tan}^{-1} \frac{X_{C_3} - X_{L_3}}{R_3}$$

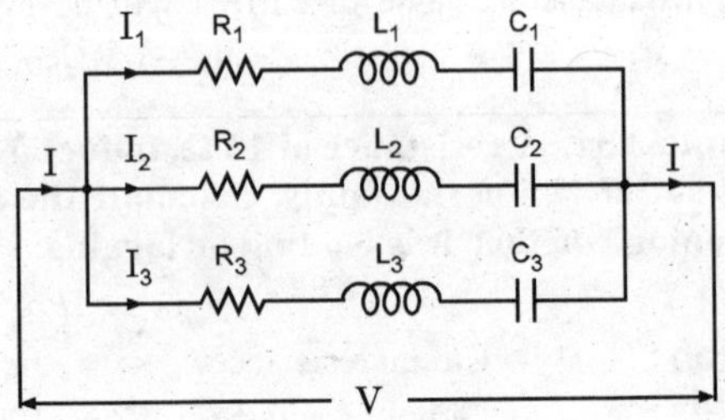

(a) Circuit Diagram

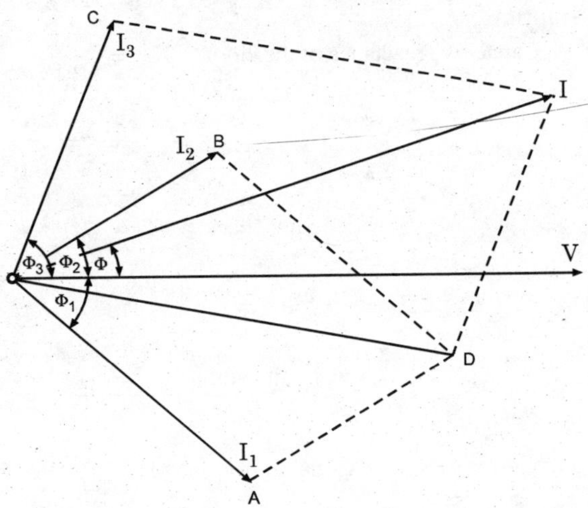

(b) Phasor Diagram

Fig. 3.18. ***Single Phase Parallel Circuit***

Resultant current, I, which is phasor sum of I_1, I_2 and I_3, can be determined either by using parallelogram law of phasors, as shown in Fig. 3.18 (*b*) or by resolving branch currents I_1, I_2, and I_3 along X-axis and Y-axis and then determining the resultant of these components analytically.

Component of resultant current I along X-axis

= Sum of components of branch currents I_1, I_2 and I_3 along X-axis

or $I \cos \Phi = I_1 \cos \Phi_1 + I_2 \cos \Phi_2 + I_3 \cos \Phi_3$

Similarly component of resultant current I along Y-axis

= Sum of components of branch currents I_1, I_2 and I_3 along Y-axis

or $I \sin \Phi = I_1 \sin \Phi_1 + I_2 \sin \Phi_2 + I_3 \sin \Phi_3$

Resultant current, $I = \sqrt{(I\cos\phi)^2 + (I\sin\phi)^2}$

$$= \sqrt{(I_1\cos\Phi_1 + I_2\cos\Phi_2 + I_3\cos\Phi_3)^2 + (I_1\sin\Phi_1 + I_2\sin\Phi_2 + I_3\sin\Phi_3)^2}$$

$$\text{and } \tan\Phi = \frac{I\sin\Phi}{I\cos\Phi}$$

$$\text{or } \Phi = \text{Tan}^{-1}\frac{I_1\sin\Phi_1 + I_2\sin\Phi_2 + I_3\sin\Phi_3}{I_1\cos\Phi_1 + I_2\cos\Phi_2 + I_3\cos\Phi_3}$$

If Φ is + ve current I will lead the applied voltage V, if Φ is – ve current I will lag behind the applied voltage.

Example 3.35. A resistance of 50 Ω, an inductance of 0.15 H and a capacitance of 100 μF are connected in parallel across a 100 V, 50 Hz, supply. Calculate *(a)* the current in each path *(b)* the resultant current *(c)* the phase angle between the resultant current and the supply voltage *(d)* power factor of the circuit.

Solution : Resistance, R = 50 Ω

Inductive reactance, $X_L = 2\pi fL = 2\pi \times 50 \times 0.15 = 47.1\ \Omega$

Capacitive reactance, $X_C = \dfrac{1}{2\pi fC} = \dfrac{1}{2\pi \times 50 \times 100} = 31.8\ \Omega$

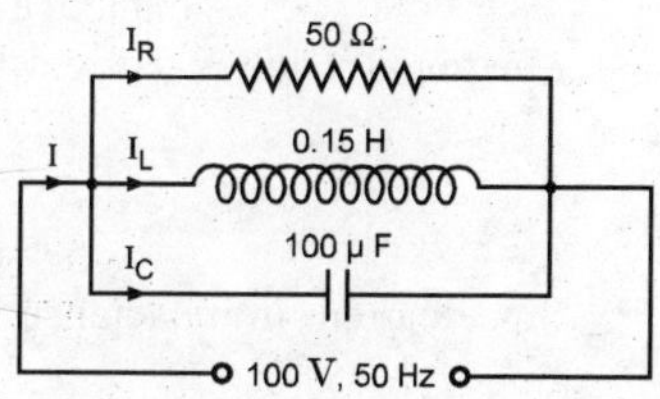

(a) Circuit Diagram

(a) Current through resistance, $I_R = \dfrac{V}{R} = \dfrac{100}{50}$

= 2 A in phase with voltage V **Ans.**

Current through inductance, $I_L = \dfrac{V}{X_L} = \dfrac{100}{47.1} = 2.12$ A

lagging behind voltage V by $\dfrac{\pi}{2}$ radians **Ans.**

Current through capacitance,

$I_C = \dfrac{V}{X_C} = \dfrac{100}{31.8}$ =3.14 A leading the voltage V by $\dfrac{\pi}{2}$ radians **Ans.**

The circuit and phasor diagrams are shown in Figs. 3.19 *(a)* and 3.19 *(b)* respectively.

Voltage phasor is taken as reference phasor and laid off along the horizontal line. The current taken by resistance is in phase with applied voltage. The current taken by inductance lags behind the voltage by $\dfrac{\pi}{2}$ radians and that taken by the capacitance leads the voltage by $\dfrac{\pi}{2}$ radians.

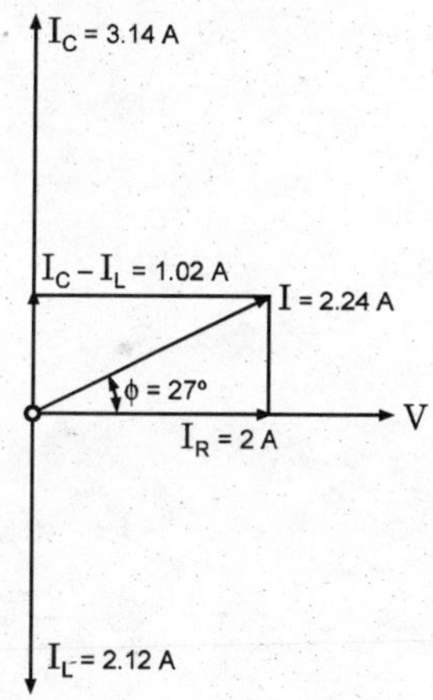

(b) Phasor Diagram

Fig. 3.19

Combining I_C and I_L which are in opposition, we get

$$I_C - I_L = 3.14 - 2.12 = 1.02 \text{ A}$$

(b) and resultant current, which is phasor sum of I_R, I_C and I_L,

$$I = \sqrt{(I_R)^2 + (I_C - I_L)^2} = \sqrt{(2)^2 + (1.02)^2} = 2.24 \text{ A} \quad \textbf{Ans.}$$

$$\textit{(c)} \text{ Phase angle, } \Phi = \text{Tan}^{-1}\frac{I_C - I_L}{I_R} = \text{Tan}^{-1}\frac{3.14 - 2.12}{2} = 27° \text{ (lead)} \quad \textbf{Ans.}$$

(d) Power factor of the circuit = Cos Φ = Cos 27° = 0.89 (lead) **Ans.**

Example 3.36. A circuit consists of resistance R = 100 ohms connected in parallel with an inductance of 638 mH. Find the expressions for instantaneous values of currents in the two branches and also for the total current if a voltage, $v = 200 \sin 100\,\pi t$ is applied across the circuit. Find also the power consumed. Sketch the voltage and current wave shapes over one cycle giving their maximum values and relative phase displacements.

Solution : Maximum value of voltage applied,

$$V_{max} = \text{Coefficient of the sine of the time angle} = 200 \text{ V}$$

$$\text{Supply frequency, } f = \frac{\text{Coefficient of time } t}{2\pi} = \frac{100\pi}{2\pi} = 50 \text{ Hz}$$

Maximum value of current through resistance,

$$I_{R\,max} = \frac{V_{max}}{R} = \frac{200}{100} = 2 \text{ A}$$

Since current in pure resistance is in phase with the applied voltage, the expression for instantaneous value is given by the expression

$$i_R = I_{R\,max} \sin \omega t = 2 \sin 100\,\pi t \quad \textbf{Ans.}$$

$$\text{Inductive reactance, } X_L = 2\pi f L = 2\pi \times 50 \times 0.638 = 200\,\Omega$$

Maximum value of current through inductance,

$$I_{L\,max} = \frac{V_{max}}{X_L} = \frac{200}{200} = 1 \text{ A}$$

Since current in inductance lags behind the voltage by $\frac{\pi}{2}$ radians, the expression for the instantaneous values of current through inductance is given by the expression

$$i_L = I_{L\,max} \sin(\omega t - \Phi) = 1 \sin\left(100\pi t - \frac{\pi}{2}\right) \quad \textbf{Ans.}$$

$$\text{Resultant current, } i_r = i_R + i_L = 2 \sin 100\pi t + 1 \sin\left(100\pi t - \frac{\pi}{2}\right)$$

$$\text{or} \quad i_r = 2 \sin 100\pi t - 1 \cos 100\pi t$$

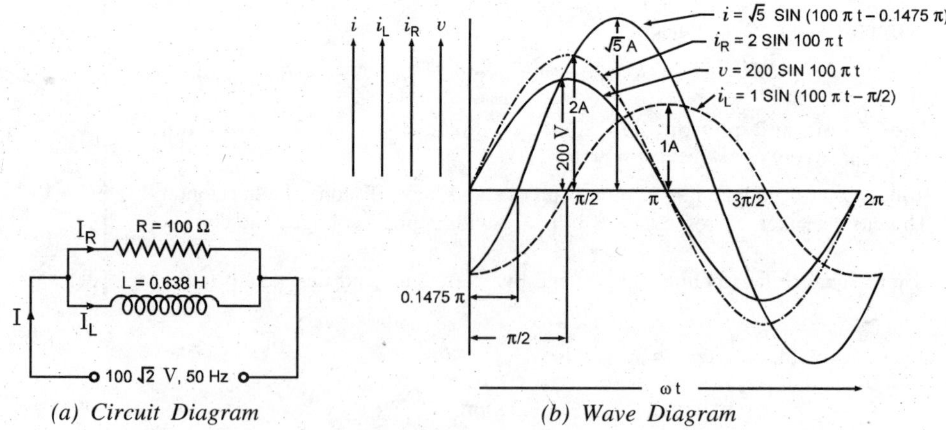

(a) Circuit Diagram (b) Wave Diagram

Fig. 3.20

Dividing both sides by $\sqrt{(2)^2 + (1)^2}$ we get

$$\frac{i_r}{\sqrt{5}} = \left[\sin 100\,\pi t \times \frac{2}{\sqrt{5}} - \cos 100\,\pi t \times \frac{1}{\sqrt{5}}\right]$$

$$\text{or} \quad i_r = \sqrt{5}\,[\sin 100\,\pi t \cos \Phi - \cos 100\,\pi t \sin \Phi]$$

$$\text{where } \cos \Phi = \frac{2}{\sqrt{5}} \text{ and } \sin \Phi = \frac{1}{\sqrt{5}}$$

or $i_r = \sqrt{5}\ \sin(100\pi t - \Phi)$ where $\Phi = \text{Tan}^{-1}\dfrac{1}{2} = 0.1475\pi$

or $i_r = \sqrt{5}\ \sin(100\pi t - 0.1475\pi)$ **Ans.**

Power consumed in resistance $= \left[\dfrac{I_{R\,max}}{\sqrt{2}}\right]^2 R = \left(\dfrac{2}{\sqrt{2}}\right)^2 \times 100 = 200$ watts

Power consumed in inductance = zero

Total power consumed, P = 200 + 0 = 200 watts **Ans.**

Example 3.37. A coil of resistance 15 Ω and inductance 0.05 H is connected in parallel with a non-inductive resistor of 20 Ω. Find (*i*) the current in each branch circuit (*ii*) the total current supplied and (*iii*) the phase angle of the combination, when a voltage of 230 V at 50 Hz is applied. Draw the relevant phasor diagram.

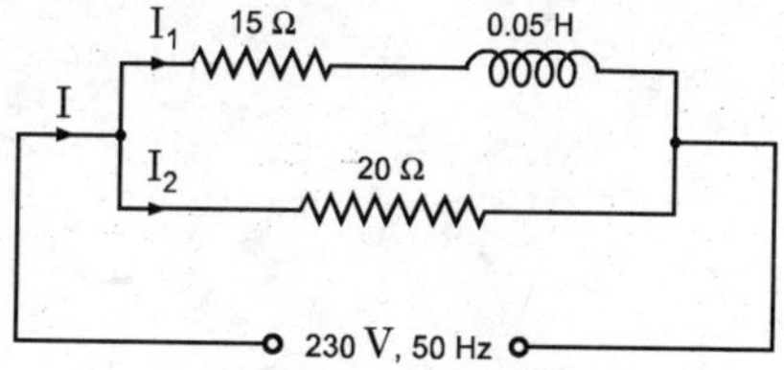

(a) Circuit Diagram

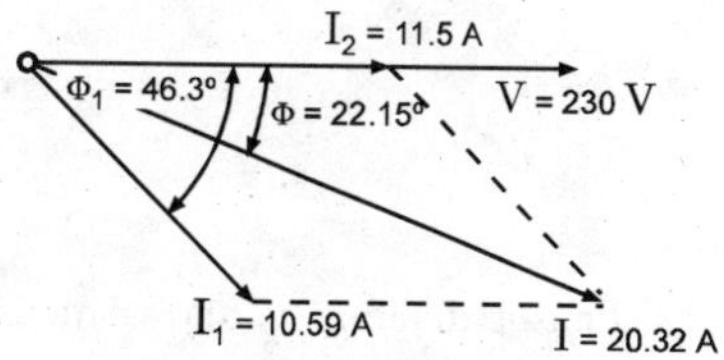

(b) Phasor Diagram

Fig. 3.21

Solution : Coil resistance, $R_1 = 15\ \Omega$

Coil inductance, $L_1 = 0.05$ H

Coil reactance (inductive), $X_{L_1} = 2\pi f L$

$= 2\pi \times 50 \times 0.05 = 15.71\ \Omega$

Coil impedance, $Z_1 = \sqrt{R_1^2 + X_{L_1}^2}$

$= \sqrt{(15)^2 + (15.71)^2} = 21.72\ \Omega$

(*i*) Current through the coil, $I_1 = \dfrac{V}{Z_1} = \dfrac{230}{21.72} = 10.59$ A **Ans.**

Phase angle of the coil, $\Phi_1 = \text{Tan}^{-1}\dfrac{X_L}{R} = \text{Tan}^{-1}\dfrac{15.71}{15}$

$= 46.3°$ (lag)

Current through non-inductive resistance of 20 Ω,

$I_2 = \dfrac{V}{R_2} = \dfrac{230}{20} = 11.5$ A in phase with voltage V **Ans.**

Circuit and phasor diagrams are shown in Figs. 3.21 (*a*) and 3.21 (*b*) respectively. Let the resultant current be of I amperes and phase angle Φ.

Resolving currents in in-phase and quadrature components we have

In-phase component, $I\cos\Phi = I_1\cos\Phi_1 + I_2 = 10.59\cos 46.3° + 11.5 = 7.32 + 11.5 = 18.82$ A ...(*i*)

Quadrature component, $I\sin\Phi = I_1\sin\Phi_1 = 10.59\sin 46.3° = 7.66$ A ...(*ii*)

Squaring both sides of Eqs. (*i*) and (*ii*) and adding them we get

$I^2(\cos^2\Phi + \sin^2\Phi) = (18.82)^2 + (7.66)^2 = 412.81$

(*ii*) or $I = \sqrt{412.81} = 20.32$ A **Ans.**

Dividing Eq. (*ii*) by Eq. (*i*) we get

$\text{Tan}\,\Phi = \dfrac{7.66}{18.82} = 0.407$

(*iii*) or Phase angle, $\Phi = \text{Tan}^{-1}\ 0.407 = 22.15°$ (lag) **Ans.**

Example 3.38. A coil in parallel with a 200 μ F capacitor is connected across a 200 V, 50 Hz, supply. The coil takes a current of 8 A and loss in the coil is 960 W. Calculate the following:

(*a*) The resistance of the coil (*b*) The inductance of the coil (*c*) The power factor of the entire circuit.

Solution : Power loss in the coil, P = 960 watts

Current through coil, $I_1 = 8$ A

(*a*) Coil resistance, $R = \frac{P}{I_1^2} = \frac{960}{(8)^2} = 15\ \Omega$ **Ans.**

Coil impedance, $Z = \frac{V}{I_1} = \frac{200}{8} = 25\ \Omega$

Coil reactance (inductive), $X_L = \sqrt{Z^2 - R^2} = \sqrt{25^2 - 15^2}$

$= 20\ \Omega$

(*b*) Coil inductance, $L = \frac{X_L}{2\pi f} = \frac{20}{2\pi \times 50} = 63.7$ mH **Ans.**

Phase angle between coil current and applied voltage V,

$$\Phi = \text{Tan}^{-1}\frac{X_L}{R} = \text{Tan}^{-1}\frac{20}{15} = 53.13°$$

Capacitive reactance, $X_C = \frac{1}{2\pi f C}$

$$= \frac{1}{2\pi \times 50 \times 200 \times 10^{-6}} = 15.915\ \Omega$$

Current through capacitor, $I_C = \frac{V}{X_C}$

$$= \frac{200}{15.915}$$

$= 12.566$ A leading applied voltage by $\frac{\pi}{2}$ radians

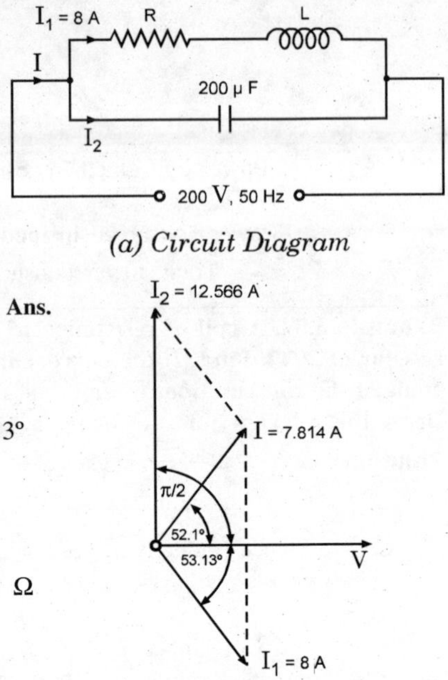

(a) Circuit Diagram

(b) Phasor Diagram

Fig. 3.22

Phasor diagram is shown in Fig. 3.22 (*b*)

Resolving currents in in-phase and quadrature components we have

$$I \cos \Phi = I_1 \cos 53.13° + I_2 \cos 90° = 8 \times 0.6 + 0 = 4.8 \text{ A} \quad ...(i)$$

$$I \sin \Phi = I_1 \sin(-53.13°) + I_2 \sin 90°$$

$$= 8 \times (-0.8) + 12.566 \times 1 = 6.166 \text{ A} \quad ...(ii)$$

Dividing Eq. (*ii*) be Eq. (*i*) we get

$$\text{Tan } \Phi = \frac{6.166}{4.8} = 1.2846$$

or $\Phi = \text{Tan}^{-1}\ 1.2846 = 52.1°$ leading

(*c*) ∴ Power factor of the entire circuit = Cos Φ = Cos 52.1° = 0.6143 (leading) **Ans.**

Example 3.39. A circuit of resistance 15 Ω and inductive reactance of 12 Ω is connected in parallel with another circuit consisting of a resistor of 25 Ω in series with a capacitor of capacitive reactance 17 Ω. The combination is energized from a 200 V, 50 Hz mains. Find the branch currents, total current and power factor of the combination.

Solution : Impedance of inductive circuit, $Z_1 = \sqrt{R_1^2 + X_L^2} = \sqrt{15^2 + 12^2} = 19.21\ \Omega$

Impedance of capacitive circuit, $Z_2 = \sqrt{R_2^2 + X_C^2} = \sqrt{25^2 + 17^2} = 30.23\ \Omega$

Current drawn by inductive circuit, $I_1 = \frac{V}{Z_1} = \frac{200}{19.21} = 10.41$ A **Ans.**

Phase angle of inductive circuit, $\Phi_1 = \text{Cos}^{-1}\frac{R}{Z} = \text{Cos}^{-1}\frac{15}{19.21} = 38.66°$ (lag)

Current drawn by capacitive circuit, $I_2 = \frac{V}{Z_2} = \frac{200}{30.23} = 6.615$ A **Ans.**

Phase angle of capacitive circuit, $\Phi_2 = \text{Cos}^{-1} \dfrac{25}{30.23} = 34.21°$ (lead)

Resolving currents in active and reactive components we get

Active component of resultant current,

$$I \cos \Phi = I_1 \cos \Phi_1 + I_2 \cos \Phi_2$$
$$= 10.41 \cos 38.66° + 6.615 \cos 34.21°$$
$$= 8.129 + 5.47 = 13.599 \text{ A} \qquad ...(i)$$

Reactive component of resultant current,

$$I \sin \Phi = I_1 \sin \Phi_1 + I_2 \sin \Phi_2$$
$$= 10.41 \sin (-38.66°) + 6.615 \sin 34.21°$$
$$= -6.503 + 3.719$$
$$= -2.784 \text{ A} \qquad ...(ii)$$

Squaring both sides of Eqs. (*i*) and (*ii*) and adding them we get

$$I^2 (\cos^2 \Phi + \sin^2 \Phi) = \sqrt{(13.599)^2 + (-2.784)^2} = 13.88 \text{ A} \quad \textbf{Ans.}$$

Power factor of the combination

$$= \frac{I \cos \phi}{I} = \frac{13.599}{13.88} = 0.98 \text{ (lagging)} \quad \textbf{Ans.}$$

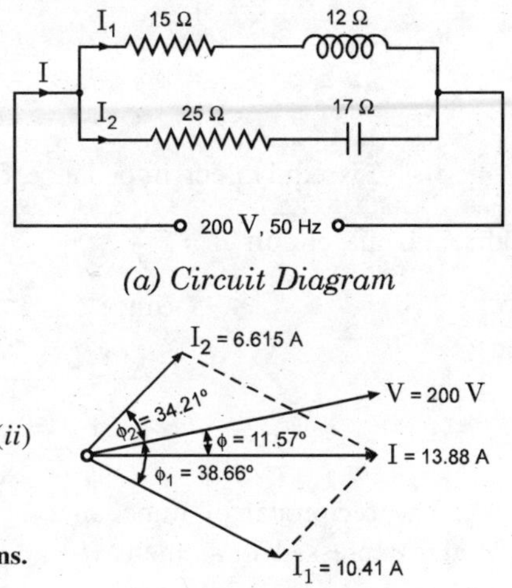

(a) Circuit Diagram

(b) Phasor Diagram

Fig. 3.23

Example 3.40. Find the active and reactive components of current taken by a series circuit consisting of a coil of inductance 0.1 H, resistance 8 Ω and a capacitor of 120 μF connected to a 240 V, 50 Hz supply mains. Find the value of the capacitor that has to be connected in parallel with the above series circuit so that the pf of the entire circuit is unity.

Solution : Inductive reactance, $X_L = 2\pi f L = 2\pi \times 50 \times 0.1 = 31.416 \ \Omega$

Capacitive reactance, $X_C = \dfrac{1}{2\pi f C} = \dfrac{1}{2\pi \times 50 \times 120 \times 10^{-6}} = 26.526 \ \Omega$

Circuit impedance, $Z = \sqrt{R^2 + (X_L - X_C)^2} = \sqrt{8^2 + (31.416 - 26.526)^2} = 9.376 \ \Omega$

Circuit current, $I = \dfrac{V}{Z} = \dfrac{240}{9.376} = 25.6$ A

Phase angle, $\phi = \text{Tan}^{-1} \dfrac{X_L - X_C}{R} = \text{Tan}^{-1} \dfrac{31.416 - 26.526}{8} = 31.435°$

Active component of current $= I \cos \phi = 25.6 \cos 31.435° = 21.84$ A **Ans.**

Reactive component of current $= I \sin \phi = 25.6 \sin 31.435° = 13.35$ A (lagging) **Ans.**

The power factor of the whole circuit will be unity if the current drawn by the capacitor, connected across the series circuit, equals the reactive (lagging) component of the series circuit.

i.e. when $2\pi f C V = I \sin \phi$

or $C = \dfrac{I \sin \phi}{2\pi f V} = \dfrac{13.35}{2\pi \times 50 \times 240} = 177 \ \mu\text{F}$ **Ans.**

3.13. ADMITTANCE METHOD

In the circuit shown in Fig. 3.18 (*a*), the branch currents

$$I_1 = \frac{V}{Z_1} = \frac{V}{\sqrt{R_1^2 + (X_{L_1} - X_{C_1})^2}}$$

$$I_2 = \frac{V}{Z_2} = \frac{V}{\sqrt{R_2^2 + (X_{L_2} - X_{C_2})^2}}$$

$$I_3 = \frac{V}{Z_3} = \frac{V}{\sqrt{R_3^2 + (X_{L_3} - X_{C_3})^2}}$$

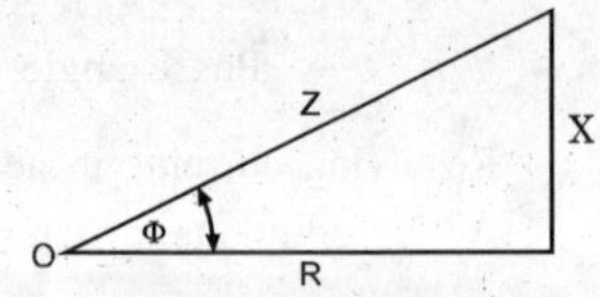

Impedance Triangle

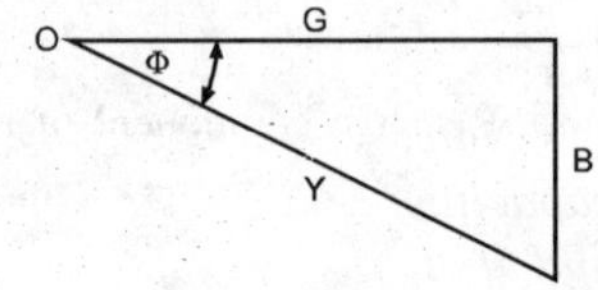

Admittance Triangle

Fig. 3.24

If Z is equivalent impedance of the circuit then total current flowing through the circuit, $I = \frac{V}{Z}$

Since $I = I_1 + I_2 + I_3$ (phasor sum)

$$\therefore \frac{V}{Z} = \frac{V}{Z_1} + \frac{V}{Z_2} + \frac{V}{Z_3} \text{ (phasor sum)}$$

$$\text{or } \frac{1}{Z} = \frac{1}{Z_1} + \frac{1}{Z_2} + \frac{1}{Z_3} \text{ (phasor sum)}$$

The reciprocal of impedance is known as admittance and is represented by English capital letter Y and expressed in siemens (S)

$$\therefore Y = \frac{1}{Z};\ Y_1 = \frac{1}{Z_1};\ Y_2 = \frac{1}{Z_2} \text{ and } Y_3 = \frac{1}{Z_3}$$

$$\text{and } Y = Y_1 + Y_2 + Y_3 \text{ (phasor sum)}$$

$$\text{Resultant current, } I = V \times \frac{1}{Z} = V \times Y$$

Just as impedance has two components, resistance R and reactance X similarly the admittance has two components known as *conductance*, G and *susceptance,* B respectively. The unit of conductance and susceptance is siemens (S)

From admittance and impedance triangles (Fig. 3.24)

$$\text{Conductance, } G = Y \cos \Phi = \frac{1}{Z} \cdot \frac{R}{Z} = \frac{R}{Z^2} = \frac{R}{R^2 + X^2}$$

$$\text{Susceptance, } B = Y \sin \Phi = \frac{1}{Z} \cdot \frac{X}{Z} = \frac{X}{Z^2} = \frac{X}{R^2 + X^2}$$

$$\text{Admittance, } Y = \sqrt{(Y \cos \Phi)^2 + (Y \sin \Phi)^2} = \sqrt{G^2 + B^2}$$

$$\text{Phase angle, } \Phi = \text{Tan}^{-1} \frac{B}{G}$$

3.14. APPLICATION OF ADMITTANCE METHOD IN SOLUTION OF SINGLE PHASE PARALLEL CIRCUITS

Determine conductance and susceptance of individual branches from the relation

$$G = \frac{R}{Z^2} \text{ and } B = \frac{X}{Z^2}$$

taking B as + ve if X is capacitive and as –ve if X is inductive. Let the conductances of the three branches of circuit shown in Fig. 3.18 (*a*) be G_1, G_2 and G_3 respectively and susceptances be B_1, B_2 and B_3 respectively. Find the equivalent conductance and susceptance of the circuit which will be equal to algebraic sum of conductances and susceptances of individual branches respectively.

For the circuit shown in Fig. 3.18 (*a*)

$$\text{Total conductance, } G = G_1 + G_2 + G_3$$

$$\text{and total susceptance, } B = B_1 + B_2 + B_3$$

$$\text{Total admittance of the circuit, } Y = \sqrt{G^2 + B^2}$$

The current in the circuit may be determined from the relation

$$I = V \times Y \text{ having phase angle } \Phi = \text{Tan}^{-1} \frac{B}{G}$$

Note: *Capacitive susceptance will be taken as + ve and inductive susceptance as –ve though capacitive reactance is taken as – ve and inductive reactance as + ve. If Φ is + ve current will lead and if Φ is – ve current will lag behind the applied voltage.*

Example 3.41. For the parallel circuit shown in the Fig. 3.25 (*a*), the parameter values are:

$$\mathbf{R_1 = 70\ \Omega\ \text{(non-inductive)}: X_C = 157\ \Omega\ \text{(at 50 Hz)}}$$

$$\mathbf{\text{Coil: } R_C = 30\ \Omega\ ;\ L_C = 0.5\ H;\ R_2 = 100\ \Omega}$$

(*i*) Determine (in magnitude and phase) the branch currents and the total current.

(*ii*) Draw phasor diagram indicating the currents and voltages across coil and condenser.

(*iii*) What current the circuit would draw if the ac source is replaced by equivalent dc source ?

Solution : Resistance of the inductive branch $= R_1 + R_C = 70 + 30 = 100\ \Omega$

Reactance of inductive branch $= 2\pi f L_C = 2\pi \times 50 \times 0.5 = 157\ \Omega$

$$\text{Conductance of inductive branch, } G_1 = \frac{R_1 + R_C}{(R_1 + R_C)^2 + (2\pi f L_C)^2} = \frac{100}{100^2 + 157^2} = 0.00288\ \text{S}$$

$$\text{Susceptance of inductive branch, } B_1 = \frac{2\pi f L_C}{(R_1 + R_C)^2 + (2\pi f L_C)^2} = \frac{157}{100^2 + 157^2} = -0.00453\ \text{S}$$

(inductive)

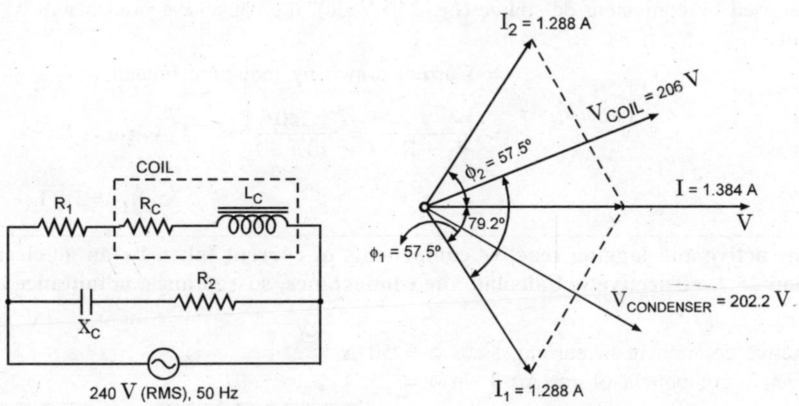

(*a*) *Circuit Diagram* (*b*) *Phasor Diagram*

Fig. 3.25

Resistance of capacitive branch, $R_2 = 100\ \Omega$

Reactance of capacitive branch, $X_C = 157\ \Omega$

$$\text{Conductance of capacitive branch, } G_2 = \frac{R_2}{R_2^2 + X_C^2} = \frac{100}{100^2 + 157^2} = 0.00288\ \text{S}$$

$$\text{Susceptance of capacitive branch, } B_2 = \frac{X_C}{R^2 + X_C^2} = \frac{157}{100^2 + 157^2} = 0.00453 \text{ S (capacitive)}$$

$$\text{Total conductance of circuit, } G = G_1 + G_2 = 0.00288 + 0.00288 = 0.00576 \text{ S}$$

$$\text{Total susceptance of circuit, } B = B_1 + B_2 = -0.00453 + 0.00453 = 0$$

$$\text{Total admittance of the circuit,} Y = G^2 + B^2 = \sqrt{(0.00576)^2 + 0^2} = 0.00576 \text{ S}$$

$$\text{Current in inductive branch, } I_1 = V \times Y_1 = 240\sqrt{(0.00288)^2 + (-0.00453)^2} = 1.288 \text{ A } \textbf{Ans.}$$

$$\text{Phase angle,} \Phi_1 = \text{Tan}^{-1}\frac{B_1}{G_1} = \tan^{-1}\frac{-0.00453}{0.00288} = -57.5^\circ \textbf{ Ans.}$$

$$\text{Current in capacitive branch, } I_2 = V \times Y_2 = 240 \times \sqrt{(0.00288)^2 + (0.00453)^2} = 1.288\text{A } \textbf{Ans.}$$

$$\text{Phase angle, } \Phi_2 = \text{Tan}^{-1}\frac{B_2}{G_2} = \text{Tan}^{-1}\frac{0.00453}{0.00288} = 57.5^\circ \textbf{ Ans.}$$

$$\text{Total current, } I = V \times Y = 240 \times 0.00576 = 1.384 \text{ A } \textbf{Ans.}$$

$$\text{Phase angle,} \Phi = \text{Tan}^{-1}\frac{B}{G} = \text{Tan}^{-1}\frac{0}{0.00576} = 0^\circ \textbf{ Ans.}$$

$$\text{Voltage across condenser} = I_2 X_C = 1.288 \times 157$$

$$= 202.2 \text{ V lagging behind } I_2 \text{ by } 90^\circ \textbf{ Ans.}$$

$$\text{Voltage across coil} = I_1 \times \sqrt{R_C^2 + (2\pi f L_C)^2} = 1.288\sqrt{30^2 + 157^2} = 206 \text{ V } \textbf{Ans.}$$

$$\text{Phase angle with current, } I_1 = \text{Tan}^{-1}\frac{157}{30} = 79.2^\circ \textbf{ Ans.}$$

(*iii*) When energised by equivalent dc source (*i.e.* 240 V dc), the capacitive branch will be open and current drawn by the circuit

= Current drawn by inductive branch

$$= \frac{V}{R_1 + R_C} = \frac{240}{70 + 30} = 2.4 \text{ A } \textbf{Ans.}$$

$$\because X_L = 2\pi f L_C = 2\pi \times 0 \times 0.5 = 0$$

Example 3.42. The active and lagging reactive components of current taken by an ac circuit from a 250 V supply are 50 A and 25 A respectively. Calculate the conductance, susceptance, admittance and power factor of the circuit.

Solution : Active component of current, $I \cos\phi = 50$ A

Lagging reactive component of current, $I \sin\phi = 25$ A

$$\text{Circuit current, } I = \sqrt{(I\cos\phi)^2 + (I\sin\phi)^2} = \sqrt{50^2 + 25^2} = 55.9 \text{ A}$$

$$\text{Admittance of the circuit, } Y = \frac{I}{V} = \frac{55.9}{250} = 0.224 \text{ S } \textbf{Ans.}$$

$$\text{Power factor of the circuit, } \cos\phi = \frac{I\cos\phi}{I} = \frac{50}{55.9} = 0.8944 \text{ (lagging) } \textbf{Ans.}$$

$$\text{Conductance of the circuit, } G = Y\cos\phi = 0.224 \times 0.8944 = 0.2 \text{ S } \textbf{Ans.}$$

$$\text{Susceptance of the circuit, } B = \sqrt{Y^2 - G^2} = \sqrt{(0.224)^2 - 0.2^2}$$

$$= 0.1 \text{ S (inductive) } \textbf{Ans.}$$

Example 3.43. A voltage $v(t) = 400 \sin 314t$ is applied to the circuit shown (Fig. 3.26). Find the currents and their phase angles w.r.t. the voltages for the three branches.

$$C = 50\ \mu F\ ;\ R_1 = 100\ \Omega\ ;\ R_2 = 50\ \Omega,\ L = 0.1\ H$$

[Pb. Technical Univ. Electrical Engineering May 2002]

Solution : Maximum value of applied voltage, V_{max} = Coefficient of the sine of time angle = 400 V

RMS value of applied voltage,

$$V_{rms} = \frac{V_{max}}{\sqrt{2}} = \frac{400}{\sqrt{2}} = 282.84\ V$$

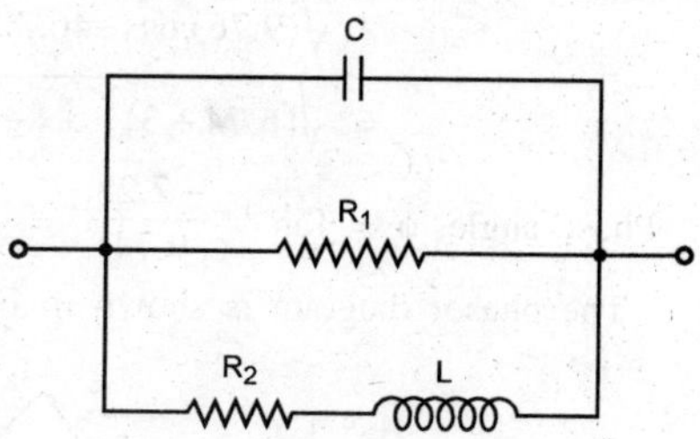

Fig. 3.26

Supply frequency,

$$f = \frac{\text{Coefficient of time } t}{2\pi} = \frac{314}{2\pi} = 50\ Hz$$

Current in capacitive branch,

$$I_1 = \frac{V}{X_C}$$

$$= 2\pi f CV = 2\pi \times 50 \times 50 \times 10^{-6} \times 282.84\ A$$

$= 4.44$ A leading applied voltage by $\frac{\pi}{2}$ radians **Ans.** *i.e.* Phase angle = 90° (leading)

Current in resistive branch,

$I_2 = \frac{V}{R_1} = \frac{282.84}{100} = 2.83$ A in phase with applied voltage **Ans.** *i.e.* phase angle = 0°

Current in R-L branch,

$$I_3 = \frac{V}{Z} = \frac{V}{\sqrt{R_2^2 + (2\pi f L)^2}} = \frac{282.84}{\sqrt{50^2 + (2\pi \times 50 \times 0.1)^2}} = \frac{282.84}{59.05} = 4.79\ A\ \textbf{Ans.}$$

Phase difference between voltage and current I_3,

$$\phi_3 = \text{Tan}^{-1}\frac{2\pi f L}{R_2} = \text{Tan}^{-1}\frac{2\pi \times 50 \times 0.1}{50} = 32.14° \text{ (lag) } \textbf{Ans.}$$

Example 3.44. Determine the rms value of current in each branch and total current of the circuit shown below (Fig. 3.27). Draw the phasor diagram. *[Pb. Technical Univ. Electrical Engineering January 2000]*

Solution : Impedance of inductive branch,

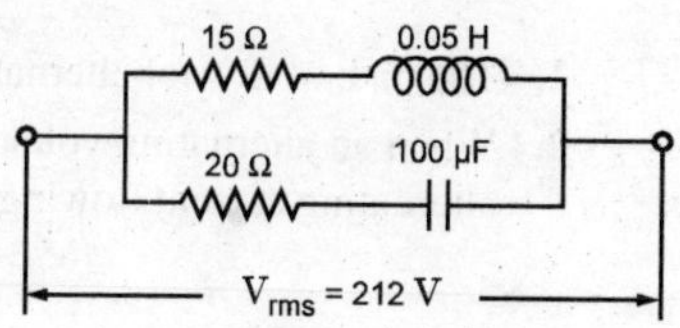

Fig. 3.27

$$Z_1 = \sqrt{R_1^2 + (2\pi f L)^2}$$

$$= \sqrt{15^2 + (2\pi \times 50 \times 0.05)^2}$$

$$= \sqrt{15^2 + 15.71^2} = 21.72\ \Omega$$

Phase angle, $\phi_1 = \text{Tan}^{-1}\frac{X_L}{R} = \text{Tan}^{-1}\frac{2\pi f L}{R} = \text{Tan}^{-1}\frac{15.71}{15} = 46.32°$ (lagging).

Impedance of capacitive branch,

$$Z_2 = \sqrt{R_2^2 + \left(\frac{1}{2\pi f C}\right)^2} = \sqrt{20^2 + \left(\frac{1}{2\pi \times 50 \times 100 \times 10^{-6}}\right)^2} = \sqrt{20^2 + (31.83)^2} = 37.59\ \Omega$$

Phase angle, $\phi_2 = \text{Tan}^{-1}\frac{X_C}{R} = \text{Tan}^{-1}\frac{31.83}{20} = 57.86°$ (leading).

Current, $I_1 = \frac{V}{Z_1} = \frac{212}{21.72} = 9.76$ A lagging behind applied voltage by 46.32° **Ans.**

$$\text{Current, } I_2 = \frac{V}{Z_2} = \frac{212}{37.59} = 5.64 \text{ A leading applied voltage by } 57.86^\circ \textbf{ Ans.}$$

Resultant current,

$$I = \sqrt{(I_1 \cos\phi_1 + I_2 \cos\phi_2)^2 + (I_1 \sin\phi_1 + I_2 \sin\phi_2)^2}$$

$$= \sqrt{(9.76 \cos(-46.32^\circ) + 5.64 \cos 57.86^\circ)^2 + (9.76 \sin(-46.32^\circ) + 5.64 \sin 57.86)^2}$$

$$= \sqrt{(6.74+3)^2 + (-7.06+4.78)^2} = \sqrt{(9.74)^2 + (-2.28)^2} = 10 \text{ A} \quad \textbf{Ans.}$$

$$\text{Phase angle, } \phi = \text{Tan}^{-1}\frac{-2.28}{9.74} = -13.2^\circ \quad \textbf{Ans.}$$

The phasor diagram is shown in Fig. 3.28.

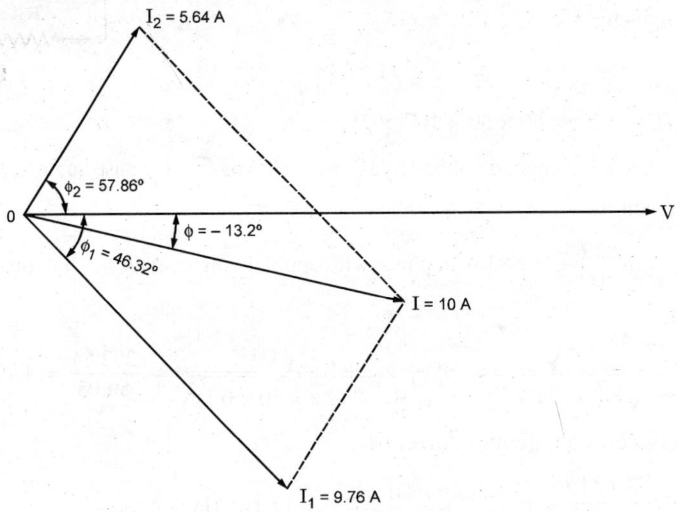

Fig. 3.28

HIGHLIGHTS

1. The path of flow of alternating current is called the ac circuit.
2. When an alternating voltage given by expression $v = V_{max} \sin \omega t$ is applied across a purely resistive circuit, the alternating current flowing through the circuit is given as

$$i = I_{max} \sin \omega t \quad \text{where } I_{max} = \frac{V_{max}}{R}$$

Power in purely resistive circuit, P $= V_{rms} \times I_{rms}$ or VI watts

Power factor of purely resistive circuit is unity.

3. When an alternating voltage given by expression $v = V_{max} \sin \omega t$ is applied across a purely inductive circuit (a coil with or without an iron core having negligible resistance), the alternating current flowing through the circuit is given as

$$i = I_{max}\left(\omega t - \frac{\pi}{2}\right)$$

where $I_{max} = \dfrac{V_{max}}{X_L}$; Inductive reactance X_L being equal to $2\pi f L$ or ωL ohms

and power absorbed by a pure inductor = 0

PF of a purely inductive circuit is zero.

4. In a purely capacitive circuit, if $v = V_{max} \sin \omega t$

$$i = I_{max} \sin \left(\omega t + \frac{\pi}{2}\right)$$

where $I_{max} = \frac{V_{max}}{X_C}$; the capacitive reactance X_C being equal to $\frac{1}{\omega C}$ or $\frac{1}{2\pi f C}$ ohms

Power absorbed, P = 0

PF is also zero.

5. An ac circuit can have three elements, namely R, L and C. Power is only consumed in R. The power consumed in elements L and C is zero.

The current through pure inductor lags behind the applied voltage by $\frac{\pi}{2}$ radians while the current through a pure capacitor leads the applied voltage by $\frac{\pi}{2}$ radians.

6. In an R-L series ac circuit, if $v = V_{max} \sin \omega t$

$$i = I_{max} \sin (\omega t - \phi)$$

$$I_{max} = \frac{V_{max}}{Z}; \; Z = \sqrt{R^2 + \omega^2 L^2}; \; \phi = \text{Tan}^{-1} \frac{\omega L}{R};$$

$$P = VI \cos \phi = I^2 R$$

Power factor, $\cos \phi = \frac{R}{Z}$ (lagging).

7. Power factor of an circuit is given by

(*i*) cosine of the phase angle between voltage and current *i.e.* $\cos \phi$

or (*ii*) the ratio of resistance to impedance *i.e.* $\frac{R}{Z}$

or (*iii*) the ratio of true power to apparent power *i.e.* $\frac{\text{True power, P}}{\text{Apparent power, S}}$

The *apparent power* of an ac circuit is given by the product of rms values of current and voltage *i.e.* S = VI volt-amperes.

The *true power* in an ac circuit is obtained by multiplying the apparent power by the power factor *i.e.*

$$P = S \cos \phi = V I \cos \phi \text{ watts.}$$

The product of apparent power VI and sine of angle between voltage and current, $\sin \phi$, is called the reactive power, Q.

$$\text{and } S = \sqrt{P^2 + Q^2}$$

8. Reciprocal of power factor of a coil is known as Q-factor of the coil. It is also called the quality factor or figure of merit of a coil.

Mathematically, Q-factor $= \frac{1}{PF} = \frac{1}{\cos \phi} = \frac{Z}{R} = \frac{\omega L}{R}$ if R is very small in comparison with inductive reactance X_L.

9. In an R-C series ac circuit, if $v = V_{max} \sin \omega t$

$$i = I_{max} \sin (\omega t + \phi)$$

$$I_{max} = \frac{V_{max}}{Z}; \; Z = \sqrt{R^2 + \left(\frac{1}{\omega C}\right)^2} = \sqrt{R^2 + X_C^2}$$

$$\phi = \text{Tan}^{-1}\frac{X_C}{R} = \text{Tan}^{-1}\frac{1}{\omega CR}$$

$$P = VI\cos\phi = I^2R$$

$$\text{Power factor, } \cos\phi = \frac{R}{Z} \text{ (leading).}$$

10. In an R-L-C series ac circuit

$$\text{Impedance, } Z = \sqrt{R^2 + (X_L - X_C)^2}$$

Circuit behaves as an inductive circuit if $X_L > X_C$; as a capacitive if $X_L < X_C$ and as a purely resistive if $X_L = X_C$

$$\text{Circuit current, } I = \frac{V}{Z}$$

$$\text{Phase angle, } \phi = \text{Tan}^{-1}\frac{X_L - X_C}{R}$$

$$\text{Power consumed, } P = VI\cos\phi = I^2R$$

$$\text{Power factor, } \cos\phi = \frac{R}{Z}$$

11. AC Parallel Circuits. An ac circuit having two or more branches, each branch having different components in series, is known as an ac parallel circuit.

AC parallel circuits can be solved by the following methods:

1. Phasor method
2. Admittance method
3. Symbolic or *j* method.

In *phasor method*, the magnitude and direction of currents flowing through different branches are determined and represented as phasors. The phasor sum of the branches currents gives the resultant current. In *admittance method*, the component values are determined as per admittance triangle which contains conductance and susceptance.

$$\text{Conductance, } G = Y\cos\phi = \frac{R}{Z^2}$$

$$\text{Susceptance, } B = Y\sin\phi = \frac{X}{Z^2}$$

$$\text{Admittance, } Y = \sqrt{G^2 + B^2}, \quad \text{where } G = G_1 + G_2 + \ldots + G_n \text{ and } B = B_1 + B_2 + \ldots + B_n$$

$$\text{Current, } I = V \times Y.$$

EXERCISES

1. Show that the instantaneous power consumed in a pure resistive circuit is not constant but it is fluctuating.
2. Derive the relationship between the voltage and current for a purely inductive circuit. Also show that the average power consumed by a circuit is zero.
[Pb. Univ. Elements of Electrical Engineering December 1988]
3. Develop the expression for the mean power consumed over a cycle of a single phase sinusoidal supply delivering power to a load comprising a resistance R in series with an inductance L.
4. Explain why the phasor of voltage across the inductor leads the current phasor by 90° and the phasor of voltage across capacitor lags its current by 90°.
5. Discuss the phase relation between emf and current when ac flows through R-L circuit.
[Pb. Technical Univ. Basic Electrical and Electronics Engineering Dec. 2007]

6. Define the terms : Impedance, Admittance, Power factor, Active power and Reactive power. **[Pb. Technical Univ. Elec. Engineering June 2000; December 2000; May 2001]**
7. Define apparent power and reactive power of an ac circuit. Explain how the true power and power factor of the circuit can be determined from them.
8. Define power factor. What is the power factor of a pure capacitor ? Differentiate between active and reactive powers. Draw a phasor diagram of a series R-L circuit.
9. Discuss the phasor relation between emf and current when ac flows through series R-C circuit. **[Pb. Technical Univ. Basic Electrical and Electronics Engineering May 2008]**
10. Discuss how do you analyse series R-C circuit. Draw its phasor diagram. **[Pb. Technical Univ. Basic Electrical and Electronics Engineering Dec. 2005]**
11. Draw the phasor diagram for given R-L-C circuit.
12. Explain inductive reactance and capacitive reactance.
13. Write the expression of currents in different branches and the total current of a circuit consisting of an inductance, a capacitance and a non-inductive resistance connected in parallel. Draw also the phasor diagram.
14. Define conductance, susceptance and admittance.
15. Distinguish between :
    (*i*) Instantaneous power and actual power.
    (*ii*) Form factor and power factor.
    (*iii*) Apparent power and reactive power.
    (*iv*) Apparent power and true power.
    (*v*) True power and reactive power.
    (*vi*) Inductance and inductive reactance.

SHORT ANSWER TYPE QUESTIONS WITH ANSWERS

Q. 1. In series circuit analysis why is current taken as reference phasor ? **[P.T.U. B.E.E.E. May 2005]**

Ans. In a series circuit, the current flowing through different components is the same, so current is taken as reference phasor.

Q. 2. What is the phase relationship between the voltage of supply and current flowing through a purely resistive circuit ?

Ans. Both supply voltage and circuit current are in phase with each other.

Q. 3. What is the phase relationship between the supply voltage and current flowing through a purely inductive circuit ?

Ans. In a purely inductive circuit, current lags behind the supply voltage by 90° or $\frac{\pi}{2}$ radians.

Q. 4. Why are the inductors usually iron cored ?

Ans. Inductors or choke coils are made of iron core because large valued flux densities can be produced in iron cores and so inductances of large value can be had. Air-cored inductors become too much bulky to provide an inductance of a required value.

Q. 5. What do you understand by power factor of a circuit ?

Ans. Power factor may be defined as the cosine of the phase angle between voltage and current. It may also be defined as the ratio of resistance to impedance or the ratio of true power to apparent power.

Q. 6. What do you understand by active and reactive components of current in an ac circuit ?

Ans. The current component which is in phase with circuit voltage (*i.e.* I cos ϕ) and contributes to active or true power of the circuit is called the active (wattfull or in-phase) component of current.

The current component which is in quadrature (or 90° out of phase) to the circuit voltage (*i.e.* I sin ϕ) and contributes to reactive power of the circuit is called the reactive (or wattless) component of current.

Q. 7. What are active and reactive powers ? Draw the power triangle. **[P.T.U. December 2003]**

Ans. The power which is actually consumed or utilized in an ac circuit is called the true or active power of the circuit. Power is consumed only in resistance. It is given by the product of the circuit voltage V, current I and power factor cos ϕ i.e. P = VI cos ϕ. It is expressed in watts.

A pure inductor and a pure capacitor do not consume any power, as in a half cycle whatsoever power is drawn from the supply source by these components, the same is returned to the supply source in the other half cycle. This power which flows back and forth (i.e. in both directions in the circuit) or reacts itself is called the reactive power. This is also known as wattless power.

The reactive power of an ac circuit is given by the product of voltage V, current I and sine of the phase angle ϕ *i.e.* Reactive power, $Q = VI \sin \phi$. It is expressed in reactive volt-amperes.

The power triangle is shown in Fig. 3.6 (*a*) page 73.

Q. 8. What is the relationship between supply voltage and circuit current in a purely capactive circuit ?

Ans. In a purely capactive circuit, current leads the applied voltage by 90° or $\frac{\pi}{2}$ radians.

Q. 9. What is the effect of frequency on inductive reactance ?

Ans. Inductive reactance of an inductor increases proportionately with the increase in supply frequency, *i.e.* $X_L \propto f$.

Q. 10. What is the effect of frequency on capactive reactance ?

Ans. Capactive reactance of a given capacitor decreases with the increase in supply frequency. Capacitive reactance is inversely proportional to supply frequency *i.e.* $X_C \propto \frac{1}{f}$.

Q. 11. What are the values of power factor for (*i*) purely resistive circuit (*ii*) purely inductive circuit and (*iii*) purely capacitive circuit ?

Ans. (*i*) Unity, (*ii*) Zero (lagging) (*iii*) Zero (leading).

Q. 12. What is the relationship between apparent power, true power and reactive power of an ac circuit ?

Ans. Apparent power = $\sqrt{(\text{True power})^2 + (\text{Reactive power})^2}$

Q. 13. In an R-L-C circuit on what factors depends whether the phase angle between voltage V and current I is lagging or leading ?

Ans. Phase angle in an R-L-C circuit is given as

$$\phi = \text{Tan}^{-1}\frac{X}{R} = \text{Tan}^{-1}\frac{X_L - X_C}{R}$$

i.e. the phase angle between voltage V and current I will depend upon whether the inductive reactance is greater or smaller than capacitive reactance. When X_L exceeds X_C voltage V will lead current I and when X_C exceeds X_L applied voltage V will lag behind the current I.

Q. 14. What is admittance ? Give units. **[P.T.U. B.E.E.E. May 2008]**

Ans. The reciprocal of impedance is known as admittance and is represented by English letter Y. Its unit is siemens or mho.

PROBLEMS

1. A 50-Hz voltage of 115 V (rms) is impressed on a 50 Ω resistance. Write the time equations for the voltage and resultant current. Let the zero point of the voltage wave be at $t = 0$ [**Ans.** 162.6 sin 314*t* and 3.25 sin 314*t*]

2. A resistance of 15 Ω is connected in series with an inductance of 0.02 H. This combination is connected across 200 V, 50 Hz supply. Calculate the following :

(*i*) Current flowing in the circuit. (*ii*) Power factor, and (*iii*) Power consumed by the circuit.

[**Ans.** (*i*) 12.3A (*ii*) 0.922 (lag) (*iii*) 226.9 watts]

3. $v = V_m \sin(\omega t + \Phi)$

$i = I_m \sin(\omega t + \theta)$

From the above expressions determine impedance; power factor of the circuit; real and reactive powers.

$$\left[\textbf{Ans. } \frac{V_m}{I_m};\ \text{Cos}\,(\Phi - \theta);\ \frac{V_m I_m}{2}\cos(\Phi - \theta);\ \frac{V_m I_m}{2}\sin(\Phi - \theta)\right]$$

4. A coil having a resistance of 6 Ω and an inductance of 0.03 H is connected across a 50 V, 60 Hz supply. Calculate (*i*) the current (*ii*) the phase angle between the current and the applied voltage, (*iii*) the power factor, (*iv*) the volt-amperes and (*v*) the power. [**Ans.** (*i*) 3.905 A (*ii*) 62.05° (*iii*) 0.4686 lag (*iv*) 195.25 VA (*v*) 91.5 W]

5. The impressed voltage and the current flowing through a circuit are given by the following equations :

$$v = 200 \sin\left(400t + \frac{\pi}{3}\right);\ i = 5 \sin\left(400t - \frac{\pi}{6}\right)$$

Draw the circuit and determine the power consumed in the circuit. **[Ans.** Zero]

6. An inductive coil takes a current of 10 A lagging 30° from 200 V, 50 Hz supply. Calculate the resistance, reactance and inductance of the coil. **[Ans.** 17.32 Ω, 10 Ω, 31.8 mH]

7. A coil is connected across a 200 V, 50 Hz supply and takes a current of 8 A. The loss in the coil is 960 W. Calculate (*i*) impedance (*ii*) resistance (*iii*) reactance and (*iv*) power factor.

[Ans. (*i*) 25 Ω (*ii*)15 Ω (*iii*) 20 Ω (iv) 0.6 (lag)]

8. In a particular R-L series circuit a voltage of 10 V at 50 Hz produces a current of 700 mA while the same voltage at 75 Hz produces a 500 mA. What are the values of R and L in the circuit ? **[Ans.** 6.9 Ω; 40 mH]

9. A 100 V, 60 W lamp is to be operated on 220 V, 50 Hz mains. What (*i*) pure resistance and (*ii*) pure inductance placed in series with the lamp, will enable it to be used without being overrun ? Which method would be more economical ? **[Ans.** (*i*) 200 Ω (*ii*) 1.04 H, Second method is more economical]

10. $v = 100 \sin 100\pi t$ and $i = 8 \sin (100\pi t + 60°)$

From these expressions find the impedance and the power factor of the circuit. **[Ans.** 12.5 Ω; 0.5 lead]

11. A voltage of 125 V at 60 Hz is applied across a non-inductive resistor connected in series with a condensor. The current is 2.2 A. The power loss in the resistor is 96.8 W and that in the condensor is negligible. Calculate the resistance and the capacitance. **[Ans.** 20 Ω ; 49.88μF]

12. A tungsten filament lamp rated at 500 W, 100 V is to be connected in series with a capacitance across 220 V, 50- Hz supply. Calculate (*a*) the value of capacitor such that the voltage and power consumed by the lamp are according to the rating of the lamp (*b*) the power factor of the current drawn from the supply (*c*) draw the phasor diagram of the circuit. **[Ans.** 81.2 μF (*b*) 0.4545 (lead)]

13. A resistance of 20 Ω, inductance of 0.2 H and capacitance of 150 μ F connected in series and are fed by a 230 V, 50 Hz supply. Find X_L, X_C, Z, Y, pf, active power and reactive power.

[Ans. 62.8 Ω; 21.2 Ω, 46.16 Ω, 0.0217 S, 0.4333(lag); 497 W, 1,030 VAR]

14. Find the impedance, current, power and power factor of the following series circuits and draw the corresponding phasor diagrams (*i*) R and L (*ii*) R and C (*iii*) R, L and C. In each case the applied voltage is 200 V, the frequency is 50 Hz; R = 10 Ω; L = 50 mH ; C = 100 μF.

[Ans. (*i*) 18.62 Ω; 10.74 A; 1.153 kW; 0.537 (lag);
(*ii*) 33.365 Ω; 5.994 A; 359 W; 0.2997 lead;
(*iii*) 18.97 Ω; 10.54 A; 1.111 kW; 0.527 lead]

15. A circuit containing a resistance of 40 Ω in parallel with an inductive reactance of 30 Ω is connected across 240 V ac mains. Calculate (*i*) the current in each branch and the total current drawn from the mains (*ii*) the power factor of each branch and power factor of the whole circuit (*iii*) the power drawn by each branch and the total power drawn from the mains. **[Ans.** (*i*) 6 A, 8 A, 10 A, (*ii*) 1.0, zero, 0.6 (lag) (*iii*) 1,440 watts, zero 1,440 watts]

16. A series combination of R and C is in parallel with a 20 Ω resistor across 50 Hz source results in a total current of 7 A, a current through the 20 Ω resistor of 5 A and a current in the R-C branch of 3 A. Determine R and C.

[Pb. Univ. Electrical Circuits and Systems December 1988]

[Ans. R = 16.667 Ω; C = 110.3 μF]

17. Two parallel circuits comprise respectively of (*i*) a coil of resistance 20 Ω and inductance 0.07 H and (*ii*) a capacitor of capacitance 60 μF in series with resistance of 50 Ω. Calculate the current in each branch and the current taken from the supply and power factor when connected to 220 V, 50 Hz supply.

[Ans. 7.4 A, 3 A, 7.8 A, 0.907 (lag)]

18. An iron choke coil takes 4 A when connected to a 20 V dc supply and takes 5 A from 65 V, 50 Hz ac supply. Determine :

(*i*) the resistance and inductance of the coil,
(*ii*) the power drawn by the coil,
(*iii*) the power factor. **[Ans.** (*i*) 5 Ω; 38.2 mH (*ii*) 125 W (*iii*) 10.3846 (lagging)]

❑❑❑

4

CHAPTER

Phasor Algebra and AC Circuits

4.1. INTRODUCTION

INSIDE THIS CHAPTER

We have seen that ac circuits cannot be solved by use of simple algebra since geometrical relations are to be taken into consideration. Though phasor diagram method is quite satisfactory for fairly simple circuits but becomes complicated when a circuit or network is made up of several branches. AC circuits can be conveniently solved by use of complex algebra. The results obtained by this method are of the same order of accuracy as those obtained by trigonometrical methods although calculations are usually much simpler.

The above system enables equations representing alternating voltages and currents and their phase relationships to be expressed in simple algebraic form. It is based upon the idea that a phasor can be resolved into two components along two axis at right angles to each other.

There are four forms or methods of representing the vector quantities viz (*i*) *symbolic form*, (*ii*) *trigonometrical form* (*iii*) *exponential form and* (*iv*) *polar form*. All of these methods of expressing vector quantities enable those operations which are carried out graphically in a phasor diagram, to be performed analytically.

4.2. SYMBOLIC FORM

A phasor, as already mentioned above, can be resolved into two components along two axis at right angles to each other. It is preferable and convenient to resolve it along X-axis and Y-axis.

In symbolic form, a phasor is resolved into two components at right angles to each other and the component lying along Y axis is designated by $+j$ or $-j$ according to its direction.

In Fig. 4.1 phasors OA_1, OA_2, OA_3 and OA_4 are shown lying in first, second, third and fourth quadrants respectively. The phasor OA_1 lying in first quadrant may be resolved into two components a_1 along X-axis (sometimes called the axis of reals) and b_1 along Y-axis (sometimes called the axis of imaginaries or j-axis). The component along Y-axis should be designated by j and should be written as jb_1. Thus we represent the phasor OA_1

$$\boldsymbol{OA_1}^* = a_1 + j\, b_1$$

The phasor, when expressed in above form, is said to be expressed in *complex form*. In electrical engineering, the horizontal component (*i.e* a_1) is called the *in-phase* or *active* component while the vertical component is called the *quadrature* or *reactive* component.

The length of line OA_1 represents, to adopted scale, the absolute value or modulus (magnitude) of the phasor and is equal to $\sqrt{a_1^2 + b_1^2}$, and its angle with the X-axis, θ_1, called the *argument* of the phasor, is given by the expression $\theta_1 = \text{Tan}^{-1} \dfrac{b_1}{a_1}$.

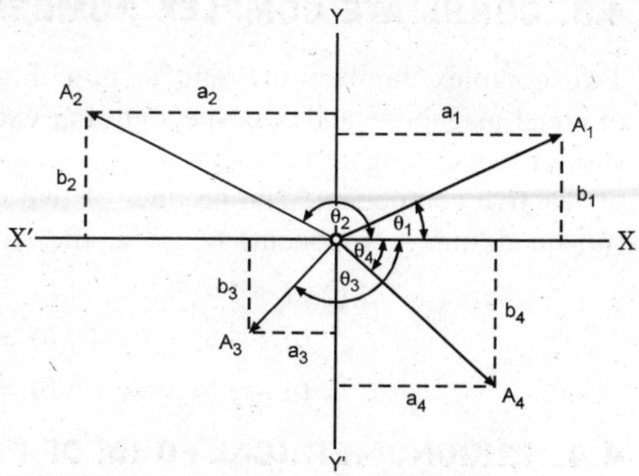

Fig. 4.1. ***Resolution of Phasors***

Similarly phasors OA_2, OA_3 and OA_4 can be resolved into two components along axis of reals and axis of imaginaries and can be expressed as

$$\boldsymbol{OA_2} = -\, a_2 + j\, b_2$$
$$\boldsymbol{OA_3} = -\, a_3 - j\, b_3$$
$$\boldsymbol{OA_4} = a_4 - j\, b_4$$

4.2.1. Significance of Operator *j*

The letter *j* used in the above expressions is an operator, the application of which to phasor *a*, or phasor component *a*, causes its rotation through 90º in the positive (counter-clockwise) direction. It is assigned the numerical value $\sqrt{-1}$. The double operation of *j* on phasor *a* rotates it in counter-clockwise direction through 180º and hence reverses its sense because $j\,j = j^2 = (\sqrt{-1})^2 = -1$. The triple operation of *j* on phasor *a* rotates it in counter-clockwise direction through 270º and so on.

For example, if phasor V lying along X-axis is operated with *j*, the phasor V will rotate through an angle of 90º in counter-clockwise direction and lie along Y-axis. If this phasor V is operated twice with operator *j* it will rotate through 180º in counter-clockwise direction from X-axis and will lie along X-axis in negative direction, as illustrated in Fig. 4.2. Application of the operator *j* thrice to this phasor V will rotate it through 270º in counter-clockwise direction from X-axis and now the phasor V will lie along Y-axis in negative direction. On operating the phasor V with *j* four times, it will rotate through 360º in counter-clockwise direction from the reference axis and so will come to the original position.

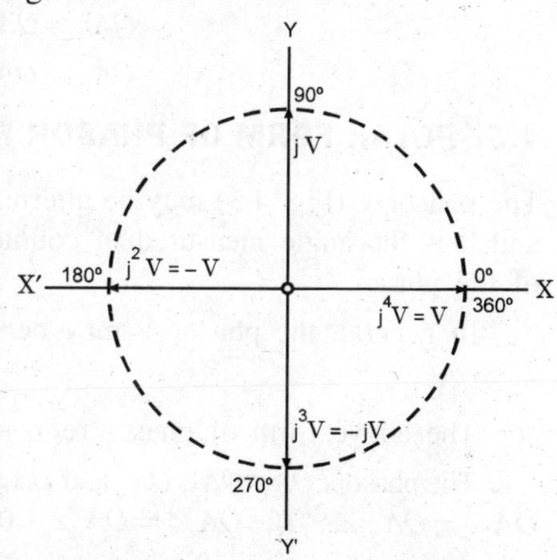

Fig. 4.2. ***Significance of Operator j***

Hence it is seen that *j* is an operator which turns phasor *a* through an angle of 90º in counter-clockwise direction, each time that it is applied and has a numerical value of $\sqrt{-1}$ so that

$$j^2 = (\sqrt{-1})^2 = -1;\; j^3 = (\sqrt{-1})^3 = -\sqrt{-1} \text{ and } j^4 = (\sqrt{-1})^4 = (-1)^2 = 1$$

---

* Symbols representing phasors are printed in bold face italics, while those representing only the magnitude are printed in ordinary italics, thus:

V, I, Z, phasors or complex numbers.

V, I, Z, magnitude or module.

Other methods of denoting phasors are by putting bar above, dot above or below letters in ordinary italics as $\bar{V}, \bar{I}, \bar{Z}, \dot{V}, \dot{I}, \dot{Z}$ or $\underset{.}{V}, \underset{.}{I}, \underset{.}{Z}$

4.3. CONJUGATE COMPLEX NUMBERS

Two complex numbers are said to be conjugate when they are of same modulus and have arguments of equal magnitude and opposite sign. For example, the numbers $(a + j\,b)$ and $(a - j\,b)$ are conjugate. The sum of two conjugate numbers gives in-phase or active component and their difference gives quadrature or reactive component. The product of two conjugate numbers gives in-phase or active component only (no quadrature component) equal to the sum of the square of their magnitudes.

i.e. $(a + j\,b) + (a - j\,b) = 2a$, active component.

$(a + j\,b) - (a - j\,b) = j\,2b$, reactive component.

$(a + j\,b) \times (a - j\,b) = (a^2 + b^2)$, active component.

4.4. TRIGONOMETRICAL FORM OF PHASOR REPRESENTATION

As may be seen from Fig. 4.3, X-component of $\mathbf{V} = V \cos \theta$ and Y-component of $\mathbf{V} = V \sin \theta$. Hence phasor V can be expressed in the trigonometrical form as

$$\mathbf{V} = V \cos \theta + j\,V \sin \theta = V (\cos \theta + j \sin \theta)$$

This is equivalent to the symbolic or rectangular form $\mathbf{V} = (a + j\,b)$ because $a = V \cos \theta$ and $b = V \sin \theta$

In general $\mathbf{V} = V (\cos \theta \pm j \sin \theta)$

The above expression represents a phasor of numerical value V and having phase angle of $\pm\,\theta$ with the reference axis, positive angle being measured in counter-clockwise direction from the reference axis and negative angle in clockwise direction from the reference axis.

The phasors OA_1, OA_2, OA_3 and OA_4 (Fig. 4.1) may be represented thus:

$$\mathbf{OA_1} = OA_1 (\cos \theta_1 + j \sin \theta_1)$$
$$\mathbf{OA_2} = OA_2 (\cos \theta_2 + j \sin \theta_2)$$
$$\mathbf{OA_3} = OA_3 (\cos \theta_3 - j \sin \theta_3)$$
$$\mathbf{OA_4} = OA_4 (\cos \theta_4 - j \sin \theta_4)$$

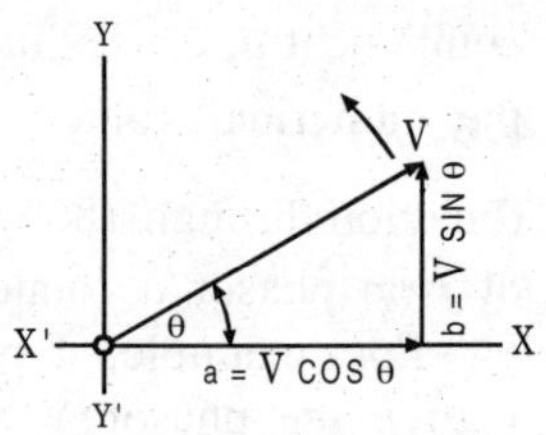

Fig. 4.3

4.5. POLAR FORM OF PHASOR REPRESENTATION

The phasor V (Fig. 4.3) may be alternatively expressed as $V\angle\theta$ where V is the magnitude of the phasor and θ is the angle measured in counter-clockwise direction from reference axis, called the *argument* of the phasor.

In general, the phasor V may be represented as

$$\mathbf{V} = V\angle\pm\theta$$

The above form of phasor representation is known as *polar form*.

The phasors OA_1, OA_2, OA_3 and OA_4 (Fig. 4.1) may be represented thus $\mathbf{OA_1} = OA_1 \angle\theta_1$; $\mathbf{OA_2} = OA_2 \angle\theta_2$; $\mathbf{OA_3}, = OA_3 \angle-\theta_3$; $\mathbf{OA_4} = OA_4\angle-\theta_4$

Note : The phasor representation in either of three mathematical forms convey the same information *i.e.* magnitude (modulus) of the phasor and its phase angle (argument). So it is possible to convert one form to the other.

Example 4.1. Write equivalent polar form of vector (3 + *j*4) and illustrate on phasor diagram.

[*Pb. Technical Univ. Basic Electrical and Electronics Engineering May 2008*]

Solution : The modulus or magnitude of the given vector is

$$\sqrt{3^2 + 4^2} = 5$$

The argument or phase angle of the given vector is

$$\theta = \text{Tan}^{-1}\frac{4}{3} = 53.13°$$

So given vector in polar form is given as $5\ \angle 53.13°$ **Ans.**

Phasor diagram is shown in Fig. 4.4.

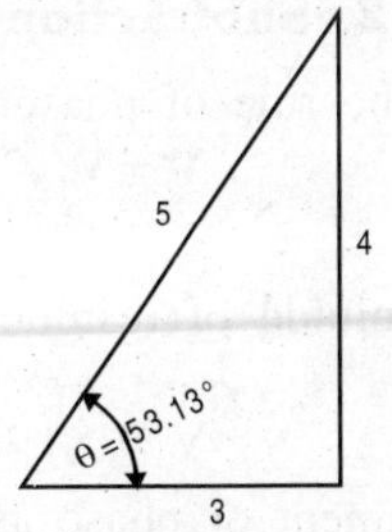

Fig. 4.4. ***Phasor Diagram***

Example 4.2. Give the voltage expressed in the symbolic form V = (5 + j 12) volts in trigonometrical and polar forms.

Solution : The modulus or magnitude of the given voltage,

$$V = \sqrt{5^2 + 12^2} = 13\ V$$

The argument or phase angle of the given voltage,

$$\theta = \text{Tan}^{-1}\frac{12}{5} = 67.38° \text{ or } 0.374\ \pi \text{ radians.}$$

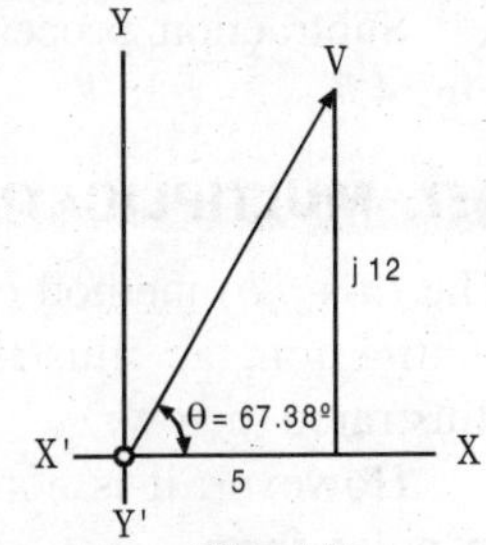

Fig. 4.5

So given voltage in

(*i*) Trigonometrical form is 13 (cos 67.38° + *j* sin 67.38°) volts **Ans.**

(*ii*) Polar form is $13\ \angle 67.38°$ volts **Ans.**

4.6. ADDITION AND SUBTRACTION OF PHASORS

The symbolic or rectangular form is most suitable form for addition or subtraction of phasors. If the phasors are given in polar or exponential form, they should be first converted to rectangular form and then addition or subtraction be carried out.

4.6.1. Addition of Phasors

For the addition of phasors in the rectangular form in-phase (or active) components are added together and quadrature components are added together.

Let the two voltage phasors V_1 and V_2 represented as $\mathbf{V_1} = a_1 + j\,b_1$, and $\mathbf{V_2} = a_2 - j\,b_2$, be added.

Then resultant voltage will be given as

$$\mathbf{V} = \mathbf{V_1} + \mathbf{V_2} = (a_1 + j\,b_1) + (a_2 - j\,b_2)$$
$$= (a_1 + a_2) + j\,(b_1 - b_2)$$

Magnitude of resultant voltage,

$$V = \sqrt{(a_1 + a_2)^2 + (b_1 - b_2)^2}$$

Argument or phase angle of the resultant voltage,

$$\theta = \text{Tan}^{-1}\frac{b_1 - b_2}{a_1 + a_2}$$

Addition process by graphic representation is illustrated in Fig. 4.6

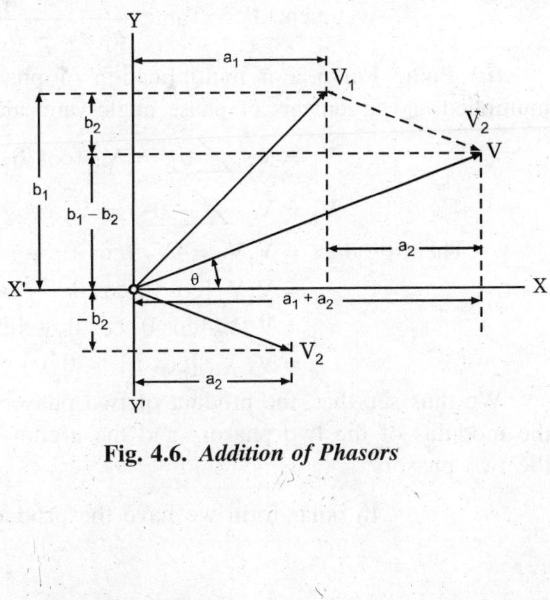

Fig. 4.6. ***Addition of Phasors***

4.6.2. Subtraction of Phasors

By the rule of phasor subtraction

$$\mathbf{V} = \mathbf{V}_1 - \mathbf{V}_2 = (a_1 + j\,b_1) - (a_2 - j\,b_2)$$
$$= (a_1 - a_2) + j\,(b_1 + b_2)$$

Magnitude of resultant voltage,

$$V = \sqrt{(a_1 - a_2)^2 + (b_1 + b_2)^2}$$

Argument or phase angle of resultant voltage,

$$\theta = \text{Tan}^{-1}\frac{b_1 + b_2}{a_1 - a_2}$$

Subtraction process by graphic representation is illustrated in Fig. 4.7.

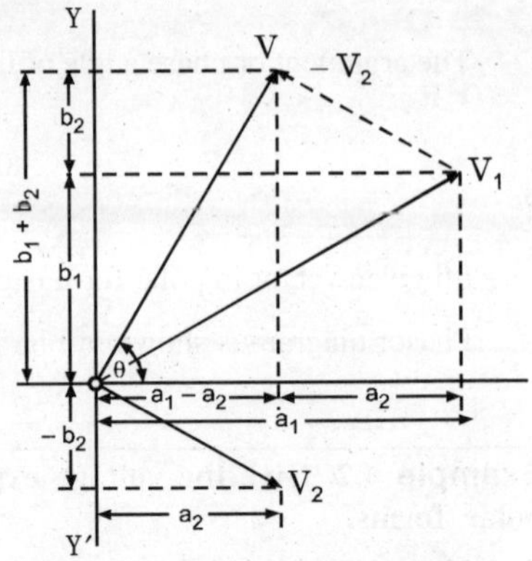

Fig. 4.7. *Subtraction of Phasors*

4.7. MULTIPLICATION AND DIVISION OF PHASORS

The $(a + j\,b)$ method of representing a phasor is particularly well suited to the process of addition and subtraction, as illustrated above, but it also lends itself readily to multiplication and division as illustrated below:

However, it is noteworthy that it is easier to multiply and divide phasors when they are expressed in polar form.

Consider two phasors given by $\mathbf{V}_1 = (a_1 + j\,b_1)$ or $V_1 \angle \theta_1$

and $\mathbf{V}_2 = (a_2 - j\,b_2)$ or $V_2 \angle -\theta_2$

4.7.1. Multiplication of Phasors

(*i*) Rectangular Form.

$$\mathbf{V}_1 \times \mathbf{V}_2 = (a_1 + j\,b_1)(a_2 - j\,b_2)$$
$$= a_1\,a_2 - j\,a_1 b_2 + j\,a_2 b_1 - j^2 b_1 b_2 = (a_1 a_2 + b_1 b_2) + j\,(a_2 b_1 - a_1 b_2)$$

$$\text{Magnitude of resultant} = \sqrt{(a_1\,a_2 + b_1\,b_2)^2 + (a_2\,b_1 - a_1\,b_2)^2}$$

$$\text{Argument, } \theta = \text{Tan}^{-1}\frac{a_2\,b_1 - a_1\,b_2}{a_1\,a_2 + b_1\,b_2}$$

(*ii*) Polar Form. For multiplication of phasors in polar forms, their magnitude or modulus are multiplied and arguments or phase angles are added as proved below:

$$\mathbf{V}_1 = V_1 \angle \theta_1 = V_1(\cos\theta_1 + j\sin\theta_1)$$
$$\mathbf{V}_2 = V_2 \angle -\theta_2 = V_2(\cos\theta_2 - j\sin\theta_2)$$

$$\text{Their product} = \mathbf{V}_1\mathbf{V}_2 = [V_1(\cos\theta_1 + j\sin\theta_1)]\,[V_2(\cos\theta_2 - j\sin\theta_2)]$$
$$= V_1V_2[\cos\theta_1\cos\theta_2 - j\cos\theta_1\sin\theta_2 + j\sin\theta_1\cos\theta_2 - j^2\sin\theta_1\sin\theta_2]$$
$$= V_1V_2[\cos\theta_1\cos\theta_2 + \sin\theta_1\sin\theta_2 + j(\sin\theta_1\cos\theta_2 - \sin\theta_2\cos\theta_1)]$$
$$= V_1V_2[\cos(\theta_1 - \theta_2) + j\sin(\theta_1 - \theta_2)]$$

We thus see that, the product of two phasors is a phasor having modulus equal to the product of the modulus of the two phasors and the argument is equal to the algebraic sum of the arguments of the two phasors.

In polar form we have the product $\mathbf{V}_1\mathbf{V}_2 = V_1V_2 \angle \theta_1 - \theta_2$

4.7.2. Division of Phasors

(i) Rectangular Form

$$\frac{\mathbf{V}_1}{\mathbf{V}_2} = \frac{a_1 + j\,b_1}{a_2 - j\,b_2}$$

Multiplying both the numerator and denominator by conjugate of $(a_2 - j\,b_2)$ *i.e.* by $(a_2 + j\,b_2)$ we have

$$\frac{\mathbf{V}_1}{\mathbf{V}_2} = \frac{a_1 + j\,b_1}{a_2 - j\,b_2} \times \frac{a_2 + j\,b_2}{a_2 + j\,b_2} = \frac{(a_1\,a_2 - b_1\,b_2) + j\,(a_1\,b_2 + a_2\,b_1)}{a_2^2 + b_2^2}$$

$$= \frac{(a_1\,a_2 - b_1\,b_2)}{a_2^2 + b_2^2} + j\,\frac{a_1\,b_2 + a_2\,b_1}{a_2^2 + b_2^2}$$

(ii) Polar Form

$$\frac{\mathbf{V}_1}{\mathbf{V}_2} = \frac{a_1 + j\,b_1}{a_2 - j\,b_2} = \frac{\mathrm{V}_1\,(\cos\theta_1 + j\sin\theta_1)}{\mathrm{V}_2\,(\cos\theta_2 - j\,sin\,\theta_2)} = \frac{\mathrm{V}_1 \angle \theta_1}{\mathrm{V}_2 \angle -\theta_2}$$

The quotient is $\dfrac{\mathbf{V}_1}{\mathbf{V}_2} = \dfrac{\mathrm{V}_1\,(\cos\theta_1 + j\sin\theta_1)}{\mathrm{V}_2\,(\cos\theta_2 - j\sin\theta_2)}$

$$= \frac{\mathrm{V}_1}{\mathrm{V}_2} \cdot \frac{(\cos\theta_1 + j\sin\theta_1)}{(\cos\theta_2 - j\sin\theta_2)} \times \frac{\cos\theta_2 + j\,\sin\theta_2}{(\cos\theta_2 + j\sin\theta_2)}$$

$$= \frac{\mathrm{V}_1}{\mathrm{V}_2}\left[\frac{[\cos\theta_1\,\cos\theta_2 - \sin\,\theta_1\,\sin\,\theta_2 + j\,(\sin\theta_1\,\cos\theta_2 + \sin\theta_2\,\cos\theta_1)}{\cos^2\,\theta_2 + \sin^2\,\theta_2}\right]$$

$$= \frac{\mathrm{V}_1}{\mathrm{V}_2}\,[\cos\,(\theta_1 + \theta_2) + j\,\sin\,(\theta_1 + \theta_2)]$$

We thus see that, when we divide one phasor by another phasor, the quotient is a phasor having modulus equal to the quotient of the modulus of the two phasors and an argument equal to the algebraic difference of the arguments of the two phasors.

In polar form we have $\dfrac{\mathbf{V}_1}{\mathbf{V}_2} = \dfrac{\mathrm{V}_1}{\mathrm{V}_2} \angle \theta_1 - (-\theta_2) = \dfrac{\mathrm{V}_1}{\mathrm{V}_2} \angle \theta_1 + \theta_2$

Example 4.3. Add, subtract and multiply the following phasors.

(3 + *j*4) and (2 + *j*3)

Solution : (*i*) Sum of phasors $(3 + j\,4)$ and $(2 + j\,3) = (3 + j\,4) + (2 + j\,3) = (5 + j\,7)$ **Ans.**

The resultant phasor lies in first quadrant having magnitude of $\sqrt{5^2 + 7^2} = 8.6$ and argument of $\tan^{-1}\dfrac{7}{5}$ or 54.5° or 0.3π radians.

(*ii*) Difference of phasors $(3 + j\,4)$ and $(2 + j\,3) = (3 + j\,4) - (2 + j\,3) = (1 + j\,1)$ **Ans.**

The resultant phasor lies in first quadrant having magnitude of $\sqrt{1^2 + 1^2} = 1.414$ and argument of $\tan^{-1}\dfrac{1}{1}$ or 45° or $\dfrac{\pi}{4}$ radians.

(*iii*) The product of phasors $(3 + j\,4)$ and $(2 + j\,3)$ $= (3 + j\,4)(2 + j\,3)$

$= (6 + j\,9 + j\,8 - 12) = (-6 + j\,17)$ **Ans.**

The resulant phasor lies in the second quadrant having magnitude of $\sqrt{(-6)^2+(17)^2}$ = 18.028 and argument of $\tan^{-1}\frac{17}{-6}$ or 109.44°.

Example 4.4. Find the resultant of (8 + j6) × (–10 – j7.5).

[*Pb. Technical Univ. Basic Electrical and Electronics Engineering Dec. 2007*]

Solution : Product of (8 + j6) and (–10 – j7.5)

$$= (8+j6)\times(-10-j7.5) = (-80-j60-j60+45)$$

$$= (-35-j120) = 125\angle{-106.26°}$$ **Ans.**

Example 4.5. Divide phasor (5 + j 6) by phasor (3j + 4)

Solution : $\frac{5+j6}{3j+4} = \frac{5+j6}{4+j3}\times\frac{4-j3}{4-j3} = \frac{20+18+j9}{4^2+3^2} = \frac{38+j9}{25} = 1.52 + j\,0.36$ **Ans.**

The resultant phasor lies in the first quadrant having magnitude of $\sqrt{(1.52)^2+(0.36)^2}$ = 1.562 and argument of $\tan^{-1}\frac{0.36}{1.52}$ or 13.32° or 0.074π radian.

Example 4.6. Determine the product of phasors 30 $\angle 45°$ and 7.5 $\angle -30°$.

Solution : $30\angle 45° \times 7.5\angle -30° = 30\times 7.5\angle 45°+(-30°) = 225\angle 15°$ **Ans.**

Example 4.7. Divide phasor 30 $\angle 45°$ by phasor 7.5 $\angle -30°$.

Solution : $\frac{30\angle 45°}{7.5\angle -30°} = \frac{30}{7.5}\angle 45°-(-30°) = 4\angle 75°$ **Ans.**

Example 4.8. Three phasors are A = 10 + j 10 ; B = 50 $\angle 120°$ and C = 8 – j 6. Evaluate $\frac{\mathbf{AB}}{\mathbf{C}}$ and express it in rectangular and polar forms.

Solution : Phasor A = (10 + j 10) have modulus of $\sqrt{10^2+10^2}$ *i.e* $10\sqrt{2}$ and argument of $\tan^{-1}\frac{10}{10}$ *i.e* 45° or $\frac{\pi}{4}$ radians.

So phasor *A* in polar form will be $10\sqrt{2}\angle 45°$

Phasor *B* in polar form is $50\angle 120°$

Phasor *C* = (8 – j 6) have modulus of $\sqrt{8^2+(-6)^2}$ *i.e.* 10 and argument of $\tan^{-1}\frac{-6}{8}$ or – 36.87°

So **C** = $10\angle -36.87°$

$$\text{Hence } \frac{\mathbf{AB}}{\mathbf{C}} = \frac{10\sqrt{2}\angle 45°\times 50\angle 120°}{10\angle -36.87°}$$

$$= \frac{10\sqrt{2}\times 50}{10}\angle 45°+120°-(-36.87°) = 50\sqrt{2}\angle 201.87°$$ **Ans.**

In rectangular form $\frac{\mathbf{AB}}{\mathbf{C}}$ will be

$$= 50\sqrt{2}\cos 201.87° + j\,50\sqrt{2}\sin 201.87° = -65.622 - j\,26.34$$ **Ans.**

4.8. POWERS AND ROOTS OF PHASORS

Powers and roots of phasors can be determined conveniently in polar form. If the phasor is not in polar form then it will be advisable to convert it first to polar form and then carry out algebraic operations.

4.8.1. Powers of Phasors

Let us find the fourth power of the phasor $5\angle 20°$, for this purpose, the phasor has to be multiplied by itself four times.

$$\text{So } (5\angle 20°)^4 = 5 \times 5 \times 5 \times 5 \angle 20° + 20° + 20° + 20° = 625 \angle 80°$$

$$\text{In general } \mathbf{A}^n = A^n \angle n\,\theta$$

4.8.2. Roots of Phasors

Let us find the one-fourth power of the phasor $5\angle 20°$

$$(5\angle 20°)^{1/4} = (5)^{1/4} \angle 20°/4 = 1.496\angle 5° \textbf{ Ans.}$$

$$\text{In general } \sqrt[n]{\mathbf{A}} = \sqrt[n]{A}\angle \theta/n$$

4.9. PHASOR ALGEBRA APPLIED TO SINGLE PHASE SERIES CIRCUITS

A simple series circuit having non-inductive resistance R and inductive reactance X_L is shown in Fig. 4.8 (*a*). Let the current phasor be taken as reference phasor and be represented as $\mathbf{I} = I + j\,0$ or $I\angle 0°$.

Since voltage drop across resistance R, $V_R = IR$ in phase with current I so it is expressed as $(IR + j\,0)$ and the voltage drop across the inductive reactance, $V_L = I X_L$ in quadrature with current and leading, therefore, it is expressed as $(0 + j\,IX_L)$. Hence the voltage applied across the circuit can be expressed as

$$\mathbf{V} = \mathbf{V}_R + \mathbf{V}_L = (IR + j\,0) + (0 + j\,IX_L) = \mathbf{I}\,(R + j\,X_L) = \mathbf{I}\,\mathbf{Z}^* = IZ\angle\Phi$$

Hence impedance **Z** can be written as $(R + jX_L)$ or $Z\angle\Phi$ where $\Phi = \text{Tan}^{-1}\dfrac{X_L}{R}$.

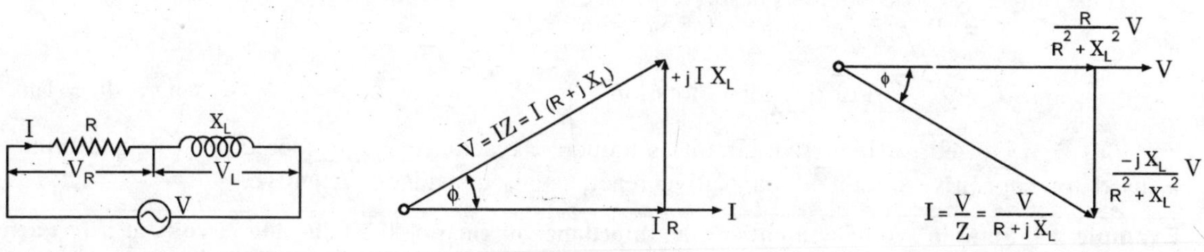

(*a*) *Circuit Diagram* (*b*) *Phasor Diagram* (*c*) *Phasor Diagram*

Fig. 4.8

Thus impedance can be written as complex quantity though it is not a complex quantity. Since impedance causes resistance and reactance voltage drops to be at right angle to each other so it is actually a complex operator.

Now let the applied voltage be taken as reference phasor and expressed as $V(1 + j\,0)$ or $V\angle 0°$ then current be determined by using impedance as phasor

$$\mathbf{I} = \frac{\mathbf{V}}{\mathbf{Z}} = \frac{V + j\,0}{R + jX_L} \times \frac{R - j\,X_L}{R - j\,X_L} = \frac{VR}{R^2 + X_L^2} - j\frac{VX_L}{R^2 + X_L^2} \quad \text{using complex notation}$$

* In the expression **V** = **IZ**, **V** and **I** differ from **Z** in that they are associated with time varying quantities, whereas **Z** is a complex number independent of time and is, therefore, not a phasor in the sense that **V** and **I** are phasors. It has, however, become the practice to refer all complex numbers used in ac circuit calculations as phasors.

and $\mathbf{I} = \frac{\mathbf{V}}{\mathbf{Z}} = \frac{V\angle 0°}{Z\angle\Phi} = \frac{V}{Z}\angle -\Phi$ where $\Phi = \text{Tan}^{-1}\frac{X_L}{R}$ using polar notation.

In Fig. 4.9 (*a*) circuit having resistance *R* and capacitive reactance X_C connected in series are shown.

Taking current *I* as reference phasor *i.e* along X-axis, as shown in phasor diagram [Fig. 4.9 (*b*)]

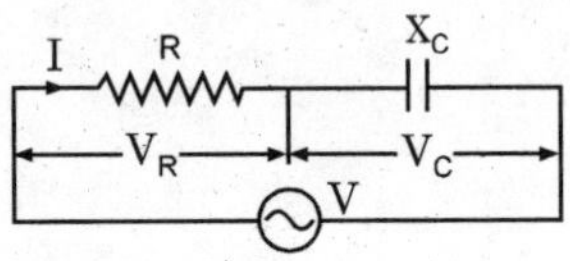

(a) Circuit Diagram

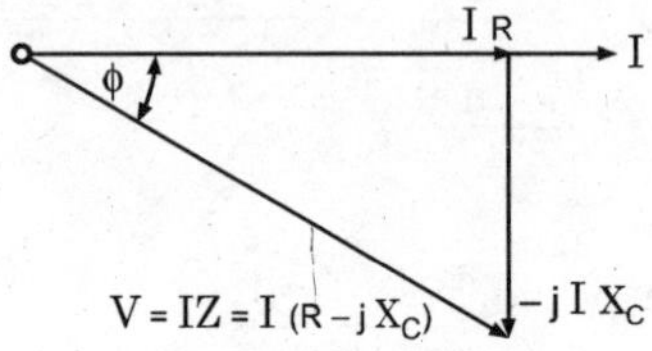

(b) Phasor Diagram

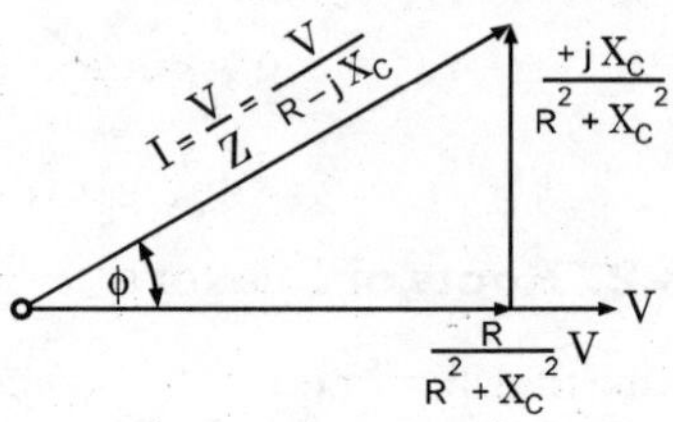

(c) Phasor Diagram

Fig. 4.9

Applied voltage $\mathbf{V} = \mathbf{V}_R + \mathbf{V}_C = (\mathbf{I}\,R + j\,0) + (0 - j\,\mathbf{I}\,X_C) = \mathbf{I}\,(R - j\,X_C) = \mathbf{I}\,\mathbf{Z} = I\,Z\angle -\Phi$

Hence impedance Z can be written as $(R - j\,X_C)$ or $Z\angle -\Phi$ where $\Phi = \text{Tan}^{-1}\frac{X_C}{R}$

Taking applied voltage as reference phasor, as shown in Fig. 4.9 (*c*), the current flowing through the circuit is given as

$$\mathbf{I} = \frac{\mathbf{V}}{\mathbf{Z}} = \frac{V + j\,0}{R - j\,X_C} \times \frac{R + j\,X_C}{R + j\,X_C} = \frac{VR}{R^2 + X_C^2} + j\frac{VX_C}{R^2 + X_C^2} \quad \text{using complex notation}$$

and $\mathbf{I} = \frac{V}{Z} = \frac{V\angle 0°}{Z\angle -\Phi} = \frac{V}{Z}\angle\Phi$ where $\Phi = \text{Tan}^{-1}\frac{X_C}{R}$ using polar notation.

Thus general expression for phasor impedance can be written as

$\mathbf{Z} = (R \pm j\,X)$ in complex notation

and $\mathbf{Z} = Z\angle \pm\Phi$ in polar notation where $\Phi = \text{Tan}^{-1}\frac{X}{R}$.

Plus sign is used for inductive circuit as inductive reactance is considered positive and minus sign is used for capacitive circuit as capacitive reactance is considered negative.

Example 4.9. State in symbolic notation, the impedance of each of the following circuits at a frequency of 50 Hz.

(*a*) A resistance of 20 Ω in series with an inductance of 0.1 H

(*b*) A resistance of 50 Ω in series with a capacitance of 40 μF

(c) Circuits (*a*) and (*b*) in series.

Solution : *For Circuit (a)*

Resistance, $R_1 = 20\ \Omega$

Reactance, $X_1 = 2\pi f L = 2\pi \times 50 \times 0.1 = 31.4\ \Omega$ (being inductive)

∴ Impedance, $\mathbf{Z}_1 = (R_1 + j\,X_1) = (20 + j\,31.4)\ \Omega$ **Ans.**

For Circuit (b)

Resistance, $R_2 = 50\ \Omega$

Reactance, $X_2 = \frac{1}{2\pi f C} = \frac{1}{2\pi \times 50 \times 40 \times 10^{-6}} = -79.6\ \Omega$ (being capacitive)

∴ Impedance, $\mathbf{Z}_2 = (50 - j\,79.6)\ \Omega$ **Ans.**

When circuits (*a*) and (*b*) are connected in series, total impedance of the circuit,

$$\mathbf{Z} = \mathbf{Z}_1 + \mathbf{Z}_2 = (20 + j\,31.4) + (50 - j\,79.6) = (70 - j\,48.2) \text{ ohms } \textbf{Ans.}$$

Example 4.10. The voltage and current of a circuit are given as $V_1 + j\,V_2$ and $I_1 - j\,I_2$ respectively. Determine the circuit impedance in symbolic notation.

Solution : Voltage, $\mathbf{V} = V_1 + j\,V_2$

Current, $\mathbf{I} = I_1 - j\,I_2$

$$\text{Impedance, } \mathbf{Z} = \frac{\mathbf{V}}{\mathbf{I}} = \frac{V_1 + j\,V_2}{I_1 - j\,I_2} = \frac{V_1 + j\,V_2}{I_1 - j\,I_2} \times \frac{I_1 + j\,I_2}{I_1 + j\,I_2}$$

$$= \frac{V_1 I_1 - V_2 I_2}{I_1^2 + I_2^2} + j\,\frac{V_2 I_1 + I_2 V_1}{I_1^2 + I_2^2} \quad \textbf{Ans.}$$

Example 4.11. A resistance of 4 Ω is connected in series with an inductive reactance of 5 Ω. Draw the impedance phasor for this circuit and express the impedance in the symbolic form and polar form.

Solution : Circuit resistance, R = 4 Ω

Circuit reactance, X = 5 Ω (inductive)

The impedance phasor of the circuit is shown in Fig. 4.10

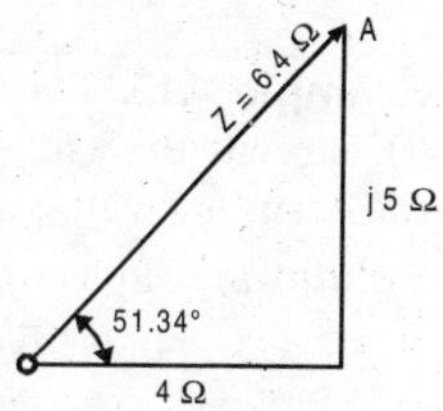

Fig. 4.10

$$\text{Magnitude of impedance, } Z = \sqrt{4^2 + 5^2} = 6.4\ \Omega$$

$$\text{Argument, } \theta = \text{Tan}^{-1}\frac{5}{4} = 51.34° \text{ or } 0.285\,\pi \text{ radians } \textbf{Ans.}$$

∴ Impedance, $\mathbf{Z} = (4 + j\,5)\ \Omega$ in symbolic form **Ans.**

or $\mathbf{Z} = 6.4 \angle 51.34°\ \Omega$ in polar form **Ans.**

Example 4.12. V = 200 ∠36.9° volts ; Z = 20 ∠–53.1 ohms. Find I.

Solution : Circuit voltage, $\mathbf{V} = 200 \angle 36.9°$ volts

Circuit impedance, $\mathbf{Z} = 20 \angle -53.1°$ ohms

$$\text{Circuit current, } \mathbf{I} = \frac{\mathbf{V}}{\mathbf{Z}} = \frac{200 \angle 36.9°}{20 \angle -53.1°} = \frac{200}{20} \angle 36.9° - (-53.1°) = 10 \angle 90° \text{ A } \textbf{Ans.}$$

Example 4.13. Determine the current flowing through the circuit having impedance of (4 + *j* 3) Ω connected across 250 V, 50 Hz supply.

Solution : Let the voltage phasor be taken as reference phasor and, therefore, be represented as

$\mathbf{V} = (250 + j\,0)$ volts

Circuit impedance, $\mathbf{Z} = (4 + j\,3)$ ohms

$$\text{Circuit current, } \mathbf{I} = \frac{\mathbf{V}}{\mathbf{Z}} = \frac{250 + j\,0}{(4 + j\,3)} = \frac{(250 + j\,0)(4 - j\,3)}{(4 + j\,3)(4 - j\,3)} = \frac{1{,}000 - j\,750}{16 + 9} = (40 - j\,30) \text{ A } \textbf{Ans.}$$

Example 4.14. A coil of resistance 10 Ω and inductance 0.1 H is connected in series with a condenser of capacitance 150 μ F across a 200 V, 50 Hz supply. Determine the following:

(*i*) Impedance (*ii*) Current (*iii*) Power factor (*iv*) Voltage across the coil; and (*v*) Voltage across the condenser.

Solution : Resistance of the circuit, R = 10 Ω

Inductive reactance of the circuit, $X_L = 2\pi f L = 2\pi \times 50 \times 0.1 = 31.42\ \Omega$

$$\text{Capacitive reactance of the circuit, } X_C = \frac{1}{2\pi f C} = \frac{1}{2\pi \times 50 \times 150 \times 10^{-6}} = 21.22\ \Omega$$

Magnitude of impedance, $Z = \sqrt{R^2 + (X_L - X_C)^2} = \sqrt{(10^2) + (31.42 - 21.22)^2} = 14.28\ \Omega$

Argument, $\theta = \text{Tan}^{-1} \dfrac{X_L - X_C}{R} = \text{Tan}^{-1} \dfrac{31.42 - 21.22}{10} = 45.57°$

(*i*) Hence impedance, $Z = (10 + j\ 10.2)$ or $14.28 \angle 45.57°\ \Omega$ **Ans.**

(*ii*) Current, $\mathbf{I} = \dfrac{\mathbf{V}}{\mathbf{Z}} = \dfrac{200 \angle 0°}{14.28 \angle 45.57°} = 14 \angle{-45.57°}$ A **Ans.**

(*iii*) Power factor = $\cos \Phi = \cos 45.57° = 0.7$ (lagging) **Ans.**

(*iv*) Voltage across the coil = $\mathbf{I}\ (R + jX_L)$

$$= 14 \angle{-45.57°} \times (10 + j\ 31.42)$$
$$= 14 \angle{-45.57°} \times 32.97 \angle 72.34°$$
$$= 14 \times 32.97 \angle{-45.57° + 72.34°}$$
$$= 461.58 \angle 26.77° \text{ volts}$$ **Ans.**

(*v*) Voltage across the capacitor = $I X_C = 14 \angle{-45.57°} \times 21.22 \angle{-90°}$

$$= 297.08 \angle{-135.57°} \text{ volts}$$ **Ans.**

Example 4.15. The circuit shown in Fig. 4.11 (*a*) is connected to a 230 V, 50 Hz supply. Find the values of (*i*) current drawn; (*ii*) voltages V_1 and V_2 and (*iii*) power factor. Draw also a phasor diagram indicating the terminal voltages V_1, V_2 and supply voltage with respect to current.

Solution : Inductive reactance corresponding to 0.04 H = $j\, 2\pi \times 50 \times 0.04 = j\ 12.566\ \Omega$

Inductive reactance corresponding to 0.01 H, = $j\, 2\pi \times 50 \times 0.01 = j\ 3.141\ \Omega$

Capacitive reactance corresponding to 100 μF capacitance

$$= \frac{1}{j 2\pi \times 50 \times 100 \times 10^{-6}} = -j\ 31.831\ \Omega$$

Total series impedance, $\mathbf{Z} = 15 + j\ 12.566 + 10 + j\ 3.141 - j\ 31.831$

$$= 25 - j\ 16.124 = 29.749 \angle{-32.82°}\ \Omega$$

(*i*) Current drawn, $\mathbf{I} = \dfrac{\mathbf{V}}{\mathbf{Z}} = \dfrac{230 \angle 0°}{29.749 \angle{-32.82°}} = 7.73 \angle 32.82°$ A **Ans.**

(*ii*) Voltage, $V_1 = \mathbf{I}\ (15 + j\ 12.566)$

$$= 7.73 \angle 32.82° \times 19.57 \angle 39.95° = 151.28 \angle 72.77° \text{ V}$$ **Ans.**

Voltage, $V_2 = \mathbf{I}\ (10 + j\ 3.141 - j\ 31.831)$

$$= 7.73 \angle 32.82° \times 30.38 \angle{-70.78°} = 234.83 \angle{-37.96°} \text{ V}$$ **Ans.**

(*iii*) Power factor = $\cos \Phi = \cos 32.82° = 0.84$ (lead) **Ans.**

The phasor diagram is shown in Fig. 4.11 (*b*).

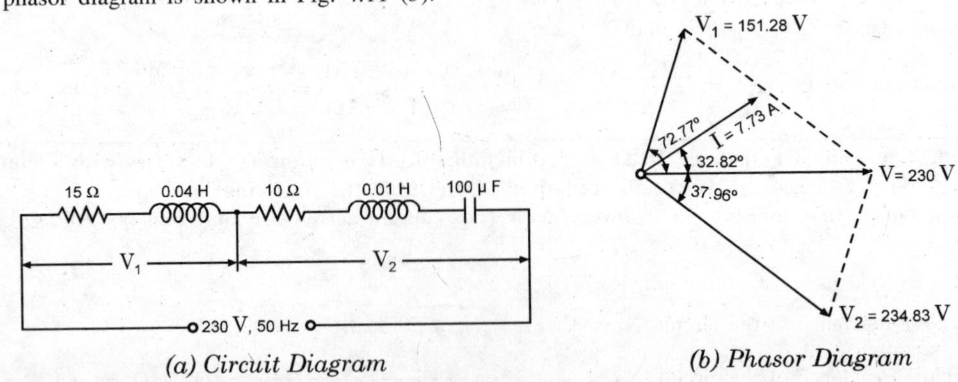

(*a*) Circuit Diagram (*b*) Phasor Diagram

Fig. 4.11

Example 4.16. In the circuit of Fig. 4.12, applied voltage is given by $(0 + j10)$ and the current is $(0.8 + j\,0.6)$ A. Determine the values of R and X and also indicate if X is inductive or capacitive.

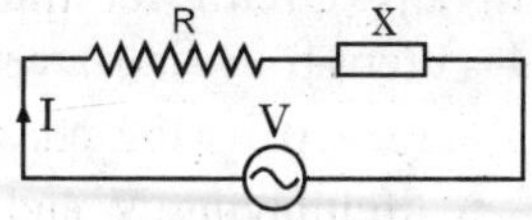

Fig. 4.12

Solution : Applied voltage, $\mathbf{V} = (0 + j\,10) = 10\angle 90^\circ$ volts

Circuit current, $\mathbf{I} = (0.8 + j\,0.6) = 1\angle 36.9^\circ$ A

Current lags behind applied voltage by

$$\Phi = (90^\circ - 36.9^\circ) \text{ i.e. } 53.1^\circ$$

Hence the circuit shown is inductive.

$$\text{Circuit impedance, } \mathbf{Z} = \frac{\mathbf{V}}{\mathbf{I}} = \frac{10\angle 90^\circ}{1\angle 36.9^\circ} = 10\angle 53.1^\circ = (6 + j\,8)\,\Omega$$

Hence R = 6 Ω and X = 8 Ω (inductive) **Ans.**

4.10. POWER DETERMINATION

If the voltage and current in a circuit are expressed in complex form, power can be determined as follows :

Let the current be expressed as $\mathbf{I} = I_1 - j\,I_2$ and voltage as $\mathbf{V} = V_1 + j\,V_2$.

From Fig. 4.13 it is evident that the real component of voltage, V_1 and real component of current I_1 are in phase. Since components of voltage and current may be treated as if they were acting alone, the power contributed by V_1 and I_1, $P_1 = V_1 I_1$.

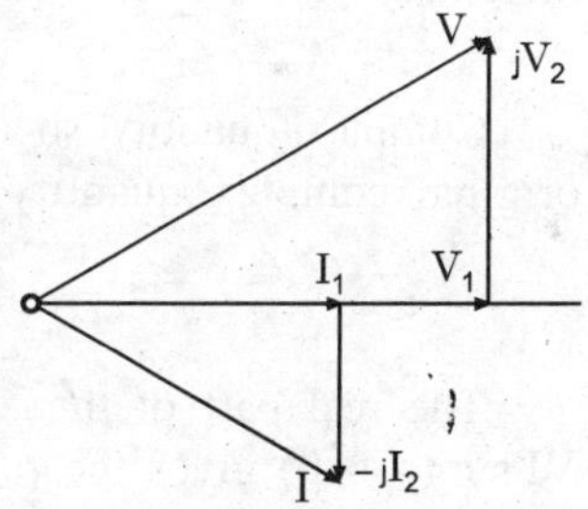

Fig. 4.13

The coefficient of V_2 is $+j$ and coefficient I_2 is $-j$, these two components of voltage and current are opposite in phase and, therefore, their product gives negative power.

$$\text{So power } P = -V_2 I_2$$

Voltage component V_1 and current component I_2 are in quadrature so power contributed by V_1 and I_2 is zero, similarly voltage component V_2 and current component I_1 are in quadrature and, therefore, contribute no power.

$$\therefore \text{ Total power in the circuit, } P = P_1 + P_2 = V_1 I_1 - V_2 I_2$$

The total power is the algebraic sum of the product of the real quantities V_1 and I_1 and the product of imaginary quantities V_2 and I_2. The sign being determined in the ordinary algebraic method. The operator j must not be included in the multiplication.

Example 4.17. Determine the power consumed in a circuit carrying a current of $(3 - j\,4)$ A and connected across $(80 + j\,40)$ volt supply mains.

Solution : Supply voltage, $\mathbf{V} = (80 + j\,40)$ volts

Circuit current, $\mathbf{I} = (3 - j\,4)$ amperes

Power consumed, $P = V_1 I_1 + V_2 I_2 = 80 \times 3 + 40\,(-4) = 80$ watts **Ans.**

Example 4.18. An ac voltage of $(160 + j\,120)$ volts is applied across a circuit and circuit current is $(5 + j6)$ amperes. Determine the power in the circuit.

Solution : Voltage applied across the circuit,

$$\mathbf{V} = (160 + j\,120) \text{ volts}$$

Circuit current, $\mathbf{I} = (5 + j\,6)$ amperes

Power consumed, $P = V_1 I_1 + V_2 I_2 = 160 \times 5 + 120 \times 6 = 1{,}520$ W **Ans.**

4.11. CONJUGATE METHOD OF POWER DETERMINATION

In any circuit let the applied voltage and current flowing be $\mathbf{V}$ = V (cos α + *j* sin α) and $\mathbf{I}$ = I (cos β + *j* sin β) respectively.

(α – β) is the phase difference between V and I so power factor of the circuit is cos (α – β)

Multiplying **V** and **I** we have

$$\mathbf{VI} = V(\cos\alpha + j\sin\alpha) \times I(\cos\beta + j\sin\beta)$$
$$= VI\,[\cos(\alpha+\beta) + j\sin(\alpha+\beta)]$$

The real part of the above expression is VI cos (α + β) and evidently it cannot be true power of the circuit, because power factor of the circuit is cos (α – β). Similarly VI sin (α + β) is not correct expression of the reactive volt-amperes. Hence to determine the power by *j* method, it is necessary to change the sign of the imaginary part of either V or I. Thus if we change the sign of imaginary part of V, leaving I as before, we have

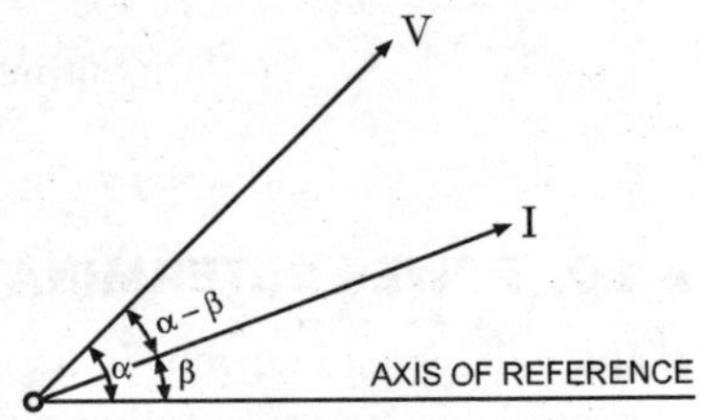

Fig. 4.14

$$\mathbf{V_C} = V(\cos\alpha - j\sin\alpha)$$

Complex quantity so obtained is known as *conjugate* of the original complex quantity.

$$\text{Now } \mathbf{V_C I} = V(\cos\alpha - j\sin\alpha) \times I(\cos\beta + j\sin\beta)$$
$$= VI\,[\cos(\alpha-\beta) - j\sin(\alpha-\beta)]$$

The real part of the above expression VI cos (α – β) gives the true power and imaginary part VI sin (α – β) gives the reactive power. It should, however, be noted that the real power obtained by the conjugate method is the same regardless of whether V or I is reversed but sign of reactive volt-amperes will depend upon it. When the conjugate of voltage is used, capacitive VARs have a positive sign and inductive VARs a negative sign. If the conjugate of the current rather than of the voltage is used the capacitive VARs have a negative sign and inductive VARs a positive sign.

Example 4.19. In an alternating circuit, the impressed voltage is given by V = (100 – *j* 50) volts, and the current in the circuit is I = (3 – *j* 4) A. Determine the real and reactive power in the circuit.

Solution :

Impressed voltage, $\mathbf{V} = (100 - j\,50)$

Conjugate of current, $\mathbf{I_C} = (3 + j\,4)$ amperes

$$\mathbf{V\,I_C} = (100 - j\,50)(3 + j\,4) = 500 + j\,250$$

True power = Real part of $\mathbf{V\,I_C}$ = 500 watts **Ans.**

Reactive power = *j* part of $\mathbf{V\,I_C}$ = 250 VAR (inductive) **Ans.**

Example 4.20. The current in a circuit is given by (5 + *j* 10) A when the applied voltage is (225 + *j* 150) V. Determine (*i*) the complex expression for the impedance stating whether it is an inductive or capacitive (*ii*) power (*iii*) phase angle between current and voltage. Draw the complete phasor diagram.

Solution :

Circuit voltage, $\mathbf{V} = (225 + j\,150)$ volts or $270.42\angle 33.7°$ V

Circuit current, $\mathbf{I} = (5 + j\,10)$ amperes or $11.18\angle 63.4°$ A

$$\text{Circuit impedance, } \mathbf{Z} = \frac{\mathbf{V}}{\mathbf{I}} = \frac{270.42}{11.18}\angle 33.7° - 63.4° = 24.2\angle -29.7°\ \Omega$$

$$\text{or } \mathbf{Z} = 24.2\,[\cos(-29.7°) + j\sin(-29.7°)] = (21 - j\,12)\ \Omega \textbf{ Ans.}$$

(*i*) Since the *j* part of the impedance expression is negative, the impedance is capacitive. **Ans.**

(*ii*) Conjugate current $\mathbf{I_C} = (5 - j\,10)$

$$\mathbf{V}\,\mathbf{I}_C = (225 + j\,150)(5 - j\,10)$$
$$= (2{,}625 - j\,1{,}500)$$

Power of the circuit, P = Real part of $\mathbf{V}\,\mathbf{I}_C$ = 2,625 W **Ans.**

Power of the circuit can also be determined as follows :

$$P = VI\cos\Phi$$
$$= 270.42 \times 11.18\cos(63.4° - 33.7°)$$
$$= 270.42 \times 11.18\cos 29.7° = 2,625 \text{ W } \textbf{Ans.}$$

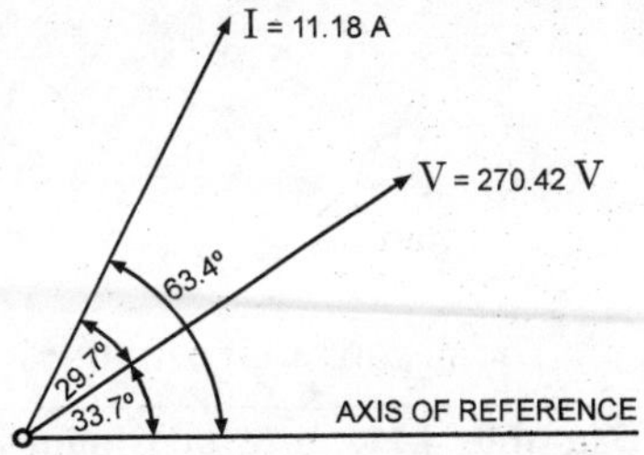

Fig. 4.15

(*iii*) Phase angle between current and voltage,

$$\Phi = \text{Argument of current} - \text{argument of voltage}$$
$$= 63.4° - 33.7° = 29.7° \text{ (leading) } \textbf{Ans.}$$

Phasor diagram is shown in Fig. 4.15.

4.12. PHASOR ALGEBRA APPLIED TO SINGLE PHASE PARALLEL CIRCUITS

When two or more circuits are connected in parallel, the current in each branch may be found in complex form and then total current in the circuit may be determined by addition of complex numbers, as detailed in Art. 4.6.

In the circuit shown in Fig. 4.16.

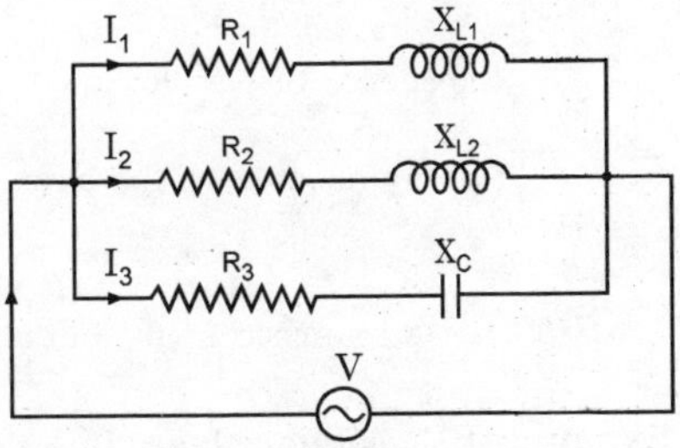

Fig. 4.16

Current in branch 1, $\mathbf{I}_1 = \dfrac{\mathbf{V}}{R_1 + jX_{L_1}}$

$$= \frac{\mathbf{V}}{R_1 + jX_{L_1}} \times \frac{R_1 - jX_{L_1}}{R_1 - jX_{L_1}}$$

$$= \mathbf{V}\left(\frac{R_1}{R_1^2 + X_{L_1}^2} - j\frac{X_{L_1}}{R_1^2 + X_{L_1}^2}\right)$$

Similarly current in branch 2, $\mathbf{I}_2 = \mathbf{V}\left(\dfrac{R_2}{R_2^2 + X_{L_2}^2} - j\dfrac{X_{L_2}}{R_2^2 + X_{L_2}^2}\right)$

and current in branch 3, $\mathbf{I}_3 = \mathbf{V}\left(\dfrac{R_3}{R_3^2 + X_C^2} + j\dfrac{X_C}{R_3^2 + X_C^2}\right)$

Total circuit current, $\mathbf{I} = \mathbf{I}_1 + \mathbf{I}_2 + \mathbf{I}_3$

4.13. EQUIVALENT PARALLEL IMPEDANCE

The ac parallel circuit may be solved in the same way as dc parallel circuit except that complex impedances rather than simple resistances are involved. With several impedances connected in parallel the equivalent impedance is expressed as

$$\frac{1}{\mathbf{Z}} = \frac{1}{\mathbf{Z}_1} + \frac{1}{\mathbf{Z}_2} + \frac{1}{\mathbf{Z}_3} + \ldots$$

Since admittance is reciprocal of impedance,

$$\mathbf{Y} = \frac{1}{\mathbf{Z}} = \mathbf{Y}_1 + \mathbf{Y}_2 + \mathbf{Y}_3 + \ldots$$

After finding equivalent impedance or admittance, current in the circuit may be determined from the relation

$$I = \frac{V}{Z} = V \times Y$$

where $$Y = \frac{1}{Z} = \frac{1}{(R \pm jX)} = \frac{R \mp jX}{(R \pm jX)(R \mp jX)} = \frac{R}{R^2 + X^2} \mp j\frac{X}{R^2 + X^2} = G \mp B$$

The negative sign is used for inductive susceptance and positive sign for capacitive susceptance.

Example 4.21. A 15 mH inductor is in series with a parallel combination of an 80 Ω resistor and 20 μF capacitor. If the angular frequency of the applied voltage is ω = 1,000 radians/second, find the admittance of the network.

Solution : Inductive reactance, $X_L = \omega L = 1{,}000 \times 15 \times 10^{-3} = 15\ \Omega$

Capacitive reactance, $X_C = \frac{1}{\omega C} = \frac{1}{1{,}000 \times 20 \times 10^{-6}} = 50\ \Omega$

$R = 80\ \Omega$

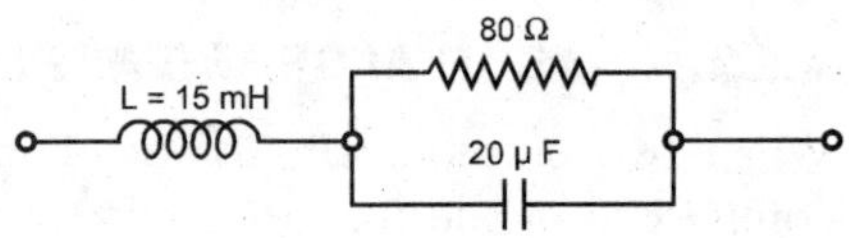

Fig. 4.17

Admittance of parallel circuit, $Y_P = \frac{1}{80} + j\frac{1}{50}$

$= (0.0125 + j\ 0.02)$ S

Impedance of parallel circuit, $Z_P = \frac{1}{Y_P} = \frac{1}{(0.0125 + j\ 0.02)} = (22.48 - j\ 35.96)\ \Omega$

Total impedance of the circuit, $Z = (0 + j\ X_L) + Z_P$

$= (0 + j15) + (22.48 - j\ 35.96) = (22.48 - j\ 20.96)\ \Omega$

Admittance of the circuit, $Y = \frac{1}{Z} = \frac{1}{22.48 - j20.96} = (0.238 + j\ 0.022)$ siemens **Ans.**

Example 4.22. An impedance (6 + j8) is connected across 200 V, 50 Hz mains in parallel with another circuit having an impedance of (8 – j6) Ω, calculate

(a) the admittance, the conductance, the susceptance of the combined circuit.
(b) the total current taken from the mains and its pf.

Solution : Admittance, $Y_1 = \frac{1}{Z_1} = \frac{1}{6 + j8} = (0.06 - j\ 0.08)$ siemen

Admittance, $Y_2 = \frac{1}{Z_2} = \frac{1}{8 - j6} = (0.08 + j\ 0.06)$ siemen

Combined admittance, $Y = Y_1 + Y_2 = (0.06 - j\ 0.08) + (0.08 + j\ 0.06)$

$= (0.14 - j\ 0.02)$ or $0.1414 \angle -8.13^\circ$ siemen **Ans.**

Conductance, G = 0.14 siemen **Ans.**

Susceptance, B = 0.02 siemen (inductive) **Ans.**

Let supply voltage be $V = 200 \angle 0^\circ$ volts

Total current $I = V\ Y = 200 \angle 0^\circ \times 0.1414 \angle -8.13^\circ = 28.28 \angle -8.13^\circ$ A **Ans.**

Power factor, $\cos \Phi = \text{Cos}\ (-8.13^\circ) = 0.99$ (lag) **Ans.**

Example 4.23. Two impedances $Z_1 = (10 + j15)$ ohms and $Z_2 = (6 - j8)$ ohms are connected in parallel and supply current is 20 A. What is power dissipation in each branch ?

[Pb. Technical Univ. Basic Electrical and Electronics Engineering Dec. 2007]

Solution : Current distribution is made by current distribution rule and accordingly

$$I_1 = \frac{Z_2}{Z_1 + Z_2} \times I = \frac{(6 - j8)}{(10 + j15) + (6 - j8)} \times (20 + j0)\ A$$

Taking current phasor as reference phasor

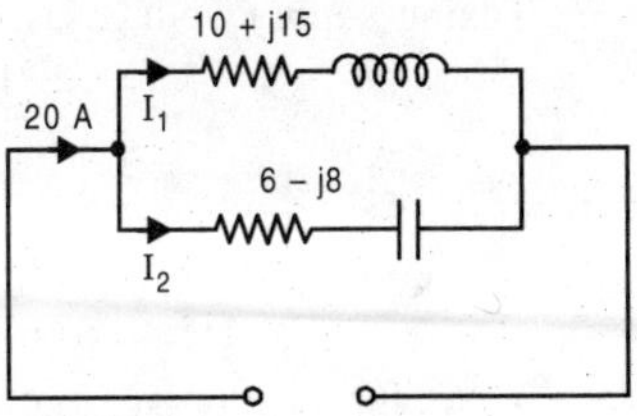

Fig. 4.18

$$= \frac{6 - j8}{(16 + j7)}(20 + j0)$$

$$= \frac{10 \angle -53.13^\circ}{17.464 \angle 23.63^\circ} \times 20 \angle 0^\circ = 11.45 \angle -76.76^\circ \text{ A}$$

and $$I_2 = \frac{\mathbf{Z}_1}{\mathbf{Z}_1 + \mathbf{Z}_2} \times \mathbf{I} = \frac{10 + j15}{(10 + j15) + (6 - j8)} \times 20 \angle 0^\circ \text{ A}$$

$$= \frac{18.028 \angle 56.3^\circ}{17.464 \angle 23.63^\circ} \times 20 \angle 0^\circ \text{ A} = 20.646 \angle 32.67^\circ \text{ A}$$

Power dissipation in inductive branch = $I_1^2 R_1 = (11.45)^2 \times 10 = 1{,}311$ W or 1.311 kW **Ans.**

Power dissipation in capacitive branch = $I_2^2 R_2 = (20.646)^2 \times 6 = 2{,}558$ W or 2.558 kW **Ans.**

Example 4.24. Consider the circuit shown in Fig. 4.19. Find the steady state current flowing through each coil. Estimate the impedance and the resistance of a single coil which will draw the same current at the same power factor from the line.

Solution : Impedance of branch I, $\mathbf{Z}_1 = R_1 + j\,2\pi f L_1 = 5 + j\,2\pi \times 50 \times 0.02 = (5 + j\,6.283)\ \Omega$

Impedance of branch II, $\mathbf{Z}_2 = 1 + j2\pi \times 50 \times 0.08 = (1 + j\,25.133)\ \Omega$

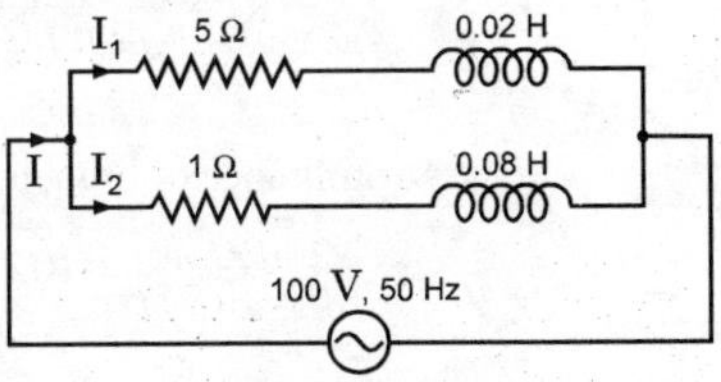

Fig. 4.19

Admittance of branch I,

$$\mathbf{Y}_1 = \frac{1}{\mathbf{Z}_1} = \frac{1}{5 + j\,6.283} = (0.0775 - j\,0.0974) \text{ S}$$

Admittance of branch II,

$$\mathbf{Y}_2 = \frac{1}{\mathbf{Z}_2} = \frac{1}{1 + j25.133} = (0.00158 - j\,0.0397) \text{ S}$$

Total admittance $\mathbf{Y} = \mathbf{Y}_1 + \mathbf{Y}_2$

$$= (0.0775 - j\,0.0974) + (0.00158 - j\,0.0397) = (0.07908 - j\,0.1371) \text{ S}$$

Current through coil I, $\mathbf{I}_1 = \mathbf{V} \times \mathbf{Y}_1 = (100 + j\,0)(0.0775 - j\,0.0974)$

$$= (7.75 - j\,9.74) \text{ or } 12.45 \text{ A} \angle -51.49^\circ$$ A **Ans.**

Current through coil II, $\mathbf{I}_2 = \mathbf{V} \times \mathbf{Y}_2 = (100 + j\,0)\ (0.00158 - j\,0.0397)$

$$= (0.158 - j\,3.97) \text{ or } 3.973 \angle -87.72^\circ$$ A **Ans.**

Admittance of a single coil which will draw the same current and at same power factor from the line $= \mathbf{Y} = (0.07908 - j\,0.1371)$ S.

Impedance of the coil, $\mathbf{Z} = \frac{1}{\mathbf{Y}} = \frac{1}{0.07908 - j0.1371} = (3.1569 + j\,5.473)$ or 6.32 Ω **Ans.**

Resistance of coil, R = 3.1569 Ω **Ans.**

Example 4.25. Two single phase load circuits, one taking 15 kW at 0.6 power factor lagging and the other 25 kW at 0.75 pf lagging, are connected in parallel to a 200 V, 50 Hz supply. Determine the total current drawn from the supply and the combined power factor of the two loads in parallel.

Solution : Current drawn by 15 kW load circuit at 0.6 power factor

$$I_1 = \frac{15{,}000}{200 \times 0.6} = 125 \text{ A}$$

Current drawn by 25 kW load circuit at 0.75 power factor,

$$I_2 = \frac{25{,}000}{200 \times 0.75} = 166.67 \text{ A}$$

Taking supply voltage as reference phasor

$$\mathbf{V} = (200 + j\,0) \text{ volts}$$

$$\Phi_1 = \text{Cos}^{-1}\,0.6 = 53.13° \text{ (lagging)}$$

$$\text{and } \Phi_2 = \text{Cos}^{-1}\,0.75 = 41.4° \text{(lagging)}$$

$$\therefore\ \mathbf{I}_1 = 125\ (\cos 53.13° - j \sin 53.13°) = 125\ (0.6 - j\,0.8) = (75 - j100)\text{A}$$

$$\text{and } \mathbf{I}_2 = 166.67(\cos 41.4° - j \sin 41.4°) = 166.67(0.75 - j\,0.6614) = (125 - j110.24)\text{ A}$$

Total current drawn from the supply,

$$\mathbf{I} = \mathbf{I}_1 + \mathbf{I}_2 = (75 - j\,100) + (125 - j\,110.24) = (200 - j\,210.24)\text{ A}$$

$$\text{Magnitude of total current, I} = \sqrt{(200)^2 + (-210.24)^2} = 290.2 \textbf{ A Ans.}$$

$$\text{Phase angle, } \Phi = \text{Tan}^{-1}\frac{-210.24}{200} = -46.4°$$

$$\text{Combined power factor} = \text{Cos } \Phi = \text{Cos } 46.4° = 0.6892 \textbf{ (lagging) Ans.}$$

Example 4.26. For the circuit shown in Fig. 4.20 calculate (*a*) the current in each branch (*b*) total current taken (*c*) total power consumed by circuit (*d*) reactive power consumed/generated (*e*) power factor of the circuit (*f*) phasor diagram of the circuit. [*Pb. Technical Univ. Basic Electrical Engineering June* 2003]

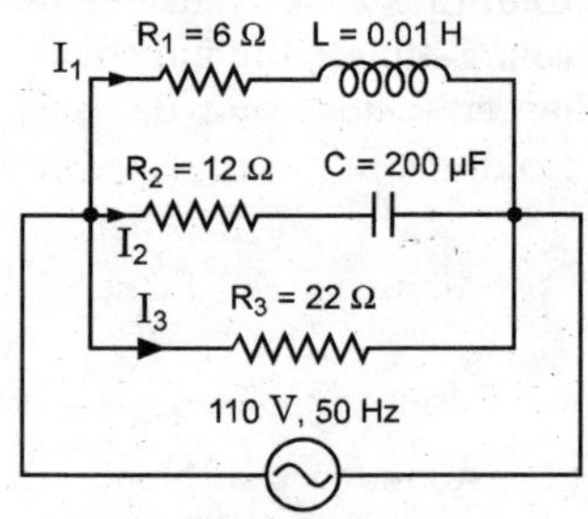

Fig. 4.20

Solution : Impedance of branch 1,

$$\mathbf{Z}_1 = \text{R}_1 + j\,2\,\pi f \text{L} = 6 + j\,2\,\pi \times 50 \times 0.01 = (6 + j\,3.14)\ \Omega$$

Impedance of branch 2,

$$\mathbf{Z}_2 = \text{R}_2 + \frac{1}{j\,2\pi f \text{C}} = 12 + \frac{1}{j \times 2\pi \times 50 \times 200 \times 10^{-6}}$$

$$= (12 - j15.92)\ \Omega$$

Admittance of branch 1,

$$\mathbf{Y}_1 = \frac{1}{\mathbf{Z}_1} = \frac{1}{(6 + j3.14)} = (0.1308 - j0.0685)\text{ S} = 0.1477\ \angle -27.64°\text{ S}$$

$$\text{Admittance of branch 2, } \mathbf{Y}_2 = \frac{1}{\mathbf{Z}_2} = \frac{1}{12 - j15.92} = (0.03 + j0.04)\text{ S} = 0.05\ \angle 53.13°\text{ S}$$

$$\text{Admittance of branch 3, } \mathbf{Y}_3 = \frac{1}{\mathbf{Z}_3} = \frac{1}{22} = 0.04545\text{ S}\ \angle 0°$$

$$\text{Current in branch 1, } \mathbf{I}_1 = \mathbf{V}\,\mathbf{Y}_1 = 110\ \angle 0° \times 0.1477\ \angle -27.64° = 16.25\ \angle -27.64° = (14.4 - j7.54)\text{ A}$$

$$\text{Current in branch 2, } \mathbf{I}_2 = \mathbf{V}\,\mathbf{Y}_2 = 110\ \angle 0° \times 0.05\ \angle 53.13° = 5.5\ \angle 53.13° = (3.3 + j4.4)\text{ A}$$

$$\text{Current in branch 3, } \mathbf{I}_3 = \mathbf{V}\,\mathbf{Y}_3 = 110 \times 0.04545 = 5\ \angle 0° = (5 + j0)\text{ A}$$

(*a*) Thus current in branch 1, I_1 = 16.25 A lagging behind applied voltage by 27.64° **Ans.**

Current in branch 2, I_2 = 5.5 A leading the applied voltage by 53.13° **Ans.**

Current in branch 3, I_3 = 5 A in phase with applied voltage **Ans.**

(*b*) Total admittance of the circuit,

$$\mathbf{Y} = \mathbf{Y}_1 + \mathbf{Y}_2 + \mathbf{Y}_3 = [(0.1308 - j\,0.0685) + (0.03 + j0.04) + 0.04545)]\text{ S}$$

$$= (0.2063 - j0.0285) = 0.208\ \angle -7.866°\text{ S}$$

$$\text{Total current taken, I} = \mathbf{V} \times \mathbf{Y} = 110\ \angle 0° \times 0.208\ \angle -7.866° = 22.91\ \angle -7.866°\text{ A } \textbf{Ans.}$$

i.e. Total current drawn, I = 22.91 A lagging behind the applied voltage by 7.866° **Ans.**

(*e*) Power factor of the circuit, *pf* = Cos ϕ = Cos 7.866° = 0.99 (lagging) **Ans.**

(*c*) Total power consumed by the circuit,

$$\text{P} = \text{VI}\cos\phi = 110 \times 22.91 \times 0.99 = 2{,}496\text{ W or } 2.496\text{ kW } \textbf{Ans.}$$

(*d*) Reactive power, Q = VI sin ϕ = 110 × 22.91 × sin 7.866° = 344.9 VAR (lagging) **Ans.**

(*f*) Phasor diagram is shown in Fig. 4.21. Voltage phasor has been taken as reference phasor.

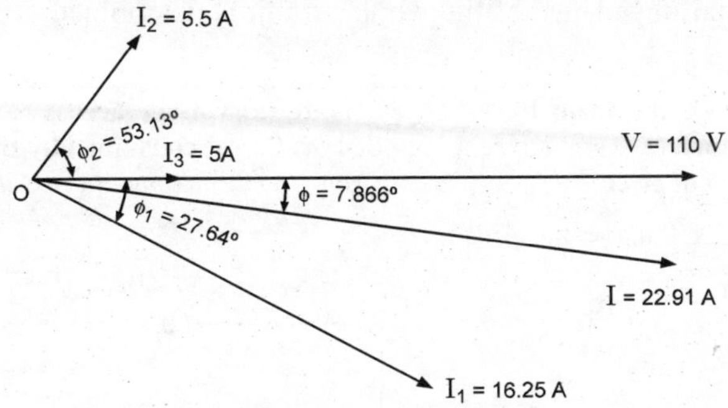

Fig. 4.21. ***Phasor Diagram***

Example 4.27. For the circuit of Fig. 4.22 calculate the current supplied by the voltage source and voltage across the current source. *[Pb. Technical Univ. Electrical Engineering May 2001]*

Solution : The current source of $2.83\angle 45°$ A is converted into its equivalent voltage source. The circuit so obtained is shown in Fig. 4.23.

$$\mathbf{Z}_{in} = 50 \parallel j\,25 = \frac{50 \times j25}{50 + j25} = \frac{1{,}250\angle 90°}{55.9\angle 26.56°} = 22.36\angle 63.44° = 10 + j\,20\ \Omega$$

$$\mathbf{V}_2 = 2.83\angle 45° \times 22.36\angle 63.44°\ \text{V} = 63.28\angle 108.44°\ \text{V} = (-20 + j\,60)\ \text{V}$$

Current supplied by voltage source,

$$\mathbf{I} = \frac{100\angle 0° - (-20 + j60)}{(10 + j20) + 50}\ \text{A} = \frac{120 - j60}{60 + j20}\ \text{A} = \frac{134.16\angle -26.56°}{63.25\angle 18.43°} = 2.12\angle -45°\ \text{A} \quad \textbf{Ans.}$$

Voltage across current source, $V_1 = \mathbf{I}\,Z_{in}$

$$= 2.12\angle -45° \times 22.36\angle 63.44° = 47.43\angle 18.44°\ \text{V} \quad \textbf{Ans.}$$

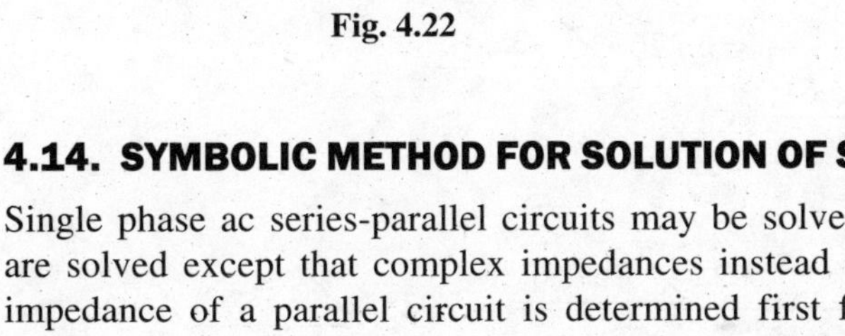

Fig. 4.22

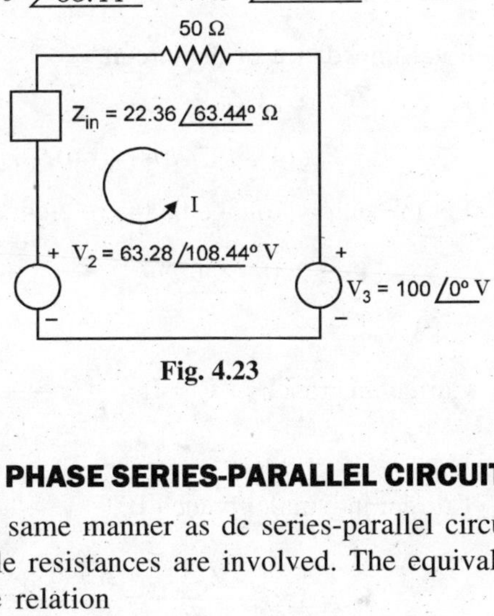

Fig. 4.23

4.14. SYMBOLIC METHOD FOR SOLUTION OF SINGLE PHASE SERIES-PARALLEL CIRCUITS

Single phase ac series-parallel circuits may be solved in the same manner as dc series-parallel circuits are solved except that complex impedances instead of simple resistances are involved. The equivalent impedance of a parallel circuit is determined first from the relation

$$\frac{1}{\mathbf{Z}_{eq}} = \frac{1}{\mathbf{Z}_1} + \frac{1}{\mathbf{Z}_2} + \frac{1}{\mathbf{Z}_3} + \ldots$$

then it is added (in complex form) to the impedance of the circuit connected in series with the combined parallel circuit and thus total impedance of the whole circuit is determined. The problems can be solved as discussed before.

Example 4.28. For the circuit shown in Fig. 4.24, calculate (*a*) voltage across each branch (A), (B) and (C) (*b*) current through each branch (*c*) Total current supplied (*d*) Power consumed by the circuit (*e*) Power factor of the circuit (*f*) Phasor diagram of circuit. [*Pb. Technical Univ. Electrical Engineering December 2002*]

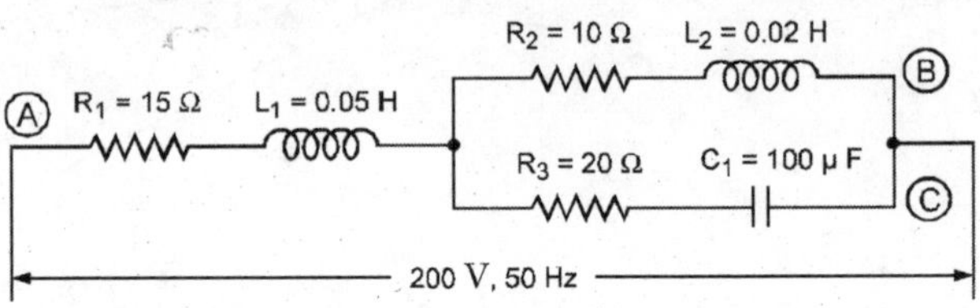

Fig. 4.24

Solution : Impedance of branch B of parallel circuit,

$$\mathbf{Z_B} = R_2 + j\,2\,\pi f L_2$$
$$= 10 + j\,2\,\pi \times 50 \times 0.02$$
$$= (10 + j\,6.283)\ \Omega$$
$$= 11.81\ \angle 32.14°\ \Omega$$

Impedance of branch C of parallel circuit,

$$\mathbf{Z_C} = R_3 - \frac{j}{2\pi f C} = 20 - \frac{j}{2\pi \times 50 \times 100 \times 10^{-6}} = (20 - j\,31.83)\ \Omega = 37.592\ \angle -57.86°\ \Omega$$

Impedance of series branch A,

$$\mathbf{Z_A} = R_1 + j\,2\pi f L_1 = 15 + j\,2\pi \times 50 \times 0.05 = (15 + j\,15.708)\ \Omega = 21.72\ \angle 46.32°\ \Omega$$

Equivalent impedance of parallel circuit,

$$\mathbf{Z_{BC}} = \frac{\mathbf{Z_B Z_C}}{\mathbf{Z_B + Z_C}} = \frac{11.81\angle 32.14° \times 37.592\angle -57.86°}{(10 + j\,6.283) + (20 - j31.83)}\ \Omega$$

$$\text{or}\quad \mathbf{Z_{BC}} = \frac{11.81 \times 37.592\angle -25.72°}{30 - j25.547} = \frac{11.81 \times 37.592\angle -25.72°}{39.4\angle -40.42°} = 11.268\ \angle 14.7°\ \Omega$$

$$= (10.9 + j\,2.86)\ \Omega$$

Total impedance of the circuit,

$$\mathbf{Z} = \mathbf{Z_A} + \mathbf{Z_{BC}}$$
$$= (15 + j\,15.708) + (10.9 + j\,2.86) = 25.9 + j\,18.568\ \Omega = 31.868\ \angle 35.64°\ \Omega$$

Let the supply voltage be the reference phasor

i.e., $\mathbf{V} = 200\ \angle 0°\,\text{V}$

Current in branch A, $\mathbf{I_A} = \mathbf{I} = \dfrac{\mathbf{V}}{\mathbf{Z}} = \dfrac{200\ \angle 0°}{31.868\ \angle 35.64°} = 6.276\ \angle -35.64°\ \text{A}$

Current in parallel branch B,

$$\mathbf{I_B} = \mathbf{I} \times \frac{\mathbf{Z_C}}{\mathbf{Z_B + Z_C}} = \frac{6.276\ \angle -35.64° \times 37.592\ \angle -57.86°}{39.4\ \angle -40.42°} = 5.988\ \angle -53.08°\ \text{A}$$

Current in parallel branch C,

$$\mathbf{I_C} = \mathbf{I} \times \frac{\mathbf{Z_B}}{\mathbf{Z_B} + \mathbf{Z_C}} = \frac{6.276 \angle -35.64^\circ \times 11.81 \angle 32.14^\circ}{39.4 \angle -40.42^\circ} = 1.88 \angle 36.92^\circ \text{ A}$$

(*a*) Voltage across branch A, $\mathbf{V_A} = \mathbf{I_A} \times \mathbf{Z_A} = 6.276 \angle -35.64^\circ \times 21.72 \angle 46.32^\circ$

$= 136.315 \angle 10.68^\circ$ V **Ans.**

Voltage across branch B, $\mathbf{V_B}$ = Voltage across branch C, $\mathbf{V_C}$

$= \mathbf{I} \times \mathbf{Z_{BC}} = 6.276 \angle -35.64^\circ \times 11.268 \angle 14.7^\circ$

$= 70.72 \angle -20.94^\circ$ V **Ans.**

(*b*) Current through branch A, I_A = 6.276 A **Ans.**

Current through branch B, I_B = 5.988 A **Ans.**

Current through branch C, I_C = 1.88 A **Ans.**

(*c*) Total current supplied, I = 6.276 A **Ans.**

(*d*) Power consumed by circuit $= V \times I \cos \phi = 200 \times 6.276 \cos 35.64^\circ = 1{,}020$ W **Ans.**

(*e*) Power factor of the circuit $= \cos \phi = \cos(-35.64^\circ) = 0.8127$ (lag) **Ans.**

Phasor diagram is shown in Fig. 4.25.

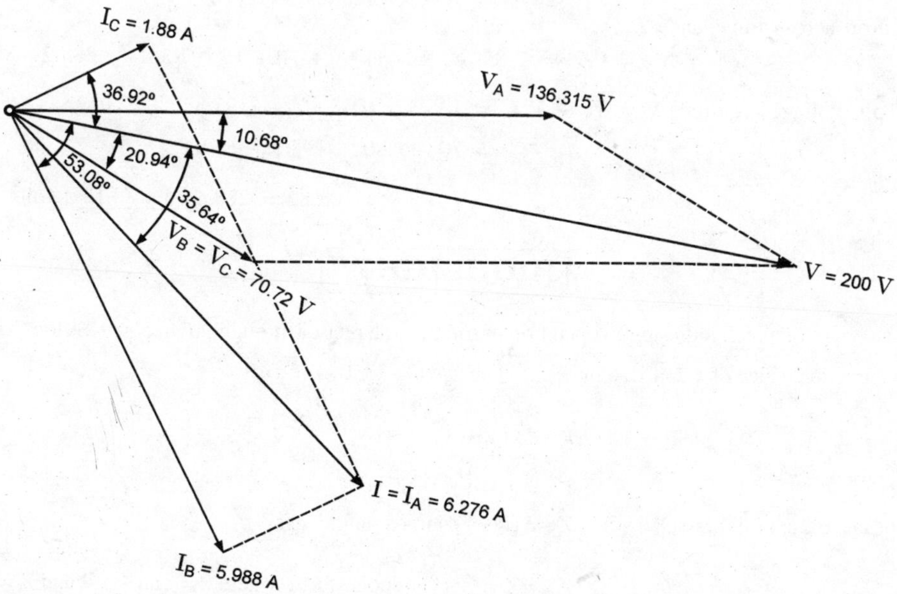

Fig. 4.25. ***Phasor Diagram***

Example 4.29. In the circuit shown in Fig. 4.26 determine what 50 Hz voltage must be applied across AB in order that a current of 10 A may flow in the capacitor.

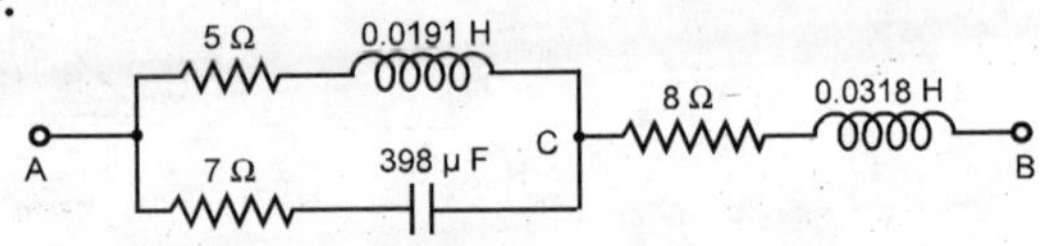

Fig. 4.26

Solution : Impedance for branch I of parallel circuit,

$$\mathbf{Z}_1 = 5 + j\,2\pi \times 50 \times 0.0191$$

$$= (5 + j\,6) \text{ or } 7.81 \angle 50.2^\circ \quad \Omega$$

Impedance of branch II of parallel circuit,

$$\mathbf{Z}_2 = 7 + \frac{1}{j2\pi \times 50 \times 398 \times 10^{-6}}$$

$$= (7 - j\,8) \text{ or } 10.63 \angle -48.8^\circ \quad \Omega$$

Impedance of series branch

$$\mathbf{Z}_3 = 8 + j\,2\pi \times 50 \times 0.0318$$

$$= (8 + j10) \text{ or } 12.8 \angle 51.34^\circ \quad \Omega$$

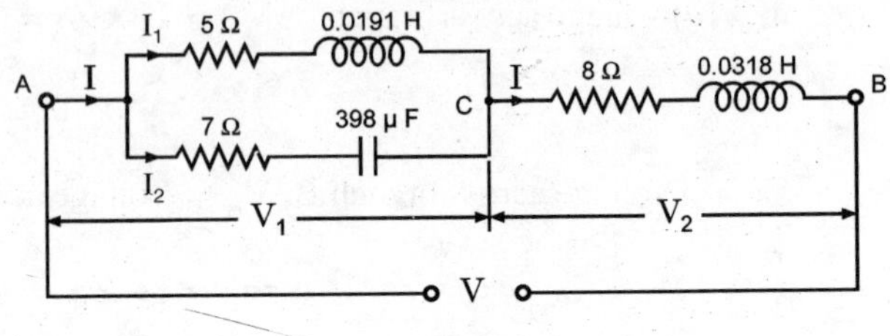

Fig. 4.27

Let the current in impedance Z_2 be the reference phasor

So $\mathbf{I}_2 = (10 + j\,0)$ or $10 \angle 0^\circ$ **A**

Voltage drop across parallel branch,

$$\mathbf{V}_1 = \mathbf{I}_2\,\mathbf{Z}_2 = 10 \angle 0^\circ \times 10.63 \angle -48.8^\circ = 106.3 \angle -48.8^\circ \text{ volts}$$

$$\text{Current in impedance } Z_1,\ \mathbf{I}_1 = \frac{\mathbf{V}_1}{\mathbf{Z}_1} = \frac{106.3 \angle -48.8^\circ}{7.81 \angle 50.2^\circ}$$

$$= \frac{106.3}{7.81} \angle -48.8^\circ - 50.2^\circ = 13.61 \angle -99^\circ = (-2.13 - j\,13.44) \text{ A}$$

$$\text{Total current, } \mathbf{I} = \mathbf{I}_1 + \mathbf{I}_2 = -2.13 - j\,13.44 + 10 + j\,0 = (7.87 - j\,13.44) \text{ A}$$

Voltage drop across impedance Z_3,

$$\mathbf{V}_2 = \mathbf{I}\,\mathbf{Z}_3 = (7.87 - j\,13.44)\,(8 + j10) = (197.36 - j\,28.82) \text{ volts}$$

$$\text{Supply voltage required, } \mathbf{V} = \mathbf{V}_1 + \mathbf{V}_2 = 106.3 \angle -48.8^\circ + (197.36 - j\,28.82)$$

$$= 70 - j\,80 + 197.3 - j\,28.82$$

$$= 267.3 - j\,108.82 = 288.6 \angle -22.15^\circ \text{ volts } \textbf{Ans.}$$

HIGHLIGHTS

1. j is an operator which indicates an operation of rotating the phasor through 90° in counter-clockwise direction. It has numerical value of $(\sqrt{-1})$ so that

$$j^2 = (\sqrt{-1})^2 = -1; \quad j^3 = (\sqrt{-1})^3 = -\sqrt{-1}$$

and $$j^4 = (\sqrt{-1})^4 = 1$$

2. Impedance can be expressed as $\mathbf{Z} = (\text{R} \pm j\text{X})$ in rectangular form

$$\mathbf{Z} = \text{Z} \angle \pm\theta \text{ in polar form where } \theta = \text{Tan}^{-1}\frac{\text{X}}{\text{R}} \text{ and } \text{Z} = \sqrt{\text{R}^2 + \text{X}^2}$$

$$\mathbf{Z} = \text{Z}\,(\cos\theta \pm j\sin\theta) \text{ in trigonometrical form}$$

3. In symbolic method or j-method of solution of ac circuits, the quantities are represented in complex, triangular or polar form and then the solution is obtained.

If applied voltage is represented as

$$\mathbf{V} = V \pm j0 \text{ or } V\angle 0^\circ$$

and I is a current lagging behind the voltage by angle ϕ, then it is represented as

$$\mathbf{I} = (a - jb) \text{ or } I(\cos\phi - j\sin\phi) \text{ or } I\angle -\phi$$

For addition and subtraction, complex form is used while for multiplication and division, polar form is used.

4. Apparent power in complex form is given as

$$S = \text{Phasor voltage} \times \text{conjugate of phasor current} = \mathbf{V}\,\mathbf{I_C}$$

EXERCISES

1. What is the use of operator j ? Explain briefly.
2. Explain the method of representing a phasor by symbolic or j notation.
3. Explain how alternating quantities may be represented by complex numbers ?
4. Is impedance a phasor quantity, if not then how it is expressed in phasor or complex form ?
5. Explain the conjugate method of determining power in ac circuits.
6. Write short note on j operator.

SHORT ANSWER TYPE QUESTIONS WITH ANSWERS

Q. 1. What are the values of j^2 and j^4 ?

Ans. $j^2 = -1$ and $j^4 = +1$

Q. 2. What is meant by complex power ?

Ans. The product $S = V.I_C$ is called the complex power where I_C is the conjugate of I.

Q. 3. Why is impedance represented by a complex number ?

Ans. In an expression $\mathbf{V} = \mathbf{IZ}$, the impedance $\mathbf{Z}$ is not associated with time varying quantities and is not a phasor in the sense that **V** and **I** are phasors. However, it is represented as a complex number. It is represented as

$\mathbf{Z} = R \pm jX$ +ve sign for inductive reactance and –ve sign for capacitive reactance.

Q. 4. How is complex impedance dependent on frequency ?

Ans. Complex impedance has two components viz. real or horizontal component and imaginary or vertical components. The reactance is represented by imaginary component. The inductive and capacitive reactances are given as

Inductive reactance, $X_L = \omega L = 2\pi f L$ where f is supply frequency in Hz.

Capacitive reactance, $X_C = \dfrac{1}{\omega C} = \dfrac{1}{2\pi f C}$ where f is supply frequency in Hz.

Since imaginary component of impedance depends upon the supply frequency so the impedance.

PROBLEMS

1. Give the current expressed in the polar form $\mathbf{I} = 10\angle 53.1^\circ$ amperes in symbolic and trigonometrical forms.

   [**Ans.** $(6 + j\,8)$A; $10(\cos 53.1^\circ + j \sin 53.1^\circ)$A]

2. Divide phasor $(5 + j\,6)$ by phasor $(4 - j\,3)$. [**Ans.** $0.08 + j1.56$]

3. Add phasors $30\angle 45^\circ$ and $7.5\angle 30^\circ$. [**Ans.** $27.7 + j\,17.46$]

4. Determine $\sqrt[5]{243} \angle 120°$. [**Ans.** $3 \angle 24°$]

5. Determine the current flowing through a circuit having impedance of $(5 + j12)\ \Omega$ connected across 230 V, 50 Hz supply. [**Ans.** $17.69 \angle -67.38°$ A]

6. A coil of resistance 20 Ω and inductance 0.1 H is connected in series with 50 μF capacitor across a 100 V, 50 Hz supply. Estimate (*i*) current in the circuit (*ii*) power (*iii*) power factor (*iv*) voltage across the capacitor.
[**Ans.** (*i*) 2.636 A (*ii*) 138.9 W (*iii*) 0.5271 leading (*iv*) 167.8 V]

7. Two impedances $\mathbf{Z}_1 = (8 + j\,6)$ ohms and $Z_2 = (3 - j\,4)$ ohms are in parallel. If the total current of this combination is 25 A, find the power taken by each impedance.Draw and explain the complete phasor diagram.
[**Ans.** 1000 watts, 1500 watts]

8. An impedance of $(2 + j\,6)$ ohms is connected in series with two impedances $(10 + j\,4)$ ohms and $(12 - j\,8)$ ohms which are in parallel. Calculate the magnitude and power factor of the main current drawn from the supply when combined circuit is supplied at 200 volts. [**Ans.** 18.6 A; 0.831 (lagging)]

9. Impedances $\mathbf{Z}_A = (8 + j\,6)$ ohms and $\mathbf{Z}_B = (6 + j\,8)$ ohms are connected in parallel and the combination is in series with an impedance $\mathbf{Z}_C = (8 - j\,6)$ ohms. If the current in $\mathbf{Z}_A = (10 + j\,0)$ amperes, find the magnitude and phase of (*i*) voltage across the parallel branch (*ii*) current in the impedance $\mathbf{Z}_B$ (*iii*) voltage across the impedance $\mathbf{Z}_C$ (*iv*) total voltage across the whole circuit. Draw the phasor diagram for the circuit.

[**Ans.** (*i*) $100 \angle 36.9°$ V(*ii*) $10 \angle 16.2°$ A (*iii*) $140\sqrt{2} \angle -45°$ V (*iv*) $234 \angle -36.1°$ V]

10. In a series-parallel circuit the parallel branches A and B are in series with C. The impedances are:

$$\mathbf{Z}_A = (4 + j\,3),\ \mathbf{Z}_B = (4 - j\,16/3) \text{ and } \mathbf{Z}_C = (2 + j\,8) \text{ ohms}$$

If the current $\mathbf{I}_C = (25 + j\,0)$, draw the complete phasor diagram determining the power (active and reactive power) for each branch and the whole circuit.

[**Ans.** (1,600 – j 1,200), (900 + j 1,200), (1,250 – j 5,000), (3,750 – j 5,000)]

❑❑❑

5

CHAPTER

Resonance in R-L-C Circuits

5.1. INTRODUCTION

INSIDE THIS CHAPTER

Resonance is identified with engineering situations which involve energy-storing elements subjected to a forcing function of varying frequency. Specifically, *resonance* is the term employed for describing the steady-state operation of a circuit or system at that frequency for which the resultant response is in time phase with the source function despite the presence of energy-storing elements. Resonance cannot take place when only one type of energy-storing element is present, e.g. capacitance or spring. There must exist two types of independent energy-storing elements capable of interchanging energy between them—for example, inductance and capacitance or mass and spring. Thus resonance is a phenomenon found in any system involving two independent energy-storing elements, be they electrical, mechanical, pneumatic, hydraulic or whatever.

If we have an ac circuit having a resistance R, an inductance L, and a capacitance C, connected in series (Fig. 5.1) and apply a small voltage V from a source that can keep the magnitude of V constant but can vary its frequency, we find that the magnitude of current drawn from the source of supply varies with the variation in frequency of supply source. There will be a value of frequency at which the current is maximum. *Electrical resonance* is said to exist when this condition is reached.

In this chapter we will discuss this phenomenon (more precisely called the *series* or *voltage resonance*) and also the situation when a constant voltage of variable frequency is applied to a *parallel circuit*. The simplest parallel circuit encountered in practice is a coil having resistance R and inductance L connected in parallel with a capacitor C. The resonant condition in this case is called *parallel resonance,* and sometimes *anti-resonance*. The latter name is prompted due to the fact that at resonance the input current to the parallel circuit is *minimum*.

Under the conditions of resonance, such a network becomes purely resistive in its effects, and the voltage and the current in the network are in phase. For this to occur, the inductive reactance and the capacitive reactance should be made equal.

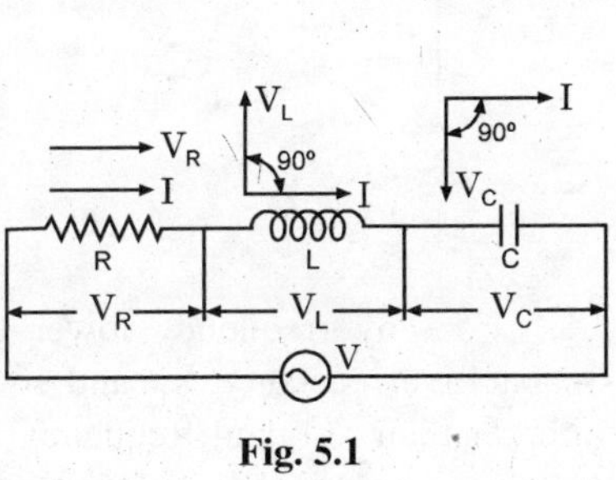

Fig. 5.1

5.2. SERIES OR VOLTAGE RESONANCE

Consider an ac circuit containing a resistance R, an inductance L and a capacitance C connected in series, as shown in Fig. 5.1.

Impedance of the circuit, $Z = \sqrt{R^2 + (X_L - X_C)^2} = \sqrt{R^2 + \left(\omega L - \frac{1}{\omega C}\right)^2}$

If for some frequency of applied voltage, $X_L = X_C$ in magnitude then

(*i*) Net reactance is zero *i.e.* $X = 0$.

(*ii*) Impedance of the circuit, $Z = R$.

(*iii*) The current flowing through the circuit is maximum and in phase with the applied voltage. The magnitude of the current will be equal to $\frac{V}{R}$.

(*iv*) The voltage drop across the inductance is equal to the voltage drop across capacitance and is maximum.

(*v*) The power factor is unity.

(*vi*) The power expended = VI watts.

When this condition exists, the circuit is said to be in resonance and the frequency at which it occurs is known as *resonant frequency*.

If resonant frequency is denoted by f_r, then

$$X_L = \omega L = 2\pi f_r L \text{ and } X_C = \frac{1}{\omega C} = \frac{1}{2\pi f_r C}$$

Since for resonance $X_L = X_C$

$$\therefore \; 2\pi f_r L = \frac{1}{2\pi f_r C}$$

$$\text{or} \quad f_r = \frac{1}{2\pi\sqrt{LC}}$$

$$\text{or } \omega_r = \sqrt{1/LC} \qquad \text{...(5.1)}$$

From the above expression it is obvious that the value of resonant frequency depends on the parameters of the two energy-storing elements.

Phasor diagrams for the series R-L-C circuit shown in Fig. 5.1 at three different frequencies *i.e.* (*a*) $f < f_r$ (*b*) $f = f_r$ and (*c*) $f > f_r$ with L and C kept constant are shown in Figs. 5.2 (*a*), 5.2 (*b*) and 5.2 (*c*) respectively.

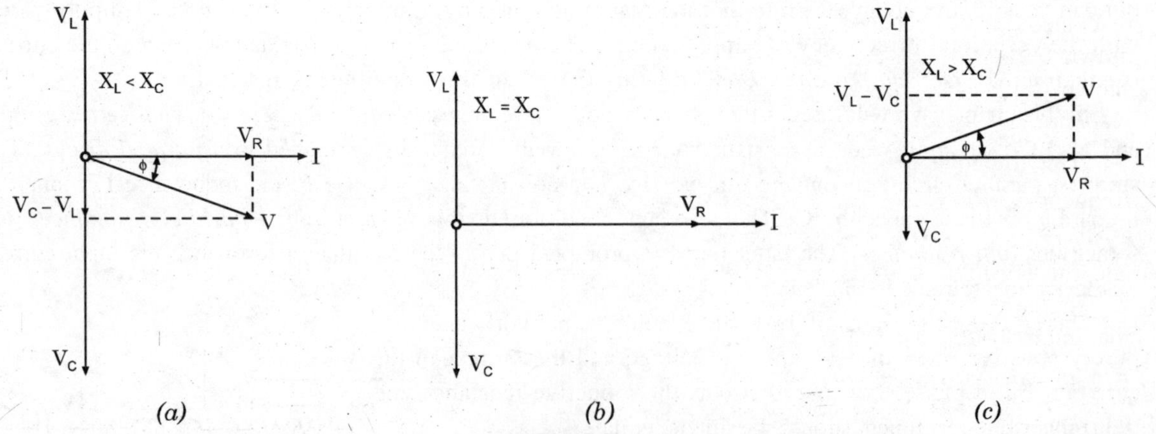

Fig. 5.2

For any frequency lower than resonant frequency f_r, inductive reactance X_L is lesser than the capacitive reactance X_C and so the circuit behaves as a capacitive circuit. Similarly for any frequency higher than resonant frequency inductive reactance is larger than capacitive reactance and so the circuit behaves as an inductive circuit. When the frequency of applied voltage is equal to the resonant frequency, the inductive reactance is equal to the capacitive reactance, the voltage drop across inductor is equal to voltage drop across capacitor in magnitude, but opposite in phase and, therefore, the circuit current I is in phase with the applied voltage *i.e.* the circuit behaves as a resistive circuit.

When the circuit is in resonance, the current is too large and will produce large voltage drop across inductance and capacitance, which will be equal in magnitude but opposite in phase and each may be several times greater than the applied voltage. If resistance R would have not been present in the circuit, such a circuit, would act like a *short circuit* to currents of frequency to which it resonates.

Since in this resonance the voltage is maximum, it is called the *voltage resonance*. The series resonant circuit is also called an *acceptor circuit* because such a circuit accepts currents at one particular frequency but rejects currents of other frequencies. Such circuits are used in radio receivers.

5.2.1. Graphical Representation of Resonance in an R-L-C Series Circuit

The circuit can be made resonant in two different ways namely (*i*) by varying L and C parameters (either one or both) at a constant supply frequency or (*ii*) by varying the supply frequency *f* with parameters L and C constant. In our study of series resonance phenomenon we shall keep the applied voltage to the circuit and parameters L and C constant and allow the frequency to vary.

The circuit *resistance* R is independent of supply frequency and, therefore, remains constant. It has been represented by a straight line, parallel to the X-axis (or frequency axis) in Fig. 5.3. *Inductive reactance* X_L, being equal to ωL, increases in direct proportion to the supply frequency and is represented by a straight line passing through the origin (as X_L is regarded positive, so it lies in the first quadrant). *Capacitive reactance*, being equal to $\frac{1}{\omega C}$, decreases inversely with the increase in frequency and is represented by a rectangular hyperbola lying in the fourth quadrant, below the frequency axis (capacitive reactance is considered negative).

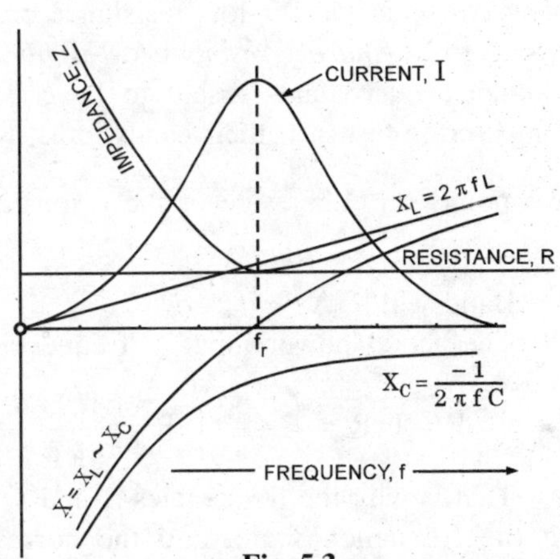

Fig. 5.3

The *net reactance* is the difference of inductive reactance X_L and capacitive reactance X_C and the curve drawn between the net reactance ($X_L \sim X_C$) and frequency will be a hyperbola (not rectangular) as shown in Fig. 5.3. The frequency at which the reactance curve crosses the frequency axis is called the *resonant frequency*, f_r (or f_o).

The impedance of the circuit, Z, being equal to $\sqrt{R^2 + (X_L \sim X_C)^2}$ is minimum at resonant frequency f_r.

At frequencies lower than resonant frequency f_r the impedance Z is large and capacitive as $X_C > X_L$ and the power factor is leading and at frequencies higher than resonant frequency f_r, the impedance Z is again large but inductive as $X_L > X_C$ and the power factor is lagging. The power factor has the maximum value of unity at resonant frequency.

5.2.2. Resonance Curve

Current varies inversely with the variation in impedance and, therefore, it is maximum at resonant frequency when impedance is minimum and decreases with the variation in frequency on both sides of the resonant frequency (as impedance Z is large), as illustrated in Fig. 5.4. The curve drawn between circuit current and the frequency of the applied voltage is called the *resonance curve* and its shape depends upon the value of circuit resistance R, as shown in the figure. For smaller values of R, the resonance curve is sharply peaked, but for larger values of R, the curve is flat (Fig. 5.4).

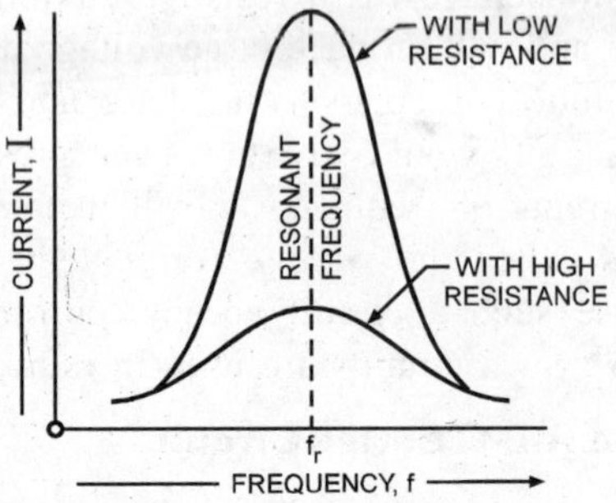

Fig. 5.4. *Resonance Curve*

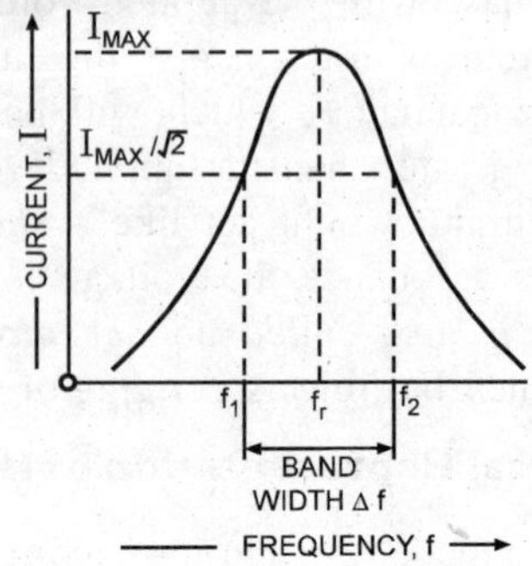

Fig. 5.5.

5.2.3. Selectivity and Bandwidth

We have seen that for low resistance circuit the resonance curve is sharply peaked and such a circuit is said to be *sharply resonant* or *highly selective.* On the other hand, high resistance circuit have flat resonance curve and is said to have poor selectivity. Selectivity of different resonant circuits are compared in terms of their bandwidths, which are given by the bands of frequencies which lie between two points on either side of the resonant frequency where current is $\frac{1}{\sqrt{2}}$ times of maximum current I_{max}

Band width, $\Delta f = f_2 - f_1$ ...(5.2)

The actual power input at frequencies f_1 and f_2

$$P = I^2 R = \left(\frac{I_{max}}{\sqrt{2}}\right)^2 R = \frac{I_{max}^2 R}{2} = \frac{1}{2} P_{max} = \frac{1}{2} \times \text{power input at resonance} \qquad ...(5.3)$$

That is why the frequencies f_1 and f_2 at the limits of the bandwidth are called the *half-power points* on the frequency scale, and the corresponding value of the bandwidth is termed as *half-power bandwidth* (B_{hp}) or –3 dB bandwidth.

The noteworthy points regarding half-power points are

1. The current is $\frac{I_{max}}{\sqrt{2}}$; 2. Impedance is $\sqrt{2}$ R or $\sqrt{2}\, Z_{min}$;

3. $P_1 = P_2 = \frac{P_{max}}{2}$; 4. The circuit phase angle $\Phi = \pm 45°$ or $\frac{\pi}{4}$ radian;

5. Tangent of the circuit phase angle at the off-resonance frequencies f_1 and f_2, $Q = \tan\theta = \tan 45° = 1$

5.2.4. Quality Factor of a Series Resonant Circuit

The Q-factor of an R-L-C series circuit can be defined in any of the following ways.

It may be given as the voltage magnification that the circuit produces at resonance.

We have seen that at resonance, the circuit current is maximum and is equal to $\frac{V}{R}$ or supply voltage, $V = I_{max} R$ and

Voltage across inductance or capacitance $= I_{max} X_L = I_{max} X_C$

$$\therefore \text{ Voltage magnification} = \frac{I_{max} X_L}{I_{max} R} = \frac{X_L}{R} = \frac{\omega_r L}{R} = \frac{1}{\omega_r RC}$$

or Q-factor at resonance, $$Q_r = \frac{\omega_r L}{R} = \frac{2\pi f_r L}{R} = \frac{2\pi L}{R} \times \frac{1}{2\pi\sqrt{LC}} = \frac{1}{R}\sqrt{\frac{L}{C}} \quad ...(5.4)$$

$$\because f_r = \frac{1}{2\pi\sqrt{LC}}$$

$$\text{Also } Q_r = \frac{2\pi f_r L}{R} = \text{Tan } \Phi \quad ...(5.5)$$

where Φ is the power factor angle of the coil.

In case of series resonance, higher value of Q-factor means not only higher voltage magnification, but also a higher selectivity of the tuning coil and that is why it is necessary that the coil be of high inductance and low resistance. In fact, Q-factor of a series resonant circuit may be defined as the ratio of resonant frequency to bandwidth

$$i.e. \quad Q_r = \frac{f_r}{\Delta f} = \frac{\frac{1}{2\pi\sqrt{LC}}}{R/2\pi L} = \frac{1}{R}\sqrt{\frac{L}{C}}, \text{ as before} \quad \text{(Refer to Eq. 5.4)}$$

Example 5.1. A series R-L-C circuit has R = 10 Ω, L = 0.1 H and C = 8 μF. Determine (*i*) resonant frequency (*ii*) Q-factor of the circuit at resonance and (*iii*) the half-power frequencies.

Solution : R = 10 Ω; L = 0.1 H; C = 8 μF = 8×10^{-6} F

(*i*) Resonant frequency, $$f_r = \frac{1}{2\pi}\sqrt{\frac{1}{LC}} = \frac{1}{2\pi}\sqrt{\frac{1}{0.1 \times 8 \times 10^{-6}}} = 177.94 \text{ Hz } \textbf{Ans.}$$

(*ii*) Q-factor of the circuit at resonance $$= \frac{1}{R}\sqrt{\frac{L}{C}} = \frac{1}{10}\sqrt{\frac{0.1}{8 \times 10^{-6}}} = 11.18 \textbf{ Ans.}$$

(*iii*) Lower half-power frequency, $$f_1 = f_r - \frac{R}{4\pi L} = 177.94 - \frac{10}{4\pi \times 0.1} = 170 \text{ Hz } \textbf{Ans.}$$

Upper half-power frequency, $$f_2 = f_r + \frac{R}{4\pi L} = 177.94 + \frac{10}{4\pi \times 0.1} = 185.9 \text{ Hz } \textbf{Ans.}$$

Example 5.2. A circuit consists of a 115 Ω resistor, a 0.024 μF capacitor and a coil of inductance L. If the resonant frequency of the circuit is 1,000 Hz, determine inductance and bandwidth.

[*Pb. Technical Univ. Basic Electrical and Electronics Engineering May 2009*]

Solution : R = 115 Ω, C = 0.024 μF = 0.024×10^{-6} F; f_r = 1,000 Hz

Since $$f_r = \frac{1}{2\pi}\sqrt{\frac{1}{LC}}$$

$$L = \frac{1}{(2\pi f_r)^2 C} = \frac{1}{(2\pi \times 1{,}000)^2 \times 0.024 \times 10^{-6}} = 1.055 \text{ H } \textbf{Ans.}$$

$$\text{Bandwidth, BW} = \frac{R}{2\pi L} = \frac{115}{2 \times \pi \times 1.055} = 17.34 \text{ Hz } \textbf{Ans.}$$

Example 5.3. A resistance of 24 Ω, a capacitor of 150 μF and an inductor of 0.16 H are connected in series with each other. A supply of 240 V, 50 Hz is applied to the ends. Calculate (*a*) the current in the circuit (*b*) the pd across each element (*c*) the frequency to which the supply would need to be changed so that the current would be at unity power factor and find the current at this frequency.

[*Pb. Technical Univ. Basic Electrical and Electronics Engineering May 2010*]

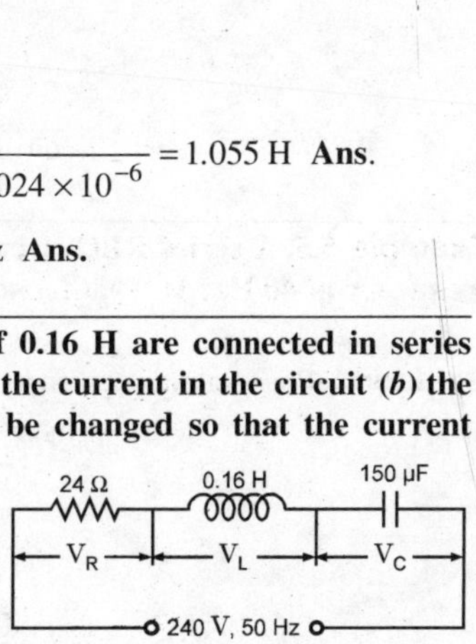

Fig. 5.6

Solution : R = 24 Ω

Inductive reactance, $X_L = 2\pi f L = 2\pi \times 50 \times 0.16$
$= 50.27$ Ω

$$\text{Capacitive reactance, } X_C = \frac{1}{2\pi f C} = \frac{1}{2\pi \times 50 \times 150 \times 10^{-6}} = 21.22\ \Omega$$

$$\text{Circuit impedance, } Z = \sqrt{R^2 + (X_L + X_C)^2} = \sqrt{24^2 + (50.27 - 21.22)^2} = 37.68\ \Omega$$

$$(a)\ \text{Circuit current, } I = \frac{V}{Z} = \frac{240}{37.68} = 6.37\text{ A} \quad \textbf{Ans.}$$

(*b*) Voltage across resistor, $V_R = IR = 6.37 \times 24 = 152.88$ V **Ans.**

Voltage across inductor, $V_L = IX_L = 6.37 \times 50.27 = 320.2$ V **Ans.**

Voltage across capacitor, $V_C = IX_C = 6.37 \times 21.22 = 135.2$ V **Ans.**

$$(c)\ \text{Resonant frequency, } f_r = \frac{1}{2\pi}\sqrt{\frac{1}{LC}} = \frac{1}{2\pi}\sqrt{\frac{1}{0.16 \times 150 \times 10^{-6}}} = 32.5\text{ Hz} \quad \textbf{Ans.}$$

$$(d)\ \text{Current at resonant frequency, } I_{max} = \frac{V}{R} = \frac{240}{24} = 10\text{ A} \quad \textbf{Ans.}$$

Note : Current would be drawn at unity power factor when the frequency becomes equal to resonant frequency.

Example 5.4. For a series R-L-C circuit the inductor is variable. Source voltage is $200\sqrt{2}$ sin $100\pi t$. Maximum current obtainable by varying the inductance is 0.314 A and the voltage across the capacitor then is 300 V. Find the circuit element values. [*Pb. Technical Univ. Basic Electrical and Electronics Engineering May 2010*]

Solution :

$$\text{Supply voltage, V(rms)} = \frac{200\sqrt{2}}{2} = 200\text{ V}$$

$$\text{Frequency, } f = \frac{100\pi}{2\pi} = 50\text{ Hz}$$

Maximum current in a series R-L-C circuit will flow under resonant conditions.

Since under resonant condition current

$$I_{max} = \frac{V}{R}$$

$$\text{So circuit resistance, } R = \frac{V}{I_{max}} = \frac{200}{0.314} = 636.6\ \Omega \quad \textbf{Ans.}$$

$$\text{Voltage across the capacitor, } V_C = \frac{I_{max}}{2\pi f C}$$

$$\text{So circuit capacitance, } C = \frac{I_{max}}{2\pi f V_C} = \frac{0.314}{2\pi \times 50 \times 300} = 3.333\ \mu\text{ F} \quad \textbf{Ans.}$$

$$\text{Since under resonant condition, } X_L = X_C$$

$$\text{or } 2\pi f L = \frac{1}{2\pi f C}$$

$$\text{or Circuit inductance, } L = \frac{1}{4\pi^2 f^2 C} = \frac{1}{4\pi^2 \times 50^2 \times 3.333 \times 10^{-6}} = 3.04\text{ H} \quad \textbf{Ans.}$$

Example 5.5. A series RLC circuit with R = 250 Ω, and L = 0.6 H results in a leading phase angle of 60° at a frequency of 40 Hz. At what frequency will the circuit resonate ?

[*Pb. Technical Univ. Basic Electrical and Electronics Engineering Dec. 2006*]

Solution : Phase angle is given by

$$\text{Tan } \phi = \frac{X}{R} = \frac{X_C - X_L}{R} \qquad \because \text{ Phase angle is leading}$$

$$= \frac{(1/2\pi f C) - 2\pi f L}{R}$$

$$\text{or } \frac{1}{2\pi f C} = R \tan \phi + 2\pi f L$$

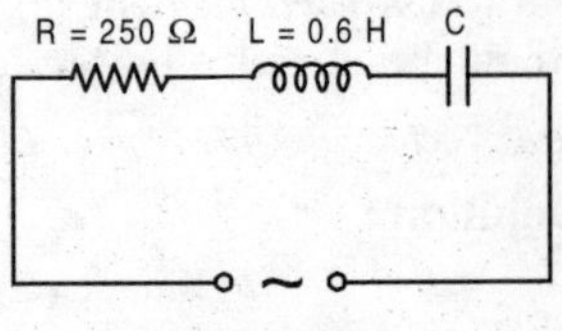

Fig. 5.7

$$= 250 \times \tan 60^\circ + 2\pi \times 50 \times 0.6$$
$$= 433 + 188.5 = 621.5$$

$$\text{or } C = \frac{1}{2\pi f \times 621.5} = \frac{1}{2\pi \times 40 \times 621.5} = 6.4\ \mu\text{F}$$

$$\text{Resonant frequency, } f_r = \frac{1}{2\pi\sqrt{LC}} = \frac{1}{2\pi\sqrt{0.6 \times 6.4 \times 10^{-6}}} = 81.2 \text{ Hz} \quad \textbf{Ans.}$$

5.3. PARALLEL OR CURRENT RESONANCE

When an inductive reactance and a capacitive reactance are connected in parallel, as shown in Fig. 5.8, conditions may reach under which *current resonance* (also known as *parallel or anti-resonance*) will take place.

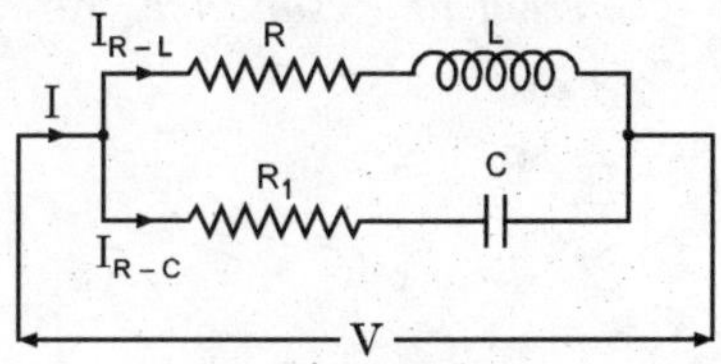

Fig. 5.8

Consider a practical case of a coil in parallel with a condenser, as shown in Fig. 5.8. Let the coil be of resistance R ohms and inductance L henrys and the condenser of resistance R ohms and capacitance C farads.

Such a circuit is said to be in electrical resonance when the reactive (or wattless) component of line current becomes zero. The frequency at which this happens is known as *resonant frequency*.

Circuit will be in electrical resonance if reactive component of R-L branch current, $I_{R\text{-}L} \sin \phi_{R\text{-}L}$ = Reactive component of R-C branch current, $I_{R\text{-}C} \sin \Phi_{R\text{-}C}$

$$\text{Now since } I_{R\text{-}L} = \frac{V}{\sqrt{R^2 + (\omega_r L)^2}} \text{ and } \sin \phi_{R\text{-}L} = \frac{X_L}{Z_{R\text{-}L}} = \frac{\omega_r L}{\sqrt{R^2 + (\omega_r L)^2}}$$

$$I_{R\text{-}C} = \frac{V}{Z_{R\text{-}C}} = \frac{V}{\sqrt{R_1^2 + \left(\frac{1}{\omega_r C}\right)^2}} \text{ and } \sin \phi_{R\text{-}C} = \frac{X_C}{Z_{R\text{-}C}} = \frac{1/\omega_r C}{\sqrt{R_1^2 + \left(\frac{1}{\omega_r C}\right)^2}}$$

$$\therefore \frac{V}{\sqrt{R^2 + (\omega_r L)^2}} \times \frac{\omega_r L}{\sqrt{R^2 + (\omega_r L)^2}} = \frac{V}{\sqrt{R_1^2 + \left(\frac{1}{\omega_r C}\right)^2}} \times \frac{1/\omega_r C}{\sqrt{R_1^2 + \left(\frac{1}{\omega_r C}\right)^2}}$$

$$\text{or } \frac{\omega_r L}{R^2 + (\omega_r L)^2} = \frac{1/\omega_r C}{R_1^2 + (1/\omega_r C)^2}$$

$$\text{or } \frac{\omega_r L}{R^2 + \omega_r^2 L^2} = \frac{\omega_r C}{\omega_r^2 R_1^2 C^2 + 1}$$

$$\text{or } L(\omega_r^2 R_1^2 C^2 + 1) = C(R^2 + \omega_r^2 L^2)$$

$$\text{or } \omega_r^2 LC(R_1^2 C - L) = CR^2 - L$$

$$\text{or } \omega_r = \frac{1}{\sqrt{LC}} \sqrt{\frac{CR^2 - L}{CR_1^2 - L}}$$

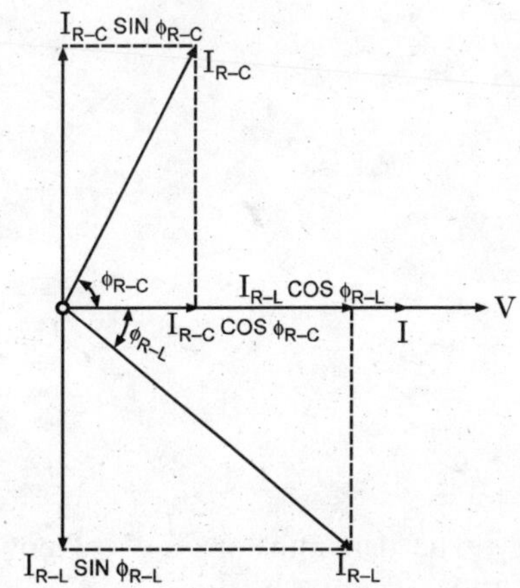

Fig. 5.9

or Resonant frequency, $$f_r = \frac{1}{2\pi}\omega_r = \frac{1}{2\pi\sqrt{LC}}\sqrt{\frac{CR^2 - L}{CR_1^2 - L}} \qquad ...(5.6)$$

If resistance of capacitor is negligible *i.e.* $R_1 = 0$, as is usually the case,

Resonant frequency, $$f_r = \frac{1}{2\pi\sqrt{LC}}\sqrt{1 - \frac{CR^2}{L}} = \frac{1}{2\pi}\sqrt{\frac{1}{LC} - \frac{R^2}{L^2}} \qquad ...(5.7)$$

If resistance of coil R is zero

Resonant frequency, $$f_r = \frac{1}{2\pi}\sqrt{\frac{1}{LC}} \qquad ...(5.8)$$

Current at Resonance. Assuming $R_1 = 0$, as usually, at resonant frequency

$$I_{R\text{-}C} = \frac{V}{Z} = \frac{V}{\sqrt{(R_1)^2 + \dfrac{1}{(2\pi f_r C)^2}}} = 2\pi f_r CV \qquad ...(5.9)$$

$$\Phi_{R\text{-}C} = \text{Cos}^{-1}\frac{R}{Z} = 90^\circ \text{ or } \frac{\pi}{2} \text{ radians}$$

Active component of R-C branch current, $I_{R-C} \cos \Phi = 0$

Reactive component of R-C branch current $I_{R-C} \sin \phi = 2\pi f_r CV$

R-L branch current, $$I_{R\text{-}L} = \frac{V}{Z} = \frac{V}{\sqrt{R^2 + (2\pi f_r L)^2}}$$

$$\Phi_{R\text{-}L} = \text{Cos}^{-1}\frac{R}{Z}$$

Active component of R-L branch current, $= I_{R\text{-}L} \cos \Phi_{R\text{-}L} = \frac{V}{Z}\cdot\frac{R}{Z} = \frac{VR}{Z^2}$

Reactive component of R-L branch current $= I_{R\text{-}L} \sin \Phi_{R\text{-}L} = \frac{V}{Z}\cdot\frac{2\pi f_r L}{Z}$

At resonance $I_{R\text{-}L} \sin \Phi_{R\text{-}L} = I_{R\text{-}C} \sin \Phi_{R\text{-}C}$

or $$\frac{V}{Z}\cdot\frac{2\pi f_r L}{Z} = 2\pi f_r CV$$

or $$\frac{L}{Z^2} = C$$

or $$Z = \sqrt{\frac{L}{C}} \qquad ...(5.10)$$

Line current, $I = I_{R\text{-}L} \cos \Phi_{R\text{-}L} + I_{R\text{-}C} \cos \Phi_{R\text{-}C}$

$$= \frac{V}{Z}\cdot\frac{R}{Z} + 0 = \frac{VR}{Z^2} = VR \times \frac{C}{L} = \frac{V}{L/CR} \qquad ...(5.11)$$

The denominator $\frac{L}{CR}$ is known as *effective or equivalent* or *dynamic impedance* of the parallel circuit at resonance.

5.3.1. Important Points About Current or Parallel Resonance

1. Net susceptance is zero *i.e.* $\frac{1}{X_C} = \frac{X_L}{Z^2}$ or $\omega_r C = \frac{\omega_r L}{Z^2}$ or $Z = \sqrt{\frac{L}{C}}$.
2. The admittance is equal to conductance.
3. Reactive or wattless component of line current is zero, hence circuit power factor is unity.
4. Impedance is purely resistive, maximum in magnitude and is equal to L/CR.
5. Line current is minimum and is equal to $\frac{V}{L/CR}$ in magnitude and is in phase with the applied voltage.
6. Frequency is equal to $\frac{1}{2\pi}\sqrt{\frac{1}{LC} - \frac{R^2}{L^2}}$ Hz.

Note : *Parallel resonant circuit is sometimes called the rejector circuit because at resonant frequency the line current is minimum or it almost rejects it.*

Since in parallel resonant circuits circulating current between the branches is many times the line current, such type of resonance is sometimes known as current resonance.

An inductive coil of inductance L connected in parallel with a capacitance C is called the tank circuit.

5.3.2. Graphical Representation of Current or Parallel Resonance

Now we will discuss the effect of variation in frequency on the susceptances of the two parallel branches. The variations are illustrated in Fig. 5.10.

Inductive susceptance, being equal to $\frac{1}{\omega L}$ or $\frac{1}{2\pi f L}$, decreases inversely with the increase in frequency and is represented by a rectangular hyperbola lying in the fourth quadrant, below the frequency axis (inductive susceptance being considered negative). *Capacitive susceptance,* being equal to ωC or $2\pi f C$, increases in direct proportion to the supply frequency and is represented by a straight line passing through origin. As capacitive susceptance is considered positive, so it lies in first quadrant. The net susceptance B is the difference of capacitive susceptance and inductive susceptance and the curve drawn between net susceptance and frequency of the applied voltage is a hyperbola (not rectangular) as illustrated in Fig. 5.10.

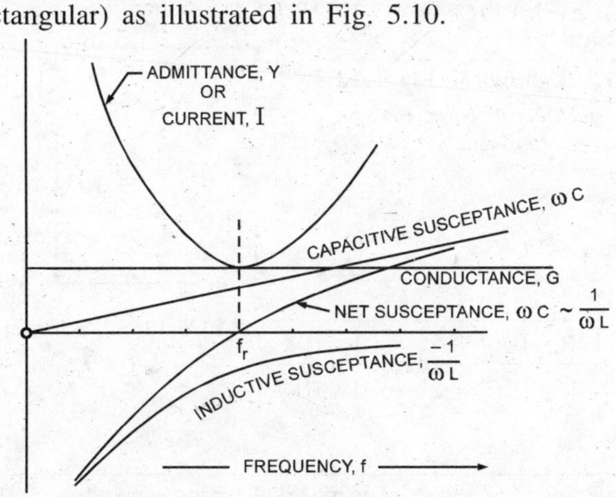

Fig. 5.10

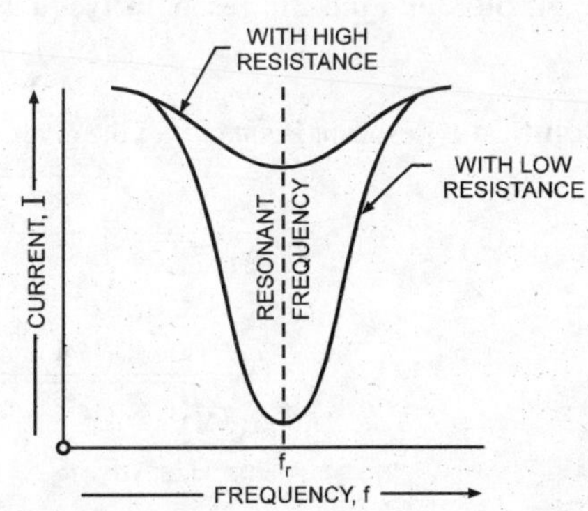

Fig. 5.11

The frequency at which the net susceptance curve crosses the frequency axis is called the *resonant frequency*. At this point impedance is maximum or admittance is minimum and is equal to G, consequently line current is minimum. Obviously at frequency lower than resonant frequency the inductive susceptance is more than capacitive susceptance, therefore, the circuit is inductive and the line current lags behind the applied voltage. But for frequencies exceeding resonant frequency capacitive susceptance predominates, therefore, the circuit is capacitive and the line current leads the applied voltage.

If the resistance is relatively low the current will fall considerably at resonant frequency and if the resistance is high, decrease in current will be less pronounced, as illustrated in Fig. 5.11.

5.3.3. Bandwidth in Case of a Parallel Resonant Circuit

The bandwidth in case of a parallel circuit is defined in the same way as in case of a series circuit. In this case also there are upper and lower half-power points where power absorbed is half of that at resonant frequency.

At bandwidth frequencies, the net susceptance B is equal to the conductance G. So at frequency f_1 the net susceptance $B_{L_1} - B_{C_1} = G$ and at frequency f_2, $B_{C_2} - B_{L_2} = G$. Thus admittance $Y = \sqrt{G^2 + B^2}$ $= \sqrt{2}\,G$ and phase angle $\Phi = \text{Tan}^{-1}\dfrac{B}{G} = \text{Tan}^{-1} 1 = 45°$ or $\dfrac{\pi}{4}$ radian.

5.3.4. Q-Factor or Current Magnification Factor

Q-factor of a parallel circuit is defined as the ratio of the circulating current to the line current or as the current magnification.

$$\therefore \quad \text{Q-factor} = \frac{\text{Circulating current between L and C}}{\text{Line current}}$$

Since from Eq. (5.9), the circulating current,

$$I_C = 2\pi f_r CV$$

and from Eq. (5.11), the line current,

$$I = \frac{V}{L/CR} = VCR/L$$

$$\therefore \text{Q-factor} = \frac{I_C}{I} = \frac{2\pi f_r CV}{VCR/L} = \frac{2\pi f_r L}{R} \quad \text{but } f_r = \frac{1}{2\pi}\sqrt{\frac{1}{LC}} \text{ neglecting R}$$

$$\text{So Q-factor} = 2\pi \times \frac{1}{2\pi}\sqrt{\frac{1}{LC}} \times \frac{L}{R} = \frac{1}{R}\sqrt{\frac{L}{C}}.$$ It is the same as for series resonance.

Example 5.6. Find the resonant frequency for the circuit shown in Fig. 5.12.

[Pb. Technical Univ. Basic Electrical and Electronics Engineering May 2005]

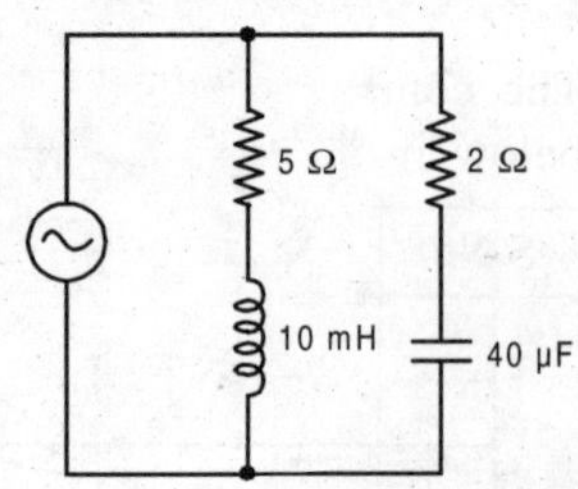

Fig. 5.12. *A Resonant Circuit*

Solution : Resonant frequency of the given circuit

$$f_r = \frac{1}{2\pi\sqrt{LC}}\sqrt{\frac{CR^2 - L}{CR_1^2 - L}}$$

$$= \frac{1}{2\pi\sqrt{10 \times 10^{-3} \times 40 \times 10^{-6}}} \times \sqrt{\frac{40 \times 10^{-6} \times 5^2 - 10 \times 10^{-3}}{40 \times 10^{-6} \times 2^2 - 10 \times 10^{-3}}}$$

$= 263$ Hz **Ans.**

Example 5.7. A coil of 10 Ω resistance and 0.1 H inductance is connected in parallel with a capacitor of 100 μF. Calculate the frequency at which the circuit will act as a non-inductive resistance of R ohms. Find also the value of R, input current at resonant frequency and the ratio of the circulating current to the supply current at resonant frequency. The applied voltage is 100 V (rms).

Solution : Capacitance, C = 100 μF = 1 × 10^{-4} F

The circuit will act as a non-inductive resistance of R ohms at resonant frequency, which is given as

$$f_r = \frac{1}{2\pi}\sqrt{\frac{1}{LC} - \frac{R^2}{L^2}} = \frac{1}{2\pi}\sqrt{\frac{1}{0.1 \times 1 \times 10^{-4}} - \frac{(10)^2}{(0.1)^2}} = 47.7 \text{ Hz } \textbf{Ans.}$$

$$\text{Dynamic resistance (or impedance)} = \frac{L}{CR} = \frac{0.1}{1 \times 10^{-4} \times 10} = 100\ \Omega \textbf{ Ans.}$$

$$\text{Input current at resonant frequency, } I = \frac{V}{L/CR} = \frac{100}{100} = 1.0 \text{ A } \textbf{Ans.}$$

The ratio of circulating current to line current at resonant frequency,

$$\frac{I_C}{I} = \frac{2\pi f_r L}{R} = \frac{2\pi \times 47.7 \times 0.1}{10} = 3 \textbf{ Ans.}$$

Example 5.8. A coil of 20 Ω resistance and 0.2 H inductance is connected in parallel with a capacitor of 100 μF capacitance. Find the frequency of resonance and the effective impedance at resonance.

[*Pb. Technical Univ. Basic Electrical and Electronics Engineering December 2008*]

Solution : Resonant frequency, $f_r = \frac{1}{2\pi}\sqrt{\frac{1}{LC} - \frac{R^2}{L^2}} = \frac{1}{2\pi}\sqrt{\frac{1}{0.2 \times 100 \times 10^{-6}} - \frac{20^2}{0.2^2}} = 31.83 \text{ Hz}$ **Ans.**

$$\text{Effective impedance at resonance, } Z = \frac{L}{CR} = \frac{0.2}{100 \times 10^{-6} \times 20} = 100\ \Omega \textbf{ Ans.}$$

Example 5.9. An inductive circuit of resistance 2 Ω and inductance 0.014 H is connected to a 250 V, 50 Hz supply. Which capacitance if placed in parallel will produce resonance ?

Solution : At, resonance, $Z = \sqrt{L/C}$

or $C = \frac{L}{Z^2}$

$$\text{Impedance of inductive circuit, } Z = \sqrt{R^2 + (2\pi f L)^2} = \sqrt{2^2 + (2\pi \times 50 \times 0.014)^2} = 4.832\ \Omega$$

$$\text{Required capacitance } C = \frac{L}{Z^2} = \frac{0.014}{(4.832)^2} = 600\ \mu\text{F} \textbf{ Ans.}$$

5.4. COMPARISON BETWEEN VOLTAGE RESONANCE AND CURRENT RESONANCE

The comparative results of voltage resonance and current resonance studied so far are summarized below in the tabular form.

S.No.	Particulars	Voltage Resonance	Current Resonance
1.	Impedance under resonant conditions	Z = R, the minimum	$Z = \frac{L}{CR}$, the maximum
2.	Magnitude of current under resonant condition	I = V/R, the maximum	$I = \frac{V}{L/CR}$, the minimum
3.	Power factor at resonance	Unity	Unity
4.	Resonant frequency	$f_r = \frac{1}{2\pi\sqrt{LC}}$	$f_r = \frac{1}{2\pi}\sqrt{\frac{1}{LC} - \frac{R^2}{L^2}}$
5.	Magnification occurs of	Voltage	Current
6.	Magnification	ωL/R	ωL/R
7.	Bandwidth	R/2πL Hz	R/2πL Hz

HIGHLIGHTS

1. An R-L-C series circuit is said to be in electrical resonance when

$$X_L = X_C \quad \textit{i.e.} \text{ the net reactance } X = 0$$

Under resonant conditions

(i) Circuit impedance, Z = R

(ii) The current flowing through the circuit is maximum and is in phase with the applied voltage and will be equal to V/R.

(iii) The voltage drop across inductance is equal to the voltage drop across capacitance and is maximum.

(iv) The power factor is unity.

(v) The power expended, P = VI watts and

(vi) The resonant frequency, $f_r = \dfrac{1}{2\pi\sqrt{LC}}$ Hz

Q-factor of a series resonant circuit is given as

$$\text{Q-factor} = \frac{\omega_r L}{R} = \frac{2\pi f_r L}{R} = \frac{1}{R}\sqrt{\frac{L}{C}}$$

$$\text{Bandwidth, BW} = \frac{R}{2\pi L} = \frac{f_r}{\text{Q-factor}}$$

2. A parallel ac circuit is said to be in resonance when it draws no reactive current.

Under resonant condition of parallel ac circuit

$$\text{Impedance, Z} = \frac{L}{CR}, \text{ the maximum}$$

$$\text{Current, I} = \frac{V}{L/CR}, \text{ the minimum}$$

$$\text{Power factor} = \text{Unity}$$

$$\text{Resonant frequency, } f_r = \frac{1}{2\pi}\sqrt{\frac{1}{LC} - \frac{R^2}{L^2}} \simeq \frac{1}{2\pi\sqrt{LC}} \quad \text{if R is negligibly small}$$

$$\text{Q-factor} = \frac{1}{R}\sqrt{L/C}$$

$$\text{Bandwidth} = \frac{R}{2\pi L} \text{ Hz}$$

EXERCISES

1. Draw and explain the phasor diagram of RLC series circuit and give the condition for resonance in this circuit. **[Pb. Technical Univ. Basic Electrical and Electronics Engineering December 2009]**
2. Discuss various characteristics of a series R-L-C circuit. Derive mathematical expressions in support of your discussion. **[Pb. Technical Univ. Basic Electrical and Electronics Engineering May 2009]**
3. Discuss the resonance in R-L-C circuit. What is the condition for it ? **[Pb. Univ. Elements of Electrical Engineering-II December 1988]**
4. An alternating voltage V is connected to a circuit containing resistor, inductor and capacitor in series. The supply frequency is gradually varied. Explain the observations.
5. Derive the conditions of resonance in an R-L-C series circuit. **[Pb. Technical Univ. Basic Electrical and Electronics Engineering December 2006]**
6. State and explain the condition of resonance in a series R-L-C circuit. Draw the phasor diagram also.
7. Define the Q-factor for the series resonant circuit and express it in terms of the circuit parameters.

8. Discuss the R-L-C parallel resonant circuit.

[Pb. Technical Univ. Basic Electrical and Electronics Engineering December 2008]

9. What are the features of resonance in parallel circuits ?

10. Explain 'series resonance' and 'parallel resonance'. Mention their similarities and dissimilarities.

SHORT ANSWER TYPE QUESTIONS WITH ANSWERS

Q. 1. What is meant by resonant frequency ? **[P.T.U. B.E.E.E. May 2010]**

Ans. The frequency of supply at which two reactances (inductive reactance and capacitive reactance) are equal is called resonant frequency.

Q. 2. What is series resonance ?

Ans. When in an R-L-C series circuit the inductive reactance equals capacitive reactance, the circuit is said to be in resonance, called the series resonance.

Q. 3. Why is the series resonance called the voltage resonance ?

Ans. Since in series resonance the voltage across inductance and capacitance is maximum, it is called the voltage resonance.

Q. 4. Why is the series resonant circuit also called the acceptor circuit ?

Ans. Series resonant circuit accepts currents at one particular frequency but rejects currents of other frequencies and, therefore, series resonant circuit is called the acceptor circuit.

Q. 5. What do you understand by the term bandwidth of a resonant circuit ?

Ans. The frequency band over which the magnitude of current is equal to $\frac{1}{\sqrt{2}}$ times the peak current is referred to as bandwidth of the circuit.

Q. 6. What is the condition of resonance in an R-L-C series circuit and also in an R-L-C parallel circuit ?

[P.T.U. December 2001]

Ans. Impedance of an R-L-C series circuit (shown in Fig. 5.1 page 127) is given as

$$Z = \sqrt{R^2 + (X_L - X_C)^2} = \sqrt{R^2 + \left[\omega L - \frac{1}{\omega C}\right]^2}$$

If for some frequency of applied voltage $X_L = X_C$ in magnitude then (*i*) net reactance is zero (*ii*) impedance of the circuit Z = R (*iii*) current flowing through the circuit is maximum and in phase with the applied voltage, the magnitude of current will be equal to V/R. (*iv*) voltage drop across the inductance is equal to the voltage drop across capacitance and is maximum (*v*) power factor is unity and (*vi*) power expended = VI watts.

When this condition exists, the series circuit is said to be in resonance and the frequency at which it occurs is known as resonance frequency, which is derived as follows :

At resonance frequency (say f_r)

$$X_L = 2\pi f_r L \text{ and } X_C = \frac{1}{2\pi f_r C}$$

Since for resonance $X_L = X_C$,

$$2\pi f_r L = \frac{1}{2\pi f_r C}$$

$$\text{or } f_r = \frac{1}{2\pi\sqrt{LC}}$$

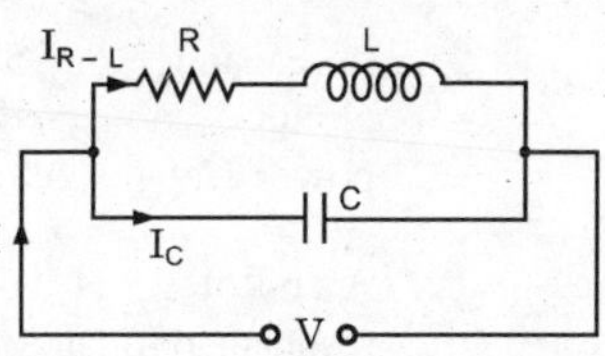

Fig. 1

An R-L-C parallel circuit shown in Fig. 1 is said to be in electrical resonance when the reactive (or wattless) component of line current is zero *i.e.*, when reactive component of R-L branch current is equal to current drawn by capacitance C.

When an R-L-C parallel circuit is in resonance, then (*i*) net susceptance is zero *i.e.*, $\frac{1}{X_C} = \frac{X_L}{Z^2}$ or $Z = \sqrt{L/C}$, (*ii*) admittance is equal to conductance (*iii*) reactive or wattless component of line current is zero *i.e.*, circuit power factor is unity (*iv*) impedance is purely resistive, maximum in magnitude and is equal to L/CR (*v*) line current is minimum and is equal to $\frac{V}{L/CR}$ in magnitude and in phase with applied voltage and (*vi*) frequency is equal to $\frac{1}{2\pi}\sqrt{\frac{1}{LC}-\frac{R^2}{L^2}}$ Hz.

Q. 7. What is parallel resonance and why it is also known as current resonance ?

[P.T.U. B.E.E.E. December 2006]

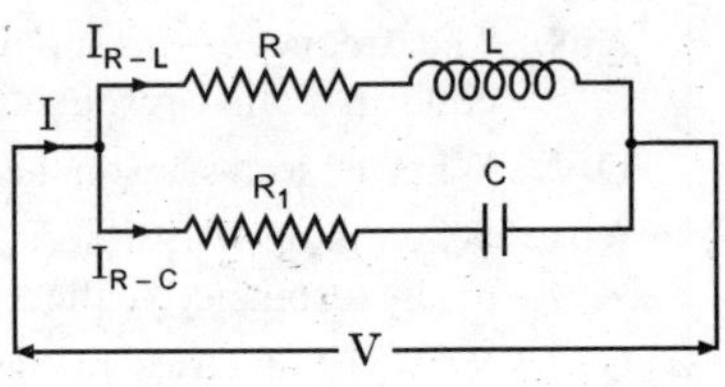

Fig. 2

Ans. When an inductive reactance and a capacitive reactance are connected in parallel, as shown in Fig. 2, conditions may reach under which *current resonance* (also known as *parallel or anti-resonance*) will take place.

Such a circuit is said to be in electrical resonance when the reactive (or wattless) component of line current becomes zero. The frequency at which this happens is known as *resonant frequency*.

Since in parallel resonant circuit, circulating current between the branches is many times the line current, such type of resonance is called the current resonance.

Q. 8. Why is parallel resonant circuit also called the rejector circuit ?

Ans. Parallel resonant circuit is also called the rejector circuit because at resonant frequency the line current is minimum or it almost rejects it.

Q. 9. What is meant by Q-factor of series resonant circuit ?

Ans. The Q-factor of a series resonant circuit may be given as the voltage magnification that the circuit produces at resonance.

$$\text{Q-factor at resonance, } Q_r = \frac{\omega_r L}{R} = \frac{2\pi f_r L}{R} = \frac{1}{R}\sqrt{\frac{L}{C}}$$

Q. 10. What is meant by Q-factor of a parallel resonant circuit ?

Ans. Q-factor of a parallel resonant circuit is defined as the ratio of the circulating current to the line current at resonance or as the current magnification.

Q-factors for series resonance and parallel resonance are the same.

PROBLEMS

1. A resistance of 15 Ω, choke coil of 0.05 H and the capacitor of 300 μF are connected in series. If the circuit is connected to 250 V, 50 Hz, power supply, find (*i*) circuit impedance (*ii*) power factor (*iii*) absorbed power (*iv*) resonant frequency and (*v*) Q-factor of the coil.
[**Ans.** (*i*) 15.84 Ω (*ii*) 0.947 (lag) (*iii*) 3,735 W (*iv*) 41 Hz (*v*) 1.056]
2. A coil of 1.5 H inductance and 5 Ω resistance and a capacitor of 5 μF are connected in series. Calculate the resonant frequency and the power consumed if it is connected to a 200 V ac mains. [**Ans.** 58 Hz; 8,000 W]
3. A circuit comprises an inductance of 26 μH associated with resistance of 0.1 Ω in parallel with a capacitance of 0.16 μF. Find the ratio of the circulating current to the supply current at resonant frequency. [**Ans.** 127.5]
4. An ac voltage of 200 V and variable frequency is applied to a coil of 20 Ω resistance and 0.3 H inductance connected in parallel with a capacitance of 200 μF. Determine the parallel resonant frequency and the input current at this frequency. [**Ans.** 20.5 Hz; 2.667 A]
5. An inductive circuit of resistance 2 Ω and inductance 0.01 H is connected to 250 V, 50 Hz supply. What capacitance placed in parallel will produce resonance ? Find the total current taken from the supply and the current in the branch circuits. [**Ans.** 721 μF; 36.05 A; 125 A; 67. $\angle -57.52°$ A; 56.63 $\angle 90°$ A]

□□□

CHAPTER 6

Three Phase Systems

6.1. INTRODUCTION

INSIDE THIS CHAPTER

The type of alternating currents and voltages discussed so far in the book are termed as *single phase currents and voltages* as they consist of a single alternating current and voltage waves. Single phase systems involving single phase currents and voltages are quite satisfactory for domestic applications. Even the motors employed in domestic applications are mostly single phase, for example, motors for mixers, coolers, fans, air-conditioners, refrigerators. However, the single phase system has its own limitations and, therefore, has been replaced by polyphase system. For supplying power to electric furnaces 2-phase system is generally employed. Six phase system is usually employed in connection with converting machinery and apparatus. For general supply three-phase system is universally used. For generation, transmission and distribution of electric power 3-phase system has been universally adopted. The two-phase supply and six-phase supply are obtained from 3-phase supply.

Polyphase system means the system which consists of numerous (poly means numerous or multiple) windings or circuits (phase* means winding or a circuit).

A polyphase system is essentially a combination of several single phase voltages having same magnitude and frequency but displaced from one another by equal angle (electrical), which depends upon the number of phases and can be determined from the following relation

$$\text{Electrical displacement} = \frac{\text{360 electrical degrees}}{\text{Number of phases}} \qquad ...(6.1)$$

The above relation does not hold good for two phase windings, which are displaced by 90 electrical degrees apart.

A supply system is said to be symmetrical when several voltages of the same frequency have equal magnitude and are displaced from one another by equal time angle. 3-phase, 3-wire or 4-wire supply system will be symmetrical when the line-to-line voltages are equal in magnitude and displaced in phase by 120 electrical degrees with respect to each other. Further in a four-wire system the voltage with respect to

* Each emf of a polyphase system is referred to as a *phase* emf.

neutral of the three phase wires should be equal to one another in magnitude and displaced in phase by 120° with respect to each other.

A 3-phase supply will be unbalanced when either of the three-phase voltages are unequal in magnitude or the phase angle between these phases is not equal to 120°.

A load circuit is said to be *balanced* when the loads (impedances) connected in various phases are same in magnitude as well as in phase. Any three phase load in which the impedances in one or more phases differ from the impedances of other phases is called an *unbalanced three-phase load**.

In case one phase out of three phases of a 3-phase supply connected to the 3-phase load is not available, then such a condition is called the *single phasing*.

6.1.1. Merits and Demerits of Polyphase System Over Single Phase System

The advantages of a polyphase system over single phase system are enumerated below:

(*i*) In a single phase circuit the power delivered is pulsating. Even when current and voltage are in phase, the power is zero twice in each cycle, and when the current leads or lags behind the voltage, the power is negative twice and zero four times during each cycle. This is not objectionable for lighting and small motors but with large motors it causes excessive vibrations. In a polyphase system the total power delivered is constant if loads are balanced though the power of any one phase or circuit may be negative. So polyphase system is highly desirable particularly for power loads.

(*ii*) The rating of a given machine increases with the increase in number of phases. For example output of a 3-phase motor is 1.5 times the output of a single phase motor of same size.

(*iii*) Single phase induction motors have no starting torque and so it is necessary to provide these motors with an auxiliary means of starting, but in case of three-phase motors except synchronous motors there is no need of providing an auxiliary means for starting.

(*iv*) Power factor of a single-phase motor is lower than that of a polyphase motor of the same rating (output and speed). The efficiency of a polyphase motor is also higher than that of a single-phase motor.

(*v*) Three phase-system requires 3/4th weight of copper of that required by single-phase system to transmit the same amount of power at a given voltage and over a given distance.

(*vi*) Rotating magnetic field can be set up by passing polyphase currents through stationary coils.

(*vii*) Polyphase system is more capable and reliable than single-phase system.

(*viii*) Parallel operation of polyphase alternators is simple as compared to that of single-phase alternators because of pulsating reaction in single phase alternators.

However, three-phase operation is not practical for domestic applications where motors are usually smaller than 1 kW and where lighting circuits supply most of the load.

6.1.2. Commonly Used Polyphase System

Although there are several polyphase systems such as two-phase, three-phase but 3-phase system is invariably adopted because of its inherent advantages over all other polyphase systems.

The demand for two-phase system has almost disappeared because that does not have any advantage that is not equalled or surpassed by three-phase system in generation, transmission or utilization. As already mentioned, 3-phase system is universally employed for generation, transmission and distribution of electric power. Two-phase supply and six-phase supply, when required, is obtained from 3-phase supply.

---

* Normally, unbalanced load is not employed on 3-phase, 3-wire ungrounded star-connected system because unbalanced loading may cause different voltage drops in lines. Consequently the three phase voltages are different (or unbalanced) in magnitude as well as in phase and it is quite possible that one phase voltage may exceed the line voltage. Such a condition is undesirable, since some loads may operate inefficiently due to lowering of the voltage and the other equipment may get damaged due to overvoltage.

A knowledge of 3-phase power system is, therefore, essential for understanding of power technology. Fortunately, the basic circuit technique used in solving single phase circuits are directly applicable to 3-phase circuits because three phases are identical and one phase represents the behaviour of all the three phases.

In this chapter we will discuss only 3-phase systems.

6.2. GENERATION OF 3-PHASE EMF

When three coils fastened rigidly together and 120º (electrical) apart rotate about the same axis in a uniform magnetic field, the induced emf in each of them will have a phase difference of 120º or 2/3π radians.

Consider three identical coils a_1a_2, b_1b_2 and c_1c_2 mounted on the same axis but displaced from each other by 120º rotating in counter-clockwise direction in a bipolar magnetic field, as shown in Fig. 6.1 (*a*). Here a_1, b_1 and c_1 are the start terminals and a_2, b_2 and c_2 are the finish terminals of the three coils. When the coil a_1a_2 is in position AB the induced emf in this coil is zero and is increasing in positive direction, the coil b_1b_2 is 120º behind coil a_1a_2 so the emf induced in this coil is approaching its maximum negative value and the coil c_1c_2 is 240º behind coil a_1a_2 so the emf induced in the coil has passed its positive maximum value and is decreasing.

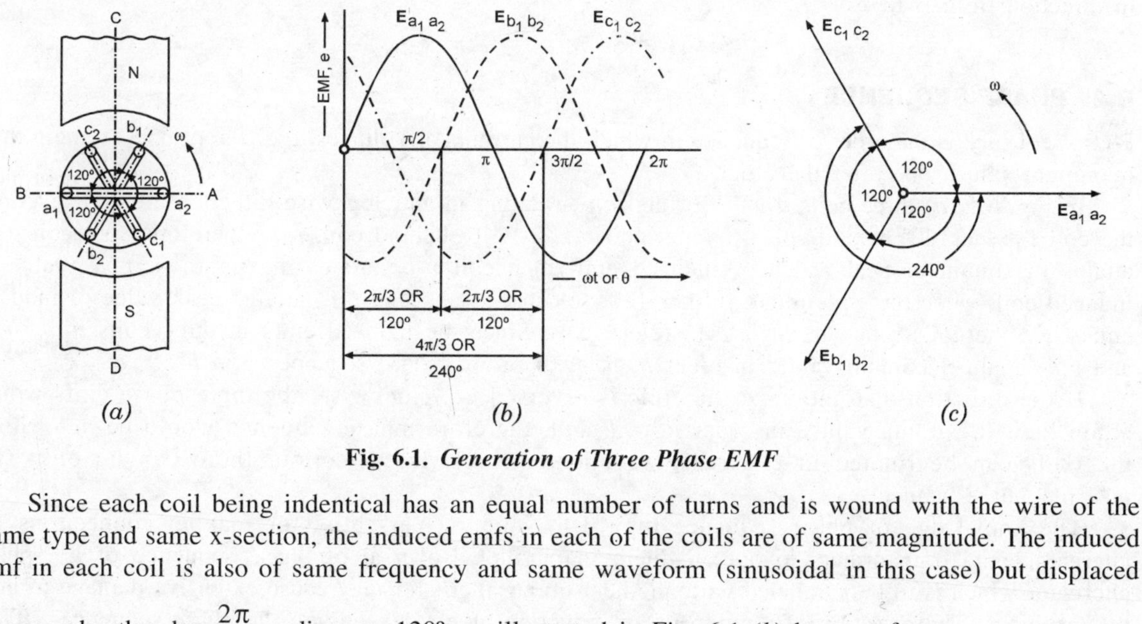

(*a*) (*b*) (*c*)

Fig. 6.1. ***Generation of Three Phase EMF***

Since each coil being indentical has an equal number of turns and is wound with the wire of the same type and same x-section, the induced emfs in each of the coils are of same magnitude. The induced emf in each coil is also of same frequency and same waveform (sinusoidal in this case) but displaced from each other by $\frac{2\pi}{3}$ radians or 120º, as illustrated in Fig. 6.1 (*b*) by waveforms.

Accordingly, the instantaneous values of the emfs induced in coils a_1a_2, b_1b_2 and c_1c_2 may be given as

$$\left.\begin{aligned} e_{a_1a_2} &= E_{max} \sin \omega t \\ e_{b_1b_2} &= E_{max} \sin\left(\omega t - \frac{2\pi}{3}\right) \text{ or } E_{max} \sin(\omega t - 120^\circ) \\ e_{c_1c_2} &= E_{max} \sin\left(\omega t - \frac{4\pi}{3}\right) \text{ or } E_{max} \sin(\omega t - 240^\circ) \text{ or } E_{max} \sin(\omega t + 120^\circ) \end{aligned}\right\} \quad ...(6.2)$$

if $t = 0$ corresponds to the instant when the voltage or emf of coil a_1a_2 passes through zero and increases in positive direction.

6.3. DOUBLE SUBSCRIPT NOTATION

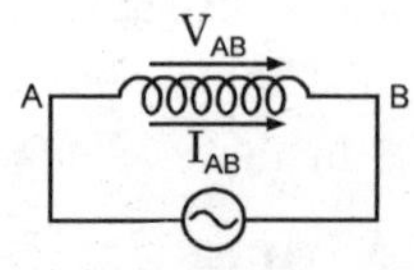

Fig. 6.2

The solution of problems involving circuits and systems containing a number of voltages and currents is simplified and less susceptible to error if the voltage and current phasors are designated by some systematic notation. The double subscript notation is a very useful concept from this point of view. In this notation two letters are placed at the foot of the symbol for voltage or current, the order in which the subscripts are written, indicates the direction in which the voltage acts or current flows. For example, if a voltage in a circuit acts in such a direction as to cause a current to flow from A to B, the positive direction of voltage is from A to B, and the voltage may be represented by V_{AB} or E_{AB}, the order of the subscripts denoting that the voltage or emf is acting from A to B. If the voltage is indicated by V_{BA} or E_{BA} it means that point B is positive w.r.t. point A (during its positive half-cycle) and the voltage causes the current to flow from B to A *i.e.* V_{BA} or E_{BA} indicates that voltage or emf acts in a direction opposite to that in which V_{AB} or E_{AB} acts.

$$\text{So } \mathbf{V}_{BA} = -\mathbf{V}_{AB} \qquad ...(6.3)$$

Similarly I_{AB} indicates that current is flowing from A to B but I_{BA} indicates that current is flowing in direction from B to A

$$i.e.\ \mathbf{I}_{BA} = -\mathbf{I}_{AB}$$

6.4. PHASE SEQUENCE

Phase sequence is the order or sequence in which the currents or voltages in different phases attain their maximum values one after the other.

In Fig. 6.1 (*a*) three coils a_1a_2, b_1b_2 and c_1c_2 rotating in anticlockwise direction are shown. Since the coil b_1b_2 is 120° behind coil a_1a_2 and coil c_1c_2 is 240° behind coil a_1a_2, therefore, first coil a_1a_2 attains maximum or peak value of induced emf, then coil b_1b_2 attains maximum or peak value of induced emf, when the coils rotate further 120° (electrical) and coil c_1c_2 attains peak value of induced emf when the coils rotate through 240° (electrical). Since the induced emfs in three coils a_1a_2, b_1b_2 and c_1c_2 attain maximum values in order *a*, *b*, *c*, therefore, phase sequence is *a b* c.

If the direction of rotation of the coils is reversed *i.e.* clockwise, the three-phase emfs would attain their maximum values in order *a*, *c*, *b* and, therefore, phase sequence would be *acb*. Since the coils can be rotated in clockwise or counter-clockwise direction, there can be only two possible phase sequences.

The knowledge of phase sequence of a 3-Φ supply is essential for making connections for alternators and transformers to operate them in parallel. Reversal of phase sequence of a 3-phase alternator which is to be paralleled with another similar alternator may cause extensive damage to both the machines. The direction of rotation of an induction motor depends on the phase sequence. If the phase sequence is reversed by interchanging any two terminals of the 3-Φ supply, the motor would rotate in the opposite direction.

The phase sequence of the voltage applied to the load, in general, is determined by the order in which three phase lines are connected. In the case of 3-phase unbalanced loads, the effect of reversal of phase sequence, in general causes a completely different set of values of the currents. So while working on such systems it is essential to specify the phase sequence clearly to avoid unnecessary confusion.

6.5. INTERCONNECTION OF THREE PHASES

Each coil of three phases has two terminals (one 'start' and another 'finish') and if individual phase is connected to a separate load circuit, as illustrated in Fig. 6.3, we get a non-interlinked three-phase, six-wire system. Then, by Ohm's law, the current in individual phase will be

$$I_P = E_P/Z_P \quad ...(6.4)$$

where I_P is the phase current, E_P is the phase emf or voltage.

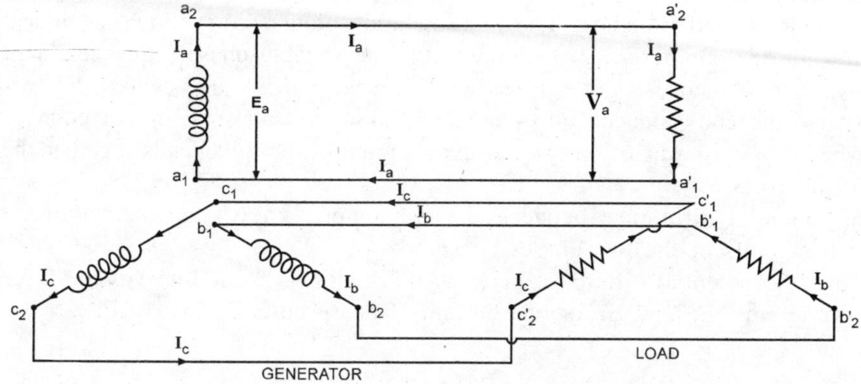

Fig. 6.3. ***Non-interlinked Three-Phase System***

In such a system, each circuit will require two conductors, therefore, 6 conductors in all.

Hence such a system will be very complicated and expensive.

In practice non-interlinked 3-phase, 6-wire system is not employed but the phases are interconnected to give 3 or 4-wire, 3-phase system. The two principal methods of connecting the separate phases of three-phase system are the star (or wye) connections and the delta (or mesh) connections.

The interconnections must be carried out in such a way that if closed circuits are formed the sum of instantaneous emfs in them must be zero in order that there shall be no circulating current in these circuits.

6.6. STAR OR WYE (Y) CONNECTED SYSTEM

This system is obtained by joining together similar ends, either the start or the finish, the other ends are joined to the line wire, as shown in Fig. 6.4 (*a*). The common point N at which similar (start or finish) ends are connected is called the *neutral or star point.* Ordinarily only three wires are carried to the external circuit giving 3-phase 3-wire star-connected system but sometimes a fourth wire, known as *neutral wire*, is carried to the neutral point of the external load circuit giving 3-phase, 4-wire star-connected system. The voltage between any line and the neutral point *i.e.* voltage across the phase winding is called the *phase voltage*; while the voltage between any two outers is called the *line voltage*. The neutral point is usually earth-connected.

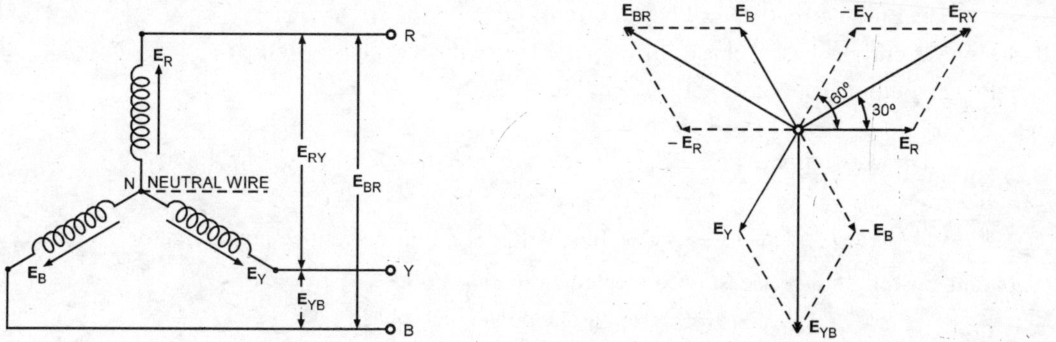

(a) Connection Diagram *(b) Phasor Diagram of Line and Phase Voltages*

Fig. 6.4. ***Three-Phase Star-Connected System***

In Fig. 6.4 (*a*) positive directions of emfs are taken from star point outwards. The arrowheads on emfs and currents indicate the positive direction and not their directions at any instant. The three phases be either numbered as 1, 2, 3, or a, b, c or R, Y, B. In Fig. 6.4 (*a*) the three-phases are numbered as usually done; R, Y and B indicate the three natural colours red, yellow and blue respectively.

It is worthwhile to state here that there are only two possible phase sequences, namely RYB and RBY. By convention sequence RYB is taken as positive and RBY as negative.

In Fig. 6.4 (*b*) the emfs induced in 3-phases are shown by phasors. In star connection, there are two windings between each pair of outers and due to joining of similar ends together the emf induced in them are in opposition.

Hence the potential difference between the two outers, known as *line voltage,* is the phasor difference of phase emfs of the two phases concerned.

For example, the potential difference between outers R and Y or line voltage E_{RY} is the phasor difference of phase emfs E_R and E_Y or phasor sum of phase emfs E_R and $(-E_Y)$.

i.e. $\mathbf{E}_{RY} = \mathbf{E}_R - \mathbf{E}_Y$ (phasor difference)

or $\mathbf{E}_{RY} = \mathbf{E}_R + (-\mathbf{E}_Y)$ (phasor sum)

Since phase angle between phasors E_R and $(-E_Y)$ is 60º,

From phasor diagram shown in Fig. 6.4 (*b*)

$$E_{RY} = \sqrt{E_R^2 + E_Y^2 + 2\,E_R\,E_Y \cos 60^\circ}$$

Let $E_R = E_Y = E_B = E_P$ (phase voltage) (Assuming balanced system)

$$\text{then line voltage } E_{RY} = \sqrt{E_P^2 + E_P^2 + 2E_P\,E_P \times (1/2)} = \sqrt{3}\,E_P$$

Similarly potential difference between outers Y and B

$$\text{or line voltage } \mathbf{E}_{YB} = \mathbf{E}_Y - \mathbf{E}_B = \sqrt{3}\,E_P \text{ in magnitude}$$

and potential difference between outers B and R

$$\text{or line voltage, } \mathbf{E}_{BR} = \mathbf{E}_B - \mathbf{E}_R = \sqrt{3}\,E_P \text{ in magnitude}$$

In balanced star (Y) system E_{RY}, E_{YB} and E_{BR} are equal in magnitude and are known as *line voltages*. Let each of them be equal to E_L

$$\therefore\ E_L = \sqrt{3}\,E_P \text{ (in balanced star-connected system)} \qquad ...(6.5)$$

Since in star-connected system each line conductor is connected to separate phase, so current flowing through the lines and phases is same

i.e. Line current, I_L = Phase current, I_P ...(6.6)

If the phase current has a phase difference of Φ with the phase voltage

$$\text{Power output per phase} = E_P\,I_P \cos \Phi \qquad ...(6.7)$$

$$\text{Total power output, } P = 3\,E_P I_P \cos \Phi = 3\,\frac{E_L}{\sqrt{3}}\,I_L \cos \Phi = \sqrt{3}\,E_L\,I_L \cos \Phi \qquad ...(6.8)$$

i.e. Power = $\sqrt{3}$ × line voltage × line current × power factor

Apparent power of 3-phase star-connected system

= 3 × apparent power per phase

$$= 3\,E_P\,I_P = 3\,\frac{E_L}{\sqrt{3}} \times I_L = \sqrt{3}\,E_L\,I_L \qquad ...(6.9)$$

It should be noted that in a balanced star-connected system

(*i*) Line voltages are 120° apart.

(*ii*) Line voltages are 30° ahead of the respective phase voltages.

(*iii*) Line voltages are $\sqrt{3}$ times of phase voltages.

(*iv*) Line currents are equal to phase currents.

(*v*) The angle between line currents and the corresponding line voltages is (30° ± Φ); +ve for lagging currents and –ve for leading currents.

(*vi*) True power output = $\sqrt{3}\ E_L\ I_L \cos\Phi$, where Φ is the angle between the respective phase current and phase voltage (not between line current and line voltage).

(*vii*) Apparent power = $\sqrt{3}\ E_L I_L$

(*viii*) In balanced system, the potential of neutral or star point is zero because potential at neutral or star point,

$$\mathbf{E_{NR} + E_{NY} + E_{NB}} = 0$$

Example 6.1. The phase voltage of a 3-phase star-connected alternator is 6,000 V. Calculate the line voltages when (*i*) the phases are correctly connected (*ii*) the connections to one of the phases are reversed.

Solution : (*i*) When the phases are correctly connected *i.e.* similar ends are connected together, as shown in Fig. 6.4 (*a*) all line voltages will be equal in magnitude and will be given as

$$\text{Each line voltage, } E_L = \sqrt{3}\ E_P = \sqrt{3} \times 6{,}000 = 10{,}392 \text{ V } \textbf{Ans.}$$

(*ii*) When the connections to one of the phases are reversed, as shown in Fig. 6.5 (*a*), voltage in phase R is in reverse direction. The voltage phasor diagram for such a case is shown in Fig. 6.5 (*b*).

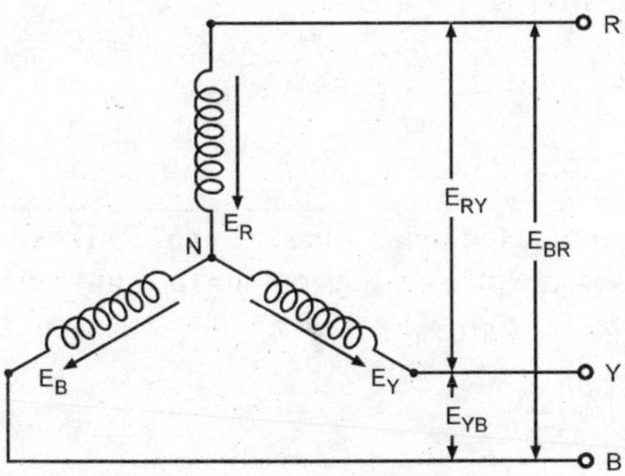

(*a*) Connection Diagram

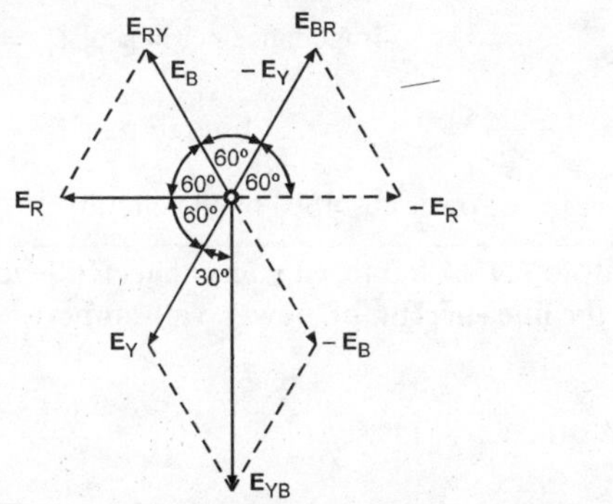

(*b*) Phasor Diagram of Line and Phase Voltages

Fig. 6.5

It should be noted that E_R has been drawn in reverse direction, so the angles between the three phase voltages are of 60°, as shown in 6.5 (*b*)

$$\text{So} \quad \mathbf{E_{RY}} = \mathbf{E_R} - \mathbf{E_Y} = 2E_P \cos 60° = E_P = 6{,}000 \text{ V } \textbf{Ans.}$$

$$\mathbf{E_{YB}} = \mathbf{E_Y} - \mathbf{E_B} = 2E_P \times \cos 30° = 2 \times 6{,}000 \times \frac{\sqrt{3}}{2} = 10{,}392 \text{ V } \textbf{Ans.}$$

$$\text{and } \mathbf{E_{BR}} = \mathbf{E_B} - \mathbf{E_R} = 2E_P \cos 60° = 6{,}000 \text{ V } \textbf{Ans.}$$

Example 6.2. A star-connected 3-phase load has a resistance of 6 Ω and an inductive reactance of 8 Ω in each branch. A line-to-line voltage of 220 V is impressed through a 3-Φ auto-transformer. Find the voltage across each branch, line voltages and line currents and total active power.

Solution : Line voltage, V_L = Line-to-line voltage = 220 V **Ans.**

$$\text{Phase voltage, } V_P = \frac{V_L}{\sqrt{3}} = \frac{220}{\sqrt{3}} = 127 \text{ V } \textbf{Ans.}$$

$$\text{Impedance per phase, } Z_P = \sqrt{R_P^2 + X_P^2} = \sqrt{6^2 + 8^2} = 10 \ \Omega$$

$$\text{Phase currents, } I_P = \frac{V_P}{Z_P} = \frac{127}{10} = 12.7 \text{ A}$$

$$\text{Total active power, } P = 3 \times \text{power of each phase}$$
$$= 3I_P^2 \ R_P = 3 \times (12.7)^2 \times 6 = 2{,}904 \text{ watts } \textbf{Ans.}$$

Total active power can also be determined from the formula $P = \sqrt{3}\ V_L\ I_L \cos \Phi$

$$\text{where } \cos \Phi = \frac{R}{Z} = \frac{6}{10} = 0.6$$

$$\text{Line current, } I_L = I_P = 12.7 \text{ A } \textbf{Ans.}$$

$$\text{and } \quad P = \sqrt{3}\ V_L\ I_L \cos \Phi$$
$$= \sqrt{3} \times 220 \times 12.7 \times 0.6 = 2{,}904 \text{ watts, as before } \textbf{Ans.}$$

Example 6.3. Three similar coils each of resistance 20 Ω and an inductance of 0.5 H are connected in star across a three-phase supply source of 400 V, 50 Hz. Calculate the line current and power absorbed by the circuit.

[Pb. Technical Univ. Basic Electrical Engineering, December 2002]

Solution :

$$\text{Supply line voltage, } V_L = 400 \text{ V}$$

$$\text{Phase voltage, } V_P = \frac{V_L}{\sqrt{3}} = \frac{400}{\sqrt{3}} = 231 \text{ V}$$

$$\text{Resistance per phase, } R_P = 20 \ \Omega$$

$$\text{Inductive reactance per phase, } X_L = 2\pi f L = 2\ \pi \times 50 \times 0.5 = 157.08 \ \Omega$$

$$\text{Impedance per phase, } Z_P = \sqrt{R_P^2 + X_L^2} = \sqrt{20^2 + 157.08^2} = 158.35 \ \Omega$$

$$\text{Line current, } I_L = \text{Phase current, } I_P = \frac{V_L}{Z_P} = \frac{231}{158.35} = 1.458 \text{ A } \textbf{Ans.}$$

$$\text{Power absorbed by the circuit, } P = 3I_P^2 R_P = 3 \times (1.458)^2 \times 20 = 127.6 \text{ W } \textbf{Ans.}$$

Example 6.4. A balanced star-connected load of (8 + *j*6) Ω/phase is connected to a 3-phase, 230 V, 50 Hz supply. Find the line current, pf, power, volt-amperes and reactive power. Draw the phasor diagram for the above circuit.

[Pb. Technical Univ. Electrical Engineering, June 1999]

Solution :

$$\text{Phase voltage, } V_P = \frac{\text{Line voltage, } V_L}{\sqrt{3}} = \frac{230}{\sqrt{3}} = 132.8 \text{ V}$$

$$\text{Impedance per phase, } Z_P = \sqrt{R_P^2 + X_P^2} = \sqrt{8^2 + 6^2} = 10 \ \Omega$$

$$\text{Phase current, } I_P = \frac{V_P}{Z_P} = \frac{132.8}{10} = 13.28 \text{ A}$$

$$\text{Line current, } I_L = I_P = 13.28 \text{ A } \textbf{Ans.}$$

$$\text{Power factor, } \cos \phi = \frac{R_P}{Z_P} = \frac{8}{10} = 0.8 \text{ (lagging) } \textbf{Ans.}$$

$$\text{Power, } P = \sqrt{3}\ V_L I_L \cos \phi$$
$$= \sqrt{3} \times 230 \times 13.28 \times 0.8 = 4{,}232 \text{ watts or } 4.232 \text{ kW } \textbf{Ans.}$$

$$\text{Total volt-amperes, } S = \sqrt{3}\ V_L\ I_L = \sqrt{3} \times 230 \times 13.28 = 5{,}290 \text{ VA } \textbf{Ans.}$$

$$\text{Reactive power, } Q = \sqrt{3}\ V_L\ I_L \sin \phi = \sqrt{3} \times 230 \times 13.28 \sin \cos^{-1}(0.8)$$
$$= \sqrt{3} \times 230 \times 13.28 \times 0.6 = 3{,}174 \text{ VAR or } 3.174 \text{ kVAR } \textbf{Ans.}$$

Phasor diagram for currents and voltages is shown in Fig. 6.6. (*b*)

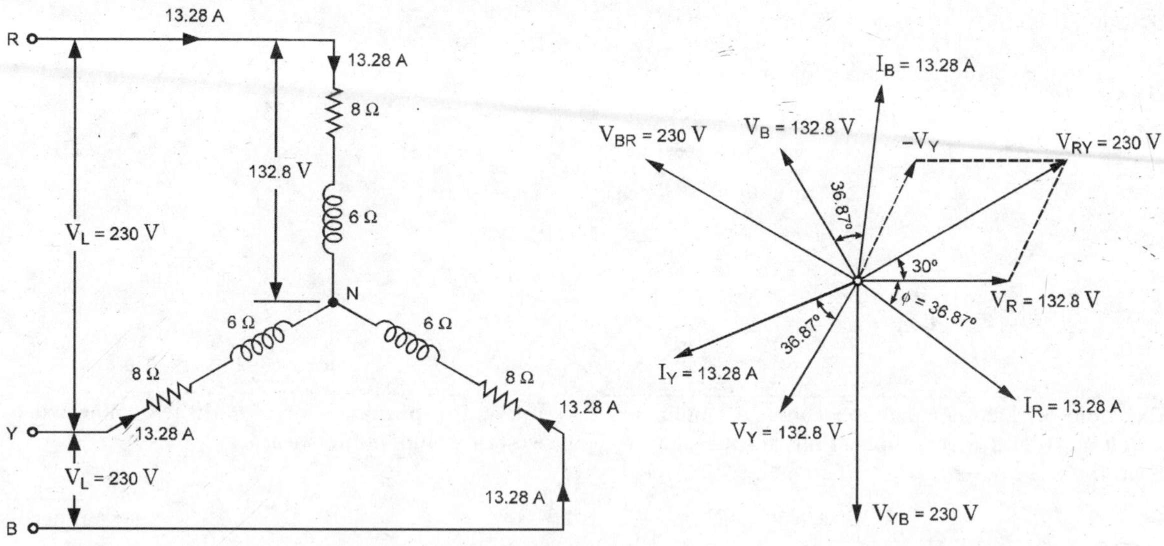

(a) Circuit Diagram *(b) Phasor Diagram For Currents and Voltages*

Fig. 6.6

Example 6.5. Each phase of a star-connected load consists of a resistance of 100 Ω in parallel with a capacitance 31.8 μF. Calculate the line current, power absorbed, total kVA and power factor of the load when connected to a 416 V, 3ϕ, 4 wire 50 Hz supply. *[Pb. Technical Univ. Basic Electrical Engineering December 2003]*

Solution : Phase voltage, $V_P = \frac{V_L}{\sqrt{3}} = \frac{416}{\sqrt{3}} = 240.2$ V

Taking phase voltage as reference phasor,

$$\mathbf{V_P} = 240.2 \angle 0° \text{ V}$$

Admittance per phase,

$$\mathbf{Y_P} = \frac{1}{R} + j\omega C$$

$$= \frac{1}{100} + j\, 2\pi \times 50 \times 31.8 \times 10^{-6}$$

$$= (0.01 + j0.01) \text{ or } 0.01414 \angle 45° \text{ S}$$

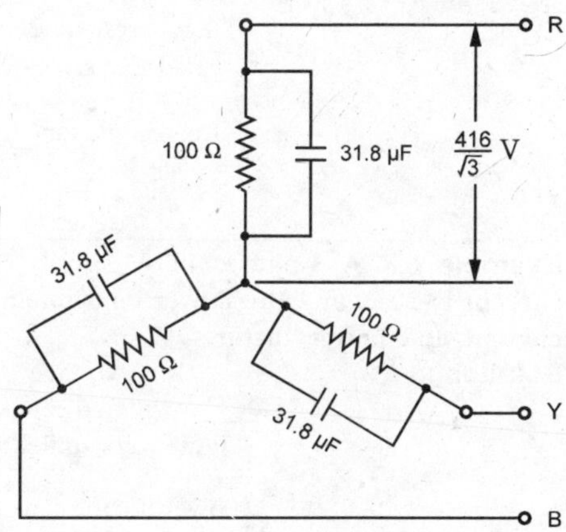

Fig. 6.7

Line current, $\mathbf{I_L} = \mathbf{I_P} = \mathbf{V_P}\,\mathbf{Y_P}$

$$= 240.2 \angle 0° \times 0.01414 \angle 45°$$

$$= 3.397 \angle 45° \text{ A}$$

So line current, I_L = 3.397 A **Ans.**

Power factor, cos ϕ = Cos 45° = 0.707 (leading) **Ans.**

Power absorbed, $P = \sqrt{3}\, V_L I_L \cos\phi = \sqrt{3} \times 416 \times 3.397 \times 0.707$ = 1,730 watts or 1.73 kW **Ans.**

Total kVA, $S = \frac{\sqrt{3} V_L I_L}{1{,}000} = \frac{\sqrt{3} \times 416 \times 3.397}{1{,}000} = 2.45$ **Ans.**

Example 6.6. A balanced 3-phase star-connected load draws a current of 25 A at a lagging power factor of 0.8 when connected to a 3-phase 400 V (line-to-line) system. Calculate: true power, reactive power and voltage across each phase of load.

Solution :

Line voltage, $V_L = 400$ V

Voltage across each phase of load $= V_P = \frac{V_L}{\sqrt{3}} = \frac{400}{\sqrt{3}} = 231$ V **Ans.**

Line current, $I_L = 25$ A

Power factor, $\cos \Phi = 0.8$ (lag)

$$\sin \Phi = \sqrt{1 - \cos^2 \Phi} = \sqrt{1 - 0.8^2} = 0.6$$

$$\text{True power, } P = \sqrt{3}\, V_L I_L \cos \Phi$$

$$= \sqrt{3} \times 400 \times 25 \times 0.8 = 13{,}856 \text{ W or } 13.856 \text{ kW}$$ **Ans.**

$$\text{Reactive power, } Q = \sqrt{3}\, V_L I_L \sin \Phi$$

$$= \sqrt{3} \times 400 \times 25 \times 0.6 = 10{,}392 \text{ VAR or } 10.392 \text{ kVAR}$$ **Ans.**

Example 6.7. Three equal star-connected inductors take 9 kW at a power factor of 0.8 when connected to a 960 V, 3-Φ, 3-wire supply. Find the per phase load resistance and inductance.

Solution :

Line voltage, $V_L = 960$ V

$$\text{Phase voltage, } V_P = \frac{960}{\sqrt{3}} = 554.3 \text{ V}$$

Total power drawn = 9 kW = 9,000 watts

Power factor, $\cos \Phi = 0.8$ (lag) $\quad\because$ load is inductive

Total power taken, $P = 3\, V_P\, I_P \cos \Phi$

$9{,}000 = 3 \times 554.3 \times I_P \times 0.8$

$$\text{or phase current, } I_P = \frac{9{,}000}{3 \times 554.3 \times 0.8} = 6.766 \text{ A}$$

$$\text{Load impedance, } Z_L = \frac{V_P}{I_P} = \frac{554.3}{6.766} = 81.92\ \Omega$$

Load resistance, $R = Z \cos \Phi = 81.92 \times 0.8 = 65.54\ \Omega$ **Ans.**

Load reactance, $X_L = Z \sin \Phi = 81.92 \times 0.6 = 49.15\ \Omega$

$$\text{Load inductance, } L = \frac{X_L}{2\pi f} = \frac{49.15}{2\pi \times 50} = 0.156 \text{ H}$$ **Ans.**

(assuming supply frequency of 50 Hz)

Example 6.8. A 3-phase, 400 V supply is connected to a 3-phase star-connected balanced load. The line current is 20 A and the power consumed by the load is 12 kW. Calculate the impedance of the load, phase current and power factor.

Solution :

Line voltage, $V_L = 400$ V

Line current, $I_L = 20$ A

Power consumed, P = 12 k W = 12,000 W

$$\text{Power factor, } \cos \Phi = \frac{P}{\sqrt{3}\, V_L\, I_L} = \frac{12{,}000}{\sqrt{3} \times 400 \times 20} = 0.866$$ **Ans.**

Phase current, $I_P = I_L = 20$ A **Ans.**

$$\text{Phase voltage, } V_P = \frac{V_L}{\sqrt{3}} = \frac{400}{\sqrt{3}} = 231 \text{ V}$$

$$\text{Load impedance, } Z_P = \frac{V_L}{I_P} = \frac{231}{20} = 11.55\ \Omega$$ **Ans.**

6.7. MESH OR DELTA (Δ) CONNECTED SYSTEM

When the starting end of one coil is connected to the finishing end of another coil, as shown in Fig. 6.8 (*a*), the *delta or mesh connection is obtained.* The direction of emfs in the coils have been taken as positive from start end to finish end in Fig. 6.8 (*a*).

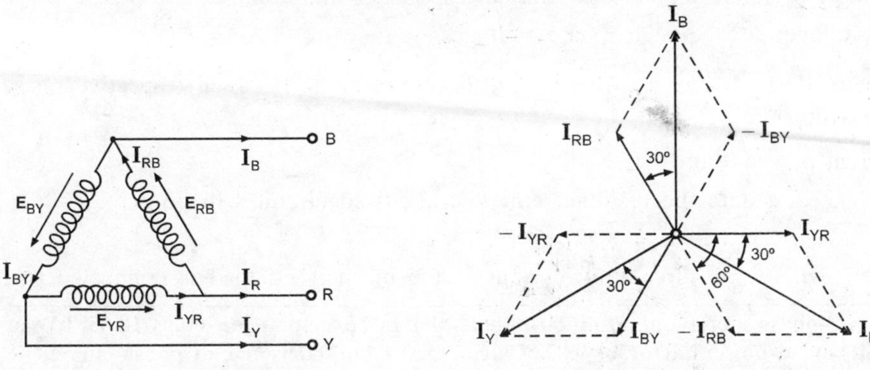

(a) Connection Diagram *(b) Phasor Diagram*

Fig. 6.8. ***Three Phase Delta-Connected System***

From Fig. 6.8 (*b*) it is obvious that line current is phasor difference of phase currents of two phases concerned. For example the line current in red outer I_R will be equal to the phasor difference of phase currents I_{YR} and I_{RB}. The current phasors are shown in Fig. 6.8 (*b*).

From Figs. 6.8 (*a*) and 6.8 (*b*) we have

$$\text{Line current, } \mathbf{I}_R = \mathbf{I}_{YR} - \mathbf{I}_{RB} = \mathbf{I}_{YR} + (-\mathbf{I}_{RB})$$

Since phase angle between phase current phasors $\mathbf{I}_{YR}$ and $-\mathbf{I}_{RB}$ is 60°,

$$\therefore\ I_R = \sqrt{I^2{}_{YR} + I^2{}_{RB} + 2\, I_{YR}\, I_{RB} \cos 60°}$$

Assuming balanced load

The phase current in each winding is equal and let each be equal to I_P

$$\therefore \text{ Line current } I_R = \sqrt{I^2{}_P + I^2{}_P + 2\, I_P\, I_P \times \frac{1}{2}} = \sqrt{3}\, I_P \quad ...(6.10)$$

$$\text{Similarly line current } \mathbf{I}_Y = \mathbf{I}_{BY} - \mathbf{I}_{YR} = \sqrt{3}\, I_P \text{ in magnitude}$$

$$\text{and line current } \mathbf{I}_B = \mathbf{I}_{RB} - \mathbf{I}_{BY} = \sqrt{3}\, I_P \text{ in magnitude}$$

Since in delta-connected system, only one phase is included between any pair of line outers, potential difference between the line outers, called the *line voltage*, is equal to phase voltage.

$$\textit{i.e.} \text{ Line voltage, } E_L = \text{Phase voltage, } E_P \quad ...(6.11)$$

$$\text{Power output per phase} = E_P\, I_P \cos \Phi \text{ where } \cos \Phi \text{ is the power factor of the load}$$

$$\text{Total power output} = 3\, E_P\, I_P \cos \Phi = 3\, E_L \frac{I_L}{\sqrt{3}} \cos \Phi = \sqrt{3}\, E_L\, I_L \cos \Phi \quad ...(6.12)$$

i.e. Total power output = $\sqrt{3}$ × line voltage × line current × power factor

Apparent power of 3-phase delta-connected system

$$= 3 \times \text{apparent power per phase} = 3\, E_P\, I_P$$

$$= 3\, E_L \frac{I_L}{\sqrt{3}} = \sqrt{3}\, E_L\, I_L \quad ...(6.13)$$

It should be noted that in balanced delta connected system

(*i*) Line currents are 120° apart.

(*ii*) Line currents are 30° behind the respective phase currents.

(*iii*) Line currents are $\sqrt{3}$ times phase currents.

(*iv*) The angle between line currents and corresponding line voltages is (30º ± Φ) as in star system.

(*v*) Line voltages are equal to phase voltages.

(*vi*) Power = $\sqrt{3}\, E_L\, I_L \cos \Phi$, where Φ is the phase angle between respective phase current and phase voltage.

(*vii*) Apparent power = $\sqrt{3}\, E_L\, I_L$

(*viii*) In balanced system the resultant emf in the closed circuit will be zero.

i.e. $\mathbf{E}_{RY} + \mathbf{E}_{YB} + \mathbf{E}_{BR} = 0$

Hence there will be no circulating current in the mesh if no load is connected to the lines.

Example 6.9. A 3-phase motor operating on a 400 V balanced system develops 18.65 kW at an efficiency of 0.87 per unit and a power factor of 0.82. Calculate the line current and phase current if the windings are delta-connected.

Solution :

$$\text{Motor output} = 18.65 \text{ kW} = 18{,}650 \text{ watts}$$

$$\text{Motor efficiency, } \eta = 0.87$$

$$\text{Motor input, P} = \frac{\text{Motor output}}{\eta} = \frac{18{,}650}{0.87} = 21{,}437 \text{ watts}$$

$$\text{Line voltage, } V_L = 400 \text{ V}$$

$$\text{Power factor, } \cos\Phi = 0.82$$

$$\text{So line current, } I_L = \frac{P}{\sqrt{3}\, V_L \cos\Phi} = \frac{21{,}437}{\sqrt{3} \times 400 \times 0.82} = 37.73 \text{ A} \quad \textbf{Ans.}$$

$$\text{Phase current, } I_P = \frac{I_L}{\sqrt{3}} = \frac{37.73}{\sqrt{3}} = 21.8 \text{ A} \quad \textbf{Ans.}$$

Example 6.10. Three coils each of resistance 4 Ω and inductive reactance 3 Ω are connected in delta across 400 V, 50 Hz supply. Find the current in each coil, line current, active, reactive and apparent power.

Solution :

$$\text{Resistance in each phase, } R_P = 4\ \Omega$$

$$\text{Inductive reactance in each phase, } X_P = 3\ \Omega$$

$$\text{Impedance in each phase, } Z_P = \sqrt{R_P{}^2 + X_P{}^2} = \sqrt{4^2 + 3^2} = 5\ \Omega$$

$$\text{Phase voltage, } V_P = V_L = 400 \text{ V}$$

$$\text{Current in each coil} = I_P = \frac{V_P}{Z_P} = \frac{400}{5} = 80 \text{ A} \quad \textbf{Ans.}$$

$$\text{Line current, } I_L = \sqrt{3}\, I_P = \sqrt{3} \times 80 = 138.6 \text{ A} \quad \textbf{Ans.}$$

$$\text{Power factor, } \cos\Phi = \frac{R_P}{Z_P} = \frac{4}{5} = 0.8$$

$$\sin\Phi = \frac{X_P}{Z_P} = \frac{3}{5} = 0.6$$

$$\text{Active power, P} = \sqrt{3}\, V_L\, I_P \cos\Phi = \sqrt{3} \times 400 \times 138.6 \times 0.8 = 76{,}800 \text{ W or } 76.8 \text{ kW} \quad \textbf{Ans.}$$

$$\text{Reactive power, Q} = \sqrt{3}\, V_L\, I_L \sin\Phi = \sqrt{3} \times 400 \times 138.6 \times 0.6$$

$$= 57{,}600 \text{ VAR or } 57.6 \text{ kVAR} \quad \textbf{Ans.}$$

$$\text{Apparent power} = \sqrt{3}\, V_L\, I_L = \sqrt{3} \times 400 \times 138.6 = 96{,}000 \text{ V A or } 96 \text{ kVA} \quad \textbf{Ans.}$$

Example 6.11. A delta-connected 3-phase load has a resistance of 6 Ω and an inductive reactance of 8 Ω in each branch line. Line voltage is 230 V, 50 Hz. What are the rms values of current and voltage in each branch ? Calculate the total power consumed by the circuit and the power factor. Draw the phasor diagram.

[Pb. Technical Univ. Electrical Engineering December 2001]

Solution :

Resistance in each phase, $R_P = 6\ \Omega$

Inductive reactance in each phase, $X_P = 8\ \Omega$

Impedance in each phase, $Z_P = \sqrt{R_P^2 + X_P^2}$

$= \sqrt{6^2 + 8^2} = 10\ \Omega$

Phase voltage, V_P = Line voltage

= 230 V **Ans.**

RMS value of phase current, $I_P = \dfrac{V_P}{Z_P}$

$= \dfrac{230}{10} = 23$ A **Ans.**

Circuit power factor, $\cos\phi = \dfrac{R_P}{Z_P} = \dfrac{6}{10} = 0.6$ (lag) **Ans.**

Total power consumed in the circuit, $P = 3\, V_P I_P \cos\phi$

$= 3 \times 230 \times 23 \times 0.6 = 9{,}522$ W or 9.522 kW **Ans.**

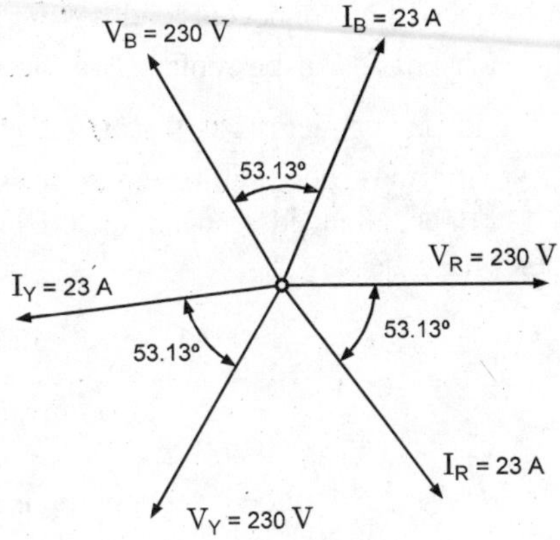

Fig. 6.9

6.8. COMPARISON BETWEEN STAR AND DELTA SYSTEMS

1. In a star-connected system similar ends are joined together whereas in a delta-connected system dissimilar ends are joined together.
2. In a star-connected system phase voltage is equal to $\dfrac{1}{\sqrt{3}}$ times the line voltage, but in a delta-connected system phase voltage is equal to line voltage, therefore, star-connected alternator needs less number of turns per phase and less insulation than a delta-connected alternator for the same line voltage. These are the reasons that 3-phase alternators are usually star-connected.
3. In a star-connected system, phase current is equal to line current *i.e.* $I_P = I_L$ but in a delta-connected system, the phase current is $\dfrac{1}{\sqrt{3}}$ times of line current *i.e.* $I_P = \dfrac{I_L}{\sqrt{3}}$.
4. It is possible to carry neutral wire to the load in star-connected system only *i.e.* only star-connected system can give 3-phase 4-wire arrangement and hence only star-connected system can be used for lighting as well as power loads *i.e.* for an unbalanced load. Lighting loads are connected between neutral and on any one of the outers and power circuits can be connected to the lines.
5. Neutral point of a star-connected system can be connected to earth, so relays and protective devices can be provided in star-connected system for protection of the system against ground faults and it is usually done in practice.
6. Delta-connections are mostly used in transformers for operating small low-voltage 3-phase motors and are best suited for rotary convertors. Most of the three-phase induction motors are delta-connected, 3-phase loads are also usually delta-connected as it provides flexibility for addition or removal of a load on a single phase in case of an unbalanced load.

6.9. CONVERSION OF BALANCED LOAD SYSTEM FROM STAR TO DELTA AND VICE-VERSA

Any balanced star-connected system can be completely replaced by an equivalent delta-connected system or vice versa because of their relationships between phase and line voltages and currents. For example, a balanced star-connected load having an impedance of magnitude Z with a power factor cos Φ (or $Z\angle\Phi$) in each phase can be replaced by an equivalent delta-connected load having an impedance of magnitude 3Z and power factor cos Φ (*i.e.* $3Z\angle\Phi$) in each phase.

This may be established as follows :

For a balanced star-connected load, let the line voltage be V_L, line current be I_L and impedance per phase be Z

$$\text{Phase voltage, } V_P = \frac{V_L}{\sqrt{3}}$$

$$\text{Phase current, } I_P = I_L$$

$$\text{So impedance per phase, } Z_S = \frac{V_P}{I_P} = \frac{V_L/\sqrt{3}}{I_L} = \frac{1}{\sqrt{3}}\frac{V_L}{I_L} \quad ...(6.14)$$

Now in equivalent delta-connected system for the same line values of voltage and current as in case of star-connected system

$$\text{Phase voltage, } V_P = V_L$$

$$\text{Phase current, } I_P = \frac{I_L}{\sqrt{3}}$$

$$\text{So impedance per phase, } Z_D = \frac{V_P}{I_P} = \frac{V_L}{I_L/\sqrt{3}} = \sqrt{3}\frac{V_L}{I_L} \quad ...(6.15)$$

Comparing Eqs. (6.14) and (6.15) we have

$$Z_S = \frac{1}{3} Z_D \text{ or } Z_D = 3Z_S \quad ...(6.16)$$

Example 6.12. A 3-phase star-connected 1,000 V alternator supplies power to 500 kW delta-connected induction motor. If the motor power factor is 0.8 lagging and its efficiency 0.9, find the current in each alternator and motor phase.

Solution :

$$\text{Line voltage, } V_L = 1,000 \text{ V}$$

$$\text{Motor output} = 500 \text{ kW}$$

$$\text{Motor input, } P = \frac{\text{Motor output}}{\text{Motor efficiency}} = \frac{500}{0.9} = 555.55 \text{ kW}$$

$$\text{Motor power factor, } \cos\Phi = 0.8 \text{ (lagging)}$$

$$\text{Line current, } I_L = \frac{\text{P in watts}}{\sqrt{3}\,V_L\cos\Phi} = \frac{555.55\times 1,000}{\sqrt{3}\times 1,000\times 0.8} = 401 \text{ A}$$

$$\text{Current in star-connected alternator phase, } I_P = I_L = 401 \text{ A } \textbf{Ans.}$$

$$\text{Current in delta-connected motor phase, } I_P = \frac{I_L}{\sqrt{3}} = \frac{401}{\sqrt{3}} = 231.5 \text{ A } \textbf{Ans.}$$

Example 6.13. A 3-phase star-connected alternator supplies a delta-connected load, each phase of which has a resistance of 10 Ω and reactance of 10 Ω. Calculate the current supplied by the alternator and the output of the alternator in kW and kVA, neglecting the losses in the line between the alternator and the load. The line voltage is 400 V.

Solution :

$$\text{Impedance per phase, } Z_P = \sqrt{R_P^2 + X_P^2} = \sqrt{10^2 + 10^2} = 14.14\ \Omega$$

$$\text{Power factor, } \cos\Phi = \frac{R_P}{Z_P} = \frac{10}{14.14} = 0.707 \text{ lag}$$

$$\text{Line voltage, } V_L = 400 \text{ V}$$

$$\text{Voltage across each phase of load, } V_P = V_L = 400 \text{ V} \qquad \because \text{ Load is } \Delta\text{-connected}$$

$$\text{Current in each phase of load, } I_P = \frac{V_P}{Z_P} = \frac{400}{14.14} = 28.284 \text{ A}$$

$$\text{Current supplied by the alternator} = \text{Line current of load} = \sqrt{3} \times I_P = \sqrt{3} \times 28.284 = 49 \text{ A } \textbf{Ans.}$$

$$\text{Output of alternator} = \text{Input to load} \qquad \because \text{ line losses are negligible}$$

$$= \sqrt{3}\, V_L I_L \cos\Phi$$

$$= \sqrt{3} \times 400 \times 49 \times 0.707 = 24{,}000 \text{ W or } 24 \text{ kW } \textbf{Ans.}$$

$$\text{Output of alternator in kVA} = \frac{\text{Output in kW}}{\text{Load power factor}} = \frac{24}{0.707} = 33.94 \textbf{ Ans.}$$

Example 6.14. Define phase sequence in a 3-phase system. Three 100 Ω resistors are connected in (*a*) star and (*b*) delta across 440 V, 50 Hz, 3-phase lines.

Calculate the line currents and phase currents and power factor taken from supply mains in each case.

[*Pb. Technical Univ. Basic Electrical and Electronics Engineering May 2005*]

Solution : (*a*) **When the resistors are connected in star**

$$\text{Phase voltage, } V_P = \frac{V_L}{\sqrt{3}} = \frac{440}{\sqrt{3}} = 254 \text{ V}$$

$$\text{Phase current, } I_P = \frac{V_P}{R} = \frac{254}{100} = 2.54 \text{ A } \textbf{Ans.}$$

$$\text{Line current, } I_L = I_P = 2.54 \text{ A } \textbf{Ans.}$$

$$\text{Power consumed, } P = 3I_P^2 \times R = 3 \times (2.54)^2 \times 100 = 1{,}936 \text{ W } \textbf{Ans.}$$

(*b*) **When the resistors are connected in delta**

$$\text{Phase voltage, } V_P = V_L = 440 \text{ V}$$

$$\text{Phase current, } I_P = \frac{V_P}{R} = \frac{440}{100} = 4.4 \text{ A } \textbf{Ans.}$$

$$\text{Line current, } I_L = \sqrt{3}\, I_P = \sqrt{3} \times 4.4 = 7.62 \text{ A } \textbf{Ans.}$$

$$\text{Power consumed, } P = 3I_P^2 \times R = 3 \times (4.4)^2 \times 100 = 5{,}808 \text{ W } \textbf{Ans.}$$

Example 6.15. A delta-connected balanced 3-phase load is supplied from a 3-phase, 400 V supply. The line current is 20 A and the power taken by the load is 10 kW. Find (*i*) impedance in each branch (*ii*) the line current, power factor and power consumed if the same load is connected in star.

Solution : (*i*) **When load is delta-connected**

$$\text{Phase voltage, } V_P = \text{Line voltage, } V_L = 400 \text{ V}$$

$$\text{Phase current, } I_P = \frac{I_L}{\sqrt{3}} = \frac{20}{\sqrt{3}} = 11.55 \text{ A}$$

$$\text{Phase impedance, } Z_P = \frac{V_P}{I_P} = \frac{400}{11.55} = 34.64 \ \Omega \textbf{ Ans.}$$

$$\text{Power factor, } \cos\Phi = \frac{\text{Power taken by the load}}{\sqrt{3}\, V_L I_L} = \frac{10{,}000}{\sqrt{3} \times 400 \times 20} = 0.7217$$

(*ii*) **When load is star-connected**

$$\text{Phase voltage, } V_P = \frac{V_L}{\sqrt{3}} = \frac{400}{\sqrt{3}} = 231 \text{ V}$$

$$\text{Phase current, } I_P = \frac{V_P}{Z_P} = \frac{231}{34.64} = 6.667 \text{ A}$$

Line current, $I_L = I_P = 6.667$ A **Ans.**

Power factor, $\cos \Phi = 0.7217$, as before, **Ans.**

Power consumed $= \sqrt{3}\, V_L I_L \cos \Phi = \sqrt{3} \times 400 \times 6.667 \times 0.7217 = 3{,}333$ W or 3.333 kW **Ans.**

Example 6.16. Three similar resistors connected in star draw a line current of 5 A from a 400 V, 3-phase mains. To what value should the line voltage be changed to obtain the same line current with the resistors connected in delta ?

Solution : When star-connected

$$\text{Phase voltage, } V_P = \frac{V_L}{\sqrt{3}} = \frac{400}{\sqrt{3}} = 231 \text{ V}$$

$$\text{Phase current, } I_P = I_L = 5 \text{ A}$$

$$\text{Phase resistors, } R_P = \frac{V_P}{I_P} = \frac{231}{5} = 46.2\ \Omega$$

When delta-connected

Line current, $I_L = 5$ A, the same as in star-connection

$$\text{Phase current, } I_P = \frac{I_L}{\sqrt{3}} = \frac{5}{\sqrt{3}} = 2.887 \text{ A}$$

$$\text{Phase voltage, } V_P = I_P R_P = 2.887 \times 46.2 = 133.33 \text{ V}$$

Line voltage, $V_L = V_P = 133.33$ V, one-third of line voltage for star-connection **Ans.**

Example 6.17. Three 100 Ω resistors are connected in star across a 440 V, 50 Hz, 3-phase supply. One of the phase is disconnected. What will be the line current and total absorbed power.

[*Pb. Technical Univ. Basic Electrical and Electronics Engineering Dec. 2005*]

Solution : After disconnecting one of the resistors, the circuit is no longer three-phase but converted into single phase circuit, having two resistors, each of 100 Ω connected in series across supply of 440 V. [See Fig. 6.10 (*b*)].

$$\text{Line current, } I_L = \frac{V_L}{2R_P} = \frac{440}{2 \times 100} = 2.2 \text{ A } \textbf{Ans.}$$

$$\text{Power absorbed, } P = I_L^2 \times 2R_P = (2.2)^2 \times 2 \times 100 = 968 \text{ Watts } \textbf{Ans.}$$

Example 6.18. A 3-wire, 3-phase supply feeds a load consisting of three equal resistors. By how much is the load reduced if one of the resistor be removed when the load is in (*i*) star (*ii*) delta ?

Solution : Let the line voltage of 3-phase supply be V_L volts and resistance of each phase be of R ohms.

(*i*) **In Star**

$$\text{Power consumed, } P = 3 \times \frac{V_P^2}{R} = 3 \times \left(\frac{V_L}{\sqrt{3}}\right)^2 \times \frac{1}{R} = \frac{V_L^2}{R}$$

After disconnection or removal of one of the resistances, the circuit is no longer 3-phase but converted into single phase circuit, having two resistors, each of R ohms connected in series across supply of V_L volts, as shown in Fig. 6.10 (*b*).

$$\text{Hence total power consumed} = 2\frac{(V_L/2)^2}{R} = \frac{V_L^2}{2R}$$

$$\therefore \text{ Reduction in load} = \frac{\frac{V_L^2}{R} - \frac{V_L^2}{2R}}{\frac{V_L^2}{R}} \times 100 = 50\% \textbf{ Ans.}$$

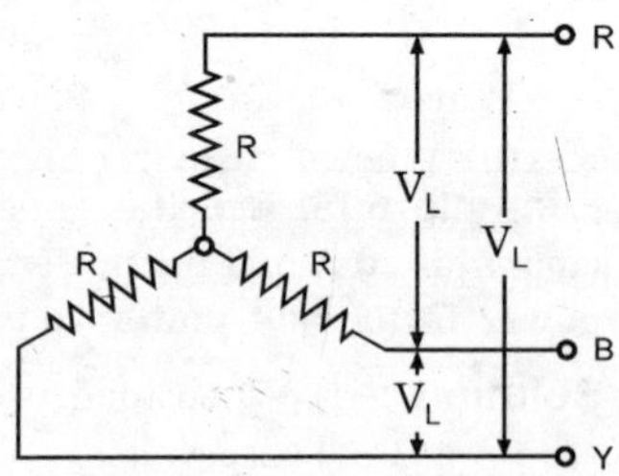

(*a*)

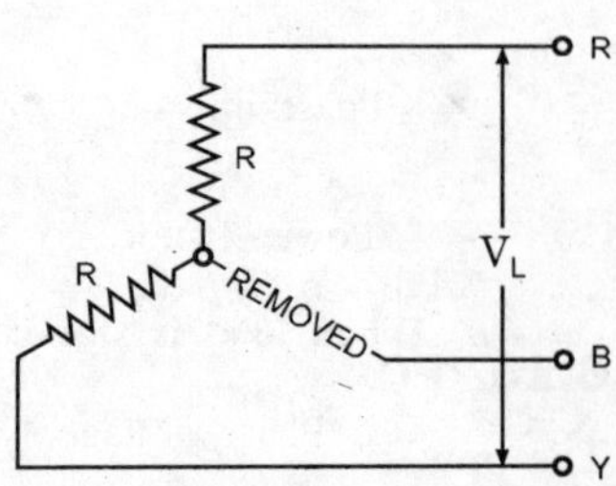

(*b*)

Fig. 6.10

(ii) In Delta Power consumed $= 3 \times \dfrac{V_P^2}{R} = 3 \times \dfrac{V_L^2}{R} = 3\dfrac{V_L^2}{R}$

After removal of one of the resistance, each of the remaining two resistances is connected across supply of V_L volts.

$$\therefore \text{ Power consumed} = 2 \times \frac{V_L^2}{R} = \frac{2V_L^2}{R}$$

$$\text{So reduction in load} = \frac{\dfrac{3V_L^2}{R} - \dfrac{2V_L^2}{R}}{\dfrac{3V_L^2}{R}} \times 100$$

$$= 33.33\ \% \ \textbf{Ans.}$$

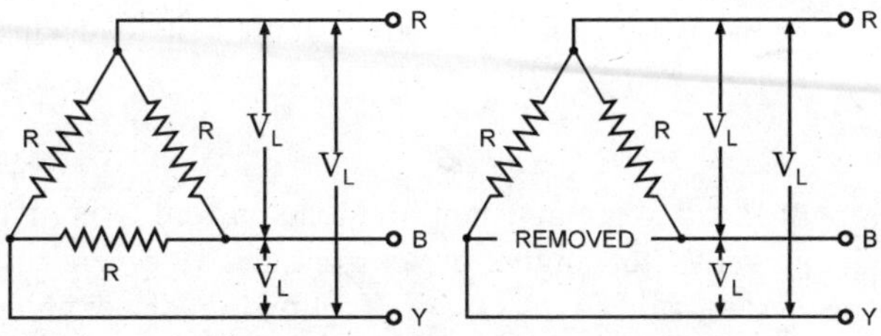

(a) (b)

Fig. 6.11

Example 6.19. Three 100 Ω non-inductive resistances are connected in (a) star (b) delta across a 400-V, 50 Hz, 3-phase mains. Calculate the power taken from the supply system in each case. In the event of one of the three resistances getting open circuited, what would be the value of the total power taken from the mains in each of the two cases.

Solution : (a) **When connected in star**

$$\text{Phase voltage, } V_P = \frac{V_L}{\sqrt{3}} = \frac{400}{\sqrt{3}} = 231 \text{ V}$$

$$\text{Phase current, } I_P = \frac{V_P}{R_P} = \frac{231}{100} = 2.31 \text{ A}$$

Power factor, $\cos \Phi = 1$ $\because$ circuit is pure resistive

Power taken, $P = 3V_PI_P \cos \Phi = 3 \times 231 \times 2.31 \times 1 = 1{,}600$ watts **Ans.**

(b) **When connected in delta**

Phase voltage, $V_P = V_L = 400$ V

$$\text{Phase current, } I_P = \frac{V_P}{R_P} = \frac{400}{100} = 4 \text{ A}$$

Power taken, $P = 3\ V_P\ I_P \cos \Phi = 3 \times 400 \times 4 \times 1 = 4{,}800$ watts **Ans.**

When one of the resistances is disconnected.

(a) **When connected in star**

After disconnecting one of the resistances, the circuit is no longer 3 phase but converted into single phase circuit, having two resistances, each of 100 Ω connected in series across supply of 400 volts.

$$\text{Hence line current, } I_L = \frac{V_L}{2R_P} = \frac{400}{2 \times 100} = 2 \text{ A}$$

Power absorbed, $P = V_LI_L \cos \Phi = 400 \times 2 \times 1 = 800$ watts **Ans.**

(b) **When connected in delta**

PD across each resistance, $V_L = 400$ V

$$\text{Current in each resistance, } I = \frac{400}{100} = 4 \text{ A}$$

Power absorbed in each resistance $= I^2 R = (4)^2 \times 100 = 1{,}600$ watts.

Power absorbed in both resistances $= 1{,}600 \times 2 = 3{,}200$ watts **Ans.**

6.10. POWER FACTOR IMPROVEMENT

The power factor for lighting and heating loads supplied from 3-phase power supply ranges from 0.95 to unity but the power factor for motor loads ranges from 0.5 to 0.9. Power factor for single phase motors may be as low as 0.4 (lagging) and for electric welding may be even 0.2 or 0.3 (lagging).

6.10.1. Disadvantages of Low Power Factor

The current for a given load supplied at a constant voltage will be higher at lower power factor and low at higher power factor. For example if load P is supplied at terminal voltage V and at power factor cos Φ by a 3-phase balanced system then load current is given by

$$I_L = \frac{P}{\sqrt{3}\, V \cos \Phi}$$

If P and V are constant, the load current I_L is inversely proportional to power factor, cos Φ *i.e.* lower the power factor, higher the current and vice-versa. The higher current due to poor power factor affects the system and results in the following disadvantages:

(*i*) Rating of generators and transformers are proportional to their output current hence inversely proportional to power factor, therefore, large generators and transformers are required to deliver same load but at a low power factor.

(*ii*) The x-sectional area of the bus-bar and the contact surface of switchgear is required to be enlarged for the same power to be delivered but at a low power factor.

(*iii*) For the same power to be transmitted but at low power factor, the transmission line or the distributor or cable will have to carry more current. Thus more conductor material is required for transmission line, distributors and cables to deliver the same load but at a low power factor.

(*iv*) Copper losses are proportional to the square of the current hence inversely proportional to the square of the power factor *i.e.* more copper losses incur at low power factor, which results in poor efficiency.

(*v*) Low lagging power factor results in large voltage drop in generators, transformers, transmission lines and distributors which results in poor regulation. Hence extra regulating equipment is required to keep the voltage drop within permissible limits.

Thus we see that the low power factor leads to a high capital cost for the alternators, switchgears, transformers, transmission lines, distributors and cables etc.

6.10.2. Causes of Low Power Factor

(*i*) All ac motors (except overexcited synchronous motors and certain types of commutator motors) and transformers operate at lagging power factor. The power factor decreases with the decrease in load. For example an induction motor has a reasonable higher power factor of 0.85 at full load, 0.8 at 75% of full-load, 0.7 at half-full load, 0.5 at 25% of full-load and as low as 0.1 on no load.

(*ii*) Arc lamps operate at low power factor (lagging) due to typical characteristic of the arc.

(*iii*) Industrial heating furnaces such as arc and induction furnaces operate on very low lagging power factor.

6.10.3. Methods of Power Factor Improvement

The following methods are usually adopted for improving the power factor:

(i) By Use of Induction Motors With Phase Advancers. The power factor of an induction motor falls mainly due to its exciting current drawn from the ac supply mains, because exciting current lags behind the voltage by $\frac{\pi}{2}$. It may be improved by equipping the set with an ac exciter or phase advancer which supplies this exciting current to the motor circuit at slip frequency. Such an exciter may be mounted on the same shaft as the main motor or may be suitably driven from it. Use of phase advancer is not generally economical in conjuction with motors below 150 kW but above this size, phase advancers are frequently employed.

(ii) By Use of Static Capacitors. Power factor can be improved by connecting the capacitors in parallel with the equipment operating at lagging power factor such as induction motors, fluorescent tubes. Static capacitors have small losses, require almost no maintenance and are very convenient in the small sizes. The capacitors draw current leading the supply voltage by $\frac{\pi}{2}$ and neutralize the quadrature or wattless component of current drawn by the equipment across which these are connected. Capacitors can be connected either in star or in delta, as shown in Fig. 6.12 (*a*) and (*b*) respectively.

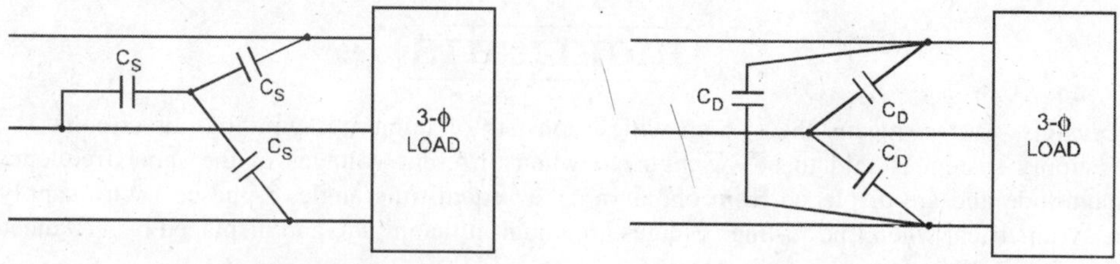

(a) Star-Connected Capacitors *(b) Delta-Connected Capacitors*

Fig. 6.12

These capacitors remain connected permanently in parallel with the equipment and are across the supply mains whenever the equipment is switched on. Since the static capacitors are not variable, they tend to overcompensate on light loads, unless arrangements for automatic switching off the capacitor banks are made.

Whenever the load is operated with a fixed wattless component of current the static condensers are the best devices to improve the power factor. This is the cheapest method of power factor improvement.

(iii) By Use of Synchronous Condensers. An overexcited synchronous motor operating on no load is called the *synchronous condenser* or *synchronous phase advancer* and behaves like a capacitor, the capacitive reactance of which depends on the motor excitation. Power factor can be improved by using synchronous condensers like shunt capacitors connected across the supply.

The advantage of synchronous condenser over static capacitor is that power factor can very easily be controlled by variation of the field excitation.

The disadvantages of synchronous condensers are (*i*) except in size above about 500 kVA, the cost is higher than that of static capacitors of the same rating, (*ii*) possibility of their falling out of synchronism with resulting in interruption of supply, and (*iii*) increase of short-circuit currents when the fault occurs near the synchronous condenser.

6.10.4. Advantages of Power Factor Improvement

If the power factor of the supply or power station is raised to unity, the current for the same amount of power to be supplied is reduced to minimum. This results in reduction of transmission line copper losses and reduction of voltage drop in transmission line and in alternator windings as copper losses are directly proportional to the square of supply current and voltage drop is directly proportional to the supply current.

The advantages of good power factor are that (*i*) load output of a given plant is better utilized (*ii*) line losses are reduced to minimum, therefore, efficiency of the plant is improved (*iii*) voltage drop or voltage regulation is considerably reduced.

Example 6.20. A 3-phase, 5 kW induction motor has power factor of 0.7 at full load. What should be done to improve the power factor of the load to 0.9 ?

Solution : Motor input, P = 5 k W = 5,000 watts

$\text{Cos } \Phi_1 = 0.7;$ $\quad$ $\text{Tan } \Phi_1 = 1.0202;$

$\text{Cos } \Phi_2 = 0.9$ and $\text{Tan } \Phi_2 = 0.4843$

Leading kVAR to be supplied by the condensers to be connected to improve the power factor

$$= P (\tan \Phi_1 - \tan \Phi_2) = 5{,}000\ (1.0202 - 0.4843) = 2{,}680$$

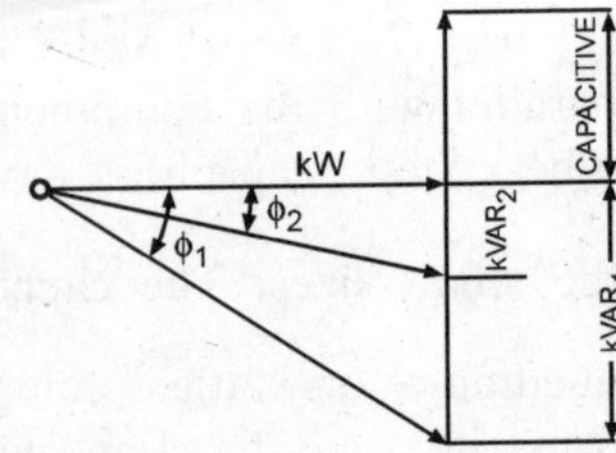

Hence three condensers having ratings of $\frac{2{,}680}{3}$ *i.e.* 893.33 VA must be connected across the supply terminals to raise the power factor to 0.9 (assuming capacitors loss free). **Ans.**

Fig. 6.13

HIGHLIGHTS

1. *Polyphase system* means the system which consists of numerous windings or circuits.
2. A supply system is said to be *symmetrical* when different voltages of the same frequency have equal magnitude and are displaced from one another by equal time angle. 3-phase, 3-wire supply system will be symmetrical when line-to-line voltages are equal in magnitude and displaced by 120 electrical degrees w.r.t. each other.
3. A 3-phase supply will be *unbalanced* when either of three phase voltages are unequal in magnitude or the phase angles between these phases is not equal to 120º.
4. A load circuit is said to be balanced when the loads (impedances) connected in different phases are same in magnitude as well as in phase.
5. For a balanced system, the phasor sum of the three phase voltages is always zero.
6. In case one phase out of three phases of a 3-phase supply connected to the three phase load is not available, then such a condition is called the single phasing.
7. Commonly used polyphase system is 3-phase system.
8. 3-phase system is advantageous over single phase system and have the advantages of constant power, higher rating (1.5 times), economy in power transmission and superiority of 3-phase induction motors etc.
9. *Phase sequence* is the order in which the currents or voltages in different phases attain their maximum values.
10. The three phase system is connected in two ways namely star connection and delta connection.
11. Star-connected system is obtained by joining together similar ends, either the start or the finish, the other ends are joined to the line wires. The common point at which similar ends are connected is called the neutral or star point.
12. In a balanced star-connected system
    (*i*) Line voltages are 120º apart.
    (*ii*) Line voltages are 30º ahead of the respective phase voltages.
    (*iii*) Line voltages are $\sqrt{3}$ times of phase voltages.
    (*iv*) Line currents are equal to phase currents.
    (*v*) The angle between line currents and the corresponding line voltages is (30º ± Φ); + ve for lagging currents and –ve for leading currents.
    (*vi*) True power output = $\sqrt{3}\ E_L I_L \cos \Phi$, where Φ is the angle between the respective phase current and phase voltage (not between line current and line voltage).
    (*vii*) Apparent power = $\sqrt{3}\ E_L I_L$.
    (*viii*) In balanced system, the potential of neutral or star point is zero because potential at neutral or star point,
    $$E_{NR} + E_{NY} + E_{NB} = 0.$$
13. When the starting end of one coil is joined to the finishing end of another coil, the delta or mesh connection is obtained. The direction of emf in the coils is taken as positive from start to finished ends.
14. In balanced delta connected system
    (*i*) Line currents are 120º apart.
    (*ii*) Line currents are 30º behind the respective phase currents.
    (*iii*) Line currents are $\sqrt{3}$ times phase currents.
    (*iv*) The angle between line currents and corresponding line voltage is (30º ± Φ) as in star system.
    (*v*) Line voltages are equal to phase voltages.

(*vi*) Power = $\sqrt{3}\ E_L I_L \cos \Phi$, where Φ is the phase angle between respective phase current and phase voltage.

(*vii*) Apparent power = $\sqrt{3}\ E_L I_L$.

(*viii*) In balanced system the resultant emf in the closed circuit will be zero.

$$i.e.\ \mathbf{E}_{RY} + \mathbf{E}_{YB} + \mathbf{E}_{BR} = 0.$$

Hence there will be no circulating current in the mesh if no load is connected to the lines.

EXERCISES

1. What do you understand by the following terms:
(*i*) Three phase balanced supply. (*ii*) Three phase balanced load. (*iii*) Three phase unbalanced supply. (*iv*) Three phase unbalanced load. (*v*) Single phasing.

2. State the advantages of three phase system over single phase system.
[Pb. Technical Univ. Basic Electrical Engineering June 2003]

3. Explain with the help of neat diagram how three phase emf can be generated.

4. What are the two ways of connecting a 3-phase system ? Draw their phasor diagrams and write down the relationship between phase and line voltages and currents for these systems.

5. Deduce the relationship between the phase and line voltages and currents in a 3-phase star-connected circuit. Draw phasor diagram to establish it.

6. Derive the relationship between line current and phase current in a 3-phase delta-connected balanced load.

7. Show that the total power in a 3-phase balanced load is P = $\sqrt{3}$ VI cos ϕ where V is the rms line voltage, I is the rms line current and ϕ is the phase angle between phase voltage and phase current.

8. Compare 3-phase star- and delta-connected systems.

9. Show that power consumption, when three identical impedances are connected in delta across a balanced 3-phase supply, is three times that when the same impedances are connected in star across the same 3-phase supply.

10. Discuss the practical importance of power factor.

11. It is normally desired that the power factor should be high, why ? Explain.

12. What are the causes and disadvantages of poor power factor. Enumerate the different methods of power factor improvement ? Discuss any one of them.

13. What are the causes of low power factor in supply system ? Discuss its harming effects. How can the power factor be improved ?

14. Write short notes on the following :
(*i*) Advantages of 3-phase system over single phase system.
(*ii*) Double subscript notation.
(*iii*) Interconnection of 3-phases.
(*iv*) Power factor improvement.

SHORT ANSWER TYPE QUESTIONS WITH ANSWERS

Q. 1. What do you mean by three-phase balanced load ?

Ans. Any three-phase load will be balanced when the loads (impedances) connected in three phases are same in magnitude as well as in phase.

Q. 2. What is three-phase unbalanced load ?

Ans. Any three-phase load will be unbalanced if impedances in one or more phases differ from the impedance (s) of the remaining phase (s).

Q. 3. Why is an unbalanced load not normally used on a 3-phase 3-wire system ? Why line/phase voltages equal in such situations ? **[P.T.U. 1999]**

Ans. Normally, unbalanced load is not employed on 3-phase, 3-wire ungrounded star-connected system because unbalanced loading may cause different voltage drops in lines. Consequently the three phase voltages are different (or unbalanced) in magnitude as well as in phase and it is quite possible that one phase voltage may exceed the line voltage. Such a condition is undesirable, since some loads may operate inefficiently due to lowering of the voltage and the other equipment may get damaged due to over-voltage.

Q. 4. What is phase sequence in a 3-phase system ? How is given phase sequence reversed ? **[P.T.U. 1999]**

Ans. Phase sequence is the order or sequence in which the currents or voltages in different phases attain their maximum values one after the other.

The given phase sequence can be reversed by interchanging any two terminals of the supply.

Q. 5. Why interconnection of 3-phase system is necessary ?

Ans. Non-interlinked three-phase system will require 6 wires while interconnected 3-phase system will require 3 or 4 wires. So non-interlinked system will be very complicated and expensive.

Q. 6. Differentiate between star and delta connections. **[P.T.U. June 2003]**

Ans. In star-connected system similar (either start or finish) terminals of the three phases are connected together to provide star or neutral point [refer Fig. 6.4 (*a*)] while in delta-connected system start terminal of one phase is connected to the finish terminal of the second phase and start terminal of second phase is connected to finish terminal of the third phase and so on to provide a closed circuit in 3-phase system [refer Fig. 6.8 (*a*)].

Q. 7. Give the relation between phase and line values of voltage and current for star connection. **[P.T.U. B.E.E.E. December 2009]**

Ans. In star-connection,

$$\text{Line current, } I_L = \text{Phase current, } I_P$$

$$\text{Line voltage, } V_L = \sqrt{3} \text{ times phase voltage.}$$

Q. 8. Discuss, why the stator windings of alternator are generally star-connected ? **[P.T.U. December 2000]**

Ans. In a star-connected system phase voltage is equal to $\frac{1}{\sqrt{3}}$ times the line voltage, but in a delta-connected system phase voltage is equal to line voltage, therefore, star-connected alternator needs less number of turns per phase and less insulation than a delta-connected alternator for the same line voltage.

Neutral point of a star-connected system can be connected to earth, so relays and protective devices can be provided in star-connected system for protection of the system against ground faults and it is usually done in practice. These are the reasons that 3-phase alternators are usually star-connected.

Q. 9. Why does star-connected alternator need less number of turns per phase as compared to that required by delta-connected one to provide the same line voltage ?

Ans. In a star-connected system phase voltage is equal to $\frac{1}{\sqrt{3}}$ times the line voltage, but in a delta-connected system phase voltage is equal to line voltage, therefore, star-connected alternator needs less number of turns per phase than a delta-connected alternator for the same line voltage.

Q. 10. Why delta-connected system cannot be employed for supplying lighting as well as power loads ?

Ans. Delta-connected system cannot provide 3-phase 4-wire system (*i.e.* neutral wire cannot be carried) which is essential for supplying lighting as well as power loads.

Q. 11. Discuss, why the phase spread of three-phase winding is 60° and not 120° ? **[P.T.U. B.E.E.E. May 2010]**

Ans. The phase spread is to be taken in two dimensions. Total spread available in two dimensions is 180° and, therefore, the spread of three phase winding is $\frac{180°}{3}$ *i.e.* 60°.

Q. 12. Name two electrical equipments in each case having unity power factor and lagging power factor. **[P.T.U. June 2003]**

Ans. Electrical equipments operating

at unity power factor: Incandescent lamps, electric heater.

at lagging power factor: Arc lamps, induction furnaces.

PROBLEMS

1. Given a balanced 3-Φ, 3-wire system with Y-connected load for which line voltage is 230 V and impedance of each phase is (6 + *j* 8) Ω. Find the line current and power absorbed. [**Ans.** 13.28 A; 1,058 W]
2. Three impedances, each of resistance 12 Ω and inductive reactance 5 Ω are connected in star-connection across a 3-phase, 440 V, 50 Hz supply. Find the line current and the total power consumed in the circuit. [**Ans.** 19.54 A; 13,747 W]
3. Find (*i*) line current (*ii*) power factor and (*iii*) reactive power in the circuit shown in Fig. 1. [**Ans.** (*i*) 23.1A (*ii*) 0.8 (leading) (*iii*) 9.6 kVAR]
4. Three equal star-connected inductors take 8 kW at a power factor of 0.8 when connected across a 460 V, 3-phase, 3-wire supply. Find the circuit constants of the load per phase. Draw the phasor diagram showing phase and the line voltages and currents. [**Ans.** 16.93 Ω; 12.7 Ω]
5. A 3-phase, 415-V, 50 Hz ac source supplies a balanced star-connected load of 7.5 kW. If the current in the lines is 15 A, determine the resistive and reactive components of impedance. Draw phasor diagram showing the line voltages, line currents and phase voltages. [**Ans.** 11.11 Ω ; 11.47 Ω]

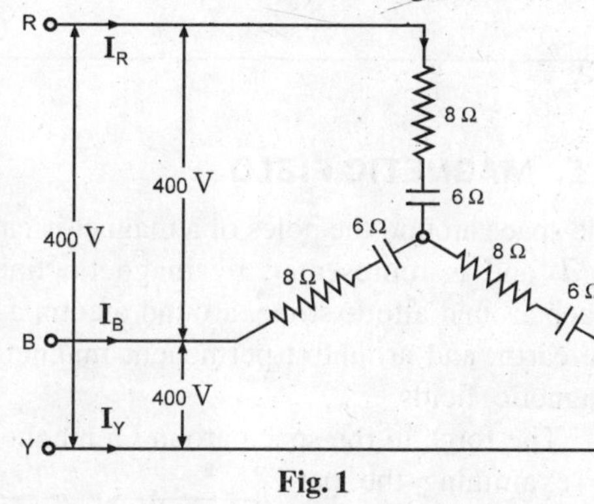

Fig. 1

6. A balanced 3-phase star-connected load takes a power of 5 kW at 0.8 pf lagging when connected to a 400 V, 3-phase supply. Calculate the line current and impedance per phase of the load. [**Ans.** 9.02 A; 25.6 Ω]
7. A 3-phase, 400 V, 50 Hz, 10 hp motor is delta-connected and operates at a power of 0.8 (lag) and efficiency of 80%. What current will it draw from the supply ? (1 hp = 735 W) [**Ans.** 16.576 A]
8. Three identical coils of resistance 8 Ω and inductive reactance 6 Ω are connected in delta across 400 V mains. Determine consumed power, power factor and line current. [**Ans.** 38.4 kW, 0.8 (lag); 69.28 A]
9. A load consisting of three identical impedances **Z** = $10\angle{-45^\circ}$ Ω in Δ is connected to a 3-phase, 220 V source. Determine phase and line currents and also the total power drawn by the load. [**Ans.** 22 A; 38.1 A; 10.266 kW]
10. A star-connected alternator supplies a delta-connected load. The impedance of the load branch is (8 + *j* 6) ohms/phase. The line voltage is 230 V. Determine (*a*) the current in the load branch (*b*) power consumed by the load (*c*) power factor of the load (*d*) reactive power of the load. [**Ans.** (*a*) 23 A (*b*) 12.696 kW (*c*) 0.8 (lag) (*d*) 9.522 kVAR]
11. Three coils are connected in delta to a 3-phase, three-wire, 415 V, 50 Hz supply and take a line current of 5 A at a pf of 0.8 lagging. Calculate the resistance and inductance of the coils. If the coils are star-connected to the same supply, calculate the line current and total power. [**Ans.** 115 Ω; 0.2746 H; 1.667 A; 958.3 watts]
12. Three identical coils connected in star serve as a 3-phase balanced load to a 3-phase, 50 Hz balanced supply. The line current is 28 A and the load kVA and kW are 22 and 13 respectively. Calculate the load kVAR, phase voltage and reactance of each coil. What will be new values of line current and power consumed if the coils are connected in delta ? [**Ans.** 17.748 kVAR; 261.9 V; 5.546 Ω; 84.0 A; 39 kW]

❑❑❑

7
CHAPTER

Magnetic Circuits

7.1. MAGNETIC FIELD

The space around the poles of a magnet is called the *magnetic field,* and is represented by magnetic lines of force. The space around a lode stone, around a compass needle, around the earth, and around a permanent magnet are examples of magnetic fields.

The force in the space around a magnet can be pictured by examining the pattern made by iron fillings, sprinkled on a card board placed over the magnet (Fig. 7.2). Each little splinter of iron acts like a compass needle, attracting other fillings, at its ends and repelling those lying parallel to it. These chains of fillings led to the assumptions that the region (field) around a magnet contains invisible "lines of force". The total number of lines of force surrounding a magnet, as illustrated in Fig. 7.2, is called the total *magnetic flux.*

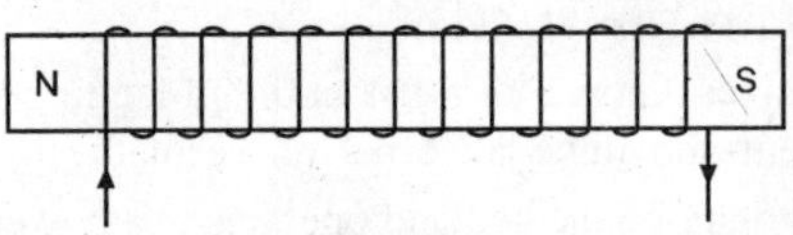

Fig. 7.1

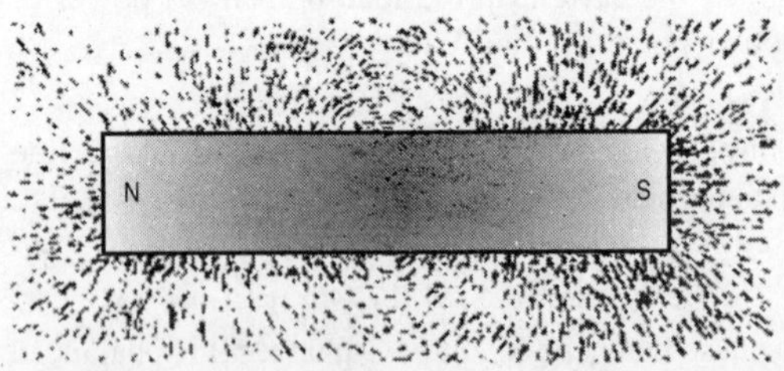

Fig. 7.2. ***Pattern of Iron-Fillings Around a Bar-Magnet***

INSIDE THIS CHAPTER

7.2. LINES OF MAGNETIC FLUX

Inspite of the fact that lines of magnetic flux have no physical existence, they do form a very convenient and useful basis for explaining various magnetic effects and for calculating their magnitudes.

Magnetic line of force is defined as the curve along which a unit N-pole would move if free to do so, when placed in a magnetic field. The tangent to these curves at any point gives the direction of magnetic field at that point.

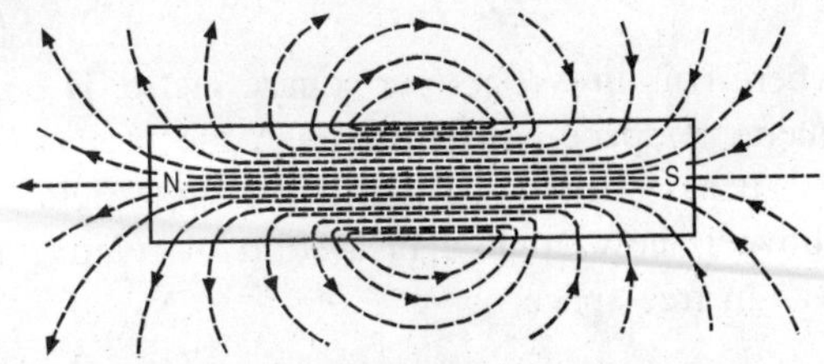

Fig. 7.3. ***Field Around a Bar Magnet Represented By Lines of Force***

Lines of magnetic flux are assumed to have the following properties :

(*i*) Each line of magnetic flux forms a closed loop, as shown by dotted lines in Fig. 7.5. This means that a line of flux emerging from any point at N-pole end of a magnet passing through the surrounding space back to the S-pole end and is then assumed to continue through the magnet to the point at which it is emerged at the N-pole end.

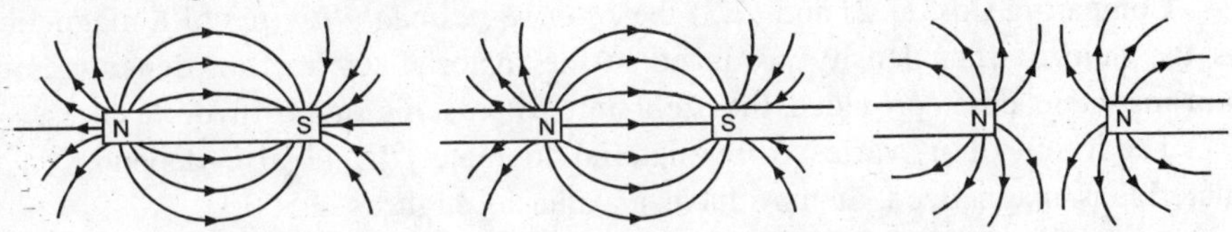

Fig. 7.4

(*ii*) The lines of magnetic flux never intersect.

(*iii*) Lines of magnetic flux tend to contract lengthwise and repel another sideways.

(*iv*) Lines of force emanate from and terminate on a surface normally.

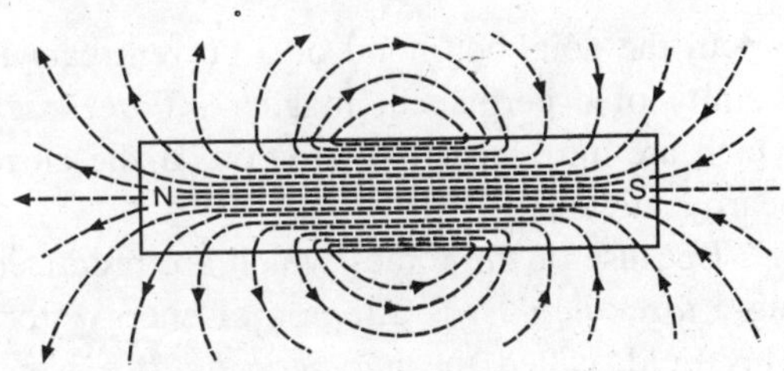

Fig. 7.5

7.3. LINES OF INDUCTION

Lines of magnetic flux travel from N-pole to S-pole externally, and continue to travel through the magnet and finally they reach the N-pole again forming closed curves, as illustrated in Fig. 7.5. The portion of the curves within the magnetic material are called the *lines of induction*.

7.4. MAGNETIC FLUX

Magnetic flux is the total number of lines of force comprising the magnetic field. In SI unit system unit north pole *i.e.* north pole of strength of one weber is supposed to radiate out the flux of one weber. It is represented by symbol ϕ and is measured in webers.

$\therefore$ Flux emanated from north pole of strength m webers = m Wb.

7.5. MAGNETIC FLUX DENSITY

Magnetic flux density is defined as the magnetic flux passing per unit area through any material through a plane at right angles to the direction of flux. This is also known as *magnetic induction.* It is represented by English letter B and is measured in Tesla (weber/m^2)

$$i.e.\ \mathrm{B} = \frac{\phi}{a} \qquad ...(7.1)$$

Magnetic flux density is a scalar quantity.

7.6. RELATION BETWEEN MAGNETIC FIELD INTENSITY (H) AND INDUCTION DENSITY (B)

At any point in a magnetic field, field strength or magnetic field intensity H is the force that maintains the magnetic flux and produces a particular value of flux density B at that point. Thus the field intensity H is the cause and the flux density B is the effect. Thus the flux density can be assumed to be proportional to the field intensity in a magnetic field, *i.e.* in free space

$$\mathbf{B} = \mu_0 \mathbf{H} \quad ...(7.2)$$

where B is in webers per square metre, H is in newtons per weber (or ampere-turns/metre) and μ_0 is a magnetic space constant, called the *permeability of free space,* and is equal to $4\pi \times 10^{-7}$ H/m.

It has been found that if the same field intensity H, which was applied in free space as mentioned above is now applied in a solid of relative permeability μ_r, the flux density B will be μ_r times that was in free space

$$i.e. \quad \mathbf{B} = \mu_0\mu_r \ \mathbf{H} \quad ...(7.3)$$

$$\text{or} \quad \mathbf{B} = \mu\, \mathbf{H} \qquad \text{where } \mu = \mu_0\mu_r \quad ...(7.4)$$

Comparing Eqs. (7.2) and (7.3) the *relative permeability* (μ_r) of a magnetic material may be defined as the ratio of flux density produced in the material to the flux density produced in vacuum or in a non-magnetic core, provided the field intensity is the same in both cases.

The value of μ_r varies from material to material and also depends on the degree to which the material is magnetized. It may have a value as high as 2,500.

7.7. MAGNETIC EFFECTS OF ELECTRIC CURRENT

In the neighbourhood of a current carrying conductor, effects are produced very like those in the vicinity of a permanent magnet : Forces act on magnets, iron, or other current carrying conductors which are introduced. In addition, if the current is changing in magnitude, voltages are induced in the nearby circuits.

Because of the forces which are produced, the neighbourhood of conductor is regarded as possessing a *force field.* It is a region of energy storage possessing the ability to produce forces and do work. This field, called the *magnetic field,* exists simultaneously with the electric field.

7.8. MAGNETIC FIELD DUE TO A CURRENT CARRYING CONDUCTOR

In 1819 it was discovered by a Danish Physicist, Hans Christian Oersted that an electric current is always accomplished by certain magnetic effects.

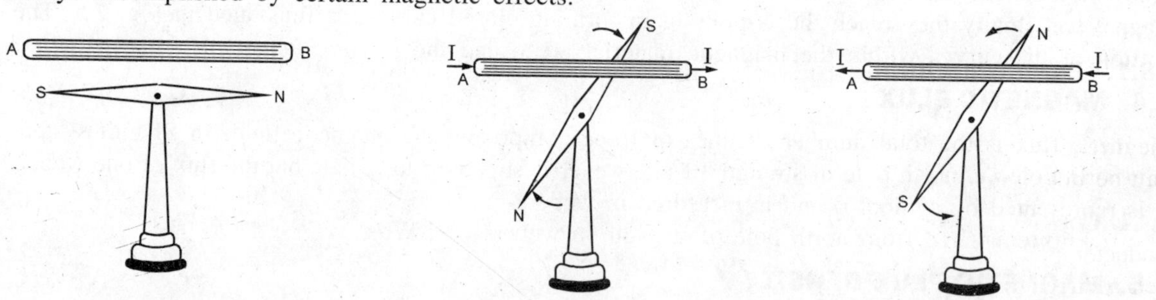

No Current Flow (*a*) *Current Flowing From A to B* (*b*) *Current Flowing From B to A* (*c*)

Fig. 7.6

Oersted found that when current is passed through a conductor placed above the magnetic needle, the needle turns in a certain direction, as shown in Fig. 7.6. He also found that when the direction of flow of current is reversed the magnetic needle also deflects in opposite direction.

Further investigation showed that the field around the current carrying conductor consists of lines of force, which encircles the conductor. It can be proved experimentally by passing a current carrying conductor AB in the card board and plotting the field with the help of magnetic needle on it, as shown in Figs. 7.7 (*a*) and 7.7 (*b*).

From Figs. 7.7 (*a*) and 7.7 (*b*) it is observed that when the current is passed through conductor in upward direction, the direction of lines of force is counter-clockwise direction (observed from the top

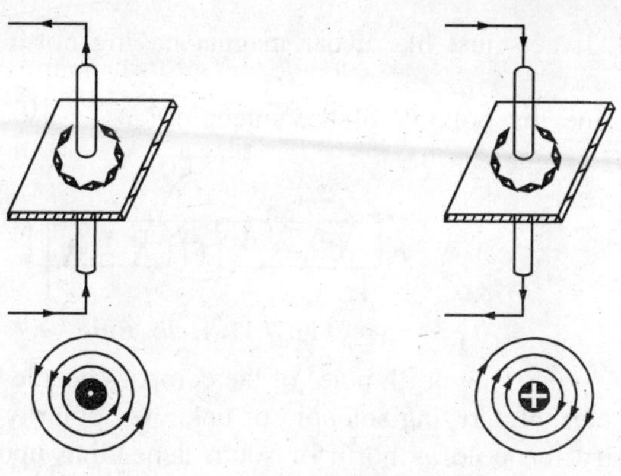

(a) (b)

Fig. 7.7

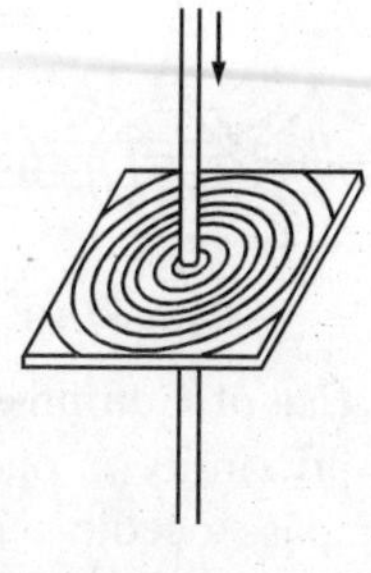

Fig. 7.8. *Magnetic Field Around A Current Carrying Conductor*

of the conductor) and when the current is passed through the conductor in downward direction, the direction of lines of force is clockwise (observed from the top of the conductor).

The properties of the lines of magnetic induction around a current carrying conductor are summarized as below:

(*i*) Lines of magnetic induction are circles, symmetrical about, and concentric with, the axis of the conductor.

(*ii*) The spacing between the lines of induction decreases as we move closer to the conductor (Fig. 7.8).

(*iii*) The direction of lines of magnetic induction depends on the direction of flow of current through the conductor.

(*iv*) Magnetic induction or flux density depends upon the strength (or magnitude) of the current flowing through the conductor.

7.8.1. Determination of Direction of Magnetic Field Around a Current Carrying Conductor

The direction of lines of force (magnetic field) around a straight current carrying conductor may be determined by any of the following rules:

1. Cork Screw Rule. If the right handed cork screw is held with its axis parallel to the conductor pointing the direction of flow of current and the head of the screw is rotated in such a direction that the screw moves in the direction of flow of current then the direction in which the head of screw is rotated, will be the direction of magnetic lines of force.

(a) Cork Screw Rule

2. Right Hand Rule. If the current carrying conductor is held in right hand by the observer so that it is encircled by fingers stretching the thumb at right angle to the fingers in the direction of flow of current then finger tips will point the direction of magnetic lines of force, as shown in Fig. 7.9 (*b*).

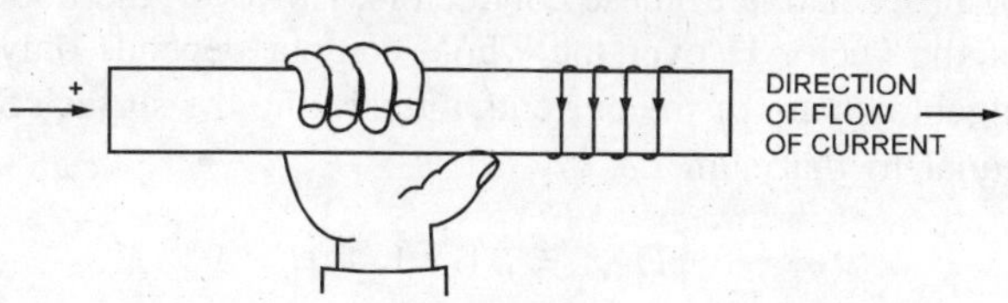

(b) Right Hand Rule

Fig. 7.9

7.9. SOLENOID

The current carrying wire wound spirally in the form of helix about an axis, as shown in Fig. 7.10, is known as *solenoid or coil.* Magnetic field produced due to current carrying solenoid is fairly uniform over a small region in the middle of the coil. It acts just like a bar magnet having north and south poles.

There are several methods used to determine the polarity of the solenoid.

Fig. 7.10. *Solenoid*

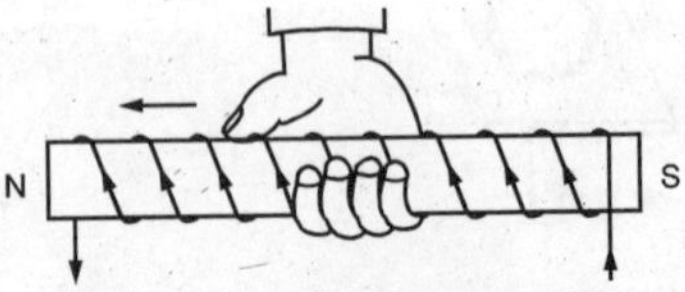

Fig. 7.11. *Helix Rule*

1. By Use of Compass Needle. If one of the poles (say north pole) of the compass needle be brought into close proximity to one of the poles of the current carrying solenoid of unknown polarity the action of the compass needle will immediately classify the pole as north or south depending upon whether the needle is repelled or attracted.

2. Helix Rule. If the helix is held in right hand in such a manner that the finger tips point in the direction of flow of current and thumb is outstretched longitudinally along the coil, it will point towards north pole (Fig. 7.11).

7.10. BASIC LAW OF MAGNETIC FIELD

A magnetic field always accompanies with electric currents. Knowing the magnitude and paths of the currents, the corresponding magnetic fields can be computed. In doing so, account must be taken not only of the currents in the conductors of the electric circuit but also of the intermolecular currents that cause magnetization of the medium surrounding the current carrying conductors. In practice, it is necessary only when there are ferromagnetic substances (such as iron, steel, cast iron, nickel, cobalt and certain alloys) near the conductors.

Let us imagine a magnetic field with currents I_1, I_2 and I_3, an arbitrary closed contour l and an arbitrary surface S supported by it (Fig. 7.12). At least a part of these currents intersect the surface: some in one direction and others in the other direction. Let us choose an arbitrary direction of transverse of the contour l as positive and consider all currents as positive if they intersect the surface S in the same direction as the end of a corkscrew handle moving along the contour l in a positive direction. Thus the current I_1 is negative and current I_2 is positive.

The vector of magnetic field intensity **H** at any point of the contour P depends on the currents both intersecting and not intersecting the surface S, on the shape and dimensions of the conductors carrying these currents, and on the position of point P relative to these conductors. However, the line integral of the vector **H** over the whole contour depends only on the algebraic sum of the currents intersecting the surface S, and is equal to that sum *i.e.*

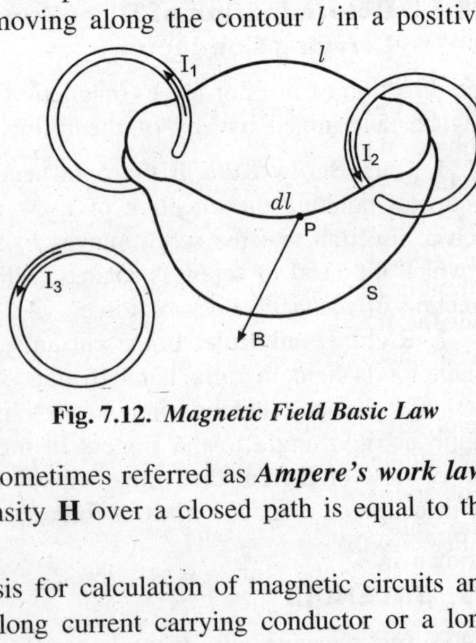

Fig. 7.12. *Magnetic Field Basic Law*

$$\oint \mathbf{H}\, dl = \Sigma I = I_2 - I_1 = I \qquad ...(7.5)$$

The above relation is known as *Ampere's circuital law*, sometimes referred as **Ampere's work law**, which states that the line integral of the magnetic field intensity **H** over a closed path is equal to the current enclosed.

The above law is very comprehensive and provides a basis for calculation of magnetic circuits and makes it possible in some cases (such as in the cases of a long current carrying conductor or a long solenoid) to determine readily the strength of the magnetic field.

7.11. BIOT-SAVART LAW

The quantitative expression for magnetic induction (or flux density) was derived experimentally by Biot and Savart. Taking the deflection of a compass needle as a measure of the intensity of magnetic field of a current, varying in magnitude and shape, the two scientists concluded that any current element projects into space a magnetic field, the magnetic flux density of which, dB, is directly proportional to the length of the element dl, the current I, and the sine of the angle θ between the direction of the current and the vector joining a given point of the field and the current element, and is inversely proportional to the square of the distance of the given point from the current element, r

$$i.e., \quad dB \propto \frac{I\,dl \sin\theta}{r^2} \quad \text{or} \quad dB = K\frac{I\,dl \sin\theta}{r^2}$$

where K is a proportionality constant, which depends upon the magnetic properties of the medium and the system of units employed. In SI system of units $K = \frac{\mu_0 \mu_r}{4\pi}$.

$$\text{So } dB = \frac{\mu_0 \mu_r}{4\pi} \cdot \frac{I\,dl \sin\theta}{r^2} \quad ...(7.6)$$

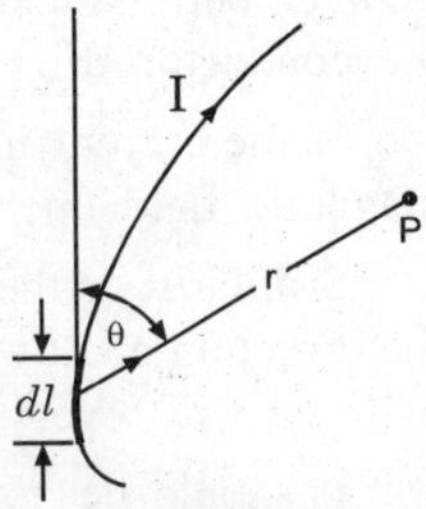

Fig. 7.13

7.12. FORCE ON A CURRENT CARRYING CONDUCTOR LYING IN THE MAGNETIC FIELD

In Fig. 7.14 (*a*) a uniform magnetic field between the two opposite poles is shown.

In Fig. 7.14 (*b*) the cross section of a conductor carrying current in inward direction placed between the two magnets, the field being temporarily removed, is shown. By applying the right-hand thumb rule, the direction of the field around the conductor is found to be clockwise.

If the current carrying conductor shown in Fig. 7.14 (*b*) is placed in the magnetic field shown in Fig. 7.14 (*a*), the resultant magnetic field would be similar to that shown in Fig. 7.14 (*c*).

The lines of force above the conductor are strengthened, since they are in same direction but the lines of force below the conductor are weakened because the two fields below the conductor are opposite in directions and hence tend to destroy each other. Magnetic lines like rubber bands have a tendency to strengthen out and, therefore, a force is experienced on the conductor in the downward direction, as shown in Fig. 7.14 (*c*).

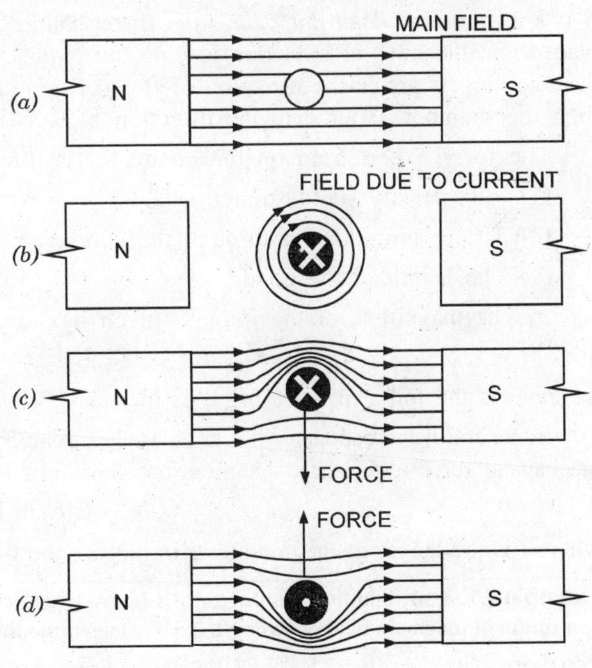

Fig. 7.14

If the direction of current is reversed in the conductor, as shown in Fig. 7.14 (*d*), the direction of force experienced is reversed. In this case the lines of force above the conductor are weakened while those below the conductor are strengthened.

Hence it is observed that when a current carrying conductor is placed at right angle to the direction of magnetic field, a mechanical force is experienced on the conductor in a direction perpendicular to both the direction of magnetic field and flow of current.

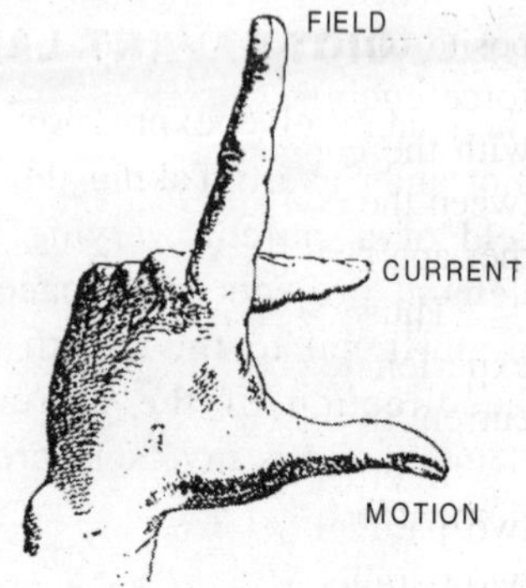

Fleming's Left Hand
Fig. 7.15 (*a*)

The direction of this force can be determined by applying *Fleming's Left Hand Rule* which states that if the thumb, forefinger and middle finger of the left hand are stretched in such a way that they are at right angles to each other mutually and forefinger points towards the direction of the magnetic field, middle finger towards the direction of the flow of current then thumb will point the direction of force acting on the conductor [Fig. 7.15 (*a*)].

If the current in the conductor is reversed, keeping the direction of magnetic field unchanged the direction of force will reverse.

Similarly, if the direction of the magnetic field is reversed, keeping the direction of flow of current in the conductor unchanged, the direction of force will reverse.

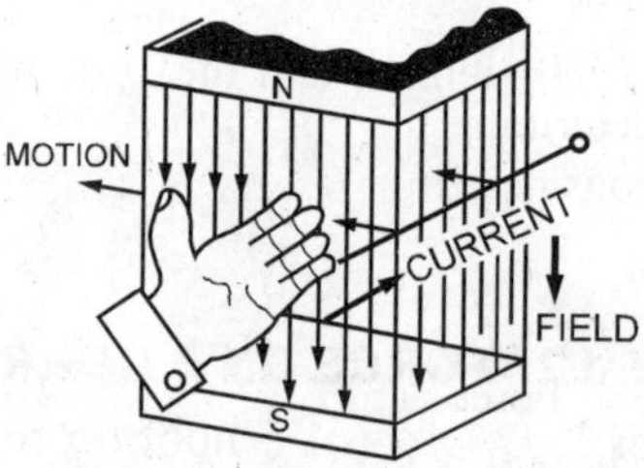

Fig. 7.15 (*b*)

It should be noted that no force is exerted on a conductor when it lies parallel to the magnetic field.

Figure 7.15 (*b*) shows another method of finding the direction of force acting on a current carrying conductor placed in a magnetic field. It is known as *left hand flat palm rule*. If the palm of the left hand is so placed that the lines of induction leaving the N-pole enter the palm and the four fingers are in the direction of flow of current in the conductor, then the thumb will indicate the direction of force acting on the conductor.

The force experienced on the conductor is directly proportional to

(1) Flux density (field strength), B

(2) The current flowing through the conductor, I and

(3) The length of the conductor, l

∴ The magnitude of the force is given by,

$$F = BIl \quad ...(7.7)$$

where F is the force in newtons, B is in teslas (Wb/m^2), I is in amperes and l is in metres.

In general if the conductor lies at an angle θ with a magnetic field of flux density B webers/metre^2, then mechanical force experienced on a current carrying conductor is given by

$$F = BIl \sin\theta \text{ newtons} \quad ...(7.8)$$

where l is the length of the conductor in metres and I is the current carried by the conductor in amperes.

Example 7.1. A straight long conductor of 1 metre length carrying a current of 60 amperes is placed at right angle to a uniform magnetic field of strength 2.5 T. Determine the mechanical force acting on the conductor.

Solution : Given that B = 2.5 T, I = 60 A and l = 1 m

Force acting on the conductor, $F = B I l = 2.5 \times 60 \times 1 = 150$ N **Ans.**

7.13. FORCE BETWEEN TWO PARALLEL CURRENT CARRYING CONDUCTORS

When two parallel conductors carrying current in the same direction (say in downward direction), as shown in Fig. 7.16 (*a*) the lines of force encircle each conductor in the same direction and the resultant field is an envelop of acting like elastic bands tending to pull the conductors together.

When two parallel conductors carry current in the opposite directions, as shown in Fig. 7.16 (*b*), the lines of force encircle but not concentric either with another or with the conductors. The lines of force are crowded between the conductors tending to repel the conductors further apart.

Thus we see that the force of attraction or repulsion is experienced between two parallel conductors carrying current in the same or in opposite directions respectively.

In order to determine the magnitude of force between two parallel current carrying conductors let us consider two parallel conductors A and B each of length l metres and placed d metres apart in air.

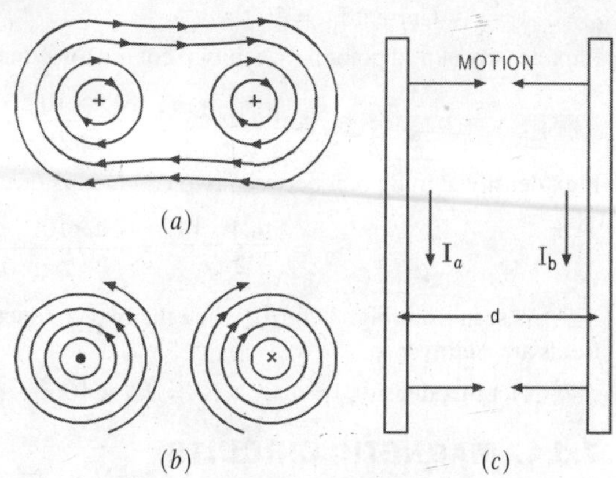

Fig. 7.16

Let the current flowing through the conductors A and B be I_a and I_b respectively.

Field produced due to one of the two conductors experiences a force on another conductor, therefore, to determine the force between two parallel current carrying conductors one of them (say conductor A) be considered placed in the magnetic field due to another one (due to conductor B.)

Magnetic field due to conductor B, $H = \dfrac{I_b}{2\pi d}$ N/Wb

Force acting on conductor A, $F = B\, I_a\, l$

where B is the flux density of the field due to conductor B

or $F = \mu_0 \mu_r\, H \times I_a \times l$ Since $B = \mu_0 \mu_r\, H$

$$= 4\,\pi \times 10^{-7} \times 1 \times \frac{I_b}{2\pi d} \times I_a \times l = \frac{2 I_a I_b\, l}{d} \times 10^{-7} \text{ N or } \frac{2 I_a I_b}{d} \times 10^{-7} \text{ N/m} \qquad ...(7.9)$$

7.13.1. Definition of Ampere

In above expression if $I_a = I_b = 1$ A, $d = 1$ m and $l = 1$ m, it follows that the force per metre length of conductor is 2×10^{-7} N. This value forms the basis for the definition of the ampere.

Thus, the ampere is defined as that current which if maintained in two straight parallel conductors of infinite length, of negligible circular cross section, and placed 1 metre apart in a vacuum, would produce between these conductors a force of 2×10^{-7} N/m *of length.*

Example 7.2. Two parallel wires each of 3 metre length have a separation of 4 mm. Calculate the force exerted on each of these wires when they carry a current of 5 A in the same direction and opposite direction. Assume $\mu_0 = 4\pi \times 10^{-7}$ H/m.

Solution : Current flowing through wires, I = 5 A

Length of each conductor, l = 3 m

Separation between wires, d = 4 mm = 0.004 m

$$\text{Force experienced, } F = \frac{\mu_0 I^2}{2\pi d} \times l = \frac{4\pi \times 10^{-7}}{2\pi} \times \frac{5^2 \times 3}{0.004} = 3.75 \times 10^{-3} \text{ N } \textbf{Ans.}$$

Force experienced between wires will be of attraction when they carry current in the same direction and will be of repulsion when they carry current in the opposite direction.

Example 7.3. Two long parallel wires A and B, 12 cm apart carry currents of 750 A and 500 A respectively in opposite directions. Determine the flux density at the midpoint of the perpendicular line between the wires.

Solution : Current I_A = 750 A

Current $I_B = 500$ A

Flux density at mid point between two conductors, due to conductor A,

$$B_1 = \frac{\mu_0 \mu_r I_A}{2\pi r} = \frac{4\pi \times 10^{-7} \times 1 \times 750}{2\pi \times 0.06} = 2.5 \times 10^{-3} \text{ T}$$

Flux density at mid point between two conductors due to conductor B,

$$B_2 = \frac{\mu_0 \mu_r I_B}{2\pi r} = \frac{4\pi \times 10^{-7} \times 1 \times 500}{2\pi \times 0.06} = 1.667 \times 10^{-3} \text{ T}$$

As seen from Fig. 7.16 (*b*), when the two conductors carry current in opposite directions, the fields are additive, so

Net flux density, $B = B_1 + B_2 = 2.5 \times 10^{-3} + 1.667 \times 10^{-3} = 4.167 \times 10^{-3}$ T **Ans.**

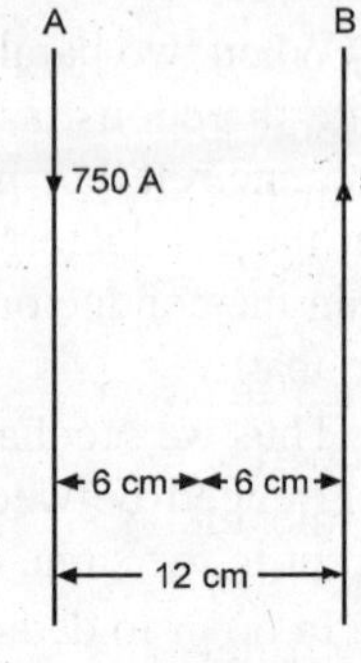

Fig. 7.17

7.14. MAGNETIC CIRCUITS

Practically all electric power machinery (*e.g.* transformers, generators, motors) depend for their operation upon the magnetism produced by the magnetic circuits. The closed path followed by magnetic flux is called a *magnetic circuit* (Fig. 7.18) just as the closed path followed by current is called an electric circuit. There is a lot of similarity between the magnetic and electric circuits, and will be shown later on during discussion.

A magnetic circuit consists of a structure composed for the most part of high permeability magnetic material. The presence of high permeability material causes the magnetic flux to be confined to the paths defined by the structure, much as currents are confined to the conductors of an electric circuit. A simple example of a magnetic circuit is shown in Fig. 7.18. The core is assumed to be composed of magnetic material whose permeability is much greater than that of the surrounding air. The core is of uniform cross-section and is excited by a winding having N turns and carrying a current of I amperes. This winding develops a magnetic field in the core, as illustrated in the figure. The magnetic field can be visualized in terms of flux lines, which form closed loops interlinking with the winding. According to basic law of magnetic field, called the *Ampere's circuital law* (sometimes referred to as *Ampere's work law*) the line integral of **H** around a closed path is equal to the net current enclosed by that path *i.e.*

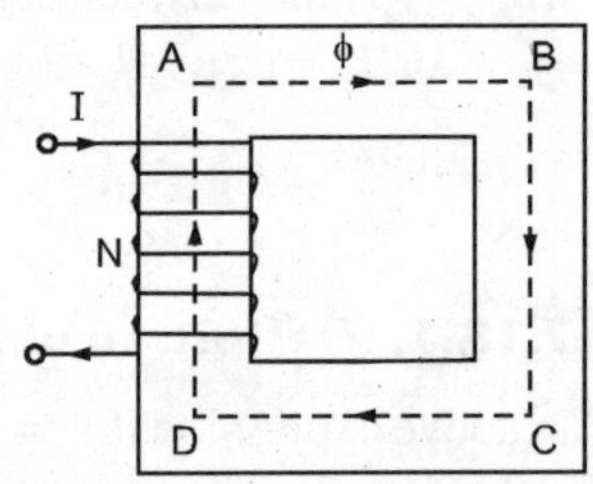

Fig. 7.18. *Simple Magnetic Circuit*

$$\oint \mathbf{H}\, dl = \Sigma I \qquad ...(7.10)$$

The above law is very comprehensive and provides a basis for calculation of magnetic circuits and makes it possible in some cases (such as in the cases of a long current carrying conductor or a long solenoid) to determine readily the strength of the magnetic field.

7.14.1. Magnetic Circuit Concepts

The flux producing ability of the coil in Fig. 7.18 or of a coil on any other magnetic circuit is proportional to the number of turns N and the current I. The product NI is called the *magnetomotive force* (mmf) and determines the amount of flux developed in the magnetic circuit.

MMF of the magnetic circuit is the magnetic potential difference that tends to force flux around the magnetic circuit and is analogous to the electromotive force (emf) in an electric circuit.

$$\text{MMF} = \text{NI ampere-turns} \qquad ...(7.11)$$

The resulting flux ϕ in the magnetic circuit, besides being dependent on the mmf, is also a function of the opposition of the iron to carrying flux. This opposition is called the *reluctance* S of the magnetic circuit.

As in the case with resistance in the electric circuit, reluctance S of the magnetic circuit is directly proportional to length l, inversely proportional to cross-sectional area a, and dependent on the nature of material of the magnetic circuit. The reluctance of the magnetic circuit is given as

$$S = \frac{l}{\mu a} \text{ ampere-turns/weber} \quad ...(7.12)$$

when the flux ϕ is constant over the length and uniform over the area. The quantity μ expresses the property of the magnetic material called its *permeability*. Permeability is a measure of the receptiveness of the material of having magnetic flux developed in it. For free space, the permeability μ_0 equals $4\pi \times 10^{-7}$ H/m in SI system. The relative permeability of magnetic material μ_r may range up to thousands.

Total flux developed in the circuit is given as

$$\phi = \frac{\text{MMF}}{\text{S}} \text{ webers} \quad ...(7.13)$$

The above Eq. (7.13) is sometimes referred to as *Ohm's law* for the magnetic circuit. It serves to emphasize the mathematical analogy between the magnetic circuit and the electric circuit. Analogous quantities in the two circuits are listed in Art. 7.14.4.

Magnetic circuits differ from electric circuits in one important respect. The reluctance of the magnetic circuit containing iron or ferromagnetic material depends upon the flux carried by it. With the increase in flux, a large change in mmf is required to develop the same change in flux.

7.14.2. Terminology

1. MMF of the magnetic circuit is defined as the magnetic potential that drives or tends to drive flux around the magnetic circuit and is analogous to the emf in an electric circuit. It is measured in ampere-turns (AT) or simply in amperes.

2. Magnetic Field Intensity. MMF per unit length (along the path of magnetic flux) is called the magnetic field intensity (H) and is given as

$$\text{H} = \frac{\text{MMF}}{\text{Length}} = \frac{\text{NI}}{l} \text{ AT/m}$$

3. Reluctance. It is the name given to that property of material which opposes the creation of magnetic flux in it. It, in fact, measures the resistance offered to the passage of magnetic flux through a material and is analogous to resistance in an electric circuit. It is measured in ampere-turns / Wb or amperes / Wb.

4. Permeability. It is a measure of receptiveness of the material of having magnetic flux developed in it.

Every substance possesses a certain power of conducting magnetic flux. For example, iron is a better conductor for magnetic flux than air (or vacuum). Permeability of a material (μ) is thus its conducting power for magnetic flux. It is equal to the ratio of flux density B and the magnetic field strength H *i.e.* $\mu = \frac{\text{B}}{\text{H}}$.

5. Relative Permeability. Relative permeability of a material is defined as the ratio of the number of lines of magnetic flux per unit area in the given medium to the number of lines of magnetic flux per unit area, when medium is replaced by vacuum.

Relative permeability of free space is unity and of magnetic material may range up to thousands.

7.14.3. Determination of Ampere-Turns

In any magnetic circuit, flux created is given as

$$\phi = \frac{\text{MMF}}{\text{Reluctance}} = \frac{\text{AT}}{l/\mu a} = \frac{\text{AT}}{l/\mu_0 \mu_r a}$$

$$\text{or AT required} = \frac{\phi}{a} \times \frac{1}{\mu_0 \mu_r} \times l = \frac{B}{\mu_0 \mu_r} \times l = H \times l \text{ ampere-turns} \qquad ...(7.14)$$

Hence for determination of AT for a magnetic circuit

(*i*) First find field strength H in each part of the magnetic circuit, (*ii*) find the length of various parts of magnetic circuit, (*iii*) find the number of ampere-turns required for the various parts of magnetic circuit from the relation AT = Hl where l is the length of the part in metres and lastly, (*iv*) find total number of ampere-turns for the whole series magnetic circuit by adding ampere-turns determined for various paths in magnetic circuit.

This will be more clear to the students from examples given in this chapter.

As both permeability and reluctance may change from one operating condition to another. Hence direct numerical application of the reluctance concept and Eqs. (7.13) and (7.14) is rare. For quantitative analysis, graphical methods are generally used because they can easily be adapted to the non-linearities involved.

7.14.4. Comparison Between Magnetic and Electric Circuits

	Magnetic Circuit	Electric Circuit
	SIMILARITIES	
Circuit	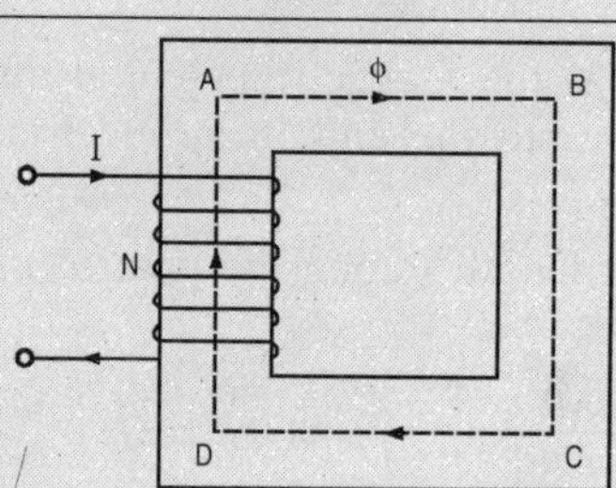	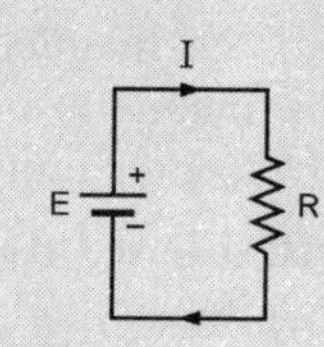
1. Definition	The closed path followed by magnetic flux is called a magnetic circuit	The closed path followed by an electric current is called an *electric circuit*
2. Driving force	**MMF** is required to establish flux ϕ in the magnetic circuit and is measured in ampere-turns (AT) or amperes	**EMF** is required to cause flow of current in an electric circuit and is measured in volts
3. Response	Flux, $\phi = \frac{\text{Driving force}}{\text{Magnetic reluctance}} = \frac{AT}{S}$ webers	Current, $I = \frac{\text{Driving force}}{\text{Electric resistance}} = \frac{E}{R}$ amperes
4. Impedance	Reluctance, $S = \frac{l}{\mu_0 \mu_r a} = \frac{l}{\mu a}$ AT/Wb	Resistance, $R = \rho \frac{l}{a}$ ohms
(*a*) For series circuits	$S = S_1 + S_2 + S_3 + ...$	$R = R_1 + R_2 + R_3 + ...$
(*b*) For parallel circuits	$S = \frac{1}{\frac{1}{S_1} + \frac{1}{S_2} + \frac{1}{S_3} + ...}$	$R = \frac{1}{\frac{1}{R_1} + \frac{1}{R_2} + \frac{1}{R_3} + ...}$
5. Admittance	Permeance $= \frac{1}{\text{Reluctance}}$ in Wb/AT	Conductance $= \frac{1}{\text{Resistance}}$ in siemens
6. Proportionality constant	Reluctivity $= \frac{1}{\text{Permeability}} = \frac{1}{\mu}$ in m/H	Resistivity, $\rho = \frac{1}{\text{Conductivity}}$ in ohm-metre

7. Density	Flux density, B = μ × magnetic field intensity = μH is tesla or Wb/m^2	Current density, $J = \frac{\text{Electric field intensity}}{\rho}$ in A/m^2
8. Field intensity	Magnetic field intensity, $H = \frac{MMF}{l} = \frac{NI}{l}$ in AT/m	Electric field intensity $= \frac{E}{l}$ in volts/metre
9. Drop	MMF drop = φ × reluctance = φS	Voltage drop = I × Resistance = IR

DISSIMILARITIES

1. Flux does not actually flow in a magnetic circuit.	Current does flow in an electric circuit.
2. Permeability does not largely vary from material to material and there is hardly any material which can act as an insulator to the magnetic flux.	Conductivity varies largely from material to material so much so that some materials are insulators to electric current and some are very good conductors.
3. For a particular temperature the permeability depends upon the flux density (or total flux).	For a particular temperature, conductivity (or resistivity) is constant and independent of current strength.
4. Flux can pass through air.	Current would not flow through the air until an arc is struck.
5. Residual flux persists after removal of mmf.	The current is reduced to zero after removal of source of emf.
6. There is no waste of energy due to reluctance in the magnetic circuit and, therefore, energy is required only to create the magnetic flux but not to maintain it.	In an electric circuit, resistance causes heat to be generated resulting in waste of energy and, therefore, energy is required as long as the current flows.

7.14.5. Magnetic Circuits With Air Gaps

Energy-conversion devices which incorporate a moving element have necessarily airgaps in their magnetic circuits. Airgaps are also provided in the magnetic circuits to avoid saturation. A magnetic circuit with an airgap is shown in Fig. 7.19. An airgap is nothing else but a volume of air between two magnetic surfaces. The length of airgap l_g equals the distance between the two magnetic surfaces. The area of x-section of any one of the surfaces gives the airgap area, a_g. When the airgap length l_g is much smaller than the dimensions of the adjacent core faces, the magnetic flux φ is constrained essentially to reside in the core and the airgap and is continuous throughout the magnetic circuit. Thus the configuration shown in Fig. 7.19 can be analysed as a magnetic circuit with two series components, a magnetic or iron core of permeability μ and mean length l_i and an airgap of permeability μ_0 and length l_g. Since the permeability of air is constant, the airgap is a linear part of the magnetic circuit and the flux density in the airgap is proportional to the mmf across the airgap. The necessary mmf is calculated separately for the airgap and the iron portions and then added to determine the total mmf.

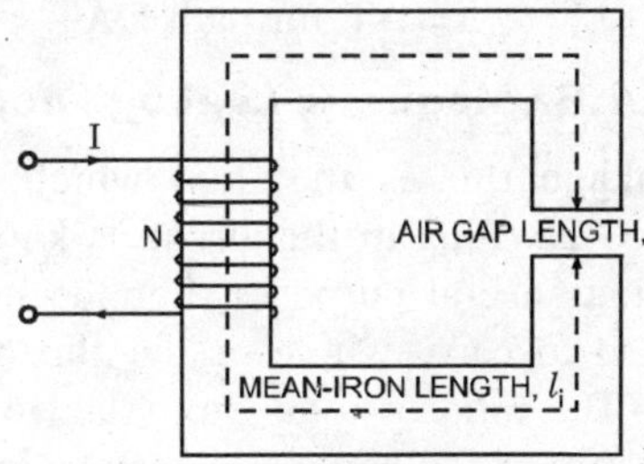

Fig. 7.19. *Magnetic Circuit With Air Gap*

7.14.6. Composite Magnetic Circuits

Consider a circular ring made from different materials of lengths l_1, l_2 and l_3, cross-sectional areas a_1, a_2 and a_3 and relative permeability μ_{r1}, μ_{r2} and μ_{r3} respectively with a cut of length l_g known as airgap. The total reluctance is the arithmetic sum of individual reluctances as they are joined in series.

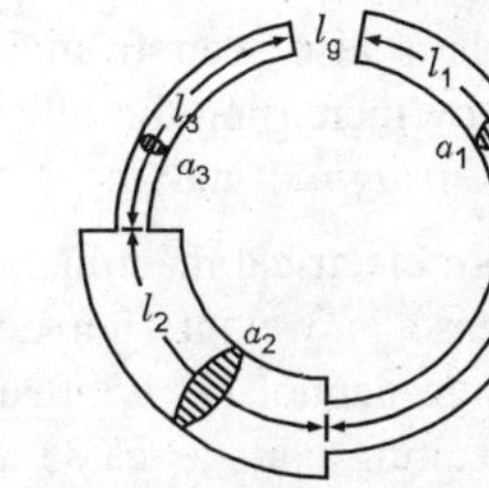

Fig. 7.20. *Composite Magnetic Circuit*

$$\therefore \text{ Total reluctance, } S = \frac{l_1}{\mu_0 \mu_{r1} a_1} + \frac{l_2}{\mu_0 \mu_{r2} a_2} + \frac{l_3}{\mu_0 \mu_{r3} a_3} + \frac{l_g}{\mu_0 a_g}$$

$$\text{Total mmf} = \phi \times S = \phi\left[\frac{l_1}{\mu_0 \mu_{r1} a_1} + \frac{l_2}{\mu_0 \mu_{r2} a_2} + \frac{l_3}{\mu_0 \mu_{r3} a_3} + \frac{l_g}{\mu_0 a_g}\right]$$

or Total ampere-turns required $= H_1 l_1 + H_2 l_2 + H_3 l_3 + H_g l_g$ ...(7.15)

= Sum of ampere-turns required for individual parts of the magnetic circuit.

7.14.7. Parallel Magnetic Circuits

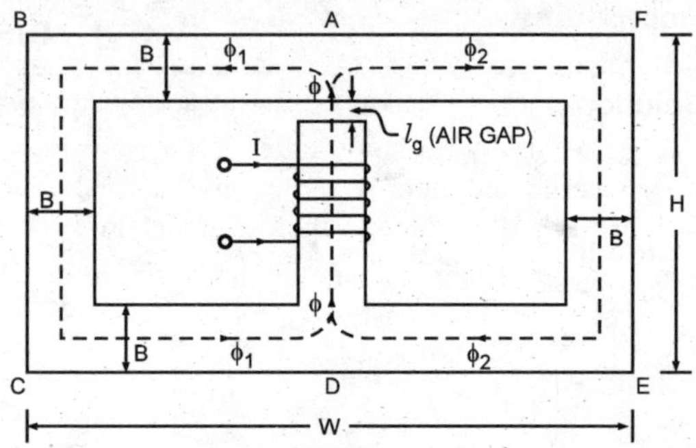

Fig. 7.21

In series circuits, all parts of the magnetic circuit carry same flux and total ampere-turns required to create a given flux is the arithmetic sum of the ampere-turns required for individual parts of the circuit.

But if the various paths of the magnetic circuit are in parallel, as shown in Fig. 7.21, the ampere-turns required for the combination is equal to the ampere turns required to create the given flux in one path.

For example for the circuit shown in Fig. 7.21 paths ABCD and AFED are in parallel, so ampere-turns required to create flux ϕ_1 in path ABCD is equal to ampere-turns required to create flux ϕ_2 in path AFED and also equal to the ampere-turns required for both of the paths.

Hence total ampere-turns required for magnetic circuit shown in Fig. 7.21

= AT for path DA + AT for path ABCD

= AT for path DA + AT for path AFED

7.14.8. Magnetic Leakage and Fringing

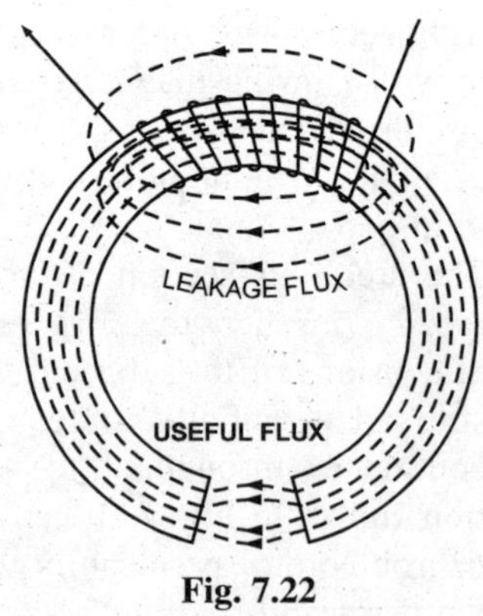

Fig. 7.22

Leakage flux is the flux, which follows a leaking path, as shown in Fig. 7.22. Flux in the airgap is known as *useful flux* which is utilized for various useful purposes. For the purpose of calculations the iron is supposed to carry whole of the flux throughout its entire length.

The ratio of total flux (flux in the iron path) to the useful flux (flux in the air) is known as *leakage factor*.

$$\textit{i.e.} \text{ Leakage factor} = \frac{\text{Total flux}}{\text{Useful flux}}$$

$$\text{or Leakage factor} = \frac{\text{Flux in the iron path}}{\text{Flux in the airgap}}$$

It is also seen from Fig. 7.22 that the useful flux passing across the gap tends to bulge outwards, thereby increasing the effective area of gap and reducing the flux density in the gap. This effect is referred to as *fringing*; and the longer the airgap, the greater is the fringing.

In electrical machines (such as in generators and motors), magnetic leakage is undesirable as it causes increase in weight (not decrease in their power efficiency) and cost of manufacture. Though magnetic leakage cannot be avoided completely but can be reduced to the minimum by placing the magnetizing or exciting coils as close as possible to the airgap or to the points in the magnetic circuit where the flux is to be utilized for useful purposes.

The value of leakage factor for modern electric machinery is about 1.2.

Example 7.4. What will be the current passing through the ring shaped air-cored coil when number of turns is 800 and ampere-turns are 3,200. [*Pb. Technical Univ. Basic Electrical and Electronics Engineering May 2006*]

Solution : Ampere-turns, AT = 3,200

Number of turns, N = 800

Current through the ring shaped air-cored coil,

$$I = \frac{AT}{N} = \frac{3,200}{800} = 4 \text{ A} \textbf{ Ans.}$$

Example 7.5. An electromagnet has an gap of 4 mm and flux density in the gap is 1.3 Wb/m². Determine the ampere-turns for the gap.

Solution : Magnetising force, $H_g = \frac{B}{\mu_0 \mu_r} = \frac{1.3}{4\pi \times 10^{-7} \times 1} = 1.035 \times 10^6$ AT/m $\because \mu_r$ for air = 1

Ampere-turns required for the gap $= H_g l_g = 1.035 \times 10^6 \times 4 \times 10^{-3} = 4,140$ **Ans.**

Example 7.6. A ring of mean diameter 30 cm is wound with 200 turns of copper wire carrying a current of 2 A. The x-section of the magnetic material of the ring is 12 cm² and its relative permeability is 1,000. Determine the flux through it.

Solution : Length of iron ring, $l = \pi D = \pi \times 0.3 = 0.942$ m

Area of x-section of iron path, $a = 12 \text{ cm}^2 = 0.0012 \text{ m}^2$

Relative permeability, $\mu_r = 1,000$ and $\mu_0 = 4\pi \times 10^{-7}$ H/m

$$\text{Flux through iron path, } \phi = \frac{NI}{l/\mu_r \mu_0 a} = \frac{200 \times 2}{0.942/1,000 \times 4\pi \times 10^{-7} \times 0.0012}$$

$$= 0.64 \text{ mWb } \textbf{Ans.}$$

Example 7.7. An iron ring having a mean diameter of 25 cm and cross-sectional area 2 cm² is uniformly wound with 400 turns and carries a current of 5 A. The permeability of iron is 450. Calculate (*i*) magneto motive force (*ii*) reluctance (*iii*) flux (*iv*) flux density produced.

Solution : Mean iron length, $l = \pi D = \pi \times 0.25 = 0.25 \pi$m

Cross-sectional area of iron, $a = 2 \text{ cm}^2 = 0.0002 \text{ m}^2$

(*i*) MMF = N × I = 400 × 5 = 2,000 AT **Ans.**

$$(ii) \text{ Reluctance, } S = \frac{l}{\mu_0 \mu_r a}$$

$$= \frac{0.25\pi}{4\pi \times 10^{-7} \times 450 \times 0.0002} = 0.694 \times 10^7 \text{ AT/Wb } \textbf{Ans.}$$

$$(iii) \text{ Flux, } \phi = \frac{MMF}{S} = \frac{2,000}{0.694 \times 10^7} = 0.288 \text{ mWb } \textbf{Ans.}$$

$$(iv) \text{ Flux density, } B = \frac{\phi}{a} = \frac{0.288 \times 10^{-3}}{0.0002} = 1.44 \text{ T } \textbf{Ans.}$$

Example 7.8. An iron ring has a mean diameter of 25 cm and a cross-sectional area of 4 cm². It is wound with a coil of 1,200 turns. An air gap of 1.5 mm width is cut in the ring. Determine the current required in the coil to produce a flux of 0.48 mWb in the air gap, if the relative permeability of iron under these conditions is 800. Neglect leakage and fringing.

Solution : Flux to be produced, $\phi = 0.48$ m Wb $= 0.48 \times 10^{-3}$ Wb

Area of x-section, $a = 0.0004 \text{ m}^2$

$$\text{Flux density required, } B = \frac{\phi}{a} = \frac{0.48 \times 10^{-3}}{0.0004} = 1.2 \text{ T}$$

Air-gap length, $l_g = 1.5 \text{ mm} = 1.5 \times 10^{-3}$ m

Length of the iron path, $l_i = \pi d - l_g = 0.25\pi - 1.5 \times 10^{-3} = 0.784$ m

$$\text{Total ampere-turns required} = \frac{B}{\mu_0} l_g + \frac{B}{\mu_0 \mu_r} l_i = \frac{1.2}{4\pi \times 10^{-7}} \times 1.5 \times 10^{-3} + \frac{1.2 \times 0.784}{4\pi \times 10^{-7} \times 800}$$

$$= 1{,}432 + 936 = 2{,}368 \text{ AT}$$

$$\text{Current required, I} = \frac{AT}{N} = \frac{2{,}368}{1{,}200} = 1.973 \text{ A } \textbf{Ans.}$$

Example 7.9. An iron ring of mean length 50 cm and relative permeability 300 has an air gap of 1 mm. If the ring is provided with a winding of 200 turns and a current of 1 A is allowed to flow through, find the flux density across the airgap.

Solution : Let the flux density through the iron as well as air gap be B Wb/m^2

$$\text{Total ampere-turns provided, AT} = N \times I = 200 \times 1 = 200$$

$$\text{Total ampere-turns required} = \text{AT required for air gap} + \text{AT required for iron ring}$$

or $$\text{Total ampere-turns provided} = \frac{B}{\mu_0} l_g + \frac{B}{\mu_0 \mu_r} l_i$$

or $$200 = \frac{B}{\mu_0}\left(l_g + \frac{l_i}{\mu_r}\right) = \frac{B}{\mu_0}\left(0.001 + \frac{0.5}{300}\right)$$

or $$\text{Flux density, B} = \frac{200 \times \mu_0}{0.001 + \frac{0.5}{300}} = \frac{200 \times 4\pi \times 10^{-7}}{0.001 + 0.001667} = 0.0942 \text{ T } \textbf{Ans.}$$

Example 7.10. An iron ring of mean circumference 80 cm is made from round iron of area 8 cm^2. It has a saw cut of 2 mm wide and is wound with 500 turns. Find the current required to produce a flux of 0.8 m Wb across the air gap. Assume a relative permeability for iron as 625 and a leakage factor of 1.25.

Solution : Area of x-section, $a = 8 \text{ cm}^2 = 0.0008 \text{ m}^2$

$$\text{Flux density in the air gap, } B_g = \frac{\phi_g}{a} = \frac{0.8 \times 10^{-3}}{0.0008} = 1 \text{ T}$$

$$\text{Ampere-turns required for air gap, } AT_g = \frac{B_g}{\mu_0} . l_g = \frac{1.0 \times 2 \times 10^{-3}}{4\pi \times 10^{-7}} = 1{,}592$$

$$\text{Flux through iron path, } \phi_i = \phi_g \times \text{leakage factor} = 0.8 \times 10^{-3} \times 1.25 = 0.001 \text{ Wb}$$

$$\text{Flux density in iron path, } B_i = \frac{0.001}{0.0008} = 1.25 \text{ T}$$

$$\text{Ampere-turns required for iron path, } AT_i = \frac{B_i}{\mu_0 \mu_r} l_i = \frac{1.25}{4\pi \times 10^{-7} \times 625} \times (0.8 - 0.002) = 1{,}270$$

$$\text{Total ampere-turns required, AT} = AT_g + AT_i = 1{,}592 + 1{,}270 = 2{,}862$$

$$\text{Current required, I} = \frac{AT}{N} = \frac{2{,}862}{500} = 5.72 \text{ A } \textbf{Ans.}$$

Example 7.11. The magnetic circuit frame shown in Fig. 7.23 is built of iron of square cross section of 3 cm. Each air gap is 2 mm wide and each of the coil is wound with 1,000 turns. The relative permeability of part A and B may be taken as 1,000 and 1,200 respectively. Calculate (*a*) reluctance of part A (*b*) reluctance of two air gaps (*c*) total reluctance of complete magnetic path.

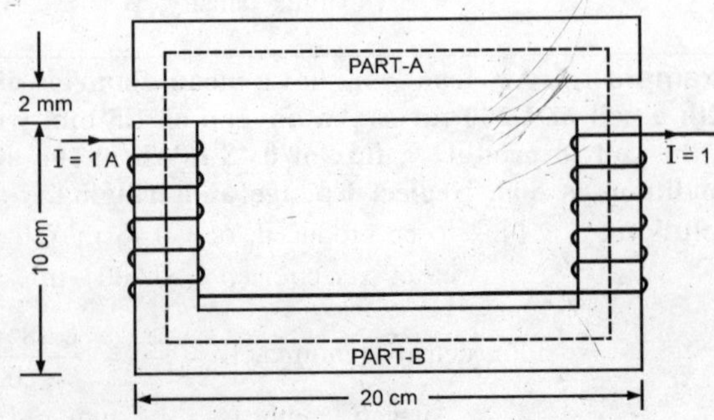

Fig. 7.23

Solution : Cross-sectional area,

$$a = (0.03)^2 = 0.0009 \text{ m}^2$$

Mean length flux path in part A,

$$l_A = 20 - (1.5 + 1.5) = 17 \text{ cm or } 0.17 \text{ m}$$

Mean length flux path in part B, $l_B = (20 - 1.5 - 1.5) + (10 - 1.5) + (10 - 1.5) = 34$ cm or 0.34 m

Relative permeability of part A, $\mu_{rA} = 1{,}000$

Relative permeability of part B, $\mu_{rB} = 1{,}200$

Length of two air gaps, $l_g = 2 \times 2 = 4$ mm $= 0.004$ m

(*i*) Reluctance of part A, $S_A = \dfrac{l_A}{\mu_0 \mu_{rA} \times a} = \dfrac{0.17}{4\pi \times 10^{-7} \times 1{,}000 \times 0.0009} = 1{,}50{,}313$ AT/Wb **Ans.**

(*ii*) Reluctance of part B, $S_B = \dfrac{l_B}{\mu_0 \mu_{rB} \times a} = \dfrac{0.34}{4\pi \times 10^{-7} \times 1{,}200 \times 0.0009} = 2{,}50{,}521$ AT/Wb

(*iii*) Reluctance of two air gaps, $S_g = \dfrac{l_g}{\mu_0 a} = \dfrac{0.004}{4\pi \times 10^{-7} \times 0.0009} = 35{,}36{,}776$ AT/Wb **Ans.**

(*iv*) Total reluctance, $S = S_A + S_B + S_g = 1{,}50{,}313 + 2{,}50{,}521 + 35{,}36{,}776$
$= 39{,}37{,}610$ AT/Wb **Ans.**

Example 7.12. For the magnetic circuit frame given is Fig. 7.23 calculate (*a*) mmf (*b*) total flux (*c*) flux density.

Solution : (*a*) Total mmf $= 2 \times 1{,}000 \times 1 = 2{,}000$ AT **Ans.**

because mmf is produced by two coils on part B each having 1,000 turns and carrying a current of 1 A

(*b*) Total flux $= \dfrac{\text{MMF}}{\text{Reluctance}} = \dfrac{2{,}000}{39{,}37{,}610} = 5.08 \times 10^{-4}$ Wb

$\because$ Reluctance $= 39{,}37{,}610$ AT/Wb, as determined in Example 7.11

(*c*) Flux density, $B = \dfrac{\phi}{a} = \dfrac{5.08 \times 10^{-4}}{0.0009} = 0.564$ Wb/m^2 (or Tesla) **Ans.**

Example 7.13. A rectangular magnetic core shown in Fig. 7.24 (*a*) has square cross section of area 16 cm². An air gap of 2 mm is cut across one of its limbs. Find the exciting current needed in a coil having 1,000 turns wound on the core to create an air-gap flux of 4 mWb. The relative permeability of the core is 2,000.

Solution : Flux to be created, $\phi = 4$ mWb
$= 0.004$ Wb

Area of x-section, $a = 16$ cm^2
$= 16 \times 10^{-4}$ m^2

Fluse density required, $B = \dfrac{\phi}{a}$
$= \dfrac{4 \times 10^{-3}}{16 \times 10^{-4}}$
$= 2.5$ T

Length of air gap $l_g = 2$ mm
$= 2 \times 10^{-3}$ m

Each side of cross-section $= \sqrt{16} = 4$ cm

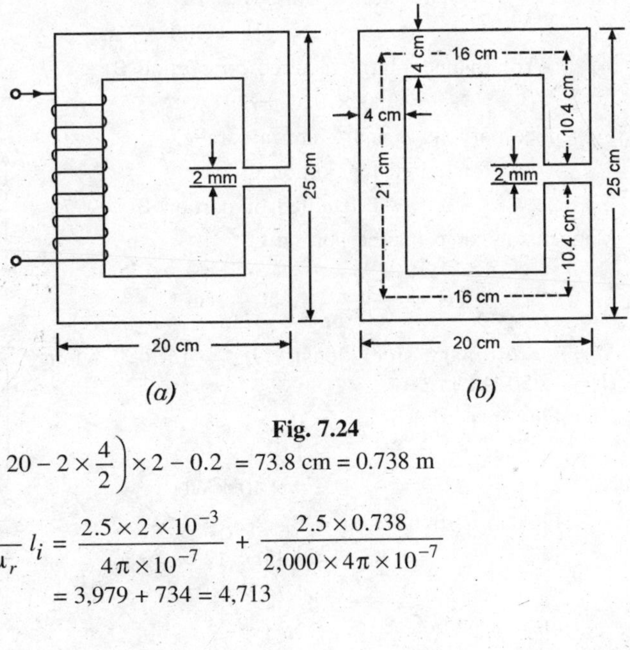

Fig. 7.24

Length of iron-path, $l_i = \left(25 - 2 \times \dfrac{4}{2} + 20 - 2 \times \dfrac{4}{2}\right) \times 2 - 0.2 = 73.8$ cm $= 0.738$ m

Total ampere-turns required, $AT = \dfrac{B}{\mu_0} l_g + \dfrac{B}{\mu_0 \mu_r} l_i = \dfrac{2.5 \times 2 \times 10^{-3}}{4\pi \times 10^{-7}} + \dfrac{2.5 \times 0.738}{2{,}000 \times 4\pi \times 10^{-7}}$
$= 3{,}979 + 734 = 4{,}713$

Number of turns on the coil $= 1{,}000$

Exciting current required, $I = \dfrac{AT}{N} = \dfrac{4{,}713}{1{,}000} = 4.713$ A **Ans.**

Example 7.14. A ring of cast steel has an external diameter of 24 cm and a square x-section of 3 cm side. Inside and across the ring an ordinary steel bar 18 × 3 × 0. 4 cm is fitted with negligible gap. Calculate the number of ampere-turns required to be applied to one half of the ring to produce a flux density of 1 Wb/m² in the other half. Neglect leakage.

The B-H characteristics area follows :

	For cast steel			For ordinary steel		
B(Wb/m²)	**1.0**	**1.1**	**1.2**	**1.2**	**1.4**	**1.45**
H(AT/m)	**900**	**1,020**	**1,230**	**590**	**1,200**	**1,650**

[*Pb. Univ. Applied Electricity April, 1974*]

Solution :

External diameter of the ring = 24 cm = 0.24 m

Internal diameter of the ring = 24 – 6 = 18 cm = 0.18 m

$$\text{Mean diameter of the ring, } D = \frac{0.24 + 0.18}{2} = 0.21 \text{ m}$$

Mean length of the ring = $\pi D = \pi \times 0.21 = 0.66$ m

$$\text{Length of upper portion of ring, } l_B = \text{length of lower portion of ring } l_A = \frac{0.66}{2} = 0.33 \text{ m}$$

Area of x-section of ring $a_A = a_B = 0.03 \times 0.03 = 9 \times 10^{-4}\ \text{m}^2$

Length of cross-bar, l_c = 18 cm = 0.18 m

Area of cross section of cross-bar, $a_c = 0.03 \times 0.004 = 12 \times 10^{-5}\ \text{m}^2$

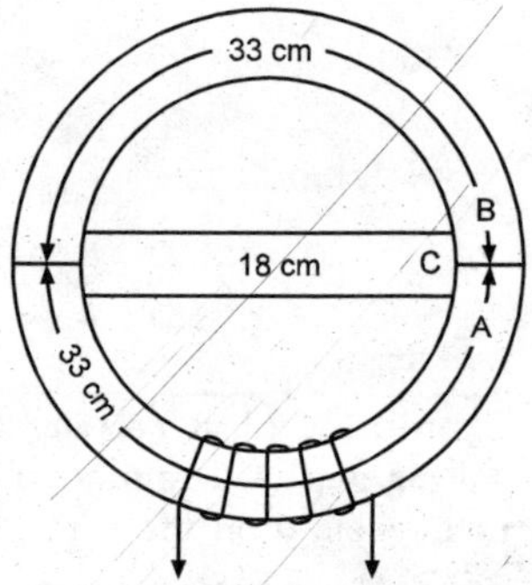

Fig. 7.25

This is a case of parallel magnetic circuit.

Flux created in A divides itself when it comes to the junction point depending upon the reluctance of portions B and C respectively.

i.e. $\phi_A = \phi_B + \phi_C$

Flux density in the upper half,

B_B = 1 T

For cast steel, against B = 1 T ;

H = 900 AT/m

∴ Total ampere-turns required for portion B

= 900 × 0.33 = 297

Since portions B and C are in parallel

∴ AT required for portion C

= AT required of portion B = 297

Magnetising force for the portion C,

$$H = \frac{297}{0.18} = 1,650 \text{ AT/m}$$

For ordinary steel density B_c = 1.45 T when H = 1,650 AT/m

Flux in portion B, $\phi_B = B_B \times a_B$

$= 1 \times 9 \times 10^{-4}$

$= 9 \times 10^{-4}$ Wb

Flux in portion C, $\phi_C = B_C \times a_C$

$= 1.45 \times 12 \times 10^{-5}$

$= 1.74 \times 10^{-4}$ Wb

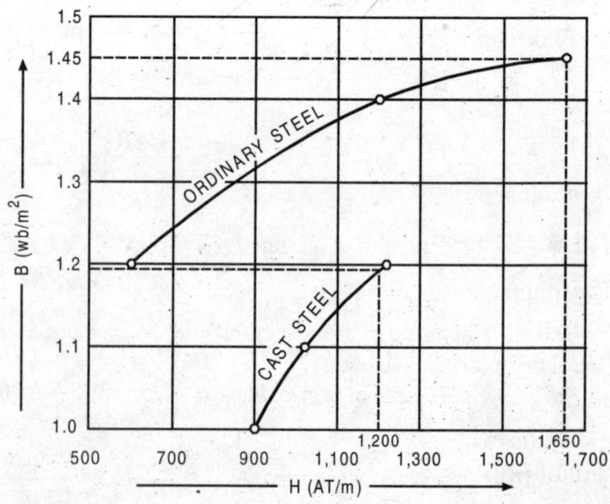

Fig. 7.26

Flux in portion A, $\phi_A = \phi_B + \phi_C$

$= 9 \times 10^{-4} + 1.74 \times 10^{-4} = 10.74 \times 10^{-4}$ Wb

$$\text{Flux density in portion A, } B_A = \frac{\phi_A}{a_A} = \frac{10.74 \times 10^{-4}}{9 \times 10^{-4}} = 1.1933 \text{ T}$$

From the B-H curve drawn between H and B for cast steel magnetising force corresponding to flux density of 1.1933 T is 1,200 AT/m.

AT required for portion A = 1200 × 0.33 = 396

Total ampere-turns required = AT for portion A + AT for portion B (or AT for portion C)

= 396 + 297 = 693 **Ans.**

Example 7.15. A cast steel magnetic structure made of a bar of section 2 cm × 2 cm is shown in Fig. 7.27. Determine the current that the 500 turn magnetizing coil on the left limb should carry so that a flux of 2 mWb is produced in the right limb. Take $\mu_r = 600$ and neglect leakage.

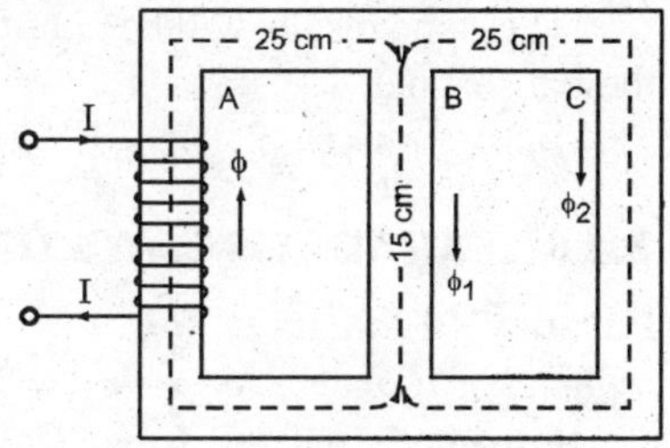

Fig. 7.27

Solution : Flux created by magnetizing coil, say ϕ, is divided at junction point into two paths depending upon the reluctances of portions B and C respectively.

$$\text{Reluctance of portion B, } S_B = \frac{l_B}{\mu_0 \mu_r \times a} = \frac{0.15}{\mu_0 \mu_r a}$$

$$\text{Reluctance of portion C, } S_C = \frac{l_C}{\mu_0 \mu_r a} = \frac{0.25}{\mu_0 \mu_r a}$$

Flux in portion C, ϕ_C = 2 mWb = 0.002 Wb (given)

$$\text{Flux in portion B, } \phi_B = \phi_C \times \frac{S_C}{S_B} = 2 \times 10^{-3} \times \frac{0.25/\mu_0\mu_r a}{0.15/\mu_0\mu_r a} = 3.33 \times 10^{-3} \text{ Wb}$$

Total flux in portion A, $\phi = \phi_B + \phi_C = 3.33 \times 10^{-3} + 2 \times 10^{-3} = 5.33 \times 10^{-3}$ Wb

$$\text{Flux density in portion A, } B_A = \frac{\phi}{a} = \frac{5.33 \times 10^{-3}}{4 \times 10^{-4}} = 13.33 \text{ Wb/m}^2$$

$$\text{AT required for portion A, } AT_A = \frac{B_A \times l_A}{\mu_0 \mu_r} = \frac{13.33 \times 0.25}{4\pi \times 10^{-7} \times 600} = 4{,}416$$

$$\text{Flux density in path B, } B_B = \frac{3.33 \times 10^{-3}}{4 \times 10^{-4}} = 8.33 \text{ Wb/m}^2$$

$$\text{AT required for portion B, } AT_B = \frac{B_B\, l_B}{\mu_0 \mu_r} = \frac{8.33 \times 0.15}{4\pi \times 10^{-7} \times 600} = 1{,}658$$

Total ampere-turns required, AT = $AT_A + AT_B$ = 4,416 + 1,658 = 6,074 **Ans.**

7.15. ELECTROMAGNETIC INDUCTION

The phenomenon by which an emf is induced in a circuit, when magnetic flux linking with it changes, is known as *electromagnetic induction.*

The method of producing emf by electromagnetic induction is discussed below :

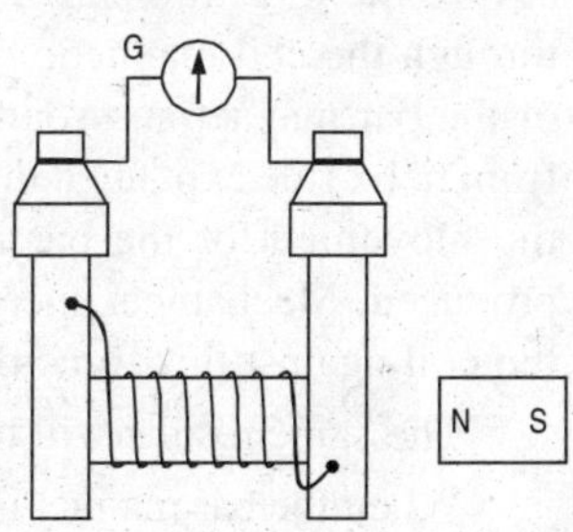

Fig. 7.28

It has been shown in Art. 7.12 that when a current carrying conductor is suitably placed in a magnetic field a force is exerted on the conductor in a direction perpendicular to both the direction of flow of current and that of magnetic field. Faraday discovered that reverse of the above is also true *i.e.* an emf can be produced by moving a conductor in a magnetic field. This will be clear from the following experiment.

When two ends of a coil are connected to a galvanometer G and a bar magnet N S is brought nearer to the coil, deflection in the galvanometer is observed. The deflection of galvanometer indicates the flow of current through the coil and this deflection will persist so long as the magnet is moved. If the magnet movement is stopped, the deflection will reduce to zero. Now if the magnet is moved away from the coil, galvanometer will again show deflection but in opposite direction.

When the magnet is moved towards the coil, linking flux with the coil increases, therefore, an emf is induced in the coil, which causes flow of current through the coil and galvanometer shows deflection. Similarly when the magnet is moved away from the coil, the linking flux decreases, therefore, an emf is induced but in direction opposite to previous one. But when the movement of the magnet is stopped, though flux links with the coil, but there is no change in linking flux, therefore, no emf is induced in the coil.

The same results will be obtained if bar magnet is kept stationary and coil is moved towards or away from the bar magnet.

Hence, for the production of an emf, the change in the magnitude of the linking flux is essential.

7.16. FARADAY'S LAWS OF ELECTROMAGNETIC INDUCTION

Faraday's First Law. *This law states that when the flux linking with the coil or circuit changes, an emf is induced in it or whenever the magnetic flux is cut by the conductor an emf is induced in the conductor.*

Faraday's Second Law. *This law states that the magnitude of emf induced is directly proportional to the rate of change of flux linkage or to the product of number of turns and rate of change of flux linking the coil.*

i.e. induced emf, $e \propto N.\frac{d\phi}{dt}$ where $N\frac{d\phi}{dt}$ is the product of number of turns and rate of change of linking flux and is termed as *rate of change of flux linkage.*

7.17. LENZ'S LAW

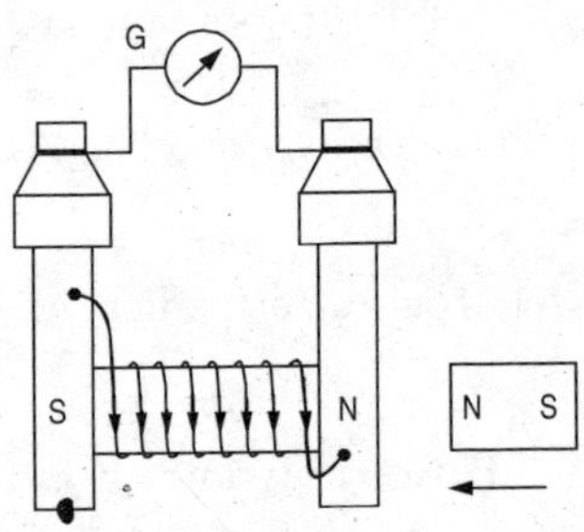

Fig. 7.29

According to Newton's third law of motion, every action causes an equal and opposite reaction. This is true not only in mechanics but also in electromagnetic induction, as discovered by Lenz. Faraday's law determines the magnitude of the induced emf but its direction is determined by Lenz's law.

According to Lenz's law, the direction of induced emf is such that the current produced by it sets up a magnetic field which tends to oppose the cause which produces it. Lenz's law is actually the application of the principle that if there is any action to change the state of the system, the system reacts in such a way as to oppose this action.

The above statement will be clarified from the following experiment.

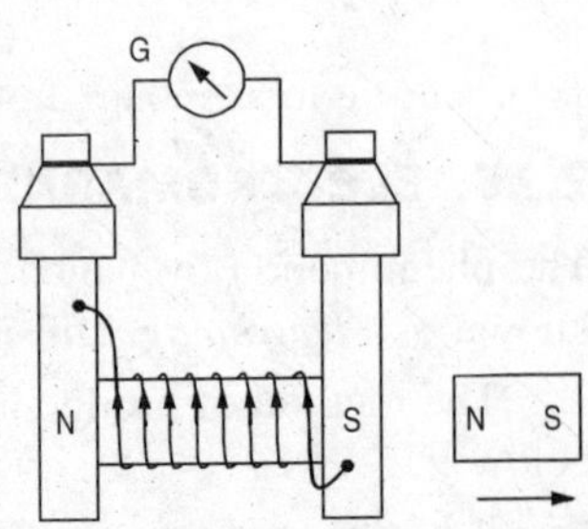

Fig. 7.30

When the bar magnet is moved towards the coil, as shown in Fig. 7.29, an emf is induced in the coil, which causes an electric current to flow through the coil in the direction shown in Fig. 7.29. Due to flow of this electric current through the coil magnetic field is produced, north pole adjacent to north pole of the bar magnet and south pole on the other end, which can be determined from helix rule explained in Art. 7.9. Now since similar poles repel each other, the movement of the bar magnet towards the coil is opposed by the field produced. Mechanical energy is expended in moving the bar magnet towards the coil against this force of repulsion and this expended energy is converted into electrical energy.

The same results will be observed when the bar magnet is moved away from the coil.

When the bar magnet is moved away from the coil, an emf is induced in the coil, current flows through it in the direction opposite to that in which it was flowing when the magnet was moved towards the coil, so magnetic field will be produced of opposite polarity *i.e.* south pole adjacent to north pole of the bar magnet. Therefore, force of attraction will be exerted between the coil and bar magnet, which will oppose the movement of the bar magnet away from the coil. Mechanical energy is expended in this case also and is converted into electrical energy.

7.18. INDUCED EMF

Let the flux linking with the coil of turns N be changed by an amount $d\phi$ in short time dt.

EMF induced, e = Rate of change of flux linkage = Number of turns × rate of change of flux = $N\frac{d\phi}{dt}$

A minus sign is required to be placed before the right hand side quantity of above expression just to indicate the phenomenon explained by Lenz's law, therefore, expression for induced emf may be written as

$$e = -N\frac{d\phi}{dt} \text{ volts} \quad ...(7.16)$$

Example 7.16. A coil has 400 turns. Find the induced voltage in it if the flux changes from 0.2 mWb to 1 mWb in 0.2 second.

Solution : Change of flux in 0.2 s, $d\phi$ = 0.001 – 0.0002 = 0.0008 Wb

Rate of change of flux, $\frac{d\phi}{dt} = \frac{0.0008}{0.2} = 0.004$ Wb/s

Induced voltage, $e = N\frac{d\phi}{dt} = 400 \times 0.004 = 1.6$ V **Ans.**

Example 7.17. A coil of resistance 100 Ω is placed in a magnetic field of 1 mWb. The coil has 100 turns and a galvanometer of 400 Ω resistance is connected in series with it. Find the average emf and the current if the coil is moved from given field to a field of 0.2 mWb in 0.1 second.

Solution : Number of turns on the coil, N = 100

Change of flux in 0.1 s, $d\phi$ = 0.001 – 0.0002 = 0.0008 Wb

Rate of change of flux, $\frac{d\phi}{dt} = \frac{0.0008}{0.1} = 0.008$ Wb/s

Induced emf, $e = N\frac{d\phi}{dt} = 100 \times 0.008 = 0.8$ V **Ans.**

Current through the coil, $I = \frac{e}{\text{Coil resistance + galvanometer resistance}}$

$$= \frac{e}{100 + 400} = \frac{0.8}{500} = 0.0016 \text{ A or } 1.6 \text{ mA} \textbf{ Ans.}$$

7.19. DYNAMICALLY INDUCED EMF

We have learnt that when the flux linking with the coil or circuit changes, an emf is induced in the coil or circuit.

EMF can be induced by changing the flux linking in two ways :

(*i*) By increasing or decreasing the magnitude of the current producing the linking flux. In this case there is no motion of the conductor or of coil relative to the field and, therefore, emf induced in this way is known as *statically induced emf.*

(*ii*) By moving a conductor in a uniform magnetic field and emf produced in this way is known as *dynamically induced emf.*

Consider a conductor of length l metres placed in a uniform magnetic field of density B Wb/m^2, as shown in Fig. 7.31 (*a*).

Let this conductor be moved with velocity v m/s in the direction of the field, as shown in Fig. 7.31 (*b*). In this case no flux is cut by the conductor, therefore, no emf is induced in it.

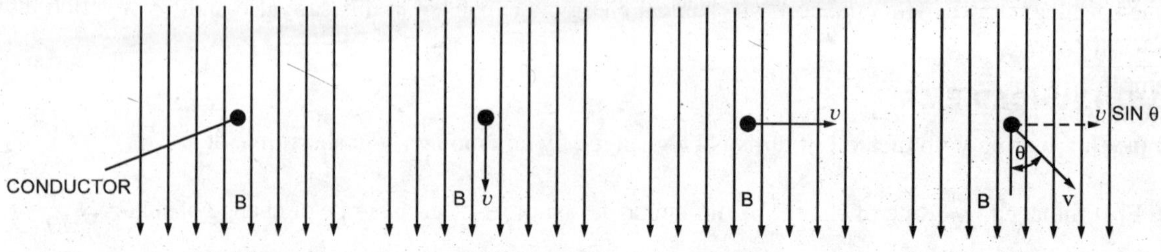

(*a*) (*b*) (*c*) (*d*)

Fig. 7.31

Now if this conductor is moved with velocity v m/s in a direction perpendicular to its own length and perpendicular to the direction of the magnetic field, as shown in Fig. 7.31 (*c*) flux is cut by the conductor, therefore, an emf is induced in the conductor.

Area swept per second by the conductor = $l \times v$ m^2/s

$$\text{Flux cut per second} = \text{Flux density} \times \text{area swept per second} = Blv$$

$$\text{Rate of change of flux, } \frac{d\phi}{dt} = \text{Flux cut per second} = Blv \text{ Wb/s}$$

$$\text{Induced emf, } e = \frac{d\phi}{dt} = Blv \text{ volts}$$

If the conductor is moved with velocity v metres per second in a direction perpendicular to its own length and at an angle θ to the direction of magnetic field, as shown in Fig. 7.31 (*d*),

The magnitude of emf induced is proportional to the component of the velocity in a direction perpendicular to the direction of the magnetic field and induced emf is given by

$$e = Blv \sin\theta \text{ volts} \qquad ...(7.17)$$

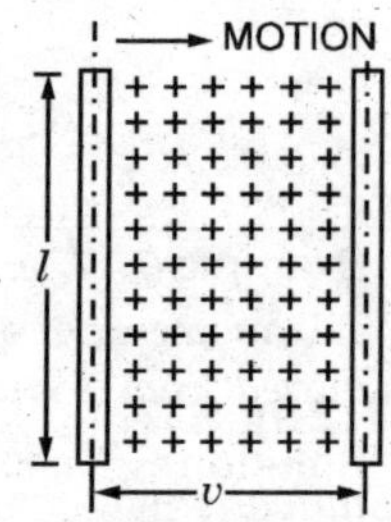

Fig. 7.32

The direction of this induced emf is given by Fleming's right hand rule.

If the thumb, forefinger and middle finger of right hand are held mutually perpendicular to each other, forefinger pointing into the direction of the field and thumb in the direction of motion of conductor then the middle finger will point in the direction of the induced emf as shown in Fig. 7.33 (*a*).

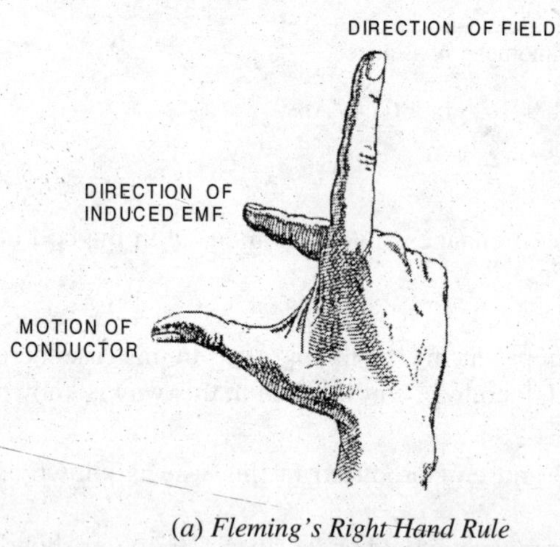

(*a*) *Fleming's Right Hand Rule*

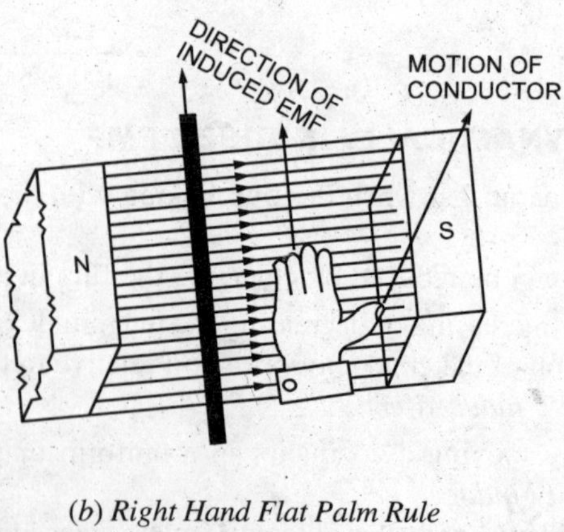

(*b*) *Right Hand Flat Palm Rule*

Fig. 7.33. ***Determination of Direction of Induced EMF***

Figure 7.33 (*b*) illustrates another way of determination of induced emf, known as *right hand flat palm* rule. This law states that if right hand is so placed in a magnetic field along the conductor that the magnetic lines of force emerging from the north pole enter the palm and the thumb points in the direction of motion of conductor, the other four fingers will give the direction of induced emf or current.

Example 7.18. Find the induced emf in a conductor of length 150 cm moving at an angle of 30° to the direction of uniform magnetic field of flux density 1.2 Wb/m² with a velocity of 60 m/s.

Solution : Flux density B $= 1.2$ T

Length of conductor, $l = 150 \text{ cm} = 1.5 \text{ m}$

Velocity of conductor, $v = 60$ m/s

Angle of movement of conductor from the direction of magnetic field,

$$\theta = 30°$$

EMF induced, $e = Blv \sin\theta = 1.2 \times 1.5 \times 60 \times \sin 30° = 108 \times 0.5 = 54$ V **Ans.**

Example 7.19. A conductor of active length 30 cm carries a current of 100 A and lies at right angles to a magnetic field of strength 0.4 Wb/m². Calculate the force in newtons exerted on it.

If the force causes the conductor to move at a velocity of 10 m/s, calculate (*i*) the emf induced in it and (*ii*) the power in watts developed by it.

Solution : Length of conductor, $l = 30 \text{ cm} = 0.3 \text{ m}$

Current flowing through conductor, I $= 100$ A

Field strength, B $= 0.4 \text{ Wb/m}^2$

Force exerted on the conductor, F $= \text{B I } l = 0.4 \times 100 \times 0.3 = 12$ N **Ans.**

Conductor velocity, $v = 10$ m/s

Induced emf, $e = \text{B } l\, v = 0.4 \times 0.3 \times 10 = 1.2$ V **Ans.**

Power developed, P $= \text{F} \times v = 12 \times 10 = 120$ watts **Ans.**

7.20. STATICALLY INDUCED EMF

Statically induced emf may be (*a*) self induced emf or (*b*) mutually induced emf.

7.20.1. Self Induced EMF

When the current flowing through the coil is changed, the flux linking with its own winding changes and due to the change in linking flux with the coil, an emf, known as *self induced emf, is induced.*

Since according to Lenz's law, an induced emf acts to oppose the change that produces it, a self induced emf is always in such a direction as to oppose the change of current in the coil or circuit in which it is induced. This property of the coil or circuit due to which it opposes any change of the current in the coil or circuit, is known as *self inductance*.

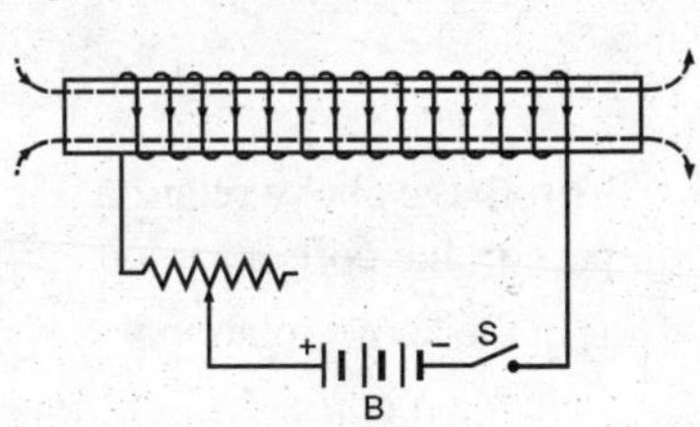

Fig. 7.34

Consider a solenoid of N turns, length l metres, area of x-section a square metres and of relative permeability μ_r. When the solenoid carries a current of i amperes, a magnetic field of flux $\dfrac{Ni}{l/\mu_0 \mu_r a}$ webers is set up around the solenoid and links with it.

If the current flowing through the solenoid is changed, the flux produced by it will change and, therefore, an emf will be induced.

$$\text{Self induced emf, } e = -N\frac{d\phi}{dt} = -N\frac{d}{dt}\left[\frac{Ni}{l/\mu_r \mu_0 a}\right] = -N\frac{N}{l/\mu_r \mu_0 a}\frac{di}{dt} = -\frac{N^2 \mu_r \mu_0 a}{l}\cdot\frac{di}{dt} \quad ...(7.18)$$

The quantity $\frac{N^2 \mu_r \mu_0 a}{l}$ is a constant for any given coil or circuit and is called *coefficient of self inductance*. It is represented by symbol L and is measured in henries.

Hence self induced emf, $e = -L\frac{di}{dt}$ where $L = \frac{N^2 \mu_r \mu_0 a}{l}$ henrys

7.20.2. Coefficient of Self Induction

The coefficient of self induction (L) can be determined from any one of the following three relations.

First Method. In case the dimensions of the solenoid are given, the coefficient of self induction may be determined from the relation

$$L = \frac{N^2 a \mu_r \mu_0}{l} \text{ henrys} \quad ...(7.19) \quad \text{[Refer to Art. 7.20.1]}$$

Second Method. In case the magnitude of induced emf in a coil for a given rate of change of current in the coil is known, self inductance of the coil may be determined from the following relation.

$$e = L\frac{di}{dt} \quad ...\text{[Refer to Art. 7.20.1]}$$

$$\text{or } L = \frac{e}{di/dt} \quad ...(7.20)$$

Third Method. In case the number of turns of the coil and flux produced per ampere of current in the coil is known, the self inductance of the coil may be determined from the following relation

$$L = \frac{N\phi}{i} \quad ...(7.21)$$

The above relation can be derived as follows:

Magnetic flux produced in a coil of N turns, length l metres, area of x-section a metres^2 and relative permeability μ_r when carrying a current of i amperes is given by

$$\phi = \frac{Ni}{l/\mu_r \mu_0 a} = \frac{N}{l/\mu_r \mu_0 a} i$$

and self inductance of the coil

$$L = \frac{N^2 a \mu_r \mu_0}{l} = \frac{N}{i} \frac{N a \mu_r \mu_0}{l} i = \frac{N\phi}{i}$$

From the above relation it is obvious that the self inductance of a coil or circuit is equal to weber-turns per ampere in the coil or circuit.

In the above relation if $N\phi$ = 1 Wb-turn and i = 1 A then L = 1 H.

Hence a coil is said to have a self inductance of one henry if a current of 1 A, when flowing through it, produces flux linkage of 1 wb-turn in it.

7.20.3. Mutually Induced EMF

Consider two coils A and B placed closed together so that the flux created by one coil completely links with the other coil. Let coil A have a battery and switch S and coil B be connected to the galvanometer G.

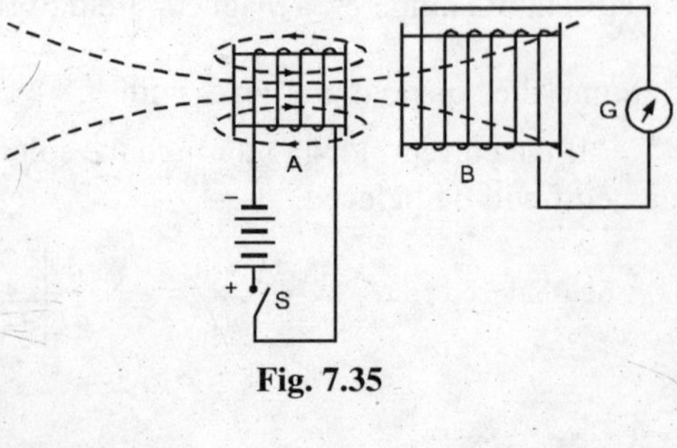

Fig. 7.35

When switch S is opened, no current flows through coil A, so no flux is created in coil A, *i.e.* no flux links with coil B, therefore, no emf is induced across coil B, the fact is indicated by galvanometer zero deflection. Now when the switch S is closed current in coil A starts rising from zero value to a finite value, the flux is produced during this period and increases with the increase in current of coil A, therefore, flux linking with the coil B increases and an emf, known as *mutually induced emf,* is produced in coil B, the fact is indicated by galvanometer deflection. As soon as the current in coil A reaches its finite value, the flux produced or flux linking with coil B becomes constant, so no emf is induced in coil B, and galvanometer pointer returns back to zero position. Now if the switch S is opened, current will start decreasing, resulting in decrease in flux linking with coil B, an emf will be again induced but in direction opposite to previous one, this fact will be shown by the galvanometer deflection in opposite direction.

Hence, whenever the current in coil A changes, the flux linking with coil B changes and an emf, known as mutually induced emf, is induced in coil B.

Consider coil A of turns N_1 wound on a core of length l metres, area of cross section a square metres and relative permeability μ_r. When the current of i_1 amperes flows through it, a flux of $\frac{N_1 i_1}{l/\mu_0 \mu_r a}$ webers is set up around the coil. Let whole of the flux produced due to flow of current in coil A be linked with the coil B having N_2 turns and placed nearby coil A.

Mutually induced emf,

$$e_m = -\text{ Rate of change of flux linkage of coil B}$$

$$= -N_2 \times \text{rate of change of flux in coil A} = -N_2 \frac{d}{dt}\left[\frac{N_1 i_1}{l/\mu_0 \mu_r a}\right] = -\frac{N_1 N_2 a \mu_0 \mu_r}{l} \cdot \frac{di_1}{dt} \quad \text{...(7.22)}$$

The quantity $\frac{N_1 N_2 a \mu_0 \mu_r}{l}$ is called the *coefficient of mutual induction* of coil B with respect to coil A. It is represented by symbol M and is measured in henrys.

Hence mutually inducd emf,

$$e_m = -M \frac{di_1}{dt} \qquad \text{where } M = \frac{N_1 N_2 a \mu_0 \mu_r}{l} \text{ henrys}$$

7.20.4. Coefficient of Mutual Induction

Mutual inductance may be defined as the ability of one coil or circuit to induce an emf in a nearby coil by induction when the current flowing in the first coil is changed. The action is also reciprocal i.e. the change in current flowing through second coil will also induce an emf in the first coil. The ability of reciprocal induction is measured in terms of the coefficient of mutual induction M.

The coefficient of mutual induction (M) can be determined from any one of the following three relations.

First Method. In case the dimensions of the coils are given, the coefficient of mutual induction may be determined from the relation

$$M = \frac{N_1 N_2 a \mu_0 \mu_r}{l} \text{ henrys} \quad \text{...(7.23) [Refer to Art. 7.20.3]}$$

Second Method. In case the magnitude of induced emf in the second coil for a given rate of change of current in the first coil is known, mutual inductance between the coil may be determined from the following relation

$$e_m = M \frac{di_1}{dt} \quad \text{[Refer to Art. 7.20.3]}$$

$$\text{or } M = \frac{e_m}{di_1/dt} \quad \text{...(7.24)}$$

Third Method. In case the number of turns of the coil and flux linking with this coil per ampere of current in another coil is known, the mutual inductance of the coil may be determined from the following relation

$$M = N_2 \frac{\phi_2}{i_1} \text{ Henry} \qquad \text{...(7.25)}$$

Example 7.20. A solenoid has 1,200 turns and carries a current of 2 A. The iron core has a length of 0.4 m and cross section of 80 cm², the relative permeability is 1,000. Calculate the self induced emf in the solenoid, if the current is switched off in 0.01 second.

Solution : Area of x-section of iron core, $a = 80 \text{ cm}^2 = 0.008 \text{ m}^2$

$$\text{Self-inductance of coil, } L = \frac{N^2 a \mu_r \mu_0}{l} = \frac{(1{,}200)^2 \times 0.008 \times 1{,}000 \times 4\pi \times 10^{-7}}{0.4} = 36.191 \text{ H}$$

$$\text{Rate of change of current, } \frac{di}{dt} = \frac{2-0}{0.01} = 200 \text{ A/s}$$

$$\text{Self induced emf, } e = L\frac{di}{dt} = 36.191 \times 200 = 7{,}238 \text{ V } \textbf{Ans.}$$

Example 7.21. A toroid has a core of square x-section 2,500 mm² in area and a mean diameter of 250 mm. The core material has a relative permeability 1,000. Calculate the number of turns wound on the core to obtain an inductance of 1 H.

Solution : Length of core, $l = \pi \times D = \pi \times 0.25 = 0.25\,\pi$ metres

Area of x-section of core, $a = 2{,}500 \text{ mm}^2 = 0.0025 \text{ m}^2$

Relative permeability of core material, $\mu_r = 1{,}000$

$\mu_0 = 4\pi \times 10^{-7}$ H/m

Self-inductance of coil, L = 1 H

$$\because L = \frac{N^2 a \mu_0 \mu_r}{l},$$

$$\text{Number of turns on the coil, } N = \sqrt{\frac{L \times l}{\mu_0 \mu_r a}} = \sqrt{\frac{1 \times 0.25\pi}{4\pi \times 10^{-7} \times 1{,}000 \times 0.0025}} = 500 \textbf{ Ans.}$$

Example 7.22. The resistance and inductance of a coil are 3 Ω and 0.1 mH respectively. What potential difference exists at the terminals of the solenoid at the instant when the current is 1 A, but increasing at the rate of 10,000 A per second ?

Solution : Resistance of the coil, R = 3 Ω

Inductance of the coil, $L = 0.1 \text{ mH} = 1 \times 10^{-4}$ H

Current flowing through the coil, I = 1.0 A

$$\text{Rate of increase of current, } \frac{di}{dt} = 10{,}000 \text{ A/s}$$

Potential difference across the solenoid,

V = Voltage drop in resistance + induced voltage in the coil

$$= iR + L\frac{di}{dt} = 1 \times 3 + 1 \times 10^{-4} \times 10{,}000 = 4 \text{ V } \textbf{Ans.}$$

Example 7.23. An iron ring of 0.15 metre diameter and 0.001 m² in cross section with a saw cut 2 mm wide is wound with 300 turns of wire. The gap flux density is 1 T. The relative permeability of the iron is 500. Determine the exciting current and inductance. Ignore leakage and fringing.

Solution : Flux density, B = 1 T

Relative permeability for iron, $\mu_r = 500$

Length of iron path $= \pi \times 0.15 - 0.002 = 0.469$ m

Total ampere-turns required, $AT = H_g l_g + H_i l_i$

$$= \frac{B}{\mu_0} l_g + \frac{B}{\mu_0 \mu_r} l_i = \frac{1}{4\pi \times 10^{-7}} \times 0.002 + \frac{1}{4\pi \times 10^{-7} \times 500} \times 0.469$$

$$= 1{,}591 + 747 = 2{,}338$$

$$\text{Exciting current, } i = \frac{AT}{N} = \frac{2{,}338}{300} = 7.76 \text{ A } \textbf{Ans.}$$

$$\text{Flux, } \phi = B \times a = 1 \times 0.001 = 0.001 \text{ Wb}$$

$$\text{Inductance, } L = \frac{N\phi}{i} = \frac{300 \times 0.001}{7.76} = 38.66 \text{ mH } \textbf{Ans.}$$

Example 7.24. A circuit has 1,000 turns enclosing a magnet circuit 20 cm^2 in cross section. With 4 A, the flux density is 1.0 Wb/m^2 and with 9 A it is 1.4 Wb/m^2. Find the mean value of the inductance between these current limits and the induced emf if the current falls from 9 A to 4 A in 0.05 second.

Solution : Flux corresponding to the current of 4 A, $\phi_1 = 1 \times 20 \times 10^{-4} = 0.002$ Wb

$$\text{Self-inductance corresponding to the current of 4 A, } L_1 = \frac{N\phi_1}{i_1} = \frac{1{,}000 \times 0.002}{4} = 0.5 \text{ H}$$

Flux corresponding to the current of 9 A, $\phi_2 = 1.4 \times 20 \times 10^{-4} = 0.0028$ Wb

$$\text{Self-inductance corresponding to the current of 9 A, } L_2 = \frac{N\phi_2}{i_2} = \frac{1{,}000 \times 0.0028}{9} = 0.31 \text{ H}$$

$$\text{Mean value of self-inductance, } L = \frac{L_1 + L_2}{2} = \frac{0.5 + 0.31}{2} = 0.405 \text{ H } \textbf{Ans.}$$

$$\text{Rate of change of current, } \frac{di}{dt} = \frac{9 - 4}{0.05} = 100 \text{ A/s}$$

$$\text{EMF induced, } e = L\frac{di}{dt} = 0.405 \times 100 = 40.5 \text{ V } \textbf{Ans.}$$

Example 7.25. A coil of 300 turns, wound on a core of non-magnetic material, has an inductance of 10 mH. Calculate (*i*) the flux produced by a current of 5 A and (*ii*) the average value of emf induced when a current of 5 A is reversed in 8 milliseconds.

Solution :

$$(i) \text{ Flux produced, } \phi = \frac{iL}{N} = \frac{5 \times 0.01}{300} = 0.1667 \text{ mWb } \textbf{Ans.} \qquad \because L = \frac{N\phi}{i}$$

$$\text{Rate of change of current, } \frac{di}{dt} = \frac{5 - (-5)}{0.008} = 1{,}250 \text{ A/s}$$

$$\text{Induced emf, } e = L\frac{di}{dt} = 0.01 \times 1{,}250 = 12.5 \text{ V } \textbf{Ans.}$$

Example 7.26. Find the inductance of a coil in which a current of 0.2 A increasing at the rate of 0.4 A per second represents a power flow of 0.4 watt.

Solution : Power flow, $p = 0.4$ watt

Current, $i = 0.2$ A

$$\text{Induced emf, } e = \frac{p}{i} = \frac{0.4}{0.2} = 2 \text{ V} \qquad \because p = ei$$

Rate of increase of current, $\frac{di}{dt} = 0.4$ A/s

Inductance of coil, $L = \frac{e}{di/dt} = \frac{2}{0.4} = 5$ H **Ans.**

Example 7.27. Two coils having 100 and 150 turns respectively are wound side by side on a closed iron circuit of section 125 cm^2 and mean length 200 cm. If the permeability of iron is 2,000, calculate (*a*) self-inductance of each coil (*b*) mutual inductance between them (*c*) the emf induced in the second coil if current in first coil changes from 0 to 5 A in 0.02 s.

Solution : Area of x-section of iron circuit, $a = 125 \text{ cm}^2 = 0.0125 \text{ m}^2$

Length of iron circuit, $l = 200 \text{ cm} = 2.0 \text{ m}$

(*a*) Self-inductance of first coil, $L_1 = \frac{N_1^2 \mu_0 \mu_r a}{l} = \frac{(100)^2 \times 4\pi \times 10^{-7} \times 2{,}000 \times 0.0125}{2.0}$

$= 157.1$ m H **Ans.**

Self-inductance of second coil, $L_2 = \frac{N_2^2 \mu_0 \mu_r a}{l} = \frac{(150)^2 \times 4\pi \times 10^{-7} \times 2{,}000 \times 0.0125}{2.0}$

$= 353.4$ m H **Ans.**

(*b*) Mutual inductance between coil, $M = \frac{N_1 N_2 \mu_0 \mu_r a}{l}$

$= \frac{100 \times 150 \times 4\pi \times 10^{-7} \times 2{,}000 \times 0.0125}{2.0} = 235.6$ m H **Ans.**

Rate of change of current in first coil, $\frac{di_1}{dt} = \frac{5-0}{0.02} = 250$ A/s

EMF induced in second coil, $e_2 = M\frac{di_1}{dt} = 235.6 \times 10^{-3} \times 250 = 58.9$ V **Ans.**

Example 7.28. Two coils A and B are wound on the same iron core. There are 500 turns on coil A and 3,000 turns on coil B. A current of 5 amperes through coil A produces a flux of 600 × 10^{-6} Wb in the core. If this current is reversed in 0.02 s, calculate the average emf induced in coils A and B.

Solution : Number of turns on coil A, $N_1 = 500$

Number of turns on coil B, $N_2 = 3{,}000$

Flux per ampere in coil A, $\frac{\phi_1}{i_1} = \frac{600 \times 10^{-6}}{5} = 120 \times 10^{-6}$ Wb/A

Self-inductance of coil A, $L_1 = N_1 \frac{\phi_1}{i_1} = 500 \times 120 \times 10^{-6} = 0.06$ H **Ans.**

Flux linking the coil B per ampere of current in coil A, $\frac{\phi_2}{i_1} = 120 \times 10^{-6}$ Wb

(Assuming flux created in coil A links the coil B completely *i.e.* $\phi_2 = \phi_1$)

Mutual inductance between the coils, $M = N_2 \frac{\phi_2}{i_1} = 3{,}000 \times 120 \times 10^{-6} = 0.36$ H

Rate of change of current in coil A, $\frac{di_1}{dt} = \frac{5-(-5)}{0.02} = 500$ A/s

Average emf induced in coil A, $e_1 = L_1 \frac{di_1}{dt} = 0.06 \times 500 = 30$ V **Ans.**

Average emf induced in coil B, $e_2 = M \frac{di_1}{dt} = 0.36 \times 500 = 180$ V **Ans.**

Example 7.29. Two coils have a mutual inductance of 0.3 H. If the current in one coil is varied from 5 A to 2 A in 0.4 s, calculate (*i*) the average emf induced in the second coil (*ii*) the change of flux linked with the second coil assuming that it is wound with 200 turns.

Solution : Mutual inductance, M = 0.3 H

Rate of change of current in one coil, $\frac{di_1}{dt} = \frac{5-2}{0.4} = 7.5$ A/s

(*i*) EMF induced in second coil , $e_2 = M\frac{di_1}{dt} = 0.3 \times 7.5 = 2.25$ V **Ans.**

(*ii*) Since mutually induced e m f, $e_2 = N_2 \frac{d\phi_2}{dt}$,

Change of flux linked with the second coil, $\frac{d\phi_2}{dt} = \frac{e_2}{N_2} = \frac{2.25}{200} = 0.01125$ Wb/s **Ans.**

Example 7.30. The self-inductance of a coil of 500 turns is 0.25 H. If 60% of the flux is linked with a second coil of 10,000 turns, calculate

(*i*) The mutual inductance of the two coils.

(*ii*) EMF induced in the second coil when current in the first coil changes at the rate of 100 A/s.

Solution : Self-inductance of the first coil, $L_1 = 0.25$ H

Number of turns on first coil, $N_1 = 500$

Flux per ampere of current in first coil, $\frac{\phi_1}{i_1} = \frac{L_1}{N_1} = \frac{0.25}{500} = 0.5$ mWb $\quad \because L_1 = N_1\frac{\phi_1}{i_1}$

Flux linked with second coil per ampere of current in first coil

$$\frac{\phi_2}{i_1} = \frac{0.6\,\phi_1}{i_1} = 0.6 \times 0.5 = 0.3 \text{ mWb}$$

(*i*) The mutual inductance of the two coils, $M = \frac{N_2\phi_2}{i_1} = 10{,}000 \times 0.3 \times 10^{-3} = 3$ H **Ans.**

(*ii*) EMF induced in the second coil, $e_2 = M\frac{di_1}{dt} = 3 \times 100 = 300$ V **Ans.**

Example 7.31. Two identical 1,000 turn coils X and Y lie in parallel planes such that 60% of the flux produced by one coil links the other. A current of 5 A in X produces in it a flux of 0.05 m Wb. If the current in X changes from 6 A to – 6 A in 0.01 s ; what is the emf induced in Y ? Calculate the self-inductance of each coil and their mutual inductance. [*Pb. Univ. Elements of Electrical Engineering April 1975*]

Solution : Flux produced per ampere of current in coil X,

$$\frac{\phi_1}{i_1} = \frac{0.05 \times 10^{-3}}{5} = 1 \times 10^{-5} \text{ wb}$$

Number of turns on coil X, $N_1 = 1{,}000$

Number of turns on coil Y, $N_2 = 1{,}000$

Self-inductance of coil X (or Y), $L = N_1\frac{\phi_1}{i_1} = 1{,}000 \times 1 \times 10^{-5} = 0.01$ H **Ans.**

Flux linking in coil Y per ampere of current in coil X,

$$\frac{\phi_2}{i_1} = \frac{0.6 \times 0.05 \times 10^{-3}}{5} = 0.6 \times 10^{-5} \text{ Wb}$$

Mutual inductance between the coils, $M = N_2\frac{\phi_2}{i_1} = 1{,}000 \times 0.6 \times 10^{-5} = 0.006$ H **Ans.**

Rate of change of current in coil X, $\frac{di_1}{dt} = \frac{6-(-6)}{0.01} = 1{,}200$ A/s

EMF induced in coil Y, $e_2 = M\frac{di_1}{dt} = 0.006 \times 1{,}200 = 7.2$ V **Ans.**

7.21. MUTUAL COUPLING IN A SIMPLE MAGNETIC CIRCUIT

The self inductance of a circuit is intimately associated with the magnetic field linking the circuit. The self inductance emf may be thought of as the emf induced in the circuit by a magnetic field produced by the circuit current.

Since a magnetic field exists in the region around the current that develops it, there is also a possibility that an emf be induced in the other circuits linked by the field. Two circuits linked by the same magnetic field are said to be *coupled* to each other. The circuit element used to represent *magnetic coupling* is shown in Fig. 7.36 (*a*) and is called *mutual inductance.* It is represented by symbol M and is measured in henrys, already stated in Art. 7.20.3. The volt-ampere relationship is one which gives the induced emf in one circuit by a current in another and is given as

$$e_2 = M\frac{di_1}{dt}$$

A similar equation can, of course, be written giving an emf e_1 induced by a current i_2. The two dots, called *polarity markings,* in Fig. 7.36 (*a*) are used for indicating the direction of the magnetic coupling between the two coils : note from the figure and from equation given above that by matching the dots, the directions of currents and voltage drops are made to correspond with those of Fig. 7.36 (*b*) and Eq. (7.23) for self inductance.

If currents are present in both coupled circuits, emfs of self inductance and mutual inductance are induced in each circuit. The self induced emf, have the directions shown in Fig. 7.36 (*c*) the mutually induced emfs, follow the pattern of Fig. 7.36 (*a*). Figure 7.36 (*b*) shows two such elements. Each coil is characterized by its own self inductance; the combination has a mutual inductance indicating the coupling between the coils. The volt-ampere relationships for this case are given below :

$$e_1 = L_1\frac{di_1}{dt} + M\frac{di_2}{dt} \quad \text{...(7.26)}$$

$$\text{and } e_2 = L_2\frac{di_2}{dt} + M\frac{di_1}{dt} \quad \text{...(7.27)}$$

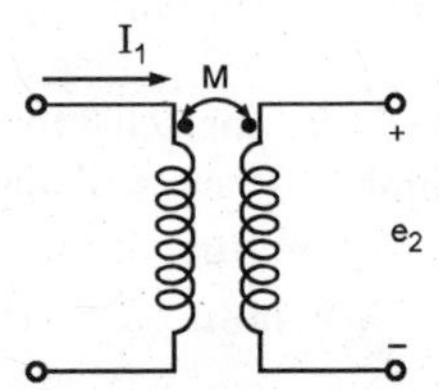

(a) Schematic Representation of Mutual Inductance

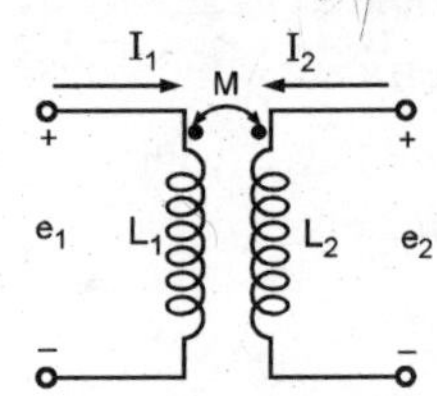

(b) Mutual Inductance With Current in Both Coils

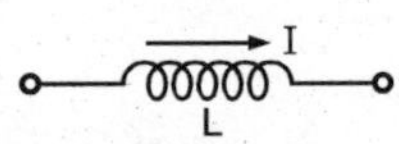

(c) Schematic Representation of Self Inductance

Fig. 7.36

Coupling between two closed circuits permits the transfer of energy between the circuits through the medium of the mutual magnetic field. This phenomenon is the basis on which transformers operate.

7.22. COEFFICIENT OF COUPLING BETWEEN TWO MAGNETICALLY COUPLED CIRCUITS

When two coils are placed near each other, all the flux produced by one coil does not link with the other coil, as assumed in Art. 7.20.3, only a certain portion (say, K) of flux produced by one coil links with other coil, K being less than unity. K is called the *coefficient of coupling.*

Flux created in coil A due to current of i_1 amperes,

$$\phi_1 = \frac{N_1 i_1}{l/\mu_0 \mu_r a} \text{ Wb}$$

$$\text{Flux linking with coil B, } \phi_2 = K\,\phi_1 = \frac{K\,N_1\,i_1}{l/\mu_0\,\mu_r\,a}\ \text{Wb}$$

$$\text{Coefficient of mutual inductance, M} = \frac{N_2\,\phi_2}{i_1} = \frac{K\,N_1\,N_2}{l/\mu_0\,\mu_r\,a} \quad ...(i)$$

$$\text{Coefficient of self inductance of coil A, } L_1 = \frac{N_1\,\phi_1}{i_1} = \frac{N_1^2}{l/\mu_0\,\mu_r\,a} \quad ...(ii)$$

$$\text{Coefficient of self inductance of coil B, } L_2 = \frac{N_2\,\phi_2}{i_2} = \frac{N_2^2}{l/\mu_0\,\mu_r\,a} \quad ...(iii)$$

Multiplying Eqs. (*ii*) and (*iii*) and taking square root of both sides we get

$$\sqrt{L_1\,L_2} = \frac{N_1\,N_2}{l/\mu_0\,\mu_r\,a} \quad ...(iv)$$

Comparing Eqs. (*i*) and (*iv*) we get

$$M = K\sqrt{L_1\,L_2}$$

$$\text{or } K = \frac{M}{\sqrt{L_1\,L_2}} \quad ...(7.28)$$

When the coils are tightly coupled *i.e.* when the flux due to one coil links with the other coil completely, the coefficient of coupling, K is unity and the coefficient of mutual inductance M is given as

$$M = \sqrt{L_1\,L_2} \quad ...(7.29)$$

When the flux due to one coil does not link with the other coil at all, the value of coefficient of coupling, K is zero.

Example 7.32. Two coils A of 1,200 turns and B of 800 turns lie near each other so that 60 percent of the flux produced in one links the other. It is found that a current of 5 A in A produces a flux of 0.25 mWb while the same current in B produces a flux of 0.15 mWb. Determine the mutual inductance and coefficient of coupling between the coils.

Solution : Number of turns on coil A, $N_1 = 1{,}200$

Number of turns on coil B, $N_2 = 800$

$$\text{Flux per ampere of current in coil A, } \frac{\phi_1}{i_1} = \frac{0.25\times10^{-3}}{5} = 0.05\times10^{-3}\ \text{Wb}$$

$$\text{Self inductance of coil A, } L_1 = N_1\frac{\phi_1}{i_1} = 1{,}200\times0.05\times10^{-3} = 0.06\ \text{H}$$

$$\text{Flux per ampere of current in coil B, } \frac{\phi_2}{i_2} = \frac{0.15\times10^{-3}}{5} = 0.03\times10^{-3}\ \text{Wb}$$

$$\text{Self inductance of coil B, } L_2 = N_2\frac{\phi_2}{i_2} = 800\times0.03\times10^{-3} = 0.024\ \text{H}$$

$$\text{Flux linking coil B per ampere of current in coil A, } \frac{\phi_2}{i_1} = \frac{0.25\times10^{-3}}{5}\times0.6 = 0.03\times10^{-3}\ \text{Wb}$$

$$\text{Mutual inductance between the coils, } M = N_2\frac{\phi_2}{i_1} = 800\times0.03\times10^{-3} = 0.024\ \text{H}\ \textbf{Ans.}$$

$$\text{Coefficient of coupling, } K = \frac{M}{\sqrt{L_1\,L_2}} = \frac{0.024}{\sqrt{0.06\times0.024}} = 0.6325\ \textbf{Ans.}$$

Example 7.33. Define coefficient of coupling and calculate it for two coils having self inductances of 60 mH and 80 mH. The mutual inductance between them is 40 mH.

Solution :

$$\text{Self inductance, } L_1 = 60 \text{ mH} = 0.06 \text{ H}$$

$$\text{Self inductance, } L_2 = 80 \text{ mH} = 0.08 \text{ H}$$

$$\text{Mutual inductance, } M = 40 \text{ mH} = 0.04 \text{ H}$$

$$\text{Coefficient of coupling, } K = \frac{M}{\sqrt{L_1 L_2}} = \frac{0.04}{\sqrt{0.06 \times 0.08}} = 0.577 \textbf{ Ans.}$$

HIGHLIGHTS

1. The space around the poles of a magnet is called the magnetic field, and is represented by magnetic lines of force.
2. Magnetic flux density is given by

$$B = \mu_0 \mu_r H \text{ Wb/m}^2$$

3. Ampere's circuital law or Ampere's work law is given by

$$\oint H\, dl = \Sigma I$$

4. According to Biot-Savart law

$$dB = \frac{\mu_0 \mu_r}{4\pi} \frac{I\, dl \sin\theta}{r^2}$$

5. Force on a current carrying conductor lying in the magnetic field is given by

$$F = BIl \sin\theta \text{ newtons}$$

6. Force between two long current carrying conductors is given by

$$F = \frac{2I_a I_b}{d} \times l \times 10^{-7} \text{ Newtons}$$

7. The closed path followed by magnetic flux is called a *magnetic circuit*. There is a lot of similarity between the magnetic and electric circuits, as mentioned in Art. 7.14.4.
8. **MMF** of the magnetic circuit is a magnetic potential that tends to force flux around the magnetic circuit and is analogous to the emf in an electric circuit.

$$\text{MMF} = \text{NI ampere-turns}$$

9. **Reluctance** is the name given to that property of material which opposes the creation of magnetic flux in it. It is measured in ampere-turns/Wb or amperes/Wb. It may be given as

$$S = \frac{l}{\mu_o \mu_r a} \text{ AT/Wb}$$

where l is the length of magnetic path, a is x-sectional area, μ_o is the permeability of the evacuated space or air and is equal to $4\pi \times 10^{-7}$ H/m, and μ_r is the relative permeability of the medium with respect to evacuated space or air.

10. Flux produced ϕ, mmf and reluctance are related as follows

$$\phi = \frac{\text{MMF}}{\text{Reluctance}} = \frac{\text{NI}}{l} \times \mu_o \mu_r a$$

11. Permeance of a magnetic circuit is the reciprocal of reluctance and is given as

$$\text{Permeance} = \frac{1}{\text{Reluctance}} = \frac{\mu_o \mu_r a}{l}$$

12. For a series magnetic circuit

Total ampere-turns required = $H_1 l_1 + H_2 l_2 + H_3 l_3 + \ldots$

where H_1 is the magnetizing force required to create flux through part l_1 of the magnetic circuit and so on.

13. For parallel magnetic circuit,

AT required = AT required for either parallel magnetic path.

14. Leakage factor is the ratio of total flux (flux in the iron path) and useful flux (flux in the air gap).
15. The curve drawn giving relationship between induction density B and magnetizing force H is known as B-H curve or magnetizing curve.
16. When the flux linking the circuit changes, an emf is induced, and its magnitude is given as

$$e = N\frac{d\phi}{dt} = \frac{N(\phi_2 - \phi_1)}{dt}$$

17. The magnitude of dynamically induced emf in a conductor moving at right angles to a uniform magnetic field is given as $e = Blv$ volts

If the conductor moves at an angle θ to the magnetic field, then

$$e = Blv \sin\theta \text{ volts}$$

The direction of dynamically induced emf is given by Fleming's right hand rule. According to Fleming's right hand rule:

If the thumb, forefinger and middle finger of right hand are held mutually perpendicular to each other, forefinger pointing into the direction of the field and thumb in the direction of motion of conductor then the middle finger will point in the direction of the induced emf as shown in Fig 1.

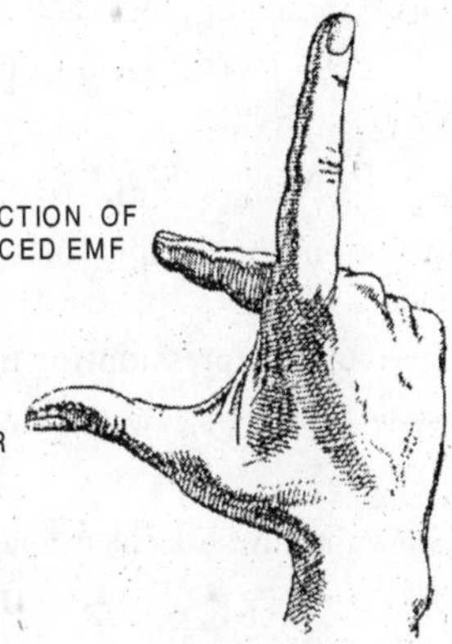

Fig. 1

18. When the current flowing through the coil is changed, the flux linking with its own winding changes and due to the change in linking flux with the coil, an emf, known as *self induced emf, is induced.* The magnitude of self induced emf, *e* in a coil of N turns is given by

$$e = N\frac{d\phi}{dt} = L\frac{di}{dt}$$

where L is known as *self inductance* of the coil and is given as

$$L = \frac{e}{di/dt} = \frac{N\phi}{i} = \frac{N^2\mu_o\mu_r a}{l} = \frac{N^2}{\text{Reluctance, S}}$$

19. Whenever the current in coil A (placed near another coil B) changes, the flux linking with coil B changes and an emf, known as mutually induced emf, is induced in coil B.

The magnitude of mutually induced emf is given as

$$e_m = M\frac{di_1}{dt} \text{ volts}$$

where $\frac{di_1}{dt}$ is the rate of change of current in coil A and M is mutual inductance of the two coils in henry and is given as

$$M = \frac{e_m}{di_1/dt} = \frac{N_2\phi_2}{i_1} = \frac{N_1N_2 a\mu_o\mu_r}{l} = \frac{N_1N_2}{\text{Reluctance, S}}$$

20. The coefficient of coupling between two coils is given as

$$K = \frac{M}{\sqrt{L_1L_2}}$$

where L_1 and L_2 are the individual inductances of the coils and M is the mutual inductance between the coils.

EXERCISES

1. Derive an expression for the electrodynamic force on a conductor carrying current in a magnetic filed.
2. State Biot-Savart law. Apply this law to derive magnetic field due to an infinitely long straight current carrying thin conductor.

3. Obtain an expression for the force between two parallel conductors carrying currents.

[Pb. Technical Univ. Basic Electrical and Electronics Enginering, May 2009]

4. Compare electric and magnetic circuits. Establish the relationship between magnetomotive force, magnetic flux and magnetic reluctance.

[Pb. Technical Univ. Basic Electrical and Electronics Engineering, December 2008]

5. Give the magnetic circuit concepts. On what factors does the magnetic reluctance depend ? Is it constant for all flux densities ?

   Give relation between mmf, reluctance and flux in a magnetic circuit.

6. Give an analogy between electric circuit and magnetic circuit.

[Pb. Technical Univ. Basic Electrical and Electronics Engineering Dec. 2004; May 2010]

7. State similarities of electric and magnetic circuits.

[Pb. Technical Univ. Basic Electrical and Electronics Engineering Dec. 2006]

8. Bring out the analogy between electric and magnetic circuits. Point out major difference if any between them. **[Pb. Technical Univ. Basic Electrical and Electronics Engineering May 2005]**

9. Derive an expression for the electrodynamic force on a conductor carrying current in a magnetic field.

10. State Faraday's laws of electromagnetic induction.

[Pb. Technical Univ. Basic Electrical and Electronics Engineering Dec. 2006]

11. State the laws of electromagnetic induction.

[Pb. Technical Univ. Basic Electrical and Electronics Engineering May 2007]

12. Derive the expression for the dynamically induced emf. On what factor does this emf depend ? Explain.

13. What do you understand by statically and dynamically induced emfs ? Write down difference between them with example.

14. Define the coefficient of inductance.

[Pb. Technical Univ. Basic Electrical and Electronics Engineering, May 2009]

15. Explain what is meant by self inductance and mutual inductance. Define the units in which each is measured.

16. For a magnetically coupled coils with self inductances L_1 and L_2 respectively and mutual inductance M, show that coefficient of coupling is given by $K = \frac{M}{\sqrt{L_1 L_2}}, K < 1$.

[Pb. Technical Univ. Basic Electrical and Electronics Engineering Dec. 2004]

17. Define mutual inductance (M) and show that

$$K = \frac{M}{\sqrt{L_1 L_2}}, K < 1$$

where L_1 and L_2 are inductances of coil 1 and coil 2 respectively and K is coefficient of coupling.

[Pb. Technical Univ. Basic Electrical and Electronics Engineering Dec. 2005]

SHORT ANSWER TYPE QUESTIONS WITH ANSWERS

Q. 1. What is the value of permeability of free space ?

Ans. The permeability for free space is $4\pi \times 10^{-7}$ H/m.

Q. 2. Give the relationship between magnetic field intensity H and magnetic flux density.

Ans. Field intensity H is the cause and flux density B is the effect. Flux density will be μ times the field intensity where μ is the permeability of the magnetic material.

Q. 3. How will you find the direction of magnetic field around a current carrying conductor ?

Ans. By Cork Screw rule or right hand rule.

Q. 4. How will you determine the polarity of a solenoid ?

Ans. By using compass needle or by using helix rule.

Q. 5. On what factors does the force experienced by a current carrying conductor placed in a magnetic field depend ?

Ans. The force experienced on a current carrying conductor placed in a magnetic field is directly proportional to (*i*) flux density (field strength), B (*ii*) the magnitude of current flowing through the conductor (*iii*) the length of the conductor and (*iv*) sine of the angle at which the conductor lies with the magnetic field *i.e.* $F = BIl \sin \theta$ newtons.

Q. 6. What is Fleming's left hand rule ?

Ans. Fleming's left hand rule is used to determine the direction of mechanical force experienced by a current carrying conductor placed in a magnetic field.

Fleming's left hand rule states that if the thumb, forefinger and middle finger of the left hand are stretched in such a way that they are at right angles to each other mutually and forefinger points towards the direction of the magnetic field, middle finger towards the direction of flow of current then thumb will point in the direction of force acting on the conductor [Fig. 7.15(*a*)].

Q. 7. On what factors does the magnetic reluctance depend ?

Ans. The reluctance of the magnetic circuit is directly proportional to length l, inversely proportional to cross-sectional area and dependent on the nature of material of the magnetic circuit.

Q. 8. Is magnetic reluctance constant for all flux densities ?

Ans. No.

Q. 9. What is meant by leakage and fringing ?

Ans. Leakage flux is the flux, which follows a leaking path, as shown in Fig. 7.22.

Useful flux passing across the gap tends to bulge outwards, as shown in Fig. 7.22, thereby increasing the effective area of gap and reducing the flux density in the gap. This effect is referred to as fringing; and the longer the air gap, the greater is the fringing.

Q. 10. What is leakage coefficient ? Give its effects in magnetic circuit. **[P.T.U. B.E.E.E. May 2009]**

Ans. Leakage flux is the flux, which follows a leaking path, as shown in Fig. 7.22 (page 194). Flux in the air gap is known as *useful flux* which is utilized for various useful purposes. For the purpose of calculations the iron is supposed to carry whole of the flux throughout its entire length.

The ratio of total flux (flux in the iron path) to the useful flux (flux in the air gap) is known as *leakage factor*.

$$i.e.\ \text{Leakage factor} = \frac{\text{Total flux}}{\text{Useful flux}}$$

$$\text{or Leakage factor} = \frac{\text{Flux in the iron path}}{\text{Flux in the air gap}}$$

In electrical machines (such as in generators and motors), magnetic leakage is undesirable as it causes increase in weight (not decrease in their power efficiency) and cost of manufacture. Though magnetic leakage cannot be avoided completely but can be reduced to the minimum by placing the magnetizing or exciting coils as close as possible to the air gap or to the points in the magnetic circuit where the flux is to be utilized for useful purposes.

The value of leakage factor for modern electric machinery is about 1.2.

Q. 11. On what factors does the force acting between two parallel current carrying conductors depend ?

Ans. Force acting between two parallel current carrying conductors depend on the magnitude of currents carried by them and spacing between the conductors.

The force experienced between two parallel conductors spaced d metres apart in air and carrying currents of I_1 and I_2 amperes respectively is given by the expression

$$F = \frac{2I_1I_2}{d} \times 10^{-7} \text{ Newtons per metre}$$

Q. 12. State Faraday's laws of electromagnetic induction. **[P.T.U. B.E.E.E. December 2008; December 2009]**

Ans. A very important effect of a magnetic field on an electric circuit is that when the flux linking the circuit changes, an emf is induced. Electromagnetic induction of emf (or voltage) is basic to the operation of transformers, generators (ac or dc) and motors (ac or dc). The effect is described by Faraday's law, which states that *the magnitude of emf (or voltage) induced is directly proportional to the rate of change of flux linkage or to the product of number of turns and rate of change of flux linking the coil.*

$$\textit{i.e.} \text{ induced emf, } e = N\frac{d\phi}{dt}$$

where $N\frac{d\phi}{dt}$ is the product of number of turns and rate of change of linking flux and is termed as rate of change of flux linkage.

Q. 13. What is indicated by the negative sign before $N\frac{d\phi}{dt}$ in expression $e = -N\frac{d\phi}{dt}$?

Ans. The negative sign before $N\frac{d\phi}{dt}$ in expression $e = -N\frac{d\phi}{dt}$ indicates the phenomenon explained by Lenz's law.

Q. 14. What is Lenz's law ?

Ans. According to Lenz's law, the direction of induced emf is such that the current produced by it sets up a magnetic field which tends to oppose the cause which produces it. Lenz's law is actually the application of the principle that if there is any action to change the state of the system, the system reacts in such a way as to oppose this action.

Q. 15. What is the difference between a statically induced emf and a dynamically induced emf ?

[P.T.U. January 2000]

Ans. When flux linking with any coil or circuit changes, an emf is induced in the coil or circuit, called the *electromagnetically induced emf.*

EMF induced in a coil or circuit by increasing or decreasing magnitude of the linking flux (no relative motion between coil or circuit and magnetic field) is known as statically *induced emf.*

The emf induced in a conductor or coil due to relative motion between it and the magnetic field, that produces flux, is known as dynamically induced emf.

Q. 16. What is the application of Fleming's right hand rule ?

Ans. Fleming's right hand rule is used in determination of direction of emf induced in a conductor moving in a uniform magnetic field.

Q. 17. What is Fleming's right hand rule ?

Ans. According to Fleming's right hand rule if the thumb, forefinger and middle finger of the right hand are held mutually perpendicular to each other, forefinger pointing into the direction of the field and thumb in the direction of motion of conductor then middle finger will point in the direction of the induced emf, as shown in Fig. 7.33 (*a*).

Q. 18. What is mutual inductance ? Give its unit. **[P.T.U. B.E.E.E. Dec. 2007]**

Ans. Mutual inductance may be defined as the ability of one coil or circuit to induce an emf in a nearby coil by induction when the current flowing in the first coil is changed. The action is also reciprocal *i.e.* the change in current flowing through second coil will also induce an emf in the first coil. The ability of reciprocal induction is measured in terms of the coefficient of mutual induction M. Its unit is Henry.

Q. 19. What is meant by coefficient of coupling between the coils ?

Ans. When the two coils are placed near each other, all the flux produced by one coil does not link with the other coil, only a certain portion (say, K) of flux produced by one coil link with the other coil, K being less than unity. K is called the coefficient of coupling.

Q. 20. Give the value of coefficient of coupling between two tightly coupled inductive coils.

Ans. Unity.

PROBLEMS

1. An electromagnet has an air gap of 5 mm and the flux density in the air gap is 1.2 Wb/m^2. Determine the ampere-turns for the gap. [**Ans.** 4,775]

2. A flux density of 1.2 Wb/m^2 is required in 2 mm air gap of an electromagnet having an iron path of one metre long. Calculate mmf required assuming relative permeability of iron 1,500. [**Ans.** 2,546 AT]

3. A magnetic circuit consists of an iron ring of mean circumference 80 cm with cross-sectional area 12 cm^2 throughout. A current of 2 A in the magnetising coil of 200 turns produces a total flux of 1.2 m Wb in the iron. Calculate (*i*) the flux density in the iron (*ii*) the absolute and relative permeability of iron (*iii*) the reluctance of the circuit. [**Ans.** (*i*) 1.0 T (*ii*) 0.002; 1,592 (*iii*) 3.334×10^6 AT/Wb]

4. A cast steel electromagnet has an air gap of length 2 mm and an iron path of length 30 cm. Find the number of ampere-turns necessary to produce a flux density of 0.8 Wb/m^2 in the air gap. Take $\mu_r = 700$. Neglect leakage and fringing. [**Ans.** 1,546]

5. Calculate the relative permeability of an iron when the exciting current taken by the 600 turns coil is 1.2 A, and the total flux produced is 1 mWb. The mean circumference of the ring is 0.5 m and the area of cross-section is 10 cm^2. [**Ans.** 552]

6. A steel ring of circular cross-sectional area of 2 cm^2 has mean diameter of 30 cm. Assuming the relative permeability of iron to be constant at 700, determine the reluctance of the ring. If a saw cut of 1 mm width is cut in the ring, what will be the total reluctance of the whole magnetic circuit ?
[**Ans.** 5.36×10^6 AT/Wb; 9.33×10^6 AT/Wb]

7. A steel ring of 25 cm mean diameter and of circular section 3 cm in diameter has an air gap of 1.5 mm length. It is wound uniformly with 700 turns of wire carrying a current of 2 A. Calculate (*i*) magnetomotive force. (*ii*) flux density (*iii*) magnetic flux (*iv*) reluctance and (*v*) relative permeability of steel ring.
Neglect magnetic leakage and assume that iron path takes about 35 percent of the total magneto-motive force.
[**Ans.** (*i*) 1,400 AT (*ii*) 0.762 T (*iii*) 0.539 mWb (*iv*) 2.6×10^6 AT/Wb (*v*) 968]

8. An iron ring of mean diameter 22 cm and cross section 10 cm^2 has an air gap 1 mm wide. The ring is wound uniformly with 200 turns of wire. The permeability of ring material is 1,000. A flux of 0.16 mWb is required in the gap. What current should be passed through the wire ?
[Pb. Univ. Elec. Engineering-1, April 1982] [**Ans.** 1.075 A]

9. An iron ring has a x-section of 4 cm^2. An air gap of 0.5 mm has been made by a cut across its section. The effective iron path is 75 cm. The ring is wound with a coil of 500 turns through which a current of 1.5 A is passed. If the total flux in the air gap is 0.3 mWb, find the relative permeability of iron.
[**Ans.** 991.2]

10. An iron ring has a mean circumferential length of 60 cm with an air gap of 1 mm and a uniform winding of 300 turns. When a current of 1 A flows through the coil, find the flux density. The relative permeability of iron is 300. Assume $\mu = 4\pi \times 10^{-7}$ H/m. [**Ans.** 0.1256 T]

11. A coil of 200 turns is linked by a flux of 8 mWb. Determine the emf induced in the coil if the flux is reversed in 0.02 second. [**Ans.** 160 volts]

12. A 0.5 m long wire moves at right angles to its length at 40 m/s in uniform magnetic field of 1 Wb/m^2. Calculate the emf induced in the conductor when the direction of motion is (*i*) perpendicular to field, (*ii*) inclined at 30° to the direction of field. [**Ans.** (*i*) 20 V (*ii*) 10 V]

13. A conductor of active length 30 cm and carrying a current of 100 A moves at an angle 30° with the direction of a uniform magnetic field of density of 0.4 Wb/m^2. Find the emf induced in it if the velocity of the conductor is 0.5 m/s. **[Ans.** 30 mV]

14. If a current through a coil of inductance 1 H is reduced from 10 A to –10 A in 0.05 s, calculate the mean value of emf induced in the coil. **[Ans.** 400 V]

15. The resistance and inductance of a coil are 3 Ω and 0.1 mH respectively. What potential difference exists at the terminals of the solenoid at the instant when the current is 1 A but is increasing at the rate of 10,000 A per second ? **[Ans.** 4 V]

16. Calculate the inductance of a toroid 25 cm mean diameter and 6.25 cm^2 circular x-section wound uniformly with 1,000 turns of wire. Hence calculate the emf induced when the current in it increases at the 100 A/s. **[Ans.** 1 mH, 0.1 V]

17. A circuit has 1,000 turns enclosing a magnetic circuit 20 cm^2 in section. With 4 A, the flux density is 1.0 Wb/m^2 and with 9 A it is 1.4 Wb/m^2. Find the mean value of the inductance between these current limits and the induced emf if the current falls from 9 A to 4 A in 0.05 s. **[Ans.** 0.16 H H,16 V]

18. Two identical coils A and B each having 1,000 turns lie in parallel planes such that 60% of the lines of force produced by one coil link the other. A current of 5 amperes in coil A produces in it a flux of 0.05 mWb. If the current in A changes from + 6 A to – 6 A in 0.01 second, what will be the magnitude of the emf induced in B ? Calculate the self and mutual inductances of each coil.

[Ans. 0.01 H; 0.006 H: 7.2 V]

19. Two coils A and B of 600 and 100 turns respectively are wound uniformly around a wooden ring of mean circumference 80 cm. The x-sectional area of the ring is 4 cm^2. Calculate (*a*) self-inductance of each coil (*b*) mutual inductance between the coils (*c*) emf induced in the coil B when a current of 2 A in coil A is reversed in 0.01 second. **[Ans.** (*a*) 226.2 μH; 6.283 μH; (*b*) 37.7 μ H; (*c*) 15.08 mV]

20. Two coils having 50 and 500 turns respectively are wound side by side on a closed iron circuit of section 50 cm^2 and mean length 120 cm. Estimate the mutual inductance between the coils if the permeability of iron is 1,000. Also find the self-inductance of each coil and the coefficient of coupling.

[Pb. Univ. Elec. Engg-1 April 1982] **[Ans.** 0.13 H; 0.013 H; 1.3 H; 1.0]

❑❑❑

8
CHAPTER

Single Phase Transformer

8.1. INTRODUCTION

INSIDE THIS CHAPTER

Electrical energy is generated at places where it is easier to get water head, oil or coal for hydroelectric, diesel or thermal power stations respectively. Then energy is to be transmitted at considerable distances for use in villages, towns and cities located at distant places. Since transmission of electrical energy at high voltages is economical, some means are required for stepping up the voltage at generating stations and stepping down the same at the places where it is to be used. Electric machine used for this purpose is "*transformer*". In our country the electrical energy is usually generated at 6.6, 11 or 33 kV, stepped up to 132, 220, 400, or 765 kV with the help of step-up transformers for transmission and then stepped down to 66 kV or 33 kV at grid substations for feeding various substations, which further step down the voltage to 11 kV for feeding distributing transformers stepping down the voltage further to 440/250 volts for the consumer uses.

Transformer is an ac machine that (*i*) transfers electrical energy from one electric circuit to another (*ii*) does so without a change of frequency (*iii*) does so by the principle of electromagnetic induction and (*iv*) has electric circuits that are linked by a common magnetic circuit. The energy transfer usually takes place with a change of voltage, although this is not always necessary. When the transformer raises the voltage *i.e.* when the output voltage of a transformer is higher than its input voltage, it is called the *step-up transformer* and when it lowers the voltage it is called the *step-down transformer*.

Since its basic construction requires no moving parts, it is often called the '*static transformer*' and it is very rugged machine requiring the minimum amount of repair and maintenance. Owing to the lack of rotating parts there are no friction or windage losses. Further, the other losses are relatively low, so that the efficiency of a transformer is high. Typical transformer efficiencies at full load lie between 96% and 97% and with extremely large capacity transformers the efficiencies are as high as 99%.

8.2. BASIC CONSTRUCTION AND WORKING PRINCIPLE OF TRANSFORMER

An elementary transformer consists of a soft iron or silicon steel core and two windings placed on it. The windings are insulated from both the core and each other. The core is built up of thin soft iron or silicon steel laminations to provide a path of low reluctance to the magnetic flux. The winding connected to the supply main is called the *primary* and the winding connected to the load circuit is called the *secondary*. The winding connected to higher voltage circuit is called the *high-voltage (hv) winding* while that connected to the lower voltage circuit is called the *low-voltage (lv) winding*. In case of a step-up transformer, low-voltage winding is the primary and high-voltage winding is the secondary while in case of a step-down transformer the high-voltage winding is the primary and low-voltage winding is the secondary.

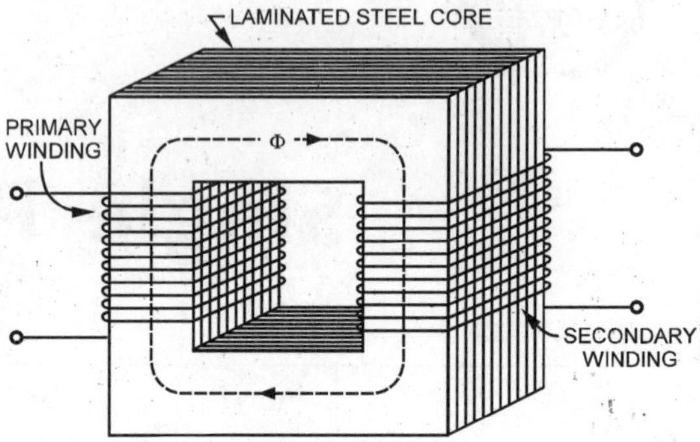

Fig. 8.1. *Simple Tarnsformer*

The action of a transformer is based on the principle that energy may be efficiently transferred by induction from one set of coils to another by means of a varying magnetic flux, provided that both the sets of coils are on a common magnetic circuit. In a transformer, the coils and magnetic circuit are all stationary with respect to one another. The emfs are induced by the variation in the magnitude of flux with time, as illustrated in Fig. 8.2.

Although in the actual construction the two windings are usually wound one over the other, for the sake of simplicity, the figures for analysing transformer theory show the windings on opposite sides of the core as in Fig. 8.1.

When the primary winding is connected to an ac supply mains, a current flows through it. Since this winding links with an iron core, current flowing through this winding produces an alternating flux ϕ in the core. Since this flux is alternating and links with the secondary winding also, so induces an emf in the secondary winding. The frequency of induced emf in secondary winding is the same as that of the flux or that of the supply voltage. The induced emf in the secondary winding enables it to deliver current to an external load connected across it. Thus the energy is transformed from primary winding to the secondary winding by means of electromagnetic induction without any change in frequency. The flux ϕ of the iron core links not only with the secondary winding but also with the primary winding, so produces self induced emf in the primary winding. This induced emf in the primary winding opposes the applied voltage and, therefore, sometimes it is known as *back emf* of the primary. In fact the induced emf in the primary winding limits the primary current in much the same way that the back emf in a dc motor limits the armature current.

8.3. TRANSFORMER ON DC

A transformer cannot operate on dc supply and never be connected to a dc source. If a rated dc voltage is applied to the primary of a transformer, the flux produced in the transformer core will not vary but remain constant in magnitude and, therefore, no emf will be induced in the secondary winding except at the moment of switching *on*. Thus the transformer is not capable of raising or lowering the dc voltage. Also there will be no self induced emf in the primary winding, which is only possible with varying flux linkage, to oppose the applied voltage and since the resistance of primary winding is quite low, a heavy current will flow through the primary winding which may result in the burning out of the primary winding. This is the reason that *dc is never applied to a transformer*.

8.4. IDEAL TRANSFORMER

For a better understanding and an easier explanation of a practical transformer, certain idealizing assumptions are made which are close approximations for a practical transformer. A transformer having these ideal properties is hypothetical (has no real existence) and is referred to as the *ideal transformer.* It possesses certain essential features of a real transformer but some details of minor significance are ignored which will be introduced step-by-step while analysing a transformer. The idealizing assumptions made are as follows:

(*i*) *No winding resistance i.e.* the primary and secondary windings have zero resistance. It means that there is no ohmic power loss and no resistive voltage drop in an ideal transformer.

(*ii*) *No magnetic leakage i.e.* there is no leakage flux and all the flux set up is confined to the core and links both the windings.

(*iii*) *No iron loss i.e.* hysteresis and eddy current losses in transformer core are zero.

(*iv*) *Zero magnetizing current i.e.* the core has infinite permeability and zero reluctance so that zero magnetizing current is required for establishing the requisite amount of flux in the core.

From the above discussion an ideal transformer is supposed to consist of two purely inductive coils wound on a loss-free core.

8.5. EMF EQUATION

When an alternating (sinusoidal) voltage is applied to the primary winding of a transformer, an alternating (sinusoidal) flux, as shown is Fig. 8.2, is set up in the iron core which links both the windings (primary and secondary windings).

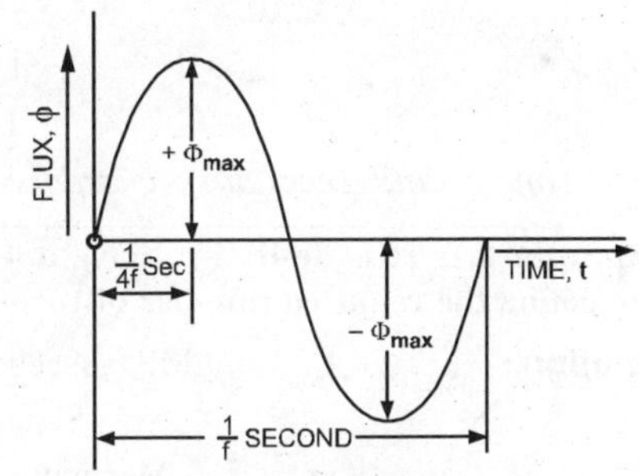

Fig. 8.2. ***Sinusoidal Variation of Flux With Time***

Let ϕ_{max} = Maximum value of flux in webers

and f = Supply frequency in hertz.

As illustrated in Fig. 8.2, the magnetic flux increases from zero to its maximum value ϕ_{max} in one-fourth of a cycle *i.e.* in $\frac{1}{4f}$ second.

So average rate of change of flux, $\frac{d\phi}{dt} = \frac{\phi_{max}}{1/4f} = 4f\phi_{max}$

Since average emf induced per turn in volts is equal to the average rate of change of flux,

Average emf induced per turn $= 4f\phi_{max}$ volts

Since flux ϕ varies sinusoidally, emf induced will be sinusoidal and form factor for sinusoidal wave is 1.11 *i.e.* the rms or effective value is 1.11 times the average value.

$\therefore$ RMS value of emf induced per turn $= 1.11 \times 4f\phi_{max}$ volts ...(8.1)

If the number of turns on primary and secondary windings are N_1 and N_2 respectively, then

RMS value of emf induced in primary, E_1 = EMF induced per turn × number of primary turns

$= 4.44 f\phi_{max} \times N_1 = 4.44 f N_1 \phi_{max}$ volts ...(8.2)

Similarly rms value of emf induced in secondary,

$E_2 = 4.44 f \phi_{max} \times N_2$ volts ...(8.3)

In an ideal transformer the voltage drops in primary and secondary windings are negligible, so

EMF induced in primary winding, E_1 = Applied voltage to primary, V_1

and terminal voltage, V_2 = EMF induced in secondary, E_2

Note: If B_{max} is the maximum allowable flux density in Wb/m^2 (or T) and *a* is the area of x-section of iron core in square metres, then in Eqs. (8.1), (8.2) and (8.3), ϕ_{max} is given as

$$\phi_{max} = B_{max}\, a \text{ webers}$$

8.6. VOLTAGE AND CURRENT TRANSFORMATION RATIOS

Referring to Eq. (8.1), it is clear that the *volts per turn* is exactly the same for both the primary and secondary windings *i.e.* in any transformer, the secondary and primary induced emfs are related to each other by the ratio of the number of secondary and primary turns. Thus,

$$\frac{V_2}{V_1} = \frac{E_2}{E_1} = \frac{N_2}{N_1} = K \qquad ...(8.4)$$

The same relationship can be derived by dividing Eq. (8.3) by Eq. (8.2).

The constant K in Eq. (8.4) is called the *voltage transformation ratio.*

For step-up transformer, $V_2 > V_1$ or voltage transformation ratio, $K > 1$.

For step-down transformer, $V_2 < V_1$ or voltage transformation ratio, $K < 1$.

In an ideal transformer, the losses are negligible, so the volt-ampere input to the primary and volt-ampere output from secondary can be approximately equated *i.e.*

$$\text{Output VA} = \text{Input VA}$$

$$\text{or } V_2 I_2 = V_1 I_1$$

$$\text{or } \frac{I_2}{I_1} = \frac{V_1}{V_2} = \frac{E_1}{E_2} = \frac{N_1}{N_2} = \frac{1}{K} \qquad ...(8.5)$$

i.e. Primary and secondary currents are inversely proportional to their respective turns.

Example 8.1. It is desired to have a 4.13 mWb maximum core flux in a transformer at 110 V and 50 Hz. Determine the required number of turns in the primary. [*Pb. Technical Univ. Electrical Engineering December, 2000*]

Solution :

EMF induced in primary, $E_1 = 110$ V

Supply frequency, $f = 50$ Hz

Maximum core flux, $\phi_{max} = 4.13 \text{ mWb} = 4.13 \times 10^{-3}$ Wb

Required number of turns on primary, $N_1 = \dfrac{E_1}{4.44 f \phi_{max}}$ ...Refer to Eq. (8.2)

$$= \frac{110}{4.44 \times 50 \times 4.13 \times 10^{-3}} = 120 \textbf{ Ans.}$$

Example 8.2. The emf per turn of a single phase 10 kVA, 2,200 / 220 V, 50 Hz transformer is 10 V. Calculate (*i*) the number of primary and secondary turns, (*ii*) the net cross-sectional area of core for a maximum flux density of 1.5 T.

Solution :

EMF per turn = 10 V

Primary induced emf, $E_1 = V_1 = 2{,}200$ V

Secondary induced emf, $E_2 = V_2 = 220$ V

Supply frequency, $f = 50$ Hz

Maximum flux density, $B_{max} = 1.5$ T

(*i*) Number of primary turns, $N_1 = \dfrac{E_1}{\text{EMF per turn}} = \dfrac{2{,}200}{10} = 220$ **Ans.**

Number of secondary turns, $N_2 = \dfrac{E_2}{\text{EMF per turn}} = \dfrac{220}{10} = 22$ **Ans.**

Maximum value of flux, $\phi_{max} = \dfrac{\text{EMF per turn}}{4.44 f} = \dfrac{10}{4.44 \times 50} = 0.045$ Wb ...Refer to Eq. (8.1)

(*ii*) Net cross-sectional area of core, $a = \dfrac{\phi_{max}}{B_{max}} = \dfrac{0.045}{1.5} = 0.03\ m^2$ **Ans.**

Example 8.3. A single phase transformer has 400 primary and 1,000 secondary turns. The net cross-sectional area of the core is 60 cm². If the primary winding be connected to a 50 Hz supply at 500 V, calculate (*i*) the peak value of the flux density in the core, and (*ii*) the voltage induced in the secondary winding.

Solution :

Primary induced emf, $E_1 = V_1 = 500$ V

Supply frequency, $f = 50$ Hz

Primary turns, $N_1 = 400$

Net cross-sectional area of core, $a = 60\ cm^2 = 0.006\ m^2$

Maximum value of flux, $\phi_{max} = \dfrac{E_1}{4.44 f N_1}$ ...Refer to Eq. (8.2)

$$= \frac{500}{4.44 \times 50 \times 400} = 0.00563 \text{ Wb}$$

(*i*) Peak value of flux density in the core, $B_{max} = \dfrac{\phi_{max}}{a} = \dfrac{0.00563}{0.006} = 0.9384$ T **Ans.**

(*ii*) Voltage induced in the secondary winding, $E_2 = E_1 \times \dfrac{N_2}{N_1}$ ...Refer to Eq. (8.4)

$$= 500 \times \frac{1{,}000}{400} = 1{,}250 \text{ V}$$ **Ans.**

Example 8.4. A 3,000/200 V, 50 Hz single phase transformer is built on a core having an effective cross section area of 150 cm² and has 80 turns in low voltage winding. Calculate (*a*) the value of maximum flux density in the core (*b*) number of turns in high-voltage winding.

Solution : Maximum value of flux, $\phi_{max} = \dfrac{\text{Induced emf in low - voltage winding}}{4.44 \times \text{supply frequency} \times \text{low - voltage winding turns}}$

$$= \frac{200}{4.44 \times 50 \times 80} = 0.01126 \text{ Wb}$$

(*i*) Value of maximum flux density in the core,

$$B_{max} = \frac{\phi_{max}}{a} = \frac{0.01126}{150 \times 10^{-4}} = 0.75 \text{ T}$$ **Ans.**

(*ii*) Number of turns on high voltage winding

$$= \frac{\text{Induced emf in hv winding} \times \text{lv winding turns}}{\text{induced emf in lv winding}}$$

$$= \frac{3{,}000}{200} \times 80 = 1{,}200$$ **Ans.**

Example 8.5. A 200 kVA, 6600/400 V, 50 Hz single phase transformer has 80 turns on the secondary. Calculate (*i*) the approximate values of the primary and secondary currents (*ii*) the approximate number of primary turns and (*iii*) the maximum value of flux.

Solution : Output = 200 kVA

(*i*) Approximate value of primary current on full-load,

$$I_1 = \frac{\text{Rated kVA} \times 1{,}000}{V_1} = \frac{200 \times 1{,}000}{6{,}600} = 30.3 \text{ A}$$ **Ans.**

Approximate value of secondary current on full-load,

$$I_2 = \frac{\text{Rated kVA} \times 1{,}000}{V_2} = \frac{200 \times 1{,}000}{400} = 500 \text{ A } \textbf{Ans.}$$

(*ii*) Approximate number of primary turns,

$$N_1 = N_2 \times \frac{E_1}{E_2} = N_2 \times \frac{V_1}{V_2} = 80 \times \frac{6{,}600}{400} = 1{,}320 \textbf{ Ans.}$$

(*iii*) Maximum value of flux in the core,

$$\phi_{max} = \frac{E_1}{4.44\, f\, N_1} = \frac{6{,}600}{4.44 \times 50 \times 1{,}320} = 0.0225 \text{ Wb or } 22.5 \text{ mWb } \textbf{Ans.}$$

Example 8.6. A single phase transformer with 10 : 1 turn-ratio and rated at 50 kVA, 2400/240 V, 50 Hz is used to step down the voltage of a distribution system. The low tension voltage is to be kept constant at 240 V. Find the value of load impedance of the *lt* side so that the transformer will be loaded fully. Find also the value of maximum flux inside the core if the *lt* side has 23 turns.

Solution : Low tension (secondary) voltage, V_2 = 240 volts

$$\text{Full-load secondary current, } I_2 = \frac{\text{Rated kVA} \times 1{,}000}{V_2} = \frac{50 \times 1{,}000}{240} = 208.33 \text{ A}$$

$$\text{Load impedance connected on secondary side, } Z = \frac{V_2}{I_2} = \frac{240}{208.33} = 1.152\ \Omega \textbf{ Ans.}$$

$$\text{Maximum value of flux in the core, } \phi_{max} = \frac{E_2}{4.44\, f\, N_2} = \frac{240}{4.44 \times 50 \times 23} = 0.047 \text{ Wb } \textbf{Ans.}$$

8.7. TRANSFORMER ON NO LOAD

When the primary of a transformer is connected to the source of ac supply and the secondary is open, the transformer is said to be at no load (*i.e.* there is no load on secondary).

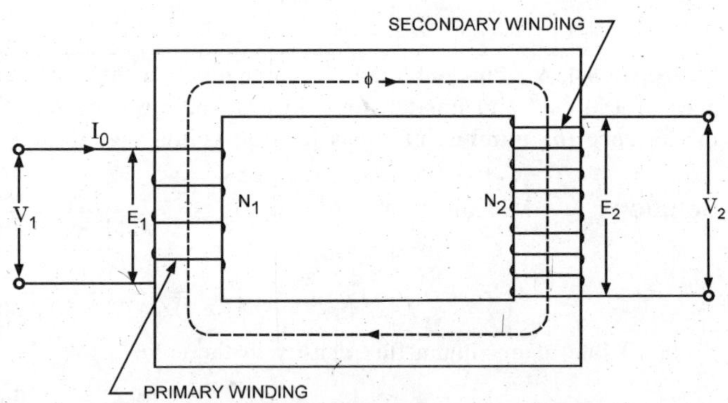

Fig. 8.3. *Transformer on No Load*

Consider an ideal transformer whose secondary side is open and the primary winding is connected to a sinusoidal alternating voltage V_1. The alternating voltage applied to the primary winding will cause flow of alternating current in the primary winding. Since the primary coil is purely inductive and there is no output (secondary being open), the primary draws the magnetising current I_m only. The function of this current is merely to magnetize the core. If the transformer is truely ideal, the magnitude of I_m should be zero by virtue of assumption (*iv*) made in Art. 8.4. Since the reluctance of the magnetic circuit is never zero, I_m has definite magnitude. The magnetising current, I_m is small in magnitude and lags behind supply voltage V_1 by 90°. This magnetising current I_m produces an alternating flux ϕ which is, at all times, proportional to the current (assuming permeability of the magnetic circuit to be constant) and, hence, in phase with it.

Let the instantaneous linking flux be given as

$$\phi = \phi_{max} \sin \omega t \qquad \text{..(8.6)}$$

The varying flux is linked with both of the windings (primary and secondary) and so induces emfs in the primary and secondary windings. The instantaneous values of induced emfs in the primary and secondary windings will be

$$e_1 = -N_1 \frac{d\phi}{dt} = -N_1 \frac{d}{dt}(\phi_{max} \sin \omega t)$$
$$= -N_1 \omega\phi_{max} \cos \omega t$$
$$= N_1\omega\phi_{max} \sin\left(\omega t - \frac{\pi}{2}\right) \quad ...(8.7)$$

Similarly, $e_2 = N_2\omega\phi_{max} \sin\left(\omega t - \frac{\pi}{2}\right)$ ...(8.8)

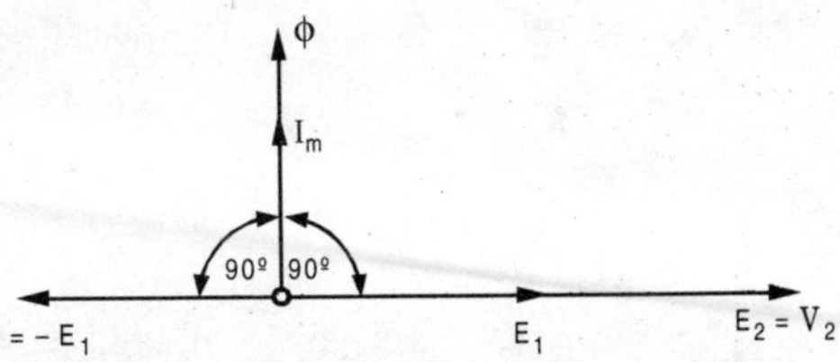

Fig. 8.4. ***No-Load Phasor Diagram For an Ideal Transformer***

Since primary winding has no ohmic resistance, (as assumed), applied voltage to primary winding is to only oppose the induced emf in the primary winding, hence instantaneous applied voltage to primary will be given by

$$v_1 = -e_1 = -N_1 \omega \phi_{max} \sin\left(\omega t - \frac{\pi}{2}\right) \quad ...(8.9)$$

Comparing Eqs. (8.6), (8.7), (8.8) and (8.9) we conclude that

(*i*) Induced emfs in primary and secondary windings, E_1 and E_2 lag behind the main flux ϕ by $\frac{\pi}{2}$, so these emfs (E_1 and E_2) are in phase with each other, as shown in Fig. 8.4 vectorially.

(*ii*) Applied voltage to the primary winding leads the main flux by $\frac{\pi}{2}$ and is in phase opposition to the induced emf in the primary winding, as shown in Fig. 8.4 vectorially.

(*iii*) Secondary voltage $V_2 = E_2$ as there is no voltage drop in secondary.

The instantaneous values of applied voltage, induced emfs, flux and magnetising current, in case of an ideal transformer, are illustrated by sinusoidal waves in Fig. 8.5.

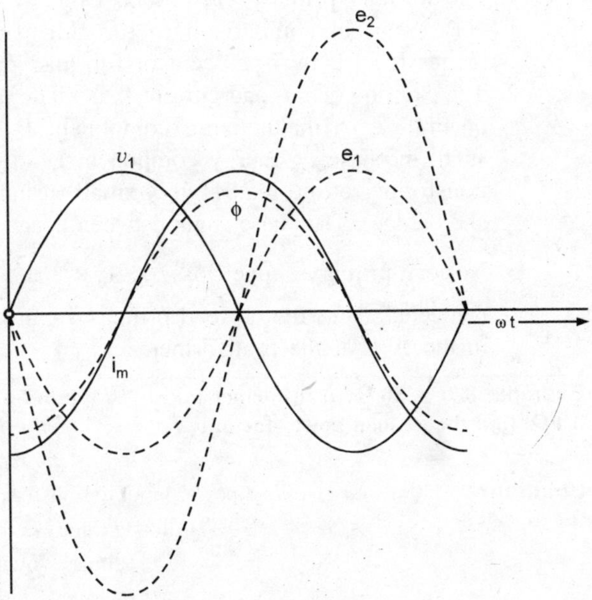

Fig. 8.5

However, when a varying flux is set up in magnetic material, there will be power loss, called the *iron or core loss.* So the input current to the primary under no-load condition has also to supply the hysteresis and eddy current losses (iron losses) occurring in the core in addition to small amount of copper loss occurring in primary winding (no copper loss occurs in secondary winding on open circuit or on no load). Hence, the no-load primary current I_0 does not lag behind applied voltage V_1 by 90° but lags behind V_1 by angle $\phi_0 < 90°$.

Input power on no load, $P_0 = V_1I_0 \cos \phi_0$ where $\cos \phi_0$ is the primary power factor under no-load condition.

As seen from phasor diagram shown in Fig. 8.6, input current to the primary I_0, called the *exciting current,* has two components (*i*) in-phase, active or energy component, I_e used to meet the iron loss in addition to small amount of copper loss occurring in the primary winding and (*ii*) quadrature component or wattless component, called the magnetizing component, I_m used to create the alternating flux in the core.

$$\text{Thus,}\quad I_e = I_0 \cos\phi_0$$
$$\text{and}\quad I_m = I_0 \sin\phi_0$$
$$\text{and}\quad \sqrt{I_e^2 + I_m^2} = I_0$$

$$\text{Angle of lag, } \phi_0 = \text{Tan}^{-1}\frac{I_m}{I_e}$$

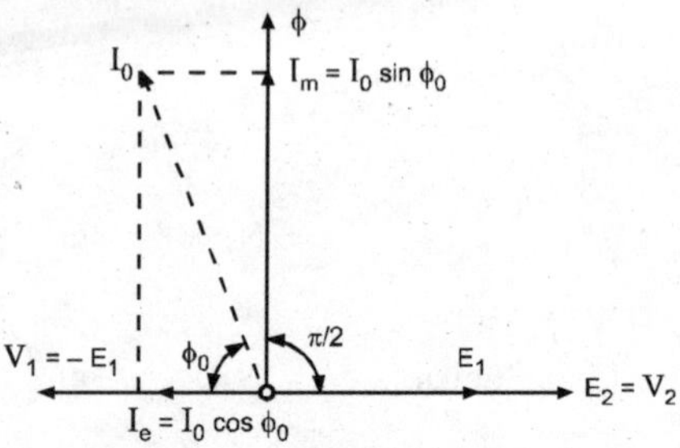

Fig. 8.6

The equivalent circuit of a transformer on no load is illustrated in Fig. 8.7, where in two components of no-load current; I_e and I_m are represented by currents drawn by a non-inductive resistance R_0 and a pure inductive reactance X_0 respectively. Both these currents are drawn at induced emf $E_1 = V_1$ for resistance-less, no-leakage primary coil; even otherwise $E_1 \simeq V_1$

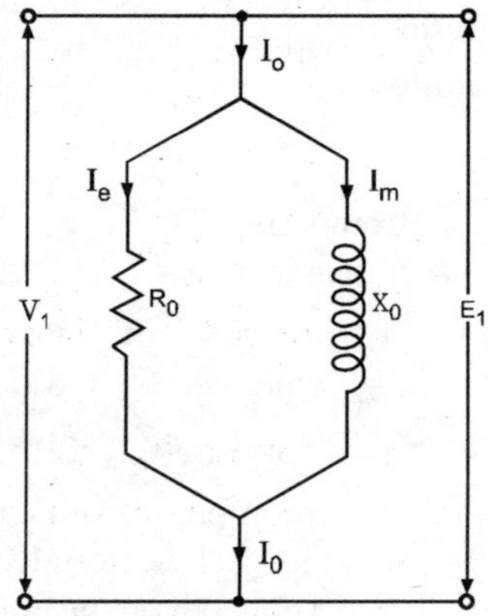

Fig. 8.7

The noteworthy points are given below:

1. The no-load primary current I_0, called the *exciting current*, is very small in comparison to the full-load primary current. It ranges from 2 to 5 per cent of full-load primary current.
2. The exciting or no-load current I_0 is made up of a relatively large quadrature or magnetizing component I_m and a comparatively small in-phase or energy component I_e so the power factor of a transformer on no load is very small (usually varies between 0.1 and 0.2 lag). The phase angle between I_0 and V_1 is about 78° to 87°.
3. No-load primary copper loss $\left(i.e.\ I_0^2 R_1\right)$ is very small and may be neglected. Thus the no-load primary input power is practically equal to the iron loss occurring in the core of the transformer.

Example 8.7. A 50 Hz transformer takes 75 W input at 1.5 A and 120 V. If the primary winding resistance is 0.4 Ω, find the no-load power factor.

[*Pb. Technical Univ. Electrical Engineering June 2000; May 2001*]

Solution :

No-load power input to transformer, $P_0 = 75$ W

No-load primary current, $I_0 = 1.5$ A

Primary voltage, $V_1 = 120$ V

Primary resistance, $R_1 = 0.4\ \Omega$

Copper loss in primary winding, $P_C = I_0^2 R_1 = (1.5)^2 \times 0.4 = 0.9$ W

No-load iron loss, $P_i = P_0 - P_C = 75 - 0.9 = 74.1$ W

$$\text{No-load power factor, } \cos\phi_0 = \frac{P_0}{V_1 \times I_0} = \frac{75}{120 \times 1.5} = 0.417 \text{ (lag)}$$ **Ans.**

Example 8.8. A 100 kVA, 2400/240 V, 50 Hz single phase transformer has an exciting current of 0.64 A and a core loss of 700 W, when its high voltage side is energised at rated voltage and frequency. Calculate the components of no-load current and no-load branch parameters.

[*Pb. Technical Univ. Basic Electrical Engineering December 2002*]

Solution :

Primary voltage, $V_1 = 2{,}400$ V

Exciting or no-load current, $I_0 = 0.64$ A

Iron loss = 700 W

$$\text{Iron-loss current, } I_e = I_0 \cos\phi_0 = \frac{\text{Iron loss}}{V_1} = \frac{700}{2{,}400} = 0.292 \text{ A}$$ **Ans.**

Magnetising component of no-load current,

$$I_m = \sqrt{I_0^2 - I_e^2} = \sqrt{(0.64)^2 - (0.292)^2} = 0.57 \text{ A } \textbf{Ans.}$$

No-load branch parameters, $R_0 = \frac{V_1}{I_e} = \frac{2,400}{0.292} = 8,220\ \Omega$ or $8.22\ k\Omega$ **Ans.**

$$X_0 = \frac{V_1}{I_m} = \frac{2,400}{0.57} = 4,210\ \Omega \text{ or } 4.21\ k\Omega \textbf{ Ans.}$$

Example 8.9. The no-load current of a transformer is 5 A at 0.25 pf when supplied at 235 V, 50 Hz. The number of turns on the primary winding is 200. Calculate (*i*) the maximum value of flux in the core, (*ii*) the core loss (*iii*) the magnetizing component.

Solution :

No-load current, $I_0 = 5$ A

No-load power factor, $\cos \phi_0 = 0.25$

Primary induced emf, $E_1 = V_1 = 235$ V

Supply frequency, $f = 50$ Hz

Number of primary turns, $N_1 = 200$

(*i*) Maximum value of flux in the core, $\phi_{max} = \frac{E_1}{4.44 f N_1} = \frac{235}{4.44 \times 50 \times 200}$

$= 0.005293$ Wb or 5.293 mWb **Ans.**

(*ii*) Core loss = Input on no load

$= V_1 I_0 \cos \phi_0 = 235 \times 5 \times 0.25 = 293.75$ W **Ans.**

(*iii*) Magnetizing component, $I_m = I_0 \sin \phi_0 = I_0 \sqrt{1 - (\cos \phi_0)^2} = 5\sqrt{1 - (0.25)^2} = 4.84$ A **Ans.**

8.8. TRANSFORMER ON LOAD

When the secondary circuit of a transformer is completed through an impedance, or load, the transformer is said to be loaded and current flows through the secondary winding and the load. The magnitude and phase of secondary current I_2 with respect to secondary terminal voltage V_2 will depend upon the characteristic of load *i.e.* current I_2 will be in phase, lag behind and lead the terminal voltage V_2 respectively when the load is non-inductive, inductive and capacitive.

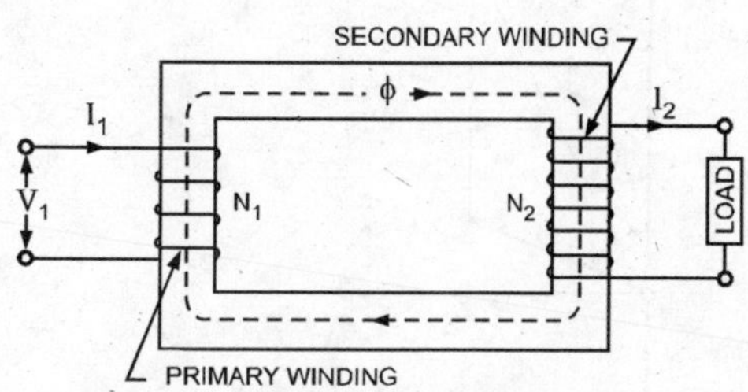

Fig. 8.8. ***An Ideal Transformer On Load***

When the transformer is on no load, as shown in Fig. 8.3, it draws no-load current I_0 from the supply mains. The no-load current I_0 sets up an mmf $N_1 I_0$ which produces flux ϕ in the core. When an impedance is connected across the secondary terminals, as shown in Fig. 8.8, current I_2 flows through the secondary winding. The secondary current I_2 sets up its own mmf and hence creates a secondary flux ϕ_2. The secondary flux ϕ_2 opposes the main flux ϕ set up by the exciting current I_0 according to Lenz's law. The opposing secondary flux ϕ_2 weakens the main flux ϕ momentarily, so primary back emf E_1 tends to be reduced. So difference of applied voltage V_1 and back emf E_1 increases, therefore, more current is drawn from the source of supply flowing through the primary winding until the original value of flux ϕ is obtained. It again causes increase in back emf E_1 and it adjusts itself as such that there is a balance between applied voltage V_1 and back emf E_1. Let the additional primary current be I_1'. The current I_1' is in phase opposition with secondary current I_2 and is called the *counterbalancing current.* The additional current I_1' sets up an mmf $N_1 I_1'$

producing flux ϕ_1' in the same direction as that of main flux ϕ and cancels the flux ϕ_2 produced by secondary mmf N_2I_2 being equal in magnitude.

$$\text{So } N_1 I_1' = N_2 I_2$$

$$\text{or } I_1' = \frac{N_2}{N_1} I_2$$

The total primary current I_1 is, therefore, phasor sum of primary counterbalancing current I_1' and no-load current I_0, which will be approximately equal to I_1' as I_0 is usually very small in comparison to I_1'.

$$\therefore I_1 = I_1' = \frac{N_2}{N_1} I_2$$

$$\text{or } \frac{I_1}{I_2} = \frac{N_2}{N_1} = K \quad \text{(transformation ratio)}$$

Hence primary and secondary currents are inversely proportional to their respective turns.

Since the secondary flux ϕ_2 produced by secondary mmf $N_2 I_2$ is neutralized by the flux ϕ_1' produced by mmf N_1I_1' set up by counterbalancing primary current I_1', the flux in the transformer core remains almost constant from no load to full load.

The phasor diagrams for a transformer on non-inductive, inductive and capacitive loads are shown in Figs. 8.9 (*a*), 8.9 (*b*) and 8.9 (*c*) respectively.

Since the voltage drops in both of the windings of the transformer are assumed to be negligible,

$$\mathbf{V_2 = E_2 \text{ and } V_1 = -E_1}$$

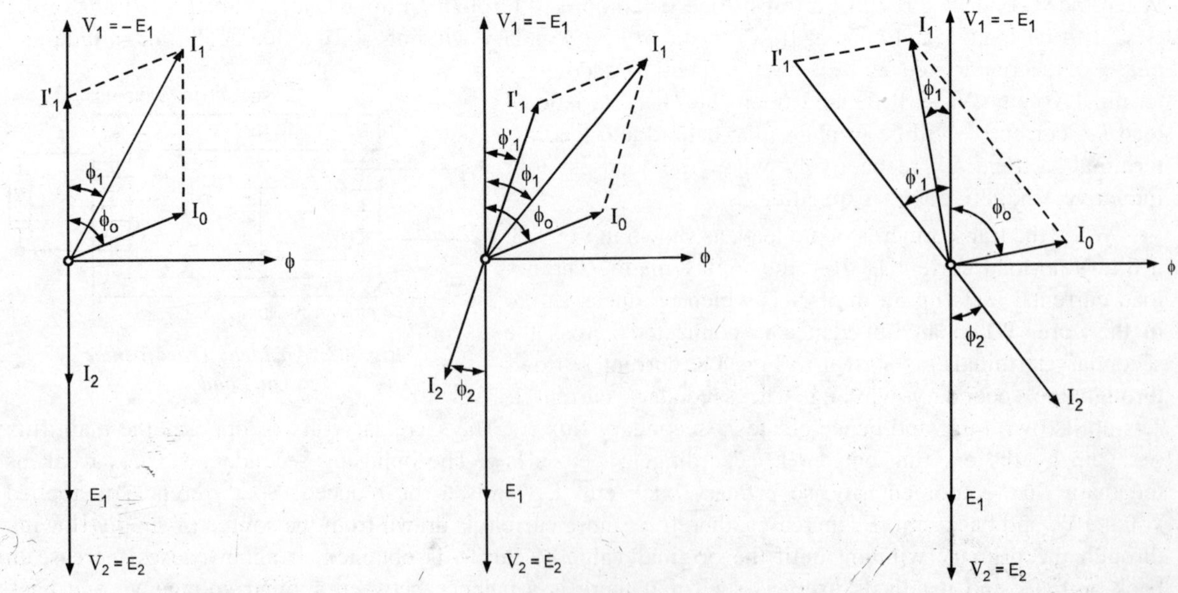

(a) For Pure Resistive Load *(b) For Inductive Load* *(c) For Capacitive Load*

Fig. 8.9. ***Phasor Diagram For an Ideal Transformer On Load***

The secondary current I_2 is in phase, lags behind and leads the secondary terminal voltage V_2 by an angle ϕ_2 for pure resistive, inductive and capacitive loads respectively.

The induced primary current I_1', also known as *counterbalancing current,* is always in opposition to secondary current I_2 and since no-load current I_0 is very small, the total primary current I_1 is almost opposite in phase to I_2 and K times the secondary current I_2, where K is transformation ratio.

Note : In phasor diagrams shown in Figs. 8.9 (*a*), 8.9 (*b*) and 8.9 (*c*) no-load current has been drawn on exaggerated scale for sake of clarity.

Example 8.10. A single phase transformer with a ratio of 440/110 V takes a no-load current of 5 A at 0.2 power factor lagging. If the secondary supplies a current of 120 A at a pf of 0.8 lagging, estimate the current taken by the primary.

Solution :

Primary emf, $E_1 = 440$ V

Secondary emf, $E_2 = 110$ V

No-load current, $I_0 = 5$ A

No-load power factor, $\cos \phi_0 = 0.2$ (lag)

$$\sin \phi_0 = \sqrt{1 - 0.2^2} = 0.98$$

Secondary current, $I_2 = 120$ A

Load power factor, $\cos \phi_2 = 0.8$ (lag)

$$\text{Transformation ratio, } K = \frac{V_2}{V_1} = \frac{110}{440} = \frac{1}{4}$$

$$\text{Primary counterbalancing current, } I_1' = KI_2 = \frac{1}{4} \times 120 = 30 \text{ A}$$

lagging behind the applied voltage V_1 by an angle ϕ_1' where

$$\phi_1' = \phi_2 = \text{Cos}^{-1}\ 0.8 = 36.87°$$

No-load current, $I_0 = 5$ A lagging behind the applied voltage V_1 by an angle

$$\phi_0 = \text{Cos}^{-1}\ 0.2 = 78.46°$$

$$\text{Primary current, } \mathbf{I_1} = \mathbf{I_1'} + \mathbf{I_0}$$

$$= (I_1' \cos \phi_1 + I_0 \cos \phi_0) + j\,(I_1' \sin \phi_1 + I_0 \sin \phi_0)$$

$$= (30 \times 0.8 + 5 \times 0.2) + j\,(30 \times 0.6 + 5 \times 0.98)$$

$$= (25 + j\,22.9) \text{ A}$$

$$\therefore \text{Magnitude of primary current, } I_1 = \sqrt{25^2 + 22.9^2} = 33.9 \text{ A } \textbf{Ans.}$$

$$\text{Phase angle, } \phi_1 = \text{Tan}^{-1}\frac{22.9}{25} = 42.49° \text{ (lag) } \textbf{Ans.}$$

8.9. RESISTANCE AND LEAKAGE REACTANCE

In preceding discussions we considered an ideal transformer, which according to our assumptions, has got no resistance in the windings and no leakage flux but in actual practice it is impossible to obtain such an ideal transformer.

In actual transformer both the windings, primary and secondary windings have finite resistances R_1 and R_2 which cause copper losses and voltage drops in them. The result is that :

(*i*) The secondary terminal voltage V_2 is less than the secondary induced emf E_2 and is equal to phasor difference of secondary induced emf E_2 and voltage drop in the secondary winding I_2R_2, if magnetic leakage is negligible, *i.e.*

$$\mathbf{V_2} = \mathbf{E_2} - \mathbf{I_2}\,R_2$$

where I_2 is the secondary current and R_2 is the secondary winding resistance.

(*ii*) Similarly the counter-emf of primary, $-E_1$ is equal to phasor difference of voltage applied to the primary winding V_1 and voltage drop in the primary winding $I_1 R_1$, provided magnetic leakage is negligible, *i.e.*,

$$-\mathbf{E}_1 = \mathbf{V}_1 - \mathbf{I}_1 \mathbf{R}_1$$

where I_1 is the primary current and R_1 is the primary resistance.

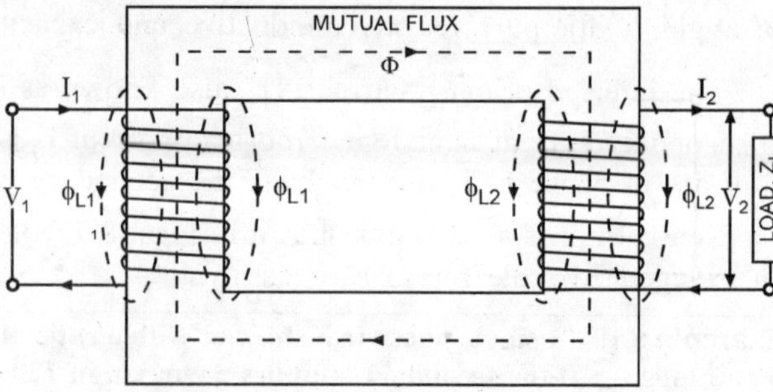

Fig. 8.10. ***Magnetic Fluxes in a Transformer***

It was previously assumed that the entire flux ϕ, developed by the primary winding, links with and cut every turn of both the primary and secondary windings. In practice, it is impossible to realize this condition. However, part of the flux set up by the primary winding links only the primary turns, as illustrated in Fig. 8.10 by flux ϕ_{L1}. Also, some of the flux set up by the secondary winding links only the secondary turns, as illustrated in Fig. 8.10 by ϕ_{L2}. These two fluxes ϕ_{L1} and ϕ_{L2} are known as leakage flux *i.e. that flux which leaks out of the core and does not link both windings.* The flux which does pass completely through the core and links both the windings is known as the *mutual flux* and is illustrated as ϕ.

The primary leakage flux ϕ_{L1} linking with the primary winding is produced by primary ampere-turns only, therefore, it is proportional to primary current, number of primary turns being fixed. On no load primary current is so small that leakage flux ϕ_{L1} produced by it can be neglected but on load primary current increases resulting in increase in ampere-turns and hence leakage flux ϕ_{L1} increases. The primary leakage flux ϕ_{L1} is in phase with I_1 and produces self induced emf E_{L1} given by $E_{L1} = 2\pi f L_1 I_1$ in primary winding, where L_1 is the self inductance of the primary winding produced by primary leakage flux ϕ_{L1}.

The self induced emf E_{L1}, due to primary leakage flux, in the primary winding must lag leakage flux ϕ_{L1} and primary current I by 90°. The emf necessary to balance this counter-emf is opposite and equal to it and, therefore, leads the primary current I_1 by 90°. As this emf, induced by the primary leakage flux, is proportional to the current and lags it by 90°, it is nothing more than a reactance voltage and is denoted by $-I_1X_1$. The component of line voltage that balances this emf is $+I_1X_1$. The effect of the primary leakage flux, therefore, is to induce an emf that opposes the flow of current to the transformer. The reactance of the primary winding, X_1 can be obtained by dividing self induced emf E_{L1} by the primary current I_1 *i.e.*

$$X_1 = \frac{E_{L1}}{I_1} = \frac{2\pi f L_1 I_1}{I_1} = 2\pi f L_1$$

Similarly secondary leakage flux ϕ_{L2} is set up by secondary ampere-turns and is proportional to secondary current I_2. On no load there is no current in secondary winding and, therefore, no leakage flux exists across the secondary winding on no load. On load leakage flux ϕ_{L2} in phase with secondary current I_2 and produces self induced emf $E_{L2} = 2\pi f L_2 I_2$ in the secondary winding where L_2 is self inductance of secondary winding due to leakage flux ϕ_{L2}. This is also a reactance voltage, and the component that balances it leads the secondary current by 90°. The secondary reactance X_2 opposes the current flowing out of the transformer and can be obtained by dividing self induced emf in secondary winding, E_{L2} by the secondary current I_2 *i.e.*

$$X_2 = \frac{E_{L2}}{I_2} = \frac{2\pi f L_2 I_2}{I_2} = 2\pi f L_2$$

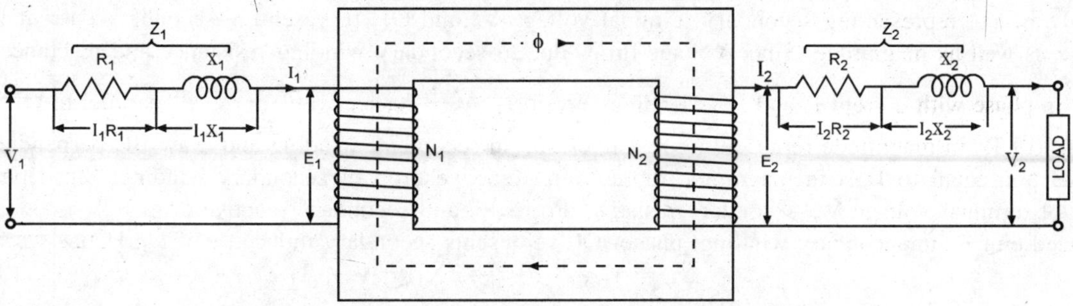

Fig. 8.11. *An Equivalent Diagram of Actual Transformer*

The effect of magnetic leakage is, thus to produce in their respective windings emfs of self inductance which are proportional to the current, and are, therefore, equivalent in effect to the addition of an inductive coil in series with each winding, the reactance of which is called the leakage reactance.

A transformer with magnetic leakage and winding resistance is equivalent to an ideal transformer (having no resistance and leakage reactance) having inductive and resistive coils connected in series with each winding as shown in Fig. 8.11.

8.10. PHASOR DIAGRAM OF ACTUAL TRANSFORMER ON LOAD

Consider a transformer shown in Fig. 8.11 having primary and secondary windings of resistances R_1 and R_2 and reactances X_1 and X_2 respectively. The impedance of primary winding is given by $\mathbf{Z}_1 = R_1 + j\,X_1$ and impedance of secondary winding is given by $\mathbf{Z}_2 = R_2 + j\,X_2$.

The phasor diagrams of above transformer on (*i*) pure resistive, (*ii*) resistive-inductive, and (*iii*) resistive-capacitive loads are shown in Fig. 8.12 (*a*), 8.12 (*b*) and 8.12 (*c*) respectively.

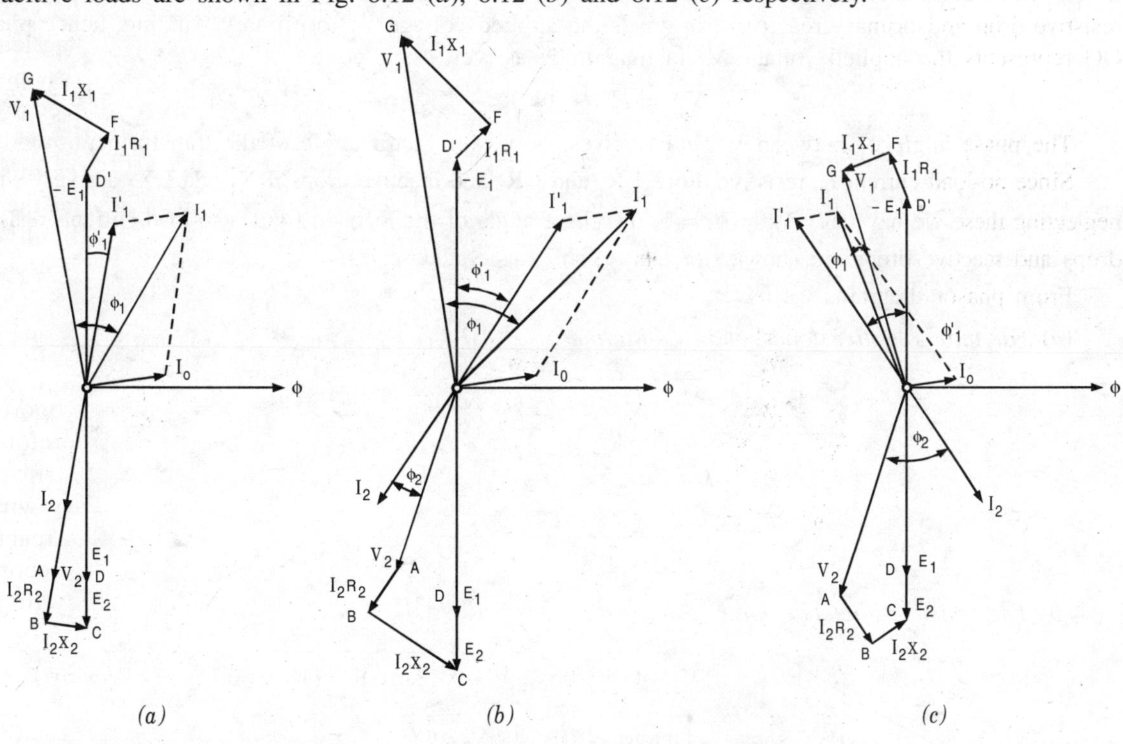

Fig. 8.12. *Phasor Diagram of an Actual Transformer*

Draw *OA* representing secondary terminal voltage V_2 and OI_2 representing secondary current I_2 in phase as well as magnitude. Since voltage drops due to secondary winding resistance and reactance are I_2R_2 in phase with current I_2 and I_2X_2 leading current I_2 by $\frac{\pi}{2}$ respectively, draw AB parallel to OI_2 and equal to I_2R_2 in magnitude representing resistive drop in secondary winding and draw BC perpendicular to AB and equal to I_2X_2 in magnitude representing reactive drop of secondary winding. Since phasor sum of terminal voltage V_2, secondary resistive drop I_2R_2 and secondary reactive drop I_2X_2 is equal to induced emf E_2 in secondary winding, phasor OC represents secondary induced emf E_2. Hence we have

$$\mathbf{E_2} = \mathbf{V_2} + \mathbf{I_2}\,(R_2 + j\,X_2) = \mathbf{V_2} + \mathbf{I_2}\,\mathbf{Z_2} \qquad ...(8.10)$$

The induced emf E_1 in primary winding is in phase with E_2 and equal to $\frac{N_1}{N_2}$ E_2 in magnitude, so take $OD = \frac{N_1}{N_2}$ OC representing E_1. Produce DO to D′ taking OD′ = OD hence representing $(-E_1)$.

The induced primary current I_1' is equal to $-\,I_2\,\frac{N_2}{N_1}$ so draw OI_1' equal to $OI_2 \times \frac{N_2}{N_1}$ by producing line I_2 O. Draw line OI_0 representing no-load current in magnitude as well as in phase. The phasor sum of induced primary current I_1' and no-load current I_0 gives primary current represented by phasor OI_1 in Fig. 8.12.

Since voltage drop due to primary winding resistance and reactance are $I_1\,R_1$ in phase with primary current I_1 and I_1X_1 leading current I_1 by $\frac{\pi}{2}$ respectively, draw D′ F parallel to OI_1 and equal to I_1R_1 in magnitude representing resistive drop in primary winding and draw FG perpendicular to D′ F equal to I_1X_1 in magnitude representing reactive drop in primary winding. As the phasor sum of $(-E_1)$, primary resistive drop and primary reactive drop gives the applied voltage V_1 to primary winding, hence phasor OG represents the applied voltage V_1 in magnitude as well as in phase.

$$i.e.\ \mathbf{V_1} = -\,\mathbf{E_1} + \mathbf{I_1}\,(R_1 + jX_1) = -\,\mathbf{E_1} + \mathbf{I_1 Z_1} \qquad ...(8.11)$$

The phase angle ϕ_1 between V_1 and I_1 gives the power factor angle of the transformer.

Since no-load current I_0, resistive drops I_1R_1 and I_2R_2 and reactive drops, I_1X_1 and I_2X_2 are very small, neglecting these we have $\phi_2 = \phi_1' = \phi_1 = \phi$, the phase angle of the load. In figure, no-load current, resistive drops and reactive drops are shown, for clarity, on exaggerated scales.

From phasor diagrams we have

(a) For pure resistive load [phasor diagram 8.12 (a)]

$$E_2 = \sqrt{(V_2 + I_2\,R_2)^2 + (I_2\,X_2)^2} \approx V_2 + I_2\,R_2 \qquad ...(8.12)$$

$$E_1 = \frac{E_2}{K}$$

$$\text{and } V_1 = \sqrt{(E_1 + I_1\,R_1)^2 + (I_1\,X_1)^2} \approx E_1 + I_1\,R_1 \qquad ...(8.13)$$

(b) For resistive-inductive load [phasor diagram 8.12 (b)]

$$E_2 = \sqrt{\left(V_2 + I_2\,R_2\cos\phi + I_2\,X_2\sin\phi\right)^2 + \left(I_2\,X_2\cos\phi - I_2\,R_2\sin\phi\right)^2}$$

$$\approx V_2 + I_2\,R_2\cos\phi + I_2\,X_2\sin\phi \qquad ...(8.14)$$

$$E_1 = \frac{E_2}{K}$$

$$\text{and } V_1 = \sqrt{(E_1 + I_1 R_1 \cos\phi + I_1 X_1 \sin\phi)^2 + (I_1 X_1 \cos\phi - I_1 R_1 \sin\phi)^2}$$

$$\approx E_1 + I_1 R_1 \cos\phi + I_1 X_1 \sin\phi \quad ...(8.15)$$

(c) For resistive-capacitive load [phasor diagram 8.12 (c)]

$$E_2 = \sqrt{(V_2 + I_2 R_2 \cos\phi - I_2 X_2 \sin\phi)^2 + (I_2 X_2 \cos\phi + I_2 R_2 \sin\phi)^2}$$

$$\approx V_2 + I_2 R_2 \cos\phi - I_2 X_2 \sin\phi \quad ...(8.16)$$

$$E_1 = \frac{E_2}{K}$$

$$\text{and } V_1 = \sqrt{(E_1 + I_1 R_1 \cos\phi - I_1 X_1 \sin\phi)^2 + (I_1 X_1 \cos\phi + I_1 R_1 \sin\phi)^2}$$

$$= E_1 + I_1 R_1 \cos\phi - I_1 X_1 \sin\phi \quad ...(8.17)$$

Example 8.11. A 230/460 V transformer has a primary resistance of 0.2 Ω and a reactance of 0.5 Ω and the corresponding values for the secondary are 0.75 Ω and 1.8 Ω respectively. Find the secondary terminal voltage when supplying *(a)* 10 A at 0.8 pf lagging *(b)* 10 A at pf 0.8 leading.

Solution : Transformation ratio, $K = \frac{V_2}{V_1} = \frac{460}{230} = 2$

(a) **When load is 10 A at 0.8 pf lagging** *i.e.* when

$$I_2 = 10 \text{ A and } \cos\phi = 0.8 \text{ lagging.}$$

$$\text{Primary current } I_1 = KI_2 = 2 \times 10 = 20 \text{ A}$$

From Eq. (8.15)

$$\text{Primary induced emf, } E_1 = V_1 - (I_1 R_1 \cos\phi + I_1 X_1 \sin\phi)$$

$$= 230 - (20 \times 0.2 \times 0.8 + 20 \times 0.5 \times 0.6) = 220.8 \text{ V}$$

$$\text{Secondary induced emf, } E_2 = KE_1 = 2 \times 220.8 = 441.6 \text{ V}$$

From Eq. (8.14)

$$\text{Secondary terminal voltage, } V_2 = E_2 - (I_2 R_2 \cos\phi + I_2 X_2 \sin\phi)$$

$$= 441.6 - (10 \times 0.75 \times 0.8 + 10 \times 1.8 \times 0.6) = 424.8 \text{ V} \textbf{ Ans.}$$

(b) **When load is 10 A at 0.8 pf leading**

From Eq. (8.17)

$$\text{Primary induced emf, } E_1 = V_1 - (I_1 R_1 \cos\phi - I_1 X_1 \sin\phi)$$

$$= 230 - (20 \times 0.2 \times 0.8 - 20 \times 0.5 \times 0.6) = 232.8 \text{ V}$$

and from Eq. (8.16)

$$\text{Secondary terminal voltage, } V_2 = KE_1 - (I_2 R_2 \cos\phi - I_2 X_2 \sin\phi)$$

$$= 2 \times 232.8 - (10 \times 0.75 \times 0.8 - 10 \times 1.8 \times 0.6) = 470.4 \text{ V} \textbf{ Ans.}$$

8.11. EQUIVALENT RESISTANCE AND REACTANCE

The two independent circuits of a transformer can be resolved into an equivalent circuit to make the calculations simple.

Let resistances and reactances of primary and secondary windings be R_1 and R_2 and X_1 and X_2 ohms respectively and let transformation ratio be K.

Resistive drop in secondary winding $= I_2R_2$

Reactive drop in secondary winding $= I_2X_2$

Resistive drop in primary winding $= I_1R_1$

Reactive drop in primary winding $= I_1X_1$

Referred To Secondary Side. Since transformation ratio is K, primary resistive and reactive drops as referred to secondary will be K times, *i.e.* $K I_1 R_1$ and $K I_1 X_1$ respectively. If I_1 is substituted equal to KI_2 then we have primary resistive and reactive drops referred to secondary equal to $K^2 I_2 R_1$ and $K^2 I_2 X_1$ respectively.

Total resistive drop in a transformer $= K^2 I_2 R_1 + I_2 R_2 = I_2 (K^2 R_1 + R_2) = I_2 R_{02}$

Total reactive drop in a transformer $= K^2 I_2 X_1 + I_2 X_2 = I_2 (K^2 X_1 + X_2) = I_2 X_{02}$

The term $(K^2 R_1 + R_2)$ and $(K^2 X_1 + X_2)$ represent the equivalent resistance and reactance respectively of the transformer referred to secondary and let these be represented by R_{02} and X_{02} respectively. Equivalent circuit referred to secondary has been shown in Fig. 8.13.

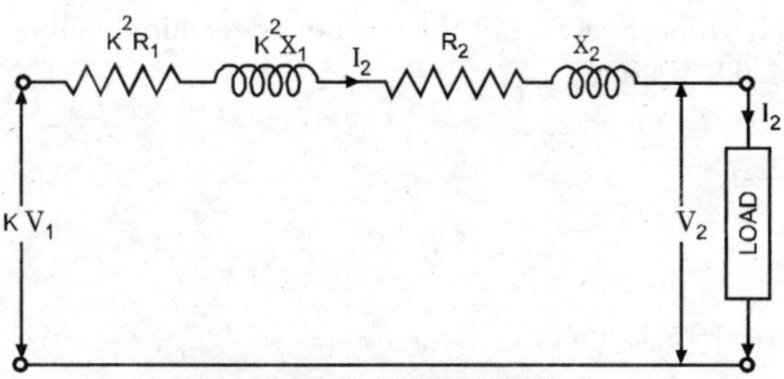

Fig. 8.13. ***Equivalent Circuit of a Transformer Referred to Secondary***

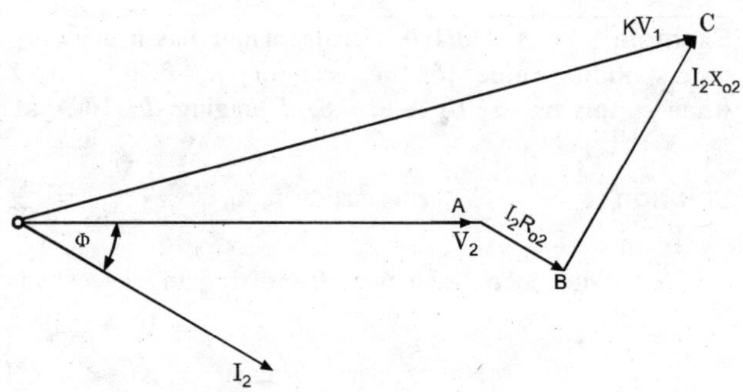

Fig. 8.14. ***Phasor Diagram***

From phasor diagram (Fig. 8.14)

$$KV_1 = \sqrt{(V_2 + I_2 R_{02} \cos\phi + I_2 X_{02} \sin\phi)^2 + (I_2 X_{02} \cos\phi - I_2 R_{02} \sin\phi)^2}$$

where V_2 is secondary terminal voltage, I_2 is secondary current lagging behind the terminal voltage V_2 by ϕ.

Since term $(I_2 X_{02} \cos\phi - I_2 R_{02} \sin\phi)$ is very small as compared to the term $(V_2 + I_2 R_{02} \cos\phi + I_2 X_{02} \sin\phi)$, neglecting the former we have

$$K V_1 = V_2 + I_2 R_{02} \cos\phi + I_2 X_{02} \sin\phi$$

$$\text{or } V_2 = KV_1 - I_2 R_{02} \cos\phi - I_2 X_{02} \sin\phi \quad ...(8.18)$$

where V_1 is applied voltage to primary winding.

If load is pure resistive, $\phi = 0$ and $V_2 = KV_1 - I_2 R_{02}$ ...(8.19)

If load is capacitive then ϕ should be taken as –ve hence we have

$$V_2 = KV_1 - I_2 R_{02} \cos\phi + I_2 X_{02} \sin\phi \quad ...(8.20)$$

Referred To Primary Side. Secondary resistive drop referred to primary

$$= \frac{I_2 R_2}{K} = \frac{I_1 R_2}{K^2} \qquad \text{since } I_2 = \frac{I_1}{K}$$

Secondary reactive drop referred to primary = $\frac{I_2 X_2}{K} = \frac{I_1 X_2}{K^2}$

Total resistive drop in the transformer referred to primary

$$= I_1 R_1 + \frac{I_1 R_2}{K^2} = I_1 \left(R_1 + \frac{R_2}{K^2} \right) = I_1 R_{01}$$

Total reactive drop in the transformer referred to primary

$$= I_1 X_1 + \frac{I_1 X_2}{K^2} = I_1 \left(X_1 + \frac{X_2}{K^2} \right) = I_1 X_{01}$$

The terms $\left(R_1 + \frac{R_2}{K^2} \right)$ and $\left(X_1 + \frac{X_2}{K^2} \right)$ represent the total resistance and reactance of the transformer referred to primary respectively. Let these be represented by R_{01} and X_{01} respectively.

Equivalent circuit referred to primary is shown in Fig. 8.15 (*a*).

From phasor diagram shown in Fig. 8.15 (*b*) we have

$$V_1 = \sqrt{\left(\frac{V_2}{K} + I_1 R_{01} \cos\phi + I_1 X_{01} \sin\phi \right)^2 + (I_1 X_{01} \cos\phi - I_1 R_{01} \sin\phi)^2}$$

Since term $(I_1 X_{01} \cos\phi - I_1 R_{01} \sin\phi)$ is very small as compared to the term $\left(\frac{V_2}{K} + I_1 R_{01} \cos\phi + I_1 X_{01} \sin\phi \right)$, neglecting the former we have

$$V_1 = \frac{V_2}{K} + I_1 R_{01} \cos\phi + I_1 X_{01} \sin\phi \qquad ...(8.21)$$

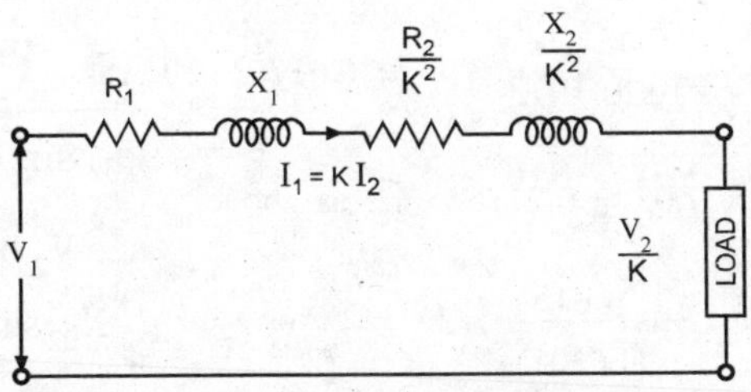

(a) *Equivalent Circuit of a Transformer Referred To Primary*

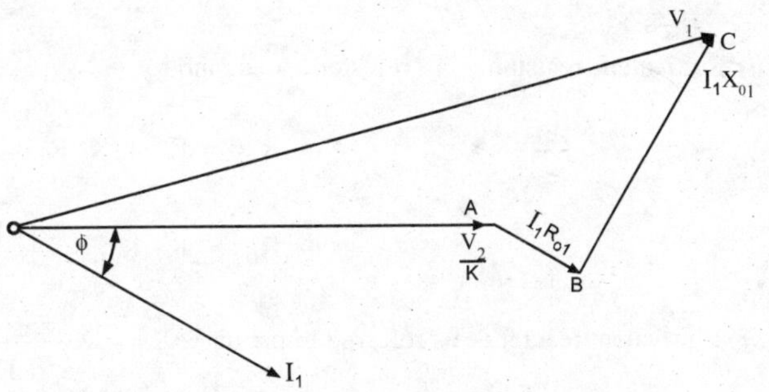

(b) *Phasor Diagram*

Fig. 8.15

If load is pure resistive *i.e.* $\phi = 0$ then $V_1 = \frac{V_2}{K} + I_1 R_{01}$ ...(8.22)

If load is capacitive then ϕ should be taken as –ve so we have

$$V_1 = \frac{V_2}{K} + I_1 R_{01} \cos\phi - I_1 X_{01} \sin\phi \qquad ...(8.23)$$

Example 8.12. A 5 kVA, 240/120 V, single phase transformer supplies rated current to a load at 120 V. Determine the magnitude of load impedance as seen from input terminals of transformer. The load impedance is (5 + *j*15) Ω.

[*Pb. Technical Univ. Basic Electrical Engineering December 2002*]

Solution : Transformation ratio, $K = \frac{V_2}{V_1} = \frac{120}{240} = 1/2$

Load impedance referred to secondary, $Z_L = (5 + j15)\ \Omega$

Load impedance referred to primary, $Z_L = \frac{Z_L}{K^2} = \frac{5 + j15}{(½)^2} = (20 + j60)\ \Omega$ **Ans.**

Example 8.13. A 50 kVA, 4,400/220 V transformer has $R_1 = 3.45\ \Omega$, $R_2 = 0.009\ \Omega$. The values of reactances are $X_1 = 5.2\ \Omega$ and $X_2 = 0.015\ \Omega$. Calculate for the transformer (*i*) equivalent resistance as referred to primary (*ii*) equivalent resistance referred to secondary (*iii*) equivalent reactance as referred to both primary and secondary (*iv*) equivalent impedance as referred to both primary and secondary (*v*) total copper loss, first using individual resistances of the two windings and secondly using equivalent resistances as referred to each side.

Solution : Transformation ratio, $K = \frac{V_2}{V_1} = \frac{220}{4,400} = \frac{1}{20}$

Full-load secondary current, $I_2 = \frac{50 \times 1,000}{220} = 227.3$ A

Full-load primary current, $I_1 = KI_2 = \frac{1}{20} \times 227.3 = 11.36$ A

Primary resistance, $R_1 = 3.45\ \Omega$

Secondary resistance, $R_2 = 0.009\ \Omega$

Primary reactance, $X_1 = 5.2\ \Omega$

Secondary reactance, $X_2 = 0.015\ \Omega$

(*i*) Equivalent resistance as referred to primary, $R_{01} = R_1 + \frac{R_2}{K^2} = 3.45 + \frac{0.009}{(1/20)^2} = 7.05\ \Omega$ **Ans.**

(*ii*) Equivalent resistance as referred to secondary,

$$R_{02} = R_2 + K^2 R_1 = 0.009 + \left(\frac{1}{20}\right)^2 \times 3.45 = 0.017625\ \Omega \text{ Ans.}$$

$$\text{or } R_{02} = K^2 R_{01} = \left(\frac{1}{20}\right)^2 \times 7.05 = 0.017625\ \Omega, \text{ as above.}$$

(*iii*) Equivalent reactance as referred to primary, $X_{01} = X_1 + \frac{X_2}{K^2} = 5.2 + \frac{0.015}{(1/20)^2} = 11.2\ \Omega$ **Ans.**

Equivalent reactance as referred to secondary, $X_{02} = K^2 X_{01} = \left(\frac{1}{20}\right)^2 \times 11.2 = 0.028\ \Omega$ **Ans.**

(*iv*) Equivalent impedance as referred to primary,

$$\mathbf{Z}_{01} = R_{01} + jX_{01} = (7.05 + j11.2)\ \Omega \text{ or } 13.23\ \angle 57.81^\circ\ \Omega \text{ Ans.}$$

Equivalent impedance as referred to secondary,

$$\mathbf{Z}_{02} = R_{02} + jX_{02} = (0.017625 + j\ 0.028)\ \Omega \text{ or } 0.0331\ \angle 57.81^\circ\ \Omega \text{ Ans.}$$

(*v*) Copper loss $= I_1^2 R_1 + I_2^2 R_2 = (11.36)^2 \times 3.45 + (227.3)^2 \times 0.009$

$= 910.4$ watts **Ans.**

Also copper loss $= I_1^2 R_{01} = (11.36)^2 \times 7.05 = 910.4$ watts **Ans.**

Also copper loss $= I_2^2 R_{02} = (227.3)^2 \times 0.017625 = 910.4$ watts **Ans.**

} The same as expected.

Example 8.14. A 10 kVA, 500/100 V transformer has the following circuit parameters referred to primary; Resistance R_{01} = 0.3 Ω; Reactance X_{01} = 5.2 Ω. When supplying power to a lagging load, the current, power and voltage measured on primary side were 20 A, 8 kW and 500 V respectively. Calculate voltage on secondary terminals under these conditions.

Solution : Primary voltage, V_1 = 500 V

Secondary voltage, V_2 = 100 V

$$\text{Transformation ratio, } K = \frac{V_2}{V_1} = \frac{100}{500} = 0.2$$

Equivalent resistance of transformer referred to secondary,

$$R_{02} = K^2 R_{01} = (0.2)^2 \times 0.3 = 0.012\ \Omega$$

Equivalent reactance of transformer referred to secondary,

$$X_{02} = K^2 X_{01} = (0.2)^2 \times 5.2 = 0.208\ \Omega$$

$$\text{Secondary current, } I_2 = \frac{I_1}{K} = \frac{20}{0.2} = 100\ \text{A}$$

$$\text{Power factor, } \cos\phi = \frac{8 \times 1{,}000}{500 \times 20} = 0.8$$

$$\text{Sin } \phi = \text{Sin } (\cos^{-1} 0.8) = 0.6$$

$$\text{Secondary terminal voltage, } V_2 = KV_1 - I_2 R_{02} \cos\phi - I_2 X_{02} \sin\phi$$

$$= 0.2 \times 500 - 100 \times 0.012 \times 0.8 - 100 \times 0.208 \times 0.6 = 86.56 \text{ V}$$ **Ans.**

8.12. EQUIVALENT CIRCUIT OF A TRANSFORMER

The equivalent circuit of any device can be quite helpful in predetermination of the behaviour of the device under various conditions of operation and it can be drawn if the equations describing its behaviour are known. If any electrical device is to be analysed and investigated further for suitable modifications, its appropriate equivalent circuit is necessary. The equivalent circuit for electromagnetic devices consists of a combination of resistances, inductances, capacitances, voltage etc. Such an equivalent circuit (or circuit model) can, therefore, be analysed and studied easily by the direct application of electric circuit theory.

As stated above, equivalent circuit is simply a circuit representation of the equations describing the performance of the device. Equations (8.10) and (8.11) describe the behaviour of the transformer under load and are helpful in arriving at the transformer equivalent circuit.

Equivalent circuit of a transformer having transformation ratio $K = \frac{E_2}{E_1}$ is shown in Fig. 8.16.

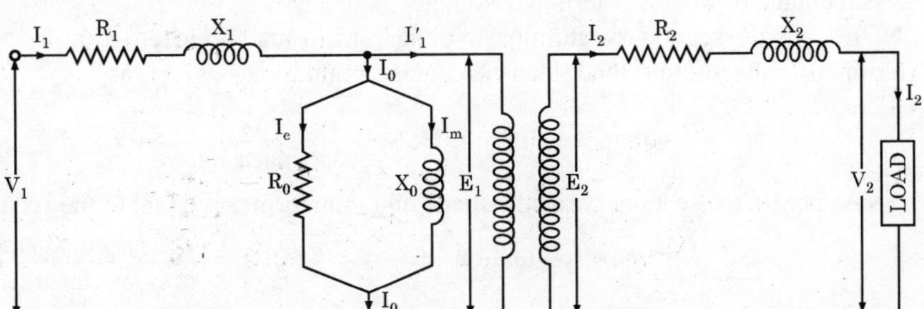

Fig. 8.16. ***Equivalent Circuit of a Transformer***

The induced emf in primary winding E_1 is primary applied voltage V_1 less primary voltage drop. This voltage causes iron loss current I_e and magnetising current I_m and we can, therefore, represent these two components of no-load current by the current drawn by a non-inductive resistance R_0 and pure reactance X_0 having the voltage E_1 or (V_1—primary voltage drop) applied across them, as shown in Fig. 8.16.

Secondary current,

$$I_2 = \frac{I_1'}{K} = \frac{I_1 - I_0}{K}$$

Terminal voltage V_2 across load is induced emf E_2 in secondary winding less voltage drop in secondary winding.

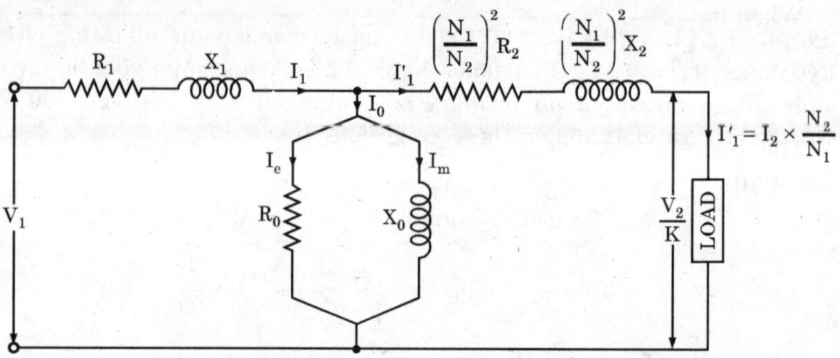

Fig. 8.17. ***Equivalent Circuit of a Transformer With All Secondary Impedances Transferred To Primary Side***

The equivalent circuit can be simplified by transferring the voltage, current and impedance to the primary side. After transferring the secondary voltage, current and impedance to primary side equivalent circuit is reduced to that shown in Fig. 8.17.

The equivalent circuit diagram can further be simplified by transferring the resistance R_0 and reactance X_0 towards left end, as shown in Fig. 8.18. The error introduced by doing so is very small and can be neglected.

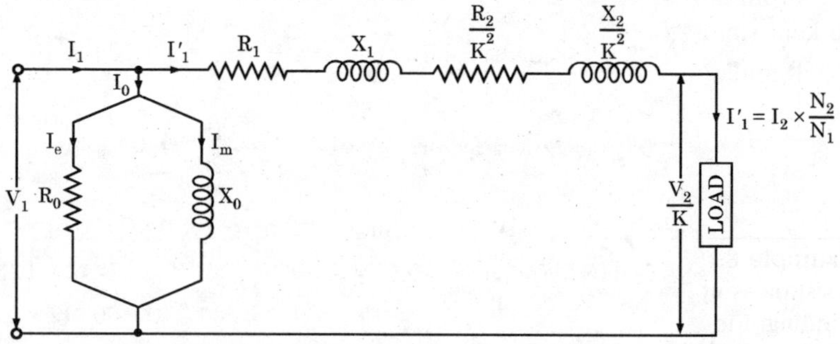

Fig. 8.18. ***Approximate Equivalent Circuit of a Transformer***

No-load current I_0 is hardly 3 to 5 per cent of the full-load rated current, the parallel branch consisting of resistance R_0 and reactance X_0 can be omitted without introducing any appreciable error in the behaviour of the transformer under loaded condition. Such a circuit is shown in Fig. 8.15 (*a*). The equivalent circuit referred to secondary side (neglecting no-load current I_0) is illustrated in Fig. 8.13.

8.13. VOLTAGE REGULATION

The way in which the secondary terminal voltage varies with the load depends upon load current, the internal impedance and the load power factor. The change in secondary terminal voltage from no load to full load, with primary voltage and frequency held constant, is termed the *inherent regulation*.

The figure of merit which determines the voltage characteristic of a transformer is termed as *voltage regulation*. It is usually expressed as a percentage of the rated terminal (secondary) voltage. It is defined as the change in magnitude of terminal (secondary) voltage when full load (rated load) of specified power factor supplied at rated voltage is thrown off (reduced to no load) with primary voltage and frequency held constant, as percentage of the rated terminal voltage.

If V_2 is the secondary (terminal) voltage at full load at specified power factor and E_2 is the secondary (terminal) voltage at no load, then percentage regulation is given as

$$\text{Voltage regulation} = \frac{E_2 - V_2}{\text{Secondary rated voltage}} \times 100$$

As per IS, the secondary rated voltage of a transformer is equal to the secondary voltage at no load *i.e.* E_2

$$\text{So percentage regulation} = \frac{E_2 - V_2}{E_2} \times 100 = \frac{\text{Voltage drop in transformer at full load}}{\text{No-load rated secondary voltage, } E_2} \times 100$$

$$= \frac{I_2 R_{02} \cos\phi + I_2 X_{02} \sin\phi}{E_2} \times 100 \qquad \text{...(8.24)}$$

When the power factor is leading, the percentage regulation is given as

$$\text{Percentage regulation} = \frac{I_2 R_{02} \cos\phi - I_2 X_{02} \sin\phi}{E_2} \times 100 \qquad ...(8.25)$$

The percentage regulation can be given as

$$\% \text{ regulation} = \frac{I_2 R_{02} \cos\phi \pm I_2 X_{02} \sin\phi}{E_2} \times 100$$

+ ve sign for lagging power factor and
– ve sign for leading power factor

$$= \left(\frac{I_2 R_{02}}{E_2} \cos\phi \pm \frac{I_2 X_{02}}{E_2} \sin\phi\right) \times 100$$

$$= (\text{PU resistance} \times \cos\phi \pm \text{PU reactance} \sin\phi) \times 100 \qquad ...(8.26)$$

Voltage regulation of a transformer, on an average, is about 4 per cent.

From the consumer's viewpoint, voltage regulation due to variations in load is undesirable and should be kept smallest possible.

Regulation will be zero if the numerator in Eq. (8.24) will be equal to zero *i.e.*

$$I_2 R_{02} \cos\phi + I_2 X_{02} \sin\phi = 0 \text{ or } \tan\phi = \frac{-R_{02}}{X_{02}} \qquad ...(8.27)$$

The –ve sign indicates that zero regulation occurs at a leading power factor.

Example 8.15. The primary and secondary windings of a 40 kVA, 6,600/250 V single phase transformer have resistances of 10 Ω and 0.02 Ω respectively. The total leakage reactance is 35 Ω as referred to the primary winding. Find full-load regulation at a pf of 0.8 lagging.

Solution : Transformation ratio, $K = \frac{V_2}{V_1} = \frac{250}{6,600} = 0.03788$

Equivalent resistance of transformer referred to secondary,

$$R_{02} = K^2 R_1 + R_2 = (0.03788)^2 \times 10 + 0.02 = 0.03435 \ \Omega$$

Equivalent leakage reactance of transformer referred to secondary,

$$X_{02} = K^2 X_{01} = (0.03788)^2 \times 35 = 0.05022 \ \Omega$$

$$\text{Secondary rated current, } I_2 = \frac{\text{Rated kVA} \times 1,000}{V_2} = \frac{40 \times 1,000}{250} = 160 \text{ A}$$

$$\text{Power factor, } \cos\phi = 0.8 \text{ and } \sin\phi = \sqrt{1 - 0.8^2} = 0.6$$

$$\text{Full load regulation} = \frac{I_2 R_{02} \cos\phi + I_2 X_{02} \sin\phi}{E_2} \times 100$$

$$= \frac{160 \times 0.03435 \times 0.8 + 160 \times 0.05022 \times 0.6}{250} \times 100 = 3.687\% \ \textbf{Ans.}$$

Example 8.16. A 5 kVA, 2,200/220 V single phase transformer has the following parameters.

HV side : $r_1 = 3.4 \ \Omega$, $x_1 = 7.2 \ \Omega$

LV side : $r_2 = 0.028 \ \Omega$, $x_2 = 0.060 \ \Omega$

The transformer is made to deliver rated current at 0.8 lagging power factor to a load connected to LV side. If the load voltage is 220 V, calculate the terminal voltage on the HV side. Also find the voltage regulation at this load.

Solution : Transformation ratio, $K = \frac{220}{2,200} = 0.1$

Equivalent resistance referred to primary,

$$R_{01} = R_1 + \frac{R_2}{K^2} = 3.4 + \frac{0.028}{(0.1)^2} = 6.2\ \Omega$$

Equivalent reactance referred to primary,

$$X_{01} = X_1 + \frac{X_2}{K^2} = 7.2 + \frac{0.060}{(0.1)^2} = 13.2\ \Omega$$

$$\text{Rated current on primary side, } I_1 = \frac{\text{Rated kVA} \times 1{,}000}{V_1} = \frac{5 \times 1{,}000}{2{,}200} = 2.273 \text{ A}$$

$$\text{Power factor, } \cos\phi = 0.8 \text{ (lagging)}$$

$$\text{Sin}\,\phi = \sqrt{(1-\cos^2\phi)} = \sqrt{1-(0.8)^2} = 0.6$$

$$\text{Terminal voltage on H V side, } V_1 = \frac{V_2}{K} + I_1 R_{01}\cos\phi + I_1 X_{o1}\sin\phi$$

$$= \frac{220}{0.1} + 2.273 \times 6.2 \times 0.8 + 2.273 \times 13.2 \times 0.6$$

$$= 2{,}200 + 11.28 + 18 = 2{,}229.28 \text{ V} \quad \textbf{Ans.}$$

$$\text{Full-load regulation} = \frac{2{,}229.28 - 2{,}200}{2{,}200} \times 100 = 1.33\% \quad \textbf{Ans.}$$

Example 8.17. A single phase transformer on full load has an impedance drop of 20 V and resistance drop of 10 V. Calculate the value of power factor when its regulation will be zero.

Solution : Regulation of a transformer is given as $\dfrac{I_2 R_{02}\cos\phi + I_2 X_{02}\sin\phi}{E_2}$.

Regulation will be zero when the numerator of the expression for regulation is zero

i.e. $$I_2 R_{02}\cos\phi + I_2 X_{02}\sin\phi = 0 \quad ...(i)$$

$$\text{Resistance drop, } I_2 R_{02} = 10 \text{ V}$$

$$\text{Impedance drop, } I_2 Z_{02} = 20 \text{ V}$$

$$\text{Reactance drop, } I_2 X_{02} = \sqrt{(I_2 Z_{02})^2 - (I_2 R_{02})^2} = \sqrt{20^2 - 10^2} = 17.32 \text{ V}$$

Substituting $I_2 R_{02} = 10$ V and $I_2 X_{02} = 17.32$ V in Eq. (*i*) we have

$$10\cos\phi + 17.32\sin\phi = 0$$

$$\text{or phase angle, } \phi = \text{Tan}^{-1}\frac{-10}{17.32} = -30^\circ \text{ or } 30^\circ \text{ (leading)}$$

$$\text{Power factor} = \text{Cos}\,\phi = \text{Cos } 30^\circ = 0.866 \text{ (leading)} \quad \textbf{Ans.}$$

Example 8.18. Calculate the regulation of a transformer in which ohmic drop is 1% and reactance drop 5 % of the voltage at the following power factors (*i*) 0.8 lagging (*ii*) 0.8 leading.

Solution :

$$\text{Percentage resistive drop} = \frac{I_2 R_{02}}{E_2} \times 100 = 1$$

$$\text{Percentage reactive drop} = \frac{I_2 X_{02}}{E_2} \times 100 = 5$$

(*i*) When power factor $\cos\phi = 0.8$ lagging; $\sin\phi = 0.6$

$$\text{Voltage regulation} = \frac{I_2 R_{02}\cos\phi + I_2 X_{02}\sin\phi}{E_2} \times 100$$

$$= \frac{I_2 R_{02}}{E_2} \times 100\cos\phi + \frac{I_2 X_{02}}{E_2} \times 100 \times \sin\phi = 1 \times 0.8 + 5 \times 0.6 = 3.8\% \quad \textbf{Ans.}$$

(*ii*) When power factor $\cos \phi = 0.8$ (leading); $\sin \phi = -0.6$

$$\text{Voltage regulation} = \frac{I_{02}\, R_{02}}{E_2} \times 100 \cos \phi + \frac{I_2\, X_{02}}{E_2} \times 100 \sin \phi = 1 \times 0.8 + 5 \times (-0.6) = -2.2\ \%\ \textbf{Ans.}$$

8.14. TRANSFORMER LOSSES

1. Iron or Core Losses. Iron loss is caused by the alternating flux in the core and consists of hysteresis and eddy current losses.

(*a*) Hysteresis Loss. The core of a transformer is subjected to an alternating magnetizing force and for each cycle of emf a hysteresis loop is traced out. The hysteresis loss per second is given by the equation

$$\text{Hysteresis loss, } P_h = \eta' (B_{max})^x f\, v \text{ joules per second or watts} \quad ...(8.28)$$

where f is the supply frequency in Hz, v is the volume of core in cubic metres, η' is the hysteresis coefficient, B_{max} is peak value of flux density in the core and x lies between 1.5 and 2.5 depending upon the material and is often taken as 1.6.

(*b*) Eddy Current Loss. We have seen that whenever the flux linkage with a closed electric circuit changes, an emf is induced in the circuit and a current flows, the value of which depends on the emf around the circuit and the resistance of the circuit. It is not necessary that the circuit be a wire and that the flux passes entirely through it. If a solid block of metal is traversed by a varying flux, metallic circuits in the block itself, which are linked by the flux, will carry current. If the magnetic circuit is made up of iron and if the flux in the circuit is variable, currents will be induced by induction in the iron circuit itself. All such currents are known as *eddy currents.*

The eddy current loss is given by expression

$$P_e = K_e (B_{max})^2 f^2 t^2 \text{ v watts} \quad ...(8.29)$$

2. Copper or Ohmic Losses. These losses occur due to ohmic resistance of the transformer windings. If I_1 and I_2 are the primary and secondary currents respectively and R_1 and R_2 are the respective resistances of primary and secondary windings then copper losses occurring in primary and secondary windings will be $I_1^2R_1$ and $I_2^2R_2$ respectively. So total copper losses will be $(I_1^2R_1 + I_2^2R_2)$. These losses vary as the square of the load current or kVA. For example if the copper losses at full load are P_c then copper losses at one-half or one-third of full load will be $\left(\frac{1}{2}\right)^2 P_c$ or $\left(\frac{1}{3}\right)^2 P_c$ *i.e.* $\frac{P_c}{4}$ or $\frac{P_c}{9}$ respectively.

Copper losses are determined on the basis of constant equivalent resistance R_{eq} determined from the short-circuit test (refer to Art. 8.16) and then corrected to 75°C (since the standard operating temperature of electrical machines is taken 75°C).

8.15. OPEN-CIRCUIT TEST (OR NO-LOAD TEST)

The purpose of this test is to determine the core (or iron or excitation) loss, P_i and no-load current I_0 and thereby the shunt branch parameters R_0 and X_0 of the equivalent circuit.

In this test, one of the windings (usually high voltage winding) is kept open circuited and the rated voltage at rated frequency is applied to the other winding, as shown in Fig. 8.19. No doubt, the core loss will be the same whether the measurements are made on *lv* winding or *hv* winding so long as the rated voltage of that winding is applied to it but in case the measurements are made on *hv* winding, the voltage required to be applied would often be inconveniently large while the current I_0 would be inconveniently small.

Either an auto-transformer or a voltage divider (VD) is used for varying the voltage applied to the low-voltage winding. Ammeter A and wattmeter W are connected to measure no-load current I_0 and input power P_0. Voltmeter V is connected to measure the applied voltage.

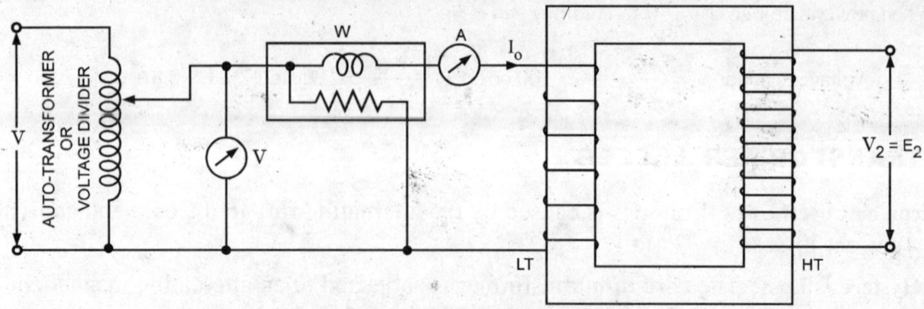

Fig. 8.19. ***Circuit Diagram For Open-Circuit Test***

Since no current flows in the open-circuited secondary, the current in the primary will be merely that necessary to magnetize the core at normal voltage. Moreover, this magnetising current is a very small fraction of the full-load current (usually 3 to 10% of full-load current) and may be neglected as far as the copper loss is concerned consequently, the test gives core loss alone practically.

With normal voltage applied to the primary, normal flux will be set up in the core and, therefore, normal iron (or core) loss will occur which are recorded by a wattmeter W.

The open-circuit test gives enough data to compute the equivalent circuit constants R_0, X_0, no-load power factor cos ϕ_0, no-load current I_0 and no-load power loss (iron loss) of a transformer.

$$\text{Iron loss, } P_i = \text{Input power on no load} = P_0 \text{ watts (say)}$$

$$\text{No-load current} = I_0 \text{ amperes}$$

$$\text{Applied voltage to primary} = V_1 \text{ volts}$$

$$\text{Angle of lag, } \phi_0 = \text{Cos}^{-1} \frac{P_0}{V_1 I_0} \quad \text{...(8.30)}$$

$$\text{No-load current energy component, } I_e = I_0 \cos \phi_0 = \frac{P_0}{V_1} \quad \text{...(8.31)}$$

$$\text{No-load current magnetizing component, } I_m = \sqrt{I_0^2 - I_e^2} \quad \text{...(8.32)}$$

$$\text{Equivalent circuit parameter, } R_0 = \frac{V_1}{I_e} = \frac{V_1^2}{P_0} \quad \text{...(8.33)}$$

$$\text{Equivalent circuit parameter, } X_0 = \frac{V_1}{I_m} = \frac{V_1}{\sqrt{I_0^2 - I_e^2}} \quad \text{...(8.34)}$$

Notes 1. Since no-load current I_0 is very small, pressure coils of wattmeter and the voltmeter should be connected such that the currents drawn by them do not flow through the current coils of the wattmeter and ammeter.

2. Since power factor at no-load is quite low (in the range of 0.1 – 0.2 lag), a low power factor wattmeter should be used to ensure accurate measurements.

3. The error due to power loss in ammeter can be eliminated by short circuiting the ammeter while reading wattmeter.

4. Sometimes a high resistance voltmeter is connected across the secondary to indicate the emf induced in the secondary (hv winding). This helps in determination of transformation ratio K.

5. It must, however, be remembered that in making this test, hv side is hot and, therefore, its terminals must be properly insulated.

8.16. SHORT-CIRCUIT TEST (OR IMPEDANCE TEST)

The purpose of this test is to determine full-load copper loss and equivalent resistance and equivalent reactance referred to metering side.

In this test, the terminals of secondary winding (usually of low-voltage winding*) are short circuited by a thick wire or strip or through an ammeter (which may serve the additional purpose of indicating secondary rated load current) and variable low voltage is applied to the primary through an auto-transformer or potential divider, as shown in Fig. 8.20. The transformer now becomes equivalent to a coil having an impedance equal to impedance of both the windings.

The applied voltage, V_s to the primary is gradually increased till the ammeter A indicates the full-load (rated) current of the metering side. Since applied voltage is very low (5 – 8% of the rated voltage), flux linking with the core is very small and, therefore, iron losses are so small that these can be neglected. Thus the power input (reading of wattmeter W) gives total copper loss at rated load, output being nil. Let the readings of voltmeter, ammeter and wattmeter be V_s, I_s and W_s respectively.

$$\text{Full-load copper loss, } P_c = I_s^2 R_{eq} = W_s \quad ...(8.35)$$

$$\text{Equivalent resistance, } R_{eq} = \frac{W_s}{I_s^2} \quad ...(8.36)$$

$$\text{Equivalent impedance, } Z_{eq} = \frac{V_s}{I_s} \quad ...(8.37)$$

$$\text{Equivalent reactance, } X_{eq} = \sqrt{Z_{eq}^2 - R_{eq}^2} \quad ...(8.38)$$

The above values are referred to the metering side (high voltage side in above case). If desired, the values could be easily determined referred to the other side, as discussed in Art. 8.11.

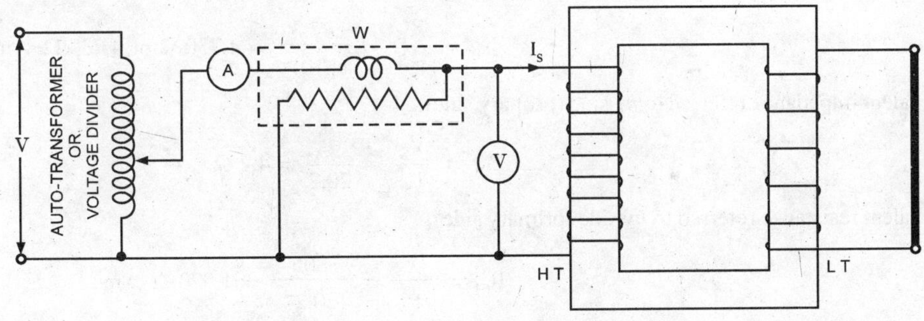

Fig. 8.20. *Circuit Diagram For Short-Circuit Test*

* Voltage required for the short-circuit test is about 5 per cent of the rated value. For a 200 kVA, 2200/220 V transformer, test on high voltage side would need the voltage (to be applied) of $2{,}200 \times \frac{5}{100}$ *i.e.* 110 V (which is standard voltage for instrument coils) and a current of $\frac{200 \times 1{,}000}{2{,}200} = 91$ A.

If the test is conducted on low voltage side of the above transformer, the voltage needed would be $220 \times \frac{5}{100} = 11$ V and the current would be $\frac{200 \times 1{,}000}{220} = 910$ A (very high). At this low voltage, high precision would not be readily obtainable with ordinary instruments.

Thus we see that if the measurements were made on low voltage side, the voltage needed would be inconveniently low, while the current would often be inconveniently large.

Example 8.19. Given below are the results conducted on a 50 kVA, 2200/220 V transformer.

Open-Circuit (LV Side) Test: 405 W, 5 A, 220 V

Short-Circuit (HV Side) Test: 805 W, 20.2 A, 95 V

Calculate the parameters of the equivalent circuit.

[Pb. Technical Univ. Basic Electrical Engineering December 2003]

Solution :

$$\text{Transformation ratio, } K = \frac{V_2}{V_1} = \frac{220}{2{,}200} = 1/10$$

$$\text{No-load current, } I_0 = 5 \text{ A}$$

$$\text{No-load input power, } P_0 = 405 \text{ W}$$

$$\text{Secondary (}lv\text{ side) voltage, } V_2 = 220 \text{ V}$$

$$\text{No-load power factor, } \cos\phi_0 = \frac{P_0}{V_2 I_0} = \frac{405}{220 \times 5} = 0.3682 \text{ (lag)}$$

$$\text{Energy-component of no-load current, } I_e = I_0 \cos\phi_0 = 5 \times 0.3682 = 1.84 \text{ A}$$

$$\text{Magnetising-component of no-load current, } I_m = \sqrt{I_0^2 - I_e^2} = \sqrt{5^2 - 1.84^2} = 4.65 \text{ A}$$

$$\text{Exciting resistance referred to } lv \text{ side, } R_0 = \frac{V_2}{I_e} = \frac{220}{1.84} = 119.5\ \Omega$$

$$\text{Magnetising reactance referred to } lv \text{ side, } X_0 = \frac{V_2}{I_m} = \frac{220}{4.65} = 47.3\ \Omega$$

Exciting resistance referred to *hv* side (primary side)

$$R_0' = \frac{R_0}{K^2} = \frac{119.5}{(1/10)^2} = 11{,}950\ \Omega \text{ or } 11.95\ k\Omega \textbf{ Ans.}$$

Magnetising reactance referred to *hv* side (primary side),

$$X_0' = \frac{X_0}{K^2} = \frac{47.3}{(1/10)^2} = 4{,}730\ \Omega \text{ or } 4.73\ k\Omega \textbf{ Ans.}$$

Equivalent impedance referred to *hv* side (primary side),

$$Z_{01} = \frac{V_S}{I_S} = \frac{95}{20.2} = 4.7\ \Omega$$

Equivalent resistance referred to *hv* side (primary side),

$$R_{01} = \frac{W_S}{I_S^2} = \frac{805}{(20.2)^2} = 1.973\ \Omega \textbf{ Ans.}$$

Equivalent reactance referred to *hv* side (primary side),

$$X_{01} = \sqrt{(Z_{01})^2 - (R_{01})^2} = \sqrt{4.7^2 - 1.973^2} = 4.266\ \Omega \textbf{ Ans.}$$

8.17. DETERMINATION OF REGULATION OF TRANSFORMER FROM OPEN-CIRCUIT AND SHORT-CIRCUIT TESTS

Percentage regulation is given as

$$\text{\%age regulation} = \frac{I_2 R_{02} \cos\phi \pm I_2 X_{02} \sin\phi}{\text{No-load rated secondary voltage}} \times 100 \qquad \text{...Refer to Eq. (8.26)}$$

Equivalent resistance R_{02} and reactance X_{02} referred to secondary can be determined from short-circuit test, as explained in Art. 8.16. I_2 and $\cos\phi$ are the load current and power factor (lagging or leading) of the load, so known. No-load secondary terminal voltage is equal to emf induced in the secondary E_2.

$$\text{So percentage regulation} = \frac{I_2 R_{02} \cos\phi \pm I_2 X_{02} \sin\phi}{E_2} \times 100$$

Note: *Open-circuit test datas are not needed for determination of voltage regulation of transformer.*

Example 8.20. A short-circuit test when performed on the hv side of a 10 kVA 2,000/400 V single phase transformer, gave the following data — 60 V, 4 A, 100 W

If the lv side is delivering full-load current at 0.8 pf lag and at 400 V, find the applied voltage to hv side.

Solution : Equivalent impedance referred to primary, $Z_{01} = \frac{V_s}{I_s} = \frac{60}{4} = 15\ \Omega$

∵ Short-circuit test has been conducted on the hv (primary) side

$$\text{Equivalent resistance referred to primary, } R_{01} = \frac{P_s}{I_s^2} = \frac{100}{4^2} = 6.25\ \Omega$$

$$\text{Equivalent reactance referred to primary, } X_{01} = \sqrt{(Z_{01})^2 - (R_{01})^2} = \sqrt{(15)^2 - (6.25)^2} = 13.64\ \Omega$$

$$\text{Full-load primary current, } I_1 = \frac{\text{KVA (rated)} \times 1{,}000}{V_1} = \frac{10 \times 1{,}000}{2{,}000} = 5\ \text{A}$$

Total voltage drop in transformer winding as referred to primary

$$= I_1 (R_{01} \cos\phi + X_{01} \sin\phi)$$

$$= 5\ (6.25 \times 0.8 + 13.64 \times 0.6) = 65.9\ \text{V}$$

Applied voltage to the hv side = Load voltage + voltage drop

$$= 2{,}000 + 65.9 = 2{,}065.9\ \text{V}$$ **Ans.**

8.18. TRANSFORMER EFFICIENCY

The rated capacity of a transformer is defined as the product of rated voltage and full-load (rated) current on the output side. The power output depends upon the power factor of the load.

The efficiency (η) of a transformer, like that of any other apparatus, is defined as the ratio of useful power output to the input power, the two being measured in same units (either in watts or killowatts).

$$\textit{i.e. } \text{Transformer efficiency, } \eta = \frac{\text{Output}}{\text{Input}} = \frac{\text{Output}}{\text{Output + losses}}$$

$$= \frac{\text{Output}}{\text{Output + iron loss + copper loss}}$$

$$= 1 - \frac{\text{Iron loss + copper loss}}{\text{Output + iron loss + copper loss}}$$

Now power output $= V_2 I_2 \cos\phi$

where V_2 is the secondary terminal voltage on load, I_2 is the secondary current at load and $\cos\phi$ is the power factor of the load.

$$\text{Iron loss, } P_i = \text{Hysteresis loss + eddy current loss}$$

$$\text{Copper loss} = I_1^2 R_1 + I_2^2 R_2 = I_1^2 R_{01} = I_2^2 R_{02}$$

8.18.1. Determination of Transformer Efficiency

The ordinary transformer has a very high efficiency (in the range of 96-99%). Hence the transformer efficiency cannot be determined with high precision by direct measurement of output and input, since the losses are of the order of only 1-4 %. The difference between the readings of output and input instruments is then so small that an instrument error as low as 0.5% would cause an error of the order of 15% in the losses. Further, it is inconvenient and costly to have the necessary loading devices of the correct current and voltage ratings and power factor to load the transformer. There is also a wastage of large amount of power (equal to that of power output + losses) and no information is available from such a test about the proportion of copper and iron losses.

The best and accurate method of determination of efficiency of a transformer would be to compute losses from open-circuit and short-circuit tests and determine efficiency as follows :

Iron loss, $P_i = W_0$ or P_0, determined from open-circuit test

Copper loss at full load, $P_c = W_s$ or P_s , determined from short-circuit test

Copper loss at a load x times full load $= I_2^2 R_{02} = x^2 P_c$

where x is the ratio of load current I_2 to secondary full-load secondary current.

$$\text{and transformer efficiency, } \eta = \frac{V_2 I_2 \cos\phi}{V_2 I_2 \cos\phi + P_i + x^2 P_c} \qquad ...(8.39)$$

In Eq. (8.39) the effect of instrument readings is confined to losses only so that overall efficiency obtained from it is far more accurate than that obtained by direct loading. Another great advantage of this method is that it is not necessary for the transformer to be loaded to its full-load rating during testing, and the kW rating of the test plant need be equal to the value of the individual transformer losses.

8.18.2. Condition For Maximum Efficiency

From Eq. (8.39)

$$\text{Transformer efficiency, } \eta = \frac{V_2 I_2 \cos\phi}{V_2 I_2 \cos\phi + P_i + x^2 P_c} = \frac{xP}{xP + P_i + x^2 P_c} \qquad ...(8.40)$$

where P is equal to the full-load (rated output in volt-amperers or kVA) × load power factor ($\cos\phi$), P_i is the total iron loss, P_c is the full-load copper loss and x is the fraction of full-load kVA at which efficiency is maximum.

Differentiating both sides of Eq. (8.40) we get

$$\frac{d\eta}{dx} = \frac{(xP + x^2 P_c + P_i)P - xP(P + 2xP_c)}{(xP + P_i + x^2 P_c)^2} = \frac{P(P_i - x^2 P_c)}{(xP + x^2 P_c + P_i)^2}$$

Efficiency η will be maximum if $\frac{d\eta}{dx} = 0$

$$\text{or } P_i - x^2 P_c = 0 \text{ or } x^2 P_c = P_i \qquad ...(8.41)$$

Hence the efficiency will be maximum when variable loss (copper loss) is equal to constant loss (iron loss).

$$\text{Output kVA corresponding to maximum efficiency} = x \times \text{Rated kVA} = \text{Rated kVA} \times \sqrt{\frac{P_i}{P_c}} \qquad ...(8.42)$$

Since from Eq. (8.41) $x = \sqrt{\frac{P_i}{P_c}}$

Output current I_2 corresponding to maximum efficiency is determined as below :

Copper loss at given load at which efficiency is maximum $= I_2^2 R_{02}$

and since for maximum efficiency to occur it is necessary that copper loss equals P_i,

$$I_2^2 R_{02} = P_i$$

$$\text{So current corresponding to maximum efficiency, } I_2 = \sqrt{\frac{P_i}{R_{02}}} \qquad ...(8.43)$$

Power transformers employed for bulk power transmission are operated continuously near about full load and are, therefore, designed to have maximum efficiency at full load. On the other hand, the distribution transformers, which supply load varying over the day through a wide range, are designed to have maximum efficiency at about three-fourths the full load.

It is important to appreciate that, since copper loss depends on current and the iron loss depends on voltage, the total loss in the transformer depends on the volt-ampere product, and not on the phase angle between voltage and current, *i.e.* is independent of the load power factor. The transformers are, therefore, rated in kilo-volt amperes (kVA) and not in kilowatts.

Example 8.21. A 4 kVA, 220/440 V, 50 Hz, single phase transformer gave the following test figures.

No-load test performed on 220 V side keeping 440 V side open: 220 V, 0.7 A, 60 W.

Short-circuit test performed short circuiting the 440 V side through an ammeter : 9 V, 6 A, 21.6 W.

Calculate (*a*) the magnetising current and the component corresponding to iron loss at normal voltage and frequency (*b*) The efficiency on full-load at unity power factor and the corresponding secondary terminal voltage (*c*) Draw the phasor diagram corresponding to full-load operation.

[*Pb. Technical Univ. Electrical Engineering January 2000*]

Solution : Primary (lv) voltage, $V_1 = 220$ V

$$\text{No-load power factor, } \cos\phi_0 = \frac{P_0}{V_1 I_0} = \frac{60}{220 \times 0.7} = 0.3896 \text{ (lagging)}$$

(*a*) Component of current corresponding to iron loss at normal voltage and frequency,

$$I_e = I_0 \cos\phi_0 = 0.7 \times 0.3896 = 0.2727 \text{ A} \quad \textbf{Ans.}$$

$$\text{Magnetising current, } I_m = \sqrt{I_0^2 - I_e^2} = \sqrt{(0.7)^2 - (0.2727)^2} = 0.6447 \text{ A} \quad \textbf{Ans.}$$

(*b*)

$$\text{Full-load secondary current, } I_2 = \frac{\text{S in kVA} \times 1{,}000}{V_2} = \frac{4 \times 1{,}000}{440} = 9.091 \text{ A}$$

$$\text{Full-load primary current, } I_1 = \frac{4 \times 1{,}000}{220} = 18.182 \text{ A}$$

$$\text{Full-load copper loss, } P_c = \left(\frac{I_1}{I_s}\right)^2 \times P_c = \left(\frac{18.182}{6}\right)^2 \times 21.6 = 198.35 \text{ W}$$

$$\text{Full-load efficiency, } \eta = \frac{\text{Output in kVA} \times \cos\phi}{\text{Output in kVA} \times \cos\phi + P_c + P_1} \times 100$$

$$= \frac{4 \times 1{,}000 \times 1}{4 \times 1{,}000 \times 1 + 198.35 + 60} \times 100$$

$$= \frac{4{,}000}{4{,}000 + 258.35} \times 100 = 93.93\% \quad \textbf{Ans.}$$

$$\text{Equivalent impedance referred to primary, } Z_{01} = \frac{V_s}{I_s} = \frac{9}{6} = 1.5\ \Omega$$

$$\text{Equivalent resistance referred to primary, } R_{01} = \frac{P_s}{I_s^2} = \frac{21.6}{6^2} = 0.6\ \Omega$$

$$\text{Equivalent impedance referred to secondary, } Z_{02} = Z_{01} \times \left(\frac{V_2}{V_1}\right)^2 = 1.5 \times \left(\frac{440}{220}\right)^2 = 6\ \Omega$$

$$\text{Equivalent resistance referred to secondary, } R_{02} = R_{01} \times \left(\frac{V_2}{V_1}\right)^2 = 0.6 \times \left(\frac{440}{220}\right)^2 = 2.4\ \Omega$$

$$\text{Equivalent reactance referred to secondary, } X_{02} = \sqrt{(Z_{02})^2 - (R_{02})^2} = \sqrt{6^2 - 2.4^2} = 5.5\ \Omega$$

Since power factor, $\cos\phi_2 = 1.0$, $\sin\phi_2 = 0$

$$\text{and } E_2^2 = (V_2 + I_2 R_{02})^2 + (I_2 X_{02})^2$$

$$\text{So secondary voltage, } V_2 = \sqrt{E_2^2 - (I_2 X_{02})^2} - I_2 R_{02}$$

$$= \sqrt{440^2 - (9.091 \times 5.5)^2} - 9.091 \times 2.4 = 415.9 \text{ V} \quad \textbf{Ans.}$$

Phasor diagram is shown in Fig. 8.21.

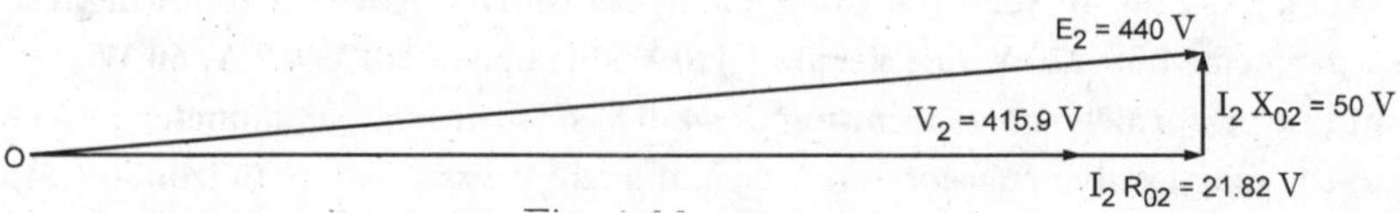

Fig. 8.21

Example 8.22. A 5 kVA, 1000/200 V, 50 Hz single phase transformer has the following test results:

Open-circuit test (Low voltage side) : 200 V, 1.2 A, 90 W

Short-circuit test (High voltage side) : 50 V, 5 A, 110 W

Calculate iron losses, copper losses, resistance of two windings of transformer and efficiency at full-load and 0.8 pf lagging. [*Pb Technical Univ. Basic Electrical Engineering June 2003*]

Solution : Full-load primary current, $I_1 = \dfrac{\text{S in kVA} \times 1{,}000}{V_1} = \dfrac{5 \times 1{,}000}{1{,}000} = 5 \text{ A}$

Iron loss, P_i = Power input on no load with rated voltage applied to the winding

= 90 W **Ans.**

Copper loss corresponding to short-circuit current of 5 A

= Wattmeter reading at short-circuit test = 110 W

Since full-load current is equal to 5 A *i.e.* equal to short-circuit current,

Full-load copper loss, P_C = 110 W **Ans.**

Resistance of the two windings of the transformer referred to the *hv* side *i.e.*

$$R_{01} = \frac{\text{Wattmeter reading at short circuit}}{\text{Short-circuit current}} = \frac{110}{5^2} = 4.4\ \Omega \quad \textbf{Ans.}$$

Efficiency of transformer at full-load and 0.8 *pf* lagging,

$$\eta = \frac{\text{Output in VA} \times \cos\phi}{\text{Output in VA} \times \cos\phi + P_i + P_C} \times 100$$

$$= \frac{5 \times 1{,}000 \times 0.8}{5 \times 1{,}000 \times 0.8 + 90 + 110} \times 100 = 95.238\% \quad \textbf{Ans.}$$

Example 8.23. The parameters of the equivalent circuit of a 150 kVA, 2400/240 V single phase transformer are as follows:

Primary resistance = 0.2 Ω ; **Secondary resistance = 2 mΩ ;**

Primary leakage reactance = 0.45 Ω ; **Secondary leakage reactance = 4.5 mΩ ;**

Core-loss resistance = 10 kΩ ; **Magnetising reactance = 1.55 kΩ**

Using the circuit referred to primary, determine the (*a*) voltage regulation (*b*) efficiency of the transformer operating at rated load with 0.8 lagging power factor. [*Pb. Technical Univ. Electrical Engineering December 2000*]

Solution : Transformer rating, S = 150 kVA

$$\text{Transformation ratio, } K = \frac{E_2}{E_1} = \frac{240}{2{,}400} = 0.1$$

$$\text{Total resistance referred to primary, } R_{01} = R_1 + \frac{R_2}{K^2} = 0.2 + \frac{2 \times 10^{-3}}{(0.1)^2} = 0.4\ \Omega$$

$$\text{Total reactance referred to primary, } X_{01} = X_1 + \frac{X_2}{K^2} = 0.45 + \frac{4.5 \times 10^{-3}}{(0.1)^2} = 0.9\ \Omega$$

$$\text{Primary full-load current, } I_1 = \frac{S \times 1{,}000}{E_1} = \frac{150 \times 1{,}000}{2{,}400} = 62.5\ \text{A}$$

$$\text{Operating power factor, } \cos\phi = 0.8\ \text{(lagging)}$$

$$\text{Sin}\ \phi = \text{Sin}\ \cos^{-1} 0.8 = 0.6$$

$$\text{Applied voltage to primary, } V_1 = \sqrt{(E_1 \cos\phi + I_1 R_{01})^2 + (E_1 \sin\phi + I_1 X_{01})^2}$$

$$= \sqrt{(2{,}400 \times 0.8 + 62.5 \times 0.4)^2\ (2{,}400 \times 0.6 + 62.5 \times 0.9)^2}$$

$$\text{or } V_1 = \sqrt{(1{,}945)^2 + (1{,}496.25)^2} = 2{,}453.9\ \text{V}$$

$$(a)\ \text{Voltage regulation} = \frac{V_1 - E_1}{V_1} \times 100 = \frac{2{,}453.9 - 2{,}400}{2{,}453.9} \times 100 = 2.1965\%.\ \textbf{Ans.}$$

$$\text{Full-load output, } P_{out} = S \times \cos\phi = 150 \times 0.8 = 120\ \text{k W}$$

$$\text{Full-load copper loss, } P_c = I_1^2\ R_{o1} = (62.5)^2 \times 0.4 = 1{,}562.5\ \text{W or } 1.5625\ \text{k W}$$

$$\text{Iron loss, } P_i = V_1\ I_e = V_1 \times \frac{V_1}{R_0} = 2{,}400 \times \frac{2{,}400}{10 \times 10^3} = 576\ \text{W or } 0.576\ \text{kW}$$

Efficiency of transformer at rated load with 0.8 lagging power factor,

$$\eta = \frac{P_{out}}{P_{out} + P_c + P_i} \times 100 = \frac{120}{120 + 1.5625 + 0.576} \times 100$$

$$= 98.25\%.\ \textbf{Ans.}$$

Example 8.24. The primary and secondary windings of a 500 kVA, 1-ph transformer have resistances of 0.4 Ω and 0.0015 Ω respectively. The primary and secondary voltages are 6,000 V and 400 V respectively and the iron loss is 3.2 kW. Calculate the efficiency on (*i*) full-load and (*ii*) half load, assuming the pf of the load to be 0.8.

Solution :

$$\text{Rated output, } S = 500\ \text{kVA}$$

$$\text{Secondary voltage, } V_2 = 400\ \text{V}$$

$$\text{Full-load (or rated) secondary current, } I_2 = \frac{S \times 1{,}000}{V_2} = \frac{500 \times 1{,}000}{400} = 1{,}250\ \text{A}$$

$$\text{Total resistance referred to secondary, } R_{02} = K^2 R_1 + R_2 = \left(\frac{400}{6{,}000}\right)^2 \times 0.4 + 0.0015 = 0.003278\ \Omega$$

$$\text{Full-load copper loss, } P_c = I_2^2 R_{02} = (1{,}250)^2 \times 0.003278 = 5{,}122\ \text{W or } 5.122\ \text{kW}$$

$$\text{Iron loss, } P_i = 3.2\ \text{kW}$$

$$(i)\ \text{Full-load output at 0.8 pf} = S \times \cos\phi = 500 \times 0.8 = 400\ \text{kW}$$

$$\text{Transformer efficiency at full-load and 0.8 pf, } \eta = \frac{\text{Output}}{\text{Output} + P_i + P_c} \times 100$$

$$= \frac{400}{400 + 3.2 + 5.122} \times 100 = 97.962\%\ \textbf{Ans.}$$

$$(ii)\ \text{Output at half load and 0.8 pf} = \frac{1}{2}\ S \times \cos\phi = \frac{1}{2} \times 500 \times 0.8 = 200\ \text{kW}$$

$$\text{Transformer efficiency at half load and 0.8 pf} = \frac{\text{Output}}{\text{Output} + P_i + \left(\frac{1}{2}\right)^2 P_c} \times 100$$

$$= \frac{200}{200 + 3.2 + \frac{1}{4} \times 5.122} \times 100 = 97.81\% \textbf{ Ans.}$$

Example 8.25. In a 25 kVA, 2000/200 V transformer, the constant and variable losses are 350 W and 400 W respectively. Calculate the efficiency on unity power factor at (*i*) full load and (*ii*) half full-load.

[Pb. Technical Univ. Electrical Engineering 1999]

Solution :

Rating of transformer, $S = 25$ kVA

Constant or iron loss, $P_i = 350 \text{ W} = 0.35 \text{ kW}$

Variable or copper loss on full-load, $P_C = 400 \text{ W} = 0.4 \text{ kW}$

Power factor, $\cos \phi = 1$

(*i*) At full load and at unity power factor

Output power, $P = S \times \text{power factor} = 25 \times 1.0 = 25 \text{ kW}$

$$\text{Efficiency, } \eta = \frac{P}{P + P_i + P_C} \times 100 = \frac{25}{25 + 0.35 + 0.4} \times 100 = 97.09\% \textbf{ Ans.}$$

(*ii*) At half full-load and unity power factor

$$\text{Output power, } P = \frac{1}{2} \times S \times \text{pf} = \frac{1}{2} \times 25 \times 1.0 = 12.5 \text{ kW}$$

$$\text{Total losses} = P_i + \left(\frac{1}{2}\right)^2 P_C = 0.35 + \left(\frac{1}{2}\right)^2 \times 0.4 = 0.45 \text{ kW}$$

$$\text{Efficiency, } \eta = \frac{12.5}{12.5 + 0.45} \times 100 = 96.53\% \textbf{ Ans.}$$

Example 8.26. In a 50 kVA, 11 kV/400 V transformer, the iron and copper losses are 500 W and 600 W respectively under rated conditions. Calculate the efficiency on unity power factor at full load. Find the load for maximum efficiency and the iron and copper losses corresponding to this load.

Solution : Full-load copper loss, $P_c = 600$ watts or 0.6 kW

Iron loss, $P_i = 500$ watts or 0.5 kW

At full load and at unity power factor

Output, P = Rated kVA × power factor = 50 × 1 = 50 kW

$$\text{Efficiency, } \eta = \frac{P}{P + P_i + P_c} \times 100 = \frac{50}{50 + 0.5 + 0.6} \times 100 = 97.85\ \% \textbf{ Ans.}$$

Let the transformer maximum efficiency occur at x times full-load

Since maximum efficiency occurs when

$$\text{Copper loss, } x^2 P_c = \text{Iron loss, } P_i$$

$$\text{so } x = \sqrt{\frac{P_i}{P_c}} = \sqrt{\frac{500}{600}} = 0.9129$$

Load in kVA corresponding to maximum efficiency

$$= x \times \text{rated kVA} = 0.9129 \times 50 = 45.645 \text{ kVA} \textbf{ Ans.}$$

Copper and iron losses will be equal at a load corresponding to maximum efficiency so each of copper and iron loss will be equal to 500 W **Ans.**

Example 8.27. A 600 kVA, 1-ph transformer has an efficiency of 92 % both at full load and half-load at unity power factor. Determine its efficiency at 60 % of full-load at 0.8 pf lag.

Solution : Let the iron loss and full-load copper loss be P_i and P_c respectively.

At full load and unity power factor

$$\text{Output, P} = 600 \times 1.0 = 600 \text{ kW}$$

$$\text{Transformer efficiency, } \eta = 92\% \text{ or } 0.92$$

$$\text{Power input} = \frac{P}{\eta} = \frac{600}{0.92} = 652.174 \text{ kW}$$

$$\text{Total losses, } (P_i + P_c) = \text{Input} - \text{output} = 652.174 - 600 = 52.174 \text{ kW} \quad ...(i)$$

At half full-load and unity power factor

$$\text{Power output} = \frac{1}{2} \times 600 \times 1.0 = 300 \text{ kW}$$

$$\text{Power input} = \frac{300}{0.92} = 326.087 \text{ kW}$$

$$\text{Total losses} = P_i + \left(\frac{1}{2}\right)^2 P_c = 326.087 - 300 = 26.087 \text{ kW} \quad ...(ii)$$

Solving Eqs. (*i*) and (*ii*) we get

$$\text{Full-load copper loss, } P_c = 34.78 \text{ kW and iron loss, } P_i = 17.39 \text{ kW}$$

At 60 % full-load and 0.8 pf lag

$$\text{Power output} = 600 \times 0.6 \times 0.8 = 288 \text{ kW}$$

$$\text{Total losses} = P_i + (0.6)^2 P_c = 17.39 + (0.6)^2 \times 34.78 = 19.91 \text{ kW}$$

$$\text{Efficiency, } \eta = \frac{288}{288 + 19.91} \times 100 = 93.53 \% \quad \textbf{Ans.}$$

Example 8.28. A 200 kVA transformer has an efficiency of 98 % at full-load. If the maximum efficiency occurs at three-quarters of full-load, calculate the efficiency at half load. Assume negligible magnetising current and pf of 0.8 at all loads.

Solution :

$$\text{Full-load output at 0.8 pf} = 200 \times 0.8 = 160 \text{ kW}$$

$$\text{Full-load efficiency, } \eta = 98 \% \text{ or } 0.98$$

$$\text{Full-load input} = \frac{\text{Output}}{\eta} = \frac{160}{0.98} = 163.265 \text{ kW}$$

$$\text{Total losses on full-load} = P_i + P_c = \text{Input} - \text{output} = 163.265 - 160 = 3.265 \text{ kW} \quad ...(i)$$

Since maximum efficiency occurs at three-quarters of full-load, so

$$\left(\frac{3}{4}\right)^2 P_c = P_i$$

$$\text{or } \frac{9}{16} P_c = P_i \quad ...(ii)$$

Solving Eqs. (*i*) and (*ii*) we get

$$\text{Full-load copper loss, } P_c = 2.09 \text{ kW and iron loss, } P_i = 1.175 \text{ kW}$$

At half-load and 0.8 pf $\text{Output} = \frac{1}{2} \times 200 \times 0.8 = 80 \text{ kW}$

$$\text{Total losses} = P_i + x^2 P_c = 1.175 + \left(\frac{1}{2}\right)^2 \times 2.09 = 1.6975 \text{ kW}$$

$$\text{Efficiency, } \eta = \frac{\text{Output}}{\text{Output} + \text{total losses}} \times 100 = \frac{80}{80 + 1.6975} \times 100 = 96.92\% \quad \textbf{Ans.}$$

Example 8.29. The efficiency of a 1,000 kVA, 110/220 V, 50 Hz single phase transformer is 98.5% at half load and 0.8 pf leading and 98.8% at full load and unity power factor. Find (*i*) iron loss, (*ii*) full-load copper loss. *[Pb. Technical Univ. Basic Electrical and Electronics Engineering December 2007]*

Solution : Let the iron loss and full-load copper loss be P_i and P_c kW respectively.

At full load and Unity Power Factor

$$\text{Output, } P_{out} = \text{kVA rating} \times \text{power factor} = 1{,}000 \times 1.0 = 1{,}000 \text{ kW}$$

$$\text{Transformer efficiency, } \eta = 98.8\% \text{ or } 0.988$$

$$\text{Power input, } P_{in} = \frac{P_{out}}{\eta} = \frac{1{,}000}{0.988} = 1{,}012.146 \text{ kW}$$

$$\text{Transformer losses, } (P_i + P_c) = P_{in} - P_{out} = 1{,}012.146 - 1{,}000 = 12.146 \text{ kW} \qquad ...(i)$$

At half full-load and 0.8 leading power factor

$$\text{Output power, } P_{out} = \frac{1}{2} \times \text{kVA rating} \times \text{power factor} = \frac{1}{2} \times 1{,}000 \times 0.8 = 400 \text{ kW}$$

$$\text{Transformer efficiency, } \eta = 98.5\% \text{ or } 0.985$$

$$\text{Power input, } P_{in} = \frac{P_{out}}{\eta} = \frac{400}{0.985} = 406.0914 \text{ kW}$$

$$\text{Transformer losses, } \left[P_i + \left(\frac{1}{2}\right)^2 P_c\right] = 406.0914 - 400 = 6.0914 \text{ kW} \qquad ...(ii)$$

Subtracting Eq. (*ii*) from Eq. (*i*), we have

$$\frac{3}{4}P_c = 12.146 - 6.0914 = 6.0546$$

$$\text{or Full-load copper loss, } P_c = 6.0546 \times \frac{4}{3} = 8.073 \text{ kW } \textbf{Ans.}$$

Subtracting P_c = 8.073 kW from Eq. (*i*), we have

$$\text{Iron loss, } P_i = 12.146 - 8.073 = 4.073 \text{ kW } \textbf{Ans.}$$

8.19. ALL-DAY EFFICIENCY (OR ENERGY EFFICIENCY)

The transformer efficiency discussed so far is the ordinary efficiency, also called *commercial efficiency,* which is defined as the ratio of power output to power input.

There are certain types of transformers whose performance cannot be judged by ordinary or commercial efficiency. For instance, distribution transformers are energized for 24 hours, but they deliver very light loads for major portion of the day. Thus iron or core loss occurs for the whole day but the copper loss occurs only when the transformer is loaded. The performance of such a transformer must be judged by its *all-day efficiency*, also called the *energy efficiency* or *operational efficiency*, which is computed on the basis of energy consumed during the whole day (24 hours).

The all-day efficiency is defined as the ratio of energy (kWh) output over 24 hours to the energy input over the same period.

$$\textit{i.e.} \text{ All-day efficiency} = \frac{\text{Output in kWh}}{\text{Input in kWh}} \qquad ...(8.44)$$

Since the distribution transformer does not supply the rated load for the whole day, the all-day efficiency of such a transformer will be less than ordinary or commercial efficiency.

For determination of all-day efficiency of a transformer, it is necessary, of course, to know how the load varies from hour to hour during the day.

Example 8.30. A 200 kVA single phase transformer is in circuit continuously. For 8 hours in a day, the load is 160 kW at 0.8 pf. For 6 hours, the load is 80 kW at unity pf and for the remaining period of 24 hours it runs on no-load. Full-load copper losses are 3.02 kW and the iron losses are 1.6 kW. Find all-day efficiency.

Solution : Full-load output = 200 kVA

$$\text{Full-load copper loss, } P_c = 3.02 \text{ kW}$$

$$\text{Iron loss, } P_i = 1.6 \text{ kW}$$

$$\text{All-day output} = (160 \times 8) + (80 \times 6) = 1{,}760 \text{ kWh}$$

$$\text{Iron loss for 24 hours} = 1.6 \times 24 = 38.4 \text{ kWh}$$

$$\text{Copper loss for 24 hours} = \left(\frac{160/0.8}{200}\right)^2 \times 3.02 \times 8 + \left(\frac{80/1.0}{200}\right)^2 \times 3.02 \times 6 = 27.0592 \text{ kWh}$$

$$\because \text{Output kVA} = \frac{\text{Output kW}}{\text{pf}} \text{ and copper loss} = P_c \left(\frac{\text{Output kVA}}{\text{Rated kVA}}\right)^2$$

$$\text{All-day input} = \text{All-day output} + \text{iron loss} + \text{copper loss}$$

$$= 1{,}760 + 38.4 + 27.0592 = 1{,}825.4592 \text{ kWh}$$

$$\text{All-day efficiency, } \eta = \frac{\text{All-day output}}{\text{All-day input}} \times 100 = \frac{1{,}760}{1{,}825.4592} \times 100 = 96.41\%$$ **Ans.**

Example 8.31. Find all-day efficiency of a transformer having maximum efficiency of 98 % at 15 kVA at unity power factor and loaded as follows:

12 hours — 2 kW at 0.5 pf lag
6 hours — 12 kW at 0.8 pf lag
6 hours — at no load.

Solution : For maximum efficiency

$$\text{Output} = 15 \times 1.0 = 15 \text{ kW}$$

$$\text{Input} = \frac{\text{Output}}{\eta} = \frac{15}{0.98} = 15.306 \text{ kW}$$

$$\text{Total losses} = \text{Input} - \text{output} = 15.306 - 15 = 0.306 \text{ kW}$$

$$\text{Full load copper loss, } P_c = \text{Iron loss} = \frac{\text{Total losses}}{2} = \frac{0.306}{2} = 0.153 \text{ kW}$$

$$\because \text{for maximum efficiency, } P_i = P_c$$

$$\text{All-day output} = (2 \times 12) + (12 \times 6) = 96 \text{ kWh}$$

$$\text{Iron loss for 24-hours} = 0.153 \times 24 = 3.672 \text{ kWh}$$

$$\text{Copper loss for 24-hour} = \left(\frac{2/0.5}{15}\right)^2 \times 0.153 \times 12 + \left(\frac{12/0.8}{15}\right)^2 \times 0.153 \times 6 = 1.04856 \text{ kWh}$$

$$\text{All-day efficiency, } \eta = \frac{\text{Output}}{\text{Output} + \text{total losses}} \times 100 = \frac{96}{96 + 3.672 + 1.04856} \times 100 = 95.31\ \%$$ **Ans.**

8.20. TRANSFORMER CONSTRUCTION

The transformer is very simple in construction and consists of *magnetic circuit* linking with two windings known as *primary* and *secondary windings*. Besides magnetic circuit and windings it consists of a suitable container for the assembled core and windings, such as a *tank*, a suitable medium for insulating the core and windings from its container such as *transformer oil,* suitable *bushings* (either of porcelain, oil-filled or

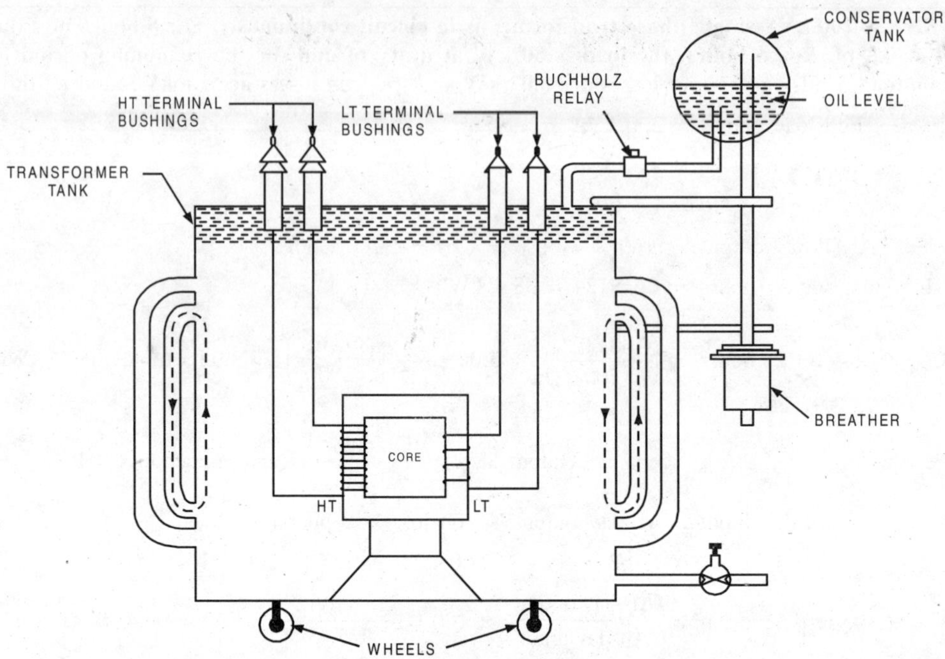

Fig. 8.22

condenser type) for insulating and bringing out terminals of the windings from the container, *temperature gauge* for measurement of temperature of hot oil or hottest spot temperature, and *oil gauge* to indicate the oil level inside the tank. Some transformers are also provided with *conservator tank* in order to slowdown deterioration of oil and keep the main tank full of oil, emergency vent to relieve the pressure inside the tank in case the pressure inside the transformer rises to a danger point and *gas operated relay* (*Buchholz relay*) in order to give alarm to indicate the presence of gas in case of some minor fault and take out the transformer out of circuit in case of serious fault.

Magnetic circuit consists of an iron core. Since core is magnetic link between the two systems in a transformer and it itself contains a lot a energy, it is not by any means the passive component it would first appear.

Ideally the steel, of which the transformer core is made, is simple in chemical composition, the hot rolled steel has 4-5% silicon and the cold rolled has 3%. Nowadays cold rolled steel is used almost exclusively inspite of its high price, because flux density 30% higher can be used. This is because of high permeability and lower losses at given flux density. Since flux is an alternating one, in order to reduce eddy current losses the core is laminated. The thickness of laminations or stampings varies from 0.35 mm to 0.5 mm. The laminations are insulated from each other by coating them with a thin coat of varnish. The iron core is rectangular in x-section for small size transformers and circular for large size ones to reduce the quantity of copper required.

As already mentioned, the electric circuit consists essentially of two windings, each being split up into two equal number of coils and the coils of two windings are arranged on each limb of iron core to reduce the magnetic leakage. In order to minimise the amount of high voltage insulation, low voltage coils are placed adjacent to the iron.

8.21. TYPES OF TRANSFORMERS

Depending upon the type of construction used, the transformers are classified into two categories viz. (*i*) *core type* and (*ii*) *shell type.*

Depending upon the type of service, in the field of power system, they are classified as (*i*) *power transformers* (*ii*) *distribution transformers*.

8.21.1. Core Type and Shell Type Transformers

In **core type construction**, as shown in Fig. 8.23, the coils are wound around the two limbs of a rectangular magnetic core. Each limb carries one half of the primary winding and one half of the secondary winding so as to reduce the leakage reactance to the minimum possible. The lv winding is wound on the inside nearer to the core while the hv winding is wound over the lv winding away from the core in order to reduce the amount of insulation materials required.

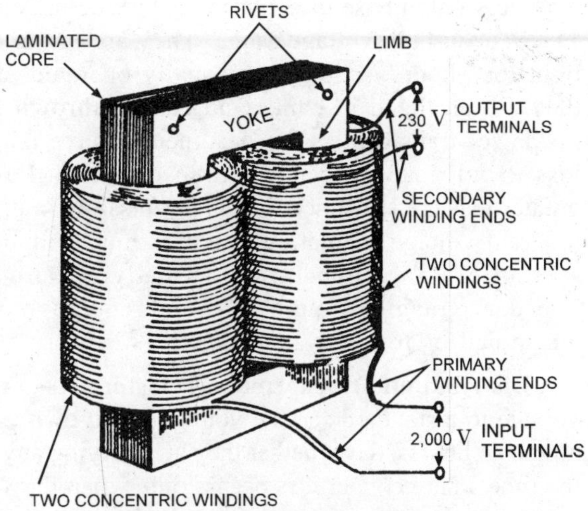

Fig. 8.23. *Core Type Transformer*

Small transformers may have cores of rectangular or square x-section with rectangular or circular coils but it is wasteful in case of large capacity transformers. In case of large sized transformers stepped cruciform core with circular cylindrical coils is employed. Such a core employs laminations of different sizes. Though the cost of manufacturing of such a cruciform core is much greater, but the circular coils that are used are easier to wind and provide more mechanical strength, especially when short-circuit occurs. Other advantages of using cruciform core are high space factor and reduced mean length of turns resulting in reduced copper loss.

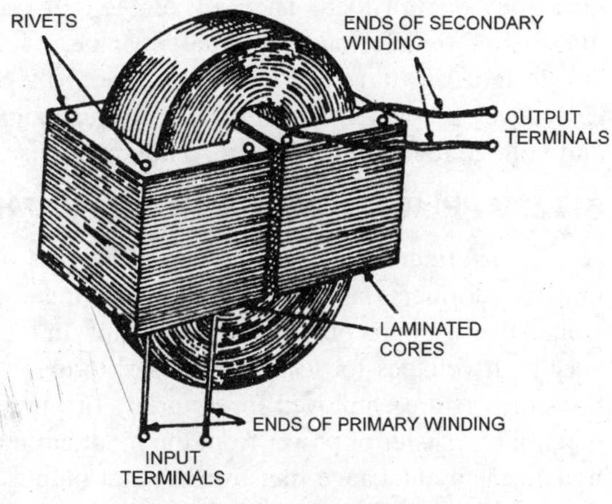

Fig. 8.24. *Shell Type Transformer*

In shell type construction, as shown in Fig. 8.24, the coils are wound on the central limb of a three limb core. The entire flux passes through the central limb and divides into two parts going to side limbs. Consequently, the x-sectional area (and hence width) of the central limb is twice of that of each of the side limbs. Sandwich type winding is used in such a construction.

If the core type and shell type transformers are compared, then we conclude that the core type transformer has a longer mean length of iron core and a shorter mean length of coil turn. The core type transformer has smaller x-section of iron and, therefore, a greater number of turns. Core type construction provides more space for insulation making it best suited for EHV requirements.

In shell type transformers advantage is gained through the core being used to protect the windings from mechanical damage. The shell type construction gives better support against electromagnetic forces between current carrying conductors. These forces are of considerable magnitude under short-circuit conditions. The shell type construction is commonly used for small transformers where a square or rectangular core cross section is suitable for economic consideration. The shell type construction needs more specialized fabrication facilities than core type, while the latter offers an additional advantage of permitting visual inspection of coils in the case of a fault and ease of repair at substation site. For these reasons, the present practice is to use core type transformers in large high voltage installations.

8.21.2. Power Transformers

The term is used to include all transformers of large size (250 kVA and above) used in generating stations and substations for transforming the voltage at each end of a power transmission line. They may be single phase or 3-phase; 3-wire delta/delta or delta/star connected; and of voltage rating 220/11 kV or in high voltage range. They are put in operation during load hours and disconnected during light load hours *i.e.* they are usually operated on approximately full load. This is possible because they are arranged in banks and can be thrown in parallel with other units or disconnected at will. So power transformers are designed to have maximum efficiency at or near full load (*i.e.* with iron loss to full load copper loss ratio of 1 : 1). Power transformers are designed to have considerable greater leakage reactance than is permissible in distribution transformers because in power transformers inherent voltage regulation is not as much important as current limiting effect of the higher leakage reactance. Power transformers usually make use of flux density of 1.5 to 1.7 T, have percentage impedance ranging from 6–18% and regulation of 6–10 per cent. They may be self oil cooled, forced air cooled or forced water cooled.

Distribution Transformers. Transformers of rating up to 200 kVA, used to step down the distribution voltage to a standard service voltage are known as *distribution transformers*. They are kept in operation all the 24 hours a day whether they are carrying any load or not. In such transformers iron loss occurs for all the time where copper loss occurs only when they are loaded. Therefore, distribution transformers should be designed with iron loss smaller in comparison to full-load copper loss (say with iron loss to full load copper loss ratio of 1 : 3). In other words, they are designed for good 'all-day efficiency' and not for highest efficiency on full load. They are of the self cooling type and are almost invariably oil-immersed. The distribution transformers are usually 3-phase, 4-wire 11 kV/440 V delta/star connected. Such transformers are designed to have an inherently good voltage regulation which is possible by arranging the coils in such a way as to have minimum leakage reactance. Such transformers have percentage impedance of 4–5% and voltage regulation of 4–8%. They make use of cold rolled steel with flux density of 1.7 T.

8.22. APPLICATIONS OF TRANSFORMERS

Transformers are extensively used like the *instrument transformers* for metering purposes. One-to-one ratio transformers (*i.e.* the transformers having equal primary and secondary voltages) are used to electrically isolate the two parts of an electrical circuit. In high-voltage laboratories the transformers are used to provide very high voltages for testing purposes, termed as *testing transformers*. In electric communication circuits transformers are employed for a variety of purposes *e.g.*, as an impedance transformation device to allow maximum transfer of power from the input circuit to the output device. In radio and television circuits input transformers, interstage transformers and output transformers are widely employed. Transformers are also employed in telephone circuits, instrumentation circuits and control circuits. Indeed the transformer is a device which plays a vital and essential role in many facets of electrical engineering.

HIGHLIGHTS

1. *Transformer* is an ac machine that (*i*) transfers electrical energy from one electric circuit to another (*ii*) does so without a change of frequency (*iii*) does so by the principle of electromagnetic induction and (*iv*) has electric circuits that are linked by a common magnetic circuit. The energy transfer usually takes place with a change of voltage, although this is not always necessary. It is a static machine and has very high efficiency (in the range of 96-99%).

   A transformer is never operated on dc. If dc supply is connected to the transformer primary or secondary, it may burn out since there is no self induced emf and heavy current will flow through it.
2. In an *ideal transformer,* winding resistance, magnetic leakage, and iron losses are neglected and core is assumed to have infinite permeability and zero reluctance.

3. The emfs induced in the transformer windings are given as

$$E_1 = 4.44 f N_1 \Phi_{max} \text{ volts}$$
$$\text{and } E_2 = 4.44 f N_2 \Phi_{max} \text{ volts}$$

4. Voltage transformation ratio, $K = \frac{E_2}{E_1} = \frac{N_2}{N_1} = \frac{V_2}{V_1} = \frac{I_1}{I_2}$

5. The current drawn by transformer on no load is called the no-load current or exciting current. It has two components (*i*) in-phase, active or *energy component* I_e ($I_0 \cos \Phi_0$) used to meet the iron loss and small amount of copper loss occurring in primary winding and (*ii*) quadrature or wattless component, called the *magnetizing component*, I_m ($I_0 \sin \Phi_0$) used to create alternating flux in the core. No-load phase angle, $\Phi_0 = \text{Tan}^{-1} \frac{I_m}{I_e}$.

6. When a transformer is loaded, the secondary ampere-turns set up a magnetic field that opposes the main field. To neutralize this effect, transformer draws extra primary current from the mains. Flux remains almost constant from no-load to full-load.

$$\text{The primary current, } \mathbf{I_1} = \mathbf{I'_1} + \mathbf{I_0} = K\ \mathbf{I_2} + \mathbf{I_0}$$

7. When a practical transformer is loaded

Induced emf in secondary winding $\mathbf{E_2} = \mathbf{V_2} + \mathbf{I_2} (R_2 + j X_2) = \mathbf{V_2} + \mathbf{I_2} \mathbf{Z_2}$

Induced emf in primary $E_1 = E_2/K$

and applied voltage to primary winding, $\mathbf{V_1} = -\mathbf{E_1} + \mathbf{I_1} (R_1 + j X_1) = -\mathbf{E_1} + \mathbf{I_1} \mathbf{Z_1}$

8. The total resistance and reactance of the transformer referred to primary are given as

$$R_{01} = R_1 + \frac{R_2}{K^2}, \quad X_{01} = X_1 + \frac{X_2}{K^2} \text{ and } Z_{01} = \sqrt{R_{01}^2 + X_{01}^2}$$

9. The total resistance and reactance of the transformer referred to secondary are given as

$$R_{02} = (K^2 R_1 + R_2),\ X_{02} = (K^2 X_1 + X_2) \text{ and } Z_{02} = \sqrt{R_{02}^2 + X_{02}^2}$$

10. Approximate voltage drop in the transformer referred to secondary
$= I_2 (R_{02} \cos \Phi \pm X_{02} \sin \Phi)$ +ve sign for lagging power factor and –ve sign for leading power factor

11. Percentage voltage regulation $= \frac{E_2 - V_2}{E_2} \times 100 = \frac{I_2 (R_{02} \cos \Phi \pm X_{02} \sin \Phi)}{\text{No-load secondary voltage, } E_2} \times 100$

12. Transformer losses are of two types namely iron loss and copper loss. Iron losses are almost constant from no-load to full-load while copper losses vary as the square of load in kVA.

13. Transformer efficiency is given as $\eta = \frac{V_2 I_2 \cos \Phi}{V_2 I_2 \cos \Phi + P_i + x^2 P_c}$

Transformer efficiency will be maximum when copper losses equal iron or core losses.

Output kVA corresponding to maximum efficiency $= \text{Rated kVA} \sqrt{\frac{\text{Iron losses}}{\text{Full-load copper losses}}}$

14. *Open-circuit lest* is performed to determine the core or iron losses, P_i, and no-load current I_0 and thereby the shunt branch parameters R_0 and X_0 of the equivalent circuit.

15. *Short-circuit test* is performed to determine the copper losses at rated load and equivalent resistance and reactance referred to metering side.

16. All-day efficiency or energy efficiency $= \frac{\text{kWh output in 24 hours}}{\text{kWh input in 24 hours}}$.

EXERCISES

1. What is a transformer ? Differentiate between step-up and step-down transformers.
2. What is an ideal transformer ?
3. Draw general schematic of a single phase transformer. Describe its working principle and deduce the expression for emf in secondary winding.
4. Derive an expression for emf equation of single phase transformer.
[Pb. Technical Univ. Basic Electrical and Electronics Engineering May 2008]
5. Derive the emf equation of single phase transformer and explain voltage transformation ratio.
6. Draw no-load phasor diagram of a transformer and derive expressions for magnetizing and core-loss components of no-load current.
7. Discuss the working of a two winding transformer. Develop its phasor diagram with an inductive load connected to secondary winding.
[Pb. Technical Univ. Basic Electrical and Electronics Engineering Dec. 2004]
8. Explain how in a transformer, as the secondary current increases, primary current also increases.
[Pb. Technical Univ. Electrical Engineering May 2002]
9. Draw the approximate equivalent circuit of a transformer referred to the primary side and indicate how it differs from the exact equivalent circuit.
10. Draw the equivalent circuit of a single phase transformer and derive expressions for all its parameters.
[Pb. Technical Univ. Electrical Engineering June 2000; December 2000; May 2001]
11. Define voltage regulation and derive the value for lagging power factor of load on the transformer.
[Pb. Technical Univ. Basic Electrical and Electronics Engineering May 2005]
12. Define voltage regulation of a transformer and mention its significance. Under what condition, will the voltage regulation be zero ?
13. State the various losses which take place in a transformer. On what factors do they depend ? Explain the steps taken to minimize these losses.
14. What are the losses in a transformer. Derive condition that efficiency at a given terminal voltage and load power factor is maximum if full-load copper loss equals iron loss in the transformer.
[Pb. Technical Univ. Basic Electrical and Electronics Engineering May 2005]
15. How short-circuit test is performed on transformer ? Draw its set-up and discuss the results obtained from the test. **[Pb. Technical Univ. Basic Electrical Engineering December 2002]**
16. Draw the connection diagrams for the open-circuit and short-circuit tests of a single phase transformer, showing all the necessary instruments. Describe briefly how would you perform the above tests.
17. Discuss how will you perform open-circuit and short-circuit tests on a single phase transformer in the laboratory. From the test results how will you find the efficiency and regulation of the transformer
18. What is all-day efficiency of a transformer ? How does it differ from ordinary efficiency ?
19. Describe the constructional features and principle of operation of a single phase transformer. Classify the transformers on the basis of (*i*) construction and (*ii*) service rendered and discuss.
20. Distinguish between
    (*i*) Step-up and step-down transformers.
    (*ii*) Voltage ratio and current ratio.
    (*iii*) All-day efficiency and power efficiency.
    (*iv*) Core and shell type transformers.
    (*v*) Power and distribution transformers.

SHORT ANSWER TYPE QUESTIONS WITH ANSWERS

Q. 1. Why does a transformer has iron core ? **[P.T.U. December 2000]**

Ans. The use of iron or steel for transformer core ensures a high permeability of the magnetic circuit and because of high permeability the magnitude of exciting current necessary to create the required flux in the core is small. The presence of steel core also causes 100% of magnetic flux created by primary to be linked with secondary.

Q. 2. Why is an iron-silicon alloy core used in a transformer ? **[P.T.U. B.E.E.E. Dec. 2005]**

Ans. An iron-silicon alloy core is used in a transformer to keep the hysteresis loss the minimum possible. Hysteresis loss occurs due to reversal of flux in the magnetic material.

Q. 3. Why is efficiency of a transformer maximum among all electrical machines ? **[P.T.U. B.E.E.E. Dec. 2004]**

Or

Why is the efficiency of a transformer maximum among electrical equipments ? Explain. **[P.T.U. 1999]**

Ans. Owing to lack of rotating parts in a transformer, there are no friction and windage losses. The other losses such as iron losses are also comparatively small because of better magnetic material and less handling of the material. So the efficiency of transformer is maximum among all electrical equipments.

Q. 4. Why the transformer is also called the static transformer ?

Ans. Since basic construction of transformer does not require any moving part, it is also called the static transformer.

Q. 5. Explain with reasons what happens when a power transformer is connected to dc supply of the same voltage rating ?

Ans. If a rated dc voltage is applied to the primary of the transformer, the flux produced in the transformer core will not vary but remain constant in magnitude and, therefore, there will be no self induced emf in the primary winding (which is only possible with varying flux linkage) to oppose the applied voltage and since the resistance of the primary winding is quite low, a heavy current will flow through the primary winding that may cause burning out of the primary winding.

Q. 6. What is the difference between an ideal and practical transformer ? **[P.T.U. January 2000]**

Ans. An *ideal transformer* is an imaginary transformer that has no winding resistance, no magnetic leakage, no iron loss and zero reluctance while a *practical transformer* is an actual transformer which operates at a particular frequency and has winding resistance, magnetic leakage, iron loss, copper loss, and magnetic reluctance.

Q. 7. What is meant by magnetising current ?

Ans. Magnetising current is quadrature or wattless component of exciting current of transformer and is used to create alternating flux in the core.

Q. 8. What do you know about the no-load current of a transformer ?

Ans. The current drawn by primary winding of a transformer on no load is called the no-load or exciting current. It is very small in magnitude in comparison to the full-load primary current (2 to 5 per cent of full-load current). It has two components—energy component (comparatively very small) to meet the iron loss in addition to small amount of copper loss occuring in the primary winding and magnetising component relatively large to create the alternating flux in the core.

Q. 9. Explain how in a transformer, as the secondary current increases, primary current also increases. **[P.T.U. May 2002]**

Or

When the load or secondary current of a transformer increases, how does the input current increases to meet with the new-conditions ? **[P.T.U. May 2001]**

Ans. When the load secondary current of a transformer increases, the secondary ampere turns (N_2I_2) create more secondary flux. This increased secondary flux opposes the main flux ϕ set up by the exciting current I_0. Thus main flux ϕ is weakened momentarily by the opposing increased secondary flux ϕ_2 and, therefore, difference of applied voltage V_1 and back emf E_1 increases, consequently, more current is drawn from the source of supply flowing through the primary winding until the original value of flux ϕ restores. The extra current drawn from the supply source increases the primary ampere-turns (N_1I_1') and compensates the extra secondary ampere-turns *i.e.* $N_1I_1' = N_2I_2$.

Thus in a transformer when the secondary current increases, primary current also increases.

Q. 10. Why the main flux in a transformer remains practically constant from no-load to full load ?

Ans. When the transformer is on no load, it draws no-load or exciting current I_0 from the supply mains. The no-load current I_0 sets up mmf N_1I_0 which produces flux ϕ in the core. When an impedance is connected across the secondary terminals, current I_2 flows through the secondary. The secondary current I_2 sets up its own mmf and hence creates a secondary flux ϕ_2 opposing the main flux ϕ. In other words, the secondary ampere-turns weaken the main flux in the core. Momentarily, self induced emf E_1 decreases due to which primary draws extra current from the supply source. This extra current drawn from the supply source increases the primary ampere-turns (N_1I_1') and compensates the secondary ampere-turns N_2I_2.

Since the secondary flux ϕ_2 produced by secondary mmf N_2I_2 is neutralized by the flux ϕ_1' produced by mmf N_1I' set up by counterbalancing primary current I_1', the flux in the transformer core remains constant from no load to full load.

Q. 11. How does the leakage flux occur in a transformer ? **[P.T.U. December 2000]**

Ans. In actual practice the total flux created does not link the primary and secondary windings but is divided into three parts namely the main flux ϕ linking both the primary and secondary windings, primary leakage flux ϕ_{L_1}, linking with primary winding only and secondary leakage flux ϕ_{L_2}, linking with secondary winding only.

The flux set up by the primary or secondary which is set up in the core linking with its own turns and not linking with the other is known as *leakage* flux.

Q. 12. What is leakage reactance of transformers ? **[P.T.U. December 2003]**

Ans. Leakage fluxes of primary and secondary produce self inductances L_1 given as $\frac{N_1\phi_{L1}}{I_1}$ and L_2 given as $\frac{N_2\phi_{L2}}{I_2}$ in the respective windings *i.e.* primary and secondary. These inductances, in turn, produce inductive reactances X_1 given as $2\pi fL_1$ and X_2 given as $2\pi fL_2$ which cause voltage drop in respective windings lagging behind the currents I_1 and I_2 respectively. The reactances produced by leakage fluxes are called the leakage reactances.

Q. 13. What is the effect of leakage flux in a transformer ? What is meant by magnetising current ? **[P.T.U. June 2000]**

Ans. The effect of leakage flux is to develop into their respective windings emfs of self inductance which are proportional to the current, and are, therefore, equivalent in effect to the addition of an inductive coil in series with each winding. The component of no-load current used to create the flux in the transformer core is known as magnetising current.

Q. 14. Draw the equivalent circuit of a transformer explaining the meaning of the parameters therein. **[P.T.U. December 2001]**

Ans. Please refer to Fig. 8.16 on page 237.

The different parameters of a transformer are :

R_0, the exciting resistance; X_0, the exciting reactance; R_1, the primary winding resistance; X_1, the primary winding leakage reactance; R_2, the secondary winding resistance and X_2, the secondary winding leakage reactance.

Q. 15. Explain with reasons what happens when the primary terminals of a single phase transformer are connected to the rated ac supply while secondary terminals are short circuited.

Ans. The transformer with secondary short circuited becomes equivalent to a coil having an impedance equal to impedance of both windings, which is quite small. When the primary terminals of the transformer are connected to the rated ac supply, heavy current will flow through primary and secondary windings that may cause burning out of these windings.

Q. 16. Is voltage regulation better or worse at 0.6 *pf* lagging than at 0.8 *pf* lagging for the same kVA output explain. **[P.T.U. May 2001]**

Ans. The voltage regulation will be the maximum and worst when power factor becomes equal to

$\cos \tan^{-1} \frac{X_{02}}{R_{02}}$ (lagging).

Since equivalent reactance of transformer is usually more than double its equivalent resistance, the voltage regulation will increase (*i.e.* become more and more worse) till power factor falls even below $\cos \tan^{-1} 2$ *i.e.* 0.45 (lagging), the voltage regulation will worsen at 0.6 pf lagging than at 0.8 pf lagging.

Q. 17. What is meant by regulation of a transformer and how is it affected by the power factor of the load supplied by it ?

Ans. The change in secondary terminal from no load to full load is termed the regulation. It is usually expressed as a percentage or a fraction of the rated no-load terminal voltage.

The voltage regulation increases with the fall in power factor of the load, if the load is inductive, and becomes maximum when the power factor of the inductive load becomes equal to $\cos\left(\tan^{-1} \frac{X_{02}}{R_{02}}\right)$.

The voltage regulation drops with the fall in power factor of the load, if the load is capacitive, and becomes zero, when the power factor of the capacitive load becomes equal to $\cos\left(\tan^{-1} \frac{-R_{02}}{X_{02}}\right)$, and then increases with the fall in power factor of load in the negative direction.

Q. 18. What factors constitute the major losses produced in a transformer ?

Ans. Eddy currents and hysteresis cause iron loss in the magnetic core and yoke and resistance of primary and secondary windings cause copper loss when the transformer is loaded.

Q. 19. On what factors does the hysteresis loss depend ? Why can this loss be assumed to be constant ? **[P.T.U. December 2000]**

Ans. Hysteresis loss depends upon the type of material, volume of the material, frequency of reversal of magnetic field and maximum value of flux density in the core of the material *i.e.*

$$P_h = \eta \, (B_{max})^{1.6} fv \text{ watts}$$

Since from no load to full load the flux linking with the core and frequency remains constant, hysteresis loss is assumed to remain constant.

Q. 20. On what factors does the eddy current loss depend ? Why this loss can be assumed to be constant ? **[P.T.U. June 2000]**

Ans. The eddy current loss depends on the maximum flux density in the core and supply frequency for a given core (*i.e.* nature of magnetic material, volume of the core, and thickness of laminations remaining unchanged). Since from no load to full load the flux linking with the core and frequency remain constant, the eddy current loss remains constant.

Q. 21. List out various losses that take place in a transformer. Which loss is independent of load ? **[P.T.U. June 2003; May 2010]**

Ans. Iron or core losses consisting of hysteresis and eddy current losses. These losses occur in the core and yoke due to alternating flux.

Copper losses occuring in primary and secondary windings due to their resistances when the transformer is loaded.

Iron losses are independent of load, and therefore, are called the constant losses.

Q. 22. How can eddy current loss be minimised ?

Ans. By using laminated iron core.

Q. 23. What is meant by equivalent circuit of a transformer?

Ans. An electrical circuit that can represent behaviour of the transformer to an approximation is called its equivalent circuit.

Q. 24. How may the iron loss be reduced to a minimum in a transformer ?

Ans. The iron loss may be minimised by using steel of high silicon content for the transformer core and by using very thin laminations (0.35 to 0.5 mm thick) insulated from each other by insulating varnish or by layers of papers.

Q. 25. What is the condition for maximum efficiency in transformer ? **[P.T.U. December 2003]**

Ans. Efficiency of a transformer will be maximum on such a load at which variable loss (copper loss) becomes equal to constant loss (iron loss).

Q. 26. Is the efficiency of a transfromer same at the same load at 0.8 *pf* lag and 0.8 *pf* lead ?

Ans. Yes, the efficiency of a transformer remains the same for a given load at a given power factor irrespective of the fact whether the *pf* is lagging or leading.

Q. 27. What is meant by all-day efficiency of a power transformer and why is it lower than commercial efficiency ?

Ans. The all-day efficiency, also known as energy efficiency or operational efficiency, is defined as the ratio of energy (kWh) output over 24 hours to the energy input over the same period

$$i.e. \quad \text{All-day efficiency} = \frac{\text{Output in kWh}}{\text{Input in kWh}}$$

There are certain types of transformers, such as distribution transformers, which remain energized for 24 hours but they supply very light loads for major portion of the day. Thus iron or core loss occurs for the whole day but the copper loss occurs only when the transformer is loaded and so the all-day efficiency of transformer is lower than its commercial efficiency.

Q. 28. Why core of a transformer is laminated ?

Ans. Transformer core is laminated to reduce eddy current loss.

Q. 29. Why *lv* winding is placed on the core and then *hv* winding in case of a core type transformer ?

Ans. Placing of *lv* winding near the core and *hv* winding around the *lv* winding in a core type transformer reduces the amount of insulation material required.

Q. 30. What is the difference between power transformers and distribution transformers ? **[P.T.U. December 2003]**

Ans. Power transformers are usually of large rating (250 kVA and above) and are installed in generating stations and substations transforming the voltage at each end of a power transmission line while distribution transformers are of small rating (upto 200 kVA) and are used to step-down the voltage to a standard service voltage (250/440 V). Power transformers are put in operation during load hours and disconnected during light load hours *i.e.* they are operated on approximately full load while distribution transformers are kept in operation all the 24 hours a day whether they are carrying any load or not. Power transformers are, therefore, designed to have maximum efficiency at a near full load (*i.e.* with iron loss to full-load copper loss ratio of 1 : 1) but distribution transformers are designed with iron loss smaller in comparison to full-load copper loss (say with iron loss to full-load copper ratio of 1 : 3).

PROBLEMS

1. The maximum flux density in the core of 250/3,000 V, 50 Hz single phase transformer is 1.2 Wb/m^2. If the emf per turn is 8 V, determine (*i*) primary and secondary turns (*ii*) area of the core. **[Ans.** (*i*) 32; 375 (*ii*) 0.03 m^2**]**
2. A single phase transformer has 400 primary and 1,000 secondary turns. The net cross-sectional area of the core is 60 cm^2. If the primary winding be connected to a 50 Hz supply at 520 V, calculate (*i*) the peak value of flux density in the core (*ii*) the voltage induced in the secondary winding. **[Ans.** 0.976T; 1,300 V**]**

3. A 200 kVA, 3300/240 V, 50 Hz single phase transformer has 80 turns on the secondary winding. Assuming an ideal transformer, calculate (*i*) primary and secondary currents on full load (*ii*) the maximum value of flux and (*iii*) number of primary turns. **[Ans.** (*i*) 60.61 A; 833.33 A (*ii*) 0.0135 Wb (*iii*) 1,100]

4. The no-load current of transformer is 5 A at 0.2 pf when supplied at 240 V, 50 Hz. The number of turns on the primary winding is 250. Determine (*i*) the maximum value of flux in the core (*ii*) the core loss and (*iii*) magnetising current. **[Ans.** (*i*) 4.32 mWb (*ii*) 240 W (*iii*) 4.899 A]

5. A 100 kVA, 2400/240 V, 50 Hz single phase transformer has no-load current of 0.64 A and a core loss of 700 W, when its high voltage side is energized at rated voltage and frequency. Calculate the two components of no-load current. If this transformer supplies a load current of 40 A at 0.8 lagging pf on its lv side, determine the primary current and its power factor. Ignore leakage impedance drop.

[Ans. 0.2917 A; 0.5697 A; 4.5838 A; 0.7618 (lag)]

6. A 10 kVA. 2,000 / 400 V single phase transformer has resistances and leakage reactances as follows: $R_1 = 50\ \Omega$; $X_1 = 12.5\ \Omega$; $R_2 = 0.2\ \Omega$; $X_2 = 0.5\ \Omega$. Determine the secondary terminal voltage when it is supplying rated current at 0.8 lagging factor. Neglect no-load current. **[Ans.** 377 volts]

7. A single phase transformer has 200 and 100 turns respectively in the secondary and primary windings. The respective resistances are 0.233 Ω and 0.067 Ω. Calculate the equivalent resistance of (*i*) the primary in terms of the secondary winding (*ii*) the secondary in terms of the primary winding and (*iii*) the total resistance of the transformer in terms of the primary. Prove the formula you use.

[Ans. (*i*) 0.268 Ω (*ii*) 0.05825 Ω (*iii*) 0.125 Ω]

8. Calculate the regulation of a transformer in which the ohmic loss is 1 % of the output and the reactance drop 5 % of the voltage, when the power factor is (*i*) 0.8 lagging (*ii*) unity, and (*iii*) 0.8 leading.

[Ans. (*i*) 3.8 % (*ii*) 0.8 % (*iii*) –2.2 %]

9. The primary and secondary windings of a 500 kVA transformer has resistances of 0.42 Ω and 0.0011 Ω respectively. The primary and secondary voltages are 6,600 V and 400 V respectively and the iron loss is 2.9 kW. Calculate the efficiency at full-load, assuming the power factor of the load to be 0.8.

[Ans. 98.273 %]

10. A 11,000 / 230 V, 150 kVA, 1-phase, 50 Hz transformer has core loss of 1.4 kW and full-load copper loss of 1.6 kW. Determine (*i*) kVA load for maximum efficiency and value of maximum efficiency at unity pf. (*ii*) the efficiency at half-full-load 0.8 pf leading. **[Ans.** 140.0311 kVA; 98.04 % (*ii*) 97.09 %]

11. A 30 kVA, single phase transformer has an iron loss of 457 W and copper loss of 125 W, when delivering half the full-load. At what percent of the full-load will the transformer have maximum efficiency ?

[Ans. 95.6%]

12. A 200 kVA single phase transformer has an efficiency of 98 % on full load. If the maximum efficiency occurs on three quarters of full-load, calculate (*i*) the iron loss and copper loss at full load (*ii*) efficiency at half load. Assume a pf of 0.8 on all loads. **[Ans.** (*i*) 1.175 kW; 2.09 kW; (*ii*) 96.92%]

13. The efficiency of a 400 kVA single phase transformer was 98.77 % when delivering full-load at 0.8 pf and 99.13 % at half load and unity power factor. Calculate the iron losses and full-load copper losses.

[Ans. 1.012 kW, 2.973 kW]

14. A 100 kVA lighting transformer has a full-load loss of 3 kW, the losses being equally divided between iron and copper. During a day, the transformer operates on full load for 3 hours, on half-load for 4 hours, the output being negligible for the remainder of the day. Calculate the all-day efficiency. **[Ans.** 92.25%]

❑❑❑

9
CHAPTER

DC Machines (Generators and Motors)

9.1. INTRODUCTION

INSIDE THIS CHAPTER

DC machine is actually an *alternating current* machine, but furnished with a special device, called the *commutator,* which under certain conditions converts ac into dc and vice versa.

The term "generator" denotes that it generates electrical energy but actually it does not. It simply converts mechanical energy supplied to it into electrical energy.

The generator operates on the principle of the production of dynamically induced emf *i.e.* whenever flux is cut by the conductor, dynamically induced emf is produced in it according to the laws of electromagnetic induction, which will cause a flow of current in the conductors if the circuit is closed.

For production of dynamically induced emf, three things are necessary, a magnetic field, a conductor and motion of the conductor with respect to the field. In dc generators the field is produced by the field magnets which are stationary. Permanent magnets are used for very small capacity machines and electromagnets are used for large machines to create magnetic flux. The conductors are situated on the periphery of the armature being rotated by the prime mover.

An electric motor is a machine which converts electrical energy into mechanical energy whereas a generator is that machine which converts mechanical energy into electrical one. As regards fundamental principles the dc motors are identical with the dc generators which have the same type of excitation, *i.e.* a machine that operates as a motor will also operate satisfactorily as a generator.

The only difference lies, however, in the mode of construction, which is due to the fact that the frame of the generator can as a rule be open but those of motors should be either partly or totally enclosed. A generator is usually placed in a suitable position and mechanical protection for the coils and armature may be reduced to minimum. Also the generator is handled by technical persons. Hence there is no risk in having the frame of the generator open, which facilitates cooling, inspection and repair. On the other hand, motors have to work in conditions of dampness, dirt, inflammable gases, chemical fumes and liability to mechanical damage and, therefore, protection must be adequate and motor frames are made flame proof, partly enclosed or totally enclosed according to the requirements of service.

9.2. SIMPLE LOOP GENERATOR

A single turn rectangular copper conductor loop A B C D rotating in clockwise direction about its own axis XX' in the uniform magnetic field is shown in Fig. 9.1. As the coil is rotated in the magnetic field by some mechanical means (not shown in the Fig.) the flux linking with the loop changes continually, therefore an emf is induced in it. The magnitude of emf induced at any instant is proportional to the rate of change of linking flux at that instant and its direction is given by Fleming's right hand rule.

While rotating at the instant the loop of wire ABCD assumes the position shown in Fig. 9.1 (*a*) *i.e.* when the loop ABCD is just parallel to the faces of field magnets N and S, the flux linking with the loop is maximum but the rate of change of linking flux is zero, as at this instant no flux is cut by the coil sides AB and CD which are just moving parallel to them. Hence induced emf is zero when the loop is in this position. This position of the loop is known as *neutral position* and let it be taken starting position and angle of rotation be measured from this position.

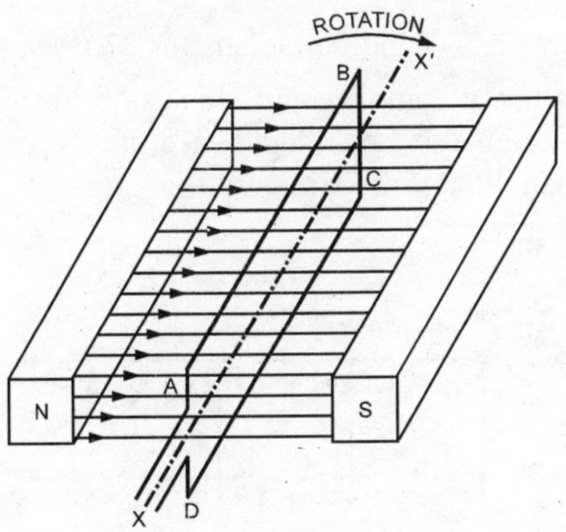

(a) Armature Coil in Neutral Plane

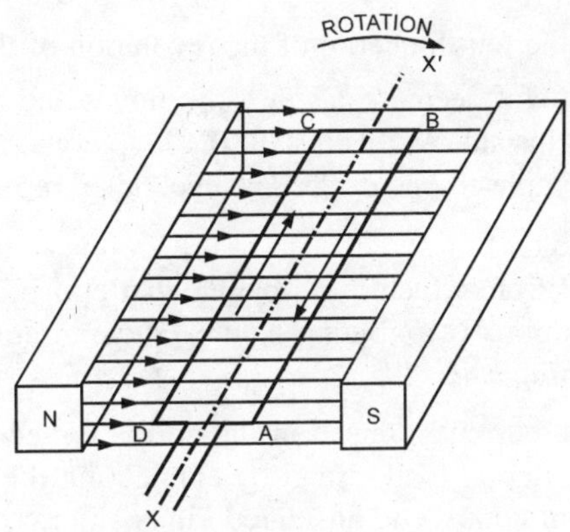

(b) Armature Coil in The 90° Position

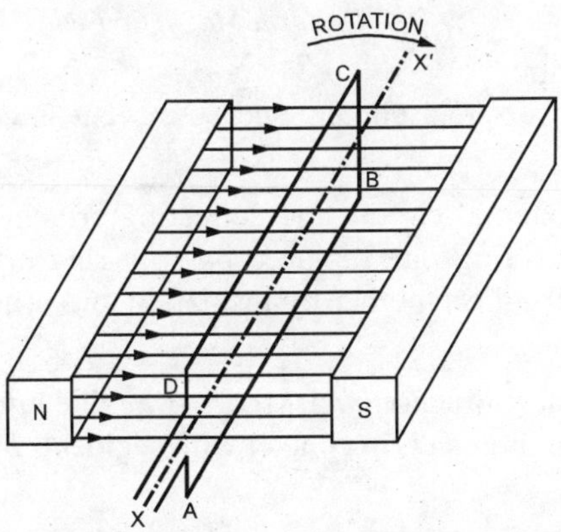

(c) Armature Coil in The 180° Position

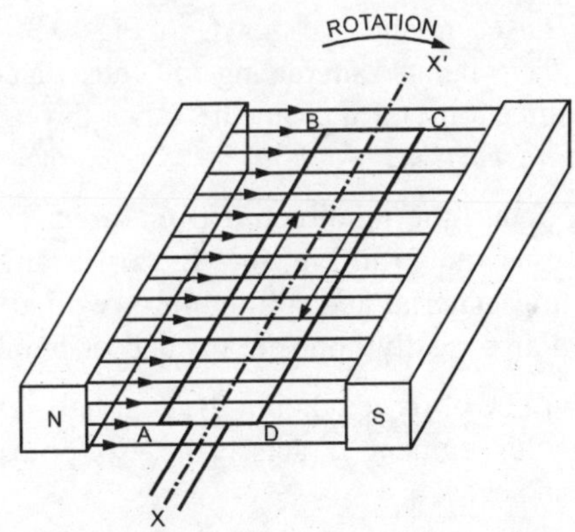

(d) Armature Coil in The 270° Position

Fig. 9.1

As the coil is turned in clockwise direction at a constant speed, the coil sides begin to cut across the field, slowly at first but at gradually increasing rate. Thus the magnitude of emf induced gradually increases as the loop moves and becomes maximum when the loop assumes the position shown in

Fig. 9.1 (*b*). During this period the loop rotates through $\frac{\pi}{2}$ radians or 90° and induced emf in the loop increases following the sine law, as shown in Fig. 9.2. The direction of emf induced in the coil, as given by Fleming's right hand rule, is from B to A and from D to C. In the position of the loop shown in Fig. 9.1 (*b*) the coil sides are moving at right angles to the field and are, therefore, cutting across the field at maximum rate, consequently, the emf induced at this instant is of maximum value. In the next quarter of the revolution of the loop *i.e.* between $\frac{\pi}{2}$ radians and π radians (or 90° to 180°), the rate at which the conductors cut across the magnetic field gradually decreases, causing the magnitude of induced emf to fall gradually with the angular movement of the loop and becomes zero at the instant the loop becomes again parallel to the faces of the field magnets but with the sides AB and CD's position interchanged with respect to zero position [Fig. 9.1 (*c*)].

In the third quarter of the revolution of loop *i.e.* between π and $\frac{3\pi}{2}$ radians (or 180° to 270°) the rate at which the conductors cut across the magnetic field hence induced emf gradually increases as the loop moves and becomes maximum at the instant, the loop assumes the position shown in Fig. 9.1 (*d*). However, the direction of emf in the loop is now from A to B and from C to D *i.e.* opposite to that in the first two quarters.

In the fourth quarter of the revolution of the loop *i.e.* between $\frac{3\pi}{2}$ and 2π radians (or 270° to 360°), the induced emf decreases as the coil moves and becomes zero when it completes 2π radians or 360° from the starting instant. At this instant the loop comes to its absolute original position and hence the loop is said to have completed one cycle. The cycle then repeats for each revolution of the armature.

The emf thus generated in the loop is of the form shown in Fig. 9.2 from which it is obvious that emf induced in armature conductors is of pulsating nature. Such an emf is known as an *alternating emf*.

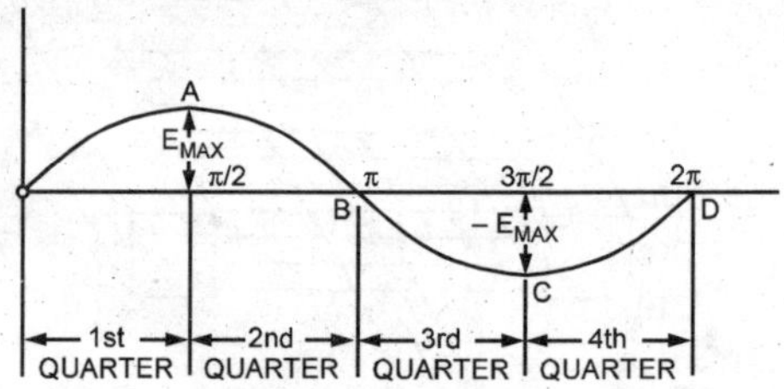

Fig. 9.2. ***Curve Showing Variation of Induced EMF in a Coil Rotated in a Uniform Magnetic Field***

The current induced in the coil is collected and conveyed to the external load circuit by connecting the coil terminals to two continuous and insulated rings, known as *slip rings or collector rings*, mounted on the generator shaft and making the two stationary brushes pressing against the slip rings, one brush bearing on each ring, as shown in Fig. 9.3. When the coil is rotated, the generated alternating emf causes a current to flow first in one direction and then in the other through the coil and external circuit. Such a current is called an *alternating current*.

To obtain the unidirectional or direct current in the external circuit, the collecting arrangement is modified as shown in Fig. 9.4. In this arrangement, the slip rings are replaced by split-ring made of a conducting material and splitted into two halves separated from each other by insulation and brushes are placed diametrically opposite instead of being side by side.

It will be observed that in the first half revolution current flows along ABMLCDA *i.e.* the brush M in contact with segment '*a*' acts as the + ve pole of the supply and brush L in contact with segment '*b*' acts as the –ve pole.

In the next half revolution, the direction of induced currents in the coil is reversed, but at the same time the positions of segments '*a*' and '*b*' are also reversed, with the result that brushes M and L again come in contact with +ve segment '*b*' and – ve segment '*a*' respectively. Thus the direction of current in the external load circuit remains the same.

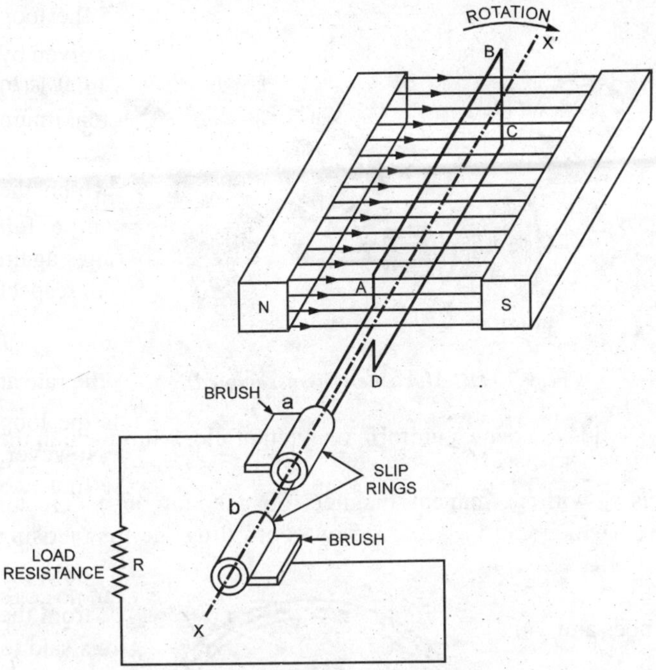

Fig. 9.3. *Elementary Generator With Brushes and Collector Rings*

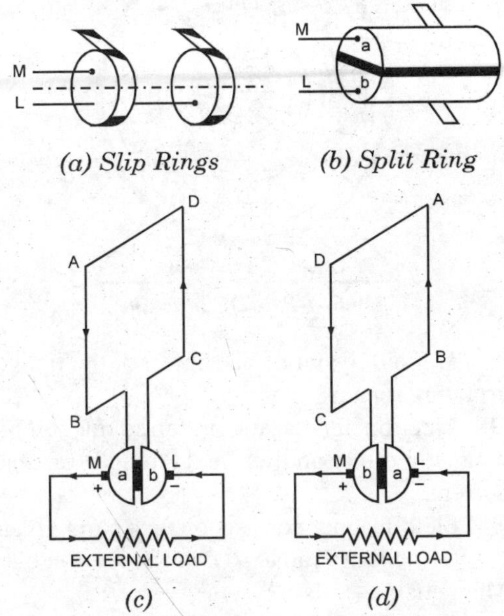

Fig. 9.4

The position of the brushes shall be so arranged that the change of segments from one segment to the other takes place when the plane of the rotating coil is perpendicular to the plane of field, since in this position the induced emf or induced current in the coil will be zero.

Thus the emf or current induced in the armature conductors of a dc generator is alternating which is rectified by the split-ring known as *commutator*.

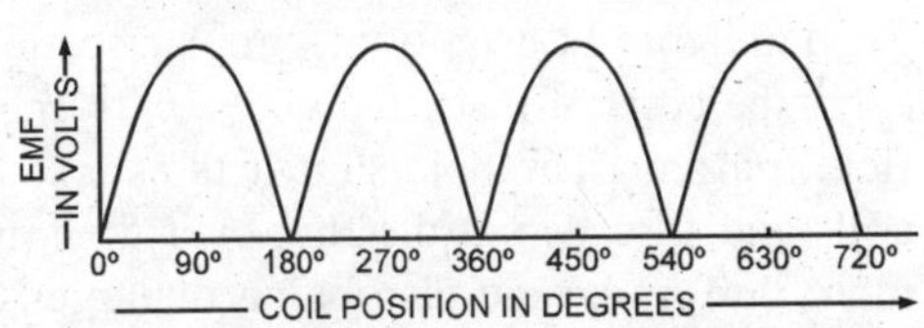

Fig. 9.5. *EMF Curve of a Simple DC Generator*

Although after rectification the current through the load circuit is always in the same direction, but it is not a steady current since the emf generated in the armature coil and applied to the brushes, varies from zero to maximum and back to zero twice each revolution. The variation in brush voltage is shown in Fig. 9.5.

A pulsating direct current such as given by a single coil generator is not suitable for most commercial purposes. However, by employing a large number of coils and commutator segments, with the coils evenly distributed around the surface of the armature the brush voltage may be made practically constant.

The voltage generated by a single turn armature coil is small. For this reason, the coils employed in commercial generators consist of several turns in series thereby increasing the magnitude of generated emf in direct proportion to the number of the turns in the coil.

9.3. CONSTRUCTIONAL DETAILS OF DC MACHINES

DC machine (whether a generator or motor) with four poles is shown in Fig. 9.6. In construction, dc machine consists of four parts mainly 1. Field magnets 2. Armature 3. Commutator 4. Brush and brush gear. Disassembled dc machine is shown in Fig. 9.7.

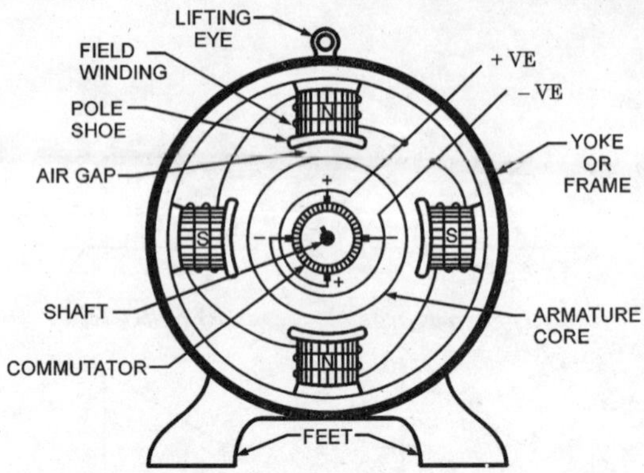

Fig. 9.6. *4-Pole DC Machine*

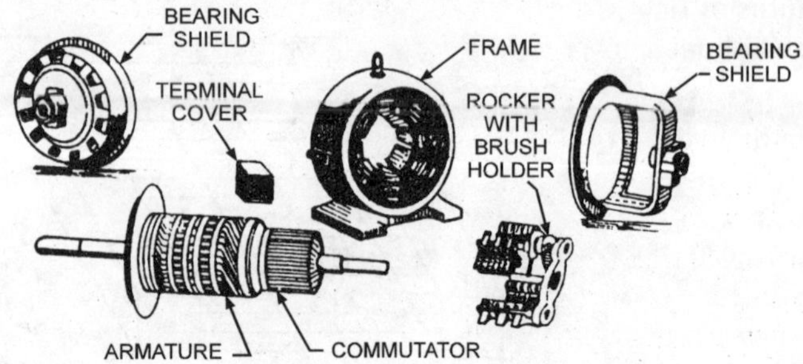

Fig. 9.7. *DC Machine (Dissassembled)*

1. Field System. The object of the field system is to create a uniform magnetic field, within which the armature rotates.

Electromagnets are preferred in comparison with permanent magnets on account of its greater magnetic effect and its field strength regulation, which can be achieved by controlling the magnetising current.

Field magnet consists of four parts given below :

(*i*) Yoke or Frame (*ii*) Pole cores (*iii*) Pole shoes and (*iv*) Magnetising coils

Cylindrical yoke is usually used which acts as a frame of the machine and carries the magnetic flux produced by the poles. In small machines, cast iron yokes are used, because of cheapness but yoke of a large machine is invariably made of fabricated steel due to its high permeability.

Pole core is usually of circular section and is used to carry the coils of insulated wires carrying the exciting (or field) current. The pole shoe acts as a support to the field coils and spreads out the flux over the armature periphery more uniformly and also being of larger cross section reduces the reluctance of the magnetic path.

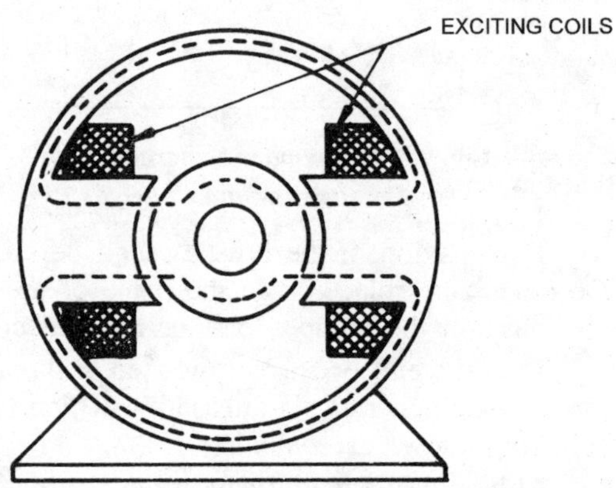

(*a*) *2-Pole DC Machine*

(*b*) *4-Pole DC Machine*

Fig. 9.8. *Complete Magnetic System of a DC Machine*

The field poles are usually formed of laminations (thin sheets of steel) and are bolted to the frame or yoke to which are also fastened the end bells with their *bearings* and the *brush rigging*. In small machines the poles are cast integral with the yoke from cast iron due to its low cost and less machining required by individual parts. In some machines the yoke and pole cores are made in single casting and laminated pole shoes are attached to the pole cores.

The object of the magnetising or field coils is to provide, under the various conditions of operation, the number of ampere-turns of excitation required to give the proper flux through the armature to induce the desired potential difference. The magnetic flux produced by the mmf developed by the field coils pass through the pole pieces, the air gap, the armature core and the yoke or frame. In Fig. 9.8 the dotted lines indicate the mean flux path through the complete magnetic circuit. It will be seen that the flux divides through two paths from each pole through the yoke.

There are several field constructions adopted according to the type of excitation. In shunt field, many turns of fine wire are used, in series field few turns of large cross sectional area are used and in compound field both shunt and series windings are used.

2. Armature. It is a rotating part of a dc machine and is built up in a cylindrical or drum shape. The purpose of armature is to rotate the conductors in the uniform magnetic field. It consists of coils of insulated wires wound around an iron and so arranged that electric currents are induced in these wires when the armature is rotated in a magnetic field. In addition, its most important function is to provide a path of very low reluctance to the magnetic flux. The armature core is made from high permeability silicon-steel stampings, each stamping being separated from its neighbouring one by thin paper or thin coating of varnish as insulation.

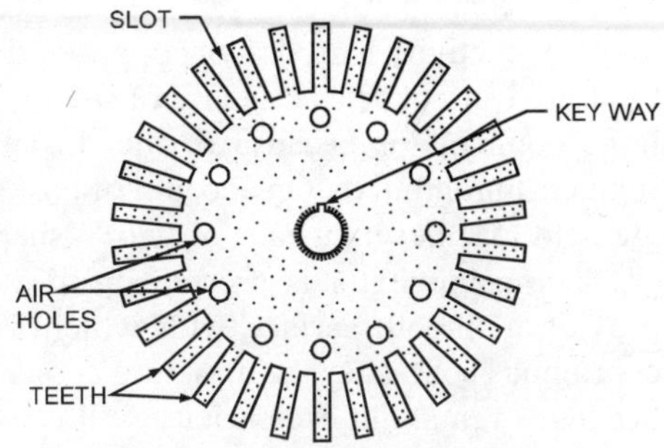

Fig. 9.9. *Armature Lamination*

A small air gap exists between the pole pieces and the armature so that there will be no rubbing in the machine. The air gap length is about 1.0 mm to 6 mm (say 1 mm for a 1 kW machine, 1.5 to 1.75 mm for medium size machines and 6 mm for 800 kW machine).

The use of high grade steel is made (*a*) to keep hysteresis loss low, which is due to cyclic change of magnetisation caused by rotation of the core in the magnetic field and (*b*) to reduce the eddy currents in the core which are induced by the rotation of the core in the magnetic field. By using stampings or laminations, the path of the eddy currents is cut into several units. The laminations must be in such a direction that they are perpendicular to the paths of eddy currents and parallel to the flux. Each lamination is about 0.3 to 0.6 mm thick.

3. Commutator. The commutator is a form of rotating switch placed between the armature and the external circuit and so arranged that it will reverse the connections to the external circuit at the instant of each reversal of current in the armature coils.

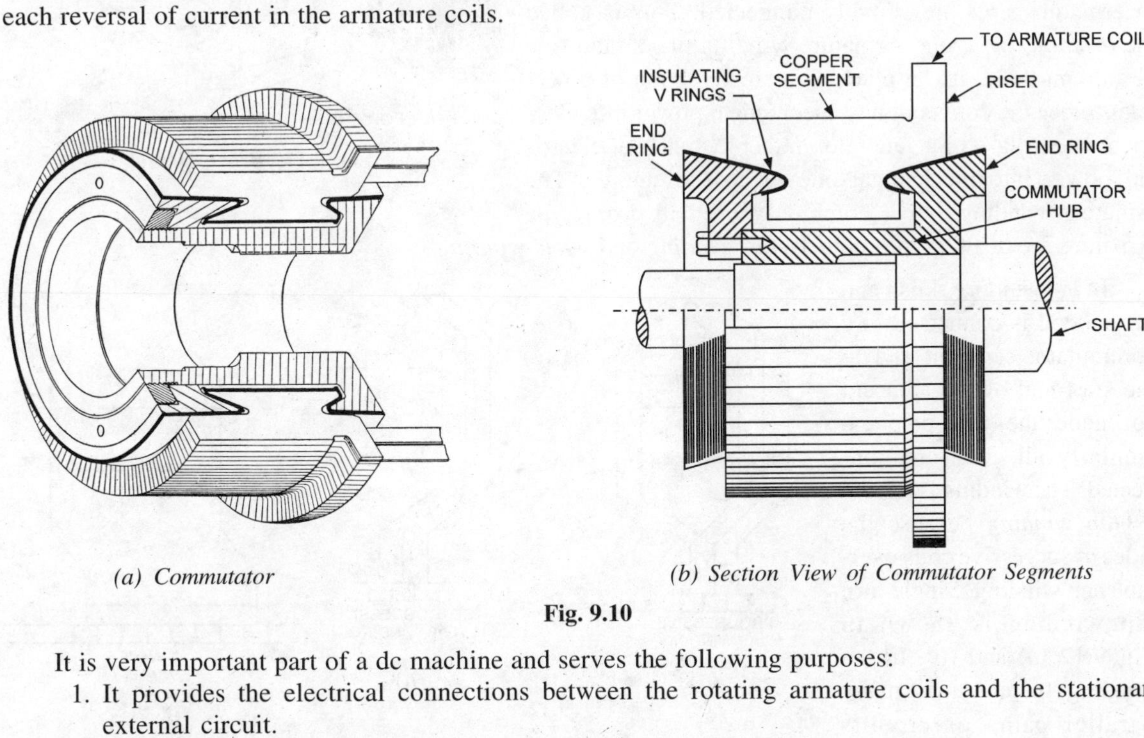

(*a*) *Commutator* (*b*) *Section View of Commutator Segments*

Fig. 9.10

It is very important part of a dc machine and serves the following purposes:

1. It provides the electrical connections between the rotating armature coils and the stationary external circuit.

2. As the armature rotates, it performs a switching action reversing the electrical connections between the external circuit and each armature coil in turn so that the armature coil voltages add together and result in a dc output voltage.
3. It also keeps the rotor or armature mmf stationary in space.

The commutator is essentially of cylindrical structure and is built up of wedge shaped segments of high conductivity hard drawn copper or drop forged copper. These segments are insulated from each other by thin layers of mica (usually of 0.5 to 1 mm thickness). Mica is to be preferred but cannot be used for large commutator because of the difficulty of obtaining large sheets, making the cost of large mica segments prohibitive. On account of cost also micanite is often used for small commutators. The segments are held together by means of two V-shaped rings that fit into the V-grooves cut into the segments.

The commutator is pressed on to the armature shaft, and the outer periphery is then machined to provide a smooth surface with which a stationary carbon (or graphite or copper) brush can maintain continuous contact as the armature and commutator rotate. Great care is taken in building the commutator because even slight eccentricity will cause the brushes to bounce, causing undue sparking.

4. Brushes. The function of brushes is to collect current from the commutator and supply it to the external load circuit (the armature of the machine being connected to the external load circuit via the commutator and brushes). The brushes are rectangular in shape and rest on the commutator. Brushes are manufactured in a variety of compositions and degrees of hardness to suit the commutation requirements. They may be classified roughly as carbon, carbon graphite, graphite, metal graphite and copper. The allowable current density at the brush contact varies from 5 A per square cm in case of carbon to 23 A per square cm in the case of copper.

The brushes are housed in brush-holders (usually of the box type) which are mounted on the brush holder studs or brackets.

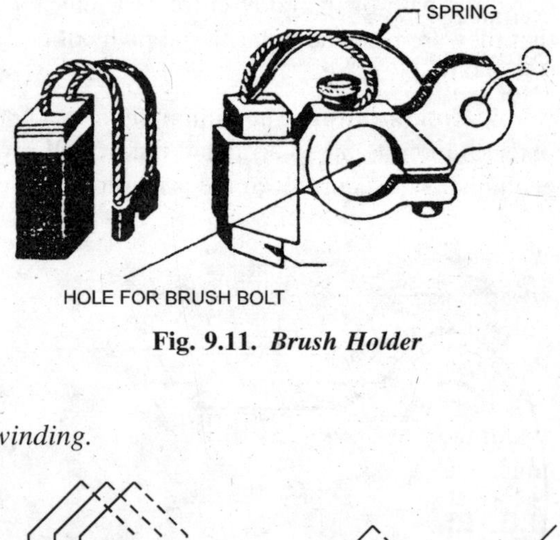

Fig. 9.11. ***Brush Holder***

5. Armature Winding. The insulated wires housed in armature slots are suitably connected. This is called the *armature winding*. Armature winding plays vital role in a dc machine. It is a place where conversion of power takes place *i.e.* conversion of mechanical power into electrical one in case of a generator and conversion of electrical power into mechanical one in case of motor. The armature windings most commonly used in drum type armature are of two types namely *lap winding* and *wave winding*.

In *lap winding* finish end of one coil is connected to a commutator segment and to the start end of the adjacent coil under the same pole and similarly all coils are connected. The winding is known as *lap winding* because the sides of successive coils overlap each winding. Single turn lap winding is shown in Fig. 9.12 (*a*) and (*b*). In lap winding there are as many parallel paths or circuits

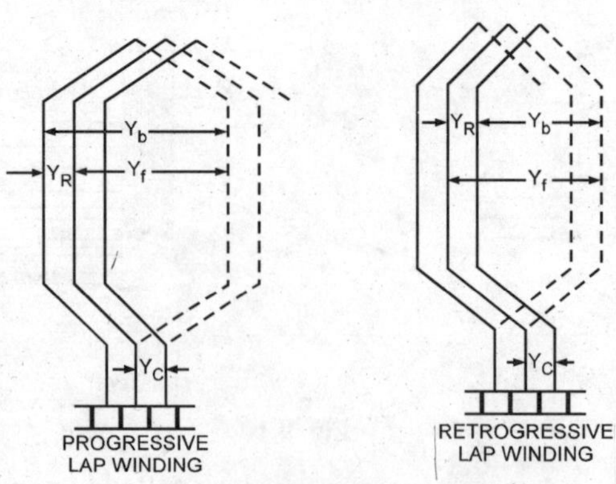

Fig. 9.12

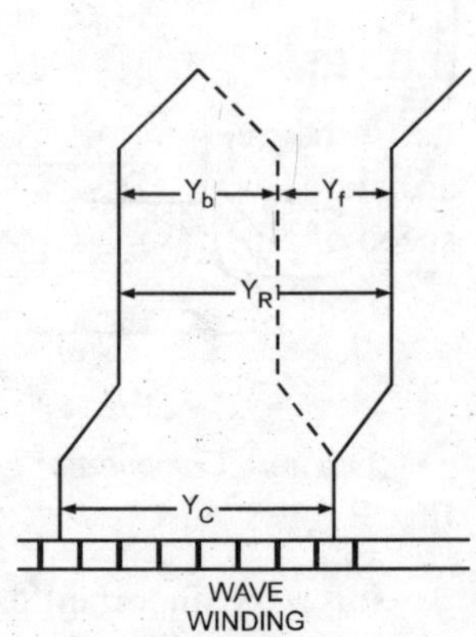

Fig. 9.13

through the winding as there are field poles on the machine. Lap winding because of large number of parallel paths and lesser number of conductors per path is suitable for large current and low voltage machines.

Wave winding is also sometimes known as *series winding*. In wave winding finish end of one coil is connected to the start of another coil as shown in Fig. 9.13. Thus in wave winding, the winding progresses, passing every N pole and S pole till it returns to the coil side from where it was started. As the winding is wavy, the winding is, therefore, called *wave winding*. The wave winding gives always two parallel paths irrespective of number of poles and, therefore, for a given number of poles and armature conductors, it gives more emf than the lap winding. Hence it is used for high voltage and low current machines.

6. Bearings. With small machines, ball bearings may be used at both ends. For larger machines, roller bearings are used at the driving end, and ball bearings may be used at the non-driving end, *i.e.* at the commutator end. Thrust bearings are used where excessive end thrust is anticipated. Sleeve bearings with ring lubrication are used for motors when very silent running is required. For large machines pedestal bearings are generally used.

7. Shaft. The shaft is made of mild steel with a maximum breaking strength. The shaft is used to transfer mechanical power from or to the machine. The rotating parts such as armature core, commutator, cooling fan etc. are keyed to the shaft.

Sectional view of a dc rotor consisting of armature shaft, armature core, armature winding and commutator is illustrated in Fig. 9.14.

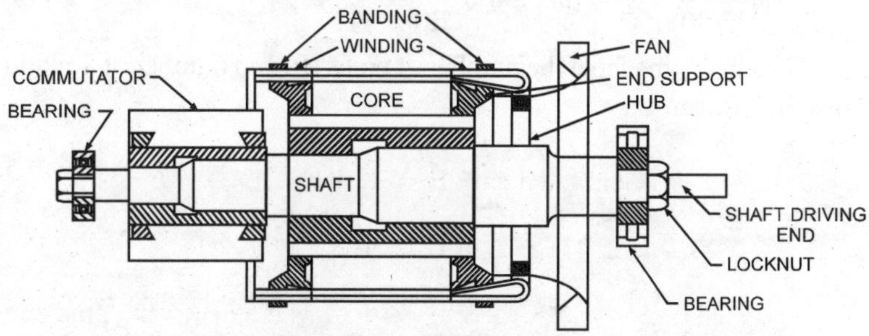

Fig. 9.14. ***Sectional View of Rotor Assembly of a DC Machine***

9.4. FUNCTION OF COMMUTATOR FOR GENERATING AND MOTORING ACTION

As mentioned in Art. 9.1, commutator is a device which under certain conditions converts ac into dc and vice versa.

The commutator is a form of rotating switch placed between the armature and external circuit and so arranged that it reverses the connections to the external circuit at the instant of each reversal of the current in the armature coil, as already explained in Art. 9.2, and thus converts induced alternating currents in armature coils into direct currents in the external load circuit.

In case of a dc motor, a direct current passes through the brushes and commutator to the armature winding; while it passes through the commutator it is converted into ac so that the group of conductors under successive field poles carry currents in opposite directions.

9.5. EMF EQUATION

Let Φ be the flux per pole in webers, Z the total number of armature conductors or coil sides on the armature, P the number of poles, A the number of parallel paths in the armature and N be the rotational speed of armature in revolutions per minute (rpm).

As will be recalled, the induced emf is proportional to the time rate of change of the magnetic flux *i.e.*

$$e = \frac{-d\phi}{dt}$$

During one revolution of armature in a P-pole generator each armature conductor cuts the magnetic flux P times, so flux cut by one conductor in one revolution= PΦ webers.

Since the number of revolutions made by the armature per minute is N so number of revolutions made per second is $\frac{N}{60}$ and, therefore, flux cut by each conductor per second

= Flux cut by one conductor per revolution × number of revolutions of armature/second

$$= \Phi P \times \frac{N}{60} \text{ webers}$$

Consequently the average emf induced in one conductor will be

$$e = \Phi P \frac{N}{60} \text{ volts}$$

The number of conductors in series between a + ve brush and – ve brush is equal to the total number of conductors divided by the number of parallel paths *i.e.* number of armature conductors per parallel path $= \frac{Z}{A^*}$.

∴ The total emf generated between the terminals,

E = Average emf induced in one conductor × number of conductors in each circuit or parallel path

$$= \Phi P \frac{N}{60} \times \frac{Z}{A} \text{ volts} = \Phi Z \frac{N}{60} \times \frac{P}{A} \text{ volts} \qquad ...(9.1)$$

For a given machine the number of poles (P) and number of armature conductors per parallel path (Z/A) are constant, therefore,

$$\text{Generated emf, } E = K\Phi N \qquad ...(9.2) \text{ where } K = \frac{PZ}{60\,A}$$

$$\text{or } E \propto \Phi N$$

$$\text{or } E \propto \Phi\,\omega \text{ where } \omega = \frac{2\pi N}{60}, \text{ the angular velocity in radians/second}$$

Thus we see that the induced emf is directly proportional to flux per pole Φ and speed N. Moreover, the polarity of the induced emf depends upon the direction of the magnetic field and the direction of rotation. If either of the two is reversed, the polarity of the induced emf *i.e* brushes is reversed, but when both are reversed, the polarity remains unchanged.

The induced emf is fundamental phenomenon to all dc machines whether they are operating as generators or motors. However, when the machine is operating as a generator, this induced emf is called the *generated emf,* E_g whereas in case of a machine operating as a motor it is called the *counter-* or *back emf,* E_b.

Example 9.1. A dynamo has a rated armature current at 250 A. What is the current per path of the armature if the armature winding is simplex wave wound or simplex lap wound ? The machine has 12 poles.

Solution : Rated armature current, I_a = 250 A

Number of poles, P = 12

With simplex wave winding

Number of parallel paths, A = 2

$$\therefore \text{Current per path, } I_c = \frac{I_a}{A} = \frac{250}{2} = 125 \text{ A} \quad \textbf{Ans.}$$

With simplex lap winding

Number of parallel paths, A = P = 12

$$\text{Current per path, } I_c = \frac{I_a}{A} = \frac{250}{12} = 20.833 \text{ A} \quad \textbf{Ans.}$$

Example 9.2. The induced emf in a dc machine when running at 500 rpm is 180 V. Calculate the induced emf while the machine is running at 600 rpm by assuming flux to be constant.

Solution : Since induced emf is proportional to the product of flux per pole and armature speed for a given machine *i.e.*

$$E \propto \Phi N$$

* A = 2 in case of wave winding and = P in case of lap winding.

$$\text{or } E \propto N \qquad \because \text{ flux } \Phi \text{ is assumed to be constant}$$

$$\text{or} \quad \frac{E_2}{E_1} = \frac{N_2}{N_1}$$

$$\text{or } E_2 = E_1 \times \frac{N_2}{N_1} = 180 \times \frac{600}{500} = 216 \text{ V } \textbf{Ans.}$$

Example 9.3. Calculate the voltage induced in the armature winding of a 4-pole, lap wound dc machine having 728 active conductors and running at 1,800 rpm. The flux per pole is 30 mWb.

[Pb Technical Univ. Electrical Engineering December 2000]

Solution :

Flux per pole, $\phi = 30 \text{ mWb} = 0.03 \text{ Wb}$

Number of armature conductors, $Z = 728$

Speed, $N = 1{,}800$ rpm

Number of poles, $P = 4$

Number of parallel paths, $A = P$ $\qquad \because$ machine is lap wound

$$\text{Voltage induced, } E = \frac{\phi Z N}{60} \times \frac{P}{A} = \frac{0.03 \times 728 \times 1{,}800}{60} \times \frac{4}{4} = 655.2 \text{ V } \textbf{Ans.}$$

Example 9.4. Calculate the voltage induced in the armature winding of a 4-pole, wave wound, dc machine having 728 conductors and running at 1,800 rpm. The flux per pole is 35 mWb.

[Pb Technical Univ. Electrical Engineering June 2000]

Solution :

Flux per pole, $\phi = 35 \text{ mWb} = 0.035 \text{ Wb}$

Number of armature conductors, $Z = 728$

Speed, $N = 1{,}800$ rpm

Number of poles, $P = 4$

Number of parallel paths, $A = 2$ $\qquad \because$ machine is wave-wound

$$\text{Voltage induced, } E = \frac{\phi Z N}{60} \times \frac{P}{A} = \frac{0.035 \times 728 \times 1{,}800}{60} \times \frac{4}{2} = 1{,}528.8 \text{ V } \textbf{Ans.}$$

Example 9.5. A 4-pole lap wound armature has 144 slots with two coil sides per slot, each coil having two turns. If the flux per pole is 20 mWb and the armature rotates at 720 rpm, what is the induced voltage ?

[Pb Technical Univ. Electrical Engineering May 2001]

Solution :

Flux per pole, $\phi = 20 \text{ mWb} = 20 \times 10^{-3}$ Wb

Number of armature conductors, Z = Number of slots × number of coil sides per slot × number of turns in each coil

$= 144 \times 2 \times 2 = 576$

Speed, $N = 720$ rpm

Number of poles, $P = 4$

Number of parallel paths, $A = P = 4$

$$\text{Induced voltage, } E = \phi \times Z \times \frac{N}{60} \times \frac{P}{A}$$

$$= 20 \times 10^{-3} \times 576 \times \frac{720}{60} \times \frac{4}{4} = 138.24 \text{ V } \textbf{Ans.}$$

9.6. ARMATURE REACTION

When the generator is supplying no load and the field winding is energized, there exists in it only the mmf of the main poles which creates the main flux. When the generator is loaded, current flows through the armature winding and a magnetizing effect or mmf that acts at right angles to the main field flux, is set up. The effect of magnetic field set up by the armature current on the distribution of the flux under main poles is known as the *armature reaction*. The effect of armature reaction is to weaken the field strength in the gap under the leading pole tips and strenghten under trailing pole tips. The magnetic field of the machine is distorted and the physical neutral line is shifted in the direction of rotation. The displacement of the neutral line depends on the magnitude of armature flux, which in turn depends on the armature or load current. As a result, the brushes have to be shifted in the direction of rotation to avoid sparking (refer to Fig. 9.15).

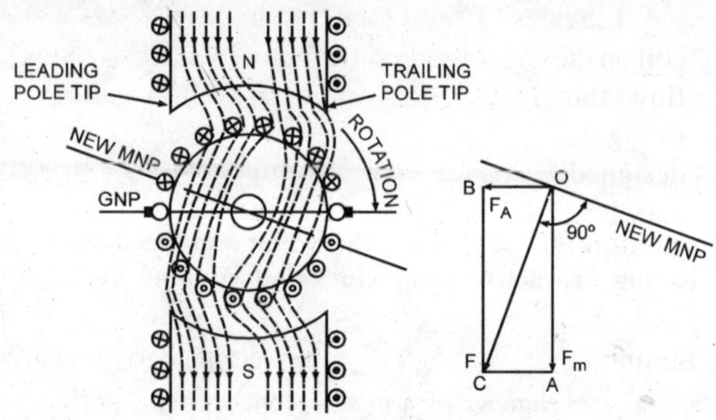

Fig. 9.15

9.7. CLASSIFICATION OF DC GENERATORS

The mmf necessary to establish the flux in the magnetic circuit of a dc generator can be obtained by means of (*i*) a permanent magnet (*ii*) field coils excited by some external source and (*iii*) field coils excited by the generator itself.

When a permanent magnet is used for establishing the flux in the magnetic circuit, the generator is known as a *permanent magnet generator*. It consists of an armature and one or several permanent magnets encircling the armature. The field developed by the poles of such machines remains fairly constant. Although these machines are very compact but in view of the low power thus generated, permanent magnet generators have not found industrial applications. Such generators are employed only in small sizes like dynamos in motor cycles etc.

9.7.1. Separately Excited DC Generators

Since the operation of a dc machine depends upon a fixed polarity of the poles which does not vary with time, the field coils need energization from a dc source.

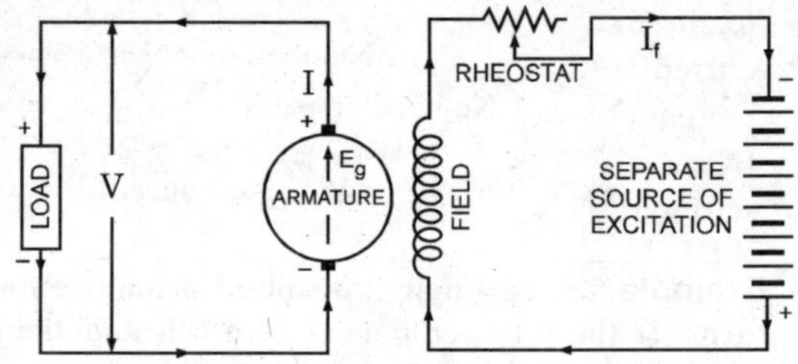

Fig. 9.16

A dc generator whose field winding is excited from an independent external dc source, such as a battery, the generator is called a *separately excited generator*. The circuit diagram is illustrated in Fig. 9.16. In this case current flowing through the armature, I_a and load, I_L is the same and the terminal voltage (voltage across the load), V is equal to the generated emf, E_g less voltage drop in armature, I_aR_a *i.e.*

$$I_a = I_L = I(\text{say}) \quad ...(9.3)$$

$$V = E_g - I_a R_a \quad \text{where } R_a \text{ is the armature resistance} \quad ...(9.4)$$

$$\text{Power developed, } P_g = E_g I_a \quad ...(9.5)$$

$$\text{Power delivered to external Load, } P_L = VI_L \quad ...(9.6)$$

9.7.2. Self-Excited DC Generators

A dc generator whose field winding is excited by the current supplied by the generator itself, is called a *self-excited generator*. In such machines the field coils are interconnected with the armature winding. The field coils may be connected either in *series* with the armature, in parallel with the armature or *partly in series and partly in parallel* with the armature. Accordingly the self-excited generators may be classified as (*i*) *series wound generators* (*ii*) *shunt wound generators* and (*iii*) *compound wound generators.*

Due to residual magnetism, some flux is always present in the poles of such machines. When the armature is rotated, a small voltage is induced in the armature winding owing to residual flux. This induced voltage causes a small current to flow in the field coils and thus increase in flux per pole. The increase in flux causes increase in induced voltage which further increases the field current and so flux per pole. These events take place rapidly and the generator builds up to the rated voltage.

1. Series Wound Generators. In a series wound generator, there is only one field winding (one field coil on the core of each salient pole) which is connected in series with armature winding so that whole current flows through the field winding as well as load. The connection diagram of a series wound generator is given in Fig. 9.17. Since the series field winding carries the full-load current, therefore, series field winding is designed with a fewer turns of thick wire or strips. The resistance of series field winding R_{se} is, therefore, very low, say of the order of 0.5 Ω. The important relations are given below :

$$I_a = I_{se} = I_L = I \text{ (say)} \quad ...(9.7)$$

$$V = E_g - I(R_a + R_{se}) \quad ...(9.8)$$

$$\text{Power developed, } P_g = E_g I \quad ...(9.9)$$

$$\text{Power delivered, } P_L = VI \quad ...(9.10)$$

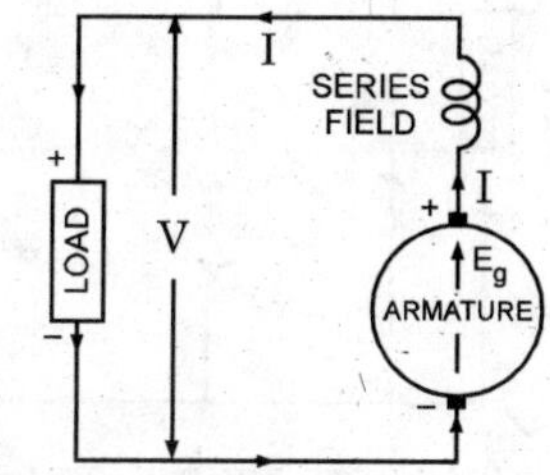

Fig. 9.17

2. Shunt Wound Generators. In a shunt wound generator also there is only one field winding, as in case of series wound generator, but in this case it is connected across the armature circuit forming a parallel or shunt circuit. The voltage across the field winding is, therefore, the same as the terminal voltage of the generator. The connection diagram is illustrated in Fig. 9.18.

In operation, total armature current I_a flowing from the +ve brush divides between two parallel paths; the external circuit and the field circuit. The load current I_L and the shunt field current I_{sh} close the circuit at the –ve brush and their sum is equal to armature current *i.e.*

$$I_a = I_L + I_{sh}$$

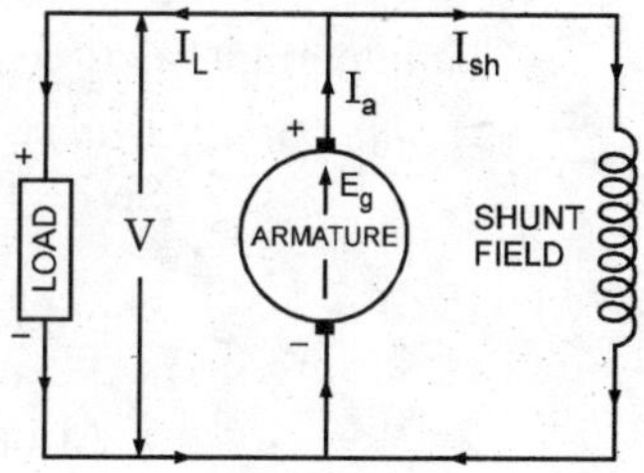

Fig. 9.18

The effective power of a generator is proportional to the current delivered to the external load circuit, therefore, it is necessary to keep shunt field current as small as possible. For shunt wound generators it is kept about 2.5 per cent of the rated armature current. Shunt field coils are, therefore, designed to produce the desired mmf by means of a relatively large number of turns of fine wire. The resistance of the shunt field winding R_{sh} will naturally be high, say of the order of 100 Ω. The important relations are given below:

$$\text{Shunt field current, } I_{sh} = \frac{V}{R_{sh}} \quad ...(9.11)$$

$$\text{Armature current } I_a = I_L + I_{sh} \quad ...(9.12)$$

$$\text{Terminal voltage, } V = E_g - I_a R_a \quad ...(9.13)$$

$$\text{Power developed, } P_g = E_g I_a \quad ...(9.14)$$

$$\text{Power delivered, } P_L = VI_L \quad ...(9.15)$$

3. Compound Wound Generators. In compound wound generators, there are two field windings (two separate coils on the core of each pole). One of them (having many turns of fine wire) is connected across the armature and the other (having few turns of thick wire) is connected in series with the armature winding. Compound generators may be connected either *short shunt*, with the shunt field winding in parallel with the armature alone, as shown in Fig. 9.19 (*a*) or *long shunt*, with the shunt field winding in parallel with both the armature and series field winding, as shown in Fig. 9.19 (*b*).

In a compound wound generator, the major portion of the excitation is usually supplied by the shunt field. The shunt field is slightly weaker and the series field is considerably weaker than those of the corresponding machine in which the entire excitation is produced by single shunt or by a single series winding.

Compound generators are also of the two types, known as *cumulative compound wound* and *differential compound wound* generators. In cumulative compound wound generators, series field assists the shunt field, as shown in Fig. 9.20 (*a*) whereas in differential compound wound generators, series field opposes the shunt field, as shown in Fig. 9.20 (*b*).

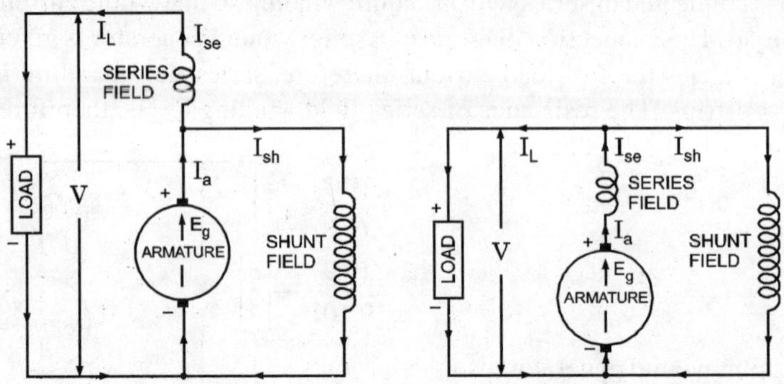

(a) Short-Shunt Compound Wound Generator *(b) Long-Shunt Compound Wound Generator*

Fig. 9.19

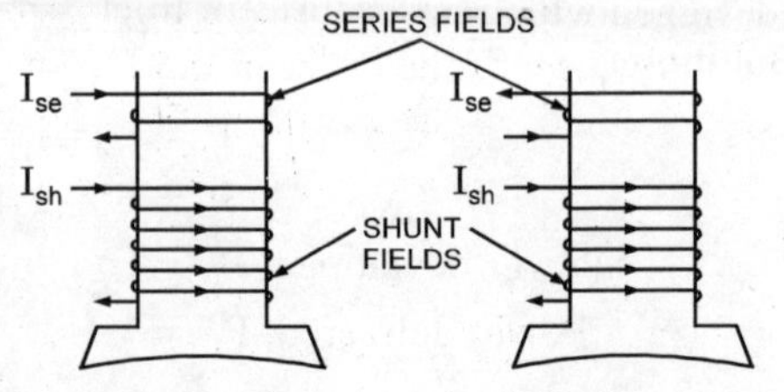

(a) Cumulative Compounding *(b) Differential Compounding*

Fig. 9.20

The important relations for short-shunt and long-shunt compound wound generators are given below:

(a) Short Shunt Compound Wound Generators

$$\text{Series field current, } I_{se} = I_L \qquad ...(9.16)$$

$$\text{Shunt field current, } I_{sh} = \frac{V + I_{se} R_{se}}{R_{sh}} \qquad ...(9.17)$$

$$\text{Armature current, } I_a = I_L + I_{sh} \qquad ...(9.18)$$

$$\text{Terminal voltage, } V = E_g - I_a R_a - I_{se} R_{se} \qquad ...(9.19)$$

$$\text{Power developed, } P_g = E_g I_a \qquad ...(9.20)$$

$$\text{Power delivered, } P_L = V I_L \qquad ...(9.21)$$

(b) Long Shunt Compound Wound Generators

$$\text{Shunt field current, } I_{sh} = \frac{V}{R_{sh}} \qquad ...(9.22)$$

$$\text{Armature current, } I_a = \text{Series field current, } I_{se} \qquad ...(9.23)$$

$$= I_L + I_{sh} \qquad ...(9.24)$$

$$\text{Terminal voltage, } V = E_g - I_a R_a - I_{se} R_{se} = E_g - I_a (R_a + R_{se}) \qquad ...(9.25)$$

$$\text{Power developed, } P_g = E_g I_a \qquad ...(9.26)$$

$$\text{Power delivered, } P_L = V I_L \qquad ...(9.27)$$

Example 9.6. A 4-pole machine running at 1,500 rpm has an armature with 90 slots having 6 conductors per slot. The flux per pole is 6 × 10⁻² Wb. Determine the induced emf as a dc generator if the coils are lap connected. If the current per conductor is 100 amperes, determine the electrical power output of the machine.

Solution :

Flux per pole, $\Phi = 6 \times 10^{-2}$ Wb

Number of armature conductors, Z = Number of slots × number of conductors per slot

$= 90 \times 6 = 540$

Number of parallel paths, A = P = 4 ∵ the machine is lap-connected.

Speed, N = 1,500 rpm

$$\text{Induced emf, } E_g = \Phi Z \frac{N}{60} \times \frac{P}{A} = 6 \times 10^{-2} \times 540 \times \frac{1{,}500}{60} \times \frac{4}{4} = 810 \text{ V } \textbf{Ans.}$$

Current per conductor, I_c = 100 A

Armature current, $I_a = I_c \times A = 100 \times 4 = 400$ A

Electrical power developed $= E_g \times I_a = 810 \times 400 = 324{,}000$ Wor 324 kW **Ans.**

Example 9.7. A 4-pole shunt generator with lap-connected armature having field and armature resistances of 50 Ω and 0.1 Ω respectively supplies a 2,400 W load at a terminal voltage of 100 V. Calculate the total armature current, the current per armature path and the generated emf.

[*Pb. Technical Univ. Electrical Engineering December 2001*]

Solution :

Power supplied, $P = 2{,}400$ W

Terminal voltage, $V = 100$ V

Load current, $I_L = \frac{P}{V} = \frac{2{,}400}{100} = 24$ A

Shunt field current, $I_{sh} = \frac{V}{R_{sh}} = \frac{100}{50} = 2$ A

Total armature current, $I_a = I_L + I_{sh}$
$= 24 + 2 = 26$ A **Ans.**

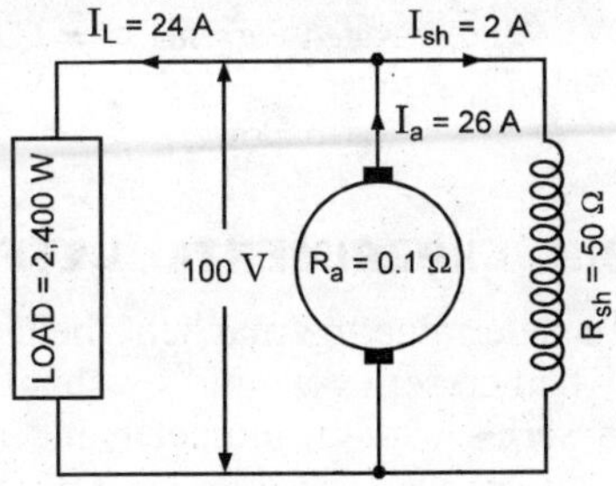

Fig. 9.21

Current per path, $I_C = \frac{I_a}{A} = \frac{26}{4} = 6.5$ A **Ans.** ∵ in lap-wound machine A = P = 4

Generated emf, $E_g = V + I_a R_a = 100 + 26 \times 0.1 = 102.6$ V **Ans.**

Example 9.8. A 4-pole shunt generator with lap-connected armature having field and armature resistances of 50 Ω and 0.1 Ω respectively supplies 60 nos 40 W lamps at 100 V. Calculate the total armature current, current per path and the generated emf. Assume a constant drop of 1 V per brush.

Solution :

Total lamp load, P_L = Number of lamps × wattage of each lamp = 60 × 40 = 2,400 W

Terminal voltage, $V = 100$ V

Load current, $I_L = \frac{P_L}{V} = \frac{2{,}400}{100} = 24$ A

Shunt field current, $I_{sh} = \frac{V}{R_{sh}} = \frac{100}{50} = 2$ A

Total armature current, $I_a = I_L + I_{sh}$
$= 24 + 2 = 26$ A **Ans.**

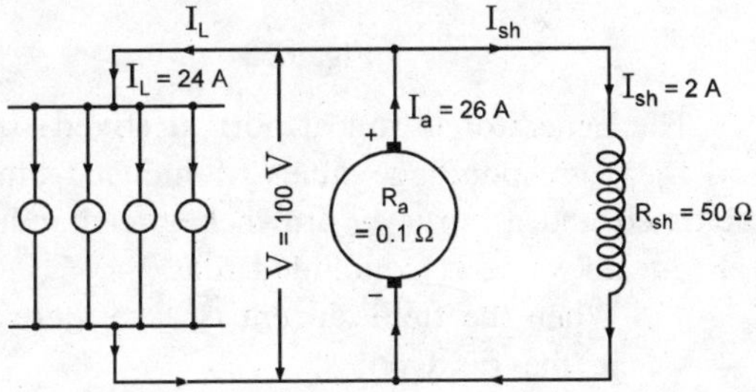

Fig. 9.22

Current per path, $I_C = \frac{I_a}{A} = \frac{26}{4} = 6.5$ A **Ans.**

Generated emf, $E_g = V + I_a R_a$ + brush drop $= 100 + 26 \times 0.1 + 2 \times 1 = 104.6$ V **Ans.**

Example 9.9. A 6-pole, 12 kW, 240 V, dc machine is wave-connected. If the same machine is lap-connected, all other things remaining same, calculate its voltage, current and power ratings.

[*Pb. Technical Univ. Basic Electrical Engineering December 2002*]

Solution : *When machine is wave-connected*

Rated power, $P = 12$ kW

Rated voltage, $V_1 = 240$ V

Voltage across each path, $V = 240$ V

Number of parallel paths, $A_1 = 2$

Rated current, $I_1 = \frac{P_1}{V_1} = \frac{12 \times 1{,}000}{240} = 50$ A

Current in each parallel path, $I_{C_1} = \frac{I_1}{A_1} = \frac{50}{2} = 25$ A

When machine is lap-connected

Number of parallel paths, $A_2 = 6$

Rated voltage, $V_2 = V_1 \times \frac{A_1}{A_2} = 240 \times \frac{2}{6} = 80$ V **Ans.**

Rated current, I_2 = Current in parallel path × number of parallel paths = 25 × 6 = 150 A **Ans.**

Rated power, $P_2 = V_2 I_2 = \frac{80 \times 150}{1,000} = 12$ kW **Ans.**

9.8. EXPERIMENTAL DETERMINATION OF MAGNETIC BEHAVIOUR OF DC MACHINE

To determine the magnetic behaviour of a dc machine experimentally, the field winding of the machine (whether series wound or shunt wound) is disconnected from the machine and, instead, is connected in series with an ammeter and rheostat, and put across dc supply main. A multi-range voltmeter V is connected across the armature terminals. The connections are shown in Fig. 9.23.

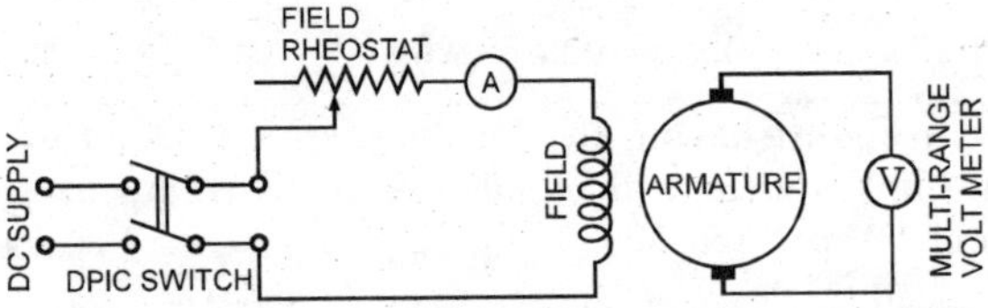

Fig. 9.23

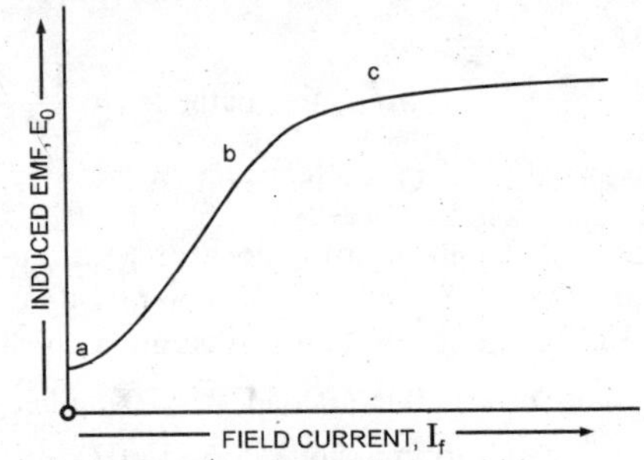

Fig. 9.24

The generator is run at normal (fixed) speed, the field current I_f is increased from zero in steps and the corresponding values of induced emf E_0 are read from the voltmeter. The observations are tabulated and a curve is drawn between induced emf E_0 and field current I_f, as shown in Fig. 9.24.

From above it is concluded that

1. When the field current is zero there is some induced emf. This is due to residual magnetism in the field poles.
2. Over a fairly wide range of field current (upto point *b* on the curve) the curve is linear.
3. After point *b* on the curve, the curve deviates from linear relationship.
4. After point *c* on the curve, the curve almost levels off. It is due to saturation effect.

9.9. EXTERNAL OR LOAD CHARACTERISTICS OF DC GENERATORS

External characteristic curve gives the relation between terminal voltage V and the load current I under the given conditions of speed and excitation. The terminal voltage V is always less than generated emf E on account of voltage drop in armature. This curve, therefore, lies below the internal characteristic. This characteristic is of utmost importance for determining the suitability of a generator for a particular service.

9.9.1. Separately Excited DC Generators

Since exciting current is independent of load or armature current, if a curve is drawn between flux per pole and load current, keeping field current constant and neglecting armature reaction, a straight line parallel to X-axis is obtained, as indicated by curve I in Fig. 9.25. But due to armature reaction, the curve of actual flux is slightly drooping as represented by curve II in Fig. 9.25.

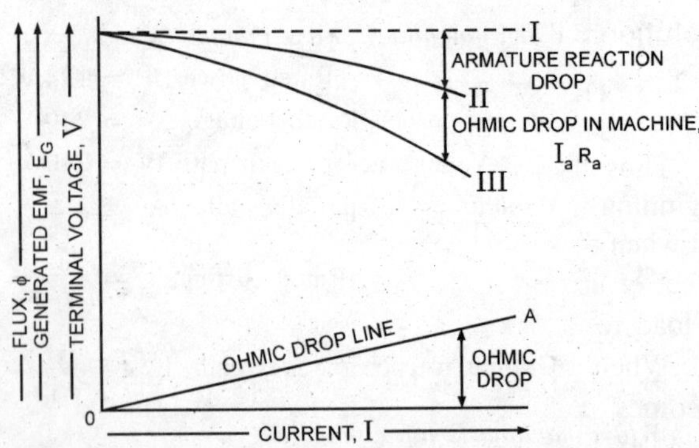

Fig. 9.25. ***Characteristics of a Separately Excited DC Generator***

The second curve also gives to some other scale the emf generated in the armature and, therefore, second curve represents the internal or total characteristic* of the separately excited dc generator. The terminal voltage on load is the emf generated less voltage drop in the armature *i.e.* $V = E - I_a R_a$, so if a line OA representing the armature resistance is drawn and its ordinates are deducted from curve II, curve III giving relation between terminal voltage and load current, known as *external characteristic* is obtained. From curve III, it is obvious that with the increase in load current, terminal voltage decreases slightly. This decrease can easily be neutralized by increasing field current slightly and, therefore, constant terminal voltage can be maintained.

9.9.2. Series Wound DC Generators

In a series wound dc generator, the armature winding, field winding and external load circuit, all are connected in series with each other, as shown in Fig. 9.17 and, therefore, the same current flows through all parts of the circuit.

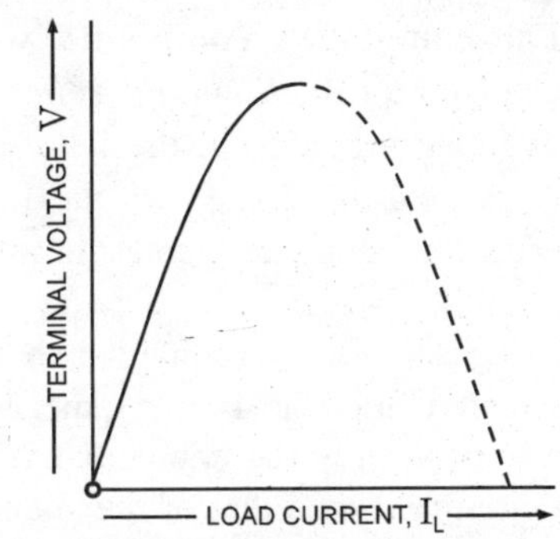

Fig. 9.26. ***Load Characteristic of Series Wound DC Generator***

At no load (the external circuit is open) no current flows through the field winding. On closing of external or load circuit, both the armature winding and the field winding carry the load current. With the increase in load current, the terminal voltage increases (the solid line in Fig. 9.26) until the magnetic circuit becomes saturated. With the further increase in load, the terminal voltage falls. This is due to increased armature reaction and larger voltage drop across the armature and field windings (the dotted line in the figure). In the case of a short circuit in the armature circuit, a fairly heavy current will be established while the terminal voltage will be zero. Because of the sharp increase in generator terminal voltage with increasing load, series wound generators are rarely used.

9.9.3. Shunt Wound DC Generators

The terminal voltage on no load is governed by the shunt field current (*i.e.* shunt field resistance), and let it be represented by OA in Fig. 9.27.

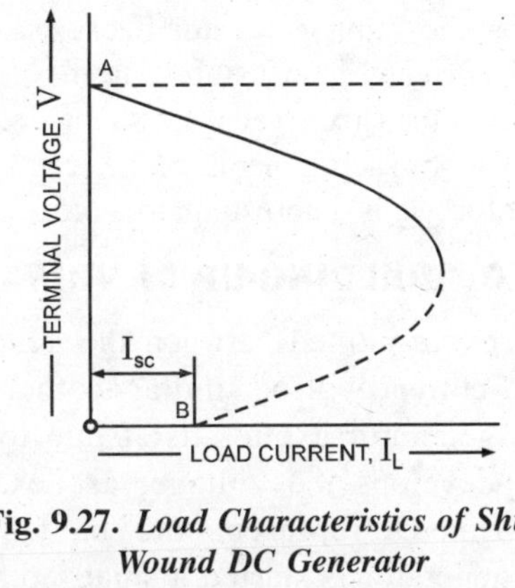

Fig. 9.27. ***Load Characteristics of Shunt Wound DC Generator***

As the generator is loaded by decreasing the resistance of the external load circuit, the terminal voltage falls. This is due to following three causes.

1. The voltage drop across the armature winding and the contact resistance of the brushes increases with the increase in armature current.
2. As the load is increased, the current in armature conductors increases, so the effect of armature reaction increases and, therefore, field is weakened. Due to weakening of field the induced emf is reduced and, therefore, terminal voltage falls.
3. The decrease in terminal voltage as a result of the first two factors causes decrease in field current. This will in turn cause the emf and, therefore terminal voltage of the generator to drop too.

Thus the effects are cumulative and terminal voltage is reduced to a much greater extent. In the beginning the effect of decrease in resistance predominates over the effect of decreased terminal voltage but when the load current reaches a certain value (much higher than the full-load current) the demagnetising effect of armature reaction and voltage drop in the armature become so important that a further decrease in load resistance causes a decrease in current rather than increase and so the characteristic turns back.

When the terminals are short circuited, there is no terminal voltage and thus the shunt winding becomes inactive but a small current is established due to a small voltage induced by the residual

* Curve giving relation between the emf actually generated in the armature and the armature current.

magnetism. Thus external characteristic meets the current axis at point B. *The shunt generators are, therefore, self-protective against accidental short circuit.*

9.9.4. Characteristics of Compound Wound DC Generators

As already mentioned in Art. 9.7, compound wound dc generators are of two types viz. *cumulative compound wound* and *differential compound wound* dc generators.

The cumulative compound wound generator combines the characteristics of the shunt and series generators. At no load the shunt field winding provides all the field flux since there is no current in the series field winding. With the increase in load on the generator, the series field adds an increasing amount of field flux to that of the shunt field.

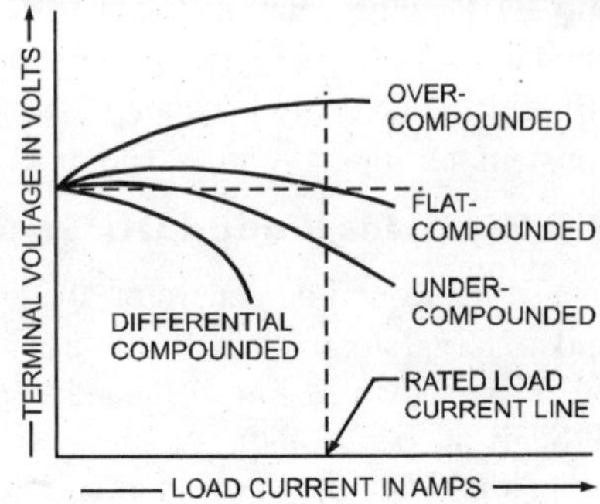

Fig. 9.28. ***External Characteristics of DC Compound Generators***

If the series excitation is such that the terminal voltage of the generator on full load is the same as that on no load, the generator is said to be level or flat-compounded*. If the series excitation becomes more prominent than that of the shunt field, the terminal voltage rises with the increase in load and the generator is said to be *over-compounded.* Similarly when the shunt field excitation plays the prominent part and full-load terminal voltage is less than no-load terminal voltage, the generator is said to be *under-compounded.*

The external characteristics of over-, under- and flat- (or level-) compound dc generators, are illustrated in Fig. 9.28.

In a **differential compound wound generator**, the induced voltage on no load would be the same as for the shunt generator because series field does not create any flux. When the generator is loaded, load current (or armature current) flows through the series field windings creating magnetic flux that opposes the flux created by shunt field. This results in decrease of induced emf and, therefore, in terminal voltage. In such generators the terminal voltage drops very rapidly with the increase in load current and, therefore it is interesting to note that a short circuit cannot cause any damage to it.

9.10. BUILDING-UP OF VOLTAGE OF A SHUNT GENERATOR AT NO LOAD

The curve plotted between the generated emf and shunt field current will be similar to that shown in Fig. 9.29. The generator excites itself due to residual magnetism and develops the voltage, as described below :

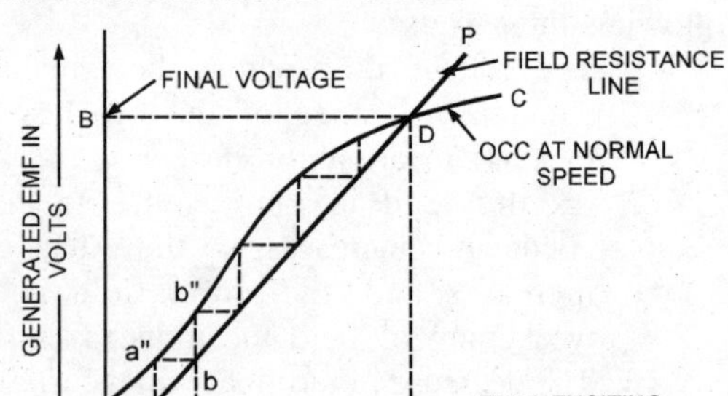

Fig. 9.29. ***Building-up of Voltage of a Shunt Generator At No Load***

Line OP represents the shunt field resistance. When the generator is started a small emf (represented by OA in the Fig. 9.29) is induced due to residual magnetism. This induced emf causes a flow of current *oa'* in the field circuit. This is obtained by drawing a horizontal line from point A meeting the field resistance line at point '*a*' and then from point '*a*' drawing perpendicular line meeting current *axis at a'. When field current is oa' the generated emf is a' a" which produces a field current ob'* which in turn produces a high voltage *b' b"*. Thus it is observed that effect is cumulative and values of induced emf and field current increase until these reach point D, the

* Even in case of a flat-compounded generator, the terminal voltage is not constant for all loads from no load to full load. As seen from Fig. 9.28, at half-full load, the terminal votlage is more than the no-load (or full-load) terminal voltage.

point of intersection of the shunt field resistance line and magnetic characteristic obtained. In Fig. 9.29 instantaneous increases in induced emf and field current are shown, but in practice flux increases gradually. OB is the maximum value of emf, that can be generated and generator cannot generate more than this voltage.

9.11. GENERATOR PERFORMANCE CURVES

In a *separately excited dc generator* the field current is supplied from a separate source and is, therefore, physically independent of armature conditions. At constant excitation and constant speed, the terminal voltage in this case drops off somewhat as load increases because of the increasing armature-resistance drop. This decrease is shown in dashed curve in Fig. 9.30.

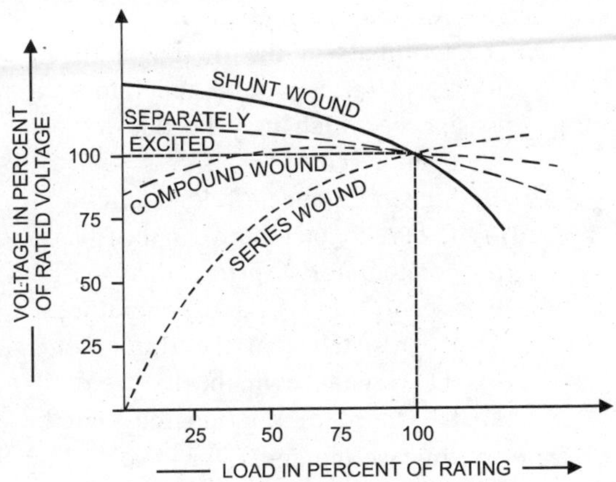

Fig. 9.30

In a *shunt wound dc generator* terminal voltage on no load depends upon shunt field resistance, as already discussed in Art. 9.9.3. The terminal voltage of a dc shunt generator will decrease as load is added because of the increase in armature resistance drop and the decrease in field current accompanying the drop in voltage. The drop in voltage with load is shown in Fig. 9.30.

In a series wound generator, the flux varies widely with load because the load current also flows through the field. The voltage-load characteristic, shown dotted in Fig. 9.30, resembles the magnetization curve but is somewhat lower because of resistance drop.

Build-up of an *unloaded compound dc generator* is the same as for a dc shunt generator since the series field carries negligible current at no load. When the generator is loaded, both the shunt and series field mmfs are present. For a cumulative compound generator, the cause of voltage drop from no load to full load in a dc shunt generator may be partially or completely compensated or overcompensated by the addition of the series field. In a *flat-compounded dc generator*, they are completely compensated, so that no-load and full-load terminal voltages are equal. If the full-load voltage is lower than the no-load voltage, the generator is *under-compounded*; if higher, *over-compounded.* The load characteristic of a flat-compounded dc generator is shown by a chained curve in Fig. 9.30.

9.12. CONDITIONS FOR SELF EXCITATION

The conditions required to be fulfilled before a series or a shunt generator excites itself are given below :

1. There must be some residual magnetism in the field system.
2. The residual magnetism must be in proper direction. The field coils should be connected with the armature in such a way that current flowing through them should increase emf induced by the residual magnetism.
3. For a series wound generator, the resistance of the external circuit should be less than the critical resistance.*
4. For a shunt wound generator, the resistance in the field circuit must be less than critical resistance** for field circuit and resistance in the load circuit must be greater than critical resistance*** for the load circuit.

---

* Critical resistance for a series wound generator is the maximum load resistance with which the generator will be able to excite. If the resistance in the external load circuit is greater than this, the generator will not excite.

** Critical field circuit resistance for a shunt wound generator is the maximum field resistance with which the generator will be able to excite. If the resistance in the field circuit exceeds this value, the generator will not excite.

***Critical load resistance for a shunt wound generator is the minimum load resistance with which the generator will be able to excite. If the resistance of the external load circuit falls below this value, the generator will fail to build up its voltage.

9.13. CAUSES OF FAILURE TO BUILD UP VOLTAGE

If a dc generator fails to build up its voltage, there may be one or more of the following reasons :

1. The residual magnetism may be zero.
2. The residual magnetism may have been reversed in polarity.
3. The terminals of the field winding may have been reversed.
4. In case of series wound generators the resistance in the circuit may be greater than the critical resistance, which may be due to :
   (*i*) open circuit;
   (*ii*) high resistance of load circuit;
   (*iii*) faulty contact between brushes and commutator; and
   (*iv*) commutator's surface dirty or greasy.
5. In case of shunt wound generator :
   (*a*) the resistance of the shunt field circuit may be greater than the critical resistance;
   (*b*) the resistance in the load circuit may be lower than the critical resistance;
   (*c*) the speed of rotation may not be equal to rated one.

Remedy. In case the generator is started up for the first time. It may be that no voltage will be built up either because the poles have no residual magnetism or the poles have retained some residual magnetism but the field winding connections are reversed so that the magnetism developed by the field winding on start has destroyed the residual magnetism and the machine cannot "build up". In both the cases, the field coils must be connected to a dc source (a storage battery) for a short while to magnetise the poles.

9.14. APPLICATIONS OF DC GENERATORS

The choice of a dc generator for a particular service depends upon its performance characteristics. Applications of various types of dc generators are given below.

1. Separately Excited DC Generators. These generators require separate source of supply and are, therefore, usually more expensive than self excited generators. On this account these are generally used where self excited generators would be relatively unsatisfactory.

2. Series Wound DC Generators. These generators have very few applications. They cannot be employed for power supply on account of their rising voltage characteristics. They can be employed as a constant current source by making use of the drooping portion of the voltage-current characteristic.

Their best application is for supplying field current for regenerative braking in dc locomotives. They are also used in series arc lighting. Another important application of these generators is as series boosters*.

3. Shunt Wound DC Generators. Such generators with field regulators are used for light and power supply purposes. These are also used for charging of batteries because they can be made to give constant output voltage.

4. Cumulative Compound Wound DC Generators. Over-compounded type dc generators are generally used for lighting and power services because overcompounding compensates voltage drop in the distribution lines and thus the voltage at the consumer terminals remains more or less constant.

5. Differential Compound Wound DC Generators. It is a constant current generator and finds an useful application as an arc welding generator where the generator is practically short circuited every time the electrode touches the metal plates to be welded.

Note: *With the advent of various power electronic devices, there is no doubt that the importance of dc generators has gone down. At present time bulk of power is generated in the form of alternating current because of technical and economical reasons. Hence the use of dc generators is very limited. They find application mainly in supplying excitation to the small and medium size alternators. For industrial applications of direct current like electrolytic*

---

* A booster is a series wound generator, which is inserted in a circuit to add a certain voltage so that the excessive drop in the line is compensated.

processes, welding processes and variable speed motor drives, the present trend is to generate ac and then convert it into dc by converters. For conversion of ac into dc the dc generator, as a part of motor (ac)-generator (dc) set, has to compete the SCR rectifiers and various other controlled power devices which usually are cheaper in cost, compact in size, relatively noise-free in operation and require minimum maintenance, though they suffer from the disadvantages of having poor power factor, harmonic generation, poor braking, etc. However, some of the important applications of dc generators include dynamometers, welding, cross-field generators, for closed-loop control systems, tachogenerators (not covered in this book) etc. Separately excited dc generators still find use for a wide output-voltage control such as in Ward-Leonard system of speed control.

9.15. WORKING PRINCIPLE OF DC MOTOR

Fig. 9.31

The principle upon which a dc motor works is very simple. If a current carrying conductor is placed in a magnetic field, mechanical force is experienced on the conductor, the direction of which is given by Fleming's left hand rule (also called *motor rule*) and hence the conductor moves in the direction of force. The magnitude of the mechanical force experienced on the conductor is given by

$$F = BI_c l_c \text{ newtons}$$

where B is the field strength in teslas (Wb/m^2), I_c is the current flowing through the conductor in amperes and l_c is the length of conductor in metres.

When the motor is connected to the dc supply mains, a direct current passes through the brushes and commutator to the armature winding; while it passes through the commutator it is converted into ac so that the group of conductors under successive field poles carry currents in the opposite directions, as shown in Fig. 9.31. Also the direction of current in the individual conductors reverses as they pass away from the influence of one pole to that of the next.

In Fig. 9.31, a 4-pole dc motor is shown when the field and armature circuits are connected across dc supply mains. Let the current in armature conductors be outwards under the N-poles (shown by dots) and inwards under S-poles (shown by crosses). By applying Fleming's left hand rule, the direction of force on each conductor can be determined, which has been illustrated in Fig. 9.31. From Fig. 9.31. it is observed that each conductor experiences a force which tends to rotate the motor armature in clockwise direction. These forces collectively produce a driving torque.

9.15.1. Commutator Action in a DC Motor

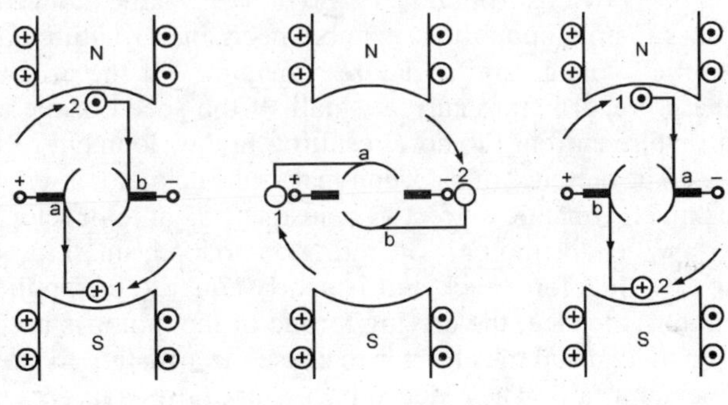

Fig. 9.32. *Commutation in a DC Motor*

In the case of a dc motor, it is necessary that the current through the coils of the armature winding be reversed as a particular coil leaves one pole (say, the north pole), crosses the neutral line and comes under the influence of next pole which is of opposite polarity (*i.e.* the south pole). The operation of the commutator, that serves the above purpose, is given below:

Consider a single turn coil, whose leads are soldered to commutator segments *a* and *b*, each carrying a brush, as illustrated in Fig. 9.32. The positive side of the supply line is connected to left hand brush and negative side to the right hand brush. In position I the line current arrives at the commutator segment a, flows through the bottom side 1 of the coil away from the reader (as shown by cross in the circle) and then through the upper side 2 of the coil towards the reader (as shown by dot in the circle), reaches the commutator segment *b* and flows again into the line through the brush. The coil will tend to rotate in clockwise direction, as determined by Fleming's left hand rule.

In position II the coil is on the magnetic neutral line; there is no contact between the commutator segments and brushes, and there is no flow of current through the coil. The coil crosses the neutral line by inertia. In case of a multi-turn coil, the remaining turns of the coil will supply the necessary torque.

In position III, the two sides of the coil, 1 and 2, have changed poles, and the current through them has reversed. The commutator segments, however, have also changed contact with the brushes. Thus the coil will continue to rotate in the same direction as before, *i.e.* clockwise.

9.16. IMPORTANCE OF BACK EMF

As already explained, when the motor armature continues to rotate due to motor action, the armature conductors cut the magnetic flux and, therefore, emfs are induced in them. The direction of this induced emf, known as *back emf,* is such that it opposes the applied voltage.

Since the back emf is induced due to the generator action, the magnitude of it is, therefore, given by the same expression as that for the generated emf in a generator

i.e. Back emf, $$E_b = \frac{\Phi ZN}{60} \times \frac{P}{A} \text{ volts} \qquad ...(9.28)$$

the symbols having their usual significance.

The equivalent circuit of a motor is shown in Fig. 9.33. The armature circuit is equivalent to a source of emf, E_b in series with a resistance, R_a put across a dc supply mains of V volts. It is evident from Fig. 9.33 that the applied voltage V must be large enough to balance both the voltage drop in armature resistance and the back emf at all times, *i.e.*

$$V = E_b + I_a R_a \qquad ...(9.29)$$

where V is the applied voltage across the armature, E_b is the induced emf in the armature by generator action, I_a is the armature current and R_a is the armature resistance.

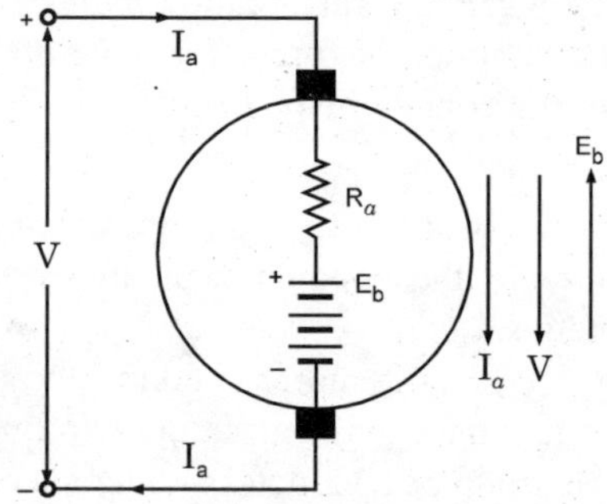

Fig. 9.33. ***Equivalent Circuit of a Motor Armature***

The Eq. (9.29) may be rewritten as $I_a = \frac{V - E_b}{R_a}$ to give armature current in terms of applied voltage V, induced emf E_b and armature resistance R_a.

As obvious from Eqs. (9.28) and (9.29) the induced emf in the armature of a motor, E_b depends among other factors upon the armature speed and armature current depends upon the back emf E_b for a constant applied voltage and armature resistance. If the armature speed is high, back emf E_b will be large and, therefore, armature current small. If the speed of the armature is low, then back emf E_b will be less and armature current I_a more resulting in development of large torque.

The presence of back emf makes the dc motor a *self regulating* machine, *i.e.* it makes the dc motor to draw as much armature current as is just sufficient to develop the required load torque. This is explained below:

When the motor is operating on no load, small torque is required to overcome the friction and windage losses, therefore, back emf is nearly equal to the applied voltage and armature current is small. When the motor is loaded, the driving torque of the motor is not sufficient to counter the increased retarding torque due to load and the effect is to cause the armature to slow down. With the decrease in the speed of armature back emf falls. The reduced back emf allows a larger current to flow through the armature. The increase in armature current results in higher electromagnetic driving torque. The motor continues to slow down till the electromagnetic torque developed matches the load torque and the steady state conditions are attained. The reverse phenomenon occurs when mechanical load on the motor falls.

When the load on the motor falls, the electromagnetic torque developed is momentarily in excess of the load requirement and, therefore, the motor armature accelerates. With the increase in armature speed, back emf increases causing armature current to decrease. The decrease in armature current causes decrease in electromagnetic torque and the steady state conditions are attained when the electromagnetic torque developed matches the load torque.

Thus it is evident that back emf E_b acts like a governor *i.e.* it makes a motor self-regulating so that it draws as much current as just required.

9.17. POWER RELATIONSHIP IN MOTOR

The Eq. (9.29) is called the *voltage equation* for a motor.

Multiplying each term of the voltage expression by I_a we get

$$V I_a = E_b I_a + I_a^2 R_a \quad ...(9.30)$$

The above equation is known as power equation of the dc motor.

The term $V I_a$ represents the power supplied to the motor armature and the term $I_a^2 R_a$ represents the power lost in the armature and, therefore, the term $E_b I_a$ must represent the power developed by the motor armature causing rotation of the armature. The power developed $E_b I_a$ is not all available at the shaft since some of it is used to overcome the mechanical or rotational losses of the motor.

The power lost in the armature, $I_a^2 R_a$ is a small portion (about 5%) of power input to the armature and the remaining portion $E_b I_a$ is converted into mechanical power within the armature.

As mentioned above, the mechanical power developed by the motor is given by the expression

$$P_m = E_b I_a = VI_a - I_a^2 R_a \quad ...(9.31)$$

Differentiating both sides with respect to armature current I_a we have

$$\frac{dP_m}{dI_a} = V - 2I_a R_a$$

Mechanical power developed by the motor will be maximum if $\frac{dP_m}{dI_a}$ is zero

$$i.e.\ V - 2 I_a R_a = 0$$

$$\text{or } I_a R_a = \frac{V}{2}$$

$$\text{and back emf, } E_b = V - I_a R_a = \frac{V}{2} \quad ...(9.32)$$

Thus the motor develops the maximum power when the armature current is such that the back emf E_b is equal to one half of the applied voltage, This is not attained in practice, since it necessitates a current well beyond the normal operating range. Also an amount of power equal to the mechanical power developed would be wasted in heating the armature, so taking other losses into account, the efficiency would be well below 50%.

9.18. CLASSIFICATION OF DC MOTORS

Similar to dc generators, the dc motors can also be classified as (1) permanent magnet, (2) separately excited, (3) series wound (4) shunt wound and (5) compound wound dc motors.

9.18.1. Permanent Magnet Motor

It consists of an armature and one or several permanent magnets encircling the armature. Field coils are usually not required. However, some of these motors do have coils wound on the poles. If they exist, these coils are intended only for recharging the magnets in the event that they lose their strength. Schematically, this motor appears as only an armature.

9.18.2. Separately Excited DC Motors

These motors have field coils similar to those of a shunt wound machine, but the armature and field coils are fed from different supply sources, as illustrated in Fig. 9.34 and may have different voltage ratings.

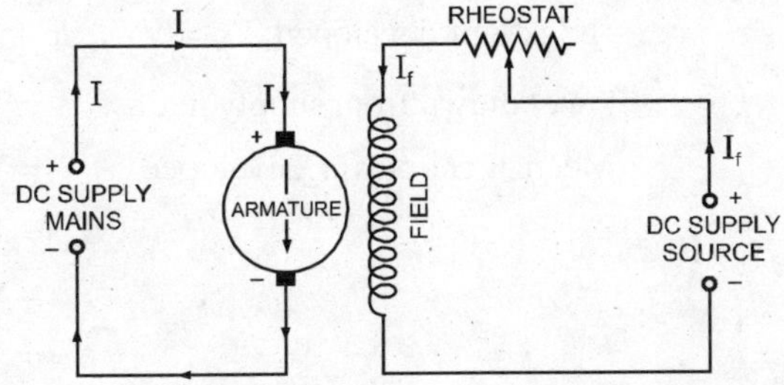

Fig. 9.34. ***Separately Excited DC Motor***

In a separately excited dc motor,

Armature current, I_a = Line current, $I_L = I$ (say) ...(9.33)

Back emf developed, $E_b = V - IR_a$ ...(9.34)

Power drawn from supply mains, $P = VI$ where V is supply voltage. ....(9.35)

Mechanical power developed, P_m = Power input to armature – power lost in armature

$= VI - I^2 R_a = I(V - IR_a) = E_b I$ ...(9.36)

9.18.3. Series Wound DC Motors

As the name implies, the field coils, consisting of few turns of thick wire, are connected in series with the armature, as illustrated in Fig. 9.35. The cross-sectional area of the wire used for field coils has to be fairly large to carry the armature current, but owing to the higher current, the number of turns of wire in them need not be large. In a dc series motor

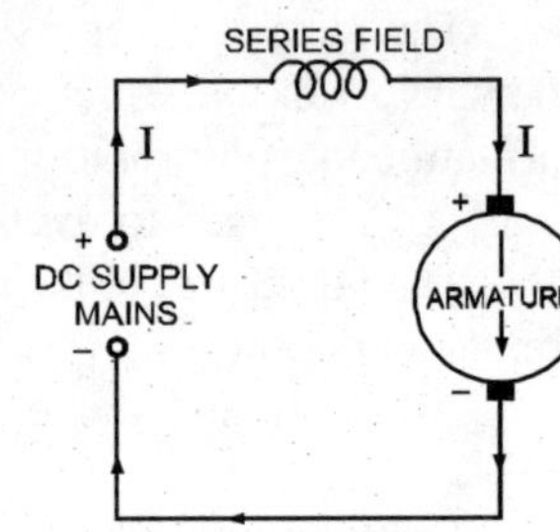

Fig. 9.35. ***Series Wound DC Motor***

Armature current, I_a = Series field current, I_{se}

= Line current, $I_L = I$ (say) ...(9.37)

Back emf developed, $E_b = V - I(R_a + R_{se})$ ...(9.38)

Power drawn from supply mains $= VI$ ...(9.39)

Mechanical power developed, P_m = Power input – losses in armature and field

$= VI - I^2 (R_a + R_{se})$

$= I[V - I(R_a + R_{se})] = E_b I$ ...(9.40)

9.18.4. Shunt Wound DC Motors

The word "shunt" means "parallel". These motors are so named because they basically operate with the field coils connected in parallel with the armature. The field winding consists of a large number of turns of comparatively fine wire so as to provide large resistance. The field current is much less than the armature current, sometimes as low as 5%.

The connection diagram is shown in Fig. 9.36. The current supplied to the motor is divided into two paths, one through the field winding and second through the armature *i.e.*,

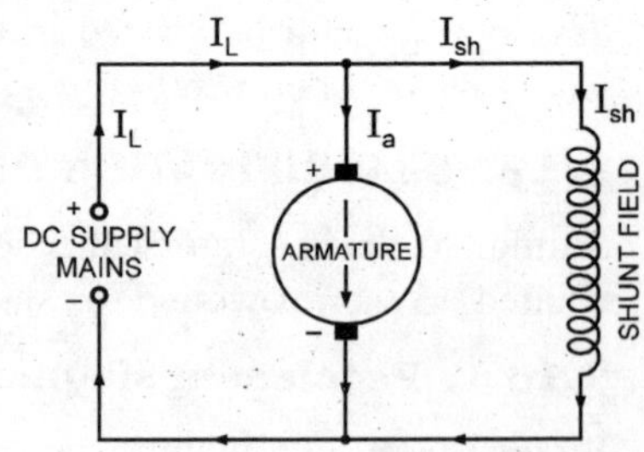

Fig. 9.36. ***Series Wound DC Motor***

Input line current, $I_L = I_a + I_{sh}$ ...(9.41)

where I_a is the armature current and I_{sh} is the shunt field current and is given by the expression

$$I_{sh} = \frac{V}{R_{sh}} \quad ...(9.42)$$

where V is the supply voltage and R_{sh} is the shunt field resistance.

Back emf developed, $E_b = V - I_a R_a$ ...(9.43)

Power drawn from supply mains, $P = VI_L$ ...(9.44)

Mechanical power developed, P_m = Power input – losses in armature and shunt field

$= VI_L - VI_{sh} - I_a^2 R_a$

$= V(I_L - I_{sh}) - I_a^2 R_a$

$= VI_a - I_a^2 R_a = I_a (V - I_a R_a) = E_b I_a$ ...(9.45)

9.18.5. Compound Wound DC Motors

A compound wound dc motor has both shunt and series field coils. The shunt field is normally the stronger of the two (*i.e.*, has more ampere-turns). Compound wound motors are of two types namely cumulative compound wound and differential compound wound motors.

Cumulative compound wound motor is one in which the field windings are connected in such a way that the direction of flow of current is same in both of the field windings, as illustrated in Fig. 9.37. In the motor of this type the flux due to series field winding strengthens the field due to the shunt field winding.

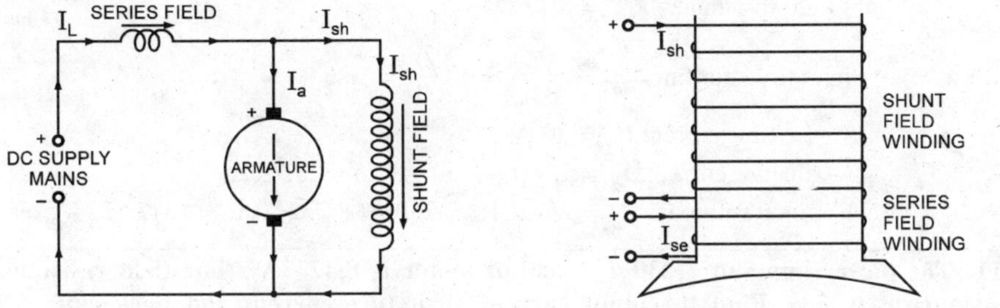

(a) Circuit Diagram *(b) Winding Connection Diagram*

Fig. 9.37. ***Cumulative Compound Wound DC Motor***

Differential compound wound motor is one in which the field windings are connected in such a way that the direction of flow of current is opposite to each other in the two field windings, as illustrated in Fig. 9.38. In this type of motor the flux due to series field winding weakens the field due to shunt field winding.

Compound wound dc motors, like compound wound dc generators, may be either long shunt connected or short shunt connected. In long shunt connected (or long shunt) compound wound motors, series field and armature are connected in series with each other and in parallel with the shunt field. In short shunt connected (or short shunt) compound wound motors, the armature and shunt field are in parallel with each other and the pair is in series with the series field. Long shunt connection sometimes results in simpler wiring. Changing from long to short shunt, or vice versa, has little effect on motor performance.

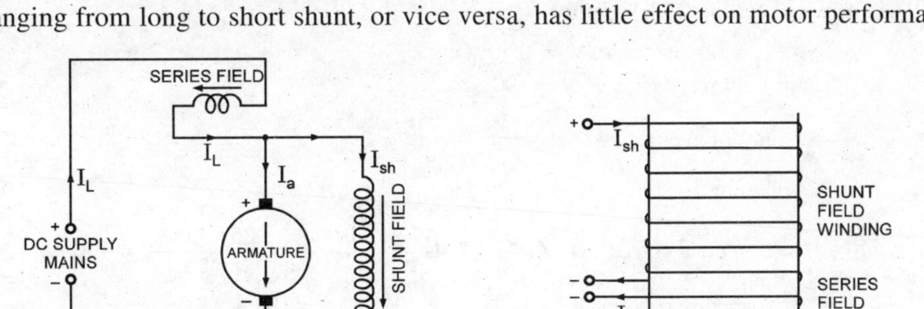

(a) Circuit Diagram *(b) Winding Connection Diagram*

Fig. 9.38. ***Differential Compound Wound DC Motor***

9.19. DIRECTION OF ROTATION OF DC MOTORS

It is evident from Fig. 9.31 that if the armature current were reversed by reversing the armature leads, but leaving the field polarity the same, torque would be developed in a counter-clockwise direction. Likewise, if the field polarity were reversed leaving the armature current as shown, torque would be developed in a counter-clockwise direction. However, if both, the direction of the armature current and field polarity were reversed, torque would be developed in a clockwise direction as before. *Hence the direction of rotation of a motor can be reversed by reversing the current through either the armature winding or the field coils. If the current through both is reversed, the motor will continue to rotate in the same direction as before.*

Example 9.10. A 250 V dc shunt motor takes 30 A current while running at full load. The resistance of motor armature and field windings are 0.1 Ω and 200 Ω respectively. Determine the back emf generated in the motor, when it runs on full load.

Solution :

Supply voltage, $V = 250$ volts

Shunt field resistance, $R_{sh} = 200\ \Omega$

Shunt field current, $I_{sh} = \dfrac{V}{R_{sh}} = \dfrac{250}{200} = 1.25$ A

Line current. $I_L = 30$ A

Armature current, $I_a = I_L - I_{sh} = 30 - 1.25 = 28.75$ A

Back emf on full load, $E_b = V - I_a R_a = 250 - 28.75 \times 0.1 = 247.125$ V **Ans.**

Example 9.11. The power input to a 230 V dc shunt motor is 8.477 kW. The field resistance is 230 Ω and armature resistance is 0.28 Ω. Find the input current, armature current and back emf.

[*Pb. Technical Univ. Electrical Engineering December 2000*]

Solution :

Power input, $P = 8.477$ kW or 8,477 watts

Supply voltage, $V = 230$ V

Shunt field current, $I_{sh} = \dfrac{V}{R_{sh}} = \dfrac{230}{230} = 1$ A

Input current, $I_L = \dfrac{P}{V} = \dfrac{8,477}{230} = 36.86$ A **Ans.**

Armature current, $I_a = I_L - I_{sh} = 36.86 - 1 = 35.86$ A **Ans.**

Back emf, $E_b = V - I_a R_a = 230 - 35.86 \times 0.28 = 219.96$ V **Ans.**

Example 9.12. A 25 kW, 250 V dc shunt generator has armature and field resistances of 0.06 Ω and 100 Ω respectively. Determine the total armature power developed when working (*i*) as generator delivering 25 kW output and (*ii*) as a motor taking 25 kW input.

Solution :

Line voltage, $V = 250$ volts

Shunt field resistance, $R_{sh} = 100\ \Omega$

Shunt field current, $I_{sh} = \dfrac{V}{R_{sh}} = \dfrac{250}{100} = 2.5$ A

Armature resistance, $R_a = 0.06\ \Omega$

(i) As generator

Load current, $I_L = \dfrac{\text{Output in kW} \times 1{,}000}{V} = \dfrac{25 \times 1{,}000}{250} = 100$ A

Armature current, $I_a = I_L + I_{sh} = 100 + 2.5 = 102.5$ A

Generated emf, $E_g = V + I_a R_a = 250 + 102.5 \times 0.06 = 256.15$ V

Power developed, $P_g = \dfrac{E_g I_a}{1{,}000} = \dfrac{256.15 \times 102.5}{1{,}000} = 26.26$ kW **Ans.**

(ii) As motor

Line current, $I_L = \dfrac{\text{Input in kW} \times 1{,}000}{V} = \dfrac{25 \times 1{,}000}{250} = 100$ A

Armature current, $I_a = I_L - I_{sh} = 100 - 2.5 = 97.5$ A

Back emf, $E_b = V - I_a R_a = 250 - 97.5 \times 0.06 = 244.15$ V

Power developed, $P_m = \dfrac{E_b I_a}{1{,}000} = \dfrac{244.15 \times 97.5}{1{,}000} = 23.8$ kW **Ans.**

9.20. SPEED EQUATION

As mentioned in Art. 9.16 the expressions for back emf developed in the armature of a dc motor are given as

$$E_b = \frac{\Phi ZN}{60} \times \frac{P}{A} \text{ volts} \qquad ...(i)$$

$$\text{and } E_b = V - I_a R_a \text{ volts} \qquad ...(ii)$$

Comparing Eqs. (*i*) and (*ii*) we get

$$\frac{\Phi ZN}{60} \times \frac{P}{A} = V - I_a R_a$$

$$\text{or } N = \frac{V - I_a R_a}{\Phi Z} \times \frac{60\,A}{P}$$

or

$$N = K \frac{V - I_a R_a}{\Phi} = K \frac{E_b}{\Phi}$$

since Z, A and P are constant for a particular machine.

For a dc motor, if initial values of speed, armature current, back emf and flux per pole are N_1, I_{a1}, E_{b1} and Φ_1 respectively and corresponding final values are N_2, I_{a2}, E_{b2} and Φ_2 respectively, then

$$N_1 \; \alpha \; \frac{E_{b1}}{\Phi_1} \qquad \text{where } E_{b1} = V - I_{a1}R_a$$

$$N_2 \; \alpha \; \frac{E_{b2}}{\Phi_2} \qquad \text{where } E_{b2} = V - I_{a2}R_a$$

$$\text{and } \frac{N_2}{N_1} = \frac{E_{b2}/\Phi_2}{E_{b1}/\Phi_1} = \frac{E_{b2}}{E_{b1}} \times \frac{\Phi_1}{\Phi_2} \qquad ...(9.46)$$

For a **dc shunt motor**, flux practically remains constant (*i.e.* $\Phi_2 = \Phi_1$) and

$$\frac{N_2}{N_1} = \frac{E_{b2}}{E_{b1}} = \frac{V - I_{a2}R}{V - I_{a1}R} \qquad ...(9.47)$$

In above expression since applied voltage V is constant and the voltage drop in armature ($I_a R_a$) is negligible in comparison to supply voltage V, therefore, speed of a dc shunt motor remains almost constant.

For a **dc series motor**, prior to saturation,

$$\Phi \; \alpha \; I_{se} \; \alpha \; I_a$$

$$\text{or } \frac{\Phi_1}{\Phi_2} = \frac{I_{a1}}{I_{a2}}$$

$$\text{and } \frac{N_2}{N_1} = \frac{E_{b2}}{E_{b1}} \times \frac{I_{a1}}{I_{a2}} \qquad ...(9.48)$$

For a **dc series motor**, after saturation

Flux Φ is independent of field current I_{se} or armature current I_a

and speed $N \; \alpha \; E_b$

$$\text{or } \frac{N_2}{N_1} = \frac{E_{b2}}{E_{b1}} \qquad ...(9.49)$$

9.21. SPEED REGULATION

The speed regulation of a dc motor is defined as the change in speed when the load on the motor is reduced from full load to zero and is expressed in percentage of rated full-load speed.

$$\therefore \text{ Percentage speed regulation} = \frac{N_0 - N_f}{N_f} \times 100 \qquad ...(9.50)$$

Example 9.13. A 500 V, 4-pole shunt motor has 720 wave-connected armature conductors. Full-load armature current is 60 A and flux per pole is 0.03 Wb. Find the full-load speed of the motor, if the armature resistance is 0.2 Ω and brush contact drop is 1 V per brush. Also calculate the speed if the above motor is lap-connected.

Solution :

Supply voltage, $V = 500$ V

Full-load armature current, $I_{af} = 60$ A

Back emf on full load, $E_{bf} = V - I_{af}R_a -$ brush contact drop $= 500 - 60 \times 0.2 - 2 \times 1 = 486$ V

Armature conductors, $Z = 720$

Flux per pole, $\Phi = 0.03$ Wb

Number of poles, $P = 4$

Full-load speed of the motor when the armature conductors are wave- connected,

$$N_f = \frac{E_b}{\Phi} \times \frac{60}{Z} \times \frac{A}{P} = \frac{486}{0.03} \times \frac{60}{720} \times \frac{2}{4} = 675 \text{ rpm } \textbf{Ans.}$$

Full-load speed of the motor when the armature conductors are lap-connected,

$$N'_f = \frac{486}{0.03} \times \frac{60}{720} \times \frac{4}{4} = 1{,}350 \text{ rpm } \textbf{Ans.}$$

Example 9.14. A 250 V shunt motor runs at 1,000 rpm at no load and takes 8 A. The total armature and shunt field resistances are respectively 0.2 Ω and 250 Ω. Calculate the speed when loaded and taking 50 A. Assume the flux to be constant.

Solution :

Supply voltage, $V = 250$ V

Shunt field current, $I_{sh} = \frac{V}{R_{sh}} = \frac{250}{250} = 1$ A

No-load armature current, $I_{a0} = I_{L0} - I_{sh} = 8 - 1 = 7$ A

No-load back emf, $E_{b0} = V - I_{a0} R_a = 250 - 7 \times 0.2 = 248.6$ V

No-load speed, $N_0 = 1{,}000$ rpm

When loaded Armature current, $I_{af} = I_{Lf} - I_{sh} = 50 - 1 = 49$ A

Back emf, $E_{bf} = V - I_{af}R_a = 250 - 49 \times 0.2 = 240.2$ V

Flux, $\Phi_f = \Phi_0$ ∵ flux is assumed to be constant

$$\text{Speed when loaded, } N_f = N_0 \times \frac{E_{bf}}{E_{b0}} \times \frac{\Phi_0}{\Phi_f} = 1{,}000 \times \frac{240.2}{248.6} \times \frac{\Phi_0}{\Phi_0} = 966.2 \text{ rpm } \textbf{Ans.}$$

Example 9.15. A 250 V dc shunt motor, on no load, runs at 1,000 rpm and takes 5 A. The field and armature resistances are 250 Ω and 0.25 Ω respectively. Calculate the speed when the motor is loaded such that it takes 41 A if the armature reaction weakens the field by 3%.

[Pb Technical Univ. Electrical Engineering May 2001]

Solution :

Supply voltage, $V = 250$ V

Shunt field current, $I_{sh} = \frac{250}{250} = 1$ A

No-load armature current, $I_{ao} = I_{Lo} - I_{sh} = 5 - 1 = 4$ A

No-load back emf, $E_{bo} = V - I_{ao} R_a = 250 - 4 \times 0.25 = 249$ V.

No-load speed, $N_o = 1{,}000$ rpm

When loaded Armature current, $I_{af} = I_{Lf} - I_{sh} = 41 - 1 = 40$ A

Back emf, $E_{bf} = V - I_{af}R_a = 250 - 40 \times 0.25 = 240$ V

Flux per pole, $\phi_f = 0.97\, \phi_0$

Since $E_b \propto \phi N$

Speed of motor when loaded, $N_f = N_o \times \frac{E_{bf}}{E_{bo}} \times \frac{\phi_o}{\phi_f} = 1{,}000 \times \frac{240}{249} \times \frac{\phi_0}{0.97\,\phi_0} = 993.67$ rpm **Ans.**

Example 9.16. A 120 V dc shunt motor has an armature resistance of 0.2 Ω and a field resistance of 60 Ω. The full-load line current is 60 A and full-load speed is 1,800 rpm. If the brush contact drop is 3 V, find the speed of the motor at half load. *[Pb Technical Univ. Electrical Engineering June 2000]*

Solution : Supply voltage, $V = 120$ V

Armature resistance, $R_a = 0.2$

Field resistance, $R_{sh} = 60\ \Omega$

Shunt field current, $I_{sh} = \frac{V}{R_{sh}} = \frac{120}{60} = 2$ A

At full load

Line current, $I_{lf} = 60$ A

Armature current, $I_{af} = I_{lf} - I_{sh} = 60 - 2 = 58$ A

Back emf, $E_{bf} = V - I_{af} R_a -$ brush contact drop $= 120 - 58 \times 0.2 - 3 = 105.4$ V

Speed, $N_f = 1{,}800$ rpm

At half full-load

Line current, $I_{lh} = \frac{60}{2} = 30$ A

Armature current, $I_{ah} = I_{lh} - I_{sh} = 30 - 2 = 28$ A

Back emf, $E_{bh} = V - I_{ah} R_a -$ brush contact drop $= 120 - 28 \times 0.2 - 3 = 111.4$ V

Since $E_b \propto N\phi \propto N \quad \because \phi$ remains constant

So speed of motor at half-full-load, $N_h = \frac{N_f \times E_{bh}}{E_{bf}} = \frac{1{,}800 \times 111.4}{105.4} = 1{,}902$ rpm **Ans.**

Example 9.17. A 25 hp, 250 V dc series motor has armature resistance of 0.1 Ω and field resistance of 0.05 Ω and brush contact drop 3 V. When the line current is 80 A, the speed is 600 rpm. Find the speed when the line current is 100 A.

Solution : Back emf when line current is 80 A,

$$E_{b1} = V - I(R_a + R_{se}) - \text{brush contact drop} = 250 - 80\,(0.1 + 0.05) - 3 = 235 \text{ V}$$

Back emf when line current is 100 A,

$$E_{b2} = 250 - 100\,(0.1 + 0.05) - 3 = 232 \text{ V}$$

Since $\Phi \propto I$ upto saturation point,

$$\because \frac{\Phi_2}{\Phi_1} = \frac{I_2}{I_1} = \frac{100}{80} = 1.25$$

Speed when line current is 100 A, $N_2 = N_1 \times \frac{E_{b2}}{E_{b1}} \times \frac{\Phi_1}{\Phi_2} = 600 \times \frac{232}{235} \times \frac{1}{1.25} = 473.9$ rpm **Ans.**

Example 9.18. A dc shunt machine connected to 250 V mains has an armature resistance (including brushes) of 0.1 Ω and the resistance of the field circuit is 125 Ω. Find the ratio of the speed as a generator to the speed as a motor, the line current in each case being 80 A.

Solution : Terminal voltage, $V = 250$ volts

Shunt field resistance, $R_{sh} = 125\ \Omega$

Shunt field current, $I_{sh} = \frac{V}{R_{sh}} = \frac{250}{125} = 2$ A

Armature resistance, $R_a = 0.1\ \Omega$

Line current, $I_L = 80$ A

As Generator Armature current, $I_{ag} = I_L + I_{sh} = 80 + 2 = 82$ A

Generated emf, $E_g = V + I_{ag} R_a = 250 + 82 \times 0.1 = 258.2$ V

As Motor Armature current, $I_{am} = I_L - I_{sh} = 80 - 2 = 78$ A

Back emf developed, $E_b = V - I_{am} R_a = 250 - 78 \times 0.1 = 242.2$ V

Since machine is shunt wound and field current is also same, flux can be assumed to be the same in both cases.

$$\frac{\text{Speed as a generator}}{\text{Speed as a motor}} = \frac{E_g}{E_b} = \frac{258.2}{242.2} = 1.066 \textbf{ Ans.}$$

Example 9.19. A belt driven 100 kW shunt generator running at 300 rpm on 220 V bus-bars continues to run as a motor when the belt breaks, then taking 10 kW. What will be its speed ? Given armature resistance = 0.025 Ω, field resistance = 60 Ω and contact drop under each brush = 1 V. Ignore armature reaction.

Solution : As Generator

$$\text{Load current, } I_L = \frac{\text{Output in kW} \times 1{,}000}{\text{Bus - bar voltage}} = \frac{100 \times 1{,}000}{220} = 454.55 \text{ A}$$

$$\text{Shunt field current, } I_{sh} = \frac{\text{Bus - bar voltage}}{R_{sh}} = \frac{220}{60} = 3.67 \text{ A}$$

Armature current, $I_a = I_L + I_{sh} = 454.55 + 3.67 = 458.22$ A

Generated emf, $E_g = V + I_a R_a$ + brush contact drop $= 220 + 458.22 \times 0.025 + 2 \times 1 = 233.46$ V

Speed, $N_g = 300$ rpm

As motor

$$\text{Line current, } I_L = \frac{\text{Input in kW} \times 1{,}000}{\text{Bus - bar voltage}} = \frac{10 \times 1{,}000}{220} = 45.46 \text{ A}$$

$$\text{Shunt field current, } I_{sh} = \frac{V}{R_{sh}} = \frac{220}{60} = 3.67 \text{ A}$$

Armature current, $I_a = I_L - I_{sh} = 45.46 - 3.67 = 41.79$ A

Back emf, $E_b = V - I_a R_a -$ brush contact drop $= 220 - 41.79 \times 0.025 - 2 \times 1 = 216.955$ V

Since shunt field current is same and armature reaction is negligible, flux is same in both cases and speed of the machine as a motor,

$$N_m = N_g \times \frac{E_b}{E_g} = 300 \times \frac{216.955}{233.46} = 278.8 \text{ rpm } \textbf{Ans.}$$

9.22. TORQUE

The measure of causing the rotation of a wheel or the turning or twisting moment of a force about the axis is called the torque.

Torque is measured by the product of force and the radius at which this force acts.

Consider a wheel of radius r metres acted by a circumferential force F newtons, as shown in Fig. 9.39. Let the force F cause the wheel to rotate at n rps.

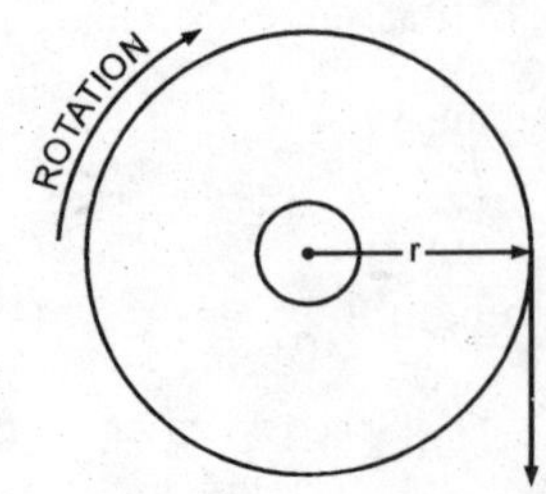

Fig. 9.39

Torque, $T = F \times r$ newton-metres

Work done per revolution $= F \times$ distance moved $= F \times 2\pi r$ joules

Work done per second $= F \times 2\pi r \times n = F \times r \times 2\pi n$ joules/second or watts

Since $F \times r$ = Torque, T and $2\pi n = \omega$, angular velocity in radians per second

Work done per second $= T\,\omega$ joules per second or watts ...(9.51)

So power developed, $P = T\omega = T \times \frac{2\pi N}{60} = 0.105$ NT watts ...(9.52)

9.23. ARMATURE TORQUE

Let T_e be the electromagnetic torque developed in newton-metres by the motor running at n rps.

$$\text{Power developed} = \text{Work done per second} = T_e\ \omega = T_e \times 2\pi n \text{ watts} \quad ...(i)$$

Electrical equivalent of mechanical power developed by the armature, as mentioned in Art. 9.17, also

$$= E_b I_a \text{ watts} \quad ...(ii)$$

Comparing Eqs. (*i*) and (*ii*) we have

$$T_e \times 2\pi n = E_b I_a$$

$$\text{or } T_e = \frac{E_b I_a}{2\pi n} = 0.159 \frac{E_b I_a}{n} \text{ Nm} \quad ....(9.53)$$

$$\text{Also } T_e = \frac{E_b I_a}{2\pi \frac{N}{60}} = 9.55 \frac{E_b I_a}{N} \text{ Nm} \quad ...(9.54)$$ where N is speed in rpm

Substituting $E_b = \Phi Z \frac{N}{60} \times \frac{P}{A}$ in expression (9.54) we have

$$T_e = 9.55 \times \Phi \times Z \times \frac{N}{60} \times \frac{P}{A} \times \frac{I_a}{N} = 0.159\ \Phi Z P \frac{I_a}{A} \text{ newton-metres} \quad ...(9.55)$$

Since Z, P and A are constant for a particular machine,

$$T = K \frac{\Phi I_a}{2\pi} \quad \text{where } K = \frac{PZ}{A}$$

The constant K for a given machine is the same for the emf equation as well as the torque equation

$$\text{or } T \propto \Phi I_a$$

Thus it may be concluded that

(*i*) the electromagnetic torque developed by the armature is proportional to the product of flux per pole and armature current.

(*ii*) the direction of electromagnetic torque developed by armature depends upon the direction of flux or magnetic field and the direction of flow of current in armature conductors. If either of the two is reversed the direction of torque developed will be reversed and, therefore, the direction of rotation. When both (the direction of field as well as that of armature current) are reversed the direction of torque (or rotation) will not change. This has been explained in Art. 9.19 also.

In case of a series wound motor flux Φ is directly proportional to armature current I_a (before saturation) because in a series wound motor field winding and armature winding currents are same, therefore,

$$T_e \propto I_a^2$$

In case of permanent magnet motors, separately excited motors and shunt wound motors the field strength *i.e.* Φ is practically constant and, therefore,

$$T_e \propto I_a$$

On the basis of the same killowatt output and speed, a dc series motor develops the highest starting torque and the dc shunt motor the least, while the cumulative compound wound dc motor falls somewhere between the first two.

9.24. SHAFT TORQUE

The armature torque is the gross torque, which is developed by the armature. Whole of this torque is not available at the pulley, since certain percentage of torque developed by the armature is lost to overcome the iron and friction losses. *Net torque i.e. gross torque less torque lost in iron and friction losses is known as shaft torque.*

If T_a is the torque developed by armature in N-m, T_f is the torque lost in iron and friction losses, and T_{sh} is the shaft torque or useful torque, then

$$T_a = \frac{E_b I_a}{\frac{2\pi N}{60}} = 9.55 \frac{E_b I_a}{N} \text{ newton metres} \qquad \text{where N is speed of armature in rpm}$$

$$T_f = \frac{\text{Iron and frictional losses in watts}}{2\pi N/60} = \frac{9.55}{N} \text{ (iron and frictional losses in watts)}$$

$$\text{Shaft torque, } T_{sh} = T_a - T_f = \frac{E_b I_a - \text{iron and frictional losses}}{2\pi N/60} \text{ newton–metres}$$

$$T_{sh} \text{ also} = \frac{\text{Output in watts}}{2\pi N/60} = \frac{9.55 \times \text{output in watts}}{N} \text{ newton-metres} \qquad ...(9.56)$$

9.24.1. Brake Horse Power (BHP)

In case of electric motors (ac or dc), the mechanical power available at the shaft in horse power is known as *brake horse power* (BHP). If T_{sh} is the shaft torque in newton metres and N is speed in rpm then

$$\text{Output in BHP} = \frac{2\pi N T_{sh}}{60 \times 735.5} \qquad ...(9.57)$$

Example 9.20. A dc shunt motor develops an open-circuit emf of 250 V at 1,500 rpm. Find the torque developed by machine for an armature current of 20 A.

[Pb Technical Univ. Basic Electrical Engineering December 2002]

Solution :

Back emf developed, E_b = Open-circuit emf = 250 V

Armature current, I_a = 20 A

Speed, N = 1,500 rpm

$$\text{Torque developed, } T = 9.55 \frac{E_b I_a}{N} = \frac{9.55 \times 250 \times 20}{1{,}500} = 31.83 \text{ Nm} \quad \textbf{Ans.}$$

Example 9.21. A dc shunt machine develops an open-circuit emf of 250 V at 1,500 rpm. Find its torque and mechanical power developed for an armature current of 50 A. State the simplifying assumptions.

Solution :

Open-circuit emf, E_0 = 250 V

Armature current, I_a = 50 A

Assuming back emf E_b equal to open-circuit emf E_0, though it may be slightly less due to armature reaction if field current remains unchanged.

Mechanical power developed, $P_{mech} = E_b I_a = 250 \times 50 = 12,500$ W or 12.5 kW **Ans.**

$$\text{Torque developed, } T_e = \frac{9.55 \times P_{mech}}{N} = \frac{9.55 \times 12{,}500}{1{,}500} = 79.583 \text{ Nm} \quad \textbf{Ans.}$$

Example 9.22. Calculate the armature torque developed by a 440 V dc shunt motor having an armature resistance of 0.25 Ω and running at 750 rpm when taking an armature current of 60 A.

Solution :

Back emf developed, $E_b = V - I_a R_a = 440 - 60 \times 0.25 = 425$ V

$$\text{Torque developed, } T_e = \frac{9.55 \times E_b I_a}{N} = \frac{9.55 \times 425 \times 60}{750} = 324.7 \text{ Nm} \quad \textbf{Ans.}$$

Example 9.23. Find the torque in N-m exerted by a 4-pole series motor whose armature has 1,200 conductors connected up in a 2-circuit winding. The motor current is 10 A and the flux per pole is 0.02 Wb.

Solution :

Flux per pole, Φ = 0.02 Wb

Total number of armature conductors, Z = 1,200

Number of poles, P = 4

Number of parallel paths, A = 2

Armature current, I_a = 10 A

$$\text{Torque developed, } T_e = 0.159 \times \Phi \times Z \times P \times \frac{I_a}{A}$$

$$= 0.159 \times 0.02 \times 1{,}200 \times 4 \times \frac{10}{2} = 76.32 \text{ Nm} \quad \textbf{Ans.}$$

Example 9.24. The armature of a 6-pole lap wound dc shunt motor takes 400 A at a speed of 350 rpm. The flux per pole is 80 m Wb, the number of turns is 600, and 3% of the torque is lost in windage, friction and iron losses. Calculate the brake horse power.

Solution : Total or gross torque developed by armature,

$$T_a = 0.159 \times \Phi \times Z \times P \times \frac{I_a}{A}$$

$$= 0.159 \times 80 \times 10^{-3} \times (600 \times 2) \times 6 \times \frac{400}{6} = 6{,}105.5 \text{ Nm} \quad \because Z = 2 \times \text{number of turns}$$

Shaft torque, T_{sh} = Total torque developed – torque lost in windage, friction and iron losses

$$= T_a - T_f = T_a\left(1 - \frac{T_f}{T_a}\right) = 6{,}105.5\,(1 - 0.03) = 5{,}922.34 \text{ Nm}$$

$$\text{Brake horse power} = T_{sh} \times \frac{2\pi \times N}{60} \times \frac{1}{735.5} = 5{,}922.34 \times \frac{2\pi \times 350}{60} \times \frac{1}{735.5} = 295.13 \textbf{ Ans.}$$

Example 9.25. A 10 hp, 230 V shunt motor has an armature circuit resistance of 0.5 Ω and a field resistance of 115 Ω. At no load and rated voltage the speed is 1,200 rpm and the armature current is 2 A. If the load is applied, the speed drops to 1,100 rpm. Determine the armature current, the line current and torque.

[*Pb Technical Univ. Electrical Engineering May 2002*]

Solution :

Supply voltage, $V = 230$ V

No-load armature current, $I_{a0} = 2$ A

No-load speed, $N_0 = 1{,}200$ rpm

Full-load speed, $N_f = 1{,}100$ rpm

No-load emf, $E_{b0} = V - I_{a0} R_a = 230 - 2 \times 0.5 = 229$ V

Since $N \alpha \frac{E_b}{\phi}$ and for a dc shunt motor ϕ is constant, $N \alpha E_b$

$$\text{or } E_{bf} = \frac{N_f}{N_0} \times E_b = \frac{1{,}100}{1{,}200} \times 229 = 209.9 \text{ V}$$

and since $E_{bf} = V - I_{af} R_a$,

$$I_{af} = \frac{V - E_{bf}}{R_a} = \frac{230 - 209.9}{0.5} = 40.2 \text{ A } \textbf{Ans.}$$

$$\text{Shunt field current, } I_{sh} = \frac{V}{R_{sh}} = \frac{230}{115} = 2 \text{ A}$$

Full-load line current, $I_{Lf} = I_{af} + I_{sh} = 40.2 + 2 = 42.2$ A **Ans.**

$$\text{Torque developed at full-load, } T_f = \frac{E_{bf} \times I_{af}}{2\pi N/60} = \frac{209.9 \times 40.2 \times 60}{2\pi \times 1{,}100} = 73.2 \text{ Nm } \textbf{Ans.}$$

Example 9.26. A 440 V shunt motor takes 105 A as armature current from the supply and runs at 1,000 rpm. The armature resistance is 0.15 Ω. If the total torque developed is unchanged, calculate the speed and armature current if the magnetic field is reduced to 70% of the initial value.

Solution :

Supply voltage, $V = 440$ V

Armature current, $I_{a1} = 105$ A

Armature resistance, $R_a = 0.15\ \Omega$

Back emf, $E_{b1} = V - I_{a1} R_a = 440 - 105 \times 0.15 = 424.25$ V

Speed, $N_1 = 1{,}000$ rpm

Flux, $\Phi_2 = 70\%$ of Φ_1 *i.e.* $0.7\ \Phi_1$

Since total torque developed remains unchanged,

$$I_{a2}\, \Phi_2 = I_{a1}\, \Phi_1$$

$$\text{or Armature current, } I_{a2} = I_{a1} \times \frac{\Phi_1}{\Phi_2} = \frac{105}{0.7} = 150 \text{ A} \quad \textbf{Ans.}$$

$$\text{Back emf, } E_{b2} = V - I_{a2} R_a = 440 - 150 \times 0.15 = 417.5 \text{ V}$$

$$\text{Speed, } N_2 = \frac{E_{b2}}{E_{b1}} \times \frac{\Phi_1}{\Phi_2} \times N_1 = \frac{417.5}{424.25} \times \frac{1}{0.7} \times 1{,}000 = 1{,}406 \text{ rpm} \quad \textbf{Ans.}$$

9.25. OPERATING CHARACTERISTICS OF DC MOTORS

The performance and, therefore, suitability of a dc motor is determined from its characteristics, known as *performance characteristics*. The important characteristics of dc motors are:

1. Torque-Armature Current Characteristic. This characteristic curve gives relation between torque developed in the armature T and armature current I_a. This is also known as *electrical characteristic.*

2. Speed-Armature Current Characteristic. This characteristic curve gives relation between speed N and armature current I_a. This is also known *as speed characteristic.*

3. Speed-Torque Characteristic. This characteristic gives relation between speed N and torque developed in armature, T. This is also known as *mechanical characteristic*. This curve may be derived from the two characteristic curves mentioned above.

The important relations to be kept in mind while discussing motor characteristics are :

(*i*) $I_a = \frac{V - E_b}{R_a}$ (*ii*) $N \propto \frac{E_b}{\phi}$ and (*iii*) $T \propto \Phi I_a$.

9.26. OPERATING CHARACTERISTICS OF DC SERIES MOTORS

1. Speed-Current Characteristic. In case of a dc series motor, the mmf due to the exciting coils increases in direct proportion to the line or armature current, so (neglecting armature reaction effects) the value of flux varies with the load current according to the ordinary magnetization curve. Owing to armature reaction, the actual curve representing the useful flux falls below the open-circuit magnetization curve, as illustrated in Fig. 9.40. With larger currents the magnetic circuit gets saturated and flux Φ tends to approach a constant value.

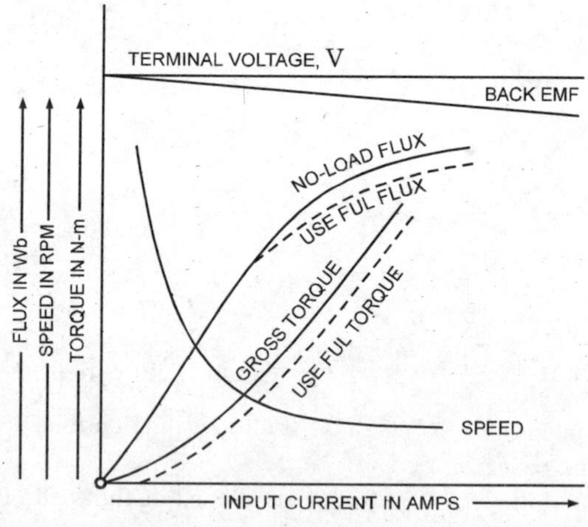

Fig. 9.40

From the speed equation, it is obvious that speed is proportional to back emf E_b and inversely proportional to flux per pole Φ. With the increase in armature current voltage drop in armature circuit and series field $[I(R_a + R_{se})]$ increases and, therefore, back emf E_b decreases, as shown in Fig. 9.40. However, under normal conditions $I(R_a + R_{se})$ drop is quite small and may be neglected. Thus if the applied voltage remains constant, speed N is inversely proportional to flux Φ. If a curve is drawn between speed and input (or line) current I, it will be a rectangular hyperbola before magnetic saturation as upto saturation point the magnetisation curve is a straight line. In this region, the speed decreases abruptly with the increase in input current. After magnetic saturation, the flux Φ tends to become constant and speed-current characteristic becomes a straight line and speed decreases slightly due to voltage drop in armature and series field, as shown in Fig. 9.40. The speed becomes zero when the input current is the normal short-circuit current of the motor *i.e.* equal to applied voltage divided by the motor resistance $(R_a + R_{se})$. Normally this current is many many times full-load value.

From speed-current characteristic curve it is obvious that the series motor is a variable speed motor *i.e.* speed varies with the variation in line current. With the decrease in load on a dc series motor, the speed increases and may become dangerously high at very light loads.

Since on no load the speed is dangerously high, the machine may get damaged due to heavy centrifugal forces set up in the rotating parts. *This is the reason that series motors are never started on no load, which is explained below.*

When the motor is connected across the supply mains without load, it draws a small current from supply mains flowing through the series field and armature, the speed tends to increase so that back emf may approach the applied voltage in magnitude. The increase in back emf weakens the armature current and hence the field current. This causes again increase in speed so in back emf. Thus the field continues to weaken and speed continues to increase until the armature gets damaged.

To start a dc series motor, mechanical load is first put and then the motor is started.

Since on no load the series motor attains dangerously high speed, which causes heavy centrifugal force resulting in the damage of machine, series motors are not suitable for the services:

(*i*) where the load may be entirely removed and

(*ii*) for driving by means of belts because mishap to the belt would cause the motor run on no load.

The motors are suitable for gear drive, because gears provide some load on account of the frictional resistances of the gear teeth in case of sudden release of load.

However, very small series motors may be used with belts, since in case of mishap to belt, the comparatively large frictional resistance would represent an appreciable load on it.

The minimum load on a dc series motor should be great enough (not below 15% of full-load) to keep the speed of the motor within limits. In case the speed becomes dangerously high the motor must be disconnected from the supply main.

2. Torque-Current Characteristic. From the expression of mechanical torque T it is obvious that torque is directly proportional to the product of flux per pole Φ and armature current I_a. Upto saturation point flux is proportional to field current and hence to the armature current, because $I_a = I_f$. Therefore, on light load mechanical torque T is proportional to the square of the armature current *i.e.*, $T \propto I_a^2$ and hence curve drawn between torque and armature current upto saturation point is a parabola, as shown in Fig. 9.40.

After saturation point flux Φ is almost independent of excitation current and so the torque is proportional to the armature current *i.e.* $T \propto I_a$. Hence the characteristic becomes a straight line. The useful (or shaft) torque is, of course, less than the total (or gross) torque developed. This is due to torque lost in iron and friction and windage losses.

From the torque-armature current curve it is evident that so long as the field of the motor is not saturated, the series motor exerts a torque proportional to the square of current i.e. starting torque is very high. Hence series motors are used where large starting torque is required for accelerating heavy masses quickly such as in hoists, electric railways, trolleys and electric vehicles.

From torque-current characteristic it is evident that series motor develops large starting torque to accelerate the heavy masses and from speed-current characteristic it is also evident that speed falls as the load increases, so series motor is automatically relieved from heavy excessive load, therefore, series motor is best suited for electric traction work.

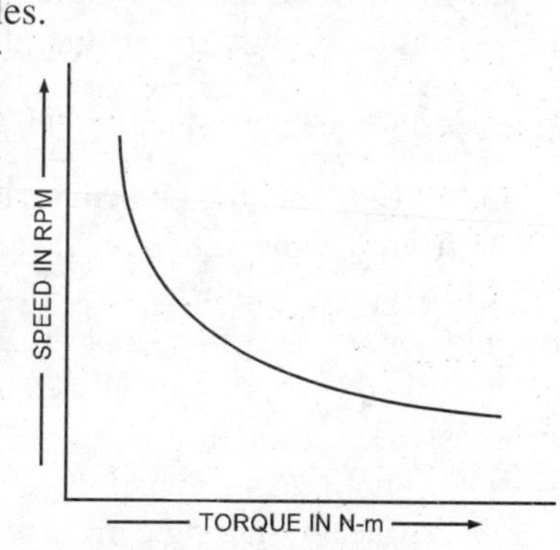

Fig. 9.41. ***Speed-Torque Characteristic of DC Series Motor***

3. Speed-Torque Characteristic. The speed-torque characteristic can be drawn with the help of above two characteristics, as illustrated in Fig. 9.41. Speed-torque characteristic, also known as mechanical characteristic, sharply falls with the increase in torque for smaller values of load. But at higher loads, the speed drops linearly but slowly with increasing torque.

Hence series motors are best suited for services where the motor is directly coupled to the load such as fans whose speed falls with the increase in load torque.

9.27. OPERATING CHARACTERISTICS OF DC SHUNT MOTORS

1. Speed - Armature Current Characteristic. If applied voltage V is kept constant, the field current will remain constant, hence flux will have maximum value on no load but will slightly decrease due to armature reaction as the load increases but for most purposes the flux is considered to be constant, neglecting armature reaction effect.

From speed equation, speed N is directly proportional to back emf E_b or $(V - I_a R_a)$ and inversely proportional to the flux Φ. Since flux is considered to be constant as mentioned above, with the increase in armature current the speed slightly falls due to increase in voltage drop in armature and the speed-armature current curve coincides with the back emf-armature current curve. Since voltage drop in armature at full load is very small as compared to applied voltage drop in speed from no-load to full-load is very small. If demagnetising effect of armature reaction is considered, the drop in speed due to voltage drop in armature is compensated for upto some extent due to decrease in flux with the increase in armature current and speed-armature current characteristic is less drooping, as shown dotted in Fig. 9.42 or even be rising if demagnetisation is high. Thus the drop in speed from no load to full load is very small and for all practical purposes the dc shunt motor is taken as a constant speed motor.

There is a slight variation in speed of the shunt motor from no load to full load and this slight variation in speed can be made up by inserting resistance in the shunt field and so reducing the flux. Therefore, shunt motors can be used for the loads which are totally and suddenly thrown off without resulting in excessive speed. Shunt motors being constant speed motors are best suited for driving of line shafts, machine lathes, milling machines, conveyors, fans and for all purposes where constant speed is required. It is not suitable for use with flywheel or with fluctuating loads or for parallel operation due to its constant speed characteristic. It is also useful where a moderate degree of speed control is required.

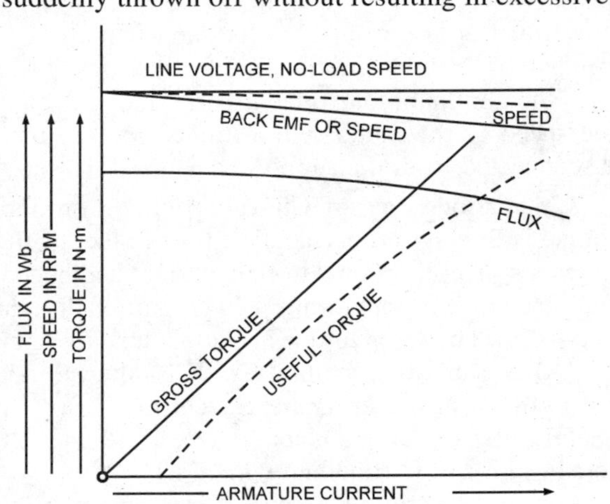

Fig. 9.42

2. Torque-Armature Current Characteristic. From the expression for the torque of a dc motor, torque is directly proportional to the product of flux per pole Φ and armature current I_a. Since in case of dc shunt motor the flux per pole Φ is considered to be constant, torque increases with the increase in load current following linear law, *i.e.* torque-armature current characteristic is a straight line passing through origin O (Fig. 9.42). But the weakening of field due to armature reaction causes the torque line to droop slightly, and the iron and friction losses cause it to be slightly lower than the line representing the electromagnetic developed torque corresponding to $T = 0.159\ \Phi ZP \frac{I_a}{A}$ Nm.

3. Speed-Torque Characteristic. This characteristic curve can be drawn from the above two characteristics and is shown in Fig. 9.43.

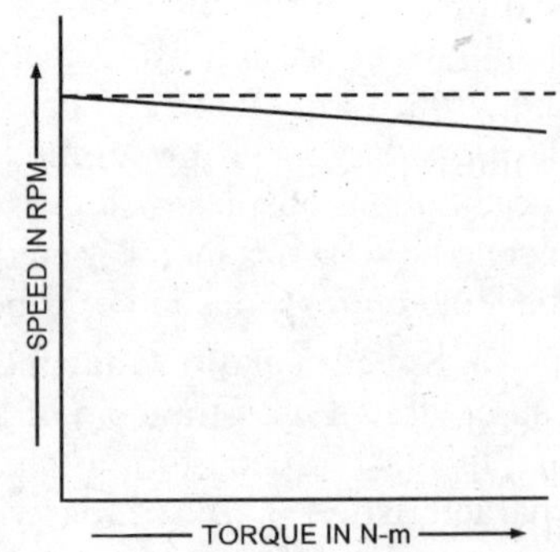

Fig. 9.43

This type of motor is used in applications requiring medium starting torque such as centrifugal pumps, blowers, fans, conveyors, boring mills, shapers, wood working machines, spinning and weaving machines, printing presses, machine tools, etc.

DC shunt motors should never be started on heavy loads because such loads need heavy starting current.

9.28. OPERATING CHARACTERISTICS OF COMPOUND WOUND DC MOTORS

As explained in Art. 9.18, the compound wound dc motors are of two types viz cumulative-compound wound motors and differential compound-wound motors.

9.28.1. Operating Characteristics of Cumulative Compound Wound DC Motors

The characteristics of cumulative compound wound motor are combination of shunt and series characteristics and lie between those of shunt and series motors.

As the load is increased, the flux due to series field winding increases and causes the torque greater than it would have been with shunt field winding alone for a given machine and for a given current. The increase in flux due to series field winding on account of increase in load causes the speed to fall more rapidly than it would have done in shunt motor. The characteristics are shown in Fig. 9.44.

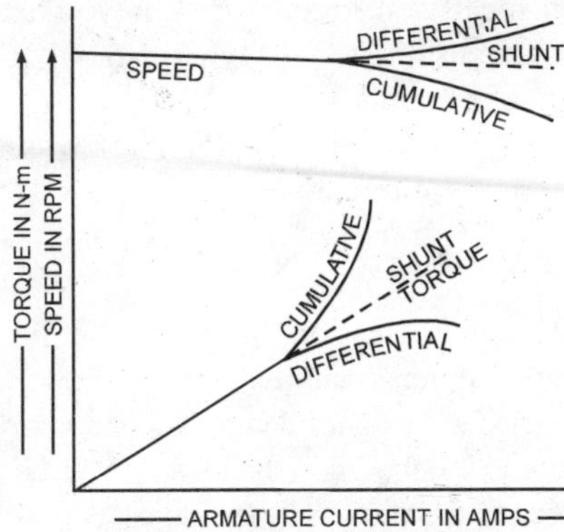

Fig. 9.44. *Characteristics of Compound Wound Motors*

Like the series motor, the cumulative compound wound motor has excellent characteristics for starting heavy loads and yet there is no danger of "running away" at light loads when used without a speed controller.

By varying the proportions of shunt field mmf and series field mmf, compound wound motors may be designed to give speed and torque characteristics anywhere intermediate between those for shunt and series motors. In many cases only enough shunt field is provided to guarantee a safe no-load speed, in which case the motor be called a *stabilised series motor*. Compound wound motors are usually designed so that there will be a drop in speed of about 20% between no-load to full-load.

Cumulative compound wound motors are used in driving machines which are subject to sudden applications of heavy loads, such as occur in rolling mills, punching and shearing machines, lifts, mine-hoist etc. This type of motor is used also where a large starting torque is required but series motor cannot be employed conveniently such as in cranes and elevators.

Another advantage of the cumulative compound wound motor for suddenly applied loads is that the motor automatically undergoes a substantial drop in speed when the load is applied. Accordingly much of its stored kinetic energy becomes available for supplying a part of the increased load thus reducing the electrical load on the motor as well as the peaks on the power system. The available kinetic energy is frequently increased by the use of a flywheel, particularly with rolling mill motors.

Cumulative compound wound motors are not practical for applications requiring adjustable speed for field control. With a weakened shunt field, the series field becomes dominating hence the changes in load may produce unstable speeds.

9.28.2. Operating Characteristics of Differential Compound Wound DC Motors

In differential compound wound motor, since the series field opposes the shunt field, the resultant flux decreases with the increase in load; thus the machine runs at a higher speed than it would do as a shunt motor. The decrease in flux with the increase in load causes the torque to be less than that of a shunt motor, as illustrated in Fig. 9.44. Such motors are rarely employed in practice as the differential arrangement causes difficulties during overloads and starting.

9.28.3. Speed-Torque Characteristics of Different Types of DC Motors

Speed-torque characteristics of dc series, shunt, separately excited, cumulatively compounded and differentially compounded motors are shown in Fig. 9.45.

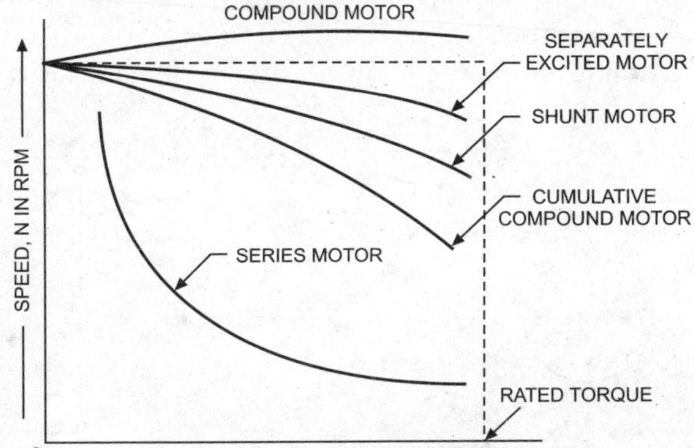

Fig. 9.45. *Speed-Torque Characteristics of Different Types of DC Motors*

9.29. SPEED CONTROL OF DC MOTORS

Speed control means intentional change of the drive speed to a value required for performing the specific work process.

One of the attractive features the dc motor offers over all other types is the relative ease with which speed control can be achieved and, therefore, dc motors are indispensable for many adjustable speed drives.The various schemes available for speed control can be deduced from the expression of speed for a dc motor which is repeated here with one modification

$$N = K \frac{V - I_a (R + R_a)}{\Phi}$$

The modification involves the inclusion of an external resistance in the armature circuit. The above expression reveals that the speed can be controlled by adjusting any one of the three factors appearing on the right hand side of the expression: (*i*) applied voltage to the armature terminals, V (*ii*) external resistance in the armature circuit, R and (*iii*) flux per pole, Φ. The first two possibilities involve adjustment affecting the armature circuit, whereas the third involves change in the magnetic field. Therefore, speed control methods are broadly classified as *armature control methods* and *field control methods.* Sometimes a combination of the two methods is employed. With armature control the speed decreases as the voltage applied to the armature terminals is reduced, whereas with field control the speed increases as the flux is reduced.

1. *Field Control Methods.* In case of *dc series motors,* the flux can be varied by any one of the following methods.

(i) Field Divertor Method. The field flux can be reduced by shunting a portion of motor current around the series field, thus reducing the excitation mmf and weakening of field. This method is illustrated in Fig. 9.46 (*a*).

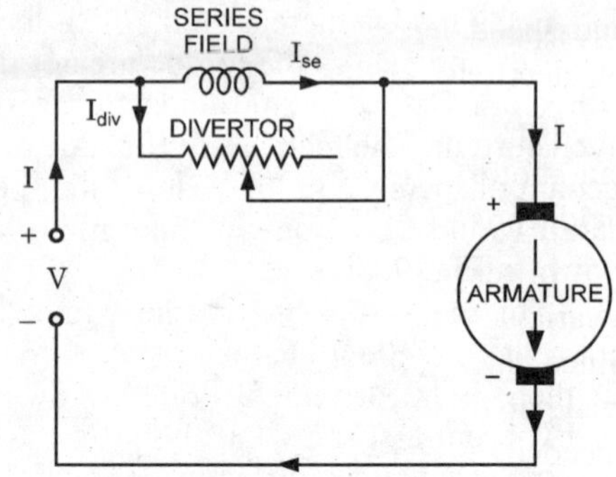

(a) Field Divertor Method of Speed Control For DC Series Motors

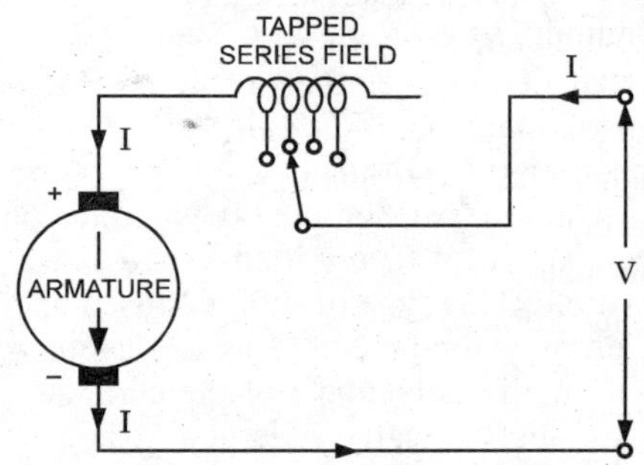

(b) Tapped Field Control For DC Series Motors

Fig. 9.46

This method provides speeds above normal because flux is reduced by this method. Lesser the divertor resistance, less the field current, less flux and, therefore, more the speed. This method is convenient as well as economical and provides the speed control range usually not exceeding 2 : 1. This method of speed control is used in electric drives in which the speed should rise sharply as soon as the load falls.

(ii) Tapped Field Control. This is another method of increasing the speed by reducing the flux and it is accomplished by reducing the number of turns of the field winding through which the current flows. In this method of speed control of dc series motors a number of tappings from the field winding are brought outside, as illustrated in Fig. 9.46 (*b*). A number of series field turns can be short circuited according to the requirement. When all field turns are in circuit, the motor runs at the lowest speed

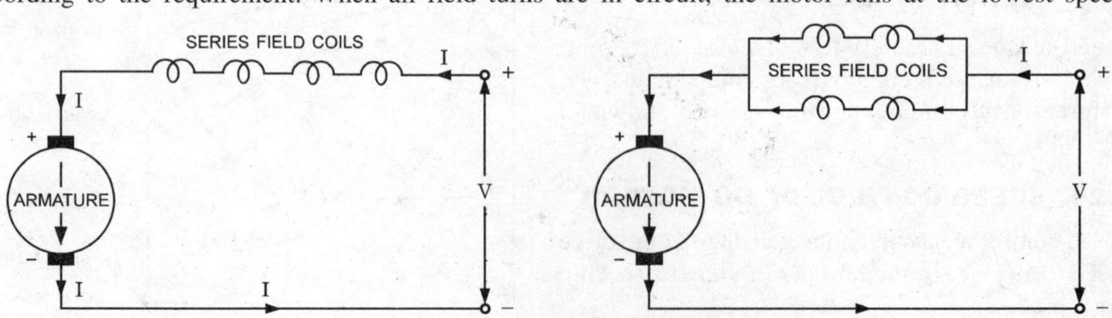

(i) *(ii)*

Fig. 9.46 (*c*)

and speed increases with cutting out some of the series field turns. This method is usually employed in electric traction.

(iii) Paralleling Field Coils Method. In this method, used in fan motors, several speeds can be obtained by re-grouping of field coils, as illustrated in Fig. 9.46 (*c*).

In case of dc **shunt motors** the flux can be varied by inserting a variable resistance in series with the field winding (refer Fig. 9.47). Since this resistance has to carry only a small current, so it is made up of slide-wire type of resistor to have continuously variable speed over the range.

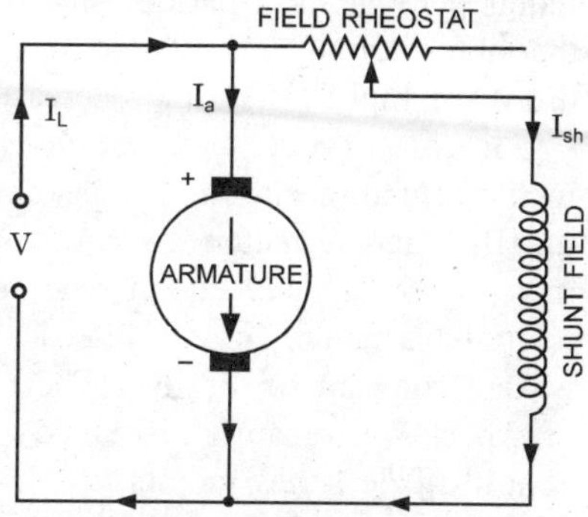

Fig. 9.47. ***Field Rheostat Control For a DC Shunt Motor***

Since the flux can be only decreased (not increased), the speeds only above normal one can be obtained by this method. The speed is minimum at the maximum value of flux and depends upon the design of the field and its saturation point. The speed is maximum at the minimum value of flux, which is governed by the demagnetizing effect of armature reaction on the field as at higher speeds the motor tends to be unstable and difficulties in commutation arise. The high speed is also restricted due to mechanical considerations as the centrifugal forces are set up at high speeds. Speed variation by this method is limited to ratio of 4 or 5 to 1. Creeping speeds cannot be obtained by this method. The power output being proportional to TN or VI_a remains constant. This method is, therefore, suitable only where power of the load remains constant. This method of speed control is very simple, convenient and most economical and is, therefore, extensively used in electric drives. The power wasted in the controlling resistance is very little, field current being very small. This method of speed control is independent of load on the motor and permits the remote control of speed.

2. ***Armature Resistance Control Methods.*** In this method of speed control, reduced speeds are obtained by inserting resistance in the armature circuit. It may be employed with series, shunt and compound wound motors. For the last two types, the series resistance is connected between the shunt field and armature, not between the line and motor.

In case of a dc *shunt motor* an increase in armature circuit resistance will cause more voltage drop in the armature circuit, so the speed will be reduced. Field current will remain unaffected, as the shunt field is directly connected across the supply voltage. For a constant torque load, the armature current remains the same so input to the motor remains the same but the output decreases in proportion to the speeds. Operating costs are, therefore, comparatively high for long time running at reduced speed. In case of fans and centrifugal pumps where the load torque decreases with the decrease in speed, the losses are considerably low and because of its low initial cost and simplicity this method may be quite convenient and economical for short-time or intermittent slowdowns. Wide range of speed (below normal one) can be obtained by this method and at the same time motor will develop any desired torque over its operating range. The main advantage of this method is that speeds below base speed down to creeping speeds of only a few rpm are easily

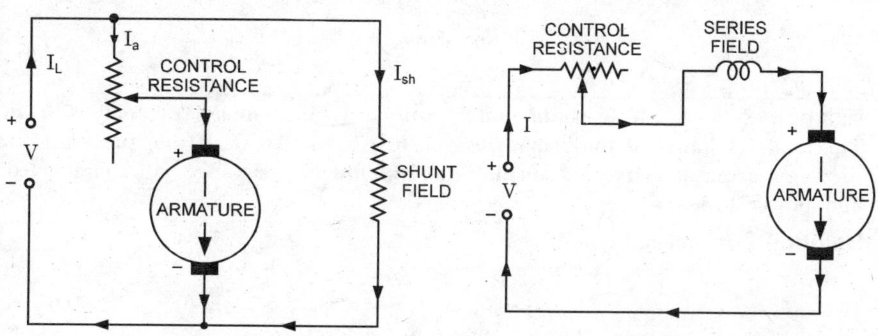

(*a*) *For DC Shunt Motor* (*b*) *For DC Series Motor*

Fig. 9.48. ***Armature-Resistance Control***

available. This method of speed control, therefore, is employed where speeds lower than rated one are required for a short period only and also occasionally as in printing machines, cranes and hoists where the motor is frequently started and stopped. This method of speed control is also employed where the load drops off rapidly with the decrease in speed, as in fans and blowers.

In case of a dc *series motor* an increase in armature circuit resistance will cause more voltage drop in the armature circuit, so speed will be reduced. The torque which is proportional to the product of the flux and armature current will also be reduced with the increase in armature circuit resistance as increase in armature circuit resistance will reduce the armature current and the flux which also depends upon the armature current. In order to obtain different speeds for constant load torque, the armature current is kept constant so the flux. The resistance in the armature circuit is increased to reduce the speed. The drawbacks of armature resistance control for machines with shunt fields are not as important in the speed control of dc series motors.

The poor speed regulation that is inherent in this method has no significance for the control of dc series motors, since the speed characteristic of a dc series motor is a rapidly drooping curve. The power loss in the control resistance for many applications of dc series motors is not too serious, since in these applications the control is utilized for a large portion of time for reducing the speed under light-load conditions and is only employed intermittently when the motor is carrying full load. The maximum range of speed control of about 3 : 1 will be available depending on the load. This method of speed control is most economical for constant torque drives. This method of speed control is employed chiefly for dc series motors driving cranes, hoists, trains etc. because such drives operate on intermittent duty.

3. *Armature Voltage Control Method.* This method of speed control requires a source of variable voltage separate from the source supplying the field current. This method avoids the disadvantages of poor speed regulation and low efficiency which are the characteristics of the armature–resistance control method but it is more expensive in initial cost. The adjustable voltage for the armature is obtained from an adjustable voltage generator or from an adjustable electronic rectifier*. This method provides a large speed range with any desired number of speed points. It is essentially a constant-torque system, because the output delivered by the motor decreases with a decrease in applied voltage and a corresponding decrease in speed.

Example 9.27. A 500 V shunt motor runs at its normal speed of 250 rpm when the armature current is 200 A. The resistance of armature is 0.12 Ω. Calculate the speed when a resistance is inserted in the field, reducing the shunt field to 80% of normal value, and the armature current is 100 A.

Solution :

$$\text{Back emf, } E_{b1} = V - I_{a1} R_a = 500 - 200 \times 0.12 = 476 \text{ V}$$

$$\text{Back emf, } E_{b2} = V - I_{a2} R_a = 500 - 100 \times 0.12 = 488 \text{ V}$$

$$\Phi_2 = 80\ \% \text{ of } \Phi_1 = 0.8\ \Phi_1$$

$$\text{Speed, } N_1 = 250 \text{ rpm}$$

$$\text{Speed, } N_2 = N_1 \times \frac{E_{b2}}{E_{b1}} \times \frac{\Phi_1}{\Phi_2} = 250 \times \frac{488}{476} \times \frac{\Phi_1}{0.8\Phi_1} = 320.4 \text{ rpm } \textbf{Ans.}$$

Example 9.28. A 240 V shunt motor runs at 1,450 rpm on full load with an armature current of 11 A. The total resistance of the armature and brushes is 0.6 Ω. If the speed is to be reduced to 1,000 rpm with the same armature current, calculate the amount of resistance to be connected in series with the armature and power lost.

Solution : In normal conditions

$$\text{Armature current, } I_{a2} = I_{a1} = 11 \text{ A}$$

$$\text{Back emf, } E_{b1} = V - I_{a1} R_a = 240 - 11 \times 0.6 = 233.4 \text{ V}$$

$$\text{Speed, } N_1 = 1{,}450 \text{ rpm}$$

* For most applications dc motors are fed by rectification of the available ac supply. It is usually necessary to insert a transformer immediately ahead of the rectifier.

With additional resistance (say of R ohms) in armature circuit

Speed, $N_2 = 1{,}000$ rpm

Back emf, $E_{b2} = V - I_{a2}(R + R_a) = 240 - 11(0.6 + R) = 233.4 - 11R$

$$\frac{E_{b2}}{E_{b1}} = \frac{N_2}{N_1} \quad \text{Assuming flux to remain constant}$$

or $$\frac{233.4 - 11R}{233.4} = \frac{1000}{1{,}450}$$

or $R = 6.585\ \Omega$ **Ans.**

Power lost $= I^2(R + R_a) = 11^2 \times (6.585 + 0.6) = 869.4$ watts **Ans.**

Example 9.29. A 200 V dc series motor runs at 500 rpm when taking a current of 25 A. The resistance of the armature is 0.5 Ω and that of field is 0.3 Ω. If the current remains constant, calculate the resistance necessary to reduce the speed to 250 rpm.

Solution : Motor input current, $I_1 = 25$ A

Back emf, $E_{b1} = V - I_1(R_a + R_{se}) = 200 - 25(0.5 + 0.3) = 180$ V

Speed, $N_1 = 500$ rpm

Let the resistance R be connected in series with the motor to reduce the speed to

$N_2 = 250$ rpm

Motor input current, $I_2 = I_1 = 25$ A

Back emf $E_{b2} = V - I_2(R + R_a + R_{se}) = 200 - 25(R + 0.5 + 0.3) = 180 - 25R$

Since $N \propto \dfrac{E_b}{\Phi}$

$$\therefore \frac{N_2}{N_1} = \frac{E_{b2}}{E_{b1}} \times \frac{\Phi_1}{\Phi_2} = \frac{E_{b2}}{E_{b1}} \quad \because \phi_2 = \Phi_1 \text{ as field current remains the same}$$

or $$\frac{250}{500} = \frac{180 - 25R}{180}$$

or $R = 3.6\ \Omega$ **Ans.**

9.30. POWER LOSSES IN DC MACHINES

The power losses in a dc machine consist of input power that is converted into heat. Power losses occurring in a dc machine are divided into (*i*) copper or electrical losses, (*ii*) iron or magnetic losses, and (*iii*) mechanical losses.

1. Copper or Electrical Losses. Copper or electrical losses include power lost in heating the shunt field, and also in heating armature circuit, which includes armature winding, brush contacts, series field and interpole field. These losses are due to power consumed in forcing a current against the resistance of windings and connections. The actual copper losses of the windings will depend upon the effective resistance of the winding for the operating conditions. The effective resistance of a winding will vary with the operating conditions of load and excitation, and its exact value even for one particular set of operating conditions is hard to determine. For these reasons it is standard practice to compute losses on the basis of the dc resistance of the winding at 75°C. These losses are further subdivided into (*a*) armature copper loss and (*b*) field copper loss.

(a) Armature Copper Loss. This is given as $I_a^2 R_a$ where I_a is the armature current and R_a is the armature resistance at the operating temperature (assumed to be 75°C). Brush contact resistance loss is also taken into account by including the brush contact resistance in the armature resistance R_a. This loss is proportional to the square of armature current and is about 30 to 40% of full-load losses.

(b) Field Copper Loss. This is further sub divided into (*i*) shunt field loss, (*ii*) interpole loss, and (*iii*) series field loss. Shunt field loss is given as $I_{sh}^2 R_{sh}$ where I_{sh} is the shunt field current and R_{sh}

is the shunt field resistance. The shunt field loss incurs in shunt and compound dc machines and is practically constant if applied voltage across the machine remains the same. Interpole loss is given as $I_a^2 R_i$ where I_a is the armature current and R_i is the interpole field resistance. Interpole loss incurs only in the machines provided with interpoles. The series field loss, which incurs in series and compound machines, is given as $I_{se}^2 R_{se}$ where I_{se} is series field current and R_{se} is series field resistance. With the increase in load on machine, armature current increases, therefore, series field current increases resulting increase in interpole and series field losses. Field copper loss is about 20 to 30% of the full-load losses.

2. Iron or Magnetic Losses. These losses are also called core losses and include the hysteresis and eddy current losses which are present in any part of the machine that is constructed of iron and subjected to variations of magnetic flux. The hysteresis loss depends upon the range of variations of the flux density and the frequency of this variation of flux. The hysteresis loss can be expressed with reasonable accuracy by following expression:

$$P_h = \eta \, (B_{max})^{1.6} fv \text{ joules/second or watts}$$

where η = Steimmetz hysteresis coefficient; v = volume of core in metres^3 and f is the frequency of the magnetic cycles per second.

The eddy current loss also depends upon the range of variation of flux density and the frequency of the variation of the flux. In addition it depends upon the thickness of the iron laminations. It can be determined with reasonable accuracy by the following expression:

$$P_e = K_e \, (B_{max})^2 f^2 vt^2$$

where K_e is a constant depending upon the electrical resistance of the magnetic material and the units employed in the expression, B_{max} is the maximum flux density attained throughout the magnetic cycle, f is the frequency of magnetic cycles per second, v is the volume of magnetic material and t is the thickness of the magnetic material laminations.

The core losses are approximately constant in a machine whose flux and speed remain constant as in case of shunt wound dc machines but are variable in case of series wound and compound wound dc machines. These losses are about 10 to 20% of full-load losses.

3. Mechanical Losses. These losses consist of power losses due to friction of the bearings, air friction or windage, as it is called, caused by the motion of the moving parts through the surrounding medium, and the friction between brushes and commutator or slip rings. These losses are constant in a machine operating at approximately constant speed and are independent of load. These losses are about 10 to 20% of full-load losses.

Usually magnetic and mechanical losses collectively termed as stray losses and the shunt field copper losses are collectively known as *constant losses.*

Example 9.30. A dc shunt motor develops 10 bhp at 600 rpm when drawing 18 A at 500 V. Calculate efficiency and torque.

Solution :

$$\text{Motor output} = 10 \times 735.5 = 7{,}355 \text{ watts or J/s}$$

$$\text{Motor input} = V \times I = 500 \times 18 = 9{,}000 \text{ watts}$$

$$\text{Motor efficiency, } \eta = \frac{\text{Output}}{\text{Input}} \times 100 = \frac{7{,}355}{9{,}000} \times 100 = 81.72\ \% \quad \textbf{Ans.}$$

$$\text{Torque, T} = \text{Output} \times \frac{60}{2\pi N} = \frac{7{,}355 \times 60}{2\pi \times 600} = 117 \text{ newton-metres} \quad \textbf{Ans.}$$

Example 9.31. A 400 V dc motor takes 5 A at no load. Its armature and field resistances are 0.5 Ω and 200 Ω respectively. Calculate the efficiency when motor takes 50 A on full load.

Solution :

$$\text{Shunt field current, } I_{sh} = \frac{V}{R_{sh}} = \frac{400}{200} = 2 \text{ A}$$

No-load line current, $I_{L_0} = 5$ A

No-load input $= VI_{L_0} = 400 \times 5 = 2{,}000$ W

No-load armature current, $I_{a_0} = I_{L_0} - I_{sh} = 5 - 2 = 3$ A

No-load armature copper loss $= I_{a_0}^2 R_a = (3)^2 \times 0.5 = 4.5$ W

Constant losses = No-load input – no-load copper losses = 2,000 – 4.5 = 1,995.5 W

On Full Load: Line current, $I_L = 50$ A

Input $= VI_L = 400 \times 50 = 20{,}000$ W

Armature current, $I_a = I_L - I_{sh} = 50 - 2 = 48$ A

Armature copper losses $= I_a^2 R_a = 48^2 \times 0.5 = 1{,}152$ W

Total losses = Constant losses + armature copper loss = 1,995.5 + 1,152 = 3,147.5 W

Output = Input – total losses = 20,000 – 3,147.5 = 16,852.5 W

$$\text{Efficiency, } \eta = \frac{\text{Output}}{\text{Input}} \times 100 = \frac{16{,}852.5}{20{,}000} \times 100 = 84.26\ \%$$ **Ans.**

Example 9.32. A shunt generator delivers 195 A at terminal pd of 250 V. The armature resistance and shunt field resistance are 0.02 Ω at 50 Ω respectively. The iron and friction losses equal 950 W. Find (*a*) emf generated (*b*) copper losses (*c*) output of the prime mover (*d*) commercial, mechanical and electrical efficiencies.

Solution : Useful output of generator $= VI_L = 250 \times 195 = 48{,}750$ watts

Shunt field current, $I_{sh} = \dfrac{V}{R_{sh}} = \dfrac{250}{50} = 5$ A

Armature current, $I_a = I_L + I_{sh}$

$= 195 + 5 = 200$ A

Shunt field copper loss $= VI_{sh} = 250 \times 5$

$= 1{,}250$ watts

Armature copper loss $= I_a^2 R_a$

$= (200)^2 \times 0.02$

$= 800$ watts

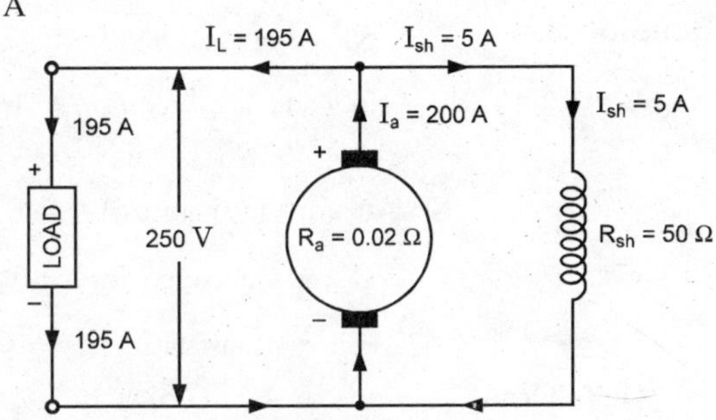

Fig. 9.49

Iron and friction (or stray) losses, $P_S = 950$ watts

(*a*) EMF generated, $E_g = V + I_a R_a = 250 + 200 \times 0.02 = 254$ A **Ans.**

(*b*) Total copper losses, $P_{copper} = I_a^2 R_a + V I_{sh} = 800 + 1{,}250 = 2{,}050$ watts **Ans.**

Electrical power developed = Useful power output + total copper loss

= 48,750 + 2,050 = 50,800 watts

Total power input = Electrical power developed + stray losses = 50,800 + 950 = 51,750 watts

(*c*) Output of prime mover = Total power supplied to generator = 51,750 W or 51.75 kW **Ans.**

$$\text{(d) Commercial efficiency, } \eta = \frac{\text{Useful power output}}{\text{Total power input}} \times 100 = \frac{48{,}750}{51{,}750} \times 100 = 94.2\%$$ **Ans.**

$$\text{Mechanical efficiency, } \eta_m = \frac{\text{Electrical power developed}}{\text{Total power input}} \times 100 = \frac{50{,}800}{51{,}750} \times 100 = 98.16\%$$ **Ans.**

$$\text{Electrical efficiency, } \eta_e = \frac{\text{Useful power output}}{\text{Electrical power developed}} \times 100 = \frac{48{,}750}{50{,}800} \times 100 = 95.96\%$$ **Ans.**

Example 9.33. A 400 V shunt generator has a full-load current of 200 A, its armature resistance is 0.06 Ω, field resistance 100 Ω; the stray losses are 2,000 W. Find the horse power of its prime mover when it is delivering full load.

Solution : On full load

$$\text{Output power, } P_o = V \times I_L = 400 \times 200 = 80{,}00 \text{ watts}$$

$$\text{Shunt field current, } I_{sh} = \frac{V}{R_{sh}} = \frac{400}{100} = 4 \text{ A}$$

$$\text{Armature current, } I_a = I_L + I_{sh} = 200 + 4 = 204 \text{ A}$$

$$\text{Armature copper loss} = I_a^2 R_a = (204)^2 \times 0.06 = 2{,}497 \text{ W}$$

$$\text{Shunt field copper loss} = V I_{sh} = 400 \times 4 = 1{,}600 \text{ W}$$

$$\text{Stray losses, } P_S = 2{,}000 \text{ W}$$

$$\text{Output power of prime mover} = \text{Input power to generator} = VI_L + I_a^2 R_a + VI_{sh} + P_S$$

$$= 80{,}000 + 2{,}497 + 1{,}600 + 2{,}000 = 86{,}097 \text{ W}$$

$$\text{Horse power of prime mover} = \frac{\text{Output of prime mover in watts}}{735.5} = \frac{86{,}097}{735.5} = 117 \textbf{ Ans.}$$

Example 9.34. A 60 kW, 250 V shunt motor takes 16 A when running light at 1,440 rpm. The resistances of armature and field windings are 0.2 ohm and 125 ohm respectively (*i*) Find the efficiency of the motor when taking 152 A. (*ii*) Also estimate the efficiency when working as generator and delivering 152 A at 250 V.

[*Pb. Technical Univ. Basic Electrical and Electronics Engineering May 2008*]

Solution :

$$\text{Shunt field current, } I_{sh} = \frac{V}{R_{sh}} = \frac{250}{125} = 2 \text{ A}$$

$$\text{No-load line current, } I_{Lo} = 16 \text{ A}$$

$$\text{No-load input} = VI_{Lo} = 250 \times 16 = 4{,}000 \text{ W}$$

$$\text{No-load armature current, } I_{ao} = I_{Lo} - I_{sh} = 16 - 2 = 14 \text{ A}$$

$$\text{No-load armature copper loss} = I_{a0}^2 R_a = 14^2 \times 0.2 = 39.2 \text{ W}$$

$$\text{Constant losses} = \text{No-load input} - \text{no-load armature copper loss} = 4{,}000 - 39.2 = 3{,}660.8 \text{ W}$$

On full load,

$$\text{Line current, } I_L = 152 \text{ A}$$

$$\text{Input} = VI_L = 250 \times 152 = 38{,}000 \text{ W}$$

$$\text{Armature current, } I_a = I_L - I_{sh} = 152 - 2 = 150 \text{ A}$$

$$\text{Armature copper losses} = I_a^2 R_a = (150)^2 \times 0.2 = 4{,}500 \text{ W}$$

$$\text{Total losses} = \text{Constant losses} + \text{armature copper losses}$$

$$= 3{,}660.8 + 4{,}500 = 8{,}160.8 \text{ W}$$

$$\text{Output} = \text{Input} - \text{total losses} = 38{,}000 - 8{,}160.8 = 29{,}839.2 \text{ W}$$

$$\text{Efficiency, } \eta = \frac{\text{Output}}{\text{Input}} \times 100 = \frac{29839.2}{38{,}000} \times 100 = 78.52\% \textbf{ Ans.}$$

As Generator,

$$\text{Line current, } I_L = 152 \text{ A}$$

$$\text{Armature current, } I_a = I_L + I_{sh} = 152 + 2 = 154 \text{ A}$$

$$\text{Armature copper loss} = I_a^2 R_a = 154^2 \times 0.2 = 4{,}743.2 \text{ W}$$

$$\text{Total losses} = \text{Constant losses} + \text{armature copper losses} = 3{,}660.8 + 4{,}743.2 = 8{,}404 \text{ W}$$

$$\text{Generator output} = VI_L = 250 \times 152 = 38{,}000 \text{ W}$$

$$\text{Generator input} = \text{Output} + \text{total losses} = 38{,}000 + 8{,}404 = 46{,}404 \text{ W}$$

$$\text{Efficiency, } \eta = \frac{\text{Output}}{\text{Input}} \times 100 = \frac{38{,}000}{46{,}404} \times 100 = 81.89\% \textbf{ Ans.}$$

9.31. APPLICATIONS OF DC MOTORS

The dc motor is often called upon to do the really tough jobs in the industry because of its high degree of flexibility and ease of control. These features cannot easily be matched by other electromechanical energy-conversion devices. The dc motor offers a wide range of control of speed and torque as well as acceleration and deceleration. The applications of the three types of dc motors are given below:

1. DC Series Motor. It is a variable speed motor *i.e.* very low speed at high torque and vice versa. However, at light or no load, the motor tends to attain dangerously high speed. The motor has a very high starting torque (upto 5 times of full-load torque). It is, therefore, used for drives requiring very high starting torque and where adjustable varying speed is satisfactory. Industrial uses are hoists, cranes, trolley cars, conveyors, electric locomotive, elevators, air-compressors, vacuum cleaners, sewing machines etc.

Loads must be positively connected, not belted. To prevent overspeed, lightest load should not be much less than 15 to 20% of full-load torque.

2. DC Shunt Motor. It is an approximately constant speed motor and has medium starting torque (usually limited to 2.5 times of full-load torque by a starting resistor).

DC shunt motor is, therefore, used essentially for constant speed applications requiring medium starting torque. May be used for adjustable speed not greater than 2 : 1 range.

Industrial applications are lathes, centrifugal pumps, reciprocating pumps, fans, blowers, conveyors, wood-working machines, machine tools, printing presses, spinning and weaving machines.

3. Compound Wound Motors. *Differential compound wound dc motors* are rarely used because of their poor torque characteristics and difficulties experienced during overloads and starting.

Cumulative compound wound dc motors are used in driving machines which are subject to sudden applications of heavy loads, such as occur in rolling mills, punching and shearing machines, lifts, mine-hoists etc. This type of motor is also used where a large starting torque is required but series motor cannot be conveniently used such as in cranes and elevators.

HIGHLIGHTS

1. *DC machine* is commutator type machine. It is actually an ac machine, but furnished with a special device, called the commutator, which under certain conditions converts ac into dc and vice versa.
2. An *electrical generator* is a machine which converts mechanical energy (or power) into electrical energy (or power). It operates on the principle of electromagnetic induction.
3. In construction a dc machine consists of four parts mainly 1. Field magnets 2. Armature 3. Commutator and 4. Brushes and brush gear.

   Field system is stationary part and it is to create a uniform magnetic field, within which the armature rotates.

   Armature is a rotating part, where mechanical energy is converted into electrical energy (in case of generator) or, conversely, electrical energy into mechanical energy (in case of motor).

   Brushes are meant for collecting current from the commutator and supply it to the external load circuit.
4. EMF equation for a generator is given as

   $$E_g = \frac{\phi Z N}{60} \times \frac{P}{A} \text{ volts; where } A = 2 \text{ in wave-wound machines, } A = P \text{ in lap-wound machines}$$
5. According to method of excitation, dc generators are classified into separately excited generators and self excited generators.

   Self excited generators are further classified in accordance how the field winding is connected w.r.t. armature. Accordingly such generators are (*i*) series wound type (*ii*) shunt wound type and (*iii*) compound wound type. Compound wound generators may be long shunt type or short shunt type.

   Compound wound dc generators can also be classified into cumulative compound wound and differential compound wound.

6. (*a*) The best application of series wound dc generators is for supplying field current for regenerative braking in dc locomotives. They are also used as series boosters and in arc lighting.
   (*b*) Cumulative compound wound dc generators are generally used for lighting and power services.
   (*c*) Differential compound generator finds an useful application as an arc welding generator.
7. As regards fundamental principles, the dc motors are identical with the dc generators which have the same type of excitation *i.e.*, a machine that operates as a motor will also operate satisfactorily as a generator.
8. When a dc motor is connected to dc supply mains, the armature continues to rotate due to motor action, the armature conductors cut the magnetic flux and, therefore, emfs are induced in them. The direction of this induced emf, known as back emf or counter emf, is such that it opposes the applied voltage.

$$\text{Back emf in a dc motor, } E_b = \frac{\phi ZN}{60} \times \frac{P}{A} \text{ volts}$$

9. Armature current of dc motor is given as

$$I_a = \frac{V - E_b}{R_a}$$

10. Voltage equation of the motor is

$$V = E_b + I_a R_a + \text{brush drop}$$

11. Power equation of a dc motor is given as

$$VI_a = E_b I_a + I_a^2 R_a \text{ in case of dc shunt motor}$$

$$\text{and } VI_a = E_b I_a + I_a^2 (R_a + R_{se}) \text{ in case of dc series motor}$$

12. Condition for maximum power developed in armature of a dc motor is

$$E_b = V/2$$

13. Similar to dc generators, the dc motors can be classified as (*i*) permanent magnet (*ii*) separately excited (*iii*) series wound (*iv*) shunt wound and (*v*) compound wound dc motors.
14. Direction of rotation of a dc motor can be reversed by reversing the current through either the armature winding or field coils not in both.
15. From emf equation $E_b \propto \phi N$

$$\text{or } N \propto \frac{E_b}{\phi} = K\frac{V - I_a R_a}{\phi}$$

$$\text{or } N = \frac{E_b}{\phi} \times \frac{60A}{Z \times P} \text{ rpm}$$

16. Speed regulation of a motor is given by

$$\text{Percentage speed regulation} = \frac{\text{No-load speed} - \text{full-load speed}}{\text{Full-load speed}} \times 100$$

17. Torque developed by armature in a dc motor is given as

$$T_a = 0.159\, \phi ZP \times \frac{I_a}{A} \text{ newton-metres}$$

$$\text{Also } T_a = 9.55 \frac{E_b I_a}{N} \text{ N-m}$$

18. Shaft or useful torque is given as

$$T_{sh} = \frac{\text{Output in watts}}{2\pi N / 60} \text{ N-m}$$

$$\text{Torque lost in iron and friction losses} = \frac{\text{Iron and friction losses in watts}}{2\pi N/60} \text{ Nm}$$

19. *Important Characteristics.*

Series motor develops high torque at load and has ability to pick up. Its speed increases with the decrease in load. DC series motors are never started on no load because their speed at no load is dangerously high. Series motors are best suited for electric traction work.

Shunt motor is a constant speed motor. Such a motor is used in applications requiring medium starting torque such as centrifugal pumps, blowers, fans, conveyors, boring mills, shapers, wood-working machines etc.

20. Speed of a dc motor is given as

$$N = K \frac{E_b}{\phi} = K \frac{V - I_a(R_a + R)}{\phi}$$

Based on above equation, the speed of a dc motor can be controlled by field control, armature resistance control or armature voltage control.

21. *In field control of dc motors*

In case of dc shunt motors, speed variation is accomplished by means of a variable resistance inserted in series with the shunt field.

In case of dc series motors, speed variation is accomplished by using a divertor across the series field winding. Field control method is more economical and convenient and gives speeds above the normal ones.

22. *In armature control of dc motors*, a variable controlling resistance is connected in series with the armature circuit.

Rheostatic control or armature control gives speeds below the normal, and is expensive and less efficient.

23. *In armature voltage control*, the speed of a dc motor can be varied by varying the voltage applied to the armature of the dc motor.

24. The total losses occurring in a dc machine are summarized below:

- Total input
  - Useful output
  - Total losses
    - Copper losses
      - Armature copper loss
      - Field copper loss
    - Iron losses
      - Hysteresis loss
      - Eddy current loss
    - Mechanical losses
      - Friction loss
      - Windage loss

25. (*a*) **DC series motors** are used for drives requiring very high starting torque and where adjustable varying speed is satisfactory. Industrial uses are hoists, cranes, trolley cars, conveyors, electric locomotives, elevators, air-compressors, vacuum cleaners, sewing machines etc.

(*b*) **DC shunt motors** are used essentially for constant speed applications requiring medium starting torque. Industrial applications are lathes, centrifugal pumps, reciprocating pumps, fans, blowers, conveyors, wood-working machines, machine tools, printing presses, spinning and weaving machines.

(*c*) **Compound Wound Motors:** Differentially compound wound dc motors are rarely used.

Cumulative compound wound dc motors are used in driving machines which are subject to sudden applications of heavy loads, such as occur in rolling mills, punching and shearing machines, lifts, mine-hoists etc. This type of motor is also used where a large starting torque is required but series motor cannot be conveniently used such as in cranes and elevators.

EXERCISES

1. What is the principle of operation of a dc generator ? Why is a commutator and brush arrangement necessary for the operation of a dc generator ?
2. How does the commutator help in the development of emf in a dc machine ?
[Pb. Technical Univ. Basic Electrical Engineering December 2002]
3. With the help of a neat diagram discuss the various parts of a dc machine.
[Pb. Technical Univ. Basic Electrical and Electronics Engineering May 2009]
4. Name the major parts of a dc machine. Draw the sketch and show the path of magnetic flux in a 4-pole machine. How can you determine the magnetic behaviour of the machine experimentally ?
[Pb. Technical Univ. Electrical Engineering May 2002]

5. Explain the principle and working of a dc generator.
[Pb. Technical Univ. Basic Electrical and Electronics Engineering May 2006]
6. Derive an expression for emf equation of dc generator.
[Pb. Technical Univ. Basic Electrical and Electronics Engineering Dec. 2007]
7. Derive an expression for the emf generated in an armature winding of a dc generator.
What is the field of application of dc shunt motor and dc series motor ?
[Pb. Technical Univ. Electrical Engineering January 2000]
8. Illustrate the methods of excitation of dc machines. [Pb. Univ. Elec. Machines-I, Dec. 1992]
9. Explain the constructional details and working principle of a dc generator. Draw the internal and external characteristics. [Pb. Technical Univ. Basic Electrical and Electronics Engineering Dec. 2006]
10. Explain how the output voltage of (*i*) dc shunt generator, and (*ii*) dc series generator vary with the increasing load current. [U.P. Technical Univ. Electrical Engineering February 2002]
11. Along with connection diagrams, draw and briefly explain the load (external) characteristics of different types of dc generators. [U.P. Technical Univ. Electrical Engineering July 2002]
12. What is meant by armature reaction in a dc generator ? Discuss external characteristics of dc generator.
[Pb. Technical Univ. Basic Electrical and Electronics Engineering Dec. 2008]
13. Discuss in detail the load characteristics of (*i*) shunt generator (*ii*) series generator.
14. Draw external characteristics of dc separately and shunt excited generator. Which one has more drooping characteristics and why ?
15. Differentiate over-, level- and differential-compoundings with the help of external characteristics of the generators.
16. A dc shunt generator fails to build up voltage when it is run at the rated speed. What may be the possible reasons for this ? Explain.
17. Give the applications of different types of dc generators with reasons and justification.
18. What is the working principle of dc motor ?
[Pb. Technical Univ. Basic Electrical and Electronics Engineering Dec. 2009]
19. Explain the principle of operation of dc motors. What is back emf in dc motors ? What is its effect ?
20. What is meant by back emf ? Is the back emf greater or lesser than the applied voltage ? Why ? By what amount do the two voltages differ ? Write voltage equation of a motor.
[Pb. Univ. Elements of Elec. Engineering May 1988]
21. Give the concept of counter-emf in a dc motor. [Pb. Univ. Elec. Machines Dec. 1992]
22. Mention different types of dc motors. How may the direction of rotation of a dc shunt motor be reversed ? What is the effect of reversing the line terminals ?
23. Explain back emf in a dc motor. Explain the effect of field current on the speed of a dc motor.
24. Derive an expression for the speed of a dc motor in terms of back emf and flux per pole.
25. Derive the emf and torque equation of a dc machine. Give applications of series and shunt motors.
[Pb. Technical Univ. Basic Electrical Engineering December 2003]
26. Derive expression of electromagnetic torque of a dc shunt motor.
[Pb. Technical Univ. Electrical Engineering 1999]
27. Explain the speed-current, torque-current and speed-torque characteristics of dc series motor.
28. Explain the speed-current, torque-current and speed-torque characteristics of a dc shunt motor.
29. Draw the torque-armature current characteristics for dc shunt and dc series motors. Why is a series motor not started at no load ? Explain. [Pb. Univ. Elec. Machines II May 1994]
30. Draw speed-torque characteristics of (*i*) dc shunt motor, (*ii*) dc series motor, (*iii*) dc cumulatively compounded motor. Mention one industrial application of each of these motors.
31. What are the important characteristics of dc motors ? Sketch all these characteristics of dc series motor.
[Pb. Technical Univ. Basic Electrical Engineering June 2003]
32. How may the speed of a dc shunt motor be controlled by armature voltage and field flux ?
33. Explain the speed control of a dc series motor by (*i*) field divertor (*ii*) variable resistance in series with the motor.

34. What are the losses taking place in a dc machine and how do they vary with load current ?

35. Explain briefly typical applications of each type of dc motor (stating the reason for choice).

36. Write short notes on the following:

(*i*) Function and action of a commutator.

(*ii*) Classification of dc generators.

(*iii*) External characteristics of dc generators

(*iv*) Applications of dc generators.

(*v*) Working principle of a dc motor.

(*vi*) Characteristics and applications of dc motors.

SHORT ANSWER TYPE QUESTIONS WITH ANSWERS

Q. 1. On what principle do the generators operate ?

Ans. The generator operates on the principle of production of dynamically induced emf.

Q. 2. Why are electromagnets preferred over permanent magnets for use in large dc machines ?

Ans. On account of its greater magnetic effect and its field strength regulation.

Q. 3. Why are cast iron yokes used in small dc machines ?

Ans. On account of low cost.

Q. 4. Why are fabricated steel yokes used in large machines ?

Ans. On account of high permeability thereby resulting in reduced weight.

Q. 5. What for field coils are provided in a dc machine ?

Ans. Field coils are provided to give, under various conditions of operation, the number of ampere-turns of excitation required to give proper flux through the armature to induce the emf.

Q. 6. What is the function of armature in a dc generator ?

Ans. Its function is to rotate the conductors in a uniform magnetic field and provide a path of very low reluctance to the magnetic flux.

Q. 7. What is the function of commutator in a dc machine (*i*) in motoring action (*ii*) in generating action ?

Ans. (*i*) To convert dc into ac (*ii*) To convert ac into dc.

Q. 8. What for brushes are employed in dc machines ?

Ans. To collect current from the commutator and supply it to the external load circuit.

Q. 9. Why is induced emf in a dc motor called the back or counter emf ?

Ans. Since induced emf in a dc motor opposes the voltage applied to the armature of a dc motor, it is called the back or counter emf.

Q. 10. Give significance of back emf in dc motors. **[P.T.U. B.E.E.E. Dec. 2008]**

Ans. The induced emf in the armature of a motor, E_b depends, among other factors, upon the armature speed (E_b or $E_g \propto N$) and armature current depends upon the back emf for a constant applied voltage and armature resistance $\left(I_a = \dfrac{V - E_b}{R_a}\right)$. If the armature speed is high, back emf E_b will be large and therefore, armature current small. If the speed of the armature is low, then back emf E_b will be less and armature current I_a more resulting in development of large torque.

Thus, the presence of back emf makes the dc motor a *self regulating* machine, *i.e.* it makes the dc motor to draw as much armature current as is just sufficient to develop the required load torque.

Q. 11. What do you understand by self excitation mode of dc machine ? Name two dc machines working in this mode. **[P.T.U. June 2003]**

Ans. The dc machine in which field winding(s) is/are excited by the current supplied by the machine itself is called the self-excitation mode of the machine. In such machines the field coils are inter-connected with the armature winding.

The self excited machines may be (*i*) dc series machine (*ii*) dc shunt machine.

Q. 12. How can the direction of rotation of a dc motor be determined ?

Ans. By applying Fleming's left hand rule.

Q. 13. How can the direction of rotation of a dc shunt motor be reversed ?

Ans. Direction of rotation of a dc shunt motor can be reversed by reversing the current flow through either the armature winding or the field winding.

Q. 14. How can the direction of rotation of a dc series motor be changed ?

Ans. By reversing the supply terminals to either armature or series field.

Q. 15. A dc shunt motor is running at 1,500 rpm at rated load torque. What happened to the motor operation if the following changes are made:

(*i*) Field winding terminals are reversed.
(*ii*) Armature winding terminals are reversed.
(*iii*) Supply terminals are reversed.
(*iv*) Both field winding and armature winding terminals are reversed. **[P.T.U. June 2003]**

Ans. (*i*) The direction of rotation of the motor will be reversed.
(*ii*) The direction of rotation of the motor will be reversed.
(*iii*) The motor will keep rotating in the same direction as before the reversal of the line terminals.
(*iv*) The direction of rotation of the motor will remain the same.

Q. 16. Why a series motor should not be fitted to a load through belt ?

Ans. Because the mishap to the belt would cause the motor to run on no load and on no load the series motor will attain dangerously high speed and thus the motor may get damaged due to heavy centrifugal forces set up in the rotating parts.

Q. 17. Why are dc series motors never started at no load ? **[P.T.U. B.E.E.E. May 2006]**

Or

Why are the dc series motors not started under no-load condition ? **[P.T.U. B.E.E.E. May 2007]**

Ans. On the no load the series motor would attain tremendously high speed and the motor may get damaged due to heavy centrifugal forces set up in rotating parts.

Q. 18. Why are series dc motors always started on load ? **[P.T.U. December 2003]**

Ans. Refer to answer of Q. 17.

Q. 19. Why a dc series motor is ideally suited for traction purposes ?

Ans. Because dc series motor develops large starting torque and slows down with the increase in load and so gets automatically relieved from heavy excessive load.

Q. 20. For what type of load dc shunt motor is suitable and why ?

Ans. DC shunt motors are suitable for loads requiring medium starting torque, approximately constant speed and which are totally and suddenly thrown off because dc shunt motors develop modern starting torque and run almost on constant speed from no load to full load.

Q. 21. What will happen if a dc shunt motor running at no load has its shunt winding opened accidently ? **[P.T.U. B.E.E.E. May 2005]**

Ans. The flux will drop to practically zero value and, therefore, the speed will increase to an extremely high value and the motor may get damaged due to heavy centrifugal forces. However, the protective device (no-volt release coil) will come into action and motor will be disconnected from the supply mains as the starter arm comes to the OFF position.

Q. 22. Why the dc motors are mostly gear motors ?

Ans. Because the low speed motors are much more expensive than high speed motors of the same kW output rating.

Q. 23. In the armature circuit of dc motor, an external resistance is introduced. How shall it affect the speed of the motor ? **[P.T.U. May 2002]**

Ans. The speed will decrease.

Q. 24. What will be the effect on the speed of a dc shunt motor if an additional resistance is introduced in the field circuit ?

Ans. Speed will increase.

Q. 25. Which method we must adopt to control the speed of a dc shunt motor above the base speed ?

Ans. Field control method.

Q. 26. While controlling the speed of a dc shunt motor what should be done to achieve a constant torque drive ?

Ans. Applied voltage should be maintained constant so as to maintain field strength constant.

Q. 27. Which losses occur in a dc machine. **[P.T.U. B.E.E.E. December 2005]**

Ans. The different types of losses that occur in a dc machine are (*i*) copper or electrical losses (*ii*) iron or magnetic losses and (*iii*) mechanical or friction and windage losses.

Copper or electrical losses include power lost in heating the shunt field, and also in heating the armature circuit, which includes armature winding, brush contacts, series field and interpole field.

Iron or magnetic losses include the hysteresis and eddy current losses which are present in any part of the machine that is constructed of iron and subjected to variations of magnetic flux.

Mechanical losses consist of power losses due to friction of the bearings, air friction or windage, as it is called, caused by the motion of the moving parts through the surrounding medium, and the friction between brushes and commutator.

Q. 28. Give two applications of series motor and shunt motor each. **[P.T.U. December 2002]**

Ans. **Applications of dc series motors.** Electric locomotives, hoists and cranes.

Applications of dc shunt motors. Lathes, reciprocating and centrifugal pumps.

Q. 29. What is the field of application of dc shunt motor and dc series motor. **[P.T.U. January 2000]**

Ans. *DC shunt motors* are almost constant speed motors, develop medium starting torque and their speeds can be adjusted by using speed control methods. So the shunt motors can be used for the loads which are totally and suddenly thrownoff without resulting in excessive speed. Being constant speed motors, they are best suited for driving of line shafts, machine lathes, milling machines, conveyors, fans and for all purposes where constant speed is required.

DC series motors are variable speed motors, develop large starting torque but attain dangerously high speed on no load. So these motors cannot be operated on no load. Series motors are best suited for electric traction, elevators, cranes, vacuum cleaners, sewing machines etc.

Q. 30. Which dc motor is used for traction purpose and why ? **[P.T.U. B.E.E.E. May 2009]**

Ans. DC series motor is ideally suited for traction purposes because dc series motor develops large starting torque and slows down with the increase in load and so gets automatically relieved from heavy excessive load.

PROBLEMS

1. A 6-pole, lap wound armature has 840 conductors and flux per pole of 0.018 Wb. The generator is run at 1,200 rpm. Calculate the emf generated. [**Ans.** 302.4 V]
2. Calculate the emf generated by a 4-pole wave wound armature with 45 slots, with 18 conductors per slot when driven at 1,000 rpm. The flux per pole is 0.02 Wb. [**Ans.** 540 V]
3. An 8-pole dc generator has 500 armature conductors and a useful flux of 0.05 wb. What will be the emf generated, if it is lap-connected and runs at 1,200 rpm ? What must be the speed at which it is to be driven to produce the same emf if it is wave-connected. [**Ans.** 500 V; 300 rpm]
4. A lap wound dc generator having 8-poles develops an emf of 500 V at 400 rpm. The armature has 144 slots and each slot contains 6 conductors. Calculate the flux per pole.
[Pb. Univ. Elements of Elec. Engineering Dec. 1986] [**Ans.** 0.0868 Wb]
5. Determine the power output of a dc armature having 1,152 lap-connected conductors carrying 150 A and rotating at 300 rpm in a 12-pole field. The flux per pole is 60 mW. [**Ans.** 51.84 kW]
6. A 440 V dc shunt generator has an armature resistance of 0.25 Ω and the resistance of the shunt field is 220 Ω. While delivering a load current of 50 A, it has a terminal voltage of 440 V. Determine the generated emf. [**Ans.** 453 V]

7. A 250 V dc shunt motor takes 41 A at full load. Resistance of motor armature and shunt field windings are 0.1 Ω and 250 Ω respectively. Find the back emf on full load. What will be its generated emf, if working as generator and supplying 41 A to load at terminal voltage of 250 V. **[Ans.** 246 V, 254.2 V]

8. A 230 V dc shunt motor takes 51 A at full load. Resistances of armature and field windings are 0.1 Ω and 230 Ω respectively. Determine (*i*) field current (*ii*) armature current and (*iii*) back emf developed at full load ? **[Ans.** (*i*) 1 A (*ii*) 50 A (*iii*) 225 V]

9. A dc series motor has an armature resistance of 0.12 Ω and field resistance of 0.08 Ω. The supply voltage across the motor terminals is 230 V. Determine the back emf and power developed by motor when line current drawn by the motor is 30 A. **[Ans.** 224 V; 6.72 kW]

10. A 250 V dc shunt machine has line current of 80 A. It has armature and field resistances of 0.1 Ω and 125 Ω respectively. Calculate power developed in armature when running as (*a*) generator (*b*) motor. **[Ans.** (*a*) 21.172 kW (*b*) 18.89 kW]

11. A 4-pole, 500 V wave wound dc shunt motor has 960 conductors. Calculate the speed of the motor if its armature current is 80 A, flux per pole is 70 mWb and armature resistance is 0.125 Ω. **[Ans.** 218.75 rpm]

12. A 220V shunt motor takes 5 A on no load and is running at 800 rpm. The resistances of armature and shunt field are 0.25 Ω and 110 Ω respectively. Determine the speed of the motor when loaded and taking 40 A from the line. **[Ans.** 768 rpm]

13. A dc machine develops an open circuit emf of 250 V at 1,500 rpm. Find the developed torque for an armature current of 20 A. **[Pb. Univ. Elec. Machines-I June, 1993] [Ans.** 31.833 Nm]

14. A 250 V, 4-pole wave wound dc series motor has 782 conductors on its armature. It has armature and series field resistance of 0.75 Ω. The motor takes a current of 40 A. Estimate its speed and gross torque developed if it has a flux per pole of 25 mWb. **[Ans.** 337.6 rpm ; 248.9 Nm]

15. A 200 V dc shunt motor running at 1,000 rpm takes an armature current of 17.5 A. It is required to reduce the speed to 600 rpm. What must be the value of resistance to be inserted in the armature circuit if the original armature resistance is 0.4 Ω ? Take armature current to be constant during this process. **[Ans.** 4.41 Ω]

16. A series motor is taking 25 A from a 220 V supply. The total resistance of the armature and field circuits is 0.5 Ω. Calculate the resistance to be connected in series with the armature to reduce the speed by 40 per cent, if the torque is proportional to the square of speed and flux is proportional to current. **[Ans.** 0.1867 Ω]

❑❑❑

10 CHAPTER

Three Phase Induction Motors

10.1. INTRODUCTION

The polyphase induction motor is, by a very considerable margin, the most widely used ac motor (almost more than 90 percent of the mechanical power used in industry is provided by 3-phase induction motors). The reasons are its low cost, simple and rugged construction, absence of commutator, good operating characteristics (reasonably good power factor, sufficiently high efficiency and good speed regulation). An induction motor of a medium size may have an efficiency as high as 90 percent and a power factor of 0.89. It is substantially a constant speed motor with a shunt characteristic; a few percent speed drop from no load to full load. The physical size of such a motor for a given output rating is relatively small as compared with other types of motors.

The distinguishing feature of such a motor is that it is a singly excited machine, although such machines are equipped with both field and armature windings. In such a machine the field* (or stator) winding is connected to an ac supply and there is no electrical connection from the armature* (or rotor) to any source of supply. Currents are made to flow in the armature (or rotor) conductors by induction which interact with the field produced by the field (or stator) winding and thereby produce a net unidirectional torque. Such motors are also called *asynchronous motors* as they run at a speed other than the synchronous speed of the rotating field developed by the stator currents.

Like other electrical machines, the asynchronous machine is reversible, *i.e.*, it can operate as both a motor and a generator. The mode of operation of the machine is determined by the speed of the rotating field in relation to the rotor.

An asynchronous machine may be considered to be a transformer in the sense that the power is transferred from the stator (primary) to the rotor (secondary) winding only by mutual induction. For this reason such a machine is often called the *induction* machine.

INSIDE THIS CHAPTER

* The terminology of the field and armature windings is seldom used for polyphase induction motors.

There is no simple and cheaper method of controlling the speed of the induction motor as is possible in case of a dc shunt motor and the torque developed by a polyphase induction motor is also not so strong as that developed by a dc shunt motor. It still finds stiff competition from the dc shunt motor in industrial applications.

10.2. CONSTRUCTION

The 3-phase induction motor is very simple in construction compared to a dc motor or a synchronous motor. The essential features of a polyphase induction motor are: a laminated stator core carrying a polyphase winding; a laminated *rotor* core carrying either a cage or polyphase winding, the latter with shaft-mounted slip rings; a stiff *shaft* to preserve the very short air gap; a *frame* to form the stator housing and carry the end covers, bearings, and terminal box. It is not possible, as in the dc machines, to use the frame as part of the magnetic circuit. The end covers receive the ball or roller bearings with their clamping plates. Non-salient pole construction is used for all polyphase induction motors.

10.2.1. Frame

It is the outer body of the motor. Its functions are to support the stator core and winding, to protect the inner parts of the machine and serve as a ventilating housing or means of guiding the coolant into effective channels. The frame may be die-cast or fabricated. Machines upto about 50 kW rating may have frame die-cast in a strong silicon-aluminium alloy, sometimes with the stator core cast in. The frames for medium and large machines are almost exclusively fabricated.

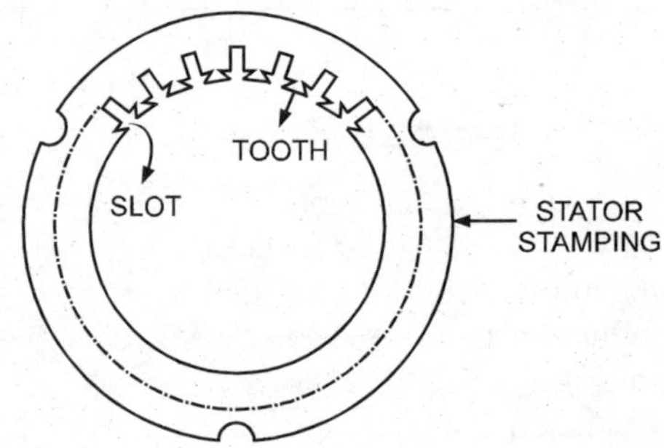

Fig 10.1

10.2.2. Stator Core

The stator of an induction motor is quite similar in construction to that of a 3-phase synchronous alternator or motor. The stator core is to carry the alternating flux which produces hysteresis and eddy current losses. In order to reduce eddy currents and hysteresis loss in the stator core it is assembled of high grade, low electrical loss, silicon steel punchings. The thickness of punchings varies from 0.35 mm to 0.65 mm. In small motors, or if low core loss is not so important slightly thicker laminations may be used. In the smaller sizes, the stator plates form complete annular rings, as shown in Fig. 10.1, but for larger machines they are prepared of segmental laminations to avoid wastage of steel. The punchings are insulated from one another by coating of varnish or oxide produced by heat treatment. The stampings are assembled under hydraulic pressure and are keyed to the frame. Careful alignment is required to prevent the rotor from rubbing against the stator, the air gap between the stator and rotor being very small. The radial ventilating ducts are provided along the length of the stator core, spaced every 5 or 7 cm. These are provided by the use of spacers placed between the laminations. The air gap between the stator and rotor is made as small as practicable (0.3 to 0.35 mm in small machines and 1.0 to 1.5 mm in high power machines) so as to make air-gap reluctance minimum. The slots are punched out on the inner periphery of the stator laminations, as shown in Fig. 10.1, to accommodate the stator winding, also sometimes known as *primary winding.* Some motors, especially of large ratings, have open type slots so that former wound coils can be used but usually the motors are provided with semi-closed type slots to make the air-gap reluctance minimum and to prevent excessive tooth loss. In large, high speed motors it is not essential to employ semi-closed slots to reduce the reluctance. The reason of it is that in large, high speed motors, pole pitch is large, therefore, the number of conductors per pole is large, so the desired flux density in the air gap can be attained without an excessive magnetizing current, even if open type slots are employed.

10.2.3. Stator or Primary or Field Winding

Motors upto 250 kW operate usually at voltages in the range 380-440 V. For 3.3 kV the minimum economical rating is 250-300 kW, and for 11 kV about 750 kW.

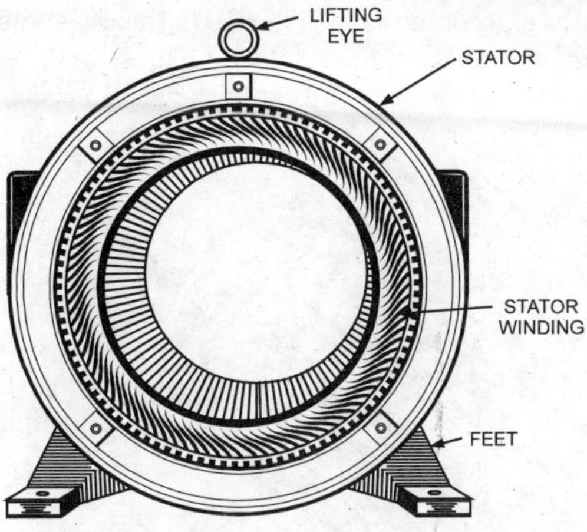

Fig. 10.2. *Stator of a 3-Phase Induction Motor*

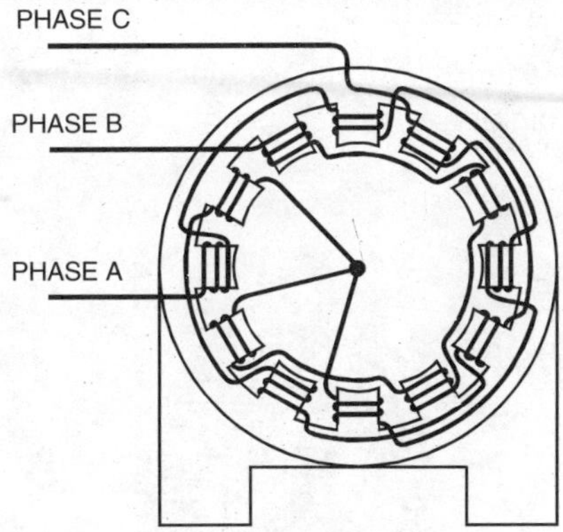

Fig. 10.3. *3-Phase, 4-Pole, Star-Connected Stator Winding*

In a polyphase induction motor the stator winding is usually 3-phase winding which is usually supplied from a 3-phase supply mains. The three phases of the winding can be connected in either star or delta depending upon the methods of starting used. The squirrel cage motors are usually started by star-delta starters and, therefore, their stators are designed for delta connections and six terminals (two from each phase) are brought out to be connected to the starter. The wound rotor motors are started by inserting resistance in the rotor circuit and, therefore, the stator winding can be connected in either star or delta as desired. It is wound for a definite number of poles as per requirement of speed. Greater the number of poles, lesser the speed or vice versa for a supply of given frequency.

Stators of a 3-phase induction motor and 3-phase, 4-pole, star-connected stator winding are illustrated in Figs. 10.2 and 10.3 respectively.

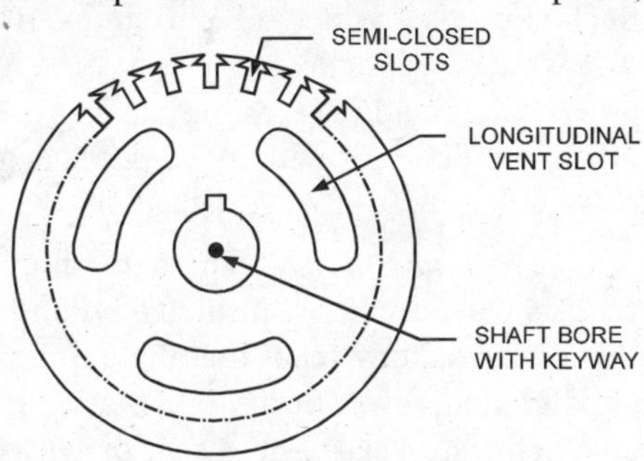

Fig. 10.4. *Rotor Stamping*

10.2.4. Rotor

The rotor comprises a cylindrical laminated iron core, much the same as that of a dc machine, with slots, around the core, carrying the rotor conductors. In general the same sheet steel laminations are employed for the rotor core as for the stator, but owing to the lower frequencies of the rotor flux, thicker laminations can be employed without excessive iron loss. The rotor stamping is shown in Fig. 10.4. Like stator, the rotor laminations are punched in one piece for small machines; in larger machines the laminations are segmented and dovetailed to a fabricated spider. If there are ventilating ducts on the stator core, an equal number of such ducts is provided on the rotor core also. Fan blades are usually employed on the ends of the rotor core to force circulating air through the motor for cooling.

The rotors employed in 3-phase induction motors, according to the type of windings used, are of two types, viz; squirrel cage rotor and the wound rotor.

1. ***Squirrel Cage Rotor.*** *Almost* 90 percent of induction motors are provided with squirrel cage rotor because of its very simple, robust and almost instructible construction. In cage construction, copper, brass, or aluminium bars are placed, as the rotor conductors, parallel or approximately parallel to the shaft (one bar in each slot) and close to the rotor surface. The conductors are not insulated from the core, since the rotor currents naturally follow the path of least resistance, *i.e.*, the rotor conductors. At both ends of the rotor, the rotor conductors are all short circuited by the continuous end rings of similar material to that of the rotor conductors. The rotor conductors and their end rings form a complete closed circuit in itself, resembling a squirrel cage, thus explaining the name. In motors with ratings upto 100 kW, the squirrel cage structure is formed by aluminium cast (under pressure) into the slots of the rotor (Fig. 10.6). In large motors the rotor bars, instead of being cast, are wedged into the rotor slots and are then welded securely to the end rings. The slots on the rotor are always not parallel to the motor shaft but are usually skewed in order to obtain a uniform torque, reduce the magnetic locking of the stator and rotor and reduce the magnetic humming noise while running.

Fig. 10.5. *General View of Squirrel Cage Rotor*

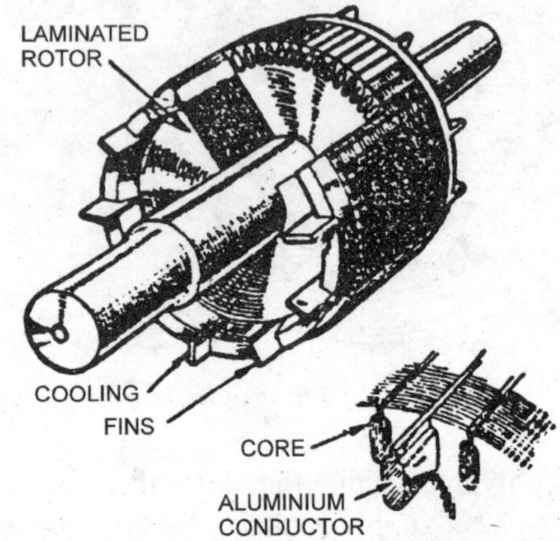

Fig. 10.6. *Cutaway View of Squirrel Cage Rotor*

The slots on rotor are either of semi-closed type or of totally closed type, because there is little difficulty in inserting the rotor bars in such slots. The advantage of semi-closed and totally closed slots is that the effective cross-sectional area of the air gap is increased, therefore, magnetizing current is reduced. Such slots also reduce the pulsations of flux in the individual teeth, therefore, tooth loss is reduced. The disadvantage of such slots is that these give higher slot inductance than the open slots, which lowers the power factor and reduces the starting and breakdown torques of the motor.

Since the rotor winding is permanently short circuited in cage construction, so there is no possibility of adding any external resistance in the rotor circuit.

2. *Wound Rotor.* *As* the name implies, such a rotor is wound with an insulated winding similar to that of the stator except that the number of slots is smaller and fewer turns per phase of a heavier conductor are used. Bar, strap, or wire is used for rotor windings, the last being used where many turns are desired. A large number of rotor turns increases the secondary voltage and reduces the current that flows through the slip rings. The secondary voltage determines the insulation that must be provided; furthermore, the voltage and current influence the value of the resistance to be employed across the slip rings. The motor operation is not influenced by the number of rotor turns, but the ratio of transformation is determined by consideration of secondary current, danger of high secondary emf at starting, and distance to secondary resistors. The standstill open-circuit slip-ring voltage is usually 100 to 400 V for small machines using hand operated gear and maximum upto 1 kV for large machines.

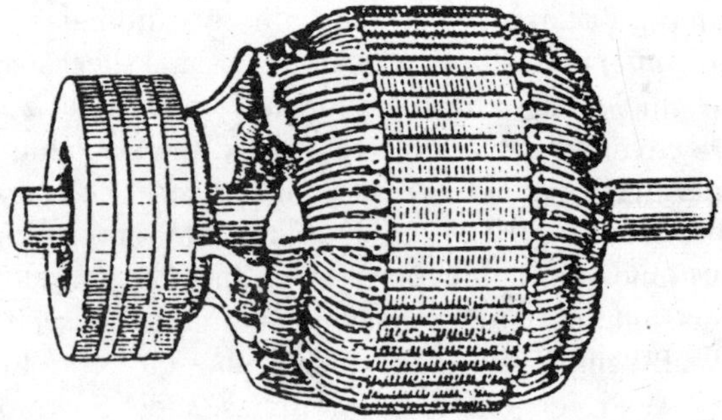

Fig. 10.7. *Rotor With Slip-rings*

The rotor is wound for the same number of poles as that of the stator. The rotor winding is always 3-phase winding even when the stator is wound for two phases. The rotor winding may be star or delta-

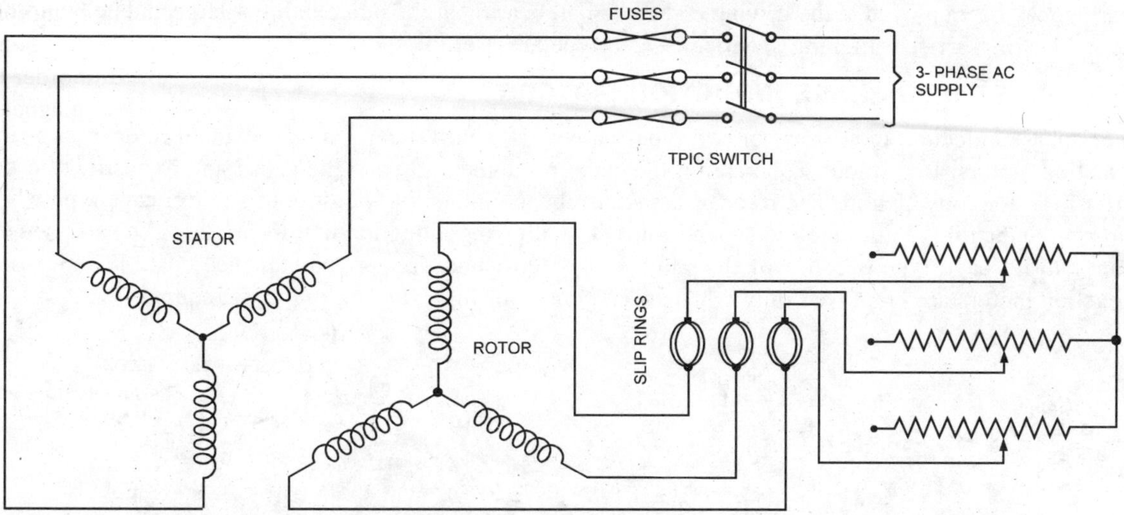

Fig. 10.8. ***3-Phase Slip-ring Induction Motor With Starting Rheostat***

connected but star-connection is usually preferred. The three finish terminals are connected together to form star-point and the three start terminals are connected to three phosphor-bronze (or brass) slip rings mounted on but insulated from the rotor shaft. The brushes, which carry the current from and to the rotor windings are held in box type holders mounted on insulated steel rods, securely bolted to the end shield. Each brush is fed forward by a lever held in tension by an adjustable spring. These brushes are further externally connected to a 3-phase star-connected rheostat for the purpose of starting and speed control (Fig. 10.8). At the time of starting, the entire resistance is included in the rotor circuit and this resistance is gradually cut out as the rotor picks up the speed. For the normal running condition the entire external resistance is cut out and the rotor windings are short circuited automatically through the slip rings by means of a metal collar which is pushed along the shaft and connects all the rings together. Some machines are provided with brush lifting and slip-ring short-circuiting arrangement for running condition. However, brush lifting and short-circuiting gear is usually not used in modern machines. The rotor is skewed in this case also. Since the connection of the wound secondary to the external terminals is made through slip rings and brushes, so wound secondary motors often are called *slip-ring induction* motors. A sectional diagram of a slip-ring induction motor is shown in Fig. 10.9.

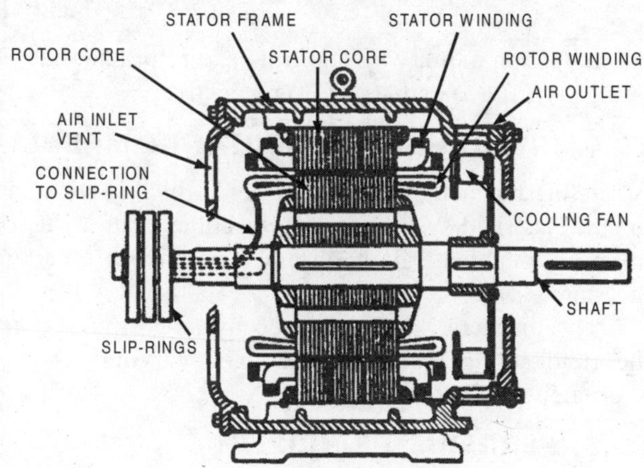

Fig. 10.9. ***Section of a Slip-Ring Induction Motor***

10.2.5. Shafts and Bearings

The rotor shaft is supported by bearings housed in the end shield. The air gap of an induction motor is kept necessarily small, therefore, the rotor shaft must be made short and stiff so that the rotor may not have any significant deflection. Even a minor deflection in the shaft would develop large irregularities in the air gap which would lead to production of an *unbalanced magnetic pull.* There is also a possibility of rotor and stator fouling with each other.

Ball and roller bearings are generally employed as their use makes accurate centring much simpler than with journal bearings. Also the overall length of the machine is reduced. In small motors, a roller bearing may be employed at the driving end and a ball bearing at the other end. For large and heavy rotors journal bearings of self aligning spherical seated type are useful.

10.3. TYPES OF 3-PHASE INDUCTION MOTORS

The 3-phase induction motors are of two types namely (*i*) squirrel cage and (*ii*) wound rotor or slip-ring induction motors. Both motors operate on the same principle and have the same stator construction but differ in rotor construction. The rotor of the squirrel cage induction motor is of squirrel cage type while slip-ring induction motor employs wound rotor. The slip-ring induction motors are less extensively used than squirrel cage type because of their higher first cost and greater maintenance cost. The slip-ring induction motors are employed only when speed control or high starting torque is required.

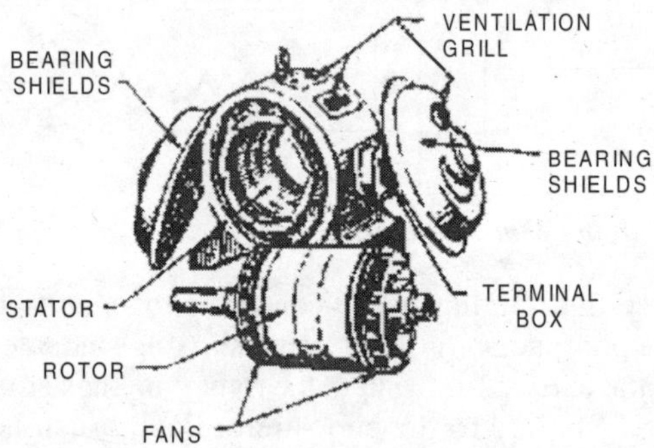

(a) Induction Motor With Squirrel Cage Rotor (Disassembled)

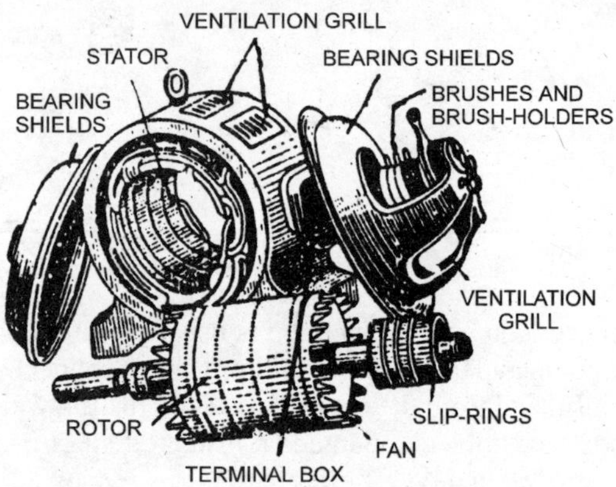

(b) Induction Motor With Wound Rotor (Disassembled)

Fig. 10.10

Induction motor with squirrel cage rotor and that with wound rotor, in disassembled form, are shown in Figs. 10.10 (*a*) and 10.10 (*b*) respectively.

10.4. INTRODUCTION TO ROTATING MAGNETIC FIELD

A rotating magnetic field is that field which is constant in magnitude but whose axis of direction rotates in space as field system of a dc machine. With a stationary field system the magnetic field is stationary in space, but if the field system is rotated at a certain speed, its magnetic field will rotate with it at the same speed.

The magnetic field produced by single phase alternating current is an *alternating magnetic field*, the field acting along a fixed axis, varying in magnitude periodically and changing its direction alternately positive and negative.

10.5. FERRARIS PRINCIPLE

The magnetic field produced by single phase alternating current is an alternating magnetic field which, by Ferraris principle, can be resolved into two rotating fields of half its amplitude and rotating in opposite directions at synchronous speed.

Consider two rotating magnetic fields OA_1 and OA_2, each having magnitude of H units and travelling in opposite directions with angular velocity ω, as shown in Fig. 10.11.

Let both the fields start travelling from axis OX at time $t = 0$.

After time t seconds the angle through which fields OA_1 and OA_2 have rotated, $\theta = \omega t$ radians.

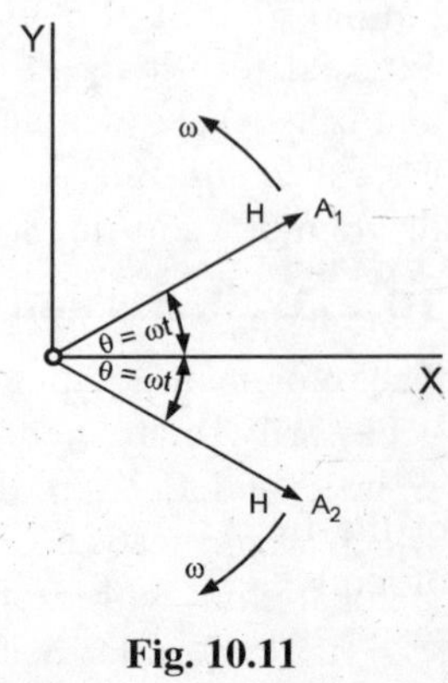

Fig. 10.11

i.e. Field represented by vector OA_1 has moved in counter-clockwise direction through an angle $\theta = \omega t$ from axis OX and field represented by vector OA_2 has moved in clockwise direction through same angle θ, because both fields are travelling with same angular velocity ω.

Resolving the magnetic fields represented by vectors OA_1 and OA_2 along axis X and axis Y, we get

X-component, $H_x = OA_1 \cos\theta + OA_2 \cos\theta = H\cos\theta + H\cos\theta = 2H\cos\theta = 2H\cos\omega t$

Y-component, $H_y = OA_1 \sin\theta - OA_2 \sin\theta = H\sin\theta - H\sin\theta = 0$

Hence resultant magnetic field is $2H\cos\omega t$ along X-axis. It is, therefore, obvious that two rotating magnetic fields, travelling in opposite directions with the same angular velocity, result in an alternating field of twice their amplitude. *Conversely an alternating field can be replaced exactly by two rotating fields of half its amplitude travelling in opposite directions at synchronous speed.*

10.6. PRODUCTION OF THREE PHASE ROTATING MAGNETIC FIELD

Consider three similar coils A, B and C displaced in space by 120°, as shown in Fig. 10.12, and connected to 3-phase ac supply. Let each of alternating fields due to currents in coils A, B and C be resolved into two components OA_1 and OA_2, OB_1 and OB_2 and OC_1 and OC_2 respectively travelling with angular velocity ω in the directions shown in Fig. 10.12, where $\omega = 2\pi f$, f being the supply frequency.

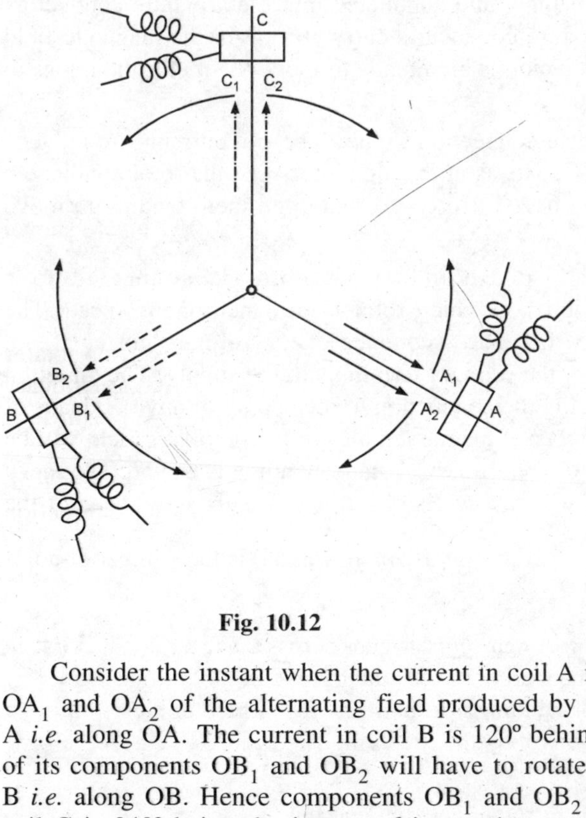

Fig. 10.12

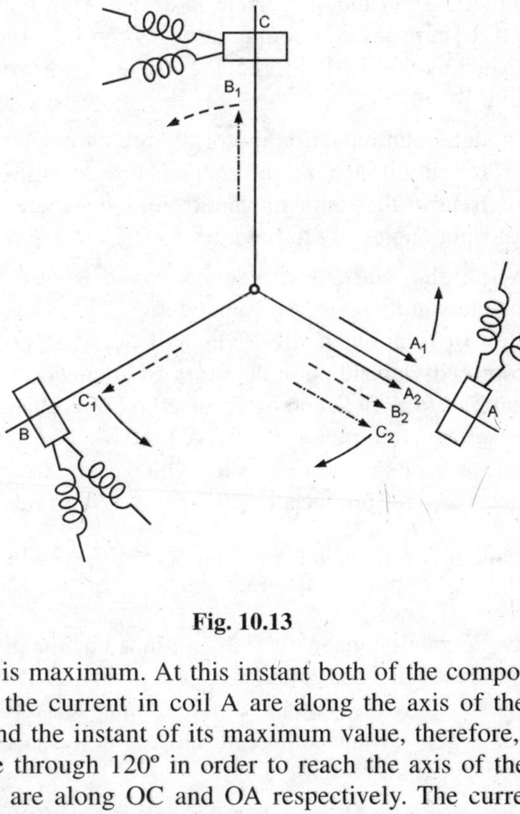

Fig. 10.13

Consider the instant when the current in coil A is maximum. At this instant both of the components OA_1 and OA_2 of the alternating field produced by the current in coil A are along the axis of the coil A *i.e.* along OA. The current in coil B is 120° behind the instant of its maximum value, therefore, each of its components OB_1 and OB_2 will have to rotate through 120° in order to reach the axis of the coil B *i.e.* along OB. Hence components OB_1 and OB_2 are along OC and OA respectively. The current in coil C is 240° being the instant of its maximum value, therefore, each of components OC_1 and OC_2 of the field produced by current in coil C will have to rotate through 240° in order to reach the axis of coil C *i.e.* along OC. Hence at this instant components OC_1 and OC_2 are along OB and OA respectively.

From Fig. 10.13. it is clear that components OA_1, OB_1 and OC_1 are rotating in counter-clockwise direction with same angular velocity ω but are 120° apart, therefore, their resultant is always zero. Components OA_2, OB_2 and OC_2 are rotating in clockwise direction with same angular velocity ω and also points in same direction, therefore, resultant gives a pure rotating field rotating at synchronous speed.

The direction of rotation is the same as the direction of phase sequence *i.e.* first current in coil A attains its maximum value, then current in coil B attains its maximum value and then current in coil C attains its maximum value, as shown in Fig. 10.14. Hence to reverse the direction of rotation of the magnetic field, it is necessary to change the phase sequence which can be done by interchanging any two of the terminals.

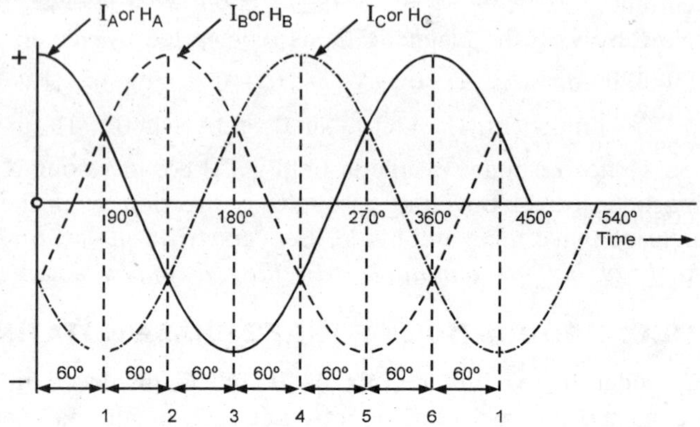

Fig. 10.14

10.7. PRINCIPLE OF OPERATION

In a direct current motor, current is drawn from the supply and conducted into the armature conductors through the brushes and commutator. When the armature conductors carry current in the magnetic field established by the field, a force is exerted on the conductors which tends to move them at right angles to the field.

Though in an induction motor, there is no electrical connection to the rotor, but currents are induced in the rotor circuit, and therefore, the same condition exists as in the dc motors *i.e.*, the rotor conductors carry current in the stator magnetic field and thereby have a force exerted upon them tending to move them at right angles to the field.

When the stator or primary winding of a 3-phase induction motor is connected to a 3-phase ac supply, a rotating magnetic field is established which rotates at synchronous speed. The direction of rotation of this field will depend upon the phase sequence of the primary currents and, therefore, will depend upon the order of connection of the primary terminals to the supply. The direction of rotation of the field can be reversed by interchanging the connection to the supply of any two leads of a 3-phase induction motor. (Refer Art. 10.6). The number of magnetic poles of the revolving field will be the same as the number of poles for which each phase of the primary or stator winding is wound. The speed at which the field produced by the primary currents will revolve is called the *synchronous speed* of the motor and is given by an expression, $N_s = \frac{120f}{P}$ where f is supply frequency and P is the number of poles on stator.

As the rotating magnetic field produced by the primary currents sweeps across the rotor conductors, an emf is induced in these conductors just as an emf is induced in the secondary winding of a transformer by the flux set up by the primary currents. Since the rotor winding is either directly shorted or closed through some external resistance, the emf induced in the secondary by the revolving field causes a current to flow in the rotor conductors.

The setting up of the torque for causing the rotor to rotate is explained below:

A section of an induction motor stator and rotor, with the magnetic field assumed to be rotating in a clockwise direction and with the rotor stationary, as at starting, is shown in Fig. 10.15. The relative motion of the rotor with respect to the stator field is anticlockwise. By applying right hand rule, the direction of induced emf or current in the rotor conductor shown is found outward. Hence the direction of the flux due to rotor current alone is anticlockwise, as shown in the Fig. Now, by applying the left hand rule, or by the effect of combined field, it is clear that the rotor conductor shown experiences a force tending to move the conductor to the right. For simplicity, only one rotor conductor is shown. However, other adjacent rotor conductors in the stator field likewise carry current in the same direction as the conductor shown and also have a force exerted upon them tending to move them towards the right. One half cycle later, the stator field direction will have reversed, but the rotor currents will have also reversed, so that the force on the rotor is still the same. Likewise rotor conductors under other stator field poles will have a force exerted upon them all tending to turn the rotor in the clockwise direction. If the developed torque is great enough to overcome the resisting torque of the load, the rotor will accelerate in the clockwise direction or in the same direction as the rotation of the stator field.

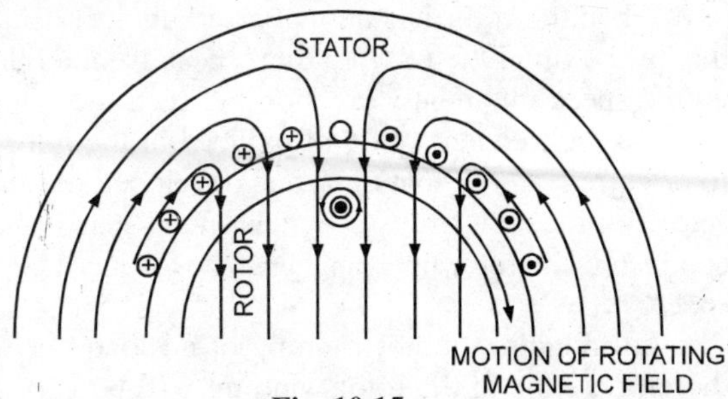

Fig. 10.15

The above facts have been illustrated diagrammatically in Fig. 10.16.

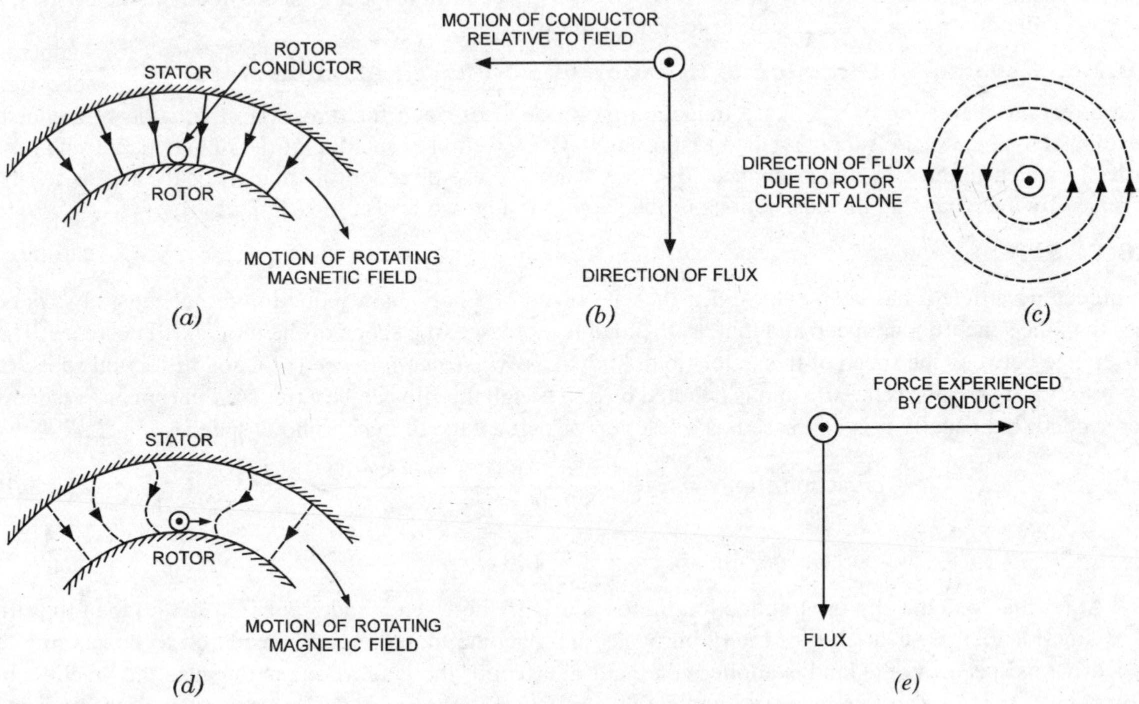

Fig. 10.16. ***Production of Torque***

This may also be explained in terms of Lenz's law. According to Lenz's law we know that the direction of induced emf would be in such a direction that it would try to oppose the very cause for which it is due. Because the cause producing the induced currents in the rotor conductors is the relative speed between the rotating magnetic field and stationary rotor conductors, therefore, they circulate in such a way that a torque is produced in the rotor tending to cause it to follow the rotating magnetic field and thus reducing the relative speed.

When the rotor is stationary and about to start, the frequency of induced emf in the rotor is equal to that of the supply fed to the stator, because the relative motion is at synchronous speed. As the motor picks up the speed, the relative motion between the rotor and the synchronously rotating magnetic field becomes less and the frequency of emf induced in the rotor decreases. The magnitude of rotor induced emf, induced rotor current and so that of torque developed depends upon the relative motion. In case the relative motion is zero, *i.e.*, the rotor runs at synchronous speed, there would be no induced emf and no current in the rotor conductors, no rotor field and hence no torque. Thus we see that an *induction motor cannot run at synchronous speed.*

An induction motor running at no load will have a speed very close to the synchronous speed and, therefore, emfs in the rotor winding will be very small. This small emf gives a small current producing a torque just sufficient to overcome the losses such as due to friction and maintain the rotor in motion. As the mechanical load is applied on the motor shaft, it must slow down because the torque developed at no load will not be sufficient to keep the rotor revolving at the no-load speed against the additional opposing torque of load. As the motor slows down, the relative motion between the magnetic field and the rotor is increased. This results in greater rotor currents and greater developed torque. Thus, as the load is increased, the motor slows down until the relative motion between the rotor and the rotating magnetic field is just sufficient to result in the development of the torque necessary for that particular load. The decrease in speed from no load to full load is usually 4 to 5 per cent in case of small and medium size induction motors and in case of large size motors it varies from 2 to 2.25 per cent. In respect of speed-load characteristics, a 3-phase induction motor is quite similar to a dc shunt motor.

10.7.1. Reversal of Direction of Rotation of a 3-Phase Induction Motor

The direction of rotation of a 3-phase induction motor depends upon the direction of rotation of the field developed by the 3-phase supply applied to the stator. The direction of rotation of the field depends upon the order in which line terminals are connected to the stator. So the direction of rotation of the motor can be reversed by interchanging the connections to the supply of any two leads.

10.8. SLIP

In preceding article it has been determined that the speed of a polyphase induction motor must always be less than the synchronous speed and that, as the load is increased, the speed of the motor will decrease. The difference between the speed of the stator field, known as *synchronous speed* (N_s), and the actual speed of the rotor (N) is known as the *slip* and is denoted by s. Though the slip can be expressed in rpm or in radians per second, but usually it is expressed as a fraction or percentage of synchronous speed.

$$\text{Fraction slip, } s = \frac{\text{Synchronous speed} - \text{ rotor speed}}{\text{Synchronous speed}} = \frac{N_s - N}{N_s} \quad ..(10.1)$$

$$\text{and percentage slip} = \frac{N_s - N}{N_s} \times 100 \quad ..(10.2)$$

At normal load the slip of an induction motor is usually between 2 and 5 per cent. At no load the slip is as small as 0.5 per cent and for many purposes the machine may then be considered to be running at synchronous speed. As the load is applied, the natural effect of the load torque is to cause the machine to slow down. As it does so, slip increases, and with it current and torque until the driving torque of the machine balances the retarding torque of the load. This determines the speed at which the machine runs on load.

The induction motor is thus a motor with substantially constant speed and plays the same role as a dc shunt motor.

Example 10.1. Find the running speed of a 4-pole induction motor working on 50 Hz supply having 3% slip.

Solution :

$$\text{Synchronous speed, } N_s = \frac{120 f}{P} = \frac{120 \times 50}{4} = 1{,}500 \text{ rpm}$$

$$\text{Slip, } s = 3\% \text{ or } 0.03$$

$$\text{Running speed of motor, } N = N_s (1 - s) = 1{,}500 (1 - 0.03) = 1{,}455 \text{ rpm } \textbf{Ans.}$$

Example 10.2. The synchronous speed of induction motor is 1,500 rpm and rotor speed is 1,440 rpm. Find slip. [*Pb. Technical Univ. Basic Electrical and Electronics Engineering May 2008*]

Solution : Synchronous speed, N_s = 1,500 rpm

Rotor speed, N = 1,440 rpm

$$\text{Slip, S} = \frac{N_s - N}{N_s} = \frac{1{,}500 - 1{,}440}{1{,}500} = \frac{60}{1{,}500} = 0.04 \text{ or } 4\% \textbf{ Ans.}$$

Example 10.3. A 12-pole, 3-phase alternator is coupled to an engine running at 500 rpm. It supplies an induction motor which has a full-load speed of 1,440 rpm. Find the percentage slip and the number of poles on the motor. [*Pb. Technical Univ. Electrical Engineering Jan. 2000; Dec. 2004*]

Solution :

$$\text{Supply frequency, } f = \frac{\text{Number of poles on alternator} \times \text{speed of alternator in rpm}}{120}$$

$$= \frac{12 \times 500}{120} = 50 \text{ Hz}$$

The number of poles even in number and to give a synchronous speed nearest higher than actual speed of motor (*i.e.* 1,440 rpm) that the motor must have is 4. **Ans.**

$$\text{Synchronous speed, } N_s = \frac{120 f}{P} = \frac{120 \times 50}{4} = 1{,}500 \text{ rpm}$$

$$\text{Slip, S} = \frac{N_s - N}{N_s} \times 100 = \frac{1{,}500 - 1{,}440}{1{,}500} \times 100 = 4\% \quad \textbf{Ans.}$$

10.9. FREQUENCY OF ROTOR CURRENT (OR EMF)

It has been shown in Art 10.7 that the rotor runs in the direction of rotating magnetic field. When the rotor is stationary, rotor conductors are being cut by the rotating flux at synchronous speed N_s $\left(N_s \text{ being equal to } \frac{120 f}{P}\right)$, therefore, the frequency of rotor current (or emf) is the same as that of supply frequency. But when the rotor starts revolving, the rate at which the rotor conductors are being cut by the rotating flux depends upon the relative speed between the rotor and the stator revolving magnetic field, called the *slip speed.* The frequency of rotor current (or emf) induced by the relative motion between rotor conductors and the stator revolving magnetic field is given by the equation:

$$\text{Rotor emf frequency, } f' = \frac{\text{Relative speed in rpm}}{120/P} = \frac{N_s - N}{120/P} \quad \text{...(10.3)}$$

$$\text{Since slip, } s = \frac{N_s - N}{N_s}$$

$$\text{or } N_s - N = sN_s = s \times \frac{120 f}{P}$$

Substituting $N_s - N = s \times \frac{120 f}{P}$ in Eq. (10.3) we have

$$\text{Rotor emf frequency, } f' = s \times \frac{120 f}{P} \times \frac{P}{120} = sf \quad \text{..(10.4)}$$

Thus the frequency of rotor current (or emf) in an induction motor is given by the product of slip s and the supply frequency f. That is why, it is also called the *slip frequency.*

Example 10.4. A 3-phase induction motor has 2 poles and is connected to 400 V, 50 Hz supply. Calculate the actual rotor speed and rotor frequency when the slip is 4 %.

Solution : Supply frequency, f = 50 Hz

Slip, s = 4 % or 0.04

$$\text{Synchronous speed, } N_s = \frac{120f}{P} = \frac{120 \times 50}{2} = 3{,}000 \text{ rpm}$$

$$\text{Rotor speed, } N = N_s(1-s) = 3{,}000(1-0.04) = 2{,}880 \text{ rpm } \textbf{Ans.}$$

$$\text{Frequency of rotor emf, } f' = sf = 0.04 \times 50 = 2 \text{ Hz } \textbf{Ans.}$$

Example 10.5. A 50 Hz, 4-pole, 3-phase induction motor has rotor current frequency of 2 Hz. Obtain slip and speed of motor. *[Pb. Technical Univ. Basic Electrical and Electronics Engineering May 2005; May 2010]*

Solution :

$$\text{Slip, } s = \frac{\text{Rotor current frequency}}{\text{Supply frequency}} = \frac{f'}{f} = \frac{2}{50} = 0.04 \text{ or } 4\% \textbf{ Ans.}$$

$$\text{Speed of motor, } N = N_s(1-s) = \frac{120f}{P}(1-s) = \frac{120 \times 50}{4}(1-0.04) = 1{,}440 \text{ rpm } \textbf{Ans.}$$

Example 10.6. A 3-phase induction motor is wound for 4 poles and is supplied from 3 phase, 50 Hz system. Calculate (*i*) the synchronous speed, (*ii*) Speed of motor when slip is 0.04 and (*iii*) the rotor frequency when the motor runs at 600 rpm. *[Pb. Technical Univ. Basic Electrical and Electronics Engineering Dec. 2005]*

Solution :

$$(i) \text{ Synchronous speed, } N_s = \frac{120f}{P} = \frac{120 \times 50}{4} = 1{,}500 \text{ rpm } \textbf{Ans.}$$

(*ii*) Speed of motor when slip, $s = 0.04$

$$N = N_s(1-s) = 1{,}500(1-0.04) = 1{,}440 \text{ rpm } \textbf{Ans.}$$

(*iii*) Slip when motor speed is 600 rpm.

$$s = \frac{N_s - N}{N_s} = \frac{1{,}500 - 600}{1{,}500} = 0.6$$

$$\text{Rotor frequency, } f' = s \times f = 0.6 \times 50 = 30 \text{ Hz } \textbf{Ans.}$$

Example 10.7. A 3-phase induction motor runs at almost 1,000 rpm at no load and 950 rpm at full load when supplied with power from a 50 Hz 3-phase line.

(*i*) How many poles has the motor ?
(*ii*) What is the percentage slip at full load ?
(*iii*) What is the corresponding frequency of rotor voltage ?
(*iv*) What is the corresponding speed of the rotor field with respect to rotor ?
(*v*) What is the corresponding speed of the rotor with respect to the stator ?
(*vi*) What is the corresponding speed of the rotor field with respect to the stator field ?
(*vii*) What is the rotor frequency at the slip of 10 per cent ?

Solution :

Supply frequency, $f = 50$ Hz
No-load speed of motor, $N_0 = 1{,}000$ rpm
Full-load speed of motor, $N_f = 950$ rpm

Since no-load speed of motor is almost 1,000 rpm, synchronous speed near 1,000 rpm is 1,000 rpm.

$$(i) \text{ Poles on the motor, } P = \frac{120f}{N_s} = \frac{120 \times 50}{1{,}000} = 6 \textbf{ Ans.}$$

$$(ii) \text{ Per cent slip on full load } = \frac{N_s - N}{N_s} \times 100 = \frac{1{,}000 - 950}{1{,}000} \times 100 = 5\% \textbf{ Ans.}$$

$$(iii) \text{ Frequency of rotor voltage, } f' = sf = 0.05 \times 50 = 2.5 \text{ Hz } \textbf{Ans.}$$

$$(iv) \text{ Speed of rotor field with respect to rotor } = \frac{120 \times f'}{P} = \frac{120 \times 2.5}{6} = 50 \text{ rpm } \textbf{Ans.}$$

(*v*) Speed of rotor with respect to stator = 950 rpm **Ans.** Since stator is stationary.

(*vi*) Rotor field and stator field are revolving at the same speed of 1,000 rpm, therefore, speed of rotor field w.r.t. stator field is always zero. **Ans.**

(*vii*) Rotor frequency at the slip of 10 % $= s' f = 0.1 \times 50 = 5$ Hz **Ans.**

Example 10.8. A 3-phase, 50 Hz induction motor has a full-load speed of 960 rpm. Calculate the speed of rotor field with respect to rotor structure, with respect to stator structure and with respect to stator field.

Solution : Supply frequency, $f = 50$ Hz

Speed of rotor field with respect to stator field is always zero because both the fields rotate synchronously in space. **Ans.**

The number of poles, even in number to give a synchronous speed nearest higher than actual speed of motor (*i.e.* 960 rpm) that the motor must have is 6, so synchronous speed,

$$N_s = \frac{120 f}{P} = \frac{120 \times 50}{6} = 1{,}000 \text{ rpm}$$

and speed of rotor field with respect to stator structure

= Speed of stator field with respect to stator structure, N_s

= 1,000 rpm **Ans.**

Speed of rotor field with respect to rotor structure $= s\,N_s = N_s - N = 1{,}000 - 960 = 40$ rpm **Ans.**

Example 10.9. A 3-phase 50 Hz induction motor has a full-load speed of 1,400 rpm. For this motor calculate slip, number of poles and speed of rotor field with respect to stator field.

[*Pb. Technical Univ. Basic Electrical Engineering June 2003*]

Solution : Supply frequency, $f = 50$ Hz

The number of poles, even in number to give a synchronous speed nearest higher than actual speed of motor (*i.e.* 1,440 rpm) that the motor has is four $\left[P \simeq \frac{120}{N} \simeq \frac{120 \times 50}{1{,}440} = 4\right]$ **Ans.**

$$\text{Synchronous speed, } N_s = \frac{120 \times f}{P} = \frac{120 \times 50}{4} = 1{,}500 \text{ rpm}$$

$$\text{Slip } s = \frac{N_s - N}{N_s} \times 100 = \frac{1{,}500 - 1{,}440}{1{,}500} \times 100 = 4\% \quad \textbf{Ans.}$$

Speed of rotor field with respect to stator field is always **zero** because both the fields rotate synchronously in space. **Ans.**

10.10. ROTOR EMF

When the rotor is at standstill, the motor is equivalent to a 3-phase transformer with secondary short circuited. So induced emf per phase E_2 in the rotor, when it is at standstill *i.e.,* at the instant of starting, is given by

$$E_2 = E_1 \times \frac{N_2}{N_1} \qquad ..(10.5)$$

where E_1 is applied voltage per phase to primary *i.e* stator winding, N_2 and N_1 are the numbers of turns per phase on rotor and stator respectively.

When the rotor starts running, the relative speed of the rotor with respect to stator flux *i.e.* slip *s* is decreased. Hence induced emf in the rotor, which drops in direct proportion with the relative speed or slip *s,* is also decreased and is given by $s\,E_2$.

Hence for slip *s*, the induced emf in the rotor is *s* times the induced emf in the rotor at standstill.

10.11. ROTOR CURRENT AND POWER FACTOR

The circuit diagram of an induction motor rotor is given in Fig. 10.17.

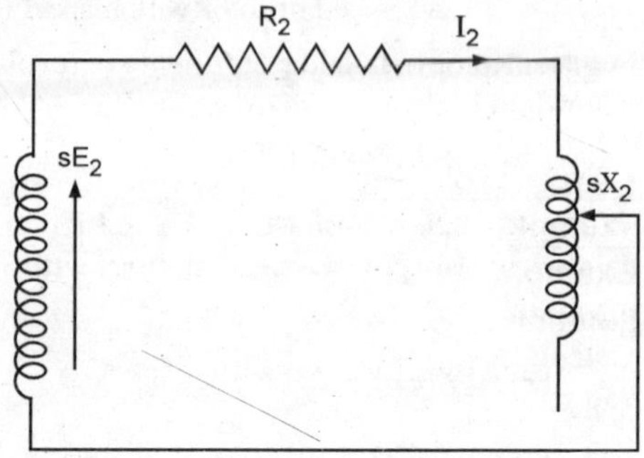

Fig. 10.17

Let the resistance and inductance per phase of rotor be R_2 and L_2 respectively and induced emf per phase in rotor at standstill be E_2.

At Standstill Induced emf per phase in rotor $= E_2$

Rotor winding resistance/phase $= R_2$

Rotor winding reactance/phase, $X_2 = 2\pi f L_2$

where f is supply frequency

$$\text{Rotor winding impedance/phase } Z_2 = \sqrt{R_2^2 + X_2^2} \quad ..(10.6)$$

$$\text{Rotor current, } I_2 = \frac{E_2}{Z_2} = \frac{E_2}{\sqrt{R_2^2 + X_2^2}} \quad ...(10.7)$$

$$\text{Power factor of rotor current, } \cos\phi_2 = \frac{R_2}{\sqrt{R_2^2 + X_2^2}} \quad ...(10.8)$$

At Slip *s*

Induced emf per phase in rotor winding $= sE_2$

Rotor winding resistance / phase $= R_2$

Rotor winding reactance / phase $= 2\pi f' L_2 = 2\pi s f L_2 = s\,(2\pi f L_2) = s\,X_2$

$$\text{Rotor winding impedance / phase} = \sqrt{R_2^2 + (sX_2)^2} \quad ...(10.9)$$

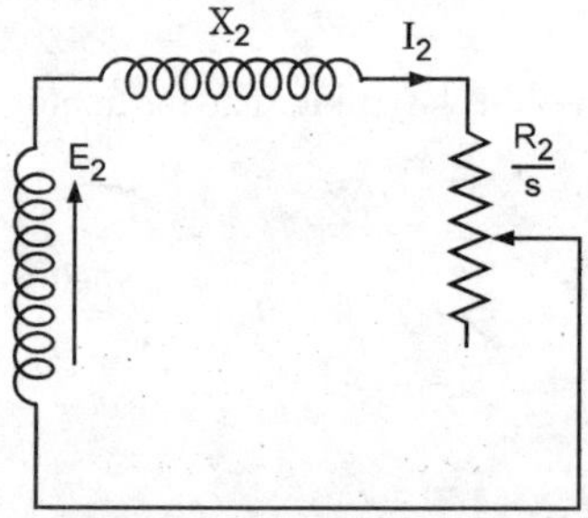

Fig. 10.18. *Equivalent Circuit of Rotor*

$$\text{Rotor current per phase, } I_2 = \frac{sE_2}{\sqrt{R_2^2 + (sX_2)^2}}$$

$$= \frac{sE_2}{\sqrt{R_2^2 + s^2 X_2^2}} \quad ...(10.10)$$

$$\text{Power factor of rotor current, } \cos\phi_2 = \frac{R_2}{\sqrt{R_2^2 + s^2 X_2^2}} \quad ...(10.11)$$

The Eq. (10.10) for rotor current I_2 may be rewritten as

$$I_2 = \frac{E_2}{\sqrt{\left(\frac{R_2}{s}\right)^2 + X_2^2}} \quad ...(10.12)$$

The above equation gives convenient form of rotor equivalent circuit, as shown in Fig. 10.18.

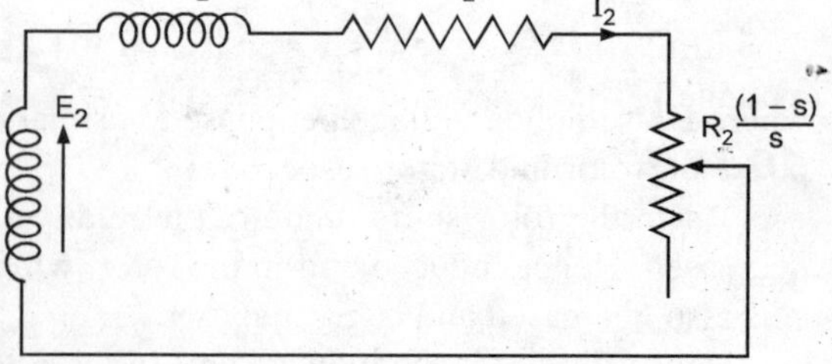

Fig. 10.19

The resistance $\frac{R_2}{s}$ is a function of slip and can be splitted into two parts R_2 and $R_2\,\frac{(1-s)}{s}$ where R_2 is rotor resistance and $R_2\,\frac{(1-s)}{s}$ is the fictitious resistance representing electrical load on the rotor. The final simplified equivalent circuit of an induction motor rotor is given in Fig. 10.19.

The power consumed by the fictitious resistance representing the rotor electrical load is given as

$I_2^2 R_2 \frac{(1-s)}{s}$ and is converted into mechanical power.

Example 10.10. In a 6-pole, 3-phase, 50 Hz motor with star-connected rotor, the rotor resistance per phase is 0.3 Ω, the reactance at standstill is 1.5 Ω per phase, and an emf between the slip rings on open circuit is 175 V. Calculate (*i*) slip at a speed of 950 rpm (*ii*) rotor emf per phase (*iii*) rotor frequency and reactance at a speed of 950 rpm.

Solution :

$$\text{Synchronous speed, } N_s = \frac{120 f}{P} = \frac{120 \times 50}{6} = 1{,}000 \text{ rpm}$$

$$\text{Rotor speed, } N = 950 \text{ rpm}$$

$$(i) \text{ Slip, } s = \frac{N_s - N}{N_s} = \frac{1{,}000 - 950}{1{,}000} = 0.05 \text{ or 5 per cent } \textbf{Ans.}$$

$$\text{Rotor emf per phase at standstill, } E_2 = \frac{175}{\sqrt{3}} = 101 \text{ volts } \textbf{Ans.}$$

$$\text{Rotor frequency at a speed of 950 rpm } f' = sf = 0.05 \times 50 = 2.5 \text{ Hz } \textbf{Ans.}$$

$$\text{Standstill rotor reactance, } X_2 = 1.5\ \Omega/\text{ phase}$$

$$\text{Rotor reactance at a speed of 950 rpm} = s X_2 = 0.05 \times 1.5 = 0.075\ \Omega \text{ per phase } \textbf{Ans.}$$

10.12. ROTOR TORQUE

The torque of an induction motor, being due to interaction of the rotor and stator fields, depends on the strength of those fields and the phase relation between them. Thus the torque developed by the rotor of an induction motor is directly proportional to (*i*) rotor current, I_2 (*ii*) stator flux per pole, ϕ and (*iii*) power factor of the rotor circuit, $\cos \phi_2$

$$i.e.\ T \propto \phi I_2 \cos \phi_2$$

Since rotor emf per phase at standstill, $E \propto \phi$

$$\therefore \text{ Torque developed, } T \propto E_2 I_2 \cos \phi_2$$

$$\text{or } T = K E_2 I_2 \cos \phi_2 \text{ where K is any constant} \quad ...(10.13.)$$

10.12.1. Torque Under Running Conditions

Substituting $I_2 = \frac{sE_2}{\sqrt{R_2^2 + s^2 X_2^2}}$ from equation (10.10) and $\cos \phi_2 = \frac{R_2}{\sqrt{R_2^2 + s^2 X_2^2}}$ from Eq. (10.11) in

Eq. (10.13) we have torque developed by an induction motor rotor,

$$T = K E_2 \frac{sE_2}{\sqrt{R_2^2 + s^2 X_2^2}} \times \frac{R_2}{\sqrt{R_2^2 + s^2 X_2^2}} = \frac{K s R_2 E_2^2}{R_2^2 + s^2 X_2^2} \quad ...(10.14)$$

It may be seen from equation (10.14) that the running torque is proportional to the square of supply voltage because rotor induced emf E_2 is proportional to supply voltage.

10.12.2. Starting Torque

At start the rotor is stationary and so the slip $s = 1.0$ and the starting torque may be obtained by substituting $s = 1.0$ in Eq. (10.14).

$$\text{Starting torque, } T_{st} = \frac{K R_2 E_2^2}{R_2^2 + X_2^2} \quad ...(10.15)$$

Generally, the stator supply voltage V is constant, so the rotating flux ϕ set up by the stator currents and rotor induced emf E_2 are also constant.

$$\therefore T_{st} = \frac{K_1 R_2}{R_2^2 + X_2^2} = \frac{K_1 R_2}{Z_2^2} \quad \text{where } K_1 \text{ is another constant} \quad ...(10.16)$$

10.12.3. Condition For Maximum Running Torque

Torque under running condition given by Eq. (10.14) is reproduced here

$$T = \frac{KR_2 sE_2^2}{R_2^2 + s^2 X_2^2}$$

Torque T for fixed supply voltage will be maximum when $\frac{sR_2}{R_2^2 + s^2 X_2^2}$ or $\frac{R_2}{\frac{R_2^2}{s} + sX_2^2}$ or $\frac{R_2}{\left(\frac{R_2}{\sqrt{s}} - X_2\sqrt{s}\right)^2 + 2R_2X_2}$ is maximum

$$\text{viz. } \frac{R_2}{\sqrt{s}} - X_2\sqrt{s} = 0$$

$$\text{or } s = \frac{R_2}{X_2}$$

Substituting $s = \frac{R_2}{X_2}$ in Eq. (10.14) we have

$$T_{max} = \frac{KR_2^2 E_2^2/X_2}{R_2^2 + R_2^2} = \frac{KE_2^2}{2X_2} \quad ...(10.17)$$

From Eq. (10.17) it is obvious that

1. Maximum torque is independent of rotor circuit resistance.
2. The slip at which maximum torque occurs is governed by the rotor resistance. As seen from above, the torque becomes maximum when rotor reactance equals the rotor resistance. So by varying the rotor circuit resistance, which is possible with wound rotors, maximum torque can be made to occur at any desired slip or motor speed.
3. Maximum torque varies inversely as standstill reactance of the rotor, hence to have maximum torque, standstill reactance *i.e.* inductance of the rotor should be kept as small as possible. This is achieved by placing the rotor conductors very close to the surface of the rotor and reducing the air gap between stator and rotor to the smallest possible value.
4. Maximum torque varies directly as the square of supply voltage.

10.12.4. Condition For Maximum Starting Torque

Starting torque T_{st} will be maximum when $\frac{R_2}{X_2} = s = 1$

$$\text{or } R_2 = X_2$$

Since the rotor resistance is not more than 1 or 2 per cent of leakage reactance otherwise efficiency will be low, so to increase the starting torque, extra resistance is required to be inserted in the rotor circuit at start and cut out gradually as the motor picks up speed. This is possible in case of wound rotor motors. This is the reason that wound rotor motors are used where heavy loads are to be accelerated at start such as cranes, lifts, elevators etc.

10.13. TORQUE-SLIP AND TORQUE-SPEED CURVES

Torque developed by an induction motor rotor is given by the equation $T = \frac{KsR_2 E_2^2}{R_2^2 + s^2 X_2^2}$.

From the above equation it is revealed that

1. When speed is synchronous *i.e.* when slip s is zero, the torque T is zero so the torque-slip curve starts from origin O.
2. When the speed is very near to synchronous speed N_s *i.e.* when slip s is very low the value of term sX_2 is very small and is negligible in comparison with rotor resistance R_2, therefore, torque T is approximately proportional to slip s, if rotor resistance R_2 is constant *i.e.* at speeds near synchronous speed the torque-speed and torque-slip curves are approximately straight lines, as shown in Figs. 10.20 (*a*) and 10.20 (*b*) respectively.
3. As the slip increases , *i.e.* as the speed drops with the increase in load, torque increases, reaches its maximum value when slip $s = \frac{R_2}{X_2}$. The maximum torque is also known as *breakdown or pull-out torque*. The slip corrresponding to breakdown torque is called the breakdown slip s_b.
4. With the further increase in slip or drop in speed due to increase in load beyond the point of

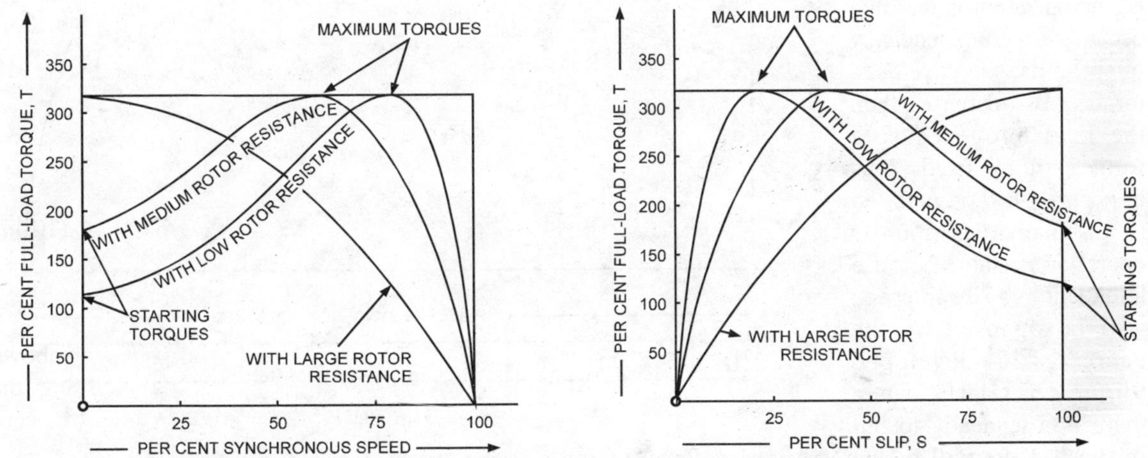

(*a*) *Torque-Speed Curves* (*b*) *Torque-Slip Curves*

Fig. 10.20

maximum torque the torque begins to decrease. The result is that the motor slows down and eventually stops. The motor operates for the values of slip between zero and that corresponding to breakdown or pull-out torque.

With higher values of slip, R_2 becomes negligible as compared to $s\,X_2$ and the torque varies according to the relation $T \propto \frac{s}{s^2 X_2^2} \propto \frac{1}{s}$ if standstill reactance X_2 is constant. It means that torque-speed curves are rectangular hyperbola with the speed or slip beyond that corresponding to breakdown or pull-out torque.

Torque-speed and torque-slip curves are shown in Figs. 10.20 (*a*) and 10.20 (*b*) respectively for different values of rotor resistance R_2.

It is seen that although maximum torque is independent of rotor resistance R_2, yet the exact location of T_{max} is dependent on it. Greater the R_2, greater is the value of slip at which maximum torque occurs.

10.13.1. Effect of Rotor Resistance Upon Torque-Slip or Torque-Speed Relationship

From the torque equation it is obvious that, for a constant supply voltage, when rotor resistance R_2 is very small compared with $s\,X_2$, the torque for a given slip is directly proportional to R_2; but when rotor resistance R_2 is large compared with s X_2, the torque for a given slip is inversely proportional to R_2.

From the Fig. 10.20 (*b*) it is seen that variation of rotor resistance does not change the magnitude of maximum torque T_{max}; merely changes the value of slip at which it occurs. Larger the rotor resistance the greater is the slip at which the maximum torque occurs.

For a given value of torque T, slip *s* is proportional to rotor resistance R_2, so addition of external resistance in the rotor circuit does not lower the torque curve but merely stretches it so that the same torque values occur at lower speeds (or higher slips) as shown in figure 10.20 (*a*) and 10.20 (*b*).

10.14. TORQUE-SPEED CURVE AND OPERATING REGION

The torque-speed curve for an induction motor is given in Fig. 10.21. From Fig. 10.21 it is evident that for any load torque, there are two operating points (say B and D). But at position B, the operation is unstable because if there is tendency of speed rise the developed torque also increases than the load torque causing further rise in speed. Thus at position B the operation is unstable. At position D operation is stable because a tendency to rise in speed will be opposed by the decrease in developed torque. Similarly when there is a tendency to fall in speed, there will be increase in developed torque to bring the motor to operating position D. Thus the region AC of the torque-speed characteristic is *unstable region* of operation while the region CF is stable region of operation.

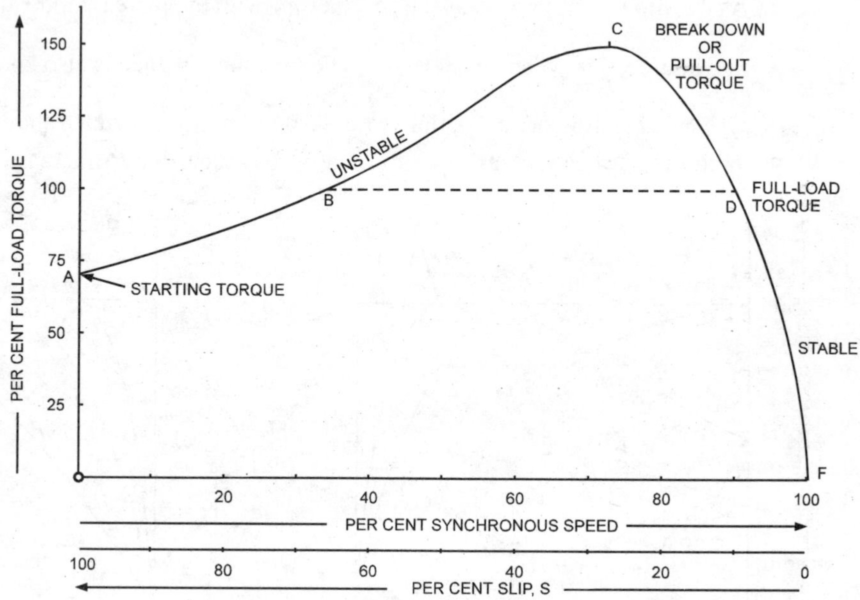

Fig. 10.21

At full load, the motor runs at a speed of N rpm. If the mechanical load applied to the motor is increased, motor speed drops till the torque developed by the motor equals the load torque. As long as the two torques are in balance, the motor will run at constant speed. However, if the load torque exceeds the breakdown torque of the motor, the motor will immediately stop.

The value of breakdown torque varies with the design of the motor but ranges from 200 to 300 per cent of full-load torque in standard squirrel cage motors.

The torque-speed characteristics of an induction motor in the operating region are quite similar to those of a dc shunt motor. Even at full load, the drop in speed from no load does not exceed 5 per cent.

10.15. POWER LOSSES IN INDUCTION MOTORS

The electrical power supplied to the stator of the motor is converted into mechanical power at the motor shaft. The various losses occuring in the machine during this energy conversion are : (*i*) stator iron loss (*ii*) stator copper loss (*iii*) rotor iron loss (*iv*) rotor copper loss and (*v*) mechanical (windage and friction) losses.

Stator iron loss (consisting of eddy current and hysteresis losses) depends on the supply frequency and the flux density in the iron core. It is practically constant.

Rotor iron loss is, however, negligible because under normal operating conditions frequency of rotor currents is always small.

Stator copper loss per phase is given as $I_1^2 R_1$ where I_1 is the stator current and R_1 is the stator resistance per phase.

Rotor copper loss per phase is given as $I_2^2 R_2$ where I_2 is the rotor current and R_2 is the rotor resistance per phase.

10.16. POWER FLOW DIAGRAM

Figure 10.22 gives a power flow diagram for a typical induction motor and clearly indicates how the electrical power supplied to the stator winding is converted into useful mechanical power available at the shaft.

Power input to the stator winding $= 3 V_1 I_1 \cos \phi_1$; where V_1 is phase supply voltage, I_1 is stator current per phase and $\cos \phi_1$ is the stator power factor.

Some of the power supplied to the stator winding is lost as heat in the iron and in the copper coils. Stator copper loss is about 3.5% and stator iron loss is about 2.5% of power input to the motor. The remaining power (about 94% of power input to the motor) is transferred magnetically across the air gap to the rotor as to a transformer secondary. Part of this rotor power received (about 3.5% of power input to motor) is lost as heat in the rotor conductors and the rest is converted into mechanical power called the mechanical power developed in the rotor. This is about 90.5% of power input to the motor. Then there are friction and windage losses (about 2% of power input to the motor), as usual and the remaining power (about 88.5 per cent of power input to the motor) is available at the motor shaft, called the mechanical power output.

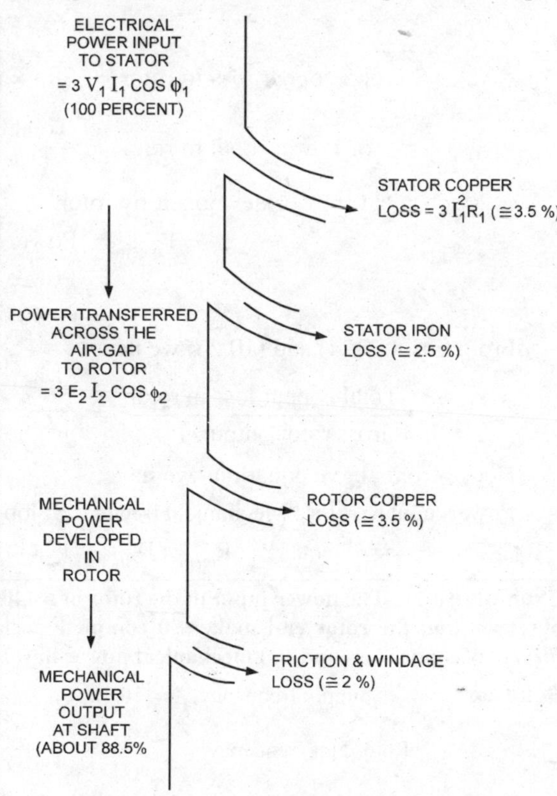

Fig. 10.22. *Power Flow Diagram For an Induction Motor*

10.17. INDUCTION MOTOR TORQUE

If T_{sh} and T_g are the shaft torque and the gross torque developed in newton metres and N is the rotational speed in rpm, then

Power output,

$$P_{out} = T_{sh} \times \frac{2\pi N}{60} \text{ watts} \qquad ...(10.18)$$

and the mechanical power developed,

$$P_{mech} = T_g \times \frac{2\pi N}{60} \text{ watts} \qquad ...(10.19)$$

$$\text{or } T_{sh} = \frac{P_{out}}{2\pi N/60} = 9.55 \frac{P_{out}}{N} \text{ Nm} \qquad ...(10.20)$$

$$\text{and } T_g = \frac{P_{mech}}{2\pi N/60} = 9.55 \frac{P_{mech}}{N} \text{ Nm} \qquad ...(10.21)$$

The difference between gross torque developed, T_g and shaft torque T_{sh} is equal to the torque lost due to friction and windage losses in the motor.

10.18. ROTOR OUTPUT

When a 3-phase induction motor is operating on a certain load, the power transferred from the stator to the rotor (*i.e.* rotor input) can be expressed in terms of the torque exerted on the rotor by the rotating flux and the synchronous speed at which this flux is rotating.

Power transferred from stator to rotor or input power to rotor or air-gap power,

$$P_2 = \frac{2\pi N_s}{60} \text{ T watts} \qquad ...(10.22)$$

where T is the torque in newton metres, exerted on the rotor by the rotating flux and N_s is the synchronous speed of rotating magnetic field in rpm.

This power transferred from the stator to the rotor (or power input to the rotor) is often referred to as the torque in synchronous watts.

Total mechanical power developed by rotor,

$$P_{mech} = \frac{2\pi N}{60} \text{ T watts} \qquad ...(10.23)$$

where N is the speed of rotor, at full load, in rpm.

Rotor copper loss = Power input to rotor – output power of rotor (or mechanical power developed by rotor)

$$= \frac{2\pi N_s T}{60} - \frac{2\pi NT}{60} = \frac{2\pi T}{60}(N_s - N) = \frac{2\pi T}{60} sN_s$$

$$= s \times \frac{2\pi N_s}{60} \times T = s \times \text{power input to rotor}$$

So total copper loss in rotor = Slip × power input to rotor ...(10.24)

$$\text{or Power input to rotor} = \frac{\text{Total copper loss in rotor}}{s} \qquad ...(10.25)$$

or Mechanical power developed by rotor,

$$P_{mech} = \text{Power input to rotor} - \text{rotor copper loss}$$

$$= \text{Power input to rotor} - \text{slip} \times \text{power input to rotor}$$

$$= (1 - s) \times \text{power input to rotor} \qquad ...(10.26)$$

From Eqs. (10.24) and (10.26) we have

$$\frac{\text{Total copper loss in rotor}}{\text{Gross rotor output, } P_{mech}} = \frac{s}{1-s} \qquad ...(10.27)$$

Hence from above equations we have

Power input to rotor : Mechanical power developed by rotor (gross rotor output power) : rotor copper loss

i.e. $P_2 : P_{mech} : P_{cu} :: 1 : (1 - s) : s$

Example 10.11. The power input to the rotor of a 440 V, 50 Hz, 3 -phase, 6-pole induction motor is 60 kW. It is observed that the rotor emf makes 90 complete cycles per minute. Calculate (*i*) the slip (*ii*) the rotor speed (*iii*) rotor copper loss and (*iv*) mechanical power developed.

Solution : Supply frequency, $f = 50$ Hz

Rotor emf frequency, $f' = \frac{90}{60} = 1.5$ Hz.

(*i*) Slip, $s = \frac{f'}{f} = \frac{1.5}{50} = 0.03$ **Ans.**

Power input to rotor, $P_2 = 60$ kW

(*ii*) Rotor speed, $N = N_s(1-s) = \frac{120f}{P}(1-s) = \frac{120 \times 50}{6} \times (1-0.03) = 970$ rpm **Ans.**

(*iii*) Rotor copper loss $= sP_2 = 0.03 \times 60 = 1.8$ kW **Ans.**

(*iv*) Mechanical power developed,

$$P_{mech} = P_2(1-s) = 60(1-0.03) = 58.2 \text{ kW } \textbf{Ans.}$$

Example 10.12. A 440 V, 50 Hz, 6-pole, 3-ϕ induction motor draws an input power of 76 kW from the mains. The rotor emf makes 120 complete cycles per minute. Its stator losses are 1 kW and rotor current per phase is 62 A. Calculate (*a*) rotor copper losses per phase (*b*) rotor resistance per phase (*c*) torque developed.

Solution : Supply frequency, $f = 50$ Hz

$$\text{Rotor emf frequency}, f' = \frac{120}{60} = 2 \text{ Hz}$$

$$\text{Slip}, s = \frac{f'}{f} = \frac{2}{50} = 0.04$$

Power input to rotor, P_2 = Power input to motor – stator losses = 76 – 1 = 75 kW

Total rotor copper losses $= sP_2 = 0.04 \times 75 = 3$ kW

(*a*) Rotor copper losses / phase $= \frac{3}{3} = 1$ kW **Ans.**

Rotor current per phase, $I_2 = 62$ A

$$\text{(b) Rotor resistance per phase, } R_2 = \frac{\text{Rotor copper loss per phase in watts}}{I_2^2} = \frac{1 \times 1{,}000}{62^2} = 0.26\ \Omega \textbf{ Ans.}$$

Rotor output power = Power input to rotor – rotor copper losses = 75 – 3 = 72 kW

$$\text{Rotor speed, } N = N_s(1-s) = \frac{120f}{P}(1-s) = \frac{120 \times 50}{6}(1-0.04) = 960 \text{ rpm}$$

$$\text{(c) Torque developed, } T = \frac{\text{Rotor output power in watts}}{2\pi N/60} = \frac{72 \times 1{,}000 \times 60}{2\pi \times 960} = 716.2 \text{ Nm } \textbf{Ans.}$$

Example 10.13. A 3-phase, 6-pole, 50 Hz induction motor develops 3,700 watts at 950 rpm. What is the stator input, if the stator loss is 300 W ?

Solution : Synchronous speed of motor, $N_s = \frac{120f}{P} = \frac{120 \times 50}{6} = 1{,}000$ rpm

Rotor speed, $N = 950$ rpm

$$\text{Slip}, s = \frac{N_s - N}{N_s} = \frac{1{,}000 - 950}{1{,}000} = 0.05$$

Mechanical power developed, $P_{mech} = 3{,}700$ W

$$\text{Power input to rotor}, P_2 = \frac{P_{mech}}{1-s} = \frac{3{,}700}{1-0.05} = 3{,}895 \text{ W}$$

Power input to stator, $P_1 = P_2$ + stator loss = 3,895 + 300 = 4,195 W **Ans.**

Example 10.14. The power input to a 500 V, 50 Hz, 6-pole, 3-phase squirrel cage induction motor running at 975 rpm is 40 kW. The stator losses are 1 kW and the friction and windage losses are 2 kW. Calculate (*i*) slip (*ii*) rotor copper loss (*iii*) brake hp and (*iv*) the efficiency.

Solution : Synchronous speed of motor, $N_s = \frac{120f}{P} = \frac{120 \times 50}{6} = 1{,}000$ rpm

Rotor speed, $N = 975$ rpm

$$(i)\ \text{Slip},\ s = \frac{N_s - N}{N_s} = \frac{1{,}000 - 975}{1{,}000} = 0.025 \text{ or } 2.5\% \ \textbf{Ans.}$$

$$\text{Power input to stator, } P_1 = 40 \text{ kW}$$

$$\text{Stator output power} = P_1 - \text{stator losses} = 40 - 1 = 39 \text{ kW}$$

$$\text{Power input to rotor, } P_2 = \text{Stator output power} = 39 \text{ kW}$$

$$(ii)\ \text{Rotor copper loss, } P_{cu} = sP_2 = 0.025 \times 39 = 0.975 \text{ kW } \textbf{Ans.}$$

$$\text{Mechanical power developed, } P_{mech} = P_2 - P_{cu} = 39 - 0.975 = 38.025 \text{ kW}$$

$$\text{Motor output} = P_{mech} - \text{friction and windage losses} = 38.025 - 2 = 36.025 \text{ kW}$$

$$(iv)\ \text{BHP} = \frac{\text{Motor output in kW}}{0.7355} = \frac{36.025}{0.7355} = 49 \ \textbf{Ans.}$$

$$(v)\ \text{Motor efficiency, } \eta = \frac{\text{Output}}{\text{Input}} \times 100 = \frac{36.025}{40} \times 100 = 90.0625\% \ \textbf{Ans.}$$

10.19. SPEED REGULATION

Like any other electrical motor, the speed regulation of an induction motor is given by

$$\%\text{ age speed regulation} = \frac{N_0 - N_f}{N_f} \times 100 \qquad ...(10.28)$$

where N_0 and N_f are no-load and full-load speeds respectively.

The change in speed of the motor from no load to full load is small *i.e.* speed regulation of an induction motor is small (say 3 to 5%).

10.20. RATINGS OF 3-PHASE INDUCTION MOTORS

Three phase induction motors are rated in terms of power output in kW, speed, voltage, frequency, phase (single or three), line current and temperature rise in a specified time. If a motor is designed to be operated on more than one voltage or at more than one speed by reconnecting the windings, a connection diagram is also often given on the name plate. Class of motors (A, B, C, D, E or F) is also to be mentioned on the name plate.

10.21. ADVANTAGES, DISADVANTAGES AND APPLICATIONS OF 3-PHASE INDUCTION MOTORS

Squirrel cage induction motors are comparatively *cheaper* than wound rotor induction motors and are *simple* and *rugged* in construction.

A cage rotor needs considerably less conductor material than a wound rotor, so copper loss in cage rotor is low and *efficiency* of a cage motor is a *little more* than that of a wound rotor motor.

Cage motor is more rugged in construction and needs no slip rings, brushes etc., therefore, its *maintenance cost is low*.

Cage motor is *explosion proof*, because the absence of slip rings and brushes eliminates risk of sparking.

Cage motor can be *cooled better* because of its bare end rings. Also there is more space for rotor fans.

A squirrel cage induction motor has very small length of overhang, so it has low rotor overhang leakage flux. This causes standstill reactance X_2 low for a cage rotor than wound rotor. So a cage motor has *more pull-out torque, greater maximum power output* and *better operating power factor* as compared to a wound rotor motor.

Other advantages of squirrel cage motors are (*i*) nearly constant speed, high overload capacity and simple starting arrangement which may be made automatic.

The disadvantages of squirrel cage induction motors are enumerated below :

(*i*) Since in cage rotors, rotor bars are permanently short circuited, it is not possible to insert any external resistance in the rotor circuit for the purpose of increasing starting torque. Consequently its *starting torque is low* with *large starting current* : — starting torque varying between 1.5 and 2 times the full-load torque with starting current 5 to 9 times the full-load current. *Power factor* is also *low* at start in case of cage motors.

(*ii*) No possibility of speed control.

(*iii*) Squirrel cage induction motors are very sensitive to fluctuations in supply voltage.

(*iv*) The total energy lost during starting of cage motor is much more than with the slip-ring induction motor and this fact is very important where frequent starting is required.

Squirrel cage induction motor is suitable for constant speed industrial drives of small power where speed control is not required and where starting torque requirements are of medium or low value, such as for printing machinery, flour mills and other shaft drives of small power.

When the squirrel cage motors are used for crane or hoist work where large starting torque is of more importance than high efficiency owing to the intermittent nature of load, the rotor resistance is increased by employing the end rings, and sometimes the bars as well, of high resistance metal such as German silver.

Advantages of wound rotor (or slip-ring) induction motors are given below :

(*i*) *High starting torque* with low starting current. By inserting an external resistance of proper value (so that rotor resistance R_2 becomes equal to its standstill reactance X_2) in each phase of the rotor circuit, maximum torque can be obtained at start.

(*ii*) *Adjustable speed.* As speed can be controlled easily in case of a wound rotor motor simply by adding external resistors in the rotor circuit. This is the reason that wound rotor induction motors are considered to be variable speed motors.

Other advantages are high overload capacity, nearly constant speed, smooth acceleration under heavy loads, no abnormal heating during starting, adoptability to automatic starting and good running characteristics after the external resistances are taken out of the rotor circuit.

The disadvantages of wound rotor induction motors are :

(*i*) Initial and maintenance costs are more than those of a squirrel cage motor, because a slip-ring induction motor needs slip rings, brushes, short-circuiting devices etc.

(*ii*) Speed regulation is poor when operated with external resistances in the rotor circuit.

(*iii*) Lower efficiency and power factor as compared with cage rotor motor.

(*iv*) Low power factor at light loads.

(*v*) Sensitivity to fluctuations in supply voltage.

Wound rotor (or slip-ring) induction motors are used for loads requiring severe starting conditions or for loads requiring speed control such as for driving line shafts, lifts, pumps, generators, winding machines, cranes, hoists, elevators, compressors, small electric excavators, printing presses, turn tables, strokers, large ventilating fans, crushers etc.

HIGHLIGHTS

1. An induction motor is simply an electric transformer whose magnetic circuit is separated by an air gap into two relatively movable portions, one carrying the primary winding and the other the secondary winding.

It essentially consists of (*i*) a stator and (*ii*) a rotor.

Rotors are of two types (*i*) squirrel-cage rotor and (*ii*) phase-wound or wound rotor.

2. *Working principle.* When a 3-ϕ supply is provided to a 3-phase wound stator of an induction motor, a rotating magnetic field is established rotating at synchronous speed. At start, stationary rotor conductors cut across the revolving magnetic field and an emf is induced in them by the electromagnetic phenomenon. Current flows through the rotor conductors, as they form a closed path, and so rotor field is developed. By interaction of stator magnetic field and rotor field, torque develops and causes the rotation of rotor in the same direction as that of revolving magnetic field.

3. The direction of rotation of a 3-phase induction motor can be reversed by interchanging the connections to the supply of any two terminals.

4. *Synchronous speed,* $N_s = \frac{120f}{P}$, f is supply frequency and P is the number of poles on stator.

5. **Slip.** The rotor never attains the synchronous speed because at that speed there would be no relative speed between rotor conductors and stator revolving field and induction phenomenon is not possible. Its speed is always less than synchronous speed.

   The difference between the speed of the stator field, known as synchronous speed (N_s) and the actual speed of the rotor (N) is known as the slip and is denoted by s

$$s = \frac{N_s - N}{N_s}$$

6. Frequency of rotor current or emf, $f' = s f$

7. Rotor emf at start, $E_2 = E_1 \times \frac{N_2}{N_1}$

8. Rotor current at start, $I_2 = \frac{E_2}{\sqrt{R_2^2 + X_2^2}}$

   Rotor current at slip s, $I_2' = \frac{s E_2}{\sqrt{R_2^2 + (s X_2)^2}}$

9. Starting torque, $T_{st} = \frac{K R_2 E^2}{R_2^2 + X_2^2} = \frac{K_1 R_2}{R_2^2 + X_2^2}$ if applied voltage is constant.

   Torque under running condition, $T = \frac{K_1 s R_2 E_2^2}{R_2^2 + s^2 X_2^2} = \frac{K_2 s R_2}{R_2^2 + s^2 X_2^2}$ if applied voltage is constant.

10. Condition for maximum torque under running condition

$$R_2 = sX_2$$

11. Slip corresponding to maximum torque, $s = \frac{R_2}{X_2}$

12. Condition for maximum starting torque, $R_2 = X_2$

13. Maximum torque, $T_{max} \propto \frac{1}{X_2}$

14. The torque-speed characteristics of an induction motor in the operating region are quite similar to those of a dc shunt motor. Even at full load, the drop in speed from no load does not exceed 5 per cent.

15. The electrical power supplied to the stator of the motor is converted into mechanical power at the motor shaft. The various losses occuring in the machine during this energy conversion are : (*i*) stator iron loss (*ii*) stator copper loss (*iii*) rotor iron loss (*iv*) rotor copper loss and (*v*) mechanical (windage and friction) losses.

    Rotor iron loss is, however, negligible because under normal operating conditions frequency of rotor currents is always small.

16. Gross torque, $T_g = \frac{P_{mech}}{2\pi N/60} = 9.55 \frac{P_{mech}}{N}$ Nm

17. Shaft torque, $T_{sh} = \frac{P_{output}}{2\pi N/60} = 9.55 \frac{P_{output}}{N}$ Nm

    where $P_{output} = P_{mech}$ – copper losses

18. Power relations

$$\text{Rotor copper loss} = s \times \text{power input to rotor}$$

$$\frac{\text{Rotor copper loss}}{\text{Gross rotor output}} = \frac{s}{1-s}$$

$$\frac{\text{Gross rotor output}}{\text{Rotor input}} = \frac{N}{N_s} = 1 - s$$

Power input to rotor : gross rotor output power : rotor copper loss

$$\textit{i.e.}\ P_2 : P_{mech} : P_{cu} :: 1 : (1-s) : s$$

19. The change in speed of the motor from no load to full load is small i.e. speed regulation of an induction motor is small (say 3 to 5%).

20. Squirrel cage induction motor is suitable for constant speed industrial drives of small power where speed control is not required and where starting torque requirements are of medium or low value, such as for printing machinery, flour mills and other shaft drives of small power.

21. Wound rotor (or slip-ring) induction motors are used for loads requiring severe starting conditions or for loads requiring speed control such as for driving line shafts, lifts, pumps, generators, winding machines, cranes, hoists, elevators, compressors, small electric excavators, printing presses, turn tables, strokers, large ventilating fans, crushers etc.

EXERCISES

1. Which is the most commonly used 3-phase ac motor ? Justify the name given for that.
2. Differentiate between squirrel cage and slip-ring induction motors.
   [Pb. Technical Univ. Basic Electrical and Electronics Engineering May 2009]
3. Sketch a layout of squirrel cage induction motor. Label all the parts and material used for each part. Explain why rotor of an induction motor can never attain synchronous speed.
   [Pb. Technical Univ. Basic Electrical Engineering June 2003]
4. Give clear explanation of the production of rotating field in a 3-phase induction motor.
   [Pb. Technical Univ. Electrical Engineering December 2000]
5. Explain how a rotating magnetic field of constant amplitude is produced when a 3-phase supply is given to a three-phase winding of induction motor.
   [Pb. Technical Univ. Basic Electrical Engineering December 2003]
6. Discuss how is rotating magnetic field produced in a 3-phase induction motor.
   [Pb. Technical Univ. Basic Electrical and Electronics Engineering Dec. 2004]
7. Explain the working principle and construction of three phase induction motor.
   [Pb. Technical Univ. Basic Electrical and Electronics Engineering, Dec. 2009]
8. Give a clear explanation of the production of torque in a 3-phase induction motor.
   [Pb. Technical Univ. Basic Electrical and Electronics Engineering, May 2010]
9. Classify 3-phase induction motors and describe briefly their construction, working and applications.
   [Pb. Technical Univ. Basic Electrical and Electronics Engineering May 2007]
10. Discuss principle of operation of a 3-phase induction motor. What is the frequency of rotor current ? Discuss. **[Pb. Technical Univ. Basic Electrical and Electronics Engineering Dec. 2005]**
11. Discuss the theory of a three-phase induction motor with the help of a diagram.
    [Pb. Technical Univ. Electrical Engineering 1999]
12. Give a clear explanation of the production of torque in a 3-phase induction motor.
    [Pb. Technical Univ. Electrical Engineering June 2000]
13. Explain the principle of working of a 3-phase induction motor and the meaning of term slip. How does the slip vary with load ? **[Pb. Technical Univ. Electrical Engineering May 2001]**

14. Explain the terms slip, slip frequency, wound rotor and cage rotor.
15. Derive an expression of rotor frequency in terms of main supply frequency and slip.
16. Derive the expression for developed torque in a 3-phase induction motor and find the condition for maximum torque.
17. Derive the condition for maximum torque of a 3-phase induction motor under running condition.
18. Draw and explain the slip-torque characteristic of a typical induction motor. Mark the starting torque and maximum torque on the diagram so drawn. How do starting and maximum torques vary with the rotor resistance ?
19. Draw the torque-speed curve of a 3-phase induction motor and indicate therein the starting torque, pull-out torque and full-load torque. What is the place of application of cage and wound rotor induction motors ?

[Pb. Technical Univ. Electrical Engineering January 2000]

20. Show that in induction motor, the rotor always run in the direction of stator field.

Sketch the torque-speed characteristic of an induction motor working at normal voltage and frequency. Indicate starting torque, pull-out torque, and normal operating region of induction motor on characteristic itself. **[Pb. Technical Univ. Basic Electrical Engineering December 2002]**

21. What are the various losses in an induction motor ? On what factors do they depend ?
22. Explain the terms air-gap power P_g, internal mechanical power developed and shaft power P_{sh}. How are these terms related with each other ? Hence show that :

$$P_g : \text{rotor ohmic loss} : P_m = 1 : s : (1 - s)$$

23. Explain the constructional details of a squirrel cage and a slip-ring rotor induction motor. Compare the merits and demerits of each type of motor.
24. Distinguish between 3-phase slip-ring and squirrel cage induction motors, from the point of view of starting torque, pull-out torque, efficiency and power factor.
25. Describe the constructional features of wound and cage rotor induction motors and hence discuss and compare salient features of their performance characteristics, merits and demerits.

[Pb. Univ. Elec. Machines II, Dec. 1994]

SHORT ANSWER TYPE QUESTIONS WITH ANSWERS

Q. 1. What is meant by an induction motor ?

Ans. An induction motor derives its name from the fact that the current in the rotor conductors is induced by the motion of rotor conductors relative to the magnetic field developed by the stator currents.

Q. 2. What are the two types of 3-phase induction motors ? Which type is generally preferred ?

Ans. Two types of 3-phase induction motors are (*i*) squirrel cage and (*ii*) wound rotor or slip-ring induction motors. Squirrel cage induction motor is generally preferred.

Q. 3. Why are induction motors called "asynchronous" motors ?

Ans. Because the rotor does not turn in synchronism with the rotating field developed by the stator currents.

Q. 4. Why closed type slots are often used for small induction motors ?

Ans. Use of closed slots makes the reluctance more uniform, reduces exciting current and gives better power factor.

Q. 5. Why the air gap between stator core and rotor of an induction motor is made very small ?

Ans. The air gap is made as small as possible so as to produce required flux with a minimum exciting current and give leakage reactance as small as possible. This results in improved power factor.

Q. 6. Why the rotor of a slip-ring induction motor should be wound for the same number of poles as its stator ?

Ans. Because the slip-ring induction motor having the same number of poles on the rotor as that of the stator will be in a position to run at slip speed.

Q. 7. Which type of induction motor develops higher starting torque ?

Ans. Slip-ring induction motor.

Q. 8. What is the necessity of short-circuited rotor conductors in a squirrel cage induction motor ?

Ans. If the rotor conductors in a squirrel cage motor are not short circuited, the rotor winding will be open circuited and when the stator is connected to **3-ϕ ac** supply, the rotating magnetic field will be set up in the air gap and an emf will be induced in the rotor but no current will flow through the rotor conductors and, therefore, no torque will be developed and the rotor will remain stationary.

Q. 9. How is a rotating magnetic field produced ? What is its importance in electrical machinery ?

[P.T.U. January 2000]

Ans. When a 3-phase wound machine is energized by 3-phase supply, a rotating magnetic field is developed that tends the rotor of the machine to move in its direction of rotation. Energy transfer phenomenon occurs due to such rotating magnetic field.

Q. 10. Explain what is rotating magnetic field and how this is made use of in electrical machinery ?

[P.T.U. December 2001]

Ans. A rotating magnetic field is that which is constant in magnitude but whose axis of direction rotates in space as field system of a dc machine. When a 3-phase winding is energized by 3-phase supply, a rotating magnetic field is developed that tends the rotor of the machine to move in its direction of rotation.

The operation of ac rotating machines (such as synchronous machines and induction machines) is based on the production of rotating magnetic field.

Q. 11. How can the direction of rotation of the 3-phase induction motor be reversed ?

Ans. The direction of rotation of a 3-phase induction motor may be reversed by interchanging the connection to the supply of any two leads of the motor.

Q. 12. What do you mean by slip in an induction motor? **[P.T.U. B.E.E.E. Dec. 2005]**

Ans. The difference between the speed of the stator field and the actual speed of the rotor expressed as a fraction of synchronous speed is known as slip of an induction motor.

Q. 13. Define terms : Slip and slip frequency in case of 3-phase induction motors. **[P.T.U. B.E.E.E. Dec. 2006]**

Ans. The difference between the speed of the stator field and the actual speed of the rotor expressed as a fraction of synchronous speed is known as slip of an induction motor.

The frequency of rotor current (or emf) in an induction motor is given by the product of slip s and the supply frequency f known as slip frequency.

Q. 14. The speed of induction motor cannot be equal to synchronous speed. Explain why ?

[P.T.U. December 2000]

Ans. An induction motor cannot run at synchronous speed. If it were possible, by some means, for the rotor to attain synchronous speed, the rotor would then be standstill with respect to rotating flux. Then no emf would be induced in the rotor, no rotor current would flow, and therefore, there would be no torque developed.

Q. 15. Explain why rotor of an induction motor can never attain synchronous speed. **[P.T.U. June 2003]**

Ans. As of Question 14.

Q. 16. Why a 3-phase induction motor cannot run at synchronous speed ?

[P.T.U. B.E.E.E. Dec. 2005; May 2010]

Ans. As of Question 14 above.

Q. 17. Show that in induction motor, the rotor always run in the direction of stator field.

[P.T.U. December 2002]

Ans. According to Lenz's law we know that the direction of induced emf would be in such a direction that it would try to oppose the very cause for which it is due. Because the cause producing the induced currents in the rotor conductors is the relative speed between the rotating magnetic field and stationary rotor conductors, therefore, they circulate in such a way that a torque is produced in the rotor tending to cause it to follow the rotating magnetic field and thus reducing the relative speed.

Q. 18. How does the slip vary with load ? **[P.T.U. May 2001]**

Ans. With the increase in load, the motor slows down until the relative motion between the rotor and the rotating magnetic field is just sufficient to result in the development of torque required for a particular load. Thus slip increases with the increase in load.

Q. 19. At what slip the torque developed in an induction motor will be maximum ?

Ans. Torque developed in an induction motor will be maximum at a slip given by the ratio of rotor resistance to rotor standstill reactance, *i.e.* $s = \frac{R_2}{X_2}$.

Q. 20. What is the condition for maximum torque at starting in a 3-phase induction motor ?

Ans. Torque developed in an induction motor at start will be maximum when rotor resistance equals rotor standstill reactance *i.e.* $R_2 = X_2$.

Q. 21. Is the maximum torque of a 3-phase induction motor dependent on the rotor resistance ?

Ans. No, it is independent of rotor resistance.

Q. 22. What is the relationship of developed torque of a 3-phase induction motor with the supply voltage ?

Ans. $T \alpha V^2$.

Q. 23. When the applied rated voltage per phase is reduced to one-half, what will be the starting torque of a squirrel cage induction motor in terms of its starting torque with full voltage ?

Ans. One-fourth of starting torque with full rated voltage.

Q. 24. What are the losses in a 3-phase induction motor ?

Ans. Stator copper loss; stator iron loss; rotor copper loss, friction and windage losses and stray load loss.

Q. 25. How is the mechanical power output represented in the circuit model of an induction motor ?

Ans. Mechanical power output per phase is represented by a pure resistive load of equivalent resistance R_L given by the relation $R_L = R_2 \left(\frac{1}{s} - 1\right)$ where R_2 is the rotor circuit resistance per phase and s is the slip.

Q. 26. The induction motor can be called a generalised transformer. Explain why ? **[P.T.U. May 2001]**

Ans. Since the two fields set up by stator (primary) and rotor (secondary) are stationary relative to each other, so the polyphase induction motor can be considered equivalent to a transformer having an air gap, separating the iron portions of the magnetic circuit carrying the primary and secondary windings. Thus the induction motor can be called a generalized transformer.

Q. 27. Is it possible to add an external resistance in the rotor circuit of a 3-phase cage induction motor ? Give reason for your answer.

Ans. No; because the rotor winding is permanently short circuited.

Q. 28. What are the advantages and disadvantages of wound rotor over squirrel cage rotor in case of induction motor ? **[P.T.U. December 2002]**

Ans. In wound rotor external resistance can be inserted and, therefore,

(*i*) high starting torque can be obtained by inserting an external resistance of proper value in each phase of the rotor circuit and

(*ii*) adjustable speed can be obtained simply by adding external resistors in the rotor circuit.

The drawbacks of wound rotor over squirrel cage rotor are initial and maintenance costs are more because wound rotor needs slip rings, brushes, short-circuiting devices etc.

Q. 29. What is meant by single phasing ?

Ans. The term 'single phasing' refers to operation with one stator terminal accidently disconnected (as by a blown fuse or a faulty contact on the run side of a star-delta switch).

Q. 30. What is the usual proportion of 3-phase induction motors, 3-phase synchronous motors, dc motors and fractional horse power motors in a typical distribution system ? **[P.T.U. January 2000]**

Ans. About 90 percent 3-phase induction motors, the rest 10 percent fractional horse power motors. 3-phase synchronous motors and dc motors both type negligible in a typical power distribution system.

Q. 31. What is the place of application of cage and wound rotor induction motors. **[P.T.U. January 2000]**

Ans. **Applications of 3-Phase Squirrel Cage Induction Motor :** Squirrel cage induction motor is suitable for constant speed industrial drives of small power where speed control is not required and where starting torque requirements are of medium or low value as for printing machinery, flour mills, saw mills, pumps, prime-movers for small generators and other shaft drives of small power.

Applications of 3-Phase Wound Rotor Motor : Wound rotor induction motor is used for loads requiring severe starting conditions or for loads requiring speed control such as for driving line shafts, lifts, pumps, generators, winding machines, cranes, hoists, elevators, compressors, small electric excavators, printing machines, turn-tables, strokers, large ventilating fans, crushers etc.

PROBLEMS

1. A motor-generator set used for providing variable frequency ac supply consists of a three-phase, 10-pole synchronous motor and a 24-pole, three-phase synchronous generator. The motor-generator set is fed from a 25 Hz, three-phase ac supply. A 6-pole, three-phase induction motor is electrically connected to the terminals of the synchronous generator and runs at a slip of 5%. Determine :

(*i*) the frequency of the generated voltage of the synchronous generator,

(*ii*) the speed at which the induction motor is running. **[Ans.** (*i*) 60 Hz (*ii*) 1,440 rpm]

2. The electromotive force in the stator of an 8-pole induction motor has a frequency of 50 Hz, and that in the rotor is 1.5 Hz. At what speed the motor is running and what is slip ? **[Ans.** 727.5 rpm, 3%]

3. A 3-ϕ induction motor is wound for 4-poles and is supplied from 50 Hz system. Calculate (*i*) the synchronous speed (*ii*) the rotor speed when slip is 4% and (*iii*) rotor frequency when rotor runs at 600 rpm.

[Ans. (*i*) 1,500 rpm (*ii*) 1,440 rpm (*iii*) 30 Hz]

4. A 4-pole, 3-phase induction motor operates from a supply whose frequency is 50 Hz. Calculate (*i*) the speed at which the magnetic field of the stator is rotating (*ii*) the speed of the rotor when the slip is 0.04 (*iii*) the frequency of rotor currents when the slip is 0.03 (*iv*) the frequency of the rotor currents at standstill.

[Ans. (*i*) 1,500 rpm (*ii*) 1,440 rpm (*iii*) 1.5 Hz (iv) 50 Hz]

5. A 3-phase induction motor has a rotor for which the resistance per phase is 0.1 Ω and reactance per phase, when stationary, is 0.4 Ω. The rotor induced emf per phase is 100 V when stationary. Calculate the rotor current and rotor power factor (*i*) when stationary (*ii*) when running with a slip of 5 percent.

[Ans. (*i*) 242.53 A; 0.2425 lag (*ii*) 49.03 A; 0.9806 lag]

6. A 6-pole, 3-phase, 50 Hz induction motor develops maximum torque of 20 kgm at a speed of 970 rpm. Determine the torque exerted by the motor at 4% slip. The rotor resistance per phase is 0.5 Ω. **[Ans.** 19.2 kgm]

7. The rotor emf of a 3-phase, 6-pole, 400 V, 50 Hz induction motor alternates at 3 Hz. Compute the speed and percentage slip of the motor. Find the rotor copper loss per phase, if the full-load input to rotor is 111.9 kW.

[Ans. 940 rpm; 6 per cent; 2.238 kW]

8. The power input to the rotor of a 400 V, 50 Hz, 6-pole, 3-phase induction motor is 80 kW. The rotor electromotive force is observed to make 100 complete alternations per minute. Calculate (*i*) the slip (*ii*) the rotor speed (*iii*) mechanical power developed (*iv*) rotor copper loss per phase (*v*) the rotor resistance per phase if the rotor current is 65 A. **[Ans.** (*i*) 3.33% (*ii*) 966.7 rpm (*iii*) 77.333 kW (*iv*) 889 watts (*v*) 0.21 Ω]

9. The power supplied to a 3-phase induction motor is 40 kW and the corresponding stator losses are 1.5 kW. Calculate (*i*) the total mechanical power developed and the rotor I^2R loss when the slip is 0.04 per unit and (*ii*) efficiency of the motor (neglecting the rotor iron loss). **[Ans.** (*i*) 36.96 kW, 1.54 kW (*ii*) 92.4%]

10. The full-load power input to a 3-phase induction motor is 50 kW, and the slip is 3.0%. Neglecting stator losses, calculate the full-load copper losses per phase and the total mechanical power developed.

[Ans. 0.5 kW; 48.5 kW]

❑❑❑

11

CHAPTER

Synchronous Machines (Generators and Motors)

11.1. INTRODUCTION

INSIDE THIS CHAPTER

A synchronous machine is an ac machine in which the rotor moves at a speed which bears a constant relationship to the frequency of currents, in the armature winding. A synchronous machine is one of the important types of electric machines. Large ac networks operating at constant frequency of 50 Hz rely almost exclusively on *synchronous generators,* also called the *alternators*, for the supply of electrical energy. Private, stand-by and peak-load plants with diesel or gas turbine prime movers also have synchronous generators. Synchronous motors provide constant speed industrial drives with the possibility of power factor correction. Synchronous machines are generally constructed in larger sizes. Small size alternators are not economical. The modern trend is to build alternators of very large sizes capable of generating 500 MVA or even more. The synchronous motor is rarely built in small sizes owing to superior performance characteristics and economical construction of induction motors.

11.1.1. Operating Principle

The operating principle of a synchronous machine is fundamentally the same as that of a dc machine, but, unlike the latter, in the synchronous machine there is no need to rectify the time varying emf which is induced in the armature winding. Consequently a synchronous machine does not require a commutator. It is, in fact, quite possible to use a dc generator as an alternator by placing a set of collector rings on the shaft and connecting these rings to the proper points on the armature winding; brushes riding on the rings can then be connected to the load. But unlike dc generator, they are to be driven at a very definite constant speed as the frequency of generated emf is determined by that speed. The latter is usually referred to as the *synchronous speed,* for which reason these machines are frequently called the *synchronous generators*.

Synchronous generators, because of absence of commutator, are comparatively simple and possess several important advantages over the dc generators.

11.1.2. Classification of Synchronous Machines

Synchronous machines, according to their applications, may be *synchronous generators* or *synchronous motors*. A synchronous generator is a synchronous machine which receives mechanical energy from a prime mover (steam turbine, hydraulic turbine or diesel engine) to which it is mechanically coupled and delivers electrical energy. A synchronous motor receives electrical energy from ac supply main and drives mechanical load.

The synchronous machines may be *single, two* or *3-phase* types. Three windings may be connected either in star or delta. Large ac machines are invariably 3-phase type.

Based on the construction of the machines, the synchronous machines may be classified as (*i*) rotating armature type and (*ii*) rotating field type.

Since it is immaterial for generation of an induced emf whether a conductor moves across a magnetic field or vice versa, synchronous generators may be constructed with either the armature or the field structure as the revolving member.

Rotating armature type alternator looks very much like a dc generator except that there are 3 slip rings in place of commutator (or 4 slip rings if it is desired to provide a connection to the generator neutral). In such generators the required magnetic field is produced by dc electromagnets placed on the stationary member called the *stator*, and the current generated is collected by means of brushes and slip rings on the revolving member, called the *rotor*. Such an arrangement is economical for the small low-voltage generators. Rotating armature type alternators are built only in small ratings upto about 200 or 250 kVA because the voltage generated is comparatively low and the current to be collected by the brushes small, no difficulty being experienced in collecting such a current. Such machines are suitable for small power plants, isolated lighting plants, where medium or small size machines are required.

Practically all medium and large machines are always constructed with revolving field. The advantages of stationary armature and revolving field system are given below:

1. It is easier to insulate stationary armature winding for very high voltage e.g. as high as 33,000 volts.
2. The load circuit can be connected directly with the fixed terminals of the stator (or armature winding) without passing through slip rings and brushes.
3. The armature winding can be more easily braced in a rigid frame to prevent any deformation which could be developed by the mechanical stresses set up due to short-circuit currents and the high centrifugal forces brought into play.
4. The armature winding is cooled more readily because the stator core can be made large enough with many air passages or cooling ducts for forced air circulation.
5. Only two slip rings are required for the supply of direct current to the rotor and since the exciting current is to be supplied at low voltage of 125 or 250 V, there is no difficulty in insulating them.
6. Since the exciting current is relatively small, the slip rings and the brush gear need be of only light construction.
7. Due to simple, light and robust construction of the rotor, higher speed of rotating dc field is possible.

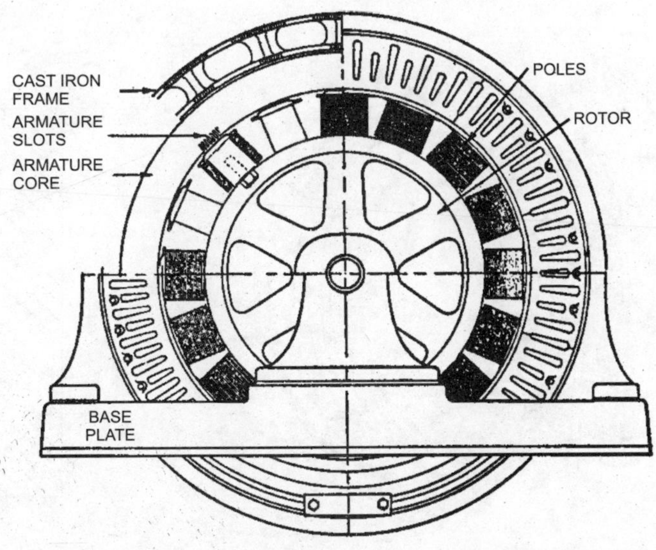

Fig. 11.1. ***Revolving Field Salient Pole Type Alternator***

Since in dc machines, the commutator makes necessary that either the armature should be rotating one, or brushes should revolve with the field, so it is convenient to have armature as a rotating member in dc machines.

Revolving field alternator with essential parts is shown in Fig. 11.1.

The synchronous machines may be classified as (*i*) salient pole machines and (*ii*) cylindrical rotor machines depending upon the type of construction used for the rotor. The salient pole construction is used for generators and motors of all ranges of output and upto all but the higher speeds. Medium- and large-sized generators for the highest speeds are of the cylindrical rotor type.

The synchronous generators, based on the type of prime movers to which they are mechanically coupled, may be classified as (*i*) turbo-generators, (*ii*) hydrogenerators and (*iii*) diesel engine driven generators.

11.2. CONSTRUCTION

Synchronous machine consists essentially of two parts namely the armature (or stator) and the field magnet system (or rotor).

1. Stator. The armature is an iron ring, formed of laminations of special magnetic iron or steel alloy (silicon steel) having slots on its inner periphery to accommodate armature conductors and is known as *stator*. The whole structure is held in a frame which may be of cast iron or welded steel plates. The field rotates in between the stator, so that flux of the rotating field cuts the core of the stator continuously and therefore causes eddy current loss in the stator core. To minimize the eddy current loss, the stator core is laminated. The laminations are stamped out in complete rings (for smaller machines) or in segments (for larger machines) and insulated from each other with paper or varnish. The stampings also have openings which make axial and radial ventilating ducts to provide efficient cooling. A general view of stator and frame is shown in Fig. 11.2.

Slots provided on the stator core are mainly of two types (*i*) open slots and (*ii*) semi-closed slots, as shown is Figs 11.3 (*a*) and 11.3 (*b*) respectively.

The open slots are more commonly used because the coils can be form-wound and insulated prior to being placed in the slots giving least expenditure and more satisfactory winding method. This type of slots also facilitate in removal and replacement of defective coils. But this type of slots have disadvantage of distributing the air gap flux into branches or tufts which tends to produce ripples in the emf wave. The semi-closed type slots are better in this respect but do not permit the use of form-wound coils. Totally closed slots are rarely used.

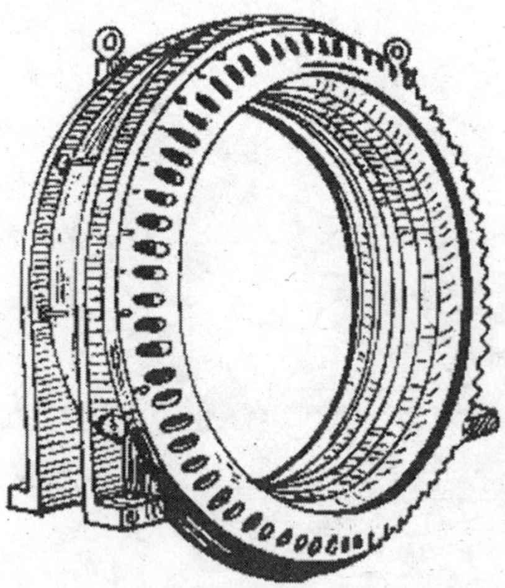

Fig. 11.2

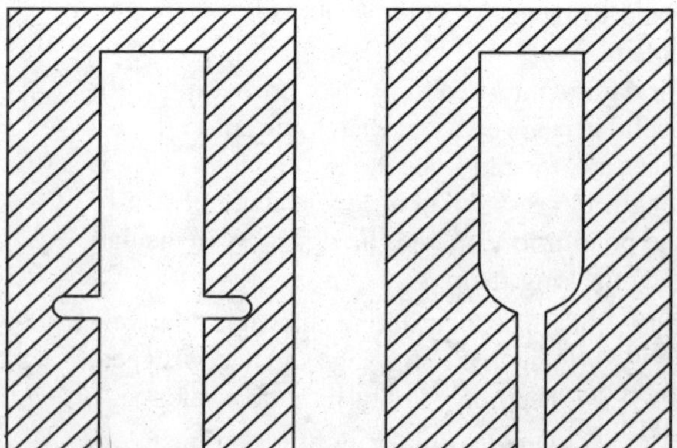

(a) Open Slot *(b) Semi-Closed Slot*

Fig. 11.3

2. Rotor. The field system is just like that of a dc generator which is excited from a separate source of 125 or 250 V dc supply. The excitation is usually provided from a small dc shunt or compound generator, *known as an exciter*, mounted on the shaft of the alternator itself. The field system of the alternator is rotated within the armature ring and is known *as rotor*. The exciting current is supplied to the rotor through two slip rings and brushes. The polarities of the field produced are alternately North and South.

The power rating of the exciter is ordinarily 0.3 to 1% of power rating of synchronous generator. The rated voltage of the exciter is usually between 125 and 250 volts.

Rotors are of two types namely (*i*) salient pole type and (*ii*) smooth cylindrical type or non-salient pole type. Both types are shown in Fig. 11.4.

Salient Pole Type. The rotor of this type is used almost entirely for low and moderate speed alternators, since it is least expensive and provides ample space for the field ampere-turns. Salient poles cannot be employed in high speed generators on account of very high peripheral speed (100 to 170 metres per second) and the difficulty of obtaining sufficient mechanical strength.

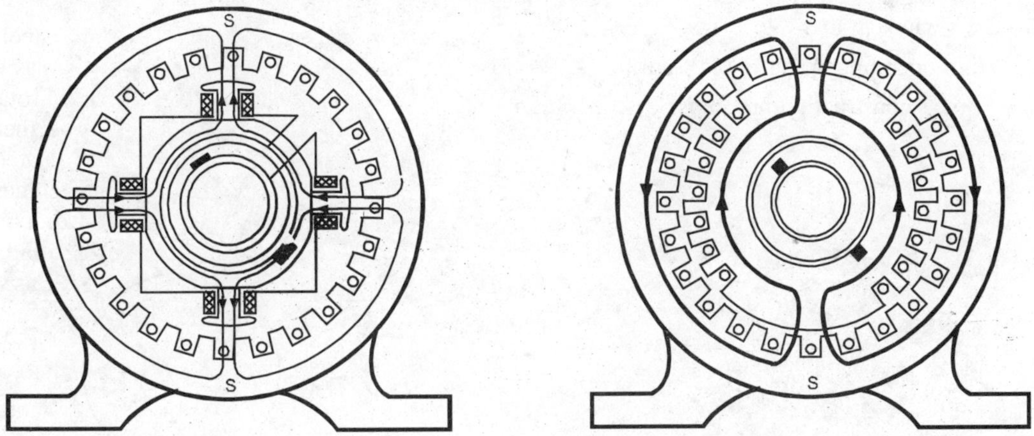

(a) Salient Pole Type Field Structure *(b) Non-Salient Pole Type Field Structure*

Fig. 11.4

The salient poles are made of thick steel laminations riveted together and are fixed to rotor by a dovetail joints. The pole faces are usually provided with slots for damper windings. These dampers are useful in preventing the hunting. The pole faces are so shaped that the radial air-gap length increases from the pole centre to the pole tips so that the flux distribution over the armature is sinusoidal and wave-form of generated emf is sinusoidal. The field coils are placed on the pole pieces and connected in series. The ends of the field windings are connected to a dc source through slip rings carrying brushes and mounted on the shaft of the field structure.

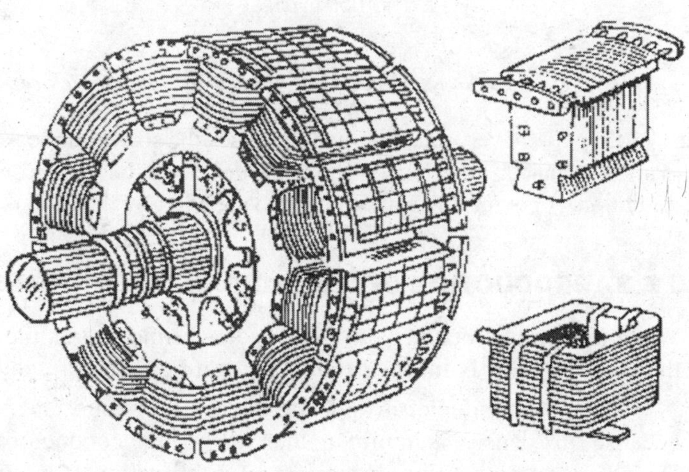

Fig. 11.5. ***Salient Pole Field Structure***

The salient pole field structure has the following special features :

(*i*) They have large diameter and short axial length.

(*ii*) The pole shoes cover about 2/3 of pole pitch.

(*iii*) These are employed with hydraulic turbines or diesel engines. Thc speed is 100 to 375 rpm.

Smooth Cylindrical or Non-Salient Pole Type. The rotors of this type are used in very high speed alternators (driven by steam turbines). To reduce the peripheral velocity, the diameter of the rotor is reduced and axial length is increased. Such rotors have two or four poles.

It consists of a cylindrical steel forging which is suitably worked mechanically and treated thermally. The forging has radial slots in which the field copper, usually in strip form, is placed. The coils are held in place by steel or bronze wedges and coil ends are fastened by metal rings. The slots over certain portions of the core are omitted to form pole faces, as shown in Fig. 11.7. Each slot is provided with a ventilation hole at the bottom. General view of an assembled smooth cylindrical rotor is shown in Fig. 11.6.

The non-salient field structure has the following special features :

(*i*) They are of small diameter and of very long axial length.

(*ii*) Less windage loss.

(*iii*) High operating speed (3,000 rpm).

(*iv*) Robust construction and noiseless operation.

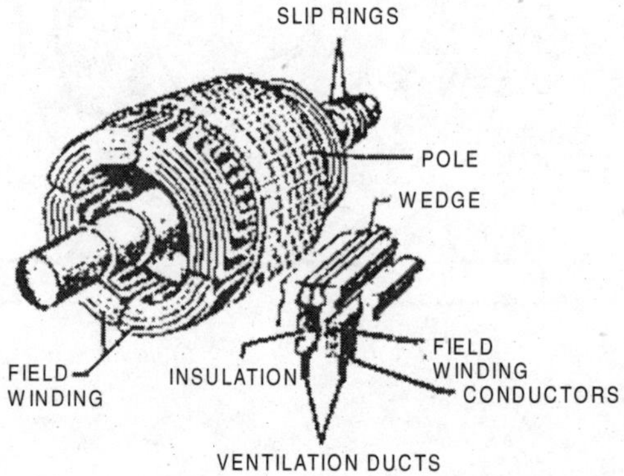

Fig. 11.6. ***Non-salient Pole (Cylindrical) Field Structure***

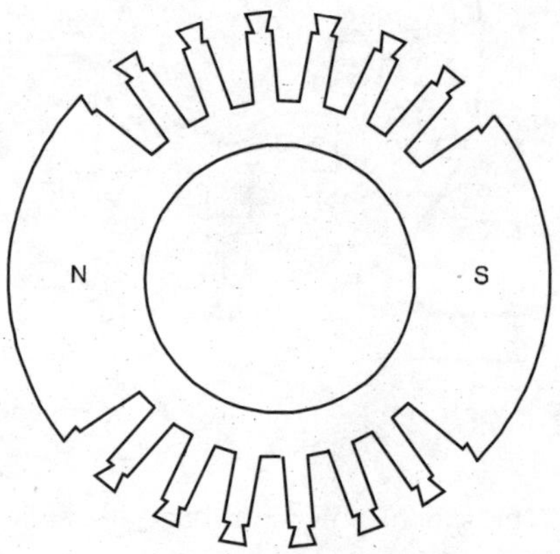

Fig. 11.7

Air gap. A very small air gap increases the stray-load loss; also the eccentricity in the air gap can result from mechanical difficulties. This also increases the synchronous reactance. A large air gap needs larger excitation. A compromise has to be made. Generally the ratio of air gap to the pole pitch is between 0.008 to 0.02.

11.3. PRODUCTION OF SINUSOIDAL ALTERNATING EMF

When the rotor is rotated by means of some prime mover, the armature conductors cut the magnetic flux, therefore, an emf is induced in the armature conductors, due to electromagnetic induction effect.

When the conductor is opposite the neutral planes, as at A, C and E, induced emf in it is minimum because flux density is minimum there. When the conductor is opposite the middle of the poles, as at B, D and F, induced emf in it is maximum, the direction of induced emf depends upon the name of the pole influencing the conductor at any given instant. Thus an alternating emf is induced in the conductors which goes through one complete cycle in an angular distance equal to twice of the pole pitch. Though the shape of the wave of alternating induced emf is not exactly the sinusoidal but taken as sinusoidal.

11.4. FREQUENCY OF INDUCED EMF

As mentioned above, an emf induced in conductor goes through one complete cycle in an angular distance equal to twice the pole pitch.

If the number of poles on rotor of an alternator is P, then P/2 cycles of emf are completed in one revolution. The number of cycles per second, *known as frequency f,* will be equal to the product of number of cycles of emf per revolution and number of revolutions made per second by the rotor.

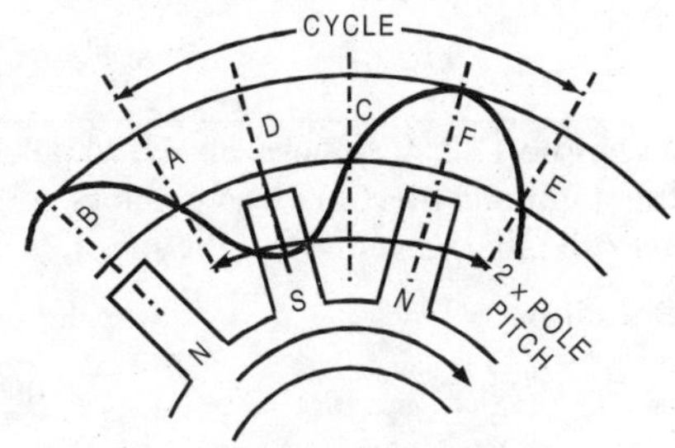

Fig. 11.8

i.e. Frequency, $f = \frac{P}{2} \times n$

where n is the number of revolutions made per second by rotor

$$\text{or } f = \frac{P}{2} \times \frac{N}{60} = \frac{PN}{120} \quad ...(11.1)$$

where N is the number of revolutions made per minute by the rotor.

Hence frequency of induced emf or current induced in stator conductors depends upon the number of poles and speed of rotor.

Example 11.1. A 6-pole ac generator is running and producing the frequency of 60 Hz. Calculate the revolutions per minute of the generator. If the frequency is reduced to 20 Hz, how many number of poles will be required if the generator is to be run at the same speed.

Solution :

Frequency generated, $f = 60$ Hz

Number of poles, P = 6

$$\text{Speed, N} = \frac{120\,f}{P} = \frac{120 \times 60}{6} = 1{,}200 \text{ rpm } \textbf{Ans.}$$

When frequency, $f' = 20$ Hz

$$\text{Number of poles required, P}' = \frac{120\,f'}{N} = \frac{120 \times 20}{1{,}200} = 2 \textbf{ Ans.}$$

11.5. EMF EQUATION

Let Z_p = Number of conductors in series per phase

P = Number of poles

ϕ = Useful flux per pole in webers

N = Rotational speed in rpm

and f = Frequency in hertz and is equal to $\frac{NP}{120}$

The flux cut by any conductor while passing from the centre of one inter polar gap to the centre of the next is ϕ webers and since during the movement, the emf wave completes half cycle *i.e.* the time taken is $\frac{1}{2f}$ seconds, therefore, the average rate of cutting the flux,

$$\frac{d\phi}{dt} = \frac{\phi}{1/2f} = 2f\phi \text{ Wb/s}$$

Hence average emf induced in each conductor = $2f\phi$ volts

Average emf per phase, E_{av}/phase = Number of conductors in series/phase arranged in one slot/pole × average emf induced per conductor

$$= Z_p \times 2f\phi = 2\,T \times 2f\phi = 4\,\phi f T \text{ volts} \qquad \because Z_p = 2\,T \text{ (turns)}$$

For distributed winding the average value of emf per phase will be K_d times the above value

$$i.e.\ E_{av} / \text{phase} = 4\,K_d\,\phi f T \text{ volts}$$

For short-pitched winding the true average value of emf per phase will be K_p time above value

$$i.e.\ E_{av}/\text{phase} = 4\, K_d\, K_p\, \phi f T \text{ volts}$$

$$\text{and } E_{rms}/\text{phase} = \text{Form factor, } K_f \times E_{av}/\text{phase} = 4\, K_f\, K_d\, K_p\, \phi f T \text{ volts}$$

$$\text{For sinusoidal wave of emf, } K_f = 1.11$$

$$E_{rms}\,/\text{ phase} = 4.44\, K_d\, K_p\, \phi f T \text{ volts} \quad ...(11.2)$$

For full-pitched and concentrated windings, $K_p = 1$ and $K_d = 1$

If the alternator is star-connected, as usually,

$$\text{Line voltage, } E_L = \sqrt{3}\ E_{rms}/\text{phase} = \sqrt{3} \times 4.44\, K_d\, K_p\, \phi f T \text{ volts} \quad ...(11.3)$$

Example 11.2. A 3-phase, 50 Hz, 20 pole salient pole alternator with star-connected stator winding has 180 slots on the stator and there are 8 conductors per slot. The flux per pole is 25 mWb and is sinusoidally distributed. The coils are full pitch. Calculate (*i*) the speed, (*ii*) the generated emf per phase and (*iii*) line emf.

Solution :

$$\text{Flux per pole, } \Phi = 25 \text{ mWb} = 0.025 \text{ Wb}$$

$$\text{Frequency, } f = 50 \text{ Hz}$$

$$\text{Number of armature conductors, } Z = 180 \times 8 = 1{,}440$$

$$\text{Number of armature conductors per phase} = \frac{1{,}440}{3} = 480$$

$$\text{Number of turns per phase, } T = \frac{480}{2} = 240$$

$$\text{Number of poles, } P = 20$$

$$(i)\ \text{Speed, } N = \frac{120 f}{P} = \frac{120 \times 50}{20} = 300 \text{ rpm } \textbf{Ans.}$$

$$(ii)\ \text{Generated emf per phase} = 4.44\, \Phi f T \text{ volts} = 4.44 \times 0.025 \times 50 \times 240 = 1{,}332 \text{ V } \textbf{Ans.}$$

$$(iii)\ \text{Line emf} = \sqrt{3} \times 1{,}332 = 2{,}307 \text{ V } \textbf{Ans.}$$

11.6. SYNCHRONOUS MOTORS

A synchronous motor has the same relationship to an alternator as a dc motor has to a dc generator *i.e.* if an alternator is supplied ac power it is capable of rotating as a motor and doing mechanical work. If the mechanical power supplied to a rotating alternator is removed while the dc field remains energized, and an ac supply is then connected across the armature terminals, torque will be developed and the alternator will continue to rotate at a speed determined by the ac supply frequency and number of poles on the synchronous machine. Changes in mechanical load within the machine's rating will not cause a change in speed.

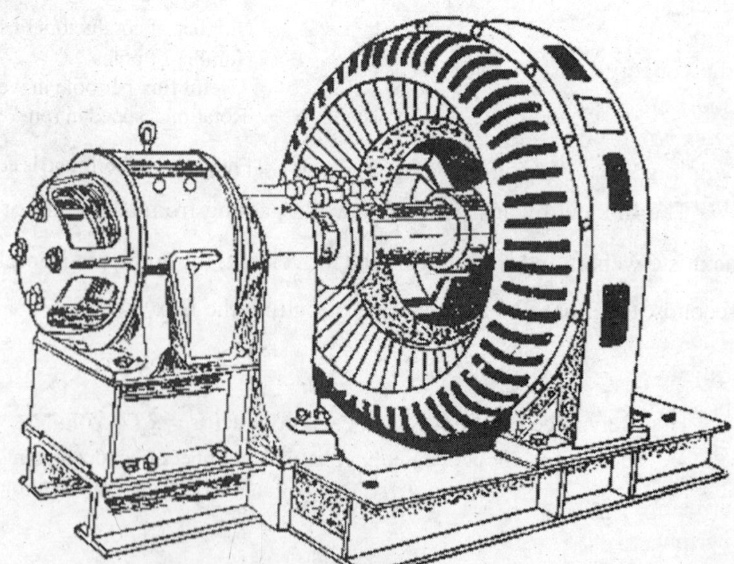

Fig. 11.9. ***Synchronous Motor With Directly Connected Exciter***

In case of a dc motor the field and armature, both, are supplied from dc supply mains, since both require direct current. But in case of a synchronous motor the field structure is to be energized by direct current, as in case of an alternator, whereas the armature winding is connected to a 3-phase ac supply mains. Such a machine, therefore, requires two sources of supply, viz., an ac source to supply the power for driving the armature and a dc source to excite the field. DC excitation may be provided either from an exciting plant dc system or, if no direct current is available, from separate dc generators. Direct-connected exciters are frequently found on high speed motors. These are overhung from specially constructed motor brackets with the exciter armature mounted on an extension of the motor shaft. On low-speed machines, belted exciters are sometimes employed, notably on air compressor drives. Separate motor-generator sets are widely used for excitation of low-speed synchronous motors. Where a group of synchronous motors is installed, it is sometimes economical to use a single motor-generator set for exciting the entire group, although greater flexibility is obtained when each motor is provided with its own exciter. Synchronous motor with a directly connected exciter is shown in Fig. 11.9.

Essentially, the construction of a polyphase synchronous motor is the same as that of an alternator. Certain features of its design may be varied to suit the particular requirements. A synchronous motor may have either a revolving armature or a revolving field, but the revolving field construction is the more common.

The essential parts of a three-phase synchronous motor are (*i*) laminated stator core with 3-phase armature winding (*ii*) rotating field structure complete with damper winding and slip rings (*iii*) brushes and brush holders and (*iv*) two end shields to house the bearings that support the rotor shaft.

The stator core and windings of a synchronous motor are similar to those of a 3-phase alternator. The leads from the stator windings terminate in a terminal box usually mounted on the side of the motor frame. The rotor of the synchronous motor usually has salient field poles connected to give alternate polarity. The number of rotor field poles must be the same as that of stator field poles. The field circuit leads are brought out to two slip rings mounted on the shaft. Carbon brushes mounted in brush holders make contact with the two slip rings. The terminals of the field circuit are brought out from the brush holders to a second terminal box mounted on the motor frame.

In the most common type of field construction, the poles are somewhat different from those of the alternator. In addition to the exciting winding, a synchronous motor often is provided with a squirrel cage winding on the rotor. The slots of this winding are in the pole faces and are parallel to the shaft. The ends of the copper or bronze bars embedded in the slots are short circuited at both ends by conducting rings. This short-circuited winding on the rotor is known as the *cage* winding, the *damper* winding, the *amortisseur* winding, or sometimes, to describe its function, the *starting* winding.

11.7. PRINCIPLE OF OPERATION OF SYNCHRONOUS MOTOR

The action of the motor may be followed by referring to Fig. 11.10, which illustrates a 3-phase, 2-pole synchronous machine. The rotor has two poles and the stator has two poles per phase. In practice it is usual to have salient poles on the rotor, but the armature winding is housed in slots in the concave periphery of the stator. In the Fig. salient type stator poles have been shown for convenience only.

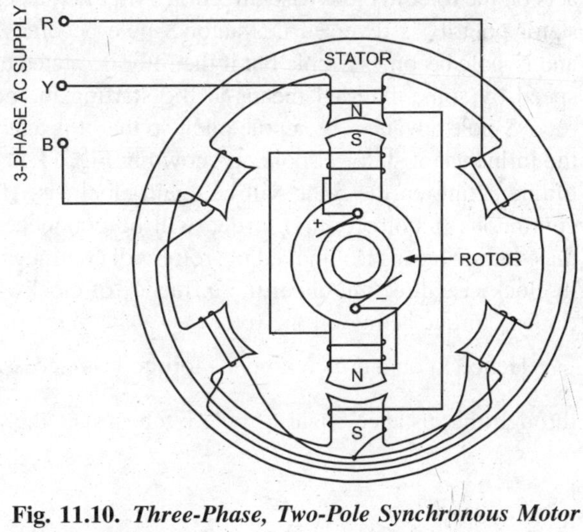

Fig. 11.10. ***Three-Phase, Two-Pole Synchronous Motor***

As the rotor is excited from dc supply, so the poles of the rotor retain the same polarity throughout but the polarity of the stator poles changes as it is connected to an ac supply mains.

First of all let us consider the rotor as stationary and in the position shown in Fig. 11.11 (*a*). At this instant the rotor S-pole is attracted to the stator N-pole and, therefore, the rotor tends to rotate in clockwise direction. After half a period $\left(i.e. \text{ after } \frac{T}{2} \text{ second whereas } T = \frac{1}{f}\right)$ the polarity of stator poles is reversed but the polarity of the rotor poles remains the same [Fig. 11.11 (*b*)]. At this instant, the rotor S-pole is repelled by the stator S-pole, being similar in nature and, therefore, rotor tends to rotate in counter-clockwise direction. Thus the torque acting on the rotor of a synchronous motor is not unidirectional but pulsating one and due to inertia of the rotor, it will not move in any direction. So the *synchronous motor has got no self-starting torque*.

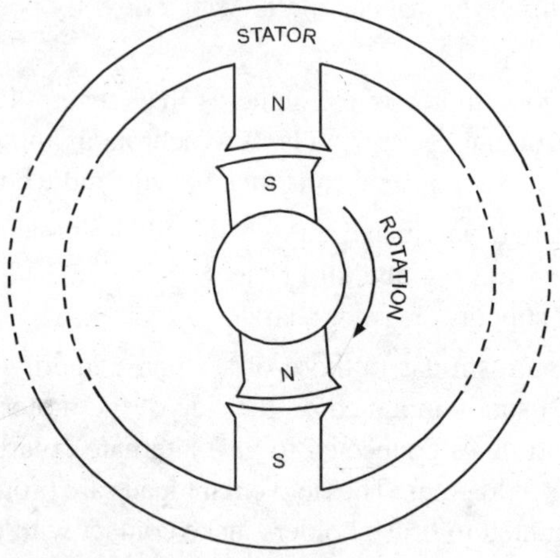

(*a*)

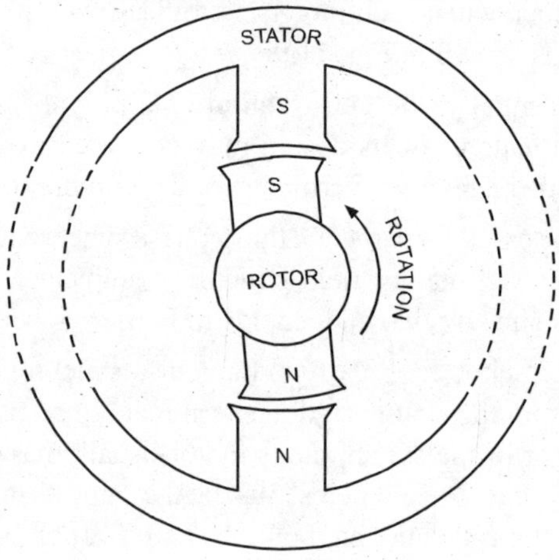

(*b*)

Fig. 11.11

Now let us consider the rotor rotating in clockwise direction, which is possible by external means, and in the position shown in Fig. 11.11 (*a*). At this instant, as mentioned above, the rotor S-pole is attracted to stator N-pole and so the torque acts on the rotor in clockwise direction. After half a period the stator polarity is reversed *i.e.* stator S-pole becomes N-pole and N-pole becomes S-pole but if the rotor is rotated at such a speed by some external means at the starting moment that rotor S-pole advances by a pole pitch so that it is again under the influence of stator N-pole, as shown in Fig. 11.11 (*c*), the torque acting on the rotor will be again clockwise. Hence a continuous (unidirectional) torque will be obtained. Now if the external means is removed the rotor will continue to rotate in clockwise direction under the influence of clockwise continuous torque acting on the rotor.

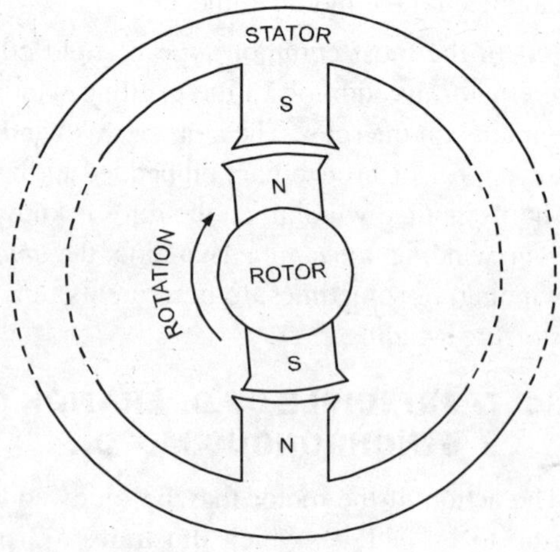

(*c*)

Fig. 11.11

Hence to obtain a continuous torque it is necessary that the rotor rotates at such a speed that it moves through the distance equal to pole pitch in half the period i.e. in $\frac{T}{2}$ or $\frac{1}{2f}$ second where f is the supply frequency. If there are P poles then to complete one revolution it should take $\frac{1}{2f}$ second or the rotor should rotate at a speed of $\frac{2f}{P} \times 60$ revolutions per minute.

Hence to obtain a continuous torque the rotor should rotate at synchronous speed given by the expression $N_s = \frac{120f}{P}$.

11.8. EFFECT OF LOAD ON A SYNCHRONOUS MOTOR

All the electric motors, whether dc or ac act as generators at the same time they are motoring. This is true because an emf is always induced opposing the applied voltage when the machine is motoring. In a dc machine, the rotating armature conductors cut magnetic flux under the stationary field poles; in an ac machine, as in an induction motor or a synchronous motor, the rotating magnetic field cuts the stationary armature conductors of the stator. In fact, it is this very opposing induced emf that limits the armature current to a value that is just enough to take care of the power requirements of the motor for a particular load.

When the load on a dc shunt motor is increased, speed of motor falls slightly. This reduces the induced emf, called the back emf, so that resultant voltage across the armature $(V - E_b)$ increases, which in turn increases the armature current, since armature current is given by the expression $I_a = \left(\frac{V - E_b}{R_a}\right)$, to supply extra torque and power required by the increased load. In the case of synchronous motor, however, the speed, as we have already discussed, remains constant irrespective of magnitude of load, therefore, method of drawing increased current from supply mains on synchronous motor cannot be same as that of a dc shunt motor.

The behaviour of a synchronous motor on a load is discussed below:

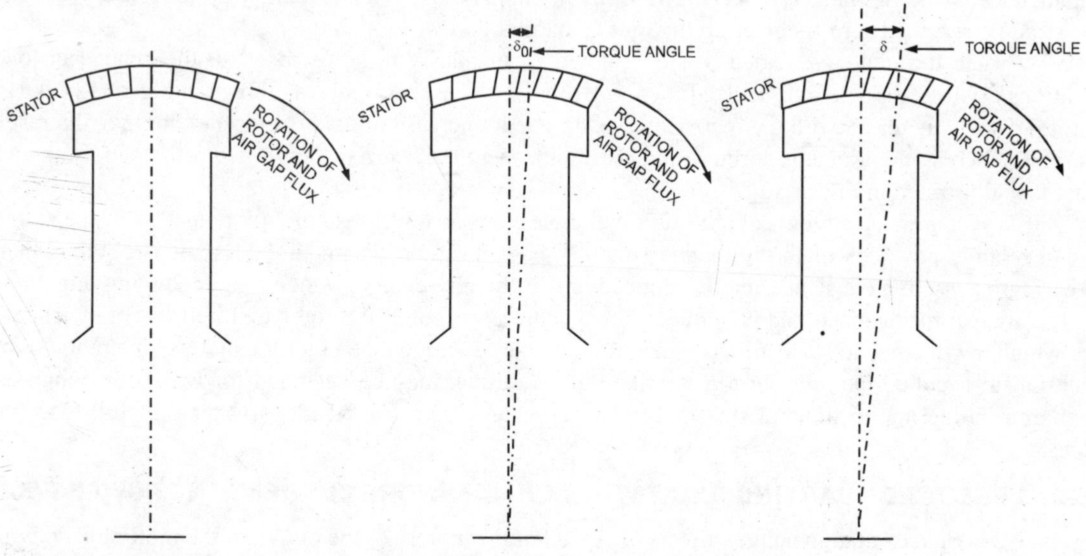

Fig. 11.12

Since the synchronous motor has got no self starting torque, first it is driven by some external means as an alternator and is synchronised with the supply. At the instant of synchronising the induced emf across the stator circuit of the motor (also called the back or counter-emf or excitation voltage) is equal to the applied line voltage V but opposite in direction, as shown in Fig. 11.13 (*a*) vectorially. The relative positions of a stator pole and dc field pole of a synchronous motor at the synchronising instant are shown in Fig. 11.12 (*a*). The centre lines of the two poles coincide. Since V = E, hence the resultant voltage is zero and so is the armature current. Motor intake is zero, as there is neither load nor losses to be met by it. Under these conditions the motor is said to be *floating*.

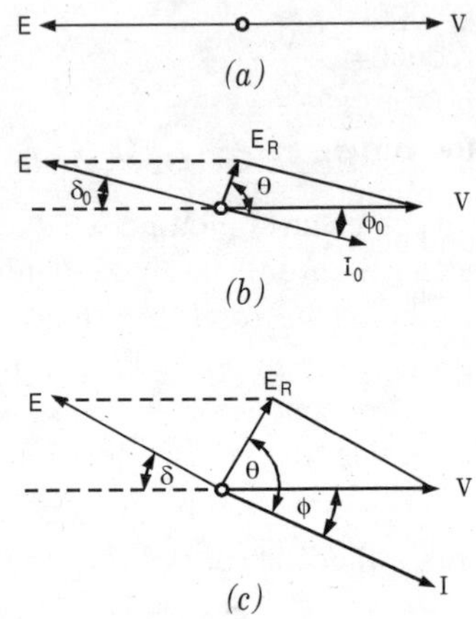

Fig. 11.13

After synchronising, the external means is removed and the motor tends to slow down. There is no change in speed; there is merely a shift in relative positions of the two poles, both of which continue to rotate at synchronous speed. The rotor pole falls back a little relative to the stator pole, as shown in Fig. 11.12 (*b*). The angular displacement between the rotor and stator poles δ is called the *torque angle*. Now the applied voltage and induced emf are no longer in direct opposition since induced emf at this instant lags behind the induced emf at the synchronising instant, say by an angle δ_0, as shown in Fig. 11.13 (*b*). The resultant voltage E_R (the resultant of applied voltage **V** and induced emf **E**) causes a current $I_0 = \frac{E_R}{Z_s}$ to flow in the stator winding; where Z_s is the synchronous impedance per phase of the armature circuit. I_0 lags E_R by an angle θ, called the *internal angle*, where $\theta = \tan^{-1}\frac{X_s}{R_e}$; X_s and R_e being the stator synchronous reactance and effective resistance respectively. Hence on no load motor draws power equal to $VI_0 \cos\phi_0$ (for one phase) from the supply mains which is sufficient to make the motor running continuously at synchronous speed.

At no load the torque angle δ_0 is very small, so the resultant voltage E_R acting across the armature circuit is small, due to which a small current is drawn from the supply mains. At no load, the torque angle δ_0 is so small that it can be assumed zero (neglecting the motor internal losses).

Now when the motor is loaded, it slows down momentarily to adjust itself to the change in loading conditions, so the rotor pole falls back a little more relative to the stator pole, as shown in Fig 11.12 (*c*). Hence the torque angle increases with the increase in load. Due to increase in load (or torque) angle δ, the resultant voltage E_R across the armature circuit increases, and, therefore, current drawn from the supply mains increases, as illustrated in Fig 11.13 (*c*).

Thus a synchronous motor is able to supply increased mechanical load, not by reduction in speed, but by shift in relative positions of the rotor and rotating magnetic field. From Fig. 11.13 it is obvious that for increasing load with a constant value of induced emf E the phase angle ϕ increases in lagging direction.

If too great mechanical load is applied to a synchronous motor, the rotor will pull out of synchronism, after which it will come to standstill. The maximum value of torque that a motor can develop without loosing synchronism is called the *pull-out torque*, like pull-in torque, may be designed for widely varying values, depending on the applications of the motor. These values range from 125% to 350% of full-load torque values.

11.9. EFFECTS OF VARYING EXCITATION ON ARMATURE CURRENT AND POWER FACTOR

The change in field excitation neither affects the speed of the motor nor the output of the motor but does affect the power factor and consequently armature current for constant supply voltage and constant input power.

The operation of a synchronous motor at a constant supply voltage demands nearly constant flux, as mentioned in the foregoing Art. Both the dc source and ac source cooperate in establishing this resultant flux. When the dc excitation is weak, the flux produced by it is weak and needs stengthening, so the stator draws the current lagging behind the supply voltage by 90° so as to magnetise the air gap to the extent needed to provide the necessary resultant flux. The magnetising component of current being a large part of the total input current results in low lagging power factor. For constant applied voltage and constant input power, the active component of current drawn from the supply mains should be constant and so because of large wattless (magnetising) component of current, armature current I is much more. With the increase in excitation to 100%, the magnetising current drawn from the 3-phase ac supply mains is reduced *i.e.* the magnetising component becomes a smaller part of the total input current and so the power factor improves. This results in reduction of armature current I drawn from 3-ϕ ac supply mains. If the excitation is further increased such that power factor becomes unity, the stator draws only energy or active current, and the dc field circuit supplies all the current necessary to magnetize the rotor field. For unity power factor, the armature current drawn will be minimum. On increasing the dc excitation further, the field produced by it becomes stronger than it is required and needs de-magnetization. So the motor draws the leading current so as to demagnetise the main field. This makes the power factor leading and lesser than unity and so the armature current increases.

The effect of variation in field excitation on armature current and power factor can also be explained as follows.

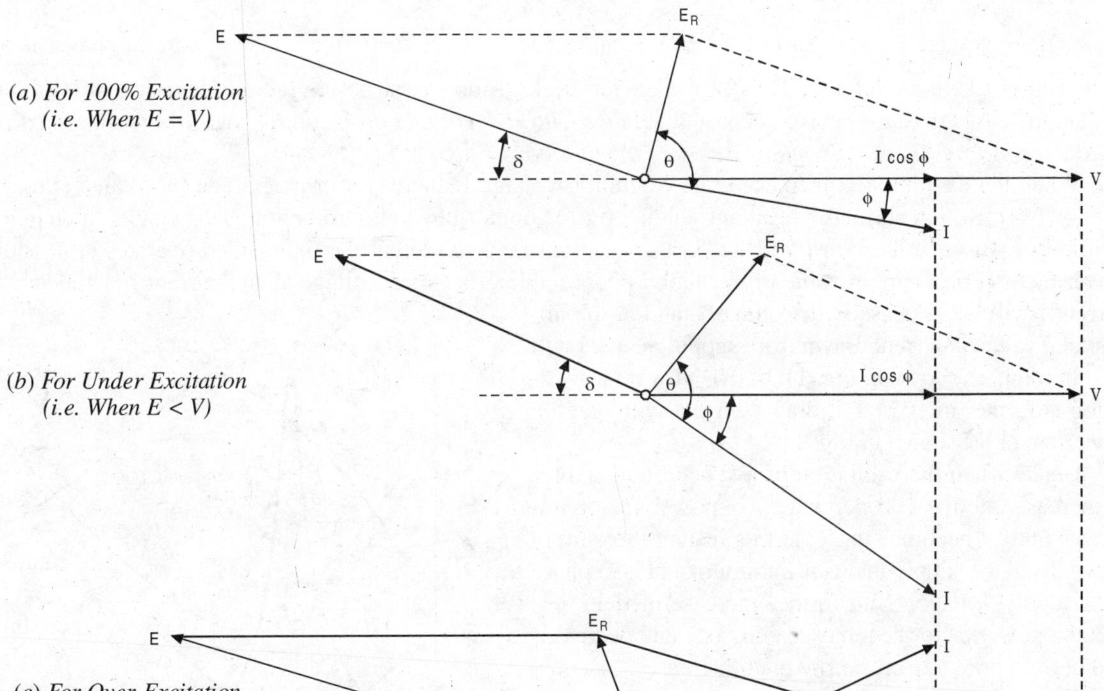

Fig. 11.14

Consider a synchronous motor loaded with a constant mechanical load and operating at a constant supply voltage. When the excitation is 100% *i.e.* when E = V, armature current I lags behind the resultant voltage E_R (phasor sum of V and E) by a fixed angle θ, given as $\theta = \tan^{-1} \frac{X_s}{R_e}$ and lags behind the applied voltage by an angle φ, power factor angle [Fig. 11.14 (*a*)]. X_s is the synchronous reactance per phase, R_e is the effective resistance of armature per phase, θ is the *internal angle* and δ is the load or torque angle. The power drawn per phase from the 3-φ ac supply mains is V I cos φ. Since applied voltage and input power or load is constant, the active component of the current drawn from the supply mains, I cos φ remains constant for constant mechanical load (neglecting internal losses). As the speed of the motor remains constant so increase in dc excitation causes increase in excitation voltage E and decrease in excitation causes reduction in excitation voltage E.

When the excitation is reduced the excitation voltage E decreases in magnitude to the value shown in Fig 11.14 (*b*) which shifts the resultant voltage E_R in clockwise direction and since the phase angle θ between the resultant voltage E_R and current I, being equal to $\tan^{-1} \frac{X_s}{R_e}$, is fixed, the armature current phasor is also shifted in clockwise direction. Thus power factor angle φ increases, power factor decreases and armature current increases because I cos φ is constant. *Hence with reduction in dc excitation synchronous motor draws more current from the supply mains at lower lagging power factor*.

When the excitation is increased, the excitation voltage E increases in magnitude to a value shown in Fig. 11.14 (*c*) which shifts the resultant voltage E_R in counter-clockwise direction. This results in change in position of phasor of current I with respect to terminal voltage so that the motor then operates at a leading power factor. For a certain value of excitation phase angle φ between voltage V and current I is zero *i.e.* the current I will be in phase with voltage V and at this instant the value of current drawn from supply mains, I will be minimum, as shown in Fig. 11.14 (*d*). Now if the excitation is further increased, it draws current with leading power factor, as shown in Fig. 11.14 (*c*).

Hence when the excitation of a synchronous motor is increased above 100 percent, first power factor improves until it becomes unity; at this instant the current drawn from the supply mains is minimum and is in phase with supply voltage. With further increase in field current the power factor becomes leading one and decreases and current drawn from supply main increases.

The variation of current and power factor of a synchronous motor with a variation of field current (excitation) and for a constant load is shown in Fig. 11.15. Because of their shape, excitation-armature current curves are called the *V-curves* of a synchronous motor.

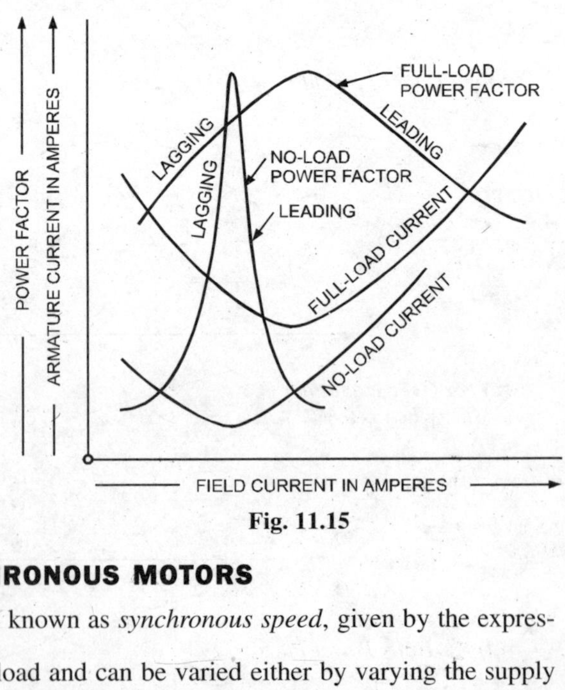

Fig. 11.15

11.10. MAIN CHARACTERISTICS OF SYNCHRONOUS MOTORS

1. Synchronous motor operates only at one speed, known as *synchronous speed*, given by the expression $N_s = \frac{120 f}{P}$. The speed is independent of load and can be varied either by varying the supply frequency or by changing the number of poles.
2. It is, inherently, not a self starting motor and, therefore, before connecting it to an ac supply it is to be run upto the synchronous speed by some external means.

3. It can be operated under a wide range of power factor both leading and lagging by controlling its excitation current.
4. On no load synchronous motor draws very little current from the supply mains to meet the internal losses. With fixed excitation the input current increases with the increase in load. After the input current reaches maximum no further increase in load is possible. If the motor is further loaded, the motor will stop.
5. If field excitation is small, the stator inductance will be more so the motor draws lagging current and, therefore, power factor is low. If the excitation is increased the stator flux is neutralized by the rotor flux and, therefore, power factor becomes unity. In this position, the current and voltage are in phase and motor current will be minimum.

If excitation is further increased rotor flux increases. It neutralizes the stator flux and causes leading current. This property makes the motor extremely useful as a power factor corrector. By taking leading current it compensates the lagging current drawn by other equipment from the supply mains and thus brings the total current drawn from the supply mains more nearly in phase with the voltage.

11.11. MERITS AND DEMERITS OF SYNCHRONOUS MOTORS

Merits. (*i*) The ease with which the power factor can be controlled. An overexcited synchronous motor having a leading power factor can be operated in parallel with induction motors and other power apparatus operating at lagging power factors, thereby improving the power factor of the supply system.

(*ii*) The speed is constant and independent of load. This characteristic is mainly of use when the motor is required to drive another alternator to generate a supply at a different frequency as in frequency changers.

(*iii*) Electromagnetic power varies linearly with the voltage.

(*iv*) These motors can be constructed with wider air gaps than induction motors, which make them better mechanically.

(*v*) These motors usually operate at higher efficiencies, especially in the low speed unity pf ranges.

Demerits (*i*) The cost per kW output is generaly higher than that of an induction motor.

(*ii*) It requires dc excitation which must be supplied from external source.

(*iii*) The synchronous motor is inherently not self-starting motor and needs some arrangement for its starting and synchronizing.

(*iv*) It cannot be used for variable speed jobs as there is no possibility of speed adjustment.

(*v*) It cannot be started under load. Its starting torque is zero.

(*vi*) It has a tendency to hunt.

(*vii*) It may fall out of synchronism and stop when overloaded.

(*viii*) Collector rings and brushes are required.

(*ix*) For some purposes synchronous motors are not desirable as for driving shafts in small workshops having no other power available for starting and in cases where frequent starting or strong starting torque is required.

11.12. APPLICATIONS OF SYNCHRONOUS MOTORS

Synchronous motors are rarely used below 40 kW output in the medium speed range because of their much higher initial cost in comparison to that of induction motors. In addition they need a dc excitation source, and the starting and control devices are usually more expensive—especially where automatic operation is required.

However, there is a kW output and speed range where the disadvantage of higher initial cost vanishes, even to the point of putting the synchronous motors to the advantage. Where low speeds and high kW outputs are involved the induction motors are no longer cheaper because they need large amounts of iron in order to restrict the air gap flux density to 0.7 T. In the synchronous machines, on the other hand, a value twice this figure is permissible because of the separate excitation.

The various classes of service for which synchronous motors are employed may be classified as (*i*) power factor correction (*ii*) voltage regulation and (*iii*) constant speed, constant-load drives.

1. They are used in power houses and substations in parallel to the bus-bars to improve the power factor. For this purpose they are run without mechanical load on them and overexcited.
2. In factories having a large number of induction motors, or other power apparatus operating at lagging power, they are employed to improve the power factor.
3. Such motors are also used to regulate the voltage at the end of transmission lines.
4. Because of the higher efficiency possible with synchronous motors, they can be employed advantageously for the loads where constant speed is required. Typical applications of high-speed synchronous motors (above 500 rpm) are drives such as fans, blowers, dc generators, line shafts, centrifugal pumps and compressors, reciprocating pumps and compressors, constant speed frequency changers, rubber and paper mills etc.

The fields of applications of low-speed synchronous motors (below 500 rpm) are drives such as reciprocating compressors when started unloaded, dc generators, centrifugal and screw type pumps, vacuum pumps, electroplating generators, line shafts, rubber and band mills, ball and tube mills, chippers, metal rolling mills etc. Flywheel is used for pulsating loads.

High power electronic convertors generating very low frequencies made the operation of synchronous motors at ultra low speeds. Synchronous motors in very large sizes (as high as 10 MW size) operating at ultra low speeds are employed to drive crushers, rotary kilns and variable speed ball mills.

HIGHLIGHTS

1. A *synchronous machine* is an ac machine in which the rotor moves at a speed which bears a constant relationship to the frequency of currents in the armature winding.

2. *Alternator* is that machine which converts mechanical power into ac electrical power at a desired frequency. This is also known as a synchronous generator.

3. An *alternator* consists of a stator and a rotor. The stator provides the armature windings whereas the rotor provides the rotating magnetic field.

The rotors are of two types :

(*i*) Salient pole type used almost entirely for low and moderate speed alternators (hydrogenerators or diesel-engine driven generators) being least expensive and providing ample space for field ampere-turns.

(*ii*) Smooth cylindrical or non-salient pole type used in very high speed alternators (turbo-generators), being of smaller diameter and robust construction.

4. The alternators may be rotating armature type or rotating field type but practically all medium and large machines are always constructed with revolving field because of its superiority over the rotating armature type.

5. The frequency of the alternating currents produced is

$$f = \frac{PN}{120} \text{ Hz}$$

6. EMF equation is E_{rms} per phase = $4.44\, K_d\, K_p\, \phi f T$ volts.

7. An alternating field can be replaced exactly by two rotating fields of half its amplitude travelling in opposite directions at synchronous speed. This is the Ferraris principle.
8. When a 3-ϕ supply is given to a 3-ϕ winding, a resultant field is set up which rotates in space at a constant speed called the synchronous speed given as

$$N_s = \frac{120\,f}{P}.$$

9. *Synchronous motor* is identical in construction with the alternator and converts ac electrical power into mechanical power.
10. When 3-phase supply is given to the stator of a 3-phase synchronous motor, a revolving field is produced. This field tries to drag the rotor with it but could not do so owing to rotor inertia. Hence starting torque is not developed. Thus a synchronous motor is inherently not self-starting and, therefore, it is necessary to provide some external means to start it and bring its speed near synchronous speed.
11. Synchronous motor operates only at one speed known as synchronous speed given as $N_s = \frac{120\,f}{P}$. It can be operated under a wide range of power factor both leading and lagging by controlling its excitation current.
12. Synchronous motors are rarely used in small ratings (below 40 kW output) in the medium speed range because of being expensive in comparison to induction motors. Where low speeds and high kW outputs are involved the synchronous motors prove economical.

    Synchronous motors are used for (*i*) constant speed, constant load drives (*ii*) power factor correction and (*iii*) voltage regulation.

EXERCISES

1. The field of a dc machine is usually kept on stator but in synchronous machines it is on rotor. Justify.
2. Give constructional differences between salient pole and cylindrical rotor synchronous machines. Why is cylindrical rotor machine used in thermal power plant and salient pole machine in hydroelectric plant ?
3. Explain the principle of operation of alternator.

   Deduce the relationship between the number of poles, the frequency and the speed of the alternator.
4. Derive the emf equation for a 3-phase synchronous generator.
5. Discuss the main constructional features of cylindrical rotor and salient pole alternators.
6. Explain the principle of operation of a 3-phase synchronous motor.
7. Explain the principle of operation of a 3-phase synchronous motor. Why is starting torque not produced in this motor.
8. Is it possible to vary the speed of a synchronous motor by varying the field excitation or by any other method ? Explain what happens when the field current is (*a*) increased (*b*) decreased.
9. The field excitation of a 3-phase synchronous motor operating on a fixed load is increased from a very low value. With the help of a V-curve or otherwise explain how the magnitude and power factor of the armature current will vary.
10. Give the constructional details of salient pole synchronous machine. Explain 'V' curve and give the applications of synchronous motor.
11. Discuss the advantages and disadvantages of synchronous motors.
12. Explain V-curve and applications of synchronous motor.
13. Explain the working principle of synchronous motor. Draw V-curves and give its applications.
14. Explain principle of operation of synchronous motor. Give its applications.
15. Three phase induction motor is self starting but three phase synchronous motor is not self starting. Explain why ? Give applications of 3-phase synchronous motor.
16. Explain the following:
    (*i*) Synchronous motor always runs at synchronous speed.
    (*ii*) Field winding of synchronous generators is placed on rotor.

17. Write short notes on the following :

(*i*) Constructional features and principle of operation of an alternator.

(*ii*) Construction and working principle of a synchronous motor.

(*iii*) Main characteristics of a synchronous motor.

(*iv*) Applications of synchronous motors.

SHORT ANSWER TYPE QUESTIONS WITH ANSWERS

Q. 1. What factors determine the number of poles of a synchronous motor ?

Ans. The number of poles of a synchronous motor is determined by supply frequency and speed.

Q. 2. A synchronous motor runs at N_S rpm at full load. What will be its speed if the load is reduced to half the full-load ?

Ans. N_s rpm.

Q. 3. Why a 3-phase synchronous motor will always run at synchronous speed ?

Ans. The synchronous motor always runs at synchronous speed because there is an interlocking action between stator and rotor fields of this motor.

Q. 4. How the power factor of a synchronous motor is changed keeping the shaft load undisturbed ?

Ans. By varying the field excitation.

Q. 5. Does the change in excitation affect the synchronous motor speed and power factor ?

Ans. Change in excitation will affect the power factor (not the speed) of the synchronous motor.

Q. 6. What is the effect of increase in excitation of a synchronous motor ?

Ans. When the excitation of a synchronous motor is increased, first power factor improves until it becomes unity and with the further increase in excitation the power factor becomes leading one and decreases.

Q. 7. What is meant by V-curves of a synchronous motor ?

Ans. The curves drawn between armature current and field current for different constant loads are known as V-curves.

Q. 8. What are the advantages of synchronous motor ?

Ans. The advantages of synchronous motor are (*i*) constant operation speed, (*ii*) higher efficiency, especially in the low speed unity pf ranges and (*iii*) capability of being operated under a wide range of power factor both lagging and leading.

❑❑❑

PART–B

ELECTRONICS ENGINEERING

12

CHAPTER

Transducers

12.1. INTRODUCTION

INSIDE THIS CHAPTER

Basically transducer is defined as a device which converts energy or information from one form to another. These are widely used in measurement work because all quantities that need to be measured, cannot be displayed as easily as others. A better measurement of a quantity can usually be made if it may be converted to another form which is more conveniently or accurately displayed. For example, the common *mercury thermometer* converts variations in temperature into variations in the length of a column of mercury. Since the variation in the length of the mercury column is rather simple to measure, the mercury thermometer becomes a convenient device for measuring temperature. On the other hand, the actual temperature variation is not as easy to display directly.

Thus the transducer is a device which provides a usable output in response to specific input measurand which may be physical or mechanical quantity, property or condition. The transducer may be mechanical, electrical, magnetic, optical, chemical, acoustic, thermal, nuclear, or a combination of any two or more of these.

Mostly quantities to be measured are non-electrical such as temperature, pressure displacement, humidity, fluid flow, speed etc., but these quantities cannot be measured directly. Hence such quantities are required to be sensed and changed into some form for easy measurement. Electrical quantities such as current, voltage, resistance, inductance and capacitance etc., can be conveniently measured, transferred and stored and, therefore, for measurement of non-electrical quantities these are to be converted into electrical quantities first and then measured. The function of converting non-electrical quantity into electrical one is accomplished by a device called the *electrical transducer.*

Basically an *electrical transducer* is a sensing device by which a physical, mechanical or optical quantity to be measured is transformed directly, with a suitable mechanism, into an electrical signal (current, voltage or frequency). The production of these signals is based upon electrical effects which may be resistive, inductive, capacitive etc., in nature. Transducers should meet the following requirements :

1. *Ruggedness.* It should be capable of withstanding overload and some safety arrangement should be provided for overload protection.

2. *Linearity.* Its input-output characteristics should be linear and it should produce these characteristics in symmetrical way.

3. *Repeatability.* It should reproduce same output signal when the same input signal is applied again and again under fixed environmental conditions *e.g.,* temperature, pressure, humidity etc.

4. *High Output Signal Quality.* The quality of output signal should be good *i.e.*, the ratio of the signal to the noise should be high and the amplitude of the output signal should be enough.

5. *High Reliability and Stability.* It should give minimum error in measurement for temperature variations, vibrations and other various changes in surroundings.

6. *Good Dynamic Response.* Its output should be faithful to input when taken as a function of time. The effect is analysed as the frequency response.

7. *No Hysteresis.* It should not give any hysteresis during measurement while input signal is varied from its low value to high value and *vice versa*.

8. *Residual Deformation.* There should be no deformation on removal of input signal after long period of application.

12.2. CLASSIFICATION OF TRANSDUCERS

The transducers may be classified in various ways such as on the basis of electrical principles involved, methods of application, methods of energy conversion used, nature of output signal etc.

1. *Primary and Secondary Transducers.* Transducers, on the basis of methods of applications, may be classified into *primary* and *secondary transducers.*

When the input signal is directly sensed by the transducer and physical phenomenon is converted into the electrical form directly then such a transducer is called the *primary transducer.*

When the input signal is sensed first by some detector or sensor and then its output being of some form other than input signal is given as input to a transducer for conversion into electrical form, then such a transducer falls in the category of *secondary transducers.*

For example, in case of pressure measurement, bourdon tube is a primary sensor which converts pressure first into displacement, and then the displacement is converted into an output voltage by an LVDT. In this case LVDT is secondary transducer.

2. *Active and Passive Transducers.* Transducers, on the basis of methods of energy conversion used, may be classified into *active* and *passive transducers.*

Self generating type transducers i.e. the transducers, which develop their output in the form of electrical voltage or current without any auxiliary source, are called the active transducers. Such transducers draw energy from the system under measurement. Normally such transducers give very small output and, therefore, use of amplifier becomes essential. Tacho-generators used for measurement of angular velocity, thermocouples used for measurement of temperature, piezoelectric crystal used for measurement of force fall in this category.

Transducers, in which electrical parameters i.e., resistance, inductance or capacitance changes with the change in input signal, are called the *passive transducers.* These transducers require external power source for energy conversion. In such transducers electrical parameters i.e., resistance, inductance or capacitance causes a change in voltage, current or frequency of the external power source. These transducers may draw some energy from the system under measurement. Resistive, inductive and capacitive transducers fall in this category.

3. *Analog and Digital Transducers.* Transducers, on the basis of nature of output signal, may be classified into analog and digital transducers.

Analog transducer converts input signal into output signal, which is a continuous function of time such as thermistor, strain gauge, LVDT, thermocouple etc.

Digital transducer converts input signal into the output signal of the form of pulses *e.g.* it gives discrete output.

4. ***Transducers and Inverse Transducers.*** Transducer, as already defined, is a device that converts a non-electrical quantity into an electrical quantity. An *inverse transducer* is a device that converts an electrical quantity into a non-electrical quantity.

12.3. STRAIN GAUGES

The strain gauge is basically a device used for measuring mechanical surface strain and is one of the most extensively used electrical transducers. Its popularity stems from the fact that it can detect and convert force or small mechanical displacements into electrical signals. Strain gauges, on the basis of operating technique used, may be classified into mechanical, optical or electrical strain gauges. The electrical resistance strain gauge, over the years, have become most extensively used device, and this is what usually meant when the term '*strain gauge*' is used.

12.3.1. Operating Principle of Resistance Strain Gauge

The working of strain gauge is based on the fact that when stress is applied on the metal conductor its resistance changes owing to change in length and x-sectional area of the conductor (Fig. 12.1). Resistance of conductor under stress is also changed due to change in resistivity of the conductor, this property is called the *piezoresistive effect*. That is why strain gauges are also called the *piezoresistive strain* gauges.

If a conductor of length l, area of cross section A is subjected to axial tension, the resistance will change because of change in length, area of x-section and resistivity of the material.

Resistance of an unstrained conductor is given by an equation

$$R = \rho \frac{l}{A} \qquad ...(12.1)$$

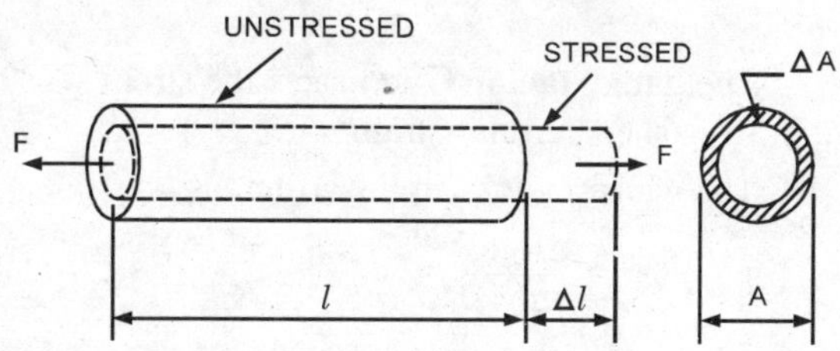

Fig. 12.1

Let under strained conditions resistance of conductor be changed by ΔR because of change in length by Δl, cross-sectional area by ΔA and resistivity by Δρ. These quantities can be related with each other by differentiating Eq. (12.1) with respect to stress *i.e.*,

$$\frac{dR}{dp} = \frac{d}{dp}\left(\rho \frac{l}{A}\right) = \frac{\rho}{A}\frac{dl}{dp} - \frac{\rho l}{A^2}\frac{dA}{dp} + \frac{l}{A}\frac{d\rho}{dp} \qquad ...(12.2)$$

Dividing both sides of Eq. (12.2) by R, we get

$$\frac{1}{R}\frac{dR}{dp} = \frac{1}{l}\frac{dl}{dp} - \frac{1}{A}\frac{dA}{dp} + \frac{1}{\rho}\frac{d\rho}{dp} \qquad ...(12.3)$$

For small variations, above equation can also be written as

$$\frac{\Delta R}{R} = \frac{\Delta l}{l} - \frac{\Delta A}{A} + \frac{\Delta \rho}{\rho} \qquad ...(12.4)$$

Under strained conditions, cross-sectional area of the conductor changes, so the diameter of the conductor is also changed.

$$\text{Area, } A = \frac{\pi}{4} D^2$$

$$\text{or} \quad \frac{\delta A}{\delta s} = 2 \times \frac{\pi}{4} D \frac{\delta D}{\delta s}$$

Dividing both sides of above expression by A, we get

$$\frac{1}{A} \cdot \frac{\delta A}{\delta s} = \frac{2 \frac{\pi}{4} D}{\frac{\pi}{4} D^2} \frac{\delta D}{\delta s} \qquad ...(12.5)$$

For small variations Eq. (12.5) can be written as

$$\frac{\Delta A}{A} = \frac{2\Delta D}{D} \quad ...(12.6)$$

Now change in diameter of wire can be related with change in length of wire by the Poisson's ratio* μ of the wire material as given below:

$$\frac{\Delta D}{D} = -\mu\frac{\Delta l}{l} \quad ...(12.7)$$

Value of μ is approximately 0.3 for most of the materials.

From Eqs. (12.6) and (12.7) we have

$$\frac{\Delta A}{A} = -2\mu\frac{\Delta l}{l} \quad ...(12.8)$$

Substituting $\frac{\Delta A}{A} = -2\mu\frac{\Delta l}{l}$ in Eq. (12.4) we get

$$\frac{\Delta R}{R} = \frac{\Delta l}{l} + 2\mu\frac{\Delta l}{l} + \frac{\Delta \rho}{\rho}$$

$$\text{or} \quad \frac{\Delta R/R}{\Delta l/l} = 1 + 2\mu + \frac{\Delta\rho/\rho}{\Delta l/l} = G, \text{ the gauge factor}$$

The **gauge factor G** indicates the strain sensitivity of the gauge in terms of the change in resistance per unit resistance per unit strain.

The gauge factor can be written as

$$G = \frac{\Delta R/R}{\Delta l/l} = \underbrace{1}_{\text{Change in resistance due to change in length}} + \underbrace{2\mu}_{\text{Change in resistance due to change in area}} + \underbrace{\frac{\Delta\rho/\rho}{\Delta l/l}}_{\text{Change in resistance due to piezoresistive effect}} \quad ...(12.9)$$

Thus the strain $\frac{\Delta l}{l}$ can be measured in terms of gauge factor, G and $\Delta R/R$.

The strain $\frac{\Delta l}{l} = \varepsilon$ and is usually measured in microstrain.

12.3.2. Types of Electrical Resistance Strain Gauges

The strain gauges are mainly of four types namely, wire strain gauges, foil strain gauges, thin film strain gauges and semiconductor strain gauges.

1. ***Wire Strain Gauges*** are normally of smaller size, and are subject to minimal leakage, and can be used in high temperature applications. Wire strain gauges are of two types namely unbonded wire strain gauges and bonded wire strain gauges.

As the name depicts, in unbonded wire strain gauges, resistance wire is not bonded on the base, instead of that strain is directly transferred to the resistance wire and so a smaller force is required for changing the wire length. Four resistance wires are generally connected in four arms of Wheatstone bridge [Fig. 12.2 (*b*)]. With no external load applied *i.e.,* at initial preload, the strains and resistances of four wires are nominally equal giving a balanced bridge *i.e.,* output voltage $v_{out} = 0$. This is generally accomplished by providing adjustable resistances in the bridge. Application of an external force increases tension in two wires and reduces it in other two wires (wires never go slack) causing corresponding variation in resistance of wires unbalancing the bridge and causing an output voltage v_{out} in proportion to the displacement. The output

* For an elastic material the Poisson's ratio μ is defined as the ratio of lateral strain to longitudinal strain *i.e.* $\mu = \frac{\text{Lateral strain}}{\text{Longitudinal strain}}$.

device can be calibrated to read the magnitude of the displacement of the armature directly. The actual displacement that can be measured is very small, of the order of 50 micron. Electrical resistance of each bridge arm is 120 to 1,000 Ω; maximum excitation voltage is 5 to 10 volts and full-scale output is 20 to 50 mV. The length of wire is 25 mm or less. The transducer will measure from ±40 g to ±2 kg, full scale.

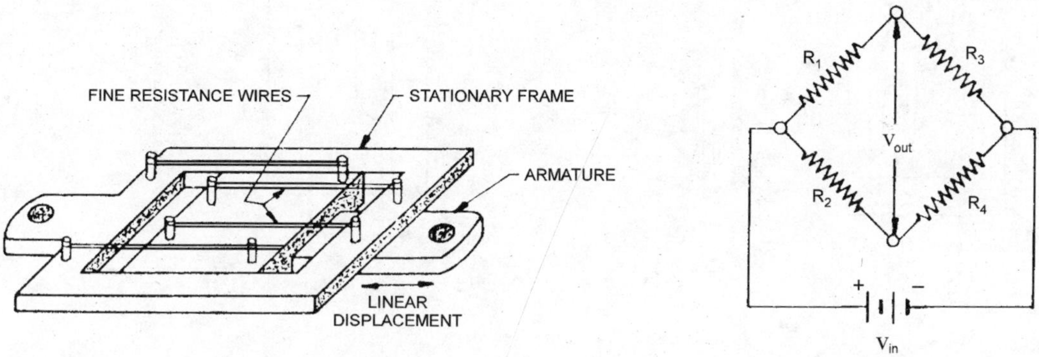

(a) (b)

Fig. 12.2. ***Unbonded Strain Gauge (Four Active Elements)***

The main advantage of such a gauge lies in its radial symmetry which assists in the cancellation of spurious signals from traverse forces. The other advantages are very low hysteresis and creep because of absence of wire binding, miniaturization of the transducing element by integration of the force-summing unit and the force-sensing components, and adaptability of high-temperature environments due to the welded construction. The main drawback of such a device is low sensitivity.

These devices are used mainly for measurement of force, pressure and acceleration rather than for measuring directly displacement or strain.

The most common type of *bonded wire strain guage* used for detecting displacement caused by large forces is *flat grid type,* as illustrated in Fig. 12.3. In flat grid type bonded wire strain gauge, fine resistance wire of diameter of 25 microns or less is wound back and forth in a grid format, with as many as loops as possible, laid side by side. Two connecting leads are soldered or welded to the ends of the grid. The bonded strain gauge is cemented with a special adhesive to the structure (specimen) whose tensile or compressive strain is to be measured. The strain of wire grid is, therefore, exactly the same as the strain of the specimen. The strain of wire grid is measured with the Wheatstone bridge connecting the gauge in one of the four arms while the remaining three arms have standard resistances of nearly equal resistance as that of gauge resistance in the unstrained condition.

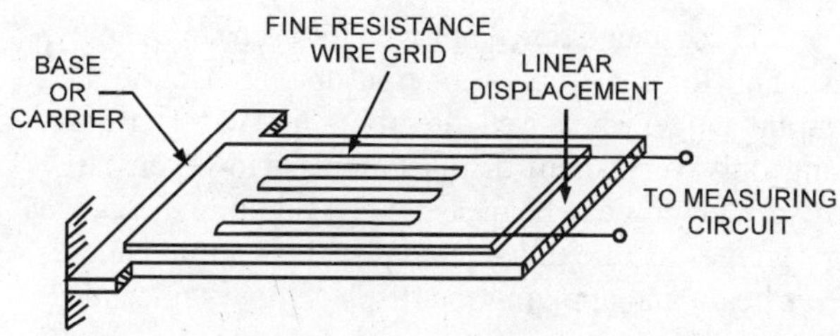

Fig. 12.3. ***Flat Grid Type Bonded Wire Strain Gauge***

Other types of bonded wire strain gauges are wrap-around, single wire, and wooven ones.

In a *wrap-around type gauge,* the resistance wire is wound on a flattened tube paper or on a paper strip, as illustrated in Fig. 12.4 (*a*). Hence for the same value resistance gauge size becomes smaller but the thickness of the gauge increases as the grid wire is in two planes. It introduces different transfer characteristics to two layers of wire resulting in larger hysteresis and creep.

In a *single wire type gauge* two or more single wires are stretched across and laid as illustrated in Fig. 12.4 (*b*), and instead of loop formed by the same wires thick copper wires are used for joining the ends, reducing the cross-sensitivity considerably. Such gauges are not very popular and are used for large gauge lengths only.

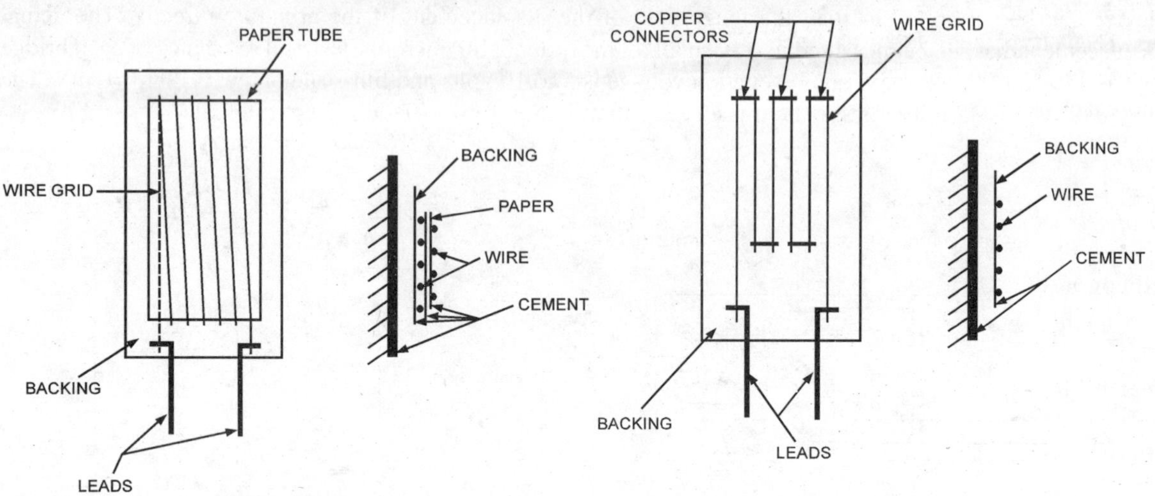

(a) Wrap-Around Type *(b) Single Wire Type*

In a *woven type gauge* a silk-insulated Eureka wire is wound as the weft on a rayon wrap. Such a gauge is useful for tests on fabrics and leather.

Bonded wire strain gauges are easy to manufacture and are relatively cheaper in cost. The size varies with its application and can be as small as 3 mm × 3 mm square but of size not more than 25 mm long and 12.5 mm wide.

These gauges are employed for both stress analysis and for construction of transducers. The bonded strain gauge, when used directly, is useful for measuring only very small displacements. However, large displacements can be measured by bonding the gauge or gauges to a flexible element such as a thin cantilever beam and applying the displacement under measurement to the end of the beam.

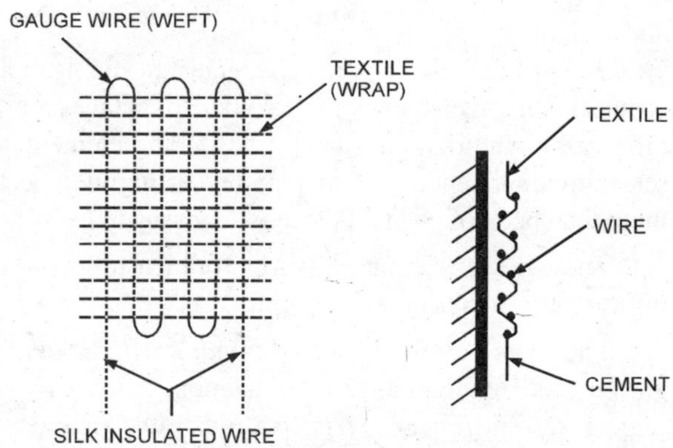

(c) Woven Type

Fig. 12.4. ***Bonded Wire Strain Gauges***

2. ***Foil Strain Gauges.*** It is basically an extension of the bonded wire strain gauge, differs in construction and has certain advantages. The bonded wire strain gauges have been completely superseded by foil strain gauges. In such gauges, metal foils are used instead of wire, as shown in Fig. 12.5. The thickness of foil varies from 2.5 microns to 6 microns. Dimensions of a single bonded foil strain gauge can be as small as 0.8 mm × 1.6 mm.

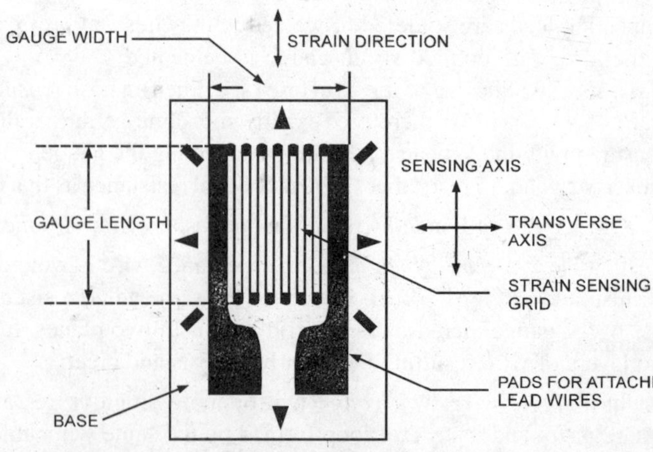

Fig. 12.5. ***Foil Strain Gauge***

Foil strain gauges with gauge resistance values of 120, 350 and 1000 Ω are common. The maximum gauge current is about 30 mA, maximum strain for ± 1 per cent linearity is about ± 0.3 per cent, and strain variation up to at least 50 kHz can be measured.

3. *Thin-Film Strain Gauges.* The thin-film strain gauge transducers have several advantages over others. They are rugged and have very low hysteresis and creep. The sensitivity is also high because of a higher gauge factor and high resistance. A resistance as high as 5 kΩ can be produced in order to permit increased input and output voltages with low power consumption. They can operate over a wide temperature range from –200°C to 400°C with good stability.

4. *Semiconductor Strain Gauges.* Semiconductor strain gauges are based on the *piezoresistive property* of doped silicon and germanium and that is why the transducers based on semiconductor gauges are often called *piezoresistive transducers.* They are made as continuous films of germanium or silicon base materials.

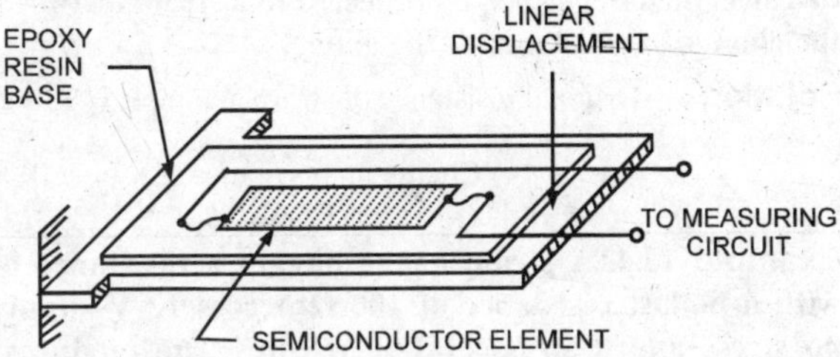

Fig. 12.6. *Semiconductor Strain Gauge*

Under strained condition, resistivity of any material changes and this is called the *piezoresistive* property of material. In metal alloy strain gauges, variation in resistance is caused, under strained condition, mainly because of change in geometrical shape and very less due to change in resistivity of the material. The gauge factor of metal alloy strain gauges is found to be about 2.0. But in semiconductor strain gauges, variation in resistance, under strained condition, is mainly due to change in resistance *i.e.,* most of the resistance variation comes from piezoresistive effect. The *p*-type gauges increase resistance with applied tensile strain while the *n*-type decrease resistance. Their big advantage is a very high gauge factor. A typical semiconductor strain gauge has a gauge factor of – 100 to + 150 with a maximum strain of 0.003 and a temperature range of – 10°C to + 70°C. In addition to the high gauge factor, other advantages are low cross sensitivity, freedom from hysteresis and creep effects, chemical inertness and good fatigue life. Unfortunately, the high gauge factor is accompanied by high temperature sensitivity, nonlinearity, and mounting difficulties. Semiconductor strain gauges can be very small, ranging from 0.7 to 7 mm, and having width typically one-tenth of the element length and, therefore, they are very useful in the measurement of highly localized strains. They permit measurement of very small strains, as small as 0.01 microstrain.

Example 12.1. What is gauge factor ? A resistance wire strain gauge uses a soft iron wire of small diameter. The gauge factor is + 5.2. Neglecting the piezoresistive effect, calculate the Poisson's ratio. [*Pb. Technical Univ. May 2008*]

Solution : From Eq. (12.9)

$$G = 1 + 2\mu + \frac{\Delta\rho/\rho}{\Delta l/l}$$

Since piezoresistive effect is negligible, the last term *i.e.* $\frac{\Delta\rho/\rho}{\Delta l/l}$ becomes negligible and the equation for gauge factor becomes,

$$G = 1 + 2\mu$$

$$\text{or Poisson's ratio, } \mu = \frac{G-1}{2} = \frac{5.2-1}{2} = 2.1 \textbf{ Ans.}$$

Example 12.2. A strain gauge is cemented to an iron post of length 1 m, which is subjected to a compressible force. The nominal resistance of the gauge is 200 Ω. The change in resistance of strain gauge is measured to be 0.7 Ω. The final length of the iron post is 999 mm. Calculate the gauge factor of the strain gauge.

Solution : Initial length of wire, l = 1 m = 1,000 mm

Final length of wire, l' = 999 mm

Increase in length of wire, $\Delta l = l' - l = 999 - 1{,}000 = -1$ mm

$$\text{Strain, } \varepsilon = \frac{\Delta l}{l} = \frac{-1}{1{,}000} = -0.001$$

Initial resistance of wire, R = 200 Ω

Change in resistance of wire, ΔR = – 0.7 Ω ∵ Under compression the resistance of strain gauge will decrease

$$\text{Gauge factor of the device, } G = \frac{\Delta R/R}{\varepsilon} = \frac{-0.7/200}{-0.001} = 3.5 \textbf{ Ans.}$$

Example 12.3. Define gauge factor of a strain gauge. A resistance strain gauge having a gauge factor of 2 is fastened to a steel member which is subjected to a strain of 10^{-6}. If the resistance of the unstrained gauge is 120 Ω, calculate the change in resistance of the gauge.

Solution : Original resistance of strain gauge, R = 120 Ω

$$\text{Change in resistance, } \Delta R = \varepsilon G R = 1 \times 10^{-6} \times 2 \times 120 = 240\ \mu\Omega \textbf{ Ans.} \qquad \because \varepsilon = \frac{\Delta R/R}{G}$$

Example 12.4. A strain gauge having a resistance of 100 Ω and a gauge factor of 2 is connected in series with a ballast resistance of 100 Ω across 12 V supply. Calculate the difference between output voltage with no stress and a stress of 140 MN/m². The modulus of elasticity is 200 GN/m².

Solution : Under normal condition *i.e.*, when no stress is applied,

Voltage across the strain gauge

$$= \text{Supply voltage} \times \frac{R}{R + R_B} = 12 \times \frac{100}{100 + 100} = 6 \text{ V}$$

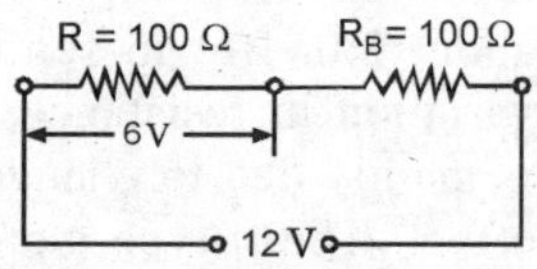

Fig. 12.7

$$\text{Strain, } \varepsilon = \frac{p}{E} = \frac{140 \times 10^6}{200 \times 10^9} = 0.0007$$

Change in resistance, $\Delta R = R \times \varepsilon \times G = 100 \times 0.0007 \times 2 = 0.14\ \Omega$

Voltage across the gauge under strained condition

$$= \frac{R + \Delta R}{R + \Delta R + R_B} \times V = \frac{100 + 0.14}{100 + 0.14 + 100} \times 12 = 6.0042 \text{ V}$$

Change in output voltage = 6.0042 – 6 = 0.0042 V = 4.2 mV **Ans.**

12.4. THERMISTOR

Thermistors are also called the *thermal resistors* and the name is derived from *thermally sensitive resistors,* as the resistance of a thermistor varies as a function of temperature. Thermistors are also called the *thermal resistors* and the name is derived from *thermally sensitive resistors,* as the resistance of a thermistor varies as a function of temperature.

Thermistors are essentially semiconductor devices, that behave as resistors with high negative temperature coefficient (usually – 0.04 per °C at room temperature of 25°C) and are at least 10 times as sensitive as the platinum resistance thermometer, as illustrated in Fig. 12.8.

Thermistor has a very nonlinear resistance-temperature relation, as obvious from the characteristic curves shown in Fig. 12.8. The resistance R of a thermistor at a temperature T can be given by the expression

$$R = \alpha e^{\beta/T} \qquad \text{...(12.10)}$$

where α and β are constants depending upon the material and manufacturing techniques used.

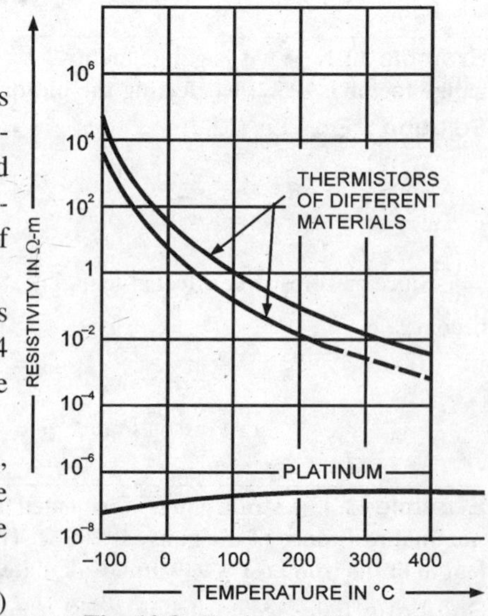

Fig. 12.8. ***Resistance-Temperature Characteristics For Thermistors and Platinum***

The approximate relationship between resistance and temperature, applying to most thermistors, can be obtained by rewriting above Eq. (12.10) for temperatures T_1 and T_0

$$R_1 = R_0 e^{\beta (1/T_1 - 1/T_0)} \qquad ...(12.11)$$

where R_1 and R_0 are resistances in ohms at absolute temperatures T_1 and T_0 respectively and β is a thermistor constant. β is expressed in Kelvin and is of the order of 4,000. The reference temperature T_0 is usually taken as 298 K or 25 °C.

β is assumed to remain constant for all practical purposes but has generally a rising tendency with temperature. The assumption that β is constant gives a relation for the resistance temperature coefficient which is given as

Temperature coefficient of resistance,

$$\alpha = \frac{dR_1}{dT} / R_1 = -\frac{\beta}{T^2} \qquad ...(12.12)$$

From Eq. (12.12) it is evident that α is not constant. It actually drops with rise in temperature.

If β = 4,000 and T = 25°C or 298 K then temperature coefficient of resistance,

$$\alpha = \frac{-4{,}000}{(298)^2} = -0.045 \ \Omega/\Omega/°C$$

Modern thermistors are manufactured from the oxides of metals like manganese, nickel, cobalt, copper, iron, zinc, aluminium, titanium, magnesium and uranium. These oxides or their sulphides and silicates are milled, mixed in appropriate proportion, pressed into desired shapes with appropriate binders, and finally sintered. The electrical terminals are embedded before sintering or baked afterwards. The resistance of *the thermistors at room temperature of* 25°C varies over a wide range, from several hundred ohms to mega ohms. Because of their high resistance they can be manufactured in very small sizes in the form of beads, discs and rods. These are available in a wide variety of shapes and sizes and commercial forms are shown in Fig. 12.9.

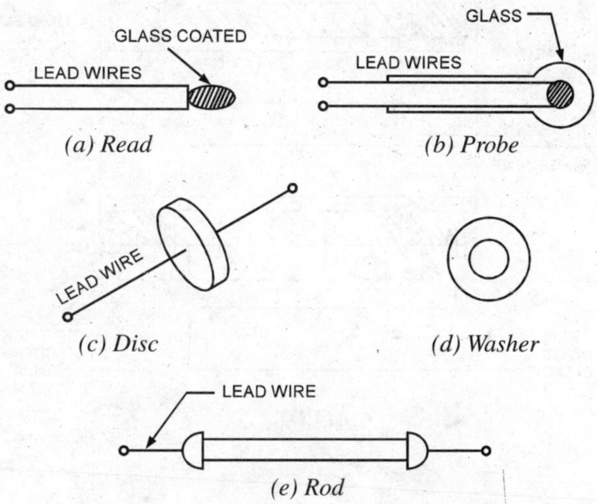

Fig. 12.9. ***Commerical Forms of Thermistors***

Because of large change of resistance per degree of temperature variation in thermistor (*i.e.* large temperature coefficient of resistance) they can provide good accuracy and resolution when used for measurement of temperatures between –100°C and +300°C. If an ammeter is used for monitoring of the current through a thermistor, temperature variations as small as ± 0.1°C can be detected. If the thermistor is instead put into a Wheatstone bridge, the measuring system can detect temperature variations as small as ± 0.005°C.

The thermistors are rugged in construction, cheaper in cost and highly sensitive devices. Although most applications of thermistors are in the field of measurement and control of temperature, they can be employed in so many other applications.

12.5. LINEAR VARIABLE DIFFERENTIAL TRANSFORMER (LVDT)

This is the most widely used inductive transducer for translating the linear motion into an electrical signal. The basic construction of an LVDT is shown in Fig. 12.10 (*a*).

LVDT is a differential transformer consisting of one primary winding P and two identical secondary windings S_1 and S_2, wound over a hollow bobbin of non-magnetic and insulating material. The secondary windings S_1 and S_2, which have equal number of turns, are arranged concentrically and placed either side of the primary winding P. A soft iron core, attached to the sensing element of which displacement is to be measured, in the shape of rod or cylinder slides freely in the hollow portion of the bobbin. In order to overcome the problem of eddy current losses in the core, nickel-iron alloy is used as core material and is slotted longitudinally. Primary winding is connected to an ac source of voltage varying from 5 to 25 V and of frequency ranging from 50 Hz to 20 kHz. When the core is moved inside the bobbin it varies coupling of primary winding to secondary windings S_1 and S_2. In null position of the core *i.e.*, in central position, coupling of primary winding to both of the secondary windings are equal and so output voltages induced in secondary windings S_1 and S_2 are equal. As the core is moved toward left from its null position, the magnetic linkages to secondary winding S_1 increases and to secondary winding S_2 decreases because the portion of the core inside secondary winding S_1 increases and that inside secondary winding S_2 decreases. Therefore, output voltage induced in secondary winding S_1 increases whereas the output voltage induced in the secondary winding S_2 decreases. The movement of the core to the right will have opposite effect. Secondary windings S_1 and S_2 are connected in series opposition so that difference of output voltages of secondary windings gives the measurement of displacement.

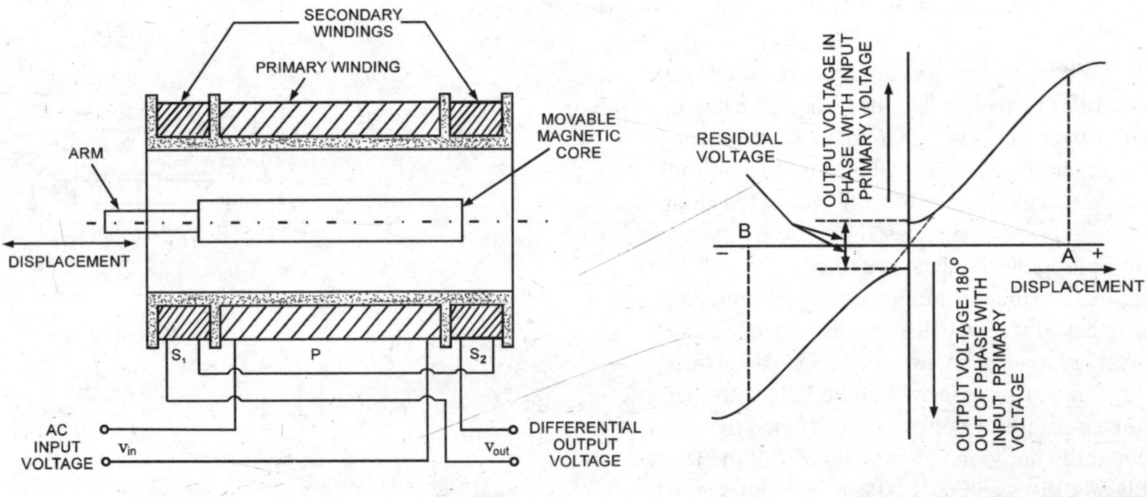

(a) LVDT

(b) Variation of Differential Output Voltage With Displacement For LVDT

Fig. 12.10

Difference of output voltages of secondary windings gives the amount of displacement following a linear law. In either direction of movement of the core, difference of output voltages increases or decreases giving the idea of direction of movement. With the movement of the core in one direction away from the null position, the differential output voltage increases and it is in phase with the input voltage of primary winding whereas the movement of the core in other direction from the null position causes again differential output voltage to increase but 180° out of phase with the input primary voltage. Now by measuring the magnitude and comparing the phase of differential output voltage with the input primary voltage, amount and direction of displacement of core may be determined.

Curve for variation of differential output voltage with displacement of core for an LVDT is illustrated in Fig. 12.10 (*b*). From this curve it is clear that this characteristic of LVDT is linear for limited range of displacement (say about 5 mm either side from the null position) and beyond this range curve starts flattening out) at both ends.

In practice, output voltage is not zero at null position and some residual voltage exists at output terminals of LVDT but usually it is less than 1% of maximum value of output voltage in linear range. This residual voltage may be either due to magnetic unbalance or due to electrical unbalance. Presence of harmonics in input supply voltage or harmonics produced in the output voltage because of saturation of iron core also contribute to residual voltage. Stray magnetic fields may also be the cause for residual voltage, which may be overcome by using a magnetic shield slotted longitudinally and fitted over the transducer.

In commercially available LVDTs normally range for displacement varies from ± 0.01 mm to ± 25 mm.

12.5.1. Merits, Demerits and Applications of LVDTs

Merits. 1. LVDT has infinite resolution as it gives stepless output and it has got no mechanical element to change output in discrete steps. Nowadays transducers are available with the resolution upto 1 micron.

2. LVDT has almost linear characteristic within its prescribed range. Linearity up to ± 0.25% can be achieved in commercially available LVDTs.
3. LVDT has high sensitivity. It usually varies from 10 mV/mm to 40 V/mm.
4. Its output is very high which, even in some cases, eliminates the need for amplification devices.
5. These devices consume very less power (less than 1 watt).
6. LVDT can be used on high frequencies upto 20 kHz.
7. Output impedance of LVDT remains constant which in case of potentiometer varies with displacement.
8. Absence of sliding contact makes LVDT a more reliable device.
9. LVDT has very low hysteresis so a good repeatability can be achieved with it.
10. LVDT is very rugged device in construction so it can tolerate shocks and vibrations without any adverse effect.
11. It is a very stable and easy to align and maintain due to simplicity of construction, small size and light weight.

Demerits. 1. These devices are sensitive to stray magnetic fields but its problem is overcome by using magnetic shields.

2. Relatively large displacements are required for appreciable differential output.
3. Sometimes, the transducer performance is affected by vibrations.

Applications. LVDTs are suitable for use in applications where the displacements are too large for strain gauges to handle. For example, LVDTs can be employed for measurement of displacements that range from a fraction of a mm to a few cm. If LVDT is to be employed for measurement of mechanical displacement greater than 25 mm, an appropriate mechanical ratioing (gearing) must be used.

Since the LVDTs can also be connected to other transducers, whose outputs are mechanical displacements, these are often employed together with other transducers for measurement of force, weight, pressure etc.

12.5.2. Measurement of Displacement by LVDT

Measurement of displacement can be accomplished by using two LVDTs in a closed loop servosystem, as illustrated in Fig. 12.11.

In this scheme two LVDTs (one input LVDT and another balancing LVDT) connected in series opposition are employed. The resultant output of the combination is fed to a servoamplifier that drives a two phase servomotor. When the cores of the two LVDTs are in their reference positions, the resultant output is zero and so no supply is delivered to the servomotor. On displacement of the core of the input LVDT from its reference position an output voltage is delivered to the servoamplifier and the servomotor rotates. The servomotor shaft is mechanically coupled to the core of the balancing LVDT. The servomotor continues rotating until the outputs of the two LVDTs are equal as in that case only no voltage is delivered to the servoamplifier. The displacement of the core of the balancing LVDT, which is a measure of the displacement under measurement, is read on the scale.

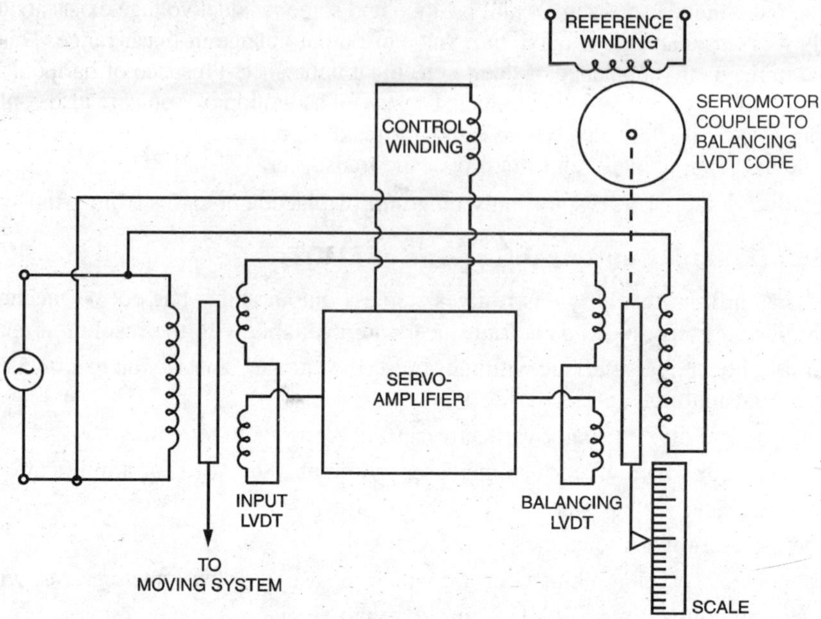

Fig. 12.11

The scheme can be employed for measuring a displacement at a considerable height from the ground.

Example 12.5. An LVDT produces an rms output voltage of 2.6 V for displacement of 0.4 μm. Calculate the sensitivity of LVDT.

Solution : Sensitivity of LVDT, $s = \dfrac{\text{RMS value of output voltage}}{\text{Displacement in } \mu\text{m}} = \dfrac{2.6}{0.4} = 6.5 \text{ V}/\mu\text{m}$ **Ans.**

Example 12.6. An LVDT has a secondary voltage of 5.0 V for a displacement of ± 12.5 mm. Determine the output voltage for a core displacement of 8.0 mm from its central position.

Solution : Sensitivity of LVDT, $s = \dfrac{5.0}{12.5} = 0.4 \text{ V/mm}$

Output voltage for a core displacement of 8.0 mm

$= 0.4 \times 8.0 = 3.2 \text{ V}$ **Ans.**

Example 12.7. An LVDT is employed for measuring the deflection of a bellows. The sensitivity of the LVDT is 60 V/mm. The bellows is deflected by 0.15 mm by a pressure of 1.2 × 10^6 N/m^2. Determine the sensitivity of the LVDT in V per N/m^2 and the pressure when the output voltage is 4.5 V.

Solution : LVDT sensitivity, $s = 60 \text{ V/mm}$

Output voltage for a deflection of 0.15 mm, $V_{out} = 60 \times 0.15 = 9 \text{ V}$

Hence the sensitivity of LVDT $= \dfrac{V_{out}}{p} = \dfrac{9}{1.2 \times 10^6} = 7.5 \times 10^{-6} \text{ V per N/m}^2$ **Ans.**

Pressure for output voltage of 4.5 V, $P = \dfrac{4.5}{7.5 \times 10^{-6}} = 6 \times 10^5 \text{ N/m}^2$ **Ans.**

12.6. DIGITAL VOLTMETER

The digital voltmeter (DVM) displays measurements of ac or dc voltages as discrete numerals instead of a pointer deflection on a continuous scale as in analog instruments. It is a versatile and accurate instrument that

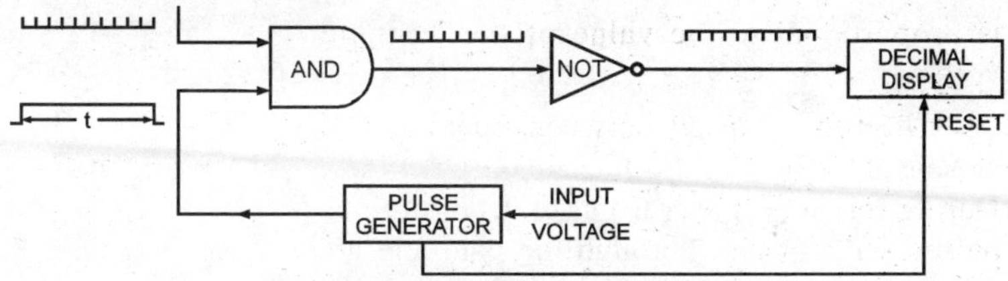

Fig. 12.12. ***Block Diagram of DVM***

is employed in many laboratory measurement applications. Because of development and perfection of IC modules, the size, power requirements and cost of the digital voltmeter has been drastically reduced and, therefore, DVMs can actively compete with conventional analog instruments, both in price and portability.

The block diagram of a simple digital voltmeter is shown in Fig. 12.12. The unknown voltage signal is fed to the pulse generator which generates a pulse whose width is directly proportional to the input unknown voltage. The output of the pulse generator is applied to one leg of an AND gate. The input signal to the other leg of the AND gate is a train of pulses. The output of the AND gate is, thus, a positive trigger train of duration t seconds and the inverter converts it into a negative trigger train. The counter, counts the number of triggers in t seconds which is proportional to the voltage under measurement. Thus, the counter can be calibrated to indicate voltage in volts directly.

Thus, we see that the DVM described above is an ADC which converts an analog signal into a train of pulses, the number of which is proportional to the input voltage. So, a digital voltmeter can be made by using any one of the analog-to-digital (A/D) conversion methods given below :

1. Ramp A/D conversion.
2. Dual slope A/D conversion.
3. Successive approximation A/D conversion.
4. Voltage-to-frequency A/D conversion.
5. Parallel A/D conversion.
6. Recirculating remainder A/D conversion.

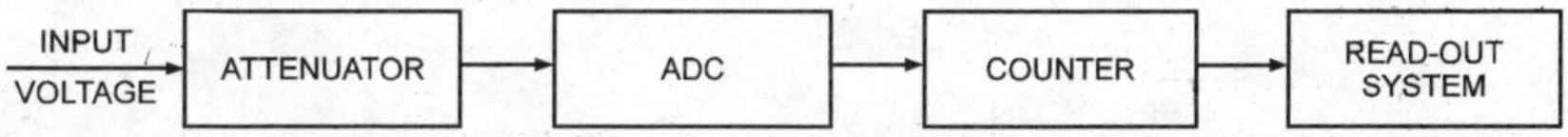

Fig. 12.13

The input range of the DVM may vary from ± 1.00000 V to ± 1,000.00 V and its limiting accuracy is as high as ± 0.005 per cent of the reading. Its resolution may be 1 part in 10^6, giving 1 μV reading of the 1 V input range. It has a high input resistance of the order of 10 MΩ and input capacitance of the order of 40 pF.

Digital voltmeter (DVM) employing ramp A/D conversion method will be discussed below :

Ramp Type DVM. The operation of a ramp type DVM is based on the measurement of the time that a linear ramp voltage takes to change from the level of the input voltage to zero voltage or vice-versa. This time interval is measured with an electronic time interval counter and the count is displayed as a number of digits on the electronic indicating tubes of the voltmeter output readout.

The operating principle and block diagram of a *ramp type DVM* are given in Figs. 12.14 (*a*) and 12.14 (*b*) respectively.

At the start of measurement a ramp voltage is initiated, this voltage can be positive-going or negative going. In the Fig. 12.14 (*a*) negative going voltage ramp is illustrated. The ramp voltage is continuously compared with the voltage under measurement (or unknown voltage). At the instant the value of ramp voltage becomes equal to the voltage under measurement, a coincidence circuit, called the *input comparator,* generates a pulse which opens a gate, as shown in Fig. 12.14 (*b*). The ramp voltage continues to fall till it reaches zero value (or ground level). At this instant another comparator, called the *ground comparator*, generates a stop pulse. The output pulse from this ground comparator closes the gate. The time duration of the gate opening is proportional to the value of the dc input voltage.

The time elapsed between opening and closing of gate is *t*, as illustrated in Fig. 12.14 (*a*). During this time interval pulses from a clock pulse oscillator pass through the gate and are counted and displayed. The decimal number indicated by the readout is a measure of the value of the input voltage.

The sample rate multivibrator determines the rate at which the measurement cycles are initiated. The sample rate circuit provides an initiating pulse for the ramp generator to start its next ramp voltage. At the same time a reset pulse is generated, which resets the counter to zero state.

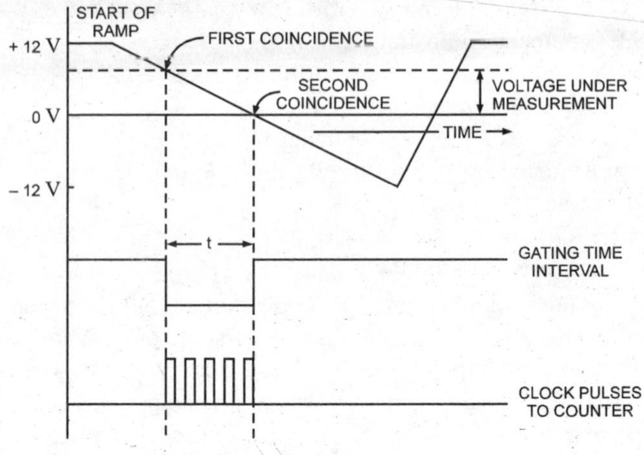

(a) Voltage-To-Time Conversion

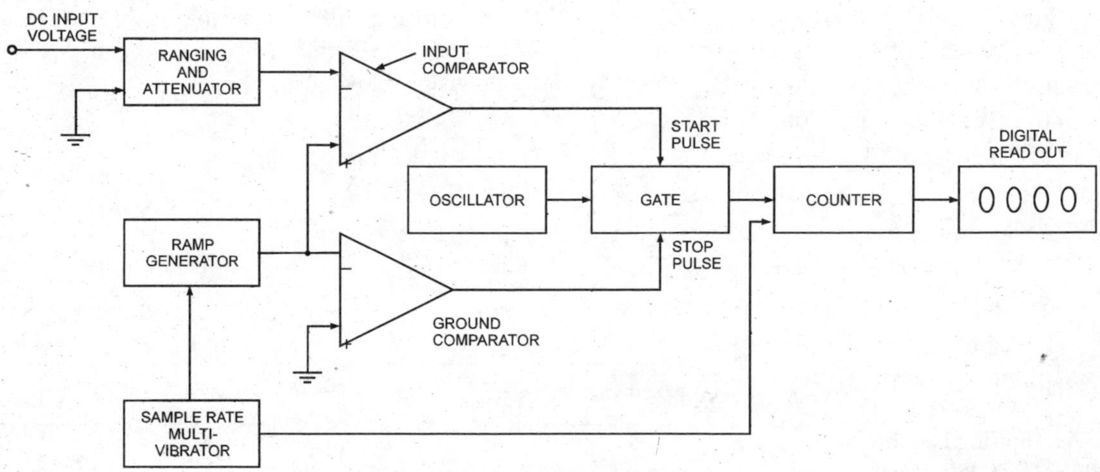

(b) Block Diagram of a Ramp Type DVM

Fig. 12.14

12.7. DIGITAL MULTIMETER

Digital multimeter (DMM) is basically a digital voltmeter and may be used for the measurement of voltage, current (dc or ac) and resistance. All quantities other than dc voltage are first converted into an equivalent dc voltage by some device.

The block diagram of a basic digital multimeter is given in Fig. 12.15.

For measurement of ac voltage the input voltage is converted into a dc voltage by means of a rectifier. A compensated attenuator is employed. Many manufacturers provide the same attenuator for both dc and ac measurements.

For measurement of resistance a constant current, depending on the range, supplied from a battery or a constant current source is passed through the resistance under measurement and the voltage developed across it is measured. The resistance value is displayed in ohms.

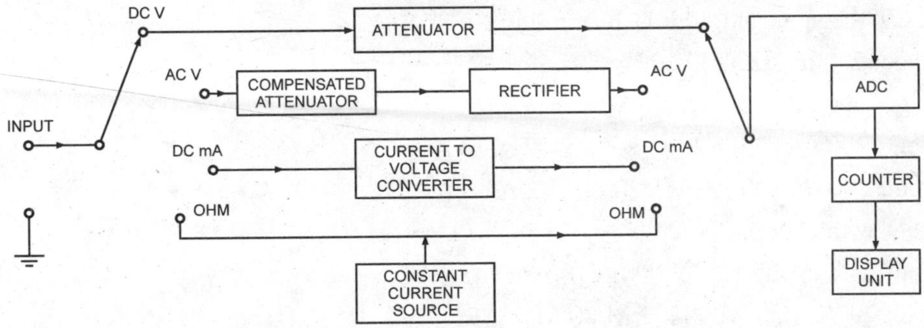

Fig. 12.15. ***Block Diagram of Digital Multimeter***

For measurement of current, the unknown current is passed through a precision resistor in many commercial digital multimeters and the voltage developed across the precision resistor is measured. The current value is displayed in mA.

For measurement of current, a current-to-voltage converter may also be used, as illustrated by block diagram in Fig. 12.16.

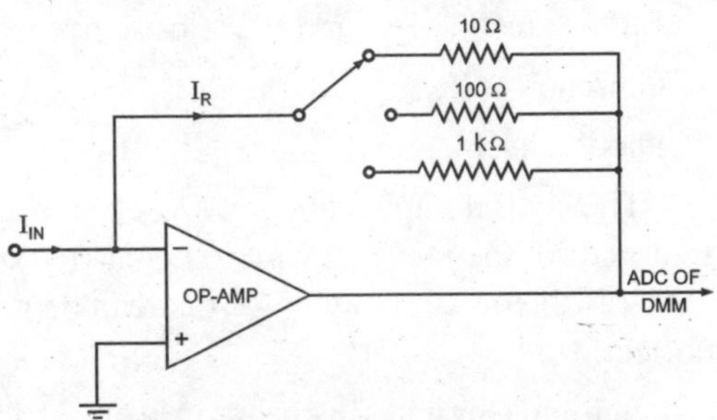

Fig. 12.16. ***Block Diagram of Current-To-Voltage Converter***

The current under measurement is applied to the summing junction at the input of the op-amp. The current in the feedback resistor I_R is equal to the input current I_{IN} because of very high input impedance of the op-amp. The current I_R causes a voltage drop across one of the resistors, which is proportional to the input current I_{IN}. Different resistors are employed for different ranges.

The performance of a digital multimeter largely depends on the A/D converter. Low-cost hand held multimeters use a single chip $3\frac{1}{2}$ digit ADC whereas high precision $5\frac{1}{2}$ and $6\frac{1}{2}$ digit multimeters make use of microprocessor controlled discrete ADCs. Digital multimeters give very high input impedance, better accuracy and resolution. Such multimeters have autoranging, autopolarity and auto zero facilities.

The hand-held (or pocket type), $3\frac{1}{2}$ digit digital multimeter DMM with auto-range and data hold function is shown in Fig. 12.17. The main features and technical specifications for the DMM shown in Fig. 12.17 are as follows:

Fig. 12.17. ***$3\frac{1}{2}$ Digit DMM With Autorange and Data-hold Function***

$3\frac{1}{2}$ digits LCD maximum reading 1999 and backlight

12 functions for DCV, ACV, DCA, ACA and OHM

Diode check

Auto power off

Display: 1999 counts and digit is 16 mm high

DC voltage: 200 m/2/20/200/600 V

AC voltage: 200 m/2/20/200/600 V

DC current: 200 μ/2000 μ/20 m/200 mA/2 A/10 A

AC current: 200 μ/2000 μ/20 m/200 mA/2 A/10 A

DC current with clamp: 200/2,000A (0.1 mV/0.1 A)

AC current with clamp: 200/2,000 A (0.1 mV/0.1 A)

Resistance: 200/2 k/20 k/200 k/2 M/20 MΩ

Temperature: 20 to 1, 000 degrees C/0 to 1,800 degrees F

Battery test: 1.5, 3 and 9 V (test current approximately 60 mA)

Continuity: Yes

Diode check: Yes

The central knob (function switch) has many positions and can be switched to any position which is required for the measurement to be carried out. For example, if the meter is switched to 200 V DC, then 200 V is the maximum voltage that can be measured. This is sometimes known as 200 V FSD (full-scale deflection).

A minus sign also appears in front of the numerical display, when required, to indicate a negative measured quantity.

For multifunction instruments, it is very important to use correct terminals and to set the function switch correctly for the meter desired application. Incorrect connection and function selection may cause damage to the circuit under test and to the instrument.

Another $3\frac{1}{2}$ digit multimeter is shown in Fig. 12.18. It is ideal for use in the field, office and home. This meter is extremely convenient for the on the go professionals that only requires to do basic measurements. The special features and technical specifications, this meter has, are as follows:

Features: Audible continuity test; Transistor test; Auto zero adjustment; Overrange indication; Low battery indication.

Technical Specifications

Maximum display : 1999

Power supply : 12 V Battery (included)

DC voltage : 200 mV to 500 V with accuracy ± (0.5% + 2 digit)

AC voltage : 1 V to 500 V with accuracy ± (1.2% + 10 digit)

DC current : 2 mA to 200 mA with accuracy ± (1% + 2 digit)

Resistance : 200 Ω to 2000 kΩ with accuracy ± (0.8% + 2 digit)

Diode test : Yes

Transistor test : Yes

Continuity test : Audible

Battery capacity check : 1.5 V (AA) and 9 V

***Fig. 12.18.** $3\frac{1}{2}$ Digit Compact, Pocket Sized DMM*

HIGHLIGHTS

1. Basically transducer is defined as a device that converts energy or information from one form to another. These are widely used in measurement work because not all quantities that need to be measured can be displayed as easily as others.
2. Basically an electrical transducer is a sensing device by which a physical, mechanical or optical quantity to be measured is transformed directly, with a suitable mechanism, into an electrical signal, current, voltage or frequency.
3. Basic requirements of transducers are ruggedness, linearity, repeatability, good dynamic response, no hysteresis and no residual magnetism.
4. Transducers may be classified in different ways such as on the basis of electrical principles involved, methods of application, methods of energy conversion used, nature of output signal like primary and secondary transducers, active and passive transducers, analog and digital transducers and transducers and inverse transducers.
5. The strain gauge is basically a device used for measuring mechanical surface strain and is one of the most extensively used electrical transducers. Its working is based on the fact that when stress is applied on the metal conductor its resistance changes owing to change in length and cross-sectional area of the conductor (Fig. 12.1). Resistance of the conductor under stress is also changed due to change in resistivity of the conductor. Strain gauges on the basis of operating technique used, may be mechanical, optical or electrical. Electrical strain gauges are used for a variety of applications, mainly because of ease in instrumentation, high accuracy and excellent reliability.
6. Thermistors are essentially semiconductor devices, that behave as resistors with high negative temperature coefficient (usually – 0.04 per ºC) and are at least 10 times as sensitive as the platinum resistance thermometer.

   Thermistors are rugged in construction, cheaper in cost and highly sensitive devices and are, therefore, well suited to temperature measurement, control and compensation.
7. LVDT is the most widely used inductive transducer for translating the linear motion into an electrical signal. LVDT is a differential transformer consisting of one primary winding P and two identical secondary windings S_1 and S_2, wound over a hollow bobbin of non-magnetic and insulating material.

   LVDTs are suitable for use in applications where the displacements are too large for strain gauges to handle. For example, LVDTs can be employed for measurement of displacements that range from a fraction of a mm to a few cm.

   Since the LVDTs can also be connected to other transducers, whose outputs are mechanical displacements, these are often employed together with other transducers for measurement of force, weight, pressure etc.
8. The digital voltmeter (DVM) displays measurements of ac or dc voltages as discrete numerals instead of a pointer deflection on a continuous scale as in analog instruments. It is a versatile and accurate instrument that is employed in many laboratory measurement applications. Because of development and perfection of IC modules, the size, power requirements and cost of the digital voltmeter has been drastically reduced and, therefore, DVMs can actively compete with conventional analog instruments, both in price and portability.

   A digital voltmeter can be made by using any one of the A/D conversion method. Thus, the DVMs can be classified on the basis of ADCs used.

   The input range of the DVM may vary from ± 1.00000 V to ± 1,000.00 V and its limiting accuracy is as high as ± 0.005 per cent of the reading. Its resolution may be 1 part in 10^6, giving 1 μV reading of the 1 V input range. It has a high input resistance of the order of 10 MΩ and input capacitance of the order of 40 pF.
9. Digital multimeter (DMM) is basically a digital voltmeter and may be used for the measurement of voltage, current (dc or ac) and resistance. All quantities other than dc voltage are first converted into an equivalent dc voltage by some device.

EXERCISES

1. What is an electrical transducer ? What are the basic requirements of a transducer ? Give the classification of transducers.
2. Define transducers. How are these classified. Give four examples of active transducers.
[Pb. Technical Univ. Basic Electrical and Electronics Engineering Dec. 2004; May 2010]
3. Define tranducer. State its classification with 4 examples of passive transducers.
[Pb. Technical Univ. Basic Electrical and Electronics Engineering Dec. 2005]
4. Explain the term gauge factor with respect to resistance strain gauge. Obtain an expression for the gauge factor in terms of Poisson's ratio μ. Discuss the merits and demerits of the various materials used for the strain gauge.
5. Show that for a wire wound resistance strain gauge, the gauge factor, G_f is given by the equation $G_f = 1 + 2\mu + \frac{\delta\rho/\rho}{\delta l/l}$ where the symbols used have their usual meanings.
6. Explain principle and operation of unbonded metal strain gauge and bonded resistance wire strain gauge.
[Pb. Technical Univ. Basic Electrical and Electronics Engineering May 2008]
7. What do you understand by piezoresistive effect ? Derive an expression to show that if the change in resistivity of a material when strained is neglected then the gauge factor (G_f) of a strain gauge equals to $1 + 2\gamma$ where γ is poisson's ratio. **[Pb. Technical Univ. Basic Electrical and Electronics Engineering Dec. 2006]**
8. Write the working principle of thermistor.
[Pb. Technical Univ. Basic Electrical and Electronics Engineering December 2009]
9. What are thermistors ? Explain the working and construction of thermistors.
10. What is transducer ? Explain the working of an LVDT.
[Pb. Technical Univ. Basic Electrical and Electronics Engineering May 2005]
11. Discuss basic principle of operation for inductive transducers. Briefly explain working of LVDT.
[Pb. Technical Univ. Basic Electrical and Electronics Engineering May 2007]
12. With the help of a neat diagram explain the principle of operation of a linear variable differential transformer transducer. List out advantages of LVDTs. **[Pb. Technical Univ. Basic Electrical and Electronics Engineering May 2006]**
13. Explain the working of LVDT. **[Pb. Technical Univ. Basic Electrical and Electronics Engineering Dec. 2007]**
14. Discuss construction, working principle, merits and demerits of LVDT.
[Pb. Technical Univ. Basic Electrical and Electronics Engineering December 2008]
15. What is LVDT ? Explain the use for measurement of displacement.
[Pb. Technical Univ. Basic Electrical and Electronics Engineering, December 2009]
16. What is digital voltmeter ? What are its advantages ? Explain its working with the help of block diagram.
17. Explain with block diagrams how a DMM can measure dc and ac signals, and various electrical parameters.
18. Write short notes on :
    (*i*) Strain gauges. **[P.T.U. December 2004]**
    (*ii*) Thermistors.
    (*iii*) LVDT.
    (*iv*) DMM.

SHORT ANSWER TYPE QUESTIONS WITH ANSWERS

Q. 1. What is transducer ? Give its classification. **[P.T.U. B.E.E.E. Dec. 2004]**

Ans. Transducer is defined as a device which converts energy or information from one form to another.

Transducers may be classified in various ways such as on the basis of electrical principles, methods of application, methods of energy conversion used, nature of output signal etc.

Transducers, on the basis of methods of application, may be classified into primary and secondary transducers.

Transducers, on the basis of methods of energy conversion used, may be classified into active and passive transducers.

Transducers, on the basis of nature of output signal, may be classified into analog and digital transducers.

Q. 2. What is the need of a transducer ?

Ans. Mostly quantities to be measured are non-electrical such as temperature, pressure, displacement, humidity, fluid flow, speed etc., but these quantities cannot be measured directly. Hence such quantities are required to be sensed and changed into some other form for easy measurement. Electrical quantities such as current, voltage, resistance, inductance and capacitance etc., can be conveniently measured, transferred and, stored and therefore, for measurement of non-electrical quantities these are to be converted into electrical quantities first and then measured. The function of converting non-electrical quantity into electrical one is accomplished by a device called the *electrical transducer.*

Q. 3. What is an electrical transducer ?

Ans. Basically an electrical transducer is a sensing device by which a physical, mechanical or optical quantity to be measured is transformed directly, with a suitable mechanism, into an electrical signal, current, voltage or frequency.

Q. 4. Distinguish between active and passive transducers.

Ans. Self generating type transducers *i.e.* the transducers, which develop their output in the form of electrical voltage or current without any auxiliary source, are called the active transducers. On the other hand the transducers in which electrical parameters (resistance, inductance or capacitance) changes with the change in input signal are called the passive transducers.

Q. 5. What is the difference between analog and digital transducers ?

Ans. Analog transducers convert input signal into output signal, which is a continuous function of time such as thermistors, strain gauge, LVDT etc. On the other hand digital transducers convert input signal into the output signal of the form of pulse *e.g.* discrete output.

Q. 6. How does inverse transducer differ from a transducer ? **[Pb. Technical Univ. 2005-06]**

Ans. Transducer is defined as a device that converts a non-electrical signal into an electrical quantity while an inverse transducer is defined as a device that converts electrical quantity into a non-electrical quantity.

Q. 7. What is strain gauge ?

Ans. The strain gauge is basically a device used for measuring mechanical surface strain and is one of the most extensively used electrical transducers. Its working is based on the fact that when stress is applied on the metal conductor its resistance changes owing to change in length and cross-sectional area of the conductor (Fig. 12.2). Resistance of the conductor under stress is also changed due to change in resistivity of the conductor. Strain gauges on the basis of operating technique used, may be mechanical, optical or electrical.

Q. 8. Where is strain gauge used ? **[Pb. Technical Univ. 2004-05]**

Ans. Electrical strain gauges are used for a variety of applications, mainly because of ease in instrumentation, high accuracy and excellent reliability.

(*i*) Strain gauge transducers are extensively used for analyzing the dynamic strains in complex structures, such as the stress and strain in bridges, automobiles, roads etc., where the output of the bridge can be recorded with the help of oscillograph calibrated to read the instantaneous stress or strains.

(*ii*) Strain gauge transducers are used for measuring tension, torque, force, stresses in structures. They are particularly useful where a minimum disturbance due to presence of transducer is required as in wind tunnel measurements.

(*iii*) Strain gauge transducers are also used for measurement of force by the strain produced in load rings.

Q. 9. Why are strain gauges called the piezoresistive strain gauges ?

Ans. Resistance of a conductor under stress is also changed due to change in resistivity of the conductor, this property is called the piezoresistive effect and the strain gauges are called the piezoresistive.

Q. 10. Define gauge factor. **[P.T.U. B.E.E.E. May 2009]**

OR

What is gauge factor ? **[P.T.U. B.E.E.E. December 2007; May 2008]**

Ans. The **gauge factor G** indicates the strain sensitivity of the gauge in terms of the change in resistance per unit resistance per unit strain.

The gauge factor can be written as

$$G = \frac{\Delta R/R}{\Delta l/l} = \underbrace{1}_{\text{Change in resistance due to change in length}} + \underbrace{2\mu}_{\text{Change in resistance due to change in area}} + \underbrace{\frac{\Delta\rho/\rho}{\Delta l/l}}_{\text{Change in resistance due to piezoresistive effect}}$$

Thus the strain $\frac{\Delta l}{l}$ can be measured in terms of gauge factor, G and $\Delta R/R$.

The strain $\frac{\Delta l}{l} = \varepsilon$ and is usually measured in micro-strain.

Q. 11. Give the advantages and limitations of foil strain gauges. **[P.T.U. B.E.E.E. December 2008]**

Ans. Strain gauges have the advantages of better sensitivity due to larger surface area and better gauge factor. But they have the limitations of low tensile strength and so heavy loads cannot be applied to it.

Q. 12. Explain why semiconductor strain gauges have high values for gauge factor. **[P.T.U. B.E.E.E. May 2009]**

Ans. Wire semiconductor devices produce large change in resistance for a small change in its strain, so its gauge factor is 50 times higher as compared to metal strain wire gauges.

Q. 13. What are thermistors ? **[U.P.S.C. I.E.S. 2001]**

Ans. Thermistors are also called the thermal resistors and the name is derived from thermally sensitive resistors, as the resistance of a thermistor varies as a function of temperature.

Thermistors are essentially semiconductor devices, that behave as resistors with high negative temperature coefficient (usually – 0.04 per °C) and are at least 10 times as sensitive as the platinum resistance thermometer.

Q. 14. List the uses of thermistors. **[P.T.U. B.E.E.E. May 2005]**

Ans. Uses of thermistors are : (*i*) Measurement of temperature (*ii*) Measurement of difference of two temperatures, (*iii*) Control of temperature, (*iv*) Temperature compensation, (*v*) Thermal conductivity measurement, (*vi*) Measurement of gas composition and (*vii*) Flow measurement.

Q. 15. Why is thermistor well suited to precision temperature measurement, control and compensation ?

Ans. Because thermistors are rugged in construction, cheaper in cost and highly sensitive devices.

Q. 16. What are uses of LVDT for measurement ? **[P.T.U. B.E.E.E. Dec. 2005]**

Ans. LVDTs are suitable for use in applications where the displacements are too large for strain gauges to handle. For example, LVDTs can be employed for measurement of displacements that range from a fraction of a mm to a few cm. If LVDT is to be employed for measurement of mechanical displacement greater than 25 mm, an appropriate mechanical ratioing (gearing) must be used.

Since the LVDTs can also be connected to other transducers, whose outputs are mechanical displacements, these are often employed together with other transducers for measurement of force, weight, pressure etc.

Q. 17. What is digital instrument ?

Ans. A digital instrument indicates the measured value in the form of decimal digits.

Q. 18. What is a digital voltmeter (DVM) ?

Ans. A digital voltmeter displays the value of ac or dc voltages being measured directly as discrete numerals in the decimal number system.

Q. 19. What are the advantages of a digital voltmeter (DVM) ?

Ans. A digital voltmeter displays the value of ac or dc voltage in decimal number system thus eliminating the errors due to parallax and observational errors. The use of DVMs increases the speed with which the readings can be taken. Also the output of DVMs can be fed to memory devices for future purpose which is not available in electronic voltmeter. On account of development of IC technology, it has been possible to reduce the size, cost of DVM. In fact DVM is less costly than analog type.

Q. 20. What is the principle of operation of a DVM ?

Ans. A DVM comprises of a clock generator, a circuit that converts the voltage to be measured into pulses and a digital frequency counter.

Q. 21. What is DMM ?

Ans. DMM (digital multimeter) is basically a digital voltmeter and may be used for the measurement of voltage, current (dc or ac) and resistance.

PROBLEMS

1. A thin circular wire of soft iron has a poisson's ratio of 0.6. Determine its gauge factor. [**Ans.** 2.2]
2. A resistance strain gauge is used to measure stress on steel. The steel is stressed to 1,200 kg/cm^2. Assume Young modulus of steel 2×10^6 kg/cm^2. Determine the percentage change of resistance of a strain gauge assuming gauge factor equal to 2.5. [**Ans.** 0.15%]

❑❑❑

13 CHAPTER

P-N Junction Diode

13.1. SEMICONDUCTOR MATERIALS

INSIDE THIS CHAPTER

We are familiar with conducting and insulating materials. *Conducting materials* (such as silver, copper, aluminium etc.) are good conductors of electricity and are characterized by a large electrical conductivity and small electrical resistivity. *Insulating materials* (such as porcelain, glass, quartz, rubber, bakelite etc.) are characterized by poor electrical conductivity and are used to block current from flowing where it is not to pass. There is another group of materials (such as germanium, silicon, boron, selenium) which are neither good conductors nor good insulators. At room temperature such materials have conductivities considerably lower than those of conductors and much higher than those of insulators. Such materials are called the *semiconductors*. The resistivity of various semiconductor materials lies in a very wide range from 10^{-4} to about 0.5 Ω-m whereas the resistivities of conductors and insulators are of the order of 10^{-8} Ω-m and 10^{12} Ω-m respectively. An even more important property of semiconductors is that *their resistance depends largely on various factors and, therefore, it can be controlled.*

The resistance of conducting materials except that of carbon, increases with the increase in temperature and resistance of insulating materials remains almost constant up to a certain temperature after which it suddenly falls off and insulating materials start conducting. Semiconductors behave in somewhat different way. The resistance of semiconductors decreases with the increase in temperature, as shown in Fig. 13.1, *i.e.* temperature coefficient of resistance of semiconductors is negative. Thus, the semiconductors behave like insulators at very low temperatures but act as conductors at high temperatures.

The resistivity of semiconductors depends on illumination and decreases in brighter surroundings (the "light" resistivity of a semiconductor is smaller than its "dark" resistivity).

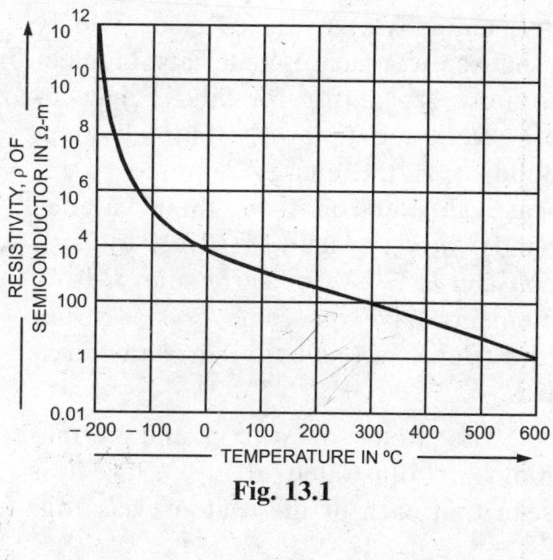

Fig. 13.1

The resistivity of semiconductors depends on the magnitude of the electric field in the semiconductor, so that the current in it is not proportional to the voltage, as with conductors that obey Ohm's law, but increases by far more than the voltage, *i.e.*, semiconductors are *nonlinear resistors*.

The semiconductors have another very important property—*the resistivity (or conductivity) of semiconductors changes considerably when even minute amounts of certain other substances, called the impurities, are added to them.* Moreover, the introduction of certain impurities (such as arsenic, antimony, gallium, indium etc.) may even change the mechanism of electrical conductivity in the semiconductor, and certain physical phenomena will begin to occur at the junction between sections of the semiconductor having different impurities. Such phenomena make semiconductor devices important as *diodes, transistors, SCRs, UJTs, varistors* and *thermistors*.

The *elemental semiconductors* (*i.e.*, semiconductors comprising a single chemical element) are germanium (Ge), silicon (Si), grey (crystalline) tin, selenium (Se), tellurium (Te), boron (B) etc. *Germanium* and *silicon* are two of the best known and widely used semiconductors.

The devices (such as diodes, transistors etc.) manufactured from semiconductors have many advantages such as compact size, low cost, light weight, rugged construction, more resistive to shocks and vibrations, instantaneous operation (no heating required), low operating voltage, high operating efficiency (no heat loss) and long life with essentially no ageing effect if operated within permissible limits of temperature and frequency.

13.2. INTRINSIC SEMICONDUCTORS

Semiconductor materials may be classified into two classes viz. *intrinsic* (or *pure*) *semiconductors* and *extrinsic* (or *impure*) *semiconductors*.

An *intrinsic semiconductor* is one which is made of the semiconductor material in its extremely pure form. A semiconductor is not truely intrinsic unless its impurity content is less than one part in 100 million parts of semiconductor. In fact, recent advances have reduced impurity levels in the pure material to 1 part in 10 billion (1:10,000,000,000). One might ask if such low impurity levels are really required. They are certainly required. It is to be noted here that the addition of one part impurity (of the proper type) per million in a wafer of silicon material can change that material from a relatively poor conductor to a good conductor of electricity.

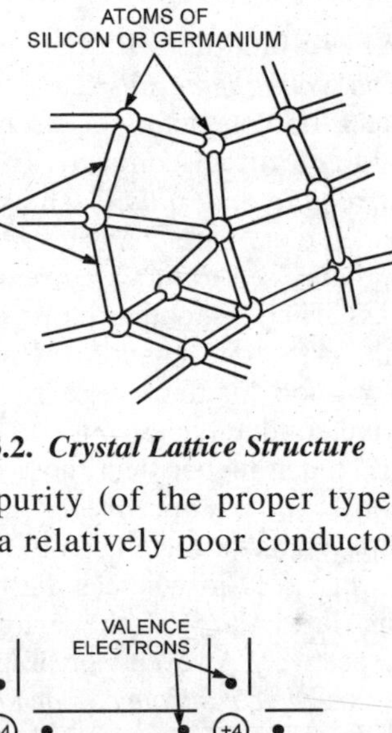

Fig. 13.2. ***Crystal Lattice Structure***

There are many semiconductor materials such as germanium, silicon, grey crystalline tin, selenium, tellurium, boron etc. available but silicon (Si) and germanium (Ge) are the two most widely used semiconductor materials in electronics. The reason for this is that energy required to break their covalent bonds (*i.e.* the energy required to release an electron from their valence bands) is very small (1.12 eV for silicon and 0.72 eV for germanium). Both elements have the same crystal structure (Fig. 13.2) and similar characteristics.

The atoms of silicon and germanium are illustrated in Fig. 13.4. It is seen that each of these atoms has four

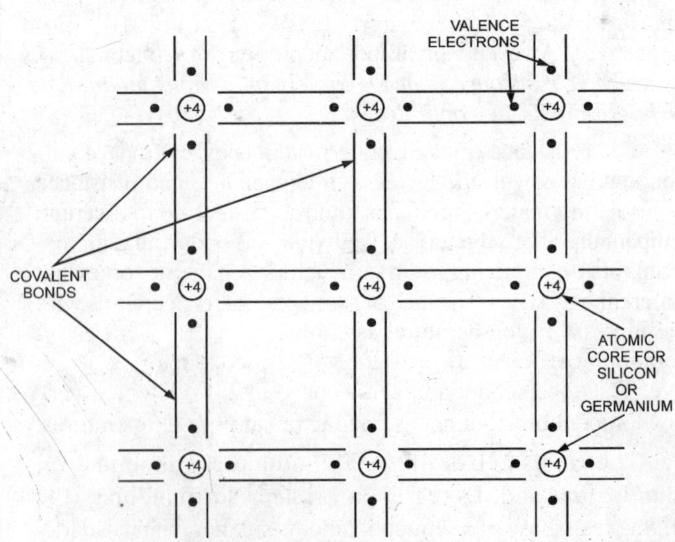

Fig. 13.3. ***Covalent Bonding***

electrons in a valence shell that can have a maximum of eight. Thus, we say that their valence shells have four electrons and four holes. *A hole is defined simply as an absence of an electron in a shell where one could exist.*

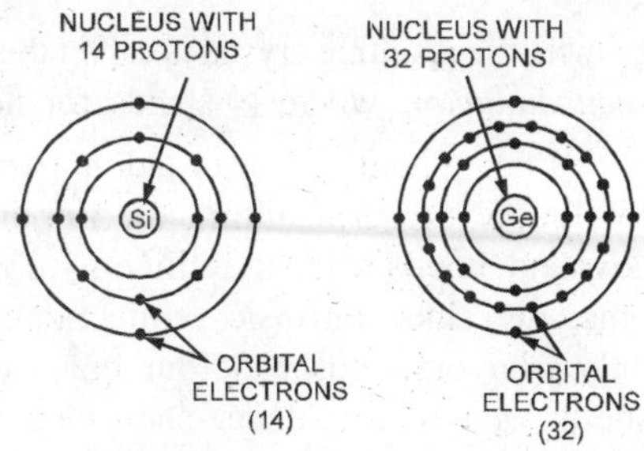

(*a*) Silicon (*b*) Germanium

Fig. 13.4. ***Bohr's Atomic Model***

Silicon is a tetravalent element and its atomic number is 14. It has, therefore, 14 protons and 14 electrons (two in first orbit, eight in second orbit and four in the third orbit), as shown in Fig. 13.4 (*a*).

Crystal lattice structure of silicon (also applicable to germanium) is shown in Fig. 13.2. Each atom is equidistant from its four neighbouring atoms. In a crystalline structure each atom shares its four valence electrons with the four adjacent atoms. This occurs because each tetravalent semiconductor atom needs a total of eight outer-shell electrons. The covalent bonds are illustrated in Fig. 13.2 by means of rods connecting the adjacent atoms. For simplified discussion, the crystal structure may be reduced to two dimensional, as illustrated in Fig. 13.3. The core represents the nucleus and all the orbiting electrons except the valence electrons. Since there are as many protons in the nucleus as there are electrons orbiting around it and there are four valence electrons, the core will have an excess +4 charge. The core of silicon contains 14 protons and 10 electrons (whereas germanium core has 32 protons and 28 electrons).

Germanium is also a tetravalent element *i.e.* it has 4 valence electrons. Its atomic number is 32. It has, therefore, 32, protons and 32 electrons [two in first orbit, eight in second orbit, eighteen in third orbit and four in fourth orbit], as shown in Fig. 13.4 (*b*). Its crystal lattice structure has already been discussed above.

For semiconductors at a temperature of absolute zero (– 273·15°C) the valence band is usually full, and there may be no electrons in the conduction band. At absolute zero temperature E_G for silicon and germanium are 1.21 eV and 0·785 eV respectively and it is difficult to get additional energy of this magnitude from applied electric field. Hence the *conductivity of intrinsic semiconductors at absolute zero temperature is zero*. However, at room temperature (300 K), the energy provided by the heat is sufficient to lift electrons from the valence band to the conduction band. Some electrons do jump the gap and go into the conduction band. So at room temperature, semiconductors are capable of conducting some electric current.

Since energy required in transferring electrons from valence band to conduction band is more in case of silicon than that in case of germanium, the conductivity of silicon will be less than that of germanium at room temperature.

13.3. EXTRINSIC SEMICONDUCTORS

Intrinsic (or pure) semiconductor by itself is of little significance as it has little current conduction capability at ordinary room temperature. However, the electrical conductivity of intrinsic semiconductor can be increased many times by adding very small amount of impurity (of the order of one atom per million atoms of pure semiconductor) to it in the process of crystallization. This process is called *doping* and the doped material is called the *impurity* or *extrinsic* semiconductor.

Germanium and silicon are tetravalent. So the impurity or doping material may be either pentavalent or trivalent. Accordingly the impurity introduced may be of two types viz. (*i*) donor type or N-type and (*ii*) acceptor type or P-type. Depending on the type of impurity added, the extrinsic semiconductors can be divided into two classes namely (*i*) N-type semiconductors and (*ii*) P-type semiconductors.

1. N-Type Extrinsic Semiconductors. When a small amount of pentavalent impurity, such as *arsenic, antimony, bismuth* or *phosphorus*, is added to a pure semiconductor crystal during the crystal growth, the resulting crystal is called the *N-type extrinsic semiconductor*, where N stands for negative.

When a pentavalent or a donor impurity is added to silicon (or germanium), the impurity atoms form covalent bonds with the silicon (or germanium) atoms; but since intrinsic semiconductor atoms have only four electrons and four holes in their valence shells, one spare valence-shell electron is produced for each impurity atom added. Each spare electron so produced enters the conduction band of pure semiconductor as a free electron. The pentavalent impurity is called the *donor type impurity* as it donates one electron to the conduction band of a pure semiconductor. Though each impurity provides only one free electron yet an extremely small amount of impurity provides enough atoms to supply millions of free electrons.

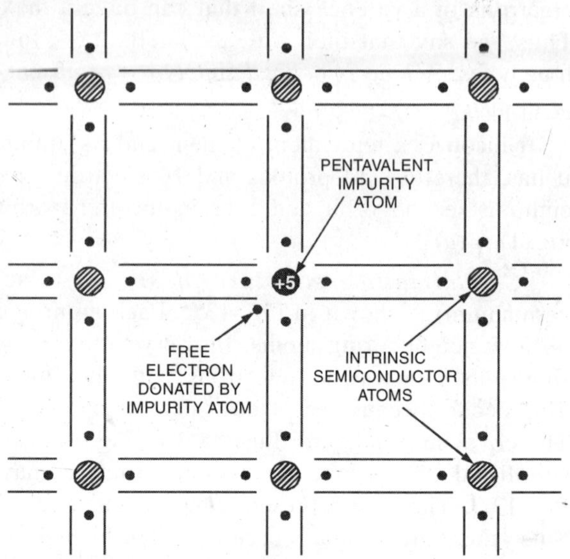

(a)

Fig. 13.5

The effect of adding donor impurity in a semiconductor is illustrated in Fig. 13.5. As shown in the figure, the unbound electron remains near its parent atom. This occurs because the impurity atom has a proton in its nucleus which attracts the unbound electron. However, due to the geometry involved, the electron is partially shielded from its parent atom and the attractive force between them is somewhat reduced. The energy required to detach this fifth electron from the atom is of the order of only 0.05 eV for silicon (0.01 eV for germanium). This energy is so small that at room temperature practically all such electrons become free. In other words, at room temperature each impurity atom donates one electron to the conduction band. These donated electrons are called *excess electrons*, because they are in excess to the electrons which are thermally generated by breaking covalent bonds. However, the number of thermally generated electron-hole pairs will be very small in comparison to the number of free electrons due to the impurity atoms. At room temperature in an intrinsic silicon there is about 1 free electron for every 10^{12} atoms (1 to 10^9 for germanium). If the dosage level were 1 in 10 million (10^7), the ratio $10^{12}/10^7$ *i.e.* 10^5 would indicate that the carrier concentration has increased by a ratio $10^5 : 1$.

Negative or N-Type Conductivity. When an electric field is applied, the excess electrons donated by impurity atoms move towards the positive terminal of the battery, as shown in Fig. 13.6. This constitutes the electric current. This type of conductivity is called the *negative* or *N-type conductivity* because the current flows through the crystal due to free electrons (negatively charged particles).

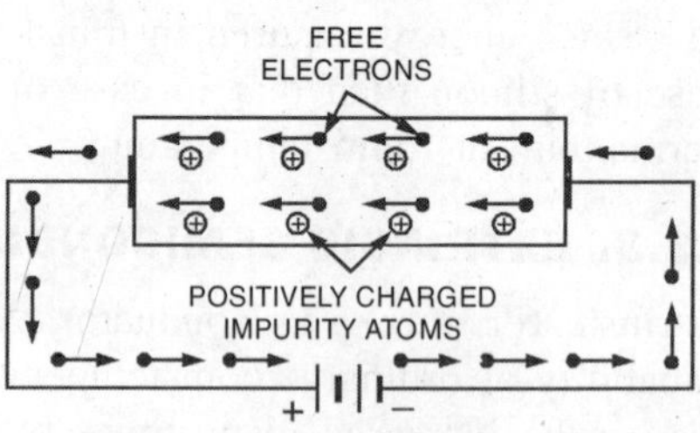

Fig. 13.6

Although N-type semiconductor has excess of electrons but it is electrically neutral. This is due to the fact that electrons are created by the addition of neutral pentavalent impurity atoms to the pure semiconductor *i.e.* there is no addition of either negative charges or positive charges.

Since pentavalent impurity donates electrons, which have negative charge, to the pure semiconductor, such an impurity is also called the *N-type impurity*.

Since current flowing through the crystal is primarily due to free electrons (negatively charged particles), such a conductor is called the *N-type semiconductor*.

In an N-type semiconductor material, the number of holes is small in comparison to parent intrinsic semiconductor because the larger number of electrons present increase the rate of recombination of electrons with the holes. Thus, the number of free electrons becomes far greater than the number of holes. That is why it is said that an N-type semiconductor has electrons as *majority* carriers, and the holes as *minority* carriers.

2. P-Type Extrinsic Semiconductors. When a small amount of trivalent impurity, such as *boron, gallium, indium* or *aluminium,* is added to a pure semiconductor crystal during the crystal growth, the resulting crystal is called the *P-type extrinsic semiconductor*, where P stands for positive. The effect of adding one of these impurities in Si or Ge is illustrated in Fig. 13.7.

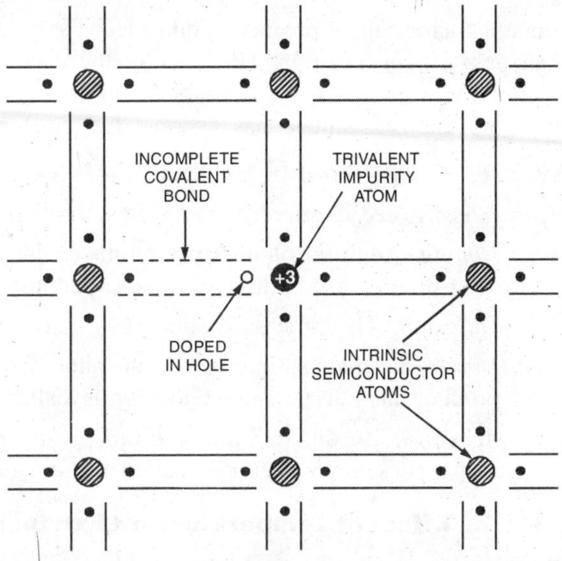

Fig. 13.7

When a trivalent impurity is added to silicon (or germanium), these impurity atoms form covalent bonds with four surrounding intrinsic semiconductor (silicon or germanium) atoms but one bond is left incomplete and gives rise to a hole, as illustrated in Fig. 13.7. Such impurities make available positive carriers because they create holes which can accept electrons. These impurities are consequently known as *acceptor* or *P-type* impurities. The amount of impurity which is required to be added to have an appreciable effect on the conductivity is very small.

Since holes can be said to have a positive charge, acceptor-doped semiconductor material is referred to as *P-type semiconductor*. The resulting P-type material is electrically neutral, for the same reasons as for the N-type material.

The noteworthy points about P-type semiconductors are :

(*i*) In P-type semiconductors, the majority carriers are holes while the minority carriers are electrons.

(*ii*) P-type semiconductor remains electrically neutral as the number of mobile holes under all conditions remains equal to the number of acceptors.

(*iii*) When an electric field is applied across a P-type semiconductor, as shown in Fig. 13.8, the current conduction is primarily due to holes. Here the holes are shifted from one covalent bond to another covalent bond. As the holes are positively charged, they are directed towards the negative terminal and constitute the hole current. The hole current flows more slowly than electron current in N-type semiconductor.

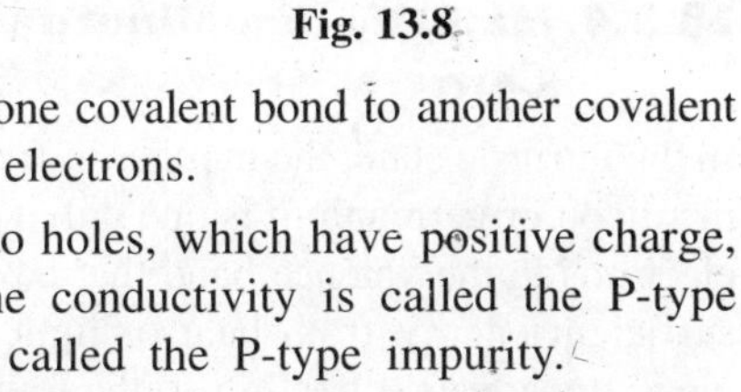

Fig. 13.8

(*iv*) In P-type conductors, the valence electrons move from one covalent bond to another covalent bond unlike N-type where current conduction is by free electrons.

Since the current flowing through the crystal is primarily due to holes, which have positive charge, such a semiconductor is called the P-type semiconductor and the conductivity is called the P-type conductivity. The impurity making P-type semiconductor is also called the P-type impurity.

Note : In the manufacturing process of an extrinsic semiconductor, before doping a pure crystal of silicon or germanium is produced. Then by controlling the amount of impurity, the properties of the semiconductor can be precisely controlled. Historically, pure germanium crystals were easier to produce than pure silicon crystals. This is why the earliest semiconductor devices were made of germanium. Eventually, manufacturing techniques improved and pure silicon crystals became available. Because of its advantages, silicon has become the most popular and useful semiconductor material.

13.3.1. Terms Used in Extrinsic Semiconductors

Various terms used in reference to extrinsic semiconductors are explained below.

1. *Doping.* Addition of impurity atoms to intrinsic semiconductor crystal is called the *doping* and the impurity used for doping is called the *dopant*.
2. *Diffusion.* The process of doping a semiconductor with impurities is known as *diffusion*.
3. *Donor.* Arsenic, antimony or phosphorus or any other pentavalent impurity used as dopant to produce an N-type semiconductor is called the *donor type impurity*.
4. *Acceptor.* Boron, gallium, indium or any other trivalent impurity used as dopant to produce a P-type semiconductor is called the *acceptor type impurity*.

13.3.2. Effect of Temperature on Extrinsic Semiconductors

Addition of a small amount of impurity (donor or acceptor) produces a large number of charge carriers in an extrinsic semiconductor, as already discussed above. In fact, this number is so large that the conductivity of an extrinsic semiconductor is many times that of an intrinsic semiconductor at room temperature.

If the temperature of an extrinsic semiconductor (say N-type) is raised, the additional thermal energy increases the thermally generated carriers. As a result, the concentration of minority carriers increases. Eventually, a temperature is reached, called the *critical temperature*, (85°C in case of Ge and 200°C in case of Si) when the number of covalent bonds, that are broken is very large, so that the number of holes becomes approximately the same as the number of electrons. The extrinsic semiconductor now behaves like an intrinsic semiconductor but with higher conductivity.

13.3.3. Charge Carriers

As discussed above, current conduction in an N-type semiconductor is due to excess of free electrons whereas in a P-type semiconductor it is due to excess of holes. One may think that N-type semiconductor has a net negative charge and P-type has a net positive charge. But this conclusion is wrong. Even though the N-type semiconductor has excess free electrons but these extra electrons were supplied by the atoms of donor impurity and each atom of donor impurity is electrically neutral. When the donor impurity is added, the term "*excess electrons*" refers to the excess with reference to the number of electrons required to fill the covalent bonds in the semiconductor crystal. The extra free electrons are there to increase the conductivity of the semiconductor, however, N-type semiconductor is electrically neutral. Similar situation occurs in P-type semiconductor which is also electrically neutral. Free electrons in N-type semiconductors and holes in P-type semiconductors are *charge carriers*.

13.3.4. Majority and Minority Charge Carriers

In the intrinsic state, the number of free electrons in silicon or germanium is due only to those few electrons in the valence band that have acquired sufficient energy from heat or light sources to break the covalent bond or to the few impurities that could not be removed. The vacancies left behind in the covalent bonding structure represent very limited supply of holes. In an

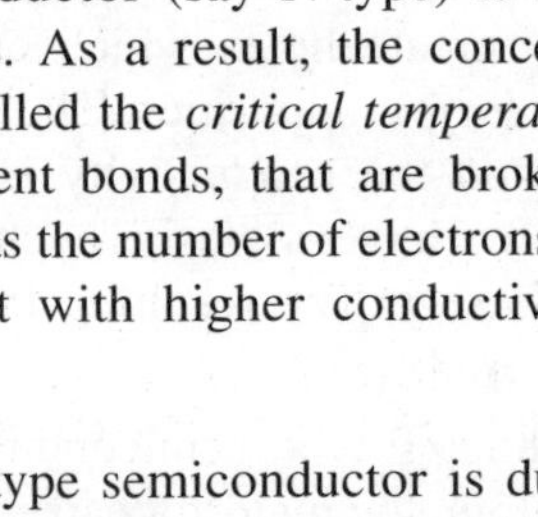

Fig. 13.9. *N-Type Material*

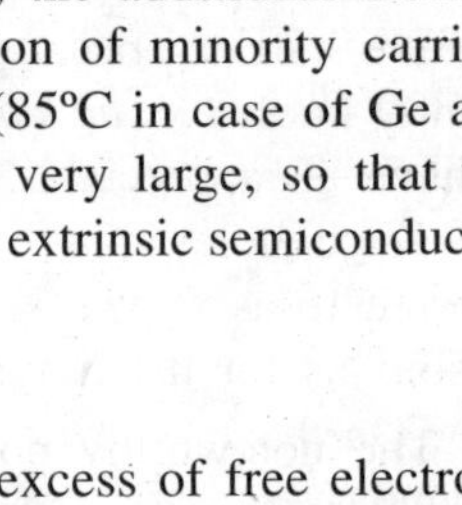

Fig. 13.10. *P-Type Material*

N-type material, the number of holes has not changed significantly from this intrinsic level and the number of electrons (both added and thermally generated) far outweighs the number of thermally generated holes. Hence in an N-type material electrons are majority carriers and responsible for flow of current.

The reverse is true for the P-type material *i.e.* P-type materials have holes as majority carriers and free electrons as minority carriers, as illustrated in Fig. 13.10.

13.4. BASIC STRUCTURE OF P-N JUNCTION

Most semiconductor devices employ one or more P-N junctions. The P-N junction is the control element for the performance of all semiconductor devices such as rectifiers, amplifiers, switching devices, linear and digital integrated circuits. The P-N junction is produced* by placing a layer of P-type semiconductor next to the layer of N-type semiconductor. The interface separating the N and P regions is referred to as the *metallurgical junction.*

Figure 13.11 represents two blocks of semiconductor material, one P-type, and the other N-type. The P-type semiconductor block has mobile holes (shown by small circles) and the same number of fixed negative acceptor ions (shown by encircled minus sign). Similarly the N-type semiconductor block has mobile or free electrons (shown by dots) and the same number of fixed donor positive ions. Normally the holes, which are the majority charge carriers in P-type of material, are uniformly distributed throughout the volume of that material. Similarly the electrons, which are the majority charge carriers in N-type of material, are uniformly distributed throughout the volume of that material. Each region is electrically neutral because each of them carries equal positive and negative charges.

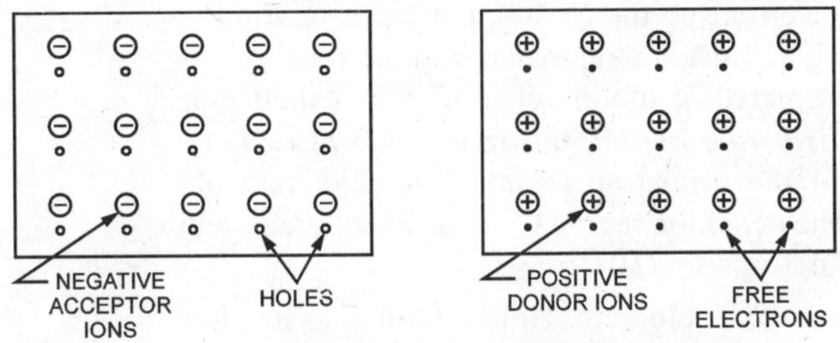

(a) P-Type Semiconductor Material *(b) N-Type Semiconductor Material*

Fig. 13.11

On the formation of P-N junction some of the holes from P-type material tend to diffuse across the boundary into N-type material and some of the free electrons similarly diffuse into the P-type material, as illustrated in Fig. 13.12. This happens due to density gradient (as concentration of holes is higher on P-side than that on N-side and concentration of electrons is higher on N-side than that on P-side). This process is known as *diffusion.*

The doping profile of an ideal uniformly doped P-N junction is depicted in Fig. 13.13.

As a result of the displacement of the charges, an electric field appears across the junction. Equilibrium is established when the field becomes large enough to restrain the process of diffusion. The general shape of the charge distribution may be as illustrated in Fig. 13.12 (*b*). The electric charges are confined to the neighbourhood of the junction, and consists of immobile ions.

We see that the free electrons crossing the junction create negative ions on the P-side by giving some atoms one more electron than their total number of protons. The electrons also leave positive ions (atoms with one fewer electron than the number of protons) behind them on the N-side. As negative ions are created on the P-side of the junction, the P-side acquires a negative potential, as illustrated in Fig. 13.12 (*e*). Similarly, the positive ions are created on the N-side and the N-side acquires a positive

* It is to be noted that a P-N junction is not formed by merging two separate P-type and N-type materials. If this were done, a polycrystalline structure would have resulted in, which would have not performed very well. The P-N junction found in the semiconductor devices is actually formed in a single monocrystalline structure by adding carefully controlled amounts of donor and acceptor impurities.

potential as illustrated in Fig. 13.12 (*d*). The negative potential on the P-side prevents the migration of any more electrons from the N-type material to the P-type material. Similarly, the positive potential on the N-side prevents any further migration of holes across the boundary. Thus, the initial diffusion of charge carriers creates a *barrier potential* at the junction.

The region around the junction is completely ionised. As a result, there are no free electrons on the N-side, nor holes on the P-side. Since the region around the junction is depleted of mobile charges it is called the *depletion region*, the *space-charge region*, or the *transition region*. The thickness of the depletion region (or layer) is of the order of 1 micron (10^{-6} m).

The electric field intensity in the neighbourhood of the junction is illustrated in Fig. 13.12 (*c*). The electrostatic potential variation in the depletion region is shown in Fig. 13.12 (*d*), and is the negative integral of the function E of Fig. 13.12 (*c*). The variation constitutes a potential-energy barrier against the further diffusion of holes across the barrier. The form of potential-energy barrier against the flow of electrons from the N-side across the junction is shown in Fig. 13.12 (*e*). It is similar to that shown in Fig. 13.12 (*d*), except that it is inverted, since the charge on electron is negative.

The necessity for the existence of a potential barrier called the *contact*, or *diffusion, potential* is now considered further. Under open-circuited conditions the net hole current must be zero. If this statement were not true, the hole density at one end of the semiconductor would continue to increase indefinitely with time, a situation which is obviously physically impossible. Since the concentration of holes in the P-side is much greater than that in the N-side, a very large diffusion current tends to flow across the junction from the P-type material to the N-type material. Hence an electric field must build up across the junction in such a direction that a

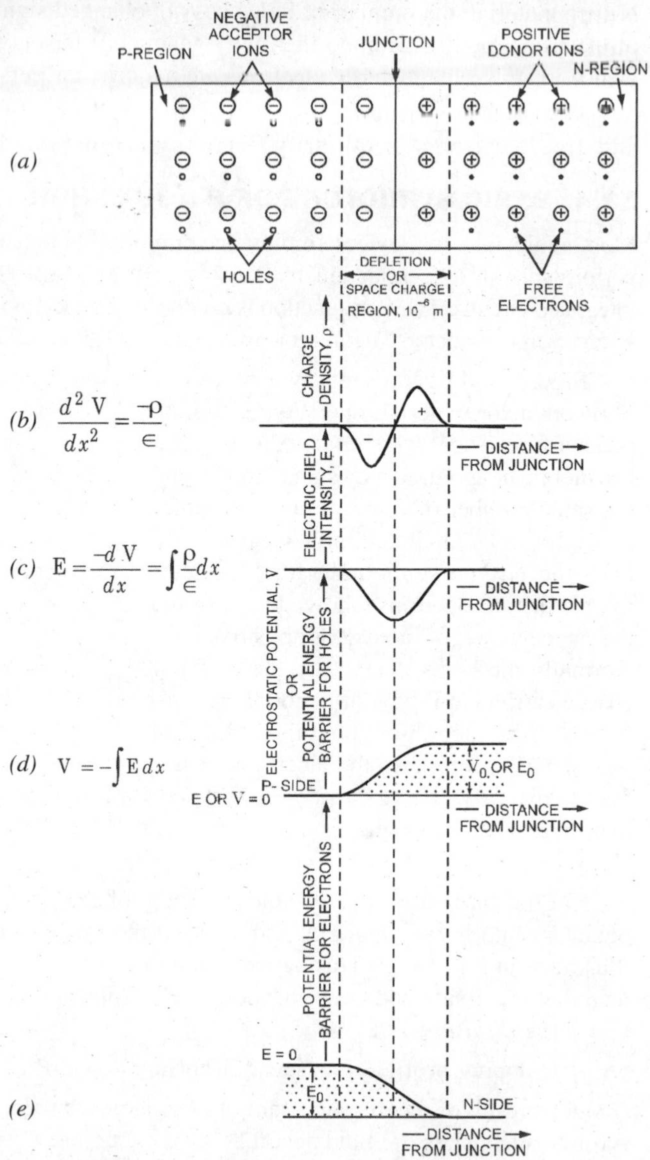

Fig. 13.12. ***A Schematic Diagram of a P-N Junction, Including The Charge Density, Electric Field Intensity, and Potential-Energy Barriers at The Junction (Not To Scale)***

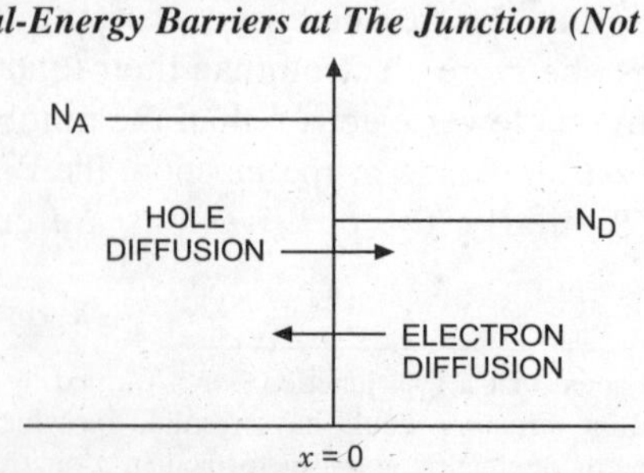

Fig. 13.13. ***Doping Profile of an Ideal Uniformly Doped P-N Junction***

drift current will tend to flow across the junction from the N-side to the P-side in order to counterbalance the diffusion current. This equilibrium condition of zero resultant hole current allows us to determine the height of the potential barrier in terms of the donor and acceptor concentrations. The magnitude of barrier potential V_0 is of the order of few tenths of a volt (0.3 V in case of Ge and 0.7 V for Si).

Barrier voltage depends on doping density, electronic charge and temperature. For a given junction, the first two factors are fixed, thus making barrier potential dependent on temperature. Increase in temperature creates more minority charge carriers leading to their increased drift across the junction. These extra minority charge carriers reduce the width of the depletion layer equivalent to reduction in potential barrier. It is found that for either silicon or germanium diodes, the barrier potential decreases 2 mV for each Celsius degree rise.

13.5. P-N JUNCTION AS A DIODE

The essential electrical characteristic of a P-N junction is that it constitutes a diode which permits the easy flow of current in one direction but restrains the flow in the opposite direction. We consider now, qualitatively, how this diode action comes about.

13.5.1. Forward Biasing

When an external field, with P-region connected to positive terminal and N-region connected to negative terminal of the battery, is applied across the junction, as shown in Fig. 13.14, the junction is said to be *forward biased.* In this circuit arrangement, the holes on the P-side being positively charged particles are repelled from the positive bias terminal and driven towards the junction. Similarly, the electrons on the N-side are repelled from the negative bias terminal and driven towards the junction. The result is that the depletion region is reduced in width, and the barrier potential is also reduced. If the applied bias voltage is increased from zero, the barrier potential gets progressively smaller and smaller until it effectively disappears and charge carriers can easily flow across the junction. Electrons from the N-side are then attracted across to the positive bias terminal on the P-side, and holes from the P-side flow across to the negative bias terminal on the N-side. Thus a majority carrier current flows. Since barrier potential is very small (0.3 V for Ge and 0.7 V for Si), a small forward voltage is sufficient to eliminate the barrier completely. Once the barrier is eliminated by the application of forward voltage, junction resistance becomes almost zero and a low resistance path is established in the entire circuit. The current, called the *forward current*, flows in the circuit.

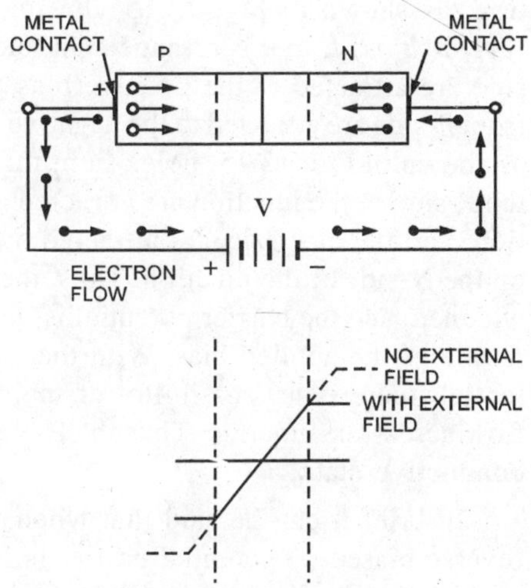

Fig. 13.14. ***Forward Biasing***

In brief it can be said that when the junction is forward biased (P-region connected to the positive terminal and N-region connected to the negative terminal of the battery), (*i*) barrier is reduced and at a forward voltage of 0.3 V in case of germanium (0.7 V in case of silicon) it is eliminated altogether, (*ii*) the junction offers low resistance, called the *forward resistance* r_f, to the flow of current and (*iii*) current flows in the circuit due to establishment of low resistance path and the magnitude of current depends upon the magnitude of applied forward voltage.

The mechanism of current flow in a forward biased P-N junction is summed up as follows :

(*i*) The electrons from the negative bias terminal continue to arrive into the N-region while the free electrons in the N-region move towards the P-N junction.

(*ii*) The electrons travel through the N-region as free electrons *i.e.* the current in N-region is by free electrons.

(*iii*) These free electrons on reaching P-N junction, combine with holes and become valence electrons. Since a hole is in the covalent bond and, therefore, when a free electron combines with a hole, it becomes a valence electron.

(*iv*) The electrons travel through P-region as valence electrons and current in this P-region is, therefore, by holes.

(*v*) These valence electrons, on reaching the left end of the crystal, flow into the positive terminal of the battery.

Thus we see that the current in the N-type region is due to movement of electrons whereas in the P-type region it is carried by the holes. However, in the external circuit (*i.e.* in connecting wires) the current is carried by electrons.

13.5.2. Reverse Applied Bias

If an external bias voltage is applied with positive terminal to N-side and negative terminal to the P-side of a P-N junction, as shown in Fig. 13.15, the junction is said to *be reverse biased*. In this arrangement, electrons from the N-side are attracted to the positive bias terminal, and holes from P-side are attracted to the negative bias terminal. Thus, as shown in Fig. 13.15, holes from the impurity atoms on the P-side of the junction are attracted away from the junction, and the electrons are attracted away from the atoms on the N-side of the junction. Thus, the depletion region is widened, and the barrier potential is increased by the magnitude of the applied bias. With the increased barrier potential, there is no possibility of majority carrier current flow across the junction. Thus the P-N junction is in a non-conductive state.

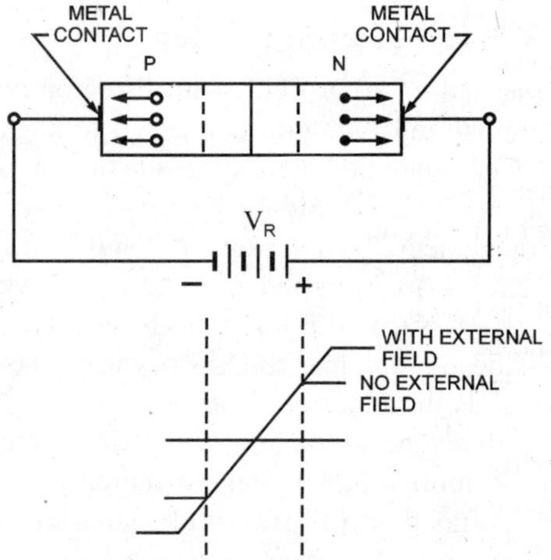

Fig. 13.15. ***Reverse Biasing***

In brief it can be said that when the P-N junction is reverse biased (*i*) potential barrier is increased and width of depletion is also increased (*ii*) the junction offers high resistance, called the reverse resistance r_r, to the flow of current (*iii*) owing to establishment of high resistance path, very small current, called the *leakage current* or *reverse current*, flows through the circuit. The reverse or leakage current is not affected by barrier and depends on temperature, material and construction of the junction.

Thus the P-N junction allows flow of large current only when forward biased and only small reverse (or leakage) current when reverse biased.

13.5.3. Volt-Ampere Characteristics of P-N Junction

The graph plotted between the potential difference across the P-N junction and the circuit current is known as volt-ampere characteristic of a P-N junction.

1. ***Forward Characteristics.*** When the external voltage is zero *i.e.* when the circuit is open, the potential barrier at the junction does not allow the flow of current and, therefore, the circuit current is zero, as shown in Fig. 13.16.

With forward bias to the P-N junction (*i.e.* P-region connected to +ve terminal and N-region connected to –ve terminal of the battery) very little current, called the *forward current*, flows until the *forward voltage exceeds* the junction barrier potential (0.3 V for Ge and 0.7 V for Si). The characteristics follow an exponential law. As the forward voltage is increased to the knee of the characteristic, the barrier potential is progressively reduced to zero, allowing more and more majority charge carriers to flow across the junction. Beyond the knee of the characteristic, the potential barrier is completely eliminated, forward current increases almost linearly with the increase in forward voltage and the P-N junction starts behaving as a resistor. If the forward voltage is increased beyond a certain value, extremely large current will flow and the P-N junction may get destroyed due to overheating.

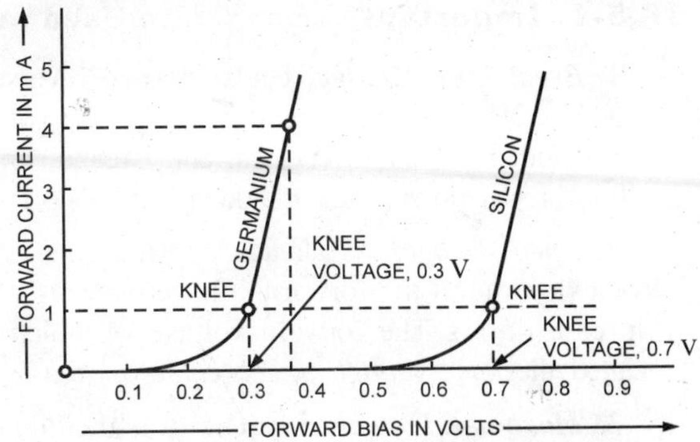

Fig. 13.16. ***Forward Characteristics of P-N Junction***

Because of the nonlinear knee of the curve, if a very small signal voltage is applied to the diode so that it only operates around the knee, the signal will be distorted. The signal must, therefore, be large enough to operate the diode over the linear part of the characteristic. Forward resistance of P-N junction is low (of the order of 75 Ω in case of Ge and 150 Ω in case of Si).

2. *Reverse Characteristics.* When a reverse bias is applied (*i.e.* when P-region is connected to the negative bias terminal and N-region is connected to the positive bias terminal), as shown in Fig. 13.15, the potential barrier at junction is increased, therefore, the junction resistance becomes very high and there is no possibility of majority carriers flowing across a reverse biased junction. But minority carriers generated on each side can still cross the junction. Electrons on the P-side are attracted across the junction to the positive bias terminal on the N-side and holes on the N-side may travel across the junction to the negative bias terminal on the P-side. This results in a very small current, as illustrated in Fig. 13.17. This current is called *reverse current* and is due to minority carriers.

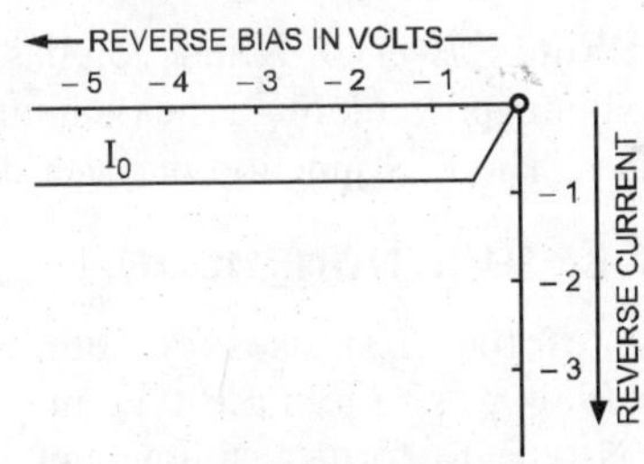

Fig. 13.17. ***Reverse Characteristics of P-N Junction***

For silicon the reverse current* is typically less than 1 μA and may be as low as 1 *n*A. In case of germanium, the reverse current may exceed 10 μA. Only a very small reverse voltage is necessary to direct all available minority carriers across the junction, and when all minority carriers are flowing across, further increase in bias voltage will not increase the current. This current is referred to as a *reverse saturation current* and is designated as I_0. Reverse resistance is very high (of the order of MΩ).

As discussed above, the reverse saturation current I_0 is made up of minority charge carriers crossing the junction. When the temperature of the semiconductor material is increased, the additional thermal energy causes more electrons to break away from atoms. This creates more hole-electron pairs and generates more minority charge carriers. So reverse saturation current I_0 increases with the increase in the junction temperature.** It has been found that I_0 approximately doubles for each 10ºC rise in temperature in case of germanium and for each 8ºC rise in temperature in case of silicon.

* There are more minority charge carriers available in germanium than in silicon, since charge carriers are more easily detached from germanium atoms.

** It can be shown that reverse saturation current I_0 depends upon electronic charge, doping density and junction area, as well as temperature. With the exception of temperature, all other factors (*i.e.* electronic charge, doping density and junction area) are constant for a given junction, I_0 is affected only by variation in temperature.

13.5.4. Important Terms Often Used in P-N Junction

1. *Breakdown Voltage.* Under normal reverse voltage, a very little reverse current flows through a P-N junction. However, on increasing the reverse voltage, a point may reach at which the junction breaks down with a sudden rise in reverse current. Breakdown voltage is defined as the reverse voltage at which P-N junction breaks down with a sudden rise in reverse current.

2. *Knee Voltage.* As already explained in Art. 13.5.3, with forward bias to P-N junction very little current flows until the forward voltage exceeds the junction barrier potential (0.3 V in case of Ge and 0.7 V in case of Si). The forward voltage, at which the current through the junction starts increasing rapidly, is called the *knee voltage* or the *cutin voltage*.

3. *Maximum Forward Current.* It is the highest instantaneous forward current that a P-N junction can conduct without damage to the junction. In case the forward current exceeds this rating, the junction will get destroyed due to overheating.

4. *Peak Inverse Voltage.* It is the maximum reverse voltage that can be applied to the P-N junction without damage to the junction. If the reverse voltage across the junction exceeds its peak inverse voltage (PIV), the junction may get destroyed owing to excessive heat.

5. *Maximum Power Rating.* It is the maximum power that can be dissipated at the junction without any damage to it.

Note : Every P-N junction has limiting values of maximum forward current, peak inverse voltage and maximum power rating. The P-N junction gives satisfactory performance when operated within these limiting values. The P-N junction may get destroyed due to excessive heat if any of these values is exceeded.

13.6. JUNCTION DIODE

One of the first uses of semiconductor materials was to join a P-type material with an N-type to form a simple but important semiconductor device called a *diode*.

The term *diode* denotes a two-electrode device. A *semiconductor diode* is simply a P-N junction with connecting leads or terminals on the two sides of the P-N junction. P-N junction is also called the crystal diode because the P-N junction is grown out of crystal. A diode is a unidirectional device permitting the easy flow of current in one direction but restraining the flow in opposite direction. A major application of diodes is in *rectification i.e.* in conversion of ac into dc. Semiconductor diode is becoming more and more popular these days due to its smaller size, cheapness, robustness and higher operating efficiency.

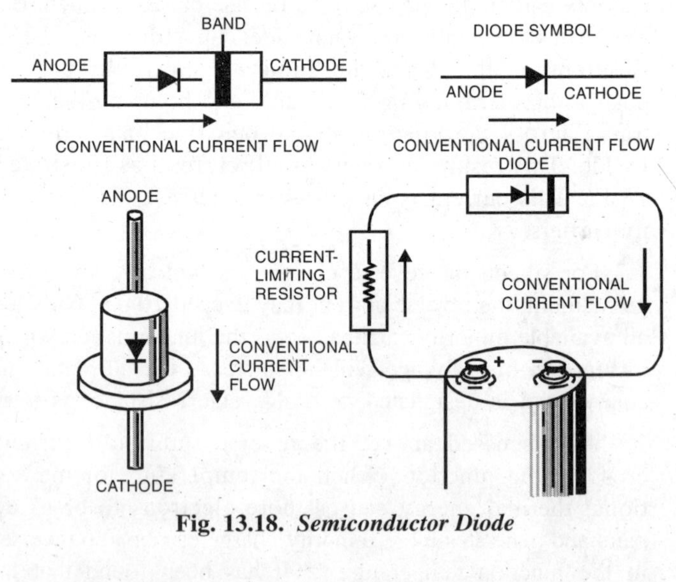

Fig. 13.18. ***Semiconductor Diode***

The P-type material of the diode is the positive electrode and is called the *anode* whereas the N-type material is the negative electrode and is called the *cathode.* The arrow symbol used for the anode represents the *direction of flow of conventional current.* The *electron current* flows in a direction *opposite* to the arrow.

The symbol for semiconductor diode is shown in Fig. 13.18. The arrow in the symbol indicates the direction of flow of conventional current. Diodes have *polarity* and, therefore, must be connected in circuits in the correct direction. Small diodes are marked with a band to indicate the cathode end. Larger diodes use the diode symbol to indicate their polarity. Diodes are available in many package configurations, depending on their ratings. Small diodes, used for low current application in circuits up to 1 A of average current, are generally packaged in a simple glass or plastic tube. Diodes for handling larger currents are packaged in metal containers and are often mounted by studs on metal plates to dissipate heat.

Diodes are rated in terms of their peak and average forward current-carrying capability and in terms of their reverse voltage or peak inverse voltage ratings. None of these values can be exceeded, otherwise the diode will get damaged. For the P-N diode junction, the reverse voltage rating applies to the instantaneous peak reverse voltage across the diode.

The diode junction can be viewed as an automatic switch that is closed when the P-type material is relatively positive and open when the P-type material is relatively negative. Remember, it is the *voltage across the diode* that is important *not* the voltage to another part of the circuit. Because the depletion region is never completely absent, it takes a small voltage across a diode to make current flow, even when the polarity is correct for flow of larger current.

Most diodes in use today make use of silicon semiconductor material; however, for some applications, germanium is also used.

The silicon junction diode is similar in appearance to the germanium diode but differs in properties. Silicon diodes have, in general, higher peak inverse voltage (PIV) and current ratings and wider temperature ranges than germanium diode.

(*i*) The peak inverse voltage ratings for the silicon can be as high as 1,000 V whereas for germanium it is near about 400 V.

(*ii*) Silicon can be operated at temperatures up to about 200°C compared with about 75–80°C for germanium.

(*iii*) The forward voltage drop of silicon diode is approximately double of that of the corresponding germanium diode. It is typically of the order of 0.7 V for commercially available silicon diodes and 0.3 V for germanium diodes.

(*iv*) The reverse current at a given temperature and voltage is approximately a hundredth of that of the corresponding germanium diode, but there is little sign of current saturation as in case of germanium. In fact, the reverse current of a silicon diode is approximately proportional to the square root of the voltage until breakdown is approached.

(*v*) The reverse current of a silicon diode, for a given voltage, practically doubles for every 8°C rise of temperature, compared with 10°C for germanium.

13.7. JUNCTION BREAKDOWN

When an ordinary P-N junction diode is reverse biased, normally only very small reverse saturation current flows. This current is due to movement of minority carriers. It is almost independent of the voltage applied. However, if the reverse bias is increased, a point is reached when the junction breaks down and the reverse current increases abruptly, as shown in Fig. 13.19. This current could be large enough to destroy the junction. If the reverse current is limited by means of a suitable series resistor, the power dissipation at the junction will not be excessive, and the device may be operated continuously in its breakdown region to its normal (reverse saturation) level. It is found that for a suitably designed diode, the breakdown voltage is very stable over a wide range of reverse currents. This quality gives the *breakdown diode* many useful applications as a *voltage reference source.*

The critical value of the voltage, at which the breakdown of a P-N junction diode occurs is called the *breakdown voltage.* The breakdown voltage depends on the width of the depletion region, which, in turn, depends on the doping level. The junction offers almost zero resistance at the breakdown point.

There are two mechanisms by which breakdown can occur at a reverse biased P-N junction : *avalanche breakdown* and *zener breakdown.*

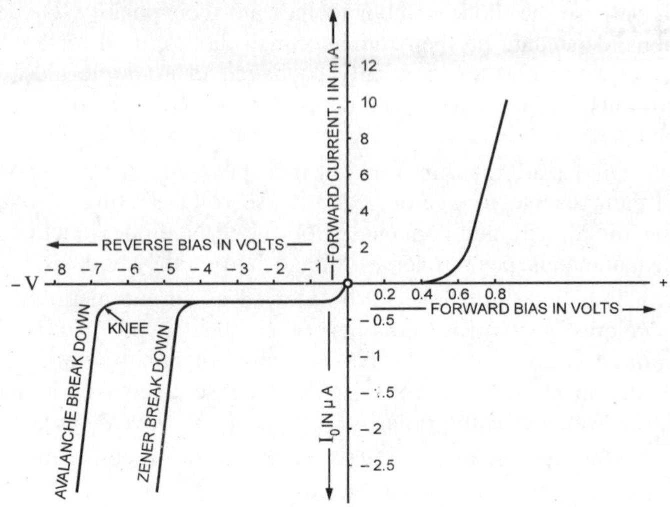

Fig. 13.19

The minority carriers, under reverse biased conditions, flowing through the junction acquire a kinetic energy which increases with the increase in reverse voltage. At a sufficiently high reverse voltage (say 5 V or more), the kinetic energy of minority carriers becomes so large that they knock out electrons from the covalent bonds of the semiconductor material. As a result of collision, the liberated electrons in turn liberate more electrons and the current becomes very large leading to the breakdown of the crystal structure itself. This phenomenon is called the *avalanche breakdown.* The breakdown region is the *'knee'* of the characteristic curve. Now the current is not controlled by the junction voltage but rather by the external circuit.

Under a very high reverse voltage, the depletion region expands and the potential barrier increases leading to a very high electric field across the junction. The electric field will break some of the covalent bonds of the semiconductor atoms leading to a large number of free minority carriers, which suddenly increase the reverse current. This is called the *zener effect.* The breakdown occurs at a particular and constant value of reverse voltage called the *breakdown voltage.* It is found that zener breakdown occurs at electric field intensity of about 3×10^7 V/m.*

Either of the two (zener breakdown or avalanche breakdown) may occur independently, or both of these may occur simultaneously. Diode junctions that break down below 5 V are caused by *zener effect.* Junctions that experience breakdown above 5 V are caused by *avalanche effect.* Junctions that break down around 5 V are usually caused by combination of two effects. The zener breakdown occurs in heavily doped junctions (P-type semiconductor moderately doped and N-type heavily doped), which produce narrow depletion layers. The avalanche breakdown occurs in lightly doped junctions, which produce wide depletion layers.

With the increase in junction temperature zener breakdown voltage is reduced while the avalanche breakdown voltage increases. The zener diodes have a negative temperature coefficient while avalanche diodes have a positive temperature coefficient. Diodes that have breakdown voltages around 5 V have zero temperature coefficient.

The breakdown phenomenon is reversible and harmless so long as the safe operating temperature is maintained.

Despite the fact that avalanche and zener constitute two types of breakdown diode, the name zener is commonly applied to both types of diodes.

---

* From basic circuit theory electric field intensity E is given as $\frac{V}{d}$, where V is the applied reverse voltage in volts and d is the width of depletion region in metres. The narrower the depletion region, the smaller the required bias to cause zener breakdown. A small reverse bias can produce a sufficiently strong electric field in a narrow depletion region.

13.7.1. Comparison of Zener and Avalanche Breakdown

S. No.	Avalanche Breakdown	Zener Breakdown
1.	This is observed in zener diodes having reverse voltage of 5 or more volts.	This is observed in zener diodes having reverse voltage of less than 5 volts.
2.	The valence bands are pushed into conduction band due to the energy imparted by colliding accelerated minority carriers.	The valance electrons are pulled into conduction band due to very intense electric field appearing across the narrow depletion region.
3.	The breakdown voltage increases with the increase in temperature *i.e.* temperature coefficient is positive.	The breakdown voltage decreases with the increase in temperature *i.e.* temperature coefficient is negative.
4.	V-I characteristics with the avalanche breakdown increase gradually. It is not as sharp as that with the zener breakdown.	V-I characteristics with the zener breakdown are very sharp.

13.8. IDEAL DIODE

It is a two terminal device and permits only *unidirectional conduction.* It conducts well in the forward direction and poorly in the reverse direction. It would have been ideal if it acted as a perfect conductor (with zero resistance or zero voltage drop across it) when forward biased and as a perfect insulator (with an infinite resistance or with no current through it) when reverse biased. The volt-ampere characteristics of such an ideal diode have been shown in Fig. 13.20 (*b*). An ideal diode acts like an *automatic switch.* The switch is closed when the diode is forward biased and is opened when reverse biased.

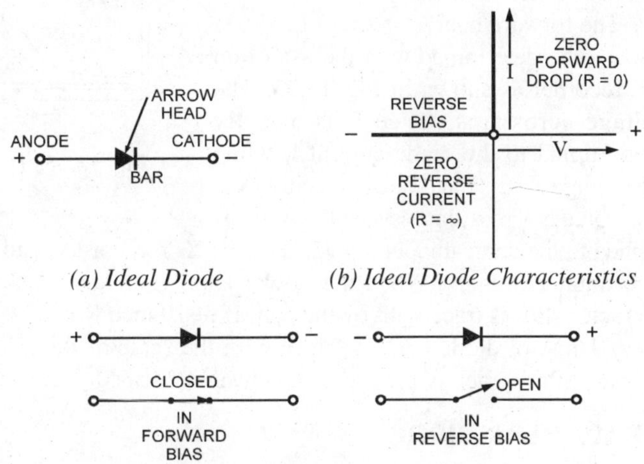

(*a*) *Ideal Diode* (*b*) *Ideal Diode Characteristics*

(*c*) *Switch Analogy*

Fig. 13.20

No diode can act as an ideal diode. An actual diode does not behave as a perfect conductor when forward biased and as a perfect insulator when reverse biased. Neither it offers zero resistance when forward biased nor infinite resistance when reverse biased. However, there are many applications in which diodes can be assumed to be nearly ideal devices, if the voltage drop across the diode when it is forward biased *i.e.*, V_D is taken into account. In cases when the circuit supply voltage V_S is much larger than the forward voltage drop V_D, V_D can be assumed constant without introducing any serious error. Also, the diode forward current I is usually so much larger than the reverse saturation current I_0 that the I_0 can just be ignored. These assumptions lead to a nearly ideal, or approximate, characteristics for germanium and silicon diodes, as illustrated in Fig. 13.21.

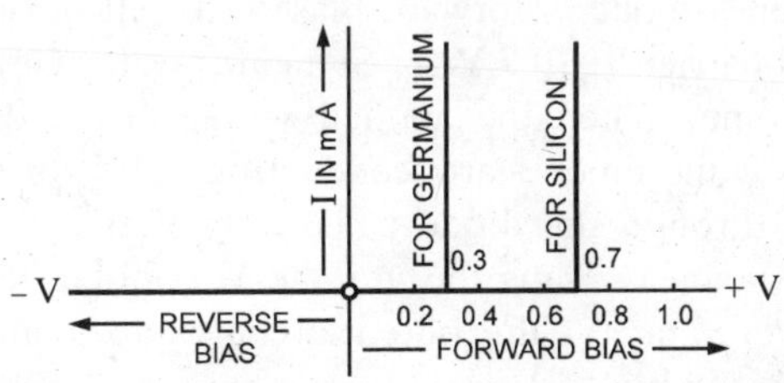

Fig. 13.21. ***Approximate Characteristics For Ge and Si Diodes***

13.9. DETERMINATION OF DIODE CHARACTERISTICS

The forward characteristics of a diode can be obtained by use of the circuit depicted in Fig. 13.22 (*a*). The diode voltage is set at a series of convenient levels, and the corresponding current levels are measured and recorded. The characteristics are then plotted from the values obtained. The reverse characteristics can be obtained in the same way, except that a very sensitive microammeter is required for measuring the diode reverse current. Also, the micro-ammeter must be connected directly in series with the diode as shown in Fig. 13.22 (*b*); otherwise the voltmeter current may introduce a serious error.

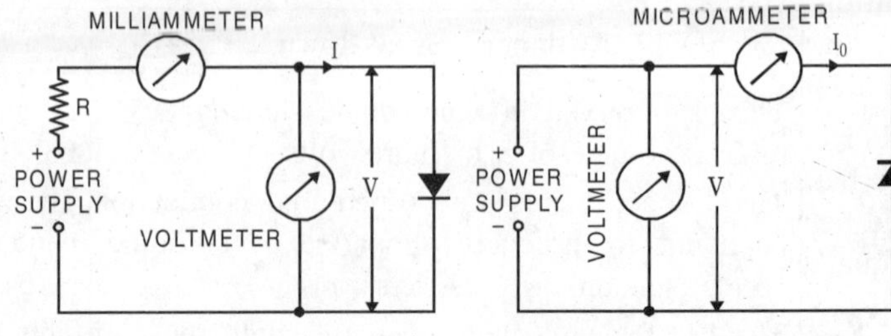

(*a*) *For Forward Characteristics* (*b*) *For Reverse Characteristics*

Fig. 13.22. ***Circuits For Determination of Diode Characteristics***

The forward characteristics of a diode can also be determined with the use of an XY recorder as shown in Fig. 13.23. The voltage across resistance R is directly proportional to the diode current I, while the voltage across the diode is of course V_D. The voltage across the resistor R *i.e.* V_R is applied to the vertical input terminals of the X-Y recorder, and voltage across the diode, V_D is applied to the horizontal terminals. When the power supply voltage is slowly increased from zero, the diode forward characteristics is traced out by the pen. If resistance R is of 1 kΩ, there is a voltage drop of 1 V across it for every 1 mA of diode current. Thus, with the scale of the XY recorder set to 1 V/cm, the (vertical) current ordinate of the graph is 1 mA/cm. A convenient scale for the (horizontal) voltage ordinate is 0.1 V/cm.

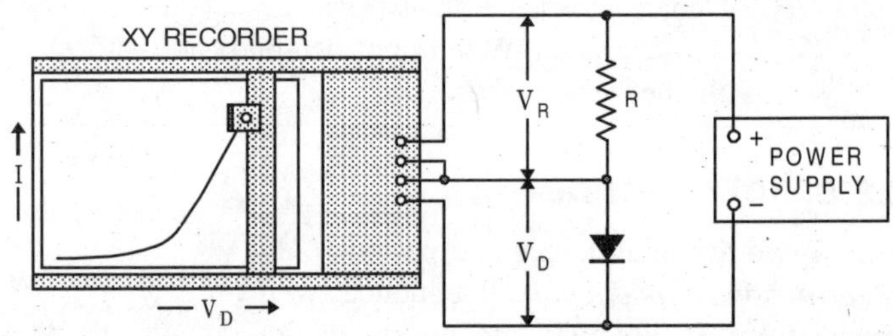

Fig. 13.23. ***Determination of Diode Forward Characteristics With The Use of XY Recorder***

13.10. REAL DIODE

As already mentioned in Art. 13.8, an actual diode neither behaves as a perfect conductor when forward biased nor as a perfect insulator when reverse biased.

When a diode is forward biased, it will not conduct till the potential barrier V_0 (0.7 V for Si and 0.3 V for Ge) at the junction is overcome. When the applied voltage exceeds the potential barrier V_0, the diode starts conducting. The forward current, I flowing through the diode causes a voltage drop in its internal resistance r'. Thus the voltage V applied across the diode is to overcome (*i*) the potential barrier V_0 and (*ii*) internal drop Ir'

$$\text{So } V = V_0 + Ir' \qquad ...(13.1)$$

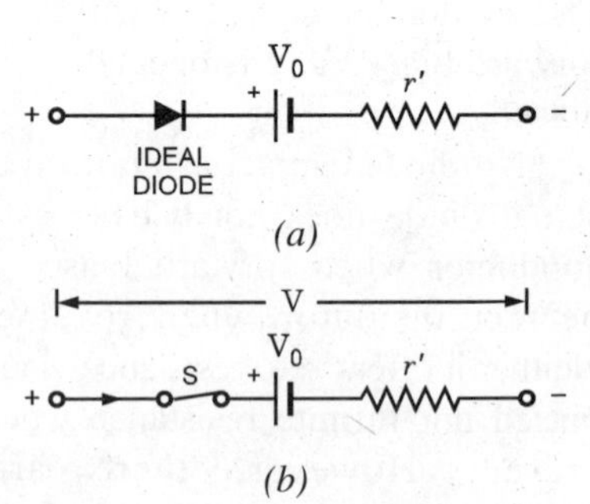

Fig. 13.24. ***Real Diode Representation Under Forward Bias Condition***

Hence, a real diode may be considered as equivalent to an ideal diode in series with a barrier potential of V_0 volts and a forward bias ac resistance r', as illustrated in Fig. 13.24 (*a*). The equivalent circuit of a real diode under forward bias condition is shown in Fig. 13.24 (*b*). This circuit shows that a real diode still acts as a switch when forward biased but the voltage required to operate this switch is V given by Eq. (13.1).

Voltage-current characteristic of a real diode under forward bias condition is shown in Fig. 13.25 which follows the Eq. (13.31). Here, the reciprocal of the slope of this characteristic represents the ac or dynamic resistance r' of the diode.

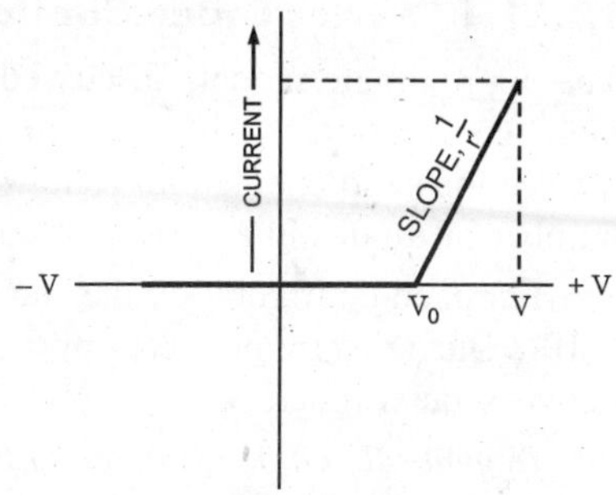

Fig. 13.25

For *large signal sources* whose voltage is much greater than the diode barrier potential V_0 (or knee voltage V_k), forward current I would be large and the ac or dynamic resistance r' of the diode will be equal to bulk resistance r_B of the diode.

For *small signal sources* whose voltage is much smaller than V_0 or V_k, forward current I would be small and the ac or dynamic resistance of r' of the diode will be equal to apparent forward or junction resistance r because for smaller values of forward current r_B is negligible.

For most applications, the ac or dynamic resistance r' of a diode can be neglected, being very small as compared to other elements and the real diode can be represented as shown in Fig. 13.26. The V-I characteristics for a real diode, neglecting its dynamic resistance become as illustrated in Fig. 13.21.

An actual or real diode does not behave as a perfect insulator when reverse biased. The reverse resistance of silicon and germanium diodes is quite large compared to the forward resistance. It may be of the order of many thousands of mega ohms. Hence an actual diode in the reverse bias condition can be thought of as equivalent to a resistor of high resistance.

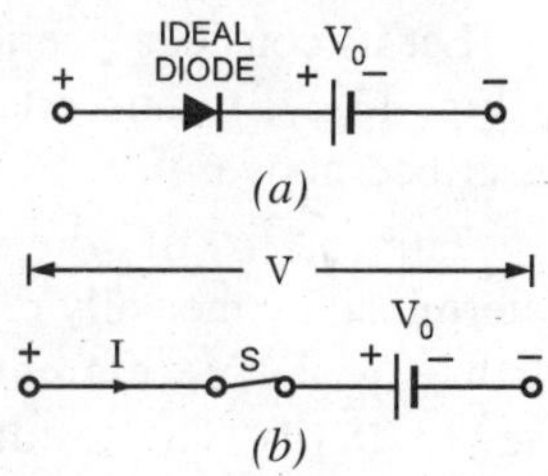

Fig. 13.26

The above discussion would be true only in case of small frequency signals. For large frequency signals the capacitive effects would have to be taken into account.

Unilateral diodes can be used for rectification of alternating currents, waveshaping of ac signals and many other applications. Waveshaping involves alteration of ac signals by passing only certain portions of the signal while blocking other portions.

Junction diodes designed for use as rectifiers must have volt-ampere characteristics close to those of an ideal diode. The reverse current should be negligible and forward current should exhibit little voltage dependence. The reverse breakdown voltage should be large and the offset or barrier potential V_0 in forward direction should be small. No diode can offer all these properties.

13.10.1. Piecewise Linear Diode Characteristic

A large-signal approximation which usually leads to a sufficiently accurate engineering solution is the **piecewise linear** representation. For instance, the piecewise linear approximation for a semiconductor diode is given in Fig. 13.25. The break point is not at the origin, and therefore, V_0 (or V_γ) is also known as the offset, or threshold, voltage. The diode behaves like an open circuit when the applied voltage is less than potential barrier V_0, and has a constant incremental resistance $r' = \frac{dV}{dI}$ when applied voltage exceeds potential barrier or threshold voltage. The equivalent circuit for this linear approximation is a constant voltage source in series with a resistor [Fig. 13.24 (*b*)]. This method models the diode with segments of straight lines : thus the name *piecewise linear model.*

13.11. INTERCONNECTION OF DIODES

Diodes can be connected in several ways such as series, parallel or series-parallel. In series arrangement, the diodes may be connected with polarities in the same direction (in forward-biased mode or in reverse-biased mode) or in opposite directions.

13.11.1. Series Diode Configurations

The approximate models discussed in preceding section are used for investigation of a number of series diode configurations with dc inputs. This establishes a foundation in diode analysis that will carry over into the sections and chapters to follow. The procedure described can, in fact, be applied to networks with any number of diodes in a variety of configurations.

For each configuration the state of each diode is first determined which diodes are "on" and which are "off"? Once determined, the appropriate equivalent can be substituted and the remaining parameters of the network determined.

In general, a diode is in the "on" state if the current caused by the applied sources is such that its direction matches that of the arrow in the diode symbol, and the diode voltage V_D exceeds V_0 (0.7 V in case of silicon and 0.3 V for germanium diode).

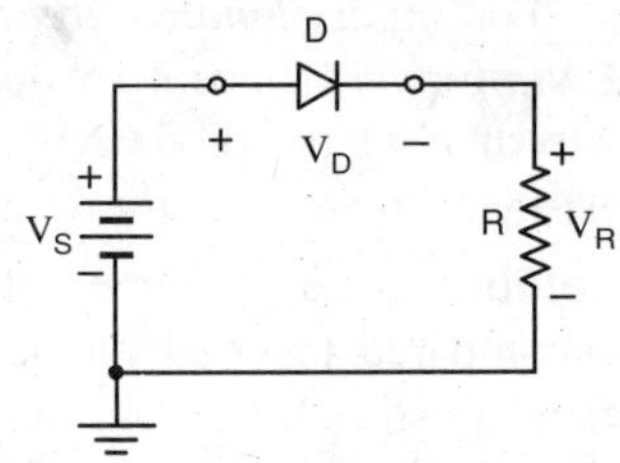

Fig. 13.27. *Series Diode Configuration*

For each configuration, mentally replace the diodes with resistive elements and note the resulting current direction. If the direction of resulting current "matches" with the arrow in the diode symbol, the diode is in conducting or "on" state. If the diode is in the "on state", diode can be replaced by a 0.7 V drop across the element or the network is redrawn with the V_0 or V_K equivalent circuit. It is preferable to include the 0.7 V drop across each "on" diode and draw a line through each diode in the "off" or open state.

Let us consider a series circuit, shown in Fig. 13.27, to understand the approach described above.

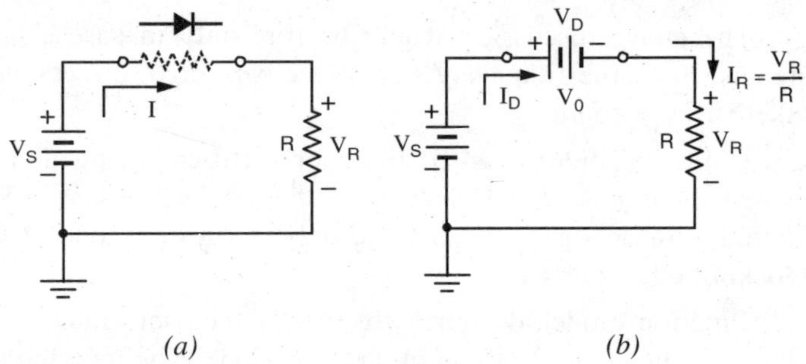

Fig. 13.28

The state of the diode is first determined by mentally replacing the diode with a resistive element as shown in Fig. 13.28 (*a*). The resulting direction of current I matches with the arrow in the diode symbol, and since V_S exceeds V_0 or V_K, the diode is in the conducting (or "on") state. The network is redrawn, as depicted in Fig. 13.28 (*b*), with the appropriate equivalent model for the forward-biased semiconductor diode. Note for future reference that the polarity of V_D is the same as would result if in fact the diode were a resistive element. The various relations between different voltages and current are given below :

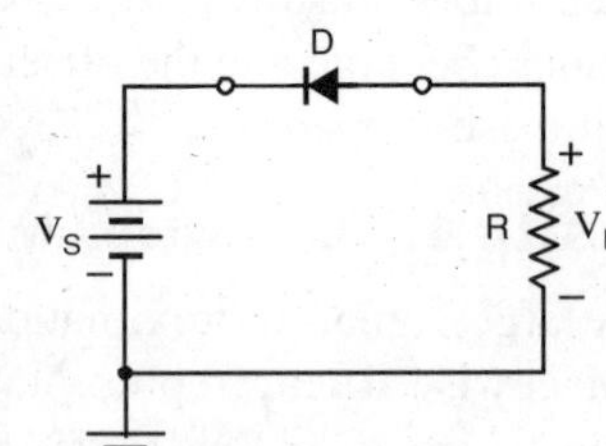

Fig. 13.29

Voltage drop across diode, $V_D = V_0 (\text{or } V_K)$ ...(13.2)

Voltage drop across resistor R, $V_R = V_S - V_D = V_S - V_0$ ...(13.3)

Diode current, $I_D = I_R = \frac{V_R}{R}$ ...(13.4)

Now let us consider the diode connected in reverse-biased mode, as shown in Fig. 13.29. If the diode is replaced mentally with a resistive element as depicted in Fig. 13.30 (*a*), it is revealed that the direction of flow of resulting current does not match the arrow in the diode symbol. It means that the diode is in "off" state

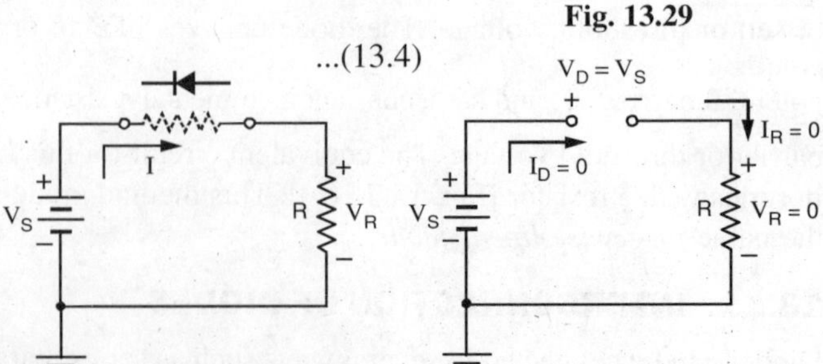

Fig. 13.30

and the resultant equivalent circuit will be as shown in Fig. 13.30 (*b*). Open circuit will not allow the current to flow in the circuit *i.e.* $I_D = I_R = 0$.

Voltage drop across resistor, $V_R = I_R R = 0 \times R = 0$ V ...(13.5)

and Voltage drop across diode, $V_D = V_S$ volts ...(13.6)

13.11.2. Parallel and Series-Parallel Configurations

The method described in the preceding section can be extended to the analysis of parallel and series-parallel configurations. For each area of application, simply the sequential series of steps applied to series diode configurations are to be matched.

Example 13.1. Determine diode voltage, diode current and voltage across 100 Ω resistor in the series diode configuration shown in Fig. 13.31.

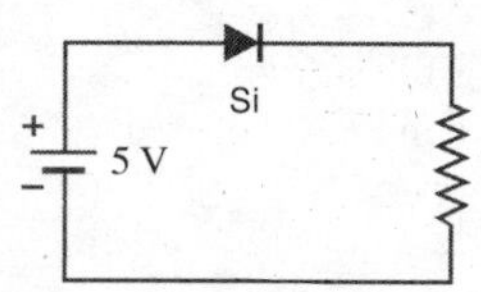

Fig. 13.31

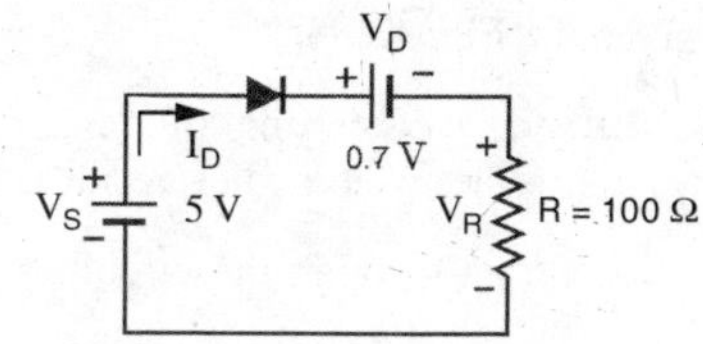

Fig. 13.32

Solution : Since the applied voltage establishes the current to flow in clockwise direction in the given circuit, it matches the arrow of the diode symbol, hence the diode is in conducting state.

Since the diode is silicon one, and the barrier potential for silicon diode is 0.7 V, so $V_0 = 0.7$ V.

The barrier potential acts in the direction opposite to that of the applied voltage and the equivalent circuit becomes as shown in Fig. 13.32. So

Diode voltage, $V_D = V_0 = 0.7$ V

Diode current, $I_D = \dfrac{V_S - V_D}{R}$ neglecting diode forward resistance

$= \dfrac{5 - 0.7}{100} = 0.043$ A or 43 mA **Ans.**

Voltage across resistor, $V_R = V_S - V_D = 5 - 0.7 = 4.3$ V **Ans.**

Example 13.2. Calculate forward current I_F for silicon diode with dynamic resistance r_d = 0.25 Ω used in the following circuit (Fig. 13.33).

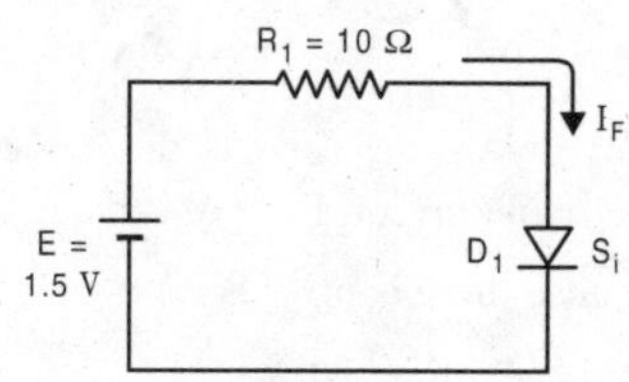

Fig. 13.33

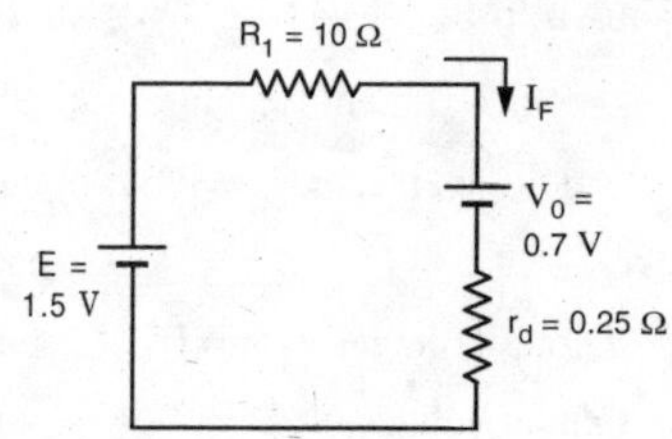

Fig. 13.34

Solution : Supply voltage, E = 1.5 V

Dynamic resistance, r_d = 0.25 Ω

Resistance, R_1 = 10 Ω

Potential barrier of silicon diode, V_0 = 0.7 V

The equivalent circuit is shown in Fig. 13.34.

Forward current, $I_F = \dfrac{E - V_0}{R_1 + r_d} = \dfrac{1.5 - 0.7}{10 + 0.25} = 0.078$ A or 78 mA **Ans.**

Example 13.3. Determine diode voltage, diode current and voltage across 100 Ω resistor in the series diode configuration shown in Fig. 13.31 with the diode polarity reversed.

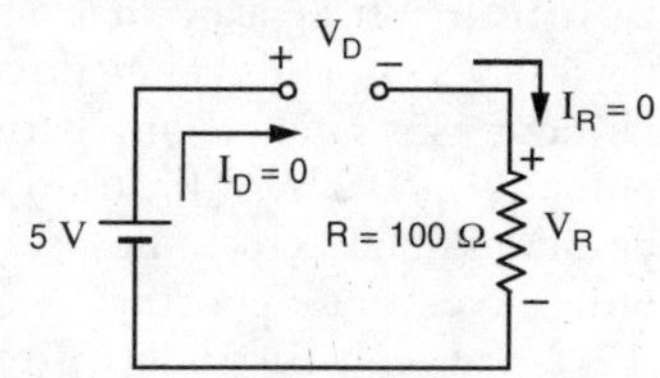

Fig. 13.35

Solution : Replacing the diode by open circuit, the equivalent circuit becomes as shown in Fig. 13.35.

Resulting current in the circuit, I = 0 ∵ the circuit is open

Diode current, I_D = Current through resistor, I_R = I = 0 A **Ans.**

Voltage across resistor, $V_R = I_R \times R = 0 \times 100 = 0$ V **Ans.**

Voltage drop across diode, $V_D = V_S = 5$ V **Ans.**

Example 13.4. Determine diode current, diode voltage and voltage across 500 Ω resistor, for the series diode configuration shown in Fig. 13.36.

Fig. 13.36

Solution : Although the applied voltage establishes a current in the same direction as that of diode arrow symbol, but the magnitude of applied voltage is insufficient, being less than potential barrier of silicon diode, to turn the silicon diode "on". The equivalent circuit is shown in Fig. 13.37.

Diode current, $I_D = 0$ A

Resistor current, $I_R = 0$ A

Voltage drop across resistor, $V_R = I_R \times R = 0 \times 500 = 0$ V **Ans.**

Diode voltage, $V_D = V_S = 0.4$ V **Ans.**

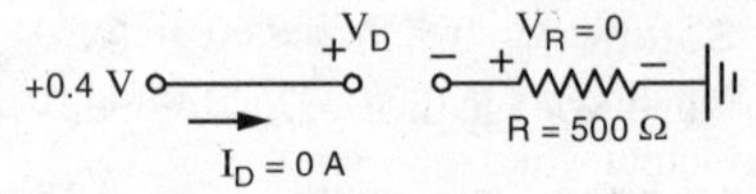

Fig. 13.37

Example 13.5. Find the voltage V_A of the circuit given in Fig. 13.38.

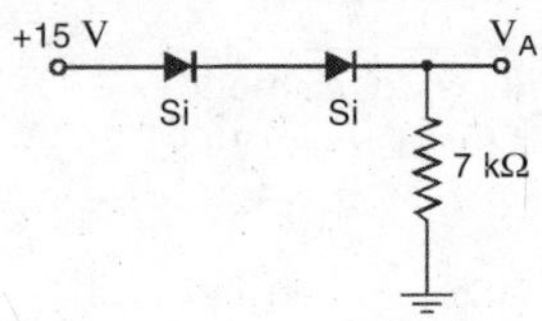

Fig. 13.38

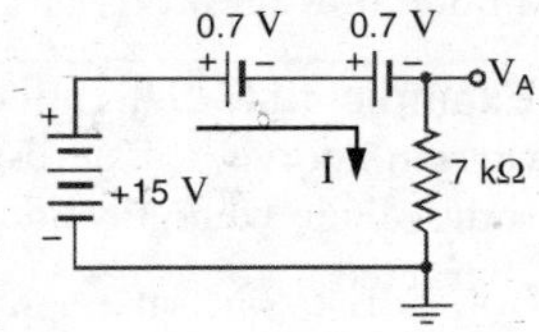

Fig. 13.39

Solution : Assuming diodes of negligible forward resistance, the given circuit is redrawn, as given in Fig. 13.39.

Current through the circuit,

$$I = \frac{V - 2V_0}{R} = \frac{15 - 2 \times 0.7}{7\ k\Omega} = 1.943\ mA$$

$\because$ barrier potential of silicon diode, $V_0 = 0.7$ V

Potential V_A = Voltage drop across 7 kΩ

= 1.943 mA × 7 kΩ = 13.6 V **Ans.**

Example 13.6. Determine diode current, diode voltage and output voltage for the circuit shown in Fig. 13.40.

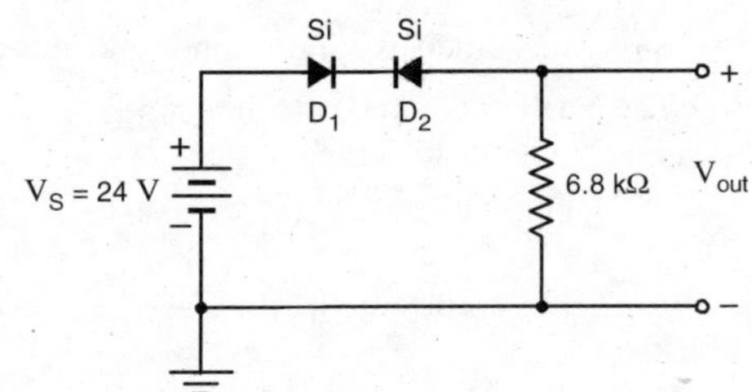

Fig. 13.40

Solution : Although the current established by the applied voltages matches the direction of symbol arrow of diode D_1 but not that of diode D_2. Thus there is an open circuit and the equivalent circuit is shown in Fig. 13.41.

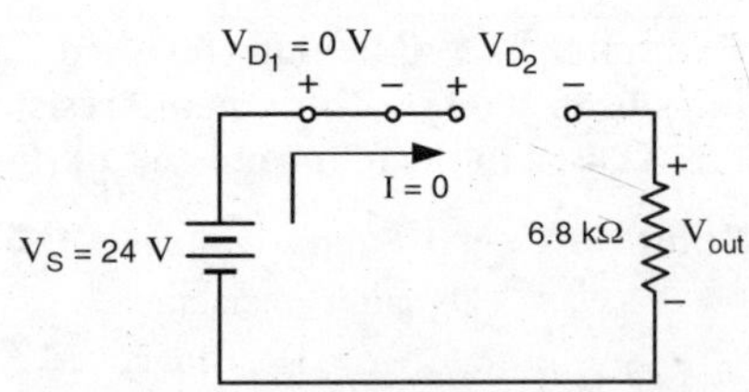

Fig. 13.41

Since the equivalent circuit shown in Fig. 13.41 is open one,

Diode current, I_0 = Output current, I_R

= Resulting circuit current, I = 0 A **Ans.**

Diode voltage, $V_{D_1} = 0$ because there is no bias

Output voltage, $V_{out} = IR = 0 \times 6.8\ k\Omega = 0$ V **Ans.**

Diode voltage, $V_2 = V_S - V_{D_1} - V_{out} = 24 - 0 - 0 = 24$ V **Ans.**

Example 13.7. Determine the voltage V_A in the circuit shown in Fig. 13.42.

Solution : It appears that on switching on the supply both the diodes turn 'on', but this is not true. When the supply voltage is switched on, germanium diode, whose barrier potential is 0.3 V, will turn on first and a level of 0.3 V will be maintained across the parallel branch. The silicon diode will never get opportunity to have 0.7 V across it and, therefore, remain in open-circuit state, as illustrated in Fig. 13.43.

So, $V_A = 15 - 0.3 = 14.7$ V **Ans.**

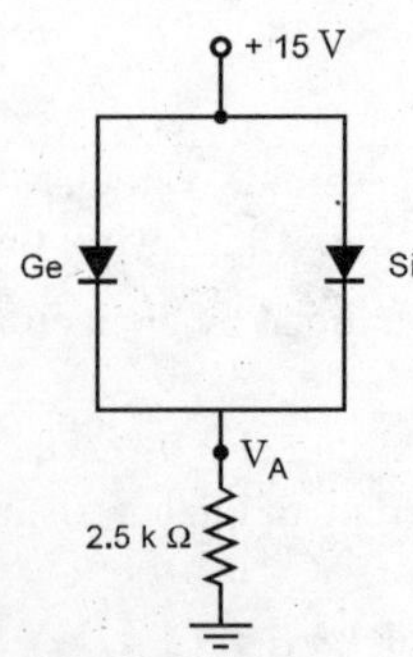

Fig. 13.42

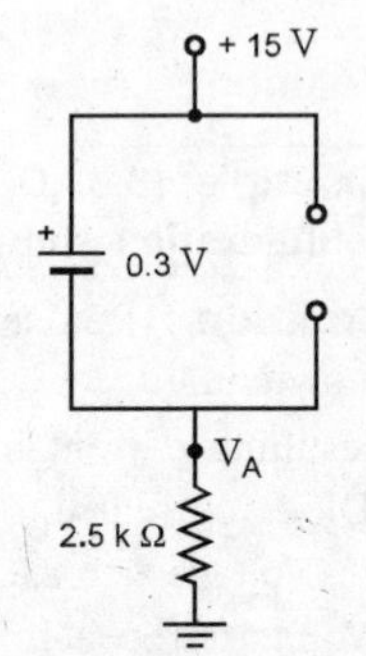

Fig. 13.43

Example 13.8. Determine V_0 and I for the following circuit (Fig. 13.44)

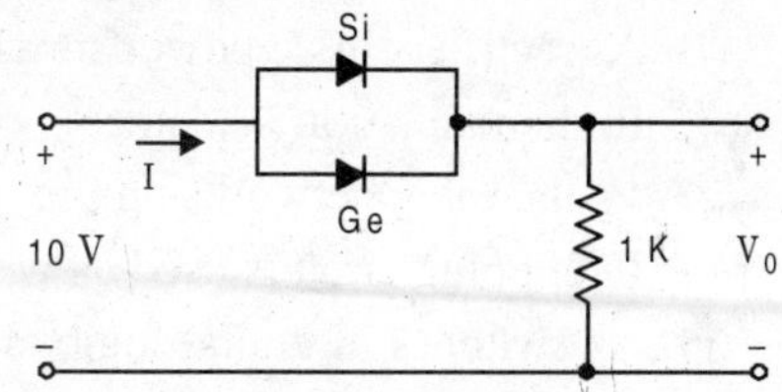

Fig. 13.44

Solution : As discussed in Example 13.7, the silicon diode will remain open circuited and the equivalent circuit is shown in Fig. 13.45.

Output voltage, $V_{out} = V_S$ – barrier potential of Ge diode

$= 10 - 0.3 = 9.7$ V **Ans.**

$$\text{Current, } I = \frac{V_{out}}{R_L} = \frac{9.7}{1 \times 10^3} = 9.7 \text{ mA} \quad \textbf{Ans.}$$

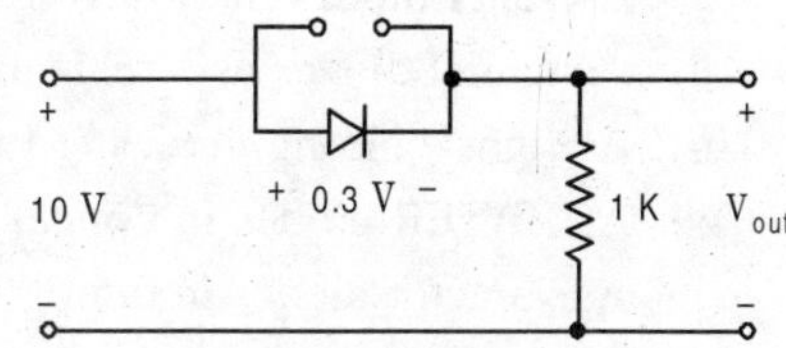

Fig. 13.45

Example 13.9. Obtain the voltage across the load (V_L) and across the diode (V_D) for symmetrical 5 V, trapezoidal waveform of the source voltage (V_i) shown in Fig. 13.46. Also draw equivalent diode characteristic circuit.

[*Pb. Technical Univ. Basic Electrical and Electronics Engineering Dec. 2004*]

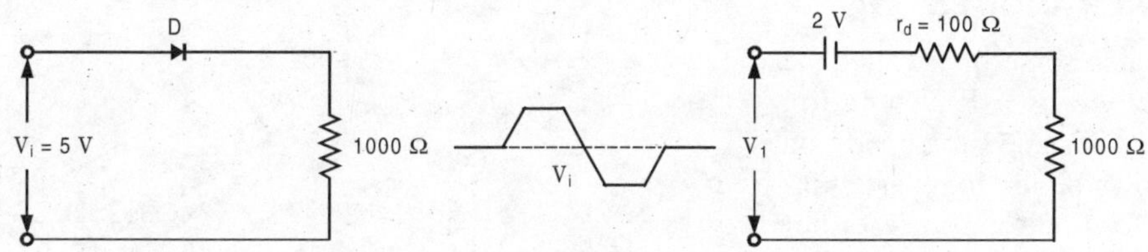

Fig. 13.46

Solution :

Input voltage, V_{in} = 5 V

Diode voltage, V_d = 2 V

During positive half cycle

Effective voltage, $V = V_{in} - V_d = 5 - 2 = 3$ V

Resistance in the circuit, $R = R_L + r_d = 1{,}000 + 100 = 1{,}100\ \Omega$

$$\text{Circuit current, } I = \frac{V}{R} = \frac{3}{1{,}100} = 2.73 \times 10^{-3} \text{ A or } 2.73 \text{ mA} \quad \textbf{Ans.}$$

Voltage across load, $V_L = I \times R_L = 2.73 \times 10^{-3} \times 1{,}000 = 2.73$ V **Ans.**

Voltage across diode, $V_d = 2 + I \times r_d = 2 + 2.73 \times 10^{-3} \times 100 = 2.273$ V **Ans.**

During negative half cycle, voltage across load is zero.

Example 13.10. A germanium diode has a saturation current of 10^{-8} A. Calculate the junction current for a forward bias of 0.4 V and 300 K. [*Pb. Technical Univ. Basic Electrical and Electronics Engineering Dec. 2005*]

Solution : Diode current is given as $I = I_0 (e^{V/\eta V_T} - 1)$ ...(*i*)

Saturation current, $I_0 = 1 \times 10^{-8}$ A

Forward bias voltage, V = 0.40 V

For germanium, η = 1

At room temperature (300°K), V_T = 26 mV = 0.026 V

Substituting above values in Eq. (*i*), we have

Junction current, $I = 1 \times 10^{-8} (e^{0.40/0.026} - 1) = 0.048$ A = 48 mA **Ans.**

13.12. P-N DIODE APPLICATIONS

An ideal P-N junction diode is a two terminal polarity sensitive device that has zero resistance (*i.e.* diode conducts) when it is forward biased and infinite resistance (*i.e.* diode does not conduct) when reverse biased. Because of this property the diode finds use in many applications as enumerated below :

(i) As rectifiers in dc power supplies.
(ii) In demodulation or detector circuits.
(iii) In clamping networks employed as dc restorers in TV receivers and voltage multipliers.
(iv) In clipping circuits used as waveshaping circuits in computers, radars, radio and TV receivers.
(v) As switches in digital logic circuits.

The same P-N junction with different doping concentrations finds special applications as follows :

(i) As zener diodes in voltage regulators, peak clippers, and in switching operation.
(ii) As tunnel diodes as a relaxation oscillator at microwave frequencies.
(iii) As light-emitting diodes (LEDs) in digital displays.
(iv) As LASER diodes in optical communications.
(v) As varactor diodes in tuning sections of radio and TV receivers.
(vi) As detectors (PIN photodiode) in optical communication circuits.

13.13. ZENER DIODES

Zener diode, also sometimes called the *breakdown diode*, is a P-N junction diode specially designed for operation in the breakdown region in reverse bias condition. The diode may use either zener breakdown mechanism or avalanche breakdown mechanism, already discussed in detail in Art. 13.7. Despite the fact that avalanche and zener constitute two types of breakdown diodes, the name zener is commonly applied to both types of diodes. The breakdown diode may be silicon or germanium one but silicon is preferred over germanium because of higher operating temperature and current capability. The knee point is also more sharp in case of a silicon diode. These diodes are analogous to gas discharge tubes in which large current appears on reaching the breakdown potential.

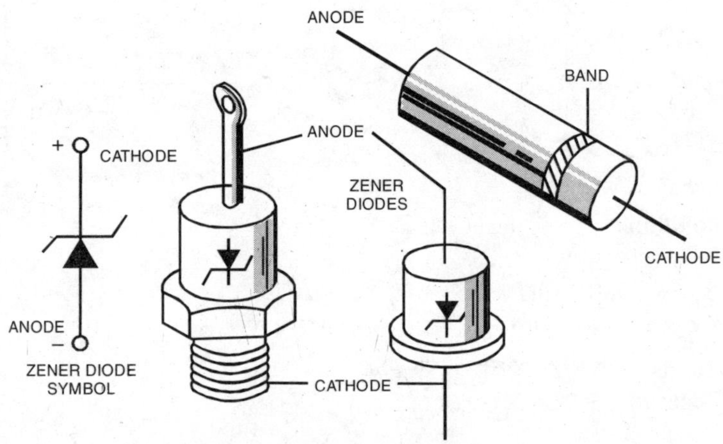

Fig. 13.47. ***Zener Diode***

Zener diodes are connected into circuits with their *cathodes* as the *positive* terminal because they operate in the reverse-bias mode. Zener diodes are packaged in the same way as conventional semiconductor diodes. Small zener diodes are marked the same way as conventional diodes with a cathode band. The larger diodes are marked with the zener diode symbol shown in the illustration (Fig. 13.47). In addition to have a voltage rating (reverse breakdown), zener diodes have a power rating that must not be exceeded.

The symbol of a zener diode is given in Fig. 13.47. This is similar to that of a normal diode except that the line representing the cathode is bent at both ends *i.e.* the bar is turned into Z-shape; stands for zener.

13.13.1. Volt-Ampere Characteristics

It has already been discussed in Art. 13.7 that when the reverse bias on a crystal diode is gradually increased, a point is reached when the junction breaks down and a reverse current increases abruptly, as illustrated in Fig. 13.48. The breakdown region is the knee of the reverse characteristic, as illustrated in Fig. 13.48. This breakdown of junction was first explained by the American scientist, C. Zener. So, the breakdown voltage is sometimes called, *zener voltage* V_Z and the sharply increased current is called the *zener current* I_Z.

Volt-ampere characteristics of typical silicon and germanium diodes are illustrated in Fig. 13.48. Breakdown diode is operated over the breakdown region and the current is limited by an external resistance. The breakdown or zener voltage V_Z depends upon the amount of doping. A heavily doped diode will have very thin depletion layer, so zener voltage will be very low. By controlling the thickness of depletion layer and doping concentration it is possible to make the zener diode to break down at a specified voltage, called the zener voltage (ranging from 2.4 V to 200 V). The forward characteristics of a zener diode are similar to those of an ordinary P-N junction diode, as illustrated.

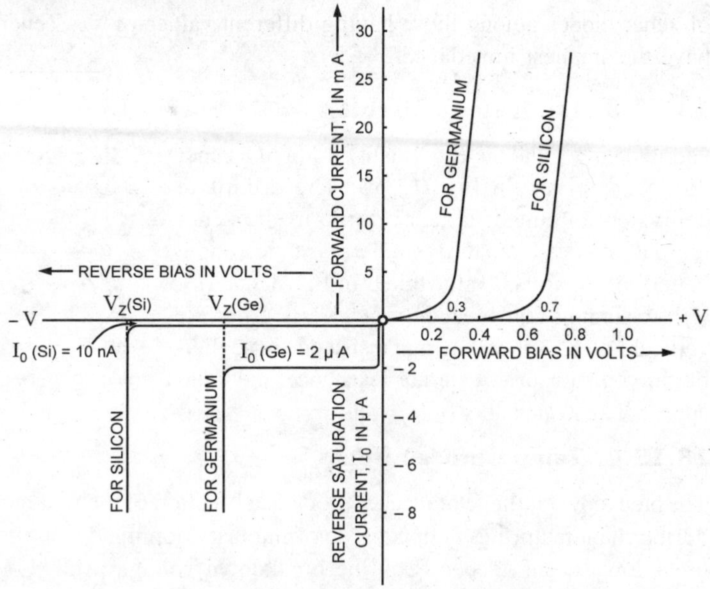

Fig. 13.48. ***Volt-Ampere Characteristics of Typical Silicon and Germanium Zener Diodes***

During the operation in the breakdown region, it does not burn out immediately. As long as the current through the diode is limited by the external circuit within permissible values, it does not burn out.

13.13.2. Zener Impedance

This is very important parameter of a zener diode. Zener impedance Z_Z is essentially the dynamic resistance of a zener diode (similar to dynamic resistance of an ordinary P-N diode). It is defined as the reciprocal of the slope of the zener curve

$$i.e.\ r_Z = \frac{\Delta V_Z}{\Delta I_Z} \quad ...(13.7)$$

where ΔV_Z and ΔI_Z are the small variations in voltage and current respectively.

It is usually measured just above the knee in zener region and is named as Z_{ZK}. When measured at test point, it is designated Z_{ZT}. It decreases with increase in zener current. For this reason, the diode should be operated with as much reverse current as possible, consistent with rating limitations.

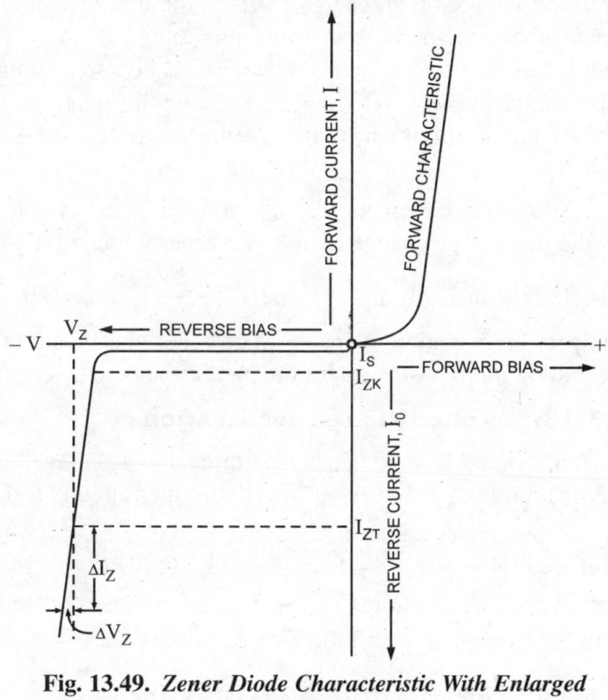

Fig. 13.49. ***Zener Diode Characteristic With Enlarged View of Breakdown Region***

Small value of zener impedance is usually desirable as it indicates steep curve where relatively large change in current, ΔI_Z occurs with a small change in voltage ΔV_Z.

Ideally the breakdown curve should be a perfect vertical line and r_Z should be zero. In practice, however, r_Z has a small value ranging from a few ohms to several hundred ohms. There is also a variation in impedance of zener diodes among those having different values of V_Z. Zener diodes with breakdown voltage near 7 V have the smallest impedances.

13.13.3. Equivalent Circuit

Circuit symbol and the equivalent circuit of a zener diode are given in Fig. 13.50. The complete equivalent circuit of a zener diode in the zener region includes a small dynamic resistance and a battery of V_Z volts, as illustrated in Fig. 13.50 (*b*).

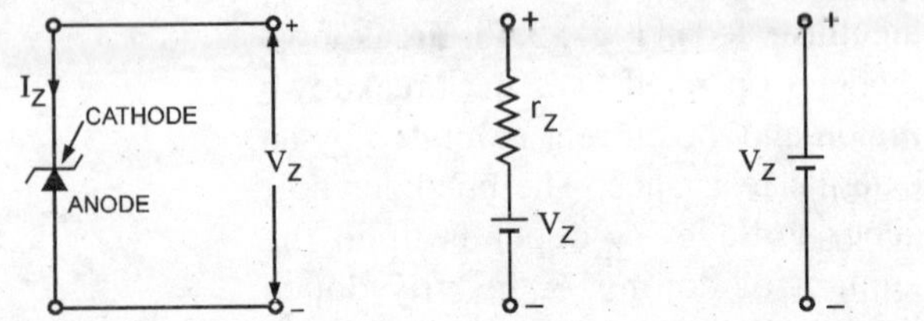

(a) Zener Diode Symbol *(b) Complete* *(c) Approximate*

Fig. 13.50. ***Equivalent Circuits of Zener Diode***

Assuming that breakdown curve is a perfect vertical line *i.e.* zener dynamic resistance is negligible in comparison to external resistance, the equivalent circuit becomes as illustrated in Fig. 13.50 (*c*) *i.e.* it appears like a *dc* battery of V_Z volts.

13.13.4. Temperature Effects

The breakdown voltage of a zener diode is a function of the width of its depletion region*, which is controlled during manufacturing by the degree of impurity doping. As discussed in Art. 13.7, the mechanism by which breakdown occurs depends on the breakdown voltage itself. For V_Z less than 5 V, the high electric field intensity across narrow depletion region (about 3×10^7 V/m) strips carriers directly from their bonds, a phenomenon usually known as zener breakdown. For V_Z exceeding 8 V, breakdown occurs as a result of collision between high-energy carriers, the mechanism known as *avalanching*. Between 5 V and 8 V, both the avalanching and zener mechanisms contribute to breakdown. The practical significance of these facts is that the breakdown mechanism determines how temperature variations affect the value of zener voltage V_Z. Low-voltage zener diodes that break down by zener mechanism have negative temperature coefficient i.e. V_Z decreases with the increase in temperature. Higher-voltage avalanche zeners have positive temperature coefficients.

The effect of temperature on zener voltage is given in terms of temperature coefficient, which is defined as the percentage change in nominal zener voltage for each degree Celsius of change in junction temperature. Mathematically it is given as $\frac{\Delta V_Z}{V_Z (t - t_0)} \times 100$ where ΔV_Z is the change in zener voltage due to temperature variation of $(t - t_0)$. The zener coefficient may be positive or negative, depending on current, becoming more positive with the increase in current.

13.13.5. Zener Diode Specifications

A zener diode is specified by its breakdown voltage V_Z, breakdown current, I_{ZK}, the maximum power dissipation $P_{Z\,(max)}$ and zener impedance measured at test point, Z_{ZT}.

Zener diodes are available for zener voltages of 2.4 to 200 V with accuracies between 5 and 20%. The preferred values are 3.3, 4.7, 5.1, 6.1, 9.1, 10, 11, 12, 13, 15, to 200 V.

Power dissipation is the product of V_Z and I_Z. The maximum power rating varies from 150 mW to 250 W. Commonly available power ratings are 0.4, 1, 3, 10, and 50 watts.

13.14. ZENER DIODE APPLICATIONS

Zener diodes have numerous applications in transistor circuitry. Some of their common applications are explained below.

1. Zener Diode Shunt Regulator (or Voltage Regulation). Voltage regulation is a measure of a circuit's ability to maintain a constant output voltage even when either input voltage or load current varies.

* Heavy doping increases conductivity, which narrows the depletion region and, therefore, reduces the voltage at which breakdown occurs.

Figure 13.51 shows how a zener diode can be used as a voltage regulator to provide a constant voltage from a source whose voltage may vary appreciably. A resistor R_S is necessary to limit the reverse current through the diode to safer value. The voltage source V_S and the resistor R_S are so selected that the diode operates in the breakdown region. The diode voltage in this region, which is also the voltage across the load R_L, is called *zener voltage* V_Z and the diode current is called the *zener current* I_Z. The series resistor R_S absorbs the output voltage fluctuations so as to maintain voltage across the load constant. Zener diode is reverse connected across the input voltage whose variations are to be regulated.

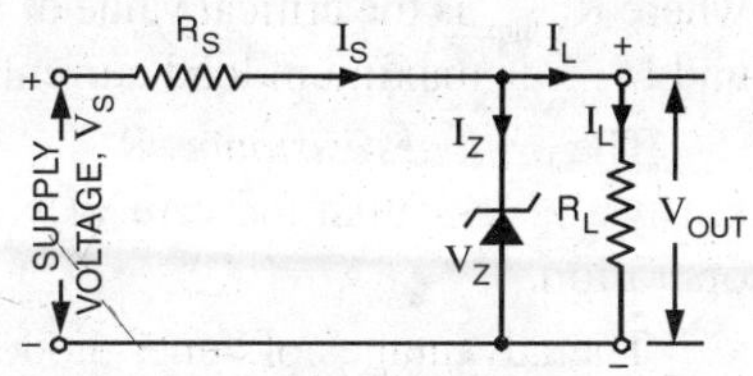

Fig. 13.51

As long as voltage across the load resistor R_L is less than the breakdown voltage V_Z, the zener diode does not conduct, the resistors R_S and R_L constitute a potential divider across V_S. At an increased supply voltage V_S, the voltage drop across load resistor (or zener diode) becomes greater than the zener breakdown voltage. It then operates in its breakdown region. The series resistor R_S limits the zener current I_Z from exceeding its rated maximum $I_{Z\,max}$. Current through resistor R_S is given as

$$I_S = \frac{V_S - V_Z}{R_S} \quad ...(13.8)$$

The current from the power supply splits at the junction of zener diode and the load resistor R_L.

$$\text{So } I_S = I_Z + I_L \quad ...(13.9)$$

When the zener diode operates in its breakdown region, the voltage across it, V_Z remains fairly constant even though the current I_Z flowing through it may vary considerably.

When the supply voltage V_S increases, the current through both the zener diode and load resistance R_L increases. At the same time, however, the zener diode resistance decreases and the current through the diode increases more than proportionately. As a result, a greater voltage drop will occur across the series resistor R_S and the output voltage V_{OUT} (voltage across the diode or load resistance R_L) will become very close to the original value. The reverse is also true. Thus a zener diode can maintain the output voltage V_{OUT} within a fraction of a volt when the supply or input voltage V_S may vary over a range of several volts. The zener diode will maintain a constant voltage across the load as long as the supply voltage is more than the zener voltage.

Let us examine the other cause of the output voltage variation. When the load resistance R_L decreases for constant input voltage V_S, load current I_L increases. This additional current is not supplied from the source of supply but the demand of additional load current is met by decrease in zener current I_Z. This keeps the voltage drop across series resistance R_S constant and so the output voltage V_{OUT}. The worst case occurs for minimum source voltage and maximum load current because the zener current reduces to a minimum. In such a case,

$$I_{S\,min} = \frac{V_{S\,min} - V_Z}{R_{S\,max}}$$

$$\text{or } R_{S\,max} = \frac{V_{S\,min} - V_Z}{I_{S\,min}} \quad ...(13.10)$$

From Eq. (13.9) $\quad I_Z = I_S - I_L$

In worst case, this may be written as

$$I_{Z\,min} = I_{S\,min} - I_{L\,max}$$

The critical point occurs when $I_{L\,max} = I_{S\,min}$. At this point, the zener current, I_Z reduces to zero, and regulation is lost.

By substituting $I_{L\,max}$ for $I_{S\,min}$ in Eq. (13.10), we have

$$R_{S\,max} = \frac{V_{S\,min} - V_Z}{I_{L\,max}} \quad ...(13.11)$$

where $R_{S\ max}$ is the critical value of series resistance, $V_{S\ min}$ is the minimum source voltage, V_Z is zener voltage and $I_{L\ max}$ is maximum load current.

The *critical resistance*, $R_{S\ max}$ is the maximum allowable series resistance. The series resistance R_S must always be less than the critical value; otherwise; breakdown operation is lost, and the regulator stops its operation.

The advantages of zener diode voltage regulators over other voltage regulators are that they are smaller, lighter, more rugged and have a longer life. They are also simpler and inherently cheaper.

Such regulators have the following drawbacks :

1. Their efficiency is low for heavy load currents because there is a considerable power loss in series resistor R_S and the diode.
2. The output voltage varies slightly due to zener impedance r_Z.
3. The output voltage cannot be choosen independently but depends upon the breakdown voltage of the zener diode.

This regulator is employed when there are small variations in load current and supply voltage.

2. Meter Protection. Figure 13.52 shows how does a zener diode protect a meter (AVO meter or multimeter) movement or circuit against damage from an accidental application of excessive voltage. In this arrangement zener diode is connected in parallel with the meter, as shown in Fig. 13.52. In case of accidental overload most of the current will pass through the zener diode and so the meter movement will be saved from damage. Two zener diodes connected, as shown in Fig. 13.53, can provide overload protection regardless of the applied polarity.

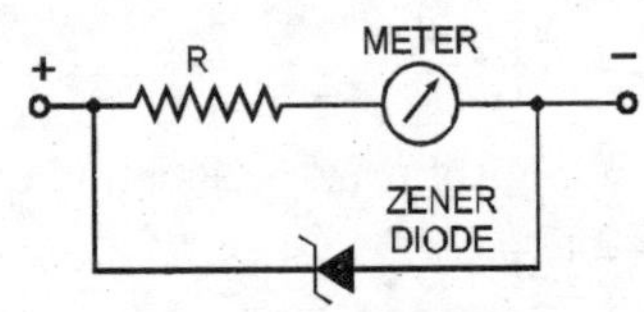

Fig. 13.52

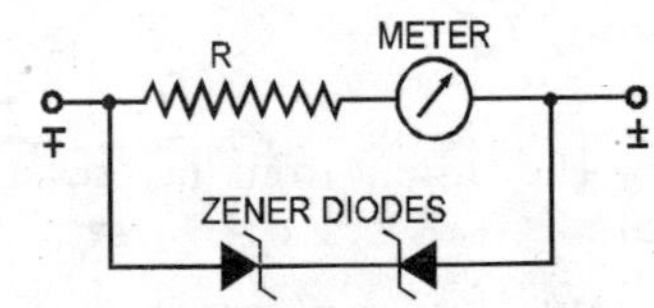

Fig. 13.53

3. Switching Operation. Zener diode can produce a sudden change from low current to high current, so it is useful for switching operations. The use of zener diode in switching operations has made possible an extremely fast performance in computer applications.

Example 13.11. For the circuit shown in Fig. 13.54, find

(*a*) the output voltage,

(*b*) the voltage drop across R_S,

(*c*) the current through zener.

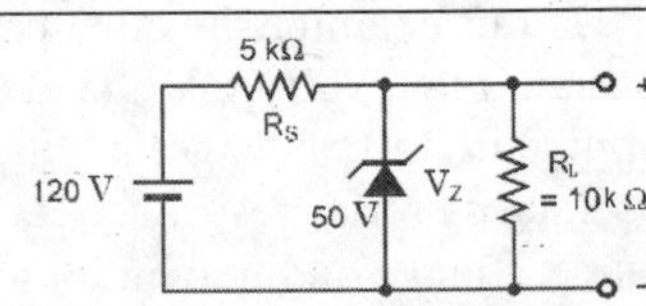

Fig. 13.54

Solution : From the figure shown in Fig. 13.55 we have

(*a*) Output voltage, $V_L = V_Z = 50$ V **Ans.**

(*b*) Voltage drop across $R_S = I_S \times R_S = V_S - V_Z = 120 - 50 = 70$ V **Ans.**

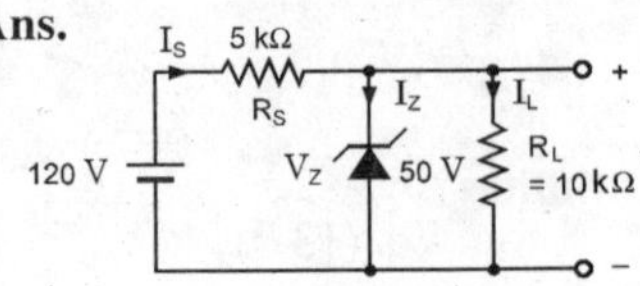

Fig. 13.55

(*c*) Load current, $I_L = \dfrac{V_L}{R_L} = \dfrac{50\ \text{V}}{10\ \text{k}\Omega} = 5$ mA

Current through R_S, $I_S = \dfrac{70\ \text{V}}{5\ \text{k}\Omega} = 14$ mA

Hence, current through zener diode is given by

$$I_Z = I_S - I_L = 14\ \text{mA} - 5\ \text{mA} = 9\ \text{mA} \quad \textbf{Ans.}$$

Example 13.12. For the zener regulator shown in Fig. 13.56, find the maximum and minimum current flowing through the zener diode.

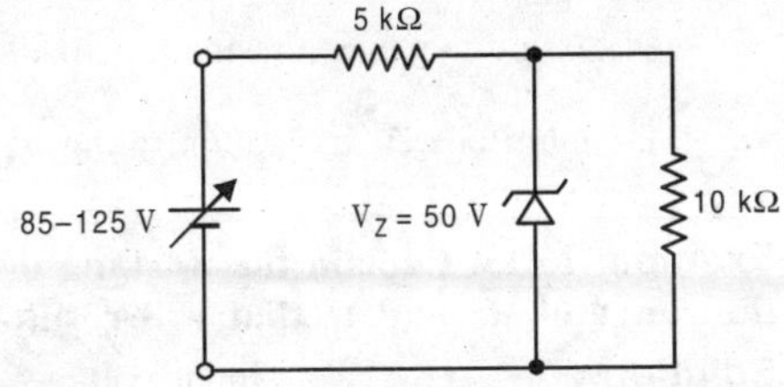

Fig. 13.56

Solution : Load current, $I_L = \dfrac{V_{out}}{R_L} = \dfrac{V_Z}{R_L} = \dfrac{50}{10\text{ k}\Omega} = 5\text{ mA}$

Zener current will be maximum when input voltage is maximum *i.e.* 125 V.

Corresponding current through series resistance of 5 kΩ

$$I_{S\,max} = \frac{V_{S\,max} - V_{out}}{R_S} = \frac{125-50}{5\text{ k}\Omega} = 15\text{ mA}$$

Corresponding zener current, $I_{Z\,max} = I_{S\,max} - I_L = 15 - 5 = 10\text{ mA}$ **Ans.**

The zener current will be minimum when input voltage is minimum *i.e.* 85 V. Corresponding current through series resistance (5 kΩ).

$$I_{S\,min} = \frac{85-50}{5\text{ k}\Omega} = 7\text{ mA}$$

Corresponding zener current, $I_{Z\,min} = I_{S\,min} - I_L = 7 - 5 = 2\text{ mA}$ **Ans.**

Example 13.13. For a zener diode network shown below in Fig. 13.57, determine V_L, V_R, I_Z and P_Z.

[*Pb. Technical Univ. Electronics Devices and Circuits 2005-06*]

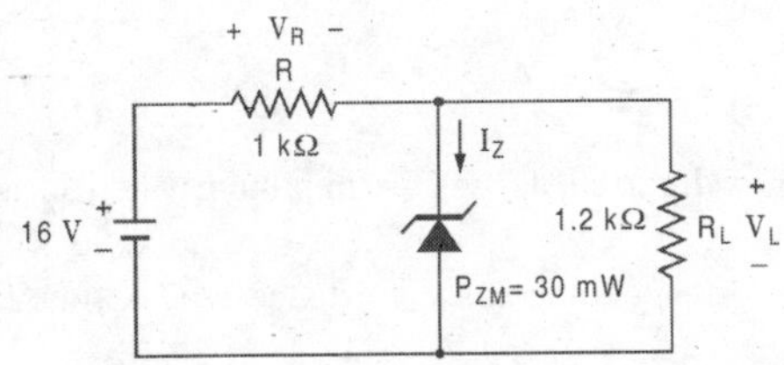

Fig. 13.57

Solution : Assume breakdown voltage of zener diode 10 V.

Voltage across the load, if zener diode is open circuited, by voltage divide rule,

$$V_L = \frac{V_S \times R_L}{R_S + R_L} = \frac{16 \times 1.2}{1 + 1.2} = 8.73\text{ V}$$

Since voltage across load, V_L is less than breakdown voltage of zener diode *i.e.* $V_L < V_Z$, the zener diode will not conduct and

$$V_L = 8.73\text{ V}$$

$$V_R = V_S - V_L = 16 = 8.73 = 7.27\text{ V}\ \textbf{Ans.}$$

$$I_Z = 0$$

$$P_Z = V_Z I_Z = V_L I_Z = 8.73 \times 0 = 0\text{ watt}\ \textbf{Ans.}$$

Example 13.14. For the circuit shown in Fig. 13.58, R_S = 20 Ω, V_Z = 18 V and R_L = 200 Ω. If V_{in} can vary from 20 V to 30 V, find

(*i*) The minimum and maximum currents in zener diode.

(*ii*) The minimum and maximum power dissipated in the diode.

(*iii*) The maximum rated power dissipation that R_S should have.

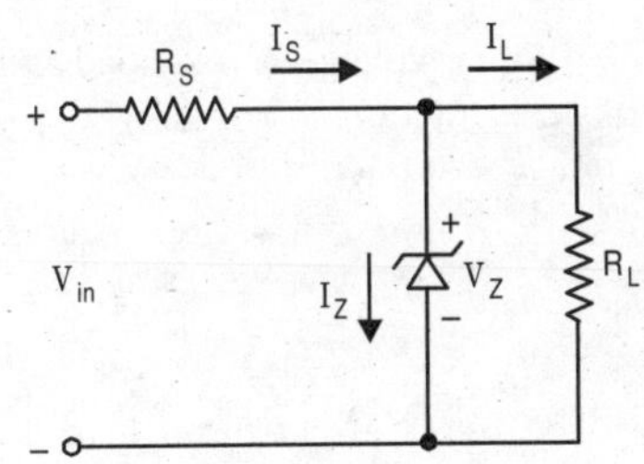

Fig. 13.58

Solution : Load current, $I_L = \dfrac{V_{out}}{R_L} = \dfrac{V_Z}{R_L} = \dfrac{18}{200} = 0.09\text{ A}$

Maximum supply current, $I_{S\,max} = \dfrac{V_{in\,max} - V_Z}{R_S} = \dfrac{30-18}{20} = 0.6\text{ A}$

(*i*) Corresponding zener current, $I_{Z\,max} = I_{S\,max} - I_L = 0.6 - 0.09 = 0.51\text{ A}$ **Ans.**

Minimum supply current, $I_{S\,min} = \dfrac{V_{in\,min} - V_Z}{R_S} = \dfrac{20-18}{20} = 0.1\text{ A}$

Corresponding zener current, $I_{Z\,min} = I_{S\,min} - I_L = 0.1 - 0.09 = 0.01\text{ A}$ or 10 mA **Ans.**

(*ii*) Minimum power dissipation in diode $= V_Z \times I_{Z\,min} = 18 \times 0.01 = 0.18\text{ W}$ **Ans.**

Maximum power dissipation in diode $= V_Z \times I_{Z\,max} = 18 \times 0.51 = 9.18$ W **Ans.**

(*iii*) Maximum power dissipation rating of $R_S = I^2_{S\,max} R_S = 0.6^2 \times 20 = 7.2$ W **Ans.**

Example 13.15. Explain the working of diode as a shunt regulator. For the circuit of Fig. 13.59, determine the range of R_L and I_L that will result in V_{R_L} being maintained at 10 V.

Solution :

Input voltage, $V_{in} = 50$ V

Series resistance, $R = 1$ kΩ

$V_Z = 10$ V

$I_{Zm} = 32$ mA

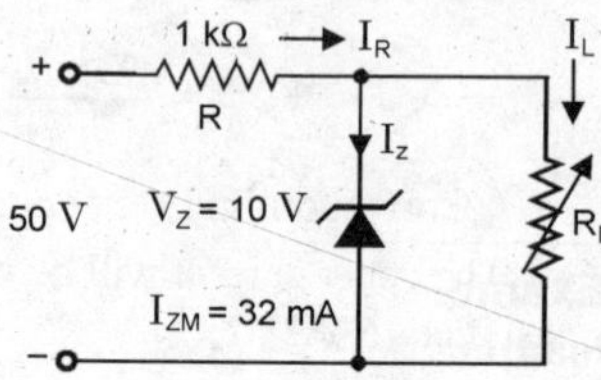

Fig. 13.59

$$\text{Supply current, } I_R = \frac{V_{in} - V_L}{R} = \frac{50 - 10}{1{,}000} = 40 \text{ mA}$$

Load current I_L will be maximum when zener current $I_Z = 0$

So $I_{L\,max} = I_R - I_{Z\,min} = 40 - 0 = 40$ mA

Corresponding load resistance will be minimum and

$$R_{L\,min} = \frac{V_L}{I_{L\,max}} = \frac{V_Z}{I_{L\,max}} = \frac{10}{40 \times 10^{-3}} = 250\ \Omega$$

Load current I_L will be minimum when I_Z is maximum *i.e.* 32 mA.

So $I_{L\,min} = I_R - I_{Z\,max} = 40 - 32 = 8$ mA

Corresponding load resistance will be maximum and

$$R_{L\,max} = \frac{V_L}{I_{L\,min}} = \frac{10}{8 \times 10^{-3}} = 1.25 \text{ k}\Omega$$

Hence range of R_L : From 250 Ω to 1.25 kΩ **Ans.** and range of I_L : From 8 mA to 40 mA **Ans.**

Example 13.16. Determine V_L, I_L, I_Z and I_R for the following circuit. (Fig. 13.60)

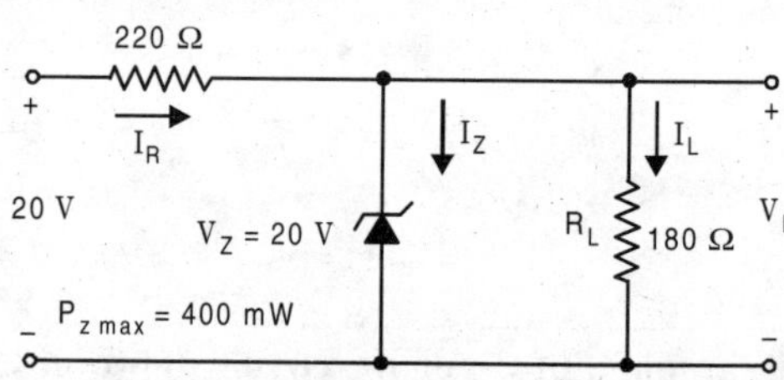

Fig. 13.60

Solution : Voltage across load if zener diode is open circuited.

$$V_L = V_{in} \times \frac{R_L}{R + R_L} = 20 \times \frac{180}{220 + 180} = 9 \text{ V}$$

Since voltage across load is less than the breakdown voltage of zener diode *i.e.* $V_L < V_Z$, the zener diode will not conduct and voltage drop across load, $V_L = 9$ V **Ans.**

Zener diode current, $I_Z = 0$ $\because$ Zener diode is open circuited

$$\text{Load current, } I_L = \frac{V_{in}}{R + R_L} = \frac{20}{220 + 180} = 0.05 \text{ A} \quad \textbf{Ans.}$$

Current through resistance R, $I_R = I_L = 0.05$ A **Ans.**

Example 13.17. Describe the physical mechanism for zener breakdown. For the circuit shown in Fig. 13.61, find the voltage drop across the resistance of 5 kΩ.

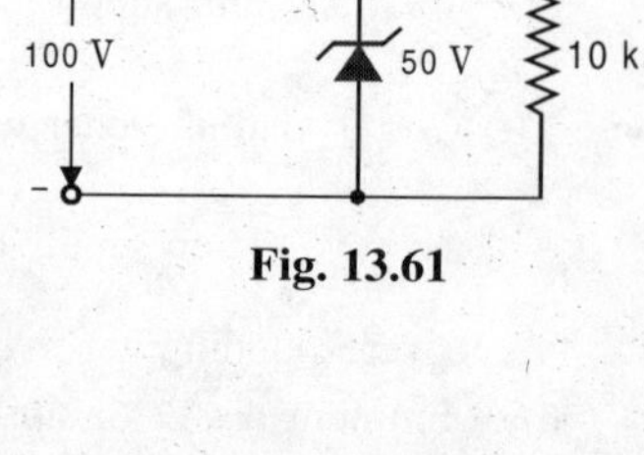

Fig. 13.61

Solution : Removing the zener diode from the circuit, we have

$$V_L = \frac{10}{10 + 5} \times 100$$

$= 66.67$ V $\quad \therefore \; V_L > V_Z$ (*i.e.* 50 V)

So zener diode is in ON state.

Thus the voltage drop across the 10 kΩ load resistor

$$V_L = V_Z$$
$$= 50 \text{ V}$$

Voltage drop across 5 kΩ resistance

$$= V_S - V_L$$
$$= 100 - 50$$
$$= 50 \text{ V} \quad \textbf{Ans.}$$

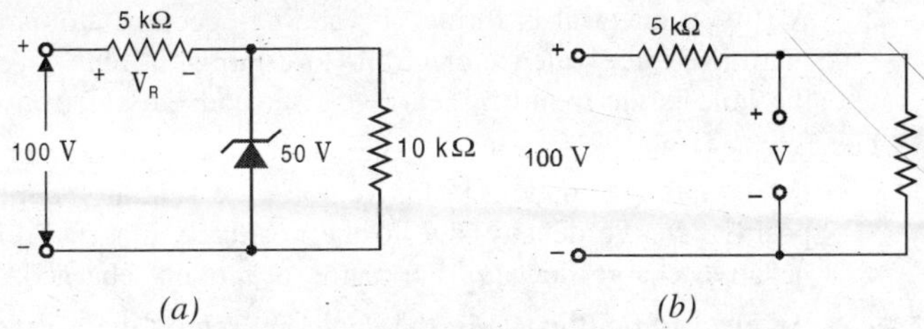

Fig. 13.62

Example 13.18. A zener diode regulator has V_Z = 12 V. The input voltage may vary from 20 to 35 V and load current from 15 to 75 mA. To hold voltage constant under all conditions, what should be the value of series resistance ? *[Pb. Technical Univ. Analog Electronics December 2005]*

Solution : Maximum value of required series resistance,

$$R_{S\,max} = \frac{V_{S\,min} - V_Z}{I_{L\,max}} = \frac{20-12}{75 \times 10^{-3}} = 106.7\ \Omega$$

Select R_S = 100 Ω, standard value **Ans.**

HIGHLIGHTS

1. A semiconductor is a material that has a resistivity somewhere between that of a good conductor and that of an insulator. The resistivity of different semiconductor materials lies in a very wide range from 10^{-4} to about 0.5 Ω-m whereas the resistivities of conductors and insulators are of the order of 10^{-8} Ω-m and 10^{12} Ω-m respectively. Most semiconductor materials used in electronics industry have negative temperature coefficients.

   The semiconductors have another very important property—*the resistivity (or conductivity) of semiconductors changes considerably when even minute amounts of certain other substances, called the impurities, are added to them.* Semiconductor materials are nonlinear resistors.

2. Intrinsic materials are those semiconductors that have very low level of impurities, whereas extrinsic materials are semiconductors that have been exposed to a doping process.

   Silicon (Si) and germanium (Ge) are the two most widely used semiconductor materials in electronics industry.

   At absolute zero temperature E_G for silicon and germanium are 1.21 eV and 0.785 eV respectively.

   Hole, strictly speaking, is not a particle; but for all practical purposes it is viewed as a positive charged particle capable of conducting current. The mobility of hole is about half that of an **electron.**

   The resistance of a semiconductor depends largely on (*a*) *temperature* — decreases with the increase in temperature, (*b*) *illumination* — decreases in brighter surroundings and (*c*) *magnitude of the electric field* — the current does not increase in proportion to the applied potential, as in case of conductors obeying Ohm's law, but increases by far more than the voltage *i.e., the semiconductors are nonlinear.*

   Conduction in a pure semiconductor consists of electron flow in the conduction band and hole flow in the valence band.

   A germanium device is seldom used at a temperature higher than 75°C, a silicon seldom at higher than 175°C.

3. Depending on the type of impurity added, the extrinsic semiconductors can be divided into two classes viz. (*i*) N-type semiconductors and (*ii*) P-type semiconductors.

   An N-type material is formed by adding donor atoms (such as arsenic, antimony, bismuth or phosphorus) that have five valence electrons to establish a high level of relatively free electrons. In an N-type material, the electron is the majority carrier and the hole is the minority carrier.

A P-type material is formed by adding acceptor atoms (such as boron, gallium, indium or aluminium) that have three valence electrons to establish a high level of holes in the material. In a P-type material, the hole is the majority charge carrier and the electron is the minority charge carrier.

4. A space charge region or depletion region is formed on either of the metallurgical junction separating the P and N junctions. This region is essentially depleted of any mobile electrons or holes. A net positive charge density, due to the positively charged donor impurity ions, exists in the N region and a negative charge density, due to the negatively charged acceptor impurity ions, exists in the P region.

   An electric field exists in the depletion region due to the net space charge density. The direction of the electric field is from the N region to the P region.

   A pd exists across the space-charge region. Under zero applied bias, the pd, known as built-in potential barrier, maintains thermal equilibrium and holds back majority carrier electrons in the N-region and majority carrier holes in the P region.

   The pd existing across the space charge region is known as the contact potential.

5. A forward-biased PN junction has the positive terminal applied to the P side and the negative to the N side. The reverse is true for a reverse-biased junction.

   When a forward bias voltage is applied across a PN junction, the potential barrier is lowered so that holes from the P-region and electrons from the N region can flow across the junction.

6. An applied reverse bias voltage (N region positive w.r.t. the P region) increases the potential barrier, increases the space-charge width, and increases the magnitude of the electric field.

7. A diode is a unidirectional device permitting the easy flow of current in one direction but restraining the flow in opposite direction. A semiconductor diode is simply a P-N junction with connecting leads or terminals on its two sides. P-N junction is also called the crystal diode because it is grown out of crystal.

8. There are two mechanisms by which breakdown can occur at a reverse biased P-N junction : avalanche breakdown and zener breakdown. The breakdown phenomenon is reversible and harmless so long as the safe operating temperature is maintained.

9. An ideal diode is a two terminal polarity sensitive device that has zero resistance when it is forward biased and infinite resistance when reverse biased.

10. The reverse saturation current I_0 will just double in magnitude for every 10-fold increase in temperature.

    The temperature coefficient for the forward voltage of a P-N junction is approximately –1.8 mV/°C for silicon diode and –2.02 mV/°C for germanium diode.

11. P-N junction diodes find use in many applications like rectifiers, demodulation circuits, clipping circuits, clamping circuits and switching circuits.

12. A zener diode is a P-N junction operated in the reverse-biased mode to take advantage of its sharply defined breakdown voltage. The direction of conduction for a zener diode is opposite to that of the arrow in the symbol, and the zener voltage has a polarity opposite to that of a forward-biased diode.

13. The zener voltage V_Z is specified at some test value of current I_{ZT}, at which the diode will exhibit some dynamic impedance Z_Z known as zener impedance and given by $Z_Z = \frac{\Delta V_Z}{\Delta I_Z}$ where ΔV_Z and ΔI_Z are the small variations in voltage and current respectively.

14. Zener diodes have a temperature coefficient, which generally is negative for V_Z below 5 V but positive above 6 V, and is expressed in percent of V_Z per °C with the change in zener voltage given by E_q. $\alpha_Z = \frac{\Delta V_Z}{V_Z(t - t_0)} \times 100$. Reference zener diodes are available with α_Z as low as 0.0005% per °C.

15. A zener diode may be used to regulate the load voltage at the value V_Z by acting as a "bypass value" to counteract line voltage or load current variations. Zener diodes are also useful in voltage reference circuits.

EXERCISES

1. Distinguish between intrinsic and extrinsic semiconductors.
2. What are junction diodes ? Explain the flow of current across forward and reverse biased junctions.
3. Discuss the formation and volt-amp characteristics of P-N junction. What happens when a P-N junction is biased in (*i*) forward direction, (*ii*) backward direction.
4. Draw the characteristics in forward and reverse bias and explain them.
5. What is a P-N junction diode and how is it formed ? Explain.

   Explain V-I characteristics of a P-N junction diode.

   [Pb. Technical Univ. Basic Electrical and Electronics Engineering Dec. 2004; May 2010]
6. Discuss working principle of P-N junction diode. Also draw its characteristics and mention some of its applications. **[Pb. Technical Univ. Basic Electrical and Electronics Engineering Dec. 2005]**
7. Explain how the process of avalanche breakdown occurs in a P-N junction diode. How is it different from zener breakdown ?

   [Pb. Technical Univ. Basic Electrical and Electronics Engineering December 2008]
8. What is difference between avalanche and zener breakdown ? How does zener diode maintain a constant voltage across the load. **[Pb. Technical Univ. Basic Electrical and Electronics Engineering Dec. 2006]**
9. Breakdown diodes with a breakdown voltage greater than 5 V generally have a positive temp. coefficient but those with lower breakdown voltage have a negative temp. coefficient. Do you agree with this statement ? Justify.
10. What is zener diode ? Explain its use as a voltage regulator.

    [Pb. Technical Univ. Basic Electrical and Electronics Engineering, December 2009]
11. Briefly describe the working of zener diode shunt regulator.

    [Pb. Technical Univ. Electronic Devices and Circuits 2003-04]
12. Write short notes on the following :
    (*i*) Intrinsic and extrinsic semiconductors.
    (*ii*) V-I characteristics of P-N junction.
    (*iii*) Zener breakdown.
    (*iv*) Working of zener diode.
    (*v*) Applications of zener diode.

SHORT ANSWER TYPE QUESTIONS WITH ANSWERS

Q. 1. What is the basis for classifying a material as a conductor, semiconductor or a dielectric ? What is the conductivity of perfect dielectric ?

Ans. Conductors possess high conductivity whereas the characteristic property of insulating materials (or dielectrics) is poor conductivity. Semiconductors occupy an intermediate position between conductors and insulators. Though there is no rigid line separating the conductors from semiconductors and semiconductors from insulators, but still according to resistivity the materials of resistivity of the order from 10^{-8} to 10^{-3}, 10^{-3} to 10^{6} and 10^{6} to 10^{18} ohm-metres may be classified as conductors, semiconductors and dielectrics respectively.

Q. 2. What is intrinsic semiconductor ?

Ans. An intrinsic semiconductor is one which is made of the semiconductor material in the extremely pure form (impurity content not exceeding one part in 100 million parts of semiconductors).

Q. 3. Why silicon and germanium are the two widely used semiconductor materials ?

Ans. Because the energy required to release an electron from their valence band (*i.e.*, to break their covalent bonds) is very small (1.12 eV for Si and 0.72 eV for Ge).

Q. 4. Which of the two semiconductor materials Si or Ge has larger conductivity at room temperature ? Why ?

Ans. Since energy required in transferring electrons from valence band to conduction band is more in case of silicon than that in case of germanium, the conductivity of Ge will be more than that of Si at room temperature.

Q. 5. What happens to the conductivity of the semiconductor with the rise in temperature? Compare with the conductivity of metals.

Ans. With the increase in temperature, the concentration of charge carriers increases resulting in increase of conductivity of semiconductors. The conductivity of metal decreases with the increase in temperature.

Q. 6. Why the temperature coefficient of resistance of a semiconductor is negative?

Ans. With the increase in temperature, the concentration of charge carriers (electrons and holes) increases. As more charge carriers are made available, the conductivity of a pure semiconductor increases *i.e.* resistivity of a pure semiconductor decreases with the rise in temperature *i.e.* semiconductors have negative temperature coefficient of resistance.

Q. 7. Differentiate between intrinsic semiconductors and extrinsic semiconductors ? Give examples.

Ans. An intrinsic semiconductor is one which is made of the semiconductor material in its extremely pure form.

When a small amount of impurity is added to a pure semiconductor crystal during the crystal growth in order to increase its conductivity, the resulting crystal is called the extrinsic semiconductor.

Q. 8. Give reasons — Doping is done in semiconductors.

Ans. Intrinsic (or pure) semiconductor by itself is of little significance as it has little current conduction capability at ordinary room temperature but if very small amount of impurity (of the order of one atom per million atoms of pure semiconductor) is added to it in the process of crystallization, the electrical conductivity is increased many times.

Q. 9. Describe the difference between P-type and N-type semiconductor materials.

Ans. When a small amount of trivalent impurity (such as boron, gallium, indium or aluminium) is added to a pure semiconductor crystal during crystal growth, the resulting crystal is called the P-type semiconductor.

When a small amount of pentavalent impurity (such as arsenic, antimony, bismuth or phosphorus) is added to a pure semiconductor crystal during crystal growth, the resulting crystal is called the N-type semiconductor.

Q. 10. What do you mean by donor and acceptor impurities?

Ans. Donor impurities (such as arsenic, antimony, bismuth or phosphorus) when added to a pure semiconductor lattice, form N-type extrinsic semiconductor. The pentavalent impurities are called the donor type impurities as such impurities donate electrons to the lattice.

Acceptor impurities (such as boron, gallium, indium or aluminium) when added to a pure semiconductor lattice form P-type extrinsic semiconductor. The trivalent impurities are called the acceptor type impurities because such impurities accept electrons from the lattice.

Q. 11. Explain the term doping and its need.

Ans. The electrical conductivity of intrinsic semiconductor, which has little current conduction capability at ordinary room temperature and so is of little use, can be increased many times by adding very small amount of impurity (of the order of one atom per million atoms of pure semiconductor) to it in the process of crystallization. This process is called the doping.

Q. 12. What are the majority and minority carriers in P-type semiconductor materials ?

Ans. Holes are majority carriers and free electrons are minority carriers in P-type semiconductor materials.

Q. 13. What are the charge carriers in N-type and P-type semiconductors ?

Ans. Free electrons in N-type semiconductors and holes in P-type semiconductors are the charge carriers.

Q. 14. For the same order of doping, why does N-type semiconductor exhibit larger conductivity than P-type semiconductor ?

Ans. Since the mobility of electrons is greater than that of holes, for same level of doping, N-type semiconductor exhibits larger conductivity.

Q. 15. What is the ratio of majority and minority carriers in intrinsic and extrinsic semiconductors ?

Ans.

Types of Semiconductor Material	***Ratio of Majority and Minority Carriers***
Intrinsic	Unity
Extrinsic	Very large

Q. 16. What is a P-N junction ?

Ans. The contact surface between the layers of P-type and N-type semiconductor pieces placed together so as to form a P-N junction is called the P-N junction.

Q. 17. How do the transition region width and contact pontential across a P-N junction vary with the applied bias voltage ?

Ans. When the P-N junction is forward biased, the transition region width is reduced and the contact potential is also reduced with the increase in applied bias voltage.

When the P-N junction is reverse biased, the transition is widened, and the contact potential is increased with the increase in applied bias voltage.

Q. 18. Which type of charges are present on the two opposite faces of the junction ?

Ans. Positive charge on N-side and negative charge on P-side of the junction.

Q. 19. What types of carriers are present in space charge region.

Ans. No mobile carrier is present in the space charge region.

Q. 20. Why is space region called the depletion region ?

Ans. The region around the junction is completely ionised on formation of P-N junction. As a result, there are no free electrons on the N-side nor the holes on the P-side. Since the region around the junction is depleted of mobile charges, it is called the depletion region.

Q. 21. Define cutin voltage of a p-n junction diode ?

Ans. The forward voltage, at which the current through the P-N junction starts increasing rapidly, is called the cutin voltage.

Q. 22. What do you understand by reverse saturation current of a diode ?

Ans. Reverse saturation current of a diode is due to minority carriers and is caused when the diode is reverse biased. Only a very small voltage is required to direct all available minority carriers across the junction, and when all minority carrjers are flowing across, further increase in bias voltage will not cause increase in current. This current is referred to as a reverse saturation current.

Q. 23. What is the effect of temperature on the reverse current of a P-N junction ?

Ans. Reverse current of a P-N junction increases with the increase in junction temperature.

Q. 24. Why is silicon preferred to germanium in the manufacturing of semiconductor devices ?

Ans. Silicon is preferred to germanium in the manufacturing of semiconductor devices because such devices have higher peak inverse voltage and current ratings and wider temperature range than germanium one.

Q. 25. Define peak inverse voltage ?

Ans. Peak inverse voltage is the maximum voltage that can be applied to the P-N junction without damage to the junction. If the reverse voltage across the junction exceeds its peak-inverse voltage (PIV), the junction may get destroyed owing to excessive heat.

Q. 26. Define the limitations in the operating conditions of P-N junction.

Ans. Every P-N junction has limiting values of (*i*) maximum forward current (*ii*) peak inverse voltage (PIV) and (*iii*) maximum power rating. The P-N junction provides satisfactory performance when operated within these limiting values. The P-N junction diode may get destroyed due to excessive heat if any of these values is exceeded.

Q. 27. What does the arrow head represent in a schematic symbol of a P-N junction ?

Ans. The arrowhead in the schematic symbol of a P-N junction indicates the direction of conventional current flow when the diode is forward biased.

Q. 28. What are the two mechanisms of breakdown in a P-N junction ?

Ans. Avalanche and zener breakdowns.

Q. 29. Name the breakdown mechanism in a lightly doped P-N junction under reverse biased condition.

Ans. Avalanche breakdown.

Q. 30. Name the breakdown mechanism in a heavily doped P-N junction under reverse biased condition.

Ans. Zener breakdown.

Q. 31. What is an ideal diode ?

Ans. An ideal diode is a two terminal polarity sensitive device that has zero resistance when it is forward biased and infinite resistance when reverse biased.

Q. 32. What is reverse saturation current ?

Ans. Reverse current of a diode is due to minority carriers and is caused when the diode is reversed biased. The reverse current is constant, independent of the applied reverse bias, consequently it is called the reverse saturation current.

Q. 33. Is reverse saturation current of a junction diode independent of reverse bias voltage ?

Ans. Yes.

Q. 34. Germanium is more temperature dependent than silicon. Why ?

Ans. Because the reverse saturation current in case of a germanium diode is approximately 1,000 times larger.

Q. 35. What is a zener diode ?

Ans. Zener diode is a P-N junction diode specially designed for operation in its breakdown region.

Q. 36. State salient features of zener diode. **[P.T.U. B.E.E.E. May 2009]**

Ans. A zener diode is a P-N junction operated in the reverse-biased mode to take advantage of its sharply defined breakdown voltage. The direction of conduction for a zener diode is opposite to that of the arrow in the symbol, and the zener voltage has a polarity opposite to that of a forward-biased diode.

A zener diode may be used to regulate the load voltage at the value V_Z by acting as a "bypass value" to counteract line voltage or load current variations. Zener diodes are also useful in voltage reference circuits.

Q. 37. What is zener voltage ?

Ans. The voltage at which the zener diode breaks down is called the zener voltage.

Q. 38. What happens to the series current, load current and zener current when the dc input voltage of a zener regulator increases ?

Ans. Zener current and series current increase while the load current remains unchanged.

Q. 39. Why is zener diode used as voltage regulator ?

Ans. Zener diode has the property of behaving like a dc battery in 'on' state (*i.e.* when the voltage across zener diode exceeds its zener voltage rating V_Z). In 'on state', the voltage across zener diode remains constant until the voltage across it drops less than V_Z. This property of zener diode makes its use as voltage regulator.

Q. 40. Explain how zener diode maintains constant voltage across the load ?

Ans. Zener diode has the property of behaving like a dc battery in 'on' state. If the zener diode is shunted across the load R_L and the voltage across zener diode is more than zener voltage V_Z then zener diode is in 'on' state, and any variation in voltage across the zener diode due to variations either in supply voltage or in load resistance is not able to change the output voltage. Thus zener diode maintains voltage constant across the load.

PROBLEMS

1. A zener diode shunt regulator is shown in Fig. 1. Determine
   (*i*) load voltage,
   (*ii*) voltage drop across series resistance,
   (*iii*) current through the diode.

   [**Ans.** (*i*) 12 V (*ii*) 18 V (*iii*) 51 mA]

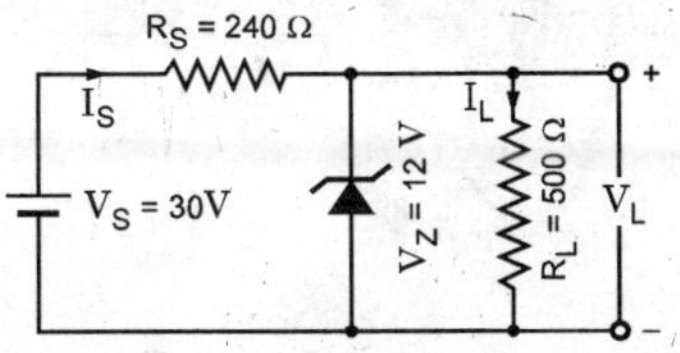

Fig. 1

2. For the circuit shown in Fig. 2 find
   (*a*) the output voltage,
   (*b*) the voltage drop across series resistance,
   (*c*) the current through zener diode.

   [**Ans.** (*a*) 50 V (*b*) 70 V (*c*) 9 mA]

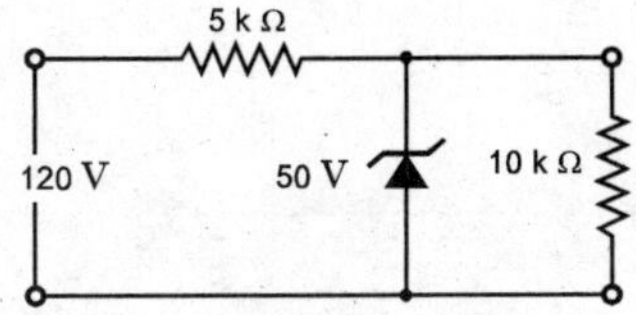

Fig. 2

3. Determine the maximum and minimum values of zener current if in the circuit shown in Fig. 13.51 the load resistance R_L = 5 kΩ, series resistance is 2 kΩ, input voltage varies between 80 V and 100 V and the output voltage is 50 V.

   [**Ans.** 15 mA and 5 mA]

4. In the circuit shown in Fig. 13.51 the series resistance is 2 kΩ, load resistance is 5 kΩ, input voltage is 60 V and output voltage is 35 V. Determine (*i*) voltage drop across series resistance (*ii*) load current (*iii*) current through series resistance and (*iv*) zener diode current .

   [**Ans.** (*i*) 25 V (*ii*) 7.0 mA (*iii*) 12.5 mA (*iv*) 5.5 mA]

5. Determine the temperature coefficient of a 7.5 V zener diode at 25°C if the nominal voltage increases to 7.8 V at a temperature of 105° C. [**Ans.** +0.05 %]

❑❑❑

CHAPTER 14

Bipolar Junction Transistor (BJT)

14.1. INTRODUCTION

INSIDE THIS CHAPTER

The transistor is a solid state device, whose operation depends upon the flow of electric charge carriers within the solid. Transistor is capable of amplification and in most respect it is analogous to a vacuum triode. The main difference between the two is that the transistor is a current controlled device whereas vacuum triode is a voltage controlled device. The transistor is only about six decades old, yet it is replacing vacuum tubes in almost all applications. The reasons are obviously its advantages over vacuum tubes such as compact size, light weight, rugged construction, more resistive to shocks and vibrations, instantaneous operation (no heating required), low operating voltage, high operating efficiency (no heat loss) and long life with essentially no ageing effect if operated within permissible limits of temperature and frequency. However, transistors, in comparison to vacuum triodes, have some drawbacks also such as loud hum noise, restricted operating temperature (upto 75ºC) and operating frequency (upto a few MHz only).

It consists of a silicon or germanium (preferably silicon because of its smaller cutoff current I_{CBO}, smaller variations in I_{CBO} due to variations in temperature and higher operating temperature) crystal in which a layer of N-type material is sandwiched between two layers of P-type material, as shown in Fig. 14.1 (*a*). Alternatively a transistor may consist of a layer of P-type material sandwiched between two layers of N-type material, as shown in Fig. 14.1 (*b*). In the former case the transistor is referred to as a P-N-P *transistor* and in latter case, as an N-P-N *transistor*. Each type of transistor has two P-N junctions—one junction between the emitter and base, called the *emitter-base junction* or simply the *emitter junction* and the other junction between the base and collector, called the *collector-base junction* or simply the *collector junction*. Thus

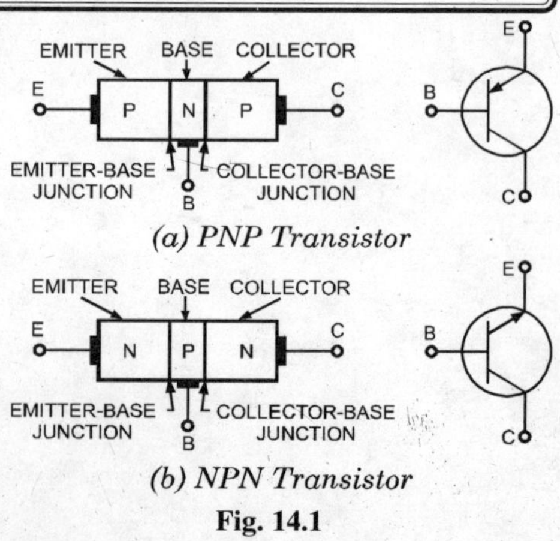

(*a*) *PNP Transistor*

(*b*) *NPN Transistor*

Fig. 14.1

a transistor is like two P-N junction diodes connected back to back. Each of these P-N junction diodes have all the characteristics discussed in earlier chapters. The two junctions give rise to three regions provided with three terminals called the *emitter*, *base* and *collector*, as shown in Fig. 14.1. The emitter, base and collector correspond in a general way to the cathode, grid and plate or anode of a vacuum triode.

14.2. TRANSISTOR TERMINALS

As mentioned above, transistor is a single crystal in which there are two P-N junctions, as shown in Fig. 14.1. The idea behind is to have first section to supply the charges (either holes or electrons) to be collected by the third section through the middle section. One side section supplying free charges is called the *emitter*, other side section collecting these charges is called the *collector* and the middle section which is formed between the emitter and collector is called the *base*.

1. *Emitter*. It is left hand section (or region) of the transistor and its main function is to supply majority charge carriers (electrons in case of NPN transistors and holes in case of P-N-P transistors) to the base. The emitter is always forward biased w.r.t. base so that it is able to supply majority charge carriers to the base. The emitter is heavily doped so that it may be able to inject a large number of charge carriers. It is of moderate size in order to maintain heavy doping without diluting it or mesh formation in it.

2. *Collector*. It is the right hand section of the transistor and its main function is to collect majority charge carriers. Collector is always reverse biased so as to remove the charge carriers away from its junction with the base. It is moderately doped to avoid the chances of mesh formation even after taking the carriers from the emitter. Large in size to withstand the temperature generated at the collector.

3. *Base*. It is the middle section of the transistor and is very lightly doped to reduce the recombination within the base so as to increase collector current and is very thin (of the order of μm) in comparison to either emitter or collector so that it may pass most of the injected charge carriers to the collector.

Base forms two junctions *i.e.* emitter junction and collector junction, each having its own barrier voltage. The forward-biased emitter-base junction offers low resistance to the emitter current whereas the collector-base junction, which is reverse biased, offers high resistance to the collector current. As the resistance of emitter-base junction is very small as compared to that of collector-base junction, therefore, the forward bias applied to the emitter-base junction is usually very small whereas the reverse bias on the collector-base junction is much large.

In most of the transistors, the collector region is made physically larger than emitter region because collector is to dissipate much power.* Due to this difference, collector and emitter are not interchangeable. However, for sake of convenience, it is customary to show emitter and collector to be of equal size.

As regards the symbols, arrowhead is always at the emitter. The direction indicates the conventional direction of current flow (from emitter to base in case of P-N-P transistor and from base to emitter in case of NPN transistor).** Generally, no arrowhead is marked for collector since its reverse leakage current is always opposite to the direction of emitter current.

14.3. TRANSISTOR ACTION

Transistors of both types (P-N-P and N-P-N) behave exactly the same way except change in biasing and majority carriers. In P-N-P transistors the conduction is by holes whereas in N-P-N transistors conduction is by electrons. However, N-P-N transistors are preferred due to their better high frequency response. Basic connections and biasing for P-N-P and N-P-N type transistors are shown in Fig. 14.2 (*a*) and (*b*) respectively. Let us consider an N-P-N transistor, which is in common use, for discussion.

* During operation of a transistor, much heat is generated at the collector junction.

** In a P-N-P transistor emitter is positive w.r.t. base while in case of an N-P-N transistor base is positive w.r.t. emitter.

The emitter is forward biased and as a result a large forward current flows across the emitter junction due to flow of majority carriers *i.e.* electrons from the emitter region enter the base region and holes from base region into the emitter region. However, since the conductivity of emitter region is larger than that of base region, the electrons from the emitter region out the number of holes from the base region. It may be assumed with good accuracy that forward current in the emitter junction flows due to the movement of electrons from the emitter region to the base region. We can, therefore, say that electrons from the emitter region due to forward bias in the emitter junction cross into the base region where electrons are in minority. Injection of electrons makes the electrons concentration on the emitter junction very large and on the collector junction the concentration of the electrons is extremely small and as such the gradient of electrons concentration is very large in base. Injected electrons diffuse into collector region due to extremely small thickness of base which is much less than the diffusion length. Most of the electrons cross into the collector region. Collector is reverse biased and creates a strong electrostatic field between base and collector. The field immediately collects the diffused electrons which enter the collector junction. Flow of electrons into the base region when confronted with the holes, a few electrons (say 1 to 5%) combined and neutralize and rest of the electrons say 95 to 99% of the injected electrons diffuse into the collector region and collected by the collector electrode. To maintain the base neutrality base electrode provides equal number of electrons which have combined with holes and results in a base current. Thus the emitter current I_E is equal to the sum of collector current I_C and base current I_B. The ratio of collector current I_C to emitter current I_E is called the α and it is measure of the possible current amplification. In a transistor α cannot be higher than unity, but practical values of 0.95 to 0.99 are attained in commercial transistors.

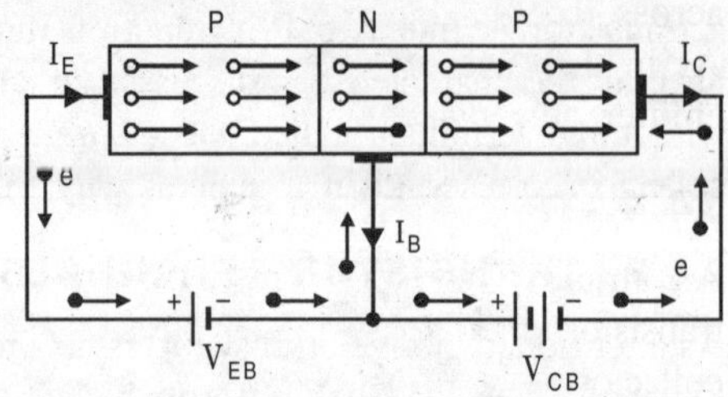

(a) Basic Connections of P-N-P Transistor

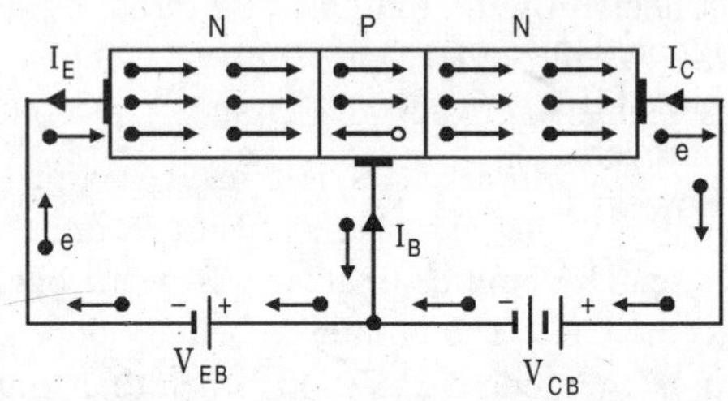

(b) Basic Connections of N-P-N Transistor

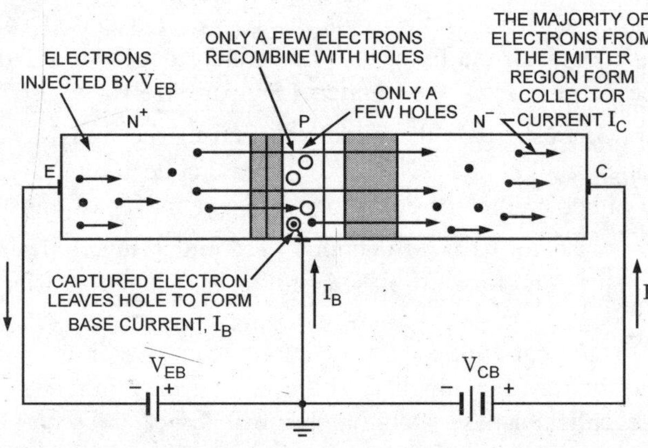

(c) Charge Carrier Movement in an N-P-N Transistor

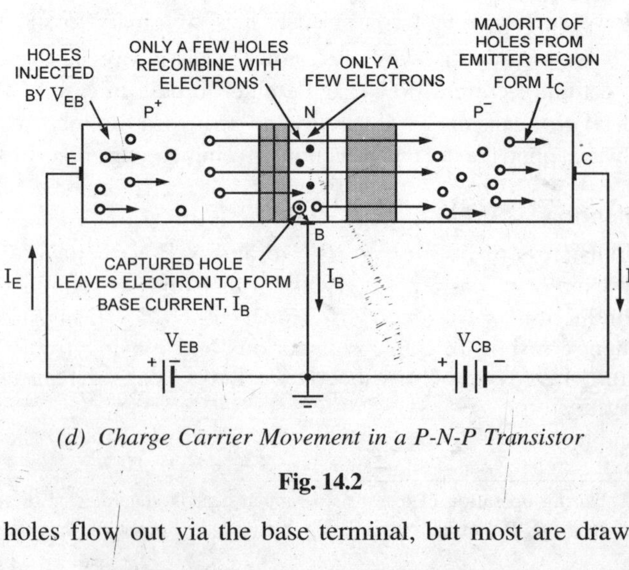

(d) Charge Carrier Movement in a P-N-P Transistor

Fig. 14.2

Operation of P-N-P Transistor. The P-N-P transistor behaves exactly the same way as an N-P-N device, with the exception that the majority charge carriers are holes. As illustrated in Fig. 14.2 (*d*), holes are emitted from the P-type emitter across the forward-biased emitter-base junction into the base. In the lightly doped N-type base, the holes find few electrons to recombine with. Some of the holes flow out via the base terminal, but most are drawn across the collector by the positive-negative electric field at the reverse-biased collector-base junction. As in the case of NPN device, the forward bias at the emitter base junction controls the collector and emitter currents.

14.4. TRANSISTOR BIASING

As already mentiond above, for normal operation of a transistor emitter-base junction is always forward biased and collector-base junction is always reverse biased.

For a moment if the biasing of emitter-base junction is ignored and only collector-base junction, which is reverse biased [Fig. 14.3 (*a*)] is considered, then due to high resistance of reverse biased collector diode, practically* no current would flow in the collector circuit.

Now if the biasing of collector-base junction is ignored and only emitter-base junction, which is forward biased [Fig. 14.3 (*b*)] is considered then practically whole of the emitter current flows through the base *i.e.* $I_B = I_E$.

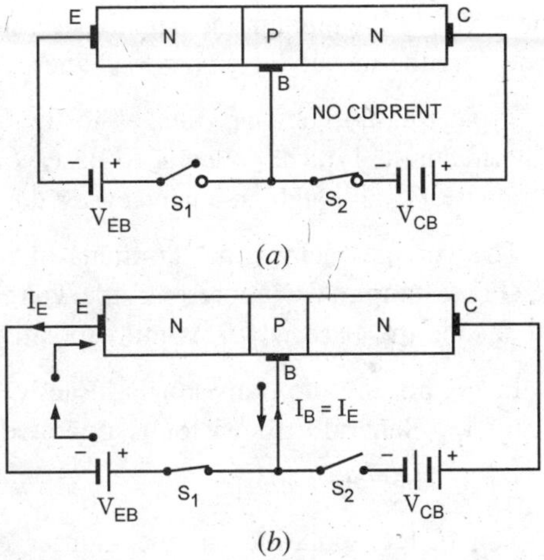

Fig. 14.3

However, if the emitter-base junction is forward biased and the collector-base junction is reverse biased it is seen that almost whole (about 95%) of the current constituted by the emitter flows through collector circuit. Thus the magnitude of collector current I_C depends upon the magnitude of emitter current I_E. No emitter current means almost no collector current. This is precisely what happens in a transistor.

14.4.1. Different Operating Conditions For a Transistor

A transistor has two junctions (emitter-base and collector-base junctions), and each of these two junctions, may be forward biased or reverse biased, therefore, there are four possible ways of biasing these two junctions. Accordingly it may operate in different conditions as listed in Table 14.1.

TABLE 14.1

Condition	*Emitter-Base (EB) Junction*	*Collector-Base (CB) Junction*	*Region of Operation*
1. Forward-Reverse (FR)	Forward-Biased	Reverse-Biased	Active
2. Forward-Forward (FF)	Forward-Biased	Forward-Biased	Saturation
3. Reverse-Reverse (RR)	Reverse-Biased	Reverse-Biased	Cutoff
4. Reverse-Forward (RF)	Reverse-Biased	Forward-Biased	Inverted

1. Forward-Reverse Biasing. As already mentioned above, for normal operation of a transistor emitter-base junction is always forward biased and collector-base junction is always reverse biased. This type of biasing is known as *FR biasing*. For this purpose a battery V_{EB} is connected between emitter and base while a battery V_{CB} is connected between the collector and base, as shown in Fig. 14.4. In Fig. 14.4 (*a*),

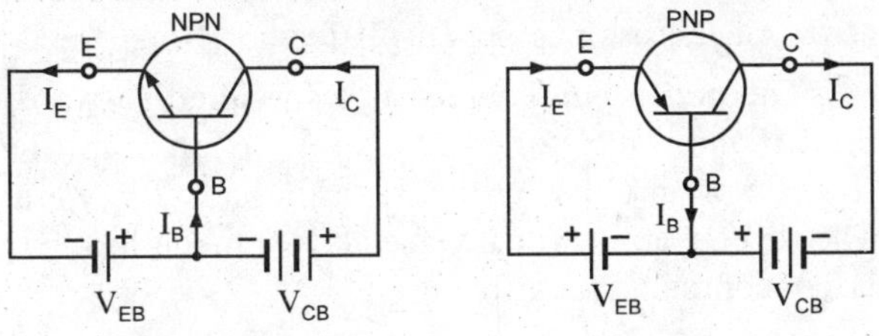

(*a*) *NPN Transistor* (*b*) *PNP Transistor*

Fig. 14.4. ***Active Region***

* In actual practice a very small reverse saturation current would flow in the collector circuit. This is called collector cut-off current and is because of minority carriers.

the emitter-base junction of N-P-N transistor is forward biased by connecting the negative terminal of battery V_{EB} to the emitter and positive terminal to the base while collector-base junction is reverse biased by connecting the positive terminal of the battery V_{CB} to the collector and negative terminal to the base. Similarly in Fig. 14.4 (*b*) the emitter-base junction of P-N-P transistor is forward biased by connecting the positive terminal of battery V_{EB} to the emitter and negative terminal to the base while collector-base junction is reverse biased by connecting the negative terminal of battery V_{CB} to the collector and positive terminal to the base.

Typical base-to-emitter voltages for both N-P-N and P-N-P transistors are 0.7 V for silicon and 0.3 V for germanium. Typical collector-to-base voltages might be anything between 3 V and 20 V for most types of transistors, although in many cases the collector-to-base voltage may exceed 20 volts even.

The voltages across the terminals of a transistor follow the usual double subscript voltage convention. For example V_{BE} represents the voltage of base with respect to the emitter. If V_{BE} is equal to 0.7 V it means the base is 0.7 V more positive.

In FR biasing the transistor is in active region and the collector current I_C, depends upon the emitter current I_E. Generally transistor is operated in this region for amplification.

For connecting batteries properly in the transistor circuits the following scheme may be remembered.

The battery polarities to the emitter and base are the same as the letter of the emitter and base impurity designations. Thus, in the P-N-P type, the emitter receives the P (positive) terminal, and the base receives the N (negative) terminal. Similarly, in N-P-N type, the emitter receives the N (negative) terminal, the base receives the P (positive) terminal. The collector receives a battery polarity opposite to its impurity type designation. Thus P-type collector receives the N (negative) terminal of the battery and the N-type collector receives the P (positive) terminal.

2. Forward-Forward Biasing. In this biasing, both collector and emitter are forward biased, as illustrated in Fig. 14.5. The transistor operates in saturation region. In this region, the collector current becomes independent of the base current. Thus the transistor acts like a close switch.

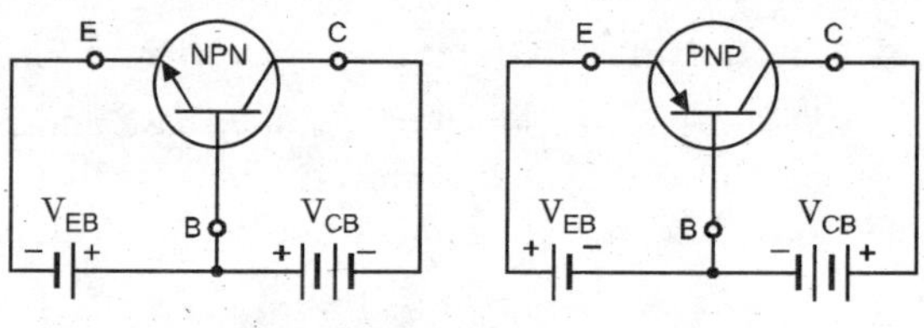

Fig. 14.5. *Saturation Region*

3. Reverse-Reverse Biasing. In this biasing, both the junctions are reverse biased. In this biasing, the transistor has practically zero current because the emitter does not emit charge carriers into the base and no charge carriers are collected by the collector except a few thermally generated minority carriers. Thus, the transistor acts like an open switch.

4. Reverse-Forward Biasing. In this biasing, the emitter junction is reverse biased while the collector junction is forward biased. As the collector is not doped to the extent as the emitter is doped, therefore, it cannot inject the majority charge carriers to the base. So, in this region, the action of transistor is poor. This region is of little importance.

The *active mode* which is also called forward active mode, is the one used if the transistor is to operate as an amplifier. Switching operations (*e.g.*, logic circuits) utilize both the *cutoff mode* and the *saturation mode*. The reverse active (or inverse active) mode has very limited application but is conceptually significant.

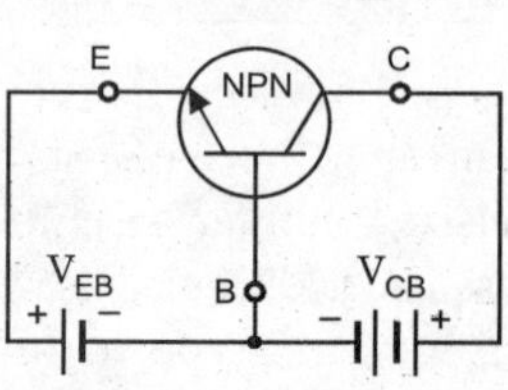

Fig. 14.6. *Cutoff Region*

14.5. IMPORTANT POINTS REGARDING WORKING OF TRANSISTORS

The noteworthy points regarding working of transistors are summarized below.

1. Current conduction in N-P-N transistor is by electrons and the conventional current flow will be in the opposite direction. Current conduction in P-N-P transistor is by holes but in external leads, the

current conduction will be by flow of electrons.

2. The collector current I_C is always less than emitter current I_E, being the difference of emitter current I_E and base current I_B.
3. The base current is only a small fraction (usually 5%) of emitter current.
4. As a standard convention, all the currents entering into the transistor are considered to be positive. Conversely currents that flow out of the transistor are taken to be negative. It means that if the actual conventional current flows in the outward direction, a negative sign is included along with its magnitude. Thus in an N-P-N transistor, the emitter current I_E is to be taken negative because it flows out of the transistor [Figs. 14.2 (*b*) and 14.4 (*a*)] while both the base current I_B and collector current I_C are to be taken positive because they flow into the transistor. Similarly in a P-N-P transistor I_E is to be taken positive and both base current I_B and collector current I_C are to be taken negative [Figs. 14.2 (*a*) and 14.4 (*b*)]. However, to avoid confusion, the actual direction of flow of current is indicated in the diagrams.
5. Emitter-base junction is always forward biased and collector-base junction is always reverse biased for active region of operation.
6. The input circuit (*i.e.* emitter-base junction), because of forward bias, offers low resistance and so needs usually very small bias (approximately 0.7 V for Si and 0.3 V for Ge). The output circuit (*i.e.* collector-base junction), because of reverse bias, offers high resistance and, therefore, needs much higher bias (3 to 20 V).
7. Transistor transfers the input signal from a low resistance circuit to a high resistance circuit, therefore, it is called the TRANsfer reSISTOR (TRANSISTOR).
8. Since both of the charge carriers (holes as well as electrons) are involved in current flow through a transistor (may be either P-N-P or N-P-N), these devices are sometimes called the *bipolar junction transistors* (BJTs).
9. Although, it is said that a transistor is a device with two PN diodes connected back to back, but it does not mean that two discrete diodes connected back to back, as shown in Fig. 14.7, can work as a transistor. In such a case, each diode has two equally doped regions, so that the overall circuit has four equally doped regions. This would not work as base region is not the same as in a transistor.

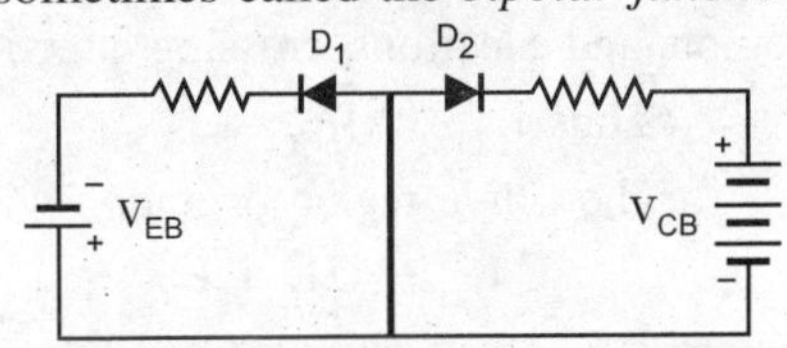

Fig. 14.7. ***Two Diodes Connected Back to Back***

The key to the transistor action is the lightly doped thin base between the heavily doped emitter and moderately doped collector.

In an N-P-N transistor, the free electrons passing through the base to the collector region have a short lifetime. As long as the base is thin, the free electrons can reach the collector. But in case of two discrete back to back connected diodes there are four doped regions instead of three and there is nothing that resembles a thin base region between an emitter and a collector.

Hence two discrete diodes connected back to back can never work as a transistor.

10. The choice of N-P-N transistor is made more often because majority charge carriers are electrons whose mobility is much more (almost double) than that of holes.

14.6. TRANSISTOR CURRENT COMPONENTS

The various current components which flow across the forward-biased emitter junction and reverse biased collector junction are indicated in Fig. 14.8. The current flowing into the emitter is referred to as the *emitter current* and designated as I_E. The emitter current I_E consists of two constituents (hole current I_{hE} due to crossing of holes from emitter into the base and electron current I_{eE} due to crossing of electrons from base into the emitter).

$$i.e.\ I_E = I_{hE} + I_{eE} \qquad ...(14.1)$$

In a commercial transistor the emitter is heavily doped and the base is very lightly doped and, therefore, electron current I_{eE} is negligibly small in comparison to hole current I_{hE}. Thus the entire emitter current I_E in a P-N-P transistor is due to flow of holes from emitter to the base. It is to be noted that the indicated direction of I_E external to the transistor is the conventional current direction.

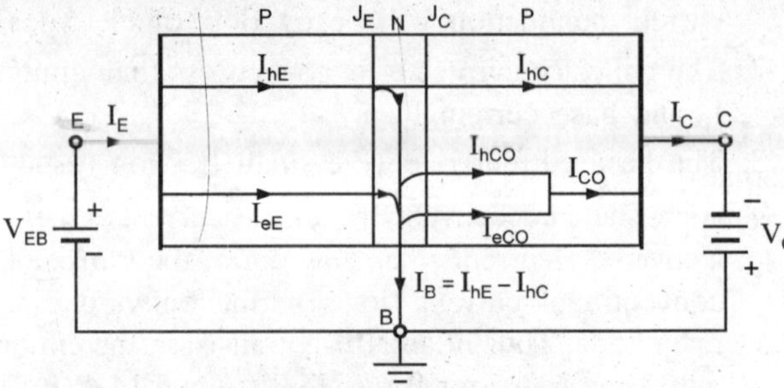

Fig. 14.8. *Current Components in a PNP Transistor For a Forward Biased Emitter Junction and a Reverse Biased Collector Junction*

All the holes crossing the emitter junction J_E do not reach the collector junction J_C because some of them combine with the electrons in the N-type base. If I_{hC} is the hole current at collector junction J_C, there must be a bulk recombination current I_{hE} less I_{hC} leaving the base, as shown in Fig. 14.8. In fact electrons enter the base region through the base lead to replenish those electrons which have been lost by recombination with the holes injected into the base across the emitter junction J_E. The holes on reaching the collector junction J_C cross this junction and enter the collector P-region.

If the emitter were open circuited, I_E would have been zero and then I_{hC} also would have been zero. Under these conditions, the base and collector would act as a reverse-biased diode and the collector current I_C would be equal to that reverse saturation current I_{CBO} (or simply I_{CO}).

When the emitter circuit is closed *i.e.* when $I_E \neq 0$, then from Fig. 14.8

$$\text{Collector current, } I_C = I_{hC} + I_{CO} \qquad ...(14.2)$$

For a P-N-P transistor, I_{CO} consists of holes moving across junction J_C from base region to collector region and electrons crossing junction J_C in the opposite direction.

$$\text{Thus } I_{CO} = I_{hCO} + I_{eCO} \qquad ...(14.3)$$

In the active region operation with emitter junction J_E forward biased, the collector current is given as

$$I_C = \alpha I_E + I_{CO} \qquad ...(14.4)$$

where α is the fraction of the total current I_E which constitutes I_{hC}.

It is to be noted that the indicated direction of emitter current I_E external to the transistor is the conventional current direction. Base current I_B and collector current I_C are also shown external to the transistor conventional current direction. According to the Kirchhoff's first law total current flowing into the transistor must be equal to the total current flowing out of it. So emitter current I_E is equal to the sum of collector and base currents I_C and I_B.

$$i.e.\ I_E = I_C + I_B \qquad ...(14.5)^*$$

Let us now define different parameters which relate the current components discussed above.

Emitter Efficiency. The emitter or injection efficiency denoted by γ is defined as the ratio of current of injected carriers at J_E to the total emitter current

$$i.e.\ \gamma = \frac{\text{Current of injected carriers at } J_E}{\text{Total emitter current}} = \frac{I_{hE}}{I_{hE} + I_{eE}} = \frac{I_{hE}}{I_E} \qquad ...(14.6)$$

where I_{hE} is the injected hole diffusion current at emitter junction and I_{eE} is the injected electron diffusion current at emitter junction.

Transport Factor. Tansport factor, denoted by β, is defined as the ratio of injected carrier current reaching J_C to the injected carrier current at J_E

* Some authors taking standard transistor convention, write this equation as $I_E = -(I_C + I_B)$

$$i.e.\ \beta = \frac{\text{Injected carrier current reaching } J_C}{\text{Injected carrier current at } J_E} = \frac{I_{hC}}{I_{hE}} \quad ...(14.7)$$

Large Signal Current Gain α. The term α has already been defined above. However, α can alternatively be defined from Eq. (14.4). From Eq. (14.4) α may be defined as the ratio of collector current increment from cutoff value ($I_C = I_{CO}$) to I_C and the emitter current increment from cutoff value ($I_E = 0$) to I_E. Thus, we may write

$$\alpha = \frac{I_C - I_{CO}}{I_E - 0} = \frac{I_C - I_{CO}}{I_E} \quad ...(14.8)$$

Term α is called the *large signal current gain* of a common base transistor. The parameter α is extremely important in transistor theory. It is to be pointed here that α is not constant, but varies with emitter current I_E, collector voltage V_{CB}, and the temperature. Typically α lies in the range 0.95 to 0.99.

From Eqs. (14.2) and (14.8) we have

$$\alpha = \frac{I_{hC}}{I_E} = \frac{I_{hC}}{I_{hE}} \times \frac{I_{hE}}{I_E} = \beta\gamma \quad ...(14.9) \quad \because \text{From Eq. (14.7) } \frac{I_{hC}}{I_{hE}} = \beta \text{ and from Eq. (14.6) } \frac{I_{hE}}{I_E} = \gamma$$

For an efficient transistor we would like β and γ to be very near unity *i.e.* the emitter current should be due mostly to holes ($\gamma = 1$) and most of the injected holes should eventually participate in the collector current ($\beta \simeq 1$).

The product $\beta\gamma$ is defined as the factor α, called the *current gain ratio,* which represents the emitter-to-collector current amplification. There is no real amplification between these currents, since α is lesser than unity. On the other hand, the relation between collector current and base current is more promising for amplification.

DC Current Gain. If I_{CO} is negligibly small in comparison to I_C, α approximately equals I_C/I_E. This is referred to as the *dc current gain* of the common base (CB) transistor and is denoted by α_{dc}

$$\text{Thus} \quad \alpha_{dc} = \frac{I_C}{I_E} \quad ...(14.10)$$

α_{dc} is always positive and less than unity.

Small Signal Current Gain α_{ac}. Small signal current gain α_{ac} is defined as the ratio of change in collector current to change in emitter current

$$i.e.\ \alpha_{ac} = \frac{\Delta I_C}{\Delta I_E} \text{ for constant collector-base voltage } V_{CB} \quad ...(14.11)$$

α_{ac} is always positive but less than unity (very close to unity). It also varies with I_E, V_{CB} and temperature.

Generalized Expression For Collector Current. We know that a small reverse current flows across a P-N junction diode due to *thermally generated minority carriers* that are propelled by the barrier potential. When the junction is reverse biased, this reverse current increases slightly. For moderate reverse-bias voltages, the reverse current attains its saturation value, I_S (or I_0). Because the collector-base junction of a transistor is reverse biased, there is likewise a reverse current due to thermally generated carriers. Of course this "reverse" current, in the context of a transistor, is in the same direction as the main (collector) current flowing through the device due to the injection of minority carriers into the base. The total collector current is, therefore, the sum of these two components: the injected minority carriers and the thermally generated minority carriers

$$i.e.\ I_C = \alpha I_E + I_{CO}$$

The reverse current I_{CBO} is given by the volt-ampere relationship of equation $I = I_0(e^{V/\eta V_T} - 1)$ with I_0 replaced by $-I_{CO}$ and V by V_C where symbol V_C represents the voltage drop across junction J_C from the P-side to the N-side and the complete expression for collector current I_C may be written as

$$I_C = \alpha I_E + I_{CO}(1 - e^{V_C/\eta V_T}) \quad ...(14.12)$$

If V_C is negative and is of large magnitude in comparison with that of V_T, Eq. (14.12) reduces to Eq. (14.4).

The physical interpretation of Eq. (14.12) is that the P-N junction diode current crossing the collector junction J_C is supplemented by a fraction α of the current I_E flowing in the emitter.

Example 14.1. For a transistor if I_E = 1 mA and α = 0.98, determine the value of I_B.

[Pb. Technical Univ. Basic Electrical & Electronics Engg. May 2006]

Solution :

Emitter current, $I_E = 1$ mA

Current amplification factor, $\alpha = 0.98$

Collector current, $I_C = \alpha I_E = 0.98 \times 1$ mA $= 0.98$ mA

Base current, $I_B = I_E - I_C = 1.0 - 0.98 = 0.02$ mA **Ans.**

14.7. CURRENT AMPLIFICATION FACTORS

There are two current amplification factors, the alpha factor (α) and the beta factor (β) defined below :

The alpha factor (α) is also called the *current amplification factor* and is the ratio of collector current (I_C) to emitter current (I_E).

$$\text{In general, } \alpha = \frac{\Delta I_C}{\Delta I_E} \quad V_{CB} \text{ remaining constant} \qquad ...(14.13)$$

The beta factor (β) is the current gain factor (*also called the transport factor*) of a common emitter circuit and is defined as the ratio of collector current (I_C) and base current (I_B).

$$\text{In general, } \beta = \frac{\Delta I_C}{\Delta I_B} \qquad ...(14.14)$$

14.8. RELATIONSHIP BETWEEN α AND β

$$\text{As defined above } \beta = \frac{\Delta I_C}{\Delta I_B}$$

$$\text{and } \alpha = \frac{\Delta I_C}{\Delta I_E} = \frac{\Delta I_C}{\Delta I_C + \Delta I_B} \qquad \because \Delta I_E = \Delta I_B + \Delta I_C$$

$$\text{or } \frac{1}{\alpha} = \frac{\Delta I_C + \Delta I_B}{\Delta I_C} = 1 + \frac{\Delta I_B}{\Delta I_C} = 1 + \frac{1}{\beta} = \frac{1+\beta}{\beta}$$

$$\text{Hence } \alpha = \frac{\beta}{1+\beta} \quad \text{and} \quad \beta = \frac{\alpha}{1-\alpha} \qquad ...(14.15)$$

$$\text{Thus, if } \alpha = 0.98 \qquad \beta = \frac{0.98}{0.02} = 49$$

$$\text{and if } \alpha = 0.99 \qquad \beta = \frac{0.99}{0.01} = 99$$

i.e. a small variation in α corresponds to a large variation in β. It is, therefore, better to determine β experimentally and calculate therefrom the corresponding value of α by means of expression

$$\alpha = \frac{\beta}{1+\beta} \qquad ...(14.16)$$

Example 14.2. In a common base configuration of a PNP transistor the current amplification factor is 0.988. Determine the base current if emitter current is 1.2 mA. Neglect collector leakage current.

Solution :

Current amplification factor, $\alpha = 0.988$

Emitter current, $I_E = 1.2$ mA

Collector leakage current, $I_{CO} \simeq 0$

From Eq. (14.4)

$$\text{Collector current, } I_C = \alpha I_E + I_{CO} = 0.988 \times 1.2 + 0 = 1.1856 \text{ mA}$$

$$\text{Base current, } I_B = I_E - I_C = 1.2 - 1.1856 = 0.0144 \text{ mA} = 14.4\ \mu\text{A} \quad \textbf{Ans.}$$

Example 14.3. For a given transistor $\alpha_{dc} = 0.98$, $I_E = 2$ mA. Calculate β_{dc} and I_B.

Solution :

$$\text{Emitter current, } I_E = 2 \text{ mA}$$

$$\alpha_{dc} = 0.98$$

$$\beta_{dc} = \frac{\alpha_{dc}}{1-\alpha_{dc}} = \frac{0.98}{1-0.98} = 49 \quad \textbf{Ans.}$$

$$\text{and base current, } I_B = \frac{I_C}{\beta_{dc}} = \frac{\alpha_{dc} I_E}{\beta_{dc}} = \frac{0.98 \times 2}{49} = 0.04 \text{ mA or } 40\ \mu\text{A} \quad \textbf{Ans.}$$

Example 14.4. For a PNP transistor the base current and collector current are respectively 45 μA and 5.45 mA. Determine (*i*) values of α, β and I_E; (*ii*) Base current required to make collector current of 10 mA.

Solution :

$$\text{Base current, } I_B = 45\ \mu\text{A} = 45 \times 10^{-6} \text{ A} = 0.045 \text{ mA}$$

$$\text{Collector current, } I_C = 5.45 \text{ mA}$$

$$(i) \text{ Emitter current, } I_E = I_B + I_C = 0.045 + 5.45 = 5.495 \text{ mA} \quad \textbf{Ans.}$$

$$\text{Current amplification factor, } \alpha = \frac{I_C}{I_E} = \frac{5.45}{5.495} = 0.9918 \quad \textbf{Ans.}$$

$$\text{Current gain factor, } \beta = \frac{I_C}{I_B} = \frac{5.45}{0.045} = 121 \quad \textbf{Ans.}$$

(*ii*) Base current required to make collector current I_C of 10 mA,

$$I_B = \frac{I_C}{\beta} = \frac{10}{121} = 0.0826 \text{ mA or } 82.6\ \mu\text{A} \quad \textbf{Ans.}$$

14.9. TRANSISTOR CIRCUIT CONFIGURATIONS

As already mentioned, a transistor is a three-terminal device (having three terminals namely emitter, base and collector) but we require four terminals—two for the input and two for the output for connecting it in a circuit. Hence one of the terminals of the transistor is made common to the input and output circuits. Thus there are three types of configurations for operation of a transistor. These configurations are *common base* (CB), *common emitter* (CE) and *common collector* (CC) configurations. The term 'common' is used to denote the electrode that is common to input and output circuits. Since the common terminal is usually grounded, these configurations are also referred to as *grounded base*, *grounded emitter* and *grounded collector* configurations, as shown in Figs. 14.9 (*a*), 14.9 (*b*) and 14.9 (*c*) respectively. Each configuration has its own merits and demerits. It is to be remembered that regardless of the circuit configurations, the emitter is always forward biased and the collector is always reverse biased.

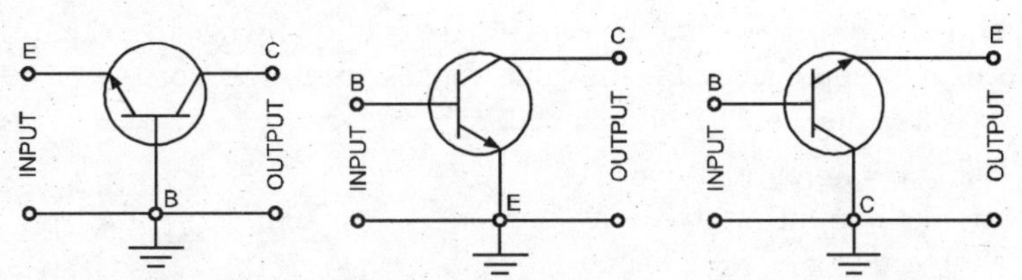

(*a*) *Common Base Configuration* (*b*) *Common Emitter Configuration* (*c*) *Common Collector Configuration*

Fig. 14.9. ***Different Circuit Configurations For NPN Transistor***

14.10. COMMON BASE (CB) CONFIGURATION

In common base (CB) configuration, input is connected between emitter and base and output is taken across collector and base, as illustrated in Fig. 14.10 for an N-P-N transistor. Thus base is common between input and output circuits.

In this arrangement (Fig. 14.10) the emitter-base junction is forward biased and collector-base junction reverse biased. The emitter current I_E flows in the input circuit and the collector current I_C in the output circuit. A change in emitter current produces a similar change in the collector current due to change in diffusion of electrons or holes across the junctions. The impedances of these two circuits are different and thus some voltage and power gains can be achieved since a current variation in the low resistance emitter-base circuit has similar current variations in the high resistance collector-base circuit.

The ratio of collector current I_C (output current) to the emitter current I_E (input current) is called the *current amplification* factor α. If there is no input ac signal then the ratio of I_C to I_E is called the dc alpha (α_{dc}).

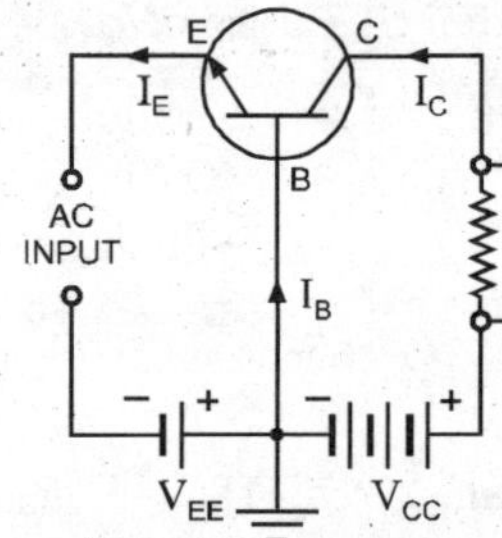

Fig. 14.10. ***Common Base NPN Transistor***

i.e. $\alpha_{dc} = \dfrac{-I_C}{I_E}$ (negative sign signifies that I_E flows out of transistor while I_C into transistor).*

If we write α_{dc} simply by α and ignore –ve sign,

$$\alpha = \frac{I_C}{I_E} \qquad ...(14.17)$$

Incidently, there is also an ac alpha (α_{ac}) for a transistor and it refers to the ratio of change in collector current to change in emitter current

$$i.e.\ \alpha_{ac} = \frac{-\Delta I_C}{\Delta I_E} \text{ for constant collector-base voltage } V_{CB}. \qquad ...(14.18)$$

α** of a transistor is a measure of its quality–higher the value of α better is the transistor in the sense that I_C approaches I_E, its value ranges from 0.95 to 0.99. The value of α can be increased by making base thin and lightly doped.

The collector current consists of two parts (*i*) the current produced by normal transistor action *i.e.* component depending upon emitter current (α I_E) which is produced by majority carriers and (*ii*) the leakage current due to movement of minority carriers across base-collector junction on account of it being reverse biased. This leakage current flows in the same direction as that of I_C and is abbreviated as I_{CBO}*** (collector-base current with emitter open).

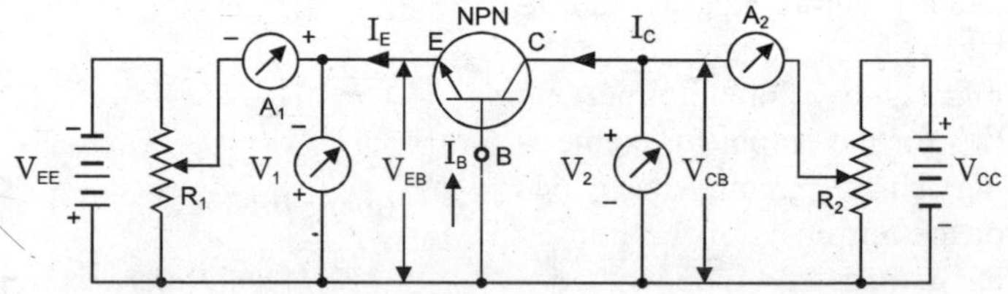

Fig. 14.11. ***Circuit Arrangement For Determination of Input and Output Characteristic Curves of a Transistor in CB Configuration***

$\therefore$ Total collector current, $I_C = \alpha\, I_E + I_{CBO}$ ...(14.19)

* In case of a P-N-P transistor emitter current I_E flows into the transistor and collector current I_C flows out of it.

** For all practical purposes α_{dc}, α_{ac} and α have the same values.

***I_{CBO} is exactly like the reverse saturation current I_S or I_0 of a reverse biased diode discussed earlier and is extremely temperature dependent as it is made up of thermally generated minority carriers.

The current I_{CBO}, sometimes referred to as I_{CO}, is usually very small and may be neglected in transistor calculations.

From Eq. (14.19)

$$\alpha = \frac{I_C - I_{CO}}{I_E} \simeq \frac{I_C}{I_E} \quad ...(14.20) \quad \text{neglecting } I_{CO} \text{ in comparison with } I_C$$

$$I_B = I_E - I_C = I_E - (\alpha I_E + I_{CO}) = (I - \alpha) I_E - I_{CO} \quad ...(14.21) \quad \because \text{ from Eq. (14.19) } I_C = \alpha I_E + I_{CO}$$

14.10.1. Characteristics of Common Base Configuration

The performance of transistors, when connected in a circuit, may be determined from their characteristic curves that relate different dc currents and voltages of a transistor. Such curves are known as *static characteristic curves*. There are two important characteristics of a transistor viz. input characteristics and output characteristics.

Both of the input and output characteristics can be determined with the same circuit arrangement shown in Fig. 14.11.

Input Characteristics. The curve drawn between emitter current I_E and emitter-base voltage V_{EB} for a given value of collector-base voltage V_{CB} is known as *input characteristic*. For determination of the input characteristics, the output voltage V_{CB} is maintained constant, and the input voltage V_{EB} is set at several convenient levels. For each level of input voltage, the input current I_E is noted. Then the curve is plotted between emitter current I_E and emitter base voltage V_{EB}, as shown in Fig. 14.12.

For a given value of V_{CB} the curve is just like that of a forward biased P-N junction because emitter-base junction in forward bias mode is similar to a P-N junction. This characteristic may be used to determine input resistance of the transistor.

From Fig. 14.12 it is seen that with the increase in V_{CB}, it conducts better, although the effect is not very significant. This is because large collector-base (reverse bias) voltages cause the depletion layer at the collector-base junction to penetrate deeper into the base of the transistor, thus reducing the distance and the resistance between the emitter-base and collector-base regions.

There exists a *cutin*, *offset* or *threshold* voltage V_γ below which the emitter current is very small. Substantial emitter current flows only when emitter-base junction is forward biased beyond the knee of the characteristic (approximately 0.7 V for Si and 0.3 V for Ge).

The emitter current I_E increases rapidly with the small increase in V_{EB} which means low dynamic input resistance because input resistance of a transistor is given as the reciprocal of the slope of its input characteristic *i.e.* mathematically

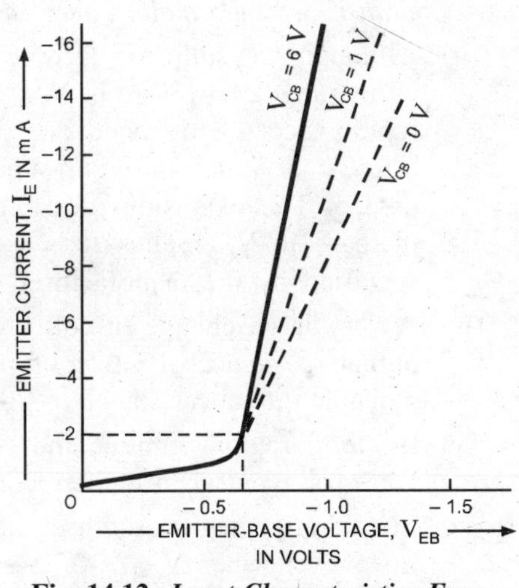

Fig. 14.12. ***Input Characteristics For Common Base NPN Transistor***

$$r_{in} = \frac{\Delta V_{EB}}{\Delta I_E} \quad ...V_{CB} \text{ constant} \quad ...(14.22)$$

Since a small increase in V_{EB} (ΔV_{EB}) causes large change in I_E (ΔI_E), r_{in} is quite low.

The value of dynamic input resistance r_{in} varies from point to point in the initial part of the characteristic because it is nonlinear. Its value over the linear part of the characteristic is about 100 Ω but for lower values of V_{EB}, it is considerably larger. This dependence of r_{in} on V_{EB} causes distortion of the signals handled by the transistors.

Output Characteristics. The curve drawn between collector current I_C and collector-base voltage V_{CB} for a given value of emitter current I_E is known as *output characteristic*.

For determination of output characteristics, the emitter current is held constant at each of several fixed levels. For each fixed level of emitter current I_E, the output voltage V_{CB} is adjusted in convenient steps, and the corresponding levels of collector current I_C are noted. From these data output characteristics can be drawn. The curves so drawn are shown in Fig. 14.13.

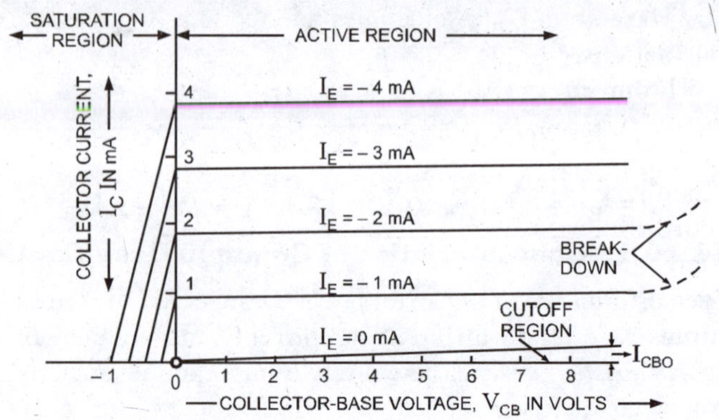

Fig. 14.13. ***Output Characteristics For Common Base NPN Transistor***

The noteworthy points regarding output characteristics of common base (CB) configuration are given below :

(*i*) The collector current I_C varies with V_{CB} only for very low voltage (below 1 V) but transistor is never operated in this region.

(*ii*) In active region (emitter forward biased and collector reverse biased) collector current I_C is almost equal to I_E and appears to remain constant when V_{CB} is increased. In fact, there is very small increase in I_C with increase in V_{CB}. This is because the increase in V_{CB} expands the collector-base depletion region and thus shortens the distance between the two depletion regions. With emitter current I_E held constant, however, the increase in I_C is so small that it is usually noticeable only for large variations in V_{CB}. I_C is slightly lesser than I_E in magnitude. *Transistor is normally operated in active region.*

(*iii*) Although the collector current I_C is practically independent of V_{CB} over the transistor operating range, however, if V_{CB} is increased beyond a certain value, I_C eventually increases rapidly because of avalanche or zener (or both) effects. This condition is known as *punch-through* or *reach-through*. When it occurs large currents can flow, possibly destroying the device. The extension of the depletion region is, of course, the direct consequence of the increase in V_{CB}. Thus it is very essential to maintain V_{CB} below the maximum safe limit specified by the manufacturer of the device.

(*iv*) A very large change in collector voltage causes a very small change in collector current *i.e.* output resistance of CB configuration is very high (of the order of few hundred kΩ)—the dynamic output resistance r_{out} being given as the ratio of ΔV_{CB} and ΔI_C for a given value of I_E.

(*v*) In *cutoff* region (emitter and collector junctions both reverse biased) small collector current I_C flows even when emitter current $I_E = 0$. This is the collector leakage current I_{CBO} or I_{CO}.

(*vi*) In saturation region (both emitter and collector junctions forward biased) collector current I_C flows even when $V_{CB} \simeq 0$. Even when the externally applied bias voltage is reduced to zero, there is still a barrier potential existing at the collector-base junction, and this assists in the flow of I_C. To stop it the collector-base junction has to be forward biased. Consequently, as illustrated in Fig. 14.13, collector current I_C is reduced to zero when V_{CB} is increased negatively.

We have already determined that a small signal voltage impressed in the low-resistance input (emitter) circuit of a transistor causes a relatively large emitter current. Almost the same amount of current will flow in the high-resistance output (collector) circuit of transistor, where the voltage may be very high. Evidently, then, both the output voltage and power can be quite large, as compared to the tiny input voltage and power present at the emitter.

CB configuration is rarely used in audio-frequency (AF) circuits because its current gain is less than unity and its input and output resistances are quite different.

Current Gain Characteristics. The current gain characteristics, also referred to as *forward transfer characteristics,* can be determined experimentally (by keeping V_{CB} constant at a convenient value, recording I_C value measured for different settings of I_E and then plotting I_C versus I_E) or from the common-base output characteristics as shown in Fig. 14.14. A vertical line is drawn through a selected value of V_{CB} and corresponding values of I_E and I_C are read off along this line. Then a graph is plotted between I_C versus I_E, and the characteristic is labelled with the V_{CB} used.

Note that for $I_E = 2$ mA, this line (representing I_C) is always less than 2 mA. Similarly, for all other values of I_E, $I_C < I_E$.

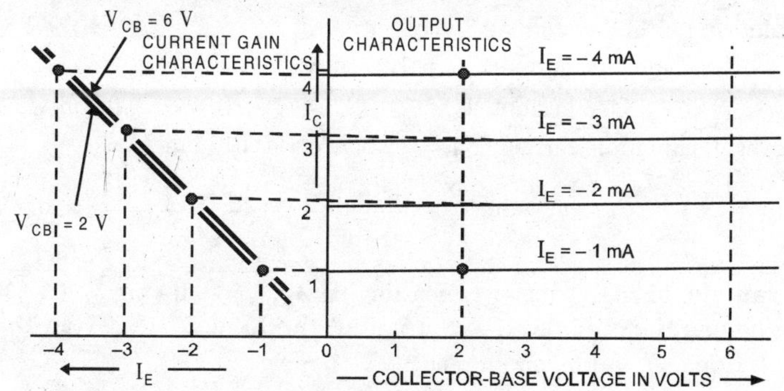

Fig. 14.14. ***Current Gain Characteristics For Common Base NPN Transistor***

Example 14.5. A transistor is connnected in CB configuration. When the emitter voltage is changed by 200 mV, the emitter current changes by 5 mA. During this variation, the collector-to-base voltage is kept fixed. Calculate the dynamic input resistance of transistor.

Solution : The dynamic input resistance for CB configuration is given as

$$r_{in} = \left.\frac{\Delta V_{EB}}{\Delta I_E}\right|_{V_{CB} = \text{constant}} = \frac{200 \text{ mV}}{5 \text{ mA}} = 40\ \Omega \textbf{ Ans.}$$

Example 14.6. In a grounded base configuration (Fig. 14.10) the voltage drop across load resistance of 4 kΩ is 3 V. Determine base current. α = 0.96.

Solution : Load resistance $R_L = 4$ kΩ

Voltage drop across load resistance, $I_C R_L = 3$ V

$$\text{Collector current, } I_C = \frac{I_C R_L}{R_L} = \frac{3\text{ V}}{4\text{k}\Omega} = 0.75 \text{ mA}$$

$$\text{Emitter current, } I_E = \frac{I_C}{\alpha} = \frac{0.75}{0.96} = 0.78 \text{ mA}$$

Base current, $I_B = I_E - I_C = 0.78 - 0.75 = 0.03$ mA **Ans.**

Example 14.7. The current gain of an N-P-N transistor $\alpha_0 = 0.98$. It is connected in CB configuration and gives reverse saturation $I_{CO} = 10$ μA. Find the base and collector currents for an emitter current of 2 mA.

Solution : Emitter current, $I_E = 2$ mA

Reverse saturation current, $I_{CO} = 10\ \mu A = 0.01$ mA

Total collector current, $I_C = \alpha I_E + I_{CBO} = 0.98 \times 2 + 0.01 = 1.97$ mA **Ans.**

Base current, $I_B = I_E - I_C = 2.0 - 1.97 = 0.03$ mA **Ans.**

Example 14.8. In a common base configuration, the emitter current is 1.6 mA. The collector current with emitter circuit open is 10 μA. Determine the total collector current.

Solution : Emitter current, $I_E = 1.6$ mA

Collector current with emitter circuit open, $I_{CO} = 10\ \mu A = 0.01$ mA

Assume $\alpha = 0.98$

Collector current, $I_C = \alpha I_E + I_{CO}$ Refer to Eq. (14.4)

$= 0.98 \times 1.6 + 0.01 = 1.578$ mA **Ans.**

Example 14.9. The reverse saturation current of an N-P-N transistor in common base circuit is 12.5 μA. For an emitter current of 2 mA, collector current is 1.97 mA. Determine the current gain and base current.

Solution : Emitter current, I_E = 2 mA

Collector current, I_C = 1.97 mA

Base current, $I_B = I_E - I_C = 2 - 1.97 = 0.03$ mA **Ans.**

Reverse saturation current, I_{CO} = 12.5 μA = 0.0125 mA

$$\text{Current gain, } \alpha = \frac{I_C - I_{CO}}{I_E} = \frac{1.97 - 0.0125}{2} = 0.979 \quad \textbf{Ans.}$$

Example 14.10. A silicon transistor has I_{CBO} = 1 μA at 27°C. The leakage current doubles every 6°C rise in temperature. If the I_E = 1 mA, find the I_B at 57°C. Given that α = 0.95.

Solution : Emitter current, I_E = 1 mA

Current amplification factor, α = 0.95

$$\text{Leakage current at 57°C, } I_{CBO_{57}} = I_{CBO_{27}} \times 2^{\frac{57-27}{6}} \quad \because I_{CBO} \text{ doubles every 6°C rise in temperature}$$

$$= 1 \times 10^{-6} \times 2^5 = 32 \times 10^{-6} \text{ A} = 0.032 \text{ mA}$$

Collector current, $I_C = \alpha I_E + I_{CBO} = 0.95 \times 1 + 0.032 = 0.982$ mA

Base current at 57°C, $I_B = I_E - I_C = 1 - 0.982 = 0.018$ mA = 18 μA **Ans.**

14.11. COMMON EMITTER (CE) CONFIGURATION

In common emitter configuration input is connected between base and emitter while the output is taken between collector and emitter, as shown in Fig. 14.15. Thus emitter is common to input and output circuits.

In this configuration, the bias voltages are applied between collector and emitter and base and emitter. Emitter-base junction is forward biased and the base is made more positive than the emitter by V_{BB}, collector-emitter junction is reverse biased and the collector is made more positive than emitter by V_{CC}. The value of V_{CC} must be greater than that of V_{BB}.

In this circuit arrangement (Fig. 14.15) the base current I_B flows in the input circuit and collector current I_C flows in the output circuit. In this arrangement current gain between the input and output sides is obtained; and since the input resistance is again less than the output resistance (though difference is not as much as in case of common base arrangement) there will be high voltage and power gains like those in equivalent vacuum tube circuits. The common emitter produces a reversal between input and output signals. Common emitter (CE) configuration is commonly used because its current, voltage and power gains are quite high and output to input impedance ratio is moderate.

The ratio of change in collector current (output current) and change in base current (input current) is called the *base current amplification factor* β

$$i.e.\ \beta = \frac{\Delta I_C}{\Delta I_B} \quad \text{...(14.23)}$$

Almost in all transistors, the base current is less than 5% of emitter current I_E so β is usually higher than 20. Usually β ranges from 20 to 100.

$$\text{Also } \beta = \frac{I_C}{I_B} = \frac{I_C}{I_E - I_C} = \frac{I_C/I_E}{1-\frac{I_C}{I_E}} = \frac{\alpha}{1-\alpha} \quad \text{...(14.24)}$$

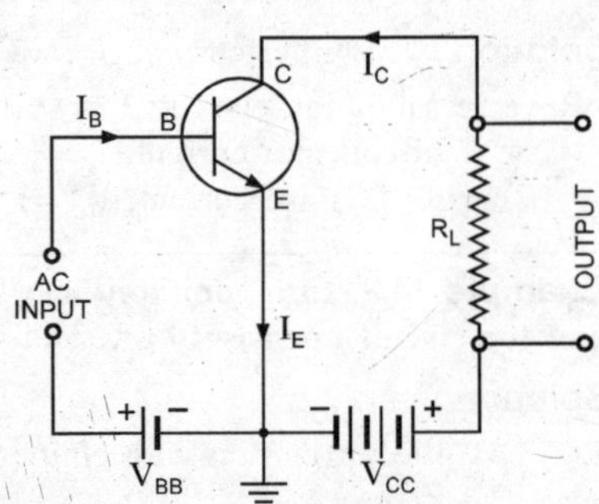

Fig. 14.15. ***Common Emitter NPN Transistor***

In CE configuration, a small collector current flows even when base current is zero (*i.e.* base lead is open). This is the collector *cutoff* current and denoted by I_{CEO}. I_{CEO} must be larger than I_{CBO}.

The current equations are $I_E = I_B + I_C$ ...(14.25)

$$\text{and } I_C = \alpha I_E + I_{CBO} = \alpha (I_B + I_C) + I_{CBO}$$

$$\text{or } I_C (1 - \alpha) = \alpha I_B + I_{CBO}$$

$$\text{or } I_C = \frac{\alpha}{1-\alpha} I_B + \frac{I_{CBO}}{1-\alpha} \quad ...(14.26)$$

and total collector current, $I_C = \beta I_B + I_{CEO}$ ...(14.27)

Comparing Eqs. (14.26) and (14.27) we have

$$\beta = \frac{\alpha}{1-\alpha}$$ as already derived in Eq. (14.24)

$$\text{and } I_{CEO} = \frac{1}{1-\alpha} I_{CBO} \quad ...(14.28)$$

Substituting the value of $I_{CEO} = \frac{1}{1-\alpha} I_{CBO}$ in Eq. (14.27) we have

$$I_C = \beta I_B + \frac{1}{1-\alpha} I_{CBO}$$

$$= \beta I_B + (\beta + 1) I_{CBO} \quad ...(14.29) \quad \because \text{ from Eq. (14.24) } \frac{1}{1-\alpha} = \beta + 1$$

14.11.1. Characteristics of Common Emitter Configuration

The circuit used for determination of transistor common emitter characteristics is shown in Fig. 14.16. The input voltage is applied between base and emitter terminals, and the output is taken from collector and emitter terminals. Input and output voltages and currents are measured by voltmeters and ammeters as shown.

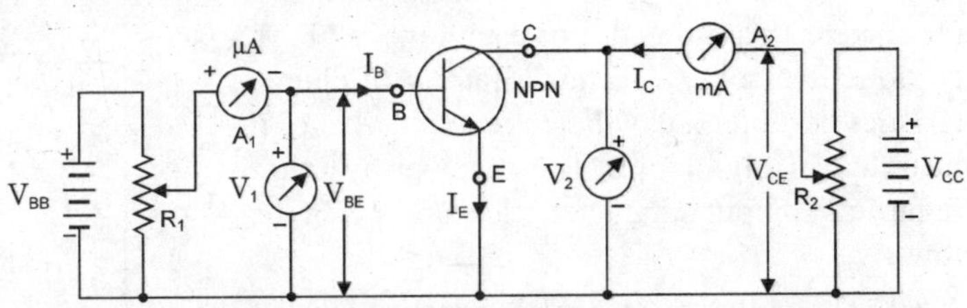

Fig. 14.16. *Circuit Arrangement For Determination of Input and Output Characteristics of a Transistor in CE Configuration*

Input Characteristics. The curve drawn between base current I_B and base-emitter voltage V_{BE} for a given value of collector-emitter voltage V_{CE} is known as the *input characteristic*. For determination of input characteristics, collector-emitter voltage V_{CE} is held constant and base current I_B is recorded for different values of base-emitter voltage V_{BE}. Now the curves are drawn between base current I_B and base-emitter voltage V_{BE} for different levels of V_{CE}, as shown in Fig. 14.17. The noteworthy points regarding input characteristics are given below :

(*i*) The input characteristics of CE transistors are quite similar to those of a forward biased diode because the base-emitter region of the transistor is a diode and it is forward biased.

(*ii*) In comparison to common base (CB) arrangement base (or input) current increases less rapidly with the increase in base-emitter voltage, V_{BE}. This indicates that input resistance is larger in common emitter configuration than that in common base configuration. Due to initial nonlinearity of the curve, input resistance varies from point to point in the initial part of the characteristic. Its value over the linear part of the curve is of the order of few hundred ohms.

(*iii*) An increment in value of V_{CE} causes the input current I_B to be lower for a given level of V_{BE}. This is because the higher levels of V_{CE} provide greater collector-base junction reverse bias, causing greater depletion region penetration into the base and thus reducing the distance between the collector-base and emitter-base regions. As a result more of the charge carriers from the emitter flow across the collector-base junction, and few flow out through the base lead. Usually effect of change in V_{CE} on the input characteristic is ignored.

The ratio of change in base-emitter voltage (ΔV_{BE}) to the resulting change in base current (ΔI_B) at constant collector-emitter voltage (V_{CE}) is known as *dynamic input resistance* (r_{in}). It is calculated as the reciprocal of the slope of the input characteristic at a given V_{BE}.

$$i.e., \quad r_{in} = \left.\frac{\Delta V_{BE}}{\Delta I_B}\right|_{V_{CE} = \text{constant}} \qquad ...(14.30)$$

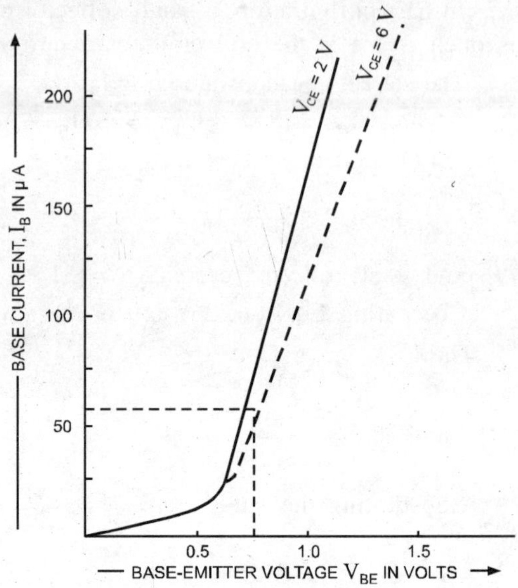

Fig. 14.17. ***Input Characteristics For Common Emitter NPN Transistor***

Output Characteristics. Output characteristic for a common emitter transistor is the curve drawn between collector current I_C and collector-emitter voltage V_{CE} for a given value of base current I_B. For determination of common emitter output characteristics, base current I_B is maintained constant at several convenient levels. At each fixed level of I_B, collector-emitter voltage V_{CE} is adjusted in steps, and the corresponding values of collector current I_C are noted. For each level of I_B, I_C is plotted versus V_{CE}. Thus a family of characteristics is obtained which are typically as illustrated in Fig. 14.18. The noteworthy points regarding output characteristics are given below :

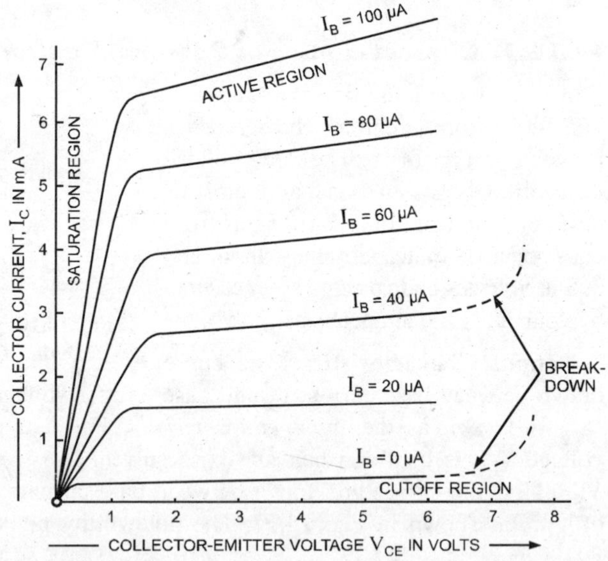

Fig. 14.18. ***Output Characteristics For Common Emitter NPN Transistor***

(*i*) The collector current I_C varies with V_{CE} for V_{CE} between 0 and 1 V and then becomes almost constant and independent of V_{CE}. The transistors are always operated above 1 V.

(*ii*) Output characteristic in CE configuration has some slope while CB configuration has almost horizontal characteristics. This indicates that output resistance in case of CE configuration is less than that in CB configuration.

(*iii*) In active region (collector junction reverse biased and emitter junction forward biased), for small values of base current I_B the effect of collector voltage V_C over I_C is small but for large values of I_B this effect increases.

The shape of the characteristic is same as in CB configuration but with the difference that collector current I_C is larger than input current (base current I_B). Thus the current gain for this configuration is larger than unity. The transistor must always be operated in the active region, when employed as an amplifying device.

(*iv*) With low values (ideally zero) of V_{CE} the transistor is said to be operated in saturation region and in this region base current I_B does not cause a corresponding change in collector current I_C.

(*v*) With much higher V_{CE}, the collector-base junction completely breaks down and because of this avalanche breakdown collector current I_C increases rapidly and the transistor gets damaged.

(*vi*) In cutoff region, small amount of collector current I_C flows even when base current $I_B = 0$. This is called I_{CEO}. Since main current I_C is zero, the transistor is said to be cutoff.

(*vii*) Moderate output to input impedance ratio makes this configuration an ideal one for coupling between various transistor stages.

The ratio of change in collector-emitter voltage (ΔV_{CE}) to the change in collector current (ΔI_C) at constant base current is known as *dynamic* ouput resistance.

It is calculated as the reciprocal of the slope of output characteristic at a given V_{CE}.

i.e., $$r_{out} = \left.\frac{\Delta V_{CE}}{\Delta I_C}\right|_{I_B = \text{constant}} \quad ...(14.31)$$

The output resistance of CE configuration is less than that of CB configuration as the slope of the ouput characteristic is more in this case. Its value is of the order of 50 kΩ.

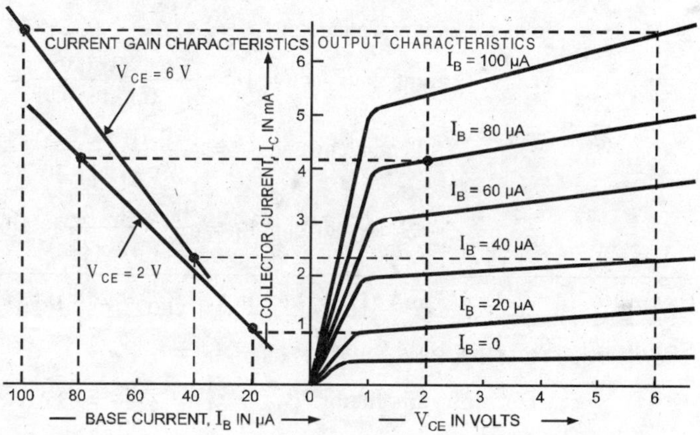

Fig. 14.19. ***Current Gain Characteristics For Common Emitter NPN Transistor***

The output characteristics may also be used to determine the dc current gain β and ac current gain β_0 as follows

$$\text{DC current gain, } \beta = \frac{I_C}{I_B} \quad ...(14.32)$$

$$\text{AC current gain, } \beta_0 = \left.\frac{\Delta I_C}{\Delta I_B}\right|_{V_{CE} = \text{constant}} \quad ...(14.33)$$

Current Gain Characteristics. Like the common-base current gain characteristics, they may be obtained experimentally, or determined from the output characteristics. For experimental determination of current gain characteristics, V_{CE} is kept constant at a convenient value and value of collector current I_C is observed and recorded for different values of base current I_B and then a graph is plotted between I_C and I_B.

The current gain characteristics can be obtained from the output characteristics as shown in Fig. 14.19, and detailed in Art. 14.10 except that I_C is plotted against I_B rather than I_C against I_E.

Example 14.11. A change of 250 mV in base-emitter voltage causes a change of 1 mA in the base current. Determine the dynamic input resistance.

Solution : Change in base-emitter voltage, $\Delta V_{BE} = 250$ mV

Change in base current, $\Delta I_B = 1$ mA

$$\text{Dynamic input resistance, } r_{in} = \frac{\Delta V_{BE}}{\Delta I_B} = \frac{250 \times 10^{-3}}{1 \times 10^{-3}} = 250\ \Omega \textbf{ Ans.}$$

Example 14.12. Increase in collector-emitter voltage from 5 V to 10 V causes increase in collector current from 5 mA to 5.8 mA. Determine the dynamic output resistance.

Solution : Change in collector-emitter voltage, $V_{CE} = 10 - 5 = 5$ V

Change in collector current, $\Delta I_C = 5.8 - 5.0 = 0.8$ mA

Dynamic output resistance, $r_{out} = \dfrac{\Delta V_{CE}}{\Delta I_C} = \dfrac{5}{0.8 \times 10^{-3}} = 6.25$ kΩ **Ans.**

Example 14.13. In a CE configuration, the collector supply voltage is 10 V. When a resistor $R_C = 1$ kΩ is connected in the collector circuit, the voltage drop across it is 0.5 V. For $\alpha = 0.98$, determine the collector-emitter voltage and the base current.

Solution : Collector supply voltage, $V_{CC} = 10$ V

Voltage drop across load resistance $= I_C R_C = 0.5$ V

Collector-emitter voltage, $V_{CE} = V_{CC} - I_C R_C = 10 - 0.5 = 9.5$ V **Ans.**

Current gain factor, $\beta = \dfrac{\alpha}{1-\alpha} = \dfrac{0.98}{1-0.98} = 49$

Collector current, $I_C = \dfrac{I_C R_C}{R_C} = \dfrac{0.5}{1 \times 10^3} = 0.5$ mA

Base current, $I_B = \dfrac{I_C}{\beta} = \dfrac{0.5 \text{ mA}}{49} = 10.2\ \mu$A **Ans.**

Example 14.14. $I_E = 5$ mA, $I_C = 4.95$ mA, $I_{CEO} = 200\ \mu$A. Calculate β_{dc} and leakage current I_{CBO}.

Solution : Base current, $I_B = I_E - I_C = 5 - 4.95 = 0.05$ mA $= 50\ \mu$A **Ans.**

Current gain factor, $\beta_{dc} = \dfrac{I_C}{I_B} = \dfrac{4.95}{0.05} = 99$ **Ans.**

Current amplification factor, $\alpha_{dc} = \dfrac{\beta_{dc}}{1+\beta_{dc}} = \dfrac{99}{1+99} = 0.99$

Collector-to-base leakage current, $I_{CBO} = I_{CEO}(1 - \alpha_{dc}) = 200(1 - 0.99) = 2\ \mu$A **Ans.**

Example 14.15. In an NPN transistor, $\alpha = 0.995$, $I_E = 10$ mA and $I_{CO} = 0.5\ \mu$A. Determine the values of I_C, I_B, β and I_{CEO}.

Solution : Emitter current, $I_E = 10$ mA

Collector-to-base leakage current, $I_{CO} = 0.5\ \mu$A $= 0.0005$ mA

Collector current, $I_C = \alpha I_E + I_{CO} = 0.995 \times 10 + 0.0005 = 9.9505$ mA **Ans.**

Base current, $I_B = I_E - I_C = 10 - 9.9505 = 0.0495$ mA $= 49.5\ \mu$A **Ans.**

Current gain factor, $\beta = \dfrac{\alpha}{1-\alpha} = \dfrac{0.995}{1-0.995} = 199$ **Ans.**

Collector current with base lead open, $I_{CEO} = \dfrac{I_{CBO}}{1-\alpha} = \dfrac{0.5 \times 10^{-6}}{1-0.995} = 100\ \mu$A **Ans.** Refer to Eq. (14.28)

Example 14.16. The collector leakage current in a transistor is 300 μA in CE arrangement. If the transistor is now connected in CB arrangement, what will be the leakage current. Given that $\beta = 100$.

Solution : I_{CEO} = Collector leakage current in CE arrangement = 300 μA

Current gain factor, $\beta = 100$

Current amplification factor, $\alpha = \dfrac{\beta}{1+\beta} = \dfrac{100}{101} = 0.99$

Leakage current in CB arrangement $= I_{CBO} = I_{CEO}(1 - \alpha) = 300(1 - 0.99) = 3\ \mu$A **Ans.**

Example 14.17. The following measurements were made in a particular transistor when connected in CB mode. I_C = 10.525 mA, I_B = 100 μA and I_{CBO} = 5 μA. Determine α, β and I_E.

Solution : $$\beta = \frac{I_C - I_{CBO}}{I_B + I_{CBO}} = \frac{10.525 - 0.005}{0.1 + 0.005} = \frac{10.52}{0.105} = 100.19 \simeq 100 \text{ Ans.} \qquad \because I_C = \beta I_B + (1+\beta) I_{CBO}$$

$$\alpha = \frac{\beta}{1+\beta} = \frac{100}{1+100} = 0.99 \text{ Ans.}$$

Emitter current, $I_E = I_C + I_B = 10.525 + 0.1 = 10.625$ mA **Ans.**

Example 14.18. The collector and base current of an N-P-N transistor are measured as I_C = 5 mA, I_B = 50 μA and I_{CBO} = 1 μA.

(*i*) Determine α, β and I_E

(*ii*) Determine the new level of I_B required to produce I_C = 10 mA.

Solution : (*i*) Given that

$$I_C = 5 \text{ mA}$$
$$I_B = 50\ \mu\text{A}$$
$$I_{CBO} = 1\ \mu\text{A}$$

Since $I_C = \beta I_B + (\beta + 1) I_{CBO}$

$$\text{or } \beta = \frac{I_C - I_{CBO}}{I_B + I_{CBO}} = \frac{5\times10^{-3} - 1\times10^{-6}}{50\times10^{-6} + 1\times10^{-6}} = \mathbf{98} \text{ Ans.}$$

$$\alpha = \frac{\beta}{1+\beta} = \frac{98}{99} = \mathbf{0.99} \text{ Ans.}$$

$$I_E = I_B + I_C = 50\times10^{-6} + 5\times10^{-3} = \mathbf{5.05\ mA} \text{ Ans.}$$

$$(ii)\ I_B = \frac{I_C - (\beta+1) I_{CBO}}{\beta} = \frac{10\times10^{-3} - (98+1)\times1\times10^{-6}}{98} = \mathbf{101\ \mu A} \text{ Ans.}$$

Example 14.19. A transistor is connected in common emitter configuration as an amplifier. The parameters of the transistor specified are :

I_B = 25 μA, I_{CBO} = 100 nA, β = 100. Find its I_C, I_E, I_{CEO} and α, where symbols have their usual meaning.

Solution : Collector current,

$$I_C = \beta I_B + (\beta+1) I_{CBO} = 100 \times 25 \times 10^{-6} + (100+1) \times 100 \times 10^{-9}$$
$$= (2{,}500 + 10.1) \times 10^{-6} = 2{,}510.1\ \mu\text{A or } 2.51 \text{ mA } \textbf{Ans.}$$

Emitter current, $I_E = I_C + I_B = 2{,}510.1 + 25 = 2{,}535\ \mu\text{A}$ or 2.535 mA **Ans.**

Current gain, $\alpha = \frac{\beta}{\beta+1} = \frac{100}{101} = 0.99$

Current, $I_{CEO} = \frac{I_{CBO}}{1-\alpha} = \frac{100\times10^{-9}}{1-0.99} = 10\ \mu\text{A}$ **Ans.**

14.12. COMMON COLLECTOR (CC) CONFIGURATION

In this circuit arrangement, input is applied between base and collector while the output is taken across emitter and collector. Thus the collector forms the terminal common to both the input and output circuits. The circuit is essentially the same as that of CE configuration except that load resistor R_L is now placed in the emitter circuit instead of collector circuit. The operation of the circuit in terms of the currents that flow is basically the same as that for the CE configuration. In this arrangement base current I_B flows in the input circuit and emitter current I_E flows in the output circuit. So the change in emitter current ΔI_E to the change in base current ΔI_B gives the current amplification factor γ. The circuit provides nearly the same current gain as in case of CE arrangement because $\Delta I_E \simeq \Delta I_C$.

With base current I_B equal to I_{CO}, the emitter current I_E is zero so no current flows in the load resistor R_L. With the increase in input current I_B, the transistor passes through the active region and finally reaches saturation. Almost the complete supply voltage, except for a small voltage drop across the transistor appears across the load resistor R_L.

The current equations are

$$I_C = \alpha I_E + I_{CBO}$$

and $$I_E = I_B + I_C = I_B + \alpha I_E + I_{CBO}$$

or $$I_E (1 - \alpha) = I_B + I_{CBO}$$

or $$I_E = \frac{I_B}{1-\alpha} + \frac{I_{CBO}}{1-\alpha} = (\beta + 1) I_B + (\beta + 1) I_{CBO} \quad ...(14.34) \quad \because \frac{1}{1-\alpha} = \beta + 1$$

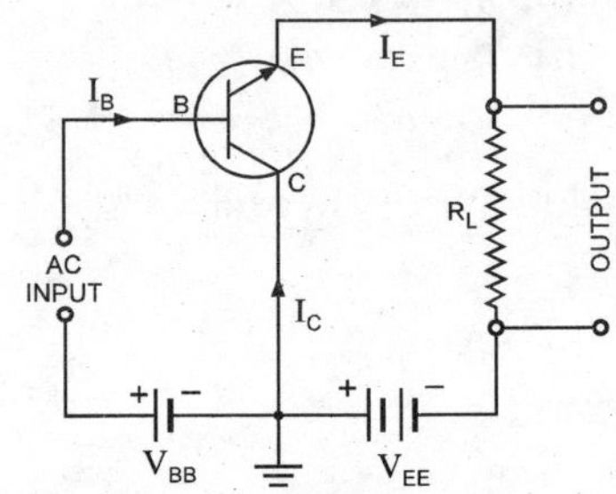

Fig. 14.20. *Common Collector NPN Transistor*

Current gain, $$\gamma = \frac{\Delta I_E}{\Delta I_B} = \frac{\Delta I_E}{\Delta I_E - \Delta I_C}$$

$$= \frac{\Delta I_E / \Delta I_C}{\frac{\Delta I_E}{\Delta I_C} - 1}$$

$$= \frac{1/\alpha}{1/\alpha - 1}$$

$$= \frac{1}{1-\alpha} = \beta + 1 \quad ...(14.35)$$

14.12.1. Characteristics of Common Collector Configuration

The circuit used for determination of transistor common collector configuration is shown in Fig. 14.21. For determination of input characteristics, V_{CE} is kept at a suitable fixed level and base current I_B is recorded for different values of collector-base voltage V_{CB}. Now the curves are drawn between base current and collector-base voltage, as shown in Fig. 14.22.

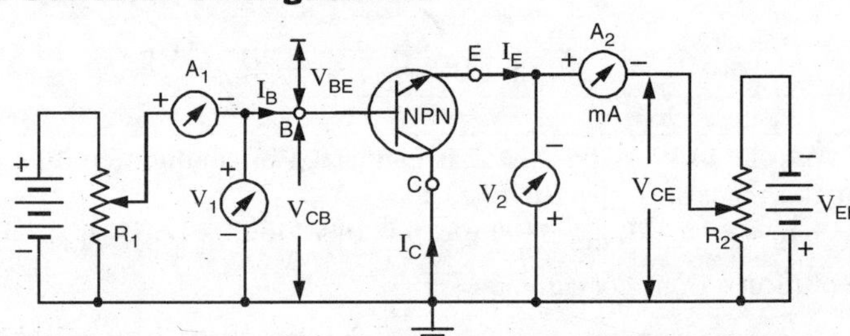

Fig. 14.21. *Circuit Arrangement For Determination of Input and Output Characteristics of a Transistor in CC Configuration*

The output characteristics are plotted between emitter current I_E and collector-emitter voltage V_{CE}. The common collector gain characteristics are plotted between emitter-current and base current. Fig. 14.23 depicts the output and current gain characteristics.

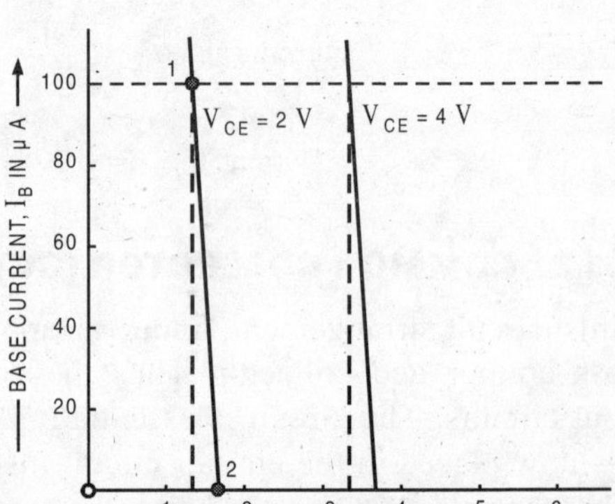

Fig. 14.22. *Input Characteristics For Common Collector NPN Transistor*

The common collector (CC) arrangement gives very high input impedance and very low output impedance and, therefore, its voltage gain is always less than unity. Hence this configuration is seldom used for amplification. However, owing to relatively high input impedance and low output impedance, this configuration is primarily used for impedance matching *i.e.* for driving low impedance from a high impedance source. This configuration is also called *emitter follower.*

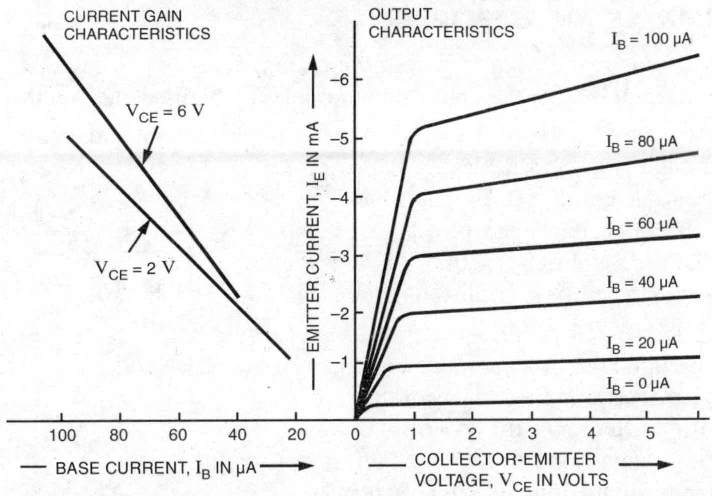

Fig. 14.23. ***Output and Current Gain Characteristics of Transistor in CC Configuration***

14.13. COMPARISON OF CHARACTERISTICS OF TRANSISTOR IN DIFFERENT CONFIGURATIONS

The essential characteristics of each of basic configurations are tabulated below :

Configuration / *Characteristics*	*Common Base*	*Common Emitter*	*Common Collector*
Input impedance	Low (about 100 Ω)	Medium (about 800 Ω)	Very high (about 750 kΩ)
Output impedance	Very high (about 500 kΩ)	High (about 50 kΩ)	Low (about 50 Ω)
Current gain	Less than unity but usually more than 0.9 (about 0.98)	High (about 80)	High (about 100)
Voltage gain	About 150	About 500	Less than unity
Leakage current	Very small (5 μA for Ge and 1 μA for Si)	Very large (500 μA for Ge and 20 μA for Si)	Very large
Output Signal phase	In phase with input	Reverse	In phase with input
Applications	For high frequency applications	For AF applications	For impedance matching

14.14. TRANSISTOR APPLICATIONS

A signal of small amplitude applied to the base is available in the "magnified form" at the collector of the transistor. This is the "amplification" provided by a transistor. The additional power required for it is obtained from an external source (dc power supply).

Bipolar junction transistor is a basic building block for almost all the electronic circuits right from a simple regulator or oscillator circuit, logic gates to a digital computer.

Some of the important transistor applications are enumerated as below :

1. Amplifiers, 2. Oscillators, 3. Switching circuits, 4. Waveshaping circuits, 5. Logic circuits, 6. Delay circuits, 7. Timers and multivibrators.

14.15. TRANSISTOR AS AN AMPLIFIER

The main utility of a transistor lies in its ability of amplifying weak signals. The weak signal is applied at the input terminals and the amplified output is obtained across the output terminals.

A transistor alone cannot perform the function of amplification and some passive components such as resistors and capacitors and a biasing battery is to be connected. Common emitter (CE) configuration, because of its high current, voltage and power gains, is much suited for most of the amplifier circuits. A common emitter N-P-N transistor amplifier circuit is given in Fig. 14.24.

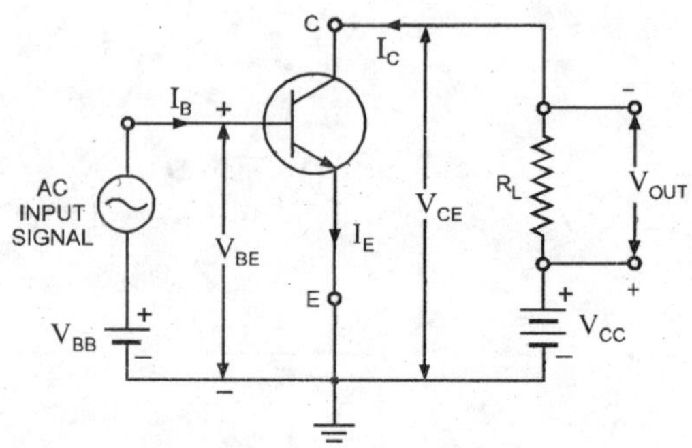

Fig. 14.24. *Common Emitter NPN Transistor Amplifier Circuit*

The weak signal is applied between emitter-base junction and output is taken across a load resistor R_L connected in series with collector supply voltage V_{CC} (Fig. 14.24). In order to obtain faithful amplification, it is necessary that the input circuit remains always forward biased regardless the polarity of the ac input signal. So a battery V_{BB} is inserted in the circuit with the polarity indicated in addition to the signal voltage. This dc voltage is called the *bias voltage* and its magnitude is such as to keep the input circuit always forward biased regardless the polarity of the input signal.

The input circuit being forward biased has low resistance and a small change ΔV_{in} in input signal voltage causes a relatively large change ΔI_E in emitter current. This causes almost the same change in collector current because of transistor action. The collector current flowing through a high load resistance R_L develops a large voltage across it. The change in output voltage across load resistance R_L may be many times the change in input signal voltage. Thus voltage amplification $A = \frac{\Delta V_{out}}{\Delta V_{in}}$ will be greater than unity, and the transistor acts as an amplifier. This is further illustrated below by considering typical circuit values.

Let the load resistance R_L be of 10 kΩ and a change of 0.1 V in input signal voltage cause a change of 0.5 mA in emitter current. This change of 0.5 mA in emitter current will also change collector current I_C by approximately 0.5 mA. This change of 0.5 mA in collector current will produce a change of $0.5 \times 10^{-3} \times 10 \times 10^3$ *i.e.* 5 V in output voltage appearing across the load resistor R_L of 10 kΩ. Thus the change of 0.1 V in input signal voltage causes a change of 5 V in the output voltage giving a voltage amplification of $\frac{5}{0.1} = 50$.

Necessity of Biasing. For most of the applications, transistors are required to operate as *linear amplifiers* (*i.e.* to amplify output voltage as a linear function of the input voltage). To achieve this, it is necessary to operate the transistor over region of its characteristic curves which are linear, parallel and equi-spaced for equal increments of the parameters. Such an operation can be ensured by proper selection of zero signal operating point and limiting the operation of the transistor over the linear portion of the characteristics. For proper selection of zero signal operating point, proper biasing *i.e.* application of dc voltages at emitter-to-base junction and collector-to-base junction is required.

If the transistor is not biased properly, it would work inefficiently and produce distortion in the output signal.

14.16. TRANSISTOR AS A SWITCH

A transistor can be employed as an electronic switch. Operating a transistor as a switch means operating it at either saturation or cutoff but nowhere else along the load line. When a transistor is saturated, it is like a closed switch from the collector to emitter. When the transistor is cutoff, it is like an open switch. The operation of a transistor as a switch is illustrated in Fig. 14.25.

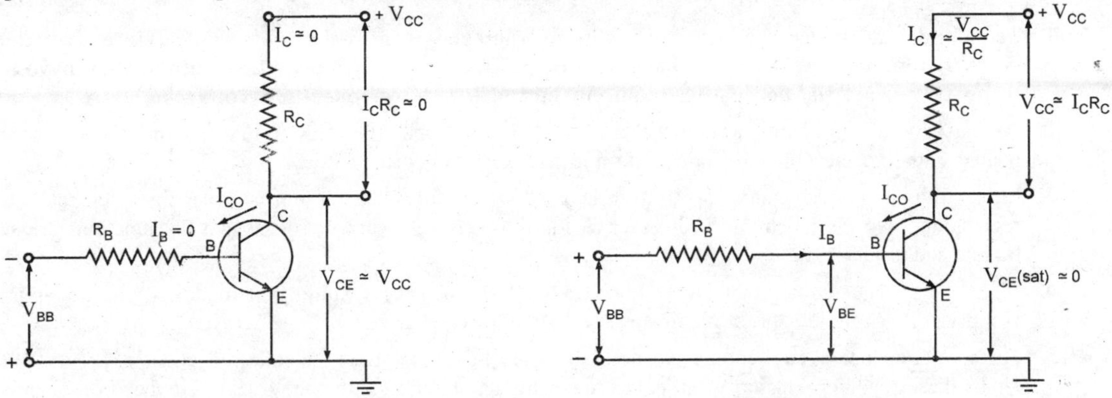

(a) Off-Biased Transistor *(b) On-Biased Transistor*

Fig. 14.25. ***Operation of a Transistor as a Switch***

14.17. STANDARD NOTATIONS FOR VOLTAGES AND CURRENTS

Standard notations or symbols are usually used for voltages and currents in transistor circuit analysis and applications.

1. Subscripts E (or *e*), B (or *b*) and C (or *c*) usually refer to emitter, base and collector quantities respectively.
2. Upper case (capital) letters with capital subscripts of the proper electrode symbol are used for dc or average values of voltages and currents.
3. Upper case (capital) letters with lower subscripts of the proper electrode symbol are used for rms values of ac voltages and currents.
4. Lower case (small) letters with lower subscripts of proper electrode symbol are used for instantaneous values of ac voltages and currents.

Notations used for various quantities (voltages and currents) in transistor circuit analysis and applications are given below in tabular form.

S.No.	*Particular*	*DC or average*	*RMS ac*	*Instantaneous ac*
1.	Emitter current	I_E	I_e	i_e
2.	Base current	I_B	I_b	i_b
3.	Collector current	I_C	I_c	i_c
4.	Collector-emitter voltage	V_{CE}	V_{ce}	v_{ce}
5.	Emitter-base voltage	V_{EB}	V_{eb}	v_{eb}
6.	Collector-base voltage	V_{CB}	V_{cb}	v_{cb}
7.	Emitter supply voltage	V_{EE}	—	—
8.	Base supply voltage	V_{BB}	—	—
9.	Collector supply voltage	V_{CC}	—	—
10.	Emitter voltage to ground	V_E	V_e	v_e
11.	Base voltage to ground	V_B	V_b	v_b
12.	Collector voltage to ground	V_C	V_c	v_c

HIGHLIGHTS

1. Semiconductor devices have the following advantages over vacuum tubes : They are (*i*) smaller in size (*ii*) lighter in weight, (*iii*) more rugged and (*iv*) more efficient. In addition, they have (*i*) no warm up period, (*ii*) no heating requirement and (*iii*) lower operating voltages.
2. A transistor may be compared loosely with a triode in that the emitter may be thought of as a non-heated cathode, the collector as a plate and the base as a grid.
3. There are two complementary bipolar transistors—NPN and PNP. Each transistor has three separately doped regions and two PN junctions. The center region (base) is very narrow, so the two PN junctions are said to be interacting junctions.
4. Since a transistor consists of either PNP or NPN slices, a very simplified model for biasing purposes comprises two diodes "pointing inward or outward" respectively.
5. For normal operation, base-emitter junction is forward biased and the base-collector junction is reverse biased. Under these conditions, majority carriers from the emitter are injected into the base where they become minority carriers. These minority carriers diffuse across the base into the B-C space charge region where they are swept into the collector.

   In this mode of operation, the current at one terminal of the transistor (collector current) is controlled by the voltage applied across the other two terminals of the transistor (base-emitter voltage). This is the basic transistor action.
6. The modes of operation of a BJT are determined by the biases applied to the two junctions. The four modes of operation are : forward active, cutoff, saturation, and inverse active. In the forward-active mode, the B-E junction is forward biased and the B-C junction is reverse biased, and the collector and base currents are related by the common-emitter current gain $\beta(I_C = \beta I_B)$. The relationship is the same for both NPN and PNP transistors, as long as the conventional current directions are maintained. In the cutoff mode of operation, the B-E and B-C junctions are both reverse biased and all currents are zero. In saturation mode, the B-E and B-C junctions are both forward biased and the collector current is no longer a function of base current.
7. The dc emitter current I_E is always the largest one of a transistor, whereas the base current I_B is always the smallest. The collector current I_C is almost as large as emitter current I_E (*i.e.* $I_C \simeq I_E$). The emitter current is always equal to the sum of the other two (*i.e.* $I_E = I_B + I_C$).
8. The collector current I_C is made up of two components : the majority component and the minority component (also known as the leakage current).

   i.e., $I_C = I_{C_{majority}} + I_{CO_{minority}}$
9. The arrow in the transistor symbol defines the direction of conventional current flow for the emitter current and thereby defines the directions for the other currents in the device.

   The arrow in the symbol of an NPN transistor points out of the device (not pointing in), whereas the arrow points to the centre of the symbol for a PNP transistor (pointing in).
10. On an average basis, as a second approximation, the base-to-emitter voltage of an operating transistor can be assumed to be 0.7 V in case of silicon device and 0.3 V in case of germanium device.
11. The quantity α relates the collector and emitter currents $\left(\alpha = \dfrac{I_C}{I_E}\right)$ and is always approximately equal to one (never greater than 1).
12. The impedance between terminals of a forward-biased junction is always relatively small, whereas the impedance between terminals of a reverse-biased junction is usually quite large.
13. The quantity beta β provides an important relationship between the base and collector currents and is usually between 50 and 400.

14. The dc beta is defined by a simple ratio of dc currents at an operating point $\left(i.e.\, \beta_{dc} = \frac{I_C}{I_B}\right)$, whereas the ac beta β_{ac} is sensitive to the characteristics in the region of interest. For most applications, however, the two are considered equivalent as a first approximation.
15. A three-terminal device needs two sets of characteristics to completely define its characteristics. The transistor input characteristic is essentially that of a forward biased junction, with some dependence upon the transistor's output voltage.
16. For a transistor used in CB configuration, the collector or output characteristics consist of practically horizontal lines of I_C versus V_{CB} with I_E as the controlling parameter.
17. The CB ac current gain, $\alpha_{ac} = \left.\frac{\Delta I_C}{\Delta I_E}\right|_{V_{CB} = \text{Constant}}$ while $\alpha_{dc} = \frac{I_C}{I_E}$ and is less than unity.
18. The reverse saturation current in a CB configuration = $I_{CBO} = I_{CO}$, while in a CE configuration it is $I_{CEO} = (\beta + 1)I_{CBO}$.
19. The collector or output characteristics for a transistor used in the CE configuration have the collector current I_C versus collector-emitter voltage curves sloping upward slightly with base current as the controlling parameter.
20. The CE ac current gain, $\beta_{ac} = \left.\frac{\Delta I_C}{\Delta I_B}\right|_{V_{CE} = \text{Constant}}$, while $\beta_{dc} = \frac{I_C}{I_B}$.

EXERCISES

1. Draw NPN and PNP transistors. Label all the currents and show the direction of flow. How are all the currents of a transistor related ?
2. What is a transistor ? Show the transistor action for the N-P-N and P-N-P transistors.
3. Draw the schematic diagram of the NPN transistor and explain its working.
4. Draw a schematic diagram of a transistor indicating the different currents. Arrive at the relationship among them.
5. Explain the action of PNP transistor.

[Pb. Technical Univ. Basic Electrical and Electronics Engineering Dec. 2007]

6. If two P-N junction diodes are connected back to back as shown in Fig. 1, will it work as a transistor with the terminals designated in the figure ? Explain.

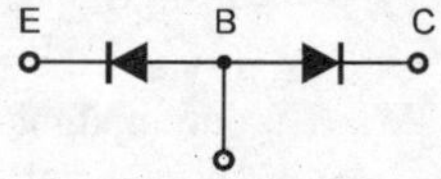

Fig. 1

7. Define with respect to BJT the following :
(*i*) I_{CO}, (*ii*) α, (*iii*) β, (*iv*) I_{CEO}, (*v*) I_{CBO}.
8. Differentiate between I_{CO} and I_{CBO}. What is the effect of temperature on I_{CBO} ?
9. Explain different components of current flowing through the structure of a N-P-N transistor. How the emitter injection efficiency and base transport factor influences the amplification factor ?
10. Enumerate types of transistor configurations and show their circuit diagrams.

[Pb. Technical Univ. Basic Electrical and Electronics Engineering May 2005]

11. Draw the input and output characteristic curves of a CB configuration for a transistor. Explain the shape of curves. Clearly indicate cutoff, active and saturation regions.
12. What are the three transistor configurations ? Compare and contrast the characteristics of these configurations. Also state any one application of each of these configurations.
13. For a transistor in CE configuration, (*i*) Derive the expression between I_C, I_B and I_{CO}, (*ii*) Explain why slope of the output characteristics is more than in CB configuration, (*iii*) cutoff condition.
14. Give BJT characteristics for CE configuration and explain the same giving various operating regions.

15. Draw a simple circuit of a common-collector amplifier and explain its operation. Obtain an expression for its voltage gain.
16. What is bipolar junction transistor (BJT) ? List applications of BJT.

[Pb. Technical Univ. Basic Electrical and Electronics Engineering, May 2010]

17. Compare the characteristics of transistor amplifiers in the three possible configurations.
18. Explain how transistor acts as an amplifier.
19. Write short notes on the following :
    (*i*) Transistor action.
    (*ii*) Comparison of characteristics of transistor in different configurations.
    (*iii*) Transistor applications.
    (*iv*) Transistor as an amplifier.

SHORT ANSWER TYPE QUESTIONS WITH ANSWERS

Q. 1. Explain why an ordinary transistor is called bipolar ? **[P.T.U. May 2008]**

Ans. Because the transistor operation is carried out by two types of charge carriers (majority and minority carriers), an ordinary transistor is called bipolar.

Q. 2. Why transistor is called current controlled device ?

Ans. The output voltage, current or power is controlled by the input current in a transistor so it is called the current controlled device.

Q. 3. What is the significance of arrowhead in the transistor symbol ?

Ans. Arrowhead is always marked on the emitter. The direction indicates the conventional direction of current flow (from emitter-to-base in case of P-N-P transistor and from base-to-emitter in case of N-P-N transistor). Generally, no arrowhead is marked for collector since its reverse leakage current is always opposite to the direction of emitter current.

Q. 4. What do you understand by transistor biasing ? Why is it necessary to bias a transistor ? **[P.T.U. May 2008]**

Ans. For normal operation, base-emitter junction should be forward biased and the collector-base junction reverse biased. The proper flow of zero signal collector current and the maintenance of proper collector-emitter voltage during the passage of signal is called the transistor biasing. The amount of bias required is significant for establishment of the operating or Q-point which is dictated by the mode of operation desired.

In case the transistor is not biased properly, it would (*i*) work inefficiently, (*ii*) produce distortion in the output signal and (*iii*) with the change in transistor parameters or temperature rise, the operating point may shift and the amplifier output will be unstable.

Q. 5. What are 'emitter injection efficiency' and 'base transport factor' and how do they influence the transistor operation ?

Ans. The ratio of current of injected carriers at emitter junction to the total emitter current is called the emitter injection efficiency. The ratio of collector current to base current is known as transport factor

$$i.e., \ \beta = \frac{I_C}{I_B}$$

The larger the value of emitter injection efficiency, the larger the injected carriers at emitter junction and this increases the collector current. The larger the β value larger the injected carriers across collector junction and hence collector current increases.

Q. 6. Which of the transistor currents is always the largest ? Which is always the smallest ? Which two currents are relatively close in magnitude ?

Ans. The emitter current I_E is always the largest one. The base current I_B is always the smallest. The collector current I_C and emitter current I_E are relatively close in magnitude *i.e.* $I_C \simeq I_E$.

Q. 7. Why silicon type transistors are more often used than germanium type ?

Ans. Because silicon transistor has smaller cutoff current I_{CBO}, small variations in I_{CBO} due to variations in temperature and high operating temperature as compared to those in case of germanium type.

Q. 8. Why the width of the base region of a transistor is kept very small compared to other regions ?

Ans. Base region of a transistor is kept very small and very lightly doped so as to pass most of the injected charge carriers to the collector.

Q. 9. Why emitter is always forward biased ?

Ans. Emitter is always forward biased w.r.t. base so as to supply majority charge carriers to the base.

Q. 10. Why collector is always reverse biased w.r.t. base ?

Ans. Collector is always reverse biased w.r.t. base so as to remove the charge carriers away from the base-collector junction.

Q. 11. What are three regions of operation of a transistor ? **[P.T.U. December 2007]**

Ans. The three regions of operation of a transistor are :

(*i*) Cutoff region, (*ii*) Active region and (*iii*) Saturation region.

Q. 12. Can a transistor be obtained by connecting two semiconductor diodes back to back ?

Ans. No. Because in case of two discrete back to back connected diodes there are four doped regions instead of three and there is nothing that resembles a thin base region between an emitter and a collector.

Q. 13. How α and β are related to each other ?

Ans. α and β are related as below :

$$\alpha = \frac{\beta}{1+\beta} \quad \text{or} \quad \beta = \frac{\alpha}{1-\alpha}$$

Q. 14. Define beta of a transistor.

Ans. The β factor of a transistor is the common-emitter current gain of that transistor and is defined as the ratio of collector current to base current.

$$i.e., \ \beta = \frac{I_C}{I_B}.$$

Q. 15. Explain why $I_{CEO} >> I_{CBO}$?

Ans. The collector cutoff current denoted by I_{CEO} is much larger than I_{CBO}. I_{CEO} is given as

$$I_{CEO} = \frac{1}{1-\alpha} I_{CBO}$$

Because α is nearly equal to unity (slightly less than unity), $I_{CEO} >> I_{CBO}$.

Q. 16. Give reasons why common emitter (CE) configuration is widely used in amplifier circuits. **[Pb. Technical Univ. May 2008]**

Ans. CE configuration is mainly used because its current, voltage and power gains are quite high and the ratio of output impedance and input impedance are quite moderate.

Q. 17. Why is CE configuration preferred for cascaded amplifiers ? **[Pb. Technical Univ. May 2007]**

Ans. CE configuration, because of its moderate output to input impedance ratio, is preferred for cascaded amplifiers.

Q. 18. Why CC configuration is called a voltage buffer ? What is other name ?

Ans. Because of its high input impedance and low output impedance, the common collector circuit finds wide application as a buffer amplifier between a high impedance source and low impedance load, that is why it is called a voltage buffer. Its other name is emitter follower.

Q. 19. What are the main purposes for which a CC amplifier may be used.

Ans. Because of its high input impedance and low output impedance, the common collector circuit finds wide application as a buffer amplifier between a high impedance source and low impedance load.

Q. 20. Compare three configurations CB, CC and CE in terms of their resistances

[Pb. Technical Univ. May 2008]

Ans.

Configuration/ Resistances	Common Base	Common Emitter	Common Collector
Input Resistance	Low (about 100 Ω)	Medium (about 800 Ω)	Very high (about 750 kΩ)
Output Resistance	Very high (about 500 kΩ)	High (about 50 kΩ)	Low (about 50 Ω)

Q. 21. What do you understand by collector reverse saturation current ? In which configuration does it have a greater value ?

Ans. When input current (I_E in case of CB configuration and I_B in case of CE configuration) is zero, collector current I_C is not zero although it is very small. In fact this is the reverse leakage current or collector reverse saturation current (I_{CBO} or simply I_{CO} in CB configuration and I_{CEO} in CE configuration). In case of CE configuration it is much more than that in case of CB configuration.

PROBLEMS

1. Determine emitter current I_E, collector current I_C for a transistor with α_{dc} = 0.97 and collector to base leakage current 10 μA. Base current is 50 μA. **[Ans.** 2.00 mA, 1.95 mA]
2. For a certain transistor collector current is 6.25 mA, base current is 100 μA and collector-to-base leakage current 5 μA. Determine emitter current and α_{dc}. **[Ans.** 6.35 m A, 0.983]
3. In a particular transistor the collector current is 5.6 mA and the emitter current is 5.75 mA. Determine α_{dc}. **[Ans.** 0.974]
4. A transistor has α = 0.98. If emitter current of the transistor is 1 mA, determine the base current and gain factor β. **[Ans.** 20 μA; 49]
5. Determine β if α = 0.99. **[Ans.** 99]
6. A bipolar junction transistor has a collector current of 4 mA and base current of 20 μA. Determine its β. **[Ans.** 200]
7. A BJT has a base current of 200 μA and emitter current of 20 mA. Determine collector current and β. **[Ans.** 19.8 mA; 99]
8. In a transistor I_C = 0.95 mA, I_E = 1 mA. Determine base current, α and β in CB configuration. **[Ans.** 50 μA; 0.95; 19]
9. In a transistor CE configuration the voltage across 5 kΩ load resistance is 5 V. Determine collector current, base current and emitter current if β = 50. **[Ans.** 1.0 mA; 20 μA; 1.02 mA]
10. In a grounded base configuration the voltage drop across load resistance of 4 kΩ is 3.2 V. Determine collector current, emitter current and base current if α = 0.96. **[Ans.** 0.8 mA; 0.8333 mA; 0.0333 mA]
11. A CE transistor amplifier has input resistance of 100 Ω and output resistance of 6 kΩ. Determine the voltage gain and power gain if current gain factor is 90. **[Ans.** 5,400; 486,000]
12. Increase in collector-emitter voltage from 4 V to 8 V causes increase in collector current from 5 mA to 6 mA. Determine the output resistance. **[Ans.** 4 kΩ]
13. A change of 200 mV in base-emitter voltage causes a change of 60 μA in the base current. Determine its input resistance. **[Ans.** 3.33 kΩ]
14. A transistor is connected in CB configuration. When the emitter voltage is changed by 200 mV, the emitter current changes by 5 mA. During this variation, the collector to base voltage is kept fixed. Calculate the dynamic input resistance of transistor. **[Ans.** 40 Ω]

❑❑❑

15 CHAPTER

Field-Effect Transistors (FETs)

15.1. INTRODUCTION

INSIDE THIS CHAPTER

We know that P-N junctions have two dominant features *viz.* the injection of minority carriers with forward bias and a variation of the depletion width W with reverse bias. These two P-N junction properties are used in two important types of transistors *viz. bipolar junction* transistor abbreviated as BJT and the junction *field effect transistor abbreviated as* JFET. The BJT discussed so far makes use of the injection of minority carriers across a forward-biased junction, while JFET discussed in this chapter depends on the control of a junction depletion width under reverse bias. FET is a majority carrier device, and is, therefore, often called a *unipolar* transistor while BJT operates by the injection and collection of minority carriers. Since in this device, both holes and electrons take part in conduction, it is called a *bipolar transistor*. Like its bipolar counterpart, the FET is a three-terminal (namely drain, source and gate) semiconductor device in which current conduction is by one type of majority carriers (electrons in case of an N-channel FET or holes in a P-channel FET). Unlike the BJT, a FET needs virtually no input (bias signal) current. Field effect devices are controlled by an electric field (or voltage) and give an extremely *high input impedance*—most important advantage over BJT devices.

There are two major categories of field-effect transistors namely (*i*) junction field-effect transistors (abbreviated JFETs or simply FETs) and (*ii*) the insulated gate field effect transistors (IGFETs), more commonly known as metal-oxide semiconductor field-effect transistors (abbreviated MOSFETs or MOSTs).

15.2. JUNCTION FIELD-EFFECT TRANSISTORS

Depending upon the majority carriers, JFETs have been classified into two types, namely (*i*) N-channel JFETs with electrons as the majority carriers and (*ii*) P-channel JFETs with holes as the majority charge carriers. N-channel JFETs are preferred because of mobility of electrons higher than that of holes. N-channel and P-channel FETs are shown in Figs. 15.1 (*a*) and 15.1 (*b*) respectively.

15.2.1. Basic Construction

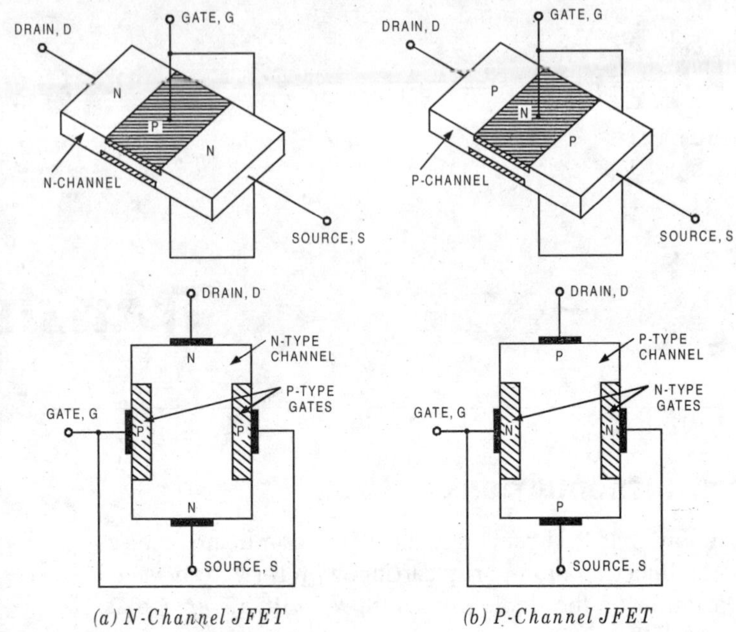

(a) N-Channel JFET *(b) P-Channel JFET*

Fig. 15.1. ***JFETs***

Its structure is quite simple. In an N-channel JFET an N-type silicon bar, referred to as the *channel*, has two smaller pieces of P-type silicon material diffused on the opposite sides of its middle part, forming P-N junctions, as illustrated in Fig. 15.1 (*a*). The two P-N junctions forming diodes or gates are connected internally and a common terminal,* called the *gate terminal*, is brought out. Ohmic contacts (direct electrical connections) are made at the two ends of the channel—one lead is called the *source terminal* S and the other *drain terminal* D.

The silicon bar behaves like a resistor between its two terminals D and S. The gate terminal is analogous to the base of an ordinary transistor (BJT). It is used to control the flow of current from source to drain. Thus, source and drain terminals are analogous to emitter and collector terminals respectively of a BJT.

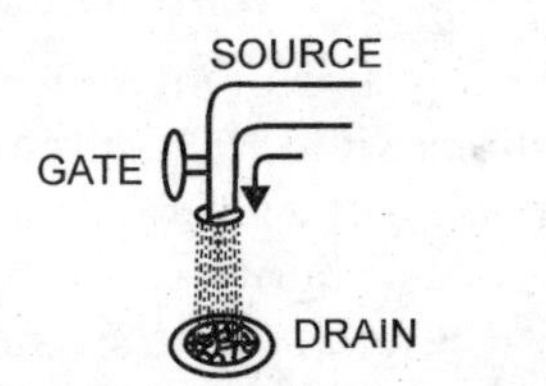

Fig. 15.2. ***Water Analogy For The JFET Control Mechanism***

Analogies are seldom perfect and at times can be misleading, but the water analogy of Fig. 15.2 does provide a sense for the JFET control at the gate terminal and the appropriateness of the terminology applied to the terminals of the device. The source of water pressure can be likened to the applied voltage from drain to source, which results in a flow of water (electrons) from the spigot (source). The "gate", through an applied signal (potential) controls the flow of water (charge) to the "drain", the drain and source terminals are at opposite ends of the N-channel as introduced in Fig. 15.1 because the terminology is defined for electron flow.

In Fig. 15.1 (*a*), the gate is P-region, while the source and the drain are N-regions. Because of this, a JFET is similar to two diodes. The gate and the source form one of the diodes, and the gate and the drain form the other diode. These two diodes are usually referred to as the *gate-source diode* and the *gate-drain diode*. Since JFET is a silicon device, it takes only 0.7 V for forward bias to get significant current in either diode.

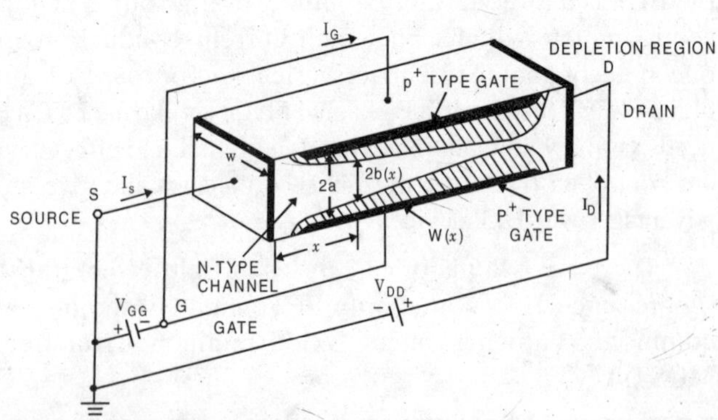

Fig. 15.3. ***Basic Structure of an N-Channel FET***

* When a manufacturer connects a separate lead to each gate, the device is called a dual-gate JFET. The main use of a dual-gate JFET is with a mixer, a special circuit used in communication equipment.

With the gate terminal not connected, and a potential applied (+ve at the drain and –ve at the source), a current called the *drain current*, I_D flows through the channel located between the two P-regions. This current consists of only majority carriers – electrons in this case.

P-channel JFET is similar in construction to N-channel JFET except that P-type semiconductor material is sandwiched between two N-type junctions, as shown in Fig. 15.1 (*b*). In this case majority carriers are holes.

The basic structure of an N-channel JFET, together with the supply voltages, is shown in Fig. 15.3.

15.2.2. Standard Notations

The noteworthy FET notations are as follows :

1. ***Source.*** The terminal, through which the majority carriers enter the channel, is called the *source* terminal S and the conventional current entering the channel at S is designated as I_S.
2. ***Drain.*** The terminal, through which the majority carriers leave the channel, is called the *drain* terminal D and the conventional current leaving the channel at D is designated as I_D.

   The drain-to-source voltage is called V_{DS}, and is positive if D is more positive than S.
3. ***Gate.*** There are two internally connected heavily doped impurity regions formed by alloying, by diffusion, or by any other method available to create two P-N junctions. These impurity regions are called the gate G. A voltage V_{GS} is applied between the gate and source in the direction to reverse bias the P-N junction. Conventional current entering the channel at G is designated as I_G.
4. ***Channel.*** The region between the source and drain, sandwiched between the two gates is called the *channel* and the majority carriers move from source to drain through this channel.

15.2.3. Schematic Symbols

The schematic symbols for N-type and P-type JFETs are shown in Figs. 15.4 (*a*) and 15.4 (*b*) respectively. The vertical line in the symbol may be thought as channel and source S and drain D connected to the line.

Note that the direction of the arrow at the gate indicates the direction in which the gate current flows when the gate junction is forward biased. Thus for the N-channel JFET, the arrow at the gate junction points into the device and in P-channel JFET, it is away from the device.

An alternative symbol for an N-channel JFET is shown in Fig. 15.4 (*c*). This symbol with its offset gate is preferred by many engineers and technicians. The offset gate points to the source end of the device, a definite advantage in complicated multistage circuit.

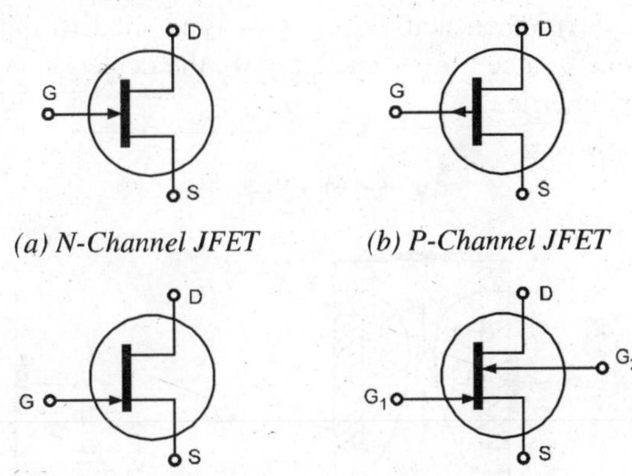

(*a*) *N-Channel JFET* (*b*) *P-Channel JFET*

(*c*) *Offset Gate Symbol For an N-Channel JFET* (*d*) *Tetrode-Connected JFET*

Fig. 15.4. ***Schematic Symbols For JFETs***

The symbol shown in Fig. 15.4 (*d*) is used where the terminals of the two gate regions are provided with separate connecting leads. In this case device is referred to as a **tetrode-connected FET.**

15.2.4. Polarity Conventions

The polarities for N-channel and P-channel JFETs are shown in Figs. 15.5 (*a*) and 15.5 (*b*) respectively. In both of the cases the voltage between gate and source is such that the gate is reverse biased. This is the normal method of connection of JFETs. The drain and source terminals are interchangeable* *i.e.* either end can be used as a source and the other end as a drain. The source terminal is always connected to that end of the drain voltage supply which provides the necessary charge carriers *i.e.* in an N-channel JFET source terminal, S is connected to the –ve end of the drain voltage supply for obtaining electrons whereas in P-channel JFET source terminal, S is connected to the +ve end of the drain voltage supply for getting holes.

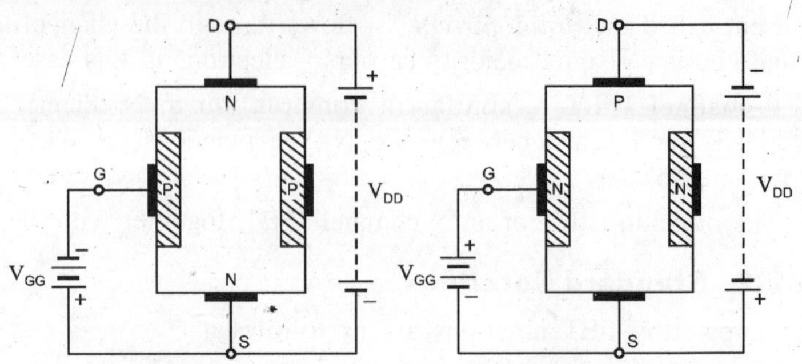

(a) N-Channel JFET *(b) P-Channel JFET*

Fig. 15.5. ***Polarity Conventions For JFETs***

Other noteworthy points are as follows :

(*i*) For N-channel JFET, I_D and V_{DD} are positive while V_{GG} is negative.

(*ii*) For P-cannel JFET, I_D and V_{DD} are negative while V_{GG} is positive.

(*iii*) Assumed positive directions of I_S, I_D and I_G are into the JFET.

From the above statements it may be concluded that the action of a P-channel JFET is complementary, that is, all voltages and currents are reversed.

15.2.5. Operation

Let us consider an N-channel JFET for discussing its operation.

(*i*) When neither any bias is applied to the gate (*i.e.* when $V_{GS} = 0$) nor any voltage to the drain w.r.t. source (*i.e.* when $V_{DS} = 0$), the depletion regions around the P-N junctions are of equal thickness and symmetrical.

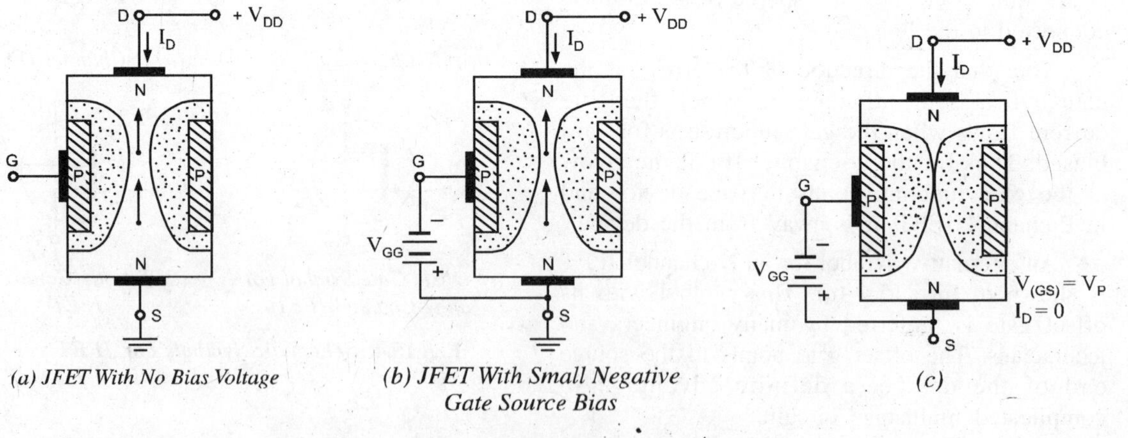

(a) JFET With No Bias Voltage *(b) JFET With Small Negative Gate Source Bias* *(c)*

Fig. 15.6

* The drain and source terminals are not interchangeable at high frequencies. Almost always, the manufacturer minimizes the internal capacitance on the drain side of the JFET *i.e.* the capacitance between the gate and the drain is smaller than the capacitance between the gate and the source.

(*ii*) When positive voltage is applied to the drain terminal D w.r.t. source terminal S without connecting gate terminal G to supply, as illustrated in Fig. 15.6 (*a*), the electrons (which are the majority carriers) flow from terminal S to terminal D whereas conventional drain current I_D flows through the channel from D to S. Due to flow of this current, there is a uniform voltage drop across the channel resistance as we move from terminal D to terminal S. This voltage drop reverse biases the diode. The gate is more negative with respect to those points in the channel which are nearer to D than to S. Hence, depletion layers penetrate more deeply into the channel at points lying closer to D than to S. Thus wedge-shaped depletion regions are formed, as shown in Fig. 15.6 (*a*), when V_{DS} is applied. The size of the depletion layer formed determines the width of the channel and hence the magnitude of current I_D flowing through the channel.

To see how the width of the channel varies with the variation in gate voltage, let us consider the situation when the gate is biased negative with respect to the source while the drain is applied with positive bias with respect to the source, as illustrated in Fig. 15.6 (*b*). Now the P-N junctions are reverse biased and depletion regions are formed. P-regions are heavily doped compared to the N-channel, so the depletion regions penetrate deeply into the channel. Since a depletion region is a region depleted of the charge carriers, it behaves as an insulator. The result is that the channel is narrowed, the resistance is increased and drain current I_D is reduced. If the negative voltage at the gate is further increased, depletion layers meet at the centre and the drain current I_D is cutoff completely. On the other hand, if the negative bias to the gate is reduced, the width of the depletion layers gets reduced causing decrease in resistance and, therefore, increase in drain current I_D. The gate-source voltage V_{GS} at which drain current I_D is cutoff completely (pinched off), is called the *pinch-off voltage* V_P. It is also to be noted that

1. The amount of reverse bias is not the same throughout the length of the P-N junction. When the drain current flows through the channel, there is a voltage drop along its length. The result is that the reverse bias at the drain end is more than that at the source end making the width of depletion layer more at the drain end than that at the source end. Thus the channel becomes narrower at the drain end in comparison to that at source end, as shown in Fig. 15.6 (*b*).

2. The channel is not completely closed at the drain end. Because in that case there will be no drain current, so there will be no voltage drop along the channel length and amount of reverse bias will become uniform and the wedge shaped depletion region will become rectangular one. The channel will open and the drain current will flow.
   However, at pinch-off voltage, the channel width is reduced to a constant minimum value to allow the flow of drain current.

3. The N-channel JFET behaves as a vacuum tube triode. The drain and source perform the same functions as the plate and cathode, respectively and, like the grid of a triode, the JFET gate controls the drain current. As is also the case with a grid, gate current is to be avoided, so the gate channel junctions are normally never forward biased.

The device is called the *field-effect transistor* (FET) because the drain current (output current) is controlled by the effect of the extension of the field associated with the depletion region developed by the reverse bias at the gate.

P-channel JFET operates in the same manner as an N-channel JFET except that channel current carriers will be the holes in place of electrons and the polarities of V_{GS} and V_{DS} are reversed.

15.3. CHARACTERISTICS OF JFETs

There are two types of static characteristics viz. (1) output or drain characteristic and (2) transfer characteristic.

1. *Output or Drain Characteristic*. The curve drawn between drain current I_D and drain-source voltage V_{DS} with gate-to-source voltage V_{GS} as the parameter is called the *drain* or *output characteristic*. This characteristic is analogous to collector characteristic of a BJT.

(a) Drain Characteristic With Shorted Gate. The circuit diagram for determining the drain characteristic with shorted gate for an N-channel JFET is given in Fig. 15.7 (*a*) and the drain characteristic with shorted gate is shown in Fig. 15.7 (*b*).

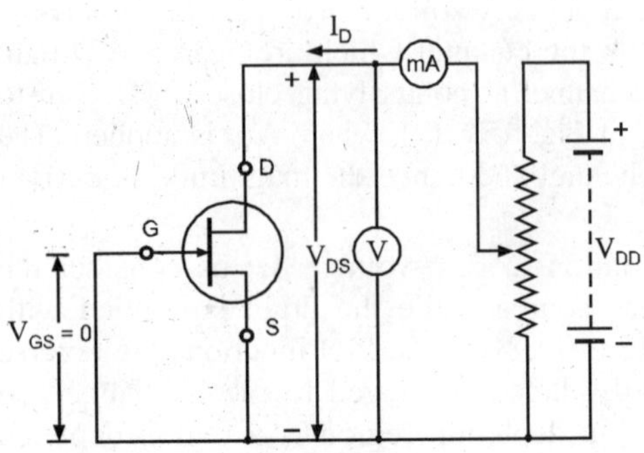

(a) Circuit Diagram For Determining Drain Characteristic With Shorted Gate For an N-Channel JFET

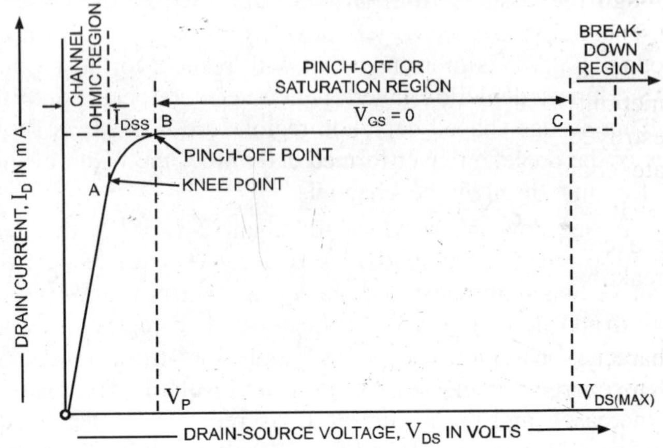

(b) JFET Drain Characteristic With Shorted Gate

Fig. 15.7

Initially when drain-source voltage V_{DS} is zero, there is no attracting potential at the drain, so no current flows inspite of the fact that the channel is fully open. This gives drain current $I_D = 0$. For small applied voltage V_{DS}, the N-type bar acts as a simple semiconductor resistor, and the drain current I_D increases linearly with the increase in V_{DS}, up to the knee point. This region (to the left of the knee point) of the curve is called the *channel ohmic region*, because in this region the FET behaves like an ordinary resistor.

With the increase in drain current I_D, the ohmic voltage drop between the source and channel region reverse biases the gate junction. The reverse biasing of the gate junction is not uniform throughout. The reverse bias is more at the drain end than that at the source end of the channel, so with the increase in V_{DS}, the conducting portion of the channel begins to constric more at the drain end. Eventually a voltage V_{DS} is reached at which the channel is *pinched off*. The drain current I_D no longer increases with the increase in V_{DS}. It approaches a constant saturation value. The value of voltage V_{DS} at which the channel is pinched off (*i.e.* all the free charges from the channel get removed), is called the *pinch-off voltage* V_P. The pinch-off voltage V_P, not too sharply defined on the curve, where the drain current I_D begins to level off and attains a constant value. From point A (knee point) to the point B (pinch-off point) the drain current I_D increases with the increase in voltage V_{DS} following a reverse square law. The region of the characteristic in which drain current I_D remains fairly constant is called the *pinch-off region*. It is also sometimes called the *saturation region* or *amplifier region*. In this region the JFET operates as a *constant current device* since drain current (or output current) remains almost constant. It is the normal operating region of the JFET when used as an amplifier. The drain current in the pinch-off region with $V_{GS} = 0$ is referred to as the *drain-source saturation current*, I_{DSS}.

It is to be noted that in the pinch-off (or saturation) region the channel resistance increases in proportion to increase in V_{DS} and so keeps the drain current almost constant and the reverse bias required by the gate channel junction is supplied entirely by the voltage drop across the channel resistance due to flow of I_{DSS} and not by the external bias because $V_{GS} = 0$.

Drain current in the pinch-off region is given by Shockley's equation

$$I_D = I_{DSS}\left(1 - \frac{V_{GS}}{V_P}\right)^2 = I_{DSS}\left(1 - \frac{V_{GS}}{V_{GS\,(OFF)}}\right)^2 \qquad ...(15.1)$$

where I_D is the drain current at a given gate-source voltage V_{GS}, I_{DSS} is the drain current with gate shorted to source and $V_{GS\,(OFF)}$ is the gate-source cutoff voltage.

If drain-source voltage, V_{DS} is continuously increased, a stage comes when the gate channel junction breaks down. At this point the drain current increases very rapidly, and the JFET may get destroyed. This happens because the charge carriers making up the reverse saturation current at the gate channel junction accelerate to a high velocity and produce an *avalanche effect.*

It will be very interesting to observe that JFET behaves as an ordinary resistor in ohmic region, as a constant current source in pinch-off (or saturation) region and as a constant voltage source in breakdown region [Fig. 15.7 (*b*)].

(*b*) Drain Characteristics With External Bias. The circuit diagram for determining the drain characteristics with different values of external bias is shown in Fig. 15.8 (*a*) and a family of drain characteristics for different values of gate-source voltage V_{GS} is given in Fig. 15.8 (*b*).

It is observed that as the negative gate bias voltage is increased

(*i*) the maximum saturation drain current becomes smaller because the conducting channel now becomes narrower.

(*ii*) Pinch-off voltage is reached at a lower value of drain current I_D than when $V_{GS} = 0$.

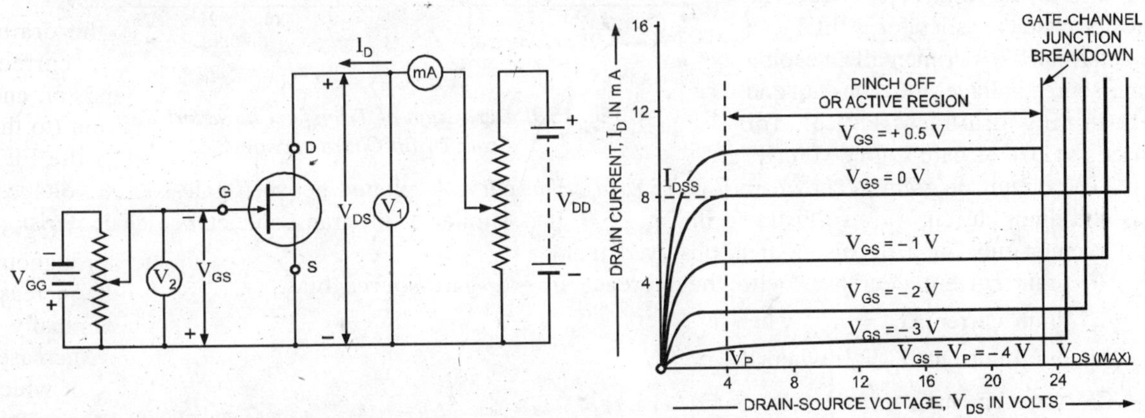

(*a*) *Circuit Diagram For Determining Drain Characteristic With External Bias For an N-Channel JFET*

(*b*) *JFET Drain-Characteristics With External Bias*

Fig. 15.8

When an external bias of, say, –1 V is applied between the gate and the source, the gate channel junctions are reverse biased even when drain current, I_D is zero. Hence the depletion regions are already penetrating the channel to a certain extent when drain-source voltage, V_{DS} is zero. Due to this reason, a smaller voltage drop along the channel (*i.e.* smaller than that for $V_{GS} = 0$) will increase the depletion regions to the point where they pinch-off the current. Consequently, the pinch-off voltage V_P is reached at a lower drain current, I_D when $V_{GS} = 0$.

(*iii*) The ohmic region portion decreases.

(*iv*) Value of drain-source voltage V_{DS} for the avalanche breakdown of the gate junction is reduced.

Value of drain-source voltage, V_{DS} for breakdown with the increase in negative bias voltage is reduced simply due to the fact that gate-source voltage, V_{GS} keeps adding to the reverse bias at the junction produced by current flow. Thus the maximum value of V_{DS} that can be applied to a FET is the lowest voltage which causes avalanche breakdown.

It is also observed that with $V_{GS} = 0$, I_D saturates at I_{DSS} and the characteristic shows $V_P = 4$ V. When an external bias of –1 V is applied, the gate channel junctions still require –4 V to achieve pinch-off. It means that a 3 V drop is now required along the channel instead of the previous 4.0 V. Obviously, this drop of 3 V can be achieved with a lower value of drain current. Similarly, when $V_{GS} = -2$ V and -3 V, pinch-off is achieved with 2 V and 1 V respectively, along the channel. These drops of 2 V and 1 V are, of course, achieved with further reduced values of drain current, I_D. It is further observed that when the gate-source bias is numerically equal to pinch-off voltage, V_P (–4 V in this case), no channel drop is required and, therefore, drain current, I_D is zero. The gate-source bias voltage required to reduce drain current I_D to zero is designated the gate-source cutoff voltage, $V_{GS\,(OFF)}$ and, as explained,

$$V_{GS\,(OFF)} = V_P$$

Hence for working of JFET in the pinch-off or active region it is necessary that the following conditions be fulfilled.

$$V_P < V_{DS} < V_{DS\,(MAX)} \quad ...(15.2)$$

$$V_{GS\,(OFF)} < V_{GS} < 0 \quad ...(15.3)$$

$$0 < I_D < I_{DSS} \quad ...(15.4)$$

2. *Transfer Characteristics.* The transfer characteristics for a JFET can be determined experimentally, keeping drain-source voltage, V_{DS} constant and determining drain current, I_D for various values of gate-source voltage, V_{GS}. The circuit diagram is shown in Fig. 15.8 (*a*). The curve is plotted between gate-source voltage, V_{GS} and drain current, I_D, as illustrated in Fig. 15.9. It is similar to the transconductance characteristic of a vacuum tube or a transistor. It is observed that

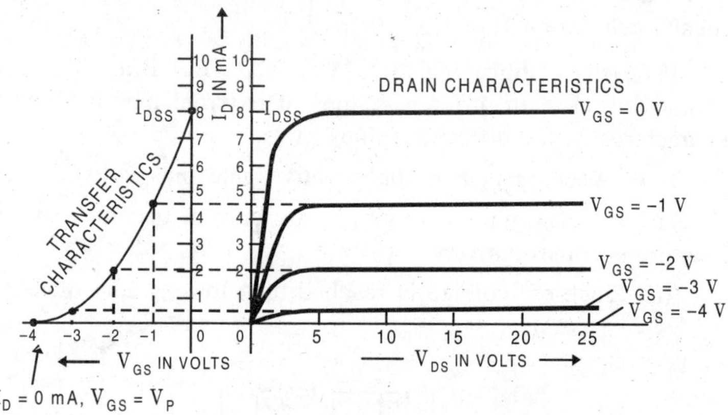

Fig. 15.9. ***Derivation of Transfer Characteristics From Drain Characteristics***

(*i*) drain current decreases with the increase in –ve gate-source bias,

(*ii*) drain current, $I_D = I_{DSS}$ when $V_{GS} = 0$

(*iii*) drain current, $I_D = 0$ when $V_{GS} = V_P$

The transfer characteristic follows Eq. (15.1)

The transfer characteristic can also be derived from the drain characteristic by noting values of drain current, I_D corresponding to various values of gate-source voltage, V_{GS} for a constant drain-source voltage and plotting them.

It may be noted that a P-channel JFET operates in the same way and have the similar characteristics as an N-channel JFET except that channel carriers are holes instead of electrons and the polarities of V_{GS} and V_{DS} are reversed.

15.4. MERITS AND DEMERITS OF JFETs

Junction field-effect transistors combine several merits of both conventional (or bipolar) transistors and vacuum tubes. Some of these are enumerated below :

1. Its operation depends upon the flow of majority carriers only. It is, therefore, a *unipolar* (one type of carrier) device. On the other hand in an ordinary transistor both majority and minority carriers take part in conduction and, therefore, ordinary transistor is sometimes called the *bipolar* transistor. The vacuum tube is another example of a unipolar device.

2. It is simpler to fabricate, smaller in size, rugged in construction and has longer life and higher efficiency. Simpler to fabricate in IC form and space requirement is also lesser.
3. It has high input impedance (of the order of 100 MΩ), because its input circuit (gate to source) is reverse biased, and so permits high degree of isolation between the input and output circuits. However, the input circuit of an ordinary transistor is forward biased and, therefore, ordinary transistor has low input impedance.
4. It carries very small current because of reverse-biased gate and, therefore, it operates just like a vacuum tube where control grid (corresponding to gate in JFET) carries extremely small current and input voltage controls the output current. This is the reason that JFET is essentially a *voltage driven* device. Ordinary transistor is a current operated device since input current controls the output current.
5. An ordinary transistor uses a current into its base for controlling a large current between collector and emitter whereas in a JFET voltage on the gate (base) terminal is used for controlling the drain current (current between drain and source). Thus an ordinary transistor gain is characterised by current gain whereas the JFET gain is characterised as the *transconductance* (the ratio of drain current and gate-source voltage).
6. JFET has no junction like an ordinary transistor and the conduction is through bulk material current carriers (N-type or P-type semiconductor material) that do not cross junctions. Hence the inherent noise of tubes (owing to high temperature operation) and that of ordinary transistors (owing to junction transitions) is not present in JFET.
7. It is relatively immune to radiation.
8. It has –ve temperature coefficient of resistance and, therefore, has better thermal stability.
9. It has high power gain and, therefore, the necessity of employing driver stages is eliminated.
10. It exhibits no offset voltage at zero drain current and, therefore, makes an excellent signal chopper.
11. It has square-law characteristics and, therefore, it is very useful in the tuners of radio and TV receivers.
12. It has got high frequency response.

The main drawback of JFET is its relative small gain-bandwidth product in comparison with that of a conventional transistor and greater susceptibility to damage in its handling. Other disadvantages of JFET are low voltage gains because of small transconductance and higher cost than that of BJTs.

15.5. FET CONFIGURATIONS

FETs (JFETs and MOSFETs) have three configurations just as BJTs have. The three configurations of FETs—the common gate, common source, and common drain configurations have been given in Fig. 15.10. Input and output terminals of different configurations have been given in Table 15.1.

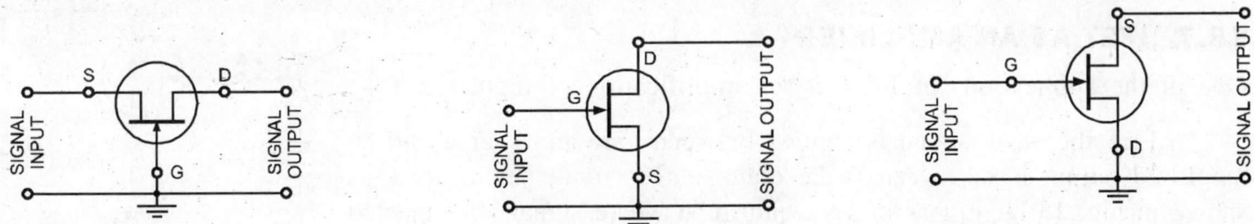

(a) Common Gate Configuration (b) Common Source Configuration (c) Common Drain Configuration

Fig. 15.10. ***N-Channel FET Configurations***

TABLE 15.1. *FET Configurations*

Configuration	*Input Terminal*	*Output Terminal*
Common gate	Source	Drain
Common source	Gate	Drain
Common drain	Gate	Source

TABLE 15.2. *FET-BJT Correspondence*

FET Configuration	*BJT Configuration*
Common gate	Common base
Common source	Common emitter
Common drain	Common collector

The three FET configurations have characteristics that are similar to the three corresponding BJT configurations, as given in Table 15.2.

15.6. JFET TEMPERATURE EFFECTS

The JFET, like other solid state devices, is subject to the effect of temperature. However, it is possible to bias the JFET such that it will exhibit a *zero temperature coefficient i.e.* drain current I_D will be independent of temperature. This is explained below :

There are two primary mechanisms underlying the temperature sensitivity of the conductivity of a JFET channel.

1. Decrease in the depletion region width at the channel-gate P-N junction with the increase in temperature. This results in increase of channel thickness.
2. Decrease in carrier mobility with the increase in temperature.

Increase in channel thickness with the increase in temperature makes drain current I_D to increase. Another way of looking into the situation is to note that $V_{GS\,(OFF)}$ increases in magnitude with the increase in temperature. So far a fixed value of V_{GS}, I_D will increase with the temperature. $V_{GS\,(OFF)}$ [V_P] has a positive temperature coefficient of about 2.2 mV/°C.

The second factor [decrease in carrier mobility with the increase in temperature] makes the channel conductivity to decrease with the increase in temperature. The result is that drain current I_D decreases (assuming V_{DS} constant) with the increase in temperature. Thus we see that I_{DSS} tends to fall as the temperature rises.

Thus we have two distinct mechanisms affecting drain current I_D as a function of temperature. Carrier mobility gives I_D a negative temperature coefficient while the effect of $V_{GS\,(OFF)}$ is to give I_D a positive temperature coefficient. Since both of these mechanisms occur simultaneously, it is possible to bias the JFET so as to exhibit a zero temperature coefficient. Thus JFETs have better thermal stability.

The value of V_{GS} that will give drain current I_D for both N- and P-channel JFETs a zero temperature coefficient is given as

$$|V_{GS}| \simeq |V_P| - 0.63 \text{ V.} \quad ...(15.5)$$

15.7. JFET AS AN AMPLIFIER

One of the applications of JFET is an amplification of input signal.

In FET the weak signal is applied between gate and source and an amplified output is taken across the drain-source circuit having load, as shown in Fig. 15.11, just as in a vacuum tube where signal is connected between grid and cathode and output is taken across a load connected in plate circuit.

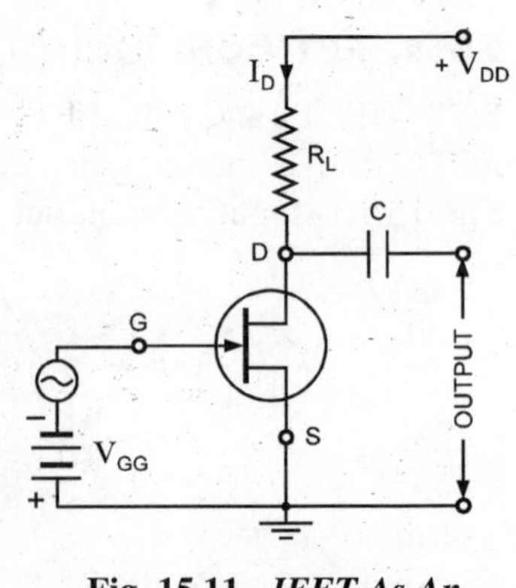

Fig. 15.11. ***JFET As An Amplifier***

For proper working of JFET it is necessary that the gate is made negative w.r.t. source (*i.e.* the input circuit is always reverse biased). This is achieved either by inserting a battery V_{GG} in the gate circuit with the polarity shown in Fig. 15.11 or by providing a biasing circuit.

On application of an alternating weak signal, the fixed reverse bias of the gate changes and a small change in reverse bias on the gate causes a large change in drain current owing to change in width of the channel. This fact makes the FET capable of amplifying the strength of weak signals. During the positive half cycle of the input signal, the reverse bias on the gate decreases causing increase in channel width and hence the drain current. During the negative half cycle of the input signal, the reverse bias on the gate increases causing drain current to decrease. The result is that the small change in voltage at the gate produces a large change in the drain current. These large variations in drain current give large output across the load R_L and thus, JFET operates as an amplifier.

15.8. APPLICATIONS OF FETs

FET has a very high input impedance (100 MΩ in case of JFETs and 10^4 to 10^9 MΩ in case of MOSFETs), the major shortcomings of an ordinary transistor *i.e.* low input impedance with consequent of loading of signal source is eliminated in FET. Hence FET is an ideal device for use in almost every application in which transistors can be used.

1. FETs are widely used as input amplifiers in oscilloscopes, electronic voltmeters and other measuring and testing equipment because of their high input impedance.
2. FETs are used in RF amplifiers in FM tuners and communication equipments for the low noise level.
3. FETs are used as voltage variable resistors (VVRs) in operational amplifiers and tone controllers etc. because it is a voltage controlled device.
4. FETs are used in mixer circuits in FM and TV receivers, and communication equipments because of their low intermodulation distortion.
5. FETs are used in low frequency amplifiers in hearing aids and inductive transducers because of the small coupling capacitors.
6. FETs are used in digital circuits in computers, LSI and memory circuits because of very small size.

15.9. METAL-OXIDE SEMICONDUCTOR FIELD-EFFECT TRANSISTORS (MOSFETs)

Metal-oxide-semiconductor field-effect transistor (MOSFET) is an important semiconductor device and is widely employed in many circuit applications. Since it is constructed with the gate terminal insulated from the channel, it is sometimes called *insulated gate* FET (IGFET). Like, a JFET, a MOSFET is also a three terminal (source, gate and drain) device and drain current in it is also controlled by gate bias. The operation of MOSFET is similar to that of JFET. It can be employed in any of the circuits covered for the JFET and, therefore, all the equations apply equally well to the MOSFET and JFET in amplifier connections. However, MOSFET has lower capacitance and input impedance much more than that of a JFET owing to small leakage current. In case of a MOSFET the positive voltage may be applied to the gate and still the gate current remains zero.

MOSFETs are of two types namely (*i*) *enhancement type* MOSFET or E-MOSFET and (*ii*) depletion enhancement MOSFET or DE-MOSFET. In the depletion-mode construction a channel is physically constructed and a current between drain and source is due to voltage applied across the drain-source terminals. The enhancement MOSFET structure has no channel formed during its construction. Voltage is applied to the gate, in this case, to develop a channel of charge carriers so that a current results when a voltage is applied across the drain-source terminals.

The enhancement-type MOSFET (E-MOSFET) is widely used in both discrete and integrated circuits. In discrete circuits, the main use is in power switching, which means turning largest currents *on* and *off.* In ICs, the main use is in digital switching, the basic process behind modern computers. Although their use is declined, depletion-mode MOSFETs (DE-MOSFETs) are still found in high-frequency front-end communications circuits as RF amplifiers.

15.10. DEPLETION-ENHANCEMENT MOSFETs (DE-MOSFETs)

In Art. 15.2.5 we have seen that when the gate is biased negative w.r.t. the source in an N-channel JFET, the depletion-region widths are increased. The increase in the depletion regions reduces the channel thickness, which increases its resistance. The net result is that drain current I_D is reduced.

If the polarity of V_{GG} were reversed so as to apply a positive bias to the gate w.r.t. source, the P-N junctions between the gate and the channel would then be forward biased. Since a forward bias reduces the width of a depletion region, the thickness of channel would increase with a corresponding decrease in channel resistance. As a result, drain current I_D would increase beyond the JFET's I_{DSS} value.

The normal operation of a JFET is in its depletion mode of operation. However, as discussed above, it is also possible to *enhance* the conductivity of the JFET channel. However, the forward bias of the silicon P-N junction is usually restricted to a maximum of 0.5 V (more conservative limit is 0.2 V) so as to limit the gate current.

Greater the drain current I_D compared to I_{DSS}, greater the conductance g_m and voltage gain is directly proportional to g_m. So, in general, the higher the g_m, the better it is. This is one of the advantages of being able to enhance the channel.

As its name suggests, the *depletion-enhancement* MOSFET (DE-MOSFET) was developed to be used in either or both the depletion and enhancement modes.

15.10.1. Construction

Figure 15.12 shows the construction of an N-channel DE-MOSFET. It consists of a lightly doped P-type substrate into which two blocks of heavily doped N-type material are diffused forming the source and drain. An N-channel is formed by diffusion between the source and drain. The type of impurity for the channel is the same as for the source and drain. Now a thin layer of SiO_2 dielectric is grown over the entire surface and holes are cut through the SiO_2 (silicon dioxide) layer to make contact with the N-type blocks (Source and Drain). Metal is deposited through the holes to provide drain and source terminals, and on the surface area between drain and source, a metal plate is deposited. This layer constitutes the gate. SiO_2 layer results in an extremely high input impedance of the order of 10^{10} to 10^{15} Ω for this device. The chip area of a MOSFET is typically 0.003 μm^2 or less which is about only 5% of the area required by a BJT.

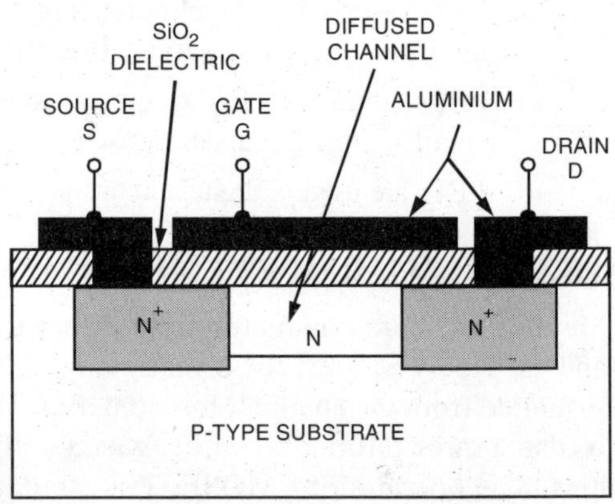

Fig. 15.12. ***N-Channel DE-MOSFET Structure***

A P-channel DE-MOSFET is constructed like an N-channel DE-MOSFET, starting with an N-type substrate and diffusing P-type drain and source blocks and connecting them internally by a P-dopped channel region.

In some cases the substrate is internally connected to the source terminal S and in others an additional terminal labelled SS is provided resulting in a four terminal device.

Silicon dioxide (SiO_2) is a type of insulator referred to as a *dielectric,* which develops opposing (as indicated by the prefix di-) electric fields within the dielectric when exposed to an externally applied field.

The fact that the SiO_2 layer is an insulating layer means that there is no direct electrical connection between the gate terminal G and the channel of a MOSFET. It is the insulating layer of SiO_2 in the MOSFET construction that accounts for the very desirable high input impedance of the device.

In fact, the input resistance of a MOSFET is usually much higher than that of a typically JFET, even though the input impedance of most JFETs is sufficiently high for applications. Because of very high input impedance, the gate current I_G is essentially zero for dc biased configurations.

15.10.2. Operation

It is obvious from Fig. 15.12 that there is no P-N junction between gate and channel. Here the diffused channel N, the insulating dielectric SiO_2 layer and the metal layer of the gate forms a parallel plate capacitor. DE-MOSFET can be operated with either a positive or a negative gate. When gate is positive with respect to the source it operates in the enhancement—or E-mode and when the gate is negative w.r.t. the source, as illustrated in Fig. 15.13, it operates in depletion-mode.

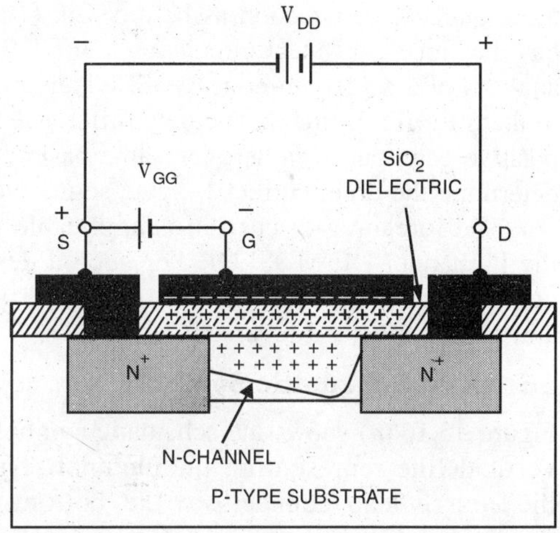

Fig. 15.13. *Depletion Mode Operation*

When the drain is made +ve with respect to source, a drain current will flow, even with zero gate potential and the MOSFET is said to be operating in E-mode. In this mode of operation gate attracts the –ve charge carriers from the P-substrate to the N-channel and thus reduces the channel resistance and increases the drain current. The more positive the gate is made, the more drain current flows.

On the other hand when the gate is made –ve with respect to the substrate, the gate repells some of the –ve charge carriers out of the N-channel. This creates a depletion region in the channel, as illustrated in Fig. 15.13, and, therefore, increases the channel resistance and reduces the drain current. The more –ve the gate, the less the drain current. In this mode of operation the device is referred to as a *depletion-mode MOSFET*. Here too much negative gate voltage can pinch-off the channel. Thus operation is similar to that of JFET.

15.10.3. Characteristics

Typical *drain characteristics*, for various levels of gate-source voltage, of an N-channel MOSFET are shown in Fig. 15.14. The upper curves are for positive V_{GS} and the lower curves are for negative V_{GS}. The bottom drain curve is for $V_{GS} = V_{GS\,(OFF)}$. For a specified drain-source voltage V_{DS}, $V_{GS\,(OFF)}$ is the gate-source voltage at which drain current reduces to a certain specified negligibly small value, as shown in Fig. 15.14. This voltage corresponds to the pinch-off voltage V_P of JFET. For V_{GS} between $V_{GS\,(OFF)}$ and zero, the device operates in *depletion-mode* while for V_{GS} exceeding zero the device operates in *enhancement mode*. These drain curves again display an ohmic region, a constant current source region and a cutoff region. MOSFET has two major applications: a constant current source and a voltage variable resistor.

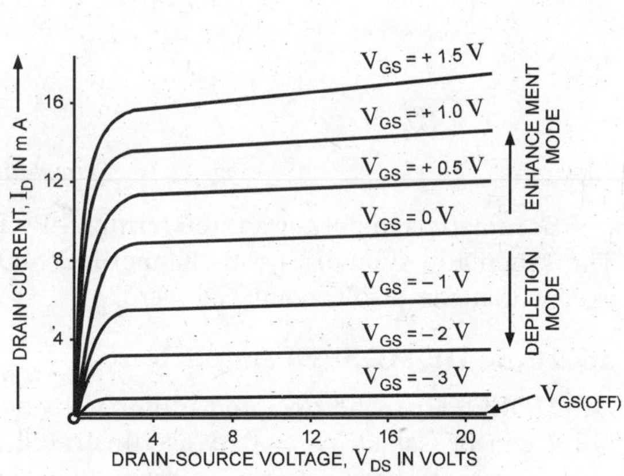

Fig. 15.14. *Drain Characteristics*

The *transfer* (or *transconductance*) characteristic for an N-channel DE-MOSFET is shown in Fig. 15.15. I_{DSS} is the drain current with a shorted gate. Since the curve extends to the right of the origin, I_{DSS} is no longer the maximum possible drain current. Mathematically, the curve is still part of a parabola and the same square-law relation exists as with a JFET. In fact, the depletion-mode MOSFET has a drain current given by the same transconductance equation as before, Eq. (15.1). Furthermore, it has the same equivalent circuits as a JFET. Because of this, the analysis of a depletion-mode MOSFET circuit is almost identical to that of a JFET circuit. The only difference is the analysis for a positive gate, but even here the same basic formulas are used to determine the drain current I_D, gate-source voltage V_{GS} etc.

The foregoing discussion is applicable in principle also to the P-channel DE-MOSFET. For such a device the sign of all currents and voltages in the characteristics [Figs. 15.14 and 15.15] must be reversed.

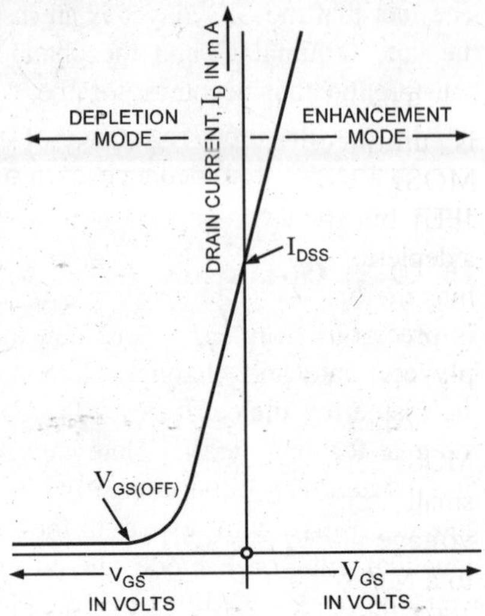

Fig. 15.15. ***Transfer Characteristic***

15.10.4. Schematic Symbols

Figure 15.16 (*a*) shows the schematic symbol for a DE-MOSFET. Just to the right of the gate is the thin vertical line representing the channel. The drain lead comes out from the top of the channel and the source lead connects to the bottom. The arrow is on the P-substrate and points to the N-material. In some applications, a voltage can be applied to the substrate for added control of drain current. For this reason, some DE-MOSFETs have four terminal leads. But in most applications, the substrate is connected to the source. Usually the substrate is connected to the source internally by the manufacturer. This results in a three terminal device whose schematic symbol is shown in Fig. 15.16 (*b*).

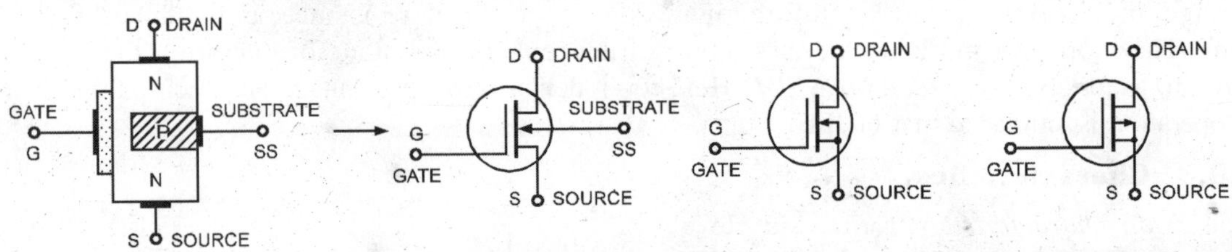

Fig. 15.16. ***Schematic Symbols For DE-MOSFETs***

Schematic symbol for a three terminal P-channel DE-MOSFET device is shown in Fig. 15.16 (*c*). The schematic symbol of a P-channel DE-MOSFET is similar to that of an N-channel DE-MOSFET, except that the arrow points outward.

15.10.5. DE-MOSFET Amplifiers

DE-MOSFET can be operated either with positive or negative gate, so its operating or quiescent point can be set at $V_{GS} = 0$ V, as illustrated in Fig. 15.17 (*b*). In Fig. 15.17 (*a*) the gate G is at ground potential for dc since ac signal source appears to be short circuited for dc and sets Q-point at $V_{GS} = 0$ V. The ac input signal causes the gate-source voltage V_{gs} to vary above and below zero developing drain current I_D and drain voltage V_{out}. The biasing circuit given in Fig. 15.17 (*a*) has $V_{gs} = 0$ V, $I_D = I_{DSS}$ and dc drain voltage

$$V_{DS} = V_{DD} - I_{DSS} R_D \quad ...(15.51)$$

The zero bias of Fig. 15.17 (*a*) is unique with depletion-mode MOSFETs. Although any of the JFET biasing method will work with a depletion-mode MOSFET, the zero-bias method shown in Fig. 15.17 (*a*) is preferred as it does the job simply and adequately.

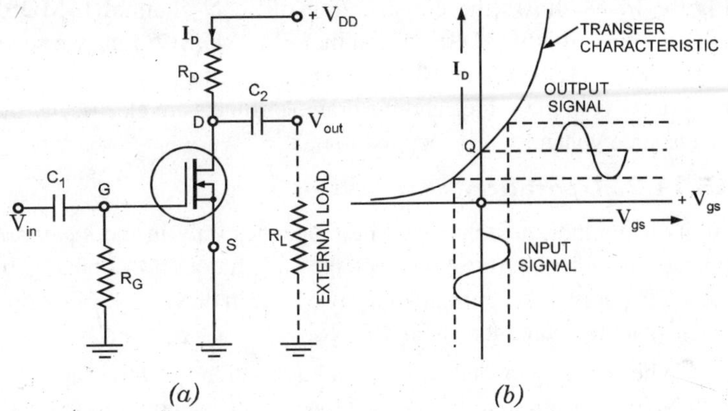

Fig. 15.17

After biasing the depletion-mode MOSFET to a Q-point, it can amplify small signals. The JFET formulas for voltage gain can be applied directly to a MOSFET amplifier. But like the JFET, the depletion-mode MOSFET has a relatively low voltage gain. MOSFETs have excellent low-noise properties, a definite advantage for any stage near the front end of a system where the signal is weak, such as a TV receiver. As with a JFET, the transconductance g_m in case of MOSFET can be controlled by varying the gate-source voltage V_{gs}. Due to the fact that a DE-MOSFET is normally an *on* device, it is also possible to use self-bias by adding a source resistor. The operation becomes the same as a self-biased JFET circuit.

15.11. ENHANCEMENT-ONLY MOSFETs (E-MOSFETs)

Although DE-MOSFET is useful in special applications, it does not enjoy widespread use. However, it played an important role in history because it was part of the evolution towards the E-mode MOSFET, a device that has revolutionized the electronic industry. E-MOSFET has become enormously important in digital electronics and computers. In the absence of E-MOSFETs, the personal computers (PCs) that are now so widespread would not exist.

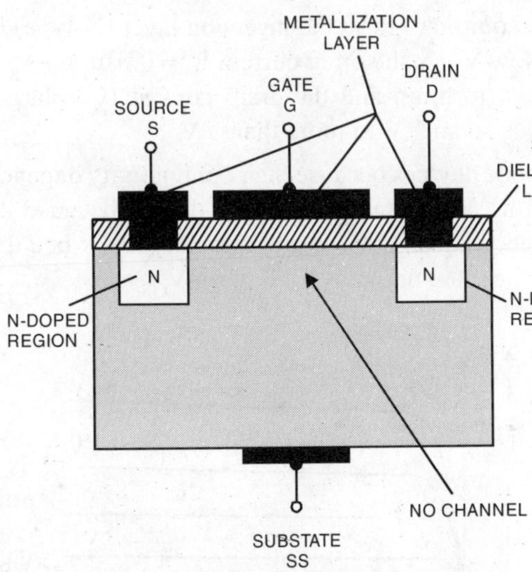

Fig. 15.18. ***A Metallic Crystal Structure***

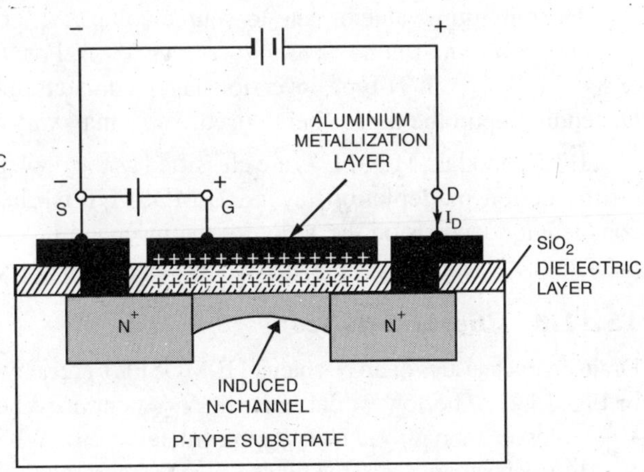

Fig. 15.19. ***Operation of N-Channel E-MOSFET***

15.11.1. Construction

Figure 15.18 shows the construction of an N-channel E-MOSFET. The main difference between the construction of DE-MOSFET and that of E-MOSFET, as we see from Figs. 15.12 and 15.18 respectively, is that in E-MOSFET substrate extends all the way to the silicon dioxide (SiO_2) and no channels are doped between the source and the drain. Channels are electrically induced in these MOSFETs, when a positive gate-source voltage V_{GS} is applied to it.

15.11.2. Operation

As its name indicates, this MOSFET operates only in the *enhancement mode* and has no depletion mode. It operates with large positive gate voltage only. It does not conduct when gate-source voltage $V_{GS} = 0$. This is the reason that it is called *normally-off* MOSFET. In these MOSFETs drain current I_D flows only when V_{GS} exceeds V_{GST} [gate-to-source threshold voltage].

When drain is applied with +ve voltage with respect to source and no potential is applied to the gate, the two N-regions and one P-substrate form two P-N junctions connected back to back with a resistance of the P-substrate. Both the junctions cannot be forward biased at the same time, so only an extremely small drain current *i.e.* a reverse leakage current flows. If the P-type substrate is now connected to the source terminal, there is zero voltage across the source-substrate junction, and the drain-substrate junction remains reverse biased. When the gate is made positive with respect to the source and the substrate, negative (*i.e.* minority) charge carriers within the substrate are attracted to the +ve gate and accumulate close to the surface of the substrate. As the gate voltage is increased, more and more electrons accumulate under the gate. Since these electrons cannot flow across the insulated layer of silicon dioxide to the gate, they accumulate at the surface of the substrate just below the gate. These accumulated minority charge carriers make an N-type channel stretching from drain to source. When this occurs, a channel is induced by forming what is termed an *inversion layer* (N-type). Now a drain current starts flowing. The strength of the drain current depends upon the channel resistance, which in turn depends on the number of charge carriers attracted to the positive gate. Thus drain current is controlled by the gate potential.

Since the conductivity of the channel is enhanced by the +ve bias on the gate, this device is called the *enhancement MOSFET* or E-MOSFET.

The minimum value of gate-to-source voltage V_{GS} that is required to form the inversion layer (N-type) is termed the *gate-to-source threshold voltage* V_{GST}. For V_{GS} below V_{GST}, the drain current $I_D = 0$. But for V_{GS} exceeding V_{GST} an N-type inversion layer connects the source to drain and the drain current I_D is large. Depending upon the device being used, V_{GST} may vary from less than 1 V to more than 5 V.

JFETs and DE-MOSFETs are classified as the depletion-mode devices because their conductivity depends on the action of depletion layers. E-MOSFET is classified as an enhancement-mode device because its conductivity depends on the action of the inversion layer. Depletion-mode devices are normally ON when the gate-source voltage $V_{GS} = 0$, whereas the enhancement-mode devices are normally OFF when $V_{GS} = 0$.

15.11.3. Characteristics

Drain characteristics of an N-channel E-MOSFET are shown in Fig. 15.20. The lowest curve is the V_{GST} curve. When V_{GS} is lesser than V_{GST}, I_D is approximately zero. When V_{GS} is greater than V_{GST}, the device *turns on* and the drain current I_D is controlled by the gate voltage. The characteristic curves have almost vertical and almost horizontal parts. The almost vertical components of the curves correspond to the ohmic region, and the horizontal components correspond to the constant current region. Thus

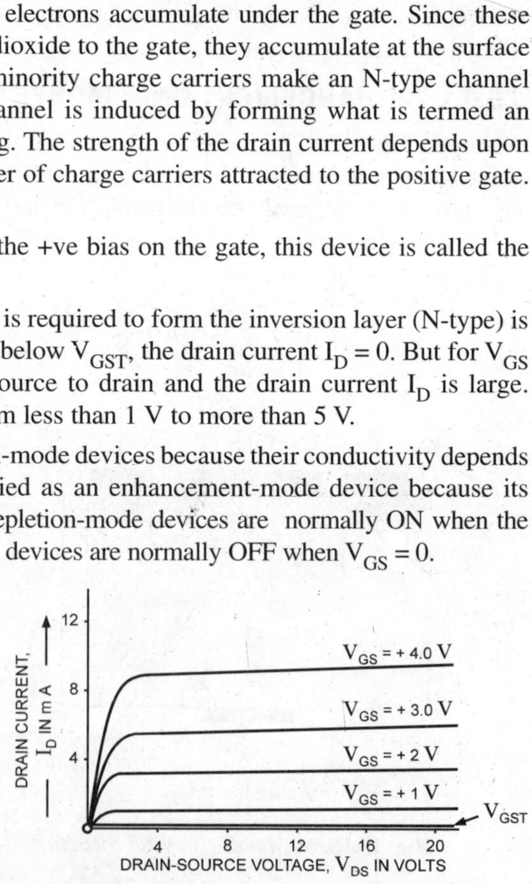

Fig. 15.20. ***Drain Characteristics***

E-MOSFET can be operated in either of these regions *i.e.* it can be used as a variable voltage resistor (VVR) or as a constant current source.

Figure 15.21 shows a typical transconductance curve. The current I_{DSS} at $V_{GS} \leq 0$ is very small, being of the order of a few nanoamperes. When the V_{GS} is made positive, the drain current I_D increases slowly at first, and then much more rapidly with an increase in V_{GS}. The manufacturer sometimes indicates the *gate-source threshold voltage* V_{GST} at which the drain current I_D attains some defined small value, say 10 μA. The current $I_{D\,(ON)}$, corresponding approximately to the maximum permissible value of drain current given on the drain characteristics and the values of V_{GS} required to give this current $V_{GS\,(ON)}$ are also usually given on the manufacturers data sheet.

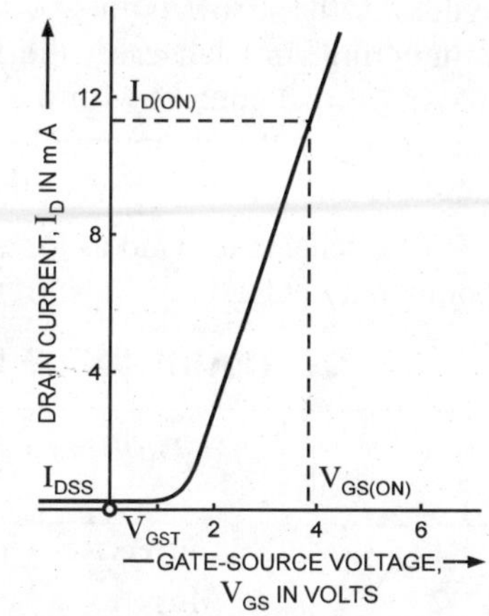

Fig. 15.21. ***Transfer Characteristic***

The equation for the transfer characteristic does not obey Eq. (15.1). However, it does follow a similar "square-law type" of relationship. The equation for the transfer characteristic of E-MOSFETs is given as

$$I_D = K(V_{GS} - V_{GST})^2 \quad \text{...(15.6)}$$

where K, typically 0.3 mA/V², is a property of the device construction. Note that I_{DSS} cannot be associated with an E-MOSFET, because no drain current flows with $V_{GS} = 0$ volt.

With JFETs and DE-MOSFETs, the value of I_{DSS} and $V_{GS\,(OFF)}$ are the key quantities required for analysis. With E-MOSFETs, the key quantities are $I_{D\,(ON)}$, V_{GST}, and $V_{GS\,(ON)}$ indicated in Fig. 15.21. These three quantities are the first items to look for on a data sheet. By substituting these quantities in Eq. (15.6), the equation can be arranged in a more useful form:

$$I_D = KI_{D\,(ON)} \quad \text{...(15.7)}$$

$$\text{where } K = \left(\frac{V_{GS} - V_{GST}}{V_{GS\,(ON)} - V_{GST}}\right)^2 \quad \text{...(15.8)}$$

The input impedance of the E-MOSFET is very high (of the order of 10^{15} Ω) as the gate of the E-MOSFET is insulated from the channel. Transconductance for this device typically ranges from 1,000 μS to 2,000 μS (or 1 to 2 mA per volt).

15.11.4. Schematic Symbols

Schematic symbols for an N-channel E-MOSFET are shown in Fig. 15.22 (*a*). For zero value of V_{GS}, the E-MOSFET is OFF because there is no conducting channel between source and drain. Each of schematic symbols shown in Figs. 15.22 (*a*) (*ii*) and 15.22 (*a*) (*iii*), has broken channel line to indicate this normally OFF condition. As we know that for V_{GS} exceeding the threshold voltage V_{GST}, an N-type inversion layer, connecting the source to drain, is created. In each of the schematic

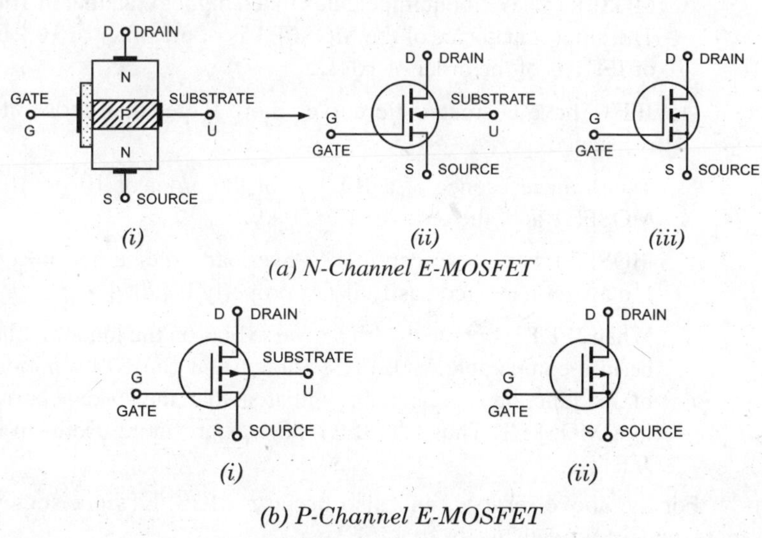

Fig. 15.22. ***Schematic Symbols For E-MOSFETs***

symbols, the arrow points to this inversion layer, which acts like an N-channel when the device is conducting. In each case, the fact that the device has an insulated gate is indicated by the gate not making direct contact with the channel. The schematic symbol shown in Fig. 15.22 (*a*) (*iii*) shows the source and substrate internally connected, while the other symbol shown in Fig. 15.22 (*a*) (*ii*) shows the substrate connection brought out separately from the source.

The schematic symbols for a P-channel E-MOSFET are shown in Fig. 15.22 (*b*). In these cases the arrow points outwards.

15.12. COMPARISON OF N-CHANNEL FETs WITH P-CHANNEL FETs

S. No.	*Particulars*	*N-Channel FETs*	*P-Channel FETs*
1.	Current carriers	Electrons	Holes
2.	Mobility	Large	Poor
3.	Noise	Less	More
4.	Transconductance	Larger	Lesser

15.13. COMPARISON OF JFETs AND MOSFETs

JFETs and MOSFETs are quite similar in their operating principles and in their electrical characteristics. However, they differ in some aspects, as detailed below :

1. In MOSFETs (DE-MOSFETs as well as E-MOSFETs), the transverse electric field induced across an insulating layer deposited on the semiconductor material controls the conductivity of the channel whereas in the JFETs the transverse electric field across the reverse-biased PN junction controls the conductivity of the channel.
2. JFETs can only be operated in the depletion mode whereas MOSFETs can be operated in either depletion or in enhancement mode. In a JFET, if the gate is forward biased, excess-carrier injunction occurs and the gate current is substantial. Thus channel conductance is enhanced to some degree due to excess carriers but the device is never operated with gate forward biased because gate current is undesirable.
3. MOSFETs have input impedance much higher than that of JFETs because of small leakage currents. The input impedance of the MOSFET is of the order of 10^{10} to 10^{15} Ω whereas the input impedance of JFET is of the order of 10^8 Ω.
4. JFETs have characteristic curves more flatter than those of MOSFETs indicating a higher drain resistance.

   The drain resistance of a JFET is of the order of 10^5 to 10^6 Ω whereas the drain resistance of a MOSFET is of the order of 1 to 50 kΩ.
5. MOSFET is very susceptible to overload voltage and needs special handling during installation. It may get damaged easily if not properly handled.
6. When JFET is operated with a reverse bias on the junction, the gate current I_G is larger than it would be in a comparable MOSFET. The current caused by minority carrier extraction across a reverse biased junction is greater, per unit area, than the leakage current that is supported by the oxide layer in a MOSFET. Thus MOSFET devices are more useful in electrometer applications than are the JFETs.

For the above reasons, and also because MOSFETs are somewhat easier to manufacture, they are more widely used than are the JFETs.

15.14. COMPARISON OF NMOS AND PMOS

1. P-channel is much easier and cheaper to produce than the N-channel device.
2. The drain resistance of P-channel MOSFET is three times higher than that of an identical N-channel device.
3. The N-channel MOSFET is smaller for the same complexity than the P-channel MOSFET.
4. The N-channel MOSFET has the higher packing density which makes it faster in switching applications due to the smaller junction areas and low inherent capacitances.
5. A P-channel MOSFET occupies a larger area than the N-channel MOSFET for the given drain current rating. This is because the electron mobility is 2.5 times of that of an hole.
6. The N-channel MOSFET has higher false turn-on possibility than the P-channel device because of the positively charged contaminants.

HIGHLIGHTS

1. The FET is a majority carrier device, and is, therefore, often called a unipolar transistor. Like its bipolar counterpart, the FET is a three-terminal (namely drain, source and gate) semiconductor device in which current conduction is by one type of majority carriers (either electrons or holes).
2. Field effect devices are controlled by an electric field (or voltage) and give an extremely high input impedance—most important advantage over BJT devices. FETs are of several types such as JFETs and MOSFETs.
3. The JFET is a voltage controlled device and has two diodes; the gate-source diode and the gate-drain diode. For normal operation, the gate-source diode is reverse biased. Then the gate voltage controls the drain current.
4. The JFET has a high input impedance (of the order of 100 MΩ), a high degree of isolation between input and output, and a low noise figure. The main disadvantage of a JFET is its relatively low gain-bandwidth product, due to the junction capacitive effects.
5. The JFET can actually be used as a voltage-controlled resistor because of unique sensitivity of the drain-to-source resistance R_{DS} to the gate-to-source voltage V_{GS}.
6. The curve drawn between drain current I_D and drain-to-source voltage V_{DS} with gate-to-source voltage V_{GS} as the parameter is called the drain characteristic.

   Maximum drain current for any JFET labelled as I_{DSS} occurs when the gate-source voltage V_{GS} is zero. The minimum drain current for a JFET occurs at pinch-off voltage defined by $V_{GS} = V_P$. The pinch-off voltage separates the ohmic and active regions for $V_{GS} = 0$. The gate-source cutoff voltage has the same magnitude as the pinch-off voltage. $V_{GS(off)}$ turns the JFET off.
7. The relationship between the drain current I_D and the gate-to-source voltage V_{GS} of a JFET is a nonlinear defined by Shockley's equation

   $$I_D = I_{DSS}\left[1-\frac{V_{GS}}{V_P}\right]^2 = I_{DSS}\left[1-\frac{V_{GS}}{V_{GS(off)}}\right]^2$$

   As the current level approaches I_{DSS}, the sensitivity of I_D to variations in V_{GS} increases significantly.
8. The curve drawn between drain current I_D and gate-to-source voltage V_{GS} for a given value of drain-to-source voltage V_{DS} is called the transfer characteristic. The transfer characteristics are the characteristics of the device itself and are not sensitive to the circuit in which the JFET is used. The drain current increases more rapidly as V_{GS} approaches zero. Because the drain current equation contains a squared quantity, JFETs are referred to as square-law devices. The normalized transconductance curve indicates that I_D equals one-quarter of the maximum when V_{GS} equals half of the cutoff

   $\left(i.e. I_D = \frac{I_{DSS}}{4} \text{ when } V_{GS} = \frac{V_P}{2}\right)$ and at a point where $I_D = \frac{I_{DSS}}{2}$, $V_{GS} \cong 0.3$ V.

9. JFETs are employed in multiplexers, chopper amplifiers, buffer amplifiers, voltage-controlled resistors, cascade amplifiers, current limiters, current sources and AGC circuits.

10. The operation of a metal-semiconductor field-effect transistor (MOSFET) is identical to that of a JFET. In the MOSFET, however, a metal-semiconductor rectifying contact, instead of a P-N junction, is used for the gate electrode.

11. The MOSFET is a field-effect transistor in which the metal gate is insulated by a very thin oxide layer from the semiconductor channel. It is acually a four-terminal device, with the substrate or body being the fourth terminal.

    There are two types of MOSFETs—the N-channel in which current is due to the flow of electrons in the inversion layer, and the P-channel in which current is due to the flow of holes in the inversion layer. Each of these devices can be either enhancement mode (E-mode) in which device is normally "off" and is turned on by applying a gate voltage, or depletion mode (DE-mode) in which the device is normally "on" and is turned off by applying a gate voltage.

12. The current in a MOSFET is due to the flow of carriers in the inversion layer between the source and drain terminals. The inversion layer charge density and channel conductance are controlled by the gate voltage, which means that the channel current is also controlled by the gate voltage.

13. The depletion-mode MOSFET, has a source, gate, and drain. The gate is insulated from the channel. Because of this, the input resistance is very high. The DE-MOSFET has limited use, mainly in RF circuits.

14. The drain characteristic curves for a DE-MOSFET are similar to those of a JFET when operating in depletion mode. Unlike JFETs, DE-MOSFETs can also operate in the enhancement mode. When operating in enhancement mode, the drain current is greater than I_{DSS} (*i.e.* $I_D > I_{DSS}$).

    DE-MOSFETs are mainly employed as RF amplifiers. They have good high frequency response, generate low levels of electrical noise, and maintain high input impedance values when V_{GS} is negative or positive.

15. The enhancement type MOSFET is normally off. When the gate voltage is equal to the threshold voltage (V_{GST}), an N-type inversion layer connects the source to the drain. When the gate voltage exceeds the V_{GST}, the device conducts heavily. Because of the thin insulating layer, MOSFETs are easily destroyed unless precautions are taken in their handling.

16. Since the E-MOSFET is primarily a switching device, it usually operates between cutoff and saturation. When it is biased in the ohmic region, it acts like a small resistance.

EXERCISES

1. Explain the working principle of a FET. Discuss also the advantages of FET over bipolar transistor.
2. Sketch and explain the basic structure of an N-channel junction field-effect transistor. Show the circuit symbol for the JFET.
3. Explain the construction and working of JFET. What is the difference between FET and bipolar transistor ?
   [Pb. Technical Univ. Basic Electrical and Electronics Engineering Dec. 2006]
4. Discuss the operation of N-channel JFET and plot its drain characteristics. Define pinch-off voltage.
   [Pb. Technical Univ. Basic Electrical and Electronics Engineering, May 2009]
5. Describe the constructional details, working principle and characteristics of field-effect transistor.
   [Pb. Technical Univ. Basic Electrical and Electronics Engineering May 2007]
6. Explain the four distinct regions of the output characteristics of a JFET.
7. Draw the structure of a JFET and explain its principle of operation with neat diagrams. Also sketch its V-I characteristics. Define pinch-off voltage and mark it on the characteristics. Explain its importance.
8. Explain the working of a JFET alongwith its construction with neat sketch. Also draw its transfer characteristic curve with explanation.
9. Explain FET as an amplifier.
   [Pb. Technical Univ. Basic Electrical and Electronics Engineering May 2006]

10. What is field-effect transistor ? List its applications.

[Pb. Technical Univ. Basic Electrical and Electronics Engineering Dec. 2005]

11. With a neat sketch, explain the working of a P-channel depletion mode MOSFET and draw the transfer characteristics.

12. Sketch the cross section of a depletion type MOSFET and explain its function.

13. What is MOSFET ? With the help of suitable diagram explain the working of different MOSFETs.

[Pb. Technical Univ. Basic Electrical and Electronics Engineering May 2005]

14. Write short note on construction and operation of N-channel JFET.

[Pb. Technical Univ. Basic Electrical and Electronics Engineering May 2005]

15. Sketch the cross-sectional view of an enhancement mode MOSFET. Explain its operation and characteristics.

16. Draw the structures of enhancement mode and depletion mode MOSFET. Show the formation of channel in these two modes.

17. What is the difference between MOSFET and FET ?

[Pb. Technical Univ. Basic Electrical and Electronics Engineering May 2006]

18. Explain the difference between the enhancement mode and depletion mode MOSFETs.

19. Write short notes on the following :

(*i*) Characteristics of JFET.

(*ii*) JFET's merits and demerits.

(*iii*) MOSFET.

(*iv*) Characteristics of MOSFET.

SHORT ANSWER TYPE QUESTIONS WITH ANSWERS

Q. 1. What is the major difference between a bipolar and a unipolar device ? Explain with example.

Ans. In a bipolar device, current conduction is by both majority carriers and minority carriers (both by holes as well as electrons), such as in a case of BJT. On the other hand, in a unipolar device, the current conduction is by only one type of majority carriers (either by electrons or by holes), such as in case of FET.

Q. 2. Explain why BJTs are called bipolar devices while FETs are called unipolar devices.

Ans. In ordinary transistors, current conduction is by both majority carriers and minority carriers (both by holes as well as electrons) and, therefore, these are called bipolar transistors. In field-effect transistors current conduction is by only one type of majority carriers (either by electrons or by holes) and, therefore, these are called unipolar transistors.

Q. 3. Define "current-controlled device" and "voltage-controlled device".

Ans. A current-controlled device is one in which a current defines the operating conditions of the device, whereas a voltage-controlled device is one in which a particular voltage defines the operating conditions.

Q. 4. Why is terminology "field effect" appropriate for this terminal device ? **[Pb. Technical Univ. May 2005]**

Ans. The device is called field-effect transistor (FET) because the drain current (output current) is controlled by the effect of extension of the field associated with the depletion region developed by the reverse bias at the gate.

Q. 5. Why I_G is effectively zero ampere for a JFET transistor ? **[Pb. Technical Univ. May 2005]**

Ans. The gate channel junction in a JFET is a P-N junction and since it is normally reverse biased, a minority charge carrier current flows. This is gate-source cutoff current, known as gate reverse current. So the gate current, I_G can be said to be effectively zero ampere.

Q. 6. What do you understand by terms channel and drain in JFET ? **[Pb. Technical Univ. May 2008]**

Ans. ***Channel.*** The region between the source and drain, sandwiched between the two gates is called the *channel* and the majority carriers move from source to drain through this channel.

Drain. The terminal, through which the majority carriers leave the channel, is called the *drain* terminal D and the conventional current leaving the channel at D is designated as I_D.

Q. 7. Why the channel of a JFET is never completely closed at the drain end ?

Ans. If the channel is completely closed in JFET, then there will be no drain current, so there will be no voltage drop along the channel length and amount of reverse bias will become uniform and wedge-shaped depletion region will become rectangular one.

Q. 8. How is drain current controlled in a JFET ?

Ans. In a JFET drain current is controlled by controlling the reverse bias given to its gate (*i.e.* V_{GS}).

Q. 9. What is meant by drain characteristic of FETs ?

Ans. The curve drawn between drain current and drain-source voltage with gate-to-source voltage as the parameter is called the drain characteristic.

Q. 10. What happens to the drain current of a P-channel JFET when a positive voltage is applied on its gate ? **[Pb. Technical Univ. December 2008]**

Ans. Whenever the positive voltage is applied to the gate of P-channel JFET, the depletion region width increases because of which channel is narrowed, the resistance is increased and drain current is reduced.

Q. 11. What is meant by transfer characteristic of FETs ?

Ans. The curve drawn between drain current and gate-source voltage for a given value of drain-source voltage is called the transfer characteristic.

Q. 12. No current flows through the channel of a JFET when $V_{GS} = 0$ and $V_{DS} = 0$, though channel is fully open. Why ?

Ans. When drain-source voltage is zero, there is no attracting potential at the drain, so no current flows inspite of the fact that the channel is fully open.

Q. 13. Define pinch-off voltage for a JFET

Ans. The value of drain-source voltage at which channel is pinched-off (*i.e.* all the free charges from the channel get removed) is called the pinch-off voltage.

Q. 14. What are the parameters that control the pinch-off voltage of JFET ?

Ans. Parameters that control the pinch-off voltage of JFET are electron charge, donor/acceptor concentration density, permittivity of channel material and half width of channel bar.

Q. 15. Why FET is called a voltage controlled device ? **[Pb. Univ. April 1993]**

Ans. In a FET, drain current is controlled by the effect of the extension of the field associated with the depletion region developed by the reverse bias on the gate, so it is called a voltage controlled device.

Q. 16. How does the FET behave (*i*) for small values of $|V_{DS}|$ and (*ii*) for large values of $|V_{DS}|$? **[Pb. Univ. May 1994]**

Ans. (*i*) FET behaves as an ordinary resistor for small value of $| V_{DS} |$ *i.e.* in ohmic region.

(*ii*) FET behaves as a constant current source for large values of $| V_{DS} |$ till breakdown occurs.

Q. 17. How does the current vary with the gate voltage in the saturation region ? **[Pb. Univ. May 1994]**

Ans. Drain current decreases with the increase in gate-source bias $| V_{GS} |$. When $V_{GS} = 0$; drain current $I_D = I_{DSS}$, drain-source saturation current and when $V_{GS} = V_P$; drain current $I_D = 0$.

Q. 18. What is meant by gate-source cutoff voltage ?

Ans. The gate-source bias voltage required to reduce the drain current to zero is designated the gate-source cutoff voltage $V_{GS\ (OFF)}$. It is equal to pinch-off voltage V_P.

Q. 19. What is meant by saturation region ?

Ans. The region of drain characteristic of a FET in which drain current remains fairly constant is called the saturation or pinch-off region.

Q. 20. What is meant by drain-source saturation current I_{DSS} ?

Ans. The drain current in pinch-off or saturation region with zero gate-source voltage (V_{GS} = 0) is referred to the drain-source saturation current I_{DSS}.

Q. 21. What is physical significance of high input impedance of an FET ? **[Pb. Technical Univ. December 2006]**

Ans. Input circuit of an FET (gate to source) is reverse biased and the input gate current is very small (of the order of few nA), so the input impedance is very high. It permits high degree of isolation between the input and output circuits.

Q. 22. Why is MOSFET called sometimes IGFET ?

Ans. MOSFET is constructed with the gate terminal insulated from the channel so it is sometimes called insulated gate FET (or IGFET).

Q. 23. What is meant by gate-to-source threshold voltage in E-MOSFET ?

Ans. The minimum value of gate-to-source voltage V_{GS} that is required to form the inversion layer is termed the gate-to-source threshold voltage V_{GST}.

Q. 24. Name the factors which make the JFET superior to BJT ?

Ans. The high input impedance, low output impedance and low noise level make JFET far superior to the BJT.

Q. 25. Why N-channel MOSFETs are preferred over P-channel ones in integrated circuits ?

Ans. The hole mobility in silicon at normal field intensities is approximately 0.05 m^2/V-s and the same for electron is about 0.13 m^2/V-s. Thus the P-channel device will have more than twice the ON resistance of an equivalent N-channel of the same geometry and under the same working conditions. In other words, P-channel device must have more than twice the area of the N-channel device to achieve the same resistance. So, N-channel MOS circuits can be smaller for the same complexities than P-channel devices. The higher packing density of the N-channel MOS also makes it faster in switching applications due to the smaller junction areas. The operating speed is limited primarily by the internal time constants (RC), and the capacitance is directly proportional to the junction cross-sections. For all these reasons, N-channel MOS are preferred over P-channel MOS.

Q. 26. Why enhancement mode MOSFETs are preferred over the depletion mode ones in the fabrication of integrated circuits ?

Ans. Since the conductivity of the channel is enhanced by the positive bias given to the gate in case of enhancement mode MOSFETs, the E-MOSFETs are preferred over the DE-MOSFETs in the fabrication of ICs.

❑❑❑

16

CHAPTER

DC Power Supplies

16.1. INTRODUCTION

INSIDE THIS CHAPTER

The electrical power is almost exclusively generated, transmitted and distributed in the form of ac because of economical consideration but for operation of most of the electronic devices and circuits, dc supply is required. Dry cells and batteries can be used for this purpose. No doubt, they have the advantages of being portable and ripple free but their voltages are low, they need frequent replacement and are expensive in comparison to conventional dc power supplies. Nowadays, almost all electronic equipment include a circuit that converts ac supply into dc supply. The part of equipment that converts ac into dc is called the *dc power supply*. In general, at the input of the power supply, there is a power transformer. It is followed by a rectifier (a diode circuit), a *smoothing filter* and then by a *voltage* regulator circuit. A block diagram of such a power supply is given in Fig. 16.1.

As obvious from the block diagram shown in Fig. 16.1, the basic power supply is constituted by four elements viz. a *transformer*, a *rectifier*, a *filter*, and a *regulator* put together. The output of the dc power supply is used to provide a constant dc voltage across the load. Let us briefly outline the function of each of the elements of the dc power supply.

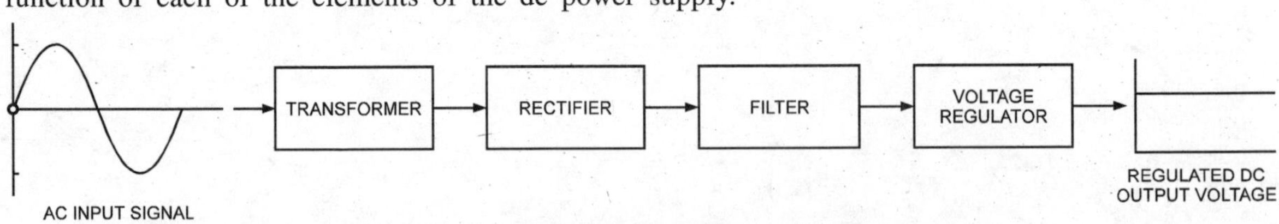

Fig. 16.1. ***Block Diagram of a DC Power Supply***

Transformer is used to step-up or step-down (usually to step-down) the supply voltage as per need of the solid-state electronic devices and circuits to be supplied by the dc power supply. It can provide isolation from the supply line—an important safety consideration. It may also include internal shielding to prevent unwanted electrical noise signal on the power line from getting into the power supply and possibly disturbing the load.

Rectifier is a device which converts the sinusoidal ac voltage into either positive or negative pulsating dc. P-N junction diode, which conducts when forward biased and practically does not conduct when reverse biased, can be used for rectification *i.e.* for conversion of ac into dc. The rectifier typically needs one, two or four diodes. Rectifiers may be either half-wave or full-wave (centre-tap or bridge) type.

The output voltage from a rectifier circuit has a pulsating character *i.e.*, it contains unwanted ac components (components of supply frequency f and its harmonics) along with dc component. For most supply purposes, constant direct voltage is required than that furnished by a rectifier. To reduce ac components from the rectifier output voltage a *filter circuit* is required. Thus *filter* is a device which passes dc component to the load and blocks ac components of the rectifier output. Filter is typically formed from reactive circuit elements such as capacitors and/or inductors and resistors.

The magnitude of output dc voltage may vary with the variation of either the input ac voltage or the magnitude of load current. So at the output of a rectifier-filter combination a voltage regulator is required, as shown in Fig. 16.1, to provide an almost constant dc voltage at the output of the regulator. The voltage regulator may be constructed from a zener diode, and/or discrete transistors, and/or integrated circuits (ICs). Its main function is to maintain a constant dc output voltage. However, it also rejects any ac ripple voltage that is not removed by the filter. The regulator may also include protective devices such as short-circuit protection, current limiting, thermal shutdown, or overvoltage protection.

16.2. HALF-WAVE RECTIFIERS

When a single rectifier unit is placed in series with the load across an ac supply, as illustrated in Fig. 16.2, it converts alternating voltage into unidirectional pulsating voltage, using one half cycles of the applied voltage, the other half cycles being suppressed because it conducts only in one direction. Unless there is an inductance or battery in the circuit, the current will be zero, for half the time. This is called *half-wave rectification*.

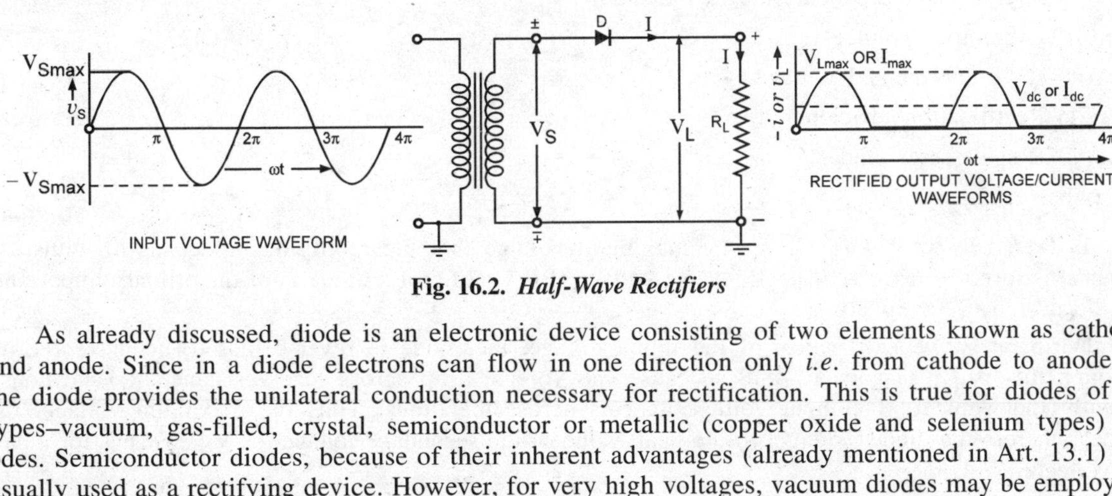

Fig. 16.2. *Half-Wave Rectifiers*

As already discussed, diode is an electronic device consisting of two elements known as cathode and anode. Since in a diode electrons can flow in one direction only *i.e.* from cathode to anode so the diode provides the unilateral conduction necessary for rectification. This is true for diodes of all types–vacuum, gas-filled, crystal, semiconductor or metallic (copper oxide and selenium types) diodes. Semiconductor diodes, because of their inherent advantages (already mentioned in Art. 13.1) are usually used as a rectifying device. However, for very high voltages, vacuum diodes may be employed.

Circuit. The half-wave rectifier circuit using a semiconductor diode with a load resistance R_L but no smoothing filter is given in Fig. 16.2. The diode is connected in series with the secondary of the transformer and the load resistance R_L, the primary of the transformer is being connected to the ac supply mains.

Working. The ac voltage across the secondary winding changes polarity after every half cycle. During the positive half cycles of the input ac voltage *i.e.* when upper end of the secondary winding is positive w.r.t. its lower end, the diode is forward biased and, therefore, conducts current. If the forward resistance of the diode is assumed to be zero (in practice, however, a small resistance exists) the input voltage during the positive half cycles is directly applied to the load resistance R_L, making its upper end positive w.r.t. its lower end. The waveforms of the output current and output voltage are of the same shape as that of the input ac voltage.

During the negative half cycles of the input ac voltage *i.e.* when the lower end of the secondary winding is positive w.r.t. its upper end, the diode is reverse biased and so does not conduct. Thus during the negative half cycles of the input ac voltage the current through and voltage across the load remains zero if the reverse current, being very small in magnitude, is neglected. Thus for the negative half cycles no power is delivered to the load.

Thus the output voltage developed across load resistance R_L (V_L) is a series of positive half cycles of alternating voltage, with intervening very small constant negative voltage levels, as shown in the Fig. 16.2. It is obvious from the figure that the output is not a steady dc, but only a pulsating dc wave. Since only half cycles of the input wave are used, it is called a *half-wave* rectifier.

Power Supply Specifications. The most important characteristics which are required to be specified for a power supply are given below :

1. The required output dc voltage.
2. The average and peak currents in the diode.
3. The peak inverse voltage (PIV) of each diode.
4. The regulation.
5. The ripple factor.

The above factors will be considered in the analysis of the circuit given below.

Analysis of Half-Wave Rectifier. Under this sub-article, the following items will be considered.

(*i*) Peak inverse voltage

(*ii*) Average and peak currents in the diode

(*iii*) DC output voltage

(*iv*) RMS current and voltage

(*v*) Rectification efficiency

(*vi*) Ripple factor, and

(*vii*) Regulation.

1. *Peak Inverse Voltage.* It is the maximum voltage that the rectifying diode has to withstand, when it is reverse biased. The peak inverse voltage (PIV) rating of a diode is of the primary importance in the design of rectification systems.

During negative half cycles of the input voltage, the diode is reverse biased, no current flows through the load resistance R_L and so causes no voltage drop across load resistance R_L and consequently the whole of the input voltage appears across the diode. Thus the maximum voltage, that appears across the diode, is equal to the peak value of the secondary voltage *i.e.* $V_{S\,max}$. Thus for a half-wave rectifier

$$PIV = V_{S\,max} \qquad ...(16.1)$$

2. *Peak Current.* Assuming the voltage across the transformer secondary sinusoidal of peak value $V_{S\,max}$, instantaneous value of the voltage applied to the rectifier is given as

$$v_s = V_{S\,max} \sin \omega t \qquad ...(16.2)$$

neglecting diode cutin voltage V_K, being very small as compared with $V_{S\,max}$

If the diode is assumed to have a forward resistance of R_F ohms and reverse resistance equal to infinity, then the current flowing through the diode (or load resistance R_L) is given as

$$i = I_{max} \sin \omega t \text{ for } 0 \le \omega t \le \pi \qquad ...(16.3)$$

$$i = 0 \text{ for } \pi \le \omega t \le 2\pi \quad ...(16.4)$$

where peak value of current flowing through diode (or load resistance R_L) is given as

$$I_{max} = \frac{V_{S\,max}}{R_F + R_L} \quad ...(16.5)$$

3. *DC Output Current.* The average or dc value of output current is given as

$$I_{dc} = \frac{1}{2\pi}\int_0^{2\pi} i\,d(\omega t) = \frac{1}{2\pi}\left[\int_0^{\pi} I_{max} \sin\omega t\, d(\omega t) + \int_{\pi}^{2\pi} 0\, d(\omega t)\right]$$

$$= \frac{1}{2\pi}\left[I_{max}\{-\cos\omega t\}_0^{\pi}\right] = \frac{I_{max}}{\pi} = 0.318\ I_{max} \quad ...(16.6)$$

Substituting the value of I_{max} from Eq. (16.5) we have

$$I_{dc} = \frac{V_{S\,max}}{\pi(R_L + R_F)} = \frac{V_{S\,max}}{\pi R_L} \text{ if } R_L >> R_F \quad ...(16.7)$$

4. *DC Output Voltage.* Average or dc value of voltage across the load is given as

$$V_{dc} = I_{dc} R_L = \frac{V_{S\,max}}{\pi(R_L + R_F)} \cdot R_L = \frac{V_{S\,max}}{\pi\left(1 + \frac{R_F}{R_L}\right)} \quad ...(16.8)$$

If $R_L >> R_F$, then $V_{dc} \simeq \frac{V_{S\,max}}{\pi}$ ...(16.9)

5. *RMS Value of Current.* RMS value of current flowing through the diode (or load resistance R_L) is given as

$$I_{rms}^2 = \frac{1}{2\pi}\int_0^{2\pi} i^2\, d(\omega t) = \frac{1}{2\pi}\left[\int_0^{\pi} I_{max}^2 \sin^2 \omega t\, d(\omega t) + \int_{\pi}^{2\pi} 0\, d(\omega t)\right] = \frac{I_{max}^2}{4}$$

or $$I_{rms} = \frac{I_{max}}{2} \quad ...(16.10)$$

Substituting the value of I_{max} from Eq. (16.5) we have

$$I_{rms} = \frac{V_{S\,max}}{2(R_F + R_L)} \quad ...(16.11)$$

6. *RMS Value of Output Voltage.* RMS value of voltage across the load is given as

$$V_{L\,rms} = I_{rms} R_L = \frac{V_{S\,max} \times R_L}{2(R_F + R_L)} = \frac{V_{S\,max}}{2\left[1 + \frac{R_F}{R_L}\right]} \quad ...(16.12)$$

If $R_L >> R_F$, then $V_{L\,rms} = \frac{V_{S\,max}}{2}$ ...(16.13)

7. *Form Factor and Peak Factor.* The form factor, defined as the ratio of rms value to average value, is given by

$$K_f = \frac{\text{RMS value}}{\text{Average value}} = \frac{I_{rms}}{I_{dc}} = \frac{V_{S\,max}/2(R_F + R_L)}{V_{S\,max}/\pi(R_F + R_L)} = \frac{\pi}{2} = 1.57 \quad ...(16.14)$$

The peak factor, defined as the ratio of peak value to rms value, is given by

$$K_p = \frac{\text{Peak value}}{\text{RMS value}} = \frac{V_{Smax}/(R_F + R_L)}{V_{Smax}/2(R_F + R_L)} = \frac{V_{Smax}}{V_{Smax}/2} = 2 \quad ...(16.15)$$

8. *Output Frequency.* The output frequency is the same as the input frequency. This can be realized by comparing output waveform with the input waveform in Fig. 16.2. Thus for an half-wave rectifier

$$f_{out} = f_{in} \quad ...(16.16)$$

9. *Rectification Efficiency.* Rectification efficiency, which is defined as the ratio of dc output power to the ac input power, is given as

$$\eta = \frac{\text{DC power delivered to the load}}{\text{AC input power from the transformer}} = \frac{P_{dc}}{P_{ac}}$$

Now dc power delivered to the load,

$$P_{dc} = I_{dc}^2 R_L = \left(\frac{I_{max}}{\pi}\right)^2 R_L \quad ...(16.17)$$

AC power input from the transformer,

P_{ac} = Power dissipated in diode junction + power dissipated in load resistance R_L

$$= I_{rms}^2 R_F + I_{rms}^2 R_L = \left(\frac{I_{max}}{2}\right)^2 R_F + \left(\frac{I_{max}}{2}\right)^2 R_L = \frac{I_{max}^2}{4}(R_F + R_L) \quad ...(16.18)$$

So rectification efficiency,

$$\eta = \frac{P_{dc}}{P_{ac}} = \frac{I_{max}^2 R_L/\pi^2}{I_{max}^2 (R_F + R_L)/4} = \frac{4}{\pi^2}\frac{R_L}{(R_F + R_L)} = \frac{0.406}{1 + \dfrac{R_F}{R_L}} \quad ...(16.19)$$

If R_F is neglected, η becomes equal to 0.406 or 40.6 per cent. Obviously it is the maximum possible efficiency for a half-wave rectifier.

10. *Ripple Factor.* The pulsating output of a rectifier can be considered to contain a dc component and ac components called the *ripples*. The ripple current is undesirable and its value should be the smallest possible in order to make the rectifier effective.

The ripple voltage or current is measured in terms of the ripple factor which is defined as the ratio of the effective value of the ac components of voltage (or current) present in the output from the rectifier to the direct or average value of the output voltage (or current).

The effective value of the load current is given as

$$I^2 = I_{dc}^2 + I_1^2 + I_2^2 + I_4^2 + \ldots = I_{dc}^2 + I_{ac}^2 \quad ...(16.20)$$

where I_1, I_2, I_4 etc. are the rms values of fundamental, second, fourth etc. harmonics and I_{ac}^2 is the sum of the squares of the rms values of the ac components.

$$\text{So ripple factor, } \gamma = \frac{I_{ac}}{I_{dc}} = \frac{\sqrt{I^2 - I_{dc}^2}}{I_{dc}} = \sqrt{\left[\frac{I_{rms}}{I_{dc}}\right]^2 - 1} = \sqrt{K_f^2 - 1} \quad ...(16.21)$$

where K_f is the form factor of the input voltage. For half-wave rectifier, form factor is given as

$$K_f = \frac{I_{rms}}{I_{av}} = \frac{I_{max}/2}{I_{max}/\pi} = \frac{\pi}{2} = 1.57$$

$$\text{So ripple factor, } \gamma = \sqrt{(1.57)^2 - 1} = 1.21$$

11. *Transformer Utilisation Factor.* Determination of transformer rating becomes necessary while designing a power supply. For this transformer utilisation factor (TUF), which is defined as the ratio of power delivered to the load and ac rating of the transformer secondary, should be known.

$$\text{So TUF} = \frac{P_{dc}}{P_{ac\,(rated)}} = \frac{I_{dc}^2 R_L}{V_{S\,rms} I_{rms}} = \frac{(I_{max}/\pi)^2 R_L}{\frac{V_{S\,max}}{\sqrt{2}} \cdot \frac{I_{max}}{2}} = \frac{2\sqrt{2}}{\pi^2} \cdot \frac{I_{max} R_L}{V_{S\,max}} \quad ...(16.22)$$

But $V_{S\,max} = I_{max}(R_F + R_L)$

$$\text{So TUF} = \frac{2\sqrt{2}}{\pi^2} \cdot \frac{I_{max} R_L}{I_{max}(R_F + R_L)} = \frac{0.286\, R_L}{R_L + R_F} \quad ...(16.23)$$

Neglecting forward resistance R_F of the diode we have TUF = 0.286.

12. *Regulation.* The variation of dc output voltage as a function of dc load current is called *regulation*. Percentage regulation is given as

$$\%\ \text{regulation} = \frac{V_{NL} - V_{FL}}{V_{FL}} \times 100$$

For an ideal power supply, output voltage should be independent of load current and the percentage regulation should be equal to zero.

For half-wave rectifier, from Eq. (16.8)

$$V_{dc} = \frac{V_{S\,max} R_L}{\pi(R_F + R_L)} = \frac{V_{S\,max}}{\pi}\left[1 - \frac{R_F}{R_F + R_L}\right] = \frac{V_{S\,max}}{\pi} - I_{dc} R_F \quad ...(16.24)$$

From above equation it seems that the half-wave rectifier functions as if it were a constant voltage source of $\frac{V_{S\,max}}{\pi}$ volts in series with an internal resistance being equal to R_F. Thus V_{dc} equals $\frac{V_{S\,max}}{\pi}$ at no load and dc output voltage decreases linearly with the increase in output dc current I_{dc}. Clearly, the effective internal resistance of the power supply is R_F. In practice, the resistance R_S of the transformer secondary is in series with the diode, and in Eq. (16.24) R_S is to be added in R_F. Thevenin's model for a half-wave rectifier is given in Fig. 16.3.

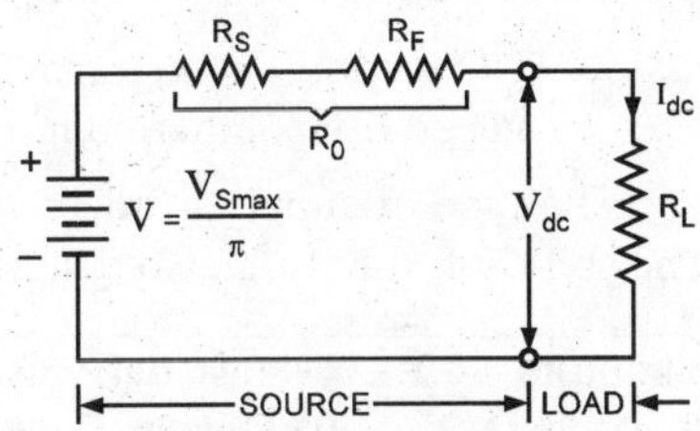

Fig. 16.3. *Thevenin's Model For a Half-Wave Rectifier*

Effect of Diode Cutin Voltage on Rectifier Output. We do not get a perfect half-wave voltage across the load resistor. Because of the barrier potential (or cutin voltage) of diode, the diode does not turn on until the ac source voltage attains approximately 0.7 V in case of silicon diode. When the peak source voltage is much greater than barrier potential (V_K) of diode, the load voltage will resemble a half-wave signal. If the peak source voltage is not much greater than diode barrier potential,

Peak value of voltage across load resistor R_L,

$$V_{out\,(max)} = V_{S\,max} - V_K \quad ...(16.25)$$

$$\text{DC output voltage, } V_{dc} = \frac{V_{S\,max} - V_K}{\pi} \quad ...(16.26)$$

$$\text{DC output current, } I_{dc} = \frac{V_{dc}}{R_L} = \frac{V_{S\,max} - V_K}{\pi R_L} \quad ...(16.27)$$

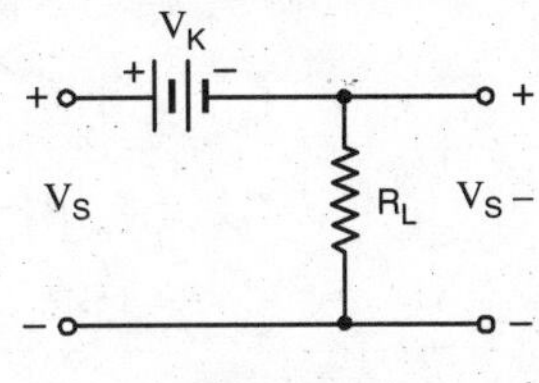

Fig. 16.4

Condition For Maximum DC Output Power in a Half-Wave Rectifier. DC output power in a half-wave single phase rectifier is given by

$$P_{dc} = I_{dc}^2 R_L = \left(\frac{I_{max}}{\pi}\right)^2 R_L = \frac{V_{S\max}^2}{\pi^2 (R_L + R_F)^2} R_L \qquad \because I_{max} = \frac{V_{S\max}}{R_L + R_F}$$

Differentiating both sides of above equation w.r.t. R_L we have

$$\frac{dP_C}{dR_L} = \frac{V_{S\max}^2}{\pi^2}\left[\frac{(R_L^2 + R_F^2 + 2R_L R_F) - R_L(2R_L + 2R_F)}{(R_L^2 + R_F^2 + 2R_L R_F)^2}\right] = \frac{V_{S\max}^2 (R_F^2 - R_L^2)}{\pi^2 (R_L^2 + R_F^2 + 2R_L R_F)^2}$$

Output will be maximum if $\frac{dP_C}{dR_L} = 0$

or $R_F^2 - R_L^2 = 0$

or $R_L = R_F$ *i.e.* when load resistance equals diode resistance.

The advantages and disadvantages of a half-wave rectifier are given below.

Advantages. Simple circuit and low cost.

Disadvantages. 1. The output current in the load contains, in addition to dc component, ac components of basic frequency equal to that of the input voltage frequency. Ripple factor is high and an elaborate filtering is, therefore, required to give steady dc output.

2. The power output and, therefore, rectification efficiency is quite low. This is due to the fact that power is delivered only half the time.
3. Transformer utilization factor is low.
4. DC saturation of transformer core resulting in magnetising current and hysteresis losses and generation of harmonics.

The type of supply available from a half-wave rectifier is not satisfactory for general power supply. This type of supply can be satisfactory for some particular purposes such as battery charging.

Example 16.1. A sinusoidal voltage of 40 V and frequency 50 Hz is applied to a half-wave rectifier. $R_L = 200\ \Omega$, $V_\gamma = 0$, $R_F = 20\ \Omega$, $R_r = \infty$. Find V_{dc}, I_{dc}, I_m, I_{rms}, P_{dc}, η, ripple factor, % regulation.

Solution : $V_{s\,rms} = 40$ V

Frequency, $f = 50$ Hz

$R_L = 200\ \Omega$

$R_F = 20\ \Omega$

$R_r = \infty$

$$I_{max} = \frac{V_{s\max}}{R_L + R_F} = \frac{40\sqrt{2}}{200 + 20} = 0.257 \text{ A or } 257 \text{ mA} \quad \textbf{Ans.}$$

$$I_{rms} = \frac{I_{max}}{2} = \frac{257}{2} = 128.5 \text{ mA} \quad \textbf{Ans.}$$

$$I_{dc} = \frac{I_{max}}{\pi} = \frac{257}{\pi} = 81.8 \text{ mA} \quad \textbf{Ans.}$$

$$V_{dc} = I_{dc} R_L = 81.8 \times 10^{-3} \times 200 = 16.36 \text{ V} \quad \textbf{Ans.}$$

$$P_{dc} = I_{dc}^2 R_L = (81.8 \times 10^{-3})^2 \times 200 = 1.338 \text{ W} \quad \textbf{Ans.}$$

$$P_{ac} = \frac{I_{max}^2}{4}(R_F + R_L) = \frac{(257 \times 10^{-3})^2}{4} \times (20 + 200) = 3.63 \text{ W}$$

$$\eta = \frac{P_{dc}}{P_{ac}} \times 100 = \frac{1.338}{3.63} \times 100 = 36.86\% \textbf{ Ans.}$$

$$\text{Ripple factor}, \gamma = \left(\frac{I_{rms}^2}{I_{dc}^2} - 1\right)^{1/2} = \sqrt{\left(\frac{128.5}{81.8}\right)^2 - 1} = 1.21 \textbf{ Ans.}$$

$$V_{dc} = \frac{V_{max}}{\pi} - I_{dc} R_F = \frac{40\sqrt{2}}{\pi} - 81.8 \times 10^{-3} \times 20 = 18 - 1.636 = 16.364 \text{ V}$$

$$\text{Regulation} = \frac{V_{NL} - V_{FL}}{V_{FL}} \times 100 = \frac{\frac{V_{max}}{\pi} - I_{dc} R_F}{I_{dc} R_F} \times 100 = \frac{18 - 16.364}{16.364} \times 100 = 9.998\% \textbf{ Ans.}$$

Example 16.2. A half-wave rectifier uses a diode with a forward resistance of 100 Ω. If the input ac voltage is 220 V and load resistance is 2 kΩ, determine the transformer utilisation factor.

Solution : Diode forward resistance, R_F = 100 Ω or 0.1 kΩ

Load resistance, R_L = 2 kΩ

$$\text{Transformer utilisation factor, TUF} = \frac{0.286}{1 + \frac{R_F}{R_L}} = \frac{0.286}{1 + \frac{0.1}{2}} = 0.2724 \textbf{ Ans.}$$

Example 16.3. What is the ripple 2 V on average of 50 V ?

Solution : RMS value of ac component, V_{rms} = 2 V

Average value of output voltage, V_{dc} = 50 V

$$\text{Ripple factor}, \gamma = \frac{V_{rms}}{V_{dc}} = \frac{2}{50} = 0.04 \textbf{ Ans.}$$

Example 16.4. A dc voltage supply provides 60 V when the output is unloaded. When connected to a load the output drops to 56 V. Calculate the value of voltage regulation.

Solution : No-load voltage, V_{NL} = 60 V

Full-load voltage, V_{FL} = 56 V

$$\text{Voltage regulation} = \frac{V_{NL} - V_{FL}}{V_{FL}} \times 100 = \frac{60 - 56}{56} \times 100 = 7.14\% \textbf{ Ans.}$$

Example 16.5. A 10 V regulated dc power supply has a regulation of 0.002. Find the magnitude of variation in output voltage. *[Pb. Technical Univ. Electronic Devices and Circuits December 2003]*

Solution : Full-load voltage, V_{FL} = 10 V

Regulation = 0.002

$$\text{Voltage regulation} = \frac{V_{NL} - V_{FL}}{V_{FL}}$$

$$\text{Variation in output voltage} = V_{NL} - V_{FL} = V_{FL} \times \text{voltage regulation} = 10 \times 0.002 = 0.02 \text{ V} \textbf{ Ans.}$$

Example 16.6. A half wave rectifier is used to supply 10 V dc to a resistive load of 400 Ω. If the crystal diode has a forward resistance of 20 Ω, determine the value of ac voltage supplied to the circuit.

Solution : Maximum value of supply voltage

$$V_{s\,max} = V_{dc} \times \pi\left(1 + \frac{R_F}{R_L}\right) = 10 \times \pi\left(1 + \frac{20}{400}\right) = 33 \text{ V} \qquad \text{Refer to Eq. (16.8)}$$

RMS value of ac voltage applied to the circuit

$$V_s = \frac{V_{s\,max}}{\sqrt{2}} = \frac{33}{\sqrt{2}} = 23.3 \text{ V} \textbf{ Ans.}$$

16.3. FULL-WAVE RECTIFIERS

In half-wave rectifiers only one half cycles of the input are utilized but in full-wave rectifiers both half cycles of the input are utilized. Alternate half cycles are inverted to give unidirectional load current. There are two types of full-wave rectifier circuits namely 1. *centre-tap rectifier* and 2. *bridge rectifier.*

16.3.1. Centre-Tap Full-Wave Rectifier

In such a rectifier, the ac input is applied through a transformer, the anodes of the two diodes D_1 and D_2 (having similar characteristics) are connected to the opposite ends of the centre-tapped secondary winding and two cathodes are connected to each other and are connected also through the load resistance R_L and back to the centre of the transformer, as shown in Fig. 16.5.

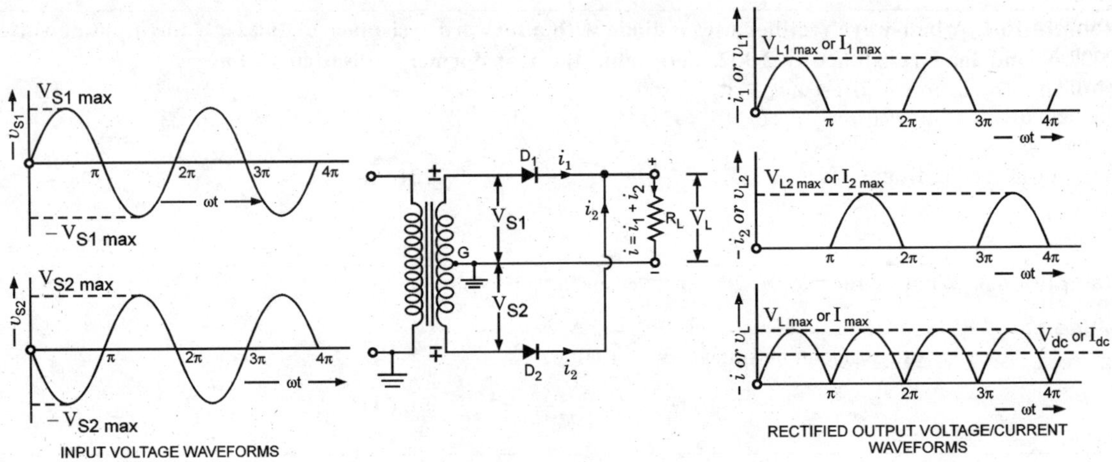

Fig. 16.5. *Centre-Tap Full-Wave Rectifier*

When the top of the transformer secondary winding is positive, say during the first half cycle of the supply, the anode of diode D_1 is positive w.r.t. cathode, and anode of diode D_2 is negative w.r.t. cathode. Thus only diode D_1 conducts, being forward biased and current flows from cathode to anode of diode D_1, through load resistance R_L and top half the transformer secondary making cathode end of load resistance R_L positive. During the second half cycle of the input voltage the polarity is reversed, making the bottom of the secondary winding positive w.r.t. centre tap and thus diode D_2 is forward biased and diode D_1 is reverse biased. Consequently during this half cycle of the input only the diode D_2 conducts and current flows through the load resistance R_L and bottom of the transformer secondary making the cathode end of the load resistance R_L positive. Thus the direction of flow of current through the load resistance R_L remains the same during both halves of the input supply voltage. Thus the circuit shown in Fig. 16.5 acts as a full-wave rectifier.

Peak Inverse Voltage. As already mentioned before, peak inverse voltage (PIV) is the maximum possible voltage across a diode when it is reverse biased. During first half cycle of the supply *i.e.* when the top of the transformer secondary winding is positive, diode D_1 conducts and offers almost zero resistance. So whole of the voltage $V_{S\,max}$ of the upper half winding is developed across the load resistance R_L. Now voltage across the non-conducting diode D_2 is the sum of voltage across the lower half of the transformer secondary and the voltage across the load resistance R_L.

Thus PIV of diode, $D_2 = V_{S\,max} + V_{S\,max} = 2V_{S\,max}$

Similarly PIV of diode $D_1 = 2\ V_{S\,max}$ ...(16.28)

16.3.2. Bridge Rectifier

In the bridge circuit four diodes are connected in the form of a Wheatstone bridge, two diametrically opposite junctions of the bridge are connected to the secondary of a transformer and the other two are connected to the load, as shown in Fig. 16.6.

When the upper end of the transformer secondary winding is positive, say during first half cycles of the input supply, diodes D_1 and D_3 are forward biased and current flows through arm AB, enters the load at positive terminal, leaves the load at negative terminal, and returns back flowing through arm DC. During this half of each input cycle, the diodes D_2 and D_4 are reverse biased and so the current is not allowed to flow in arms AD and BC. The flow of current is indicated by solid arrows in the figure. In the second half of the input cycle the lower end of ac supply becomes positive, diodes D_2 and D_4 become forward biased and current flows through arm CB, enters the load at the positive terminal, leaves the load at negative terminal and returns back flowing through arm DA. Flow of current has been shown by dotted arrows in the figure. Thus the direction of flow of current through the load resistance R_L remains the same during both half cycles of the input supply voltage.

Peak Inverse Voltage (PIV). Figure 16.6 shows a bridge rectifier circuit. Let us consider the instant the secondary voltage attains its positive peak value $V_{S\,max}$. Now diodes D_1 and D_3 are conducting whereas diodes D_2 and D_4 are non-conducting being reverse biased. The conducting diodes D_1 and D_3 have almost zero resistance (*i.e.* zero voltage drop across them). Point B has the same potential as point A and similarly point D has the same potential as point C. The entire voltage of the transformer secondary winding, V_{Smax} is developed across the load resistance R_L. The same voltage *i.e.* V_{Smax} acts across each of the non-conducting diodes D_2 and D_4. Thus

$$\text{PIV} = V_{Smax} \qquad ...(16.29)$$

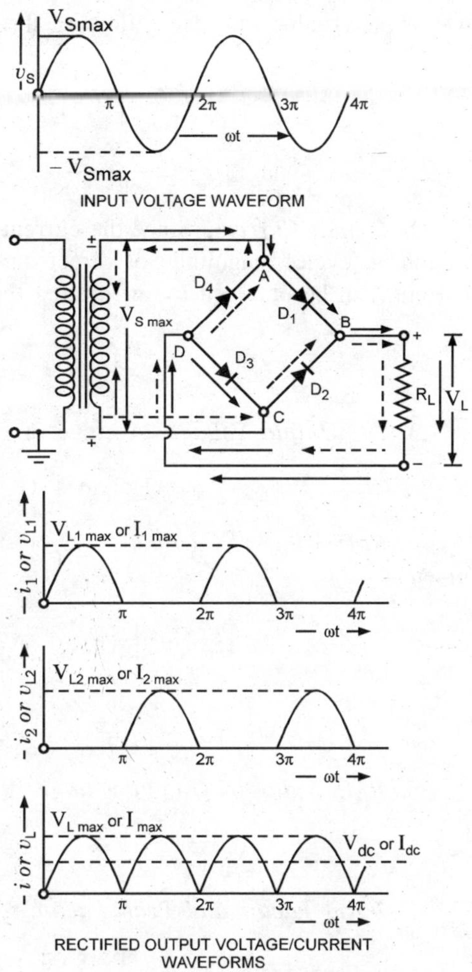

Fig. 16.6. ***Bridge Rectifier***

16.3.3. Circuit Analysis

The analysis of both of the full-wave rectifier circuits (*i.e.* centre-tap and bridge type) is the same except that (*i*) in a bridge rectifier circuit two diodes conduct during each half cycle and forward resistance becomes double *i.e.* 2 R_F and (*ii*) in a bridge rectifier circuit $V_{S\,max}$ is the maximum voltage across the transformer secondary winding whereas in a centre tap rectifier circuit $V_{S\,max}$ represents the maximum voltage across each half of the secondary winding.

1. ***Peak Current.*** Instantaneous value of voltage applied to the rectifier is given as

$$v_s = V_{S\,max} \sin \omega t \qquad ...(16.30)$$

If the diode is assumed to have a forward resistance of R_F ohms and reverse resistance equal to infinity, then current flowing through the load resistance is given as

$$i_1 = I_{max} \sin \omega t \text{ and } i_2 = 0 \text{ for first half cycle} \qquad ...(16.31)$$

$$\text{and } i_1 = 0 \text{ and } i_2 = I_{max} \sin \omega t \text{ for second half cycle} \qquad ...(16.32)$$

The total current flowing through the load resistance R_L, being the sum of currents i_1 and i_2, is given as

$$i = i_1 + i_2 = I_{max} \sin \omega t \text{ for the whole cycle} \quad ...(16.33)$$

where peak value of current flowing through the load resistance R_L is given as

$$I_{max} = \frac{V_{S\,max}}{R_F + R_L} \quad ...(16.34) \text{ in case of centre-tap rectifier}$$

and $$I_{max} = \frac{V_{S\,max}}{2R_F + R_L} \quad ...(16.35) \text{ in case of bridge rectifier}$$

2. *Output Current.* Since the current is the same through the load resistance R_L in the two halves of the ac cycle, magnitude of direct current I_{dc}, which is equal to the average value of alternating current, can be obtained by integrating the current i_1 between 0 and π or current i_2 between π and 2π.

So $$I_{dc} = \frac{1}{\pi}\int_0^{\pi} i_1 \, d(\omega t) = \frac{1}{\pi}\int_0^{\pi} I_{max} \sin \omega t \, d(\omega t) = \frac{2I_{max}}{\pi} \quad ...(16.36)$$

3. *DC Output Voltage.* Average or dc value of voltage across the load is given as

$$V_{dc} = I_{dc} R_L = \frac{2}{\pi} I_{max} R_L \quad ...(16.37)$$

4. *RMS Value of Current.* RMS or effective value of current flowing through the load resistance R_L is given as

$$I_{rms}^2 = \frac{1}{\pi}\int_0^{\pi} i_1^2 \, d(\omega t) = \frac{1}{\pi}\int_0^{\pi} I_{max}^2 \sin^2 \omega t \, d(\omega t) = \frac{I_{max}^2}{2}$$

or $$I_{rms} = \frac{I_{max}}{\sqrt{2}} \quad ...(16.38)$$

5. *RMS Value of Output Voltage.* RMS value of voltage across the load is given as

$$V_{L\,rms} = I_{rms} R_L = \frac{I_{max}}{\sqrt{2}} R_L \quad ...(16.39)$$

6. *Form Factor and Peak Factor.*

$$\text{Form factor, } K_f = \frac{\text{RMS value}}{\text{Average value}} = \frac{I_{rms}}{I_{dc}} = \frac{I_{max}/\sqrt{2}}{2I_{max}/\pi} = \frac{\pi}{2\sqrt{2}} = 1.11 \quad ...(16.40)$$

$$\text{Peak factor, } K_p = \frac{\text{Peak value}}{\text{RMS value}} = \frac{I_{max}}{I_{max}/\sqrt{2}} = \sqrt{2} \quad ...(16.41)$$

7. *Output Frequency.* With a half-wave rectifier, the output frequency equals the input frequency. But the frequency of the full-wave signal is double the input frequency. A full-wave output has twice as many cycles as the sine-wave input has. The full-wave rectifier inverts each negative half cycle, so that we get the number of positive half cycles. The effect is to double the frequency. Thus, for full-wave rectifier

$$f_{out} = 2 f_{in} \quad ...(16.42)$$

8. *Rectification Efficiency.* Power delivered to load,

$$P_{dc} = I_{dc}^2 R_L = \left(\frac{2}{\pi} I_{max}\right)^2 R_L = \frac{4}{\pi^2} I_{max}^2 R_L \quad ...(16.43)$$

$$\text{AC input power, } P_{ac} = I_{rms}^2 (R_L + R_F) = \frac{I_{max}^2}{2} (R_L + R_F) \quad ...(16.44)$$

$$\text{Rectification efficiency,} \quad \eta = \frac{P_{dc}}{P_{ac}} = \frac{\frac{4}{\pi^2} I_{max}^2 R_L}{\frac{1}{2} I_{max}^2 (R_L + R_F)} = \frac{8}{\pi^2} \frac{1}{\left(1 + \frac{R_F}{R_L}\right)} = \frac{0.812}{1 + \frac{R_F}{R_L}} \quad \text{...(16.45)}$$

In case of bridge rectifier, rectification efficiency is given as

$$\eta = \frac{0.812}{1 + \frac{2 R_F}{R_L}} \quad \text{...(16.46)}$$

9. *Ripple Factor*. Form factor of the rectified output voltage of a full-wave rectifier is given as

$$K_f = \frac{I_{rms}}{I_{av}} = \frac{I_{max}/\sqrt{2}}{2 I_{max}/\pi} = \frac{\pi}{2\sqrt{2}} = 1.11$$

Ripple factor is given as

$$\gamma = \sqrt{K_f^2 - 1} = \sqrt{(1.11)^2 - 1} = 0.482 \quad \text{...(16.47)}$$

10. *Regulation*. From Eq. (16.37).

$$V_{dc} = \frac{2}{\pi} I_{max} R_L = \frac{2 V_{S\,max} R_L}{\pi (R_F + R_L)} = \frac{2 V_{S\,max}}{\pi}\left[1 - \frac{R_F}{R_F + R_L}\right] = \frac{2 V_{S\,max}}{\pi} - I_{dc} R_F \quad \text{...(16.48)}$$

In case of a bridge rectifier,

$$V_{dc} = \frac{2 V_{S\,max}}{\pi} - 2 I_{dc} R_F \quad \text{...(16.49)}$$

11. *Transformer Utilization Factor For Centre-Tap Transformer*. In a centre-tap full-wave rectifier, there is no problem due to saturation of core because the dc currents in the two halves of the transformer secondary flow in opposite directions. The average transformer utilisation factor is found by considering the primary and secondary windings.

Transformer utilisation factor of primary

$$= \frac{P_{dc}}{\text{VA rating of primary}} = \frac{I_{dc}^2 R_L}{V_{S\,rms} I_{rms}} = \frac{\left(\frac{2 I_{max}}{\pi}\right)^2 R_L}{\frac{V_{S\,max}}{\sqrt{2}} \times \frac{I_{max}}{\sqrt{2}}}$$

$$= \frac{4 I_{max}^2 R_L / \pi^2}{\frac{I_{max}}{\sqrt{2}} (R_F + R_L) \times \frac{I_{max}}{\sqrt{2}}} = \frac{8}{\pi^2} \times \frac{R_L}{(R_F + R_L)} = \frac{8}{\pi^2} \times \frac{1}{\left(1 + \frac{R_F}{R_L}\right)} \simeq 0.812$$

Centre-tap transformer can be thought of as equivalent to two half-wave rectifiers feeding to a common load. Hence, TUF of two half secondaries can be written as

$$TUF_{(\text{Full secondary})} = 2 \times TUF_{(\text{Half-wave})} = 2 \times 0.286 = 0.572$$

The average transformer utilisation factor,

$$(TUF)_{av} = \frac{TUF_{Primary} + TUF_{Secondary}}{2} = \frac{0.812 + 0.572}{2} = 0.692$$

12. *Transformer Utilisation Factor For Bridge Rectifier*. The current flow through both of primary and secondary windings are sinusoidal. Due to this TUF of both the primary and secondary are 0.812 and the overall TUF is also 0.812. Since in case of bridge rectifier TUF is 0.812 whereas the TUF for centre-tap transformer is 0.692 *i.e.* TUF of bridge rectifier is larger than that of centre-tap rectifier, for the same dc output power, a smaller transformer can be used in the bridge circuit.

16.3.4. Effect of Diode Cutin Voltage on Rectifier Output and PIV of Diodes

1. *Centre-Tap Rectifier.* Peak value of voltage across load resistor R_L.

$$V_{out\,(max)} = V_{max} - V_K \quad ...(16.50)$$

as in case of half-wave rectifier

$$\text{DC output voltage, } V_{dc} = \frac{2}{\pi}(V_{S\,max} - V_K) \quad ...(16.51)$$

$$\text{DC output current, } I_{dc} = \frac{V_{dc}}{R_L} = \frac{2}{\pi}\frac{(V_{S\,max} - V_K)}{R_L} \quad ...(16.52)$$

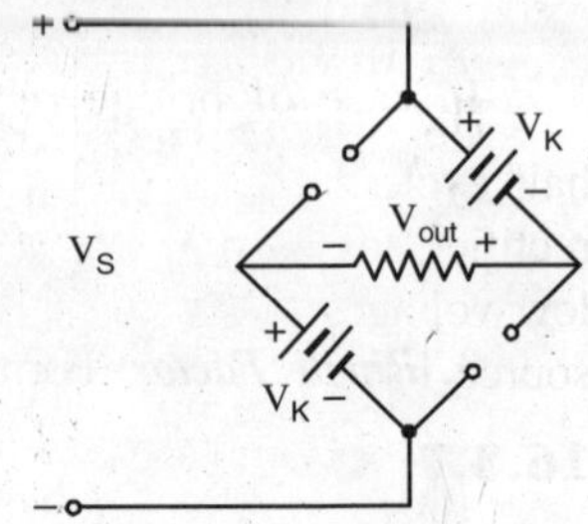

Fig. 16.7

Peak inverse voltage (PIV) of diode

$$= V_{max} + \text{voltage developed across load resistor } R_L$$

$$= V_{max} + V_{out\,(max)} = V_{max} + (V_{max} - V_K) = 2V_{max} - V_K \quad ...(16.53)$$

2. *Bridge Rectifier.* Voltage developed across load resistor R_L,

$$V_{out\,(max)} = V_{S\,max} - 2V_K \quad ...(16.54)$$

$$\text{DC output voltage, } V_{dc} = \frac{2}{\pi}(V_{S\,max} - 2V_K) \quad ...(16.55)$$

$$\text{DC output current, } I_{dc} = \frac{V_{dc}}{R_L} = \frac{2}{\pi}\frac{(V_{S\,max} - 2V_K)}{R_L} \quad ...(16.56)$$

$$\text{Peak inverse voltage (PIV) of diode} = \frac{1}{2}(V_{S\,max} + \text{voltage developed across load resistor } R_L)$$

$$= \frac{1}{2}[V_{S\,max} + (V_{S\,max} - 2V_K)] = V_{S\,max} - V_K \quad ...(16.57)$$

16.3.5. Merits and Demerits of Full-Wave Rectifiers Over Half-Wave Rectifiers

Merits. 1. The rectification efficiency of full-wave rectifier is double of that of a half-wave rectifier.

2. The ripple voltage is low and of higher frequency in case of a full-wave rectifier, so simple filtering circuit is required.
3. Higher output voltage, higher output power and higher TUF in case of a full-wave rectifier.
4. In a full-wave rectifier, there is no problem due to dc saturation of the core because the dc currents in the two halves of the transformer secondary flow in opposite directions.

Demerits. Full-wave rectifier needs more circuit elements and is costlier.

16.3.6. Merits and Demerits of Bridge Rectifiers Over Centre-Tap Rectifiers

With the availability of low-cost, highly reliable and small-sized silicon diodes bridge rectifier is becoming more and more popular in comparison to centre-tap rectifier. It has many advantages over a centre-tap rectifier, as given below.

1. No centre tap is required in the transformer secondary so in case of a bridge rectifier the transformer required is simpler. If stepping up or stepping down of voltage is not required, transformer can be eliminated even.
2. The PIV is one-half that of centre-tap rectifier. Hence bridge rectifier is highly suited for high-voltage applications.

3. Transformer utilisation factor, in case of a bridge rectifier, is higher than that of a centre-tap rectifier.
4. For a given power output, power transformer of smaller size can be used in case of the bridge rectifier because current in both (primary and secondary) windings of the supply transformer flow for the entire ac cycle.

The main drawback of a bridge rectifier is that it needs four diodes, two of which conduct in alternate half cycles. Because of this the total voltage drop in diodes becomes double of that in case of centre-tap rectifier, losses are increased and rectification efficiency is somewhat reduced. This poses a problem when low voltages are required. Another drawback of bridge rectifier is that the load resistor R_L and the supply source have no common point which may be earthed.

16.3.7. Comparison Between Half-Wave, Centre-Tap and Bridge Rectifiers

TABLE 16.1.

S.No.	*Particulars*	*Half-Wave Rectifiers*	*Full-Wave Rectifiers*	
			Centre-Tap	*Bridge*
1.	Number of diodes required	1	2	4
2.	Transformer requirement	not essential	essential by centre tap	not essential
3.	Average value of current, I_{dc}	I_{max}/π	$2\,I_{max}/\pi$	$2\,I_{max}/\pi$
4.	RMS value of current, I_{rms}	$I_{max}/2$	$I_{max}/\sqrt{2}$	$I_{max}/\sqrt{2}$
5.	Peak inverse voltage (PIV)	$V_{S\,max}$	$2\,V_{S\,max}$	$V_{S\,max}$
6.	Peak load current, I_{max}	$\frac{V_{S\,max}}{R_L + R_F}$	$\frac{V_{S\,max}}{R_L + R_F}$	$\frac{V_{S\,max}}{R_L + 2\,R_F}$
7.	DC output voltage, V_{dc}	$\frac{I_{max}}{\pi} R_L$	$\frac{2}{\pi} I_{max} R_L$	$\frac{2}{\pi} I_{max} R_L$
8.	Rectification efficiency (max)	40.6%	81.2%	81.2%
9.	Ripple factor	1.21	0.482	0.482
10.	Fundamental frequency of ripple	Supply frequency, f	$2\,f$	$2\,f$
11.	Voltage regulation	good	better	good
12.	Transformer utilisation factor	0.286	0.692	0.812
13.	Form factor	1.57	1.11	1.11
14.	Peak factor	2	$\sqrt{2}$	$\sqrt{2}$

Example 16.7. The load resistance of a centre-tapped full-wave rectifier is 500 Ω and the necessary voltage is 60 sin (100 πt). Calculate (*i*) peak, average and rms values of current ; (*ii*) ripple factor and (*iii*) efficiency of the rectifier. Each diode has an idealised I-V characteristics having slope corresponding to a resistance of 50 Ω.

Solution : Maximum value of supply voltage, $V_{S\,max} = 60$ V

Forward resistance, $R_F = 50\ \Omega$

Load resistance, $R_L = 500\ \Omega$

$$(i)\ \text{Peak current, } I_{max} = \frac{V_{S\,max}}{R_L + R_F} = \frac{60}{500+50} = 0.109 \text{ A } \textbf{Ans.}$$

$$\text{Average current, } I_{dc} = \frac{2I_{max}}{\pi} = \frac{2 \times 0.109}{\pi} = 0.0695 \text{ A } \textbf{Ans.}$$

RMS value of current, $I_{rms} = \frac{I_{max}}{\sqrt{2}} = \frac{0.109}{\sqrt{2}} = 0.077$ A **Ans.**

(*ii*) Ripple factor, $\gamma = \sqrt{\left(\frac{I_{rms}}{I_{dc}}\right)^2 - 1} = \sqrt{\left(\frac{0.077}{0.0695}\right)^2 - 1} = 0.482$ **Ans.**

(*iii*) Efficiency of rectifier, $\eta = \frac{0.812}{1+\frac{R_F}{R_L}} \times 100 = \frac{0.812 \times 100}{1+\frac{50}{500}} = 73.82\%$ **Ans.**

Example 16.8. A sine wave of RMS value 70.7 volts is applied to the primary of full-wave centre-tapped rectifier with a transformer turns-ratio of 1:2. Assuming diode drop of 0.7 V, determine the maximum PIV rating of each diode.

Solution : Primary voltage, $V_1 = 70.7$ V

Secondary voltage, $V_2 = V_{S_1} + V_{S_2} = 2V_s = 70.7 \times 2 = 141.4$ V

Supply voltage across one-half of secondary winding,

$$V_S = \frac{141.4}{2} = 70.7 \text{ V}$$

PIV of each diode $= 2V_{Smax}$ – Voltage drop across diode $= 2 \times \sqrt{2} \times 70.7 - 0.7 = 199.3$ V **Ans.**

Example 16.9. In a bridge rectifier circuit the peak value of secondary voltage is $240\sqrt{2}$ V and frequency is 50 Hz. Determine no-load dc voltage, PIV and output frequency.

Solution : Maximum supply voltage, $V_{max} = 240\sqrt{2}$ V

No-load dc output voltage $= \frac{2V_{max}}{\pi} = \frac{2 \times 240\sqrt{2}}{\pi} = 216$ V **Ans.**

PIV rating of diodes, PIV $= V_{smax} = 240\sqrt{2} = 339.4$ V **Ans.**

Output frequency, $f_{out} = 2f_{in} = 2 \times 50 = 100$ Hz **Ans.**

Example 16.10. A simple full-wave bridge rectifier circuit has an input voltage of 240 V ac rms. Assume the diodes to be ideal. Find the output dc current, dc voltage, rms values of output currents and voltages and peak inverse voltage that appears across the non-conducting diode. Assume load resistance to be 10 kΩ.

Solution : RMS value of supply voltage, $V_{rms} = 240$ V

Maximum value of supply voltage, $V_{max} = \sqrt{2}\,V_{rms} = \sqrt{2} \times 240 = 339.4$ V

Diode barrier potential, $V_K = 0$ ∵ diodes are ideal one

Diode forward resistance, $R_F = 0$ ∵ diodes are ideal one

Load resistance, $R_L = 10$ kΩ

Output dc current, $I_{dc} = \frac{2I_{max}}{\pi} = \frac{2}{\pi} \times \frac{V_{Smax}}{R_L} = \frac{2}{\pi} \times \frac{339.4}{10 \times 10^3}$

$= 0.0216$ A or 21.6 mA **Ans.**

Output dc voltage, $V_{dc} = I_{dc} \times R_L = 21.6 \times 10^{-3} \times 10 \times 10^3 = 216$ V **Ans.**

RMS value of output current, $I_{rms} = \frac{I_{max}}{\sqrt{2}} = \frac{V_{Smax}}{\sqrt{2} \times R_L} = \frac{\sqrt{2} \times 240}{\sqrt{2} \times 10 \times 10^3} = 0.024$ A or 24 mA **Ans.**

$$V_{rms} = \frac{V_{Smax}}{\sqrt{2}} = \frac{240 \times \sqrt{2}}{\sqrt{2}} = 240 \text{ V}$$

PIV $= 2V_{S\,max} = 2 \times 339.4 = 678.8$ V **Ans.**

Example 16.11. A full-wave bridge rectifier with 120 V_{rms} sinusoidal input has a load resistor of 1 kΩ.

(*i*) If silicon diodes are applied, what is dc voltage available at load.

(*ii*) Determine required PIV rating of each diode.

(*iii*) Find maximum current through each diode during conduction.

(*iv*) What is required power rating of each diode ? [*Pb. Technical Univ. Electronic Devices & Circuits 2005*]

Solution : RMS value of supply voltage to the bridge rectifier,

$$V_{S\,rms} = \text{RMS value of output voltage of transformer} = 120 \text{ V}$$

$$V_{S\,max} = \sqrt{2}\, V_{S\,rms} = \sqrt{2} \times 120 \text{ V} = 169.7 \text{ V}$$

(*ii*) PIV rating of each diode $= V_{S\,max} - V_K = 169.7 - 0.7 = 169$ V **Ans.**

(*iii*) Peak current, $I_{max} = \dfrac{V_{max} - 2V_K}{R_L} = \dfrac{169.7 - 2 \times 0.7}{1 \times 10^3} = 168.3$ mA **Ans.**

Average current, $I_{dc} = \dfrac{2I_{max}}{\pi} = \dfrac{2 \times 168.3}{\pi} = 107.14$ mA

(*i*) DC voltage available at load, $V_{dc} = I_{dc} R_L = 107.14 \times 10^{-3} \times 1 \times 10^3 = 107.14$ V **Ans.**

(*iv*) Power rating of each diode $= I_{dc}^2 R_L = (107.14 \times 10^{-3})^2 \times 1 \times 10^3 = 11.48$ W **Ans.**

Example 16.12. What is transformer utilisation factor ? Determine the rating of a transformer to deliver a 100 W of dc power to a load under full wave rectifier.

Solution : Transformer rating $= \dfrac{P_{dc}}{\text{Transformer utilisation factor}}$

$= \dfrac{100}{0.692} = 144.5$ VA **Ans.** in case of centre-tap transformer

and transformer rating $= \dfrac{100}{0.812} = 123$ VA **Ans.** in case of bridge rectifier

16.4. REGULATED POWER SUPPLY

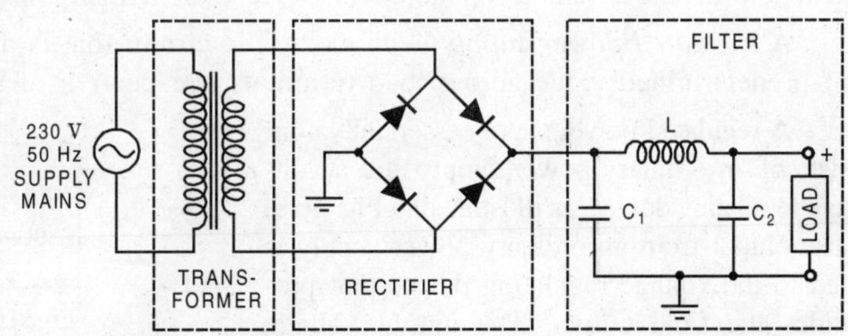

Fig. 16.8. ***Block Diagram of an Unregulated Power Supply***

Almost all electronic devices used in electronic circuits need a dc source of power to operate. The source of dc power is used to establish the dc operating points (Q-points) for the passive and active electronic devices incorporated in the system. The dc power supply is typically connected to each and every stage in an electronic system. It means that the single requirement common to all phases of electronics is the need for a supply of dc power. For portable low-power systems batteries may be used, but their operating period is limited. Thus, for long time operation frequent recharging or replacement of batteries become much costlier and complicated. More frequently, however, electronic equipment is energized by a *power supply*, derived from the standard industrial or domestic ac supply by transformation, rectification, and filtering. The combination of a transformer, a rectifier and a filter constitutes an *ordinary dc power supply*, also called an *unregulated power supply*.

The block diagram of an ordinary power supply is depicted in Fig. 16.8. Usually, a small dc voltage, in the range of 2–24 volts is required for the operation of different electronic circuits, while in India, single-phase ac supply is available at 230 V. So a small step-down transformer is used at the beginning which reduces the voltage level according to the needs. Next block is a rectifier which converts the sinusoidal ac voltage into pulsating dc. In the last there is a filter block which reduces the ripples (ac components) from the rectifier output voltage. The filter is a device which passes dc component to the load and blocks ac components of the rectifier output.

For many applications in electronics, unregulated power supply is not good enough because of the following reasons.

1. *Poor regulation*. The output voltage does not remain constant as the load varies. The internal resistance of an ordinary power supply is relatively large (more than 30 Ω). So output voltage is significantly affected by the magnitude of current drawn from the supply. The voltage drop in the internal resistance of the supply increases directly with an increase in load current.

2. *Variations in the ac supply mains*. The permissible variation in the ac supply mains voltage as per Indian Electricity Rules is 6% of its rated value. But in India, the variation in ac mains voltage is much more than this (sometimes it may vary from 180 V to 260 V). The dc output voltage being proportional to the input ac voltage, therefore, varies largely.

3. *Variations in temperature*. The dc output voltage varies with temperature, particularly if semiconductor devices are employed.

These variations in dc output voltage may cause inaccurate or erratic operation or even malfunctioning of many electronic circuits. For instance, in oscillators the frequency will shift, in transmitters output will get distorted, and in amplifiers the operating point will shift causing bias instability.

Some feedback arrangement (acting as a voltage regulator) is employed in conjunction with an unregulated power supply to overcome the above mentioned three shortcomings and also to reduce the ripple voltage. Such a system is called a *regulated power supply*.

In many applications, it is important to protect the power supply output against inadvertant short circuits that might destroy either the circuit under test or operation of the supply itself. Thus, current limiting circuits are often incorporated into the regulator design.

Power supplies are becoming steadily more sophisticated in terms of performance objectives and application strategies. A commercial power supply is typically a complex system that makes use of ICs to reduce ripple, improve regulation, and widen control options. Programmable power supplies are also available to allow remote operation that is useful in many settings.

A *regulated power supply* is an electronic circuit that is designed to provide a constant dc voltage of predetermined value across load terminals irrespective of ac mains fluctuations or load variations.

A regulated power supply essentially consists of an ordinary power supply and a voltage regulating device, as illustrated in Fig. 16.9. The output from an ordinary power supply is fed to the voltage regulating device that provides the final output. The output voltage remains constant irrespective of variations in the ac input voltage or variations in output (or load) current.

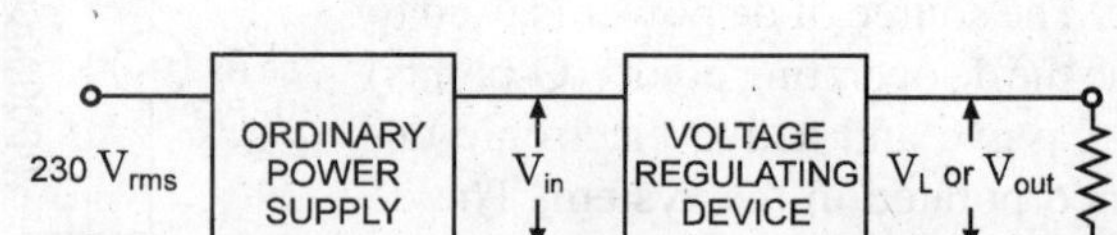

Fig. 16.9. ***Regulated Power Supply***

Figure 16.10 shows the complete circuit of a regulated power supply with a transistor series regulator as a regulating device. The ac voltage, typically 230 V_{rms} is connected to a transformer which transforms that ac voltage to the level for the desired dc output. A bridge rectifier then provides a fullwave rectified voltage that is initially filtered by a π-(or C-L-C) filter to produce a dc voltage. The resulting dc voltage usually has some ripple or ac voltage variation. A regulating circuit uses this dc input to provide a dc voltage that not only has much less ripple voltage but also remains constant even if the input dc voltage varies somewhat or the load connected to the output dc voltage changes. The regulated dc supply is available across a voltage divider.

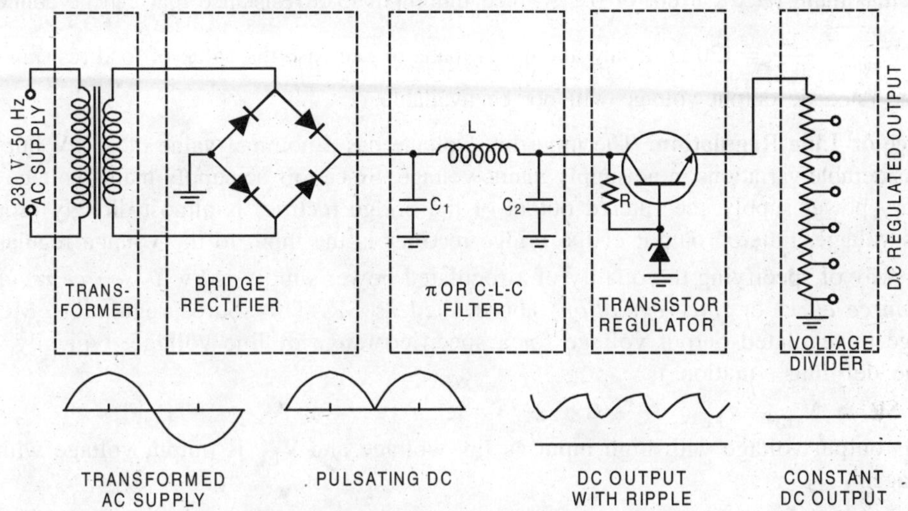

Fig. 16.10. ***Complete Regulated Power Supply Circuit***

Often more than one dc voltage is required for the operation of electronic circuits. A single power supply can provide as many voltages as are required by using a voltage (or potential) divider, as illustrated in the figure. As illustrated in Fig. 16.10, a potential divider is a single tapped resistor connected across the output terminals of the supply. The tapped resistor may consist of two or three resistors connected in series across the supply. In fact, bleeder resistor may also be employed as a potential divider.

16.5. POWER SUPPLY CHARACTERISTICS

The quality of power supply depends on different factors such as its load voltage, load current, voltage regulation, source regulation, output impedance, ripple rejection etc. Some of the characteristics of regulated power supplies are discussed below.

1. Load Regulation. The *load regulation*, abbreviated LR (also called the *load effect*), is the change in regulated output voltage when the load current changes from minimum to maximum value *i.e.*

$$LR = V_{NL} - V_{FL} \qquad ...(16.58)$$

where V_{NL} is load voltage at no load and V_{FL} is the load voltage at full load. In this equation V_{NL} occurs when the load resistance is infinite (*i.e.* out terminals are open circuited) and V_{FL} occurs when the load resistance is of the minimum value where voltage regulation is lost.

Load regulation is usually expressed in percentage and is given as

$$\%LR = \frac{V_{NL} - V_{FL}}{V_{FL}} \times 100 \qquad ...(16.59)$$

For an ideal power supply output voltage should be independent of dc load current and the percentage load regulation should be equal to zero.

2. Minimum Load Resistance. The load resistance at which a power supply delivers its full-load rated current at rated voltage is referred to as a *minimum load resistance*, $R_{L\,(min)}$.

$$i.e.\ R_{L\,(min)} = \frac{V_{FL}}{I_{FL}} \qquad ...(16.60)$$

The value of full-load current I_{FL} should never increase beyond that mentioned in the data sheet of the power supply. For instance, if a data sheet specifies that the power supply will provide an output voltage of 24 V at a maximum rated current of 0.2 A, then minimum load resistance that can be connected across the supply is $R_{min} = \frac{24}{0.2} = 120\ \Omega$. If any attempt is made to decrease the value of load resistance R_L below this value, the rated dc output voltage will not be available.

3. Source or Line Regulation. The input line voltage has a nominal value of 230 V but in practice, there are considerable variations in ac supply mains voltage. Since this ac supply mains voltage is the input to the ordinary power supply, the filtered output of the bridge rectifier is almost directly proportional to the ac mains voltage. Filtered output of the bridge rectifier is the input to the voltage regulating device.

Another way of specifying the quality of a regulated power supply is by its *source regulation* (also called the *source effect* or *line regulation*) abbreviated as SR. The source regulation, SR is defined as the change in regulated output voltage for a specified range of line voltage, typically 230 V ± 10 per cent. The defining equation is

$$SR = V_{HL} - V_{LL} \quad ...(16.61)$$

where V_{HL} is output voltage with high input ac line voltage and V_{LL} is output voltage with low input ac line voltage.

The percentage of source regulation is given as

$$\%SR = \frac{V_{HL} - V_{LL}}{\text{Nominal load voltage}} \times 100 = \frac{SR}{V_{nominal}} \times 100 \quad ...(16.62)$$

The output voltage under specific operating conditions is known as the *nominal load (or output) voltage*.

4. Output Impedance. A regulated power supply is a very stiff dc voltage source. This means that the output resistance is very small (in milliohms). Even though the external load resistance is varied, almost no change is seen in the load voltage. An ideal voltage source has an output impedance of zero. Modern regulated power supplies approach ideal voltage sources.

5. Ripple Rejection. Voltage regulators stabilize the output voltage against variations in input voltage. Ripple is equivalent to a periodic variation in the input voltage. Thus, a voltage regulator attenuates the ripple that comes in with the unregulated input voltage. Since a voltage regulator uses negative feedback, the distortion is reduced by the same factor as the gain. Ripple rejection is a measure of a power supply's ability to reject ripple voltages and is usually expressed in decibels.

16.6. VOLTAGE REGULATORS

The *regulator* may be constructed from a zener diode, and/or discrete transistors, and/or integrated circuits. All voltage regulators must have a stable voltage reference source which is provided by a special type of diode operated in reverse breakdown called a *breakdown diode*, or *zener* diode.

The primary function of a voltage regulator is to maintain a constant dc output voltage. However, it also rejects ac ripple voltage that is not removed by the filter. The regulator may also include protective functions such as short-circuit protection, current limiting, thermal shutdown, or overvoltage protection.

16.7. DISCRETE VOLTAGE REGULATORS

Basically there are two types of transistor voltage regulators viz. (*i*) series voltage regulators and the (*ii*) shunt voltage regulators. Each type of circuit can provide an output dc voltage that is regulated or maintained at a predetermined value even if the input voltage varies or the load connected to the output terminal changes.

16.7.1. Series Voltage Regulators

The basic connection of a series voltage regulator circuit is shown in the block diagram given in Fig. 16.11.

The series element controls the magnitude of the input voltage that gets to the output. The output voltage is sampled by a circuit that provides a feedback voltage to be compared to a reference voltage.

If the output voltage increases, the comparator circuit provides a control signal to cause the series control element to reduce the magnitude of the output voltage—thereby maintaining the output voltage. On the other hand, if the output voltage falls, the comparator circuit provides a control signal to cause the series control element to increase the magnitude of output voltage. Thus, the output voltage is maintained constant.

1. Transistor Series Voltage Regulator or Emitter Follower Voltage Regulator. A simple series voltage regulator using an NPN transistor and a zener diode is shown in Fig. 16.12. This circuit is called a *series regulator* because collector and emitter terminals of the transistor are in series with the load, as illustrated in the figure. This circuit is also called an *emitter follower voltage regulator* because transistor Q is connected in emitter follower configuration. Here, the transistor Q is termed a series pass transistor. The unregulated dc supply (or filtered output from the rectifier) is fed to the input terminals and regulated output voltage V_{out} is obtained across the load resistor R_L. Zener diode provides the reference voltage and the transistor acts as a variable resistor, whose resistance varies with the operating conditions (base current I_B). The principle of operation of such a regulator is based on the fact that a large proportion of the change in supply (or input) voltage appears across the transistor and, therefore, output voltage tends to remain constant.

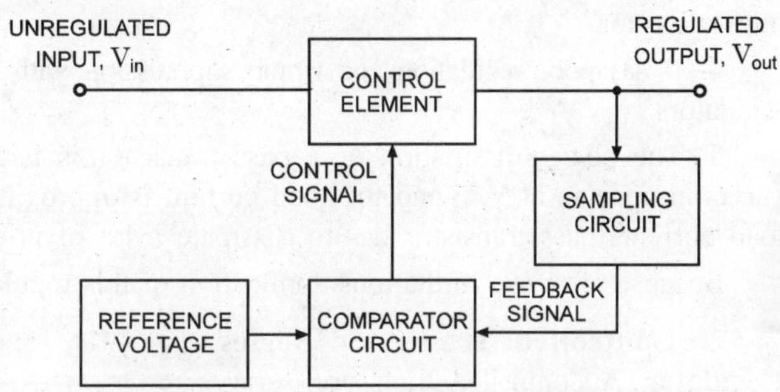

Fig. 16.11. ***Block Diagram of a Series Voltage Regulator***

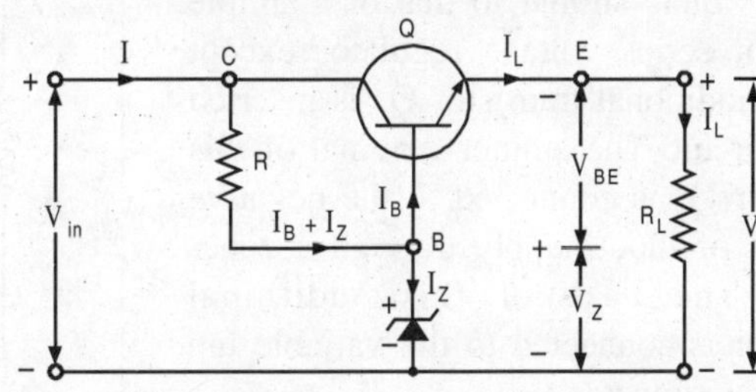

Fig. 16.12. ***Transistor Series Voltage Regulator***

Keeping in mind the polarities of different voltages we have

$$V_{out} = V_Z - V_{BE} \qquad ...(16.63)$$

The base voltage of the transistor remains almost constant being equal to that across the zener diode, V_Z.

Operation (*i*) Let the supply (or input) voltage increase which will cause the output voltage V_{out} to increase. An increase in output voltage V_{out} will result in decrease of V_{BE} because V_Z is fixed and decrease in V_{BE} will reduce the level of conduction. This will lead to increase in the collector-emitter resistance of the transistor causing an increase in collector to emitter voltage and as a result the output voltage will be reduced. Thus output voltage will remain constant. Similar explanation can be given for decrease in supply voltage.

(*ii*) Now let us consider the effect of change in load on the output voltage—say current is increased by decrease in R_L. Under such a situation the output voltage V_{out} tends to fall and, therefore, V_{BE} tends to increase. As a result the conduction level of the transistor will increase leading to decrease in the collector-emitter resistance. The decrease in the collector-emitter resistance of the transistor will cause the slight increase in input current to compensate for the decrease in R_L. Thus the output voltage being equal to $I_L R_L$ remains almost constant. Similar explanation will hold true for increase in R_L.

The *advantage* of such a circuit is that the changes in zener current are reduced by a factor β and thus the zener effect is greatly reduced and much more stabilized output is obtained.

Output voltage from a series regulator, $V_{out} = (V_Z - V_{BE})$, and maximum load current $I_{L\,(max)}$ can be the maximum emitter current that the transistor Q is capable of passing. For a 2N3055 transistor, load current I_L could be 15 A. When load current I_L is zero, the current drawn from the supply is approximately $(I_Z + I_{C\,(min)})$. The zener regulator (resistor R and zener diode form a simple zener regulator) has to supply only the base current of the transistor. The emitter follower voltage regulator is, therefore, much more efficient than a simple zener regulator.

Limitations. 1. The output voltage cannot be maintained absolutely constant because both V_{BE} and V_Z decrease with the increase in room temperature. Further, V_{BE} increases slightly with the increase in load.

2. The output voltage cannot be changed as there is no provision for it in the circuit.

3. It cannot provide good regulation at high currents because of small amplification provided by one transistor.

4. It has poor regulation and ripple suppression with respect to input variations as compared to other regulators.

5. The power dissipation of a pass transistor is large because it is equal to $V_{CC} I_C$ and almost all variation appears at V_{CE} and the load current is approximately equal to collector current. Thus for heavy load currents pass transistor has to dissipate a lot of power and, therefore, becomes hot.

Because of above limitations application of this regulator is limited to low output voltages.

2. Controlled Transistor Series Regulator. A controlled transistor series regulator is shown in Fig. 16.13. The circuit is quite similar to that of a simple transistor series voltage regulator except that an additional transistor Q_2 is inserted in the circuit. The emitter terminal of this transistor Q_2 is connected to the negative terminal of input supply through a zener diode. The base of this additional transistor is connected to the variable tap of a potentiometer.

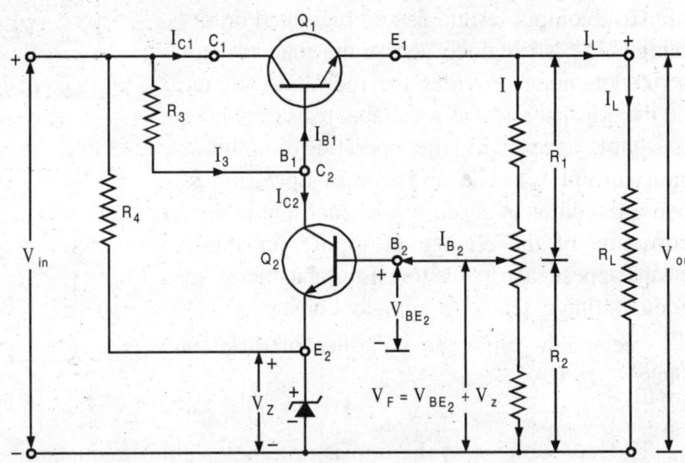

Fig. 16.13. ***Controlled Transistor Series Regulator or Negative Feedback Regulator***

This voltage regulator employs the principle of negative feedback to hold the output voltage almost constant despite variations in supply voltage and/or load current. That is why this regulator is also called a *negative feedback regulator*.

Transistor Q_1, the *control element*, is called the *pass transistor* because all the load current flows through it. Zener diode and resistor R_4 act as a *reference element.* The voltage divider (or potentiometer) consisting of resistors R_1 and R_2 samples the output voltage and delivers a negative feedback voltage to the base of transistor Q_2 and this feedback voltage ($V_F = V_{BE_2} + V_z$) controls the collector current of transistor Q_2.

Operation. Suppose the output voltage increases (due to any reason), the voltage across R_2 is also increased as it is part of the output circuit. This causes an increase in voltage ($V_{BE_2} + V_z$). As a result I_{B_2} and also I_{C_2} increases. Assuming I_3 relatively constant, I_{B_1} decreases. Decrease in base current of transistor Q_1 causes the increase in collector-emitter resistance of transistor Q_1. This causes an increase in $V_{C_1E_1}$ thereby off-setting the increase in output voltage. Thus output voltage remains constant. Reverse happens should the output voltage decrease.

The voltage V_2 provided by the potential divider R_1-R_2 must be equal to the sum of the base-emitter voltage of transistor Q_2 and the zener diode *i.e.*,

$$V_{BE_2} + V_z = V_2 = \frac{R_2}{R_1 + R_2} V_{out}$$

$$\text{or } V_{out} = \frac{R_1 + R_2}{R_2}(V_{BE_2} + V_z) = \left(1 + \frac{R_1}{R_2}\right)(V_{BE_2} + V_z) = A_f(V_{BE_2} + V_z) \quad ...(16.64)$$

Thus the regulated output voltage is equal to the closed-loop gain multiplied by the sum of zener voltage and base-emitter voltage of transistor Q_2 (the sensing element).

Like emitter follower regulator this regulator also has the drawback of excessive power dissipation. Due to this high power dissipation output of the power supply is limited to 30-40 V, as safe value of V_{CE} is 50 V.

Another drawback of this regulator is that it has no overload/short-circuit protection.

3. Controlled Transistor Series Regulator With Overload and Short-Circuit Protection. If the load resistance R_L is reduced or load terminals are shorted accidentally, a very large load current will flow in the circuit shown in Fig. 16.13. It may destroy the pass transistor Q_1, diode or possibly some other component. Fuse protection will not prove adequate because the transistor may get damaged in a very small fraction of a second.

To avoid this situation, a *current limiting circuit* is added to a series regulator, as illustrated in Fig. 16.14.

The current limiting circuit consists of a transistor Q_3 and a resistor R_5 (approximately 1 Ω) connected between base and emitter terminals of transistor Q_1. With normal load current, transistor Q_3 remains off because the voltage drop across resistor R_5 is small (less than about 0.7 V necessary for making the transistor Q_3 *on*). Under this condition, the circuit works normally, as described earlier. With the excessive load current (exceeding 0.6/1 *i.e.* 0.6 A or 600 mA) the voltage drop across R_5 becomes large enough to turn transistor Q_3 on. The collector current of transistor Q_3 flows through R_3 thereby decreasing the base voltage of transistor Q_1. This results in reduction of the conduction level of transistor Q_1. Thus further increase in load current is prevented.

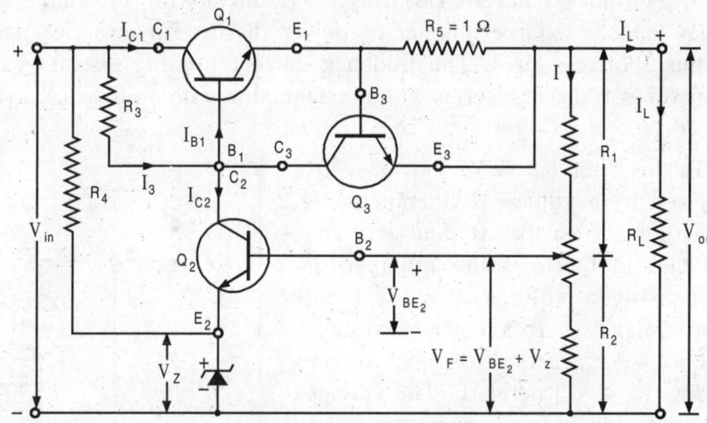

Fig. 16.14. ***Controlled Transistor Series Regulator With Short-Circuit Protection***

Figure 16.15 summarizes the current limiting. When load resistance R_L is infinite, the output voltage is regulated and has a value of V_{REG}. The load current I_L is zero for this operating condition. When R_L decreases, the load current I_L increases up to the point where R_L becomes equal to $R_{L\,(min)}$. At this minimum load resistance, I_L equals 600 mA and V_{BE} equals 0.6 V. Beyond this point, transistor Q_3 turns *on* and the current limiting sets in. Further decrease in R_L produces decrease in output voltage, and regulation is lost. When R_L is zero, the load current I_L is limited to a value between 600 mA and

700 mA. The load current with shorted load terminals is symbolized as I_{SL}. When the load terminals are shorted in Fig. 16.14, the voltage across resistor R_5 is

$$V_{BE} = I_{SL} R_5 \quad \text{or} \quad I_{SL} = \frac{V_{BE}}{R_5} \qquad ...(16.65)$$

where V_{BE} is typically between 0.6 and 0.7 V.

The minimum load resistance where regulation is lost can be estimated with the following equation

$$R_{L\,(min)} = \frac{V_{REG}}{I_{SL}} \qquad ...(16.66)$$

The exact value of $R_{L\,(min)}$ will be slightly less or greater than this.

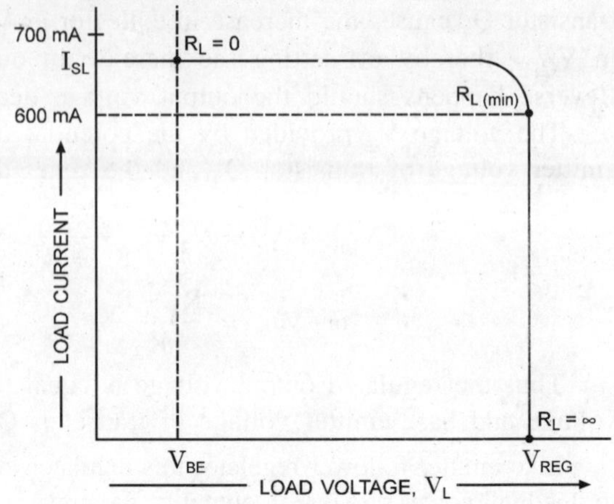

Fig. 16.15

The simple current limiting circuit also has a drawback of large power dissipation across the series pass transistor. With a short across the load, almost all the input voltage appears across the pass transistor. So the pass transistor has to dissipate approximately

$$P_v = (V_{in} - V_{BE}) I_{SL} \qquad ...(16.67)$$

where V_{BE} is the base-emitter voltage of Q_3, the *current-limiting transistor*.

4. Foldback Current Limiting. A problem with the simple current limiting circuit just discussed is that there is a large amount of power dissipation in series pass transistor Q_1 while the regulator remains short circuited. The foldback current limiting circuit is the solution of above problem. The circuit of a transistor series voltage regulator with foldback current limiting facility is illustrated in Fig. 16.16.

In this circuit base of transistor Q_3 is biased by a voltage divider network consisting of resistors R_6 and R_7. The load current I_L flows through resistor R_5, causing a voltage drop of $I_L R_5$ (approximately) across it. Thus a voltage of $(I_L R_5 + V_{out})$ acts across the voltage divider $(R_6 - R_7)$ network. The voltage applied to the base of transistor Q_3 is equal to the voltage drop across resistor R_7 and is given as

$$V_{B_3} = \frac{R_7}{R_6 + R_7} (I_L R_5 + V_{out})$$

$$= K (I_L R_5 + V_{out}) \qquad ...(16.68)$$

$$\text{where } K = \frac{R_7}{R_6 + R_7}$$

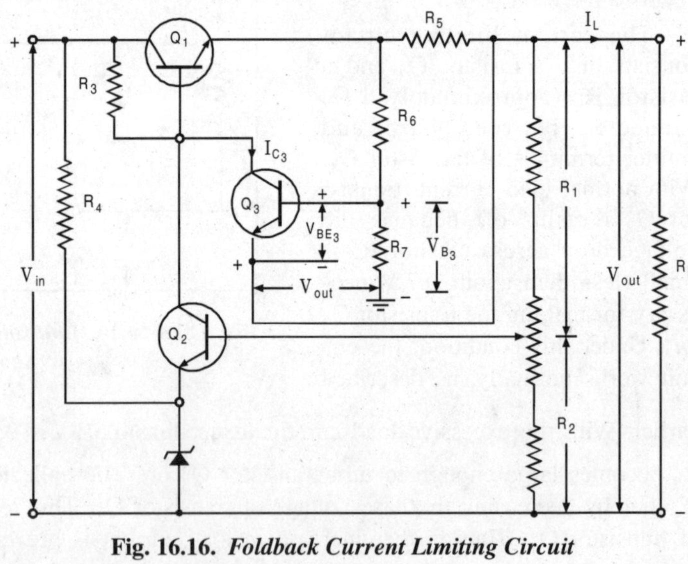

Fig. 16.16. *Foldback Current Limiting Circuit*

Emitter of transistor Q_3 is connected to the positive terminal of V_{out}. Applying Kirchhoff 's voltage law to closed mesh of Q_3 shown in the figure we have

$$V_{out} + V_{BE_3} = V_{B_3}$$

$$\text{or } V_{BE_3} = V_{B_3} - V_{out} = K(I_L R_5 + V_{out}) - V_{out} = K\, I_L R_5 + (K - 1)V_{out} \qquad ...(16.69)$$

Thus the magnitude of base drive of transistor Q_1 is by this V_{BE_3}.

Now if load resistance decreases, may be due to any reason, load current I_L will increase causing voltage drop $I_L R_5$ to increase. This causes V_{B_3} to increase. This makes transistor Q_3 *on* in a stronger way. The increased collector current I_{C_3} of transistor Q_3 flows through the resistor R_3 thereby decreasing the base voltage of transistor Q_1. This results in reduction of the conduction level of transistor Q_1. Thus further increase in load current is prevented.

From Eq. (16.69) it is obvious that V_{BE_3} in this circuit is much more than that was in circuit illustrated in Fig. 16.14 (only $I_L R_5$). It means that the increment in load current is limited by larger amount in circuit shown in Fig. 16.16.

Due to reduction in load resistance R_L, V_{BE_3} increases to a level so that transistor Q_3 gets saturated. Now collector current I_{C_3} becomes constant. Any further decrease in R_L will have no effect on I_{C_3}. The corresponding load current is $I_{L\,(max)}$ and is given as

$$I_{L\,(max)} = \frac{V_{BE_3}}{KR_5} + \frac{(1-K)}{KR_5} V_{out} \qquad ...(16.70)$$

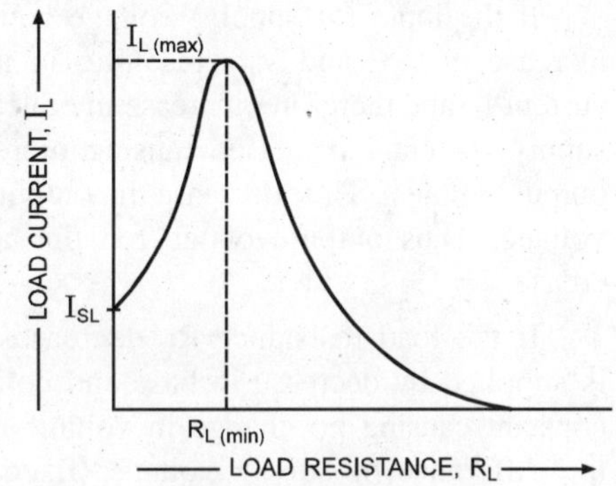

Fig. 16.17

Beyond this point V_{BE_3} also drops due to saturation. Therefore according to Eq. (16.70) load current I_L begins decreasing with decrease in R_L from $R_{L\,(min)}$, as illustrated in Fig. 16.17. When load resistance R_L is zero *i.e.* when output terminals get shorted the output voltage V_{out} becomes equal to zero. Substituting $V_{out} = 0$ in Eq. (16.70) we have

$$I_{SL} = \frac{V_{BE_3}}{KR_5} \qquad ...(16.71)$$

i.e. shorted load current I_{SL} is very much smaller than maximum load current $I_{L\,(max)}$ proving the foldback current limiting. The smaller I_{SL} limits power dissipation in pass transistor Q_1 preventing it from being damaged. This is the main advantage of this circuit.

16.7.2. Shunt Voltage Regulator

A shunt voltage regulator provides regulation by shunting current away from the load. The block diagram of such a voltage regulator is depicted in Fig. 16.18. The input unregulated voltage provides current to the load. Some of this current is shunted away by the control element to maintain the regulated output voltage across the load. If the output voltage tends to change due to change in load, the sampling circuit provides a feedback signal to a comparator circuit which then provides a control signal to vary the magnitude of current shunted away from the load. For example, when the output voltage tends to fall, the sampling circuit provides a feedback signal to the comparator circuit which then provides a control signal to draw lesser shunt current, providing more load current, thereby keeping the regulated voltage constant.

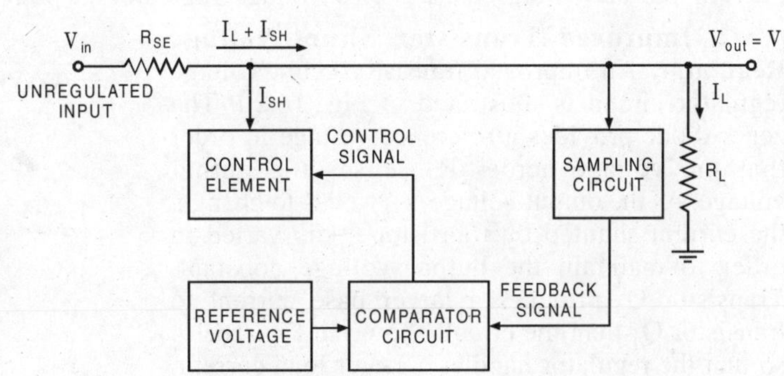

Fig. 16.18. ***Block Diagram of a Shunt Voltage Regulator***

1. Transistor Shunt Voltage Regulator. A shunt voltage regulator using an NPN transistor and a zener diode is shown in Fig. 16.19. A series resistance R_{SE} is connected in series with the unregulated (or input), supply. Zener diode is connected across the base and collector terminals of the NPN transistor and the transistor is connected across the output, as shown in Fig. 16.19.

Unregulated voltage is reduced, due to voltage drop in series resistance R_{SE}, by an amount that depends on the current supplied to the load R_L. The voltage across the load is fixed by the zener diode and transistor base-emitter voltage V_{BE}.

Output voltage is given as

$$V_{out} = V_Z + V_{BE} = V_{in} - I R_{SE} \quad ...(16.72)$$

Since both V_Z and V_{BE} remain nearly constant so output voltage V_{out} remains nearly constant. This is explained below :

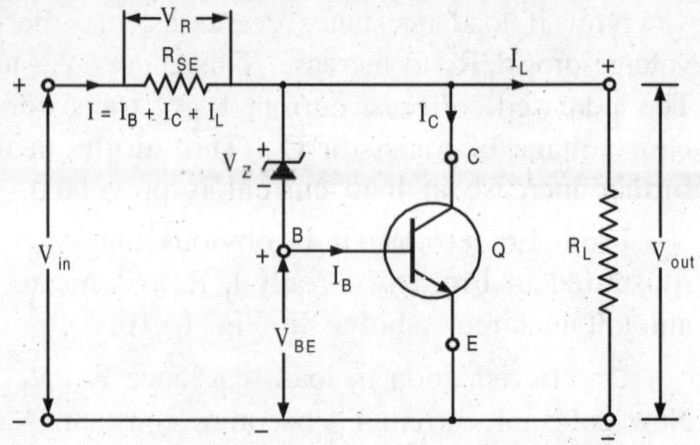

Fig. 16.19. *Transistor Shunt Voltage Regulator*

If the input (or supply) voltage increases, it causes increase in V_{out} and V_{BE} resulting in increase in base current I_B and therefore, increase in collector current I_C ($I_C = \beta I_B$). Thus with the increase in supply voltage, supply current I increases causing more voltage drop in series resistance R_{SE} and thereby reducing the output voltage. This decrease in output voltage is enough to compensate the initial increase in output voltage. Thus output voltage remains almost constant. Reverse happens should the supply voltage decrease.

If the load resistance R_L decreases, output current I_L increases and this increase in output current is supplied by decrease in base and collector currents I_B and I_C. Thus the input current I remains almost constant causing no change in voltage drop across series resistance R_{SE}. Thus output voltage V_{out} being the difference of supply voltage (fixed) and series resistor drop V_R (fixed) remains constant. Reverse happens should the load resistance increase.

Limitations. (*i*) There is considerable power loss in series resistor R_{SE}.

(*ii*) A large proportion of supply current I flows through the transistor rather than to load.

(*iii*) There are problems of overvoltage protection in the circuit.

For the above reasons, a series voltage regulator is preferred over the shunt voltage regulator.

2. Improved Transistor Shunt Voltage Regulator. An improved transistor shunt voltage regulator circuit is illustrated in Fig. 16.20. The zener diode provides a reference voltage in order that the voltage across R_1 senses the output voltage. As the output voltage V_{out} tends to change, the current shunted by transistor Q_1 is varied in order to maintain the output voltage constant. Transistor Q_2 provides a larger base current to transistor Q_1 than the circuit shown in Fig. 16.19, so that the regulator handles a larger load current. The output voltage is set by the zener voltage and that across the two transistor base emitters.

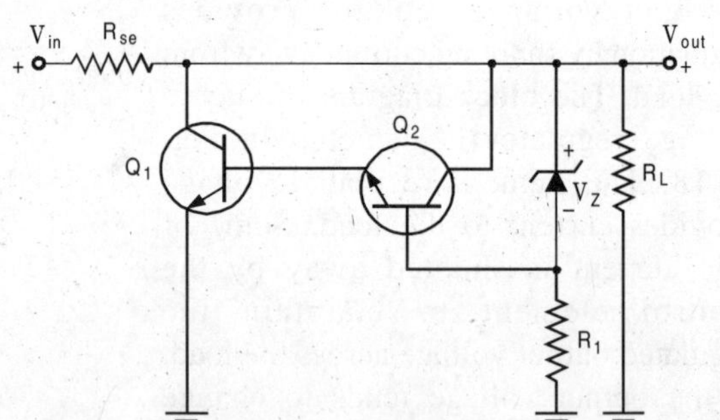

Fig. 16.20. *Improved Transistor Shunt Voltage Regulator Circuit*

$$V_{out} = V_Z + V_{BE_2} + V_{BE_1} \quad ...(16.73)$$

16.7.3. Limitations of Transistorized Power Supplies

The stabilized voltage that can be obtained from transistorized power supplies is limited to about 30 – 40 V only, because maximum safe value of V_{CE} is about 50 V. This puts a limit to the use of transistorized power supplies.

HIGHLIGHTS

1. Half-wave and full-wave rectifier circuits convert an ac signal to an approximate dc signal. A dc power supply, which is used to bias electronic circuits and systems, utilizes such circuits.
2. A half-wave rectifier is characterized by an excessive ripple, small values of rectification efficiency and transformer utilisation factor (TUF) and dc saturation of transformer core.
3. The centre-tapped and bridge type full-wave rectifiers improve on the half-wave characteristics but the three-phase, full-wave rectifier is even better.
4. Ripple factor is defined as the ratio of the effective value of the ac components of voltage (or current) present in the output from the rectifier to the direct or average value of the output voltage (or current). It is an indication of how successful is a circuit in converting ac to dc.

   Ripple factor for half-wave is 1.21 whereas for full-wave rectifier it is 0.482.
5. Rectification efficiency is defined as the ratio of dc output power to the ac input power from the transformer secondary $\left(i.e.\ \eta = \frac{P_{dc}}{P_{ac}}\right)$ and is a figure of merit to compare rectifiers.

   Rectification efficiency is 40.6% in case of half-wave rectifiers and 81.2% in case of full-wave rectifiers.
6. Transformer utilisation factor is defined as the ratio of power delivered to the load and ac rating of transformer secondary

$$i.e.\ \text{TUF} = \frac{P_{dc}}{P_{ac\ (rated)}} = \frac{I_{dc}^2 R_L}{V_{S\ rms} I_{rms}}.$$

   It is used in determination of the necessary rating of a transformer for a given dc load.

   TUF for half-wave, centre-tap and bridge rectifiers are 0.286, 0.692 and 0.812 respectively.
7. The variation of dc output voltage as a function of dc load current is called regulation. Percentage regulation is given as

$$\%\ \text{voltage regulation} = \frac{V_{NL} - V_{FL}}{V_{FL}} \times 100$$

   It is an indication of how the output voltage of a power supply varies with load.
8. The internal resistance of a power supply, given by

$$R_{in} = \frac{V_{NL} - V_{FL}}{I_{FL}}$$

   is due to the resistance of the diodes and transformer winding.
9. It is very important that the PIV rating of a diode be checked for a particular application. Simply determine the maximum voltage across the diode under reverse-bias conditions, and compare it to the name plate rating. For the typical half-wave and full-wave bridge rectifiers, it is the peak value of the applied signal. For the centre-tapped transformer full-wave rectifier, it is twice the peak value.
10. A regulated power supply is an electronic circuit designed to provide constant dc voltage of predetermined value across load terminals, which is independent of variations in load current, ac mains voltage and temperature.
11. A transformer, a rectifier circuit, a filter circuit, a voltage regulator along with some current limiting arrangement constitute a regulated power supply.

EXERCISES

1. Sketch the circuit for a half-wave rectifier. Explain its operation and derive the expression for dc current.
2. What is rectification ? Draw the circuit diagram of a half-wave rectifier and explain its working giving input and output waveforms. Derive an expression for the efficiency of half-wave rectifier.
3. Explain the terms ripple, regulation, transformer utilization factor (TUF) and PIV as referred to rectifier circuits.

4. Discuss the working of a full-wave rectifier.

[Pb. Technical Univ. Basic Electrical and Electronics Engineering December 2008]

5. Explain the working of a full-wave rectifier using centre-tapped-transformer. What is the PIV of this rectifier ?
6. Draw the circuit diagram of a full-wave rectifier circuit and explain its working.

[Pb. Technical Univ. Electronic Devices and Circuits 2004-05]

7. Draw the circuit diagram of a full-wave rectifier using two diodes and calculate (*i*) I_{dc}, (*ii*) I_{rms}, (*iii*) PIV rating of diode.
8. Draw the circuit diagram of a bridge rectifier and explain its operation with waveforms. Derive expressions for its rectification efficiency and ripple factor.
9. Explain working of P-N junction diode as full-wave rectifier.

[Pb. Technical Univ. Basic Electrical and Electronics Engineering May 2008]

10. Explain the working of a full-wave bridge rectifier. Derive the expression for ripple factor.

[Pb. Technical Univ. Basic Electrical and Electronics Engineering May 2007]

11. Explain different parameters of power supply.

    Draw the block diagram of regulated power supply and explain the function of each block.
12. Draw the circuit and explain the working of series pass transistor voltage regulator.
13. Draw shunt voltage regulator and explain its working.
14. Write short notes on the following :
    (*i*) Half-wave rectifier.
    (*ii*) Centre-tapped rectifier.
    (*iii*) Bridge rectifier.
    (*iv*) Voltage regulator.
    (*v*) Transistor series regulator.
    (*vi*) Transistor shunt regulator.

SHORT ANSWER TYPE QUESTIONS WITH ANSWERS

Q. 1. What is a dc power supply ?

Ans. The part of the equipment that converts ac into dc is called the dc power supply.

Q. 2. What is meant by rectifier ?

Ans. A rectifier is a device which converts alternating current (or voltage) into unidirectional current (or voltage).

Q. 3. Enumerate the properties that the junction diodes designed for use as rectifiers should possess.

Ans. Junction diodes designed for use in rectifiers should have the properties : (*i*) V-I characteristics as close as possible to that of an ideal diode (*ii*) Negligible forward resistance (*iii*) Negligible reverse current *i.e.* large reverse resistance (*iv*) Large reverse breakdown voltage and (*v*) Small offset voltage V_0 in the forward direction.

Q. 4. Define peak inverse voltage (PIV) of a diode. What is the difference between PIV of a center tapped FWR and bridge rectifier ?

Ans. Peak inverse voltage (PIV) is the maximum possible voltage that occurs across a diode when it is reverse biased.

PIV of a center-tapped FWR is double of that of bridge rectifier.

Q. 5. What is the importance of peak inverse voltage ?

Ans. If the applied voltage in reverse biased condition exceeds peak inverse voltage (PIV) rating of the diode, then the diode may get damaged.

Q. 6. Why half-wave rectifiers are generally not used in dc power supply ?

Ans. The type of supply available from half-wave rectifier is not satisfactory for general power supply. That is why it is generally not used in dc power supply.

Q. 7. Why diodes are not operated in the breakdown region in rectifiers ?

Ans. In breakdown region, a diode has a risk of getting damaged or burnt because the magnitude of current flowing through it increases in an uncontrollable manner. That is why diodes are not operated in the breakdown region in rectifiers.

Q. 8. What is rectification efficiency of a full-wave rectifier ?

Ans. $\eta = \dfrac{0.812}{1 + \dfrac{2R_F}{R_L}}$ *i.e.* maximum value is 81.2%

Q. 9. What is ripple factor ? **[Pb. Technical Univ. May 2005]**

Ans. The pulsating output of a rectifier can be considered to contain a dc component and ac components called the *ripples*. The ripple current is undesirable and its value should be the smallest possible in order to make the rectifier effective.

The ripple voltage or current is measured in terms of the ripple factor which is defined as the ratio of the effective value of the ac components of voltage (or current) present in the output from the rectifier to the direct or average value of the output voltage (or current).

Ripple factor is an indication of how successful is a circuit in converting ac to dc. It helps in determination of ac component in rectified output.

Q. 10. Define ripple factor in rectifiers and give its significance. **[P.T.U. B.E.E.E., December 2008]**

Ans. Refer to Answer of Q. 9.

Q. 11. What is transformer utilisation factor ?

Ans. Transformer utilisation factor is defined as the ratio of power delivered to the load and ac rating of secondary of supply power transformer.

Q. 12. The output of a 60 Hz full-wave bridge rectifier has a 60 Hz ripple. Is the circuit working properly ?

Ans. A full-wave rectifier with 60 Hz input must have lowest ripple frequency equal to twice the input frequency *i.e.* 120 Hz. If the ripple frequency is 60 Hz, it means some diodes in the circuit are not working.

Q. 13. What is a regulated power supply ?

Ans. A regulated power supply is an electronic circuit designed to provide constant dc voltage of predetermined value across load terminals, which is independent of variations in load current, ac mains voltage and temperature.

Q. 14. What is the need of voltage regulator in a power supply ? **[P.T.U. B.E.E.E. May 2009]**

Ans. Voltage regulator is provided in regulated power supply to maintain output voltage constant regardless of variations in load current and ac mains voltage.

Q. 15. What is practical importance of voltage regulation in power supply ? **[P.T.U. Dec. 2005]**

Ans. Since the output voltage available from an unregulated power supply varies with the variations in load current, ac mains voltage and temperature and variations in dc output voltage may cause inaccurate or erratic operation or malfunctioning of many electronic circuits. Unregulated power supply is not considered suitable for many of the applications in electronics. That is why voltage regulation is of utmost importance in power supplies.

Q. 16. What is a regulated power supply and what are its parts ?

Ans. A transformer, a rectifier circuit, a filter circuit, a voltage regulator along with some current limiting arrangement constitute a regulated voltage supply. Refer to Answer of Q. 13 also.

Q. 17. What do you mean by the parameter 'load regulation' ?

Ans. The load regulation is the change in regulated output voltage when the load current changes from minimum to maximum value *i.e.*

Load regulation, LR = No-load voltage, V_{NL} – Full-load voltage V_{FL}.

It is usually expressed in percentage and is given as

$$\%LR = \frac{V_{NL} - V_{FL}}{V_{FL}} \times 100$$

Q. 18. Define line and load regulation. **[P.T.U. Analog Electronics December 2007]**

Ans. The source or line regulation is defined as the change in regulated output voltage for a specified range of line voltage, typically 230 V ± 10%.

The defining equation is

$$SR = V_{HL} - V_{LL} \quad ...(i)$$

where V_{HL} is output voltage with high input ac line voltage and V_{LL} is output voltage with low input ac line voltage.

The percentage of source regulation is given as

$$\%SR = \frac{V_{HL} - V_{LL}}{\text{Nominal load voltage}} \times 100$$

$$= \frac{SR}{V_{nominal}} \times 100 \quad ...(ii)$$

The output voltage under specific operating conditions is known as the nominal load (or output) voltage.

Q. 19. What is the function of a voltage regulator ?

Ans. The primary function of a voltage regulator is to maintain a dc output voltage constant regardless of variations in load current or ac mains voltage.

Q. 20. What are the limitations of a regulator using zener diodes ?

Ans. The efficiency of zener regulated power supply becomes very low when the load current is high. The other drawback is the shift in breakdown voltage due to change in temperature unless suitable compensation techniques are not used, the reference voltage level will shift according to temperature and this may affect the voltage regulation.

Q. 21. Why pass-transistor is so called ?

Ans. Collector and emitter terminals of a transistor connected in a transistor series voltage regulator are in series with the load and whole load current pass through it, so it is called a pass-transistor.

Q. 22. Why emitter follower regulator is better than zener regulator ?

Ans. The advantage of an emitter follower regulator is that the changes in zener current are reduced by a factor β and thus the effect of zener effect is largely reduced and much more stabilised output is obtained.

Q. 23. What is the need of foldback current limiting ?

Ans. A problem with the simple current limiting, provided for protection against overload/short circuit, is that a large amount of power is dissipated in the series-pass resistor while the regulator remains short circuited. The foldback current limiting is the solution of this problem.

PROBLEMS

1. A half-wave rectifier has a load resistance of 3.5 kΩ. If the diode and secondary of the transformer have a total resistance of 800 Ω and the input voltage has an ac signal of 240 V (peak value), determine (*i*) peak, average and rms values of current flowing (*ii*) dc power output (*iii*) ac power input (*iv*) rectification efficiency and (*v*) ripple factor.

[Ans. (*i*) 55.81 mA, 17.766 mA; 27.91 mA (*ii*) 1.1 W (*iii*) 3.35 W (*iv*) 32.85% (*v*) 1.21]

2. A half-wave rectifier, having a diode of resistance 1,000 Ω and a load of 1,000 Ω rectifies an ac voltage of 310 V rms value. Calculate :

(*i*) peak, average and rms values of current (*ii*) dc power output (*iii*) ac power input (*iv*) efficiency.

[Ans. (*i*) 0.219 A; 0.07 A; 0.1096 A (*ii*) 4.868 W (*iii*) 24.025 W (*iv*) 20.26%]

3. In a power supply the dc output voltage drops from 44 V with no load to 42 V at full load. Calculate the percentage of voltage regulation. **[Ans.** 4.76%]

4. A half-wave rectifier employs a diode having a forward resistance of 10 Ω. If the input voltage to the rectifier circuit is 12 V rms, find the dc output voltage at a load of 100 mA and PIV. (Fig. 1) **[Ans.** 4.4 V; 17 V]

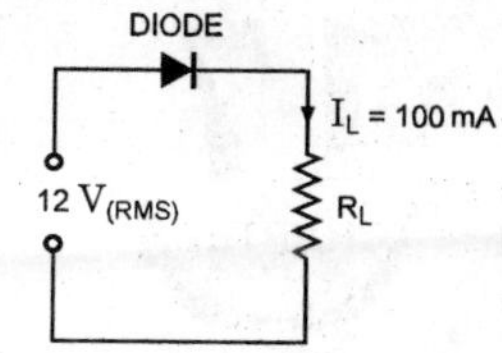

Fig. 1

5. A full-wave rectifier uses a double diode with each element having a constant forward resistance of 500 Ω. The transformer rms secondary voltage from the centre tap to each plate is 300 V and the load has a resistance of 2.5 kΩ. Determine (*i*) dc output power (*ii*) ac input power (*iii*) the rectification efficiency and (*iv*) voltage regulation from no-load to full-load.

[Ans. (*i*) 20.3 W (*ii*) 30 W (*iii*) 67.67% (*iv*) 20%]

6. A single phase full-wave rectifier uses two diodes, the internal resistance of each being 20 Ω. The transformer rms secondary voltage from centre tap to each end of secondary is 50 V and load resistance is 980 Ω. Find (*i*) the mean load current (*ii*) rms load current and output efficiency.

[Ans. (*i*) 45 mA (*ii*) 50 mA; (*iii*) 79.38%]

7. In a bridge rectifier circuit the peak value of secondary voltage is $240\sqrt{2}$ V and frequency is 50 Hz. Determine the no-load output dc voltage (*ii*) PIV and (*iii*) output frequency.

[Ans. (*i*) 216 V (*ii*) 339.4 V (*iii*) 100 Hz]

8. What is transformer utilization factor ? Determine the rating of a transformer to deliver a 100 watts of dc power to a load under full-wave rectifier.

[Ans. 144.5 W in case centre-tap transformer rectifier, and 123.2 W in case of bridge rectifier]

9. A centre-tapped transformer has a 220 V primary winding and a secondary winding rated at 12-0-12 V and is used in a full-wave rectifier circuit with a load of 100 Ω. What is the dc output voltage, dc load current and the PIV rating required for diodes ? **[Ans.** 10.8 V; 108 mA; 33.94 V]

10. A 220 V, 50 Hz ac voltage is applied to the primary of 4 : 1 stepdown transformer, which is used in bridge rectifier, having a load resistance of 1 kΩ. Assuming the diodes to be ideal, determine the following :

(*i*) DC output voltage (*ii*) DC power delivered to load
(*iii*) PIV of each diode (*iv*) Output frequency.

[Ans. (*i*) 49.52 V; (*ii*) 2.452 W; (*iii*) 77.78 V; (*iv*) 100 Hz]

11. A dc voltage supply provides 60 V when the output is unloaded. When connected to a load the output drops to 56 V. Calculate the value of voltage regulation.

[Pb. Technical Univ. Electronic Devices and Circuits 2005] **[Ans.** 7.14%]

12. A full-wave rectifier has a peak output voltage of 20 volts at 50 Hz and uses a shunt capacitor filter with C = 25 μF. The connected load is of 5 kΩ. Determine (*i*) dc load current (*ii*) dc output voltage (*iii*) ripple voltage and (*iv*) ripple factor. **[Ans.** (*i*) 3.846 m A (*ii*) 19.23 V (*iii*) 1.538 V (*iv*) 0.023]

❑❑❑

17

CHAPTER

Number Systems

17.1. INTRODUCTION

INSIDE THIS CHAPTER

We are familiar with the number system in which count is up to ten numerals. So its base is 10 and ten numerals 0, 1, 2...9 are used for expressing any arbitrary number. This number system is popularly known as the *decimal number system*. For writing any number exceeding 9, an array of numerals (digits) are used. Each position of numeral is assigned a well defined meaning. Decimal numbers are used to represent quantities outside the digital system. There are some other systems also to represent numbers. Some of the other commonly used number systems are: *binary, octal* and *hexadecimal number systems*. The *binary number system* is the most important one in digital systems, but others are also important. Binary (base-2) number system is extensively used in digital systems like digital computers which operate on binary information. Octal (base-8) system has certain advantages in digital work as it needs less circuitry to get information into and out of a digital system. Moreover, it is easier to read, record and print out octal numbers than binary numbers. Hexadecimal (base-16) number system is particularly suited for microcomputers.

17.2. DECIMAL NUMBER SYSTEM

The decimal number system, as already mentioned above, has a base of 10 and is a *position-value system*. The statement 'the decimal number system has a base of 10' implies that it contains ten unique symbols (or digits) *i.e.* 0, 1, 2, 3, 4, 5, 6, 7, 8 and 9. The ten digits do not limit us to express only ten different quantities because we use the various digits in appropriate positions within a number to indicate the magnitude of the quantity. For expressing quantities exceeding nine, two or more digits are used, and the position of each digit within the number indicates the magnitude it represents. The placement of the digits in the sequenced order from right to left is understood to carry a specific meaning. For instance consider the number 6,523 (six thousand five hundred and twenty three). The total value of this number is obtained by adding 3 unit values, 2 tens, 5 hundreds, and 6 thousands. Expressed more formally, it can be written as

$$6523 = 6 \times 10^3 + 5 \times 10^2 + 2 \times 10^1 + 3 \times 10^0$$

The figure 10 plays the role of a *base* or *radix*, hence the name *decimal number system*. The powers of 10 are incremented from zero by one as the position is moved successively toward the left from the decimal point. Furthermore, the coefficients of the powers of 10 are allowed to take on values from 0 to 9.

It will be noted that in number 6523, 3 is the *least significant digit* (LSD) whereas 6 is the *most significant digit* (MSD).

17.2.1. 9's Complements

The 9's complement of a decimal number is determined by subtracting each digit in the number from nine. For example the 9's complements of 7, 25, 321, 2435 and 57364 are 9–7 *i.e.* 2, 99–25 *i.e.* 74, 999–321 *i.e.* 678, 9999–2435 *i.e.* 7564 and 99999–57364 *i.e.* 42635 respectively.

17.2.2. 9's Complement Subtraction

The usefulness of the 9's complement stems from the fact that subtraction of a smaller decimal number from a larger one can be accomplished by adding the 9's complement of the subtrahend (in this case the smaller number) to the minuend and then adding the carry to the result (called an end-around carry). While subtracting a larger number from a smaller one, there is no carry, and the result is in 9's complement form and negative. This process has a distinct advantage in certain types of arithmetic logic. The 9's complement method of subtraction will be understood well from the following example.

Example 17.1. Perform the following subtraction by using the 9's complement method. (*a*) 2 from 7 (*b*) 08 from 14 (*c*) 235 from 572 and (*d*) 485 from 228.

Solution : *Regular Subtraction* *9's Complement Subtraction*

	Regular Subtraction	9's Complement Subtraction		
(*a*)	7	7		
	– 2	+ 7		9's complement of 2
	5	① 4		
		+ 1		Add carry to result
		5	**Ans.**	
(*b*)	14	14		
	– 08	+ 91		9's complement of 08
	6	① 05		
		+ 1		Add carry to result
		6	**Ans.**	
(*c*)	572	572		
	– 235	+ 764		9's complement of 235
	337	① 336		
		+ 1		Add carry to result
		337	**Ans.**	
(*d*)	228	228		
	– 485	514		9's complement of 485
	– 257	742		9's complement of result
		↓		
		– 257 **Ans.**		(No carry indicates that the answer is negative and in complement form)

17.2.3. 10's Complement

The 10's complement of a decimal number is equal to 9's complement plus 1. For example the 10's complements of 7, 25, 321, 2435 and 57364 are 3, 75, 679, 7565 and 42636 respectively (the 9's complements being 2,74, 678, 7564 and 42635 respectively, determined in Art. 17.2.1).

17.2.4. 10's Complement Subtraction

The 10's complement can be used to perform subtraction by adding the minuend to the 10's complement of the subtrahend and dropping the carry, as illustrated in the following example (Example 17.2).

Example 17.2. Perform the following subtraction by using the 10's complement method. (*a*) 2 from 7 (*b*) 08 from 14 (*c*) 235 from 572 and (*d*) 485 from 228.

Solution :

	Regular subtraction	*10's complement subtraction*	
(*a*)	7	7	
	– 2	+ 8	10's complement of 2
	5	× 5 **Ans.**	Drop carry
(*b*)	14	14	
	– 08	+ 92	10's complement of 08
	6	× 06 **Ans.**	Drop carry
(*c*)	572	572	
	– 235	+ 765	10's complement of 235
	337	× 337 **Ans.**	Drop carry
(*d*)	228	228	
	– 485	+ 515	10' complement of 485
	– 257	743	10's complement of result
		↓	(No carry indicates that
		– 257 **Ans.**	the answer is negative
			and in 10's complement form)

Example 17.3 Using 10's complement, subtract 72532 – 3250.

Solution :

72532	
96750	10's complement of 03250
×69282	Drop carry

i.e. 72532 – 3250 = 69282 **Ans.**

Example 17.4. Perform the subtraction with the following decimal numbers using 10's complement and 9's complement. (*i*) 5249 – 749, (*ii*) 27 – 289.

Solution : (*i*) Regular subtraction 5249 – 749 = 4500

9's complement

5249	
+ 9250	(9's complement of 0749 = 9999 – 0749 = 9250)
14499	Add carry to result
4500 **Ans.**	

10's complement

5249	
+ 9251	(10's complement of 0749)
14500	
4500 **Ans.**	Drop carry

(*ii*) Regular subtraction 27 – 289 = –262

9's complement

27	
+710	(9's complement of 289 = 999 – 289 = 710)
737	(No carry indicates that result is –ve and in 9's complement form)
–262 **Ans.**	

10's complement

27	
+ 711	(10's complement of 289)
738	(No carry indicates that result is –ve and in 10's complement form)
–262 **Ans.**	

17.3. BINARY NUMBER SYSTEM

The binary number system is simply another way to count. It is less complicated than the decimal number system because it is composed of only two digits 0 and 1. Just as the decimal system with its ten digits is a base-10 system, the binary system with its two digits is a base-2 system. These two digits namely 0 and 1 carry the same meaning as in the decimal system but different meaning is given to the position occupied by the numeral or digit. In a binary system, the base is 2, so any number can be expressed by an array of numerals representing the coefficients of power of 2 and not 10. In a binary system, the weight of each successively higher position (to the left) is an increasing power of 2.

17.3.1. Counting in Binary

To learn to count in binary, let us first look how we count in decimal number system. We start counting from 0, 1, 2, 3 and up to 9. After 9 we write 0 with 1 on its left side *i.e.* 10. Now we have two digit number. To continue counting, first digit of number *i.e.* 0 is again changed up to 9 and after that it is written 0 and second digit of number (which was earlier 1) is written 2. Thus we get number 20. This way we count the number up to 99. In further counting the number digit 99 is changed to 00 with digit 1 on the left side of these digits added. So we get number 100. This is the way of counting and writing down the numbers in decimal number system.

TABLE 17.1. ***Table Showing Equivalent Binary Numbers of Decimal Numbers***

Decimal Number	*Equivalent Binary Number*	*Decimal Number*	*Equivalent Binary Number*
0	0000	8	1000
1	0001	9	1001
2	0010	10	1010
3	0011	11	1011
4	0100	12	1100
5	0101	13	1101
6	0110	14	1110
7	0111	15	1111

A comparable situation occurs when counting in binary, except that we have only two digits. Counting is started with zero and ended with next number 1. After 1 there is no other digit in this system so in further counting, digit 1 is replaced by 0 with digit 1 on its left side *i.e.* 10. Next number follows 11. Again after number 11 digit 00 is written with digit 1 on the left side *i.e.* 100. This way we count in binary system. Table 17.1 shows the equivalent binary numbers of decimal numbers from 0 to 15. This Table 17.1 makes us more clear about binary number system.

Some terms like bit, nibble and byte are used in binary number system. *Bit* is used for a single binary digit. A binary number with four bits is called a *nibble*. When binary number has eight bits, then it is called a *byte*. Binary numbers have numerous advantages in digital computer's use.

1. An ordinary switch has two states either open or close which can be used to represent binary numbers.
2. When current flows through a wire, magnetic flux is created across the wire. Direction of flux depends upon the direction of flow of current which can be determined by right hand rule. As the current can flow either in outward or in inward direction and, therefore, flux created would be either in counter-clockwise or clockwise direction so a magnetic core can be used to represent binary numbers, because of its two state operation. In earlier computers, thousands of magnetic cores were used to store the binary instructions and data.
3. A card with punched holes can also be used to represent the binary numbers. A card has two states, one state is without punching and the other is with punched holes in it. So it is a *two-state device*. Binary data and instructions, are stored in the card with the help of a card punching machine which punches the card in a prearranged code, as shown in Fig. 17.1.

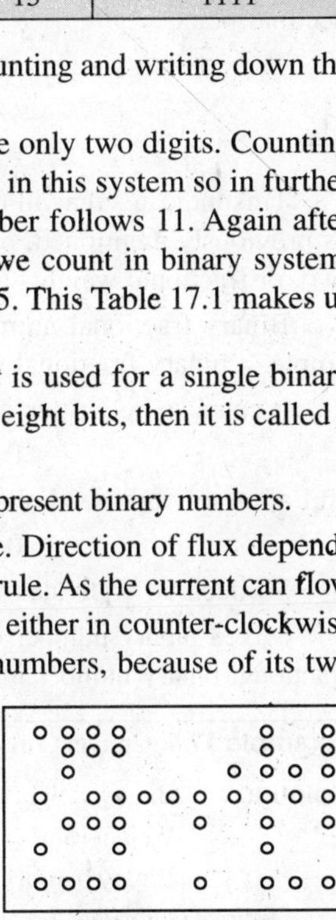

Fig. 17.1. ***Punched Card***

4. Like the punched card, a magnetic tape may also be used as a two state device. Some points of magnetic tape are magnetised while leaving other points unmagnetised. A row of the points represents data or instructions. So a magnetic tape reel can be used to store thousands of binary data and instructions.

5. Use of transistor as two state device has already been discussed. Two states of transistors *i.e. cutoff* and *saturation* are used to produce an output of a transistor either *low* or *high*.

17.3.2. Binary-To-Decimal Conversion

The binary number system is a positional system where each binary digit (bit) carries a certain weight based on its position relative to the least significant bit (LSB). The right-most bit is the LSB in a binary number and has a weight of $2^0 = 1$. Any number can be converted to its decimal equivalent simply by adding the products of each bit and its weight. For example

$$\begin{aligned} 10011 &= 1\,(2^4) + 0\,(2^3) + 0\,(2^2) + 1\,(2^1) + 1\,(2^0) \\ &= 16 + 0 + 0 + 2 + 1 = 19 \end{aligned}$$

Example 17.5. Convert $(110110)_2$ to its decimal equivalent.

Solution :

Binary weight :	2^5	2^4	2^3	2^2	2^1	2^0
Weight value :	32	16	8	4	2	1
Binary number :	1	1	0	1	1	0

So $(110110)_2 = 1 \times 32 + 1 \times 16 + 0 \times 8 + 1 \times 4 + 1 \times 2 + 0 \times 1 = (54)_{10}$ **Ans.**

The subscripts 2 and 10 identify that they are respectively binary and decimal numbers.

17.3.3. Binary Fractions

So far, we have discussed binary *integers* (whole numbers). Fractional numbers can also be represented in binary by placing bits to the right of the binary point, just as fractional decimal digits are placed to the right of the decimal point.

The column weights of a binary number are

$$2^n \ldots 2^2\, 2^1\, 2^0 \bullet 2^{-1}\, 2^{-2}\, 2^{-3} \ldots 2^{-n}$$

binary point (↑ at •)

This indicates that all the bits to the left of the binary point have weights that are positive powers of two, as previously mentioned, but all bits to the right of the binary point have weights that are negative power of two, or fractional weights ($2^{-1} = 1/2^1$ etc.).

Binary fractional number can also be converted into decimal fractional numbers. Let us take, for example, a binary fractional number 0.1001.

It can also be written as

$$0.1001 = 1\,(2^{-1}) + 0\,(2^{-2}) + 0\,(2^{-3}) + 1\,(2^{-4})$$

and after solving the above equation, we get

$$0.1001 = 1(0.5) + 0(0.25) + 0(0.125) + 1(0.0625) = 0.5625$$

So decimal equivalent of binary fractional number 0.1001 is 0.5625.

Mixed binary number can be converted into a decimal number by converting integer binary number and fractional binary number individually as described above.

Example 17.6. Convert the binary fractional number 0.1101 into its decimal equivalent.

Solution :

Binary weight	2^{-1}	2^{-2}	2^{-3}	2^{-4}
Weight value	0.5	0.25	0.125	0.0625
Binary number	1	1	0	1

Decimal equivalent $1 \times 0.5 + 1 \times 0.25 + 0 \times 0.125 + 1 \times 0.0625 = 0.8125$ **Ans.**

Example 17.7. Convert the binary number 110.111 into its equivalent decimal number.

Solution :

$$\begin{aligned} 110.111 &= 1 \times 2^2 + 1 \times 2^1 + 0 \times 2^0 + 1 \times 2^{-1} + 1 \times 2^{-2} + 1 \times 2^{-3} \\ &= 4 + 2 + 0 + 0.5 + 0.25 + 0.125 = 6.875 \text{ Ans.} \end{aligned}$$

Example 17.8. Convert the following :

$$(1011.0011)_2 = (\quad)_{10}$$

Solution : $(1011.0011)_2 = 1 \times 2^3 + 0 \times 2^2 + 1 \times 2^1 + 1 \times 2^0 + 0 \times 2^{-1} + 0 \times 2^{-2} + 1 \times 2^{-3} + 1 \times 2^{-4}$

$$= 8 + 2 + 1 + \frac{1}{8} + \frac{1}{16} = (11.1875)_{10} \textbf{ Ans.}$$

17.3.4. Decimal-To-Binary Conversion

For the conversion of a decimal number into a binary number, decimal is divided by 2 progressively and remainder is written down after each division. The remainder is then taken in a reverse order and it is the binary equivalent. This method of conversion is usually called the *double dabble method.*

Let us, for example, convert decimal number 19 into binary number to make above method clear.

Firstly divide 19 by 2

$$\begin{array}{r|l} & 9 \\ \hline 2 & 19 \\ & 18 \\ \hline & 1 \end{array} \qquad \text{First remainder} = 1$$

This division gives quotient 9 with remainder 1. Now again divide 9 by 2

$$\begin{array}{r|l} & 4 \\ \hline 2 & 9 \\ & 8 \\ \hline & 1 \end{array} \qquad \text{Second remainder} = 1$$

Second division gives quotient 4 with remainder 1. Again divide 4 by 2

$$\begin{array}{r|l} & 2 \\ \hline 2 & 4 \\ & 4 \\ \hline & 0 \end{array} \qquad \text{Third remainder} = 0$$

Third division gives quotient 2 with remainder 0. Again divide 2 by 2

$$\begin{array}{r|l} & 1 \\ \hline 2 & 2 \\ & 2 \\ \hline & 0 \end{array} \qquad \text{Fourth remainder} = 0$$

Fourth division gives quotient 1 with remainder 0. Again divide 1 by 2

$$\begin{array}{r|l} & 0 \\ \hline 2 & 1 \\ & 0 \\ \hline & 1 \end{array} \qquad \text{Last remainder} = 1$$

In this last division remainder is 1. Now write down all remainders in reverse order to get binary equivalent of decimal number 19. So binary equivalent is 10011.

Fractional decimal numbers are converted into binary numbers, multiplying it by 2. Carry in the integer position of the result is recorded and carry of the decimal position is again multiplied by 2. Again the process is repeated. The carries of integer position recorded are written downward, which gives fractional binary number equivalent of fractional decimal number.

Let us, for example, convert fractional decimal number 0.65 into fractional binary number to make the method described above clear.

0.65	$\times 2 = 1.3 = 0.3$	with a carry of 1
0.3	$\times 2 = 0.6 = 0.6$	with a carry of 0
0.6	$\times 2 = 1.2 = 0.2$	with a carry of 1
0.2	$\times 2 = 0.4 = 0.4$	with a carry of 0
0.4	$\times 2 = 0.8 = 0.8$	with a carry of 0
0.8	$\times 2 = 1.6 = 0.6$	with a carry of 1
0.6	$\times 2 = 1.2 = 0.2$	with a carry of 1
0.2	$\times 2 = 0.4 = 0.4$	with a carry of 0

Now write down all the carries downward. It gives fraction binary number 0.10100110. Thus equivalent fraction binary number of 0.65 is 0.10100110.

In above conversion we stopped after eight binary digits and, therefore, this result is approximate. We can continue multiplying by 2 until we get as many digits as necessary for our application in order to have required accuracy.

Example 17.9. Convert $(1245)_{10}$ into binary number system.

[Pb. Technical Univ. Basic Electrical Engineering and Electronics May 2008]

Solution :

2	1245	
2	622 –	1
2	311 –	0
2	155 –	1
2	77 –	1
2	38 –	1
2	19 –	0
2	9 –	1
2	4 –	1
2	2 –	0
2	1 –	0
	0 –	1

Now writing down all remainders in reverse order we get the binary equivalent of decimal number 1245. Thus

$$(1245)_{10} = (10011011101)_2$$ **Ans.**

Example 17.10. Convert the following number : show all calculation steps : $(0.0625)_{10} = (?)_2$.

Solution :

$0.0625 \times 2 = 0.125$ with a carry of 0
$0.125 \times 2 = 0.25$ with a carry of 0
$0.25 \times 2 = 0.5$ with a carry of 0
$0.5 \times 2 = 1.0$ with a carry of 1

Now writing down all the carries downward, we get the binary equivalent of decimal number 0.0625. Thus

$$(0.0625)_{10} = (0.0001)_2$$ **Ans.**

Example 17.11. Convert the fractional decimal number $(0.625)_{10}$ into a binary number.

[Pb. Technical Univ. Basic Electrical and Electronics Engineering Dec. 2005]

Solution : $(0.625)_{10}$

$0.625 \times 2 = 0.25$ with a carry of 1

$0.25 \times 2 = 0.50$ with a carry of 0

$0.50 \times 2 = 0.0$ with a carry of 1

Hence equivalent binary number of $(0.625)_{10}$ is 0.101 **Ans.**

Example 17.12. Convert a decimal number $(100.625)_{10}$ in binary number.

[*Pb. Technical Univ. Basic Electrical and Electronics Engineering Dec. 2004*]

Solution : $(100.625)_{10}$ Fraction 0.625

Integer 100

2	100
2	50 – 0
2	25 – 0
2	12 – 1
2	6 – 0
2	3 – 0
2	1 – 1
	0 – 1

$0.625 \times 2 = 0.25$ with a carry of 1

$0.25 \times 2 = 0.50$ with a carry of 0

$0.50 \times 2 = 0.0$ with a carry of 1

Hence equivalent binary number of 0.625 is 0.101

Hence equivalent binary number of 100 is 1100100

Thus $(100.625)_{10} = (1100100.101)_2$ **Ans.**

17.3.5. Binary Operations

We are familiar with the arithmetic operations such as addition, subtraction, multiplication and division in a decimal number system. Similar operations can be performed on binary numbers, infact, binary arithmetic is much simpler than decimal arithmetic because here only two digits, 0 and 1 are involved.

In binary number system, addition is the most important of the four operations of addition, subtraction, multiplication and division. By using '*complements*', subtraction can be reduced to addition. Most digital computers subtract by complements. It leads to reduction in hardware because circuitry is required only for addition operation. Similarly, multiplication is nothing but repeated addition and, finally division is nothing but repeated subtraction.

17.3.6. Binary Addition

One method of addition of binary numbers is to convert binary numbers into decimal numbers, add them and then again convert the resultant decimal number into binary number. But this method is very time consuming and, therefore, we would discuss the direct and time saving method.

There are four cases, described below for addition of binary numbers.

$$0 + 0 = 0$$
$$0 + 1 = 1$$
$$1 + 0 = 1$$
$$1 + 1 = 10$$

With the help of the above four cases, all binary numbers can be added.

Let us, for example, add two binary numbers 10010 and 11011.

Now in this case first of all least significant digits of binary numbers are added. These digits are 0 and 1. We know that addition of 0 and 1 gives us 1. So in first column we write 1

$$\begin{array}{r} 10010 \\ 11011 \\ \hline 1 \\ \hline \end{array}$$

In second column addition of 1 and 1 gives us 10 *i.e.*, 0 with carry 1

$$\begin{array}{r} 10010 \\ 11011 \\ \hline 01 \\ \hline \end{array}$$

In the third column addition of 0 and 0 with carry 1 gives us 1

```
  10010
  11011
  -----
    101
```

This way addition of binary numbers is done which gives

```
  10010
  11011
 ------
 101101
```

Example 17.13. Complete the following operation

$$00010100 + 00111001 = (\quad)_2.$$

Solution :

```
    1 1
  00010100
 +00111001
 ----------
 (01001101)₂  Ans.
```

Example 17.14. Add $(10111010)_2$ and $(101001)_2$ without converting base.

Solution : Addition :

```
   10111010
 +   101001
 ------------
 (11100011)₂   Ans.
```

17.3.7. Binary Subtraction

Again for subtraction of binary numbers, four cases are required to be considered while subtracting

$$0 - 0 = 0$$

$$1 - 0 = 1$$

$$1 - 1 = 0$$

$$10 - 1 = 1$$

Let us, for example subtract binary number 1001 from binary number 1010.

First of all least significant column is taken for subtraction. In this case 1 is to be subtracted from zero. So 1 is borrowed from the next column of binary number 1010. Now subtraction of 1 from 10 gives 1

So

```
   1010
 - 1001
 ------
      1
```

Next we subtract 0 from 0 (1 has already been borrowed) so in second column we get 0

```
   1010
 - 1001
 ------
     01
```

Again subtraction of 0 from 0 gives us 0, therefore,

```
   1010
 - 1001
 ------
    001
```

Again subtraction of 1 from 1 gives us 0 so

```
   1010
 - 1001
 ------
   0001
```

So result of binary subtraction is 0001.

Example 17.15. Subtract 0111 from 1001. [*Pb. Univ. Digital Computations 1986*]

Solution :

```
  1001      1st column : 1 – 1 = 0
– 0111      2nd column : 0 – 1 = 1 with a borrow of 1
  0010 Ans. 3rd column : 1 (after borrow) – 1 = 0 with a borrow of 1
            4th column : 0 (after borrow) – 0 = 0
```

Example 17.16. Subtract 01011 from 10110.

Solution :

```
  10110
– 01011
  01011    Ans.
```

Example 17.17. Complete the following operation :

01001111 – 0000101 = ( )$_2$.

Solution :

```
  01001111
– 0000101
 (1001010)₂    Ans.
```

17.3.8. Binary Multiplication

Multiplication is performed in binary in the same way as with decimal numbers. It involves forming the partial products, shifting each successive partial product left one place, and then adding all the partial products. The four simple rules are as follows:

Rule 1. $0 \times 0 = 0$ Rule 2. $0 \times 1 = 0$ Rule 3. $1 \times 0 = 0$ and Rule 4. $1 \times 1 = 1$

Following example will illustrate the procedure.

Example 17.18. Perform the arithmetic operation 11.01 × 10.11.

Solution :

```
     11.01
     10.11
     -----
      1101
     1101
    ××××
   1101
 ---------
 1000.1111 Ans.
```

17.3.9. Binary Division

Binary division is obtained as decimal division. Following example will illustrate the procedure.

Example 17.19. Perform the arithmetic operation indicated

$$(11101101)_2/(11101)_2 = (?)_2$$

Solution :

```
           1000.001011
      11101)11101101
            11101
            00000101000
                 11101
                00101100
                   11101
                  0011110
                    11101
                    00001   Ans. 1000.001011
```

17.3.10. Shifting a Number To Left or Right

As in a decimal number system shifting one step to the left or right corresponds respectively to multiplication or division by 10, similarly shifting binary numbers one step to the left or right corresponds respectively to multiplication or division by 2.

When binary number 11100_2 corresponding to decimal number 28 is shifted one step to the left, it becomes 111000_2 which corresponds to decimal number 56 *i.e.* it is doubled. If the given number is shifted one step to the right it becomes 1110_2 which corresponds to decimal number 14, *i.e.* it is halved.

17.3.11. Shifting The Place Point

As in a decimal number system, moving of a decimal point one place to the right or left corresponds respectively to multiplication or division by 10, similarly shifting the binary point by one place to the right or left multiplies or divides by 2. For example 1011.0_2 corresponds to $(11)_{10}$ but 10110_2 corresponds to $(22)_{10}$ while $(101.1)_2$ corresponds to $(5.5)_{10}$.

17.4. SIGNED BINARY NUMBERS

In the decimal number system, a plus (+) sign is used for denoting a positive number and a minus (–) sign for denoting a negative number. The plus sign is normally dropped, and absence of any sign means the number is positive one. This representation of numbers is called the signed number.

In binary number system, negative binary numbers are represented in three ways : Signed magnitude representation. 1's complement representation and 2's complement representation.

17.4.1. Signed Magnitude Representation

In this representation an additional bit called the sign bit is placed as the MSB to the magnitude of the numbers. The '0' is used for '+' sign and digit $(r-1)$ is used for '–' sign where r is the base (radix) of the number system and it is $(2-1)=1$ in binary system. For example in 8 bit binary number having one sign bit and seven magnitude bits, the largest positive number is 01111111 = +127 and smallest number is 11111111 = –127.

Example 17.20. Find the decimal equivalent of the following binary numbers assuming signed magnitude representation of the number. (*a*) 001000, (*b*) 1111.

Solution : (*a*) $[001000]_2 = +[01000]_2 = +8_{10}$ **Ans.** (MSB is 0 and hence sign is +ve)

(*b*) $[1111]_2 = -[111]_2 = -7_{10}$ **Ans.** (MSB is 1 and hence sign is –ve)

17.4.2. One's Complement Representation

In a digital work, two types of complements of binary numbers viz. 1's complement and 2's complement are used for complemental subtraction.

In a binary number, if each 1 is replaced by 0 and each 0 by 1, the resulting number is known as 1's complement of first number. In fact, both numbers are complement of each other. A few examples of 1's complement are given below for illustration

Binary Number	*1's Complement*
100	011
1110	0001
10101	01010
10111	01000
11011011	00100100

17.4.3. 1's Complement Subtraction

Subtraction of binary numbers can be accomplished by using 1's complement method, which *allows us to subtract using only addition*. This method of subtraction is particularly useful in arithmetic logic circuits because subtraction can be accomplished with an *adder*. Also the 1's complement of a number is easily obtained by inverting each bit in the number.

1's complement subtraction in binary can be accomplished in the same manner as 9's complement subtraction is affected in decimal number system *i.e.* subtraction of a small binary number from a larger one can be accomplished by adding 1's complement of the smaller number to the larger number and then adding the carry to the result (called an *end-around carry*). For subtracting a larger number from a smaller one, there is no carry, and the result is in 1's complement form and negative. Following example will illustrate the procedure.

Example 17.21. Perform the subtraction with the following binary numbers using 1's complement. (*i*) 11010 – 1101 (*ii*) 100 – 1100000.

Solution :

	Direct Subtraction	*1's Complement Subtraction*	
(*i*)	11010	11010	
	– 1101	+ 0010	1's complement of 1101
	1101	(1)1100	
		1	Add end around carry
		1101	**Ans.**
(*ii*)	100	100	
	– 110000	+ 001111	1's complement of 110000
	– 101100	010011	1's complement of result
		– 101100	**Ans.** Final result (No carry indicates that the answer is negative and in complement form)

17.4.4. 2's Complement Representation

The 2's complement of a binary number is obtained by adding 1 to its 1's complement *i.e.*,

2's complement = 1's complement + 1.

The following example illustrates the procedure.

Example 17.22. Find the 1's and 2's complements of the following binary numbers. 0.101; 101.0

Solution : 1's complement of binary number 0.101 is 0.010 **Ans.**

2's complement of binary number 0.101 is

0.010 + 0.001 = 0.011 **Ans.**

1's complement of binary number 101.0 is 010 **Ans.**

2's complement of binary number 101.0 is 010 + 1 = 011 **Ans.**

17.4.5. 2's Complement Subtraction

For subtracting a smaller number from a larger one by 2's complement subtraction method, the procedure applied is as follows :

1. Determine the 2's complement of the smaller number.
2. Add the 2's complement of the smaller number to the larger one.
3. Discard the carry (there is always a carry in this case).

For subtracting a larger number from smaller one by 2's complement subtraction method, the procedure applied is as follows :

1. Determine the 2's complement of the larger number.
2. Add 2's complement of the larger number to the smaller one.
3. There is no carry. The result is in 2's complement and is negative.
4. For having result in true form, take 2's complement and change the sign.

Both of the methods of complement subtraction have distinct advantages when implemented with logic circuits because they allow subtraction to be accomplished by using only addition. Both of the 1's and the 2's complements of a binary number are relatively easy to accomplish with logic circuits; and the 2's complement has an advantage over the 1's complement is that an end-around carry operation does not have to be performed.

Example 17.23. Perform the following subtraction using 2's complement method

(1) 110100 – 10101

(2) 0011.1001 – 0001.1110

Solution : *Direct subtraction* *2's complement subtraction*

	Direct subtraction	2's complement subtraction	
(1)	110100	110100	
	–10101	01011	2's complement of 10101
	11111	111111	
		11111	**Ans.** Drop end carry
(2)	0011.1001	0011.1001	
	0001.1110	0.0010	2's complement of 0001.1110
	1.1011	11.1011	
		1.1011	**Ans.** Drop end carry

Example 17.24. Perform the subtraction with the following binary numbers using 1's complement and 2's complement. (*i*) 11010 – 1101; (*ii*) 10010 – 10011.

Solution :

(*i*) *1's complement*		*2's complement*	
11010		11010	
+ 0010	(1's complement of 1101)	+ 0011	(2's complement of 1101)
11100		11101	
1	Add end around carry		Drop end carry
1101	**Ans.**	1101	**Ans.**
(*ii*) *1's complement*		*2' complement*	
10010		10010	
+ 01100	(1's complement of 10011)	+ 01101	(2's complement of 10011)
11110	(No carry indicates that result is –ve and in 1's complement form)	11111	(No carry indicates that result is –ve and in 2's complement form)
–00001	= –1 **Ans.**	–00001	= –1 **Ans.**

17.5. DOUBLE PRECISION NUMBER

For any computer, the word length is fixed. Most present day computers are 16 bit and in a 16 bit computer, that is, in a computer with 16-bit word length, only numbers from $+2^{16-1} - 1$ (*i.e.* +32,767) to -2^{16-1} (*i.e.* –32,768) can be stored in each register. To store numbers larger than these numbers, two storage locations need to be used, that is, each such number has to be stored in two registers. This is known as *double precision system.* Leaving the MSB which is the sign bit, this allows a 31-bit number length with two 16-bit registers. The format is

First word	S	High Order Bit
Second word	0	Low Order Bit

S is the sign bit and 0 is a zero.

If still larger numbers are to be stored, three registers are used to store each number. This is known as *triple precision system.*

17.6. FLOATING POINT NUMBERS

The real numbers considered so far were represented in what is known as *fixed point notation.* In such notation, the position of the decimal is fixed for a fixed number of bits. For example, a number say 725.632 can be stored in a 6 bit register as shown in Fig. 17.2.

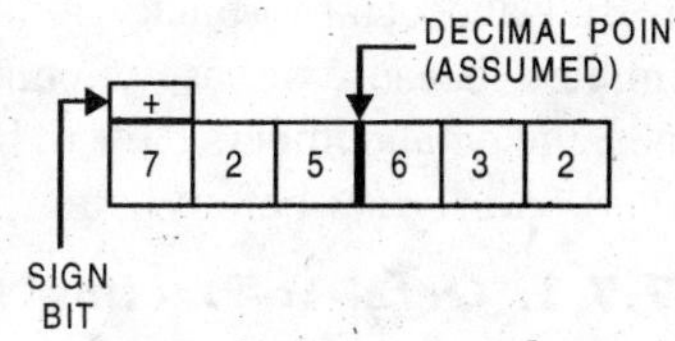

Fig. 17.2

The scientific calculations involve very large and very small numbers which cannot be expressed in fixed point notation.

In decimal system, very large and very small numbers are expressed in scientific notation, known as *floating point notation,* by stating a number (mantissa) and an exponent of 10. Examples are 3.02×10^8, 5.83×10^{-6} or 8.25×10^{-27}.

A floating point number N can be represented in the form $N = F \times b^e$, where F is any number of base *b* and *e* is exponent. For example the decimal number 55.83 can be represented as 0.5583×10^2 or 558.3×10^{-1} or 5583×10^{-2}. The number $F \times b^e$ is called a *normalized* floating point number if $1/b < F < 1$. Therefore the above decimal number 55.83 is represented 0.5583×10^2 $(1/10 < 0.5583 < 1)$ as normalized floating point number.

The floating point representation of a number have two parts, the first part is signed, fixed point number called the *mantissa* and the second part is signed exponent. For example, 55.83 is represented as

sign	*floating point number*	*sign*	*exponent*	
0	5583	0	02	$= 0.5583 \times 10^2$

Likewise binary numbers (mantissa) can also be expressed by a number and an exponent of 2. This representation of binary number is called *Floating point number.* In a 16 bit computer 10 bits are mantissa and remaining 6 bits are exponent. The mantissa is written in 2's complement form and therefore MSB is taken as sign bit. The binary point is placed after this bit (or it may be placed at the given radix point position). The remaining 6 bits are the exponent component representing 0 to 63. The binary number 100000_2 $(32)_{10}$ is added to the desired exponent for representing negative exponents. It is called the excess-32 exponent.

Example 17.25. Find out the decimal number for a floating point number 0011000000101010.

Solution :

Mantissa portion $= 0011000000 = +0.011000000$

Exponent portion $= 101010$

Actual exponent $= 101010 - 100000 = 001010 = 10_{10}$

$$\text{Number} = 0.011_2 \times 2^{10} = 11_2 \times 2^7 = 110000000_2 = 384_{10}$$

17.7. OCTAL NUMBER SYSTEM

The number system with base (or radix) eight is known as the *octal number system.* Although we can use any eight digits, it is customary to use the first eight decimal digits: 0, 1, 2, 3, 4, 5, 6, 7. There is no 8 or 9 in octal number code. These digits 0 through 7, have exactly the same physical meaning as in decimal number system. The digit positions in octal number have weights as follows.

8^4	8^3	8^2	8^1	8^0	8^{-1}	8^{-2}	8^{-3}	8^{-4}	8^{-5}

In digital systems, binary numbers are required to be entered and certain results or status signals are required to be displayed. It is highly inconvenient to handle long strings of binary numbers. It may cause errors also. Therefore, octal numbers are used for entering the binary data and displaying certain information. The octal number system requires *one-third in length* as compared to binary numbers. Thus from the users point of view, it would be comparatively much easier to handle the input and output data of a digital computer in octal form. Moreover, the printouts are more compact and easy to read. Conversion from binary-to-octal and octal-to-binary is also quick and simple. Infact, octal numbers are used to represent binary numbers because of ease of conversion and compactness. Since digital circuits can process only zeros and ones, the octal numbers have to be converted into binary form employing special circuits known as *octal-to-binary converters* before being processed by the digital circuits.

17.7.1. Octal-To-Decimal Conversion

An octal number can be easily converted to its decimal equivalent by multiplying each octal digit by its positional weight. For example

$$314_8 = 3 \times (8^2) + 1\ (8^1) + 4\ (8^0) = 192 + 8 + 4 = 204_{10}$$

Another example

$$0.734 = 7 \times (8^{-1}) + 3 \times (8^{-2}) + 4\ (8^{-3}) = 0.875 + 0.046875 + 0.0078125 = (0.9296875)_{10} \text{ or } \left(\frac{119}{128}\right)_{10}$$

Example 17.26. Convert the following :

$(1527)_8 = (\ \ )_{10}$

Solution : $(1527)_8 = 1 \times 8^3 + 5 \times 8^2 + 2 \times 8^1 + 7 \times 8^0 = 512 + 320 + 16 + 7 = (855)_{10}$ **Ans.**

Example 17.27. Convert the following numbers into desired bases : $(630.4)_8 = (?)_{10}$.

Solution : $(630.4)_8 = 6 \times 8^2 + 3 \times 8 + 0 \times 8^0 + 4 \times 8^{-1} = 384 + 24 + 0 + 0.5 = (408.5)_{10}$ **Ans.**

17.7.2. Decimal-To-Octal Conversion

A decimal integer can be converted to octal by using the same repeated division method, called the *double dabble method*, that we used in the decimal-to-binary conversion, but with a division factor of 8 rather than 2. The following examples will illustrate the procedure.

Example 17.28. Convert $(1375)_{10} = (?)_8$

Solution :

Divisor	Quotient – Remainder
8	1375
8	171 – 7
8	21 – 3
8	2 – 5
	0 – 2

So equivalent octal number of $(1375)_{10}$ is $(2537)_8$ **Ans.**

Note that first remainder becomes the least significant digit (LSD) of the total number, and the last remainder becomes the most significant digit (MSD).

Example 17.29. Convert $(0.45)_{10} = (\ \ )_8$.

Solution : $(0.45)_{10}$

$0.45 \times 8 = 0.60$ with a carry of 3

$0.60 \times 8 = 0.80$ with a carry of 4

$0.80 \times 8 = 0.40$ with a carry of 6

$0.40 \times 8 = 0.20$ with a carry of 3

$0.20 \times 8 = 0.60$ with a carry of 1

So $(0.45)_{10} = (0.34631)_8$ **Ans.**

Example 17.30. Convert the decimal number 39.75 to octal.

[*Pb. Technical Univ. Basic Electrical and Electronics Engineering May 2009*]

Solution : Integer :

$$\begin{array}{r|l} 8 & 39 \\ \hline 8 & 4-7 \\ \hline & 8-4 \end{array} \uparrow$$

Fraction : $0.75 \times 8 = 6.0$ with a carry of 6

So $(39.75)_{10} = (47.6)_8$ **Ans.**

Example 17.31. Convert the following number

$$\mathbf{(6089.25)_{10} = (\ )_8.}$$

Solution :

Integer

$$\begin{array}{r|l} 8 & 6089 \\ \hline 8 & 761-1 \\ \hline 8 & 95-1 \\ \hline 8 & 11-7 \\ \hline 8 & 1-3 \\ \hline & 0-1 \end{array} \uparrow \quad (13711)_8$$

Fraction

$0.25 \times 8 = 0$ with a carry of 2

Hence $(0.25)_{10} = (0.2)_8$

Hence $(6089.25)_{10} = (13711.2)_8$ **Ans.**

Example 17.32. Convert the following numbers to octal number :

(i) $\mathbf{(43)_{10}}$ **;** ***(ii)*** $\mathbf{(279)_{10}}$ **;** ***(iii)*** $\mathbf{(36.125)_{10}}$

Solution : (*i*)

$$\begin{array}{r|l} 8 & 43 \\ \hline 8 & 5-3 \\ \hline & 0-5 \end{array} \uparrow$$

So equivalent octal number of $(43)_{10}$ is $(53)_8$. **Ans.**

(*ii*)

$$\begin{array}{r|l} 8 & 279 \\ \hline 8 & 34-7 \\ \hline 8 & 4-2 \\ \hline & 0-4 \end{array} \uparrow$$

So equivalent octal number of $(279)_{10}$ is $(427)_8$. **Ans.**

Integer

$$\begin{array}{r|l} 8 & 36 \\ \hline 8 & 4-4 \\ \hline & 0-4 \end{array} \uparrow$$

Fraction $0.125 \times 8 = 1.0$ with a carry of 1

Thus equivalent octal number of $(36.125)_{10}$ is $(44.1)_8$ **Ans.**

17.7.3. Octal-To-Binary Conversion

Because 8 (the base of octal numbers) is the third power of 2 (the base of binary number), the conversion from octal to binary can be performed by converting each octal digit to its 3-bit binary equivalent. The eight possible digits are converted as indicated below in Table 17.2.

TABLE 17.2

Octal Digit	0	1	2	3	4	5	6	7
Binary Equivalent	000	001	010	011	100	101	110	111

Using these conversions, any octal number can be converted to binary by individually converting each digit. The following examples will illustrate the procedure.

Example 17.33. Convert the following number

$$\mathbf{(375.37)_8 = (\ )_2}$$

Solution :

$$\begin{array}{ccccccc} (375.37)_8 = & 3 & 7 & 5 & . & 3 & 7 \\ & \downarrow & \downarrow & \downarrow & & \downarrow & \downarrow \\ & 011 & 111 & 101 & & 011 & 111 \end{array}$$

Thus $(375.37)_8 = (011111101.011111)_2$ **Ans.**

Example 17.34. Convert the following : $(750.34)_8 = (\ )_2$.

Solution : $(750.34)_8 = (111\ 101\ 000.011\ 100)_2$

17.7.4. Binary-To-Octal Conversion

Conversion of a binary number to an octal number is simply the reverse of the foregoing process. The bits of the binary number are grouped into groups of three bits starting at the LSB. Then each group is converted to its octal equivalent. The following examples will illustrate the procedure.

Example 17.35. Convert the following number : show all calculation steps. $(11010111)_2 = (?)_8$

Solution :

011	010	111
↓	↓	↓
3	2	7

Hence $(11010111)_2 = (327)_8$ **Ans.**

Example 17.36. Convert the following number to octal number : $(0.101010110)_2$.

Solution : $(0.101010110)_2$ = 0.101 010 110

0.101	010	110
↓	↓	↓
5	2	6

Hence $(0.101010110)_2 = (0.526)_8$ **Ans.**

Example 17.37. Convert $(11011100.101010)_2 = (?)_8$.

Solution : 11011100.101010 011 011 100 . 101 010

011	011	100	.	101	010
↓	↓	↓			
3	3	4	.	5	2

i.e. $(110\ 11100.101010)_2 = (334.52)_8$ **Ans.**

Example 17.38. Convert 101011 into decimal system and octal system.

[*Pb. Technical Univ. Basic Electrical and Electronics Engineering December 2009*]

Solution : $(101011)_2 = 1 \times 2^5 + 0 \times 2^4 + 1 \times 2^3 + 0 \times 2^2 + 1 \times 2^1 + 1 \times 2^0 = 32 + 0 + 8 + 0 + 2 + 1 = 43$ **Ans.**

101011 ≡ 101 011

101	011
↓	↓
5	3

So octal equivalent of $(101011)_2$ is $(53)_8$ **Ans.**

17.7.5. Octal Arithmetic

Normally octal arithmetic operations using octal representation of numbers are not performed. This number system is normally employed to enter long strings of binary data into a digital system like a microcomputer. This makes the task of the operator easier. Arithmetic operations can be performed by converting the octal numbers into binary numbers and then using the rules of binary arithmetic. Following few examples are considered for illustration.

Example 17.39. Compute the results. Show all steps $(16)_8 + (72)_8$.

Solution : Using octal representation we have

$$\begin{array}{r} 16 \\ +\ 72 \\ \hline 110 \end{array} \quad i.e.\ (16)_8 + (72)_8 = (110)_8$$

Using binary representation

$$\begin{array}{rr} 16 = & 001110 \\ +\ 72 = & +111010 \\ \hline & (1001000)_2 \end{array} = (110)_8.$$ **Ans.**

Example 17.40. Add and subtract without converting the following two octal numbers 7461 and 3465.

Solution :

$$\begin{array}{r} 1\,1 \\ (7461)_8 \\ +(3465)_8 \\ \hline (13146)_8 \end{array} \quad \textbf{Ans.} \qquad \begin{array}{r} 1\,1\,1 \\ (7461)_8 \\ -(3465)_8 \\ \hline (3774)_8 \end{array} \quad \textbf{Ans.}$$

Example 17.41. Perform $(1766)_8 - (23)_8 = (?)_8$

Solution : $(1766)_8 = (001\ 111\ 110\ 110)_2$

$(23)_8 = (010\ 011)_2$

$$\begin{array}{lcccc} \text{So } (1766)_8 - (23)_8 = & 001 & 111 & 110 & 110 \\ - & & & 010 & 011 \\ \hline & 001 & 111 & 100 & 011 \\ & \downarrow & \downarrow & \downarrow & \downarrow \\ & 1 & 7 & 4 & 3 \end{array}$$

i.e. $(1766)_8 - (23)_8 = (1743)_8$ **Ans.**

Example 17.42. Write the sum of $(23.53)_{10}$ and $(23.53)_8$ in decimal.

Solution : First of all we will convert $(23.53)_8$ into its decimal equivalent and then we will add it to $(23.53)_{10}$

$(23.53)_8 = 2 \times 8^1 + 3 \times 8^0 + 5 \times 8^{-1} + 3 \times 8^{-2} = 16 + 3 + 0.625 + 0.046875 = (19.671875)_{10}$

Thus $(23.53)_{10} + (123.53)_8 = (23.53)_{10} + (19.671875)_{10} = (43.201875)_{10}$ **Ans.**

17.8. HEXADECIMAL NUMBER SYSTEM

The binary number system forms the natural choice for the two-state systems. But in this system, the numbers tend to get short rather than long. Hence to reduce the length of a given number it is quite common to use hexadecimal system. The hexadecimal system has a base of 16 : that is, it is composed of 16 digits and characters. It uses the digits 0 through 9 plus the letters A, B, C, D, E and F. Since numeric digits and alphabets both are used to represent the digits in the hexadecimal number system, this is an *alphanumeric* number system. Table 17.3 shows the relationship between hexadecimal, decimal and binary. It is important to note that hex (abbreviation for hexadecimal) digits A through F are equivalent to the decimal values 10 through 15.

From the Table 17.3, it is observed that there are 16 combinations of 4-bit binary numbers and sets of 4-bit binary numbers can be entered in the computer in the form of hexadecimal (hex) digits. These numbers are required to be converted into binary representations using hexadecimal-to-binary converter circuits before these can be processed by the digital circuits. This system is extensively used in microprocessor work.

TABLE 17.3. ***Binary and Decimal Equivalents of Hexadecimal Numbers***

Hexadecimal	*Decimal*	*Binary*
0	0	0000
1	1	0001
2	2	0010
3	3	0011
4	4	0100
5	5	0101
6	6	0110
7	7	0111
8	8	1000
9	9	1001
A	10	1010
B	11	1011
C	12	1100
D	13	1101
E	14	1110
F	15	1111

17.8.1. Counting in Hexadecimal

How do we count in hexadecimal system once we get to F ? Simply start over with another column and continue as follows

10, 11, 12, 13, 14, 15, 16, 17, 18, 19, 1A, 1B, 1C, 1D, 1E, 1F

20, 21, 22, 23, 24, 25, 26, 27, 28, 29, 2A, 2B, 2C, 2D, 2E, 2F

30, 31, 32, 33, 34, 35, 36, 37, 38, 39, 3A, 3B and so on.

With two hexadecimal digits, we can count up to FF_{16}, which is equal to 255_{10}. For counting beyond this, three hexadecimal digits are required. For instance, 100_{16} is equal to 256_{10}, 101_{16} is equal to 257_{10} and so forth. The maximum three-digit hexadecimal number is FFF_{16} which is equal to 4095_{10}.

17.8.2. Hexadecimal-To-Decimal Conversion

A hexadecimal number can be converted to its decimal equivalent by multiplying each hex digit by its weight and then taking the sum of these products. The weights of a hex number are increasing powers of 16 (from right to left). For a four-digit hex number the weights are as follows :

16^3	16^2	16^1	16^0
4096	256	16	1

For illustration let us consider some examples.

Example 17.43. Convert $(345)_{16}$ into decimal equivalent.

Solution : $(345)_{16} = 3 \times 16^2 + 4 \times 16^1 + 5 \times 16^0 = 768 + 64 + 5 = (837)_{10}$ **Ans.**

Example 17.44. Convert the following number : Hexadecimal number $(1A2B)_{16}$ to its decimal equivalent.

Solution : $(1A2B)_{16} = 1 \times 16^3 + 10 \times 16^2 + 2 \times 16^1 + 11 \times 16^0 = 4096 + 2560 + 32 + 11 = (6699)_8$ **Ans.**

Example 17.45. Convert the following : $(3A5)_{16} = (\ )_{10}$

Solution : $(3A5)_{16} = 3 \times 16^2 + 10 \times 16^1 + 5 \times 16^0 = 768 + 160 + 5 = (933)_{10}$ **Ans.**

Example 17.46. Convert the following number as indicated

$(BC64)_{16} = (\_\_\_\_\_\_\_\_)_{10}$

Solution : $(BC64)_{16} = 11 \times 16^3 + 12 \times 16^2 + 6 \times 16 + 4 = 45{,}056 + 3{,}072 + 96 + 4 = 48{,}228$ **Ans.**

Example 17.47. Perform the following conversion : $(3A.2F)_{16} = (?)_{10}$.

Solution : $(3A.2F)_{16} = 3 \times 16^1 + 10 \times 16^0 + 2 \times 16^{-1} + 15 \times 16^{-2} = 48 + 10 + 0.125 + 0.05859375$

$= (58.18359375)_{10}$ **Ans.**

17.8.3. Decimal-To-Hexadecimal Conversion

Repeated division of a decimal number by 16 will produce the equivalent hex number formed by the remainder of each division. This is similar to the repeated division by 2 for decimal-to-binary conversion and repeated division by 8 for decimal-to-octal conversion. Following examples illustrate the procedure.

Example 17.48. Convert $(4287)_{10}$ into hexadecimal number system.

[Pb. Technical Univ. Basic Electrical and Electronics Engineering May 2008]

Solution :

$$\begin{array}{r|l} 16 & 4287 \\ \hline 16 & 267 - 15 \\ \hline 16 & 16 - 11 \\ \hline 16 & 1 - 0 \\ \hline 6 & 0 - 1 \end{array} \uparrow$$

So equivalent hexadecimal number of $(4287)_{10}$ is $(10BF)_{16}$ **Ans.**

Example 17.49. Convert the decimal number $(453.75)_{10}$ into hexadecimal.

[Pb. Technical Univ. Basic Electrical and Electronics Engineering May 2007]

Solution : *Integer*

16	453
16	28 – 5
16	1 – 12
	0 – 1

i.e. $(453)_{10} = (1C5)_{16}$

Fraction

$0.75 \times 16 = 0$ with a carry of 12

i.e. $(0.75)_{10} = (0.C)_{16}$

Thus $(453.75)_{10} = (1C5.C)_{16}$ **Ans.**

Example 17.50. Convert the following bases :

$(11011.011)_{10} \rightarrow (\ )_{16}$.

Solution :

Integer Fraction

16	11011
16	688 – 3
16	43 – 0
16	2 – 11
	0 – 2

$0.011 \times 16 = 0.176$ with a carry of 0
$0.176 \times 16 = 0.816$ with a carry of 2
$0.816 \times 16 = 0.056$ with a carry of 13
$0.056 \times 16 = 0.896$ with a carry of 0
$0.896 \times 16 = 0.336$ with a carry of 14

i.e. $(11011)_{10} = (2B03)_{16}$ *i.e.* $(.011)_{10} = (02D0E)_{16}$

Thus, $(11011.011)_{10} = (2B03.02D0E)_{16}$ **Ans.**

Example 17.51. Convert the decimal number 39.75 to hexadecimal.

[Pb. Technical Univ. Basic Electrical and Electronics Engineering, December 2008]

Solution : Integer : 39

16	39
16	2 – 7
	0 – 2

Fraction : $0.75 \times 16 = 0$ with a carry of 12 *i.e.*, C

$(39.75)_{10} = (27.C)_{16}$ **Ans.**

17.8.4. Hexadecimal-To-Binary Conversion

Hex numbers can be converted into equivalent binary numbers by replacing each hex digit by its equivalent 4-bit binary number. This procedure is illustrated below.

Example 17.52. Convert the following :

$(F80B)_{16} = (\ )_2$

Solution :

F	8	0	B
↓	↓	↓	↓
1111	1000	0000	1011

Hence $(F80B)_{16} = (1111\ 1000\ 0000\ 1011)_2$ **Ans.**

Example 17.53. Convert the following number $(A6B.F5)_{16} \rightarrow (?)_2$.

Solution : Converting each hexadecimal digit to binary number of bit we have

$(A6B.F5)_{16}$ =	A	6	B	.	F	5
	↓	↓	↓		↓	↓
	1010	0110	1011		1111	0101

Hence $(A6B.F5)_{16} = (1010\ 0110\ 1011.1111\ 0101)_2$ **Ans.**

17.8.5. Binary-To-Hexadecimal Conversion

Conversion from binary to hex is just the reverse of the process above. The binary number is grouped into groups of 4-bits starting from LSB and moving towards MSB for integer part and then each group of four bits is replaced by its hex representation. Zeros are added, as required to complete a 4-bit group.

For the fractional part, the above procedure is repeated from the bit next to the binary point and moving towards the right.

Example 17.54. Convert the following binary numbers to hexadecimal.

1001011010101 ; 1011000001

Solution :

100 101 101 0101 →	0001	0010	1101	0101
	↓	↓	↓	↓
	1	2	D	5

i.e. $(100\ 101\ 1010\ 101)_2 = (12D5)_{16}$ **Ans.**

101 100 0001 →	0010	1100	0001
	↓	↓	↓
	2	C	1

i.e. $(101\ 100\ 0001)_2 = (2C1)_{16}$ **Ans.**

Example 17.55. Convert the following numbers

(1) $2D6_{16} = (\ )_2$ (2) $011010110_2 = (\ )_{16}$.

Solution : (1) Converting each hexadecimal digit to binary number of bit, we have

2	D	6
↓	↓	↓
0010	1101	0110

Hence $(2D6)_{16} = (1011010110)_2$ **Ans.**

(2) 011010110_2 →	0000	1101	0110
	↓	↓	↓
	0	D	6

i.e. $011010110_2 = (0D6)_{16}$

Example 17.56. Convert the hexadecimal number 8A3D into decimal and binary equivalent. Convert the decimal number 5796.12 into hexadecimal. [*Pb. Technical Univ. Basic Electrical and Electronics Engineering Dec. 2006*]

Solution : $(8A3D)_{16} = 8 \times 16^3 + 10 \times 16^2 + 3 \times 16 + 13 = 32{,}768 + 2{,}560 + 48 + 13 = (35{,}389)_{10}$ **Ans.**

$(8A3D)_{16} = (1000\ 1010\ 0011\ 1101)_2$ **Ans.**

Integer 5796

16	5796
16	362 – 4
16	22 – 10
16	1 – 6
	0 – 1

Fraction 0.12

$0.12 \times 16 = 1.92 = 0.92$ with a carry of 1

$0.92 \times 16 = 14.72 = 0.72$ with a carry of 14

$0.72 \times 16 = 11.52 = 0.52$ with a carry of 11

$0.52 \times 16 = 8.32 = 0.32$ with a carry of 8

Hence $(5796.12)_{10} = 16\text{A}4.1\text{EB}8$ **Ans.**

17.8.6. Conversion From Hex-To-Octal and Vice-versa

Hexadecimal numbers can be converted to equivalent octal numbers and octal numbers can be converted to equivalent hex numbers by converting the hex/octal number to equivalent binary and then to octal/hex respectively. The procedure is illustrated by the following example.

Example 17.57. Convert $(1375)_8 = (?)_2 = (?)_{16}$

Solution : $(1375)_8 \rightarrow$ 1 3 7 5

001 011 111 101

i.e. $(1375)_8 = (001011111101)_2$ **Ans.**

Now 001011111101 $\rightarrow$ 0010 1111 1101

$\downarrow$ $\downarrow$ $\downarrow$

2 15 13

$\downarrow$ $\downarrow$ $\downarrow$

2 F D

i.e. $(1375)_8 = (001011111\ 101)_2 = (2FD)_{16}$ **Ans.**

Example 17.58. Convert the following number

$(A6F.CD)_{16} = (\ )_8$.

Solution : $(A6F.CD)_{16}$

Converting to binary we have

A 6 F . C D

1010 0110 1111 . 1100 1101

Converting to octal we have

101 001 101 111 . 110 011 010

5 1 5 7 . 6 3 2

So, $(A6F.CD)_{16} = (5157.632)_8$ **Ans.**

Example 17.59. Convert the following

$(C3A.47)_H \rightarrow (\ )_8 \rightarrow (\ )_2$.

Solution : Converting to binary we have

C 3 A . 4 7

$\downarrow$ $\downarrow$ $\downarrow$ $\downarrow$ $\downarrow$

1100 0011 1010 . 0100 0111 $\rightarrow (110000111010.01000111)_2$ **Ans.**

$\downarrow$

110 000 111 010 . 010 001 110

$\downarrow$ $\downarrow$ $\downarrow$ $\downarrow$ $\downarrow$ $\downarrow$ $\downarrow$

6 0 7 2 2 1 6

Thus $(C3A.47)_H \rightarrow (6072.216)_8 \rightarrow (110000111010.01000111)_2$ **Ans.**

17.8.7. Hexadecimal Arithmetic

The rules for arithmetic operations with hex numbers are similar to the rules for decimal, octal and binary numbers. The information can be handled only in binary form in a digital circuit and it is convenient and easier to enter the information using hex numbers. Since arithmetic operations are performed by the digital circuits on binary numbers, hex numbers are to be first converted into binary numbers. Following few examples illustrate the procedure.

Example 17.60. Compute the results showing all steps :

(*i*) $(DF)_{16} + (AC)_{16}$ (*ii*) $(2B)_{16} + (84)_{16}$; (*iii*) $(C3)_{16} - (0B)_{16}$.

Solution : Using hexadecimal representation we have

(*i*)
```
  DF
+ AC
----
 18B
```
i.e. $(DF)_{16} + (AC)_{16} = (18B)_{16}$ **Ans.**

(*ii*)
```
  2B
+ 84
----
  AF
```
i.e. $(2B)_{16} + (84)_{16} = (AF)_{16}$ **Ans.**

(*iii*)
```
  C3
- 0B
----
  B8
```
i.e. $(C3)_{16} - (0B)_{16} = (B8)_{16}$ **Ans.**

Example 17.61. Solve the following : $(A6C.DF)_{16} + (1DE.9A)_{16}$.

Solution :
```
   1 1 1   1    Carry
   A 6 C . D F
 + 1 D E . 9 A
 -------------
   C 4 B . 7 9
```
i.e. $(A6C.DF)_{16} + (1DE.9A)_{16} = (C4B.79)_{16}$ **Ans.**

Example 17.62. Complete the following operation :

$$8_{16} + F_{16} = (\ )_{16}.$$

Solution :
```
   8
 + F
 ---
  17
```
Thus, $8_{16} + F_{16} = (17)_{16}$ **Ans.**

Example 17.63. Solve the following :

(*i*) $(A4F.EF)_{16}$ $+(3FD.AB)_{16}$ (*ii*) $(FA12.35)_{16}$ and $-(9BCD.EC)_{16}$

Solution :

(*i*)
```
  1 1 1 1      CARRY
   A 4 F . E F
 + 3 F D . A B
 -------------
 (E 4 D . 9 A)16  Ans.
```
(*ii*)
```
  1 1 1 1 1    BORROW
   F A 1 2 . 3 5
 - 9 B C D . E C
 ---------------
 (5 E 4 4 . 4 9)16  Ans.
```

Example 17.64. Perform indicated operation

$$FC2_{16} \times DE_{16} = ?$$

Solution :
```
    FC2
     DE
 ------
   DC9C
  CCDA
 ------
  DAA3C   Ans.
```

$2 \times 14 = 28_{10} = 1C$

$12 \times 14 = 168_{10} + 1_{10} = 169_{10} = A9$

$15 \times 14 = 210_{10} + 10_{10} = 220_{10} = DC$

$2 \times 13 = 26_{10} = 1A$

$12 \times 13 = 156_{10} + 1_{10} = 157_{10} = 9D$

$15 \times 13 = 195_{10} + 9_{10} = 204_{10} = CC$

Example 17.65. Divide $1EC87_{16}$ by $A5_{16}$.

Solution :

```
A5 ) 1 E C 8 7 (2FC
     1 4 A
     -----
       A 2 8
       9 A B
       -----
         7 D 7
         7 B C
         -----
           1 B
```

A5(165_{10}) can go in 1EC(492_{10}) twice with a remainder of A2(162_{10})

A5(165_{10}) can go in A28(2600)F(15_{10}) times with a remainder of 7D(125_{10})

A5(165_{10}) can go in 7D7(2007) C(12_{10}) with a remainder of 1B(27_{10})

2FC with a remainder of 1B. **Ans.**

HIGHLIGHTS

1. Decimal, binary, octal and hexadecimal number systems are positional weighted, which means that the values attached to the symbols depend on their location with respect to the radix point.
2. Radix or base of a number system is the number of digits or distinct symbols it uses to represent various numbers.
3. The base or radix point separates the integer and fraction parts.
4. The extreme right digit in any number is the LSD and the extreme left digit is the MSD.
5. A decimal number can be converted to a number in any other system with base *b* by repeated division by *b*. In repeated division by *b*, the remainders are read from bottom to top.
6. A decimal fraction can be converted to a fraction in any number system with base *b* by repeated multiplication by *b*. In repeated multiplication by *b*, the integers to the left of the radix point are read from top to bottom.
7. The binary number system uses digits 0 and 1. Thus the base is 2.
8. Each binary digit is abbreviated as a bit. A string of 4 bits is known as nibble and a string of 8 bits is known as byte. 1 kilo byte = 2^{10} = 1024 bits.
9. To convert a decimal number into binary, we divide the number successively by 2. The remainders when read upwards give the binary number.
10. To convert a binary number into decimal number we multiply each binary digit by its weight. The weight 2^0, 2^1, 2^2, 2^3 starting with the LSB *i.e.* bit in the extreme right position.
11. The 1's complement of a binary number is obtained by complementing each bit of the number (*i.e.* 0 for 1 and 1 for 0).
12. The 2's complement of a binary number is obtained by finding its 1's complement and adding 1 to it.
13. In binary fractions, the weights of the bits after binary point are 2^{-1}, 2^{-2}, 2^{-3} etc.
14. The rules for binary addition are 0 + 0 = 0; 0 + 1 = 1; 1 + 0 = 1; 1 + 1 = 10 and 1 + 1 + 1 = 11.
15. The rules for binary subtraction are 0 – 0 = 0; 1 – 0 = 1; 1 – 1 = 0 and 10 – 1 = 1.
16. The rules for binary multiplication are 0 × 0 = 0; 0 × 1 = 0; 1 × 0 = 0 and 1 × 1 = 1.
17. In sign magnitude representation, an additional bit called the sign bit is placed as the MSB to the magnitude of the numbers. The '0' is used for '+' sign and digit (r – 1) is used for '–' sign where r is the base (radix) of the number system and it is (2 – 1) = 1 in binary system. For example in 8 bit binary number having one sign bit and seven magnitude bits, the largest positive number is 01111111 = +127 and smallest number is 11111111 = –127.
18. Binary numbers (mantissa) can also be expressed by a number and an exponent of 2. This representation of binary number is called *Floating point number.* In a 16 bit computer 10 bits are mantissa and remaining 6 bits are exponent. The mantissa is written in 2's complement form and therefore MSB is taken as sign bit. The binary point is placed after this bit (or it may be placed at the given radix point position). The remaining 6 bits are the exponent component representing 0 to 63. The binary number 100000_2 $(32)_{10}$ is added to the desired exponent for representing negative exponents. It is called the excess-32 exponent.

19. In octal system the base is 8. The digits are 0, 1, 2, 3, 4, 5, 6, 7. In hexadecimal system the base is 16. The digits are 1, 2, 3, 4, 5, 6, 7, 8, 9, A, B, C, D, E, F.

The main advantage of octal and hexadecimal systems is their ease of conversion to and from binary. An octal number is 1/3rd the length of the corresponding binary number and a hex number is 1/4th the length of the corresponding binary number.

20. To convert an octal number into binary, replace each octal digit by its 3-bit binary equivalent. Similarly to convert a hex number into binary replace each hex digit by its 4-bit binary equivalent.

21. To convert a binary number into octal, starting from the binary point make groups of 3-bits on either side of the binary point and replace each 3-bit group by an octal digit.

Similarly, to convert a binary number into hex, starting from the binary point, make group of 4-bits on either side of the binary point and replace 4-bit group of an hex digit.

EXERCISES

1. Why binary system is preferred in digital systems ?
2. Discuss the importance of 1's and 2's complement numbers.
3. What are signed binary numbers ? How subtraction of binary numbers is carried out in 2's complement representation ? **[Pb. Technical Univ. Basic Electrical & Electronics Engg. May 2007]**
4. Give the advantages of octal and hexadecimal systems.

SHORT ANSWER TYPE QUESTIONS WITH ANSWERS

Q. 1. What is meant by radix (or base) of a number system ?

Ans. Radix or base of a number system is the number of digits or distinct symbols it uses to represent various numbers.

Q. 2. What is (*a*) octal system (*b*) hexadecimal system ?

Ans. The number systems with base (or radix) 8 and 16 are known as octal number system and hexadecimal number system respectively.

PROBLEMS

1. Perform the following subtractions by using 9's complement method :

   (*i*) 3 from 5 (*ii*) 06 from 17 (*iii*) 346 from 742 and (*iv*) 657 from 328. **[Ans.** (*i*) 2 (*ii*) 11 (*iii*) 396 and (*iv*) – 329]

2. Perform the following subtractions by using 10's complement method :

   (*i*) 2 from 7 (*ii*) 05 from 18 (*iii*) 425 from 746 and (*iv*) 725 from 436. **[Ans.** (*i*) 5 (*ii*) 13 (*iii*) 321 (*iv*) – 289]

3. Convert $(10110111)_2$ to decimal. **[Ans.** 183]
4. Convert the following : $(0.865)_{10} = (\quad)_2$. **[Ans.** $(0.865)_{10} = (0.110111)_2$]
5. Convert the following number : $(4021.25)_{10} = (?)_2$. **[Ans.** $(111110110101.01)_2$]
6. Solve the following problems given in binary form.

   (1) 110101 + 111111 (2) 10000 − 1111 (3) 1010 × 1101

   (4) 1101 ÷ 101 (5) 101010 ÷ 111

   [Ans. (1) **1110100** (2) 0001 (3) 10000010 (4) 10.10011 (5) 110]

7. Perform the arithmetic indicated $(10001.101)_2 \times (111.001)_2$. **[Ans.** 1111101.100101]
8. Perform the subtraction with the following binary numbers using 2's complement and 1's complement. Check the answer by using straight subtraction.

   (*i*) 11010 – 1101 (*ii*) 100 – 110000 **[Ans.** (*i*) 1101 (*ii*) – 101100]

9. Perform the following subtraction using 2's complement method:
(*i*) 110100 – 10101 (*ii*) 0011.1001 – 0001.1110 **[Ans.** (*i*) $(11111)_2$ (*ii*) $(1.1011)_2$]

10. Perform the subtraction using 1's complement and 2's complement method. $(11010)_2 - (10000)_2$
[Ans. $(1010)_2$]

11. Convert the following : $(2374)_8 = (\quad)_{10}$ **[Ans.** $(1,276)_{10}$]

12. Convert the following number : $(6089.25)_{10} \rightarrow (?)_8$ **[Ans.** $(13711.2)_8$]

13. Convert the decimal number 39.75 to octal. **[Pb. Technical Univ. Basic Electrical & Electronics Engg. May 2009]**
[Ans. $(47.6)_8$]

14. Convert the octal number $(345)_8$ to its equivalent binary number. **[Ans.** $(011100101)_2$]

15. Convert the following :
$(5624.37)_8 = (\quad)_2$. **[Ans.** $(101110010100.011111)_2$]

16. Convert the following binary number into octal : 0111001.0001000. **[Ans.** $(71.04)_8$]

17. Convert the following number to octal number : $(11000111.1101)_2$ **[Ans.** $(307.64)_8$]

18. Convert the following hexadecimal numbers into decimals : (*i*) A13B (*ii*) 7CA3 (*iii*) 7FD6
[Ans. $(41275)_{10}$; (*ii*) $(31907)_{10}$; (*iii*) $(32726)_{10}$]

19. Solve the following : $(7001)_8 - (5763)_8$. **[Ans.** $(1016)_8$]

20. Convert the following bases : $(11011.011)_{10} \rightarrow (?)_{16}$ **[Ans.** $(2B03.02D)_{16}$]

21. Convert the decimal number 39.75 to hexadecimal.
[Pb. Technical Univ. Electrical and Electronics Engg. Dec. 2008] **[Ans.** $(27.C)_{16}$]

22. Convert the following numbers in decimal into their hexadecimal equivalents : 17386.75; 9817.625
[Ans. 43EA.C; 2659.A]

23. Convert the following : $(FB17)_{16} = (\quad)_2$ **[Ans.** $(1111\ 1011\ 0001\ 1111)_2$]

24. Convert : $(2715)_8$ into HEX and $(1E7B)_{16}$ into octal. **[Ans.** $(5CD)_{16}$, $(17173)_8$]

25. Convert the following numbers as indicated :
(*i*) $(6089.25)_{10} = (\_\_\_\_\_)_8$ (*ii*) $(A6BF.5)_{16} = (\_\_\_\_\_)_2$ (*iii*) $(25.26)_8 = (\_\_\_\_\_)_2$.
[Ans. (*i*) $(13711.2)_8$, $(1010\ 0110\ 1011\ 1111\ 0101)_2$, $(010101.010110)_2$]

26. Convert the following number $(A6F.CD_{16}) = (?)_8$ **[Ans.** $(5157.632)_8$]

27. Convert the following binary number into octal and hexadecimal.
10110110111 and 1111110. **[Ans.** $(2667)_8$; $(5B7)_{16}$; $(176)_8$; $(7E)_{16}$]

28. Convert the following hexadecimal numbers to octal through binary. (*a*) 7FD6 (*b*) ABCD.
[Ans. (*a*) $(77726)_8$; (*b*) $(125715)_8$]

29. Convert the decimal number 749.25 to binary, Octal and hexadecimal numbers.
[Ans. $(1011101101.01)_2$; $(1355.2)_8$; $(2ED.4)_{16}$]

30. Add and subtract the following two hexadecimal numbers without converting into decimal numbers : $(4F3A)_{16}$ and $(23C1)_{16}$. **[Ans.** $(72FB)_{16}$; $(2B79)_{16}$]

31. Add the following hex numbers :
(*i*) 93 + DE; (*ii*) ABCD + EF12. **[Ans.** $(171)_{16}$; (*ii*) $(19ADF)_{16}$]

32. Perform the following : $(756)_8 - (637)_8 + (725)_{16}$. **[Ans.** 0111 0111 0100]

❑❑❑

CHAPTER 18

Logic Circuits

INSIDE THIS CHAPTER

18.1. INTRODUCTION

Logic circuit may be classified into two categories *i.e.* combinational logic circuits and sequential logic circuits. A *combinational logic circuit* contains logic gates only but does not contain storage elements. A sequential logic circuit contains storage elements in addition to logic gates.

When logic gates are connected together to give a specified output for certain specified combination of input variables, with no storage involved, the resulting network is known as *combinational logic circuit.* In combinational logic circuit, the output level is at all times dependent on the combination of input level. The block diagram of a combinational logic circuit is depicted in Fig. 18.1. It accepts input binary variables and generates output variables depending on the logical combination of logic gates.

The combinational logic circuit with memory element(s) is called the *sequential logic circuit,* as illustrated in Fig. 18.2. It consists of a combinational circuit to which memory elements are connected to form a feedback path. The memory elements are devices, capable of storing binary information within them. By block diagram, it can be said that the output(s) of sequential logic circuit is (are) dependent not only on external input(s) but also on the present state of the memory element(s). The next state of the memory element(s) is also dependent on external inputs and the present state. Thus, a sequential circuit is specified by a time sequence of inputs, outputs, and internal state.

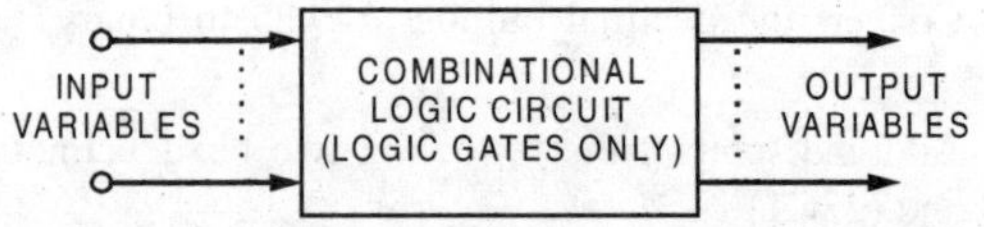

Fig. 18.1. ***Block Diagram of Combinational Logic Circuit***

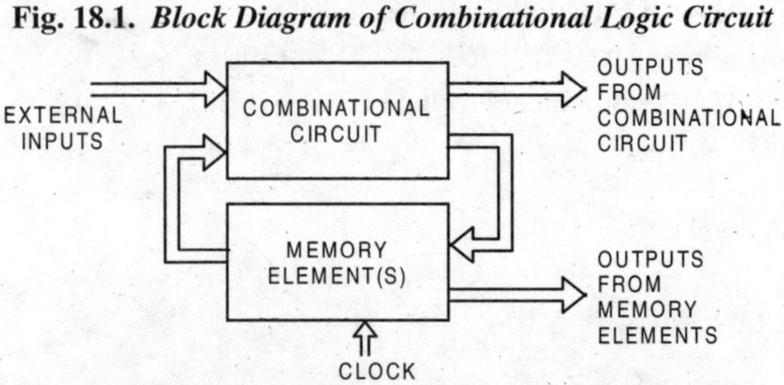

Fig. 18.2. ***Block Diagram of a Sequential Circuit***

Logic gates find wide applications in calculators and computers, digital measuring techniques, digital processing of communications, musical instruments, games and domestic appliances etc., for decision making in automatic control of machines and various industrial processes and for building more complex devices such as binary counters etc.

18.2. BOOLEAN ALGEBRA

In its basic form, logic is the realm of human reasoning that indicates us that a certain proposition (declarative statement) is true if certain conditions or premises are true. For example, 'the light is on', 'the tube is operative', 'the switch is *on*' and 'power supply is there' are the examples of propositions that can be classified as true or false. Several propositions, when combined, give rise to *propositional* or *logic functions*. For example, the propositional statement 'the light is *on*' will be true if 'the tube is operative', 'the switch is *on*' and 'power supply is there'. So this logical statement can be made as follows :

The light is *on* if and only if the tube is operative, the switch is *on* and power supply is there.

In above example the first statement is true only if the last three statements are true. The first statement *i.e.* 'the light is *on*' is then the basic proposition, and the other three statements are the conditions or premises upon which the proposition depends.

Many situations, processes, and problems that we come across in our daily lives, may be expressed in the form of propositional or logic functions. As these functions are true/false or yes/no statements, digital circuits with their two-state characteristics are quite suitable.

Boolean algebra, named after its pioneer George Boole (1815-64), is the algebra of logic applied in the design and analysis of digital systems. The rules of this algebra are based on human reasonings. It originated from the study of how we reason, what lines of reasoning are valid and what constitutes proofs etc. Boolean algebra is a mathematical system of logic in which truth functions are expressed as symbols and then these symbols are manipulated to arrive at conclusion.

Boolean functions can be performed by digital circuits. Circuits that perform complicated boolean functions can be subdivided into simpler circuits that perform simpler boolean functions. The circuits that perform the simplest boolean functions are taken as basic elements, called *logic gates* and are represented by specialised symbols. The logic gates are identified by the adjective representing the operation they perform. The most common logic gates are the OR, AND, NOT, NAND, NOR, Exlcusive OR and Exclusive NOR gates. The NAND and NOR gates are called the *universal gates*. NOT gate or inverter is the only nontrivial single input gate.

18.3. BINARY LOGIC

Binary logic presupposes two distinguishing characteristics : *two-valued variables*, and *appropriate logical operations*. Unlike the situation of ordinary numbers, the values of the variables in binary logic can be only two in number. In fact, they do not even need to be numbers. They should perhaps more properly be called *states*. One pair of terms that can be used to identify the two states that binary variables can assume is : *open* and *closed*. Another suitable pair is : *high* and *low*. Other pairs are *hot* and *cold*, *true* and *false*, *on* and *off*, *plus* and *minus*, *yes* and *no*, *up* and *down*, *excited* and *unexcited* etc. In each case the binary variable can take on at any given time only one of the two values. However, on the basis of logic, these two variables should be mutually exclusive. The examples cited obviously possess this feature. However, for binary mathematical manipulations of switching and logic functions binary variables 0 and 1 are best suited.

There are three logical operations associated with binary logic viz. AND, OR, and NOT. Two binary variables A and B, each of which can assume the value of 0 or 1, are introduced so as to facilitate the description of the three logical operations. The AND function is represented by the equation

$$Z = A \cdot B \quad ...(18.1)$$

and by definition of the logical AND operation Z = 1 if and only if A and B, both are equal to 1. Otherwise, the result of the AND logical operation is zero. This information can be represented in tabular form as below :

A	*B*	*Z*
0	0	0
0	1	0
1	0	0
1	1	1

Such a table is called a *truth table*. The dot symbol is used to identify the AND function and an inspection of truth table reveals this choice to be an appropriate one if we permit ourselves for the moment to treat the dot as multiplication. Accordingly, in the first row of the truth table we have $0 \cdot 0 = 0$. Similarly, in subsequent rows we observe $0 \cdot 1 = 0$, $1 \cdot 0 = 0$, and $1 \cdot 1 = 1$. Obviously, the logical AND cperation resembles multiplication.

The OR function is represented by the equation

$$Z = A + B \qquad ...(18.2)$$

By definition of the logical OR operation Z = 1 if A = 1 or B = 1; otherwise, Z = 0. The truth table for the OR function is given below

A	*B*	*Z*
0	0	0
0	1	1
1	0	1
1	1	1

A study of the truth table in this case reveals that the logical OR operation resembles addition. Observe that on successive rows we have $0 + 0 = 0$, $0 + 1 = 1$, $1 + 0 = 1$, and $1 + 1 = 1$. Use of the + symbol in Eq. (18.2) is, therefore, entirely appropriate. Of course, to say that the OR function resembles addition does not mean that it is addition in the sense that we use the term in ordinary arithmetic. This is obviously illustrated by the last row of the truth table. The value of the variable that represents the output of the OR function is still confined to 1, even when both A and B are at value 1.

The NOT function is represented by the equation

$$Z = \overline{A} \qquad ...(18.3)$$

The NOT function is the logical operation of *negation*. By definition if A = 0, then Z = 1. Similarly, if A = 1, then Z = 0. The NOT function acts as an *inversion* that provides the opposite state of A whatever it may be. Thus Eq. (18.3) is read as : Z equals the complement of A. It is interesting to note that the NOT operation is not the part of ordinary arithmetic or algebra.

18.4. LOGIC SYSTEMS

Logic systems, depending upon the type of excitation employed, are described below :

1. DC Logic (or Level Logic) System. In a dc logic or level logic system, a bit is represented by one of the two voltage levels—*high* or *low*. If the higher of the two voltages represents a 1 and the lower voltage represents a 0; the system is called a *positive logic system*. On the other hand, if the lower voltage represents a 1 and the higher voltage a 0, we have a *negative logic system*. For illustration, if we take + 5 V and 0 V as our logic-level voltages, then + 5 V will be designated as the *high* level voltage and 0 V as the *low* level voltage. Thus positive and negative logic can be defined as below :

Positive Logic	Negative Logic
high = 1	*high* = 0
low = 0	*low* = 1

Both positive and negative logic are used in digital systems, but positive logic is the more common.

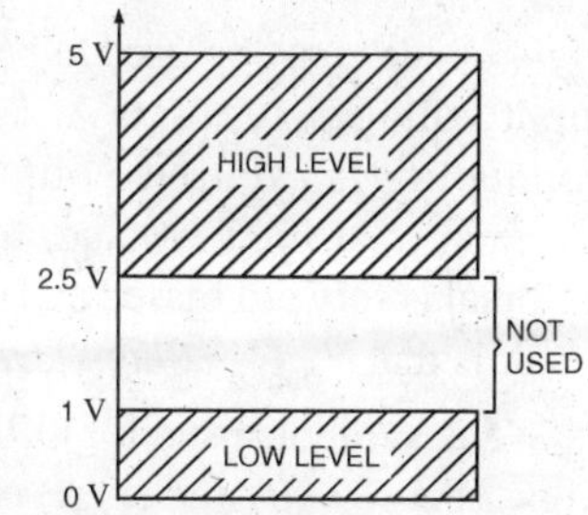

Fig. 18.3. ***DC Positive Logic***

In actuality, because of circuit variations, the 0 and 1 are not represented by the fixed voltage levels (for example + 5 V and 0 V), rather represented by voltage ranges, as illustrated in Fig. 18.3. As long as the voltage belongs to a level, it will be taken as that level and the exact value of the voltage is immaterial. For example, any voltage in the range of 2.5 to 5 V will be considered as *high* level in the positive logic system and *low* level in the negative logic system. Similarly, any voltage in the range of 0 to 1 V will be considered as *low* level in the positive logic system and *high* level in the negative logic system. The actual voltage ranges corresponding to *low* and *high* levels are not the same for all types of circuits and differ from logic family to logic family.

2. Pulse Logic System. In this logic system, a bit is recognised by the presence or absence of a pulse. It is also known as *dynamic logic system*. This can be also *dynamic positive* or *dynamic negative i.e.* state 1 may signify presence of positive pulse or negative pulse. In both cases the absence of a pulse signifies 0 level.

18.5. LOGIC GATES

The most basic elements of digital circuits are the *logic gates*. The most basic types of electronic circuit designed with active devices (such as diodes, transistors) and passive components (such as resistors) which consist of two or more inputs and one output performing very basic logic operation are known as the *logic gates*. Each gate is given several input signals to act upon. Its output may be *high* (logic-1) or *low* (logic-0) with respect to the voltage levels. The output depends upon the combination of *high* and *low* inputs and the type of gate used. Since gates have the ability to make logic decisions by producing a particular output under certain input conditions, they are commonly called *logic gates*. The interconnection of gates to provide a prescribed output is called a *logic design*.

A gate can be used alone to perform a basic logic function. The output of one gate can also be connected to the inputs of several other gates to form a big logic network. In a network one gate contributes its decision to a gate down stream to other gates to make its decision again. This way, a complex function can be performed on the basis of a few simple individual "yes" and "no" decisions. In such cases, the inputs and outputs of all gates are accessible, in the sense that external connections can be made to them. Because of their accessibility, logic circuits constructed by the interconnections of logic gates are said to be designed with discrete logic gates.

There are seven types of logic gates namely OR, AND, NOT, NOR, NAND, EX-OR and EX-NOR. First three gates *i.e.* OR, AND, NOT gates are known as *basic types of gates* while NAND and NOR gates are called the *universal gates.*

A logic circuit whose steady-state output depends only on the present steady-state inputs (and not on any prior inputs) is called a *combinational logic circuit*. To depend on previous inputs would need memory, thus a combinational logic circuit has no memory elements.

Boolean algebra allows any combinational logic circuit to be constructed with AND, OR and NOT gates. Any combinational logic circuit may also be constructed solely with NAND gates, as well as solely with NOR gates.

18.6. OR GATE

The OR gate performs logical addition, more commonly known as the OR *function*. An OR gate has two or more input signals with only one output signal. In OR gate, output voltage is *high* if any or all of the input voltages are *high*. Diodes may be used to build an OR gate, as shown in Fig. 18.4. If input voltages at points A, B and C are *low* then all diodes are non-conducting so output voltage, V_{out} is *low*. But if any of the input terminal is at *high* voltage, then diode connected with that particular terminal is forward biased and current flows through the resistance R. The result is that V_{out} is at *high level*.

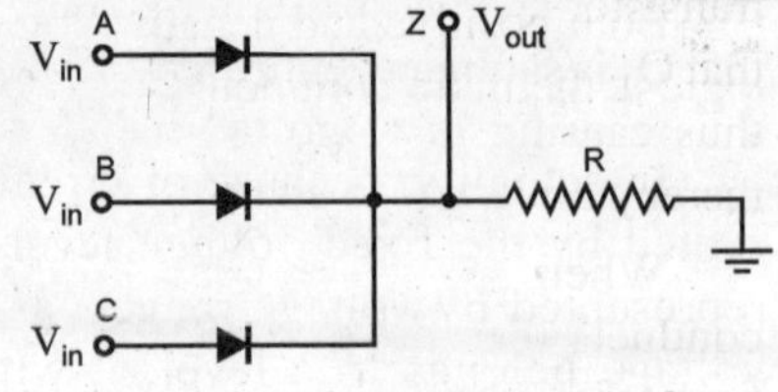

Fig. 18.4. ***Circuit Diagram of OR Gate***

The symbols used for OR gates with 2, 3 and N inputs are shown in Figs. 18.5 (*a*), (*b*) and (*c*) respectively. The inputs A, B, C....N are logic voltage levels and the output Z is a logic voltage level whose value is the result of the OR operation on A, B, C... and N. In other words, the OR gate operates in such a way that its output is *high* (logic 1) if and only if one or more inputs are *high*. The OR gate output will be *low* (logic 0) only if all inputs are *low* or at logic 0. Its logical equation is given as

$$Z = A \text{ OR } B \text{ OR } C \ldots \text{ OR } N = A + B + C + \ldots + N \qquad \ldots(18.4)$$

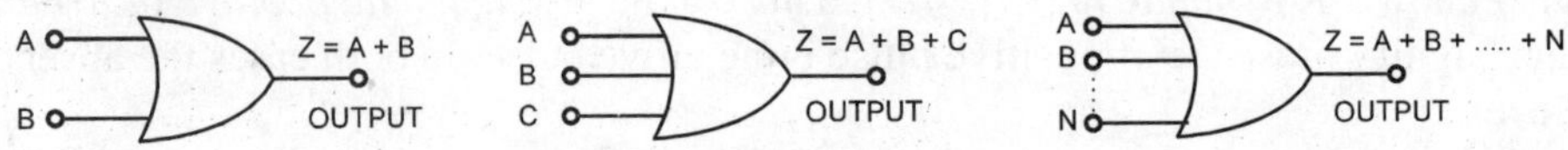

(a) Two-Input *(b) Three-Input* *(c) N-Input*

Fig. 18.5. ***OR Gate Symbols***

Equation (18.4) is known as *Boolean equation* or the *logical equation* of the OR gate.

TABLE 18.1. Truth Table For 3-Input OR Gate

Inputs			*Output*
A	*B*	*C*	*Z*
0	0	0	0
0	0	1	1
0	1	0	1
0	1	1	1
1	0	0	1
1	0	1	1
1	1	0	1
1	1	1	1

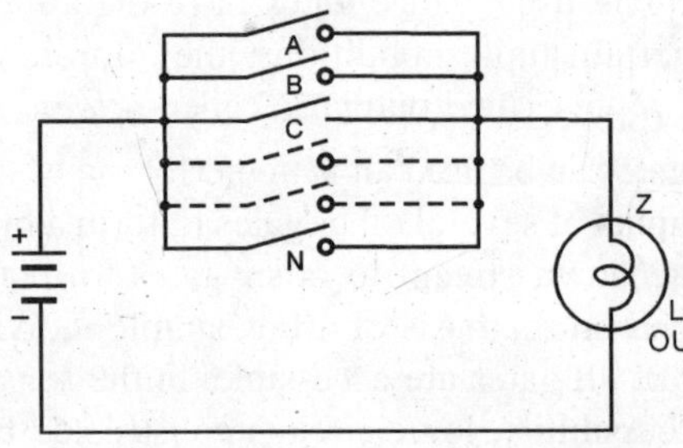

Fig. 18.6. ***Electrical Equivalent Circuit of an OR Gate***

The logical operation of the three-input OR gate is described in truth table shown in Table 18.1. The truth table can be expanded for any number of inputs; but regardless of the number of inputs, the output is *high* when any one or more of the inputs are *high*.

Incidentally, the number of rows in a truth table equals 2^n, where *n* is the number of inputs. For a 2-input OR gate, the truth table has 2^2 or 4 rows. A 3-input or gate has a 2^3 *i.e.* 8 rows and so on.

An OR gate can have as many inputs as desired, add one diode for each additional input. Six diodes result in a 6-input OR gate, twelve diodes in a 12-input OR gate. No matter how many inputs, the action of the OR gate is summarized like this; one or more *high* inputs produce a *high* output.

The OR gate can be thought of as an electrical circuit in which switches are connected in parallel with each other. Lamp will glow [output will be present] when either or all switches are closed.

18.6.1. Transistor OR Gate

Figure 18.7 depicts a possible transistor OR gate consisting of three interconnected transistors Q_1, Q_2 and Q_3 supplied from a common supply V_{CC} = +5 V. When +5 V is applied to terminal A (*i.e.* when input A is +5 V), transistor Q_1 is forward biased and so conducts. Assuming that Q_1 is saturated, entire $V_{CC} = 5$ V drops across resistor R_1 thus causing M to go to ground. This in turn, cuts off Q_3 thereby causing Z to go to V_{CC} *i.e.* +5 V.

When +5 V is applied to terminal B, transistor Q_2 conducts thereby driving M to ground *i.e.* 0 V. With no forward bias on its base, transistor Q_3 is cut off driving Z again to V_{CC} *i.e.* +5 V.

When both inputs A and B are zero both transistors Q_1 and Q_2 are cut off driving M to +5 V. Consequently, transisitor Q_3 becomes forward biased and conducts fully. In that case entire V_{CC} drops across resistor R_2 driving Z to ground (or zero).

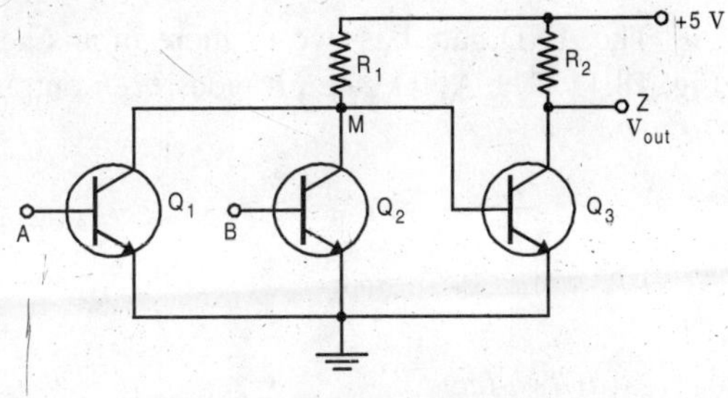

Fig. 18.7. ***Transistor OR Gate***

18.6.2. Standard Package

Figure 18.8 depicts the pinout diagram of IC7432, a TTL quad 2-input OR gate. It is a 14-pin dual-in-line package (DIP) and has four 2-input OR gates. After connecting a supply voltage of +5 V to pin 14 and a ground to pin 7, we may connect one or more of the OR gates to other TTL devices.

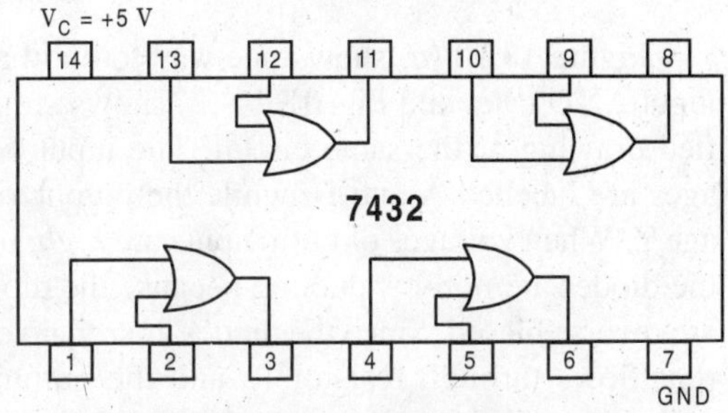

Fig. 18.8. ***Pinout Configuration of IC 7432***

18.6.3. Pulsed Operation

In many applications, the inputs of OR gate may be in the form of pulses (not in the form of single bit). The output depends upon the relationship between different pulses. Figure 18.9 illustrates two pulses A and B as inputs to an OR gate. The output Z is also shown. In the given waveform during interval t_0 both inputs are *low* so output Z is *low*. During interval t_1 input A is *high* but B is *low*, so output Z is *high*. During time interval t_2 both inputs A and B are *high* and therefore output Z is again *high*. During time interval t_3 input A is *low* but input B is *high*, so output Z is again *high.* During time interval t_4 both inputs A and B are *low*, so output Z is *low*. During time interval t_5 input A is *high* and input B is *low*, so output Z is *high* and during time interval t_6 both inputs A and B are *high* and so output Z is *high.*

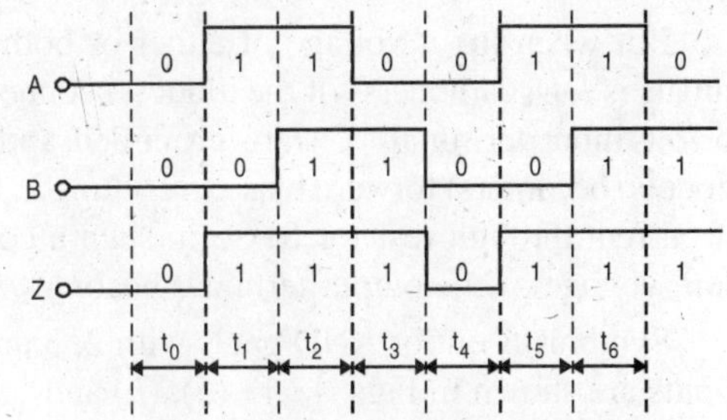

Fig. 18.9. ***Timing Diagram For OR Gate***

18.7. AND GATE

The AND gate performs logical multiplication, more commonly known as AND *function.*

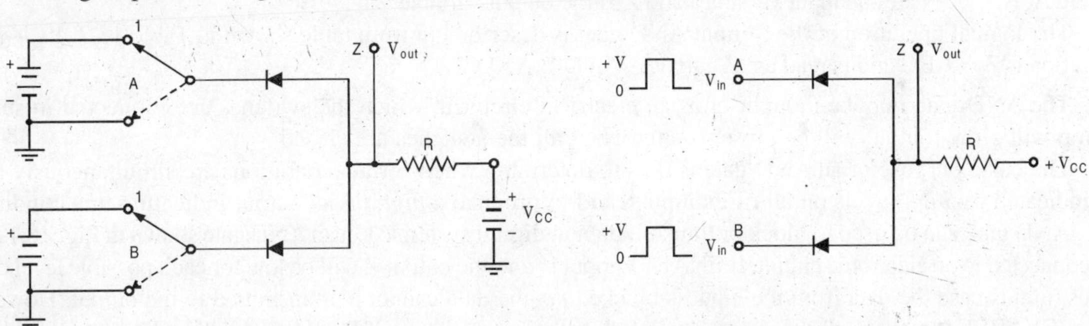

(a) Circuit Diagram *(b) Simplified Drawing*

Fig. 18.10

The AND gate has two or more inputs and a single output, as shown by the standard logic symbols in Fig. 18.11. The AND gate provides *high* output only when all inputs are *high*.

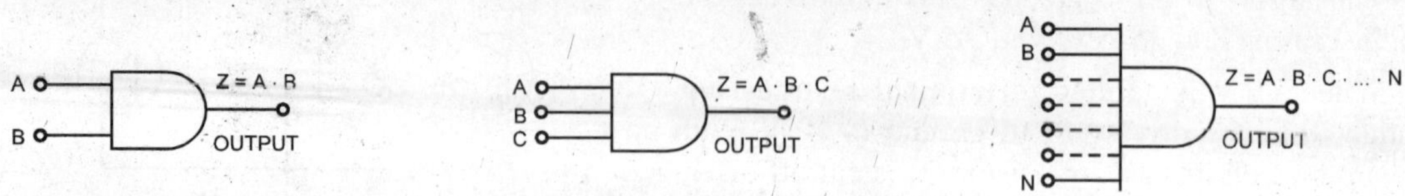

(a) Two-Input *(b) Three-Input* *(c) N-Input*

Fig. 18.11. ***AND Gate Symbols***

Figure 18.10 (*a*) shows one way to build a 2-input AND gate, and Fig. 18.10 (*b*) shows simplified drawing of the same circuit. The input voltages are labelled A and B, while the output voltage Z. When voltages of both inputs are *high*, both the diodes are non-conducting because the diodes are reverse biased. Since the diodes are *off*, no current flows through resistor R, and the output is pulled up to the supply voltage V_{CC} (+ 5 V). Thus for both inputs *high*, output is *high*.

But when input voltage of either or both terminals is *low*, cathode(s) of the diode(s) connected to *low* input terminal(s) is/are grounded and the diode(s) become(s) forward biased, resulting in flow of current through resistor R. So in such a condition, voltage of the output terminal becomes *low*.

Symbols used for AND gates with 2, 3 and N inputs are shown in Figs. 18.11 (*a*), (*b*) and (*c*) respectively. The input A, B, C ... and N are logic voltage levels and output Z is a logic voltage level whose value is the result of the AND operation on A, B, C ... and N.

The operation of the AND gate is such that the output is *high* or 1 if and only if all the inputs are *high* or 1. Mathematically, it is written as

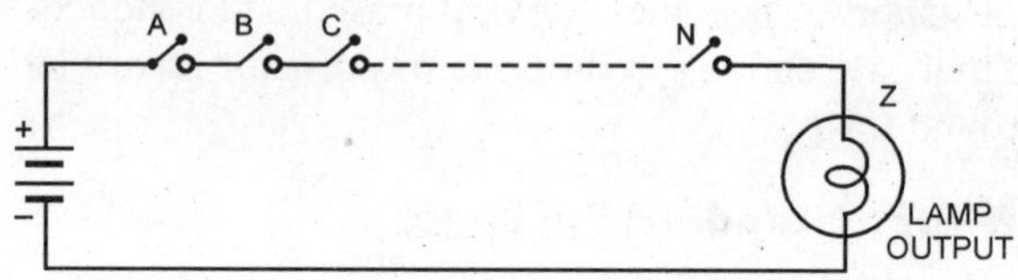

(a) Electrical Equivalent Circuit of an AND Gate

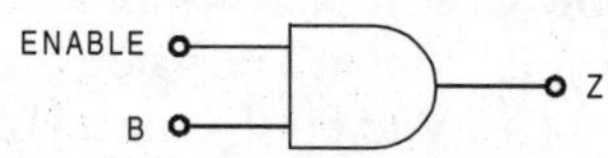

(b)

Fig. 18.12

TABLE 18.2. ***Truth Table For 3-Input AND Gate***

Inputs			*Output*
A	*B*	*C*	*Z*
0	0	0	0
0	0	1	0
0	1	0	0
0	1	1	0
1	0	0	0
1	0	1	0
1	1	0	0
1	1	1	1

$$Z = \text{A AND B AND C.. AND N} = A \cdot B \cdot C \cdot ... \cdot N = ABC... \cdot N \qquad ...(18.5)$$

where A, B, C ... N are the input variables and Z is the output variable.

The logical operation of the 3-input AND gate is described in truth table shown in Table 18.2. Its logical equation is Z = ABC and is read as 'Z equals A AND B AND C'.

The AND gate may be thought of as an electrical circuit in which the switches are connected in series. Lamp will glow (output will be present) only when all the switches are closed.

The basic purpose of an AND gate is thus to determine when certain conditions are simultaneously true, as indicated by *high* levels on all of its inputs, and to produce a *high* on its output indicating this condition.

AND gate can be used to block or transmit data in digital systems. Let an AND gate shown in Fig. 18.12 (*b*) be connected in an electronic circuit. If ENABLE input is *low,* the output Z will be low for each possible level of B. Thus, in this case the data from the input is blocked *i.e.* the data cannot be transmitted to the output. However, when ENABLE is made *high,* the output Z will follow the input at B, *i.e.* if B is *high*, Z will be *high* and if B is *low*, Z will be *low*. In this case the data at input B is enabled to be transmitted to the output.

18.7.1. Transistor AND Gate

Figure 18.13 depicts a transistor AND gate. When both transistors A and B are supplied with +5 V, the two transistors conduct. The current so produced drops the supply voltage of +5 V across R_1 thereby driving point M to ground. This, in turn cuts off Q_3 thereby causing Z to go to V_{CC} *i.e.* +5 V. It means that there is an output at Z only when there is an input at both A and B.

If either A or B is at 0 V, then Q_1 or Q_2 will be cut off and there will be no volt drop across resistor R_1. Hence point M will go to supply voltage of +5 V. Consequently Q_3 will be forward biased and conduct and whole of the supply voltage will be dropped across resistor R_2. As a result output Z will go to 0 volt.

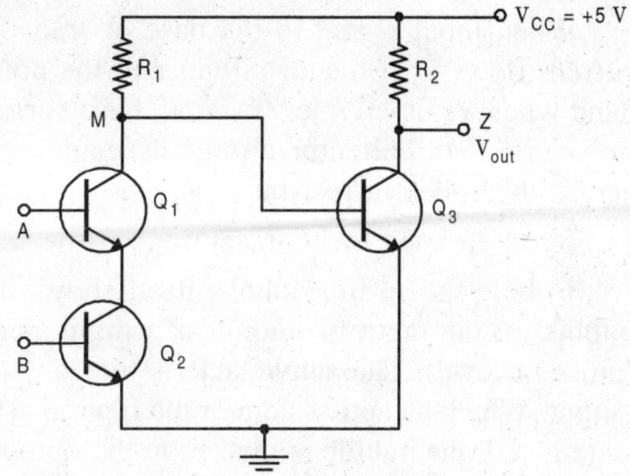

Fig. 18.13. *Transistor AND Gate*

18.7.2. Standard Package

Figure 18.14 depicts the pinout diagram of an IC7408, a TTL quad-2 input AND gate. This IC contains four 2-input AND gates. It is 14-pin DIP. After connecting a supply voltage of +5 V to pin-14 and a ground to pin-7, we can connect one or more of the AND gates to the other TTL devices.

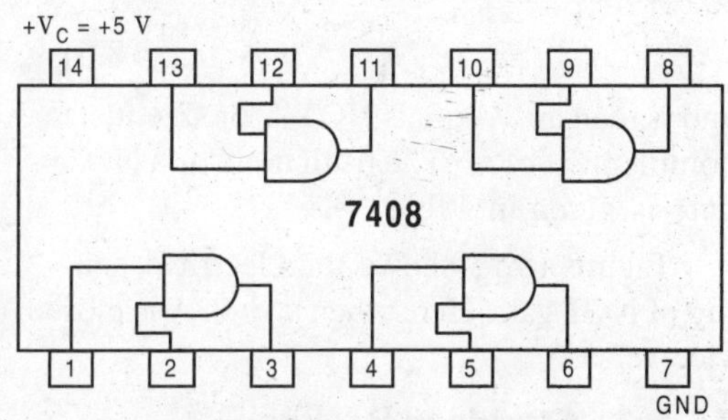

Fig. 18.14. *Pinout Configuration of IC 7408*

18.7.3. Pulsed Operation

Figure 18.15 shows two pulses A and B as inputs to an AND gate. When either input is *low* the output is *low*. Only when both inputs are *high*, the output is *high*. For the inputs, the output is *high* during time intervals t_2 and t_6 periods only. In remaining time intervals, the output is 0, as illustrated in Fig. 18.15 (*b*).

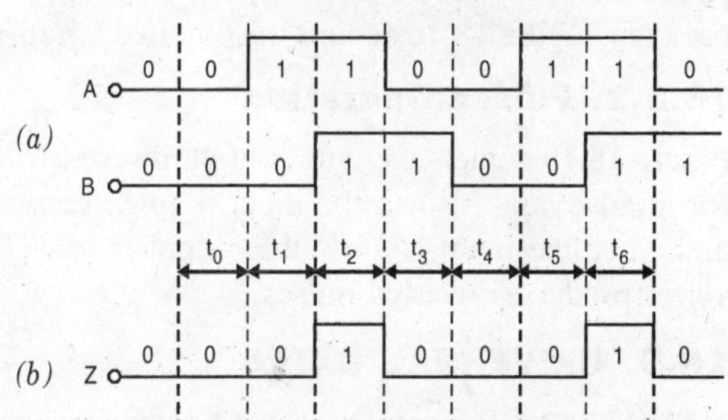

Fig. 18.15. *Timing Diagram For AND Gate*

18.8. NOT GATE OR INVERTER

The inverter (NOT circuit) performs a basic logic function called *inversion* or *complementation*. The purpose of the inverter is to change one logic level to the opposite level. In terms of bits, it changes a 1 to a 0 and vice-versa.

This gate has only one input and one output. NOT gate is called the *inverter* because output state is always opposite to the input state, so when the input is *low* signal, output signal is *high* and vice versa. It is called NOT gate because the output state is always opposite to that of input state. A transistor may be used as an inverter, as shown in Fig. 18.16 (*a*).

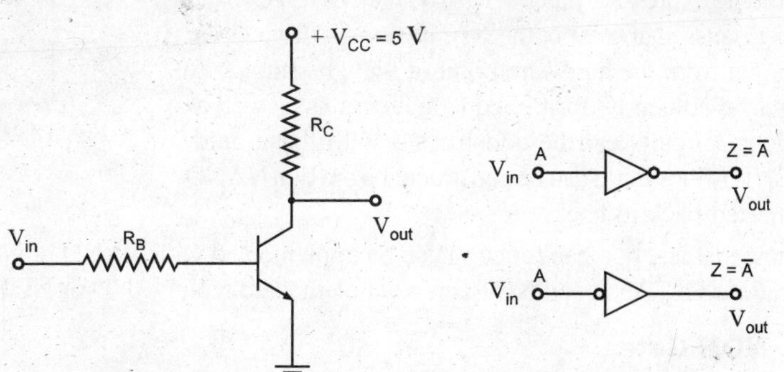

(*a*) *Transistor as an Inverter* (*b*) *Inverter Symbols*

Fig. 18.16

When input signal to the base of transistor, V_{in} is zero, base current of transistor is zero, so no current flows in collector circuit and the potential of V_{out} remains equal to V_{CC} *i.e.* 5 V. On the other hand when V_{in} is *high* level signal, base current flows through the transistor resulting in flow of saturation current in collector circuit. Because of *high* collector current there will be large voltage drop across the collector resistance R_C and so output V_{out} will be at *low* level.

Symbols used for representing inverters are shown in Fig. 18.16 (*b*).

Bubble shown in symbols, itself shows that output is a complement of input. The placement of the bubble on the input or output of a logic element is determined by the *active state* of the input signal (pulse or level). The active state is the state (1 or 0) when the signal is considered to be present on the output. When the active state of the input is a 0, the bubble is placed on the input. When the active output state is a 0, the bubble is placed on the output.

The logic equation of the NOT gate is written as

$$Z = \text{NOT}\ A = \overline{A} \quad \ldots(18.6)$$

and is read as 'Z equals NOT A' or 'Z equals complement of A'. The truth table of a NOT gate is given in Table 18.3.

TABLE 18.3. ***Truth Table For NOT Gate***

Input A	*Output* Z
0	1
1	0

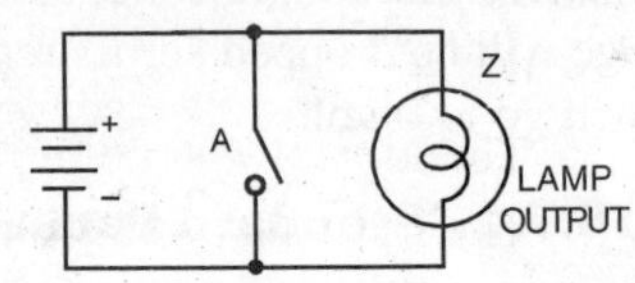

Fig. 18.17. ***Electrical Analog of NOT Gate***

Figure 18.17 shows the electrical analog of NOT gate. Here when switch A is closed, (input is *high*), the lamp does not glow (*i.e.* output is *low*) and vice versa.

18.8.1. Standard Package

It contains a TTL hex inverter. It is a 14-pin dual-in-line package. Figure 18.18 depicts the pinout diagram of IC 7404.

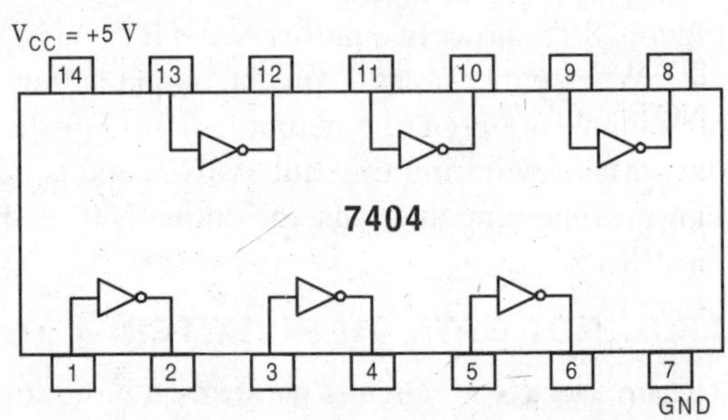

Fig. 18.18. ***Pinout Configuration of IC 7404***

18.8.2. Pulsed Operation

Figure 18.19 depicts the output of an inverter (or NOT gate) for a pulse input. When the input is *low*, the output is *high* and when the input is *high*, the output is *low*. Thus the inverter produces inverted pulses.

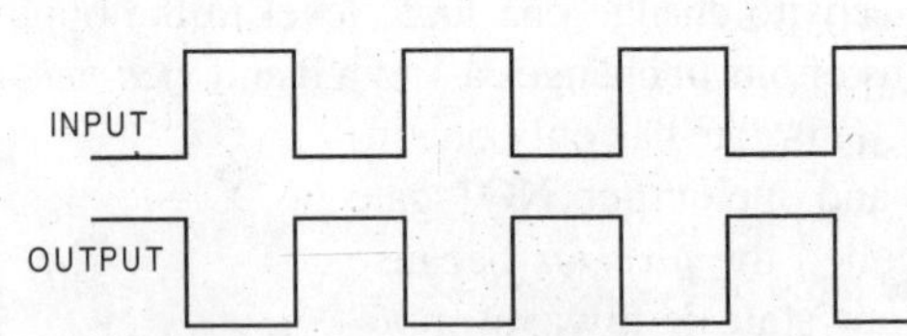

Fig. 18.19. ***Output of a NOT Gate For Pulsed Output***

18.9. UNIVERSAL GATES

NAND and NOR gates are said to be universal gates because any digital system can be implemented with these gates. Any boolean or logic expression can be realized with AND, OR and NOT gates discussed above but the NAND and NOR gates are more common from the hardware point of view because they are readily available in IC form. Sequential circuits as well as combinational circuits can be constructed with these gates because the flip-flop circuit can be constructed from two NAND gates connected back to back.

To show that any boolean function can be implemented with NAND or NOR gates, we need to show that the logic operations OR, AND and NOT can be implemented with NAND or NOR gate.

18.9.1. NOR Gate

The term NOR is a contraction of NOT-OR and implies an OR function with an inverted output. The output of this gate is *high*, only when all input voltages are *low*. A standard logic symbol for a 2-input NOR gate and the equivalent OR gate followed by an INVERTER are shown in Fig. 18.20. The standard symbol for the NOR gate is the same as the OR gate symbol except that it has a small circle on the output. This small circle represents the inversion operation. Thus, the NOR gate operates like an OR gate followed by an INVERTER, so that the circuits shown in Figs. 18.20 (*a*) and (*b*) are equivalent.

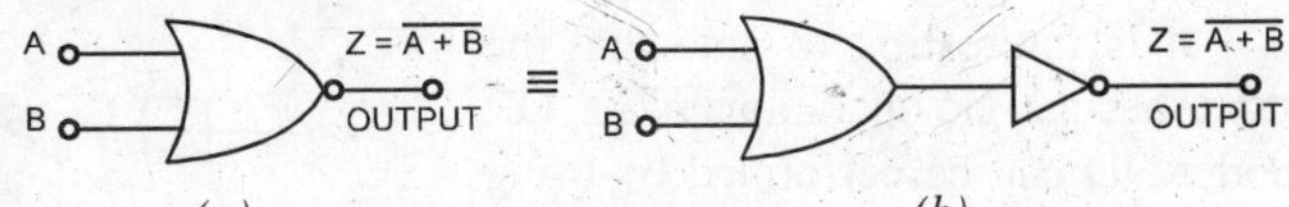

Fig. 18.20. ***Standard NOR Gate Logic Symbol***

TABLE 18.4. ***Truth Table For NOR Gate***

Inputs		*Output*
A	*B*	*Z*
0	0	1
0	1	0
1	0	0
1	1	0

The operation of this gate can be described as below :

The output of the OR gate, Z′ can be written from Eq. (18.4)

$$Z' = A + B + C + \ldots N \qquad \ldots(18.7)$$

and the output of the NOT gate can be written from Eq. (18.6)

$$Z = \overline{Z}' = \overline{A + B + C + \ldots N} \qquad \ldots(18.8)$$

The logical operation represented by Eq. (18.8) is known as the NOR operation.

The truth table for 2-input NOR gate is given in Table 18.4. From Table 18.4 we find that NOR-gate output is the exact inverse of the OR-gate output for all possible input conditions. Where as an OR-gate output goes *high* when any input is *high*, the NOR gate output goes *low* when any input is *high*. This same operation can be extended to NOR gates with more than two inputs.

The logic equation of NOR gate is given as $Z = \overline{A + B}$ and is read as 'Z equals NOT (A OR B)'.

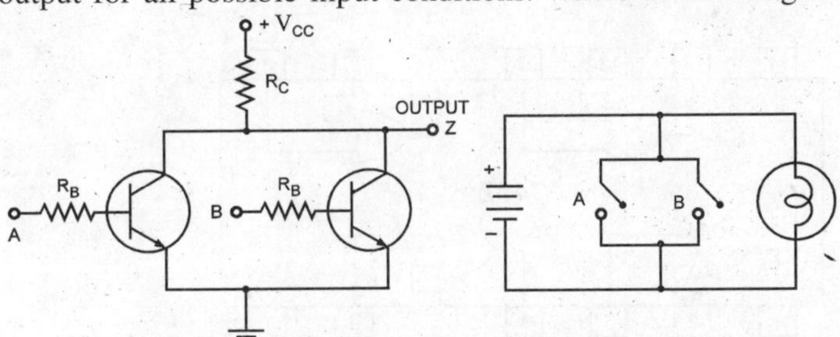

(a) Circuit For a NOR Gate *(b) Electrical Analog of NOR Gate*

Fig. 18.21

The electrical circuit and electrical analog of NOR gate are shown in Figs. 18.21 (*a*) and (*b*) respectively. As seen from the Fig. 18.21 (*a*), the output Z is 1 only when both transistors are cut off *i.e.* when A = 0 and B = 0. For any other condition of input such as A = 1, B = 0, or A = 0, B = 1 or A = 1, B = 1 one or both transistors operate in saturation and as a result the output Z is *low* (*i.e.* 0).

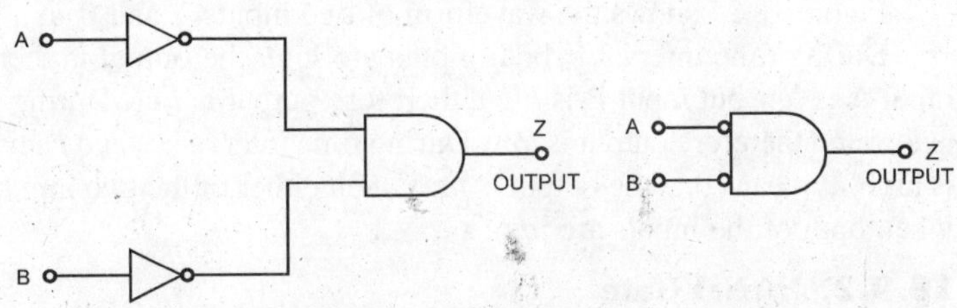

(a) AND Gate With Inverted Input Signals *(b) Inverted-Input AND Gate*

Fig. 18.22

From Fig. 18.21 (*b*) it is evident that the lamp does not glow *i.e.* the output is zero when either of the two inputs A or B is *high* (*i.e.* either of switches A or B is closed). The output is *high* (the lamp glows) only when both of the inputs are *low* (*i.e.* both of switches A and B are opened).

In Fig. 18.22 (*a*) AND gate with input signals inverted is shown. It also acts as a NOR gate. Figure 18.22 (*b*) shows the symbol for inverted input AND gate.

If we look at Table 18.5 (truth table for AND gate with its inverted inputs) we find that, it is same as Table 18.4 (truth table for NOR gate). So AND gate with its inputs inverted can be used as NOR gate.

TABLE 18.5. ***Truth Table For Inverted Input AND Gate***

Inputs		*Output*
A	*B*	*Z*
0	0	1
0	1	0
1	0	0
1	1	0

It is interesting to note that the three basic logic operations NOT, OR and AND can be performed by using only the NOR gates, as illustrated in Fig. 18.23.

NOT logic operation can be performed by tying together the two inputs, as shown in Fig. 18.23 (*a*).

OR logic operation can be performed by using two NOR gates, as shown in Fig. 18.23 (*b*). The final output is inverted and is given as Z = A + B, which is a logic function of a normal OR gate.

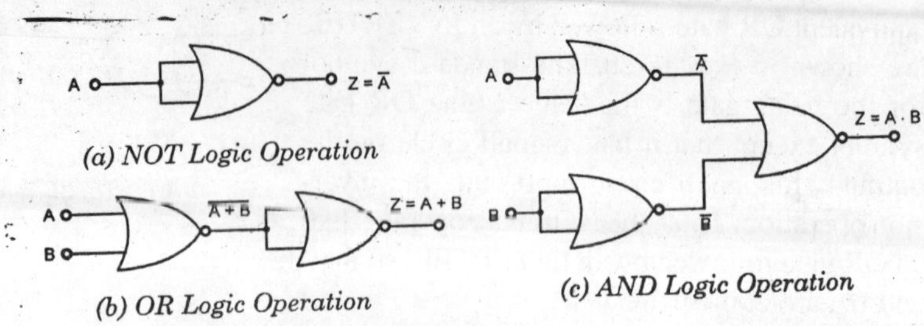

(*a*) *NOT Logic Operation*

(*b*) *OR Logic Operation*

(*c*) *AND Logic Operation*

Fig. 18.23

AND operation can be performed by using three NOR gates, as illustrated in Fig. 18.23 (*c*). The inputs have, thus, been inverted before they are applied to the NOR gate. The output $\overline{\overline{A} + \overline{B}}$ equals A·B.

Standard Package. Figure 18.24 depicts the pinout diagram of an IC 7402, a TTL quad 2 input NOR gate. This IC contains four 2-input NOR gates inside a 14-pin dual-in-line package (DIP).

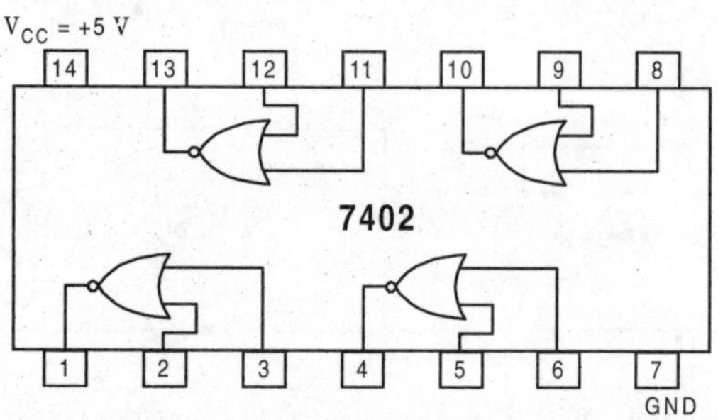

Fig. 18.24. ***Pinout Diagram of IC7402***

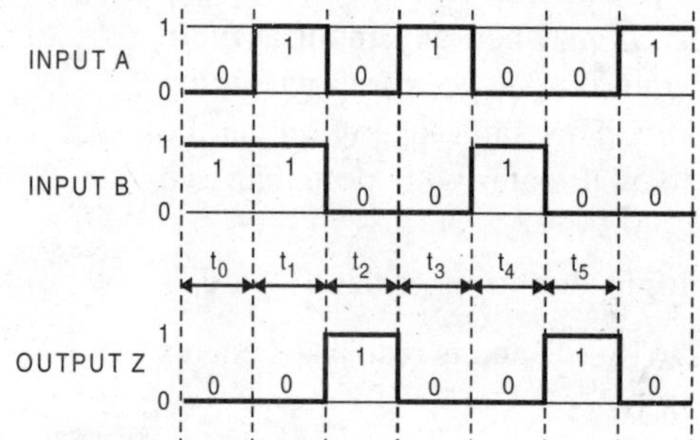

Fig. 18.25. ***Timing Diagram For NOR Gate***

Figure 18.25 shows the waveform of two inputs A and B to a NOR gate.

During time interval t_1 both inputs are *high* the output is therefore, *low*. During time intervals t_0 and t_4 input A is *low* but input B is *high*, therefore output is *low*. During time interval t_3 input A is *high* but input B is low and therefore output is *low*. During time intervals t_2 and t_5, both inputs A and B are *low*, therefore output is high. It means that the output is *low* when either of the two inputs A or B is *high* and the output is *high* only when both of the inputs are *low*.

18.9.2. NAND Gate

The term NAND is a contraction of NOT-AND and implies an AND function with complemented (inverted) output. In this gate, output is *low* only when all inputs are *high*. A standard logic symbol for a 2-input NAND gate and the equivalent AND gate followed by an INVERTER are shown in Fig. 18.26. The standard symbol for the NAND gate is the same as the AND gate symbol except for small circle on its output. Once again this small circle denotes the inverse operation. Thus the NAND gate operates like an AND gate followed by an INVERTER, so that the circuits shown in Fig. 18.26 (*a*) and (*b*) are equivalent.

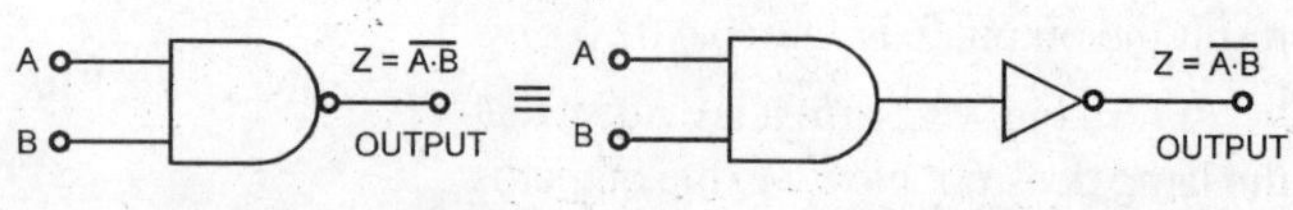

(*a*) (*b*)

Fig. 18.26. ***Standard NAND Gate Logic Symbol***

The operation of this gate can be described as below :

The output of the AND gate, Z′ can be written from Eq. (18.5)

$$Z' = A \cdot B \cdot C \ldots \cdot N \qquad \ldots(18.9)$$

and the output of the NOT gate can be written from Eq. (18.6)

$$Z = \overline{Z'} = \overline{A \cdot B \cdot C \cdots\cdot N} \quad ...(18.10)$$

The logical operation represented by Eq. (18.10) is known as NAND operation.

TABLE 18.6. ***Truth Table For NAND Gate***

Inputs		*Output*
A	*B*	*Z*
0	0	1
0	1	1
1	0	1
1	1	0

The truth table for 2-input NAND gate is given in Table 18.6. From Table 18.6 we find that NAND gate output is the exact inverse of the AND gate for all possible input conditions. The NAND gate output goes *low* only when all the inputs are *high* while the AND gate output goes *high* only when all the inputs are *high*.

The logic equation for NAND gate is given as $Z = \overline{A.B}$ and is read as 'Z' equals NOT (A AND B).

The electrical circuit and electrical analog of NAND gate are shown in Fig. 18.27 (*a*) and (*b*) respectively. Here a diode AND circuit is connected to a transistor NOT circuit that gives the NAND circuit. It can be seen from the Fig. 18.27 (*a*) that the point X would be driven to ground when either of the diodes D_1 or D_2 or both D_1 and D_2 conduct. Under such conditions, the transistor is cut off and, therefore, output goes *high*. Output is *low* only when both input voltages are *high* (at + V_{CC}) so that X is at + V_{CC} and transistor operates in saturation.

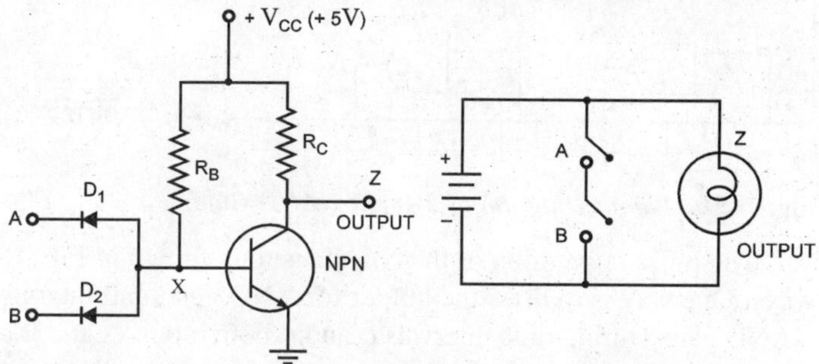

(*a*) *Circuit For NAND Gate* (*b*) *Electrical Analog of NAND Gate*

Fig. 18.27

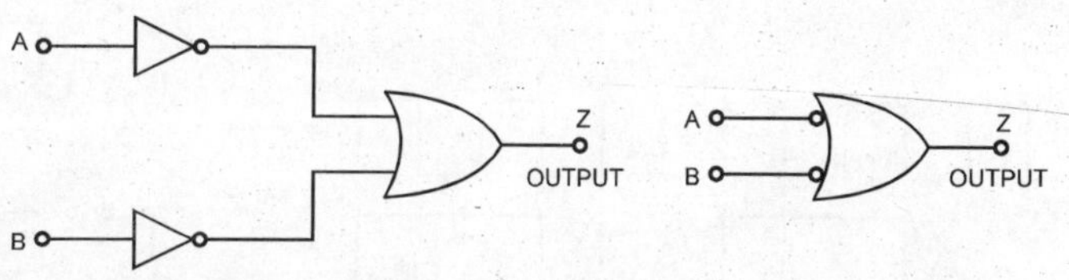

(*a*) *OR Gate With Inverted Input Signals* (*b*) *Inverted Input OR Gate*

Fig. 18.28

From Fig. 18.27 (*b*) it is evident that the lamp does not glow (*i.e.* the output is zero) only when both of the switches A and B are closed (*i.e.* both the inputs are *high*). The lamp glows (*i.e.* the output is *high*) when either or both of the switches are open (either or both of the inputs are *low*).

TABLE 18.7. ***Truth Table For Inverted Input OR Gate***

Inputs		*Output*
A	*B*	*Z*
0	0	1
0	1	1
1	0	1
1	1	0

In Fig. 18.28 (*a*) an OR gate with its inputs inverted is shown. It also acts as a NAND gate.

If we look at Table 18.7 (truth table for OR gate with its inputs inverted), we find that it is the same as truth table for NAND gate (Table 18.6). So an OR gate with its inputs inverted can be used as a NAND gate.

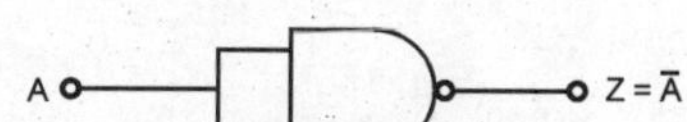

(*a*) *NOT Logic Operation*

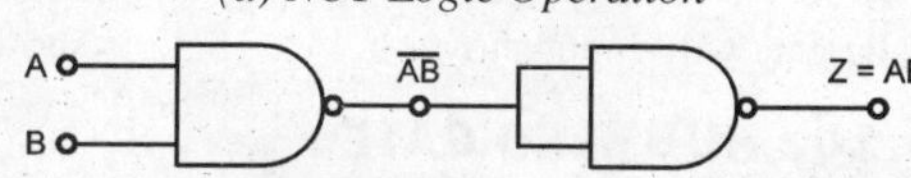

(*b*) *AND Logic Operation*

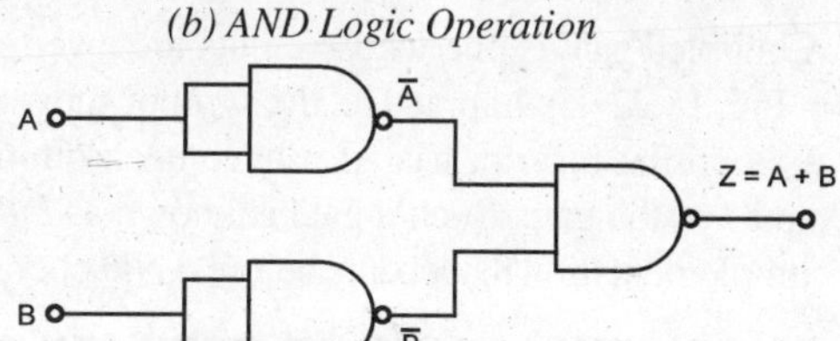

(*c*) *OR Logic Operation*

Fig. 18.29

The three logic operations NOT, OR and AND can be performed by using only the NAND gates, as illustrated in Fig. 18.29.

NOT logic operation can be performed by tying its two inputs together, as illustrated in Fig. 18.29 (*a*).

AND logic operation can be performed by using two NAND gates, as shown in Fig. 18.29 (*b*).

OR logic operation can be performed by three NAND gates, as shown in Fig. 18.29 (*c*).

Standard Package. Figure 18.30 depicts the pinout diagram of an IC7400, a TTL, quad 2-input NAND gate. It contains four 2-input NAND gates inside a 14-pin DIP.

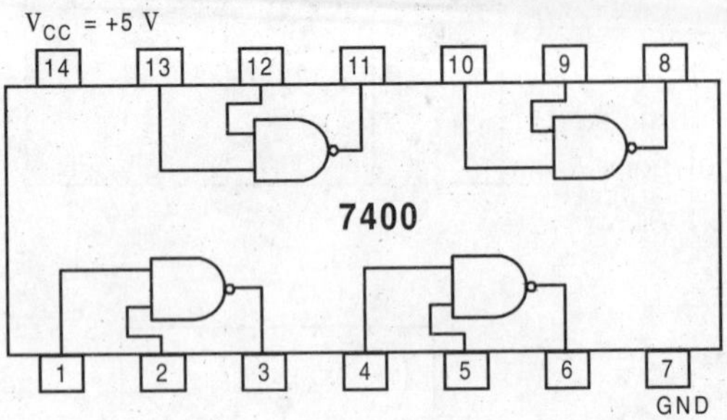

Fig. 18.30. ***Pinout Diagram of 2-Input NAND Gate***

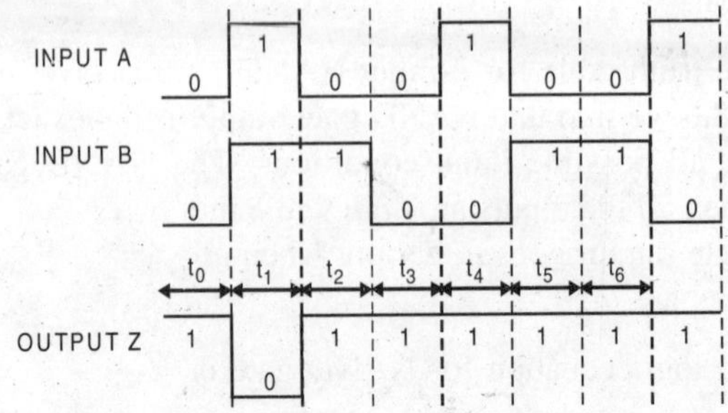

Fig. 18.31. ***Timing Diagram For NAND Gate***

The pulsed operation of the NAND gate is shown in Fig. 18.31. The output can be determined in several ways. One way is to draw the output for AND gate, and then invert it. Another way is by using truth table of NAND gate. During time intervals t_0 and t_3 both inputs A and B are *low*, the output Z is therefore, high. During time interval t_1 both inputs are *high*, so output is *low*. During time intervals t_2, t_5 and t_6, input A is low but input B is *high* therefore output is *high*.

Example 18.1. Draw the output from a 3-input NAND gate.

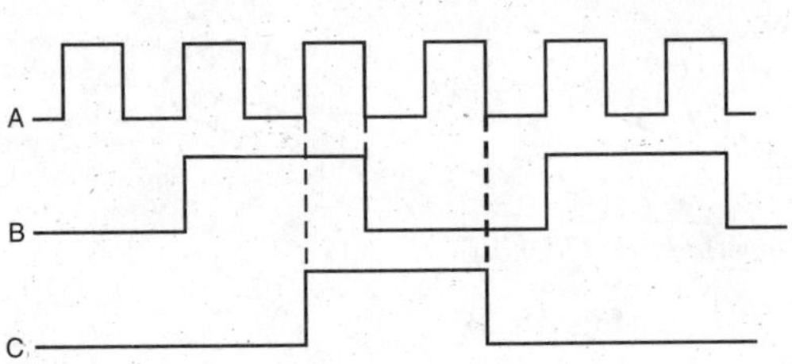

Fig. 18.32

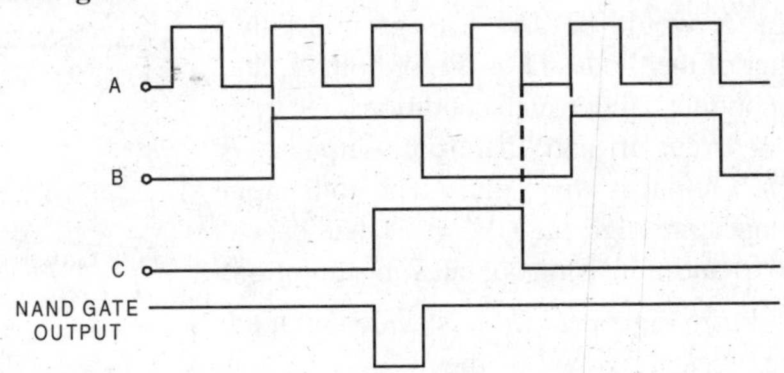

Fig. 18.33

Solution : NAND gate output goes *low* only when all the inputs go *high*, otherwise output is *high*. The output of a 3-input NAND gate with input shown in Fig. 18.32 is shown in Fig. 18.33.

18.10. BUBBLED GATES

A bubbled gate is one whose inputs are inverted *i.e.*, it is a *negated* gate. AND gate with inverted inputs is shown in Fig. 18.22 (*a*). In practice, the symbol shown in Fig. 18.22 (*b*) is widely used rather that shown in Fig. 18.22 (*a*). As seen the inverter triangles have been omitted and the two small circles or bubbles have been placed on the inputs of the gate. Such a gate is known as *bubbled AND gate*. The bubbles act as a reminder of the inversion or complementation that occurs before AND operation of the inputs. Similarly bubbled OR gate is shown in Fig. 18.28.

18.11. EXCLUSIVE-OR GATE (EX-OR GATE)

Two special logic circuits that occur quite often in digital systems are the *exclusive-OR* and *exclusive-NOR* circuits.

The circuit of an exclusive-OR gate is shown in Fig. 18.34 (*a*). As obvious, there are two inputs and one output. The exclusive-OR operation is denoted by ⊕. Hence the expression for the output of the circuit is given as

$$Z = A \oplus B = \overline{A}B + A\overline{B} \quad ...(18.11)$$

The symbol and equivalent switching circuit of an exclusive-OR gate are shown in Figs. 18.34 (*b*) and (*c*) respectively.

It is observed that output of the exclusive-OR gate is *high* if its either input, *but not both*, is *high*. It means that when the two inputs are different (one is *high* and another *low*), the output is *high* and when the two inputs are the same (either *low* or *high*) the output is *low*. This circuit is abbreviated as EX-OR gate.

The truth table for EX-OR gate is shown in Table 18.8.

An EX-OR gate has only two inputs.* The two inputs are combined so that the output, $Z = \overline{A}B + \overline{B}A$. The output expression for EX-OR gate, in short, is given as

$$Z = A \oplus B$$

where the symbol ⊕ represents the EX-OR gate operation.

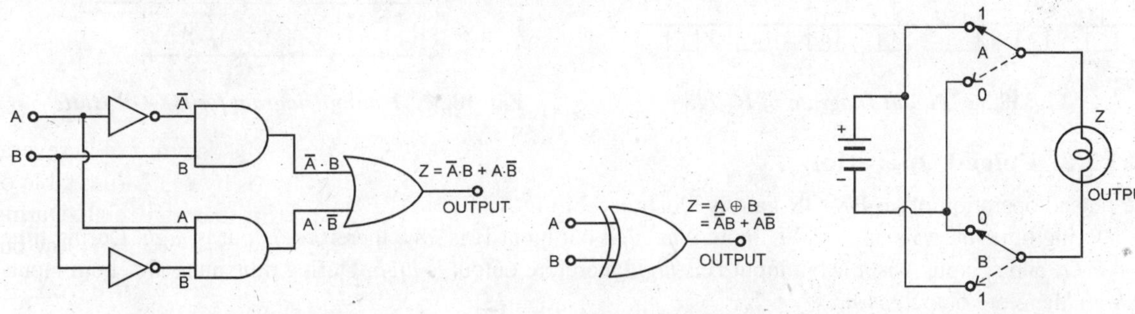

(a) Circuit of Exclusive-OR Gate *(b) Exclusive-OR Gate Symbol* *(c) Electrical Analog of Exclusive-OR Gate*

Fig. 18.34. ***Exclusive-OR Gate***

This circuit is also called an *inequality comparator* or *detector* because it produces output only when the *two inputs are different*.

The main characteristic property of an EX-OR gate is that it can perform modulo-2 addition. It should be noted that the same EX-OR truth table applies when adding two binary digits (bits). A 2-input EX-OR circuit is therefore sometimes called a *modulo-2 adder* or a *half adder*. The name half adder refers to the fact that possible carry-bit, resulting from an addition of two preceding bits, has not been taken into account. The name Exclusive-OR is derived from the fact that its output is a 1, only when exclusively one of its input is a 1 (it excludes the condition when both the inputs are 1).

TABLE 18.8. ***Truth Table For EX-OR Gate***

Inputs		***Output***
A	***B***	***Z***
0	0	0
0	1	1
1	0	1
1	1	0

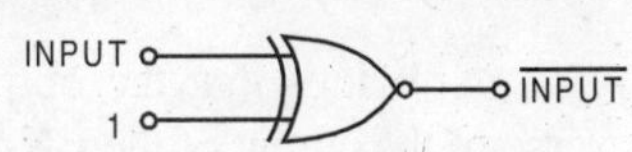

Fig. 18.35. ***EX-OR Gate as an Inverter***

Three or more variable EX-OR gates do not exist. When more than two variables are to be EX-ORed, a number of two-input EX-OR gates will be used. The EX-OR of a number of variables assumes a 1 state only when an odd number of input variables assume a 1 state.

An EX-OR gate can be used as an inverter by connecting one of its two input terminals to logic 1 and feeding the input sequence to be inverted to the other terminal as shown in Fig. 18.35. If the input bit is a 0, the output is $0 \oplus 1 = 1$, and if the input bit is a 1, the output is, $1 \oplus 1 = 0$. In fact, it can be said that an EX-OR gate can be used as a *controlled inverter*, that is, one of its inputs can be used to decide whether the signal at the other input will be inverted or not.

This circuit finds applications, obviously, where two digital signals are to be compared.

* There are no 3 or 4-input EX-OR gates.

18.11.1. Standard Package

Figure 18.36 shows the pinout diagram of an IC 7486 which is a quad 2-input EX-OR gate. It contains four 2-input EX-OR gate in a 14-pin DIP.

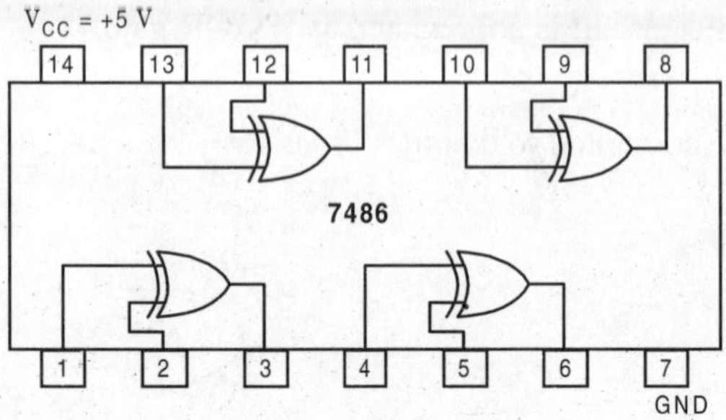

Fig. 18.36. *Pinout Diagram of IC 7486*

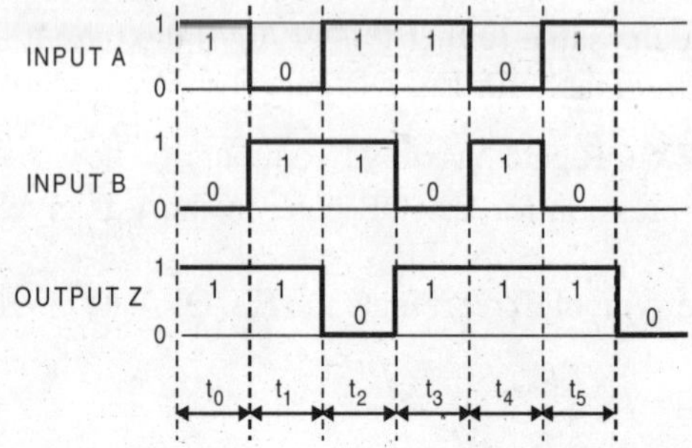

Fig. 18.37. *Timing Diagram for EX-OR Gate*

18.11.2. Pulsed Operation

The pulsed operation of an EX-OR gate is illustrated in Fig. 18.37.

During time intervals t_0, t_3 and t_5 input A is *high* but input B is *low*, therefore output is *high*. During time intervals t_1 and t_4 input A is *low* but input B is *high*, therefore output is *high*. During time interval t_2 both inputs are *high*, therefore output is *low*.

18.12. EX-NOR GATE

The exclusive-NOR circuit, abbreviated as EX-NOR, operates just opposite to the EX-OR circuit. The circuit, symbol and truth table of an EX-NOR circuit are given in Figs. 18.38 (*a*) and (*b*) and Table 18.9 respectively. This gate has *high* output if and only if its both inputs are same (either 0 or 1) and its output is *low* if inputs are different. The output equation of this gate is given as

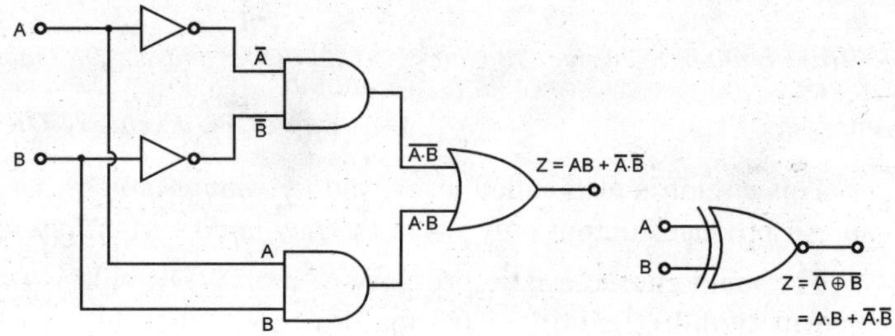

(*a*) Circuit (*b*) Symbol

Fig. 18.38. *Exclusive-NOR Gate*

$$Z = \overline{A \oplus B} = AB + \bar{A}\,\bar{B} \qquad ...(18.12)$$

It is obvious that the output of the EX-NOR circuit is the exact inverse of the output of the EX-OR circuit. So the symbol for this circuit can be obtained by adding a small circle at the output of the EX-OR symbol [Fig. 18.34 (*b*)].

The EX-NOR gate is also called a *coincidence gate*, because its output is a 1 only when its inputs coincide (either 0, 0 or 1, 1). It can be used as an *equality detector* because its output is a 1 only when its inputs are equal.

TABLE 18.9. ***Truth Table For EX-NOR Gate***

Inputs		Output
A	*B*	*Z*
0	0	1
0	1	0
1	0	0
1	1	1

Three or more variable EX-NOR gates do not exist. When more than two variables are to be EX-NORed, a number of two-input EX-NOR gates can be used. The EX-NOR of a number of variables assumes a 1 state, only when even number (including 0) of input variables assumes a 0 state.

Fig. 18.39. *EX-NOR Gate as an Inverter*

An EX-NOR gate can be used as an *inverter* by connecting one of its two input terminals to logic 0 and feeding the input sequence to be inverted to the other terminal as shown in Fig. 18.39.

If the input bit is a 0, the output is, $\overline{0 \oplus 0} = 1$, and if input is a 1, the output is $\overline{0 \oplus 1} = 0$. In fact, it can be said that an EX-NOR gate can be used as a *controlled inverter*, that is, one of its inputs can be used to decide whether the signal at the other input will be inverted or not. The EX-NOR gate can be used as a *comparator too*.

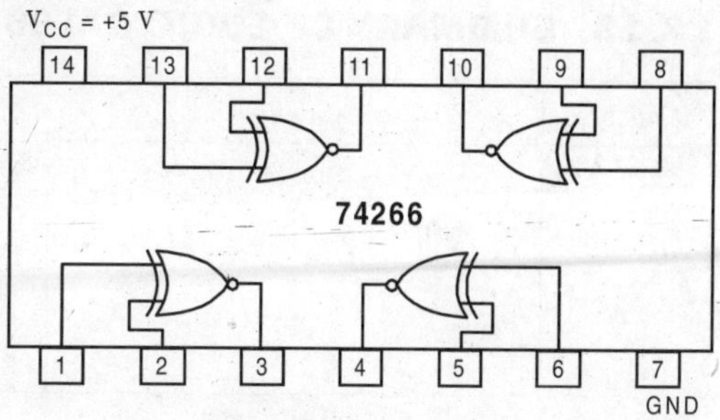

Fig. 18.40. ***Pinout Diagram of IC 74266***

18.12.1. Standard Package

Figure 18.40 shows the pinout diagram of an IC 74266, a TTL quad-2 input EX-NOR gate. It contains four 2-input EX-NOR gates in a 14-pin DIP.

18.12.2. Pulsed Operation

The pulsed operation of an EX-NOR gate is illustrated in Fig. 18.41. During time intervals t_0 and t_2 input A is *low* but input B is *high*, therefore output is *low*. During time intervals t_1 and t_5 input A is *high* but input B is *low* and therefore output is again *low*. During time interval t_3 both inputs are *low* and during time interval t_4 both inputs are *high* therefore output is *high*.

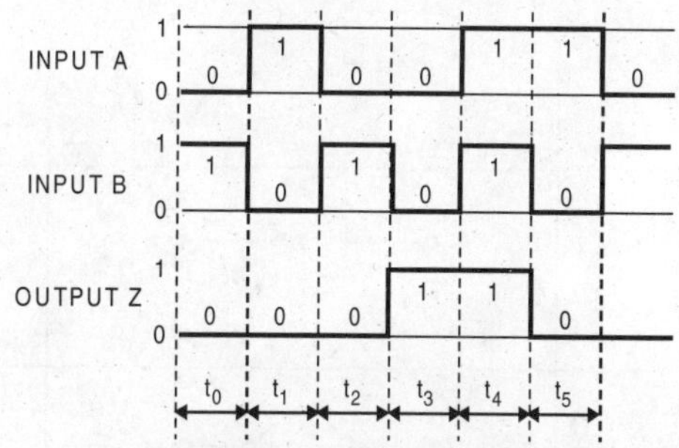

Fig. 18.41. ***Timing Diagram For EX-NOR Gate***

Example 18.2. Find the output waveforms for 2-input EX-OR and EX-NOR gates for the input (Fig. 18.42). Justify your answer.

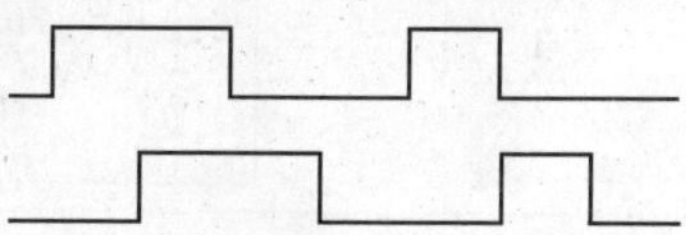

Fig. 18.42

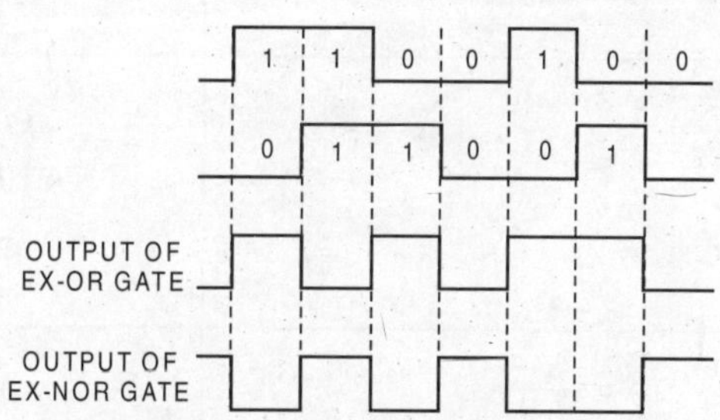

Fig. 18.43

Solution : The output of EX-OR gate goes *high* whenever inputs are different while the output of EX-NOR gate goes *high* when both input are same. The output waveforms for 2-input EX-OR and EX-NOR gates are shown in Fig. 18.43.

Example 18.3. Predict the output of the circuit (Fig. 18.44)

Solution : The given circuit consists of 2 EX-NOR gates. Final output Y becomes *high* when there are odd numbers of 1's input. Thus out of 3-inputs if any one input is *high* or if all the three inputs are *high* then the output Y will be *high*. It is illustrated in Fig. 18.45.

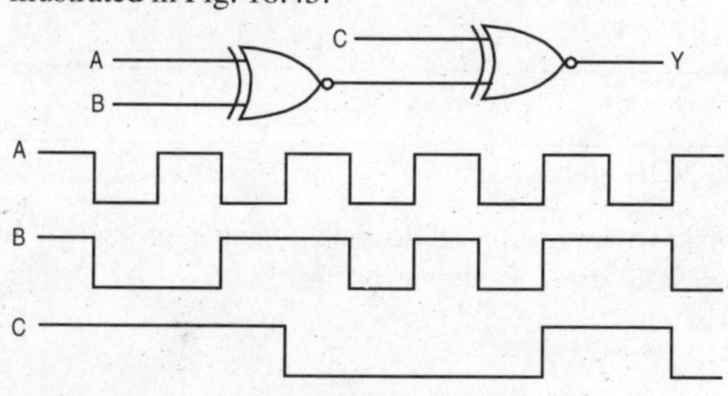

Fig. 18.44

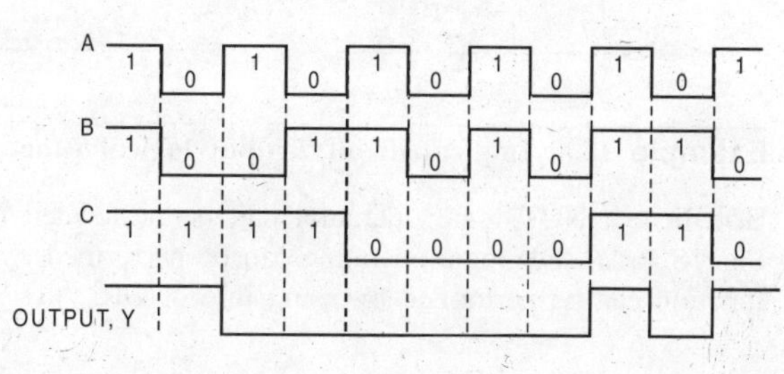

Fig. 18.45

18.13. SUMMARY OF LOGIC GATES

Gate	*Symbol*	*Function*	*Truth Table*
OR	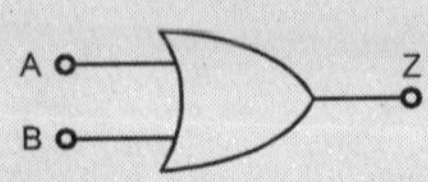	Z = A OR B $= A + B$	*Inputs* *A B* | *Output Z* 0 0 | 0 0 1 | 1 1 0 | 1 1 1 | 1
AND	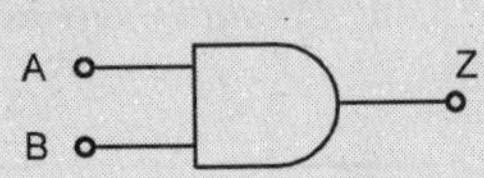	Z = A AND B $= A \cdot B$	*Inputs* *A B* | *Output Z* 0 0 | 0 0 1 | 0 1 0 | 0 1 1 | 1
NOT	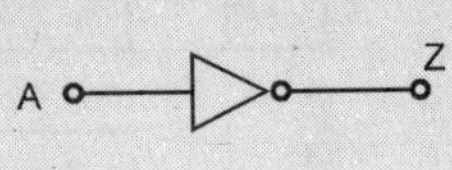	Z = NOT A $= \overline{A}$	*Inputs* *A* | *Output Z* 0 | 1 1 | 0
NOR	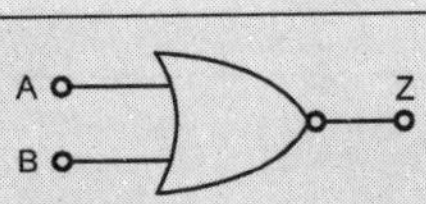	Z = A NOT OR B = A NOR B $= \overline{A + B}$	*Inputs* *A B* | *Output Z* 0 0 | 1 0 1 | 0 1 0 | 0 1 1 | 0
NAND	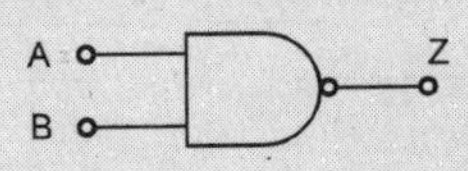	Z = A NOT AND B = A NAND B $= \overline{A \cdot B}$	*Inputs* *A B* | *Output Z* 0 0 | 1 0 1 | 1 1 0 | 1 1 1 | 0
EX-OR	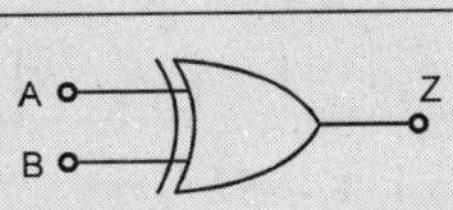	Z = A EX-OR B $= A \oplus B$ $= A\overline{B} + \overline{A}B$	*Inputs* *A B* | *Output Z* 0 0 | 0 0 1 | 1 1 0 | 1 1 1 | 0
EX-NOR	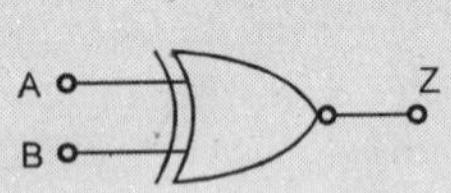	$Z = \overline{\text{A EX-OR B}}$ = A EX-NOR B $= \overline{A \oplus B} = AB + \overline{A}\,\overline{B}$	*Inputs* *A B* | *Output Z* 0 0 | 1 0 1 | 0 1 0 | 0 1 1 | 1

Example 18.4. Implement all 2-input logical functions using NOR gates.

Solution : NOT logic operation can be performed by tying its two inputs of NOR gate together as shown in Fig. 18.46 (*a*). OR logic operation can be performed by using two NOR gates as shown in Fig. 18.46 (*b*) AND logic function can be performed by using three NOR gates as shown in Fig. 18.46 (*c*).

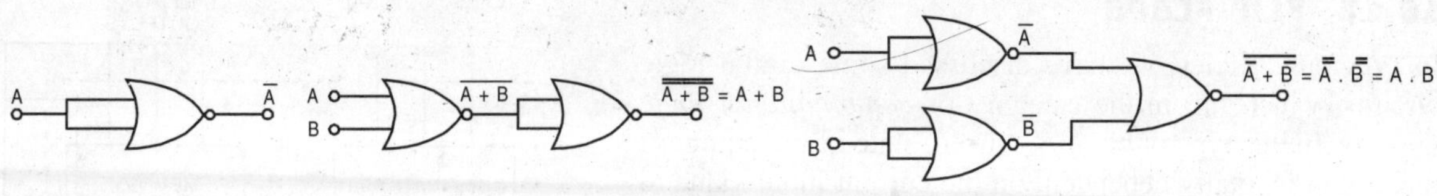

(a) *NOT Logic Operation* (b) *OR Logic Operation* (c) *AND Logic Operation*

Fig. 18.46

Example 18.5. Realise (*a*) 2-input NOR gate using 2-input NAND gates and (*b*) 2-input NAND gate using 2-input NOR gates.

Solution : (*a*) Realization of 2-input NOR gate using 2-input NAND gate is illustrated in Fig. 18.47.

(*b*) Realization of 2-input NAND gate using 2-input NOR gate is shown in Fig. 18.48.

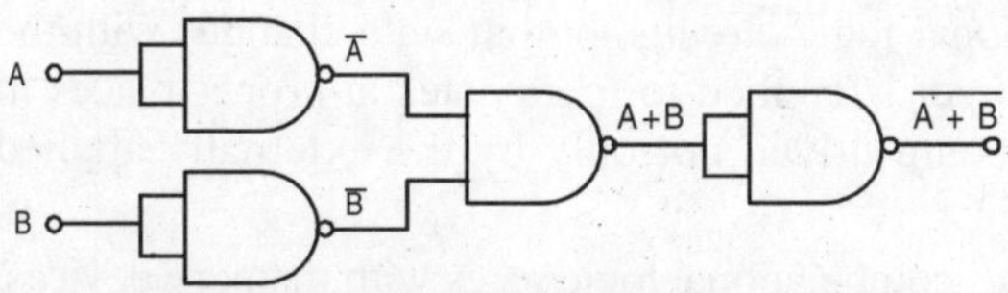

Fig. 18.47. ***Realization of 2-Input NOR Gate Using 2-Input NAND Gates***

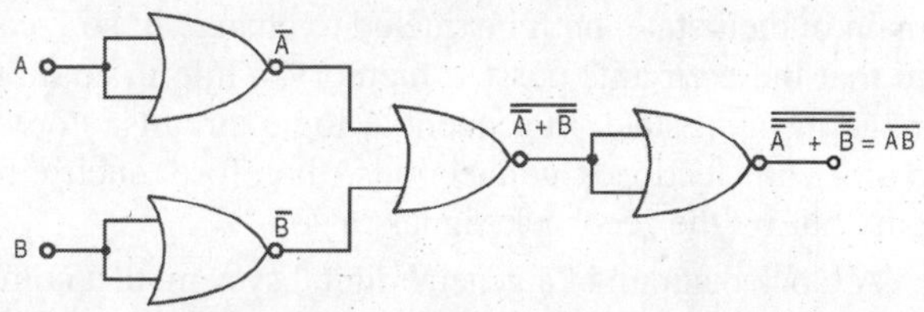

Fig. 18.48. ***Realization of 2-Input NAND Gate Using 2-Input NOR Gates***

Example 18.6. Using NOR gates only, realise EX-OR gate.

Solution : EX-OR gate using NOR gates only is shown in Fig. 18.49.

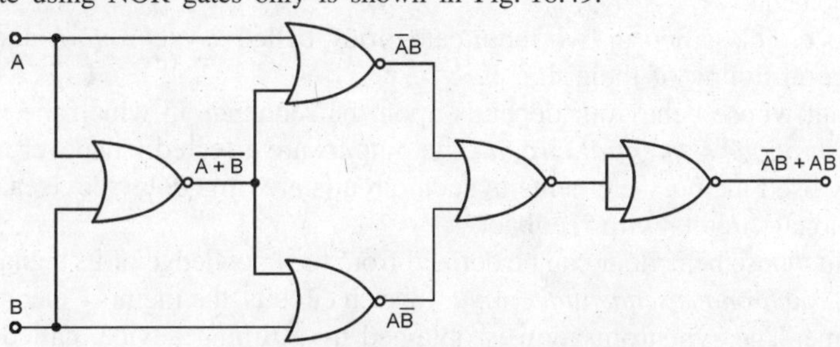

Fig. 18.49. ***Realization of EX-OR Gate Using NOR Gate Only***

Example 18.7. What are universal gates and why are they so called ? How can OR and XOR gates be realized using NAND gates only ? [*Pb. Technical Univ. Basic Electrical and Electronics Engineering Dec. 2006*]

Solution : OR logic operation can be performed by three NAND gates as shown below (Fig. 18.50)

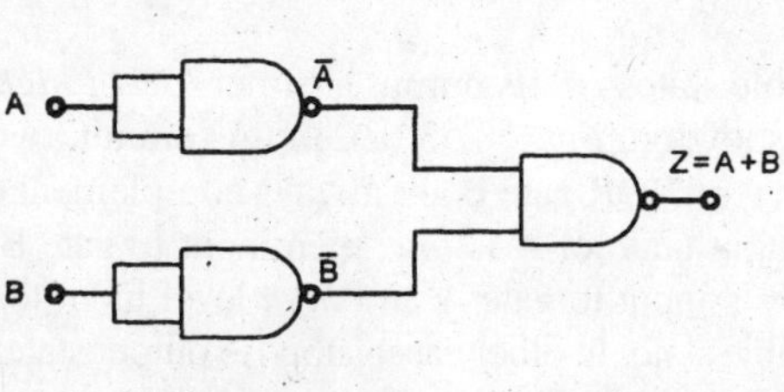

Fig. 18.50

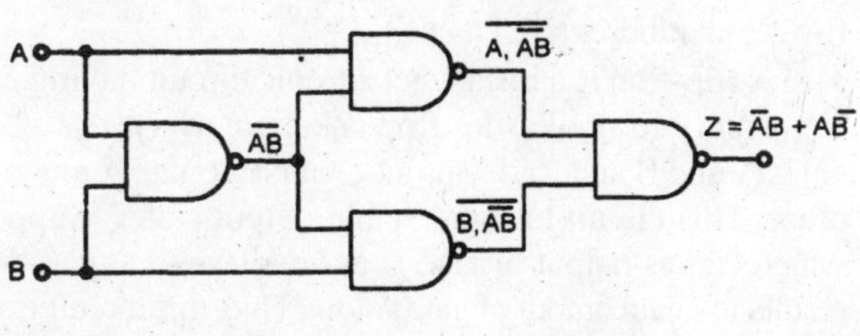

Fig. 18.51

EX-OR gate can be had by using four NAND gates as shown (Fig. 18.51).

18.14. FLIP-FLOPS

In previous articles, we have discussed some digital logic circuits, which fall in the category of *combinational logic circuits*. In this type of circuits, current state of the output of the circuit is totally dependent on the current input value to the circuit.

Other category of digital logic circuits is called the *sequential logic circuits*. In this type of circuits, current state of the output depends not only on the current input values but also upon the sequence of logic values that lead to the current input values.

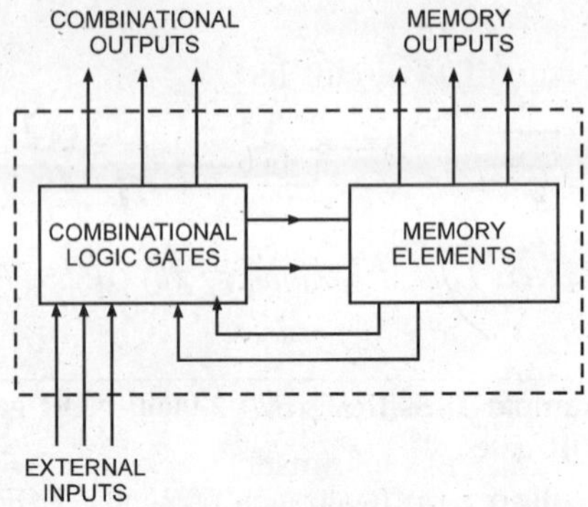

Fig. 18.52. ***Block Diagram of General Digital System***

In sequential logic circuits, output states of the circuits remain in their state, until instructed to change it, so it can be said that these circuits possess memory, while in combinational logic circuits, output state changes with the change in input state. In sequential logic circuits, stored output is applied to logic gates at proper places in circuits with feedback connections, therefore, such circuits are driven not only by the externally applied inputs but by the feedback signals also.

A block diagram of a general digital system, that combines combinational logic gates with memory devices, is shown in Fig. 18.52. The combinational portion accepts logic signals from external inputs and from the outputs of the memory elements. The combinational logic circuit operates on these inputs to generate various outputs, some of which are used to determine the binary values to be stored in the memory elements. The outputs of some of the memory elements, in turn, go to the inputs of the logic gates in the combinational circuits. This process indicates that the external outputs of a digital system are a function of both its external inputs and the information stored in its memory elements.

Sequential circuits are classified in two main categories, called *asynchronous* and *synchronous sequential circuits* depending on timing of their signals.

A sequential circuit whose behaviour depends upon the sequence in which the input signals change is referred to as an *asynchronous sequential circuit*. The outputs are affected whenever there is a change in the inputs. The commonly used memory elements in such circuits are time delay devices. These circuits may be regarded as combinational circuits with feedback.

A sequential circuit whose behaviour can be defined from the knowledge of its signals at discrete instants of time is referred to as a *synchronous sequential circuit*. In such circuits, the memory elements are affected only at discrete instants of time. The synchronisation is obtained by a timing device, called a *system clock* which generates a periodic train of clock pulses. The outputs are affected only with the application of a clock pulse. The use of asynchronous circuits because of tedious and difficult design, is limited. Sequential circuits have gained considerable domination and wide popularity and are referred to as *clocked-sequential circuits.*

The most important memory element is the *flip-flop*, which is made of an assembly of logic gates. Even though a logic gate, by itself, has no storage capability, several can be connected together in ways that permit information to be stored. There are several different gate arrangements that are used to produce these flip-flops (abbreviated as FF).

A flip-flop is a bistable electronic circuit having two stable states *i.e.* its output is either *low* or *high*. Most elementary form of a flip-flop is known as a *latch* which is shown in Fig. 18.53 (*a*). In this circuit, two NOR gates A and B are used. Output Q of NOR gate A and output $\overline{Q}$ of NOR gate B are always complement of each other. This circuit has two stable output states. Suppose output state of A is *low*, then input to gate B is *low* which causes output of gate B at *high* level. Output of gate B is input to gate A and *high* level input to gate A results in again output of gate A *low*. Thus this circuit is in a stable state. In other case, suppose output state of gate A is *high* then again output of gate B is at *low* level which also acts as an input to gate A, gives output of gate A at *high* level. This is the second stable state of the circuit.

From the above discussion the following points are noteworthy.

1. The outputs Q and $\overline{Q}$ are always complementary.
2. The circuit has two stable states; in one of the stable state Q = *high* or 1 which is referred to as a 1 state (or set state) whereas in the other stable state Q = 0 which is referred to as the 0 state (or reset state).
3. If the circuit is in 1 state, it continues to remain in this state and similarly if it is in 0 state, it continues to remain in this state. This property of the circuit is referred to as *memory*, *i.e.* it can store 1-bit of digital information.

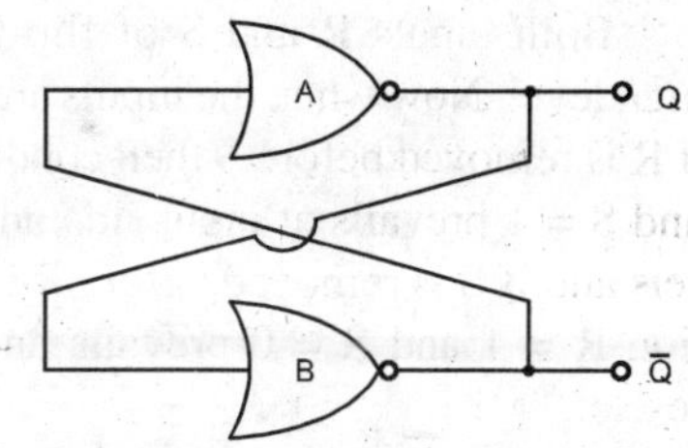

Fig. 18.53(*a*). *NOR Latch Circuit*

Since this information is locked or latched in this circuit, this circuit is known as a *latch*. Latch circuit can also be formed with NAND gates, as shown in Fig. 18.53(*b*). The behaviour of this latch circuit is the same as explained for the latch circuit shown in Fig. 18.53 (*a*).

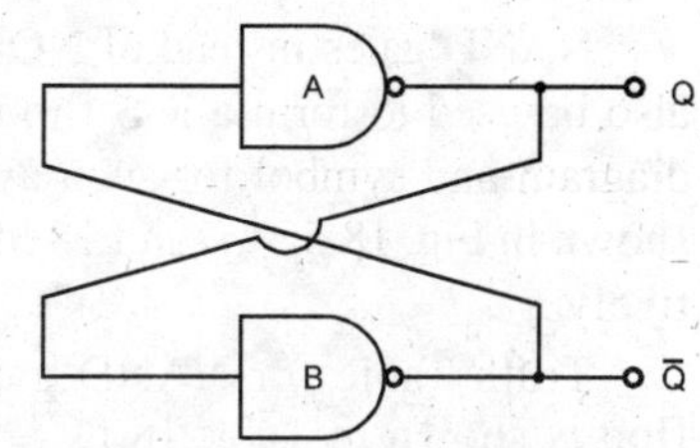

Fig. 18.53(*b*). *NAND Latch Circuit*

In the latch circuits shown in Figs. 18.53 (*a*) and (*b*), there is no possibility of entering the desired digital information to be stored in it. Infact, when the power is *switched on*, the circuit switches to one of the stable states (Q = 1 or 0) and it is not possible to predict the state.

Flip-flop is used in storage circuit, counter stages, shift register and many other computer applications. Some of the commonly used flip-flops are described in the forthcoming articles.

18.14.1. R-S Flip-Flop

Some modification is made in latch circuit in order to have control states of the flip-flop, as shown in Fig. 18.54. There are two inputs to the flip-flops called S (set) and R (reset). This is called R-S flip-flop.

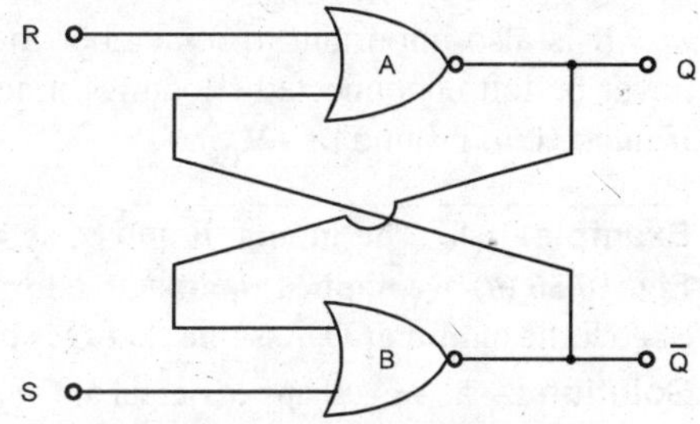

Fig. 18.54. *Circuit Diagram of R-S Flip-Flop Using NOR Gates*

Now output state of the flip-flop for all four combinations of the input signals R and S would be considered.

Case 1. When R = 0 and S = 0. Since a 0 at the input of a NOR gate has no effect on its output, the flip flop simply remains in its present state.

Case 2. When R = 0 and S = 1 with output $\overline{Q}$ = 0. Since signals at both inputs of gate A are *low* and signals at both inputs of gate B are *high* so output Q of gate A is *high* and output $\overline{Q}$ of gate B is *low*. This is a stable condition of flip-flop and is said to set the flip-flop. It changes only with the change in logical inputs of flip-flop.

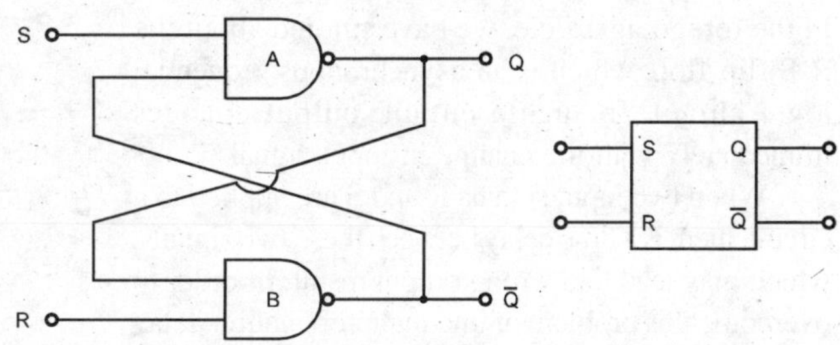

(a) Circuit Diagram *(b) Symbol*

Fig. 18.55. *R-S Flip-Flop Using NAND Gates*

Case 3. When R = 1 and S = 0 with output Q = 0. Since signals at both inputs of gate A are *high* and signals of both inputs of gate B are *low*, therefore, output Q of gate A is *low* and output $\overline{Q}$ of gate B is *high*. This is another stable condition of flip-flop and is said to reset the flip-flop. It does not change for the same logical inputs.

Case 4. When R = 1 and S = 1. In this case, the input signals R and S of the flip-flop force the output of both NOR gates to the same level. But as we know that output $\overline{Q}$ is always complement of output Q so output state of the flip-flop is unpredictable.

Both inputs R and S of flip-flop are at *high* level. Now when the inputs are removed, if R is removed before S then condition R = 0 and S = 1 prevails at input side and flip-flop sets and if S is removed before R then condition R = 1 and S = 0 prevails and flip-flop resets.

Table 18.10 is a truth table for R-S flip-flop.

NAND gates instead of NOR gates may also be used to form a R-S flip-flop. Circuit diagram and symbol for such a flip-flop are shown in Fig. 18.55 (*a*) and 18.55 (*b*) respectively.

Truth Table for NAND gate R-S flip-flop is shown in Table 18.11.

TABLE 18.10. ***Truth Table For R-S Flip-Flop***

R	*S*	*Q*	*Action*
0	0	Last value	No change
0	1	1	Set
1	0	0	Reset
1	1	?	Forbidden

TABLE 18.11. ***Truth Table For NAND Gate R-S Flip-Flop***

R	*S*	*Q*	*Action*
0	0	?	Forbidden
0	1	0	Reset
1	0	1	Set
1	1	Last value	No change

NAND flip-flop and the NOR flip-flop operate basically in the same manner, except for the difference— the NOR flip-flop inputs are normally 0 and must be pulsed to 1 state (active *high*) to change the state of the flip-flop outputs, the NAND flip-flop inputs are normally 1 and must be placed to the 0 state (active *low*) to change the flip-flop output state.

It is also important to remember that, TTL gate inputs are quite noise-sensitive and, therefore, should never be left unconnected (floating). Each input must be connected either to the output of a prior circuit, or if unused, to ground or $+V_{CC}$.

Example 18.8. The inputs $\overline{R}$ and $\overline{S}$ as depicted in Fig. 18.56 (*a*) are applied to an active low R-S latch. Sketch the output at Q. Assume that Q is initially low.

Solution : The waveshape of output at Q is depicted in Fig. 18.56 (*b*).

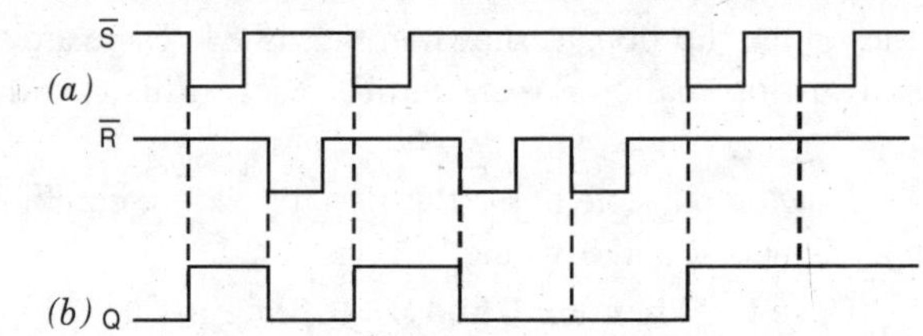

Fig. 18.56. ***Waveforms—Example 18.8.***

18.14.2. Clocked R-S Flip-Flop

In the foregoing article, we have studied about the R-S flip-flop, which is an asynchronous sequential logic circuit. In such a circuit, output changes immediately with the change in input signal.

When two input signals R and S are applied to a circuit, there is a time delay between these two signals, which may lead to a wrong output result. In order to overcome this problem of unequal propagation delay time of input signals, the circuit of R-S flip-flop is modified, as shown in Fig. 18.57. In this circuit, two AND gates and a clock pulse train input are also added to the R-S flip-flop circuit. This circuit is called the *clocked R-S flip-flop circuit*.

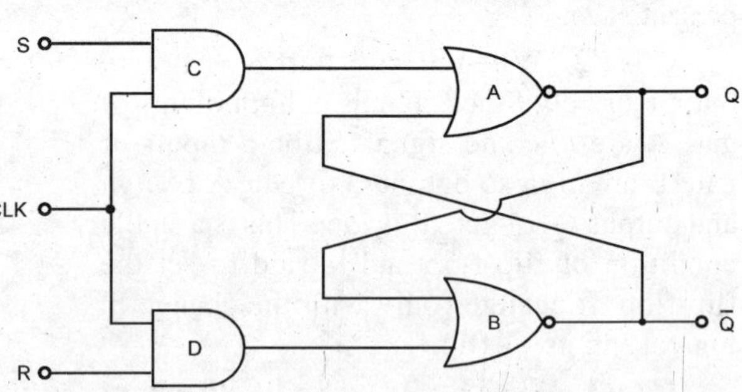

Fig. 18.57. ***Circuit Diagram of Clocked R-S Flip-Flop***

Clock supplies the *low* and *high* levels of signal to AND gates C and D regularly. When clock signal is at *low* level, AND gates C and D are disabled, so at this time, input signals R and S cannot be transferred to the NOR gates A and B, and therefore, NOR gates A and B, simply behave as a simple latch circuit, holding the previous values of signal at output. The use of the clock train pulses accordingly serves the important function of synchronising the operations that take place in digital sequential circuits. They avoid problems posed due to unequal propagation delay time associated with gates. When the clock signal is present (or at *high* level) AND gates C and D are enabled, so at this time, input signals R and S are transferred to NOR gates and output Q is changed accordingly.

In summary, it can be said that there are three inputs R, S and clock signal (CLK) which influence the output of flip-flop. Table 18.12 is a truth table for clocked R-S flip-flop. In this Table, subscript *n* is used to indicate that input R_n and S_n are applied to the circuit at *n*th instant of time. Q_n denotes the output at the *n*th instant of time and Q_{n+1} denotes the output after the next pulse of clock from the *n*th instant. Symbol * in truth table denotes for any level of signal.

TABLE 18.12. ***Truth Table For Clocked R-S Flip-Flop***

R_n	S_n	*CLK*	Q_{n+1}
*	*	0	Q_n
0	0	1	Q_n
1	0	1	0
0	1	1	1
1	1	1	?

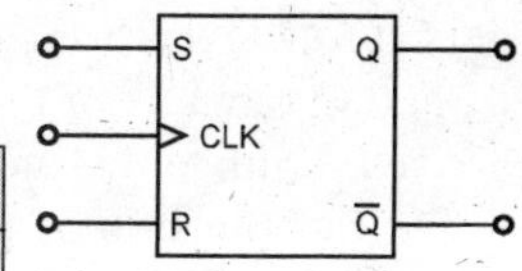

Fig. 18.58. ***Symbol of Clocked R-S Flip-Flop***

Now signal is applied to the input R and S when clock signal (CLK) is *low*, so output does not change with the change in input. When CLK signal becomes *high* this input signal changes the output of the flip-flop according to the truth Table 18.12. So in this way input signal is applied to the flip-flop at the same time and problem of unequal propagation time of the input signal is eliminated.

Clocked R-S flip-flop circuit is a synchronous sequential logic circuit because output state of the circuit changes at discrete clocked instant of time. Symbol of a clocked R-S flip-flop is shown in Fig. 18.58.

Example 18.9. The waveshapes for S, R and EN for clocked R-S latch are depicted in Fig. 18.59 (*a*). The latch is initially reset. Sketch the output at Q.

Solution : The Q output waveform is shown in Fig. 18.59 (*b*). When S is *high* and R is *low,* a *high* on the EN input *sets* the latch. When S is *low* and R is *high,* a high on the EN input resets the latch.

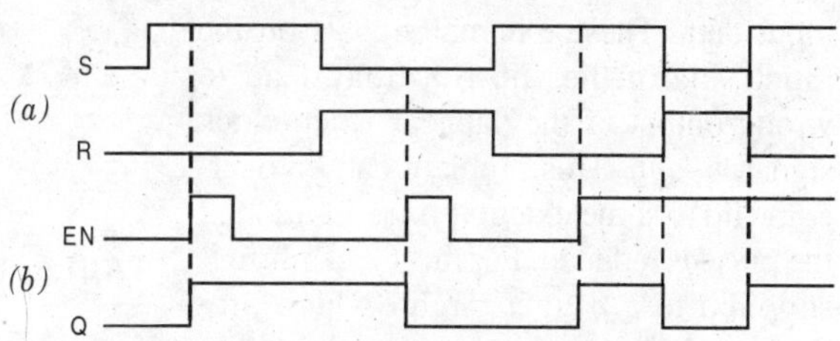

Fig. 18.59. ***Waveforms—Example 18.9***

18.14.3. D Flip-Flop

In a clocked R-S flip-flop two input signals are required to drive the flip-flop which is a disadvantage with many digital circuits. In some events, both input signals become *high* which is again an undesirable condition. So these drawbacks of clocked R-S flip-flop are overcome in D flip-flop. It is nothing but an R-S flip with an inverter in the R input. The added inverter reduces the number of inputs from two to one. Sometim:s, this type of flip-flop is called a gated D-latch. The logic diagram, circuit diagram and symbol for D flip-flop are shown in Figs. 18.60 (*a*), (*b*) and (*c*) respectively.

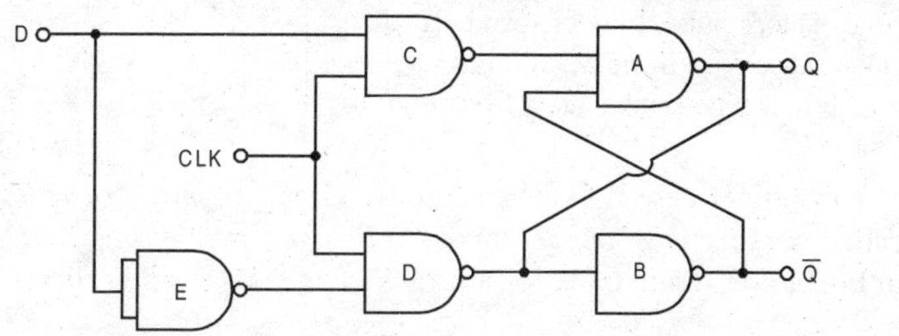

(a) Logic Diagram With NAND Gates

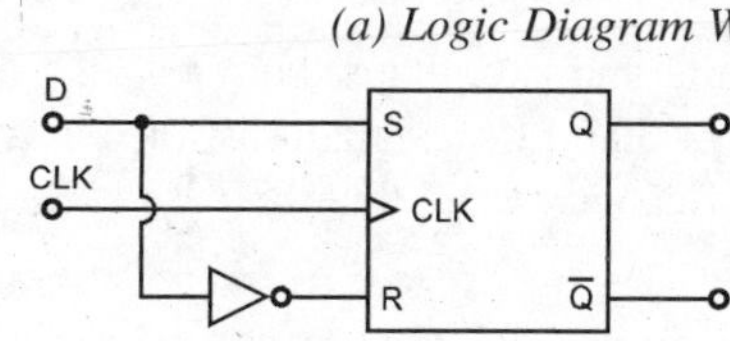

(b) Circuit Diagram

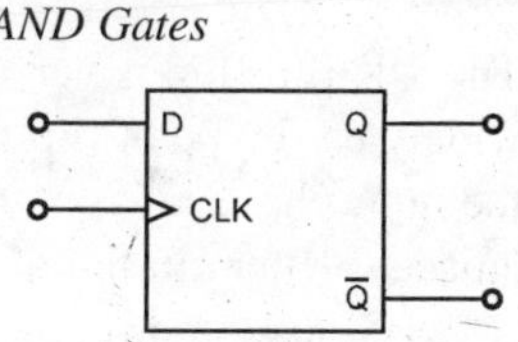

(c) Symbol

Fig. 18.60. ***D Flip-Flop***

Now, it can be seen that only single data bit, D is required to drive the flip-flop. When clock signal is at *low* level, data bit D is prevented to reach at output Q until clock signal becomes *high* at next pulse. With the help of truth table of clocked R-S flip-flop, it is easy to understand that when data bit D_n is *high*, output Q_{n+1} gets at *high* level and when data bit D_n is *low*, output Q_{n+1}

gets at *low* level. So D flip-flop transfers the data bit D to Q as it is, and Q remains in the same state until the next pulse of clock arrives. Truth table for a D-type flip-flop is given in Table 18.13. Since the transfer of data from the input to the output is delayed, the flip-flop is named delay (D) flip-flop. The D-type flip-flop is either used as a delay device or as a latch to store 1-bit of binary information.

TABLE 18.13. ***Truth Table For D-Type Flip-Flop***

Input D_n	*Output* Q_{n+1}
0	0
1	1

Example 18.10. The inputs shown in Fig. 18.61 (*a*) are applied to the gated D latch which is initially reset. Sketch the Q output waveform.

Solution : The Q output waveform is shown in Fig. 18.61 (*b*). When D is *high* and EN input is *high,* Q goes *high.* When D is *low* and the EN input is *high,* Q goes *low.* When the EN input is *low,* the state of the latch remains uneffected by the D input.

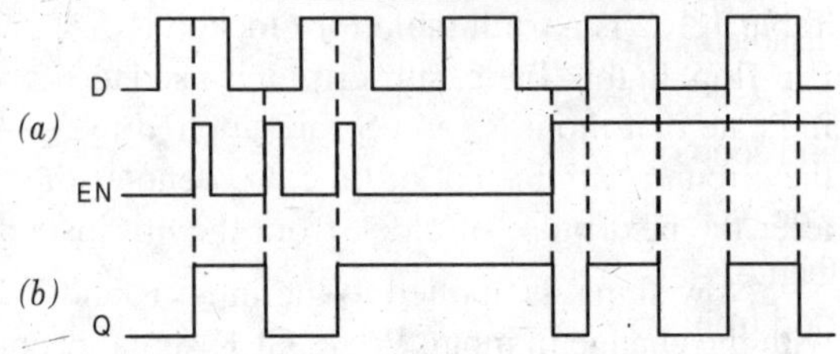

Fig. 18.61. ***Waveforms—Example 18.10***

18.14.4. Edge-Triggered D Flip-Flop

In a D flip-flop, sometimes presence of some external signals like noise is found on the input side along with the relevant input data. These external signals on the input side of the flip-flop may lead to wrong output, if they appear when clock signal is *high.* This problem can be overcome up to some extent if time duration of the pulse is reduced. For this clock pluse is supplied to a trigger circuit which produces a short duration pulse to make flip-flop enable. Trigger circuit responds at the edge of clock pulse. Either it responds when clock pulse goes from *low* to *high* level or it responds when pulse goes from *high* to *low* level.

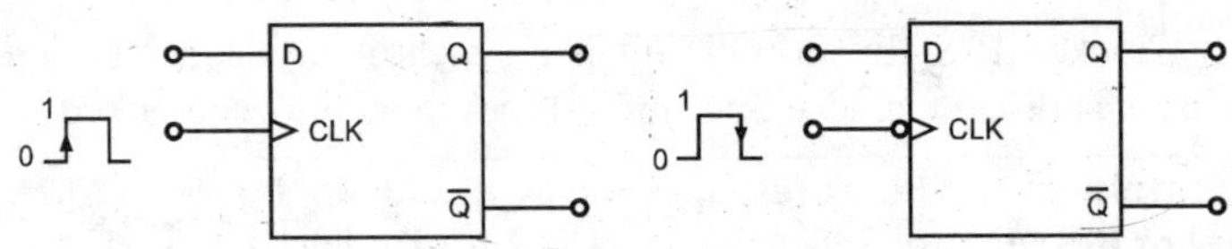

(a) Positive or Leading *(b) Negative or Trailing*

Fig. 18.62. ***Edge Triggered D Flip-Flop***

TABLE 18.14. ***Truth Table For Positive Edge-Triggered D Flip-Flop***

Inputs		*Outputs*		*Comments*
D	***CLK***	***Q***	$\bar{Q}$	
1	↑	1	0	SET (Stores a 1)
0	↑	0	1	RESET (Stores a 0)

↑ = Clock Transition *low* To *high*

In former case it is called *positive edge triggering* or *leading edge triggering*. In the second case it is called *negative edge triggering* or *trailing edge triggering*. Symbols for edge triggered D flip-flop are shown in Fig. 18.62. Truth table for a positive edge-triggered D flip-flop is shown in Table 18.14.

18.14.5. Edge-Triggered J-K Flip-Flop

The J-K flip-flop is very versatile and is perhaps the most widely used type of flip-flop. The J and K designations for the inputs have no known significance except that they are adjacent letters in the alphabet.

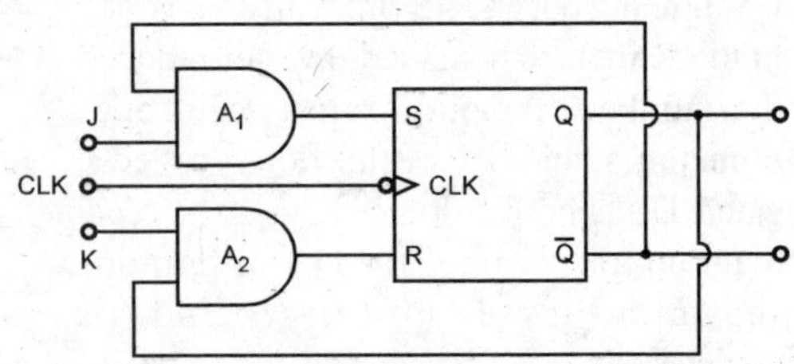

Fig. 18.63. ***Circuit Diagram of Trailing Edge Triggered J-K Flip-Flop***

The functioning of J-K flip-flop is identical to that of the R-S flip-flop in RESET, SET, and no change conditions of operation. The difference is that the J-K flip-flop has no invalid state as does the R-S flip-flop. Therefore, the J-K flip-flop is a very versatile device that finds wide application in digital devices such as counters, registers, arithmetic logic units, and other digital systems.

In J-K flip-flop clocked R-S flip-flop along with two AND gates are used. In this circuit J and K are used as input terminals and replace input terminals S and R of the S-R flip-flop respectively. Input terminal S is supplied by AND gate A_1 to which input signals are J and flip flop output $\overline{Q}$. Input terminal R is supplied by gate A_2 to which K and Q are the input signals. The purpose of additional gates A_1 and A_2, is to provide a convenient means by which the signals that are made to appear at S and R terminals are dependent not only on the input signals but also on the state of the flip-flop as well.

Output state of flip-flop remains the same when the clock signal is *low*. But when the clock pulse arrives, input signals J and K change the output state of flip-flop. Let us consider all four combinations of input signals.

Case 1. When clock signal is *high* and both input signals J and K are *low* then both AND gates A_1 and A_2 are disabled and there is no change in output state of the flip-flop.

Case 2. When input J is held at logic 1 and input K is held at logic 0. In this case AND gate A_2 is disabled and condition of AND gate A_1 depends on the output state Q_n of the flip-flop. If Q_n is *high* (or $\overline{Q}$ is *low*) then AND gate A_1 is also disabled and output state of the flip-flop Q_{n+1} remains the same as before. If Q_n is *low* (or $\overline{Q}$ is *high*) then AND gate A_1 is enabled and *high* logic signal goes to S terminal which changes the output state Q_{n+1} of flip-flop from *low* to *high*. So *high* signal at J and *low* signal at K input terminals set the flip-flop irrespective of the output state of the flip-flop before arrival of the clock pulse.

Case 3. When input J is held at logic 0 and input K is held at logic 1. In this case AND gate A_1 is disabled and condition of AND gate A_2 depends on the output state Q_n of the flip-flop. If Q_n is *low* then AND gate A_2 is also disabled and output state of flip-flop Q_{n+1} remains the same as before. If Q_n is *high*, then AND gate A_2 is enabled and *high* logic signal goes to R terminal, which changes the output state Q_{n+1} of flip-flop from *high* to *low*. So *low* signal at J and *high* signal at K input terminals reset the flip-flop irrespective of output state of flip-flop before the arrival of clock pulse.

TABLE 18.15. ***Truth Table For J-K Flip-Flop***

J	*K*	Q_{n+1}
0	0	Q_n
1	0	1
0	1	0
1	1	$\overline{Q}_n$

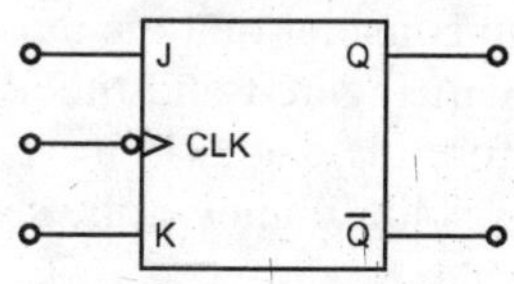

Fig. 18.64. ***Symbol For Trailing Edge Triggered J-K Flip-Flop***

Case 4. When both input signals, J and K are held at logic 1. In this case condition of both AND gates depends on the output state of the flip-flop before arrival of clock pulse. Suppose Q_n is *high* then it makes AND gate A_2 enabled, which sends *high* level signal to the input terminal R at the arrival of the clock pulse and flip-flop gets reset.

Now suppose Q_n is *low* (or $\overline{Q}_n$ is *high*) then it makes AND gate A_1 enabled, which sends *high* level signal to the input terminal S at the arrival of clock pulse and the flip-flop gets set.

Hence it can be said that in this case, output state of the flip-flop toggles when fresh clock pulse arrives each time.

Table 18.15 is a truth table for J-K flip-flop. Symbol for trailing edge triggered JK flip-flop is shown in Fig. 18.64.

Logic diagram of clocked J-K flip-flop is shown in Fig. 18.65.

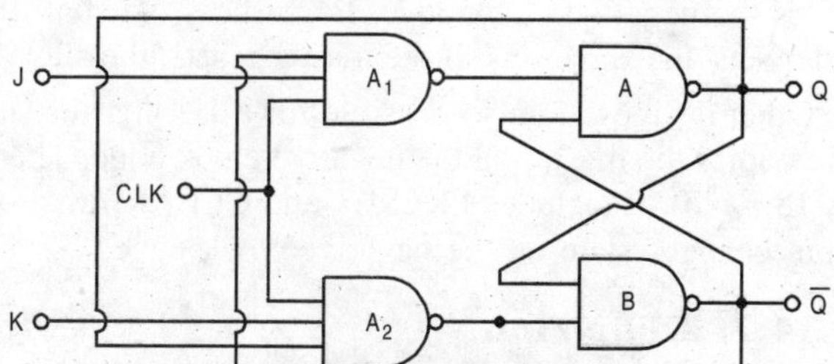

Fig. 18.65. ***Logic Diagram of Clocked J-K Flip-Flop (Negative Edge Triggered)***

Example 18.11. The inputs shown in Fig. 18.66 (*a*) are applied to the positive edge triggered JK flip-flop. Draw the Q output waveform.

Solution : The Q output waveform is shown in Fig. 18.66 (*b*).

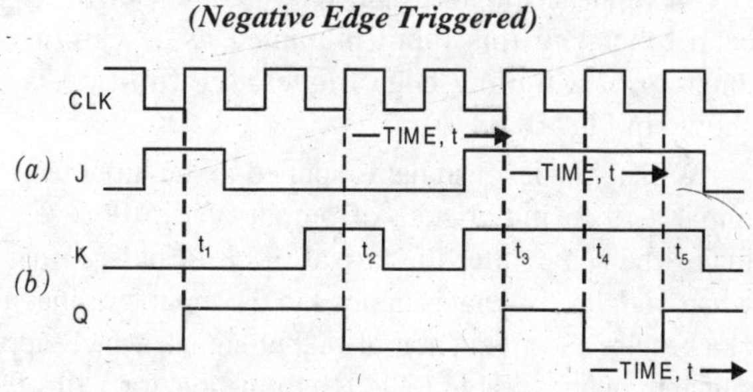

Fig. 18.66. ***Waveforms—Example 18.12***

When J is *high* and K is *low*, the rising clock edge sets Q to *high* (as shown at instant t_1). When J is *low* and K is *high*, the rising clock edge resets Q to *low* (as shown at instant t_2). When both J and K are *high*, the flip-flop toggles at each rising clock edge (instants t_3, t_4, t_5).

18.14.6. Pulse Triggered (Master-Slave J-K Flip-Flop)

The second class of flip-flops is the *pulse-triggered* or *master-slave*. The term *pulse-triggered* means that the data are entered into the flip-flop on the leading edge of the clock-pulse. The inputs must be set up prior to the clock pulse's leading edge, but the output is postponed until the trailing edge of the clock. A major restriction of the pulse-triggered flip-flop is that the data inputs must not change while the clock pulse is *high* because the flip-flop is sensitive to any change of input levels during this time.

As with the edge-triggered flip-flops, there are three basic types of pulse-triggered flip-flops : R-S, D, and J-K. The J-K is by far the most commonly available in IC form.

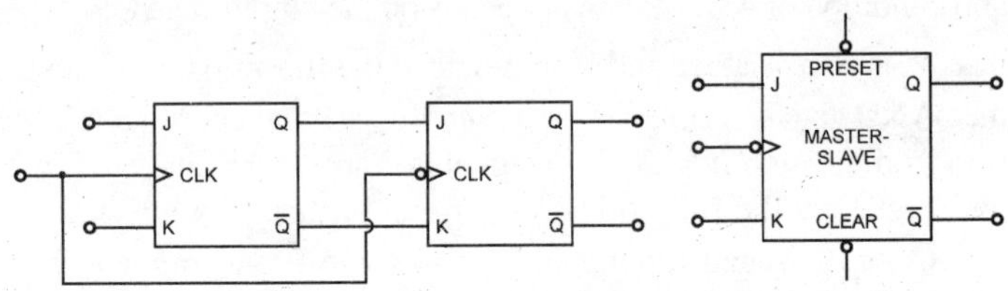

(a) Circuit (b) Symbol

Fig. 18.67. ***Master-Slave J-K Flip-Flop***

Figure 18.67 (*a*) shows one way to build a J-K master slave flip-flop. It provides another way to avoid racing. For understanding the working of the flip-flop consider that the master is positive-edge triggered and the slave, negative-edge triggered. Thus the master responds to its J and K inputs before the slave. If J = 1 and K = 0, the master sets on the positive edge. The *high* Q output of the master drives the J input of the slave, so when the negative clock edge hits, the slave sets, copying the action of the master.

If J = 0 and K = 1, the master resets on the leading edge of the clock. The *high* $\overline{Q}$ output of the master goes to the K input of the slave. Therefore, the arrival of the clock's trailing edge forces the slave to reset. Again, the slave has copied the master.

If the master's J and K inputs are both *high*, it toggles on the positive clock edge and the slave then toggles the negative clock edge. Regardless of what the master does, therefore, the slave copies it; if the master sets, the slave sets; if the master resets, the slave resets.

Other designs are also possible for a J-K master-slave flip-flop. The basic idea remains the same : the slave copies the master on the negative clock edge. The symbol used for a J-K master-slave flip is shown in Fig. 18.67 (*b*). It includes PRESET and CLEAR functions. The bubble on the clock input indicates that the output changes state on the negative clock edge.

18.14.7. T Flip-Flop

This flip-flop is basically a J-K flip-flop. In this circuit input terminals J and K are connected with each other and this input is named as T. Circuit diagram of a trailing edge triggered T flip-flop is shown in Fig. 18.68 (*a*).

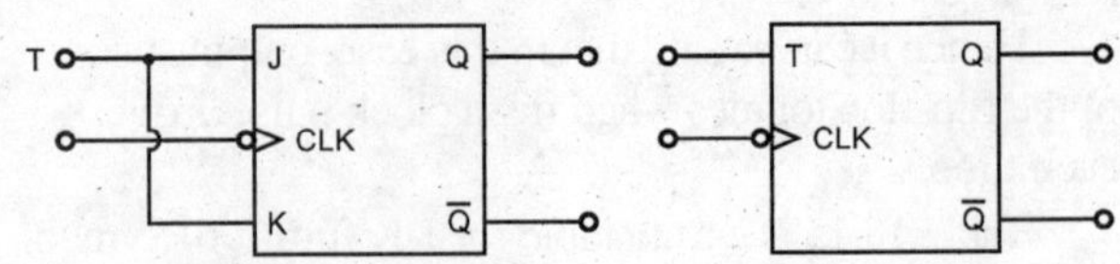

(a) Circuit Diagram (b) Symbol

Fig. 18.68. ***Trailing Edge Triggered T Flip-Flop***

When *low* level signal is applied to the input terminal T, then initial state of output of flip-flop remains the same, after the arrival of clock pulse. But when *high* level signal is applied to the input terminal T, then output of flip-flop toggles after arrival of every new clock pulse. So frequency of output signal, which appears at output terminal of T flip-flop is half of the clock signal frequency. Table 18.16 is a truth table for T flip-flop. From truth table it is clear that if T = 1 it acts as a toggle switch. For every clock pulse, the output Q changes.

An R-S flip-flop cannot be converted into a T-type because R = S = 1 is not permitted.

TABLE 18.16. ***Truth Table For T Flip-Flop***

Input T_n	*Output* Q_{n+1}
0	Q_n
1	$\overline{Q_n}$

In essence the T-flip-flop can be treated as a frequency divider or more generally as a device which takes the input frequency at the clock terminal and divides it by 2.

The waveform is shown in Fig. 18.69.

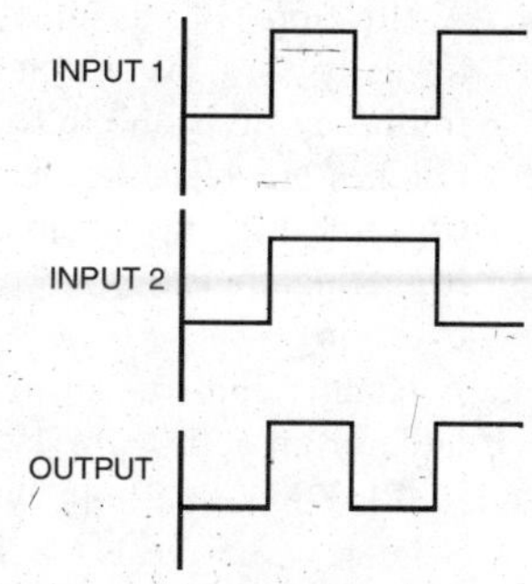

Fig. 18.69

HIGHLIGHTS

1. Logic gates are the fundamental building blocks of digital systems. OR, AND and NOT gates are the basic types of gates. The interconnection of gates to perform different logical operations is known as logic design.
2. A truth table lists all possible combinations of inputs and the corresponding outputs.
3. In a positive logic system, the higher of the two voltage levels represents logic 1 and the lower of the two voltage levels represents logic 0. On contrary in a negative logic system, the higher of the two voltage levels represents logic 0 and lower of the two voltage levels represents logic 1. Both positive and negative logic are used in digital systems, but positive logic is the more common.
4. An OR gate is *any* or *all* gate. It has two or more inputs but only one output. In an OR gate if any of the input is *high*, the output is *high*. If all the inputs are *low*, the output is *low*. It is written as $Z = A + B = A \text{ OR } B$.
5. An AND gate is an *all* or *nothing* gate. It has two or more inputs but only one output. All inputs must be *high* to provide a *high* output. It is written as $Z = A.B = A \text{ AND } B$.
6. A NOT gate (or inverter gate) has only one input and one output. It complements its input *i.e.* its output is always the opposite of the input. Thus, $Z = \overline{A} = \text{NOT } A$.
7. A NOR gate is effectively an OR gate followed by an inverter. It has two or more inputs and one output. All inputs must be *low* to give a *high* output. It, therefore, also acts as a negative AND gate. It is written as $Z = A \text{ NOR } B = \overline{A + B}$.
8. A NAND gate is effectively an AND gate followed by an inverter. It has two or more inputs and one output. All inputs must be *high* to get a *low* output. It, therefore, acts as a negative OR gate. It is written as $Z = A \text{ NAND } B = \overline{A.B}$
9. NOR and NAND gates are called universal gates. Any circuit of any complexity can be realized by using only NOR gates or only NAND gates.
10. An EX-OR gate is an anticoincidence gate. It is an inequality detector. It produces *high* output, only when its two inputs are different. The equation is $Z = A \text{ EX-OR } B = A \oplus B = \overline{A}B + A\overline{B}$.
11. An EX-NOR gate is a coincidence gate. It is an equality detector. It produces *high* output, only when its two inputs are the same. The equation is $Z = \overline{A \oplus B} = A \odot B = AB + \overline{A}\,\overline{B}$.
12. EX-OR and EX-NOR gates do not have more than two inputs. When three or more variables are to be EX-ORed or EX-NORed, a number of two input EX-OR or EX-NOR gates may be used.
13. Logic circuits can be classified into two categories *i.e.* combinational logic circuit and sequential logic circuit. In combinational logic circuits, the output depends upon the present input only, not on the previous output. There is no feedback element in combination circuit. A combinational logic circuit does not have a memory. A sequential logic circuit has a memory and its output depends not only on external input(s) but also on the present state of the memory elements.

14. A flip-flop (FF) is the basic memory element; it can store a 0 or a 1. A flip-flop is known more formally as a bistable multivibrator. It has two stable states. FFs are used for data storage counting, frequency division etc.
15. Latches are bistable elements whose state depends on asynchronous inputs. On the other hand, edge triggered FFs are bistable elements with synchronous inputs whose state depends on inputs only at the triggering transition of clock pulse. Edge triggering can be positive edge triggering or negative edge triggering.
16. A latch is constructed using two cross-coupled NOR gates or NAND gates. A latch may be active *high* latch or an active-*low* latch.
17. R-S latch can be built with NOR gates or NAND gates. The NOR gate R-S latch is an active *high* R-S latch while the NAND gate R-S latch is an active *low* R-S latch.
18. A clocked R-S latch has a clock input in addition to R and S inputs. A state change can occur only when clock signal becomes high.
19. In a D latch a common D input drives the R and S inputs, S input directly and R input through an inverter. A clocked D latch uses a clock signal to enable and disable the latch.
20. FFs often have PRESET and CLEAR inputs also. A CLEAR signal is the same as reset signal. A PRESET is equivalent to set the flip before the computer run.
21. The J-K flip-flop is very versatile and is perhaps the most widely used type of flip-flop. Edge triggered JK FF is used in digital counters.
22. The master-slave FF is made up of two FFs—a master and a slave. Master-slave FFs are called pulse-triggered FFs, because the time duration required for its output to change state is equal to the width of one clock pulse.
23. A T flip-flop is obtained by connecting the J and K inputs of JK edge triggered FF. In essence the T flip-flop can be treated as a frequency divider or more generally as a device which takes the input frequency at the clock terminal and divides it by 2.

EXERCISES

1. What is the difference between combinational and sequential circuits ?

   [Pb. Technical Univ. Basic Electrical and Electronics Engineering, December 2008]
2. What are different logic gates ? Give their truth tables.

   [Pb. Technical Univ. Basic Electrical and Electronics Engineering, December 2008]
3. Define and briefly explain the following :

   (*i*) AND gate (*ii*) OR gate (*iii*) NOT gate.
4. Show that the NOR gate is a universal building block.
5. Draw the circuits, symbols and truth tables for the following logic gates.

   (*a*) AND (*b*) OR (*c*) EX-OR (*d*) NAND (*e*) NOR.
6. What do you mean by universal gates ? Implement basic gates in terms of universal gate.
7. What are universal gates ? Why are they so called ?

   [Pb. Technical Univ. Basic Electrical and Electronics Engineering, May 2009]
8. Prove that the NAND and NOR gates are universal gates.
9. What is XOR gate ? Draw its truth table.

   [Pb. Technical Univ. Basic Electrical and Electronics Engineering, May 2009]
10. Make EX-OR gate using minimum number of gates.
11. Draw the logic symbol and construct the truth table for each of the following gates :

    (*i*) Two-input NAND gate; (*ii*) Two-input EX-OR gate; (*iii*) Two-input EX-NOR gate; (*iv*) Two-input NOR gate.
12. Give the PIN diagram and its description for ICs 74X series.

    [Pb. Technical Univ. Basic Electrical and Electronics Engineering May 2006]

13. Describe the working of S-R latch using NOR gates. Draw state table.
14. Explain the working of R-S flip-flop.
[Pb. Technical Univ. Basic Electrical and Electronics Engineering May 2006]
15. Write truth table for the flip-flops and explain the working of R-S flip-flops.
[Pb. Technical Univ. Basic Electrical and Electronics Engineering Dec. 2004]
16. Write short note on R-S flip-flops and its truth table.
[Pb. Technical Univ. Basic Electrical and Electronics Engineering May 2005]
17. Describe in detail the operation of R-S flip-flop with waveform.
[Pb. Technical Univ. Basic Electrical and Electronics Engineering May 2008]
18. What are flip-flops ? How are invectors used to represent S-R flip-flops ? Discuss.
[Pb. Technical Univ. Basic Electrical and Electronics Engineering Dec. 2004]
19. Give the logic diagram of clocked R-S flip-flop. Discuss its working.
[Pb. Technical Univ. Basic Electrical and Electronics Engineering, December 2008]
20. Draw the equivalent circuit and truth table of R-S flip-flop.
[Pb. Technical Univ. Basic Electrical and Electronics Engineering, December 2009]
21. What is J-K flip-flop ? Give truth table for J-K flip-flop.
[Pb. Technical Univ. Basic Electrical and Electronics Engineering May 2005]
22. Give the truth table of J-K flip-flop.
[Pb. Technical Univ. Basic Electrical and Electronics Engineering May 2006]
23. Draw the circuit of J-K flip-flop and explain its working with the help of truth table.
[Pb. Technical Univ. Basic Electrical and Electronics Engineering May 2007]
24. Describe in detail the operation of J-K flip-flop with waveform.
[Pb. Technical Univ. Basic Electrical and Electronics Engineering Dec. 2007]
25. What is J-K flip-flop ? Discuss its working. What is race around condition ?
[Pb. Technical Univ. Basic Electrical and Electronics Engineering, May 2009]
26. Compare R-S flip-flop with D flip-flop.
[Pb. Technical Univ. Basic Electrical and Electronics Engineering Second Semester 2005-06]
27. Draw the logic symbols and explain the operation of a D type and a T type flip-flops.
28. What is T-flip-flop ? Give its symbol and draw waveform of T-flip-flop.
[Pb. Technical Univ. Basic Electrical and Electronics Engineering Dec. 2005; May 2010]

SHORT ANSWER TYPE QUESTIONS WITH ANSWERS

Q. 1. What is Boolean algebra ?

Ans. Boolean algebra is a mathematical system of logic in which truth functions are expressed as symbols and then these symbols are manipulated to arrive at conclusion.

Q. 2. What are the basic logic elements ?

Ans. Basic logic elements are NOT gate, AND gate, OR gate and the flip-flop.

Q. 3. What is a truth table ?

Ans. Truth table is a table that gives outputs for all possible combinations of inputs to a logic circuit.

Q. 4. Define positive logic and negative logic.

Ans. If the higher of the two voltages represents a 1 and the lower voltage represents a 0, the logic is called a positive logic. On the other hand, if the lower voltage represents a 1 and the higher voltage a 0, we have a negative logic.

Q. 5. What is pulse logic system ?

Ans. A logic system in which a bit is recognised by the presence or absence of a pulse is called a pulse or dynamic logic system.

Q. 6. What is an inverter ?

Ans. An inverter is a logic gate whose output is the inverse or complement of its input.

Q. 7. What are the universal logic gates ?

Ans. Universal gate is a gate that can perform all the basic logical operations such as NAND and NOR gates.

Q. 8. What are basic logic gates and what are universal logic gates ?

Ans. Basic logic gates are AND gate, OR gate, and NOT (INVERTER) gate. Universal logic gates are NAND and NOR gates.

Q. 9. What is the speciality of NAND and NOR gates ?

Ans. The speciality of NAND and NOR gates is that they are universal gates and can perform all the basic logical operations.

Q. 10. Why NAND-NAND realization is preferred over AND-OR realization ?

Ans. NAND-NAND realization needs only one type of gate (NAND), that minimizes IC package counter.

Q. 11. Why is a two-input NAND gate called universal gate ?

Ans. NAND gate is called universal gate because any digital system can be implemented with the NAND gate. Sequential and combinational circuits can be constructed with these gates because element circuits like flip-flop can be constructed from two NAND gates connected back-to-back. NAND gates are common in hardware because they are easily available in the ICs form. A NAND gate is in fact a NOT-AND gate. It can be obtained by connecting a NOT gate in the output of an AND gate.

Q. 12. What is XOR gate ? Draw its truth table. **[P.T.U. May 2009]**

Ans. XOR gate is an inequality comparator or detector and produces output only when the inputs are different. It means that XOR gate is a logic gate that produces *high* output when the two inputs are different (one is *high* and the other is *low*) and *low* output when the two inputs are the same (either *low* or *high*). Truth table is given as Table 18.8.

Q. 13. Draw symbols of NAND gate and NOR gate. **[P.T.U. B.E.E.E. May 2008]**

Ans. Symbols of NAND gates and NOR gates are given below.

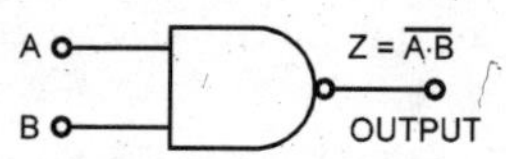

Fig. 1. Symbol of NAND Gate

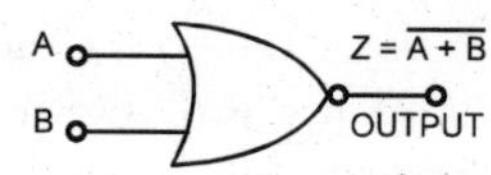

Fig. 2 Symbol of NOR Gate

Q. 14. What are universal gates and why are they called so ? **[P.T.U. B.E.E.E. Dec. 2006]**

Ans. Refer to Answers of Questions 7 and 11.

Q. 15. What is a combinational circuit ?

Ans. In a combinational circuit, the output depends upon present input(s) only *i.e.* not dependent on the previous input(s). The combinational circuit has no memory element. It consists of logic gates only.

Q. 16. Write two characteristics of combinational circuits.

Ans. The two characteristics of combinational circuits are :

(*i*) In combinational circuits, the output exists as long as the input exists.

(*ii*) A combinational circuit will always respond in the same fashion to input function, when we apply signal to the input terminal of the combinational logic circuit.

Q. 17. Differentiate between latch and flip-flop.

Ans. A flip-flop (FF) is a bistable electronic circuit having two stable states *i.e.* its output is either *low* or *high.* Most element form of a flip-flop is known as latch, as shown in Fig. 18.53. In latch circuits there is no possibility of entering the desired digital information to be stored in it. Some modification is made in latch circuit in order to have control states. This modified circuit is known as flip-flop (FF).

Q. 18. What is an excitation table ?

Ans. Excitation table gives an information about what should be the flip-flop inputs if the outputs are specified before and after the clock pulses.

Q. 19. What is a state table ?

Ans. State table consists of complete information about present state, next state, and outputs of a sequential circuit.

Q. 20. How can a D flip-flop be implemented using a J-K flip-flop.

Ans. D flip-flop can be implemented by connecting an inverter between inputs J and K, as shown in the Fig. 3.

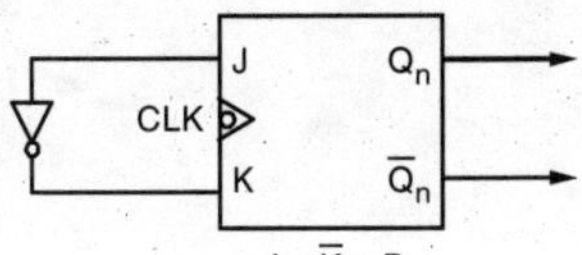

Fig. 3

Q. 21. What is race-around conditions in J-K flip-flop ? **[P.T.U. B.E.E.E. Dec. 2006]**

Ans. When both inputs J and K are *low,* the circuit is inactive with J *low* and K *high*, the circuit is reset when a positive clock edge strikes. Flip-flop remains in reset state with J = 1 and K = 0. The circuit is set at arrival of reset positive clock pulse when both inputs are *high* (*i.e.* J = 1 and K = 1), flip-flop will toggle. This is known as race round condition.

Q. 22. Give the waveform of J-K flip-flop.

Ans. Operation of J-K flip-flop is represented by the following waveform.

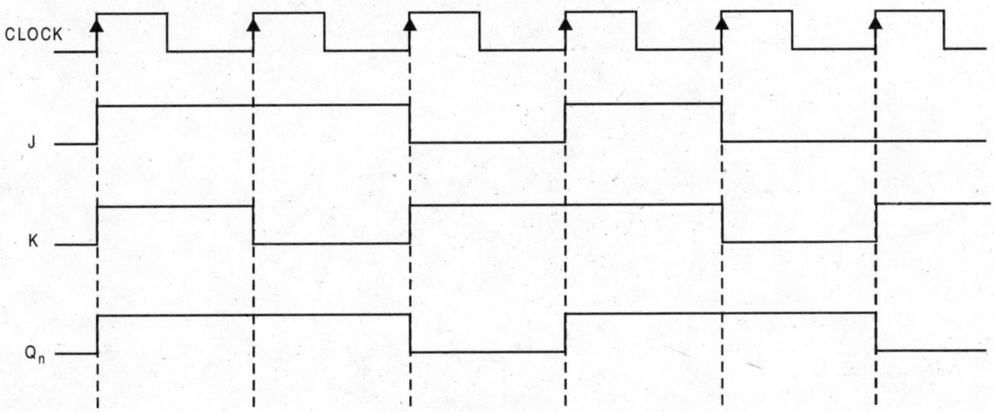

Fig. 4

❑❑❑

EXPERIMENTS

EXPERIMENT-1

Objective : To verify Ohm's law and its limitations.

Apparatus Required : 12 V battery or regulated power supply-1 no; PMMC ammeter 0-2 A-1 no; PMMC voltmeter 0-15 V-1 no; Rheostat 10 Ω, 2 A-1 no; Resistors of different values (say, 5 Ω, 10 Ω each of current rating 2 A)- 2 nos; powdered carbon and connecting wires.

Theory : Refer to Art. 1.5.

Connection Diagram :

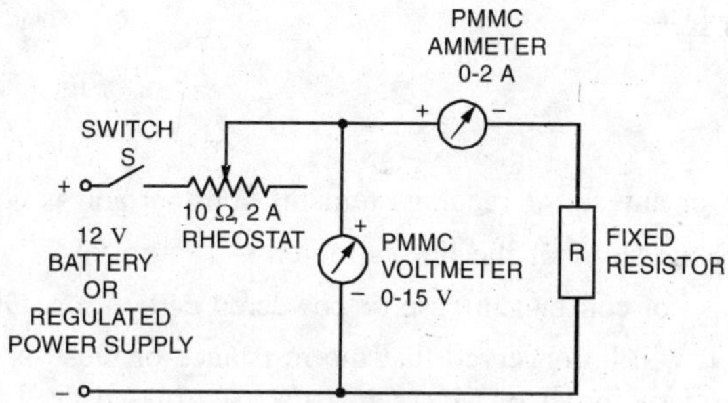

Fig. 1

Procedure : Rheostat, PMMC ammeter, PMMC voltmeter and fixed resistor R are connected to 12 V battery or regulated power supply through switch S as shown in Fig. 1.

With the rheostat at its maximum and R of 10 Ω, the switch S is closed and simultaneous readings of ammeter and voltmeter are noted. This is repeated with various currents obtained by adjusting the rheostat; the current is kept low and readings are taken faster to avoid an appreciable rise in temperature.

The above procedure is repeated with a fixed resistor of 5 Ω. The results are tabulated and graph is drawn between voltage V and current I.

The above procedure is then repeated with a nonlinear element, such as powdered carbon.

Observations :

S. No.	*R [Fixed Resistor, Nonlinear Element or P-N Junction Diode]*	*Voltmeter Reading V in volts*	*Ammeter Reading I in Amperes*	*Calculated Value of R* $R = \frac{V}{I}$
1	Resistor of 10 Ω			
2				
3				
4				
5	Resistor of 5 Ω			
6				
7				
8				
9	Powdered carbon			
10				
11				
12				

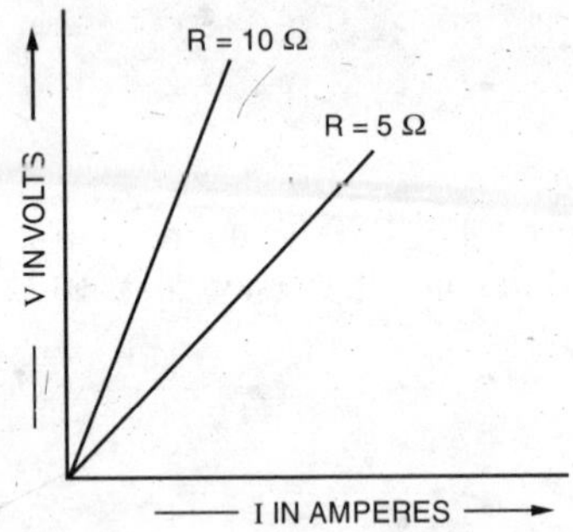

Fig. 2

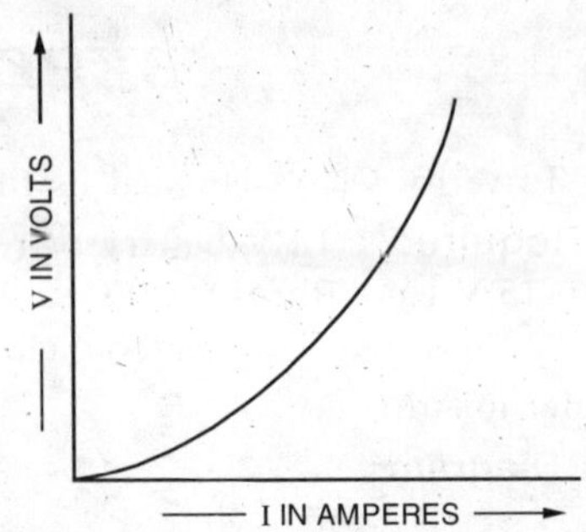

Fig. 3

Conclusion :

1. The ratio of V and I for any set of readings remains constant and is equal to R.
2. For higher value of R the slope of the line is more.
3. The ratio of V and I is not constant in case of powdered carbon, *i.e.* Ohm's law does not hold good.

Note : In Experiment 12 it will be observed that the resistance of the P-N junction diode is small when forward biased and very high (of the order of MΩ) when reversed biased *i.e.* Ohm's law does not hold good in case of electronic devices.

EXPERIMENT-2A

Objective : Verification of Kirchhoff 's current law.

Apparatus Required : 24 V battery or regulated power supply–1 no, PMMC ammeters 0–2 A — 3 nos., Rheostats 10 Ω, 2 A — 3 nos. and connecting wires.

Theory : According to Kirchhoff 's current law, in any network of wires carrying currents, the algebraic sum of all currents meeting at a point (or junction) is zero or the sum of incoming currents towards any point is equal to the sum of outgoing currents away from that point.

Connection Diagram :

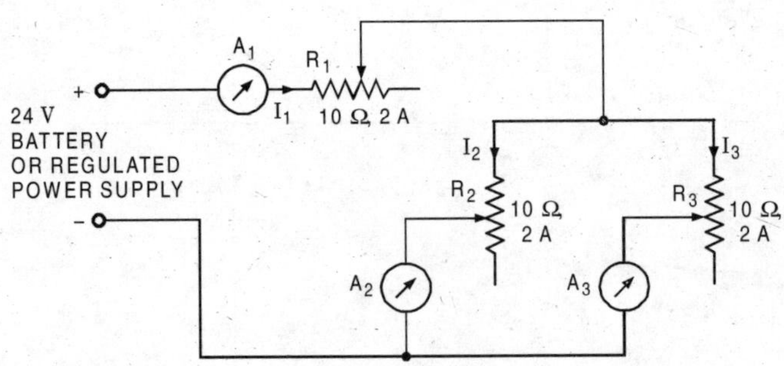

Fig. 4

Procedure: Three rheostats R_1, R_2 and R_3 and ammeters A_1, A_2 and A_3 are connected to 24 V battery or regulated power supply as shown in Fig. 4. The three rheostats are set to their maximum values, supply is switched on, and the readings of ammeters A_1, A_2 and A_3 are noted. The process may be repeated by varying either of rheostats R_1, R_2 or R_3. The current is kept low (not more than 2 A in any case) and readings are taken faster to avoid an appreciable rise in temperature.

Observations :

S. No.	*Readings of Ammeter* A_1, I_1	*Readings of Ammeter* A_2, I_2	*Readings of Ammeter* A_3, I_3	$I_2 + I_3$
1.				
2.				
3.				
4.				
5.				

Conclusion: It will be found that current I_1 is equal to the sum of currents I_2 and I_3. Hence Kirchhoff's current law is verified.

EXPERIMENT-2B

Objective: Verification of Kirchhoff's voltage law.

Apparatus Required: 24 V battery or regulated power supply–1 no., PMMC voltmeters 0–24 V — 3 nos., Rheostats 10 Ω, 2.5 A — 2 nos. and connecting wires.

Theory: According to Kirchhoff's voltage law, in any closed circuit or mesh, the algebraic sum of emfs acting in the circuit or mesh is equal to the algebraic sum of the products of the currents and resistances of each part of the circuit or mesh.

Connection Diagram:

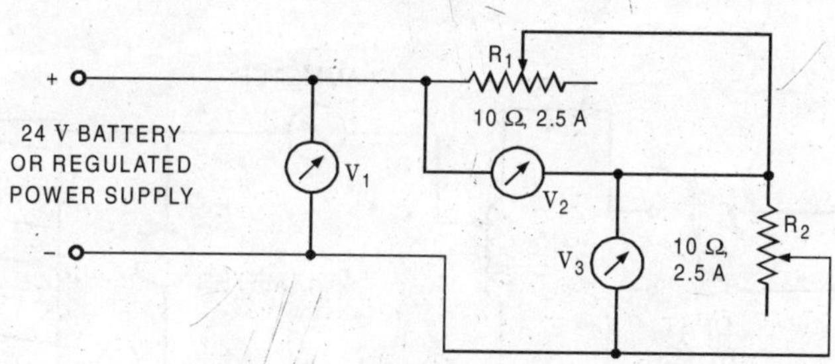

Fig. 5

Procedure: Two rheostats R_1, R_2 and three voltmeters V_1, V_2 and V_3 are connected to 24 V battery or regulated power supply as shown in Fig. 5. Two rheostats are set to their maximum values, supply is switched on, and the readings of voltmeters V_1, V_2 and V_3 are noted. The process may be repeated by varying either of the rheostats R_1 or R_2.

Observations:

S. No.	*Readings of Voltmeter* V_1 *in Volts*	*Readings of Voltmeter* V_2 *in Volts*	*Readings of Voltmeter* V_3 *in Volts*	$V_2 + V_3$
1.				
2.				
3.				
4.				
5.				

Conclusion: It will be found that voltage V_1 is equal to the sum of voltages V_2 and V_3. Hence Kirchhoff's voltage law is verified.

EXPERIMENT-3

Objective : To find voltage-current relationship in a R-L series circuit and to determine the power factor of the circuit.

Apparatus Required : Variac or 1-phase auto-transformer 1 no., Moving iron voltmeter 0-250 V-1 no; Moving iron ammeter (0-5 A)-1; Choke coil-1 no; Double pole single throw (DPST) switch-1 no. and connecting wires.

Theory : Refer to Art. 3.5 and Fig. 3.4 (*a*), (*b*) and (*c*).

From phasor diagram shown in Fig. 3.4 (*c*), we have

$$V = \sqrt{V_R^2 + V_L^2}$$

$$\text{or } IZ = \sqrt{(IR)^2 + (IX_L)^2}$$

$$\text{or } IZ = I\sqrt{R^2 + X_L^2}$$

$$\text{or Impedance, } Z = \sqrt{R^2 + X_L^2}$$

$$\text{or Inductive reactance of coil, } X_L = \sqrt{Z^2 - R^2}$$

$$\text{and Inductance of coil, } L = \frac{X_L}{2\pi f} = \frac{\sqrt{Z^2 - R^2}}{2\pi f} \qquad \because X_L = 2\pi f L$$

$$\text{Power factor of coil, } \cos\phi = \frac{R}{Z} \text{ (lagging)} \qquad \because \text{Circuit is inductive.}$$

Connection Diagram :

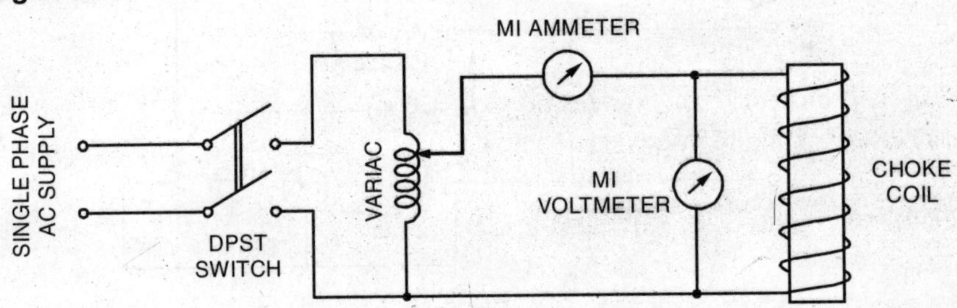

Fig. 6

Procedure: Variac, ammeter, voltmeter, choke coil are connected through double pole single throw switch (DPST switch), to single phase ac supply mains as shown in Fig. 6.

The DPST switch is closed and simultaneous readings of ammeter and voltmeter are noted. This is repeated with different currents obtained by adjusting the variac.

The coil resistance is measured as explained in Experiment no. 1 or by a multimeter.

The results are tabulated and value of inductance is determined from the relation

$$L = \frac{\sqrt{Z^2 - R^2}}{2\pi f} = \frac{X_L}{2\pi f}$$

and power factor is determined from the relation

$$\text{Pf, } \cos\phi = \frac{R}{Z}$$

Observations and Results:

Supply frequency, f = 50 Hz

Resistance of coil = Say, R ohms

S. No.	Voltmeter Reading, V in Volts	Ammeter Reading I in Amperes	Impedance of Coil $Z = \frac{V}{I}$ in ohms	Inductive Reactance of coil, $X_L = \sqrt{Z^2 - R^2}$ in Ohms	Inductance of coil $L = \frac{X_L}{2\pi f}$ in Henry	pf, $\cos\phi = \frac{R}{Z}$
1						
2						
3						
4						
5						

Conclusion: Current I increases directly in proportion to applied voltage V.

Note : If required, the coil with movable core may be included in the circuit so as to vary the coil inductance and the observations may be made with different positions of core in the coil.

EXPERIMENT-4

Objective : To find voltage-current relationship in an R-L series circuit and to determine the power factor of the circuit.

Apparatus Required : Variac or 1 phase auto-transformer-1 no; Moving iron voltmeter 0-250 V-3 nos; Moving iron ammeter 0-5 A-1 no; Variable resistor-one no; Inductor-1 no; DPST switch-1 no. and connecting wires.

Theory : Refer to Art. 3.5 and Fig. 3.4 (*a*), (*b*), (*c*).

From phasor diagram shown in Fig. 3.4 (*c*), we have

$$V = IZ$$

$$\text{or} \quad \frac{V}{I} = Z, \text{ constant}$$

$$\text{and pf of circuit, } \cos\phi = \frac{V_R}{V}$$

Connection Diagram :

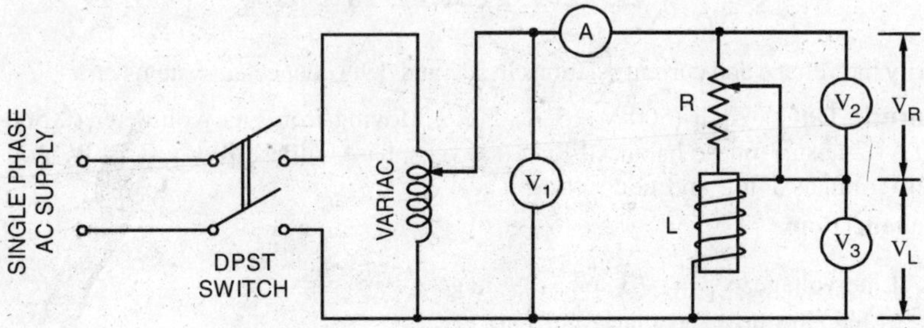

Fig. 7

Procedure: Variac, ammeter A, variable resistor R, inductor L, voltmeter V_1 (for measuring supply voltage), voltmeter V_2 (for measuring voltage across variable resistor R, V_R), and voltmeter V_3 (for measuring voltage drop across inductor L, V_L) are connected to single phase ac supply through DPST switch, as shown in Fig. 7.

DPST switch is closed and simultaneous readings of voltmeters V_1, V_2 and V_3 and ammeter A are noted. This is repeated with different currents obtained by adjusting the variac.

Value of R is changed and another set of readings are noted. The results are tabulated and value of power factor is determined.

Observations and Results:

S. No.	*Ammeter Reading, I in Amperes*	*Reading of Voltmeter V_1 (Supply Voltage V) in Volts*	*Reading of Voltmeter V_2 (Voltage across Resistor R, V_R in Volts)*	*Reading of Voltmeter V_3 (Voltage across Inductor L, V_L) in volts*	*Ratio of V and I i.e. $\frac{V}{I}$*	*Power Factor of circuit $\cos\phi = \frac{V_R}{V}$*
1						
2						
3						
4						
5						
6						
7						
8						
9						
10						
11						
12						

Conclusion: 1. For fixed value of resistor R, ratio of V and I *i.e.* $\frac{V}{I}$ remains constant and is equal to circuit impedance Z.

2. For fixed value of resistor R, power factor of the circuit remains the same and is equal to $\frac{R}{Z}$ or $\frac{V_R}{V}$.

EXPERIMENT-5

Objective : To verify the voltage and current relations in star and delta connected systems.

Apparatus Required: TPIC switch (500 V, 15 A)-1 no., Moving iron type voltmeter (0-500 V)-2 nos., Moving iron ammeter (0-10 A)-2 nos., 3-phase balanced load (say a 3-phase, 440 V, 50 Hz, 7.5 kW induction motor), Board containing six open terminals-1 no. and connecting leads.

Theory: In star-connections

$$\text{Line voltage, } V_L = \sqrt{3} \text{ phase voltage} = \sqrt{3}\, V_P$$
$$\text{Line current, } I_L = \text{Phase current, } I_P$$

In delta-connections

$$\text{Line voltage, } V_L = \text{Phase voltage, } V_P$$
$$\text{Line current, } I_L = \sqrt{3} \text{ phase current } I_P$$

Connection Diagram:

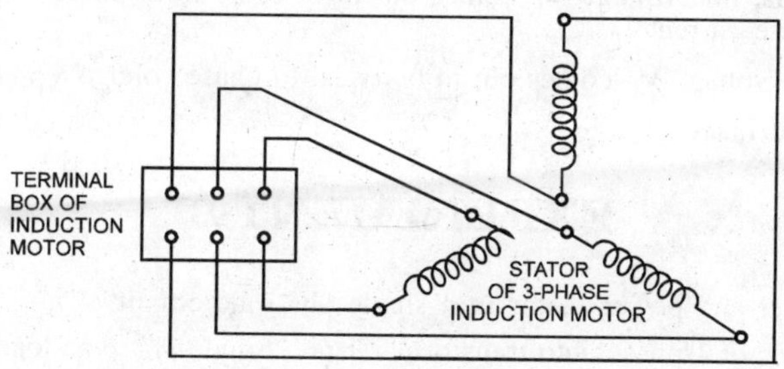

(*a*)

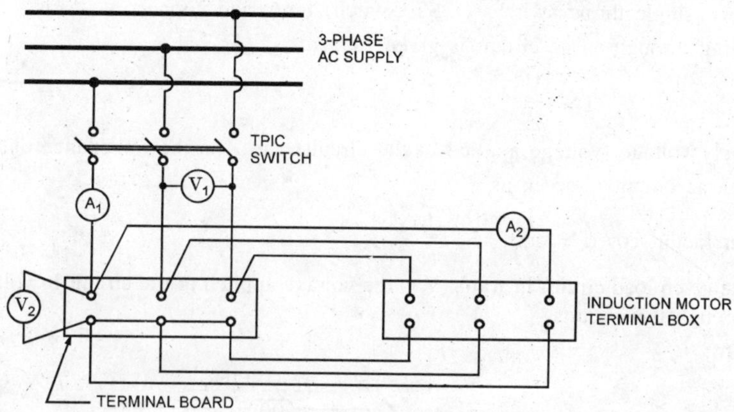

(*b*)

Fig. 8

Procedure: The connections of the terminal box of the induction motor to the stator phase windings of the motor are shown in Fig. 8 (*a*). First of all the connections in terminal box are made, as shown in Fig. 8 (*b*). The motor is started and put on load and readings of both voltmeters and ammeters are noted for different loads on the motor.

Now the TPIC switch is made off and the connections in terminal board are changed, as illustrated in Fig. 8 (*c*) for delta connections. The motor is started and put on load and the readings of both ammeters and voltmeters are noted for different loads on the motor.

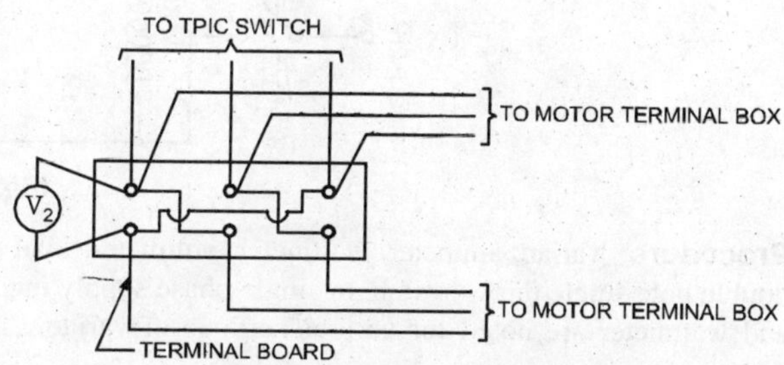

Fig. 8 (*c*) *Delta-Connections*

Connections	*S. No.*	*Voltmeter Reading V_1, Line Voltage in Volts, V_L*	*Voltmeter Reading V_2, Phase Voltage in Volts, V_P*	*Ammeter Reading A_1, Line Current in Amperes, I_L*	*Ammeter Reading A_2, Phase Current in Amperes, I_P*	$\frac{V_L}{V_P}$	$\frac{I_L}{I_P}$
Star	1 2 3						
Delta	1 2 3						

Results: In star connections, line voltage V_L comes out to be $\sqrt{3}$ times phase voltage V_P and line current I_L comes out to be equal to phase current I_P.

In delta connections, line voltage V_L comes out to be equal to phase voltage V_P and line current I_L comes out to be $\sqrt{3}$ times phase current I_P.

EXPERIMENT-6

Objective: To measure power and power factor in a single phase ac circuit.

Apparatus Required: Variac or 1-phase auto-transformer-1 no., Single-phase ac load [or lamps and choke coils], Moving-iron voltmeter [0-250 V]-1 no., Moving iron ammeter [0-5 A]-1 no., Dynamometer type wattmeter (250 V, 5 A)-1 no., Double pole, single throw switch (DPST Switch)-1 no. and connecting leads.

Theory: Current flowing through an ac circuit is given as

$$I = \frac{V}{Z}$$

where V is the ac supply voltage (voltage applied to the circuit) and Z is the impedance of the circuit in ohms.

Power factor of an ac circuit is given as

$$\text{Power factor, } \cos\phi = \frac{P}{VI}$$

where P is power of the given load circuit in watts, V is the voltage applied to the circuit in volts and I is the current in amperes flowing through the circuit.

Connection Diagram:

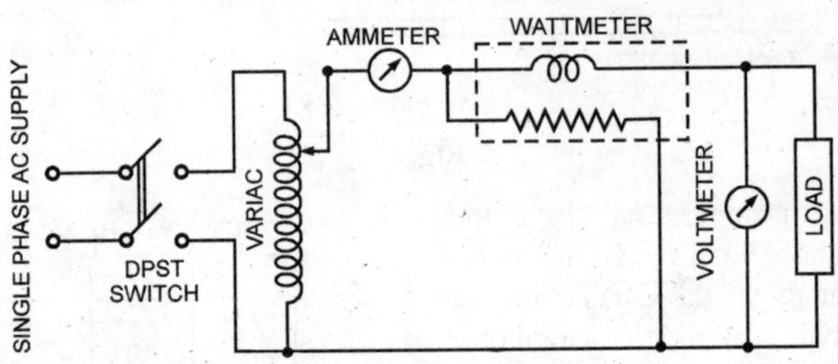

Fig. 9

Procedure: Variac, ammeter, wattmeter, voltmeter, load (lamps in series with choke coils) are connected through double pole single throw switch, to single phase supply mains as shown in Fig. 9. The readings of ammeter, voltmeter and wattmeter are noted for various settings of variac.

Observations:

S.No.	*Voltmeter Reading, V in Volts*	*Ammeter Reading, I in Amperes*	*Wattmeter Reading, P in Watts*	*Power Factor of the Circuit,* $\cos\phi = \frac{P}{VI}$
1.				
2.				
3.				
4.				

Conclusion:

1. Current I increases directly in proportion to applied voltage V.
2. Power factor of the circuit is same through out for a given load.

Note: If required, the readings of the instruments can be recorded with the different loads (by varying the number of lamps and chokes connected in series).

EXPERIMENT-7A

Objective: To verify the series resonance in ac circuits.

Apparatus Required: AC voltmeters–2 nos., AC ammeter–1 no., Power factor meter–1 no., Adjustable non-inductive resistor–1 no., Inductor–1 no., Variable capacitor–1 no., DPIC switch–1 no. and connecting wires.

Theory: Refer to Art. 5.2.

Connection Diagram: Connection diagram is shown in Fig. 10.

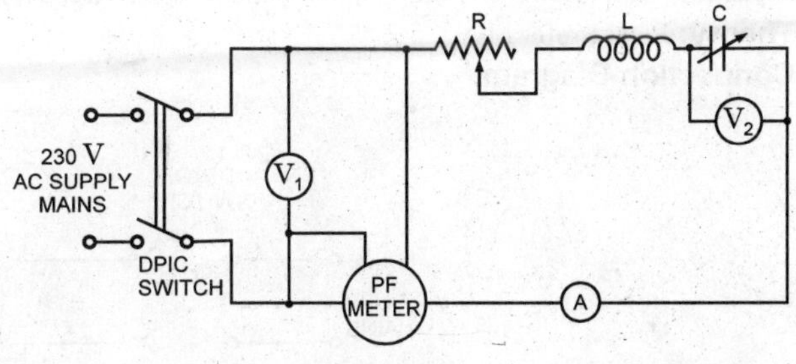

Fig. 10

Procedure: An adjustable non-inductive resistor R, an inductor L and a variable capacitor C are connected in series; a power factor meter is also included, as well as voltmeters and an ammeter, as shown in Fig. 10. With R adjusted to a value, such that the current I = V/R does not exceed the safe value and with a small capacitor, the supply is switched on and all readings are recorded. The capacitance is gradually increased, readings being noted at intervals; if, on approaching resonance, the capacitor voltage tends to reach a dangerous value, the switch must be opened and the capacitance should be increased beyond the resonant point, now the switch may be closed, and the observations continued. The results are tabulated, and curves are plotted for current, condenser voltage and power factor versus capacitance (Fig. 11).

Observations: R = ... ohms; Supply voltage, V_1 ...volts (ac); L = ... henries

Capacitance C in μF	*Capacitor voltage, V_2 in Volts*	*I in Amperes*	*Power Factor in Fraction*

$$\text{Resonant frequency, } f_r = \frac{1}{2\pi\sqrt{LC}}$$

$$\text{and Q-factor} = \frac{1}{R}\sqrt{\frac{L}{C}}$$

where C is the value of capacitance corresponding to maximum current.

Conclusions: 1. The circuit current rises from zero, when the capacitance is zero (reactance is infinite), to a maximum value V/R at resonance, beyond which it decreases, tending to a constant value as capacitance C becomes very large.

2. The capacitor voltage equals the supply voltage when its value is zero. It is because of zero current and no potential drop across either R or L. This increases at first to a maximum value near resonance, and then falls gradually to zero as capacitance C tends to infinity.

3. The power factor varies in a similar way, as the circuit current, *i.e.* it rises first to a maximum value of unity at resonance point and then falls. At resonance, the circuit behaves like a pure resistor.

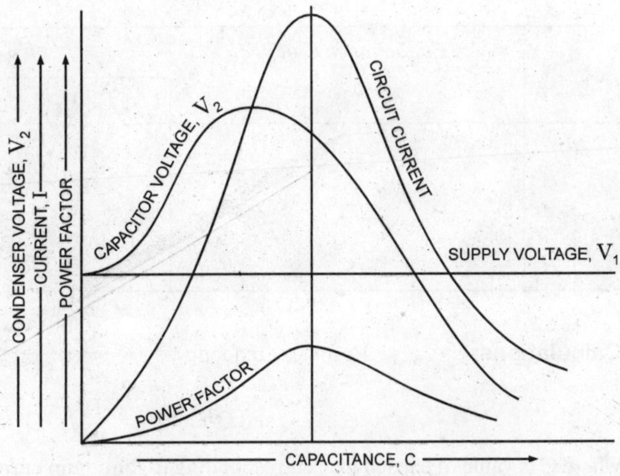

Fig. 11. ***Curves Illustrating Resonance in R-L-C Series Circuit***

EXPERIMENT-7B

Objective: To verify the parallel resonance in ac circuits.

Apparatus Required: AC ammeters–3 nos., Power factor meter–1 no., Resistor–1 no., Inductor–1 no., Variable capacitor–1 no., DPIC switch–1 no. and connecting wires.

Theory: Please refer to Art. 5.3.

Connection Diagram:

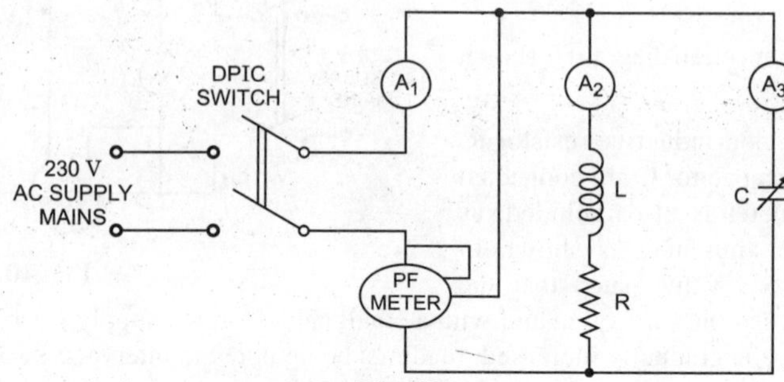

Fig. 12

Procedure: A coil of resistance R and inductance L is connected in parallel with a variable capacitor C, and supplied with alternating current through a DPIC switch as shown in Fig. 12. Three ammeters are connected in the circuit to indicate the currents in the main and branch circuits and also a power factor meter is included in the circuit. With the capacitance adjusted to a low value, the main switch is closed and the ammeter readings are noted; the capacitance is now gradually increased, the readings being taken at regular intervals and noted.

Curves are plotted of charging current I_3, current through inductive coil I_2 and the main current I_1, all to a base of capacitance as shown in Fig. 13.

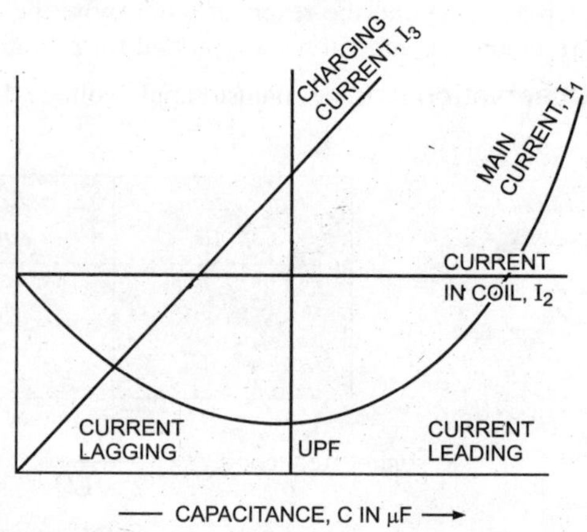

Fig. 13

Observations: R = ... Ohms L = ... Henrys

S.No.	Capacitance C in μF	Currents			Power Factor in Fraction
		I_1	I_2	I_3	
1.					
2.					
3.					
4.					
5.					

Calculations:

$$\text{Resonant frequency, } f_r = \frac{1}{2\pi}\sqrt{\frac{1}{LC} - \frac{R^2}{L^2}}$$

$$\text{and Q-factor} = \frac{1}{R}\sqrt{L/C}$$

where C is value of capacitance corresponding to minimum current.

Conclusion: The main branch current I_1 first decreases up to the point when it is in phase with voltage, and then increases again. At the minimum point a condition of maximum current resonance occurs, both the branch currents I_2 and I_3 being greater than the main current.

EXPERIMENT-8

Objective: To observe the B-H loop of ferromagnetic core material on CRO.

Apparatus Required: CRO-1 no., Transformer-1 no.; Variac-1 no.; Resistors-2 nos., Capacitor-1 no. and connecting wires.

Theory: The curve drawn giving relationship between induction density B and magnetising force H is known as *B-H curve* or *magnetising curve.* Figure 14 shows the general shape of B-H curve of a magnetic material. In general it has four distinct regions *oa*, *ab*, *bc* and the region beyond *c*. During the region *oa* the increase in flux density is very small, in region *ab* the flux density B increases almost linearly with the magnetising force H, in region *bc* the increase in flux density is again small and in region beyond point *c*, the flux density B is almost constant. The flat part of the B-H curve corresponds to *magnetic saturation* of the material. The nonlinearity of the curve indicates that the relative permeability $\mu_r \left(\frac{B}{\mu_0 H}\right)$ of a magnetic material is not* constant but depends very largely upon the flux density B.

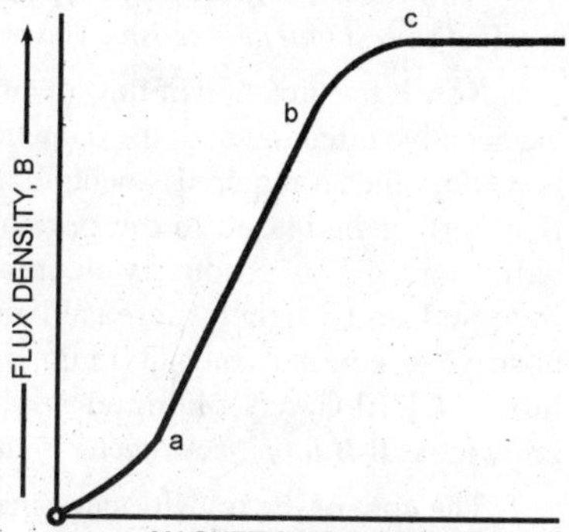

Fig. 14.

The B-H curve depends on the material only and not upon the dimensions of a specific piece. Magnetising or B-H curves for silicon alloy (electrical sheet) steel, cast steel and cast iron etc., are shown in Fig. 15. In making magnetic calculations, values of H and μ_r corresponding to working flux density B are to be taken. The use of B-H curves permits the calculations of magnetic circuits with a fair degree of accuracy.

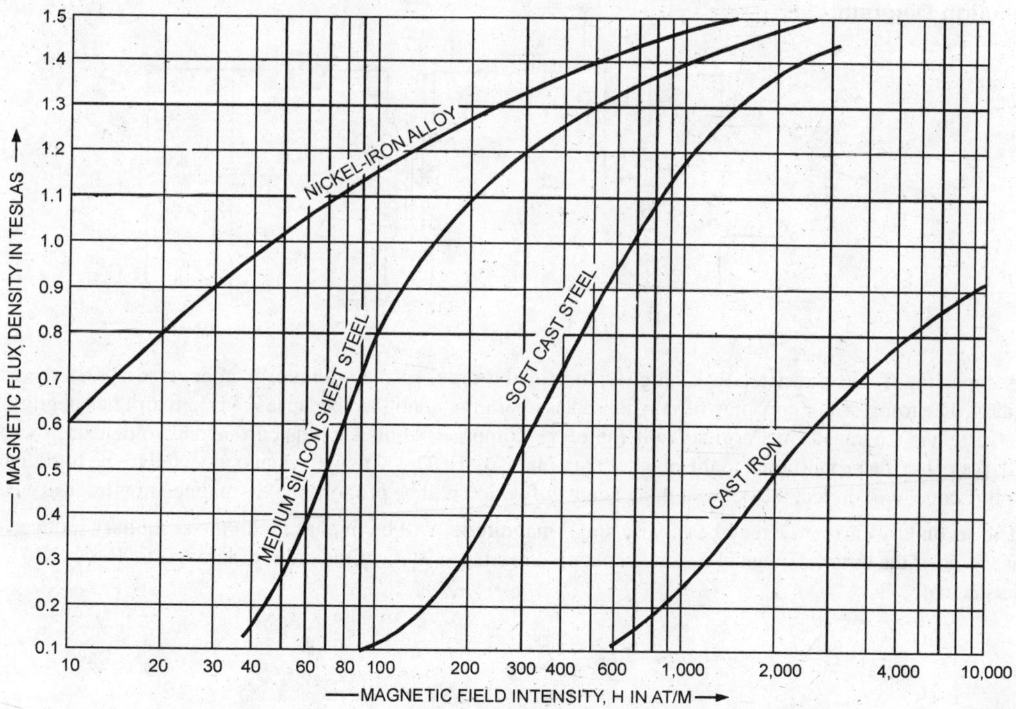

Fig. 15. ***Magnetization Curves of Typical Ferromagnetic Materials***

* Relative permeability μ_r for all non-magnetic materials (*e.g.* brass, air etc.) is constant; being equal to 1. Thus for non-magnetic materials $B = \mu_0 H$. Since μ_0 is constant, $B \propto H$ or B-H curve for a non-magnetic material is a straight line passing through origin.

Let a core of specimen of iron be wound with a number of turns of a wire and current be passed through the solenoid. A magnetic field of intensity H proportional to the current flowing through the solenoid is produced. Let magnetising force H is increased from zero to a certain maximum value and then gradually reduced to zero. If the values of flux density B in the core corresponding to various values of magnetising force H are determined and B-H curves are drawn for increasing and decreasing values of magnetising force H then it will be observed that B-H curve obtained for decreasing values of H lies above that obtained for increasing values of H.

While decreasing the magnetising force H, when H is brought to zero the induction density B, is represented by OC and is called as *residual magnetism*. The power of retaining the residual magnetism is called the *retentivity* of the material.

Now if the direction of flow of current is reversed, the magnetising force H is reversed. Let the current be increased in the negative direction until the induction density B becomes zero. At this instant *i.e.* when B = 0, the demagnetising force H = OD, which is required to neutralize the residual magnetism, and is known as *coercive force*. If the demagnetising force H is further increased to the previous maximum value and again gradually decreased to zero, reversed and further increased in original or positive direction to the maximum value, a closed loop ACDEFGA is obtained which is usually known as *B-H loop or magnetic cycle.*

The core of the transformer carries alternating current and, therefore, undergoes magnetic reversal during each cycle. Magnetic cycle or B-H loop of a transformer core can be observed on an oscilloscope by using the circuit given in Fig. 17.

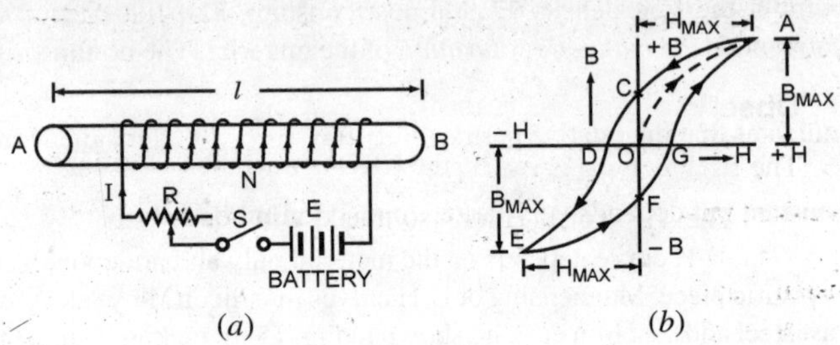

Fig. 16

Connection Diagram:

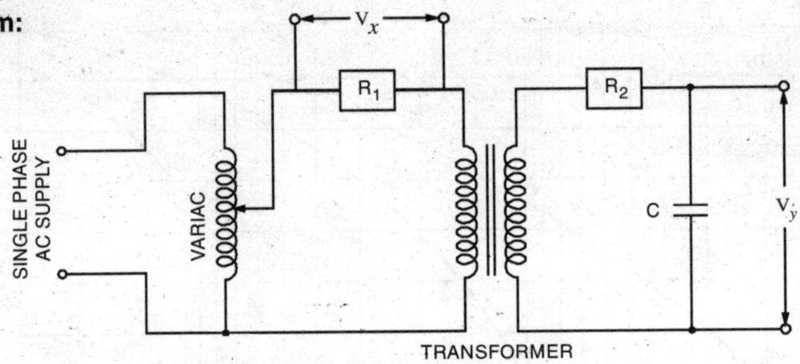

Fig. 17

A capacitor C and a resistor R_2 of high value (of the order of mega ohms) is connected across the transformer secondary. The transformer now may be considered as operating at no load because R_2-C network has negligible loading effect. Current in the transformer primary will, therefore, comprise of only exciting current and voltage drop across resistor R_1 will, therefore, be proportional to the magnetising intensity H. This voltage, when connected to the horizontal input of the oscilloscope, will deflect the beam and the beam deflection will be proportional to magnetising intensity H.

The secondary current i_2 (may be of very small magnitude) will be in phase with the secondary induced voltage e_2 because R_2 is of the order of mega ohms and capacitive reactance X_C is of the order of kilo ohms ($R_2 >> X_C$) and $Z_2 \simeq R_2$ and thus $i_2 \alpha e_2$

$$\text{or } i_2 = \frac{e_2}{R_2}$$

Also, Voltage across capacitor, $V_y = \frac{1}{C}\int i_2\, dt = \frac{1}{C}\int \frac{e_2}{R_2} dt = \frac{1}{CR_2}\int e_2\, dt$

Substituting $e_2 = N_2 \frac{d\phi}{dt}$ in above expression, we have

$$V_y = \frac{1}{CR_2}\int \frac{N_2\, d\phi}{dt} = \frac{N_2}{CR_2}\int d\phi = k\phi$$

Thus, V_y at any instant is proportional to the flux in the core and is in phase with it.

V_y represents the flux density B in the transformer core because $B = \frac{\phi}{A_i}$ and core cross section A_i is constant.

Thus V_y, when connected to the vertical input of the oscilloscope, will deflect the beam in proportion to flux density B.

Procedure: The voltage drop across R_1 is connected to the horizontal input of the oscilloscope and voltage drop across capacitor C is connected to the vertical input of the oscilloscope as shown in Fig. 17.

The horizontal and vertical gains of the CRO are adjusted so as to observe the B-H loop on the screen.

Observations and Result:

The B-H loop observed on the CRO is traced on the tracing paper.

The shape of B-H loop so traced will be as shown in Fig. 18.

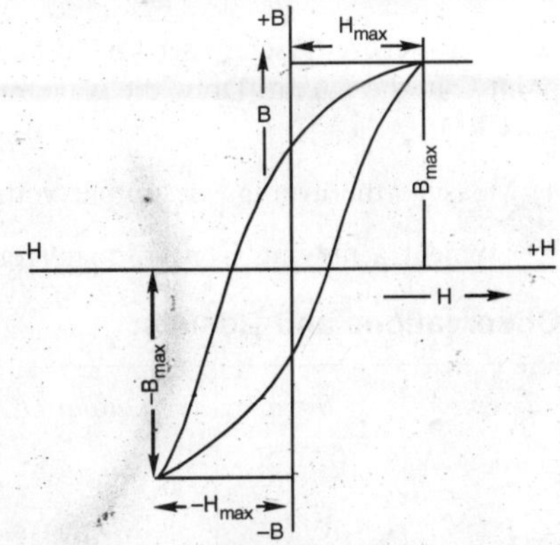

Fig. 18

EXPERIMENT-9

Objective: To use a bridge rectifier for full-wave rectification of ac supply and to determine the relationship between rms and average values of rectified voltage.

Apparatus Required: Bridge rectifier-1 no; Cathode ray oscilloscope-1 no; Moving iron voltmeter- 1 no; Moving coil voltmeter-1 no; Resistor-1 no; Variac-1 no; DPST switch; and connecting leads.

Theory: Refer to Art. 16.3.2.

From circuit analysis, we have

$$V_{rms} = \frac{V_{max}}{\sqrt{2}} \qquad ...(i)$$

$$V_{av} = \frac{V_{max}}{\pi / 2} \qquad ...(ii)$$

From Eqs. (*i*) and (*ii*), we have

$$V_{rms} = \frac{\pi}{2\sqrt{2}} V_{av} = 1.11\, V_{av}$$

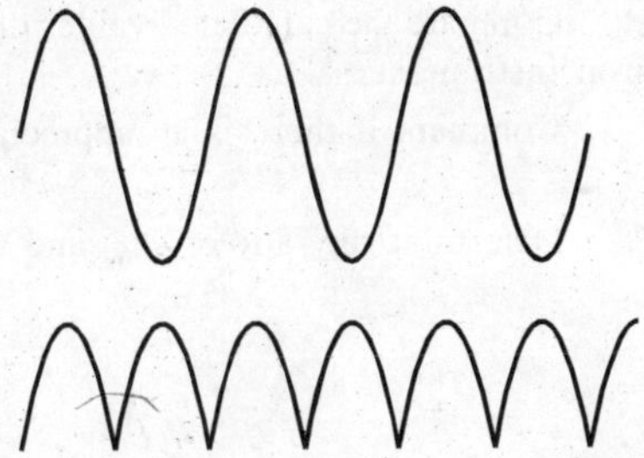

Fig. 19

The input and output waveforms for the bridge rectifier are shown in Fig. 19.

Connection Diagram:

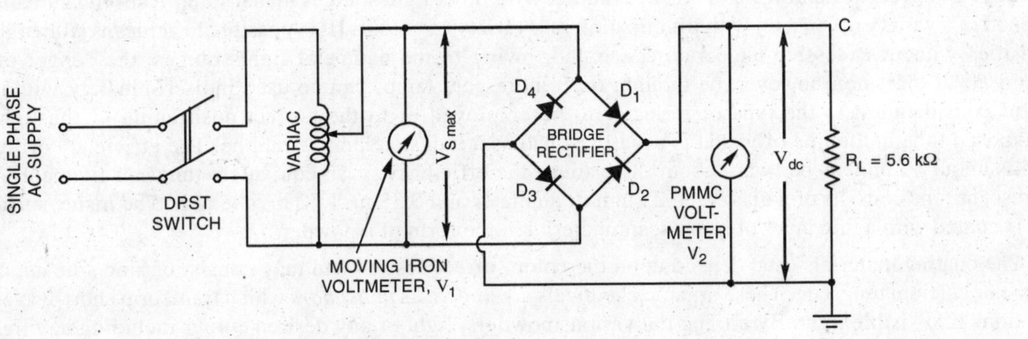

Fig. 20

Procedure: 1. Make the connections as shown in Fig. 20.

2. Switch on the supply through variac and measure the rms transformer secondary voltage. Use the appropriate terminals on variac for V_s as close as to 12 V as you can get.

3. With the oscilloscope set for dc coupling and the vertical sensitivity on 10 V/division, view the waveform at point C in the circuit. Draw the waveform on the grid provided. Observe the peak value of rectified output voltage on CRO.

4. Measure the average dc output voltage across R_L by voltmeter V_2.

5. Connect a moving iron voltmeter on the output and note its reading to measure rms value.

Observations and Results:

S. No.	Input Voltage V_1 in Volts	Output DC Voltage V_{av}, Moving Coil Voltmeter reading in Volts	Output Voltage V_{rms}, Moving Iron Voltmeter reading in Volts	V_{max} From CRO	Calculated Value of $V_{av} = 0.637\ V_{max}$	Calculated Value of $V_{rms} = \frac{V_{max}}{\sqrt{2}}$
1						
2						
3						

Calculate V_{av} and V_{rms} for each set of observation and tabulate the values.

Compare the calculated values of average and rms voltages with those observed in the moving coil and moving iron instruments.

Comment if there is an appreciable difference between the calculated and observed values.

Determine the ratio of V_{rms} and V_{av} *i.e.* $\frac{V_{rms}}{V_{av}}$, which will be found as 1.11, since supply voltage is sinusoidal.

EXPERIMENT-10

Objective: To measure the minimum operating voltage, current drawn, power consumed, and the power factor of a fluorescent tube.

Apparatus Required: Fluorescent tube with choke and starter 40 W-1 no.; Moving iron type voltmeter 0-250 V-1 no.; Moving iron ammeter 0-0.5 A-1 no.; Dynamometer type wattmeter 0-250 V; 0-0.5 A-1 no. and connecting leads.

Theory: Fluorescent lighting has a great advantage over other light sources in many applications. The tubes can be obtained in a variety of lengths, with illumination in a variety of colours. It is possible to achieve quite high lighting intensities without excessive temperature rise and, owing to the nature of light sources, the danger of glare is minimised. It must not, however, be thought that fluorescent lamps can be used indiscriminately without giving careful consideration to the type of reflector to be employed or to the correct positioning of the light source. An exposed tube in the line of vision is usually as bad as a tungsten lamp would be. The efficiency of fluorescent lamp is about 40 lumens per watt, about three times the efficiency of an equivalent tungsten filament lamp. The fluorescent tube consists of a glass tube 25 mm in diameter and 0.38 m–1.52 m in length. The inside surface of the tube is coated with a thin layer of fluorescent material in the form of a powder.

The coating materials used depend upon the colour effect desired and may consist of zinc silicate, cadmium, silicate or calcium tungstate. These organic chemicals are known as *phosphors* which transform short-wave invisible radiations into visible light. By mixing the various powders, light of any desired colour including day light can be obtained. The tube contains small quantity of argon gas at a pressure of 2.5 mm of mercury and one or two drops of mercury. It is provided with two electrodes coated with electron emissive material. A starting switch is provided in the circuit, which puts the electrodes directly across the supply mains at the time of starting, so that electrodes may get heated and emit sufficient electrons. A stabilising choke is connected in series with it, which acts as a ballast in running condition and provides a voltage impulse for starting. A capacitor is connected across the circuit to improve the power factor.

Fluorescent tubes are available in the following sizes :

Length	*Wattage*	*Length*	*Wattage*	*Length*	*Wattage*
38 cm	14 W	100 cm	25 W	152 cm	65 W
46 cm	15 W	122 cm	40 W	152 cm	80 W
61 cm	20 W				

The fluorescent lamp in common with most electrical devices may cause a certain amount of radio-interference. This interference may be caused due to one of the following factors :

1. Direct radiation from the bulb to the antenna. This effect diminishes rapidly as the radio is separated from the lamp. Thus, for example, at a radius of 2.8 metres interference from this cause is negligible.
2. Line radiation from the electric supply line to the antenna.
3. Line feedback from the lamp through the line to the radio.

The radio-interference effect may be reduced to a minimum by connecting a small capacitor (0.05 μF) across starter terminals, as shown in Fig. 21.

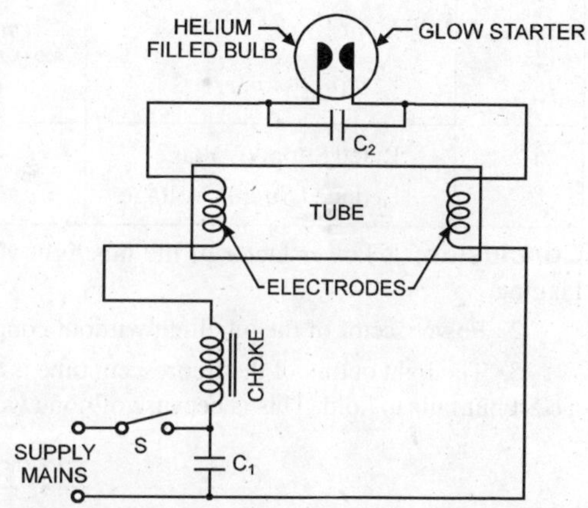

Fig. 21

The connections of a fluorescent tube incorporating a glow type starter are shown in Fig. 21. The glow type starter is a voltage operated device and consists of two bimetallic electrodes enclosed in a glass bulb filled with a mixture of helium and hydrogen. Normally the contacts are open. When the supply is switched on, the potential across bimetallic electrodes causes a small glow discharge at a small current not enough to heat up the tube electrodes (filament). This discharge is enough, however, to heat the bimetallic strips of the switch causing them to bend and make contact. The result is a large current through the electrodes, their temperature being raised to incandescence and the gas in the immediate neighbourhood is ionised. After one or two seconds the bimetallic strips cool down and the contacts open. This opening of contacts in series with the choke causes a momentary high voltage, which is sufficient to start the discharge in the main tube. The starter ceases to glow as the voltage is now too low. A small capacitor is placed inside the starter to suppress arcing and radio-interference.

Connection Diagram:

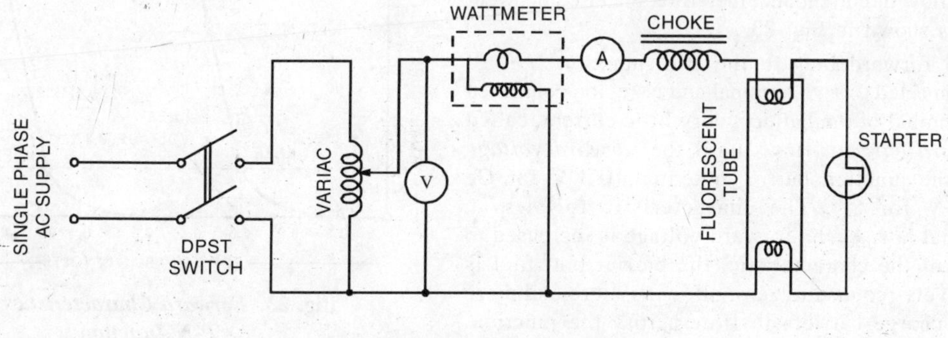

Fig. 22

Procedure: 1. Connections are made as shown in Fig. 22.

2. Supply is switched on through variac by DPST switch and the supply is adjusted to 230 V. The fluorescent tube should start glowing. Ammeter, voltmeter and wattmeter readings are noted.

3. Power factor of the fluorescent tube is determined from the relation

$$\text{Cos } \phi = \frac{W}{VI}$$

4. Power supply is switched off.

5. Reduced voltage is switched on. Applied voltage is increased gradually through variac till the fluorescent tube starts glowing (without flicker). Readings of ammeter, voltmeter and wattmeter are noted and recorded.

Observations:

S. No.	Input Voltage, V in Volts	Ammeter Reading, I in Amperes	Wattmeter Reading, P in Watts	Power Factor, $\cos \phi = \frac{W}{VI}$
1.	Rated Supply Voltage			
2.	Reduced Supply Voltage			

Conclusion: 1. Power factor of the tubelight with condenser connected across the supply is found to be about 0.95 lagging.

2. Power factor of the tubelight without condenser is found to be about 0.5 lagging. This is owing to choke.

3. The light output of the fluorescent tube is a function of supply voltage. At reduced supply voltage, the tube clicks a start but fails to hold. This is because of non availability of required holding voltage across the tube.

EXPERIMENT-11

Objective: To verify the rating of compact fluorescent lamp (CFL).

Repeat Experiment-10 with CFL in place of fluorescent tube.

EXPERIMENT-12

Objective: To obtain the characteristcs of a P-N junction diode.

Apparatus Required : P-N junction diode-1 no., IC regulated power supply-1 no., Current limting resistor-1 no., Milliammeter-1 no., Microammeter-1 no., Voltmeter 0–2 V-1 no., Voltmeter 0–10 V-1 no. and connecting leads.

Theory : When the external voltage is zero *i.e.* when the circuit is open, the potential barrier at the junction does not allow the flow of current and, therefore, the circuit current is zero, as shown in Fig. 23.

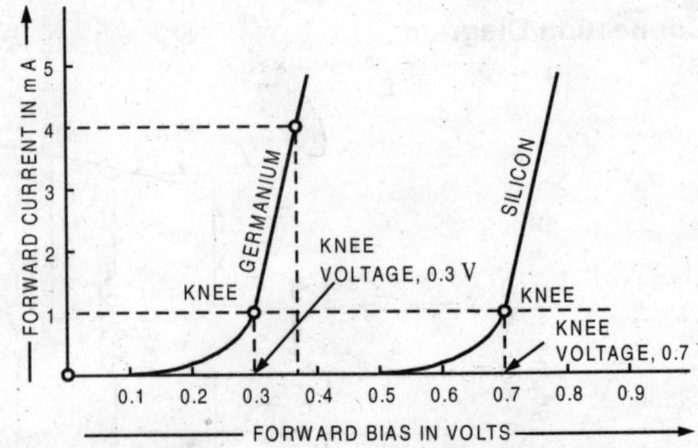

Fig. 23. ***Forward Characteristics For P-N Junction***

With forward bias to the P-N junction (*i.e.* P-region connected to + ve terminal and N-region connected to –ve terminal of the battery) very little current, called the *forward current*, flows until the *forward voltage exceeds* the junction barrier potential (0.3 V for Ge and 0.7 V for Si). The characteristic follows an exponential law. As the forward voltage is increased to the knee of the characteristic, the barrier potential is progressively reduced to zero, allowing more and more majority charge carriers to flow across the junction. Beyond the knee of the characteristics, the potential barrier is completely eliminated, forward current increases almost linearly with the increase in forward voltage and the P-N junction starts behaving as a resistor. If the forward voltage is increased beyond a certain value, extremely large current will flow and the P-N junction may get destroyed due to overheating.

Because of the nonlinear knee of the curve, if a very small signal voltage is applied to the diode so that it only operates around the knee, the signal will be distorted. The signal must, therefore, be large enough to operate the diode over the linear part of the characteristic. Forward resistance of P-N junction is low (of the order of 75 Ω in case of Ge and 150 Ω in case of Si).

Reverse Characteristics. When the reverse bias is applied (*i.e.* when P-region is connected to the negative bias terminal and N-region is connected to the positive bias terminal), the potential barrier at junction is increased, therefore, the junction resistance becomes very high and there is no possibility of a majority carriers flowing across a reverse-biased junction. But still minority carriers generated on each side can cross the junction. Electrons on the P-side are attracted across the junction to the positive potential on the N-side and holes on the N-side may travel across the junction to the negative potential on the P-side. This results in a very small current, as illustrated in Fig. 24. This current is called *reverse current* and is due to minority carriers.

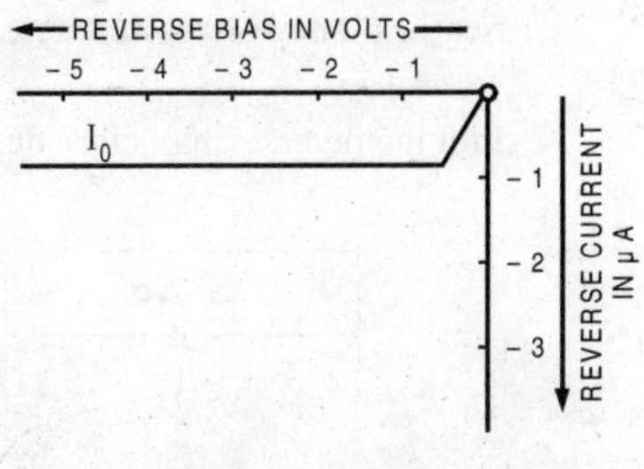

Fig. 24. ***Reverse Characteristics For P-N Junction***

For silicon the reverse current is typically less than 1 μA and may be as low as 1 *n*A. In case of germanium, the reverse current may exceed 10 μA. Only a very small reverse voltage is necessary to direct all minority carriers across the junction, and when all minority carriers are flowing across, further increase in bias voltage will not increase the current. This current is referred to as a *reverse saturation current* and is designated as I_0. Reverse resistance is very high (of the order of MΩ).

As discussed above, the reverse current I_0 is made up of minority charge carriers crossing the junction. When the temperature of the semiconductor material is increased, the additional thermal energy causes more electrons to break away from atoms. This creates more hole-electron pairs and generates more minority charge carriers. So reverse saturation current I_0 increases with the increase in the junction temperature. It has been found that I_0 approximately doubles for each 10°C rise in temperature in case of germanium and for each 8°C rise in temperature in case of silicon.

PROCEDURE :

1. Connect the P-N junction diode as shown in Fig. 25(*a*).
2. Set the diode voltage at a series of convenient levels, and measure and record the corresponding current levels in Table 1.
3. Plot the characteristics from the values obtained.
4. For reverse characteristics, connect the P-N junction diode, as shown in Fig. 25(*b*).
5. Resistance R_1 be dispensed with, as diode itself offers high resistance in reverse bias. Since very small leakage reverse current is expected, microammeter is used to measure current, and it must be directly connected in series with the diode as shown in Fig. 25(*b*), otherwise the voltmeter current may introduce a serious error.

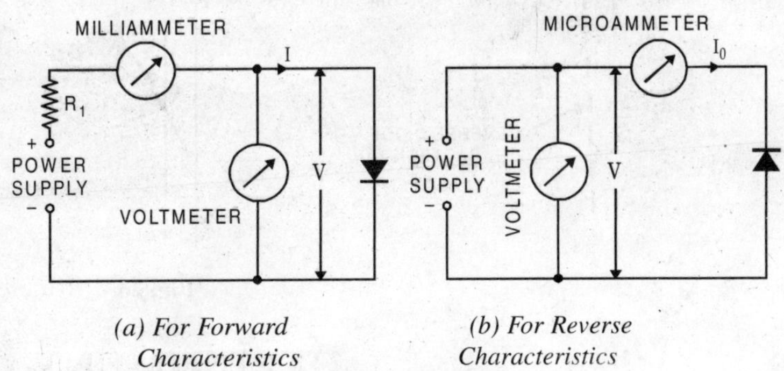

(*a*) *For Forward Characteristics*

(*b*) *For Reverse Characteristics*

Fig. 25. ***Circuits For Determination of Diode Characteristics***

TABLE 1

S. No.	Voltage V in Volts	Current I in mA
1.		
2.		
3.		

6. Set the diode voltage at a series of convenient levels, and measure and record the corresponding current levels in Table 2.
7. Plot the reverse characteristics of P-N junction diode.

TABLE 2

S. No.	Voltage V in Volts	Current I_0 in μA
1.		
2.		
3.		

CONCLUSION : Draw the V-I characteristic curve for both of the biasing (forward biasing and reverse biasing) on graph paper choosing suitable scales separately for forward and reverse biasing.

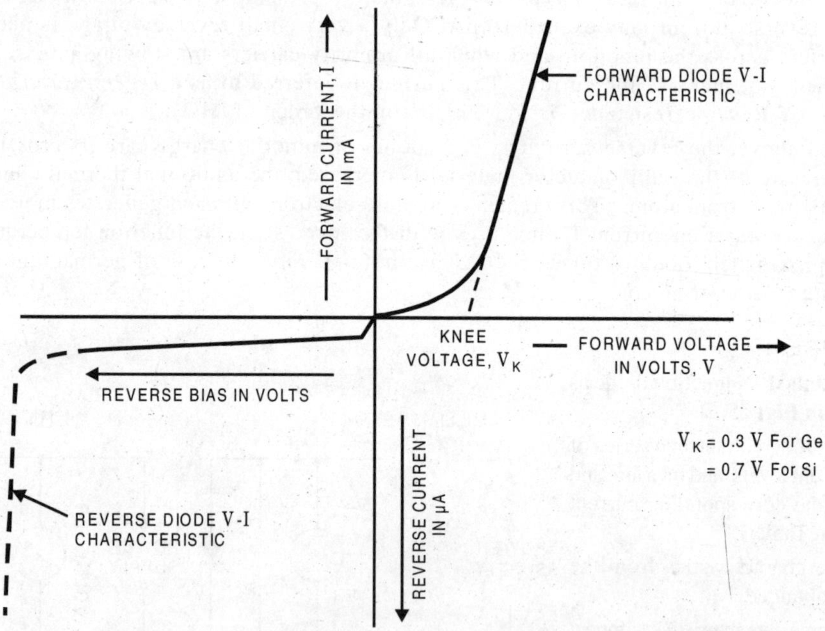

Fig. 26

EXPERIMENT-13

Objective: To verify the truth tables of logic gates.

Apparatus Used: Digital trainer kit with built-in 5 V dc power supply and

IC 7408 → 2-INPUT AND Gate Quad pack

IC 7432 → 2-INPUT OR Gate Quad pack

IC 7404 → Inverter HEX pack

IC 7400 → 2-INPUT NAND Gate Quad pack

IC 7402 → 2-INPUT NOR Gate Quad pack

IC 7486 → 2 INPUT EX-OR Gate Quad pack

IC 74266 → 2 INPUT EX-NOR Gate Quad pack.

Digital multimeter–1 no., LED indicator–1 no., Logic probe–1 no.

Theory: The most basic elements of digital circuits are logic gates. Gates are electronic circuits that have several inputs to act upon. It has one unique output (high or low) depending upon the inputs and type of gate. Each gate is defined by a list of all possible inputs and corresponding output called the *truth table.*

Procedure:

1. Set the corresponding logic gate IC with +5 V to pin 14 and ground to pin 7 of the gate with the help of switches.
2. Give suitable 'HIGH or 'LOW' input to the input pins of the IC.
3. Observe the corresponding output using digital multimeter or LED indicator as 'HIGH' or 'LOW' and verify the truth table for the Gate being tested.
4. Repeat steps 1, 2 and 3 for different logic gate ICs as OR, AND, NOT, NOR, NAND, EX-OR and EX-NOR.

Conclusion: All the logic gates of 74 series verify the truth table for given gate as mentioned in Fig. 27.

Gate	*Symbol*	*Function*	*Truth Table*		
			Inputs		**Output**
			A	**B**	**Z**
OR	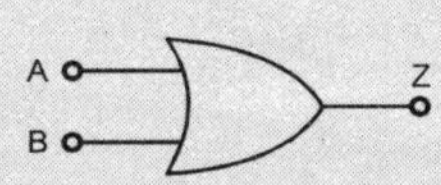	Z = A OR B = A + B	0	0	0
			0	1	1
			1	0	1
			1	1	1
			Inputs		**Output**
			A	**B**	**Z**
AND	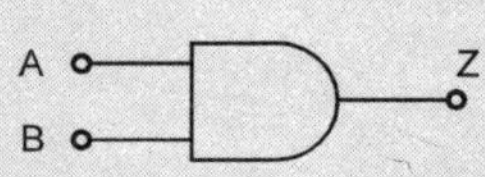	Z = A AND B = A·B	0	0	0
			0	1	0
			1	0	0
			1	1	1
			Input A		**Output Z**
NOT	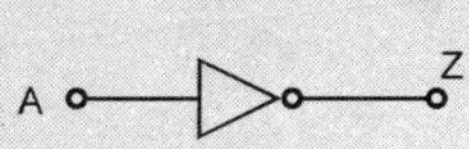	Z = NOT A $= \overline{A}$	0		1
			1		0
			Inputs		**Output**
			A	**B**	**Z**
NOR	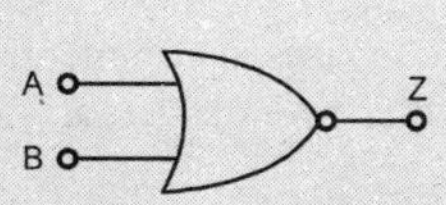	Z = A NOT OR B = A NOR B $= \overline{A + B}$	0	0	1
			0	1	0
			1	0	0
			1	1	0

Gate	*Symbol*	*Function*	*Truth Table*: Inputs *A*	*B*	*Output Z*
NAND	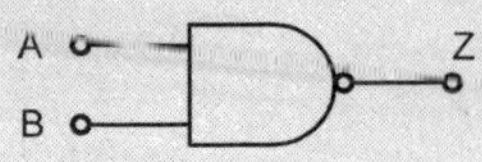	$Z = $ A NOT AND B $= $ A NAND B $= \overline{A \cdot B}$	0	0	1
			0	1	1
			1	0	1
			1	1	0
EX-OR	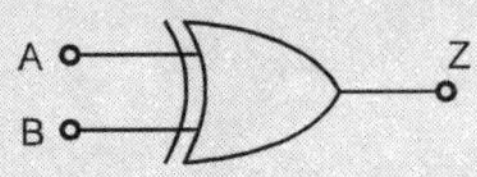	$Z = $ A EX-OR B $= A \oplus B = A\overline{B} + \overline{A}B$	0	0	0
			0	1	1
			1	0	1
			1	1	0
EX-NOR	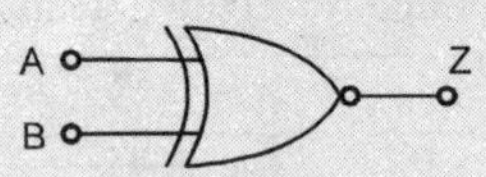	$Z = \overline{A \text{ EX-OR } B} = $ A EX-NOR B $= \overline{A \oplus B} = AB + \overline{A}\,\overline{B}$	0	0	1
			0	1	0
			1	0	0
			1	1	1

Fig. 27

EXPERIMENT-14

Objective: To connect the following measuring instruments to measure current, voltage and power in ac/dc circuits.

(*a*) Moving coil instruments.

(*b*) Moving iron instruments.

(*c*) Dynamometer instruments.

(*d*) Multimeter—both digital and analog type.

Apparatus Required: Moving coil ammeter-1 no.; Moving coil voltmeter-1 no.; Moving iron ammeter-1 no.; Moving iron voltmeter-1 no.; Dynamometer wattmeter-1 no.; Analog type multimeter-1 no. and Digital multimeter-1 no. and connecting leads.

Theory: 1. Ammeters and voltmeters, except electrostatic voltmeters, operate on the same principle. The ammeter carries the current to be measured or a definite fraction of it and this current or its definite fraction produces the deflecting torque whereas voltmeter carries the current proportional to the voltage to be measured which produces the deflecting torque.

Ammeter is of very low resistance and is connected in series with the load while voltmeter is of very high resistance and is connected across the load.

2. Moving iron instruments are widely used in laboratories and switchboards at commercial frequencies because these are cheaper in cost, robust in construction and can be manufactured with required accuracy.

There are two general types of moving iron instruments *i.e.* (*i*) the *attraction type* and (*ii*) the *repulsion type*. The attraction type instruments operate on the principle of attraction of a single piece of soft iron into a magnetic field and repulsion type instruments operate on the principle of repulsion of two adjacent iron pieces magnetised by the same magnetic field.

An attraction type instrument will usually have a lower inductance than the corresponding repulsion type instrument, and voltmeters will, therefore, be accurate over a wider range of frequency and there is a greater possibility of using shunts with ammeters. On the other hand, repulsion type instruments are more suitable for economical production in manufacture and a nearly uniform scale is more easily obtained; they are, therefore, much more common than the attraction type.

These instruments can be used both in dc as well as in ac circuits. This is due to the fact that their deflecting torque depends on the square of the current. These instruments read rms values and have cramped scales.

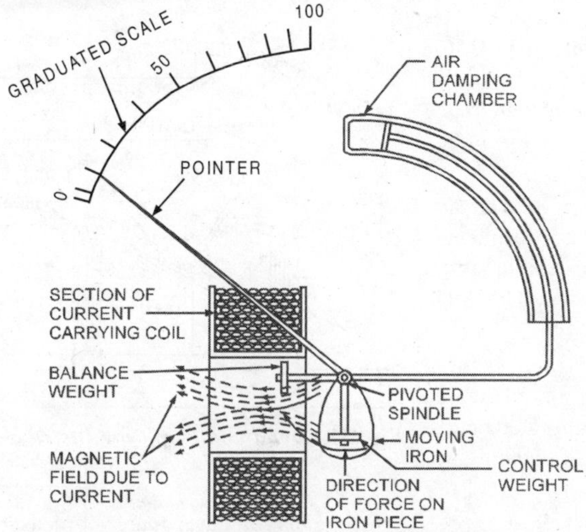

Fig. 28. ***Attraction Type Moving Iron Instrument***

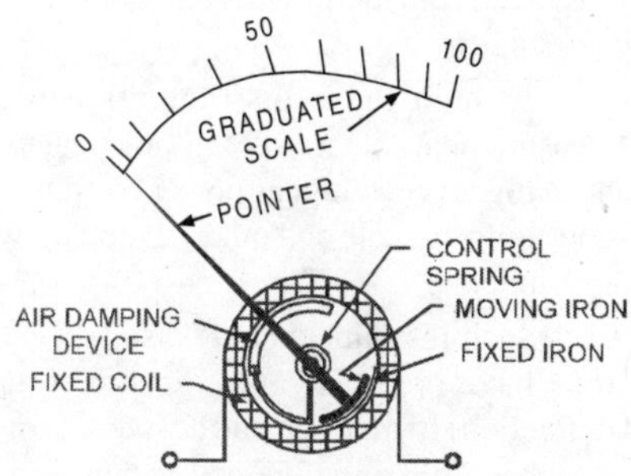

Fig. 29. ***Repulsion Type Moving Iron Instrument***

3. There are two types of moving coil instruments namely the *permanent magnet type* and *dynamometer type.*

Permanent magnet moving coil (PMMC) *instrument* is the most accurate and useful for dc measurements. In fact, were it not for the need of other instruments for ac measurements, it may be doubted whether any other type of indicating instruments would have survived except where cheapness is main requirement, in which case the simple moving iron instrument holds its place. *Dynamometer type moving coil instruments* can be used both on dc as well on ac. Their calibration is the same for both dc as well as ac and so these instruments are very useful as transfer instruments.

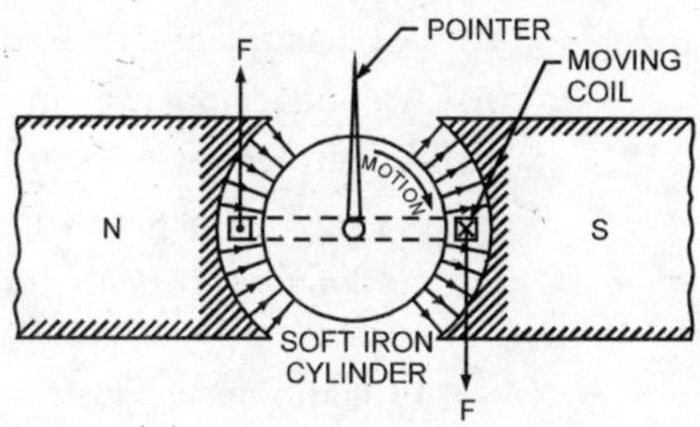

(a)

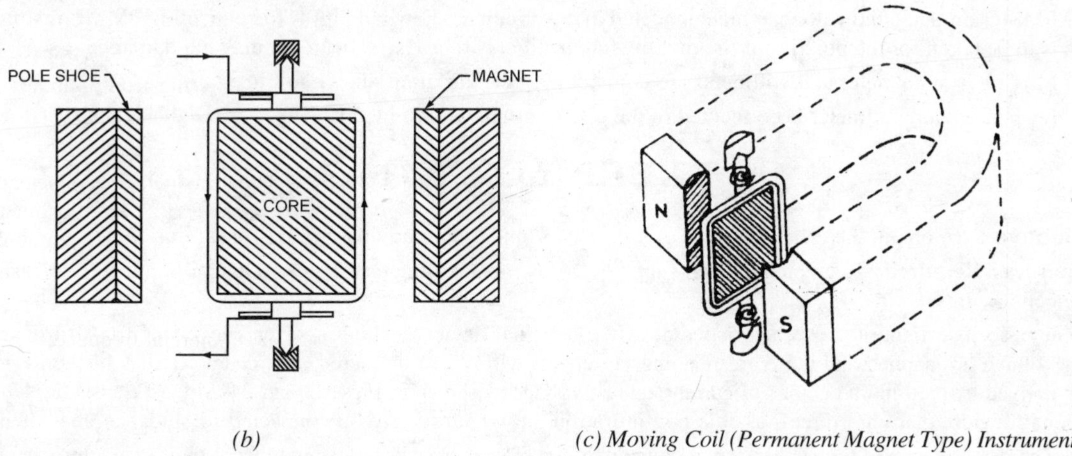

(b) *(c) Moving Coil (Permanent Magnet Type) Instrument*

Fig. 30

A permanent magnet type moving coil instrument is shown in Fig. 30. It consists of permanent powerful magnet with soft iron pole pieces. A cylindrical iron core is mounted between the two poles of the magnet giving very narrow air gap in which the sides of a pivoted light rectangular coil lies. The rectangular coil is wound of many turns from fine wire on light aluminium or copper former and acts as moving element. The purpose of using core is to make the field uniform and to reduce the reluctance of the magnetic circuit. A low reluctance helps to retain permanence of the magnet for a longer period.

These instruments can be used only in dc circuits because deflecting torque varies as the current. Their scales are uniform.

4. Dynamometer type instruments are very similar to PMMC instruments except that the permanent magnetic field is replaced by a coil (usually two fixed air-cored coils are used) which carry the current to be measured (or a definite fraction of it or proportional to the voltage to be measured).

Though these instruments can be used both, for dc and ac measurements but most of its practical uses are in ac as these instruments have higher cost, higher power consumption, lower torque-weight ratio, non-uniform scale and other several drawbacks in comparison to PMMC instruments.

Such instruments may also serve as *transfer instruments* because they have precision grade accuracy (approximately 5% of full-scale deflection) both for dc and ac measurements and can be calibrated on dc and then used directly on ac establishing a direct means of equating ac and dc measurements of voltage and current.

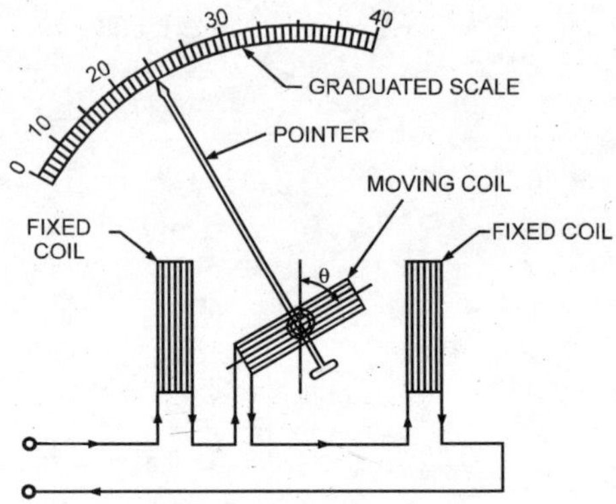

Fig. 31. ***Dynamometer Type Instrument***

5. Electronic multimeter (or VOM) is one of the most versatile general purpose instruments capable of measuring dc and ac voltages as well as current and resistances. The solid-state electronic multimeter (or VOM) generally consists of the following elements:

(*i*) A balanced bridge dc amplifier and a PMMC meter.

(*ii*) An attenuator in input stage to select the proper voltage range.

(*iii*) A rectifier for converting of an ac input voltage to proportionate dc value.

(*iv*) An internal battery and additional circuitry for providing the capability of resistance measurement.

(*v*) A *function switch* for selecting various measurement functions of the meter such as voltage, current or resistance.

In addition, the instrument is usually provided with a built-in power supply for operation on ac mains and, in most cases, one or more batteries for operation as a portable test instrument.

6. Refer to Art. 12.7.

PMMC ammeter and voltmeter are connected in the circuit as shown in Fig. 1. In connecting PMMC instruments care is to be taken for the polarity of the instrument terminals otherwise instrument may get damaged.

Moving iron ammeter and voltmeter are connected in the circuit as shown in Fig. 6.

Dynamometer wattmeter is connected in the circuit as shown in Fig. 22.

EXPERIMENT-15A

Objective : To obtain the characteristics of a transistor under common base (CB) configuration.

Apparatus Required: Regulated power supplies-2 nos., DC voltmeters-2 nos., DC milliammeters-2 nos., and connecting wires.

Theory: To investigate the characteristics of a two-terminal device (a diode), several levels of forward or reverse bias voltage are applied and the corresponding currents that flow are measured. The characteristics of the device are then derived by plotting the graphs of current against voltage as shown in Figs. 23 and 24. Since a transistor is a three-terminal device, there are three possible configurations in which it may be connected to study its characteristics.

The performance of transistors, when connected in a circuit, may be determined from their characteristic curves. There are two important characteristics of a transistor viz input characteristic and output characteristic.

In a common-base (CB) configuration, input is connected between base and emitter and output is taken across collector and base, as shown in Fig. 32 for an N-P-N transistor. The curve drawn between emitter current I_E and emitter-base voltage V_{EB} for a given value of collector-base voltage V_{CB} is known as the input characteristic (Fig. 33). For a given value of V_{CB} the curve is just like that of a forward biased P-N junction because emitter-base junction in forward bias mode is similar to a P-N junction diode.

The curve drawn between collector current I_C and collector-base voltage V_{CB} for a given value of emitter current is known as output characteristic (Fig. 34).

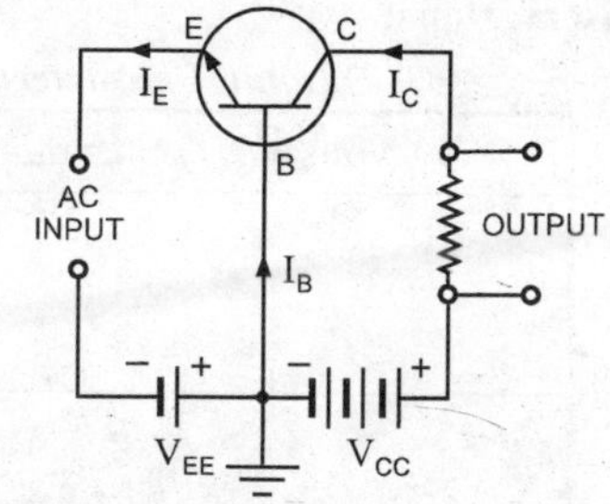

Fig. 32. ***Common Base NPN Transistor***

Connection Diagram: The connection diagram for determination of input and output characteristics of a common base N-P-N transistor is shown in Fig. 35. Voltmeters and ammeters are connected to measure input and output voltages and currents.

Procedure: For determination of input characteristics, the output voltage V_{CB} is maintained constant, and the input voltage V_{EB} is set at several convenient levels. For each level of input voltage, the input current I_E is recorded. I_E is then plotted versus V_{EB} to give the input characteristic. The procedure is repeated for other convenient values of V_{CB}.

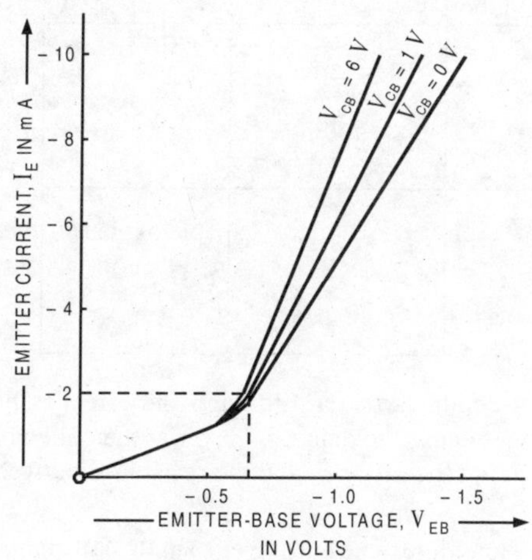

Fig. 33. ***Input Characteristics For Common Base NPN Transistor***

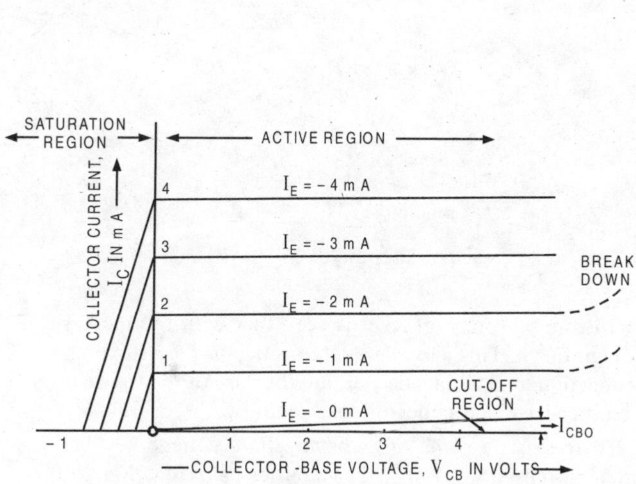

Fig. 34. ***Output Characteristics For Common Base NPN Transistor***

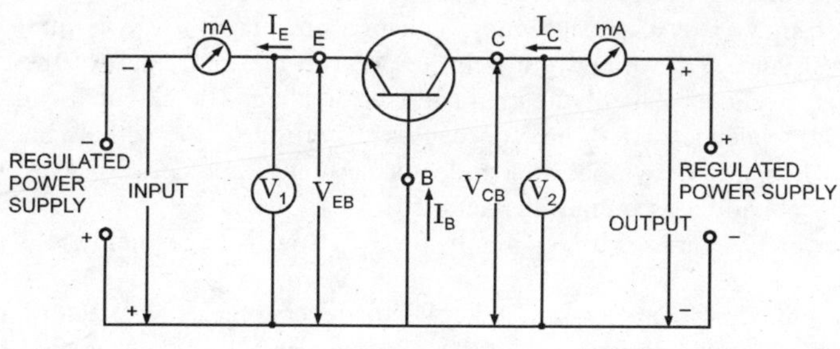

Fig. 35

For determination of output characteristics, the emitter current I_E is held constant at each of several fixed values. For each value of I_E, the output voltage V_{CB} is adjusted in convenient steps, and the corresponding values of collector current I_C are recorded. In this way, a table of values is obtained from which a family of output characteristics may be plotted.

Observations:

Input Characteristics

V_{CB} in volts	V_{EB} in volts	I_E in mA
0	– 0.5	
	– 1.0	
	– 1.5	
1	– 0.5	
	– 1.0	
	– 1.5	
6	– 0.5	
	– 1.0	
	– 1.5	

Output Characteristics

I_E in mA	V_{CB} in volts	I_C in mA
0	1	
	2	
	3	
	4	
– 1	1	
	2	
	3	
	4	
– 2	1	
	2	
	3	
	4	
– 3	1	
	2	
	3	
	4	
– 4	1	
	2	
	3	
	4	

Conclusions : From Fig. 33 it is seen that with the increase in V_{CB}, it conducts better, although the effect is not very significant. This is because large collector-base (reverse-bias) voltages cause the depletion layer at the collector-base junction to penetrate deeper into the base of the transistor, thus reducing the distance and the resistance between the emitter-base and collector-base regions.

There exists a *cutin, offset* or *threshold* voltage V_{EB} below which the emitter current is very small. Substantial emitter current flows only when emitter-base junction is forward biased beyond the knee of the characteristic (approximately 0.7 V for Si and 0.3 V for Ge)

The emitter current I_E increases rapidly with the small increase in V_{EB} which means low input resistance because input resistance of a transistor is given as the reciprocal of the slope of its input characteristic.

Since a small increase in V_{EB} (ΔV_{EB}) causes large change in I_E (Δ I_E), R_{IN} is quite low.

The value of input resistance R_{IN} varies from point-to-point in the initial part of the characteristic because it is nonlinear. Its value over the linear part of the characteristic is about 100 Ω but for lower values of V_{EB}, it is considerably larger. This dependence of R_{IN} on V_{EB} causes distortion of the signals handled by the transistors.

Output Characteristics: The curve drawn between collector current I_C and collector-base voltage V_{CB} for a given value of emitter current I_E is known as output characteristic.

(*i*) The collector current I_C varies with V_{CB} only for very low voltage (below 1 V) but transistor is never operated in this region.

(*ii*) In active region (emitter forward biased and collector reverse biased) collector current I_C is almost equal to I_E and appears to remain constant when V_{CB} is increased. In fact, there is very small increase in I_C with increase in V_{CB}. This is because the increase in V_{CB} expands the collector-base depletion region and thus shortens the distance between the two depletion regions. With emitter current I_E held constant however, the increase in I_C is so small that it is usually noticeable only for large variations in V_{CB}. I_C is slightly lesser than I_E in magnitude.

Transistor is normally operated in active region.

(*iii*) Although the collector current I_C is practically independent of V_{CB} over the transistor operating range. However, if V_{CB} is increased beyond a certain value, I_C eventually increases rapidly because of avalanche or Zener (or both) effects. This condition is known as *punch-through* or *reach-through*. When it occurs large currents can flow, possibly destroying the device. The extension of the depletion region is, of course, the direct consequence of the increase in V_{CB}. Thus it is very essential to maintain V_{CB} below the maximum safe limit specified by the manufacturer of the device.

(*iv*) A very large change in collector voltage causes a very small change in collector current *i.e.* output resistance of CB configuration is very high (of the order of few hundred kΩ)–the output resistance R_{OUT} being given as the ratio of ΔV_{CB} and ΔI_C for a given value of I_E

(*v*) In *cutoff* region (emitter and collector junctions both reverse biased) small collector current I_C flows even when emitter current $I_E = 0$. This is the collector leakage current I_{CBO} or I_{CO}.

(*vi*) In saturation region (both emitter and collector junctions forward biased) collector current I_C flows even when $V_{CB} \simeq 0$. Even when the externally applied bias voltage is reduced to zero, there is still a barrier potential existing at the collector-base junction, and this assists in the flow of I_C. To stop it the collector-base junction has to be forward biased. Consequently, as illustrated in Fig. 34, collector current I_C is reduced to zero when V_{CB} is increased negatively.

EXPERIMENT-15B

Objective: To obtain the charcteristics of a transistor under common emitter (CE) configuration.

Apparatus Required: Regulated power supplies-2 nos., DC voltmeters-2 nos., DC milliammeter-1 no., DC microammeter-1 no. and connecting wires.

Theory: In common emitter configuration input is connected between base and emitter while the output is taken between collector and emitter, as shown in Fig. 36. Thus emitter is common to input and output circuits.

In this configuration, the bias voltages are applied between collector and emitter (V_{CE}) and base and emitter (V_{BE}). Emitter base junction is forward biased and the base is made more positive than the emitter by V_{BE}, collector-emitter junction is reverse-biased and the collector is made more positive than emitter by V_{CE}. The value of V_{CE} must be greater than that of V_{BE}.

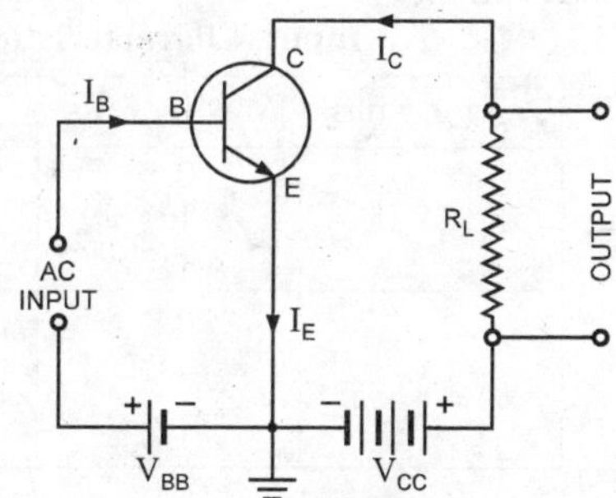

Fig. 36. ***Common Emitter NPN Transistor***

In this circuit arrangement (Fig. 36) the base current I_B flows in the input circuit and collector current I_C flows in the output circuit. In this arrangement current gain between the input and output sides is obtained; and since the input resistance is again less than the output resistance (though difference is not as much as in case of common base arrangement) there will be high voltage and power gains like those in equivalent vacuum tube circuits. The common emitter produces a reversal between input and output signals.

Connection Diagram: The connection diagram for determination of input and output characteristics of a common emitter (CE) N-P-N transistor is shown in Fig. 37.

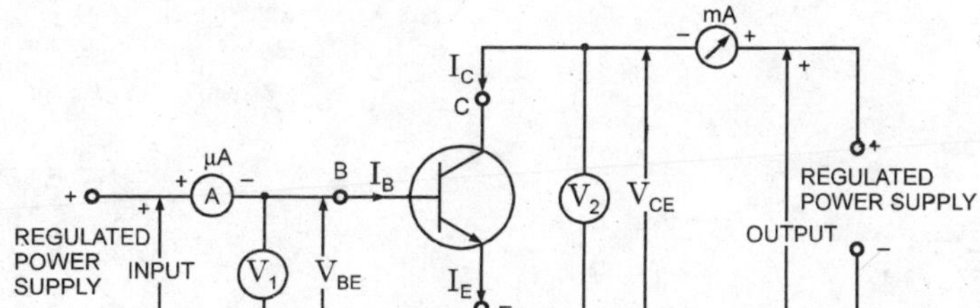

Fig. 37

Procedure: For determination of input characteristics, V_{CE} is held constant and the values of base current I_B are recorded for different values of input voltage V_{BE}, I_B is then plotted versus V_{BE}, as illustrated in Fig. 38, to give input characteristic. The procedure is repeated for another convenient value of V_{CE}.

For determination of output characteristics, base current I_B is held constant at each of several fixed values. For each value of I_B, the output voltage V_{CE} is adjusted in steps, and the corresponding values of collector current I_C are recorded. For each value of I_B, I_C is plotted versus V_{CE} to give output characteristics, as illustrated in Fig. 39.

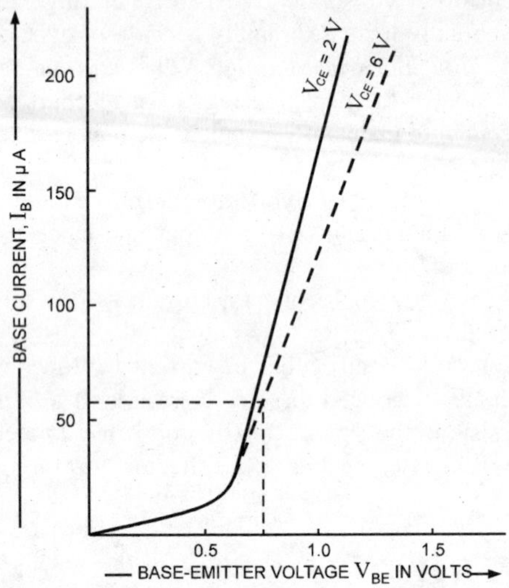

Fig. 38. ***Input Characteristics For Common Emitter NPN Transistor***

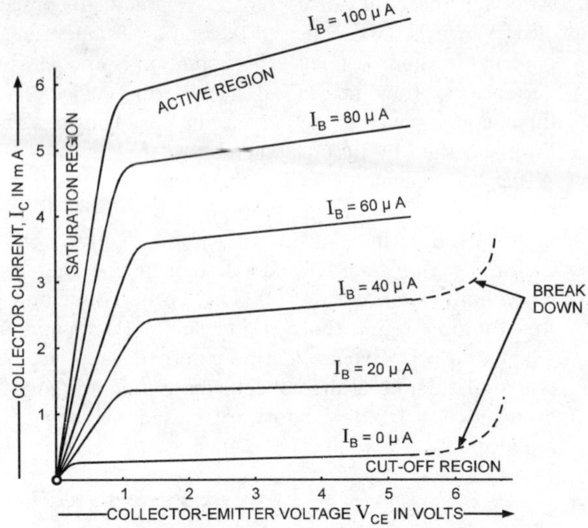

Fig. 39. ***Output Characteristics For Common Emitter NPN Transistor***

Observations:

Input Characteristics

V_{CE} in volts	V_{BE} in volts	I_B in μA
2V	0.5 1.0 1.5	
6V	0.5 1.0 1.5	

Output Characteristics

I_B in μA	V_{CE} in volts	I_C in mA
0	1 2 3 4 5	
20	1 2 3 4 5	
40	1 2 3 4 5	
60	1 2 3 4 5	
80	1 2 3 4 5	
100	1 2 3 4 5	

Conclusions:

Input Characteristics: The curve drawn between base current I_B and base-emitter voltage V_{BE} for a given value of collector-emitter voltage V_{CE} is known as the input characteristic.

(*i*) The input characteristics of CE transistors are quite similar to those of a forward biased diode because the base-emitter region of the transistor is a diode and it is forward biased.

(*ii*) In comparison to common base (CB) arrangement base (or input) current increases less rapidly with the increase in base-emitter voltage, V_{BE}. This indicates that input resistance R_{IN} is larger in common emitter configuration than that in common base configuration. Due to initial nonlinearity of the curve, input resistance varies from point-to-point in the initial part of the characteristic. Its value over the linear part of the curve is of the order of few hundred ohms.

(*iii*) An increment in value of V_{CE} causes the input current I_B to be lower for a given level of V_{BE}. This is because the higher levels of V_{CE} provide greater collector-base junction reverse bias, causing greater depletion region penetration into the base and thus reducing the distance between the collector-base and emiter-base regions. As a result more of the charge carriers from the emitter flow across the collector-base junction, and few flow out through the base lead. Usually effect of change in V_{CE} on the input characteristic is ignored.

Output Characteristics: Output characteristic for a common emitter transistor is the curve drawn between collector current I_C and collector-emitter voltage V_{CE} for a given value of base current I_B.

(*i*) The collector current I_C varies with V_{CE} for V_{CE} between 0 and 1V and then becomes almost constant and independent of V_{CE}. The transistors are always operated above 1V.

(*ii*) Output characteristics in CE configuration have some slope while CB configuration has almost horizontal characteristics. This indicates that output resistance in case of CE configuration is less than that in CB configuration.

(*iii*) In active region (collector junction reverse biased and emitter junction forward biased), for small values of base current I_B the effect of collector voltage V_C over I_C is small but for large values of I_B this effect increases.

The shape of the characteristics is same as in CB configuration but with the difference that collector current I_C is larger than input current (base current I_B). Thus the current gain for this configuration is larger than unity. The transistor must always be operated in the active region, when employed as an amplifying device.

(*iv*) With low values (ideally zero) of V_{CE} the transistor is said to be operated in saturation region and in this region base current I_B does not cause a corresponding change in collector current I_C.

(*v*) With much higher V_{CE}, the collector base junction completely breaks down and because of this avalanche breakdown collector current I_C increases rapidly and the transistor gets damaged.

(*vi*) In cutoff region, small amount of collector current I_C flows even when base current $I_B = 0$. This is called I_{CEO}. Since main current I_C is zero so the transistor is said to be cutoff.

EXPERIMENT-16

Objective: To perform open-and short-circuit tests on a single phase transformer and calculate its efficiency.

Apparatus Required: Single phase transformer (preferably 1 kVA with one winding rated at 230 V)-1 no., Variac-1 no., Low range voltmeter-1 no., Low range ammeter-1 no., Low range low power factor wattmeter-1 no., Wattmeter, ammeter and voltmeter of normal ranges (depending upon the rating of transformer under test) one each, DPIC switch-1 no. and connecting leads.

Theory: Open-circuit test is performed to determine the iron loss and the transformation ratio. Since during this test no current flows in the open-circuited secondary, the current in the primary is merely that necessary to magnetise the core at normal voltage. Moreover, the magnetising current is very small fraction of the full-load current (usually 3 to 10 per cent of full-load current) and may be neglected as far as copper loss is concerned, consequently, this test gives this core loss alone practically.

$$\text{The transformation ratio, } K = \frac{\text{Secondary turns}}{\text{Primary turns}} = \frac{\text{Secondary voltage on open circuit}}{\text{Voltage applied to primary}} = \frac{V_2}{V_1}$$

The short-circuit test is performed to determine the full-load copper loss. In this test since the terminals of the low voltage winding are short circuited, therefore, the transformer becomes equivalent to a coil having an impedance equal to the impedance of both the windings. The applied voltage is very small as it is adjusted to cause flow of rated current in the windings, so the flux linking with the core is very small and, therefore, iron losses are negligible. The power drawn from the supply, therefore, represents full-load copper loss.

Transformer efficiency at any given load is given as

$$\eta = \frac{\text{Output}}{\text{Output} + \text{iron loss} + \text{copper loss at given load}} \times 100$$

where Output = Voltmeter reading in open-circuit test, V_1 × ammeter reading during short-circuit test, I_S × pf

Iron loss, P_i = Power input in open-circuit test with rated voltage = P_o

and Copper loss, P_c = Power input in short-circuit test

Connection Diagrams: See Figures 8.19 and 8.20

Procedure: 1. Open-circuit Test. The transformer is connected, as shown in Fig. 8.19 on the low voltage side. The ammeter A is a low range ammeter and W is a low range and low power factor wattmeter. The high voltage side is kept open circuited. AC supply at rated voltage is switched on to the transformer through DPIC switch. High range voltmeter V_2 is connected across the secondary of the transformer. Readings of voltmeters V_1 and V_2, ammeter A and wattmeter W are noted and tabulated as below.

2. Short-circuit test. The transformer is connected as shown in Fig. 8.20, the low voltage winding is short circuited by connection having negligible resistance and good contacts. The voltage applied to the primary is increased in steps so that ammeter A carries 0.25, 0.5, 0.75 time rated current, rated current, and 1.5 times rated current.

The efficiency of the transformer is calculated at different loads as mentioned above and a curve is plotted between efficiency and load current.

Observations:

(*i*) Open-circuit test:

Primary voltmeter reading, V_1 = ...

Secondary voltmeter reading, V_2 = ...

Input power on no load, P_0 = Wattmeter reading, W = ...

$$\text{Transformation ratio, } K = \frac{V_2}{V_1} = \text{..........}$$

(*ii*) Short-circuit test: Rated primary current,

$$I_f = \frac{\text{Rated kVA} \times 1{,}000}{\text{Rated primary voltage}} = \text{..........}$$

Power factor, cos φ = 0.8 (lag)

S. No.	*Ammeter Reading in Amperes, I_S*	*Wattmeter Reading in Watts, P_S*	*Transformer Efficiency,* $\eta = \frac{V_1 I_S \cos\phi}{V_1 I_S \cos\phi + P_0 + P_S} \times 100 = \frac{V_1 I_S \times 0.8}{V_1 I_S \times 0.8 + P_0 + P_S} \times 100$
1.			
2.			
3.			
4.			
5.			

Results: 1. During short-circuit test, power input, P_S varies as the square of the input current, I_S *i.e.* copper loss varies as the square of the load current.

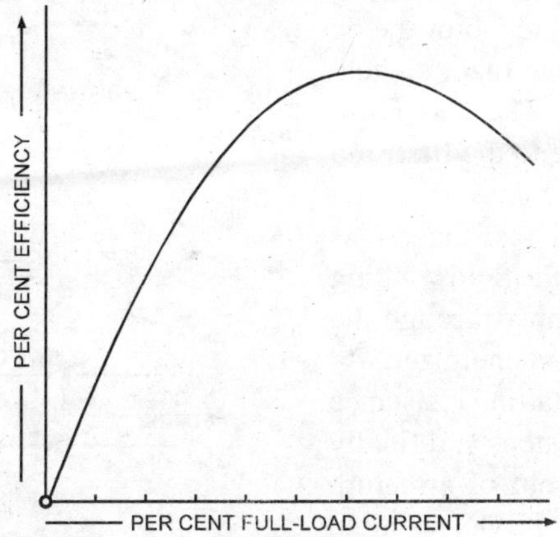

Fig. 40

2. The curve plotted between the transformer efficiency and load current [Fig. 40] shows that the efficiency becomes maximum when copper loss equals iron loss.

EXPERIMENT-17A

Objective: To start and reverse the direction of rotation of a dc motor.

Apparatus Required: DC shunt motor-one no; Three-point starter-one no; DPIC switch-1 no; and connecting leads.

Theory: Starters For DC Shunt and Compound Motors : Motor starters are generally manufactured in convenient sizes and styles for use as auxiliaries with dc shunt and compound motors. Their primary function is to limit the current in the armature circuit during the starting or accelerating period. They are always rated on the basis of output power and voltage of the motors with which they are to be employed. The simplest type of starter consists of an additional resistance (a rheostat) inserted in series with the armature alone (not with the motor as a whole), as illustrated in Fig. 41.

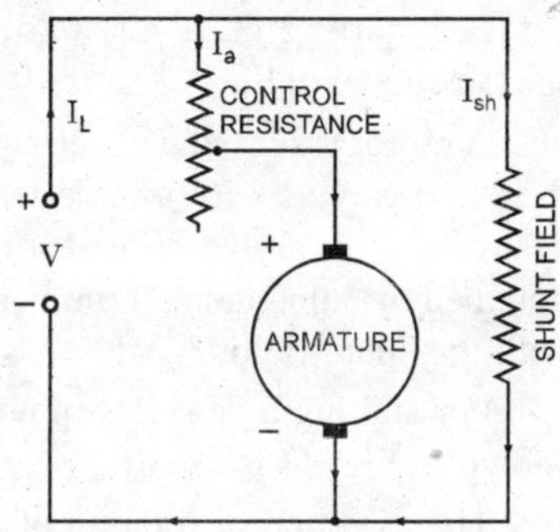

Fig. 41

This starter is, however, modified to include a few protective devices, such as no-volt release, overload release etc. There are two standard types of motor starters for dc shunt and compound wound motors, the three-point type and the four-point type starters. The four-point starter is employed when wide range of speed by shunt field control is required. When no (or little) speed control is required, either type of starter may be used.

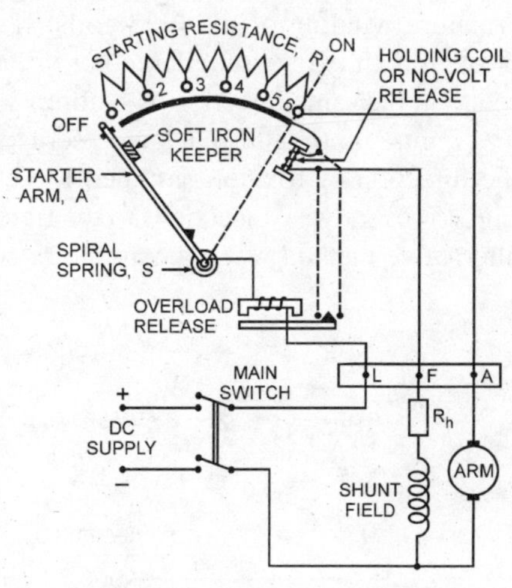

Fig. 42. ***Three-Point Starter***

In this context it is pertinent to note that while starting dc shunt and compound motors, it is advantageous to keep the field excitation at its maximum value. A large field current, therefore, a higher value of flux will result in a low operating speed and in higher motor torque for a particular value of starting current because motor torque is proportional to the product of flux per pole and armature current. Thus for a given load torque, the motor will accelerate quickly and take less lime to reach the lower operating speed from the starting instant. This will result in less heating of the armature during starting. Thus the rheostat, in series with the shunt field winding, should be at zero resistance position at the time of starting of the dc shunt and compound motors.

Operation : When the motor is at rest, the starter arm is in the OFF position due to action of strong spiral spring S. For starting the motor the dc supply is switched on by closing the main switch, keeping starter arm in OFF position. The starter arm (or handle) is then turned clockwise to the first stud and brass arc (or strip). As soon as it comes in contact with first stud, whole of the starting resistance R is inserted in series with the armature, the field winding is directly connected across the supply through the brass arc and the holding coil is also energized. As the starter arm is turned further the starting resistance is cut out of the armature circuit in steps and finally entire starting resistnace R is cut out of armature circuit. When the starter arm reaches the ON position, it is held against the action of spiral spring S by the force of attraction between holding coil magnet and soft iron keeper attached to the starter arm. The starter arm should not be held for an unduly long time in an intermediate position as it is likely to burn out the starting resistor.

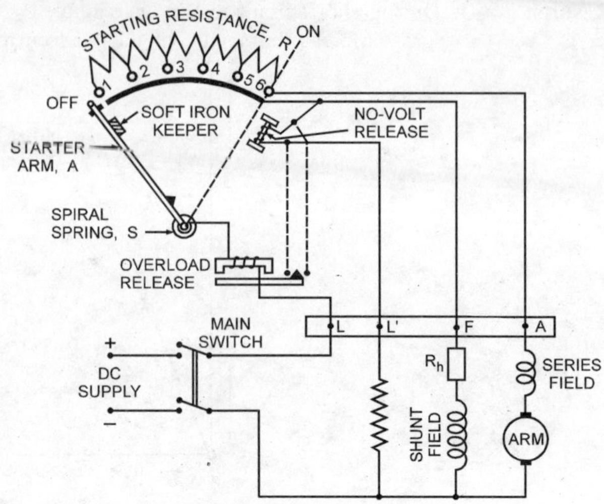

Fig. 43. *Four Point Starter*

For starting dc compound motor, four-point starter, shown in Fig. 43 is used.

A motor is started with a four-terminal starter in the same way as with a three-terminal starter. Any desired speed, above normal, of the motor can be obtained by adjustment of the field rheostat in series with the shunt field.

For starting a dc series motor face plate starter shown in Fig. 44 is used.

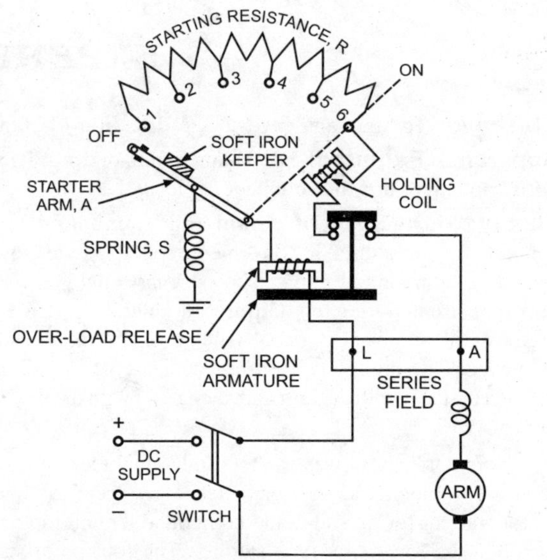

Fig. 44. *Face-Plate Starter For DC Series Motor*

The direction of rotation of a motor can be reversed by reversing the current through either the armature winding or the field coils. If the current through both is reversed, the motor will continue to rotate in the same direction as before.

Figure 45 (*a*) illustrates a dc series motor with the direction of rotation assumed to be clockwise. Fig. 45 (*b*) shows connections to the field winding reversed giving reversed rotation and Fig. 45 (*c*) shows the alternative method when the armature connections at the brushes are interchanged.

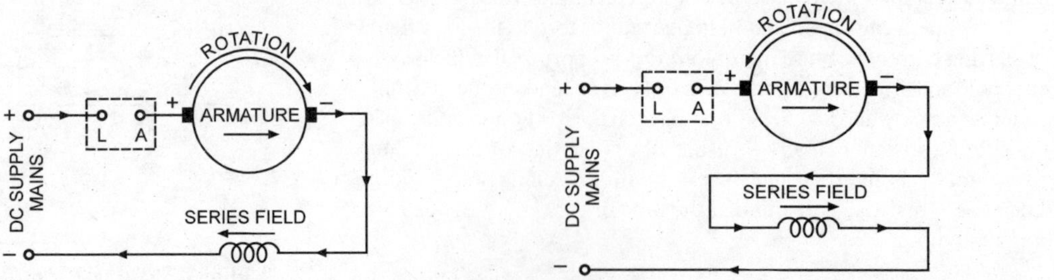

(*a*) *Normal Rotation*

(*b*) *Reversed Rotation With Field Connections Reversed*

Fig. 45

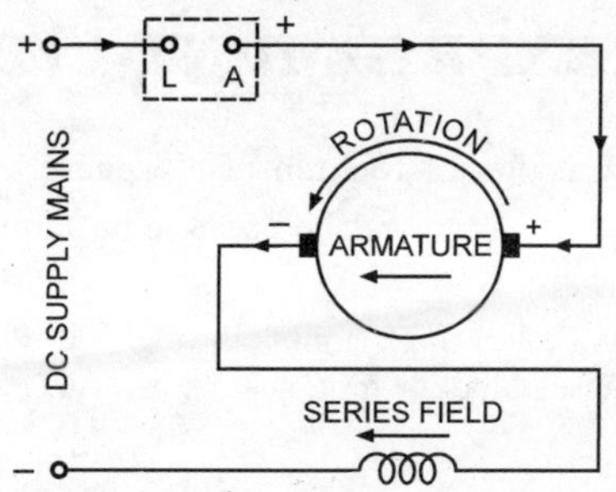

Fig. 45. (*c*) *Reversed Rotation With Armature Connections Reversed*

Figs. 46 (*a*), 46 (*b*) and 46 (*c*) illustrate the methods of reversing the direction of rotation of a dc shunt motor.

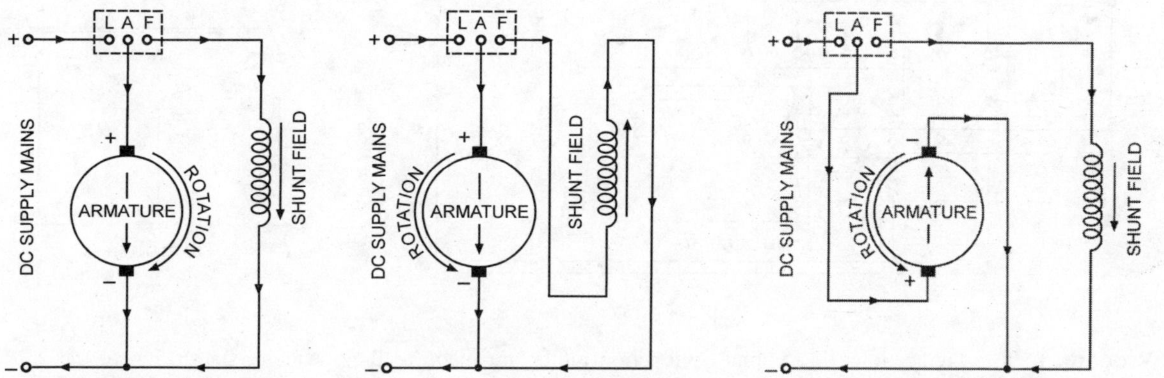

(*a*) *Normal Rotation*

(*b*) *Reversed Rotation With Field Connections Reversed*

(*c*) *Reversed Rotation With Armature Connections Reversed*

Fig. 46

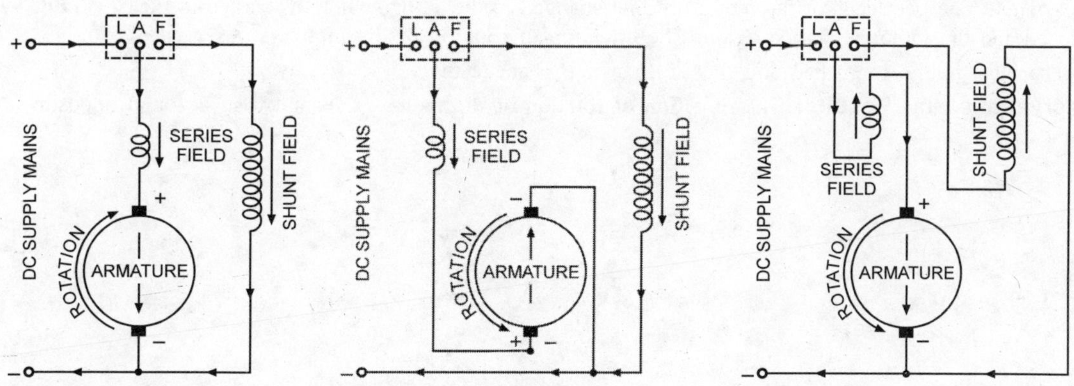

(*a*) *Normal Rotation*

(*b*) *Reversed Rotation With Armature Connections Reversed*

(*c*) *Reversed Rotation With Field Connections Reversed*

Fig. 47

The direction of rotation of a dc compound motor may be conveniently reversed by reversing the connections to the armature, as illustrated in Fig. 47 (*b*). The reversal of the field connections involve both shunt and series field windings, as shown in Fig. 47 (*c*).

EXPERIMENT-17B

Objective: To start and reverse the direction of rotation of a 3-phase induction motor.

Apparatus Required: Three phase induction motor-1 no., Star-delta starter-1 no., TPIC switch-1 no., Tools such as screw driver, plier, test pen etc. and connecting leads.

Theory: Motor is connected to the 3-phase ac supply mains through star-delta starter and TPIC switch, as shown in Fig. 48. The direction of rotation of a 3-phase induction motor can be reversed by interchanging any two terminals at the TPIC switch, as shown in Fig. 49.

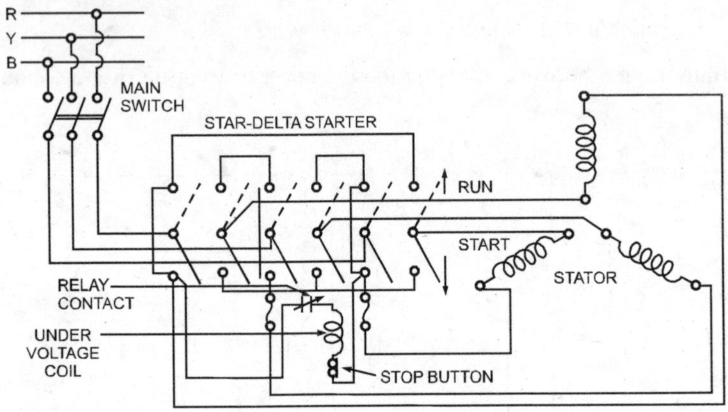

Fig. 48

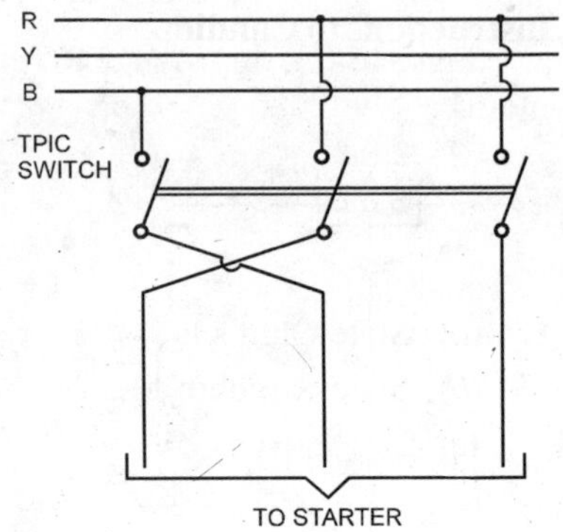

Fig. 49

Procedure: The connections of a 3-phase induction motor are made to the star-delta starter and to the TPIC switch, as shown in Fig. 48. The TPIC switch is closed and the motor is started by taking the lever of the starter to the start (star) position and then with a jerk to the run position (or delta connections). The direction of the rotation of the motor is observed. Say, it is in clock-wise direction.

Now the motor is stopped by pushing the stop button and supply to the motor is removed by opening the TPIC switch. The two leads of the motor are interchanged to the TPIC switch, as shown in Fig. 49. TPIC switch is closed and the motor is started again. The direction of rotation of the motor is observed.

The push button is pushed and the TPIC switch is made off.

Observations and Results: The direction of rotation of the motor in second case is found opposite to that in first case.

❑❑❑

EXAMINATION PAPERS

B.TECH

Examination, May-2012

Basic Electrical and Electronics Engineering

Time : 3 Hours *Maximum Marks : 60*

Instructions to Candidates:

1. Section-A is ***Compulsory***.
2. Attempt any ***Five*** questions from Sections B and C.
3. Select at least ***Two*** questions from Sections B and C.

SECTION-A **(10 × 2 = 20 Marks)**

1. (*a*) State Ohm's law in reference to dc circuits.

(*b*) State Kirchhoff's current law for linear bilateral dc circuits.

(*c*) Draw phasor diagram of a series R-L-C ac circuit.

(*d*) Explain the term form factor of sine wave and square wave.

(*e*) Prove that power on both primary as well as secondary remains same.

(*f*) Draw circuit diagram long-shunted dc motor and write its V-I equations.

(*g*) Explain in brief the term transducer.

(*h*) Draw and explain V-I characteristic of PN junction diode.

(*i*) Draw circuit diagram of field-effect transistor.

(*j*) Explain any truth table for NAND operation.

ANSWERS/SOLUTION

1. (*a*) Refer to Answer of Q. 4 page 49.

(*b*) Refer to Art. 1.23(1).

(*c*) Refer to Fig. 3.16 page 103.

(*d*) The form factor is defined as the ratio of effective value to the average or mean value of a periodic wave *i.e.*,

$$\text{Form factor, } k_f = \frac{\text{RMS or effective value}}{\text{Average or mean value}}$$

For Sinusoidal Wave

$$\text{RMS value} = \frac{\text{Peak value}}{\sqrt{2}}$$

$$\text{Average value} = \frac{\text{Peak value}}{\pi/2}$$

$$\text{So, form factor} = \frac{\dfrac{\text{Peak value}}{\sqrt{2}}}{\dfrac{\text{Peak value}}{\pi/2}} = 1.11$$

For Square Wave

$$\text{RMS value} = \text{Peak value}$$

$$\text{Average value} = \text{Peak value}$$

$$\text{Form factor} = \frac{\text{Peak value}}{\text{Peak value}} = 1$$

(*e*) Since power losses are negligible, the power on both primary as well as secondary remains same.

(*f*) Circuit diagram for long shunted dc motor is given in Fig. 1.

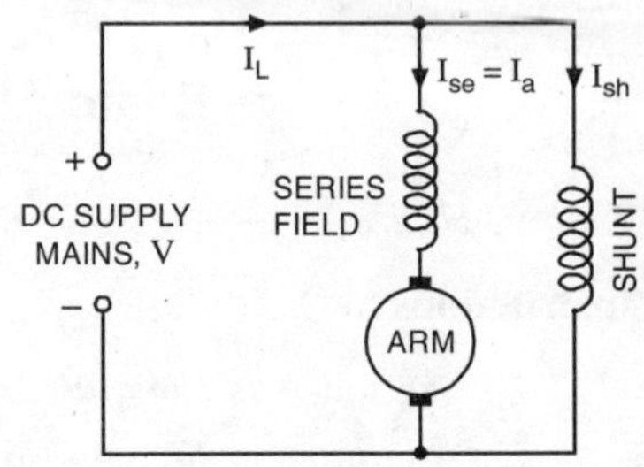

Fig. 1

$$\text{Series field current, } I_{se} = \text{Armature current, } I_a$$

$$\text{Line current, } I_L = I_a (\text{or } I_{se}) + I_{sh}$$

where I_{sh} is shunt field current

$$\text{Back emf developed, } E_b = V - I_a(R_a + R_{se})$$

where R_a and R_{se} are the resistances of armature and series field respectively.

(*g*) Basically transducer is defined as a device which converts energy or information from one form to another. These are widely used in measurement work because all quantities that need to be measured, cannot be displayed as easily as others. A better measurement of a quantity can usually be made if it may be converted to another form which is more conveniently or accurately displayed. For example, the common *mercury thermometer* converts variations in temperature into variations in the length of a column of mercury. Since the variation in the length of the mercury column is rather simple to measure, the mercury thermometer becomes a convenient device for measuring temperature. On the other hand, the actual temperature variation is not as easy to display directly.

Thus the transducer is a device which provides a usable output in response to specific input measurand which may be physical or mechanical quantity, property or condition. The transducer may be mechanical, electrical, magnetic, optical, chemical, acoustic, thermal, nuclear, or a combination of any two or more of these.

(*h*) Refer to Art. 13.5.3.

(*i*) Refer to Art. 15.2.5.

(*j*) Refer to Table 18.6 page 537.

SECTION-B (8 Marks Each)

2. Derive an expression for resistance at any temperature, if resistance at $t = 0°$, is not known.

3. A 25 kVA loss-less transformer has 500 turns on the primary and 40 turns on the secondary winding. The primary is connected to 3000 V, 50 Hz mains determine

   (*i*) primary and secondary currents in at full load;

   (*ii*) the secondary emf and

   (*iii*) the maximum flux in the core. No-load current can be neglected.

4. A voltage, $e(t) = 150 \sin (2\pi f)t$, 50 Hz, is applied to series circuit consisting of 10 Ω resistance, 0.0318 henry inductance. Determine

   (*i*) expression for current $i(t)$,

   (*ii*) phase angle between voltage and current,

   (*iii*) power factor,

   (*iv*) active power consumed,

   (*v*) maximum value pulsating energy.

5. Explain characteristics of dc shunt motor and its speed control methods.

ANSWERS/SOLUTION

2. Refer to Arts. 1.9 and 1.10.

3.
$$\text{Primary induced emf, } E_1 = V_1 = 3{,}000 \text{ V}$$
$$\text{Primary turns, } N_1 = 500$$
$$\text{Secondary turns, } N_2 = 40$$

$$(ii) \text{ Secondary emf, } E_2 = E_1 \times \frac{N_2}{N_1} = 3{,}000 \times \frac{40}{500} = 240 \text{ V}$$

$$(i) \text{ Primary current, } I_1 = \frac{\text{Rated kVA} \times 1{,}000}{E_1} = \frac{\text{Rated kVA} \times 1{,}000}{V_1}$$
$$= \frac{25 \times 1{,}000}{3{,}000} = 8.333 \text{ A} \quad \textbf{Ans.}$$

$$\text{Secondary current, } I_2 = \frac{25 \times 1{,}000}{240} = 104.167 \text{ A} \quad \textbf{Ans.}$$

$$(iii) \text{ Maximum flux in the core, } \phi_{max} = \frac{E_1}{4.44 f N_1} = \frac{3{,}000}{4.44 \times 50 \times 500} = 0.027 \text{ Wb or } 27 \text{ mWb} \quad \textbf{Ans.}$$

4.
$$e(t) = 150 \sin (2\pi f)t = 150 \sin (2\pi \times 50)t = 150 \sin 100\pi t$$
$$\text{Resistance, } R = 10 \ \Omega$$
$$\text{Inductance, } L = 0.0318 \text{ H}$$
$$\text{Inductive reactance, } X_L = 2\pi f L = 2\pi \times 50 \times 0.0318 = 10 \ \Omega$$
$$\text{Phase angle, } \phi = \text{Tan}^{-1} \frac{X_L}{R} = \text{Tan}^{-1} \frac{10}{10} = \frac{\pi}{4} \text{ (lagging)}$$
$$\text{Circuit impedance, } Z = \sqrt{R^2 + X_L^2} = \sqrt{10^2 + 10^2} = 14.142 \ \Omega$$
$$\text{Maximum value of current, } I_{max} = \frac{E_{max}}{Z} = \frac{150}{14.142} = 10.6 \text{ A}$$

(*i*) So expression for current $i(t)$ is

$$i(t) = I_{max} \sin (100\pi t - \phi) = 10.6 \sin \left(100\pi t - \frac{\pi}{4}\right) \quad \textbf{Ans.}$$

(*ii*) Phase angle between voltage and current,

$$\phi = \frac{\pi}{4} = 45° \text{ (lagging)} \quad \textbf{Ans.}$$

$$(iii) \text{ Power factor} = \text{Cos } \phi = \text{Cos } 45° = 0.707 \text{ (lagging)} \quad \textbf{Ans.}$$

$$(iv) \quad \text{Active power consumed} = EI \cos \phi = \frac{E_{max}}{\sqrt{2}} \times \frac{I_{max}}{\sqrt{2}} \times 0.707$$
$$= \frac{150}{\sqrt{2}} \times \frac{10.6}{\sqrt{2}} \times 0.707 = 562 \text{ W} \quad \textbf{Ans.}$$

$$(v) \text{ Maximum value of pulsating energy} = \frac{1}{2} E_m I_m = \frac{1}{2} \times 150 \times 10.6 = 795 \text{ VA} \quad \textbf{Ans.}$$

5. Refer to Arts. 9.27 and 9.29.

SECTION-C **(8 Marks Each)**

6. Explain in detail operation of LVDT and list its field of application.

7. A full wave centre-tap rectifier uses two crystal diodes each having a forward resistance of 25 Ω. The rms value of secondary voltage fed between centre tap to each end of secondary is 48 V and the load resistance is 1 kΩ. Determine:

(*i*) dc output voltage,

(*ii*) dc output power,

(*iii*) rectification efficiency,

(*iv*) peak inverse voltage.

8. Explain Zener diode and any one of its application.

9. Explain in detail J-K flip-flop.

ANSWERS/SOLUTION

6. Refer to Art. 12.5.

7. Secondary voltage across one-half of the secondary winding,

$$V_s = 48 \text{ V}$$

Diode forward resistance, $R_F = 25\ \Omega$

Load resistance, $R_L = 1\ k\Omega = 1{,}000\ \Omega$

(*i*) DC output voltage, $V_{dc} = I_{dc}R_L = \frac{2}{\pi} I_{max} R_L = \frac{2}{\pi} \times \frac{V_{s\,max}}{R_f + R_L} \times R_L$

$$= \frac{2}{\pi} \times \frac{48\sqrt{2}}{25 + 1{,}000} \times 1{,}000 = 42.16 \text{ V} \textbf{ Ans.}$$

$$I_{dc} = \frac{V_{dc}}{R_L} = \frac{42.16}{1{,}000} = 0.04216 \text{ A}$$

Output power $= V_{dc}I_{dc} = 42.16 \times 0.04216 = 1.778$ W **Ans.**

AC input power, $P_{ac} = V_{s\,rms} \times I_{rms} = V_{s\,rms} \times \frac{V_{s\,rms}}{R_F + R_L} = \frac{48 \times 48}{25 + 1{,}000} = 2.2478$ W

Rectification efficiency, $\eta = \frac{P_{dc}}{P_{ac}} \times 100 = \frac{1.778}{2.2478} \times 100 = 79.1\%$ **Ans.**

Peak inverse voltage, PIV $= 2V_{s\,max} = 2 \times 48\sqrt{2} = 135.76$ V **Ans.**

8. Refer to Arts. 13.3 and 13.14.

9. Refer to Arts. 18.14.5 and 18.14.6.

B.TECH

Examination, December-2012

Basic Electrical and Electronics Engineering

Time : 3 Hours *Maximum Marks : 60*

Instructions to Candidates:

1. Section-A is ***Compulsory*** consisting of ***ten*** questions carrying ***two*** marks each.
2. Section B and C have ***four*** questions each.
3. Attempt any ***five*** questions from Sections B and C carrying ***eight*** marks each.
4. Select atleast ***two*** questions from Sections B and C.

SECTION-A

1. Answer briefly :

(*a*) What will happen if the length of a conductor is doubled and its area of cross section is also doubled ?

(*i*) Resistance will increase four times.

(*ii*) Resistance remains unchanged.

(*iii*) Resistance decreases to four times.

(*iv*) Change at random.

(*b*) Give concept of work, power and energy in brief.

(*c*) Instantaneous current is given by the relation $i = 20 \sin 314t$, find rms and average value of ac ?

(*d*) What is the theory of shunt ?

(*e*) State similarities of electric and magnetic circuits.

(*f*) Why cannot a 3-phase induction motor runs at synchronous speed ?

(*g*) Give applications of strain gauge.

(*h*) Draw symbolic representation of BJT and FET.

(*i*) Convert $(4287)_{10}$ into hexadecimal number system.

(*j*) Draw symbols of EX-OR and NAND gate.

ANSWERS/SOLUTION

1. (*a*) Let the new length $l' = 2l$ and new area of cross section $a' = 2a$

So Resistance of new conductor, $R' = \rho\frac{l'}{a'} = \rho\frac{2l}{2a} = \rho\frac{l}{a} = R$

Thus *resistance remains unchanged.*

(*b*) Refer to Art. 1.12.

(*c*) Maximum value of current (sinusoidal), $I_{max} = 20$ A

$$\text{RMS value, } I_{rms} = \frac{I_{max}}{\sqrt{2}} = \frac{20}{\sqrt{2}} = 10\sqrt{2} \text{ A} \quad \textbf{Ans.}$$

$$\text{Average value, } I_{av} = \frac{2}{\pi} I_{max} = \frac{2}{\pi} \times 20 = 12.732 \text{ A} \quad \textbf{Ans.}$$

(*d*) In the electric circuit, *shunt* is a branch conductor joining the main circuit at two points and forming a parallel or derived circuit, so that the current is divided, a part passing through the branch and a part through the main circuit. The division of the current between the shunt and main circuit depends upon the ratio of the resistances of the two (*i.e.*, shunt and main circuit). Lower the ratio of the shunt resistance and the main circuit resistance, a greater portion of the current flows through the shunt.

(*e*) Refer to Art. 7.14.4.

(*f*) Refer to Answer of Q. 14 page 341.

(*g*) Refer to Answer of Q. 8 page 381.

(*h*) Refer to Figs. 14.1 and 15.4.

(*i*) Refer to Solution of Example 17.48.

(*j*) Refer to Figs. 18.34(*b*) and 18.26(*a*).

SECTION-B **(8 Marks Each)**

2. (*a*) State and explain Kirchhoffs laws.

(*b*) Illustrate Ohm's laws.

3. A wire of 100 Ω resistance is cut into how many equal pieces, so that when they are connected in parallel resultant is 1 Ω ?

4. (*a*) Derive an expression for emf equation of single phase transformer.

(*b*) An ac has frequency 50 Hz and rms current 25 Amp. Write equation of instantaneous current and find:

(*i*) Current at time 0.0025 seconds.

(*ii*) time at which current is 14.14 amp.

5. A 3 phase induction motor is wound for 4 poles and is supplied from 3 phase, 50 Hz system. Calculate :

(*a*) the synchronous speed.

(*b*) Speed of the motor when slip is 0.04 and

(*c*) The rotor frequency when motor runs at 600 rpm.

ANSWERS/SOLUTION

2. (*a*) Refer to Art. 1.23.

(*b*) Refer to Art. 1.5.

3. Let the wire of 100 Ω resistance be cut into n equal pieces, so that when they are connected in paralle, resultant is 1 Ω.

Resistance of each piece of wire, $r = \frac{100}{n}\ \Omega$

When these n equal pieces are connected in parallel, the resultant resistance,

$$R = \frac{r}{n} = \frac{100}{n^2} \text{ which equals to } 1\ \Omega$$

$$\text{So } \frac{100}{n^2} = 1$$

$$\text{or } n = \sqrt{100} = 10 \textbf{ Ans.}$$

4. (*a*) Refer to Art. 8.5.

(*b*) RMS value of current, $I_{rms} = 25$ A

Supply frequency, $f = 50$ Hz

The equation for instantaneous value of alternating current (assumed sinusoidal) with θ as zero when time, t is zero, is given as

$i = I_{max} \sin 2\pi ft = 25\sqrt{2} \sin 2\pi \times 50t = 35.355 \sin 100\pi t$ **Ans.**

Current at time 0.0025 second.

$$= 35.355 \sin 100 \times \pi \times 0.0025 = 35.355 \sin \frac{\pi}{4} = 25 \text{ A} \quad \textbf{Ans.}$$

Let the instantaneous value of current be 14.14 A for the first time, t seconds after the instant the current is zero and becomes positive.

$$\text{So,} \quad 14.14 = 35.355 \sin 100\pi t$$

$$\text{or} \quad t = \frac{1}{100\pi} \sin^{-1} \frac{14.14}{35.355} = \frac{1}{100\pi} \sin^{-1} 0.4$$

$$= 0.075 \text{ seconds or } 75 \text{ ms} \quad \textbf{Ans.}$$

5. Refer to Solution of Example 10.6.

SECTION-C

6. What is LVDT ? Discuss its principle and applications with neat diagram.

7. (*a*) Discuss working principle of a P-N junction diode. Also show its characteristics and mention its important applications.

(*b*) A germanium diode has a saturation current of 10^{-8} A. Calculate for the junction current for a forward bias of 0.4 volts and 300°K.

8. Describe the operation of RS flip-flop with suitable waveform.

9. What are universal gates and why they are called so ? How can OR and XOR gates be realized using NAND gates only ?

ANSWERS/SOLUTION

6. Refer to Arts. 12.5, 12.5.1 and 12.5.2.

7. (***a***) Refer to Arts. 13.5, 13.5.1, 13.5.2, 13.5.3 and 13.5.4.

(*b*) Junction current is given as

$$I = I_0 (e^{V/\eta V_T} - 1)$$

$$= 10^{-8} (e^{0.4/1 \times 0.026} - 1) \qquad \because V_T = 26 \text{ mV} = 0.026 \text{ V and } \eta \text{ for germanium} = 1$$

8. Refer to Arts. 18.14.1 and 18.14.2.

9. Refer to Arts. 18.9, 18.9.2 and Solution of Example 18.7.

B.TECH

Examination, May-2013

Basic Electrical and Electronics Engineering

Time : 3 Hours *Maximum Marks : 60*

Instructions to Candidates:

1. Section-A is ***Compulsory*** consisting of ***ten*** questions carrying ***two*** marks each.
2. Sections B and C have ***four*** questions each.
3. Attempt any ***five*** questions from Sections B and C carrying ***eight*** marks each.
4. Select atleast ***two*** questions from Sections B and C.

SECTION-A

1. Answer briefly :

(*a*) What is the function of Zener diode ?

(*b*) Define Octal and Hexadecimal number systems.

(*c*) Define mutual inductance.

(*d*) What is 2's complement ?

(*e*) What is the basic function of transformer ?

(*f*) What is the theory of shunt ?

(*g*) What are the various losses that take place in a transformer ?

(*h*) What is difference between a sensor and a transducer ?

(*i*) What is difference between unilateral and bilateral circuit ?

(*j*) Explain the term resistance. On what factors does it depend ?

ANSWERS/SOLUTION

1. (*a*) Refer to Answers of Q. 36 page 416.

(*b*) Refer to Answer of Q. 2 page 524.

(*c*) Refer to Answer of Q. 18 page 216.

(*d*) The 2's complement of a binary number is obtained by adding 1 to its 1's complement *i.e.*,

$$\text{2's complement} = \text{1's complement} + 1.$$

(*e*) Transformer is an ac machine that (*i*) transfers electrical energy from one electric circuit to another (*ii*) does so without a change of frequency (*iii*) does so by the principle of electromagnetic induction and (*iv*) has electric circuits that are linked by a common magnetic circuit. The energy transfer usually takes place with a change of voltage, although this is not always necessary. When the transformer raises the voltage *i.e.* when the output voltage of a transformer is higher than its input voltage, it is called the *step-up transformer* and when it lowers the voltage it is called the *step-down transformer.*

(*f*) Refer to Answer of Q. 1(*d*) of Examination Paper December 2012.

(*g*) Eddy currents and hysteresis cause iron loss in the magnetic core and yoke and resistance of primary and secondary windings cause copper loss when the transformer is loaded.

(*h*) *Sensor* is an input device which makes contact with the physical quantity under measurement, receives energy from the measured medium and produces an output depending in some way on the quantity under measurement, known as measurand.

Sensors may have a non-electrical input and output such as an orifice plate, a spring, a manometer or may have electrical input and output such as a rectifier. In case the input device or the primary sensing element has a non-electrical input and output, then it is converted into electrical signal by means of a transducer. The *transducer* is defined as a device which, when actuated by one form of energy, is capable of converting it into another form of energy.

(*i*) An electric circuit, whose characteristics or properties are same in either direction (*e.g.* a distribution or transmission line), is called the *bilateral circuit.* The distribution or transmission line can be made to perform its function equally well in either direction.

An electric circuit, whose characteristics or properties change with the direction of its operation (*e.g.* a diode rectifier), is called the *unilateral circuit.* A diode rectifier cannot perform rectification in both directions.

(*j*) Resistance may be defined as that property of a substance which opposes (or restricts) the flow of an electric current (or electrons) through it.

The resistance of a wire depends upon its length, area of x-section, type of material, purity and hardness of material of which it is made of and the operating temperature.

Resistance of a wire is

(*a*) directly proportional to its length, l *i.e.* $R \propto l$

(*b*) inversely proportional to its area of x-section, a

$$i.e. \quad R \propto 1/a$$

Combining above two facts we have $R \propto l/a$

$$\text{or } R = \rho \frac{l}{a}$$

where ρ (rho) is a constant depending upon the nature of the material and is known as the *specific resistance or resistivity of* the material of the wire.

SECTION-B (8 Marks Each)

2. What is concept of Root Mean Square value and Peak value in alternating current ?

3. Draw and describe power triangle, explaining active, reactive and apparent power.

4. Discuss the operation of PNP Transistor in CE and CC configuration.

5. Discuss the advantages and disadvantages of positive and negative feedback.

ANSWERS

2. Refer to Arts. 2.4.7 and 2.7.2.

3. Refer to Art. 3.6.

4. Refer to Arts. 14.6, 14.11, 14.12.

5. When a part or fraction of output is combined to the input, feedback is said to exist. Thus the process of combining a fraction of output energy (voltage or current) back to the input is called the *feedback.* When the feedback voltage (or current) is so applied as to increase the input signal (*i.e.,* when the feedback is in phase with input signal), it is called the *positive* or *direct* or *regenerative feedback.* Since positive feedback causes excessive distortion and instability, it is seldom used in amplifiers. However, because of its capability of increasing the power of the original signal it is used in oscillator circuits.

When the feedback voltage (or current) is so applied as to weaken the input signal (*i.e.,* when the feedback is in phase opposition to the input signal), it is called *negative, inverse* or *degenerative feedback.* Negative feedback reduces the amplifier gain but it has numerous advantages (such as gain stability, reduction in nonlinear distortion, reduction in noise, increase in bandwidth or improvement in frequency response, increase in input impedance and decrease in output impedance) and is frequently used in amplifier circuits. The reduction in gain due to negative feedback, can always be compensated by increasing the number of stages.

SECTION-C

6. Discuss various logic gates and explain how one type of logic gate can be created by combination of other logic gates.
7. Differentiate between :
   (*a*) Current and Power
   (*b*) Potential and Potential Difference
   (*c*) Active and Passive components.
8. What is Strain Gauge ? Discuss its applications.
9. Explain the working of a dc machine with the help of its neat sketch showing various parts.

ANSWERS

6. Refer to Arts. 18.5, 18.6, 18.7, 18.8, 18.9, 18.9.1, 18.9.2, 18.10, 18.11 and 18.12.
7. (*a*) Electric current may be defined as the time rate of net motion of an electric charge across a cross-sectional boundary. A random motion of electrons in a metal does not constitute a current unless there is a net transfer of charge with time.

   Power is defined as the rate of doing work or the amount of work done in unit time.

   When an electric current of I amperes flows through a conductor of resistance R ohms, then the power dissipated in the conductor is given as I^2R watts.

   (*b*) *Potential* is a term, which is defined to indicate the electrical state of a body with respect to a standard reference. The potential of earth is regarded as the reference potential and is not affected by the charge given to it or taken away from it *i.e.*, the potential of earth remains the same and is referred to as zero potential.

   Any charged conductor, which is capable of sending electricity to the earth when connected to it, is said to be at positive potential. On the other hand, any charged conductor, which receives electricity from the earth when connected to it, is said to be at a negative potential.

   The *potential difference* (pd) between two points in an electric circuit is that difference in their electrical state which tends to cause flow of electric current between them.

   (*c*) The components which supply energy to the network are known as *active components.* The voltage sourcs like batteries, dc generators, ac generators and current sources like photoelectric cells, metadyne generators fall under the category of active components. Most of the semiconductor devices like transistors are treated as current sources.

   The components which dissipates or store energy are known as *passive components.* Resistors, inductors and capacitors falls under the category of passive components. The resistor is the only component which dissipates electrical energy. The inductors and capacitors are the components which store energy, the inductor stores energy by virtue of current flowing through it whereas the capacitor stores energy by virtue of pd across it.

8. The strain gauge is basically a device used for measuring mechanical surface strain and is one of the most extensively used electrical transducers. Its working is based on the fact that when stress is applied on the metal conductor its resistance changes owing to change in length and cross-sectional area of the conductor (Fig. 12.2). Resistance of the conductor under stress is also changed due to change in resistivity of the conductor. Strain gauges on the basis of operating technique used, may be mechanical, optical or electrical.

   Electrical strain gauges are used for a variety of applications, mainly because of ease in instrumentation, high accuracy and excellent reliability.

   (*i*) Strain gauge transducers are extensively used for analyzing the dynamic strains in complex structures, such as the stress and strain in bridges, automobiles, roads etc., where the output of the bridge can be recorded with the help of oscillograph calibrated to read the instantaneous stress or strains.

   (*ii*) Strain gauge transducers are used for measuring tension, torque, force, stresses in structures. They are particularly useful where a minimum disturbance due to presence of transducer is required as in wind tunnel measurements.

   (*iii*) Strain gauge transducers are also used for measurement of force by the strain produced in load rings.

9. Refer to Arts. 9.1 and 9.2.

B.TECH

Basic Electrical and Electronics Engineering

Time : 3 Hours *Maximum Marks : 60*

Instructions to Candidates:

1. Section-A is ***Compulsory***.
2. Attempt any ***five*** questions from Sections B and C.
3. Selecting atleast ***two*** questions from Sections B and C.

SECTION-A **(2 Marks each)**

1. (*a*) Define and express Peak, RMS, Average value of alternating current and form and peak factors related to them.

(*b*) What are rectangular and polar forms ?

(*c*) Which kind of core is preferred in the medium power transformers and why ?

(*d*) Compare the three phase induction motor characteristics with single phase induction motor.

(*e*) Convert the $(6789)_{10}$ number into the hexadecimal number.

(*f*) Distinguish between Zener breakdown and Avalanche Breakdown.

(*g*) Briefly explain the principle of operation of LVDT.

(*h*) A 20 Ω resistor is connected in series with an inductor, a capacitor and ammeter across a 25 V variable frequency supply. When the frequency is 400 Hz, the current is at its maximum value of 0.5 A and potential difference across the capacitor is 150 V. Calculate

(*i*) The capacitance of the capacitor.

(*ii*) The resistance and inductance of the inductor.

(*i*) A three phase induction motor is wound for four poles and is supplied from a 50 Hz system. Calculate:

(*i*) Synchronous speed

(*ii*) The speed of the rotor when the slip is 4%.

(*j*) A germanium diode draws 40 mA with a forward bias of 0.25 V. The junction is at room temperature of 293°K. Calculate the reverse saturation current of the diode.

ANSWERS/SOLUTION

1. (*a*) Refer to Arts. 2.4.7, 2.7, 2.7.1, 2.7.2, 2.9.1 and 2.9.2.

(*b*) Refer to Arts. 4.2 and 4.5.

(*c*) Medium power transformers may have cores of rectangular or square cross section with rectangular or circular coil. This is because of lower cost of manufacturing.

(*d*) The performance characteristics of a single phase induction motor, in general, are of the same nature as those for polyphase induction motor. However, the performance of a single phase induction motor, in every respect, is somewhat inferior to that of a 3-phase induction motor for the given frame size. This is due to the presence of backward rotating field. It has a lower breakdown torque at larger slip and increased power losses. Further, it has greater power (true as well as apparent) input because of their consumption in the backward rotating field. Even the copper losses occuring in the stator winding of a single phase induction motor are higher because all the current flows through a single winding. As a consequence, efficiency is lowered and temperature rise is increased.

Further, the speed regulation of a single phase induction motor tends to be poorer than that for a polyphase motor. The normal slip of a single phase induction motor under load conditions is rather greater than that of the corresponding 3-phase motor and, therefore, power factor of a single phase motor tends to be lower. For a given power and speed rating, a single phase motor must therefore, have a larger frame size than that of a 3-phase motor.

The forward field and the rotor's backward reaction field and also the backward field and the rotor's forward reaction field move in opposite directions with relative speeds of double the synchronous speed and develop second harmonic pulsating torques with zero average value. As a consequence single phase motors tend to be somewhat noisier than three-phase ones, which have no such pulsating torque.

(*e*)

16	6789
16	424 – 5
16	26 – 8
16	1 – 10
	0 – 1

So $(6789)_{10} = (1A85)_{16}$ **Ans.**

(*f*) Refer to Art. 13.7.

(*g*) Refer to Art. 12.5.

(*h*) Voltage drop across capacitor, $V_C = 150$ V

Current through capacitor, I = 0.5 A

(*i*) Capacitor, $C = \dfrac{I}{2\pi f V_C}$ $\qquad \because I = 2\pi f C V_C$

$$= \frac{0.5}{2\pi \times 400 \times 150} = 1.3263\ \mu\text{F} \textbf{ Ans.}$$

(*ii*) Inductance of inductor, $L = \dfrac{1}{(2\pi f)^2 \times C}$ $\qquad \because$ At resonance $2\pi f L = \dfrac{1}{2\pi f C}$

$$= \frac{1}{(2\pi \times 400)^2 \times 1.3263 \times 10^{-6}} = 0.11936\ \text{H} \textbf{ Ans.}$$

Total resistance of the circuit $= \dfrac{V}{I} = \dfrac{35}{0.5} = 50\ \Omega$

So the resistance of inductor $= 50 - 20 = 30\ \Omega$ **Ans.**

(*i*) (*i*) Synchronous speed, $N_s = \dfrac{120 f}{P} = \dfrac{120 \times 50}{4} = 1{,}500$ rpm

Slip, s = 4% or 0.04

(*ii*) Motor speed, $N = N_s(1 - s) = 1{,}500\,(1 - 0.04) = 1{,}440$ rpm **Ans.**

(*j*) Applied forward bias, V = 0.25 V

Forward current, I = 40 mA = 0.04 A

Volt equivalent at room temperature of 293°K

$$= \frac{293}{11{,}600} = 0.02526\ \text{V}$$

As diode current is given as $\qquad I = I_0(e^{V/\eta V_T} - 1)$

or $I_0 = \dfrac{I}{e^{V/V_T} - 1} = \dfrac{0.04}{e^{0.25/0.02526} - 1} = 2.0\ \mu\text{A}$ **Ans.** $\because$ For Ge, $\eta = 1$

SECTION-B **(8 Marks Each)**

2. (*a*) What is the effect of temperature on resistance ? Define temperature coefficient. How does it vary with temperature ?

   (*b*) A coil has resistance of 18 Ω when its mean temperature is 20°C and of 20 Ω when its mean temperature is 50°C. Find its mean temperature rise when its resistance is 21 Ω and the surrounding temperature is 15°C.

3. Calculate the impedance and power factor of parallel L-C circuit. What do you understand by parallel circuit resonance ? Show graphically and mathematically. Define magnification factor and compute in this case.

4. (*a*) Draw the phasor diagram and equivalent circuit diagram of a transformer under different types of loads. Derive emf equation of a transformer.

   (*b*) A circular iron ring, having a cross-sectional area of 10 cm^2 and a length of (4π) cm in iron, has an air gap of 0.4π mm made a saw cut. The relative permeability of iron is 10^3 and permeability of free space is $4\pi \times 10^{-7}$ H/m. The ring is wound with a coil of 2,000 turns and carrier 2 mA. Determine the air gap flux neglecting leakage and fringing.

5. (*a*) Give construction and working details of three phase induction motor. Establish relation for torque under running conditions. Give its torque-speed characeristics. Why do we prefer it for industrial use ?

   (*b*) A three phase motor operating of a 400 V system is developing 20 kW at an efficiency of 0.87 pu and a power factor of 0.82. Calculate

   (*i*) the line current

   (*ii*) the phase current is the windings are delta-connected.

ANSWERS

2. (*a*) Refer to Arts. 1.9 and 1.10.

   (*b*) Refer to Solution of Example 1.14.

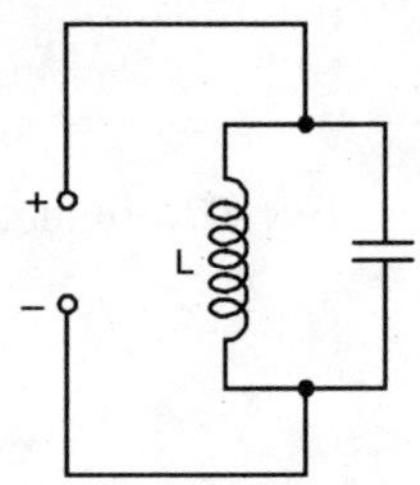

Fig. 1

3. Susceptance of inductive branch, $B_L = \dfrac{1}{2\pi f L}$

Susceptance of capacitive branch, $B_C = 2\pi f C$

Admittance of the parallel circuit $= B_C - B_L = 2\pi f C - \dfrac{1}{2\pi f L}$

Impedance of parallel circuit $= \dfrac{1}{B} = \dfrac{1}{B_C - B_L} = \dfrac{1}{2\pi f C - \dfrac{1}{2\pi f L}} = \dfrac{2\pi f L}{4\pi^2 f^2 LC - 1}$

Power factor of the parallel L-C circuit is zero.

Refer to Arts. 5.3, 5.3.1, 5.3.2 and 5.3.4.

4. (*a*) Refer to Arts. 8.10 and 8.5.

   (*b*) Let the flux density through the iron as well as air gap be B Wb/m^2

Total ampere turns provided, AT $= N \times I = 2{,}000 \times 2 \times 10^{-3} = 4$

Total ampere-turns required $= AT_g + AT_i$

$$\text{or } 4 = \frac{B}{\mu_0} l_g + \frac{B}{\mu_0 \mu_r} l_i$$

$$= \frac{B}{\mu_0}\left(l_g + \frac{l_i}{\mu_r}\right) = \frac{B}{4\pi \times 10^{-7}}\left[0.4\pi \times 10^{-3} + \frac{4\pi \times 10^{-2} - 0.4\pi \times 10^{-3}}{1,000}\right]$$

$$\text{or Flux density, } B = \frac{4 \times 4\pi \times 10^{-7}}{0.4396\pi \times 10^{-3}} = 3.64 \times 10^{-3} \text{ T}$$

$$\text{Air-gap flux} = B \times a_g = 3.64 \times 10^{-3} \times 10 \times 10^{-4} = 3.64 \times 10^{-6} \text{ Wb}$$ **Ans.**

5. (*a*) Refer to Arts. 10.2, 10.2.1, 10.2.2, 10.2.3, 10.2.4, 10.2.5, 10.7, 10.13 and 10.21.

Relation for torque under running condition is derived as follows:

When a 3-phase induction motor is operating, on a certain load, with slip *s*, then

Induced emf per phase in rotor $= s\,E_2$ volts where E_2 is standstill rotor induced emf

Rotor impedance per phase, $\mathbf{Z}_2 = R_2 + jsX_2$

where R_2 and X_2 are rotor resistance and standstill reactance per phase

$$\text{Rotor current per phase, } I_2 = \frac{sE_2}{Z_2} = \frac{sE_2}{\sqrt{R_2^2 + s^2X_2^2}}$$

$$\text{Rotor circuit power factor, } \cos\phi_2 = \frac{R_2}{\sqrt{R_2^2 + s^2X_2^2}}$$

Total rotor copper loss = Number of phases × rotor induced emf per phase × rotor current per phase × rotor circuit power factor

$$= 3 \times sE_2 \times \frac{sE_2}{\sqrt{R_2^2 + s^2X_2^2}} \times \frac{R_2}{\sqrt{R_2^2 + s^2X_2^2}}$$

$$= \frac{3R_2 s^2 E_2^2}{R_2^2 + s^2X_2^2} \quad ...(i)$$

Also,

$$\text{Total rotor copper loss} = s \times \text{power input to rotor} = \frac{s \times T \times 2\pi N_s}{60}$$

$$\text{So } s \times T\,\frac{2\pi N_s}{60} = \frac{3R_2 s^2 E_2^2}{R_2^2 + s^2X_2^2}$$

$$\text{or Torque developed by rotor, } T = \frac{3 \times 60}{2\pi N_s} \cdot \frac{R_2 sE_2^2}{R_2^2 + s^2X_2^2} \quad ...(ii)$$

(*b*) (*i*) Line current,

$$I_L = \frac{\text{Input}}{\sqrt{3}V_L \cos\phi}$$

$$= \frac{\text{Output}/\eta}{\sqrt{3}V_L \cos\phi} = \frac{20,000/0.87}{\sqrt{3} \times 400 \times 0.82} = 40.46 \text{ A}$$ **Ans.**

(*ii*) Phase current,

$$I_P = \frac{I_L}{\sqrt{3}} = \frac{40.46}{\sqrt{3}} = 23.36 \text{ A}$$ **Ans.**

SECTION-C (8 Marks Each)

6. (*a*) Define gauge factor. Explain why semiconductor strain gauges have high values for the gauge factor.

(*b*) How thermistors are used as temperature transducer ? Give its characteristics and types.

7. (*a*) Explain the working of full-wave rectifier using center tap transformer. Compute its ripple factor, efficiency.

(*b*) Give the various modes of operation and input/output characteristics of Field-effect transistors.

8. (*a*) Name the Universal gates ? Justify their name by building all the basic gates from them.

(*b*) What is JK flip-flop ? Discuss its working. What is race around condition ?

9. Write detailed short note on the following :

(*a*) LVDT performance and its improvement.

(*b*) Zener diode and its applications.

(*c*) Regions of transistor operation, as an amplifier and a switch.

ANSWERS

6. (*a*) Refer to Answers of Q. 10 page 381-382 and Q. 12 page 382.

(*b*) Refer to Art. 12.4.

7. (*a*) Refer to Arts. 16.3.1 and 16.3.3.

(*b*) Refer to Arts. 15.2.5 and 15.3.

8. (*a*) Refer to Arts. 18.9.1 and 18.9.2

(*b*) Refer to Arts. 18.4.5 and 18.4.6.

9. (*a*) Refer to Arts. 12.5 and 12.5.1.

(*b*) Refer to Arts. 13.13, 13.13.1 and 13.4.

(*c*) Refer to Arts. 14.15 and 14.16.

B.TECH

Examination, May-2014

Basic Electrical and Electronics Engineering

Time : 3 Hours *Maximum Marks : 60*

Note: (*i*) Question number 1 is ***Compulsory.***

(*ii*) Attempt ***five*** questions from part A and part B with at least two questions each from Part A and Part B.

(*iii*) Assume missing data if any and clearly indicate it.

1. (*a*) A coil of stranded copper wire having a resistance of 12 Ω at 25°C is embedded in the core of a large transformer supplied at 230 V. After the transformer has been in service for several hours, the resistance of the coil is found to be 13.4 ohms. What is the temperature of the core ? Also find power rating of the resistance. Assume temperature coefficient of wire as 0.00125/°C. **(10 × 2 = 20)**

(*b*) A series RL circuit with R = 30 ohm and L = 15 H is supplied from a constant voltage source of 60 V and is switched on at $t = 0$. Determine the current in the circuit, voltage across resistor and voltage across inductor.

(*c*) A voltage $v(t) = 170 \sin (377t + 10°)$ is applied to a circuit. It causes a steady state current to flow which is described by $i(t) = 14.14 \sin (377t - 20°)$. Determine the variation of instantaneous power. Also find the average power delivered to circuit and power factor.

(*d*) What is full form of LVDT ? Give its two applications.

(*e*) Sketch circuit symbol of PNP transistor, Zener Diode, Field-effect Transistor, Crystal diode.

(*f*) Do the following conversions :

$(101010011.1101)_2$ to $(.......)_8$

(1684)16 to $(.........)_2$

(*g*) Draw the schematic of JK flip flop using universal logic gates.

(*h*) What are two different types of three phase induction motors ? Give an application of each type.

(*i*) Name the machine used at major power generating stations in India.

(*j*) Define current gain of common emitter configuration of bipolar junction transistor. What is the approximate range of current gain ?

ANSWERS/SOLUTION

1. (*a*) Let the temperature of the core be t_2°C after being in service for several hours.

$$R_2 = R_1[1 + \alpha_1(t_2 - t_1)]$$

$$13.4 = 12[1 + 0.00125(t_2 - 25)]$$

$$\text{or } t_2 = \frac{\frac{13.4}{12} - 1}{0.00125} + 25 = 118.3 \text{ °C } \textbf{Ans.}$$

$$\text{Power rating of resistance, P} = \frac{V^2}{R} = \frac{230^2}{12} = 4{,}408 \text{ W } \textbf{Ans.}$$

(*b*) At the instant of switching

$$\text{Current in the circuit, } i(0) = \frac{V}{R}\left[1 - e^{-\frac{R}{L}t}\right] = \frac{60}{30}[1 - e^0] = 0$$

$$\text{Voltage across resistor} = i(0) \times R = 0 \times R = 0 \textbf{ Ans.}$$

$$\text{Voltage drop across inductor} = V = 60\text{ V} \textbf{ Ans.}$$

(*c*)

$$v(t) = 170 \sin(377t + 10°)$$

$$i(t) = 14.14 \sin(377t - 20°)$$

$$V_{max} = 170 \text{ V}$$

$$I_{max} = 14.14 \text{ A}$$

$$\text{Phase angle, } \phi = 10 - (-20°) = 30° \text{ (lagging)}$$

$$\text{Variation of instantaneous power} = \frac{1}{2} V_{max} I_{max} \cos(2\omega t - \phi)$$

$$= \frac{1}{2} \times 170 \times 14.14 \cos(2 \times 377t - 30°)$$

$$= 1{,}202 \cos(754t - 30°) \textbf{ Ans.}$$

$$\text{Average power delivered, P} = V_{rms} \times I_{rms} \times \cos\phi$$

$$= \frac{170}{\sqrt{2}} \times \frac{14.14}{\sqrt{2}} \cos 30° = 1{,}041 \text{ W} \textbf{ Ans.}$$

$$\text{Power factor} = \text{Cos } 30° = 0.866 \text{ (lagging)}$$

(*d*) LVDT stands for Linear Variable Differential Transformer. LVDTs are suitable for use in applications where the displacements are too large for strain gauges to handle. For example, LVDTs can be employed for measurement of displacements that range from a fraction of a mm to a few cm. If LVDT is to be employed for measurement of mechanical displacement greater than 25 mm, an appropriate mechanical ratioing (gearing) must be used.

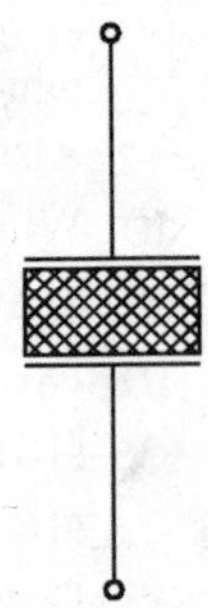

Fig. 1

(*e*) Refer to Figs. 14.1(*a*), 13.47 and 15.4.

Circuit symbol for crystal diode is given in Fig. 1.

(*f*) $(101010011.1101)_2 \rightarrow$ 101 010 011 . 110 100

↓ ↓ ↓ ↓ ↓

5 2 3 6 4

So $(101010011.1101)_2 = (523.64)_8$ **Ans.**

$(1684)_{16} \rightarrow$ 1 6 8 4

↓ ↓ ↓ ↓

0001 0110 1000 0100

So $(1684)_{16} = (1011010000100)_2$ **Ans.**

(*g*) Refer to Art. 18.14.5.

(*h*) Two types of 3-phase induction motors are (*i*) squirrel cage and (*ii*) wound rotor or slip-ring induction motors.

Applications of 3-Phase Squirrel Cage Induction Motor : Squirrel cage induction motor is suitable for constant speed industrial drives of small power where speed control is not required and where starting torque requirements are of medium or low value as for printing machinery, flour mills, saw mills, pumps, prime-movers for small generators and other shaft drives of small power.

Applications of 3-Phase Wound Rotor Motor : Wound rotor induction motor is used for loads requiring severe starting conditions or for loads requiring speed control such as for driving line shafts, lifts, pumps, generators, winding machines, cranes, hoists, elevators, compressors, small electric excavators, printing machines, turn-tables, strokers, large ventilating fans, crushers etc.

(*i*) Synchronous generators are used at major power generating stations in India.

(*j*) The ratio of change in collector current (output current) and change in base current (input current) is called the *base current amplification factor* β

$$i.e.\ \beta = \frac{\Delta I_C}{\Delta I_B} \qquad ...(i)$$

Almost in all transistors, the base current is less than 5% of emitter current I_E so β is usually higher than 20. Usually β ranges from 20 to 100.

$$\text{Also } \beta = \frac{I_C}{I_B} = \frac{I_C}{I_E - I_C} = \frac{I_C/I_E}{1-\frac{I_C}{I_E}} = \frac{\alpha}{1-\alpha} \qquad ...(ii)$$

PART-A

2. Determine the current delivered by the source in the circuit shown in Fig. 2. **(8)**
3. Show that emf induced in transformer winding is proportional to flux present in core, number of turns and frequency of source applied. **(8)**
4. Sketch the labelled constructional diagram of dc generator. Clearly mention the material used for each of the part. **(8)**
5. Show that for star connections of three phase circuit, the line voltage is equal to 1.732 times the phase voltage whereas line current is equal to phase current. Also draw phasor diagram to support your answer. **(8)**

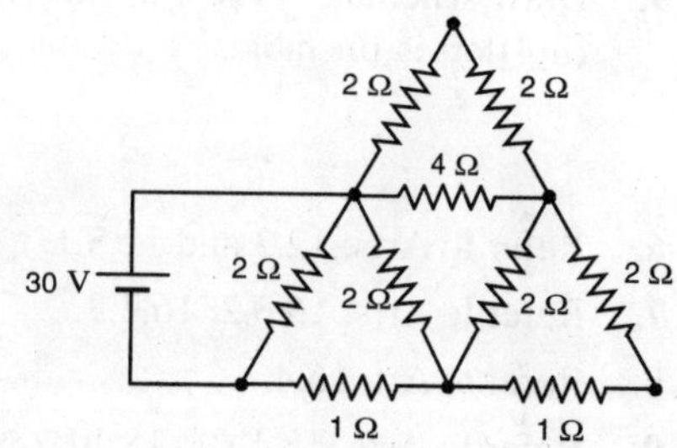

Fig. 2

ANSWERS/SOLUTION

2. The given circuit can be solved simply by simplification, as illustrated in Figs. 3 (*a*), (*b*), (*c*), (*d*), (*e*) and (*f*).

Thus effective resistance of the circuit, $R = \frac{58}{55}\ \Omega$.

Current supplied by the source, $I = \frac{V}{R} = \frac{30}{58/55} = \frac{30 \times 55}{58} = 28.45$ A **Ans.**

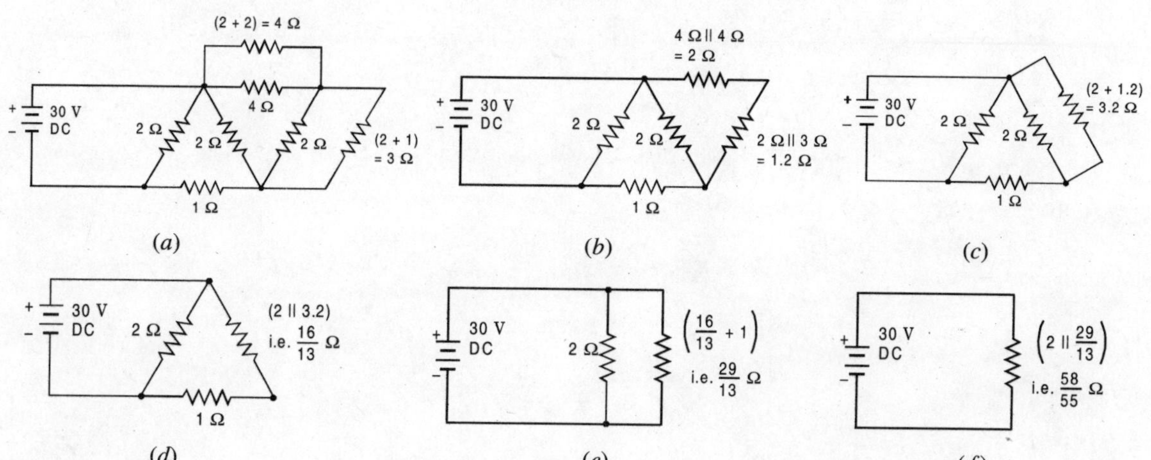

Fig. 3

3. Refer to Art. 8.5.
4. Refer to Art. 9.3.
5. Refer to Art. 6.6.

PART-B

6. **What is strain gauge ? What is its operating principle ?** **(8)**
7. **Discuss the operation of bridge configuration of full-wave rectifier feeding resistive load. Obtain the average value and rms value of output voltage. Sketch the waveforms of load voltage, voltage across one diode.** **(8)**
8. **Draw functional diagram of regulated power supply. What are the considerations of regulated power supply ? Give its applications.** **(8)**
9. **Draw schematic of Sr Flip-flop and D type flip-flop. Write their truth tables. Under what conditions SR flip-flop is unstable.** **(8)**

ANSWERS

6. Refer to Arts. 12.3 and 12.3.1.
7. Refer to Arts. 16.3.2, 16.3.3.
8. Refer to Art. 16.4.
9. Refer to Arts. 18.14.1, 18.14.2 and 18.14.3.

❑❑❑

INDEX

INDEX

A

B

C

D

E

F

G

H

I

J

Q

R

S

❑❑❑

Win Prizes

ATTENTION READERS

Win Prizes

We request for your frank assessment regarding various aspects of the book as given below :

BASIC ELECTRICAL AND ELECTRONICS ENGINEERING (PTU)

By : J.B. GUPTA

Please Fill up the Sheet in Neat and Clean Writing

(i) What topics of your syllabus, important for your examination point of view, are not covered in this book ?

....

....

....

....

(ii) What are the chapters wherein the treatment of the subject matter is not systematic or organised or updated ?

....

....

....

....

(iii) Have you come across misprints/mistakes/factual inaccuracies in the book ? Please specify the chapters and the page numbers.

....

....

....

....

(iv) What other books on the same subject have you found/heard better than the present book ? Please specify in terms of price and quality.

....

....

....

....

(v) Further suggestions and comments for the improvement of this book.

....

....

....

....

(PS : If need be please attach a separate sheet to write your views)

OTHER DETAILS

(i) Who recommended you this book ? (Please tick (✓) in the box near the option relevant to you.

☐ Teacher ☐ Friends ☐ Bookseller

....

....

....

....

(ii) Name of the recommending teacher, his designation and address:

....

....

....

....

(iii) Name and address of the bookseller you purchased the book from :

....

....

....

....

(iv) Name and address of your institution (Please mention the University or Board as the case may be)

....

....

....

....

(v) Your name and complete address :

....

....

....

....

(vi) Course you are enrolled in :

....

....

....

....

The best assessment will be awarded each month : The award will be in the form of our publications as decided by the Editorial Board.

Please mail the filled up coupon at the vearliest to :

S.K. Kataria & Sons®
4885/109, Prakash Mahal, Dr. Subhash Bhargav Lane,
Opposite Delhi Medical Association, Daryaganj, New Delhi-110002
Phones: +91-11-23243489, 43551243; Mobile: +91-9871775858
e-mail: katariabook@gmail.com; katariabooks@yahoo.com
Website: www.skkatariaandsons.com